Integrative Paths to the Past

Paleoanthropological Advances in Honor of F. Clark Howell

Edited by
Robert S. Corruccini
Southern Illinois University

Russell L. Ciochon
University of Iowa

Advances in Human Evolution Series

Prentice Hall, Englewood Cliffs, New Jersey 07632

Library of Congress Cataloging-in-Publication Data
Corruccini, Robert S.
Integrative paths to the past : paleoanthropological advances in honor of F. Clark Howell / **Robert S. Corruccini, Russell L. Ciochon.**
p. cm.
Includes bibliographical references and index.
ISBN 0-13-706773-9
1. Human evolution. 2. Man, Prehistoric. 3. Anthropology, Prehistoric. 4. Howell, F. Clark (Francis Clark) I. Ciochon, Russell L. II. Title.
GN281. C597 1994
573.2—dc20 93-36013

Acquisitions editor: *Nancy Roberts*
Editorial/production supervision : *Edie Riker*
Cover design: *Tommy Boy*
Production coordinator: *Kelly Behr/Mary Ann Gloriande*
Editorial assistant: *Pat Naturale*

A Paramount Communications Company
Englewood Cliffs, New Jersey 07632

Printed in the United States of America

10 9 8 7 6 5 4 3 2 1

ISBN 0-13-706773-9

Prentice-Hall International (UK) Limited, *London*
Prentice-Hall of Australia Pty. Limited, *Sydney*
Prentice-Hall Canada Inc., *Toronto*
Prentice-Hall Hispanoamericana, S.A., *Mexico*
Prentice-Hall of India Private Limited, *New Delhi*
Prentice-Hall of Japan, Inc., *Tokyo*
Simon & Schuster Asia Pte. Ltd., *Singapore*
Editora Prentice-Hall do Brasil, Ltda., *Rio de Janeiro*

Contents

List of Contributors

Susan C. Antón
Laboratory for Human Evolutionary Studies
Department of Anthropology
University of California
Berkeley, CA 94720

Brenda Benefit
Department of Anthropology
Southern Illinois University
Carbondale, IL 62901-4502

Noel T. Boaz
International Institute for Human Evolutionary Research
George Washington University Virginia Campus
20101 Academic Way, Suite 326
Ashburn, VA 22011

Raymonde Bonnefille
Laboratoire de Geologie du Quaternaire
C.N.R.S., Luminy - Case 907
13288 Marseille Cedex 9, France

Francis H. Brown
Department of Geology and Geophysics
University of Utah
Salt Lake City, UT 84112-1183

Russell L. Ciochon
Departments of Anthropology and Pediatric Dentistry
University of Iowa
Iowa City, Iowa 52242-1322

J. Desmond Clark
Department of Anthropology
University of California
Berkeley, CA 94720

Ron Clarke
Department of Anatomy
University of the Witwatersrand Medical School
7 York Road, Parktown, Johannesburg 2193
Republic of South Africa

Robert S. Corruccini
Department of Anthropology
Southern Illinois University
Carbondale, IL 62901-4502

Dorothy Dechant Boaz
Laboratory for Human Evolutionary Studies
Department of Anthropology
University of California
Berkeley, CA 94720

Dennis A. Etler
Laboratory for Human Evolutionary Studies
Department of Anthropology
University of California
Berkeley, CA 94720

John G. Fleagle
Department of Anatomical Sciences
School of Medicine, Health Sciences Center
State University of New York
Stony Brook, NY 11794-8081

Leslie G. Freeman
Department of Anthropology
University of Chicago
Haskell M-135, 5836-46 Greenwood Avenue
Chicago, Il 60637

Walter Carl Hartwig
Department of Anatomical Sciences
School of Medicine, Health Sciences Center
State University of New York
Stony Brook, NY 11794-8081

Jean de Heinzelin
Institut Royal des Sciences Naturelles de Belgique
Rue Vautier 29
B-1040 Brussels, Belgium

Andrew Hill
Department of Anthropology
Yale University
P.O. Box 2114 Yale Station
New Haven, CT 06520

Susan G. Keates
School of Anthropology
60 Banbury Road
University of Oxford
Oxford OX2 6PN England

Richard G. Klein
Department of Anthropology
Stanford University
Stanford, CA 94305-2145

Pelaji S. Kyauka
Laboratory for Human Evolutionary Studies
Department of Anthropology
University of California
Berkeley, CA 94720

Li Tianyuan
Hubei Archeological Institute
67 Donghu Road
Wuhan, Hubei
People's Republic of China 43077

Monte McCrossin
Department of Anthropology
University of California
Berkeley CA 94720

Henry M. McHenry
Department of Anthropology
University of California
Davis, CA 95616-8522

Geoffrey G. Pope
Department of Anthropology
William Paterson College
Wayne, NJ 07470

Naomi Porat
Institute of Earth Sciences
Hebrew University of Jerusalem
Jerusalem, Israel 91940

Yoel Rak
Department of Anatomy
Sackler School of Medicine
Tel-Aviv University
Ramat-Aviv, 69978 Tel-Aviv, Israel

Kathy Schick
Anthropology Department and CRAFT
(Center for Research into the Anthropological Foundations of Technology)
Indiana University
Bloomington, Indiana 47405

Henry Schwarcz
Department of Geology
McMaster University
Hamilton, Ontario L8S 4L9, Canada

Phillip V. Tobias
Department of Anatomy and Human Biology
University of the Witwatersrand Medical School
7 York Road, Parktown, Johannesburg 2193
Republic of South Africa

Nicholas Toth
Anthropology Department and CRAFT
(Center for Research into the Anthropological Foundations of Technology)
Indiana University
Bloomington, Indiana 47405

Russell H. Tuttle
Department of Anthropology
University of Chicago
1126 East 59th Street
Chicago, IL 60637

Elisabeth Vrba
Department of Geology and Geophysics
Kline Geology Laboratory
Yale University
P.O. Box 6666
New Haven, CT 06511-8130

Bernard Wood
Department of Human Anatomy and Cell Biology
University of Liverpool
P.O. Box 147
Liverpool L69 3BX England

Major Career Accomplishments of F. Clark Howell

1951 The place of Neanderthal man in human evolution. *Am. J. Phys. Anthropol. 9*:379-416.

1952 Pleistocene glacial ecology and the evolution of "classic Neanderthal" man. *Southw. J. Anthropol. 8*:337-410.

1953 (June): Ph.D., University of Chicago (Anthropology).

1953 Appointed Instructor, Department of Anatomy, Washington University (School of Medicine), St. Louis, Missouri.

1955 Joined the Anthropology Department, University of Chicago.

1955 The age of the australopithecines of southern Africa. *Am. J. Phys. Anthropol. 13*:635-661.

1956 Participation (American Representative) in International Neandertal Centenary Congress (Dusseldorf).

1957 The evolutionary significance of variation and varieties of "Neanderthal" man. *Quart. Rev. Biol. 32*:330-347.

1957-1958 Excavation of Isimila, an Acheulian prehistoric occupation site in the Iringa highlands, central Tanganyika, East Africa.

1958 Upper Pleistocene men of the southwest Asian Mousterian. In *Hundert Jahre Neanderthaler, 1856-1956* (G.H.R. Von Koenigswald, ed.). Utrecht: Kemink en Zoon N.V. pp. 185-198.

1959 Survey of fossiliferous lower Pleistocene beds in the Omo River region, southern Ethiopia; participation in 4th Pan-African Congress on Prehistory and Quaternary geology (Leopoldville).

1959 The Villafranchian and human origins. *Science 130*:831-844.

1960 European and northwest African Middle Pleistocene hominids. *Curr. Anthropol. 1*:195-232. (see also: More on Middle Pleistocene hominids. *Ibid., 2*:117-120. 1961).

1961 Isimila. A Paleolithic site in Africa. *Sci. Am. 205*:118-129.

1961 Organized (with F. Bourlière) and participated in symposium *African Ecology and Human Evolution* (held at Burg Wartenstein, Austria).

1961-1963 Initiated and directed excavations at prehistoric sites (Acheulian) of Torralba and Ambrona (Soria), Spain.

1961-1963 Vice-President, American Association of Physical Anthropologists.

1962 Promoted to Professor of Anthropology, Department of Anthropology, University of Chicago.

1962 Early Man and Pleistocene stratigraphy in the Circum-Mediterranean Regions (arranger and editor). *Quaternaria 6* (549 pages).

1963 Acheulian hunter-gatherers in sub-Saharan Africa (with J.D. Clark). In *African Ecology and Human Evolution*, F. Clark Howell and F. Bourlière, eds. Viking Fund Publications in Anthropology *36*. Chicago: Aldine, pp. 458-532.

1964-1967 Executive Board, American Anthropological Association.

1965 *Early Man*. Life Nature Library. New York: Time, Inc., 200 pages.

1965 Organized (with W.W. Bishop and J.D. Clark) and participated in the symposium *Systematic Investigation of the African Later Tertiary and Quaternary* (held at Burg Wartenstein, Austria).

1966 Observations on the earlier phases of the European Lower Paleolithic. *Am. Anthropol.* (Paleoanthropology issue, J. D. Clark and F. C. Howell, eds.), *68*(Pt. 2, No. 2):88-201.

1966-1969 Professor and Chairman, Department of Anthropology, University of Chicago.

1967 Recent advances in human evolutionary studies. *Quart. Rev. Biol. 42*:471-513.

1967-1973 Organized, directed and participated in interdisciplinary studies in the earth sciences, vertebrate paleontology, and paleoanthropology of the U.S.A. contingent of the International Omo Research Expedition, Omo basin, southern Ethiopia.

1968 Omo Research Expedition, 1967. *Nature 219*:567-572.

1969 Senior Scientific Advisor to Metro-Goldwyn-Mayer for TV Special, *The Man-Hunters.*

1969 Remains of Hominidae from Pliocene/Pleistocene formations in the lower Omo basin, Ethiopia. *Nature 223*:150-152.

1969 Became Member, Board of Trustees (Science and Grants Committee), L.S.B. Leakey Foundation (Los Angeles).

1970 Assumed position as Professor, Department of Anthropology, University of California, Berkeley.

1971 Pliocene/Pleistocene Hominidae in Eastern Africa: Absolute and relative ages. In *Calibration of Hominoid Evolution* (W.W. Bishop and J.A. Miller, eds.). Edinburgh: Scottish Academic Press, pp. 331-368.

1972 Member, (Assembly of Behavioral and Social Sciences), National Research Council, National Academy of Sciences.

1972 Affinities of the Swartkrans 847 cranium (with R.J. Clarke). *Am J. Phys. Anthropol. 37*:319-336.

1973 Deciduous teeth of Hominidae from Pliocene/Pleistocene of the lower Omo basin, Ethiopia (with Y. Coppens). *J. Hum. Evol.* (R.A. Dart Memorial Issue) 2:461-472.

1973 Archaeological occurrences of Early Pleistocene age from the Shungura Formation, lower Omo valley, Ethiopia (with H.V. Merrick, J. de Heinzelin, P. Haesaerts). *Nature 242*:572-575.

1973 Co-organizer of Wenner-Gren Foundation workshop conference, *Stratigraphy, Paleoecology and Evolution in the Lake Rudolf Basin,* held September 8-20, 1973, in Nairobi, Kenya, and in the field in Kenya and southern Ethiopia.

1974 Inventory of remains of Hominidae from Pliocene/Pleistocene formations of the lower Omo basin, Ethiopia (1967-1972) (with Y. Coppens). *Am. J. Phys. Anthropol. 40*:1-16.

1975 Chairman of the Paleoanthropology Delegation of the CSCPRC of the National Academy of Sciences to the People's Republic of China.

1976 *Earliest Man and Environments in the Lake Rudolf Basin. Stratigraphy, Paleoecology and Evolution* (Y. Coppens, F. C. Howell, G. L. Isaac and R.E.F. Leakey, eds.). Chicago: University of Chicago Press. With chapters on Carnivora, Mammalian faunas, remains of *Camelus*, Geological formations and overview of Hominidae from the Omo succession.

1977 Distinguished Lecturer, American Anthropological Association.

1978 Hominidae. In *Evolution of African Mammals* (H.B.S. Cooke and V.J. Maglio, eds.). Cambridge: Harvard University Press, pp. 154-248.

1978 Obervations on problems of correlation of late Cenozoic hominid-bearing formations in the north Rudolf basin (with F.H. Brown and G.G. Eck). In *Geological Background to Fossil Man.* (W.W. Bishop, ed.). London: Geological Society of London, pp. 473-498.

1978 Overview of the Pliocene and earliest Pleistocene of the lower Omo basin, southern Ethiopia. In *Early Hominids of Africa.* (C.J. Jolly, ed.). London: Duckworth, pp. 85-130.

1980 *Early Man.* Revised (2nd edition) 200 pages. Life Nature Library. New York: Time/Life, Inc.

1980-1983 Director, Ambrona Research Project, renewed investigation of an earlier Acheulian human occupation site in Soria province (Spain).

1985 John Simon Guggenheim Fellow.

1985 *Les faunes Plio-Pléistocènes de la basse Vallée de l'Omo (Éthiopie). Tome 1. Perissodactyles. Artiodactyles (Bovidae).Travaux de Paléontologie Est-Africaine* (Y. Coppens and F.C. Howell, eds.). Paris: Éditions du C.N.R.S.

1986 Variabilité chez *Homo erectus*, et le problème de la présence de cette espèce en Europe. *L'Anthropologie 90*:447-481.

1987 Senior Scientific Advisor to NGS and WQED/Pittsburgh for TV special, *Mysteries of Mankind.*

1987 Depositional environments, archaeological occurrences and hominids from Members E and F of Shungura Formation (Omo basin, Ethiopia) (with P. Haesaerts and J. de Heinzelin). *J. Hum. Evol. 16*:665-700.

1987 *Les faunes Plio-Pléistocènes de la basse vallée de l'Omo (Éthiopie). Tome 2. Les Éléphantidés (Mammalia-Proboscidea). Travaux de Paléontologie Est-Africaine.* (Y Coppens and F. C. Howell, eds.). Paris: Éditions du C.N.R.S.

1987-1988 Study tour of Miocene and Pliocene-Pleistocene fossiliferous localities and their fossil vertebrates, at the invitation of the Yunnan Provincial Museum (Kunming), P.R.C.

1988 *Les faunes Plio-Pléistocène de la basse vallé de l'Omo (Éthiopie). Tome 3. Cercopithecoidea. Travaux de Paléontologie, Est-Africaine.* Preface by F. Clark Howell. (Y. Coppens and F. C. Howell, eds.) Paris: Éditions du C.N.R.S.

1988-1989 Co-director, Yarimburgaz Research Project of cave excavations in Turkey.

1990 Foreign Member, Académie des Sciences, Paris.

1992 Honorary Degree, conferred by President Hanna H. Gray of the University of Chicago.

Career Timeline of F. Clark Howell

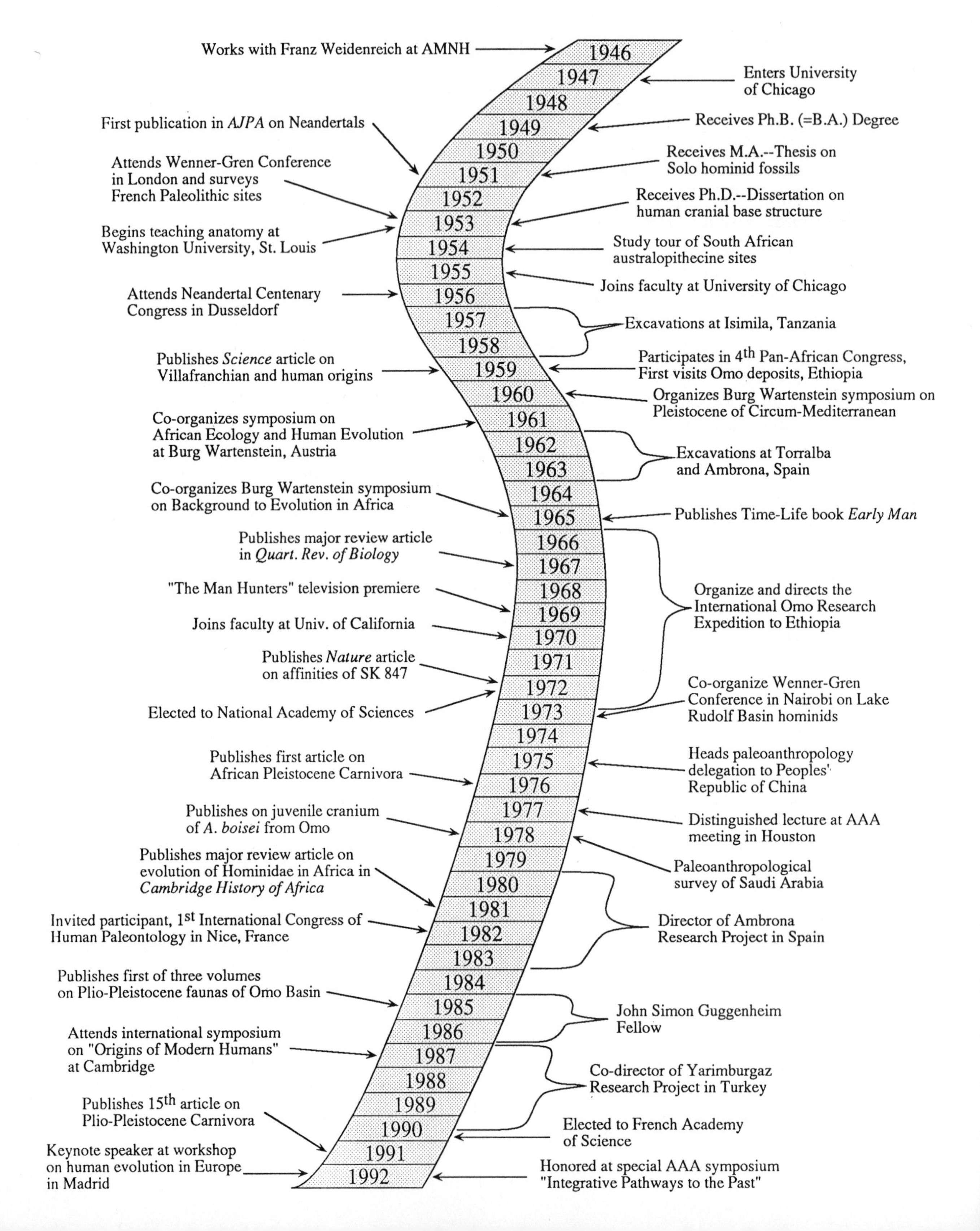

Preface

Professor F. Clark Howell retired from the University of California at Berkeley in 1991. Considering his incontestable importance to paleoanthropology and to many of our careers, we felt a special honorary volume was very overdue to acknowledge his positive influence over four decades of contributions to the field of paleoanthropology. In fact as the present volume appears, it is the 40th anniversary of Clark Howell's Ph.D. from the University of Chicago and the formal start of his paleoanthropological career.

The idea of a Festschrift is to unite students, colleagues, and associates of the honoree as authors of chapters in a book that well serves the specialty. The title for this endeavor was based on the strongly integrative path that Howell has been instrumental in bringing to studies of the human past. In order to reflect properly upon Howell's career and contributions, we sought writings concerning all the diversity of geological, paleontological, anatomical, environmental, and cultural contexts that have typified Howell's work.

We see this volume going well beyond the typical Festschrift because of its thematic concentration and determination to convey the most current thought of leading practitioners in the field. In other words, we did not seek a collection of unrelated, convenience-driven writings. Our goal was to compile an overview of the state of the art in the various subdivisions of paleoanthropology. Only time will tell if we have succeeded, but the preliminary activities, especially the symposium held in honor of Howell at the 1992 American Anthropological Association meetings in San Francisco on December 4, 1992, have been most gratifying.

In 1946 Clark Howell was discharged from the U.S. Navy. Making his way to New York, he met such eminent paleontologists as Franz Weidenreich, George Gaylord Simpson, and Ralph von Koenigswald at the American Museum of Natural History. He entered the University of Chicago in 1947 and in just six years completed bachelor's and doctorate degrees under the direction of Sherwood Washburn. By this time he already had to his credit eight publications and a very influential presentation at the American Association of Physical Anthropologists.

In the 1950s Howell travelled extensively, visiting major hominid sites in Europe and Africa, leading his first paleolithic excavation at Isimila, participating in the most important congresses of the day, and examining for the first time the Omo Basin in Ethiopia. He recognized immediately the stratigraphic importance of the volcanic tuffs there. Although organizing many conferences in the 1960s and 1970s, leading the excavation at Ambrona in Spain, and heading the first paleoanthropological delegation to the People's Republic of China in 1975 among many other honors, the Omo Research Expedition Howell led between 1966 and 1974 was perhaps his most important accomplishment. So many data were recovered for integrating the hominid Plio-Pleistocene fossil record with archeology, vertebrate paleontology, and geology that years have been required for the completion of work describing those findings. For example, Howell, the vertebrate paleontologist, has published no less than 15 papers on the non-hominid faunas. Nevertheless, Howell has been able to coordinate that project while moving on to further projects and sites. Always, whatever data were required to interpret the human evolutionary career, Clark Howell has made it his goal to study and combine such data into a coherent picture of the past. His influence on a whole generation of colleagues and students is obvious.

The Authors

Geoffrey Pope received his Ph.D. degree from the University of California at Berkeley under Howell's direction. He has held positions at the University of Illinois and William Paterson College. He has worked extensively in Asia on the stratigraphy of hominid sites and currently is continuing a series of excavations in China. That work is reflected in this volume's review chapter on early cultural and biological evolution in Asia that he has co-authored with Susan G. Keates of Oxford University. Pope opens this Festschrift with an exposition on Howell's perspective and diverse influences on the field.

John Fleagle is Professor of Anatomical Sciences at the State University of New York at Stony Brook. He edits the journal *Evolutionary Anthropology* as well as carrying out a wide variety of further editorial assignments, and has authored three major textbooks. He has researched the functional anatomy and paleontology of diverse primates and hominids from the Eocene to the Recent, conducting paleontological fieldwork in Egypt, Ethiopia, and Argentina and writing extensively on the early evolution of higher primates in Africa and South America. He served as a faculty colleague of Howell's at the University of California at Berkeley while a Visiting Professor. Fleagle's chapter conveys the latest thinking about higher primate origins, for which he is well disposed having been the co-organizer of a major Wenner-Gren sponsored conference about early Anthropoids at Duke University in 1992.

Fleagle first met Howell in person at a Wenner-Gren dinner lecture in New York in the late 1970s. "Other than that the topic was neandertals, the food was excellent, and the libations were abundant, I remember little else of the occasion. My real acquaintance with Clark has come from working with him on the Science and Grants Committee of the L.S.B. Leakey Foundation since 1985, and from a delightful semester I spent teaching in Berkeley in the spring of 1986. During that semester, Clark and I cotaught a course in Primate Evolution and I had the opportunity to experience Clark's tremendous knowledge and insight into all aspects of paleoanthropology (and most everything else in this world) as well as the extraordinary resources of books, reprints, and fossil casts that he has put together and made available to his students and colleagues. Since that time I have had the pleasure of working with Clark on a regular basis through the Leakey Foundation. He is a person of remarkable generosity and open mindedness whose wisdom and insights I have come to appreciate more and more each year."

Dennis Etler began his tenure as the resident "Asian hand" in Clark Howell's laboratory at U.C. Berkeley in 1985 and completed his doctoral dissertation, a detailed description and analysis of middle and upper Pleistocene east Asian hominids, in 1993. Consistent with Howell's broad-based approach, Etler's research interests range all of the way from the Asian fossil primate record, to Chinese Paleolithic archeology and modern human variation in China. Etler has been very active in forging collaborative research ties with Chinese paleoanthropologists and Paleolithic archeologists. He helped initiate the first joint Sino-American Paleolithic excavations at the site of Donggutuo in the Nihewan Basin as well as collaborative projects with the Yunnan Provincial Museum on the fossil apes of the Yuanmou Basin. He worked with the Hubei Archeological Institute on the middle Pleistocene hominids of Yunxian presented in this Festschrift, in collaboration with Li Tianyuan. Etler accompanied Clark Howell on his 1988 trip to the Yuanmou Basin of Yunnan, Howell's first visit to China since his trail-blazing National Academy of Sciences-sponsored visit in 1975. Etler observes "I am constantly amazed by the breadth of knowledge that Clark displays about all regions of the world. He is as much at home in China as in Europe or Africa. Clark

Clark Howell with Kenneth Oakley (middle) and Phillip Tobias (right) at the University of Chicago in the mid 1950s.

Howell (seated) with the Abbé Breuil (far right) and Ralph von Koenigswald (far left) at the Neandertal Centenary in Düsseldorf in 1956.

has done an incredible amount 'behind the scenes' to promote international cooperation within the discipline. His influence will be felt far and wide for generations to come."

Etler's chapter with Ciochon reviews past primate diversity on a continent by continent basis. They argue for a reinterpretation of the emphasis on Africa and Europe as earlier primate homelands by emphasizing the old and describing the new paleoprimatological discoveries in Asia.

Walter Hartwig began his graduate study with F. Clark Howell in 1986 and completed his doctoral thesis on the comparative morphology of New World monkey crania in 1993. He was drawn to study in Howell's laboratory at Berkeley because of the opportunity to research both the archeology and physical anthropology of early humans under a single mentor. Before long, however, the many doors of inquiry Howell opened led him to research the natural history of South American primates. "It might seem strange," Hartwig says, "studying New World monkeys under a paleoanthropologist, but my experience only further reflects the boundless intellect of F. Clark Howell. His network of colleagues, his personal library, and his influence on students range as widely as his cosmopolitan scholarship."

Monte McCrossin is a doctoral candidate at Berkeley and one of the last of Howell's students there. He has conducted extensive fieldwork at the Middle Miocene locality of Maboko Island as well as in the Western Rift valley. His chapter with Benefit details the increased understanding of early hominoid and cercopithecoid phylogeny emanating from analysis of such extinct genera as *Kenyapithecus and Victoriapithecus.*

Brenda Benefit is Assistant Professor of Anthropology at Southern Illinois University. With her spouse Monte McCrossin she has directed the recent continuation of fieldwork at Maboko Island as well as other researches on cercopithecoid systematics.

Andrew Hill is Professor of Anthropology at Yale University as well as Curator of Anthropology in the Peabody Museum of Natural History at Yale. Hill's research focuses on the evolutionary history of the Ethiopian fauna and environments over the last 25 million years, particularly that of hominids. He directs the Baringo Paleontological Research Project, which explores a 16 million year long fossiliferous sequence in the Tugen Hills of Kenya, and he also works on late Miocene sites in the Arabian peninsula. He is concerned with problems in taphonomy and site formation, particularly modern land surface bone assemblages and hyaena dens. Andrew Hill first met Clark Howell over 20 years ago while a student of Bill Bishop at London University, just beginning to deal with African fossil sites. He has been very much influenced by Howell's multidisciplinary approach, his encouragement of paleoanthropological field

Howell, Glen Cole, and Maxine Kleindienst at Isimila in 1957 (courtesy of Glen Cole).

Howell in 1959 at the 4th Pan-African Congress on Prehistory. The photo was taken on a tour outside of Kinshasa, Zaire. Bill Bishop and Maxine Kleindienst are to the left of Howell, Ray Inskeep and Louis Leakey visible to the right (courtesy of J. Desmond Clark).

studies, and his recognition that new hominid specimens are very important in formulating ideas about evolution. But most important was "Clark's insistence on the importance of context, and the need to embed human evolution in the framework of the changing African fauna and environment."

Bernard Wood is Derby Professor of Anatomy and Chairman of the Department of Human Anatomy and Cell Biology at The University of Liverpool. He qualified as a Medical Doctor in 1970 but had already begun to conduct research into hominid evolutionary history while a medical student. He joined the Koobi Fora Research Project in 1968 at Richard Leakey's invitation and served as its coordinator for several years. His research interests embrace the taxonomic and functional analysis of early hominids with particular emphasis on the evolution of the masticatory apparatus.

Wood first met Clark Howell in 1968 at Louis and Mary Leakey's house in Langata, Kenya. "I was a medical student on my first visit to Africa; he had just returned from a successful field season at the Omo and was proudly displaying the hominid teeth and jaws that his team had recovered that year. Despite my being a neophyte, Clark took the trouble to explain the significance of the finds to me in a way that did much to foster my interest in hominid paleontology. When relationships between the research groups working at the Omo and Koobi Fora were at a low ebb, it was resolved that the groups could be drawn closer together if some joint research was undertaken. To this end I joined Clark in the analysis of the Omo ulna. The resulting publication has the distinction of being one of the least-cited papers in the field of early hominid paleontology, but its importance for me was far greater for it marked the beginning of a professional relationship and a personal friendship which I value greatly." Wood's chapter synthesizes aspects of our current understanding of the biology of the earliest hominids.

Phillip V. Tobias in 1951 joined the Department of Anatomy at the University of the Witwatersrand Medical School under Raymond A. Dart. He succeeded Dart to the Chair of Anatomy in 1959 and was Head of the Department for 32 years (1959-1990). Professor Tobias has studied and described hominid fossils from Indonesia, Israel, Libya, Kenya, Tanzania, South Africa, Zambia, Zimbabwe and Namibia. His work on the Olduvai hominids led to three major monographs. In 1966 he initiated a long-term excavation at the dolomitic limestone cave deposit of Sterkfontein, near Johannesburg; this has continued for over a quarter of a century, making it the most sustained excavation of a single site in the world.

Tobias first spent time with Howell when Howell paid his first visit to the Department of Anatomy at the University of the Witwatersrand in Johannesburg in the early 1950s. "He came to examine some of the specimens in our department as well as in the Transvaal Museum, Pretoria. I remember asking him what he thought of the '*Telanthropus*' mandible from Swartkrans. He looked at it and with typical caution said 'I am not sure what it is but it is not an australopithecine jaw.' Of course, years later, it turned out to be a mandible most probably of *Homo erectus*. Subsequently, we met on many occasions. In 1956, I was a guest in Chicago of the Department of Anthropology headed by Sol Tax and Sherry Washburn and, of course, Clark was the most vigorous young member of Sherry's team. Subsequently we met again on several visits to Berkeley as well as on the Research and Grants Committee of the L.S.B. Leakey Foundation of which he was later Co-Chairman and I for many years a member."

"Clark's impressive work on the Neandertalers and his excavations in Spain were among his remarkable earlier achievements. It struck me that he, of all people, most deserved to find an early hominid—yet it seemed to elude him. I was very sad when no hominid skeletal material came into his hands during his outstanding excavation in Spain, although the faunal and cultural remains were stunning. His work in East Africa, and especially in the Omo, was another major chapter in his research career and there he had the good fortune to discover hominid teeth in those ancient deposits at the northern end of Lake Turkana. One of his masterly achievements was his great chapter on the Hominidae in the Cooke and Maglio book on African vertebrates. That was a fine achievement, following, one might say, in the tradition of Le Gros Clark's masterful synthesis, *The Fossil Evidence for Human Evolution*."

Ron Clarke writes "In the field of paleoanthropology, there are people whose expertise covers physical anthropology or paleontology or geology or archeology or human anatomy, but there are few indeed whose expertise covers all of those specialties. One of the few is F. Clark Howell, and so it was in the appropriate environment of Louis Leakey's Centre for Prehistory and Palaeontology that I first met him in the mid 1960s. It was indeed "The Centre" because from that group of buildings next to Nairobi's National Museum, expeditions set out to various parts of East Africa on archeological and paleontological field trips and brought their new discoveries back to the Centre. I was employed there by Louis Leakey as his archeological and paleontological assistant and my specialties were the cleaning, reconstruction and casting of hominid and other fossils. Thus it was one day in 1968 Clark Howell came back to Nairobi from the Omo in Ethiopia and handed to me two magnificent mandibles with the request that I clean and cast them for him. They were no ordinary mandibles. The one (Locality L7A-125) was so massive that it stretched one's credulity to the limit, yet its incisors were crammed into a minute space between the canines. I and Phillip Tobias had just recently created a mandible to fit the OH5 (*Zinjanthropus*) cranium. It had been so massive that some people had thought it an exaggeration. Yet here now was Clark's mandible to demonstrate the validity of our reconstruction. The second mandible, from Locality 74A-21, was less massive but very deep in the corpus and with a fourth premolar that looked like a huge molar.

"I was fortunate the following year (1969) to be sent by Louis Leakey to South Africa to look for the first time at the *Australopithecus* and *Paranthropus* fossils. In these latter specimens, I saw the same morphology, though somewhat less massive, that I had seen on Clark's mandibles and the crania were morphologically akin to OH5. Yet there was one supposed *Paranthropus* cranium that to my eye did not belong in that genus. It was SK 847 that, with its prominent nose, steeply rising frontal, light supraorbital torus and low cranial base, clearly belonged to the genus *Homo*. I immediately showed my discovery to Clark Howell who was also visiting the Transvaal Museum at that time, and to Bob Brain. They had no hesitation in agreeing with my observation and when I picked up the other known *Homo* palatal fragment, SK 80, to compare with the SK 847 palate, I found that they fitted together. In other words, two specimens that had been classified into two separate genera were actually one individual!"

Clarke counts himself fortunate indeed that Howell was there and that he and Bob Brain co-authored with Clarke a paper for *Nature* on this important fossil. It was for Clarke the beginning of direct involvement in research on early hominid craniofacial anatomy. Currently he is with the Palaeoanthropology Research Unit of the Department of Anatomy, University of the Witwatersrand, and is responsible for reconstructing much of our knowledge of australopithecine cranial morphology.

Clark Howell
with his wife Betty
at Torralba in 1961.

Yoel Rak received his B.A. from the Hebrew University in Jerusalem, began graduate work in the Anatomy Department of Tel Aviv University, and in 1981 received a doctorate from the Anthropology Department of the University of California, Berkeley. "It was Professor Clark Howell who brought me into the anthropology lab and introduced me to early hominids. With Professor Howell's enthusiastic support, I became engaged in the study of the morphology and architecture of the australopithecine face and ultimately wrote my dissertation on that topic." Since 1981 he has held a teaching position in anatomy and human evolution at the Sackler Faculty of Medicine of Tel Aviv University and a research position at the Institute of Human Origins in Berkeley. His interests include East and South African Plio-Pleistocene hominids, as well as the European and Middle Eastern Neandertals and early *Homo sapiens*.

Pelaji S. Kyauka is Principal Curator, National Museums of Tanzania (Natural History). Currently a graduate student in the Department of Anthropology, Laboratory for Human Evolutionary Studies, University of California at Berkeley, he plans to finish by the end of 1993. Born and educated in Tanzania, Kyauka obtained a B.Sc. (Hons) in 1976 and M.Sc. (Biol.) in 1982, at The University of Dar es Salaam. Research interests include human paleontology and growth and development of the early hominids.

Henry M. McHenry is Professor of Anthropology at the University of California, Davis. He received his Ph.D. in 1972 from Harvard University. Although he has occasionally raised his gaze and studied the craniodental fossils, most of his work has concentrated on the postcranium of the Plio-Pleistocene hominids. He writes that "When I was a graduate student at Harvard, one of my professors enjoyed referring to people with this phrase 'he has a lot to be modest about.' Perhaps he enjoyed the nervous chuckle this evoked among the students. Clark Howell is aberrant in this respect. He is modest but he has little to be modest about. He is, as I overheard one English colleague exclaim when encountering him unexpectedly at the British Museum (Natural History), 'the great man.' But he seems to be utterly without pretention. For example, he referred to his magnificent 94 page chapter in the Maglio volume (Howell 1978) as 'just a literature review' when in fact it is probably the most useful and insightful document of the decade. When he showed me the beautifully preserved Omo ulna (Locality 40), the first thing he pointed out was the tiny bit of styloid process that was missing." Russell Ciochon and Donald Johanson spent three weeks excavating at L40 looking for that missing styloid.

Russell H. Tuttle is Professor of Anthropology, of Evolutionary Biology, of the Biological and Social Sciences in the College, and in the Morris Fishbein Center for the History of Science and Medicine at the University of Chicago. He conducts field and experimental laboratory studies pertaining to the evolution of human and nonhuman primate morphology, locomotion, and other behavior. He is also engaged in paleoanthropology, particularly the evolution of bipedalism and of the human hand, and the history of theories on hominoid evolution and of social prejudice in physical anthropology.

Of his relationship to Clark Howell he says: "In 1964, fresh from the Department of Anthropology of the University of California, Berkeley, I came to the University of Chicago, and soon joined Clark Howell to co-teach the Human Career and, along with Clark, Al Dahlberg, Chuck Merbs, Ron Singer, and Charles Oxnard, to advise graduate students in a wide range of primatological, paleoanthropological and human biological research problems. Clark was a remarkably accessible, inclusive, and encouraging colleague for students and junior faculty members. I particularly recall a resplendent dinner party at his home in 1968, after which he, Al Dahlberg and I retired to the kitchen, where we lingered over his recently collected hominid treasures from Omo, Ethiopia. We never cease to regret having lost him to my alma mater."

Francis H. Brown received his Ph.D. in Geology and Geophysics from the University of California, Berkeley. He is Professor of Geology and Geophysics and Dean of the College of Mines at the University of Utah. Brown has worked in East Africa on various sites since 1966. He began dating of the Turkana Basin sequence in 1966, and still is not done. His chapter in this Festschrift is a detailed history of the efforts to radiometrically date the interbedded volcanics of the Turkana Basin. More recently his work has concentrated on correlation of sites over large regions of East Africa through analysis of glass from volcanic ash layers. In this way, stratigraphic ties between the Turkana Basin, the Middle Awash Valley, the Ethiopian Rift Valley, the Lake Albert region, and the Baringo Basin have been achieved.

Press release photo of Howell in 1969 holding the Omo specimen L7A-125 with the cranium of Olduvai Hominid 5 (courtesy of University of Chicago).

Howell working on the L894 *Homo habilis* specimen at Omo camp in 1973 (photo by Hank B. Wesselman).

Frank Brown first met Clark Howell in 1966 when Howell was on a one year visiting position at the University of California, Berkeley. Brown was in the first year of his graduate work in Geology and Geophysics at Berkeley. At that time Howell was formulating plans for work in the Lower Omo Basin, Ethiopia, and asked the young Frank Brown to go to southern Ethiopia alone in the Fall of 1966 to investigate whether there were datable volcanics. This first reconnaissance visit began Brown's life-long research in the Turkana Basin.

Henry P. Schwarcz is Professor of Geology and member of the Department of Anthropology at McMaster University, Hamilton, Ontario, Canada. His interests include dating of archeological sites by ESR and uranium-series methods, and isotopic paleodiet studies. He received his Ph.D. in Geology from Cal Tech in 1960 and specialized in isotopic geochemistry until the mid 1970s when he became interested in applications of geochemistry to archeology. In 1991 he was the recipient of the Geological Society of America's Archaeological Geology Award.

Schwarcz became acquainted with Clark Howell when working with him at Ambrona in the 1980s, attempting to date the site using U-series techniques. Later, Schwarcz turned to Howell for assistance in establishing the ESR dating lab at McMaster. Since then the two scientists have collaborated on application of ESR dating to several of the sites which Howell has worked on, including Torralba and Yarimburgaz.

Naomi Porat is a post-doctoral fellow at the Geological Survey of Israel in Jerusalem. She did her Ph.D. on the petrography of potsherds at the Hebrew University of Jerusalem, and then spent two years at McMaster doing research on ESR dating. She is now developing methods of dating Quaternary deposits in Israel.

Jean de Heinzelin has retired from the Royal Institute of Natural Sciences of Belgium in Brussels. He has written many classic monographs in geology, paleontology, and archeology. He states that "Clark has been a close friend of mine since the early fifties although more than often we lived in completely different spheres and did not meet for long times. Despite this, we never deviated from our mutual understanding. His invitation to participate to the Omo Research Expedition (1967-1973) I consider as a personal blessing, which led to the main professional reward in my career. I add, incidentally, that I was able to trace the genealogy of Clark and me up to the year 892, with astounding similarities."

Noel T. Boaz is Founder and Director of the International Institute for Human Evolutionary Research and Research Professor of Anthropology at the George Washington University. He has worked in African paleoanthropology for the past twenty years, having begun work on Clark Howell's Omo Research Expedition for his doctoral thesis in 1973. He has subsequently directed paleoanthropological research projects in Libya, Zaire, and Uganda. His current interests involve the investigation of paleoclimatic change and hominid evolution, a topic he is pursuing during the 1993-94 academic year as Meyerhoff Visiting Professor at the Weizmann Institute of Science in Rehovot, Israel.

Elisabeth Vrba is Professor of Geology and Geophysics, Adjunct Professor of Biology and Curator of Vertebrate Paleontology at the Peabody Museum, at Yale University. Before that she was for many years Head of the Department of Palaeontology and Palaeoanthropology and Deputy Director at the Transvaal Museum, Pretoria, South Africa, and directed the excavation project at the Kromdraai australopithecine site. She has a longstanding interest in evolutionary theory in general and human evolution in particular. She studied African antelope fossils associated with hominid finds for the past 20 years and still continues these studies. Her current emphasis is on paleoclimate change and the Plio-Pleistocene record in Africa.

Dorothy Dechant Boaz received her Ph.D. degree in Anthropology under the supervision of F. Clark Howell at the University of California, Berkeley. She first met Professor Howell in 1972, as an undergraduate volunteer in his laboratory, and participated in the Omo Research Expedition to Ethiopia in 1973. She has been a member of the International Sahabi Research Expedition to Libya, and the Semliki Research Expedition to Zaire. She was co-Founder and Assistant Director of the Virginia Museum of Natural History. Academic positions held include Lecturer in Gross Anatomy at New York University Medical School, Research Associate at New York University and the Virginia Museum of Natural History, and Research Scientist at the American Museum of Natural History. She currently holds the position of Research Associate at the University of California, Berkeley.

Raymonde Bonnefille is with the Laboratoire de Géologie du Quaternaire of the Centre National de la Recherche Scientifique in Marseille, France. She is a world-renowned expert on the palynology and paleoclimates of East African Plio-Pleistocene sites. She first met Clark Howell in 1969 in southern Ethiopia when she joined the French contingent of the Omo Research Expedition. Due to Howell's long-standing interest in paleoenvironmental reconstruction, she became an indispensable member of the Omo research team and participated in numerous subsequent expeditions.

Kathy Schick is co-director of CRAFT (Center for Research into the Anthropological Foundations of Technology) and a member of the Anthropology faculty at Indiana University. A Paleolithic archeologist, she has concentrated on the behavioral and geological processes of site formation and is especially interested in the origins and development of hominid technology through time. She began working with Clark Howell as a graduate student at Berkeley and was a member of his Ambrona research project in Spain.

Nicholas Toth is co-director of CRAFT (Center for Research into the Anthropological Foundations of Technology), Associate Professor of Anthropology and Adjunct Professor of Biology at Indiana University. He has specialized in the manufacture and use of early stone age tools and the behavioral and adaptive inferences that can be made from the paleolithic record. He began working with Clark Howell as a graduate student and was a member of the Ambrona research project.

J. Desmond Clark is one of the preeminent researchers in the world on earlier paleolithic cultures, especially in Africa. He has also conducted projects in India and China. He was a colleague of Howell at Berkeley for decades before they both recently retired but even prior to that time they co-organized symposia and co-authored some major works concerning the Acheulian. In 1992 J. D. Clark delivered the Distinguished Lecture to the American Anthropological Association.

Howell visits Koobi Fora in 1973 during the Wenner-Gren conference on the Lake Rudolf Basin (photo by Bernard Wood)

Howell excavating at Omo locality L345 in 1973 (photo by Hank B. Wesselman).

Richard G. Klein lectured on archeology and human evolution at the University of Chicago for many years before accepting a new professorship at Stanford University. His research focuses on the analysis of animal bones from stone age archeological sites, mostly in southern Africa. Clark Howell supervised his graduate work at the University of Chicago between 1962 and 1966 "and impressed upon me that paleoanthropology depends on the successful integration of archeological, human paleontological, and natural scientific evidence."

Leslie G. Freeman is Professor of Anthropology at the University of Chicago, where he received his Ph.D. in 1964. President of the Institute for Prehistoric Investigations in Chicago, he also serves as research director for the like named institute in Santander, Spain, and is Corresponding Academician of the Reial Academia de Belles Arts de Sant Jordi (Barcelona), one of Spain's royal academies. Freeman has been active in many paleolithic excavations in Spain and in the analysis of cave art at Altamira and other sites, as well as in studies of Spanish romanesque art and symbolism of the medieval Church.

"Clark Howell was my most impressive teacher at Chicago" writes Freeman. "I came to the University to train as a social anthropologist. During my first year (1959-60) as a graduate student, one of the core courses, called the Human Career, taught by Clark, the late François Bordes, Bob Braidwood, Art Jelinek, and Bob Adams, made an unforgettable impression on many students. It impressed me so strongly that I spent my summers excavating with Jelinek, Bill Ritchie, and others just for the fun of it, and with no idea that paleolithic archeology would become my career. I planned to do my doctoral research among the Aymara in Peru, starting sometime in 1963.

"At the annual Anthropology party in spring of 1962 Betty Howell said 'Les, Clark tells me you're coming to Spain with us this summer.' It was news to me! When I got near enough to ask Clark he said yes, he hoped I would join his crew, but he had simply become so busy he had forgotten to tell me. I wasn't quite sure whether I should be overjoyed or insulted. On the other hand, in those days, we didn't feel at all comfortable about contradicting our professors as students do today. So I agreed. Fortunately for me! Excavating with Clark, and discussing our finds and interpretations over supper (followed, of course, by cognac) was the best possible introduction to the real substance of paleolithic studies. Many of the most famous prehistorians of the day came through to see Clark's excavations, and of course he introduced me

Howell with Wu Rukang and Eric Delson in China at the IVPP in 1975 (courtesy of Eric Delson).

Clark Howell in his Berkeley office, 1986 (courtesy of Hank B. Wesselman).

to all of them. I got so involved with what one could learn about the paleolithic in Spain that I didn't return at the end of field season but stayed on to learn still more. I have been learning there ever since." Freeman gives a detailed review and defense of the seminal Torralba/Ambrona project in this volume.

Susan Antón is currently a Ph.D. candidate at U.C., Berkeley. She entered the program in 1987. "I bear the title of Clark's last graduate student. I am an NSF and Javitts graduate fellow, completing a dissertation on primate masticatory muscle architecture. I met Clark in 1984 while a sophomore at Berkeley and have resided in his lab ever since. Considering Clark a benign friendly force it was not until I ventured to Europe for fieldwork that I truly appreciated the significance of the Howell sphere of influence." Antón's published research includes work in craniofacial growth and development particularly in relation to artificial cranial deformation and congenital conditions, paleopathology, and biomechanics.

The present volume's editors, Robert S. Corruccini and Russell L. Ciochon have been "partners in crime" for over 20 years in primate biology projects resulting in two conferences, two books including 1983's *New Interpretations of Ape and Human Ancestry*, and 14 journal articles and book chapters. They began working together in 1971 at U.C., Berkeley. Ciochon has participated in paleoanthropological fieldwork in Burma, Vietnam, and China as well as in East Africa beginning with the Omo expedition in 1971, while Corruccini has conducted extensive dental anthropological fieldwork in India and in many museums. Currently Corruccini is at Southern Illinois University and Ciochon at the University of Iowa.

Acknowledgments

We thank the American Anthropological Association and its Biological Anthropology and General Anthropology sections for sponsoring the 1992 symposium bearing the same name as this volume, and Mr. and Mrs. Gordon Getty and the Leakey Foundation for hosting the dinner afterwards. Richard Nisbett compiled this volume's Index and Stephanie Coon assisted with final proof-reading. For suggestions, historical information, and photographs we thank C.K. Brain, Glen Cole, Dennis Etler, Karla Savage, Phillip Tobias, Russ Tuttle, M.J. Tyler, Hank Wesselman, John Yellen, the Editorial Board of Prentice Hall's *Advances in Human Evolution* Series, and last but not at all the least, F. Clark Howell.

RSC, RLC, June 1993

CHAPTER 1

The Howellian Perspective

Its Development and Influence on the Study of Human Evolution and Behavior

Geoffrey G. Pope

> Paleoanthropology, broadly conceived, is concerned with investigations of the biological relationships and the evolutionary relationships of the Hominidae, and of the development among the Hominidae of capacities and capabilities for culture. The roots of the discipline reach into evolutionary biology and vertebrate paleontology on the one hand, and into traditional prehistoric archeology on the other.—F. Clark Howell, 1972.

The totality and scope of F. Clark Howell's scientific contributions cannot be summarized or even adequately addressed in a single chapter. It is with a profound awareness of that fact that I present what can only be an outline of the influence and impact of a scientist who is to a large extent responsible for founding the multi- and interdisciplinary science of paleoanthropology. The perspicacity of Clark Howell's vision of modern paleoanthropology is evident not only in the previous characterization, but also in the fact that this 1971 assessment of the field was in fact first published four years previously (1967). More than one generation later, this description of our objectives still constitutes the best summation of what it is that paleoanthropologists attempt to accomplish. Clark Howell has always believed that paleoanthropological investigations should be comprehensive, multidisciplinary, hopefully new, and above all, useful. This encompasses what can rightly be called the "Howellian" perspective. I would not be surprised to see the adjective Howellian used in the description of future works that are encyclopedic, comprehensive, and integrative in their treatment of a particular subject. Mayr (1982:856), in describing Darwin's contribution, wrote the following: "In most cases the development of major new concepts has not been due to individual discoveries, but rather to the novel integration of previously established facts."

This is also an accurate description of why it is that Clark Howell as exemplified by his work in Europe and Africa is one of the most highly regarded and respected paleoanthropologists in the world. His researches at localities such as Torralba, Ambrona, and especially the Omo Basin remain as icons of superbly executed multidisciplinary scientific investigations. Mary Leakey once remarked that Clark

Howell had never made a "big find." Another colleague, putting himself in the same category, remarked that like his own name, Clark Howell's name would fade because he did not make a "tangible" discovery (Day in Lewin 1987). Yet it is a testimony both to Howell's research methods and his tremendous capacity for international collegial organization that the Omo Basin research still remains the primary "yardstick" with which other Plio-Pleistocene paleontological and archeological localities must be compared and calibrated. There Howell and the team which he co-led not only established the chronological, paleoenvironmental, and fossil "facts," but he and his colleagues both integrated their findings (Howell 1968, 1976, 1978a; Howell et al. 1978a,b; Coppens et al. 1976) and produced a series of publications detailing the hominid fossils (Howell and Coppens 1973, 1974a, 1976a; Howell and Wood 1974), archeology (Howell et al. 1973a), mammalian paleontology-biostratigraphy (Howell and Eck 1972; Howell and Coppens 1974 b,c, 1976b; Howell and Petter 1976; Howell and Grattard 1976; Howell et al. 1978 a,b) and geology (Howell et al. 1973b, 1978a) of the Omo. There is little prospect of Clark Howell's name or contributions fading in even the distant future of this new science.

Like all paleoanthropologists, Howell's science has always been primarily concerned with the past. However, with the rise of intense debate over modern human origins, paleoanthropological evidence has begun to have a direct impact on the interpretation of modern human behavior, an objective that has traditionally been the purview of the social sciences. Unfortunately, modern interpretations about such subjects as the structure, function, and very nature of extant human cultures have rarely paid more than lip service to the relevant evolutionary "facts" that the Howellian approach to the past can add to an understanding of human behavior. In the future, it is a Howellian type of integration, drawing on perhaps a dozen interrelated and disparate fields, that will have to characterize future explorations and explanations of human behavior.

"Mega-explanations," nomothetic pronouncements and prime-mover postulations of the roots of human behavior, have been proposed repeatedly in both paleoanthropology and all of the social sciences. They range from psychoanalysis through bipedalism, tool use, behaviorism, and Marxism to sociobiology. All have been or will eventually be abandoned as inadequate characterizations of the vast variation of human biology, behavior, and culture. It is this author's belief that a principal reason for the inadequacies of previous explanations has been a lack of awareness of the evidence for the prehistoric evolution of human biology and behavior (but see Barkow, 1989, for one of the few integrative approaches to human behavior). Such evidence is absolutely crucial to any convincing explanation of human nature, experience, and life. The most relevant and reliable evidence about human evolution will be exactly that which has been assembled in a Howellian way.

THE GROWTH OF THE HOWELLIAN PERSPECTIVE

As a signalman in the Pacific Theater (1944–1946), Clark used his time at sea to read everything he was able to acquire. During this period W.W. Howells' "Mankind So Far" (1944) would turn out to be one of the most influential acquisitions of the young seaman. Later, in the 1970s, Clark was to tell me that he still read constantly, but "Never fiction, there's no time for that." His interest in the past began even earlier, when he was a junior in high school in Kansas. First H.G. Wells's "A Short History of the World" initiated an interest in history and later "Men of the Old Stone Age" by Osborne led him toward the contemplation of prehistory.

After the end of World War II and his discharge in 1946, Clark made his way to the American Museum of Natural History in New York "... to see if someone like me could have a career in anthropology." There he had the chance to meet and spend a few weeks with George Gaylord Simpson, Ralph Von Koenigswald, and Franz Weidenreich. The exposure at one location and one time to the most prominent of evolutionary taxonomists, a veteran fieldworker, and a preeminent anatomist had a tremendous influence on Howell. "The experience had a profound impact on my own hope for a future career in anthropology" (Howell 1992). It is my opinion that Weidenreich had an especially important impact on Clark Howell, not only in terms of the encyclopedic thoroughness of Clark's subsequent anatomical descriptions and comparisons, but also in his taciturn manner and attitude toward teaching. Clark later described Weidenreich as the kind of person who did not teach so much as he allowed you to "observe the master." I have little doubt that this observation influenced Clark's own teaching style for decades to come. Clark's distaste for formal classroom teaching is well known and self-acknowledged. And yet he has taught so many, so well.

His experiences at the American Museum of Natural History also fanned his initial interest in the Solo hominids that were the subject of his master's thesis (University of Chicago, 1951). This in turn led to an interest in Neanderthals (Howell 1957, 1962a) to which the Solo fossils were once commonly compared. In the short space of six years (attending class 12 months a year, except one year when his mother was ill) at the University of Chicago, he went from navy signalman to the recipient of a Ph.D. (1953). By this time, Clark had published one abstract, five papers, and three reviews largely concerned with Neanderthals and other paleoanthropological subjects. After graduation, he embarked on his first study tour in Europe and conducted excavations at Abri Pataud in France. With the exception of the year he married Betty (1955) and thirty years later, 1985, he traveled abroad on at least a yearly basis. His firsthand familiarity with nearly all of the international fossil hominid material provided the bases of his later and highly influential synthetic works ("The Villafranchian and Human Origins" 1959, 1962b; "Early Man" 1965, revised 1970, 1980; "Recent Advances in Human Evolutionary Studies" 1967, revised and appended, 1972; "Hominidae" 1978; "El genero humano" 1982, "Evolution of the Hominidae in Africa" 1981; "Variabilite chez Homo erectus, et le probleme de la presence de cette espece en Europe" 1986), and continues to be indicative of his broad interests in the human past. These works have set a standard for thoroughness that subsequent works by others have yet to equal.

Another early and important influence on Clark was his exposure in the Spring Seminar Series at Chicago organized by S. L. Washburn. These talks brought the new synthesis of evolutionary biology, paleontology and ecology to the young Howell. Paleontologists such as George G. Simpson, T. T. Patterson, and Alfred Romer and biologists such as Theodosius Dobzhansky, Sewell Wright, and T. Dale Stewart were devising, integrating, and expanding the synthesis on which current work is based and F. Clark Howell was there to absorb it.

Soon after this, as an instructor of anatomy at Washington University in St. Louis, Clark showed the first indication of the immense breadth of his interests, publishing papers whose subjects ranged from paleolithic archeology to mammalian paleontology. His geographic interests not only encompassed Europe, but also expanded into the Middle East and Africa.

In 1959, the year that "Zinjanthropus" was discovered, major papers on paleoanthropology of the Levant and the "Villafranchian and Human Origins" appeared in the *Transactions of the American Philosophical Society* and in *Science*, respectively. Throughout the 1960s Clark's work became increasingly more synthetic. In addition to editing a multidisciplinary volume with F. Bourliere (1963), he published (with Maitland Edey) the immensely influential volume "Early Man," which many of my generation read as students. It was republished in various revised editions throughout the sixties and seventies.

Howell's meticulous and encyclopedic approach to paleoanthropology and his often unspoken, but always evident, conviction that nearly every natural science can directly contribute to an understanding of hominid evolution has fostered an intellectual climate that nurtures the multidisciplinary, but especially the interdisciplinary, approach that has been the hallmark of his life's work. In 1960 he coauthored "Human Evolution and Culture" with Sherry Washburn. In 1991 he published a paper entitled "The integration of archeology and paleontology: With reference to the origin of anatomically modern humans." I cannot think of any other titles that more precisely describe what paleoanthropologists (and it is hoped social scientists) will be doing in the year 2000. Few workers in paleoanthropology are respected as both archeologists and human paleontologists. Clark Howell is one of those few.

A major part of Howell's continuing work expanded in the course of what can only be described as a "Golden Age" (of science if not always collegial harmony) of paleoanthropology at U.C. Berkeley. In the 15 years between 1970 and 1985 leading scholars in all of the major subdisciplines of paleoanthropology were assembled at a single institution. Some of these leading figures included Sherry Washburn and Phyllis Dolhinow (primatology), Donald Savage (primate paleontology), Vincent Sarich and the late Allan Wilson (biochemistry), the late Glynn Isaac and Desmond Clark (paleolithic archeology), Garnis Curtis and Richard Hay (geology and geochemistry), Tim White and Donald Johanson (human paleontology), and Clark Howell. It is safe to say that although many prestigious groups are now concentrated at other universities, never has there been a more preeminent assemblage of paleoanthropologists and earth scientists at a single location.

The extraordinary opportunity that Berkeley provided could not help but reinforce Clark's multidisciplinary-encyclopedic approach. As a graduate student at Berkeley, there was no formal required curriculum, but Clark made it known that in addition to classes in anthropology, geology, and paleontology; biology and anatomy were very "useful." A number of controversies in human evolutionary studies punctuated Clark's Berkeley years. Sarich's biomolecular work (Sarich 1974) was accomplishing the demise of *Ramapithecus* as a basal hominid; the almost evangelical cladists were rearranging the taxonomy of almost every paleontological group, especially primates (Delson 1977); morphometricians were making computers an integral part of paleoanthropology (Corruccini and McHenry 1980; Corruccini 1978; Oxnard 1973); the single species theory was in its besieged death throes (Wolpoff 1973); *Homo habilis*, and later *Australopithecus afarensis* (Tobias 1991; Johanson et al. 1978) were slowly being "born" and accepted; and punctuated equilibrium was being heralded (Eldridge and Gould 1972) and denounced (Cronin et al. 1981) as the dominant mode of human evolution. But the most vitriolic controversy (at least at Berkeley) erupted over the age of the KBS Tuff (cf. Johanson and Eddey 1981; Johanson and Shreeve 1989). Most of the rifts created during this period have closed, but I think it is important to emphasize that the overall effect on students was to reenergize an interest in human evolution that endless hours of class and laboratory work can sometimes drain. Of all these controversies the KBS debate most visibly exercised Clark. Although he at first attempted to act as a mediating force in the disagreement over the age of the KBS Tuff (Lewin 1987), when he learned from a colleague that the KBS Tuff had never been "walked out," he displayed a look of dismay and irritation that I have never seen before or since. He later told Roger Lewin: "I felt Richard wasn't capable of making certain judgments, unless he had all the evidence" (1987:251). Not only does this characterize the way in which Clark reaches conclusions, but the entire KBS controversy is an excellent example of what happens when one conducts fieldwork in a non-Howellian manner. The core of the Howellian research approach has always been a meticulous and methodical concentration on simple basic field facts. As for the other debates, he followed them with concerted, but stoic interest and waited for the facts to come in.

In the Berkeley years, Clark's students, myself included, soon came to realize that it was nearly impossible to pin him down on any grandiose characterization or nomothetic pronouncement about the past. I remember once giving a seminar presentation about the "Neanderthal problem" and its possible solutions. These included "genetic swamping," "cold adaptations," "*in situ* evolution," and perhaps 10 other theories that I had scoured from more than a century of the scientific literature. When I concluded, I asked him which interpretation he thought most likely. "None of the above" he answered and then proceeded to spend the rest of the hour elaborating on "other opinions" ("not necessarily my own") before dismissing the seminar.

Clark's lack of dogma was sometimes frustrating to graduate students who had come to study at the "holy of holies" (as Yoel Rak once put it) and learn "the truth." I think that for Clark, the most important thing has always been the consolidation of useful knowledge and the establishment of cautious deductions. Howellian "dogma" (and there really is no such animal), when presented in an almost legalistic terminology, indicates what can reasonably be said on the bases of our always imperfect state of knowledge. Many scholars in paleoanthropology use the term " heuristic." Clark Howell has always been committed to this approach. From the study of carnivores to archeology he has never thought of another way of doing things. He employs this approach whether assessing the date of a particular find or summarizing what one can reasonably say about the geological, archeological, or paleontological evidence from a particular geographic region.

Clark Howell is not only a synthesizer of "facts" and reasonable, cautious conclusions about the past, but also an organizer of the propensities and energies of his colleagues. On more than one occasion he has expressed his opinion that successful research efforts (with the exception of funding) depended primarily on finding the right person with the right knowledge for the right job. This goal he realized in all his expeditions and research projects. Nowhere was this more evident than in the fruition of the Omo Research Project. His ability to "cope" with the often eccentric personalities of paleoanthropology is legendary. He has always brought a universal respect for anybody who assiduously pursued a career in the poorly paid and once fledgling discipline of modern paleoanthropology. Although he, like most of us, has occasionally been swept up in the personal and sometimes bitter conflicts that are the public trademarks of our field, the overall moderating influence of the way he treats colleagues, students, and research patrons (whom he frequently refers to as "angels") remains instrumental as a voice of moderation in what can sometimes be an acrimonious field. In an address to an audience assembled at the Senckenberg Museum commemorating "100 Years of *Pithecanthropus*" (1991, Howell in press), he noted the public scorn that Eugene Dubois had endured for his scientific views. Howell was derisive of those who had denigrated a man who had done so much, so early in the history of the study of human origins. For myself, this talk brought to mind Clark's story of his own early solitary days in the Omo when his perseverance and patient persuasion enabled him to first explore the now famous fossil localities.

His ability to synthesize disparate kinds of data and his appreciation of the construction of regional pictures has always been apparent. Nowhere was this emphasis more apparent than in his retort to Conkey's (1987) criticism of processual archeology, which Clark summarized as an "almost overtly hostile" treatment of "analytic-synthetic efforts" (Howell 1988:373). In this same review, Howell went on to emphasize the now obvious observation that processual archeology will make little progress in the absence of adequate site-by-site, environmental, and regional analyses and the inclusion of biological data about the populations being studied.

The specific site "reanalyses" of Ambrona and Torralba (Klein 1987; Binford 1987) occasioned his most vehement response to one of the rare criticisms of his researches. It was one of the principal founders of the processual school, L. Binford, who provoked a response in which Howell (1989) dissected and destroyed at every level (in my opinion) what he characterized as the "revisionist" interpretations of Binford and even those of his former student R. Klein (see also Freeman, this volume). Howell pointed to the use of simply wrong geological and taphonomic data and the convoluted statistical analyses often employed by Binford in the denigration of other people's work. Howell further decried the dismissal of these sites, and others' "potential to inform" (p. 594). His response to the attacks on Ambrona are far too detailed to be summarized here, but it is certain that this single book review which appeared in the 1989 *Journal of Human Evolution* constitutes the longest and most comprehensive review ever published there.

Restrained caution has always been a hallmark of Clark Howell's work. As a student of Asian paleoanthropology, I will always be appreciative of the contrast between Movius's (1944) characterization of the Far East as an "isolated backwater" of "cultural retardation" and Clark's understated observation that "the cultural associations [with *Homo erectus*]are surely different in eastern Asia" (Howell 1978:225).

Recently, one of my own students wrote to Clark inquiring about the age of a particular archeological discovery. Clark immediately wrote back that he certainly did not question the "presumed/ potential age of the specimens." In frustration, my student called me and asked what that meant. "Exactly what it says," I said. Such are the nature of "Howellisms." Yet in the explicit combination of our presumptions and potential meanings of the fossil and archeological record, Clark Howell's work has encouraged a critical and cautious approach that is the mark of useful observational science. Clark continues as a subtle and wonderful enigma of stoic practicality and inspiration who has always stayed well informed on both the less concrete and theoretical developments in human evolutionary research and the "hard facts."

He once had been obliged to read my predoctoral statement on the then newly polarizing "paradigm" of sociobiology. In the end, he admitted that he was unwilling to finish reading it because he did not do "that sort of thing." On another occasion, as he attempted to begin a lecture on primate evolution, a cacophonous political demonstration erupted outside the classroom window. After patiently waiting for several minutes, the always cool Professor Howell raised his voice slightly and said "Anyone that has really thought about things, sees that the really important things are biological." He then proceeded, without interruption, to read the 30 or so odd anatomical features of the primates.

This kind of epiphany is completely characteristic of the diversity of intellectual descendants that Clark has contributed to the field. It is safe to say that none of them is as diverse as Clark himself, but their ongoing individual contributions permeate the proliferating subfields of paleoanthropology. Although diversified in our geographic, substantive, and theoretical interests, we share a common recognition of the relevance of a multitude of disciplines to the study of paleoanthropology. Like Clark himself, few of us have attempted to tie our paleoanthropological research in with sociocultural anthropology. I think the time for doing exactly this has now arrived and a major reason is exactly the kind of foundation that Clark's work has established.

Unlike Washburn, Isaac, and a few others, Howell never emphasized and usually did not include in his works the implications that paleoanthropological behavior holds for modern human social behavior. As mentioned earlier, my time at Berkeley saw the rise of the newly proclaimed "paradigm" of sociobiology. This interpretation of the roots of human behavior quickly became a flashpoint among graduate students and faculty alike. But Clark Howell remained steadfastly concerned with the hard evidence of anatomy, geology, geochronology and paleontology. During this same period, Owen Lovejoy gave a lecture detailing his now well-known Provisioning Theory. When Lovejoy called for questions after his talk, one student who obviously did not like the allegedly sexist implications of the theory asked what his theory had to do with modern humans. Lovejoy replied simply, "Nothing." Clark did not attend the talk, but I suppose this answer would have met with Howell's approval, if only because it averted the task of dealing with the more speculative aspects of paleoanthropology.

Yet it is this aspect that I wish to raise in this chapter. There is, in my opinion, a considerable need to point out to our colleagues in the social and biological sciences a number of the discoveries and issues that have direct and profound implications for our understanding of modern human behavior. This is especially true for the majority of anthropologists who work in sociocultural anthropology. Additionally, what we have found out about human evolution is also of immediate importance not only to psychologists and neuroanatomists, but also to the engineers of the future, ranging from city planners to world leaders. The implication of what paleoanthropology has to say is also immediately relevant to the inhabitants of the present. Although Clark Howell may not approve of this attempt to proselytize the social importance of paleoanthropological research, his approach to the past is the very ground from which this perspective springs. All human cultures have origin myths, and paleoanthropology has succeeded in dispelling most of them, including most of the ones that paleoanthropologists consistently create. The scenarios and interpretations that have endured and will continue to endure are only those that withstand multidisciplinary examination.

PALEOANTHROPOLOGY AND THE SOCIAL SCIENCES

Archeology

One important potential link between paleoanthropology and cultural anthropology and other social sciences is paleolithic archeology, which is a subfield of both paleoanthropology and cultural anthropology. Unlike social scientists, archeologists have no direct access to the subject of their study (Trigger 1989). Trigger points out that even historical studies with the benefit of written records cannot, for instance, agree on the cause(s) of the rise and fall of the Roman Empire. In archeology and human paleontology, the problem of assuming uniformitarian social and resulting physical-cultural residues is complicated by the problem of determining when the biologically based capacity for culture reached a level of human complexity. The degree and temporal depth to which we can project modern behavioral complexity and resilience into the past has been the subject of concerted criticism for the past two decades and has taken on the dichotomous polarity so familiar to paleoanthropology. The time depth that one attributes to modern human behavior has a direct impact on the evolutionary significance and malleability of modern human behavior. This in turn directly affects our views of what has been variously called "human nature," "psychic unity," or what I prefer to call "cultural universals."

The "Binfordian School of Hominid Nature"(as it can be rightly called) strongly emphasizes the interpretation that the statistical analysis of taphonomic data indicates that recognizably modern human behavior is a geologically recent phenomenon confined largely to the latter half of the Late Pleistocene coincident to the middle and Upper Paleolithic (40 KYA), a small fraction of the known fossil record of the Hominidae. Such a demonstration of the recency of recognizably sapient human behavior has important implications for attempts to discuss the biological or social basis of modern human social behavior as it has been investigated in the context of the polarized nature–nurture continuum. Viewed in its most simplistic terms, the time depth of any particular human behavior should be directly related to the biological and genetic substrates that have evolved to support and replicate a particular behavior in successive generations. Behavioral patterns with less antiquity may be assumed to be more the result of sociocultural dynamics that are the theoretical focus of processual archeology.

Although the naivete of processual archeological studies of paleolithic data has now been abandoned, a belief in the recent advent of sapient behavior lingers (Klein 1992; Binford 1987) and has been adopted by neuroscientists (Kimura 1992). Opponents of this stance have labeled the residue of the Binfordian conclusion as "Behavioral Creationism" for its insistence that modern behavior is a distinct, archeologically detectable milestone that seems to have few demonstrable stages of development in the prior archeological record.

Indications of the lack of time depth for human (see later discussion) behavior have been based to a large extent on Binford's reanalysis of a number of internationally known localities (Torralba, Ambrona, Zhoukoudian, and Neanderthal sites), which concludes on the basis of the co-occurrence of artifactual and faunal elements that sapient behavior (planning, efficient exploitation of resources, and human social behavior) is not indicated by the evidence (Binford 1987; Binford and Ho 1985; Binford and Stone 1986).

As noted previously, major criticism of the methodology of the Binfordian approach can be found in its highly questionable use of simple statistical analysis drawn from inadequate databases and the employment of standards of behavior based on ethnographic residues of highly specialized groups that are unrepresentative of the vast majority of modern humanity as a whole. What has emerged is a tautological orientation in which sapient behavior has been equated with indications of modern behavior represented by the artifactual residues left by highly specialized groups, which are themselves highly affected by historically recent contact with agriculturally based or dependent societies. In short the capability for culture has been confused with the actualistic paleocultural residues (cf. Mellars 1988 for a discussion).

The Binfordian viewpoint has also encouraged the notion that few ethnographic examples of sufficiently pristine hunter-gatherer cultures exist that have relevance to the interpretation of the paleolithic record. This self-reifying viewpoint has tended to cast serious doubt on the importance that a hunting and gathering way of life played in the development of the biological bases of human behavior. One of the most disturbing aspects of this viewpoint is that it justifies the already common tendency of most sociocultural anthropologists and sociologists to ignore the past.

Human Paleontology

Social scientists should be aware that, based on evidence from endocranial casts, paleoneurologists are in agreement that so far as it is possible to tell, an essentially modern asymmetrically organized brain exhibiting the major speech association areas dates to at least 2 million years ago (Falk 1987; Holloway and de la Coste-Lareymondie 1982). Furthermore, there is also archeological and experimental evidence (Toth 1985) that the uniquely human characteristic of handedness has a minimum and similar antiquity of at least 1.5 MYA. The same is true of recognizably symmetric tools. Taken together these observations indicate that recognizably human behavior including the *sine qua non* of humanity, language, has been in place for well over one million years.

There is, not surprisingly, also some disagreement that essentially human behavior reaches back into the Early Pleistocene (cf. Mellars 1988), but it has come from archeologists and linguists who fail to discuss the anatomical evidence as it is known from hominid cranial endocasts and modern neuroanatomy (Klein 1992; Cavalli-Sforza et al. 1988). It is interesting to note that those who have supported the idea of replacement and a recent origin of truly sapient behavior have done so on the basis primarily of one kind of evidence (i.e., biochemical, linguistic, archeological, chronological, or anatomical). In other words, their approaches to the question of modern human origins have been decidedly non-Howellian in the breadth of their disciplinary and geographic scope. Where truly pan-geographic and interdisciplinary approaches have been employed workers are in agreement that the replacement model for the origin of human complexity is untenable (cf. Pope 1988; Frayer et al. 1993).

Where the social sciences have dealt with the biology, anatomy, and antiquity of the brain, they have done so only with lip service or more disturbingly in a distorted vein that ignores what we know from the fossil, archeological, and primatological evidence. The psychologist J. Jaynes's (1978) theory of the bicameral mind, espouses the notion of evolutionarily recent interhemispheric communication as an explanation of religion and sophisticated communication. This theory displays an appalling ignorance of the evidence for vertebrate, but especially hominid, evolution. Even the more mainstream fields of biological psychology, which have achieved stunning successes in the past two decades, continue to explore the brain/mind dichotomy with little or no reference to the paleoanthropological data that is capable of at least establishing the antiquity, adaptive importance, and variation of the human brain. Such questions as sexual dimorphism, hominid paleoecology, and the antiquity of biobehavioral diversity directly impinge on attempts to understand the ethnographic and psychological present. At the very least, an awareness of such basic paleoanthropological knowledge of hominid bipedalism has multiple implications about how the brain might be organized for communication, perception, analysis, reproduction, and many other aspects of the mind that have been the objective of study of the social sciences.

The Human Past and the Human Present

Knowledge of hunter and gatherer biology and behavior, and ethological, ethnographic, and ethnoarcheological studies growing directly or indirectly out of problems raised by paleoanthropology have played very little part in the work of modern social scientists. The following examples suffice to emphasize the great number of important conclusions that need to be incorporated in fields such as psychology, sociology, and even civil engineering.

Child Development

It is now clear that if any higher primate is raised without the benefit of physical contact with other primates, it will exhibit markedly abnormal behavior as an adult (Harlow 1959; Harlow and Harlow 1961). Human beings raised in orphanages are significantly physiologically disadvantaged due to relative lack of visual and more importantly tactile contact with others of their own species. This disadvantage is automatically imposed on the occupants of orphanages by the inadequate numbers of caretakers at such institutions. Children raised without the benefit of physical contact have a much greater chance of experiencing emotional and learning problems as adults (Skeels 1966). Conversely it is also well established that enriched environments result in larger brains in many animals (Kolb and Wishaw 1990). The determination of what constitutes an enriched environment for young hominids may be intuitive; but with reference to what we know of our hunter and gatherer past, we know that movement exploration and interaction with others of our own species formed the daily environment of maturing hominids for millions of years. Furthermore it is now evident that such cultural universals as music and the appreciation of visual "art" (defined here as the mixture of color and form ranging from body adornment to painting) is mediated by specific areas of the brain. One of the interesting things about music and art is that they frequently reinforce and mediate group cohesion. They also frequently serve as avenues for the transmission of cultural information (education). It is no coincidence that almost any form of human social organization (i.e., tribes, clubs, teams, and entire nations) is associated with particular kinds of music (anthems), symbols (insignia), and colors.

The combination of an interspecific comparison of mammals and a knowledge of human paleontology makes it clear that there is a strong correlation between the length of maturation and intelligence (defined here as behavioral plasticity). Humans not only have the longest maturation periods of any other primates, but they also continue to play throughout their adult lives. The adaptive function of play has often been pointed to as the rehearsal of adult roles. While this is true in nonhuman primates, humans continue to play long after it serves any practical value for the practice and development of adult skills. This extraordinary extension of our childhood is paralleled by the extraordinary extension of human mating beyond fertility and procreation. Allan Dundes (1980) has maintained that organized sports (ie., football) are an expression of male sexuality. While there are serious flaws with this argument, it is true that (at least in Western culture) game terms are often used to describe sexuality. What analyses such as Dundes's miss is a biological component of human behavior that emphasizes sex as play and play as sex. Such a component has evolved as the result of millions of years of selection. The exact selection pressure that allowed this uniquely species-specific human emphasis to evolve remains unknown, but it seems likely that factors such as imprinting and group cohesion may have been the "prime movers" for our unique development of these traits. Enjoyment seems to be an integral part of truly effective learning.

I think much of what we already know about child development suggests very strongly that education should be enjoyable and as physical as possible. There can be little doubt that children, like other mammals, go through a developmental period when they are highly susceptible to imprinting. It is also well known that imprinting can be induced in adult humans, especially by deprivation and isolation. However, adult humans also normally enter or initiate periods of imprinting during the course of their natural lives. Adults "falling in love" alter their behavior in a way that can accurately be described as exhibiting the qualities and emotional states of childhood. The "pet names" and terms of endearment that lovers use are highly reminiscent and frequently identical to those normally seen in infant hominids. The biological importance and strength of enjoyment is accurately summarized in the aphorism that "Time heals all wounds." In fact, it seems that human long-term memory is very good at both recalling enjoyable memories and forgetting the details of unenjoyable experiences.

Sexuality

Monogamy and permanent pair bonding are the rarest forms of reproduction and social structure among primates. It is also arguably rare in modern humans, but among primatologists it is known to represent the exception rather than the rule in all but one nonhuman primate family (Hylobatidae). Until very recently in human history, most cultures tolerated and/or encouraged polygynous marriages with the number of wives being limited only by economic constraints. In all cultures the institution has two universal characteristics: It is public and it is economic. It usually signifies a publicly recognized change in sexual availability and activity. Nearly as common is the association of marriages with public forms of food sharing and gift giving. The function of marriage as a cohesive force in societies has been long noted. Furthermore the "primal horde" once prophetized by early anthropologists and social philosophers has never been discovered. We cannot say at this point if a polygynous mating system similar to that for chimpanzees ever characterized early hominids, but it may one day be possible to indicate that the formal recognition of sociosexual bonds is an integral part of human social organization, which is of considerable antiquity.

While the universal practice of marriage strongly suggests that this institution will continue as an integral part of human society, the rising divorce rates in both Western and non-Western cultures may underline another part of our biological past. From paleoanthropological data it is clear that current human longevity is probably more than twice what it was for 4 million years and that this technological change is influencing a biology that is not geared toward unusually prolonged (by any mammalian standards) pair bonding. Sociobiologists have of course other explanations for both divorce and marriage, but the simplest explanation may lie in the differences between a slowly evolved biology and a rapidly imposed technology.

Most human sexual activity is not related to reproduction, a fact that is unique among mammals and almost certainly associated with the lack of visually displayed estrus in human females (Washburn and McCown 1972; Lovejoy 1981). It is clear that among humans copulation serves a social function far beyond simple reproduction. Pair bonding and cooperation have often been implicated with our strange form of sexuality. At the same time it seems on the basis of paleoanthropological data that a nuclear or extended family group dates back to nearly 4 million years ago (Johanson and Edey 1981).

Paleoanthropologists are less certain about the ultimate antiquity of the sexual division of labor, but anatomical sexual dimorphism (body size) dates back at least to the time of the first australopithecines (Johanson et al. 1978). One thing that seems reasonably certain is that the division of labor along sexual lines with males as hunters and females as gatherers is old enough that consistent biological differences in the brain are manifest in the early postnatal years (Kimura 1992). Such differences are not only mediated by sex hormones, but also are consistent with what we know from archeological and ethnographic studies of hunters and gatherers.

Diet and Nutrition

Studies of the dentition of fossil and extant primates reveal that human beings are specifically adapted to an omnivorous diet, which represents a significant departure from the adaptations of other higher primates and other fossil hominids (Bunn 1981; Kay and Grine 1988). There is still considerable controversy about whether early hominids were primarily hunters or scavengers. However, there is a consensus that consistent meat eating and daily food sharing are unique (among primates) and integral parts of the pattern of extant humans and their fossil ancestors.

The diet of most hunters and gatherers as it is known from the archeological and ethnographic record almost always contains far fewer calories derived from meat in comparison with vegetable resources. The high incidence of contemporary stress-related diseases is mediated not only by our past dietary adaptations as hunters and gatherers, but also by the fact that modern stress often takes the form of problems that cannot be resolved through physical activity that utilizes stored cholesterol (Campbell 1974, 1976). Thus, although the instinctive mammalian "fight or flight" response remains as part of our mammalian legacy, fat and cholesterol accumulate to contribute to the cardiovascular diseases that are so prevalent in "developed societies."

Aggression

All primates and most mammals become aggressive as the result of competition for scarce resources (Eibl-Eibesfelt 1972). Whether in competition for mates, nutrients, or space, all vertebrates rely on at least periodic aggression for survival. Among humans aggression is known in all cultures whether in a ritualized or physical form. Nomothetic pronouncements declaring our ancestors to be territorial "killer apes" (Ardrey 1961, 1966, 1976) have not been supported by a single piece of evidence drawn from the

Plio-Pleistocene paleoanthropological evidence. There is widespread agreement that with the exception of important anatomical differences associated with bipedalism and a large brain, we are "naked" but not "killer" apes (Morris 1967).

Language and Communication

Over the last few decades we have spent large amounts of money trying to teach apes to speak (Premack and Premack 1983; Gardner et al. 1989). The overwhelming and inescapable conclusion of these attempts has been that ape linguistic abilities are probably not as great as those of a four-year-old hominid (Terrace 1982, 1983). Such attempts have helped to clarify the nature of language and underscore the fundamental differences between human language and nonhuman primate communication. In immature hominids language develops in a steplike process in which the increasing complexity of linguistic abilities requires no reward but is in fact a reward in itself. On the basis of what we have learned from ape language projects and neuroanatomy, it is surprising in retrospect that anybody ever doubted Chomsky's (1972) contention that there was a "language organ" in the brain.

Ape linguistic studies have also highlighted the fact that one of the unique features of human communication is that it is not only limbic, but also nonlimbic. It is the nonlimbic component that qualifies it as unique. Despite claims that chimpanzees can mislead and even lie, it is by no means clear that they are actually deceiving their trainers. Free living nonhuman primates do have specific referent vocalizations for predators (Seyfarth et al. 1980) and may therefore be said to employ a semantic system, but such supposedly symbolic behavior (cf. Savage-Rumbaugh et al. 1983) is in fact limbic in that what is communicated is the current emotional state of the animal. Humans on the other hand are experts at both concealing and exaggerating their limbic states. But most importantly they are capable of communicating in a nonlimbic fashion, which simply conveys information that is independent of their current emotional state. Although ape linguistic studies have demonstrated that apes develop longer and slightly more complex "sentences" as training progresses, the communications always relate to their desires or opinions (feelings) about their environment or themselves. Eventually all of these animals reach a level at which the information content of their sentences ceases to increase. This is fundamentally different from the continually expanding linguistic abilities of immature hominids who not only continue to increase the length and complexity of their linguistic communications, but who also eventually develop the consistent use of nonlimbic communication. It may be that nonlimbic communication represents a watershed in human evolution that allowed early hominids to label and remember environmental facts at an unprecedented level of complexity.

Although the selection pressures that transformed an apelike primate into a hominid are still widely debated, it is almost certain that the exploitation of an increasingly wider range of resources by early hunter and gatherer groups was critical to the hominization process. A hunter and gatherer way of life may not only account for the development of language, but also for the differences in male and female brains (Kimura 1992). Although perhaps not "politically correct," there is no doubt about the fact that females have superior linguistic abilities and males test higher on tests emphasizing abilities related to spatial analysis (Kimura 1992). Such innate predispositions are wholly consistent with the picture we have developed of the past and present division of labor in hunting and gathering societies. The linguistic superiority of females may be a direct manifestation of the fact that females were selected for their ability to label and remember the location and current status of collectible resources. There is also some indication that females (in comparison with males) make greater use of landmarks in traveling over a given landscape. Males, on the other hand, appear to rely more on distance and direction (Kimura 1992). Males are also better at assessing trajectories and locating targets. These differences are so consistent with a sexual division between hunting and gathering that we should view our interpretation of prehistoric hominid ecology as essentially confirmed. In the process we should also lay to rest once and for all the notion that "man the hunter" accounted for the development of most of the unique features of human behavior such as language and tool manufacture and dependency. "*Homo* the divider of tasks," on the basis of the totality of evidence, seems a much more accurate and useful way of characterizing the hominization process.

Drawing together what we have learned in paleoanthropology and other relevant social and biological sciences not only offers a consistent and logical picture of the human present, but it also provides powerful lessons for the human future. The manifestation of our evolutionary history in our daily lives is both subtle and profound. But above all it is something of which those who plan our future and manage our present should be made aware. For example, the culturally ubiquitous existence of special areas such as parks and nonagricultural gardens in industrial societies is not just a cultural luxury. It is a biological necessity for our usual species-specific behavior. Architects eventually came to realize the importance of "green belts," but few of them are aware of paleoanthropological data that strongly suggest that they should concentrate on designing housing projects that incorporate the space and even the colors (Pope 1984) that shaped our brains and bodies for millions of years. In the United States, new "housing projects" have become a synonym for future slums and high crime areas. Recently corporations have begun to rediscover the value of exercise and its relation to mental performance. "In corpo, in sano" is hardly a new or revolutionary concept, but the reasons for its validity are confirmed in what we have recently learned about our evolutionary past. A knowledge of the modern integrated science that is evolutionary biology cannot claim to be able to solve the massive problems of poverty, racism, sexism, and repression, but it cannot help but be of value to social and economic planners charged with managing the human condition. The same is true in the individual management of our own lives.

CONCLUSION

The central premise of these short recollections of the past and speculations on the future is that integrated multidisciplinary work of the kind exemplified by F. Clark Howell represents a central direction for anthropology as a whole. Furthermore, the social sciences need to incorporate much more paleoanthropological work in their attempts to understand the origins and modern manifestations of human behavior. For myself, as an anthropologist it is unthinkable that a psychologist could believe that an understanding of "the mind" is possible without detailed knowledge of the environmental and social conditions in which the human brain evolved. There have been recent attempts to include at least some human evolutionary studies in psychology texts and articles, but frequently the authors have no way of knowing whether they are citing widely accepted or vigorously debated anthropological theory (cf. Kolb and Whishaw 1990; Bridgeman 1988; Kimura 1992). Nowhere is this more observable than in texts that cite the so-called "Eve Theory" or punctuated equilibrium as theories supported by most paleoanthropologists. The latter theory is often cited by neurobiologists and psychologists as a justification for interspecific comparison of species that are essentially static.

Among the social sciences, a belief that recognizably sapient behavior is a geologically recent phenomenon dangerously over-emphasizes the possibility that human behavior is almost solely the product of our current environment. Many largely environmental explanations of both "normal" and "aberrant" human behavior are also largely unconcerned or unfamiliar with the nature of hunter and gatherer environments that directly shaped our biology as a whole. Data capable of providing useful conclusions that can resolve such controversies derive directly from paleoanthropology and its various subdisciplines. Continuing to ignore such evidence can only hinder attempts to understand the societies and personalities of the present.

As one who has adopted a Howellian perspective, I feel the time may be fast approaching when data from modern social science research can be much more thoroughly integrated that from human evolutionary studies. Howell's work has provided a template for such an integration. The recent debate over modern human origins may have provided the impetus for an expanded integration. Just as the debate over the validity of *Australopithecus afarensis* resulted in a concrete anatomical definition of what we mean by hominids, the current debate over the origin of modern *Homo sapiens* has set the stage for developing archeological, anatomical, and behavioral definitions of what it is we mean by human. Such a consensus will demand a thoroughly Howellian integration of all the evidence. Ernst Mayr concluded his insightful "The Growth of Biological Thought" with the following observation:

"Yet, when it comes to developing a truly comprehensive science of science, it can be done only by comparing the generalizations derived from the physical sciences with those of the biological and social sciences, and by attempting to integrate all three branches. I rather suspect the raw material for such comparisons and for an integration is already available and that it is only necessary that someone adopts this as the objective of his research" (1982:857–858).

F. Clark Howell's work has already begun this process and continues to expand the rapidly maturing science of human evolutionary studies. He has brought to the study of "bones and stones" an analytical organization that will endure as the signature of that amazing Howellian way of patiently shedding light on the past of our species and on the future of our science.

SUMMARY

The life work of F. Clark Howell in paleoanthropology has resulted in the development of an encyclopedic integrative approach to human evolutionary studies that also has broad implications for both the future of paleoanthropology and the social sciences as a whole. The term "Howellian" is proposed as an apt description of studies that integrate such diverse subfields of paleoanthropology as archeology, human paleontology, paleoenvironmental reconstruction, and geochronology. Furthermore, the modern social sciences can also benefit greatly by adopting a multidisciplinary approach of the kind that F. Clark Howell brought to the study of the origin and evolution of the Hominidae. This kind of approach and the cautious conclusions which have resulted from it should be employed in any attempts to understand the biological bases of modern human behavior.

LITERATURE CITED

Ardrey, R. T. (1961) *African Genesis*. New York: Dell.

Ardrey, R. T. (1966) *The Territorial Imperative: A Personal Inquiry into the Animal Origins of Property and Nations*. New York: Atheneum.

Ardrey, R. T. (1976) *The Hunting Hypothesis*. New York: Atheneum.

Barkow, J. H. (1989) *Darwin, Sex and Status*. Toronto: University of Toronto.

Binford, L. H. (1987) Were there elephant hunters at Toralba? In M. H. Nitecki and D. V. Nitecki (eds): *The Evolution of Human Hunting*. New York: Plenum, pp.47–106.

Binford, L. R., and Ho, C. K. (1985) Taphonomy at a distance: Zhoukoudian, "The cave home of Beijing man"? *Curr. Anthropol.* 26:413–442.

Binford, L. R., and Stone, N. M. (1986) Zhoukoudian: A closer look. *Curr. Anthropol.* 27:453–475.

Bridgeman, B. (1988) *The Biology of Mind and Behavior*. New York: John Wiley and Sons.

Bunn, H. T. (1981) Archeological evidence for meat-eating by Plio-Pliestocene hominids from Koobi Fora and Olduvai Gorge. *Nature* 291:574–577.

Campbell, B. (1974) *Human Evolution*. Chicago: Aldine.

Campbell B. (1976) *Humankind Emerging*. Boston: Little, Brown.

Cavalli-Sforza, L. L., Piazza, P., Menozzi, P., and Mountain, J. (1988) Reconstruction of human evolution: Bringing together genetic, archeological, and linguistic data. *Proc. Nat. Acad. Sci.* 85: 6002-6.

Chomsky, N. (1972) *Language and the Mind*. New York: Harcourt, Brace, Jovanovich.

Conkey, M. (1987) Interpretations and problems of hunter and gatherer studies: Some thoughts on the European Upper Paleolithic. In O. Soffer (ed.): *The Pleistocene Old World: Regional Perspectives*. New York: Plenum, pp. 63–72.

Coppens, Y., Howell, F. C., and Leakey, R. E. F. (1976) *Earliest Man and Environments in the Lake Rudolf Basin: Stratigraphy, Paleoecology and Evolution*. Chicago: University of Chicago Press.

Corruccini, R. S. (1978) Morphometric analysis: Uses and Abuses. *Yearb. Phys. Anthropol.* 21:134–150.

Corruccini, R. S., and McHenry, H. M. (1980) Cladometric analysis of Pliocene hominids. *J. Hum. Evol.* 9:209-221.

Cronin, J. E., Boaz, N. T., Stringer, C. B., and Rak, Y. (1981) Tempo and mode in human evolution. *Nature* 292: 113–122.

Delson, E. (1977) Reconstruction of hominid phylogeny: A testable framework based on cladisitic analyses. *J. Hum. Evol.* 6:263–278.

Dundes, A. (1980) *Interpreting Folklore*. Bloomington: Indiana University Press.

Eibl–Eibesfelt, I. (1972) *Love and Hate. The Natural History of Behavior Patterns*. New York: Holt, Rhinehart and Winston.

Eldridge, N., and Gould, S. J. (1972) Punctuated equilibria: An alternative to phyletic gradualism. In J. Schopf (ed.): *Models in Paleobiology*. San Francisco: Freeman and Cooper.

Falk, D. (1987) Hominid paleoneurology. *Ann. Rev. Anthropol.* 16: 13–30.

Frayer, D. W., Wolpoff, M. H., Thorne, A. G., Smith, F. H., and Pope, G. G. (1993) The fossil evidence for modern human origins. *Am. Anthropol.* 95:14–50.

Gardner, R. A., Gardner, B. T., and van Cantfort, T. T. (1989) *Teaching Sign Language to Chimpanzees*. Albany: State University of New York Press.

Harlow, H. (1959) Love in infant monkeys. *Sci. Amer.* 200: 68–74.

Harlow, H., and Harlow, M. (1961) A study of animal affection. *Nat. Hist.* 70: 48–55.

Holloway, R. L., and de la Coste–Lareymondie, M. C. (1982) Brain endocast asymmetry in pongids and hominids: Some preliminary findings on the paleontology of cerebral dominance. *Am. J. Phys. Anthropol.* 58:101–110.

Howell, F. C. (1957) The evolutionary significance of variation and varieties of Neanderthal man. *Quart. Rev. Biol.* 32:330–347.

Howell, F. C. (1959) The Villafranchian and human origins. *Science* 130:831–844.

Howell, F. C. (1962a) Neanderthal Man. *Encyclopedia Britannica* 16:179–180.

Howell, F. C. (1962b) The Villafranchian and human origins, In W. W. Howells (ed.): *Ideas on Human Evolution. Selected Essays*, Cambridge Harvard Univ. Press, pp. 396–421.

Howell, F. C., (1967) Recent advances in human evolutionary biology. *Quarterly Review of Biology*. 42:471–513.

Howell, F. C. (1968) Omo research expedition. *Nature* 219:567–572.

Howell, F. C. (1972) Recent advances in human evolutionary studies. In S. L. Washburn and P. Dolhinow (eds.): *Perspectives on Human Evolution*, Vol. 2. Holt, Rhinehart and Winston, New York pp. 51–128.

Howell, F. C. (1976) Overview of the Pliocene and earlier Pleistocene of the lower Omo Basin, southern Eithiopia. In G. L. Isaac and E. R. McCown (eds.): *Human Origins: Louis Leakey and the East African Evidence*. Menlo Park: W. A. Benjamin, pp. 227–268.

Howell, F. C. (1978) Overview of the Pliocene and earliest Pleistocene of the lower Omo basin, southern Ethiopia. In C. J. Jolly (ed.): *Early Hominids of Africa*. London: Duckworth, pp. 85–130.

Howell, F. C. (1988) The Pleistocene Old World: Regional perspectives. *J. Hum. Evol.* 17: 370–374.

Howell, F. C. (1989) Book review of the Evolution of Human Hunting. *J. Hum. Evol.* 18: 583–594.

Howell, F. C. (1991) The integration of archeology and paleontology: With reference to the origin of anatomically modern humans. In M. H. Nitecki (ed.): *Origin of Anatomically Modern Humans*. Chicago: University of Chicago Press.

Howell, F. C. (1992) *Paleoanthropology*. Wenner–Gren Foundation Report for 1991, 50th Anniversary Issue, pp. 6–17.

Howell, F. C. (in press) *Thoughts on Eugene Dubois and the 'Pithecanthropus' saga. Centenerary of the discovery of* Pithecanthropus erectus. *Dubois – The* Homo erectus *problem*. Frankfurt: Courier Forschungsinstitut Senckenberg.

Howell, F. C., Boaz, N. T., Brown, F. H., and de Heinzelin, J. (1978b) Stratigraphic interpretation of the Omo Shungura and Lake Turkana fossil suid record. *Science* 202:1309.

Howell, F. C., and Bourliére, F. (1963) *African Ecology and Human Evolution. Proc. Viking Fund Publication in Anthropology* 36. Chicago: Aldine.

Howell, F. C., Brown, F. H., Chavaillon, J., Coppens, Y., Haessaerts, P., and de Heinzelin, J. (1973b) Situation stratigraphique des localitiés a Hominides des gisments plio–pleistocenes de l' Omo en Ethiopie. *C. R. Acad. Sci. Paris* 276:2781–2784, 2879–2882.

Howell, F. C., Brown, F. H., and Eck, G. G. (1978a) Observations on problems of correlation of late Cenozoic hominid–bearing formations in the north Rudolf Basin. In W. W. Bishop (ed.): *Geological Background to Fossil Man*. London: *Geological Society of London*, pp. 473–498.

Howell, F. C., and Coppens, Y. (1973) Deciduous teeth of Hominidae from Pliocene/ Pleistocene of the lower Omo basin, Ethiopia. *J. Hum. Evol.* 2:461–472.

Howell, F. C., and Coppens, Y. (1974a) Inventory of Hominidae from Pliocene/Pleistocene formations of the lower Omo basin, Ethiopia (1967–1972). *Am. J. Phys. Anthropol.* 40:1–16.

Howell, F. C., and Coppens, Y. (1974b) Les Faunes de mammiferes fossiles des formations Plio–Pleistocene de l'Omo en Ethiopie (Proboscidea, Perissodactyla, Artiodactyla). *C. R. Acad. Sci. Paris* 278–D:2275–2278.

Howell, F. C., and Coppens, Y. (1974c) Les Faunes de mammiferes fossiles des formations Plio–Pleistocene de l'Omo en Ethiopie (Tubulidentata, Hyracoidea, Lagomorpha, Rodentia, Chiroptera, Insectivora, Carnivora, Primates). *C. R. Acad. Sci. Paris* 278–D:2421–2424.

Howell, F. C., and Coppens, Y. (1976a) An overview of Hominidae from the Omo succession, Ethiopia. In Y. Coppens, F. C. Howell, G. L. Isaac, and R. E. F. Leakey (eds.): *Earliest Man and Environments in the Lake Rudolf Basin. Stratigraphy, Paleoecology and Evolution*. Chicago; University of Chicago Press, pp. 522–532.

Howell, F. C., and Coppens, Y. (1976b) Mammalian faunas of the Omo Group: Distributional and biostratigraphic aspects. In Y. Coppens, F. C. Howell, G. L. Isaac, and R. E. F. Leakey (eds.): *Earliest Man and Environments in the Lake Rudolf Basin. Stratigraphy, Paleoecology and Evolution*. Chicago: University of Chicago Press, pp. 177–192.

Howell, F. C., and Eck, J. (1972) New fossil *Cercopithecus* material from the lower Omo basin, Ethiopia. *Folia Primatol.* 18:325–355.

Howell, F. C., and Edey, M. (1965) *Early Man. Life Nature Library*. New York: Time, Inc.

Howell, F. C., and Grattard, J. L. (1976) Remains of *Camelus* from the Shungura Formation, Omo valley. In Y. Coppens, F. C. Howell, G. L. Isaac, and R. E. F. Leakey (eds.): *Earliest Man and Environments in the Lake Rudolf Basin. Stratigraphy, Paleoecology and Evolution*. Chicago: University of Chicago Press, pp. 268–274.

Howell, F. C., Merrick, H., de Heinzelein, J., and Haesaerts, P. (1973a) Archaeological occurrences of Early Pleistocene age from the Shungura Formation, lower Omo valley, Ethiopia. *Nature* 242:572–575.

Howell, F. C., and Petter, J. (1978) Carnivora from the Omo Group formations, southern Ethiopia. In Y. Coppens, F. C. Howell, G. L. Isaac, and R. E. F. Leakey (eds.): *Earliest Man and Environments in the Lake Rudolf Basin. Stratigraphy, Paleoecology and Evolution*. Chicago: University of Chicago Press, pp. 314–331.

Howell, F. C., and Washburn, S. L. (1960) *Human Evolution and Culture*. The University of Chicago Centennial, Vol II. Chicago: University of Chicago Press.

Howell, F. C., and Wood, B. (1974) An early hominid ulna from the Omo basin, Ethiopia. *Nature* 249:174–176.

Jaynes, J. (1978) *The Origin of Consciousness in the Breakdown of the Bicameral Mind*. Boston: Houghton Mifflin.

Johanson, D. C., and Edey, M. (1981) *Lucy, The Beginnings of Humankind*. New York: Simon & Schuster.

Johanson, D. C., and Shreeve, J. (1989) *Lucy's Child, The Discovery of a Human Ancestor*. New York: William Morrow.

Johanson, D. C., White, T. D., and Coppens, Y. (1978) A new species of the genus *Australopithecus*. *Kirtlandia* 28:1–14.

Kay, R., and Grine, F. E. (1988) Tooth morphology, wear and diet in *Australopithecus* and *Paranthropus*. In F. Grine (ed.): *Evolutionary History of the "Robust" Australopithecines*. New York: Aldine de Gruyter.

Kimura, D. (1992) Sex differences in the brain. *Sci. Amer.* 267:118–125.

Klein, R. G. (1987) Reconstructing how early people used animals. In M. H. Nitecki and D. V. Nitecki (eds.): *The Evolution of Human Hunting*. New York: Plenum, pp.11–46.

Klein, R. G. (1992) The archeology of modern human origins. *Evol. Anthropol.* 1:5–14.

Kolb, B., and Whishaw, Q. (1990) *Fundamentals of Human Neuropsychology*. New York: W. H. Freeman and Company.

Lewin, R. (1987) *Bones of Contention*. New York: Simon & Schuster.

Mayr, E. (1982) *The Growth of Biological Thought*. Cambridge: Belknap.

Mellars, P. (1988) Major issues in the emergence of modern humans. *Curr. Anthropol.* 30:349–385.

Morris, D. (1967) *The Naked Ape*. New York: McGraw–Hill.

Movius, H. L. (1944) Early Man and Pleistocene stratigraphy in southern and eastern Asia. *Pap. Peabody Mus. Harvard* 9:1–125.

Lovejoy, C. O. (1981) The origin of man. *Science* 211:341–350.

Oxnard, C. E. (1973) Functional inferences from morphometrics: Problems posed by uniqueness and diversity among the primates. Syst. *Zool.* 22:409–424.

Pope, G. G. (1984) Social biology, toward a bio–cultural paradigm. *Media IKA* 10:89–115.

Pope, G. G. (1988) Recent advances in Far Eastern paleoanthropology. *Ann. Rev. Anthropol.* 17:43–77.

Premack, D., and Premack, A. J. (1983) *The Mind of an Ape*. New York: W. W. Norton.

Sarich, V. M. (1974) Just how old is the hominid line? *Yearb. Phys. Anthropol.* 17:98–112.

Savage–Rumbaugh, S., Ronski, M. A., and Hopkins, W. D. et al. (1983) Symbol aquisition and use by *Pan troglodytes, Pan paniscus, Homo sapiens*. In P. G. Helne and L. A. Marguardt (eds.): *Understanding Chimpanzees*. Cambridge: Harvard University Press, pp. 266–310.

Seyfarth, R. M., Cheney, D. L., and Marler, P. (1980) Vervet monkey alarm calls. *Science* 210:801–803.

Skeels, H. M. (1966) Adult status of children with contrasting early life experiences. *Monog. Soc. Res. Child Dev.* 31:1–65.

Terrace, H. S. (1982) Why Koko can't talk. *The Sciences* 22:8–10.

Terrace, H. S. (1983) "Apes who talk": Language or the projection of language by their teachers. In J. DeLuce and H. T. Wagner (eds.): *Language in Primates*. New York: Springer–Verlag, pp. 19–42.

Tobias, P. V. (1991) *Olduvai Gorge*, Vol. 4: *The Skulls, Endocasts and Teeth of* Homo habilis. Cambridge: Cambridge University Press.

Toth, N. (1985) Archeological evidence for preferential right–handedness in the lower and middle Pleistocene, and its possible implications. *J. Hum. Evol.* 14:607–614.

Trigger, B. G. (1989) *A History of Archeological Thought*.

Washburn, S. L., and McCown, E. R. (1972) Evolution of human behavior. *Soc. Biol.* 19:163–170.

Wolpoff, M. H. (1973) The evidence for two australopithecine lineages in South Africa. *Yearb. Phys. Anthropol.* 17:113–139.

CHAPTER 2

Anthropoid Origins

John G. Fleagle

ANTHROPOIDS TODAY

Living primates are a very diverse group of mammals, containing approximately 15 families, 60 genera, and 200 species. Another 100 species, 50 genera, and 5 families are known only from the fossil record. Within the Order Primates, the anthropoids—New World monkeys, Old World monkeys and hominoids—are clearly the most distinctive natural group that stands apart from other taxa (Figure 2-1).

Biomolecular studies (e.g., Miyamoto and Goodman 1990) invariably find anthropoids to be a distinctive clade among primates, and there are numerous soft anatomical features of the placenta that separate anthropoids from other primates (e.g., Luckett 1975). In addition, there are many osteological characteristics of anthropoids that distinguish this group and potentially can be used to identify fossil anthropoids and track the evolution of the group in the fossil record. Living anthropoids share numerous craniodental features, including a fused metopic suture, postorbital closure, an anterior accessory chamber of the middle ear, a cranial blood supply through the promontory artery, a fused mandibular symphysis, and a last lower premolar with two subequal cusps (Figure 2-2). All of these characteristics have been identified as shared derived features of the group by some workers. However, while this suite of characters is unique to anthropoids, various of these features are shared with other extant and fossil primates—a situation that should enable a clear reconstruction of anthropoid origins, or at least identification of the sister taxon. However, as discussed later, the difficulty lies in the fact that anthropoids share different features with different groups of living and fossil primates—thus muddling the question of anthropoid origins.

If our knowledge of primate evolution came from only the living species, there would probably be little debate over either anthropoid origins or the basic outline of anthropoid evolution. Living anthropoids can be clearly divided into two main groups—New World monkeys (platyrrhines) and Old World higher primates (catarrhines), the latter being further divided into Old World monkeys (cercopithecoids)

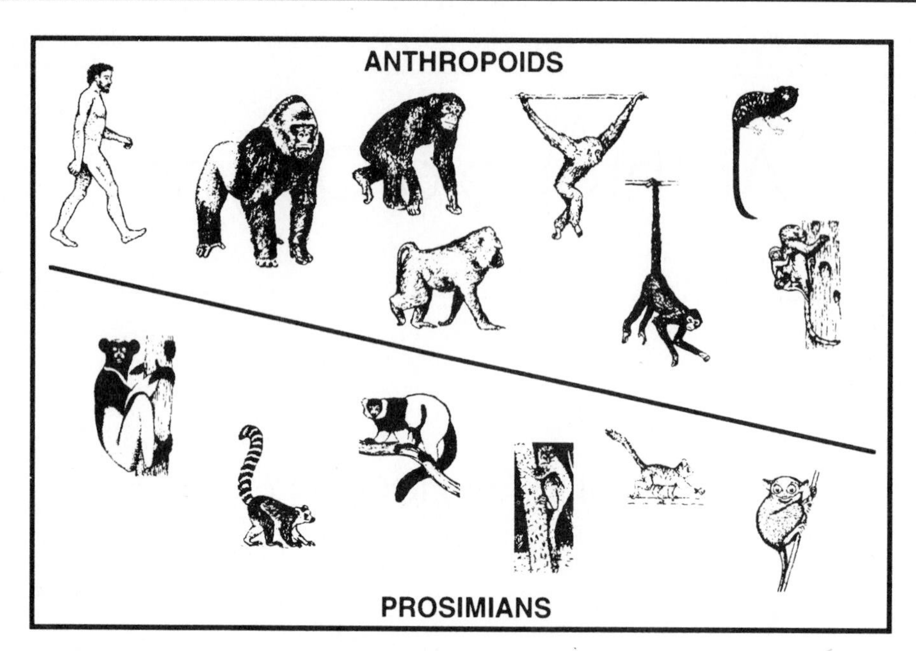

FIGURE 2-1 The major division among extant primates is between prosimians and anthropoids.

and apes and humans (hominoids). Aside from occasional suggestions that some features shared by platyrrhines and catarrhines may have been acquired independently and a persistent difficulty in identifying any derived feature to unite all platyrrhines, there is near universal acceptance of this scheme.

Moreover, in the absence of a fossil record, there would be little doubt that the sister taxon of anthropoids is the genus *Tarsius* (Figure 2-3; Cartmill 1980, in press; Martin 1990). The consensus from molecular phylogeny is that *Tarsius* is the sister taxon of anthropoids (Miyamoto and Goodman 1990), and there are metabolic similarities; numerous soft anatomical features including loss of a rhinarium; many details of placentation; and many features of the anatomy of the eye, including loss of a tapetum and presence of a retinal fovea that unite tarsiers and anthropoids. In the cranium, tarsiers share with anthropoids a cranial blood supply from the promontory artery, an anterior accessory cavity of the middle ear, a fused metopic suture, a reduced nasal turbinate system and absence of the spheno–ethmoid recess, an apical interorbital septum, and some degree of postorbital closure (Cartmill 1980, in press; Ross in press). On the basis of these similarities, tarsiers and anthropoids are commonly joined in the infraorder Haplorhini (Pocock 1918; Martin 1990). In the face of all of these shared derived features uniting tarsiers and anthropoids, it is perhaps surprising that there is any doubt at all about the origin of anthropoids.

THE PROBLEM

The confusion surrounding anthropoid origins comes from two sources. First, there is the nature of tarsiers, themselves. Although tarsiers indeed share a large number of presumably derived features with anthropoids, in many other anatomical features the extant genus *Tarsius* is a very odd primate by any measure. It has a dental formula of 2.1.3.3/1.1.3.3., with caninelike upper central incisors, and very simple premolars and molars with extraordinarily well-developed cusps. Obviously the loss of the lower central incisors precludes this genus itself from the ancestry of any species with two incisors in each quadrant. The cranium has a number of very striking features, probably associated with the gigantic orbits, including overlap of pterygoid plates with the petrosal bulla, a ventrolateral entrance of the internal carotid artery, and a gutterlike glenoid fossa that are unknown in any other extant primate. In the postcranial skeleton,

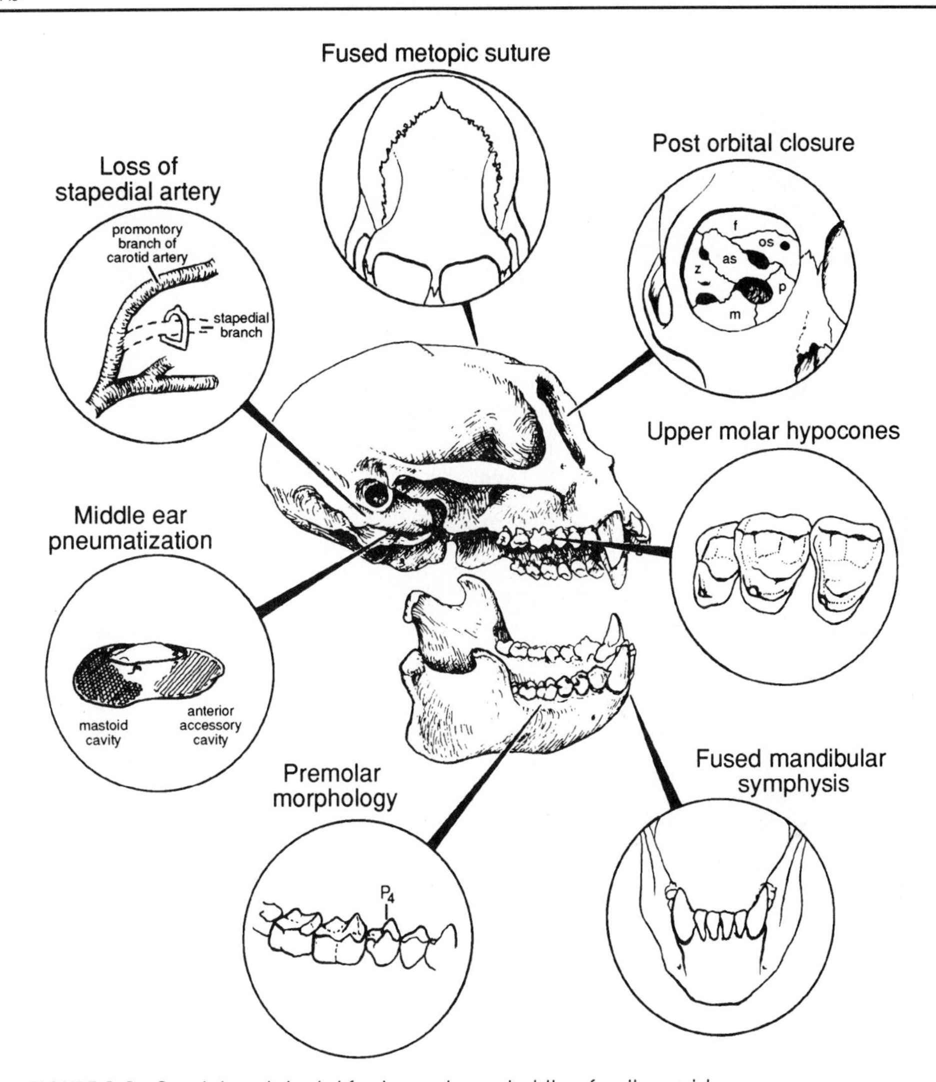

FIGURE 2-2 Cranial and dental features characteristic of anthropoids.

tarsiers have extraordinarily large hands, a fused tibia and fibula, and extremely long ankle bones that gave rise to the generic name. In addition, tarsier chromosomes are unusual among primates, and there is continued debate over the phylogenetic significance of the tarsier karyotype (Dutrillaux and Rumpler 1988). The striking anatomical specializations of this genus would seem to preclude it from any place in the direct ancestry of later, more generalized anthropoids, and many seem to feel that such strikingly unique features indicate a very early divergence of this genus from all other living primates, the anthropoid similarities notwithstanding.

However, the evidence that provides the greatest challenge to the status of tarsiers as the sister taxon of anthropoids is the information provided by the fossil record that expands our knowledge of both early anthropoids and fossil "prosimians."

The evidence from the fossil record that bears directly on the issue of anthropoid origins concerns three separate issues—the evolutionary record and nature of tarsierlike primates, the fossil record and nature of other "fossil prosimians," and the nature of early anthropoids. It is not unfair to say that as presently known and generally interpreted, all of this evidence tends to weaken rather than strengthen

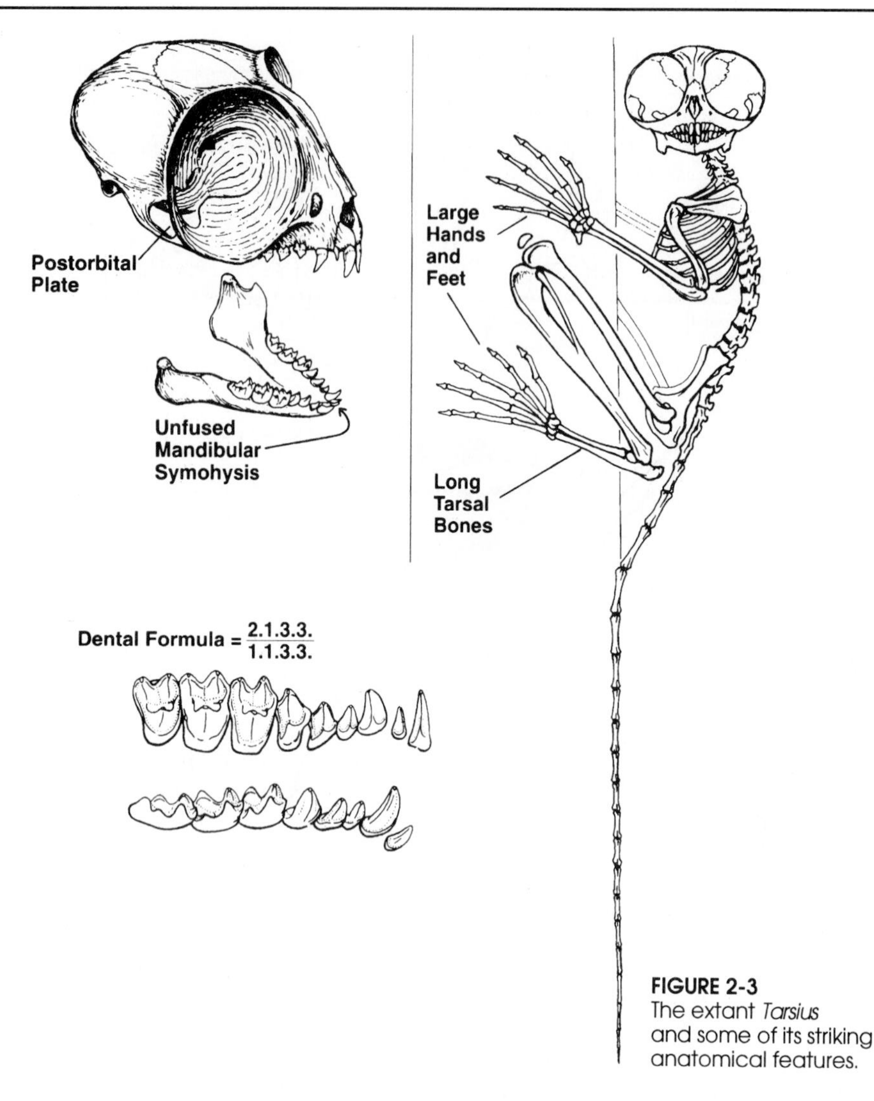

FIGURE 2-3
The extant *Tarsius* and some of its striking anatomical features.

the case for *Tarsius* as the sister taxon of anthropoids. Thus, before addressing current views of anthropoids, it is necessary to summarize first the recent advances and current interpretations concerning primate origins and the early evolution of prosimians (nonanthropoids) as well as new discoveries of early anthropoids.

PRIMATE ORIGINS

Until most recently, any consideration of preanthropoid evolution involved consideration of three groups of early tertiary mammals known primarily from North America and Europe: the predominantly Paleocene Plesiadapiformes, and two groups of predominantly Eocene prosimians—the "lemurlike" adapids and the "tarsierlike" omomyids. In the last decade, however, it has become almost uniformly accepted that only the latter groups can realistically be considered primates in any true phylogenetic sense. They share many derived features with later, including living, members of the order and may well be in the direct ancestry of most later taxa (e.g., Wible and Covert 1987). In contrast, plesiadapiforms are, at best, only the most primatelike of known Paleocene mammals. There are very few, if any, shared features to link them only with primates to the exclusion of other mammalian orders such as Dermoptera (flying lemurs) or Scandentia (tree shrews) and all known plesiadapiforms are either too specialized to be ancestral to any later primates, or in the case of *Purgatorius*, too generalized or poorly known to share any particular features with primates to the exclusion of many other orders. Moreover, not only do many plesiadapiforms seem to show more convincing shared features with other orders, such as Dermoptera (e.g., Kay et al. 1992), there are other mammalian orders, including tree shrews and megabats, that have been purported to have equally convincing shared features with primates (Wible and Covert 1987). At this time, all that seems certain is that primates, plesiadapiforms, tree shrews, and bats all seem to share a common ancestry relative to other orders, but the branching sequence among these groups, usually grouped in the superorder Archonta, is unresolved (see Simmons et al. 1991).

FOSSIL PROSIMIANS

Thus, with the elimination of the plesiadapiforms, the early Eocene adapids and omomyids are both the earliest prosimians and the earliest primates. They differ from other mammals and resemble later primates in virtually all aspects of their cranial and skeletal anatomy, including diagnostic features of the auditory region (e.g., MacPhee and Cartmill 1986; Dagosto 1988). Adaptively they share with plesiadapiforms dental features suggestive of more frugivorous-faunivorous or frugivorous-folivorous habits and especially resemble later primates in their hands (fingernails rather than claws), and eyes (large orbits with a postorbital bar) (Cartmill 1992). Both adapids and omomyids appear more or less abruptly at the Paleocene-Eocene boundary in Europe and North America where they are abundant members of early Eocene faunas. Ecologically early prosimians seem to replace the disappearing plesiadapiforms (Maas et al. 1988), but their geographic and phylogenetic origins are obscure. Not surprisingly, current hypotheses look to more poorly known continents as the ultimate source areas, either India, which was beginning to dock with Asia after a long period of isolation (Krause and Maas 1990); Africa (Sige et al. 1990; Gingerich 1990); or Asia (Szalay and Li 1986; Ciochon and Etler this volume). While both adapids and omomyids share many general features with later primates, and there is increasing evidence of remarkable dental similarity among the earliest members of the two groups (Rose and Bown 1991), the two groups seem to be clearly distinct and identifiable from the earliest Eocene, at least among North American and European members.

Adapids

As a group, the adapids are larger, probably largely diurnal, and more frugivorous and folivorous than the Omomyiformes. In dental anatomy, they are closer to the primate common ancestor in some respects—many have complete dental formula (2 • 1 • 4 • 3); and they usually have small incisors and larger, sometimes sexually dimorphic, canines. Many taxa have a fused mandibular symphysis, a feature aligning them with anthropoids. Adapids have traditionally been divided into two geographically distinct groups, the North American notharctines and the European adapines (no one seems to know what to do with the late Miocene sivaladapines (Gingerich 1984). However, many current authorities recognize three distinct groups (at either family or subfamily levels) among the Eocene adapids—the mostly North America notharctines and two groups among the European genera—the earlier, more diverse cercamoniines (Rose et al. in press) or protoadapines (Franzen 1987) and the late, abundant, but less speciose adapines. It is also evident that the geographic distinctions are not exact—the earliest notharctine is also found in Europe and there is a late cercamoniine from North America.

The North American notharctines are a very abundant and relatively well-known group of fossil primates, consisting of a large number of time-successive species (Gingerich and Simons 1977). They flourished in the earliest Eocene but are unknown after the middle Eocene. They are generally recognized as preserving the ancestral morphology for the adapid radiation.

Like notharctines, cercamoniines appear in the earliest Eocene, but unlike the notharctines, they are considerably more generically diverse and lasted to the end of the Eocene (e.g., Gingerich 1977; Franzen 1987). Most taxa are known only from teeth—a few from skulls and limbs. Like notharctines, with whom they are frequently grouped, they seem to retain many features that are primitive for both adapids and primates as a group. In addition to the European genera, there is one cercamoniine (*Mahgarita*) from the latest Eocene of Texas (Szalay and Wilson 1976) and several poorly known species from the Eocene of Pakistan (Russell and Gingerich 1987). Most recently, several new primates have been described from North Africa that have cercamoniine affinities (Hartenberger and Marandat 1992).

The better known adapines appear later in Europe and are in most aspects of cranial and skeletal anatomy (except dental formula) more specialized than the cercamoniines or notharctines (Dagosto 1983; Franzen 1987; Godinot 1984).

In addition to the groups previously described, and the late Miocene sivaladapines from Asia, there are various African and Asian genera that are frequently linked with adapids, but not clearly with any of the accepted subfamilies (e.g., Sudre 1975; Gheerbrant and Thomas 1992).

Although adapids as a group show striking phenetic similarities to extant prosimians in both craniodental and postcranial anatomy (see Gregory 1920; Tattersall and Schwartz 1985; Gingerich and Martin 1981; Rose and Walker 1985), attempts to find unique shared derived features linking modern strepsirhines with either adapids as a whole or specific lineages have yielded remarkably little detailed evidence. For example, the tooth comb characteristic of all lemurs and lorises was not present in any of the Eocene adapids and many features they share in common, such as enlarged stapedial or ring like tympanic are commonly regarded as the primitive condition for primates or as difficult to interpret precisely in most fossils. Indeed the best evidence linking adapids uniquely with strepsirhines (and presumably farther from anthropoid ancestry than Tarsius) are lateral flare on the talus (found in all adapids) and details of the wrist of Adapis (but not notharctines), both of which have been disputed (Beard et al. 1988; Beard and Godinot 1988).

Omomyids

Like adapids, omomyids are known primarily from North America and Europe, where they appear in the earliest Eocene. Compared with adapids, omomyids are generally smaller, and contain many more faunivorous-frugivorous and probably nocturnal species. Most omomyids are more dentally derived from either the primitive primate or anthropoid condition in having large procumbent incisors, and reduction of other anterior teeth is common. Many taxa have large orbits suggestive of nocturnal habits. Cranially and skeletally, they seem to share some features with other Eocene prosimians and some with tarsiers and anthropoids (Dagosto 1985; Szalay and Dagosto 1980; Beard et al. 1988). As in adapids, there are different, but not totally distinct, radiations of omomyids in North America and Europe, with a few poorly known taxa from other continents.

North American omomyids appear to have been much more diverse than their adapid contemporaries, both adaptively as determined by body size and reconstructed dietary habits (Strait 1991) and also in number of genera with a few more sympatric and synchronic genera in faunas containing only one or two adapids. Traditionally, they have been divided into two subfamilies, the more primitive, smaller, and more faunivorous anaptomorphines and the more advanced, more frugivorous and often larger omomyines, which also survive much later. Several genera of North American omomyids (Tetonius, Shoshonius) have been shown to have striking cranial similarities to the extant tarsier (Beard et al. 1991; Beard and MacPhee in press). However, recent studies have demonstrated that both the phylogenetic and morphological views of these two subfamilies are overly simplistic (Rose and Bown 1991; Bown and Rose 1991). While there are cranial similarities to tarsiers, and omomyids as a group have been traditionally allied with *Tarsius* (e.g.,

Gingerich 1981), several recent studies have noted that this is a typological view of a very diverse group. Several taxa show dental proportions and cranial features more comparable to those of adapines and/or anthropoids (Rosenberger 1986; Covert and Williams 1991; Beard and MacPhee in press).

European omomyids, the microchoerines, are much less diverse both taxonomically and adaptively. There are only four genera, probably derived from an anaptomorphine ancestry. Two taxa, *Necrolemur* and *Pseudoloris* show cranial, dental, and postcranial similarities to tarsiers (Simons and Russell 1960; Rosenberger 1985; Beard and MacPhee in press). There are also a number of poorly known taxa from Africa and Asia that have been identified as omomyids, including some that precede the appearance of omomyids in Europe and North America (Sige et al. 1990; Gingerich 1990; Szalay and Li 1986; Simons et al. 1986). However, the affinities of these have been debated.

Just as adapids have traditionally been allied with strepsirhines because of general phenetic similarities, omomyids have traditionally been linked with the genus *Tarsius* on the basis of their small size, large orbits, and elongate tarsals (Gingerich 1981). Similarly, just as the adapid-strepsirhine link has been subject to more careful phylogenetic scrutiny—so has the omomyid-tarsier link. Moreover, in light of the similarities between extant tarsiers and anthropoids, most attempts to interpret the phylogenetic affinities of omomyids and tarsiers have in one way or another borne on the issue of anthropoid origins, either in an attempt to identify haplorhine features common to the three or to distinguish the position of omomyids and tarsiers relative to anthropoids (e.g., Rosenberger and Szalay 1980). Thus on a broad level, some authorities have argued that many of the morphological similarities between omomyids and tarsiers are characteristics of small, large-eyed (presumably nocturnal), partly faunivorous primates and that these paleogene prosimians were no more similar to tarsiers than to similar-sized galagos (MacPhee and Cartmill 1986).

There are only a few features that seem to link omomyids as a group with both tarsiers and anthropoids—shapes of the ankle and elbow (e.g., Beard et al. 1988). Many have suggested that omomyids, tarsiers, and anthropoids also share an apical interorbital septum and reduced nasal turbinals, but to my knowledge, this has never been documented.

Although attempts to link omomyids as a group with tarsiers (and anthropoids) have fared poorly, there have been many more attempts to link specific omomyid taxa with tarsiers or anthropoids. Thus, Simons and Russell (1960) and Rosenberger (1985; also Rosenberger and Dagosto 1992) argued that the microchoerine *Necrolemur* was uniquely related to *Tarsius* independent of other taxa. Most recently, Beard et al. (1991; also Beard and MacPhee in press) have argued that *Shoshonius, Tetonius,* and *Necrolemur* are all uniquely related to *Tarsius* on cranial and basicranial anatomy. Most significant for understanding anthropoid origins is that in many of these studies different omomyid taxa show different similarities with tarsiers and the morphological features they share with tarsiers are not features that tarsiers share with anthropoids. Thus, if tarsiers are derived from one or more of the omomyids, tarsier-anthropoid similarities must have arisen in parallel and vice versa.

Some authors have also identified similarities between specific omomyid taxa and anthropoids. Following Simons (1961), Rosenberger (Rosenberger and Szalay 1980; Rosenberger and Dagosto 1992) has noted that *Ourayia* appears to lack enlarged orbits and has relatively broad, spatulate incisors. Covert and Williams (1991) noted that *Washakius* had similar sized lower incisors rather than an enlarged central incisor as in most omomyids, and Beard and MacPhee (in press) found basicranial similarities between *Rooneyia* and anthropoids. Most of these authors used these similarities to argue that omomyids were a relatively diverse group of primates from which anthropoids could have arisen.

EARLY ANTHROPOIDS

The fossil record of early anthropoids has increased dramatically in the last decade with new finds from the Fayum and many new sites in Oman and North Africa (e.g., Rasmussen and Simons 1992; Simons

1992; Thomas et al. 1989; Godinot in press) and hints from Asia. Early anthropoids were first found in the Fayum of Egypt in the early part of this century. Because of the large number of specimens and completeness of much of the material, the Fayum primate fauna still provides the baseline for evaluating other early anthropoid finds. The exact ages of the primate-bearing deposits from the Fayum remain a source of debate, but all probably date from no older than late Eocene to no younger than early Oligocene, or between 35 and 31 MYA (Rasmussen et al. 1992; Kappelman 1992; Gingerich 1993). Fayum anthropoids are currently placed in three major groups—propliopithecines, parapithecids and oligopithecines (Figure 2-4).

Propliopithecines

When it was first described over 80 years ago, *Propliopithecus* was identified as an early ape, probably ancestral to the European *Pliopithecus*, generally recognized as an ancestral gibbon. Likewise, the closely related *Aegyptopithecus* was originally regarded as a fossil ape. However, with increased information about the cranial and postcranial anatomy of these taxa, it has become more widely recognized that the propliopithecines are better considered generalized ancestral catarrhines. With the notable exception of their dental formula, they lack virtually all of the diagnostic features that unite extant Old World monkeys and apes and more closely resemble platyrrhines in most aspects of their cranial and postcranial anatomy (Fleagle and Kay 1983, 1987). Indeed, studies of propliopithecines demonstrate that what were previously

FIGURE 2-4 A phylogeny of early anthropoids.

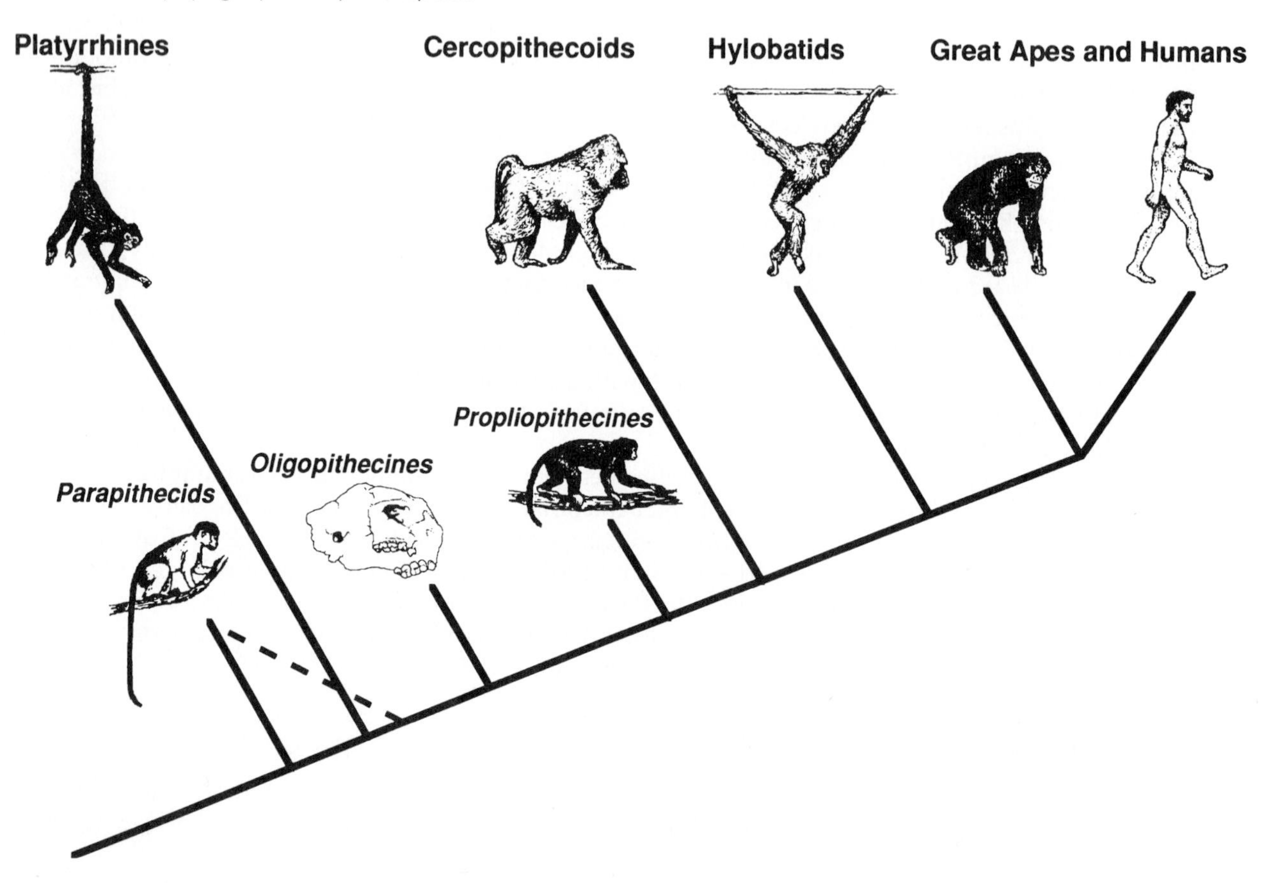

regarded as platyrrhine cranial and skeletal features are also primitive for catarrhines and thus anthropoids as a group, and that the acquisition of catarrhine features was a slow and mosaic process (e.g., Fleagle 1986).

Parapithecids

Parapithecids were also first described in the early part of this century, but the true phylogenetic position of this group and even the correct dental formula of some genera remain to be firmly established (e.g., Kay and Simons 1983; Simons 1987; Fleagle and Kay 1987). While parapithecids clearly show some striking autapomophies, overall they share many of the "platyrrhine" features of the cranium and postcranial skeletal found in propliopithecines and are more primitive than propliopithecines in other features, including premolar number (3 rather than 2), mandible shape, and several postcranial features. Moreover in some features, including premolar shape and several aspects of femoral structure, they appear to be even more primitive than platyrrhines and resemble prosimians. There is little doubt that parapithecids are more primitive than either propliopithecids or any group of living Old World anthropoid, but authorities differ in their views as to whether parapithecids precede or follow the platyrrhine-catarrhine divergence. As with the propliopithecines, the parapithecids demonstrate that the earliest Old World higher primates were clearly platyrrhine-like in much of their morphology and that in many features, platyrrhines almost certainly preserve the primitive anthropoid condition.

Oligopithecines

The genus *Oligopithecus* was first described in 1962 by Simons, who identified it as a catarrhine on the basis of its two premolars. However, the group remained poorly known until the recovery of abundant new remains and several new genera in the late 1980s from a new quarry near the base of the Jebel Qatrani Formation (Simons 1990, 1992; Rasmussen and Simons 1992). The new material confirmed the morphological paradox that had been recognized from the fragmentary remains (e.g., Gingerich 1980). It showed that these taxa were definitely anthropoid in having postorbital closure, a platyrrhine-like ear structure, and a catarrhine dental formula, but at the same time many aspects of the dentition were strikingly similar to European adapids, particularly cercamoniines. These morphological similarities have thus led Rasmussen and Simons (1992) to argue that anthropoids are derived from an adapid ancestry. However, the fact that oligopithecines have been linked with propliopithecines (in the family Propliopithecidae) on the one hand and adapids on the other has again called into question the role of parapithecids in early anthropoid phylogeny.

ANTHROPOID ORIGINS

The increased fossil record of early anthropoids has given tremendous insight into some aspects of early anthropoid evolution and remarkably little into others. By and large, we know a lot more about the early radiation of anthropoids into an extraordinary array of unsuspected creatures and the differentiation of catarrhines from a primitive ancestor, and relatively little about either the origin of platyrrhines or the origin of anthropoids as a group. Thus, the fossil record shows that platyrrhine-like morphological features were widespread in Africa 35 million years ago and the sequence in which catarrhines acquired their distinctive features (e.g., Fleagle 1986). At the same time, the fact that all of the Fayum taxa appear to be clearly anthropoid in their postorbital closure and documented features of basicranial anatomy still leaves a considerable gap between them and any of the known Eocene prosimians. Particularly perplexing are the similarities oligopithecines show to propliopithecines in premolar morphology and number on the one hand and their apparently more primitive adapid–like molars. In contrast, parapithecids seem to be more primitive than oligopithecines in premolar number and shape (as well as various postcranial features), but have more catarrhine–like molars on the other. Which derived features have evolved in parallel and which seemingly primitive features are secondary simplifications? Both parapithecids and oligopithecines have postorbital closure, and it seems unlikely that they evolved this anthropoid feature independently. However, it is not clear what their common ancestor, and hence the common ancestor of all known later anthropoids, looked like. Moreover, the one feature shared by all of the early anthropoids, postorbital closure, is the one feature that is apparently not present in any of the fossil prosimian groups usually put forth as ancestral anthropoids. As a result, for all we know about the early radiation of the group, we are still far less clear about the morphological features of the dentition that most likely characterized the common ancestor of these early anthropoids and hence where to look for the origin of the group.

Thus, at present there are four major hypothesis, each with different strengths and weaknesses, concerning which group of known nonanthropoids is closest to anthropoid origins.

Tarsier Origin

As previously noted, tarsiers (Figure 2-5) share many similarities in reproductive anatomy, eye structure, and cranial anatomy, as well as phenetic biochemical similarities with anthropoids not found in other living primates. Moreover, the features of postorbital closure and development of an anterior accessory chamber of the middle ear that unite tarsiers and anthropoids are unique features among primates or even among mammals rather than similarities that appear to have evolved in numerous groups (Ross in press). Although the cranial similarities linking tarsiers with anthropoids have been questioned (Simons and Rasmussen 1989; Beard and MacPhee in press), there seems little reason to do so on morphological grounds (Cartmill in press). Indeed, perhaps the most striking result from all the new remains of fossil primates that have been recovered in recent decades has been the demonstration that postorbital closure is the main feature distinguishing anthropoids from prosimians. Only tarsiers among all known fossil and extant prosimians share any evidence of this trait with anthropoids.

FIGURE 2-5 A phylogeny of euprimates showing *Tarsius* as the sister group of anthropoids.

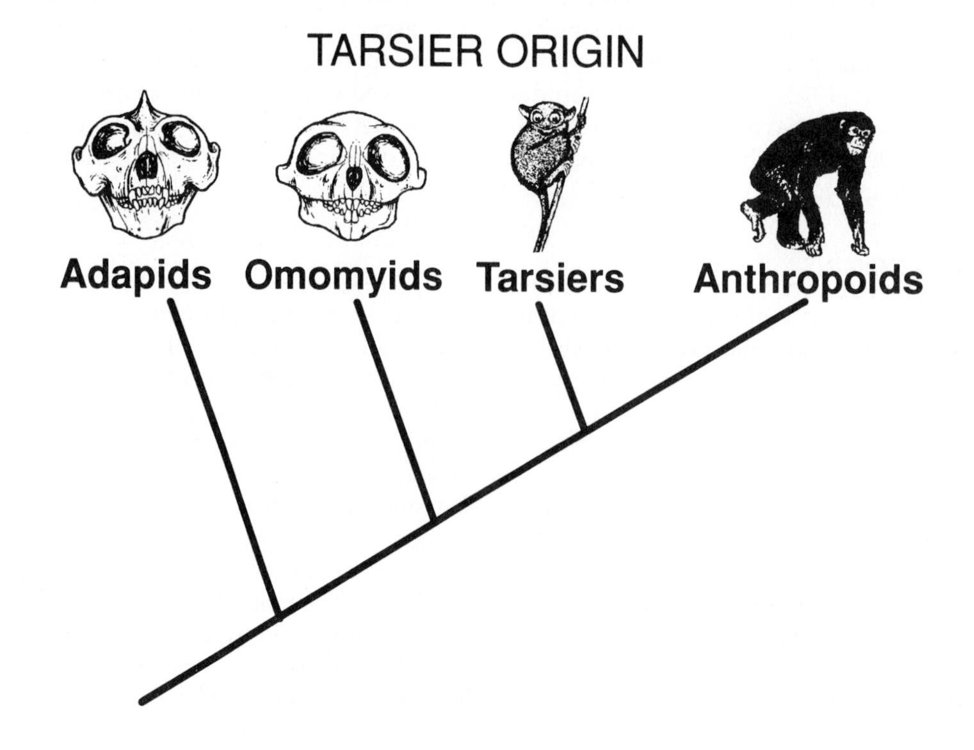

Nevertheless, the argument that one or more Eocene omomyids without postorbital closure share unique cranial features with tarsiers suggests to that many tarsier-anthropoid cranial similarities may indeed be parallelisms (Beard et al. 1991; Beard and MacPhee in press). Likewise, similarities between adapids and early anthropoids support an other–than–tarsier origin for anthropoids (Rasmussen in press).

Omomyid Origin

Many authorities have argued that the sister taxon of anthropoids is not the genus *Tarsius*, but a more generalized omomyid (Figure 2-6). In their view both tarsiers and anthropoids are descended from some common tarsiiform or omomyid ancestry (e.g., Rosenberger and Szalay 1980; Rosenberger 1986; Rosenberger and Dagosto 1992). In this view, the haplorhine features uniting *Tarsius* and anthropoids were presumably present in the omomyid ancestor of anthropoids, but not the unique features that are characteristic of *Tarsius*. This view has the advantage of being in accordance with notions of a haplorhine-strepsirhine dichotomy among primates, with molecular studies linking tarsius with anthropoids, and with studies that argue for the origin of *Tarsius* from particular omomyid taxa.

As many have pointed out, even though it would be nice to be able to apply a haplorhine-strepsirhine division to fossil prosimians, the division of fossil prosimians into haplorhines and strepsirhines is not as clear as among extant primates. It may well be that both omomyids and adapids as well as tarsiers would cluster with anthropoids and the living strepsirhines are the odd group (Rasmussen 1986). Moreover, the biggest weakness of this scenario is that the cranial features that most clearly link tarsiers with anthropoids (postorbital closure and anterior accessory chamber of the middle ear) are not present in known omomyids, including the taxa that share various other cranial features with *Tarsius*. Thus, except for a few postcranial similarities (Beard et al. 1988) there are very few derived features (possibly an apical interorbital septum and reduced nasal region) that link generalized omomyids with anthropoids. Moreover, if tarsiers are indeed derived from any of the taxa most frequently proposed (*Tetonius*, *Shoshonius*, and *Necrolemur*), then the unique tarsier-anthropoid cranial features have evolved independently in tarsiers

FIGURE 2-6 Euprimate phylogeny with generalized omomyids as the sister group of anthropoids.

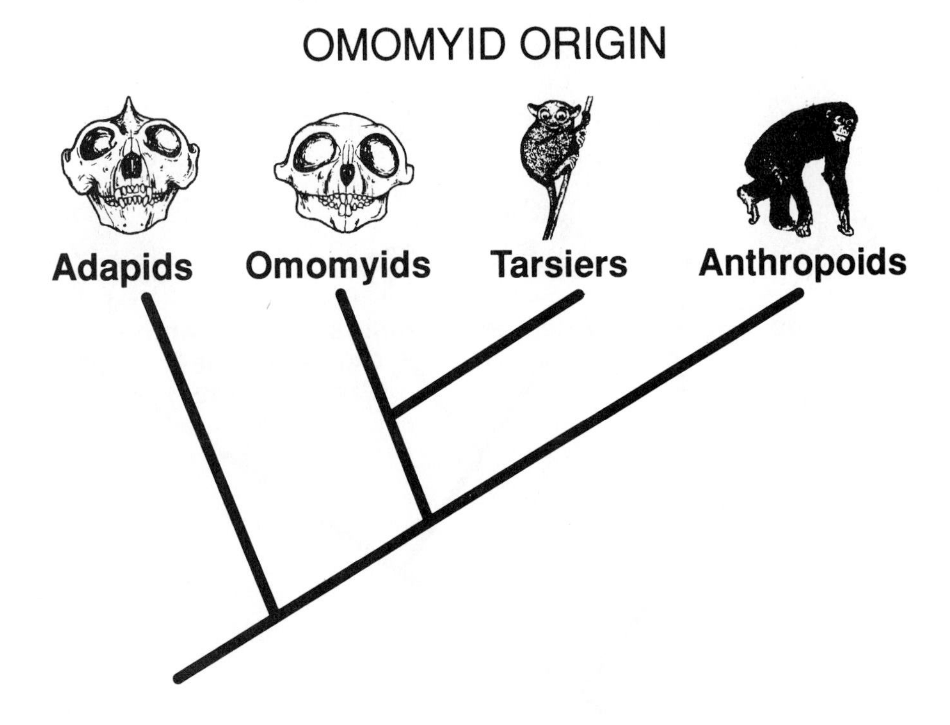

and anthropoids and the morphological evidence linking these two groups as haplorhines becomes substantially weaker and very difficult to identify among fossils.

Adapid (Cercamoniine or Protoadapine) Origin

Largely on the basis of similarities in the anterior dentition (small incisors, large canines, and frequently fused mandibular symphysis), many workers have long argued for an adapid-anthropoid relationship (e.g., Gingerich 1980; see Figure 2-7). Supporters of an adapid-anthropoid relationship have noted that omomyids tend to be too specialized dentally (but see Covert and Williams 1991) and that many of the haplorhine features do not necessarily exclude adapids as they are not clearly strepsirhine (Rasmussen 1986). The striking dental similarities between the oligopithecines and cercamoniines have strengthened the adapid-anthropoid link and have led Rasmussen (in press; see also Rasmussen and Simons 1988, 1992) to predict that some of the taxa currently recognized as cercamoniines on the basis of dental remains are actually anthropoids.

One major weakness of the adapid-anthropoid hypothesis is that there are still a number of possibly derived features that adapids share with strepsirhines (Beard et al. 1988). In addition, the features that adapids share with anthropoids are almost certainly primitive primate features, whereas the anthropoid features shared by tarsiers (postorbital closure) and perhaps omomyids (apical interorbital septum, reduced nasal region) are almost certainly derived features. Finally, as with omomyids, there is no indication among any of the known fossils that any adapid ever even approximated the one diagnostic anthropoid feature—postorbital closure.

Ancient or Other Origin

The seemingly endless frustration and difficulty of deriving anthropoids from any known group of Eocene prosimians from Europe or North America has led many to argue that the divergence of anthropoids from any known group of prosimians was an ancient event, and the anthropoid ancestor was an unknown paleogene group inhabiting some relatively unknown continent, more probably Africa or perhaps Asia

FIGURE 2-7 The protoadapine hypothesis of origins of anthropoids.

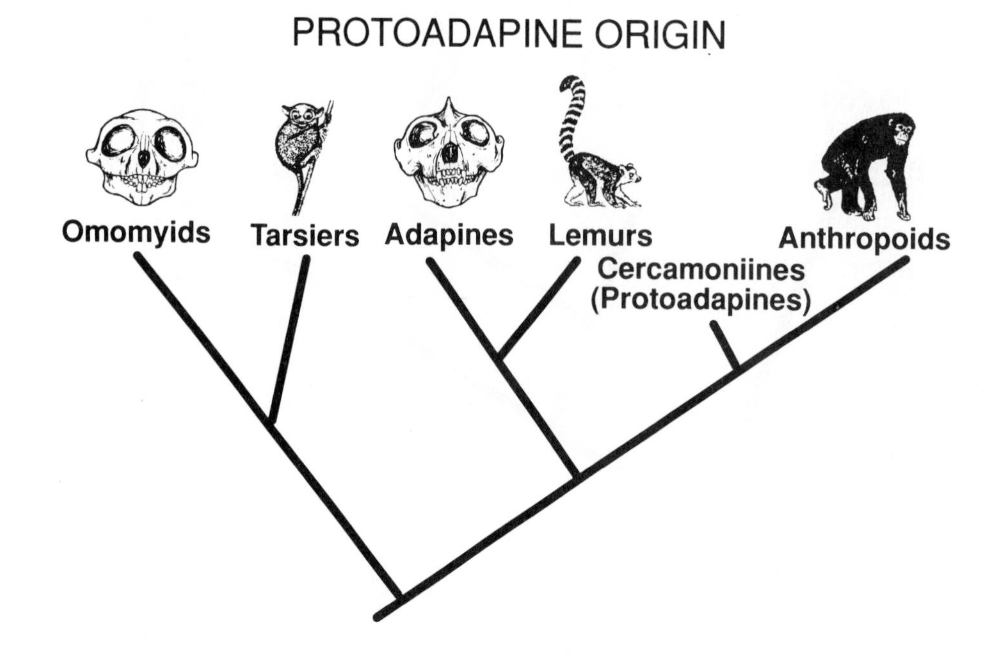

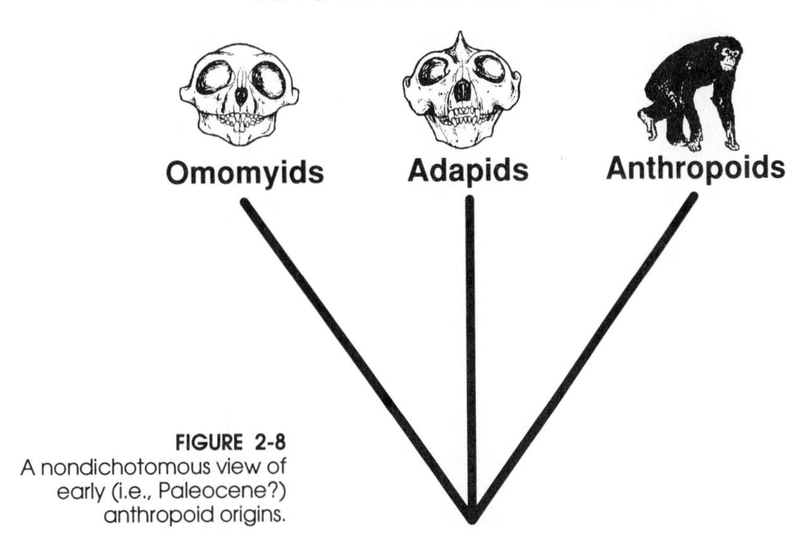

FIGURE 2-8
A nondichotomous view of early (i.e., Paleocene?) anthropoid origins.

(see Conroy 1981; Beard and MacPhee in press; see Figure 2-8). This position is consistent with the fact that the earliest appearance of anthropoids is in Africa and that many of the Eocene/Oligocene mammals that first appear with anthropoids, such as hyraces and proboscideans, also seem to have no clear Holarctic ancestors. In addition, many recent phylogenetic studies find that adapids, omomyids, and extant prosimians cluster as a group relative to anthropoids, suggesting that the anthropoid lineage is other than the known Eocene "prosimians" of the northern hemisphere.

Nevertheless, arguing for a nonadapid, nonomomyid, nontarsier origin for anthropoids is little more than a claim for ignorance regarding the origin of the group. It is certainly not unreasonable to expect that the ancestors of this group are not among currently known primate taxa, given how little we know about the fossil record from many parts of the world, but it is not very satisfying. Furthermore, a relatively ancient split among adapids, omomyids, tarsiers, and anthropoids does not mean that one of these three is not more closely related to anthropoids than the others, only that we might expect identification of their correct phylogenetic relationships more difficult to reconstruct as each would have had even more opportunity to develop parallelisms and autapomorphies (see also Martin 1993).

Indeed, recent finds from North Africa (Godinot and Mahboubi 1992; Hartenberger and Marandat 1992; see Godinot in press) are providing indications of anthropoid primates much older than the Fayum taxa and there are suggestions of others from Asia (Culotta 1992). However only when these taxa are better known will we have a clearer idea of what they really tell us about the antiquity and origins of anthropoids.

SOLVING ANTHROPOID ORIGINS

Anthropoids certainly had nonanthropoid ancestors. The origin of this group, just like the evolution of any other group of organisms is a biological phenomenon that took place. It is also almost certain that most of the hypotheses regarding anthropoid origins from tarsiers, omomyids, adapids, or some unknown group are not correct. Anthropoids most probably had a single origin from a single "prosimian" ancestor. How will we get past the current impasse and resolve the apparently contradictory and mutually exclusive scenarios implied by the current evidence?

It seems most unlikely that phylogenetic analyses alone will resolve this problem. In recent years (and months), there have been a number of very thoughtful and thorough studies of anthropoid origins examining many anatomical regions and broad samples of extant and fossil anthropoids. These studies are invaluable for clarifying precisely which alternative morphological changes are indicated by different phylogenies and providing a quantitative estimate of how much "morphological change" is required by different phylogenies. However, they have not resolved the problem of anthropoid origins, for several obvious reasons. First, the results of any analysis are obviously very dependent on the coding of characters, *a priori* determination of character homology, and the nature of the samples being studied (e.g., Cartmill 1982, in press). As a result, whether by "intent," different personal histories, or whatever, different studies often reach different conclusions from analyses of the same material. Developmental studies and extensive comparative samples can help clarify some of the homology problems, but only partially.

More significant is the obvious fact that there are no clear answers. Virtually all phylogenetic analyses find 30–50 percent homoplasy in the most parsimonious solutions. With so many features clearly evolving and reevolving or reverting many times, it is difficult to have much confidence in the most parsimonious solution when it is clear that parsimony is not obviously a common result of phyletic evolution by natural selection. Moreover, many of the solutions seem oddly illogical in their results. For example, in a recent study of primate dental evolution, Kay and Williams (1992) found that the most parsimonious solution grouped parapithecids with *Aegyptopithecus* as catarrhines rather than as primitive anthropoids near the platyrrhine/catarrhine dichotomy. However, many of the features that define the anthropoid node in this solution, are missing in parapithecids! The incongruity arises because parapithecids show other features found only among more advanced catarrhines even though they lack many of the anthropoid features. Thus, while we now have a heightened awareness of the morphological problems surrounding anthropoid origins, there is no obvious solution.

As trite as it sounds, in my opinion, resolution can only come from the recovery of new fossils that bridge the present gap between current anthropoids and prosimians and clarify which of the apparent prosimian-anthropoid similarities reflect real phylogenetic similarities and which are parallelisms and convergences. As Szalay has argued, we need to know the actual morphological transformations involved in anthropoid origins to clarify many of the morphological problems that underlie the current impasse. There are many specific inconsistencies that need to be addressed. For example, it is particularly striking that no fossil prosimians show even a hint of postorbital closure. Yet all early anthropoids show a completely walled off orbit. Where did the anthropoid condition come from? Or the tarsier condition, for that matter? Oligopithecines show a dental formula and premolar morphology similar to propliopithecines, but molar similarities to adapids, suggesting that catarrhines are derived from an adapid ancestry. But what does this imply about parapithecids which have three premolars that are more primitive than those of any other anthropoids, but good anthropoid or even catarrhine molar features? Although there are clear dental similarities between oligopithecines and adapids, adapid skeletal anatomy is very clearly not similar to that of anthropoids, but very similar to strepsirhines.

Will new fossils resolve these issues or just generate new unanticipated problems? Undoubtedly some of both. Certainly the new fossil anthropoids from Egypt (Simons 1992) are vastly different from anything anyone had any reason to anticipate in early anthropoid evolution. It seems that as we get closer to the base of the anthropoid radiation, the tree gets very bushy. It is hard to sort out the relationships when you don't even know who the characters are. Perhaps early parapithecids will help sort out phylogeny among three-premolared anthropoids and relationships between parapithecids and propliopithecines, but the new north African Eocene primates may even document an anthropoid radiation that precedes the one documented in the Fayum. In addition, the skeletal anatomy of many early anthropoids and fossil prosimians remains largely undocumented (but see Fleagle and Kay 1988; Gebo et al. in press; Dagosto and Gebo in press). This is another sort of evidence that will not eliminate current contradictions in the dental and cranial evidence but may throw support to one hypothesis over others. There is much more to be learned about early anthropoid evolution beyond which prosimians are our closest relatives.

THE PROBLEM OF ORIGINS

To a relatively small group of anthropologists, the question of anthropoid origins seems a particularly vexing and frustrating issue without a satisfactory solution. Has a unique combination of events in paleobiogeography, taphonomy, geological history, modern geomorphology, and just random factors concerning location of finds caused this to be a unique problem in primate evolution? Are the problems we face in understanding anthropoid origins characteristic of problems faced in attempts to unravel the origins of other groups? Since I have spent a number of years worrying about anthropoid origins with no satisfying answers, I would like to think that this is a particularly difficult issue. However, what little I know of other issues in primate and human evolution suggests that this is not the case at all. There are remarkably few groups of primates whose origin is understood. For some, such as adapids and omomyids, strepsirhines, and hominids, we don't have any very plausible ancestors or scenarios. For others, such as platyrrhines, Old World monkeys, and living great apes we have a pretty good idea of the group that probably includes the ancestors, but the details of the evolutionary transformations and the exact sister taxa are elusive. Then, there are groups like anthropoids, gibbons, the genus *Homo*, and anatomically modern humans for which we seem to have more plausible ancestors and different evolutionary scenarios than we need. What is the problem here? Or, as some have legitimately asked (Patterson 1981), has the fossil record contributed anything to our understanding of evolutionary relationships?

Of course, the fossil record has contributed immensely to our understanding of the history of the order Primates by providing documentation of the existence in the past 60 million years of hundreds of species that would otherwise be unknown and by and large unimaginable. In addition, it has provided evidence of both the pattern (e.g., Bown and Rose 1987) and details of evolutionary transformations that could not be reconstructed otherwise (e.g., Fleagle 1986). Students of human evolution spent all too many years trying to explain the evolution of the human features of big brain, small canines, bipedal posture, and tool use as some type of coordinated adaptive package (e.g., Darwin 1871; Washburn 1960) without realizing that they were only part of a single adaptive complex if one was comparing living humans to other primates such as chimpanzees. The fossil record demonstrates quite clearly that these "uniquely human" features evolved one by one, in what was previously an unpredictable sequence (e.g. McHenry 1975). We still may not know where the earliest hominids came from, or why hominids evolved these features that distinguish us from apes, but at least we know the proper sequence of events that must be explained.

Similarly, living catarrhines are distinguished from platyrrhines by a reduced number of premolars, a tubular ectotympanic bone lateral to the eardrum, a frontal–sphenoid connection on the side of the skull, loss of the entepicondylar foramen, and a trochlea on the humerus that is different in Old World monkeys and apes. From the distribution of these features in extant anthropoids there is no way to tell how or when these distinguishing catarrhine features were acquired by the two groups of modern catarrhines or which humerus shape is primitive for the group. However, from the fossil record, we can reconstruct the sequence in which the features were evolved in more primitive catarrhines and also learn that the humeral articulations of both Old World monkeys and hominoids are derived relative to the primitive catarrhine condition (e.g., Fleagle 1983, 1986). The value of fossils, for clarifying phylogenetic issues through preservation of intermediate forms with previously unknown combinations of features has been well documented in many groups (Donoghue et al. 1989).

For better or for worse, we humans are remarkably unimaginative compared with evolution by natural selection. We do a very poor job of imagining anything we haven't previously encountered. Thus, looking back from our present-day perspective, we tend to reconstruct fossil hominids to look like living hominids and fossil apes to look like living apes. Indeed, all too recently we were quite content to believe that hominoids have had an extraordinarily boring evolutionary history in which the few extinct hominoids known from the fossil record were seen as more or less modern apes that happened to have lived in the Miocene. Yet, even as it became clear that many were not very similar to modern apes, the first interpretations were that they must be like Old World monkeys, the only other group of living catarrhines. Only when confronted by an extraordinary diversity of forms and distinctly novel combinations of features, have we come to appreciate the true diversity of ape evolution (Ciochon and Corruccini 1983; Fleagle 1988).

However, it is these same valuable contributions of the fossil record—addition of previously unknown taxa and unique combinations of diagnostic features—that cause so much "trouble" in reconstructing the phylogeny of extant groups by showing that the taxa we have today are just pruned branches of what was originally a much more bushy tree and that the clusters of characters that seem to sort living taxa so readily tend to scatter much more widely as the other branches are grafted back. In addition, there is the simple geometric problem that the number of possible phylogenies increases dramatically with each new taxon. Thus, in the case of anthropoid origins, the fossil record has not provided a series of intermediate forms between tarsiers and humans that we might have anticipated from our knowledge of living primates, but rather a whole array of other prosimians with various anthropoid- *or* (not and) tarsier-like features and a plethora of early anthropoids, none of which have been recognized as particularly tarsier-like. We don't know less about anthropoid origins than we did without a fossil record; we just have fewer tidy stories and many more plausible hypotheses than we had earlier—as the Clark Howell dictum maintains, "The more you know, the harder it is."

ACKNOWLEDGMENTS

I am grateful to the editors for the invitation to contribute to a volume in honor of Clark Howell. Much of this chapter is a direct outcome, indeed a summary, of a workshop and conference on anthropoid origins that was held at Duke University in May 1992. It owes a great debt to all of the participants at the workshop (especially Tab Rasmussen and Chris Beard) whose contributions and ideas I have tried to represent accurately and fairly. In addition, my own views on this subject have benefited from many years of collaborative research and discussion with Elwyn Simons, Rich Kay, Alfred Rosenberger, Tom Bown, and Ken Rose. I thank Luci Betti for drafting the figures, Todd Rae for editorial and bibliographic assistance, and Kaye Reed for editorial assistance and general encouragement. This work was funded in part by NSF BNS 9012154 and grants from the LSB Leakey Foundation and the Wenner–Gren Foundation.

LITERATURE CITED

Beard, K. C., Dagasto, M., Gebo, D. L., and Godinot, M. (1988) Interrelationships among primate higher taxa. *Nature* 331:712-714.

Beard, K.C., and Godinot, M. (1988) Carpal anatomy of *Smilodectes gracilis* (Adapiformes, Notharctinae) and its significance for lemuriform phylogeny. *J. Hum. Evol.* 17:71-92.

Beard K. C., and MacPhee, R. D .E. (in press) Cranial anatomy of *Shoshonius* and the antiquity of Anthropoidea. In J. Fleagle and R. Kay (eds.): *Anthropoid Origins*. New York: Plenum Press.

Beard, K. C., Krishtalka, L., and Stucky, R. K. (1991) First skulls of the early Eocene primate *Shoshonius cooperi* and the anthropoid-tarsier dichotomy. *Nature* 349:64-67.

Bown. T. M., and Rose, K. D. (1987) Patterns of dental evolution in early Eocene Anaptomorphine primates (Omomyidae) from the Bighorn Basin, Wyoming. *J. Paleontol.* 61 (5.II, suppl.).

Bown, T. M., and Rose, K. D. (1991) Evolutionary relationships of a new genus and three new speciens of Omomyid primtes (Willwood Formtation, Lower Eocene, Bighorn Basin, Wyoming). *J. Hum. Evol.* 20:465-480.

Cartmill, M. (1980) Morphology, function, and evolution of the anthrpoid postorbital septum. In R. L. Ciochon and A. B. Chiarelli (eds.): *Evolutionary Biology of the New World Monkeys and Continental Drift.* New York: Plenum Press, pp. 243-274.

Cartmill, M. (1982) Assessing tarsier affinities: Is anatomical description phylogenetically neutral? *Geobios Mémoire Spécial* 6: 279-287.

Cartmill, M. (1992) New views on primate origins. *Evolutionary Anthropology* 1:105-111.

Cartmill, M. (in press) Cranial anatomies and antinomies in the problem of anthropoid origins. In J. Fleagle and R. Kay (eds.): *Anthropoid Origins.* New York: Plenum Press.

Ciochon, R. L., and Corruccini, R. S. (1983) *New Interpretations of Ape and Human Ancestry*. New York: Plenum Press.

Conroy, G. C. (1981) Review of Evolutionary Biology of the New World Monkeys and Continental Drift. *Folia Primatol* 36:155-156.

Covert, H. H., and Williams, B. (1991) The anterior lower dentition of *Washakius insignis* and adapid-anthropoidean affinities. *J. Hum. Evol.* 21:463-467.
Culotta, E. (1992) A new take on anthropoid origins. *Science* 256:1516-1517.
Dagosto, M. (1983) Postcranium of *Adapis parisiensis* and *Leptadapis magnus* (Adapiformes, Primates): Adaptational and phylogenetic significance. *Folia Primatol.* 41:49-101.
Dagosto, M. (1985) The distal tibia of Primates with special reference to the Omomyidae. *Int. J. Primatol.* 6:45-75.
Dagosto, M. (1988) Implications of postcranial evidence for the origin of euprimates. *J. Hum. Evol.* 17:35-56.
Dagosto, M., and Gebo, D. L. (in press) Postcranial anatomy and the origin of the Anthropoidea. In J. Fleagle and R. Kay (eds.): *Anthropoid Origins.* New York: Plenum Press.
Darwin, C. (1871) *Descent of Man and Selection According to Sex.* London: John Murray.
Donoghue, M.J., Doyle, J., Gauthier, J., Kluge, A., and Rowe, T. (1989) The importance of fossils in phylogeny reconstruction. *Ann. Rev. Ecol. Syst.* 20:431-460.
Dutrillaux, B., and Rumpler, Y. (1988) Absence of chromosomal similarities between tarsiers *(Tarsius syrichta)* and other primates. *Folia Primatol.* 50:130-133.
Fleagle, J.G. (1983) Locomotor adaptations of Oligocene and Miocene hominoids and their phyletic implications. In R.L. Ciochon and R.S. Corruccini (eds.): *New Interpretations of Ape and Human Ancestry.* New York: Plenum Press, pp. 301-324.
Fleagle, J. G. (1986) The fossil record of early catarrhine evolution. In B. Wood, L. Martin, and P. Andrews (eds): *Major Topics in Primate and Human Evolution.* Cambridge: Cambridge University Press, pp. 130-149.
Fleagle, J. G. (1988) *Primate Adaptation and Evolution.* San Diego: Academic Press.
Fleagle, J. G., and Kay, R. F. (1983) New interpretations of the phyletic position of Oligocene hominoids. In R. L. Ciochon, and R.S. Corruccini (eds.): *New Interpretations of Ape and Human Ancestry.* New York: Plenum Press, pp. 181-210.
Fleagle, J. G., and Kay, R. F. (1987) The phyletic position of the Parapithecidae. *J. Hum. Evol.* 16:483-531.
Franzen, J. L. (1987) Ein neuer Primate aus dem Mitteleozan der Grube messel (Deutschland, S–Hessen). *Cour. Forsch.–Inst. Senckenberg* 91:151-187.
Gebo, D. L., Simons, E. L., Rasmussen, D. T., and Dagosto, M. (in press) Eocene anthropoid postcrania from the Fayum, Egypt. In J. G. Fleagle and R. F. Kay (eds.): *Anthropoid Origins,* New York: Plenum Press.
Gheerbrant, E., and Thomas, H. (1992) The two first possible new adapids from the Arabian Peninsula (Taqah, early Oligocene of Sultanate of Oman). *Abstracts of the 14th Congress of the International Primatological Society.* Strasbourg: Sicop Press, pp. 349-350.
Gingerich, P. D. (1977) Radiation of Eocene Adapidae in Europe. *Geobios Mém. Spécial* 1:165-182.
Gingerich, P. D. (1980) Eocene Adapidae: Paleobiogeography and the origin of South American Platyrrhini. In R. L. Ciochon, A. B. Chiarelli (eds.): *Evolutionary Biology of the New World Monkeys and Continental Drift.* New York: Plenum Press, pp. 123-138.
Gingerich, P. D. (1981) Early Cenozoic Omomyidae and the evolutionary history of tarsiiform primates. *J. Hum. Evol.* 10:345-374.
Gingerich, P. D. (1984) Primate evolution. In T. D. Broadhead (ed): *Mammals: Notes for a short course.* Knoxville: University of Tennessee, Dept. of Geological Sciences, pp. 167-181.
Gingerich, P. D. (1990) African dawn for primates. *Nature* 346:411.
Gingerich, P. D. (1993) Oligocene age of the Gebel Qatrani Formation, Fayum, Egypt. *J. Hum. Evol.* 24:207-218.
Gingerich, P. D., and Martin, R. (1981) Cranial morphology and adaptations in Eocene Adapidae II: The Cambridge skull of *Adapis parisiensis. Am. J. Phys. Anthropol.* 56:235-257.
Gingerich, P. D., and Simons, E. L. (1977) Systematics, phylogeny, and evolution of early Eocene Adapidae (Mammalia, Primates) in North America. *Contr. Mus. Paleontol. Univ. Mich.* 24:245-279.
Godinot, M. (1984) Un nouveau genere temoignant de la diversite des Adapines (Primates, Adapidae) a l'Eocene terminal. *C.R. Acad. Sci. Paris,* Ser. 2 299:1291-1296.
Godinot, M. (in press) Early North African primates and their significance for the origin of Simiiformes (=Anthropoidea). In J. G. Fleagle and R. F. Kay (eds.): *Anthropoid Origins.* New York: Plenum Press.
Godinot, M., and Mahboubi, M. (1992) Earliest known simian primate found in Algeria. *Nature* 357:324-326.
Gregory, W. K. (1920) On the structure and relation of *Notharctus,* an American Eocene primate. *Mem. Am. Mus. Nat. Hist.* n.s. 351:243.
Hartenberger, J.-L., and Marandat, B. (1992) A new genus and species of an early Eocene primate from North Africa. *Hum. Evol.* 7:9-16.
Kappelman, J. (1992) The age of the Fayum primates as determined by paleomagnetic reversal stratigraphy. *J. Hum. Evol.* 22:495-503.
Kay, R. F., and Simons, E. L. (1983) Dental formulae and dental eruption patterns in Parapithecidae (Primates, Anthropoidea). *Am. J. Phys. Anthropol.* 62:363-375.
Kay, R. F., and Williams, B. A. (1992) Dental evidence for anthropoid origins. *Am. J. Phys. Anthropol.* Supplement 14:98.
Kay, R. F., Thewissen, J. G. M., and Yoder, A. D. (1992) Cranial anatomy of *Ignacius graybullianus* and the affinites of the Plesiadapiformes. *Am. J. Phys. Anthropol.* 89:477-498.
Krause, D. W., and Maas, M. (1990) The biogeographic origins of the late Paleocene-early Eocene mammalian immigrants to the western interior of North America. In T. M. Bown, and K. D. Rose (eds.): *Dawn of the age of mammals in the northern part of the Rocky Mountain Interior, North America.* Boulder: Geological Society of America, pp. 71-105.

Luckett, W. (1975) Ontogeny of the fetal membranes and placenta: Their bearing on primate phylogeny. In W. Luckett, and F. S. Szalay (eds.): *Phylogeny of the primates: A multidisciplinary approach.* New York: Plenum Press, pp. 157-182.

Maas, M., Krause, D. W., and Strait, S. G. (1988) The decline and extinction of Plesiadapiformes (Mammalia: ?Primates) in North America: Displacement or replacement? *Paleobiology* 14:410-431.

MacPhee, R. D. E., and Cartmill, M. (1986) Basicranial structures and primate systematics. In D. R. Swindler and J. Erwin (eds.): *Comparative primate biology,* Vol. 1: *Systematics, evolution and anatomy.* New York: Alan R. Liss, pp. 219-275.

Martin, R. (1990) *Primate Origins and Evolution: A Phylogenetic Reconstruction.* Princeton, NJ: Princeton University Press.

Martin, R. D. (1993) Primate Origins: Plugging the Gaps. *Nature* 363:223-234.

McHenry, H. (1975) Fossils and the mosaic nature of human evolution. *Science* 190:425-431.

Miyamoto, M., and Goodman, M. (1990) DNA systematics and evolution of primates. *Ann. Rev. Ecol. Syst.* 21:197-220.

Patterson, C. (1981) Significance of fossils in determining evolutionary relationships. *Ann. Rev. Ecol. Syst.* 12:195-223.

Pocock, R. I. (1918) On the external characters of the lemurs and of *Tarsius. Proc. Zool. Soc. Lond.* 1918:19-53.

Rasmussen, D. T. (1986) Anthropoid origins: A possible solution to the Adapidae-Omomyidae paradox. *J. Hum. Evol.* 15:1-12.

Rasmussen, D. T. (in press) The different meanings of a tarsioid-anthropoid clade and a new model of anthropoid origin. In J. G. Fleagle and R. F. Kay (eds.): *Anthropoid Origins.* New York: Plenum Press.

Rasmussen, D. T., and Simons, E. L. (1988) New specimens of *Oligopithecus savagei,* early Oligocene primate from the Fayum, Egypt. *Folia Primatol.* 51:182-208.

Rasmussen, D. T., and Simons, E. L. (1992) Paleobiology of the oligopithecines, the earliest known anthropoid primates. *Int. J. Primatol.* 13:477-508.

Rasmussen, D. T., Bown, T. M., and Simons, E. L. (1992) The Eocene-Oligocene transition in continental Africa. In D. Prothero and W. Berggren (eds.): *Eocene-Oligocene Climatic and Biotic Evolution.* Princeton, NJ: Princeton University Press, pp. 548-566.

Rose, K. D., and Bown, T. M. (1991) Additional fossil evidence on the differentiation of the earliest euprimates. *Proc. Natl. Acad. Sci. USA* 88:98-101.

Rose, K. D., and Walker, A. (1985) The skeleton of early Eocene *Cantius,* oldest lemuriform primate. *Am. J. Phys. Anthropol.* 66:73-89.

Rose, K. D., Godinot, M., and Bown, T. M. (in press) The early radiation of euprimates and the initial diversification of Omonyidae. In J. G. Fleagle and R. F. Kay (eds.): *Anthropoid Origins.* New York: Plenum Press.

Rosenberger, A. L. (1985) In favor of the *Necrolemur*-tarsier hypothesis. *Folia Primatol.* 45:179-194.

Rosenberger, A. L. (1986) Platyrrhines, catarrhines, and the anthropoid transition. In B. Wood, L. Martin, and P. Andrews (eds.): *Major Topics in Primate and Human Evolution.* Cambridge: Cambridge University Press, pp. 66-88.

Rosenberger, A. L., and Dagosto, M. (1992) New craniodental and postcranial evidence of fossil tarsiiforms. In S. Matano, R. Tuttle, H. Ishida, and M. Goodman (eds.): *Topics in Primatology,* Vol. 3: *Evolutionary Biology, Reproductive Endicrinology, and Virology.* Tokyo: University of Tokyo Press, pp. 37-51.

Rosenberger, A. L., and Szalay, F. S. (1980) On the tarsiiform origins of Anthropoidea. In R. L. Ciochon and A. B. Chiarelli (eds.): *Evolutionary Biology of the New World Monkeys and Continental Drift.* New York: Plenum Press, pp. 139-157.

Ross, C. (in press) The craniofacial evidence for anthropoid and tarsier relationships. In J. G. Fleagle and R. F. Kay (eds.): *Anthropoid Origins.* New York: Plenum Press.

Russel, D., and Gingerich, P. D. (1987) Nouveaux primates de l'Eocene du Pakistan. *C.R. Acad. Sci. Paris,* Ser. 2 304:209-214.

Sige, B., Jaeger, J.–J., Sudre, J., and Vianey–Liaud, M. (1990) *Altiatlasius koulchii* n. gen. et sp., primate omomyide du Paleocène supérieur du Maroc, et les origines des euprimates. *Paleontographica Abt.* A 214:31-56.

Simmons, N. B., Novacek, M. J., and Baker, R. J. (1991) Approaches, methods, and the future of of the Chiropteran monophyly controversy: A reply to J. D. Pettigrew. *Syst. Zool.* 40:239-243.

Simons, E. L. (1961) The dentition of *Ourayia:* Its bearing on relationships of omomyid prosimians. *Postilla* 54:1-20.

Simons, E. L. (1962) A new Eocene primate genus *Cantius,* and a revision of some allied European lemuroids. *Bull. Brit. Mus. (Nat. Hist.) Geol.* 7:1-30.

Simons, E. L. (1987) New faces of *Aegyptopithecus* from the Oligocene of Egypt. *J. Hum. Evol.* 16:273-289.

Simons, E. L. (1990) Discovery of the oldest known anthropoidean skull from the Paleogene of Egypt. *Science* 247:1567-1569.

Simons, E. L. (1992) Diversity in the early Tertiary anthropoidean radiation in Africa. *Proc. Natl. Acad. Sci. USA,* 89:10743-10747.

Simons, E. L., Bown, T. M., Rasmussen, D. T. (1986) Discovery of two additional prosimian primate families (Omomyidae, Lorisidae) in the African Oligocene. *J. Hum. Evol.* 15:431-438.

Simons, E. L., and Kay, R. F. (1988) New material of *Qatrania* from Egypt with comments on the phylogenetic position of the Parapithecidae (Primates, Anthropoidea). *Am. J. Primatol.* 15:337-347.

Simons, E. L., and Rasmussen, D. T. (1989) Cranial morphology of *Aegyptopithecus* and *Tarsius* and the question of the tarsier-anthropoid clade. *Am. J. Phys. Anthropol.* 79:1-23.

Simons, E. L., and Russell, D. (1960) Notes on the cranial anatomy of *Necrolemur. Breviora* 127:1-14.

Strait, S. G. (1991) Dietary Reconstruction in Small–Bodied Fossil Primates. Ph.D. Thesis, SUNY at Stony Brook.

Sudre, J. (1975) Un prosimien du Paleogene ancien du Sahara Nord–Occidental: *Azibius trerki* n.g.n.sp. *C.R. Acad. Sci. Paris,* Ser. D 280:1539-1542.
Szalay, F. S., and Dagosto, M. (1980) Locomotor adaptations as reflected in the humerus of Paleogene primates. *Folia Primitol.* 34:1-45.
Szalay, F. S., and Li, C.–K. (1986) Middle Paleocene euprimate from southern China and the distribution of primates in the Paleogene. *J. Hum. Evol.* 15:387-397.
Szalay, F. S., and Wilson, J. (1976) Basicranial morphology of the early Tertiary tarsiiform *Rooneyia* from Texas. *Folia Primatol.* 25: 288-293.
Tattersall, I., and Schwartz, J. (1985) Evolutionary relationships of living lemurs and lorises (Mammalia, Primates) and their potential affinities with European Eocene Adapidae. *Anthropol. Papers Am. Mus. Nat. Hist.* 60:1-110.
Thomas, H., Roger, J., Sen, S., Bourdillon de Grissac, C., and Al–Sulaimani, Z. (1989) Découverte de vertébrés fossiles dans L'Oligocène inférieur du Khofar (Sultanat d'Oman). *Géobios* 22:101-120.
Washburn, S. (1960) Tools and human evolution. *Sci. Amer.* 203:63-75.
Wible, J., and Covert, H. (1987) Primates: Cladistic diagnosis and relationships. *J. Hum. Evol.* 16:1-22.

CHAPTER 3

Reinterpreting Past Primate Diversity

Russell L. Ciochon
Dennis A. Etler

INTRODUCTION

Primate paleontology is an important focus of anthropological research and through its study we can gain a better appreciation of both human evolutionary biology and the circumstances leading to the emergence of our family—the Hominidae. The study of primate paleontology entails not only the investigation of the "hard evidence" of actual fossil remains; consideration is also given to analysis of the nature of past primate diversity, including an assessment of primate adaptive radiations, the divergence times of major primate lineages, primate distribution and dispersal patterns in time and space, and the ecological adaptations of past primate populations. If we are to place our origins in proper perspective, we must, therefore, be concerned not only with identifying our immediate and processional ancestors, but also with developing an appreciation of the true extent of primate diversity in the past, for it is from amidst this diversity that we emerged.

Primate paleontologists are faced with a number of challenges. The fossil record is tantalizingly incomplete. Often only the barest minimum of evidence is preserved from which conclusions can be drawn. Generally speaking most primate fossils consist of isolated teeth or fragmentary upper and lower jaws. Relatively complete cranial remains are few and far between. Diagnostic postcrania are even rarer. To make matters worse, primate teeth are notoriously labile; that is, they are subject to quick morphological change, rampant parallelism, convergence, and reversals in evolutionary trends. When more complete fossil remains of particular anatomical regions of extinct species are recovered, preconceived notions of past primate adaptation are often found to be false and the picture is frequently more complicated than previously thought. Given these circumstances, it is not surprising that the study of primate paleontology is replete with wide swings of the pendulum. Groups have fallen in and out of favor as potential ancestors as new evidence is accumulated and old evidence is reevaluated in line with advances in molecular biology and the earth sciences. We feel that primate paleontology has entered into such a period. Old paradigms are falling as new fossil evidence, some from areas of the world poorly sampled in the past, comes to light.

In addition to the problems just enumerated, the study of primate paleontology has also been plagued by an overemphasis on tracing the lineal descent of living species. While this is a natural and unavoidable tendency, it inevitably leads to the notion that past primate diversity can best be seen within the context of modern primate systematics; hence, the ubiquitous application of neontological terms to paleontological taxa—for example, the common use of the term "hominoid" to describe fossils that predate the divergence of monkeys and apes. In fact, most primate lineages have gone extinct without issue and the majority of fossil primates are certainly phylogenetically tangential to living forms. It may be assumed, for instance, that at any given time in the past, since at least the late Eocene, primates were as diverse as they are now. A conservative count of extant primate taxa indicates at least 15 families and perhaps 60 genera (Fleagle 1989). Many of these extant primate families are thought to have emerged only within the last 15-20 million years (particularly amongst the Anthropoidea). If we extrapolate these figures back to the early Miocene, it becomes apparent that much of the diversity we encounter has little or nothing to do with living primate taxa. As we go further back in the past, this becomes more and more the case.

Another set of problems relates to long-standing arguments within paleoprimatology. Over the last two decades, positions on a number of important issues became ossified and many crucial questions seemed irresolvable. Plesiadapiforms were accepted or rejected as primates more on theoretical grounds relating to nuances of systematic philosophy than on the hard fossil evidence. Anthropoids were derived without question from either the adapiform or omomyiform Eocene euprimates. The center of hominoid evolution was seen as Africa, with Europe and Asia peripheral recipients of successive hominoid dispersals. Our understanding of primate evolution has been limited to a very real extent by exigencies of the fossil record but also by long-held biases and assumptions that the meager fossil record engendered and tended to reinforce. Given the finite resources available for the study of primate paleontology and the lack of access to many areas of the world potentially rich in fossil primates, this limited focus was unavoidable. It has, nevertheless, tended to skew our appreciation of past primate diversity. As a consequence, certain assumptions have became entrenched in the literature, and a number of decisive questions relating to primate origins and evolution have been left unresolved. These include (1) when and where the earliest representatives of the primate order first appeared; (2) what the systematic relationship between extant lower and higher primates is; (3) how the present distribution of primate taxa can best be explained; (4) when, where and under what circumstances the Anthropoidea evolved; and (5) how the evolution of the modern hominoids can be understood in light of the fossil evidence, molecular biology, and biogeography. Primate paleontology has until recently been unable to adequately address many of these issues.

In this chapter, we attempt to reconstruct the singular evolutionary history of that portion of the primate record most germane to human origins by synthesizing the results of recent advances in the study of primate paleontology conducted in Africa, Europe, and Asia with modern studies of primate molecular biology, primate systematics, and primate neontology. A number of alternate hypotheses that run counter to the prevailing wisdom, such as it may be, concerning primate origins and the evolution of the primate taxa to which we humans belong will be presented.

A TERMINOLOGICAL NOTE

Throughout this chapter, we follow a convention that applies the prefix "proto-" to members of a stem lineage and its collateral sister lineages. A stem lineage is taken to mean the succession of taxa in direct procession to a particular crown group. We in turn define a crown group as consisting of all descendants both living and extinct of the last common ancestor of any given extant clade (Wible and Covert 1987; see Figure 3-1.) For example, the extant lesser and greater apes and humans constitute a holophyletic clade given the taxonomic name Hominoidea. In our usage the term "hominoid" is used only in reference to descendants of the last common ancestor of the extant gibbon, orangutan, gorilla, chimpanzee and humans. Members of the stem lineage leading to the hominoid stem species and collateral lineages diverging from it are termed "protohominoids." In this fashion the hominoid stem in combination with its crown group are referred to as the hominoid closed descent community (Ax 1985; see Figure 3-1).

Another source of terminological confusion rests with nomina used to identify modern hominoid clades. Given the widely held belief, to which we subscribe, that humans are more closely allied to the African rather than Asian apes, the traditional dichotomy between a family of great apes, Pongidae, and the human family, Hominidae, can no longer be justified. We find the current fashion of using the term "hominid" to refer to the great ape/human clade to be equally unacceptable since this usage creates

Figure 3-1 Terms used in this chapter for phylogeny reconstruction are after Ax (1985). The loose application of taxonomic terms derived from neontological systematics to the primate fossil record has tended to obscure rather than elucidate an understanding of past primate diversity. We utilize a number of terms introduced by Hennig (1966) and refined by Ax (1985) to describe the diachronic process of phylogeny reconstruction within the synchronic Linnaean taxonomic framework. In this fashion, the last common ancestor of an extant clade (i.e., its stem species) serves as a convenient phylogenetic marker around which phylogeny reconstruction can be organized. Members of lineages between two stem species, both those along the stem lineage itself and those diverging from it along collateral lineages, assume the prefix proto- to the colloquial term applied to that lineage's crown group (see text for definitions). In the example given, members of the hominoid stem lineage and associated collateral lineages are termed "protohominoids" rather than the more commonly used but ambiguous term "hominoid." In the same fashion, members of the anthropoid stem and collateral lineages that fall between the primate and anthropoid stem species are best called "protoanthropoids" (see text for elaboration). It is important to emphasize that protolineages do not share in all the derived features that define their living descendants as members of a holophyletic clade.

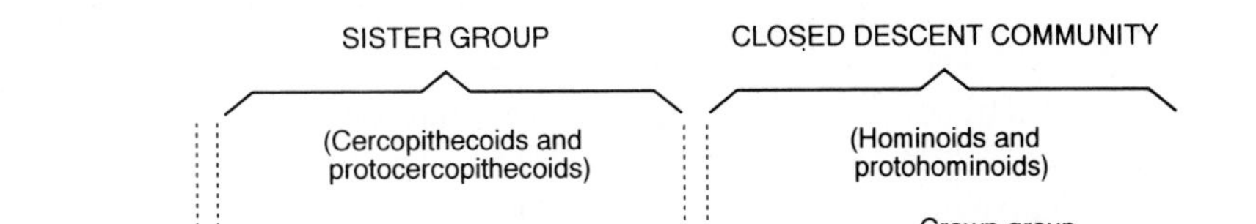

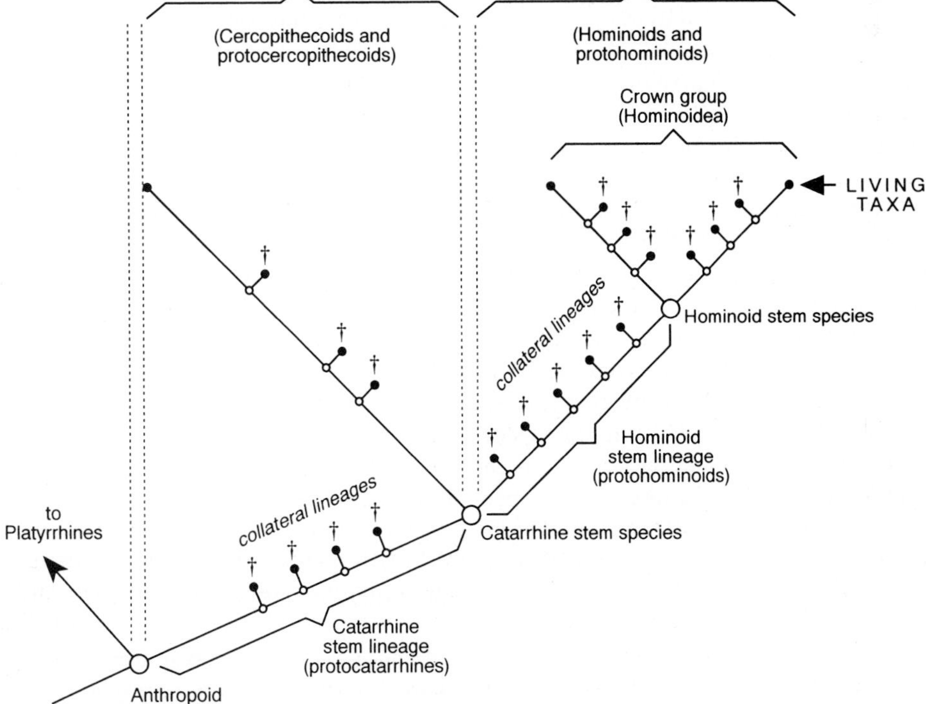

considerable confusion even for the sophisticated lay reader. We will, therefore, continue to use the term "hominid" to refer to members of an exclusive human clade. We will, however, refrain from using the term "pongid" except in the narrow sense of the clade of Asian apes represented by *Sivapithecus, Gigantopithecus,* and the modern *Pongo.* The modern African great apes can be viewed as collateral protohominids, part of the hominid closed descent community but not of the hominid crown group. In this fashion the chimp and gorilla are treated as living fossils or phylogenetic relics.

MODERN PRIMATES

As a prelude to discussing the past diversity of primates, it is useful to review the living groups of primates and their distributions. The order Primates has historically been divided into two major subgroupings. The lower primates or prosimians comprise the lemurs of Madagascar, the lorises of Africa and Asia, and the tarsiers of island Southeast Asia. The higher primates or anthropoids comprise the New World monkeys of South and Central America, the Old World monkeys of Africa and Asia, the lesser apes (the gibbon and siamang of Southeast Asia), the great apes (the orangutan of island Southeast Asia and the chimpanzee and gorilla of equatorial Africa), and humans (*Homo sapiens* and their fossil forebears).

Within the last two decades, an alternate interpretation of primate systematics has gained considerable support. It divides the order into the subordinal taxa Strepsirhini and Haplorhini (Szalay and Delson 1979). The former include the living lemurs and lorises and their supposed fossil progenitors the adapiforms. The latter include the living tarsiers and anthropoids and their purported fossil antecedents the omomyiforms (a review in support of this position is given by Andrews 1988). Although this scheme has gained a sizeable number of adherents we consider these subordinal taxa to be artificial constructs that do not reflect true evolutionary relationships (*vide infra*).

PRIMATE ORIGINS

The earliest mammals considered by some to be members of the primate order, the plesiadapiforms—often referred to as "archaic primates"—first appear abruptly in the fossil record of central North America in the early Paleocene (about 63 MYA) (Szalay and Delson 1979). From the mid-1960s to quite recently, a growing consensus developed which maintained that the first "euprimates" or "primates of modern aspect" evolved from early forms of these rather rodentlike plesiadapiforms (see Szalay et al. 1987 for an endorsement of this position). The idea that plesiadapiforms are primordial primates has, however, always been controversial (cf. Cartmill 1974), and many researchers are becoming more and more hesitant in ascribing primate affinities to them (Covert 1986; Wible and Covert 1987). This trend of thought has been reinforced by the recent recovery of postcranial remains associated with the paromomyid plesiadapiforms *Phenacolemur* and *Ignacius.* These new fossils convincingly demonstrate that paromomyids share derived features that ally them to the modern order of flying lemurs, the Dermoptera, and not primates (Beard 1990; Martin 1990; but see also Krause 1992). Other members of the plesiadapiforms, in particular the genus *Plesiadapis* itself, have been shown to share derived basicranial features that link them to the Dermoptera as well (Kay et al. 1990). Another set of fossils epitomized by the genus *Ekgmowechashala* from the late Oligocene of South Dakota, which has previously been considered highly derived tarsiiforms (Szalay and Delson 1979), have more recently been assigned to the family Plagiomenidae?Dermoptera (McKenna 1990) further demonstrating that dental features that have in the past been used to link plesiadapiforms and other specialized fossils (e.g., picrodontids, mixodectids, etc.) with the order Primates can be more convincingly seen as parallelisms or convergences rather than shared derived features indicative of a common ancestry.

There is little question that the dermopterans should not be considered primates. They are, however, candidates for sister group status with the primates, as are tree shrews (Scandentia) (McKenna 1966) and fruit bats (Megachiroptera) (Pettigrew 1986). All these groups, primates and microchiropterans included, have been placed within the supraordinal grouping Archonta (see Novacek 1990, 1992 for recent reviews of supraordinal mammalian clades), which may very well represent the initial radiation of eutherian mammals into an arboreal mileau toward the end of the Cretaceous (Sussman 1991). The plesiadapiforms, therefore, are best seen as protodermopterans and well outside the primate closed descent community.

It may be asked by the nonspecialist at this juncture, how could so many have been so wrong for so long? The problem lies primarily in the sparseness of the fossil record and the inevitable reliance on dental features for phylogeny reconstruction. The difficulty in attributing early Cenozoic dental remains to specific mammalian clades should not be underestimated. The eutherian adaptive radiation was just beginning during the transition from the Cretaceous to Tertiary periods, and many of the derived dental features that distinguish modern higher mammalian taxa from one another were incipient at best. The degree of parallelism and convergence in derived dental features among divergent lineages seems to have been considerable—witness the historic difficulty in sorting plesiadapiforms from true primates. As Kielan-Jaworowska et al. (1979:224) so aptly state in regard to early Cenozoic mammalian history, "Similarities and differences in comparative odontology cannot unequivocally be considered the result of common inheritance or indicative of 'phyletic distance.'"

The fossil record of plesiadapiforms, restricted to North America and Europe, has in the past directed attention to these continents (which were linked by a northern land bridge during the early Tertiary) as the likely source area for the later radiation of modern primates (i.e., euprimates). If, as Beard and Kay et al. suggest, the plesiadapiforms are read out of primate ancestry, there is no further need to look toward America or Europe as the center of origin of the primate order.

WHAT WERE THE FIRST REAL PRIMATES?

If we delete the plesiadapiforms from the primate fold the first primate can best be seen as the last common ancestor of all living primates species the lemurs, lorises, tarsiers and anthropoids—and early prosimian primates known in the fossil record from the Eocene of North America and Europe (i.e., the adapiforms and omomyiforms). In our opinion there has been an unfortunate tendency to subsume all nonanthropoid Eocene primates into one or another family level taxon—the Adapidae or Omomyidae (Szalay and Delson 1979; Fleagle 1988). We believe that continued use of these nomina to encompass the full diversity of Paleogene "lower primates" tends to obscure rather than illuminate the true evolutionary relationships of the various extinct prosimian lineages. We advocate use of the higher level taxonomic nomina Adapiformes (i.e., paraphyletic protolemuriforms), Omomyiformes (i.e., paraphyletic prototarsiiforms) and Simiiformes (i.e., paraphyletic protoanthropoids) and the erection of holophyletic family-level taxa to accommodate this diversity. There is some consensus, for instance, that the Eocene adapiforms of North America represent a single evolving lineage, the notharctines, which we would reelevate to the family level as Notharctidae. Some of the adapiforms of the European middle and late Eocene can be united into the Adapidae (s.s.). The omomyiforms can be tentatively divided into the primitive Euroamerican Anaptomorphidae, the primarily North American Omomyidae (*s.s.*) and the exclusively European Microchoeridae. Many Asian fossils previously attributed to the Adapidae and Omomyidae should almost certainly be removed from those taxa. The Miocene forms *Sivaladapis* and *Sinoadapis*, for instance, merit a family-level taxon, the Sivaladapidae. Many other Paleogene prosimian fossils from Asia are too scanty to be placed with assurance in any given family-level taxon.

The adapiforms and omomyiforms appear abruptly in the fossil record of both North America and Europe at the beginning of the Eocene (Maas et al. 1988). If the antecedent plesiadapiforms are not included within Primates, the adapiforms and omomyiforms would represent the earliest well-documented members of the order. The question then arises, when and where did these two groups initially evolve? It should be emphasized that the extensive Paleocene deposits of Western North America and Europe have not yielded a trace of material that can be directly linked to either of these early primate groups (Krause and Maas 1990). There is, moreover, increasing dental evidence that the differentiation of the adapiforms and omomyiforms from each other had not proceeded very far by the early Eocene (Rose and Bown 1991). It is, therefore, reasonable to assume that they shared an extra Euroamerican common ancestry sometime in the middle to late Paleocene.

Since the 1920s, paleontologists have looked to the adapiforms and omomyiforms as the ancestral stocks from which all living primates evolved. Yet there is no consensus regarding which of these Eocene taxa is the most closely related to which living primates (see varying interpretations in Aiello 1986; Covert 1986; Gingerich 1984; Rasmussen 1986; Szalay et al. 1987; Schwartz and Tattersall 1987; Beard et al. 1988; Simons and Rasmussen 1989). The position of the anthropoids in this regard has historically been the most vexing problem of all in primate phylogenetics and will be addressed next.

AN EARLY DIVERGENCE HYPOTHESIS

Based on a variety of new evidence, to be discussed later, we argue that omomyiforms and adapiforms are both prosimian taxa more closely related to each other and the living prosimians (including *Tarsius*) than either is to the Anthropoidea. This contention is supported in part by an analysis of the blood proteins of all living primate groups by Sarich and Cronin (1976) that revealed an unresolved trichotomy (apparent three-way split) for the three living monophyletic primate clades—lemurs plus lorises, tarsiers, and anthropoids. The molecular clock, moreover, indicates that this divergence occurred perhaps as early as 60 MYA. More recent studies of mitochondrial DNA evolution in primates (Hasegawa et al. 1990) reinforce Sarich and Cronin's earlier work and conclude that lemurs and tarsiers constitute a clade distinct from the anthropoids (see also Corruccini this volume). This analysis of higher level primate systematics based on molecular studies is still a minority view. Contrary views of primate phylogeny, based on the nucleotide sequencing of primate betaglobin gene clusters and earlier immunodiffusion tests, that support the sister group relationship of tarsiers and anthropoids are given in Koop et al. (1989), Dene et al. (1976), and Goodman et al. (1978).

Hasegawa et al. (1990), however, place the divergence of anthropoids from prosimians at the beginning of the Cenozoic and the divergence of the lineages leading to the lemurs and tarsiers soon thereafter. Thus, biochemically, it can be argued that the higher primates do not share any appreciable common ancestry with any extant lower primate group prior to their origin as a distinct lineage.

The molecular evidence previously cited, which supports the monophyly of the prosimians, is corroborated by recent work on the pedal, manual, basicranial and narial morphology of living and fossil primates. These studies demonstrate that (1) the adapiforms constitute a nested set of taxa that share pedal features derived from a last common ancestor and from which the radiation of the holophyletic lemuriform primates is most parsimoniously derived (Gebo 1986; Beard et al. 1988; Beard and Godinot 1988; Covert 1988; Dagosto 1988); (2) the omomyiforms likewise consist of a nested set of taxa, some of which are more closely linked than others with the tarsiers by the presence of basicranial synapomorphies (Beard et al. 1991); (3) the simiiform hand lacks specialized grasping features seen in adapiforms and later lemuriforms and may better approximate the euprimate morphotype than the more derived prosimian hand (Godinot 1992; Godinot and Beard 1991); and (4) the omomyiforms and adapiforms were structurally strepsirhine (Beard 1988), suggestive of the fact that tarsiids and anthropoids (i.e., the neontological haplorhines) either form a clade to the exclusion of known omomyiforms (see also MacPhee and Cartmill

Figure 3-2 (Opposite page) Alternative phylogenetic models for primate origins: Model 1 (after Rasmussen 1986) suggests that adapiforms have a closer relationship to both tarsiiforms and anthropoids than to lemuriforms. This model is contravened by derived pedal features that link adapiforms to lemuriforms (Beard et al. 1988). Model 2 posits omomyiforms as the sister group of anthropoids while Model 3 has anthropoids as the sister group of the more exclusive clade of tarsiids. Models 2 and 3 are variants of the strepsirhine/haplorhine subordinal division, which has gained widespread support over the last two decades (see critique of these two models in text). Model 4, championed by Gingerich (1984), has anthropoids as derivative of adapiforms based primarily on his interpretation of the paleontological evidence. Model 5, our preference, maintains the integrity of the Prosimii and hypothesizes that any similarities between tarsiers and anthropids are due to convergence. It also suggests that anthropoid origins preceded the differentiation of the prosimians from one another.

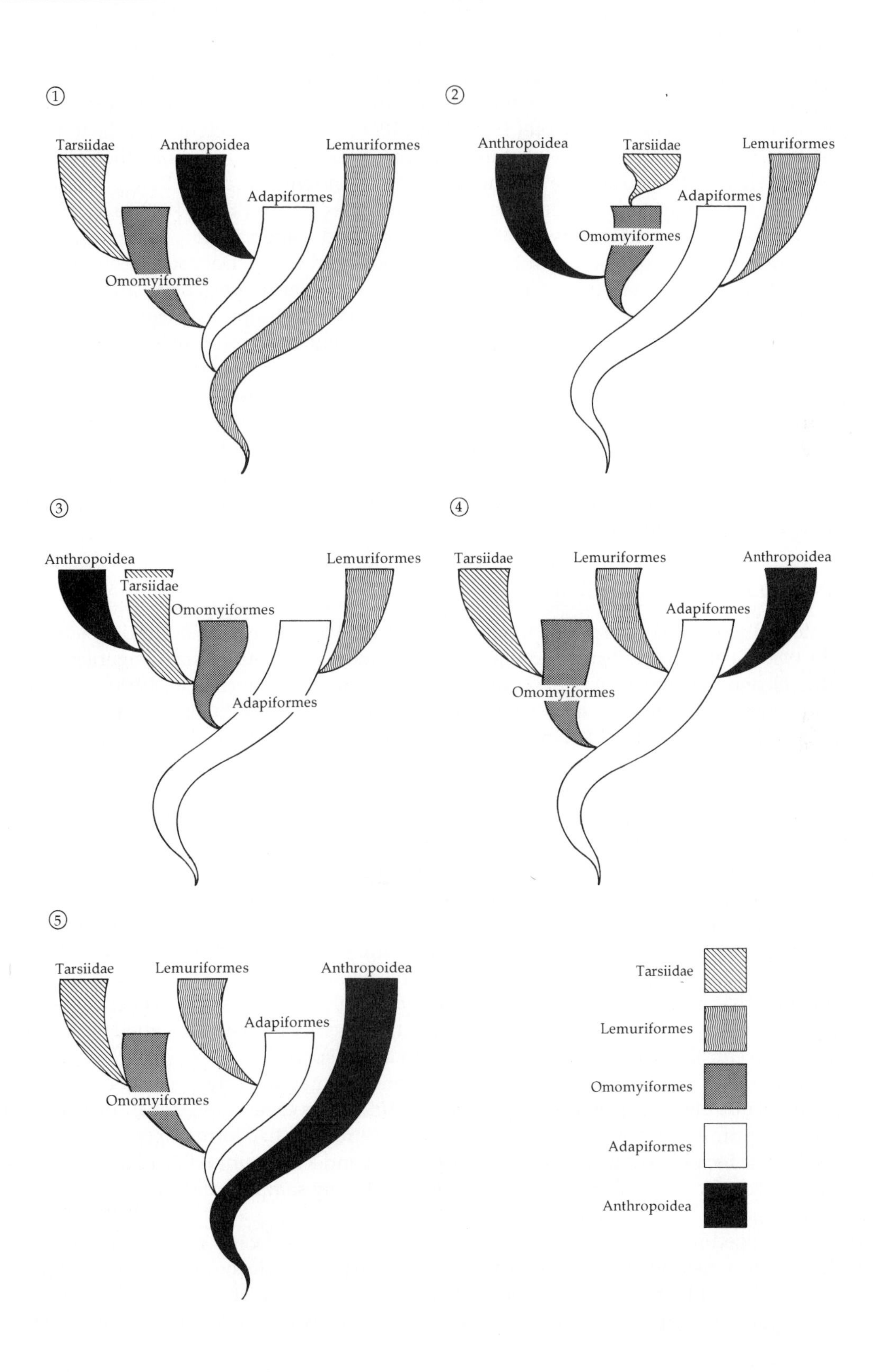
①
Tarsiidae
Anthropoidea
Lemuriformes
Adapiformes
Omomyiformes
②
Anthropoidea
Tarsiidae
Lemuriformes
Adapiformes
Omomyiformes
③
Anthropoidea
Tarsiidae
Omomyiformes
Lemuriformes
Adapiformes
④
Tarsiidae
Lemuriformes
Anthropoidea
Adapiformes
Omomyiformes
⑤
Tarsiidae
Lemuriformes
Anthropoidea
Adapiformes
Omomyiformes
Tarsiidae
Lemuriformes
Omomyiformes
Adapiformes
Anthropoidea

1986) or alternatively, the haplorhine condition evolved independently in tarsiids and anthropoids (see also Simons and Rasmussen 1989). We suggest that the latter hypothesis is more likely. The strepsirhine condition, while best seen as primitive for therians (Beard 1988), is nevertheless anatomically more complex than the haplorhine condition (Hofer 1980; Beard 1988). We, therefore, consider the independent loss of strepsirhinism in the Tarsiidae and Anthropoidea to be as or more likely than the acquisition of haplorhinism as a shared derived feature linking them in a monophyletic clade. Many other soft- and hard-body characters thought to be haplorhine synapomorphies (cf. Andrews 1988) may likewise be parallelisms (see discussions in Rasmussen 1986; Schwartz and Tattersall 1987) or nonhomologous anatomical features (Simons and Rasmussen 1989). In addition Ford (1988), Dagosto (1990), and Godinot (1992) demonstrate that prosimian (including omomyiform and tarsier) postcrania are more derived relative to the primitive euprimate pattern than are the postcrania of anthropoids. It now seems likely that the anthropoids are not derived from the Eocene radiation of either adapiforms or omomyiforms as classically conceived (see also Ford 1988) but have separate and deeper roots within the initial radiation of the Order Primates. While Dagosto (1990) rejects the holophyly of the Prosimii in favor of the derivation of anthropoids from some aspect of the prosimian radiation, a position that is acknowledged to entail a considerable number of evolutionary reversals, our line of reasoning would tend to support Prosimii as a valid holophyletic taxon and see the anthropoids as independently derived from a postcranially primitive ancestral primate morphotype (cf. Ford 1988; Godinot 1992). These considerations will be amplified in the discussion to follow.

EARLY ASIAN AND AFRICAN ANCESTORS?

Recently a spate of discoveries of purported fossil primates from Paleocene and Eocene deposits in China (Szalay and Li 1986; Qi et al. 1991; Beard and Wang 1991; Culotta 1992), south Asia (Gingerich and Sahni 1979, 1984; Russell and Gingerich 1980, 1987) and southeast Asia (Suteethorn et al. 1988), coupled with earlier, still controversial discoveries from China and Mongolia (Zdansky 1930; Woo and Chow 1957; Chow 1961, 1964; Dashzeveg and McKenna 1977; Tong 1979; Gingerich et al. 1991) have refocused attention on Asia as a possible center of early primate evolution (see Szalay and Li 1986; Beard and Wang 1991 for reviews of the Chinese specimens). Much of this material is very fragmentary, of doubtful affinity, or still under study. Of particular note, however, is the recovery of a remarkable suite of primitive euprimates from middle Eocene fissure fills at Shanghuang, Jiangsu province, China. Preliminary accounts (Culotta 1992) suggest that a variety of protoanthropoid, adapiform and tarsiid taxa are represented. The purported protoanthropoids are of very small size, lack a fused mandibular symphysis, and retain a generally primitive dental morphology coupled with certain derived features of the molar dentition that link them preferentially with later undoubted anthropoids. This new and hitherto unknown primate diversity in China during the Eocene hints at the possibility that Asia played a much more prominent role in early primate evolution than previously thought. As the early Paleogene fossil record of Chinese primates becomes better known, the likelihood increases that Asia is as good a candidate as any for occupying a central role in early primate evolution and dispersal.

Africa, as well, has until recently been a void when it comes to early primate evolution. The discovery of a true late Paleocene primate from Morocco, *Altiatlasius koulchii* (Sige et al. 1990; Gingerich 1990) and a suite of primitive primates including an early anthropoid-like form, *Algeripithecus minutus*, from a middle Eocene locality in Algeria (Sudre 1975; Godinot and Mahboubi 1992) again highlights the fact that the fossil record of early Tertiary primates has been barely sampled and that many more finds are surely to be made.

Although indulged in here, emphasis on the place of discovery of early primate specimens, be it Europe, North America, Asia or Africa, while inevitable, tends to perpetuate the notion that one or another region of the modern world served as homeland for the primate radiation. This is certainly mistaken as it envisions a static paleogeography. The Africa or Eurasia of 60 MYA was very different from the Africa or

Eurasia of today. We must seek to understand the distribution of primates in the past based on the paleogeography of the time in question. It may be more reasonable to think of north Africa during the early Paleogene as part of a sub-Tethyan biozone that linked together the southern Tethys littoral from the Atlas mountains through the Afroarabian peninsula via an island-hopping sweepstakes route to the northern shores of the Indian subcontinent and on to what is today southeast and east Asia. The lower molar morphology of *Altiatlasius,* for instance, shows clear phenetic similarities to the early Eocene Pakistani primate *Kohatius coppensi* (Russell and Gingerich 1980, 1987), which has been previously linked to early omomyiforms. If these resemblances have phylogenetic significance the biogeographic implications would be intriguing.

While north Africa as well as Europe and North America were certainly major theaters of early primate evolution, the placement of primates in Asia during the early Paleogene has some interesting ramifications. Asia is well placed geographically to have been a center of dispersal for adapiform and omomyiform primates to North America via the Bering land bridge and Europe via a southern land corridor across the Turgai Strait, the Paleogene epicontinental sea that separated Europe from Central Asia along what are now the Ural Mountains. In fact the recent discovery of the omomyid (*sensu stricto*) primate *Asiomomys changbaicus* from late Eocene deposits in the northeastern Chinese province of Jilin (Wang and Li 1990; Beard and Wang 1991) demonstrates the existence of just such a northerly primate dispersal route between Asia and North America during the early Tertiary as omomyids (*s.s.*) have previously been known only from the Eocene deposits of western North America. The search for the specific adapiform and omomyiform stocks from which the lemuriforms and tarsiiforms evolved may, therefore, best be directed toward Asia. The possibility exists that the direct ancestors of both these radiations is not represented among prosimian fossils previously known from Europe or North America.

While the fossil evidence supporting the hypothesis that the three surviving primate macroclades (Lemuriformes, Tarsiiformes and Anthropoidea) emerged in Asia from lineages that diverged from the stem primate stock is still far from persuasive, other evidence of a biogeographical nature can be marshalled to lend it support. In order to understand the origins of the modern primate groups and their current distribution it is, therefore, necessary to first discuss the distribution of the modern prosimian primates and the paleogeography of the Earth during the early Tertiary.

ISLAND HOPPING TO MADAGASCAR

One of the most perplexing problems of paleoprimatology has been to explain the presence of the lemuriform primates on the island of Madagascar. Given the overall phenetic similarity between the adapiforms of Eocene North America and Europe and modern lemurs, it has generally been thought that the latter evolved directly from the former and rafted to the island of Madagascar from Africa sometime in the Eocene. The adapiforms and modern strepsirhines, however, while sharing a few uniquely derived postcranial features in common are generally linked to one another by the retention of primitive features (Cartmill and Kay 1978; Rasmussen 1986). Those traits that suggest an ancestor/descendant relationship between notharctids and modern strepsirhines are generally primitive retentions (Krishtalka 1989) while derived pedal features linking adapids (*sensu stricto*) to modern strepsirhines demonstrate at best an indirect relationship between the two groups (Beard et al. 1988). This perception is reinforced by the lack of the derived tooth-comb in adapiforms (seen in some fashion in all extant strepsirhines but apparently lost in some late Pleistocene/Holocene subfossil forms) and the relatively late occurrence of adapiforms in the fossil record.

Another scenario can, however, be put forward. We postulate that the lemuriform primates (i.e., lemurs + lorises) originated in Asia as part of the initial adapiform radiation. The lemurs may have subsequently dispersed to Madagascar via an island-hopping sweepstakes route between India and Madagascar that appeared early in the Tertiary and the lorisoids may have dispersed from Asia to Africa along a sub-Tethyan dispersal route in the late Eocene. Recent paleogeographic reconstructions of the placement of India and Madagascar off the eastern coast of Africa at the time in question demonstrate the feasibility of the former possibility (Briggs 1989; Courtillot 1990; Courtillot et al. 1988; Duncan and Pyle 1988), while the discovery of a possible lorisid from early Oligocene deposits in the Fayum district of Egypt (Simons et al. 1986) supports the latter. As regards the first possibility, a review of evidence supporting the proximity of Madagascar, India, and south Asia during the early Tertiary is very informative. The Deccan flood basalts of western India have recently been dated to the Cretaceous/Tertiary boundary (Duncan and Pyle 1988). The massive accumulation of basalt that these beds represent are thought to be associated with a stationary mantle hotspot of which the most recent manifestation is the volcanic island of Reunion located a few hundred kilometers due east of northern Madagascar. The placement of the Indian subcontiment over this hotspot at the K/T boundary combined with a simultaneous drop in eustatic sea level leading to the exposure of a land bridge linking southern Tibet (which during the time in question was part of the Tethyan littoral zone) with the subcontinent would have opened up a dispersal route from Asia to Madagascar.

Of particular note in this respect is the occurrence in southern China and the Indian subcontinent of late Miocene prosimians, the sivaladapids, that do not bear any special resemblance to the earlier adapiforms and omomyiforms of North America and Europe but which share a number of derived traits that link them to the lorises and lemurs of today. This enigmatic group of prosimians is in its total morphological pattern unlike either the extant lemuriforms or earlier adapiforms. Dentally the sivaladapids of India (Gingerich and Sahni 1979, 1984) and Yunnan, China (Pan 1988; Pan and Wu 1986; Wu and Pan 1985a) preserve a primitive incisal and canine morphology, lacking entirely the derived toothcomb of the lemuriforms. They differ fundamentally from adapiforms, however, in their premolar and upper molar morphology. Adapiforms are characterized by (1) a trend toward expansion of the upper molars by the elaboration of either cingula or protocone derived hypocones and, in some species, the presence of a pericone; (2) an upper P4 that is generally submolariform; (3) an upper P3 that tends to have a protocone; and (4) upper and lower P2 and P1 that tend toward reduction and eventual loss. Sivaladapids, however, have (1) tritubercular upper molars without hypocones; (2) a highly molarized upper P4; (3) an unicuspid upper P3, subcaniniform upper P2 and lower premolars that are progressively higher cusped and caniniform anteriorly; and (4) a lower P2 that is extremely caniniform and vertically implanted. Many of the traits distinguishing sivaladapids from adapiforms are highly reminiscent of character states seen in lemurs and lorises. The possibility, therefore, exists that sivaladapids diverged early on from the adapiform stem lineage and have close ties to the lemuriform radiation. The possibility that the lemurs of Madagascar have an Asian origin deep in the Paleogene cannot, therefore, be discounted. The biogeographical distribution of the lower primates, that is, the tarsiers in insular southeast Asia; the modern lemurs on Madagascar; and lorises in southeast Asia, India, and Africa can perhaps be best understood as the end product of the dispersal of early prosimians out of Asia.

EARLY ANTHROPOIDS

A major conundrum in the study of primate evolution has been the origin of the higher primates or Anthropoidea. As previously argued, the ancestry of this group can be linked with the emergence of the earliest primates rather than either the Eocene adapiforms or omomyiforms. If the initial differentiation of the living primates (lemurs, tarsiers, and anthropoids) was a Paleocene event, as some molecular evidence suggests, rather than an Eocene event, the adapiforms would appear too late in the fossil record to be ancestral to both the modern lemurs and anthropoids as some have contended (Gingerich 1984). The same would apply to the relationship between omomyiforms, tarsiers, and anthropoids (Rosenberger and Szalay 1980). In this regard the recent discoveries by Qi et al. (1991) in China and Godinot and Mahboubi (1992) in Algeria indicate the widespread geographical dispersion of protoanthropoids during the middle

Eocene and the likelihood that protoanthropoid roots are as deep as those of other known primate clades. de Bonis et al. (1988) and Simons (1989, 1990, 1992) have, meanwhile, demonstrated that protocatarrhine primates ancestral to the Oligocene primate fauna of Egypt and the Afro-Arabian peninsula occur in what are probably late Eocene deposits of Algeria (*Biretia piveteaui*) and the Fayum (*Catopithecus browni, Proteopithecus sylviae, Plesiopithecus teras, Serapia eocaena, Arsinoea kallimos*). These fossils show clear similarities to both the previously known *Oligopithecus* (Simons 1971) and *Propliopithecus* respectively and have been placed by Simons in a new catarrhine primate subfamily, the Oligopithecinae. What is of particular interest in documenting the pandemic distribution of early anthropoids from east Asia through north Africa is the striking structural similarity between the lower M2 of *Oligopithecus* and related forms newly discovered from the Fayum (Simons 1992) and like material from the Eocene of China (*Hoanghonius*) (Rassmusen and Simons 1988) and Thailand (Suteethorn et al. 1988). One feature in particular seen in the Oligopithecinae and a diverse array of Asian fossil primates, including *Hoanghonius*, is the presence of a twinned entoconid and hypoconulid on lower molars.

The presence of catarrhine primates in late Eocene deposits of Africa necessitates an even earlier divergence of the platyrrhines from the anthropoid stem lineage, a circumstance consistent with the previous analysis and divergence dates for extant primate clades given in Hasegawa et al. (1990). As anthropoid origins get pushed back further into the early Eocene, how do a number of Asian fossils that have in the past been considered early anthropoids from middle Eocene (45 MYA) sediments of the Pondaung Formation in upper Burma fit into the picture? Two genera are known, *Amphipithecus* (meaning "near" or "both sides" ape) and *Pondaungia* (named for the Pondaung Formation). At present we consider the protoanthropoid status of *Pondaungia* to be unsubstantiated (Ciochon and Holroyd in press). *Amphipithecus*, however, is a somewhat better candidate for designation as a collateral protoanthropoid, although many of its features, thought previously to have been anthropoid synapomorphies, are now better seen as convergences or parallelisms (Ciochon and Holroyd in press). *Amphipithecus* does, however, exhibit a peculiar combination of features indicating that it was at or across the evolutionary transition from a "lower primate" to a "higher primate" grade of organization (Ciochon et al. 1985; Ba Maw et al. 1979; Ciochon 1985). In many ways it seems to precociously anticipate later trends in anthropoid evolution. For one thing, *Amphipithecus* was a gibbon-size animal, probably weighing about 20 pounds—relatively large in comparison to most lower primates alive in the Eocene or even today. The lower jaw is deep, both absolutely and in relation to the height of the teeth, and this depth extends the full length of the jaw. In lower primates and in some newly discovered early anthropoids, the jaw is rather shallow and lessens in height toward the front.

The cusps on the occlusal surfaces of the teeth of *Amphipithecus* are relatively flat, a trend found in fruiteaters. Many prosimian teeth, instead, have a very crested cutting surface, useful for a diet of insects and leaves. In this respect, *Amphipithecus* resembles the 35-million-year-old anthropoids from the Jebel Qatrani Formation of Egypt's Fayum province but differs from more primitive anthropoids now known from the fossil record of China and Morocco. Again *Amphipithecus* appears to be too precocious an ancestor for later anthropoids, but its status as a protoanthropoid remains justifiable.

The anthropoid-like features of *Amphipithecus* from Burma, as well as the other evidence previously reviewed, suggest that a radiation of protoanthropoid primates had occurred in Asia by at least 45 MYA if not considerably earlier. The exact placement of *Amphipithecus* in this radiation is at present difficult to determine.

Regardless of the status of *Amphipithecus*, accumulating evidence suggests that by the middle Eocene anthropoids were dispersed around the southern periphery of the Paleotethys seaway that extended from east Asia clear across to the Strait of Gibraltar (see Figure 3-3). As previously indicated, paleogeographic reconstructions of this region during the early Tertiary indicate that northern Africa was linked through the Arabian Peninsula with India and southern China (Briggs 1989), creating what may be termed the sub-Tethyean biozone. The ancient Tethys sea at this time was an eastern extension of the paleo-Mediterranean covering what is today the Middle East. It was along the southern margin of the Tethys that primates and other Asian mammals may have dispersed into Africa or vice versa. One line of evidence that could indicate a dispersal of anthropoids into Africa is that the explosive adaptive radiation of anthropoids took place there (Fleagle and Kay 1983, 1987; Simons 1992) rather than in Asia, suggesting that Africa was the recipient of the protoanthropoid expansion, not its ultimate source. Once in Africa it can be hypothesized that these protoanthropoids expanded rapidly, filling a plethora of vacant adaptive niches, with some populations giving rise both to the late Eocene and earliest Oligocene (40-35 MYA) Fayum anthropoids of Egypt, and other populations likely crossing the then-narrow equatorial Atlantic Ocean by island hopping along a series of volcanic islands that were present in the formative Atlantic (see Figure 3-3). The tectonic activity associated with seafloor spreading and continental drift (Ciochon and Chiarelli 1980) would have rendered this hypothesis feasible.

From the Oligocene on, the center of primate evolution in the Old World appears to shift to Africa. During the Oligocene, there is no fossil record of higher primates throughout Eurasia. Does this mean that the protoanthropoids of China and Burma left no descendants? This could indeed be the case, but it just as likely could reflect the simple fact that no fossil evidence of Oligocene higher primates has yet been discovered in Asia. Negative evidence, such as the lack of Asian Oligocene primates, is not a compelling argument for their certain absence. Indeed until recently there was only one published Oligocene site in the whole continent of Africa that had yielded higher primates—that was the well-known Fayum Depression in Egypt. The Afroarabian peninsula has, however, recently yielded the first Oligocene primate fauna (including an omomyiform, oligopithecid, and propliopithecid) from the Sultanate of Oman, an area intermediate between the African and Asian landmasses (Thomas et al. 1988). As a point of fact, Oligocene fossil localities are much less frequently found than Eocene or Miocene localities on all the world's continents (Savage and Russell 1983).

ORIGINS OF THE MODERN HOMINOIDS

By early Miocene times (22-18 MYA) in eastern Africa, an interesting array of protohominoid primates, sometimes called "dental apes" or "formative apes" (Corruccini et al. 1976) as well as the first Old World monkeys (*Victoriapithecus*) (von Koenigswald 1969) had evolved. These protohominoids and early Old World monkeys (protocercopithecoids) are likely descendants of the Oligocene Fayum protocatarrhines, the propliopithecids (Benefit and McCrossin 1991). The protohominoids of the East African early Miocene, such as *Proconsul, Dendropithecus, Limnopithecus,* and *Micropithecus* to name only a few, are dentally rather apelike yet decidedly monkeylike or primitive in their postcranial skeletons (Andrews 1992). *Proconsul* itself seemingly lacked a tail (Ward et al. 1992) and has a number of postcranial adaptations thought to be derived in the direction of later hominoids (Walker and Pickford 1983; Rose 1989). Their lower molar cusp pattern, which is also associated with modern apes and humans, the so-called Y5 or "dryopithecine" pattern, is, however, now generally recognized to be the primitive condition within the Catarrhini. As none of these catarrhines had the full capability to move through their environment using the below-branch movements characteristic of all modern hominoids, and hence of their likely common ancestor, these early Miocene higher primates are best referred to as "formative apes" or protohominoids (Harrison 1989).

The protohominoids of eastern Africa were first discovered in the 1930s in part through the efforts of the British paleontologist Albert Hopwood (1933) and later by Louis Leakey (Le Gros Clark and Leakey 1951). For more than four decades it was thought that these early Miocene "dental apes" were the direct ancestors of the modern hominoids and had only an eastern African distribution. In 1978, the latter perception was forever laid to rest when the first of several important primate discoveries was reported by Chinese paleontologist Li Chuankui from the early Miocene (19-17 MYA) site of Sihong in the Xiacaowan Formation in northern Jiangsu province of eastern China (see review in Etler 1989). A partial upper jaw containing three teeth was discovered, which Li christened *Dionysopithecus* (Li 1978). The surprising thing about *Dionysopithecus* is that it was a virtual carbon copy of a small primate upper jaw described from the early Miocene locality of Napak in Uganda (also in 1978) named *Micropithecus* (Fleagle and Simons 1978). This was the first indication of a link between East African early Miocene localities and the similar aged Sihong localities of eastern China. In 1983, paleoanthropologists Gu Yumin and Lin Yipu described a dental series of a medium sized protohominoid from Sihong, which they named *Platodontopithecus* (Gu and Lin 1983, reviewed in Etler 1989). The placement of the cusps of the upper and lower molars, the weakly expressed occlusal crests, and the occurrence of molar cingula link *Platodontopithecus* with the dental patterns evident in the ubiquitous East African protohominoid *Proconsul africanus*; best represented by the nearly complete skull discovered by the Louis and Mary Leakey in 1948 (Le Gros Clark and Leakey 1951). Finally, in 1985 Lei Ciyu of the Geological Museum of Nanjing described a number of primate taxa

Figure 3-3 Idealized geographic rendition of the world during the Paleogene. This map is meant merely to help the reader visualize how continental land masses and epicontinental seas were distributed early in the Cenozoic (40-65 MYA). During the Paleocene and Eocene, Europe and North America were united by a northerly land bridge as were North America and northwestern Asia. Much of central and eastern Europe was flooded by epicontinental seas, as were portions of central North America. The Ural Mountains, which today serve to divide Europe from Asia, had not yet risen and were replaced by a body of water, the Turgai Strait, linking the Arctic Ocean with the Tethys seaway. The Tethys was in turn a major extension of the Mediterranean that originally extended clear across the southern flank of Tibet into what is today the Bay of Bengal. Early in the Paleocene India was still adrift in the Indian Ocean. A series of island arcs followed in India's wake, opening up a possible island hopping sweepstakes route for the introduction of primates into Madagascar. Similar routes may have opened up elsewhere leading to the possible dispersal of primates from Africa to South America and from Asia to Africa during various periods of the early Cenozoic.

collected from the Sihong localities, including one specimen given the nomen "*Pliopithecus*" *wongi* (Lei 1985, reviewed in Etler 1989). Its occlusal pattern bears a striking resemblance to *Dendropithecus macinnesi* from the early Miocene site of Rusinga Island in Kenya (Andrews and Simons 1977).

These important discoveries by Chinese paleontologists confirm the presence of an early Miocene protohominoid or formative ape fauna in eastern China (Etler 1989), elements of which have also been reported from the Indian subcontinent (Barry et al. 1986; Bernor et al. 1988). Based on the apparent absence of any Oligocene higher primate fossils in Asia, it is most parsimonious to conclude that the primate fauna of Sihong documents a dispersal of protohominoids out of Africa into Asia about 20 MYA (Barry et al. 1986; Bernor et al. 1988). It is interesting to note that the long enigmatic central Asiatic fossil primate "*Kansupithecus*" (Bohlin 1946; Conroy and Bown 1976) discovered by Bohlin in the 1930s is the same size as *Dionysopithecus* and may actually represent the same taxon.

From the protohominoid radiation of eastern Africa and eastern Asia, the common ancestor of all living hominoids (gibbon, siamang, orangutan, gorilla, chimpanzee, and human) evolved. Based on a comparative analysis of modern hominoid taxa, the middle Miocene stem hominoid can be described hypothetically as a medium-sized ape with a short and broad face, broad interorbital region, smooth forehead lacking pronounced toral development, a primitive subnasal pattern, molar teeth with thin enamel lacking strong cingulae, and a thorax and upper limb skeleton adapted for below-branch suspensory behaviors (Ciochon 1983). The fossil primate that most closely approximates this morphotype is the Eurasian genus *Dryopithecus. Dryopithecus* has historically and conventionally been seen as more closely allied to the great apes than the lesser apes or hylobatids (Gregory and Hellman 1926; Begun 1992a,b). While the age of the dryopithecids (13-9 MYA) postdates the date hypothesized for gibbon divergence (15 MYA) they appear to be not far removed from what the stem hominoid would have been like.

It is ironic that *Dryopithecus*, which was the first fossil hominoid to be found and described (see review in Simons and Pilbeam 1965; Begun 1987), should play such a potentially pivotal role in hominoid evolution. In fact, for many years *Dryopithecus* (*s.s.*) was one of the least well-known Miocene hominoids, being represented by a mere handful of teeth and jaws and an occasional limb bone. Recent discoveries of *Rudapithecus hungaricus* (Kretzoi 1975), a form closely related to or possibly congeneric with *Dryopithecus* has, however, shed new light on this middle Miocene hominoid and allows for a better assessment of its evolutionary relationships (Begun 1992b). Fragmentary cranial remains described by Kordos (1987, 1988) and Begun (1992b), for instance, show *Rudapithecus* to conform closely to the cranial morphotype of the hypothetical common ancestor of the living hominoids previously described. Analysis of postcranial remains attributed to *Rudapithecus* (Morbeck 1983; Begun 1987) and the microstructure of its dental enamel (Xirotris and Henke 1981) likewise conform to the pattern one would expect in the common ancestor of the modern apes and humans.

Dentally *Dryopithecus* (including *Rudapithecus*), while more progressive than the protohominoids, presents a picture of overall primitiveness *vis à vis* the extant Hominoidea. *Dryopithecus* is characterized by narrow, high-crowned, buccolingually thick upper and lower incisors; tall buccolingually compressed canines with a thick distal cingula; reduced disparity in size between the lower premolar cusps; broadened lower P3s with small metaconids; elongated lower P4s with high trigonids; elongated lower molars with tall, periphalized cusps, broad basins and high dentine horns, and reduced M3s (Begun 1992b). While Begun (1992b) does not state so explicitly, there is nothing about this suite of dental features that would preclude them from being considered primitive features of the stem hominoid rather than the stem great ape (see also Andrews 1992). Begun (1992b) also discusses the craniofacial anatomy of *Dryopithecus* based on new specimens recovered over the years from Rudabánya. Many of these features are unequivocally primitive for the Hominoidea (i.e., broad interorbital distance, ethmoidal frontal sinus [seen also in *Proconsul*], poorly developed supraorbital torus and shallow supraorbital sulcus, prominent glabellar region) while others, which relate primarily to the subnasal and premaxillary regions of the face, are derived only slightly if at all in the direction of the modern great apes. Here again we would argue that most of the enumerated features indicate that *Dryopithecus* is not far removed from the stem hominoid.

Dryopithecus (= *Rudapithecus*) thus seems to be primitive enough dentally and cranially to be phenetically close to the common ancestor of all living apes yet shows signs of the derived upper body adaptation that distinguishes the modern hominoids from their protohominoid forebears the proconsulids (see Figure 3-4). It is interesting to note that the distributional range of mid-Miocene dryopithecids has recently been extended into central Asia (Xue and Delson 1988), placing them in a fortuitous position to give rise to the subsequent radiation of Asian late Miocene hominoid primates. Dryopithecid-like material has also been tentatively described from the early middle Miocene site of Sihong in eastern China (Lei 1985, reviewed in Etler 1989) and has been reconfirmed to occur at the southern Chinese middle Miocene site of Kaiyuan (Woo 1957, 1958; Kelley and Pilbeam 1986b; Begun 1987).

THE HOMINOID RADIATION

The stage is now set for a discussion of the radiation of modern hominoids, including the Asiatic lesser apes (gibbon and siamang) and great apes (orangutan and *Gigantopithecus)* to be followed by a discussion of the origin and evolution of the African great apes (gorilla and chimpanzee) and hominids.

The branching order and timing of the modern hominoid radiation has been inferred through a large number of biomolecular studies of the DNA and blood proteins of the living hominoids (Cronin et al. 1984; Gonzalez et al. 1990; Goodman et al. 1989; Hasegawa and Kishino 1991; Horai et al. 1992; Koop et al. 1986; Miyamoto et al. 1987; Ruvolo et al. 1991; Sibley and Ahlquist 1984; Sibley et al. 1990). According to Cronin et al. (1984) the line leading to the lesser apes (gibbon and siamang) first diverged about 15 MYA. This is followed by the divergence of the orangutan lineage at about 12 MYA. We believe that the preponderance of fossil and neontological evidence supports these divergence times. A study that indicates an earlier divergence of the orangutan clade at 17 MYA (Sibley and Ahlquist 1984) is flawed by reliance on the misidentification of early Miocene material from east Africa as belonging to a species of *Sivapithecus* (Leakey and Walker 1985), a position which the original authors subsequently abandoned (Leakey and Leakey 1986). The final divergence of the gorilla, chimpanzee, and human lines has long been viewed as a threeway split (trichotomy) and has been dated to 5-7 MYA. Recently, however, there has been an increasing number of molecular and morphological studies that support an exclusive relationship between the chimpanzee and humans (Ruvolo et al. 1991; Horai et al. 1992; Begun 1992a). A divergence of the gorilla at approximately 7.5 MYA and the chimpanzee at approximately 5 MYA seems reasonable in light of this new evidence. Knowledge of the cladistic branching order and temporal sequence of divergence for extant hominoids can be utilized as bounded constraints on our interpretations of the fossil evidence for the hominoid radiation precluding the possibility of certain evolutionary scenarios and lending support to others. The biomolecular evidence, however, can tell us nothing about the geographic location of the inferred splits or about the phylogenetic placement of extinct Tertiary hominoids. For this we must rely on interpretations of the fossil evidence.

GIBBON ANCESTRY

The living representatives of the lesser apes or Hylobatidae are the highly acrobatic dwellers of the high canopy forest, the gibbon and siamang. These two lesser apes are closely related and probably diverged only about 2 MYA (Groves 1972). Gibbons and siamangs occur throughout the tropical forested regions of mainland Southeast Asia and island Southeast Asia. They range from the southern provinces of China all the way to Borneo. In the Pleistocene, they had an even wider range extending up into central China (Gu 1986) and throughout Java and other populated islands of Indonesia. Today they are extinct in these regions because of habitat destruction and hunting by *Homo sapiens*. The fossil record of gibbons and siamangs in the middle and late Pleistocene of Southeast Asia is well known. Their remains are found in limestone cave sites throughout central and southern China, Vietnam, Laos, Thailand, Sumatra, Java, and Borneo. Most of these identifications have been based on isolated teeth or fragmentary jaws. The most complete Pleistocene gibbon is a partial skull known from the limestone karst cave site of Tham Khuyen, Lang San province, Vietnam (Ciochon and Olsen 1986). This middle Pleistocene cave site (probably 300,000 to 500,000 years old) has also produced remains of *Gigantopithecus* and the orangutan, as well as 30 other mammalian species. The fossil record of the gibbon and siamang prior to the Pleistocene, however, is a matter of great conjecture (Fleagle 1984). Since the distribution of the lesser apes is solely in Southeast Asia and since their closest hominoid relative, the orangutan, also has a Southeast Asian distribution and an Asian Miocene fossil record, it is logical to conclude that the lesser apes had an Asian origin. But which pre-Pleistocene Asian fossils are potential lesser ape ancestors? One candidate is *Laccopithecus robustus* from the late Miocene (7 MYA) site of Shihuiba in Lufeng county, Yunnan, Southern China (Wu and Pan 1984, 1985b). This fossil ape is known from a nearly complete face, 14 upper and lower jaws, 8 tooth rows, 67 isolated teeth, and one phalanx. The first remains of *Laccopithecus* were discovered in 1976 by Chinese paleoanthropologists and named in 1984 by Wu Rukang and Pan Yuerong. Shihuiba is a lignite (coalbearing) site that represents the accumulated remains of a swampy tropical forest habitat (Badgley et al. 1988). This is responsible for the relatively complete and numerous remains of this fossil ape.

Laccopithecus is a larger ape than the modern gibbon and siamang. It has a short and broad face, broad interorbital distance and smooth forehead lacking pronounced toral development. In these respects it resembles the hypothetical stem hominoid and would be an appropriate ancestor for the lesser apes. Unfortunately, *Laccopithecus* has no uniquely hylobatid features in its teeth or cranial region. In fact, its teeth bear a superficial resemblence to protocatarrhine crouzeliine pliopithecids known from middle Miocene deposits of Europe. They differ substantially, however, from specimens of the genus *Pliopithecus* known from Europe and China (Harrison et al. 1991). *Laccopithecus* is also sexually dimorphic in canine size, based on a good sample of fossils from Shihuiba (Pan et al. 1989); modern gibbons and siamangs exhibit greatly reduced male/female canine size differences. Finally, except for a single finger bone, no associated skeletal remains have been identified for *Laccopithecus*. It is ironic, however, that the single proximal phalanx of *Laccopithecus* may point to the fact that it is in actuality the long sought after hylobatid ancestor. The key feature used to designate hominoids from other catarrhines is the presence of a derived arm-swinging anatomy in the upper limb skeleton (Ciochon 1983). Without the confirmatory evidence of this anatomy, it is not possible to conclude that a fossil primate has the basic suspensory adaptation of the modern hominoids. In fact the protocatarrhine *Pliopithecus*, once thought to be a likely gibbon ancestor, lacked just such a locomotor adaptation, thereby foreclosing its ancestral hylobatid status (Ciochon and Corruccini 1977). The finger of *Laccopithecus*, however, has been studied and shown to have unique features that indicate a full-fledged suspensory adaptation (Meldrum and Pan 1988). *Laccopithecus* may then be the long sought after but elusive early hylobatid ancestor.

Biogeography and the molecular clock also support *Laccopithecus* as an early hylobatid. It certainly occupies the correct temporal and geographical location for such an ancestor. Based on molecular evidence (Cronin et al. 1984) and the late first occurrance of the derived upper body anatomy of modern hominoids in the fossil record, it can be safely assumed that the lineage of modern lesser apes had its origin about 15 MYA most likely, as previously discussed, from some aspect of the Dryopithecidae. At 7 MYA *Laccopithecus* is well suited temporally to be on the line leading from a dryopithecid ancestor to modern gibbons and siamangs (see also Tyler 1991). Additionally, *Laccopithecus* comes from a locality in Yunnan province, southern China, which today is an important center of gibbon diversity with 5 species living in its tropical forests (Zhang et al. 1981).

Paleoanthropologists have long had a problem with deciphering the ancestry of the hylobatids. Although numerous fossil apes have often been proposed as potential ancestors of the gibbon, there have always been a few features, such as the retention of primitive, protohominoid, or even protocatarrhine postcranial traits; a specialized dental morphology; or a lack of specializations seen in the modern gibbon, which were cause to dispute the ostensible ancestor (Fleagle 1984). In many cases, small size alone led to the identification of a fossil primate as a gibbon ancestor irrespective of its age or biogeographic distribution. How then can we explain the features that distinguish *Laccopithecus* from the modern hylobatids, in particular its robustness, various primitive dental features, and its high level of sexual dimorphism? If we assume that gibbons are descended from a moderately large fossil hominoid akin to an early form of *Dryopithecus*, it could be that the hylobatid ancestor underwent a period of phyletic dwarfing (reduction in size with concomitant allometric changes in dental and cranial anatomy) during the late Miocene and Pliocene (7.5 MYA to 1.8 MYA) as it adapted to a specialized brachiating lifestyle high in the tropical forest canopy of Southeast Asia. Changes in diet and socionomic structure of the hylobatid ancestor as it entered into this new niche may have led to the reduction in sexual dimorphism and postcanine dental size so evident in living gibbons and siamangs. This would explain why we cannot find any wholly appropriate Miocene ancestors for the lesser apes. By positing a period of phyletic dwarfing, enhancement of a brachiation adaptation and associated changes in dietary preferences and socionomic structure in the Pliocene, we arrive at a portrait of the Miocene ancestor of the "lesser apes" more robust and sexually dimorphic than its Pleistocene and recent descendants. This could explain why pre-Pliocene gibbon/siamang ancestors have been difficult to identify. It also could mean that since *Laccopithecus* possessed the appropriate arm-swinging postcranial anatomy, it is, after all, an appropriate Miocene ancestor of the living lesser apes.

ORANGUTAN ANCESTRY

The only extant Asiatic great ape, the orangutan, *Pongo pygmaeus*, has a history of discovery and a distribution similar to that of the gibbon and siamang (Gu et al. 1987; Kalke 1972). Today the orangutan is found only in island Southeast Asia on Sumatra and Borneo. This represents a relict distribution since in the middle to late Pleistocene, fossil evidence of orangutans is known from limestone caves in the southern provinces of China (Gu et al. 1987), throughout peninsular Southeast Asia (Vietnam, Laos, Thailand, Burma) and island Southeast Asia especially on Sumatra, Java, and Borneo (Hooijer 1948). Some island forms of orangutan, especially those known from the late Pleistocene caves of Sumatra discovered by Eugene Dubois in 1888, provide evidence that insular varieties of the orangutan evolved in some isolated regions of island Southeast Asia.

Today the orangutan occurs only in mostly uninhabited tropical forests in Sumatra and Borneo. Human hunting and habitat destruction are responsible for this dramatic decrease in the range of the orangutan from middle Pleistocene times until the modern era. Since the Pleistocene history of the orang is better known than that of any other great ape, what can we say about its Pliocene and Miocene ancestry?

Throughout the history of the search for fossil primates in Asia, paleontologists have often claimed to have found the Miocene ancestors of *Pongo pygmaeus*. One of the earliest fossil primates discovered anywhere in the world was an isolated upper canine described by British paleontologist Hugh Falconer in 1837 from the Siwalik beds of northern India. Falconer successfully argued that this canine was Miocene fossil evidence of the orangutan. Unfortunately, there is no way to substantiate this claim today because Falconer's fossil canine was misplaced (or lost) after his death in 1867. From the 1880s through the early part of the 1900s, numerous fossil apes were discovered from the Miocene Siwaliks of India with various claims being made that they were Miocene evidence of the orangutan lineage. Unfortunately, the critical cranial evidence to bolster these claims was lacking.

In 1980, this evidence was recovered from the fossiliferous Siwalik deposits of the Potwar Plateau in southern Pakistan (Pilbeam 1982). The specimen, GSP (Geological Survey Pakistan) 15000, consists of a relatively complete cranium and intact lower jaw of *Sivapithecus indicus* from later Miocene (8 MYA) deposits of the Nagri Formation (Pilbeam 1982). This fossil skull shares with the orangutan a unique dish-shaped face, narrow interorbital width, oval-shaped orbits, derived nasal region, and flaring cheek bones (Preuss 1982; Andrews and Cronin 1982; Ward and Brown 1986; Brown and Ward 1988). Neither the African great apes nor Asiatic lesser apes display these uniquely derived facial characteristics.

This confirmation that the orangutan of Southeast Asia is most likely descended from the Asian Miocene ape, *Sivapithecus*, has some interesting implications for the evolution of the modern African hominoids (gorilla, chimpanzee, humans). Pilbeam (1985, 1986) suggests that the genus *Sivapithecus* should be restricted to hominoids from India, Pakistan, and Turkey, which all share the derived craniofacial features of the orangutan lineage. The earliest evidence of this lineage is known from the Chinji Formation of the Siwaliks circa 12 MYA (Raza et al. 1983), a date fully in accord with the orang divergence time as documented by the molecular clock. Since the African great ape/human clade shares a different set of craniofacial features such as a broad and short face, broad interorbital region, squareshaped orbits, and differently configured subnasal morphology (Ward and Kimbell 1983), it is clear that it split from the clade leading to orangutans prior to the appearance of *Sivapithecus* (*s.s.*). It is interesting to note at this juncture that the known postcrania of *Sivapithecus*, though clearly possessing the derived hominoid upper body complex, differs considerably from that of the extant orangutan, being more generalized and lacking in derived features unique to the living arboreal form. In fact, *Sivapithecus* shows postcranial similarity to the gorilla and is in some ways more primitive than any living great ape (Pilbeam et al. 1990). There is hence the distinct possibility that some of the postcranial similarities between extant hominoids are the result of parallelisms rather than possession of uniquely shared features (Andrews 1992). Until more conclusive evidence of the postcrania of late Miocene hominoids is available, it is, however, premature to speculate as to what their locomotor adaptations may have been. Although all living hominoids share an upper body adapted to suspensory locomotion, their total locomotor patterns probably differ considerably from the basal adaptation of their last common ancestor (see Fleagle et al. 1981).

AFRICAN APE ANCESTRY

The current biomolecular and paleontological evidence for the evolution of the African great apes indicates that the splits leading to the chimpanzee, gorilla and human clades took place 5-7.5 MYA in Africa (Hill and Ward 1988; Horai et al. 1992). The earliest fossil evidence we have of this partitioning are extremely fragmentary dental and gnathic remains that predate the hominid *Australopithecus afarensis* known from the 3-4 MYA time range in Ethiopia. No clear fossil evidence for the chimpanzee or gorilla has ever been found (Hill and Ward 1988), although a partial maxilla from late Miocene deposits in the Samburu Hills of Kenya is thought by some to show affinities to the gorilla (Ishida et al. 1984). There is little doubt that the earliest evidence of the human line, typified by fossil forms found at Hadar in Ethiopia and Laetoli in Tanzania, did evolve in Africa. However, where did the common ancestor of the African great apes and humans evolve? Since the orangutan lineage is known to have split off from the great ape stem lineage about 12 MYA, there are at least 5 million years of evolution of the stem Africa great ape to account for prior to the successive splits leading to the gorilla, chimpanzee, and human lines. Could the earliest members of the African great ape/human clade have evolved in Eurasia and dispersed into Africa? If we follow the logic of allopatric speciation and parsimony, the lineage leading to the African great apes would have had its origin in Eurasia at the same time that the lineage leading to the Asian great apes had its origin. The other alternatives are that at the time of divergence between the Asian and African great apes the common ancestral species had a pandemic distribution across much of Africa and southern Eurasia with the extreme peripheries of its distribution in east Asia and Africa producing the ancestors of the living great apes, or the divergence took place in Africa with the orangutan subsequently dispersing into Asia. The first alternative contradicts fossil evidence that indicates that late Miocene hominoids were regionally distributed and represented by numerous species (Kelly and Pilbeam 1986b). The second alternative lacks substantiation as no African orangutan or possible ancestor has yet been found (Kelly and Pilbeam 1986a *contra* Leakey and Walker 1985). The probable Asian ancestor of the orangutan is, however, known (*Sivapithecus*) and from a time and place we would expect based on the molecular evidence and biogeography. In other words logic dictates that the Asian apes are derived from the radiation of late middle to late Miocene Eurasian hominoids and that the stem African great ape had a Eurasian origin as well.

In opposition to this line of reasoning, some researchers have posited a number of early middle Miocene higher primate taxa from East Africa, Turkey, and Arabia as members of the great ape/human clade based on their sharing apparently derived dental features (i.e., thick molar enamel, reduced molar cingula, mediodistally elongated upper premolars) with late Miocene hominoids such as *Sivapithecus* and the extant great apes (Andrews and Martin 1987; Pickford 1986). If these fossil primates are true great apes the long sought after African ancestors of the modern large-bodied hominoids would at long last be found. It seems likely, however, that the same logic that led to the acceptance of "*Ramapithecus*" as a basal hominid is at work here as well. By our reckoning many of these fossils (i.e., *Afropithecus*, "*Heliopithecus*," *Kenyapithecus*, and various specimens from Europe and Anatolia attributed to the genus *Griphopithecus*), some of which were previously identified with "*Ramapithecus*" or *Sivapithecus*, appear too early in the fossil record to be members of the great ape/human clade. Moreover, there is substantial evidence that these taxa retain to varying degrees primitive cranial and postcranial features that preclude their being exclusive ancestors of the modern hominoids (Begun 1987; Benefit and McCrossin this volume). *Dryopithecus* (*s.s.*), which occurs at approximately the same time, retains primitive dental features (i.e. high dentin horns, thin molar enamel, and molar cingula) but postcranial characters clearly derived in the direction of modern hominoids (Begun 1987). Since dental characters such as degree of enamel thickness and cingula expression are subject to considerable homoplasy and the mediodistal elongation of the upper premolars appears to have evolved independently in orangutans, gorillas, *Gigantopithecus*, and robust australopithecines, we feel it is most parsimonious to view *Afropithecus*, *Kenyapithecus*, and similar forms from the middle Miocene of Arabia, Turkey, and Europe as precocious thick-enameled protohominoids (more closely related to early Miocene proconsulids than the great ape/human clade) evolving in parallel with the late Miocene and extant great apes to which they have been compared (Figure 3-4).

Although we can apparently pick up the orangutan lineage in Asia soon after its postulated divergence at 12 MYA, at this point there is no fossil evidence for the stem African great ape in Asia or Africa at that time. If, however, we look later in time for evidence of modern African hominoids, we find the earliest occurrence at the late Miocene sites of Rain Ravine and Xirochori in Macedonian Greece (de Bonis and Melentis 1977, de Bonis et al. 1990). Another late Miocene hominoid from Lufeng in Yunnan province, southern China (Wu et al. 1983, 1984, 1985, 1986; Wu 1985, 1987) lacks many of the derived features noted for the orangutan and its late Miocene ancestor *Sivapithecus* and may relate to the origins of both Asian and African great apes (see later discussion).

The material from Greece (*Graecopithecus = [Ouranopithecus] macedoniensis*) has been shown to have an African ape subnasal morphology (de Bonis and Melentis 1987). In many ways, it shows dental resemblances to both the preceding dryopithecines and the later gorilla in terms of molar tooth size and cuspal proportions. In other features (i.e., the relatively small size of canines relative to cheek teeth and thick molar enamel), it shares traits common to all late Miocene great apes and early hominids. If it is acknowledged that the modern African apes are derived over their immediate Miocene ancestors, especially in secondary enamel thinning (Martin 1985) and C/P3 hypertrophy, *Graecopithecus* could well make for a viable African ape/human ancestor.

It has been generally assumed in the past that the common chimpanzee or perhaps the pygmy chimpanzee served as an appropriate model for the African great ape/human morphotype (Zihlman et al. 1978). This supposition, however, must be called into question as it is the gorilla that appears to retain a set of primitive anterior dental features *vis-à-vis* the chimpanzee and humans (Begun 1992a). In other instances it is probable that early australopithecines more closely approximate the African great ape/human craniodental morphotype than do any of the living great apes. Recent evidence suggests that the

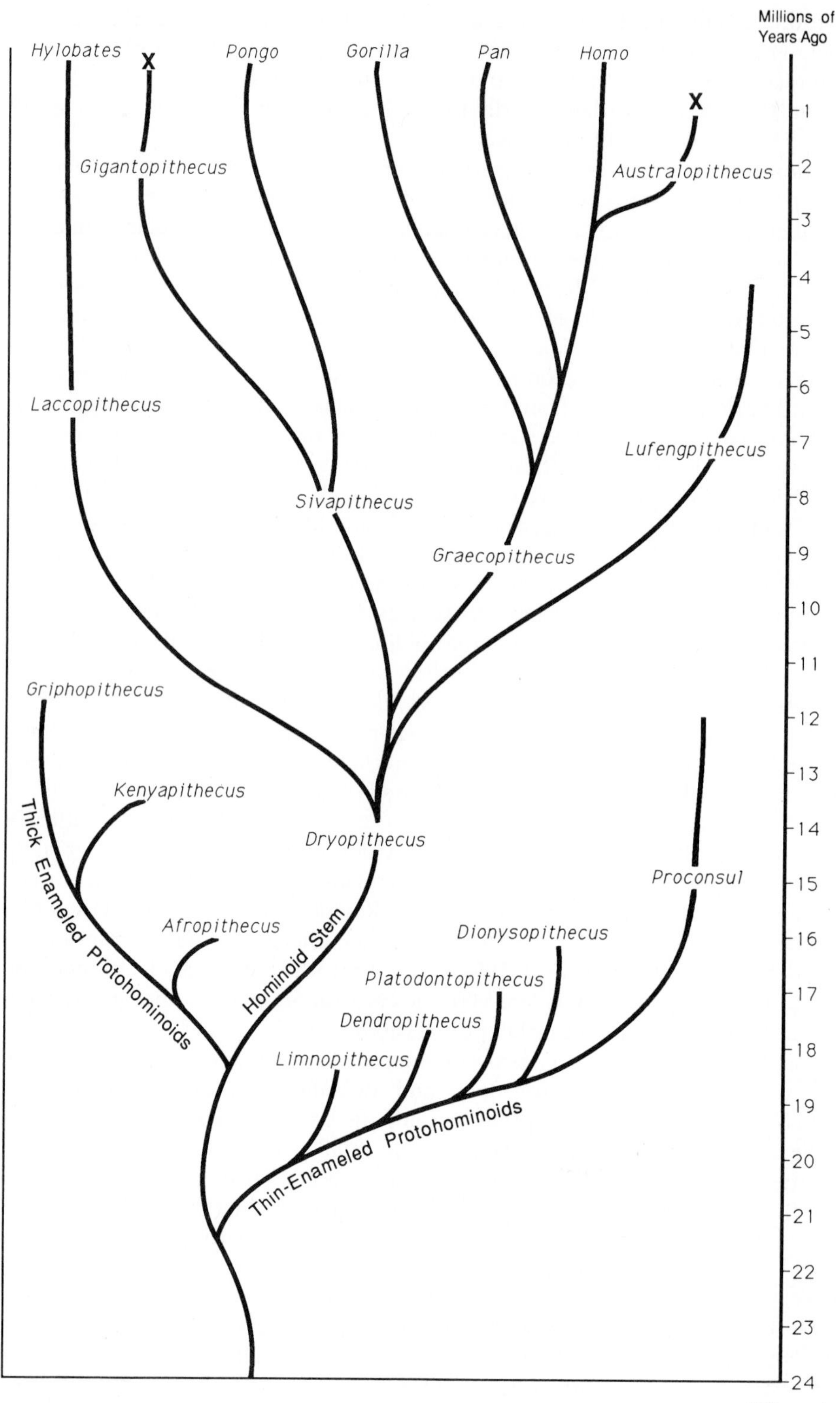

Millions of Years Ago
Hylobates
X
Pongo
Gorilla
Pan
Homo
X
Gigantopithecus
Australopithecus
Laccopithecus
Lufengpithecus
Sivapithecus
Graecopithecus
Griphopithecus
Kenyapithecus
Thick Enameled Protohominoids
Dryopithecus
Proconsul
Afropithecus
Hominoid Stem
Dionysopithecus
Platodontopithecus
Dendropithecus
Limnopithecus
Thin-Enameled Protohominoids
1
2
3
4
5
6
7
8
9
10
11
12
13
14
15
16
17
18
19
20
21
22
23
24
Millions of Years Ago

FIGURE 3-4 (Opposite page) Phylogenetic tree depicting the course of hominoid evolution from the inception of the hominoid stem lineage perhaps 24 MYA. There seems to have been a diverse array of protohominoids spread throughout much of the Old World, 15-20 MYA. The various protohominoid lineages go extinct in Africa and Eurasia by approximately 12 MYA, perhaps due to competition from the rapidly expanding Old World monkeys, the cercopithecoids. The modern hominoids, including the gibbon, most likely evolved in Eurasia from an early dryopithecid-like ancestor, which quickly diverged into a number of lineages, three of which survive today (i.e., hylobatids, pongids, and hominids (*sensu lato*)). A subsequent reintroduction of apes into Africa approximately 8-10 MYA may have set the stage for an adaptive radiation of modern African hominiods, including the gorilla, chimpanzee, and early human ancestors, the australopiths.

process of enamel thinning seen in the African great apes was initiated only 4-5 MYA (Cecchi and Pickford 1989). Earlier dentognathic specimens attributed to cf. *Australopithecus* (Hill and Ward 1988) could, therefore, predate the chimpanzee/human divergence node and represent a stem species ancestral to both.

It should be expected that the African great ape/human morphotype will possess a mosaic of features that are differentially expressed in its living and fossil descendants. The recent recovery of a nearly complete face of *Graecopithecus* from the new late Miocene Macedonian locality of Xirochori (de Bonis et al. 1990) referred to previously confirms this prediction. In its orbital, interorbital, and supraorbital structure and nasal and subnasal morphology, the new specimen clearly aligns with the African apes. The brow ridge structure of the new *Graecopithecus* skull, however, is reminiscent of that seen in early australopithecines (Asfaw 1987), indicating that in this regard the forwardly projecting supraorbital torus of the extant African great apes is derived over the ancestral African ape/human morphotype.

The Lufeng site in the southern Chinese province of Yunnan has been mentioned previously in reference to the primitive hylobatid *Laccopithecus*. The site is actually better known for the remarkable series of discoveries, beginning in the mid-1970s, of several crania of a late Miocene great ape (see review in Etler 1984; Wu et al. 1985, 1986). In 1988, Chinese paleoanthropologist Wu Rukang named this material *Lufengpithecus* based on six relatively complete skulls, numerous lower jaws, nearly 1000 isolated teeth, and a partial upper limb skeleton including a proximal humerus, partial scapula (the glenoid fossa, acromion and scapular borders), clavicle, and proximal radius. This large hominoid is very sexually dimorphic (Wu and Wang 1987; Kelley and Etler 1989; Kelley and Xu 1990; Wood and Xu 1991) and has a suite of primitive great ape craniofacial features. However, its intact shoulder joint and radius exhibit all of the suspensory characteristics of a fullfledged modern hominoid (Xiao 1981; Wu et al. 1986; Lin et al. 1987).

Lufengpithecus is an ideal prototype ancestor for all the modern great apes in that it lacks the uniquely derived craniofacial features of either the modern Asian orangutan or the African chimpanzee and gorilla; yet it shares other derived features seen exclusively within the great ape/human clade (Kelly and Etler 1989). The fact that it bears no special resemblance to Asian hominoids (*Sivapithecus*, orangutan, *Gigantopithecus*) or African hominoids (gorilla, chimpanzee, *Australopithecus*, *Homo*) indicates that *Lufengpithecus* could be phenetically close to the last common ancestor of both branches of the modern large-bodied hominoids (Kelly and Etler 1989 *contra* Schwartz 1990). Dentally, *Lufengpithecus* retains a number of primitive dryopithecid-like traits including narrow, high-crowned lower incisors; slender saberlike male lower canines; and, apparently, a greater degree of dentin separation wear than in later great apes. In other features, such as the possession of relatively large, thickly enameled molars, the dentition of *Lufengpithecus* is derived in the direction of other late Miocene large-bodied hominoids. Its possession of a derived great ape subnasal region and a gorilla-like shoulder joint clearly aligns *Lufengpithecus* with the great apes rather than more primitive hominoids. *Lufengpithecus*, therefore, seems to closely approximate the stem great ape morphotype prior to the split between the Asian and African clades. *Lufengpithecus*, however, appears too late in the fossil record (approximately 7-8 MYA) to be the phylogenetic ancestor we are looking for. It may, however, be an Asian collateral descendant of the stem great ape lineage from which all modern large-bodied hominoids evolved (Figure 3-4), further reinforcing the idea that the cladogenetic event separating the Asian and African great apes took place in Eurasia rather than Africa. The possibility that some fossil apes of Yunnan bear a more direct relationship to the African apes cannot, however, be ruled out. In fact the recent recovery of a juvenile hominoid cranium from the early Pliocene (4 MYA) site of Butterfly Ridge in the Yuanmou basin of central Yunnan by Zhang Xingyong, a paleontologist with the Yunnan Provincial Museum in Kunming, may bolster just such an argument. This skull, while not yet fully described (Zhang et al. 1988), has been seen by the authors and in some respects is reminiscent of a young chimpanzee.

If the African stem hominoid did indeed evolve in Eurasia and disperse into Africa, when would this event have occurred? The Miocene apes of Africa from 20 MYA until 12 MYA all appear to be protohominoids of the proconsulid or afropithecid (= kenyapithecid?) variety (Hill and Ward 1988; Andrews 1992). As the African Old World monkeys evolved and radiated into a variety of niches, these protohominoids became much less diverse as a result of competition with their highly successful cousins. After 12 MYA, the protohominoids go extinct. Unfortunately, the fossil record of hominoids in Africa during the period 12 MYA until 5 MYA is restricted to a few isolated teeth and gnathic fragments (Hill and Ward 1988; Hill, this volume). Perhaps modern hominoids dispersed into Africa about 10 MYA when Eurasian hipparionine horses and murid rodents first appear on the African continent (Bernor et al. 1987; de Bruijn 1986; Flynn and Sabatier 1984). The occurrence of *Graecopithecus* in southeastern Europe at this time is highly suggestive of just such a scenario. *Graecopithecus* is both well suited morphologically (it can be compared to gorillas, chimpanzees [Dean and Delson 1992], and early hominids [de Bonis et al. 1981, 1990]) and well placed in time and space to be representative of the Eurasian ape that first entered Africa to give rise to the radiation of the modern African great apes and early hominids.

Based on these considerations, we can now envision the emergence of the Hominidae as part and parcel of an adaptive radiation of Eurasian apes subsequent to their dispersal into the Ethiopian biogeographic zone of Africa between 6 and 10 MYA. Human evolution need no longer be thought of as the outcome of a unique series of events taking place solely in Africa in splendid isolation from the main theater of hominoid evolution occurring in Eurasia during the late Miocene. Rather, it can be seen as the culmination of trends inherent in the phylogenesis of the modern Hominoidea.

One final chapter in our discussion of past primate diversity concerns the origin and distribution of the giant extinct ape, *Gigantopithecus.* Based on dental and mandibular characters it is possible to place *Gigantopithecus* among the Asiatic great apes. In fact, since *Gigantopithecus* survived well into the middle Pleistocene as a contemporary of the orangutan and *Homo erectus,* it is most appropriate to view *Gigantopithecus* as the fourth great ape and the only modern ape to have gone extinct (see Ciochon et al. 1990). *Gigantopithecus* can trace its origin back to the late Miocene of India, where a primitive species of the genus is known from approximately 6 MYA (Simons and Chopra 1969). *Gigantopithecus* enters the southern Chinese faunal realm by at least the late Pliocene and persists in southern China and Vietnam into the middle Pleistocene. The recent discovery of a diverse primate fauna from what are possibly early Pliocene beds in the Yuanmou basin of Yunnan in southern China that includes at least one but possibly more hominoid taxa akin to material known from Lufeng points to a hitherto unknown diversity among Asian hominoids well into the late Neogene and possibly even into the Quaternary. The southern Chinese fauna of 3-4 MYA can be postulated to have contained late surviving forms of *Lufengpithecus* in addition to ancestral forms of *Gigantopithecus, Pongo,* and *Hylobates.* The late Neogene hominoid fauna of southern China and southeast Asia could very well have contained a phylogenetically diverse group of hominoids representing at least three distinct hominoid families: the Hylobatidae (*Hylobates*), Dryopithecidae (*Lufengpithecus),* and the Pongidae (*s.s.*)(*Gigantopithecus, Pongo*). At the same time the hominoid fauna of Africa is represented by only one family: the Hominidae (*s.s.*) (*Australopithecus* and hypothesized ancestral forms of *Pan* and *Gorilla*). The greater phyletic diversity of hominoids in southeast Asia speaks to that area of the world as a more likely center for modern hominoid origins and dispersal than Africa.

SYSTEMATIC OVERVIEW

The broad outline of primate evolution as it relates to human evolutionary biology and as conceived in this contribution runs as follows: The initial differentiation of Archonta most likely began in the late Cretaceous. The archontan radiation apparently was the result of the first wave of eutherian mammals to enter into a primarily arboreal mode of life. Once this econiche was invaded, the grand order quickly diversified into a number of distinct lineages. Among these were the plesiadapiforms; their close allies the mixodectids and microsyopsids; and the ancestors of the scandentians, dermopterans, megachiropterans, microchiropterans and primates. Looking at the archontans of the Paleocene synchronically, they can be thought of as a paleontological order of mammals that subsequently diverged to produce the neontological orders with which we are now familiar.

Until recently, it was generally accepted that the plesiadapiforms were archaic primates and as such the sister group of the euprimates or primates of modern aspect. Consistency with the method of phylogeny reconstruction utilized herein (see Figure 3-1), however, requires that the last common ancestor of the living primates be considered the stem primate species. Clades diverging prior to the primate stem species but subsequent to the last common ancestor of the primates and their living sister group (candidates of which include Scandentia, Dermoptera, and Megachiroptera) are collateral primate lineages whose members would be protoprimates. The question then arises, are plesiadapiforms protoprimates or a collateral lineage of a sister group to the primate closed descent community? New fossil evidence reviewed in this chapter supports the latter supposition.

The primates seem to have diverged very early in the Paleocene into two closed descent communities, the crown groups that today constitute the suborders Prosimii and Anthropoidea. The apomorphies which subsequently separated them may, however, have emerged sequentially over a considerable amount of time; for example, the protolemuriform may not have had the tooth-comb or other derived features of living lemuriforms and the protoanthropoid certainly lacked many of the dental, mandibular, and narial characters of its later descendants. The stem lineages of both Prosimii and Anthropoidea may not have, therefore, evolved the autapomorphies characterizing their later members until some time after the establishment of their respective clades. During this period of initial differentiation both clades may have produced a number of collateral lineages that did not share in the derived traits linking together their living members. In other words, the evolution of the synapomorphies uniting living taxa of a clade may occur at a node after the initial cladogenetic event defining the stem lineage. The descent communities of collateral lineages, moreover, do not enter into the direct ancestry of the living taxa. The clear recognition and systemization of these processes may help create more realistic phylogenies.

The surviving lineages of the initial period of primate cladogenesis are the lemuriforms, tarsiers, and anthropoids. It seems reasonable biogeographically, and more and more so paleontologically, that Asia was a primary center for the archontan and later primate radiations. Most living archontans (scandentians, dermopterans and megachiropterans) as well as the extant lineages of primates with greatest phyletic depth (i.e., *Tarsius*, the Malagasy lemuroids, and the Afro-Asian lorisoids) are today distributed in Asia or around its periphery. The Eocene prosimians of Europe and North America (i.e., the "adapids" and "omomyids") most likely represent paraphyletic collateral assemblages of protolemuriforms and prototarsiiforms, respectively, that dispersed into the Euroamerican paleobiogeographic zone late in the Paleocene and do not directly bear on extant primate origins.

The protoanthropoid lineage appears to have arisen along the sub-Tethys littoral in either north Africa or east Asia perhaps diversifying into a number of collateral anthropoid clades. Its defining node would be the primate stem species that produced the anthropoid and prosimian closed descent communities. We think that current evidence supports the contention that the stem lineage leading to modern anthropoids evolved first in Asia, dispersing into Africa in the middle Eocene with subsequent differentiation into Platyrrhini and Catarrhini through the anthropoid stem species.

At this point the main theaters of higher primate evolution seem to shift to Africa and South America. The prosimian primates of Europe and North America go extinct. In Asia, post-Eocene primates are apparently restricted to lorisoids, sivaladapids (perhaps closely tied into lemuriform origins), and tarsiids. Protoanthropoids probably go extinct in Asia during the Oligocene.

The radiation of modern anthropoids seems to have taken place in Africa during the late Eocene and early Oligocene. By the middle-late Oligocene, platyrrhines are established in South America. Protocatarrhines (e.g. propliopithecids) are clearly recognizable within the primate fauna of the Fayum, although at least one possible collateral anthropoid lineage (the protoanthropoid? *Oligopithecus* and allies) and one possibly protoplatyrrhine collateral lineage (i.e., parapithecids) are also represented.

The recent recovery of remarkably complete early Miocene "hominoid" remains from Kenya (i.e., *Afropithecus*) exhibits a set of autapomorphies that may be linked uniquely with Oligocene propliopithecids, suggesting that at least one collateral catarrhine lineage persisted in Africa well into the Miocene. Protocercopithecoid, protohominoid, and oreopithecoid lineages are also represented in the Miocene of East Africa.

It is now apparent that one or more protohominoid lineages and at least one protocatarrhine lineage (i.e., pliopithecids) dispersed out of Africa into both Europe and south and east Asia by the early-middle Miocene, with one of the former giving rise to the lesser and greater ape clades that began to differentiate approximately 15 MYA.

From about 14-6 MYA the major events in primate evolution as postulated in this review include (1) the radiation of Eurasian hominoids (i.e., hylobatids, dryopithecids, and the Asian and ancestral African great apes); (2) the extinction of African protohominoids due to competition from radiating cercopithecoids; and (3) the introduction of what is today the African great ape/human stock into Africa from Eurasia between 8-10 MYA.

As regards the last rather controversial point, it can be demonstrated by (1) the presence of proto-African great apes in Eurasia and (2) the absence of modern great ape ancestors in Africa prior to 10 MYA. We argue that the Miocene apes of Eurasia represent two distinct lineages, one leading to the orangutan and the other leading to the African apes and humans, while the Miocene apes of Africa prior to 10 MYA are represented exclusively by protohominoids or collateral catarrhines. The first point has been demonstrated by the identification of ancestral or collateral great ape lineages outside of Africa which possess derived features of the African ape/hominid clade. The second point is harder to demonstrate since it rests on negative evidence, that is, the expectation that no modern hominoids will be found in Africa prior to approximately 10 MYA. The continuing discovery of late-occurring protohominoids in Africa such as at Ngorora (Hill and Ward 1988) and Namibia (Conroy et al. 1992) to the exclusion of more modern-looking great apes does, however, confirm this expectation to a certain extent.

Our final conclusion concerns the emergence of the Hominidae itself. Our reading of the fossil record leads us to believe that early hominids (*s.s.*) such as *Australopithecus afarensis* were part and parcel of an explosive adaptive radiation of modern hominoids in Africa, which began approximately 10 MYA with the introduction of Eurasian hominoids into the Ethiopian biozone. Seeing the evolution of early hominids in the context of a dispersal event and subsequent adaptive radiation is more in keeping with the known processes of biological evolution than to see them emerging as the result of a unique series of adaptive shifts among a group of isolated and autochthonous African hominoids.

In summation, we hope that at least some of the views expressed in this chapter have more than just challenged long held assumptions but contributed in some measure to a better understanding of the scope of primate evolution and past primate diversity.

ACKNOWLEDGMENTS

This chapter could not have been written without the hospitality and assistance of many colleagues in China, including Gu Yumin, Huang Wanpo, Jia Lanpo, Li Chuankui, Li Kunsheng, Lu Qingwu,

Pan Yuerong, Qi Guoqin, Qiu Zhanxiang, Wei Qi, Wu Rukang, Xu Qinghua, Yi Guangyuan, Zhang Xingyong, Zhang Yinyun, Zhou Mingzhen, Zhou Quoxing and many others too numerous to mention. We also acknowledge the staff of the Institute of Archeology in Hanoi for their continued support and collaboration.

Profs. J. D. Clark, F. C. Howell, and T. D. White made much of this research possible and we thank them for their support. We acknowledge the artistic abilities of Douglas Beckner (Figure 3-1) and W.W. Thomson (Figures 3-2 to 3-4). Nhan Thi Le typed, edited and key-coded the entire manuscript, and Elizabeth Ramsey helped with corrections. Richard Nisbett and Pat Holroyd carefully read and commented on several versions of this chapter. Finally, we gratefully thank the Henry Luce Foundation, the L. S. B. Leakey Foundation, the National Geographic Society, the National Science Foundation, the Lowie Fund of the University of California, Berkeley, and the Carver Scientific Research Initiative Grants Program at the University of Iowa for support of travel and research related to the writing of this chapter.

This work is dedicated to F. Clark Howell for introducing both of us to the field of primate paleontology.

LITERATURE CITED

Aiello, L. C. (1986) The relationships of the Tarsiiformes: A review of the case for the Haplorhini. In B. Wood, L. Martin, and P. Andrews (eds.): *Major Topics in Primate and Human Evolution.* Cambridge: University of Cambridge Press, pp. 47-65.

Andrews, P. J. (1988) A phylogenetic analysis of the Primates. In M. J. Benton (ed.): *The Phylogeny and Classification of the Tetrapods,* Vol. 2: *Mammals.* Systematics Association Special Volume No. 35B. Oxford: Clarendon Press, pp. 143-175.

Andrews, P. J. (1992) Evolution and environment in the Hominoidea. *Nature* 360:641-646.

Andrews, P. J., and Cronin, J. E. (1982) The relationships of *Sivapithecus* and *Ramapithecus* and the evolution of the orangutan. *Nature* 297:541-546.

Andrews, P. J., and Martin, L .(1987) The phyletic position of the Ad Dabtiyah hominoid. *Bull. Br. Mus. Nat. Hist.* 41:383-393.

Andrews, P. J and Simons, E. L. (1977) A new African Miocene gibbon-like genus *Dendropithecus* (Hominoidea, Primates) with distinctive postcranial adaptations: Its significance to origin of Hylobatidae. *Folia Primatol.* 28:161-170.

Asfaw, B. (1987) The Belohdelie frontal: New evidence of early hominid cranial morphology from the Afar of Ethiopia. *J. Hum. Evol.* 16:611-624.

Ax, P. (1985) Stem species and the stem lineage concept. *Cladistics* 1:279-287.

Badgley, C., Qi, G., Chen, W., and Han, D. (1988) Paleoecology of a Miocene, tropical, upland fauna: Lufeng, China. *Nat. Geog. Res.* 4:178-195.

Ba Maw, Ciochon, R. L., and Savage, D. E. (1979) Late Eocene of Burma yields earliest anthropoid primate *Pondaungia cotteri. Nature* 282:65-67.

Barry, J. C., Jacobs, L. L., and Kelley, J. (1986) An early middle Miocene catarrhine from Pakistan with comments on the dispersal of catarrhines into Eurasia. *J. Hum. Evol.* 15:501-508.

Beard, K. C. (1988) The phylogenetic significance of strepsirhinism in Paleogene primates. *Int. J. Primatol.* 9:83-96.

Beard, K. C. (1990) Gliding behavior and paleoecology of the alleged primate family Paromomyidae (Mammalia, Dermoptera). *Nature* 345:340-341.

Beard, K. C. and Godinot, M. (1988) Carpel anatomy of *Smilodectes gracilis* (Adapiformes, Notharctinae) and its significance for lemuriform phylogeny. *J. Hum. Evol.* 17:71-92.

Beard, K. C. and Wang, B. (1991) Phylogenetic and biogeographic significance of the tarsiiform primate *Asiomomys changbaicus* from the Eocene of Jilin Province, People's Republic of China. *Am. J. Phys. Anthropol.* 85:159-166.

Beard, K. C., Dagosto, M., Gebo, D. L., and Godinot, M. (1988) Interrelationships among primate higher taxa. *Nature* 331:712-714.

Beard, K. C., Krishtalka, L., and Stucky, R. K. (1991) First skulls of the early Eocene primate *Shoshonius cooperi* and the anthropoid-tarsier dichotomy. *Nature* 349:64-67.

Begun, D. R. (1987) A Review of the Genus *Dryopithecus.* Unpublished Ph.D. Dissertation, University of Pennsylvania.

Begun, D. R. (1992a) Miocene fossil hominoids and the chimp-human clade. *Science* 257:1929-1933.

Begun, DR (1992b) Phyletic diversity and locomotion in primitive European hominids. *Am. J. Phys. Anthropol.* 87:311-340.

Benefit, B. R., and McCrossin, M. L. (1991) Ancestral facial morphology of Old World higher primates. *Proc. Nat. Acad. Sci. USA* 88:5267-5271.

Bernor, R. L., Brunet, M., Ginsburg, L., Mein, P., Pickford, M., Rogl, F., Sen, S., Steininger, F., and Thomas, H. (1987) A consideration of some major topics concerning Old World Miocene mammalian chronology, migrations and paleogeography. *Geobios* 20:431-439.

Bernor, R. L., Flynn, L. J., Harrison, T., Hussain, S. T., and Kelley, J. (1988) *Dionysopithecus* from southern Pakistan and the biochronology and biogeography of early Eurasian catarrhines. *J. Hum. Evol.* 17:339-358.

Bohlin, B. (1946) The fossil mammals from the Tertiary deposit of Taben-Buluk, western Kansu. II. Simplicidentata, Carnivora, Artiodactyla, Perissodactyla and Primates. *Palaeontol. Sin., new ser. C* (8b):1-256.

Briggs, J. C. (1989) The historic biogeography of India: Isolation or contact? *Syst. Zool.* 38:322-332.

de Bruijn, H. (1986) Is the presence of the African family Thryonomyidae in the Miocene deposits of Pakistan evidence of fauna exchange? *Palaeontol. Proceedings B* 89:125-134.

Brown, B. and Ward, S. C. (1988) Basicranial and facial topography in *Pongo* and *Sivapithecus*. In J. H. Schwartz (ed.): *Orang-utan Biology*. Oxford: Oxford University Press, pp. 247-260.

Cartmill, M. (1974) Rethinking primate origins. *Science* 184:436-443.

Cartmill, M. and Kay, R. F. (1978) Craniodental morphology, tarsier affinities, and primate suborders. In D. J. Chivers and J. A. Joysey (eds): *Recent Advances in Primatology*, Vol. 3. *Evolution*. London: Academic Press, pp. 205-214.

Cecchi, J. M., and Pickford, M, (1989) A new non-destructive technique for determining enamel prism structure in fossil mammal teeth. *C. R. Acad. Sci. Paris, Series 2* 308:1651-1654.

Chow, M. (1961) A new tarsioid primate from the Lushi Eocene, Honan. *Vert. PalAs.* 5:1-5.

Chow, M. (1964) A lemurid primate from the Eocene of Lantian, Shensi. *Vert. PalAs.* 8:260-262.

Ciochon, R. L. (1983) Hominoid cladistics and the ancestry of modern apes and humans: A summary statement. In R. L. Ciochon and R. S. Corruccini (eds.): *New Interpretations of Ape and Human Ancestry*. New York: Plenum Press, pp. 781-843.

Ciochon, R. L. (1985) Fossil ancestors of Burma. *Nat. Hist.* 10:26-36.

Ciochon, R. L., and Chiarelli, A. B. (1980) *Evolutionary Biology of the New World Monkeys and Continental Drift*. New York: Plenum.

Ciochon, R. L., and Corruccini R.S. (1977) The phenetic position of *Pliopithecus* and its phylogenetic relationship to the Hominoidea. *Syst. Zool.* 26:290-299.

Ciochon, R. L., and Corruccini, R. S. (1983) *New Interpretations of Ape and Human Ancestry*. New York: Plenum.

Ciochon, R. L., and Holroyd, P. (in press) The Asian origin of Anthropoidea revisited. In J. G. Fleagle and R. F. Kay (eds.): *Anthropoid Origins*. New York: Plenum.

Ciochon, R. L., and Olsen, J. (1986) Paleoanthropological and archeological research in the Socialist Republic of Viet Nam. *J. Hum. Evol.* 15:623-631.

Ciochon, R. L., Olsen, J., and James, J. (1990) *Other Origins: The Search for the Giant Ape in Human Prehistory*. New York: Bantam.

Ciochon, R. L., Savage, D. E., Tint, T., and Maw, B. (1985) Anthropoid origins in Asia? New discovery of *Amphipithecus* from the Eocene of Burma. *Science* 229:756-759.

Colbert, E. H. (1937) A new primate from the Upper Eocene Pondaung formations of Burma. *Amer. Mus. Novit.* 951:1-18.

Conroy, G. C., and Bown, T. M. (1976) Anthropoid origins and differentiation: The Asian question. *Yearb. Phys. Anthropol.* 18:1-6.

Conroy, G. C., Pickford, M., Senut, B., Van Couvering, J., and Mein, P. (1992) *Otavipithecus namibiensis*, first Miocene hominoid from southern Africa. *Nature* 356:144-147.

Corruccini, R. S., Ciochon, R. L., and McHenry, H. M. (1976) The postcranium of Miocene hominoids: Were dryopithecines merely "dental apes"? *Primates* 17:205-223.

Courtillot, V. (1990) What caused the mass extinction? A volcanic eruption. *Sci. Amer.* 263:85-92.

Courtillot, V., Feraud, G., Maluski, H., Vandamme, D., Moreau, M. G., and Besse, J. (1988) The Deccan flood basalts and the Cretaceous/Tertiary boundary. *Nature* 333:843-846.

Covert, H. H. (1986) Biology of early Cenozoic primates. In D. R. Swindler and J. Erwin (eds.): *Comparative Primate Biology*, Vol. 1. *Systematics, Evolution and Anatomy*. New York: Alan R. Liss, pp. 335-359.

Covert, H. H. (1988) Ankle and foot morphology of *Cantius mckennai*: Adaptations and phylogenetic implications. *J. Hum. Evol.* 17:57-70.

Cronin, J. E. (1983) Apes, humans and molecular clocks: A reappraisal. In R. L. Ciochon and R. S. Corruccini (eds.): *New Interpretations of Ape and Human Ancestry*. New York: Plenum Press, pp. 115-149.

Cronin, J. E., Sarich, V. M., and Ryder, O. (1984) Molecular evolution and the speciation of the lesser apes. In H. Preuschoft, D. J. Chivers, W. Y. Brockelman and N. Creel (eds.): *The Lesser Apes*. Edinburgh, Edinburgh University Press, pp. 467-485.

Culatto, E. (1992) A new take on anthropoid origins. *Science* 256:1516-1517.

Dagosto, M. (1988) Implications of postcranial evidence for the origin of euprimates. *J. Hum. Evol.* 17:35-56.

Dagosto, M. (1990) Models for the origin of the anthropoid postcranium. *J. Hum. Evol.* 19:121-139.

Dashzeveg, D. T., and McKenna, M. C. (1977) Tarsioid primate from the early Tertiary of the Mongolian People's Republic. *Acta Palaeontol. Polonica* 22:119-137.

Dean, D., and Delson, E. (1992) Second gorilla or third chimp. *Nature* 359:676-677.

de Bonis, L. G., and Melentis, J. (1977) Les primates hominoides du Vallesien de Macedoine (Grece): Etudes de la machiore inferieure. *Geobios* 10:849-885.

de Bonis, L. G., and Melentis, J. (1987) Interet de l'anatomie nasomaxillaire pour la phylogenie des hominidae. *C. R. Acad. Sci. Series 2* 304:767-769.

de Bonis, L. G., Bouvrain, G., Geraads, D., and Koufos, G. (1990) New hominid skull material from the late Miocene of Macedonia in northern Greece. *Nature* 345:712-715.

de Bonis, L. G., Jaeger, J. J., Coiffait, B., and Coiffait, P. E. (1988) Decouverte du plus ancien primate Catarrhinien connu dans l'Eocene superieur d'Afrique du Nord. *C.R. Acad. Sci. Paris Series 2* 306:929-934.

de Bonis, L. G., Johanson, D. C., Melentis, J., and White, T. (1981) Variations metriques de la denture chez les Hominides primitif: comparison entre *Australopithecus afarensis et Graecopithecus macedoniensis. C. R. Acad. Sci. Paris Series 2* 292:373-376.

Delson, E. (1985) The earliest *Sivapithecus*? *Nature* 318:107-108.

Dene, H., Goodman, M., Prychodko, W., and Moore, G. W. (1976) Immunodiffusion systematics of the primates. *Folia Primatol.* 25:35-61.

Duncan, R. A., and Pyle, D. G. (1988) Rapid eruption of the Deccan flood basalts at the Cretaceous/Tertiary boundary. *Nature* 333:841-843.

Etler, D. A. (1984) The fossil hominoids of Lufeng, Yunnan Province, The People's Republic of China: A series of translations. *Yearb. Phys. Anthropol.* 27:1-55.

Etler, D. A. (1989) Miocene hominoids from Sihong, Jiangsu Province, China. In A. Sahni and R. Gaur (eds.): *Perspectives in Human Evolution.* Delhi: Renaissance Publishing House, pp. 113-151.

Fleagle, J. G. (1975) A small gibbon-like hominoid from the Miocene of Uganda. *Folia Primatol.* 24:1-15.

Fleagle, J. G. (1984) Are there any fossil gibbons? In D. J. Chivers, H. Preschoft, N. Creel, and W. Brockelman (eds.): *The Lesser Apes: Evolutionary and Behavioral Biology.* Edinburgh: Edinburgh University Press, pp. 431-477.

Fleagle, J. G. (1988) *Primate Adaptation and Evolution.* San Diego: Academic Press.

Fleagle, J. G., and Kay, R. F. (1983) New interpretations of the phyletic position of Oligocene hominoids. In R. L. Ciochon and R. S. Corruccini (eds.): *New Interpretations of Ape and Human Ancestry.* New York: Plenum Press, pp. 181-210.

Fleagle, J. G., and Kay, R. F. (1987) The phyletic position of the Parapithecidae. *J. Hum. Evol.* 16:483-531.

Fleagle, J. G., and Simons, E. L. (1978) *Micropithecus clarki,* a small ape from the Miocene of Uganda. *Am. J. Phys. Anthropol.* 49:427-440.

Fleagle, J. G., Stern, J. T., Jungers, W. L., Susman, R. L., Vangor, A. K. and Wells, J. P. (1981) Climbing: A biomechanical link with brachiation and with bipedalism. In M. H. Day (ed.): *Vertebrate Locomotion. Symp. Zool. Soc. Lond.* 48:359-375.

Flynn, L. J., and Sabatier, M. (1984) A muroid rodent of Asian affinity from the Miocene of Kenya. *J. Vert. Paleontol.* 3:160-165.

Ford, S. M. (1988) Postcranial adaptations of the earliest platyrrhines. *J. Hum. Evol.* 17:155-192.

Gebo, D. L. (1986) Anthropoid origins—the foot evidence. *J. Hum. Evol.* 15:421-430.

Gingerich, P. D. (1984) Primate evolution: evidence from the fossil record, comparative morphology and molecular biology. *Yearb. of Phys. Anthropol.* 27:57-72.

Gingerich, P. D. (1990) African dawn for primates. *Nature* 346:411.

Gingerich, P. D. and Sahni, A. (1979) *Indraloris* and *Sivaladapis*: Miocene adapid primates from the Siwaliks of India and Pakistan. *Nature* 279:415-416.

Gingerich, P. D., and Sahni, A. (1984) Dentition of *Sivaladapis nagrii* (Adapidae) from the late Miocene of India. *Int. J. Primatol.* 5:63-79.

Gingerich, P. D., Dashzeveg, D., and Russell, D. (1991) Dentition and systematic relationships of *Altanius orlovi* (Mammalia, Primates) from the early Eocene of Mongolia. *Geobios* 24:637-646.

Godinot, M. (1991) Toward the locomotion of two contemporaneous *Adapis* species. *Z. Morph. Anthropol.* 78:387-405.

Godinot, M. (1992) Early euprimate hands in evolutionary perspective. *J. Hum. Evol.* 22:267-283.

Godinot, M., and Beard, K. C. (1991) Fossil primate hands: A review and an evolutionary inquiry emphasizing early forms. *Hum. Evol.* 6:307-354.

Godinot, M., and Mahboubi, M. (1992) Earliest known simian primate from Algeria. *Nature* 357:324-326.

Gonzalez, I. L., Sylvester, J. E., Smith, T. F., Stambolian, D., and Schmickel, R. D. (1990) Ribosomal RNA gene sequences and hominoid phylogeny. *Mol. Biol. Evol.* 7:203-219.

Goodman, M., Hewett-Emmett, D., and Beard, J. M. (1978) Molecular evidence on the phylogenetic relationships of *Tarsius.* In D. J. Chivers and K. A. Joysey (eds.): *Recent Advances in Primatology,* Vol. 3. New York: Academic Press, pp. 215-224.

Goodman, M., Kopp, B. F., Czelusniak, J., Fitch, D. H. A., Tagle, D.A., and Slightom, J. L. (1989) Molecular pylogeny of the family of apes and humans. *Genome* 31:3416-335.

Gregory, W. K., and Hellman, M, (1926) The dentition of *Dryopithecus* and the origin of man. *Am. Mus. Anthropol. Papers.* 28:1-123.

Groves, C. P. (1972) Systematics and phylogeny of gibbons. *Gibbon & Siamang* 1:1-89.

Gu, Y. (1986) Preliminary research of the fossil gibbon of Pleistocene China. *Acta Anthropol. Sin.* 5:208-219.

Gu, Y., and Lin, Y. (1983) First discovery of *Dryopithecus* in East China. *Acta Anthropol. Sin.* 2:305-314.

Gu, Y., Huang, W., Song, P., Guo, X., and Chen, D. (1987) The study of some fossil orang-utan from Guangdong and Guangxi. *Acta Anthropol. Sin.* 6:272-283.

Harrison, T. (1987) The phylogenetic relationships of the early catarrhine primates: A review of the current evidence. *J. Hum. Evol.* 16:41-80.

Harrison, T., Delson, E., and Guan, J. (1991) A new species of *Pliopithecus* from the middle Miocene of China and its implications for early catarrhine zoogeography. *J. Hum. Evol.* 21:329-361.

Hasegawa, M., Kishino, H., Hayasaka, K., and Horai, S. (1990) Mitochondrial DNA evolution in Primates: Transition rate has been extremely low in the lemur. *J. Mol. Evol.* 31:113-121.

Hasegawa, M., and Kishino, H. (1991) DNA sequence analysis and evolution of the Hominoidea. In M. Kimura and N. Takahata (eds.): *New Aspects of the Genetics of Molecular Evolution.* Tokyo: Springer/Verlag, p. 303.

Hennig, W. (1966) *Phylogenetic Systematics.* Urbana: University of Illinois Press.

Hill, A., and Ward, S., (1988) Origin of the Hominidae: The record of African large hominoid evolution between 14 my and 4 my. *Yearb. Phys. Anthropol.* 31:49-83.

Hofer, H. O. (1980) The external anatomy of the oro-nasal region of primates. *Z. Morph. Anthropol.* 71:233-249.

Hooijer, D. A. (1948) Prehistoric teeth of man and the orangutan from central Sumatra with notes on the fossil orangutan from Java and southern China. *Zool. Meded. Rijks Mus. Nat. Hist.* 29:175-301.

Hopwood, A. T. (1933) Miocene primates from Kenya. *Linn. Soc. (Zool.)* 38:437-464.

Horai, S., Satta, Y., Hayasaka, K., Kondo, R., Inoue, T., Ishida, T., Hayashi, S., and Takahata, N. (1992) Man's place in Hominoidea revealed by mitochondrial DNA geneology. *J. Mol. Evol.* 35:32-43.

Ishida, H., Pickford, M., Nakaya, Y., and Nakano, Y. (1984) Fossil anthropoids from Nachola and Samburu Hills, Samburu district, northern Kenya. In H. Ishida, S. Ishida and M. Pickford (eds.): *Study of the Tertiary Hominoids and their Palaeoenvironments in East Africa: 2.* African Studies Monograph, Supplementary Issue No. 2. Kyoto: Kyoto University Press. pp. 73-86.

Kalke, H. D. (1972) A review of the Pleistocene history of the orang-utan (*Pongo* Lacepede 1799). *Asian Perspectives* 15:5-13.

Kay, R., Thorington, R. W., Jr., and Houde, P. (1990) Eocene plesiadapiform shows affinities with flying lemurs not primates. *Nature* 345:342-344.

Kelley, J., and Etler, D. A. (1989) Hominoid dental variability and species number at the late Miocene site of Lufeng, China. *Am. J. Primatol.* 18:15-34.

Kelley, J., and Pilbeam, D. R. (1986a) Kenyan find not early Miocene *Sivapithecus. Nature* 321:475-476.

Kelley, J., and Pilbeam, D. R. (1986b) The dryopithecines: Taxonomy, comparative anatomy, and phylogeny of Miocene large hominoids. In D. R. Swindler and J. Erwin (eds.): *Comparative Primate Biology,* Vol. 1: *Systematics, Evolution, and Anatomy.* New York: Alan R. Liss, pp. 361-411.

Kelley, J., and Xu, Q, (1991) Extreme sexual dimorphism in a Miocene hominoid. *Nature* 352:151-153.

Kielan-Jaworowska, Z., Bown, T. M., and Lillegraven, J. A. (1979) Eutheria. In J. A. Lillegraven, Z. Kielan-Jaworowska and W. A. Clemens (eds.): *Mesozoic Mammals: The First Two-Thirds of Mammalian History.* Berkeley: University of California Press, pp. 221-258.

Koop, B. F., Goodman, M., Xu, P., Chan, and Slighton, J. L. (1986) Primate beta-globin DNA sequences and man's place among the great apes. *Nature* 319:234-238.

Koop, B. F., Tagle, D. A., Goodman, M., and Slightom, J. L. (1989) A molecular view of primate phylogeny and important systematic and evolutionary questions. *Mol. Biol. Evol.* 6:580-612.

Kordos, L (1987) Description and reconstruction of the skull of *Rudapithecus hungaricus* Kretzoi (Mammalia). *Ann. Hist.-Nat. Mus. Nat. Hungarici* 78:77-88.

Kordos, L. (1988) Comparison of early primate skulls from Rudabanya (Hungary) and Lufeng (China). *Anthropol. Hungarica* 20:9-22.

Krause, D. W. (1992) Were paromomyids gliders? Maybe, maybe not. *J. Hum. Evol.* 21:177-188.

Krause, D. W., and Maas, M. C. (1990) The biogeographic origins of late Paleocene-early Eocene mammalian immigrants to the western interior of North America. In T. M. Bown and K. D. Rose (eds.): *Dawn of the Age of Mammals in the Northern Part of the Rocky Mountain Interior, North America,* Boulder: Geological Society of America, Special Paper 243, pp. 71-105.

Kretzoi, M. (1975) New ramapithecines and *Pliopithecus* from the lower Pliocene of Rudabanya in northeastern Hungary. *Nature* 257:578-581.

Krishtalka, L. (1989) New skull of the Eocene primate *Notharctus* and the origin of anthropoids and lemuriforms. *J. Vert. Paleontol.* 9 (Supp. to no. 3):28A.

Krishtalka, L., Stucky, R. K., and Beard, K. C. (1990) The earliest fossil evidence for sexual dimorphism in primates. *Proc. Nat. Acad. Sci. USA* 87:5223-5226.

Leakey, R. E., and Leakey, M. G. (1986) A new Miocene hominoid from Kenya. *Nature* 324:143-146.

Leakey, R. E., and Walker., A (1985) New higher primates from the early Miocene of Buluk, Kenya. *Nature* 318:173-178.

Le Gros Clark, W. E., and Leakey, L. S. B. (1951) The Miocene Hominoidea of East Africa. In *Fossil Mammals of Africa. Brit. Mus. Nat. Hist.* 1:1-117.

Lei, C. (1985) Study on the mid-Miocene apes discovered in Jiangsu, China. *Acta Geol. Sin.* 59:17-24.

Li, C. (1978) A Miocene gibbon-like primate from Shihung, Kiangsu Province. *Vert. PalAs.* 16:187-192.

Li, C. and Ting, S. (1983) *The Paleogene Mammals of China.* Carnegie Museum of Natural History Bulletin 21, pp. 1-98.

Lin, Y., Wang, S., Gao, Z., and Zhang, L. (1987) The first discovery of the radius of *Sivapithecus lufengensis* in China. *Geol. Rev.* 33:1-4.

Maas, M. C., Krause, D. W., and Strait, S. G. (1988) The decline and extinction of Plesiadapiformes (Mammalia: ?Primates) in North America: Displacement or replacement. *Paleobiology* 14:410-431.

MacPhee, R. D. E., and Cartmill, M. (1986) Basicranial structures and primate systematics. In D. Swindler and J. Erwin (eds.): *Comparative Primate Biology,* Vol. 1: *Systematics, Evolution and Anatomy.* New York: Alan R. Liss, pp. 219-275.

Martin, L. (1985) Significance of enamel thickness in hominoid evolution. *Nature* 314:260-263.

Martin, R. D. (1990) Some relatives take a dive. *Nature* 345:291-292.

McKenna, M. C. (1966) Paleontology and the origin of the Primates. *Folia Primatol.* 4:1-25.

McKenna, M. C. (1990) Plagiomenids (Mammalia: ?Dermoptera) from the Oligocene of Oregon, Montana and South Dakota and middle Eocene of northwestern Wyoming. In T. M. Bown and K. D. Rose (eds.): *Dawn of the*

Age of Mammals in the Northern Part of the Rocky Mountain Interior, North America. Boulder: Geological Society of America Special Paper 243, pp. 211-234.

Meldrum, D. J., and Pan Y. (1988) Manual proximal phalanx of *Laccopithecus robustus* from the latest Miocene site of Lufeng. *J. Hum. Evol.* 17:719-731.

Miyamoto, M. M., Slighton, J. L., and Goodman, M. (1987) Phylogenetic relations of humans and African apes from DNA sequences in the ψη-globin region. *Science* 238:369-373.

Morbeck, M. E. (1983) Miocene hominoid discoveries from Rudabanya. Implications from the postcranial skeleton. In R. L. Ciochon and R. S. Corruccini (eds.): *New Interpretations of Ape and Human Ancestry.* New York: Plenum, pp. 394-404.

Novacek, M. J. (1990) Morphology, paleontology, and the higher clades of mammals. In H. H. Genoways (ed.): *Current Mammalogy,* Vol. 2. New York: Plenum.

Novacek, M. J. (1992) Mammalian phylogeny: Shaking the tree. *Nature* 356:121-125.

Osborn, H. F. (1900) The geological and faunal relations of Europe and America during the Tertiary period and the theory of successive invasions of the African fauna. *Science* 11:560-574.

Pan, Y. (1988) Small fossil primates from Lufeng, a latest Miocene site in Yunnan province, China. *J. Hum. Evol.* 17:359-366.

Pan, Y., and Wu, R. (1986) A new species of *Sinoadapis* from the hominoid site of Lufeng. *Acta Anthropol. Sin.* 5:39-50.

Pan Y., Waddle, D. M., and Fleagle, J. G. (1989) Sexual dimorphism in *Laccopithecus robustus,* a late Miocene hominoid from China. *Am. J. Phys. Anthropol.* 79:137-158.

Pettigrew, J. D. (1986) Flying primates? Megabats have the advanced pathway from eye to midbrain. *Science* 231:1304-1306.

Pickford, M. (1986) Hominoids from the Miocene of East Africa and the phyletic position of *Kenyapithecus. Z. Morph. Anthropol.* 76:117-130.

Pilbeam, D. R. (1982) New hominoid skull material from the Miocene of Pakistan. *Nature* 295:232-234.

Pilbeam, D. R. (1985) Patterns of hominoid evolution. In E. Delson (ed.): *Ancestors: The Hard Evidence.* New York: Alan R. Liss, pp. 51-59.

Pilbeam, D. R. (1986) Distinguished Lecture: Hominoid evolution and hominid origins. *Am. Anthropol.* 88:295-312.

Pilbeam, D. R., Rose, M. D., Barry, J. C., and Ibrahim Shah, S. M. (1990) New *Sivapithecus* humeri from Pakistan and the relationship of *Sivapithecus* and *Pongo. Nature* 348:237-239.

Pilgrim, G. E. (1927) A new *Sivapithecus* palate and other primate fossils from India. *Mem. Geol. Surv. India (Palaeontol. India)* 14:1-26.

Preuss, T. M. (1982) The face of *Sivapithecus indicus*: Description of a new relatively complete specimen from the Siwaliks of Pakistan. *Folia Primatol.* 38:141-157.

Qi, T., Zong, G., and Wang, Y. (1991) Discovery of *Lushilagus* and *Miacis* in Jiangsu and its zoogeographical significance. *Vert. PalAs.* 29:59-63.

Qiu, Z., and Guan, J. (1986) A lower molar of *Pliopithecus* from Tongxin, Ningxia Hui Autonomous Region. *Acta Anthropol. Sin.* 5:201-207.

Rasmussen, D. T. (1986) Anthropoid origins: A possible solution to the Adapidae-Omomyidae paradox. *J. Hum. Evol.* 15:1-12.

Rasmussen, D. T. and Simons, E. L. (1988) New specimens of *Oligopithecus savagei,* early Oligocene primate from the Fayum, Egypt. *Folia Primatol.* 51:182-208.

Raza, S. M., Barry, J. C., Pilbeam, D., Rose, M. D., Shah, S. M. I. and Ward, S. (1983) New hominoid primates from the middle Miocene Chinji Formation, Potwar Plateau, Pakistan. *Nature* 305:52-54.

Rose, K. D., and Bown, T. M. (1991) Additional fossil evidence on the differentiation of the earliest euprimates. *Proc. Nat. Acad. Sci. USA* 88:98-101.

Rose, M. D. (1989) New postcranial specimens of catarrhines from the middle Miocene Chinji formation, Pakistan: Descriptions and a discussion of proximal humeral functional morphology in anthropoids. *J. Hum. Evol.* 18:131-162.

Rosenberger, A. L., and Szalay, F. S. (1980) On the tarsiiform origins of Anthropoidea. In R. L. Ciochon and A. B. Chiarelli (eds.): *Evolutionary Biology of the New World Monkeys and Continental Drift.* New York: Plenum. pp. 139-157.

Russell, D. E., and Gingerich, P. D. (1980) Un nouveau Primate omomyide dans l'Eocene du Pakistan. *C.R. Acad. Sci. Paris, Ser. D* 291:621-624.

Russell, D. E., and Gingerich, P. D. (1987) Nouveaux Primates de l'Eocene du Pakistan. *C.R. Acad. Sci. Paris, Ser.* 2 304:209-214.

Russell, D. E., and Zhai, R, (1987) *The Paleogene of Asia; Mammals and Stratigraphy.* Museum National d'Histoire Naturelle, Sciences de la Terre, Memoires, v. 52, pp. 1-488.

Ruvolo, M., Disotell, T. R., Allard, M. W., and Brown, W. M. (1991) Resolution of the African hominoid trichotomy by use of a mitochondrial gene sequence. *Proc. Nat. Acad. Sci. USA* 88:1570-1574.

Sarich, V. M., and Cronin, J. E. (1976) Molecular systematics of the primates. In M. Goodman and R. E. Tashian (eds.): *Molecular Anthropology.* New York: Plenum Press, pp. 141-170.

Savage, D. E., and Russell, D. E. (1983) *Mammalian Paleofaunas of the World.* Reading, MA and London: Addison-Wesley.

Schwartz, J. H. (1990) *Lufengpithecus* and its potential relationship to an orang-utan clade. *J. Hum. Evol.* 19:591-605.

Schwartz, J. H., and Tattersall, I. (1987) Tarsiers, adapids and the integrity of Strepsirhini. *J. Hum. Evol.* 16:23-40.

Sibley, C. G., and Ahlquist, J. E. (1984) The phylogeny of the hominoid primates, as indicated by DNA hybridization. *J. Mol. Evol.* 20:2-15.

Sibley, C. G., Comstock, J. A., and Ahlquist, J. E. (1990) DNA hybridization evidence of hominoid phylogeny: A reanalysis of the data. *J. Mol. Evol.* 30:202-236.

Sige, B., Jaeger, J. J., Sudre, J., and Vianey-Liaud, M. (1990) *Altiatlasius koulchii* n. gen. et sp., primate omomyide du

Paleocene superieur du Maroc, et les origines des euprimates. *Palaeontographica Abt.* A 214:31-56.
Simons, E. L. (1971) Relationships of *Amphipithecus* and *Oligopithecus*. *Nature* 232:489-491.
Simons, E. L. (1989) Description of two genera and species of late Eocene Anthropoidea from Egypt. *Proc. Nat. Acad. Sci. USA* 86:9956-9960.
Simons, E. L. (1990) Discovery of the oldest known anthropoidean skull from the Paleogene of Egypt. *Science* 247:1567-1569.
Simons, E. L. (1992) Diversity in the early Tertiary anthropoidean radiation in Africa. *Proc. Nat. Acad. Sci. USA* 89:10743-10747.
Simons, E. L., and Chopra, S. R. K. (1969) *Gigantopithecus* (Pongidae, Hominoidea). A new species from northern India. *Postilla* 138:1-18.
Simons, E. L., and Pilbeam, D. (1965) Preliminary revision of the Dryopithecinae (Pongidae, Anthropoidea). *Folia Primatol.* 3:81-152.
Simons, E. L., and Rasmussen, D. T. (1989) Cranial morphology of *Aegyptopithecus* and *Tarsius* and the question of the tarsier-anthropoidean clade. *Am. J. Phys. Anthropol.* 79:1-23.
Simons, E. L., Bown, T. M., and Rasmussen, D. T. (1986) Discovery of two additional prosimian primate families (Omomyidae, Lorisidae) in the African Oligocene. *J. Hum. Evol.* 15:431-437.
Sudre, J. (1975) Un Prosimien du Paleogene ancien du Sahara nordoccidental: *Azibius trerki* n.g., n. sp. *C.R. Acad. Sci. Paris* 280:1539-1542.
Sussman, R. W. (1991) Primate origins and the evolution of angiosperms. *Am. J. Primatol.* 23:209-223.
Suteethorn, V., Buffetaut, E., Helmcke-Ingavat, R., Jaeger, J., and Jongkanjanasoontorn, Y. (1988) Oldest known Tertiary mammals from South East Asia: Middle Eocene primate and anthracotheres from Thailand. *N. Jb. Geol. Palaont. Mh.* 1988:563-570.
Szalay, F. S. (1972) *Amphipithecus* revisited. *Nature* 236:179-180.
Szalay, F. S., and Delson, E. (1979) *Evolutionary History of the Primates.* New York: Academic Press.
Szalay, F. S., and Li, C. (1986) Middle Paleocene euprimate from southern China and the distribution of primates in the Paleogene. *J. Hum. Evol.* 18:387-398.
Szalay, F. S., Li, C., and Wang, B. (1986) Middle Paleocene omomyid primate from Anhui Province, China: *Decoredon anhuiensis* (Xu, 1976), new combination Szalay and Li, and the significance of *Petrolemur* (abstract). *Am. J. Phys. Anthropol.* 69:269.
Szalay, F. S., Rosenberger, A. L., and Dagosto, M. (1987) Diagnosis and differentiation of the Order Primates. *Yearb. Phys. Anthropol.* 30:75-105.
Thomas, H., Roger, J., Sen, S., and Al-Sulaimani, Z. (1988) Decouverte des plus anciens <Anthropoides> du continent arabo-africain et d'un Primate tarsiiforme dans l'Oligocene du Sultanat d'Oman. *C.R. Acad. Sci. Paris Ser. 2,* 306:823-828.
Tong, Y. (1979) A late Paleocene primate from South China. *Vert. PalAs.* 17:65-70.
Tyler, D. E. (1991) The problems of the Pliopithecidae as a hylobatid ancestor. *Hum. Evol.* 6:73-80.
von Koenigswald, G. H. R. (1969) Miocene Cercopithecoidea and Oreopithecoidea from the Miocene of East Africa. *Foss. Verts. Afr.* 1:39-51.
Walker, A. C., and Pickford, M. (1983) New postcranial fossils of *Proconsul africanus* and *Proconsul nyanzae*. In R. Ciochon and R. Corruccini (eds.): *New Interpretations of Ape and Human Ancestry.* New York: Plenum, pp. 325-352.
Wang, B., and Li, C. (1990) First Paleogene mammalian fauna from northeast China. *Vert. PalAs.* 28:165-205.
Ward, C. V., Walker, A., and Teaford, M. F. (1992) *Proconsul* did not have a tail. *J. Hum Evol.* 21:215-220.
Ward, S. C., and Brown, B. (1986) The facial skeleton of *Sivapithecus indicus.* In D. Swindler and J. Erwin (eds.): *Comparative Primate Biology,* Vol. 1. *Systematics, Evolution and Anatomy.* New York, Alan R. Liss.
Ward, S. C., and Kimbel, W. H. (1983) Subnasal alveolar morphology and the systematic position of *Sivapithecus. Am. J. Phys. Anthropol.* 61:157-171.
Wible, J. R., and Covert, H. H. (1987) Primates: Cladistic diagnosis and relationships. *J. Hum. Evol.* 16:1-22.
Woo, J. (1957) *Dryopithecus* teeth from Keiyuan, Yunnan Province. *Vert. PalAs.* 1:25-32.
Woo, J. (1958) New materials of *Dryopithecus* from Keiyuan, Yunnan. *Vert. PalAs.* 2:31-43.
Woo, J., and Chow, M. (1957) New materials of the earliest primate known in China—*Hoanghonius stehlini. Vert. PalAs.* 1:267-272.
Wood, B. A., and Xu, Q. (1991) Variation in the Lufeng dental remains. *J. Hum. Evol.* 20:291-311.
Wu, R. (1985) The cranium of *Ramapithecus* and *Sivapithecus* from Lufeng, China. In P. Andrews and J. L. Franzen (eds.): *The Early Evolution of Man with Special Emphasis on Southeast Asia and Africa.* Frankfurt: Cour. Forsch.-Inst. Senckenberg 69, pp. 41-48.
Wu, R. (1987) A revision of the classification of the Lufeng great apes. *Acta Anthropol. Sin.* 6:265-271.
Wu, R., and Pan, Y. (1984) A late Miocene gibbon-like primate from Lufeng, Yunnan Province. *Acta Anthropol. Sin.* 3:193-200.
Wu, R., and Pan, Y. (1985a) A new genus of Miocene adapid from Lufeng. *Acta Anthropol. Sin.* 4:1-6.
Wu, R., and Pan, Y. (1985b) Preliminary observation on the cranium of *Laccopithecus robustus* from Lufeng, Yunnan, with reference to its phylogenetic position. *Acta Anthropol. Sin.* 4:7-13.
Wu, R., and Wang, L. (1987) Sexual dimorphism of fossil apes in Lufeng. *Acta Anthropol. Sin.* 6:169-174.
Wu, R., and Wang, L. (1988) Single species and sexual dimorphism in *Sinoadapis. Acta Anthropol. Sin.* 7:1-8.
Wu, R., Lu, Q., and Xu, Q. (1984) Morphological features of *Ramapithecus* and *Sivapithecus* and their phylogenetic relationship-morphology and comparison of the mandibles. *Acta Anthropol. Sin.* 3:1-10.

Wu, R., Xu, Q., and Lu, Q. (1983) Morphological features of *Ramapithecus* and *Sivapithecus* and their phylogenetic relationship-morphology and comparison of the crania. *Acta Anthropol. Sin.* 2:1-10.
Wu, R., Xu, Q., and Lu, Q. (1985) Morphological features of *Ramapithecus* and *Sivapithecus* and their phylogenetic relationship-morphology and comparison of the teeth. *Acta Anthropol. Sin.* 4:197-204.
Wu, R., Xu, Q., and Lu, Q. (1986) Relationship between Lufeng *Sivapithecus* and *Ramapithecus* and their phylogenetic position. *Acta Anthropol. Sin.* 5:1-30.
Xiao, M. (1981) Discovery of fossil hominoid scapula at Lufeng, Yunnan. *J. Yunnan Prov. Mus.* 30:41-44.
Xirotiris, N. I., and Henke, W. (1981) Enamel prism patterns of European hominoids—and their phylogenetical aspects. In A. B. Chiarelli and R. S. Corruccini (eds.): *Primate Evolutionary Biology,* Stuttgart: Springer-Verlag, pp. 109-116.
Xue, X., and Delson, E. (1988) A new species of *Dryopithecus* from Gansu, China. *Kexue Tongbao* 33:449-453 [in Chinese].
Zdansky, O. (1930) Die Alttertiaren Saugetiere China nebst stratigraphischen Bemerkungen. *Palaeontol. Sin.* C 6:1-87.
Zhang, X., Zheng, L., Gao, F., Jiang, C., and Zhang, J. (1988) A preliminary study of the skull fossils of Lama ape unearthed at Hudie Hill of Yuanmou county. *J. Yunnan Univ.* (Soc. Sci.) 5:55-61,18.
Zhang, Y., Wang, S., and Quan, G. (1981) On the geographical distribution of primates in China. *J. Hum. Evol.* 10:215-221.
Zihlman, A. L., Cronin, J. B., Cramer, D. L., and Sarich, V. M. (1978) Pygmy chimpanzee as a possible prototype for the common ancestor of humans, chimpanzees and gorillas. *Nature* 275:744-746.

CHAPTER 4

Patterns, Puzzles and Perspectives on Platyrrhine Origins

Walter Carl Hartwig

INTRODUCTION

The platyrrhine origins issue incorporates several different questions. How did platyrrhines get to South America? Are platyrrhines and catarrhines each monophyletic? What do early fossil anthropoids indicate about platyrrhine divergence? Answers to questions such as these are more or less approachable given the particular field of study (i.e., plate tectonics, paleontology, comparative anatomy, molecular biology). Integrating the answers into an empirical reconstruction of the process of platyrrhine origins remains far less tangible.

Except for a few direct analyses (e.g., Erikson 1954; Hershkovitz 1972, 1977; Cronin and Sarich 1975; Orlosky and Swindler 1975), the topic of platyrrhine origins had been treated incidentally within larger systematic reviews (e.g., Gregory 1920; Cracraft 1974; Thorington 1976; Washburn and Moore 1980) until the publication of *Evolutionary Biology of the New World Monkeys and Continental Drift* (R. L. Ciochon and A. B. Chiarelli, eds.) in 1980. This volume focused diverse analytical approaches on the question of platyrrhine origins and dispersal and in its final syntheses (Ciochon and Chiarelli 1980; Delson and Rosenberger 1980) presented rigorously formulated hypotheses from what were perceived clearly as complex, interrelated databases. It is only logical, then, that this review begins with the results of that volume and the data that have accumulated subsequently.

During the 1980s research on New World monkey ecology and biology, and discovery of several new fossil forms, proceeded apace (Fleagle and Rosenberger 1990); in the meantime, additional direct treatments of platyrrhine origins were published (Cachel 1981; Rose and Fleagle 1981; Fleagle 1986; Ford 1986b; Rosenberger 1986). Related developments in geochronology (Fleagle et al. 1986a, b; MacFadden 1990), paleontology (Simons 1989, 1990) and comparative anatomy (Fleagle and Kay 1987; Rosenberger et al. 1990, and many others) provided new grounds for testing both past and present platyrrhine origins models. The following analysis seeks to place the information of the last decade in the context of the most recent and formal platyrrhine origin and dispersal scenarios. It first reviews formal models of platyrrhine origins published since 1980, describes some of the most relevant data that have accumulated in the meantime, and concludes with a perspective on database patterns that is intended to cast approaches to this problem in a new and, it is hoped, progressive light. Allegiance to pattern and parsimony as interpretive tools of evolutionary history may be the only way to shape the concatenation of the fossil record, living primate systematics, and geophysical processes into a coherent account of the intercontinental, infraordinal New World monkey radiation.

RECENT MODELS OF PLATYRRHINE ORIGINS

Three formal treatments of the platyrrhine origins issue have appeared since 1980 (Cachel 1981; Rose and Fleagle 1981; Fleagle 1986) and are analyzed here. The Ciochon and Chiarelli (1980) Maximum Parsimony model is analyzed in detail, also. Other references to the platyrrhine origins issue have been made in contexts apart from formal models or are reconsiderations of previous holistic positions (e.g., Hershkovitz 1981; Fleagle and Rosenberger 1983; Rosenberger 1986; Ford 1986b, 1988, 1990b; Fleagle 1988) and are referenced accordingly in the following discussion.

Ciochon and Chiarelli 1980

Ciochon and Chiarelli (1980) have analyzed North American, African, Asian, and vicariance (Hershkovitz 1972, 1977; Szalay 1975) models of platyrrhine origins in great detail. Drawing on these positions, a strict reading of parsimony methodology, and the suite of contributions to the 1980 volume on New World monkey evolutionary biology, they proposed a Maximum Parsimony model that is diagrammed in Figure 4-1 and discussed later. They rigorously applied criteria of model acceptance or rejection according to outlined rules of paleobiogeographic reconstruction. In the short historical context of the last decade, the Ciochon and Chiarelli (1980) model has served as an analytical clearinghouse for later considerations of the platyrrhine origins issue.

Ciochon and Chiarelli (1980) emphasized three problem areas with the North American model, areas that are weighted much differently in other versions of this hypothesis (Gingerich and Rose 1977; Delson and Rosenberger 1980; Rose and Fleagle 1981). First, while also acknowledging a formidable distance between Africa and South America in the Eocene (Funnell and Smith 1968; Ladd et al. 1973; Smith and Briden 1977; Sclater et al. 1977), the distance between North America and South America at the same time is considered a negative factor of the North American model. Second, Ciochon and Chiarelli (1980) emphasized the influence of reconstructed paleocurrents (Frakes and Kemp 1973; Berggren and Hollister 1974; Holcombe and Moore 1977; Tarling 1980) that would counter a southward rafting event. Third, they noted the absence of a likely ancestral platyrrhine in North America, given a relatively well-known Paleocene–Eocene North American mammalian fauna. For Ciochon and Chiarelli (1980), these three factors outweighed the balance of absent evidence often used to support North American models.

For a number of reasons Ciochon and Chiarelli (1980) supported an Asian origin for anthropoids, as depicted in the two prominent models of platyrrhine dispersal to South America via North America and ultimately an Asian anthropoid ancestry (Delson and Rosenberger 1980; Gingerich 1980). The recent discovery of additional Eocene *Amphipithecus* remains in Asia (Ciochon et al. 1985) has renewed interest in establishing the presence of early primates (Szalay and Li 1986) and earliest anthropoids (Ciochon et al. 1985) in this area. While explicitly supporting the biological grounds upon which the Asian origin of anthropoids are based, Ciochon and Chiarelli (1980) rejected the Asian model of platyrrhine dispersal because it invokes the same mechanisms as the North American model, and these mechanisms failed their criteria of parsimony. Though not explicitly stated, the lack of any fossil anthropoids along the proposed dispersal route works against this scenario and must be explained before the Asian model via North America can be critically accepted.

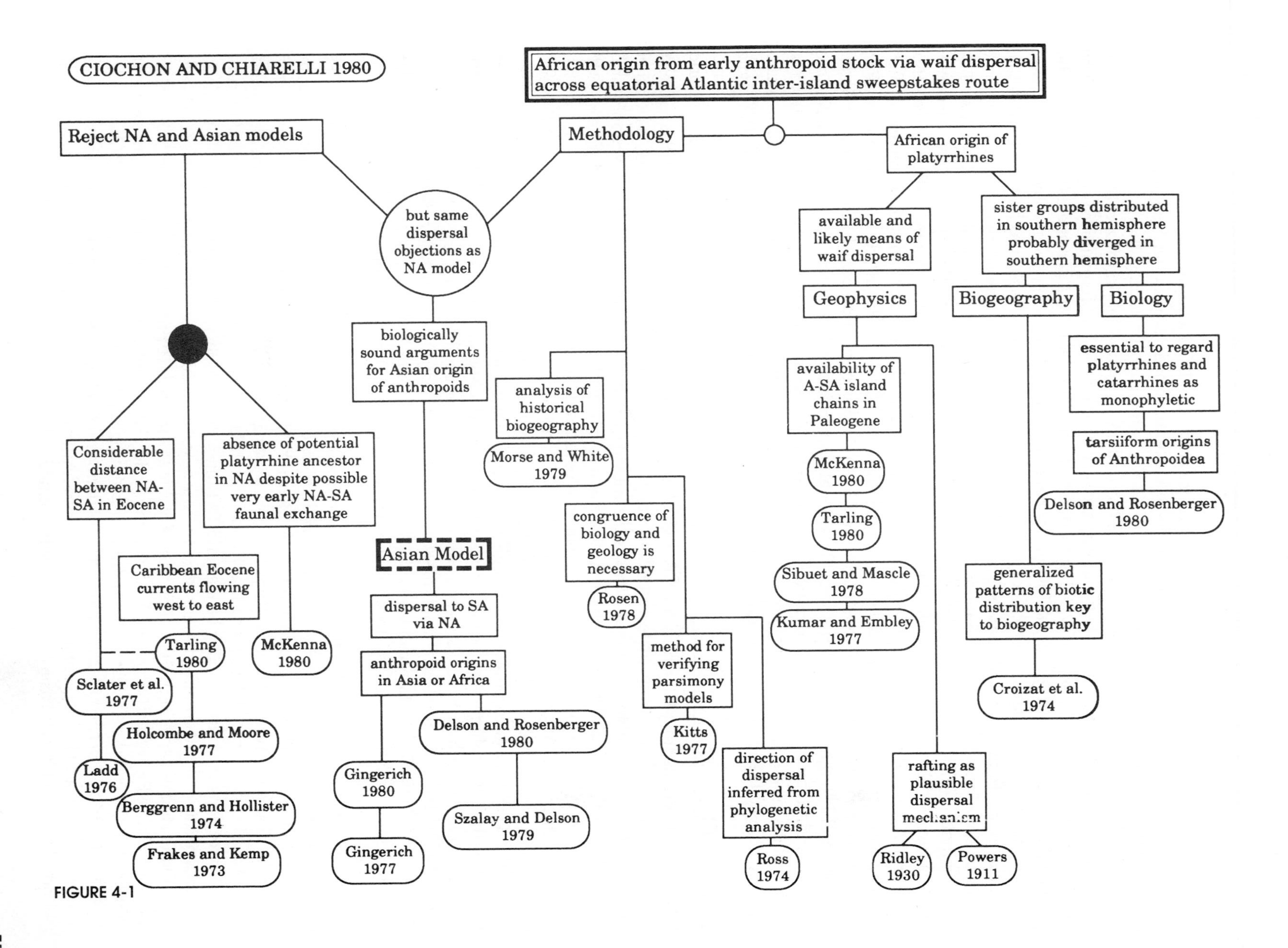
CIOCHON AND CHIARELLI 1980
African origin from early anthropoid stock via waif dispersal across equatorial Atlantic inter-island sweepstakes route
Reject NA and Asian models
Methodology
African origin of platyrrhines
but same dispersal objections as NA model
Considerable distance between NA-SA in Eocene
Caribbean Eocene currents flowing west to east
absence of potential platyrrhine ancestor in NA despite possible very early NA-SA faunal exchange
Tarling 1980
Sclater et al. 1977
Holcombe and Moore 1977
Ladd 1976
Berggrenn and Hollister 1974
Frakes and Kemp 1973
McKenna 1980
biologically sound arguments for Asian origin of anthropoids
Asian Model
dispersal to SA via NA
anthropoid origins in Asia or Africa
Gingerich 1980
Gingerich 1977
Delson and Rosenberger 1980
Szalay and Delson 1979
analysis of historical biogeography
Morse and White 1979
congruence of biology and geology is necessary
Rosen 1978
method for verifying parsimony models
Kitts 1977
direction of dispersal inferred from phylogenetic analysis
Ross 1974
available and likely means of waif dispersal
Geophysics
availability of A-SA island chains in Paleogene
McKenna 1980
Tarling 1980
Sibuet and Mascle 1978
Kumar and Embley 1977
rafting as plausible dispersal mechanism
Ridley 1930
Powers 1911
sister groups distributed in southern hemisphere probably diverged in southern hemisphere
Biogeography
Biology
generalized patterns of biotic distribution key to biogeography
Croizat et al. 1974
essential to regard platyrrhines and catarrhines as monophyletic
tarsiiform origins of Anthropoidea
Delson and Rosenberger 1980
FIGURE 4-1

In repositioning the concept of an Asian origin for anthropoids, Ciochon and Chiarelli (1980) worked the strengths of the African model into their holistic Maximum Parsimony model. Seeking to account for the greatest amount of positive evidence while at the same time excluding the greatest amount of objectionable and/or negative evidence, the Maximum Parsimony model prioritized the monophyly of platyrrhines and catarrhines and the tarsiiform ancestry of anthropoids (in qualified support of Delson and Rosenberger 1980). These two working assumptions require a southern continents staging area for platyrrhine dispersal as per the methodologies of parsimony and cladistic biogeography supported and cited by Ciochon and Chiarelli (1980). The final dimension to this model was the mechanism of the actual dispersal event. Rafting, or waif dispersal, was seen as the most logical dispersal mechanism. Objections to this means of passive transport (Simons 1976) were fully acknowledged but neutralized because they apply equally to any model of overwater dispersal (Hoffstetter 1980). In their place were anecdotal observations of actual rafting events (Powers 1911; Ridley 1930) and the documentation of *potential* equatorial Atlantic Paleogene island chains loosely connecting the South American and African mainlands (Kumar and Embley 1977; van Andel et al. 1977; Sibuet and Mascle 1978; Tarling 1980). Together these two lines of reasoning provided an available and likely means of waif dispersal from Africa to South America and brought closure to the very directional and inclusive Maximum Parsimony model (see Ciochon and Chiarelli 1980:484). This model was defended as the strongest in part because it satisfied the requirements of a southern continents origin and because the proposed deployment scenario was a "potentially viable and ultimately verifiable oceanic sweepstakes route" (Ciochon and Chiarelli 1980:478).

Cachel 1981

Figure 4-2 diagrams the platyrrhine origins argument presented by Cachel (1981) and is based almost exclusively on evidence from plate tectonics and geophysics. The purpose of the model, as Cachel explicitly stated, was not to account for all possible lines of evidence but rather to present the geophysical data as they support either Africa or North America as a source of platyrrhine dispersal to South America. In this regard, then, the only application of paleontological data to the model was the observation that early Cenozoic South American mammals show ties to North American forms (Simpson 1969, 1978) and that in some pertinent cases there is a demonstrated lack of the same to African forms (Clemens 1979). Cachel (1981) did not incorporate neontological data into the model in part because the mixed results reached in the 1980 volume supporting both North America (Delson and Rosenberger 1980; Gantt 1980; Gingerich 1980; Orlosky 1980; Perkins and Meyer 1980; Rosenberger and Szalay 1980; Wood 1980b) and Africa (Bugge 1980; Ciochon and Chiarelli 1980; Chiarelli 1980; Ford 1980; Hoffstetter 1980; Lavocat 1980; Luckett 1980; Maier 1980; Martin and Gould 1980; Sarich and Cronin 1980) as the source of deployment indicated that another database system might be more effective at eliminating one or the other continent from consideration.

From a thorough review of Paleogene structural geology, Cachel (1981) drew together a series of studies that favored North America as a more likely source of deployment for immigrant taxa to South America. In this model the preponderance of evidence held that even with conservative estimates the Paleogene distance between Africa and South America was greater than the distance between North and South America (Sclater et al. 1977), and that hypothetical island arcs between Africa and South America, even if once conjoined (Burke 1975; Lehner and DeRuiter 1977; Smith 1977), were unavailable to biotic dispersal because they were already submerged before any reasonable date of the supposed event (Thiede 1977). Other studies depicting the opening of the Atlantic as a multistage process (Le Pichon 1968; Le Pichon and Fox 1971; Le Pichon and Hayes 1971; Larson and Ladd 1973; Newton 1976; Kerr 1978; Rabinowitz and LaBrecque 1979) contrasted with earlier notions of a simple separation system (Burke and Wilson 1972; Burke et al. 1973) often cited as favorable to trans-Atlantic migration. Recalling the logistical objections of Simons (1976), Cachel (1981) thus dismissed a trans-Atlantic crossing as improbable on geophysical grounds. In the Caribbean basin tectonic activity seemed sufficiently complex (Freeland and Dietz 1971)

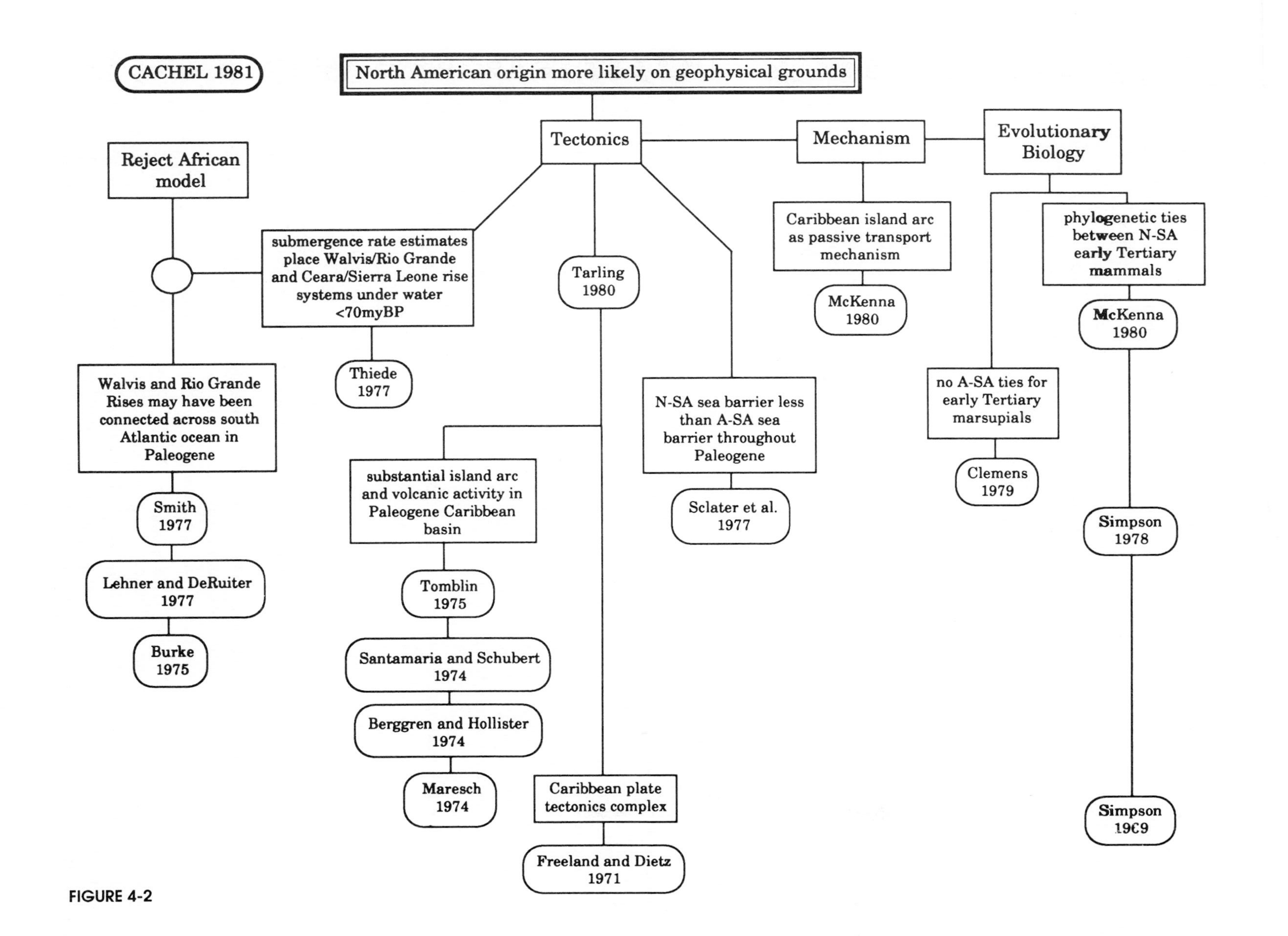

CACHEL 1981
North American origin more likely on geophysical grounds
Reject African model
Tectonics
Mechanism
Evolutionary Biology
submergence rate estimates place Walvis/Rio Grande and Ceara/Sierra Leone rise systems under water <70myBP
Thiede 1977
Walvis and Rio Grande Rises may have been connected across south Atlantic ocean in Paleogene
Smith 1977
Lehner and DeRuiter 1977
Burke 1975
Tarling 1980
substantial island arc and volcanic activity in Paleogene Caribbean basin
Tomblin 1975
Santamaria and Schubert 1974
Berggren and Hollister 1974
Maresch 1974
Caribbean plate tectonics complex
Freeland and Dietz 1971
N-SA sea barrier less than A-SA sea barrier throughout Paleogene
Sclater et al. 1977
Caribbean island arc as passive transport mechanism
McKenna 1980
no A-SA ties for early Tertiary marsupials
Clemens 1979
phylogenetic ties between N-SA early Tertiary mammals
McKenna 1980
Simpson 1978
Simpson 19€9

FIGURE 4-2

and volcanic activity sufficiently frequent (Berggren and Hollister 1974; Maresch 1974; Santamaria and Schubert 1974; Tomblin 1975) that mobile island arcs (Tarling 1980) were very likely the means of passive conveyance of platyrrhines to South America.

This model relied on criteria of likelihood narrowly defined within the perspective of structural geology, and as such was difficult to contrast with broader, more inclusive models. It is safe to say that objection to the Cachel (1981) model (e.g., Rosenberger 1986) has focused on the model's explicit and distinctly exceptional conclusion of anthropoid diphyly. The observation (Rosenberger 1986) that adequate hypothesis testing is possible only within, and not across, database systems is well worth noting here. A model built upon structural geology evidence should only be compared to the similar facets of other formal models and is not capable of measuring the validity of phylogenetic hypotheses that are based on nongeological databases.

As stated, the Cachel (1981) model discounted a trans-Atlantic dispersal scenario more than it demonstrated a North American one. For some this method of analysis is substantially different from accruing a convincing amount of positive evidence for dispersal from North America. Similar methods are often employed in paleontological analyses, where the amount of absent evidence often dwarfs the available fossil record. While the objections to a trans-Atlantic crossing raised in the Cachel (1981) model must be evaluated in their own right, they are not an adequate foundation for comment upon platyrrhine–catarrhine phylogenetic relationships. As discussed later, geophysical data do not strongly favor either continent as a point of dispersal in the early Tertiary (see the Fleagle 1986 model) or long distance oceanic crossing of any kind (Simons 1976; Hoffstetter 1980).

Rose and Fleagle 1981 and Fleagle 1986

Figure 4-3 diagrams the platyrrhine origins argument presented by Fleagle (1986). The model incorporated another post-1980 argument (Rose and Fleagle 1981) that had reached an open-ended but ultimately testable conclusion. This model was not substantially different in its conclusions from the Ciochon and Chiarelli (1980) model, but it did include updated information that somewhat restructures the argument.

In the Rose and Fleagle (1981) model, cases for both North America and Africa as areas of platyrrhine origins and dispersal were presented as plausible but untestable given what at that time was considered an insufficient fossil record for the taxa concerned. The plausibility of the North American hypothesis rested largely on the complexity of Caribbean plate tectonics (paleocurrent reconstructions and hypothesized NA–SA distances were not cited) and on one of the two polar positions on caviomorph rodent ancestry (Wood and Patterson 1970; Wood 1972, 1974, favoring North America as a point of origin). The plausibility of the African model was based on the other polar position of caviomorph ancestry (Lavocat 1969; Hoffstetter 1972, 1974, 1980). Caviomorph ancestry stands as a powerful dispersal precedent or parallel event to platyrrhine dispersal, and its resolution is clearly emphasized in the models analyzed here, despite no a priori reason to link the timing of the two events (that caviomorphs and platyrrhines first appear in South American deposits of the Deseadan Land Mammal Age reflects a stochastic fossil record more than it confirms simultaneous migration). As Fleagle (1986) observed, North American workers tended to favor a North American origin while European workers tended to favor an African one.

The Rose and Fleagle (1981) model accepted both African and North American arguments as viable pending fossil discoveries that could link South American forms to one of the two continents. Such fossil discoveries would come to light with the publication of new *Dolichocebus* teeth from the early Miocene of Argentina (Fleagle and Bown 1983) that provided for Fleagle (1986) the necessary morphological link with Fayum propliopithecids and rendered the latter as a likely sister group to platyrrhines (Fleagle and Bown 1983; see also Fleagle and Kay 1987; but see Rosenberger 1986; Rosenberger et al. 1990 for an alternative view). Combined with an interfamilial analysis of extinct and extant rodent postcrania that linked South American and African taxa (Schneider 1984), and continuing the arguments for an African ancestry of

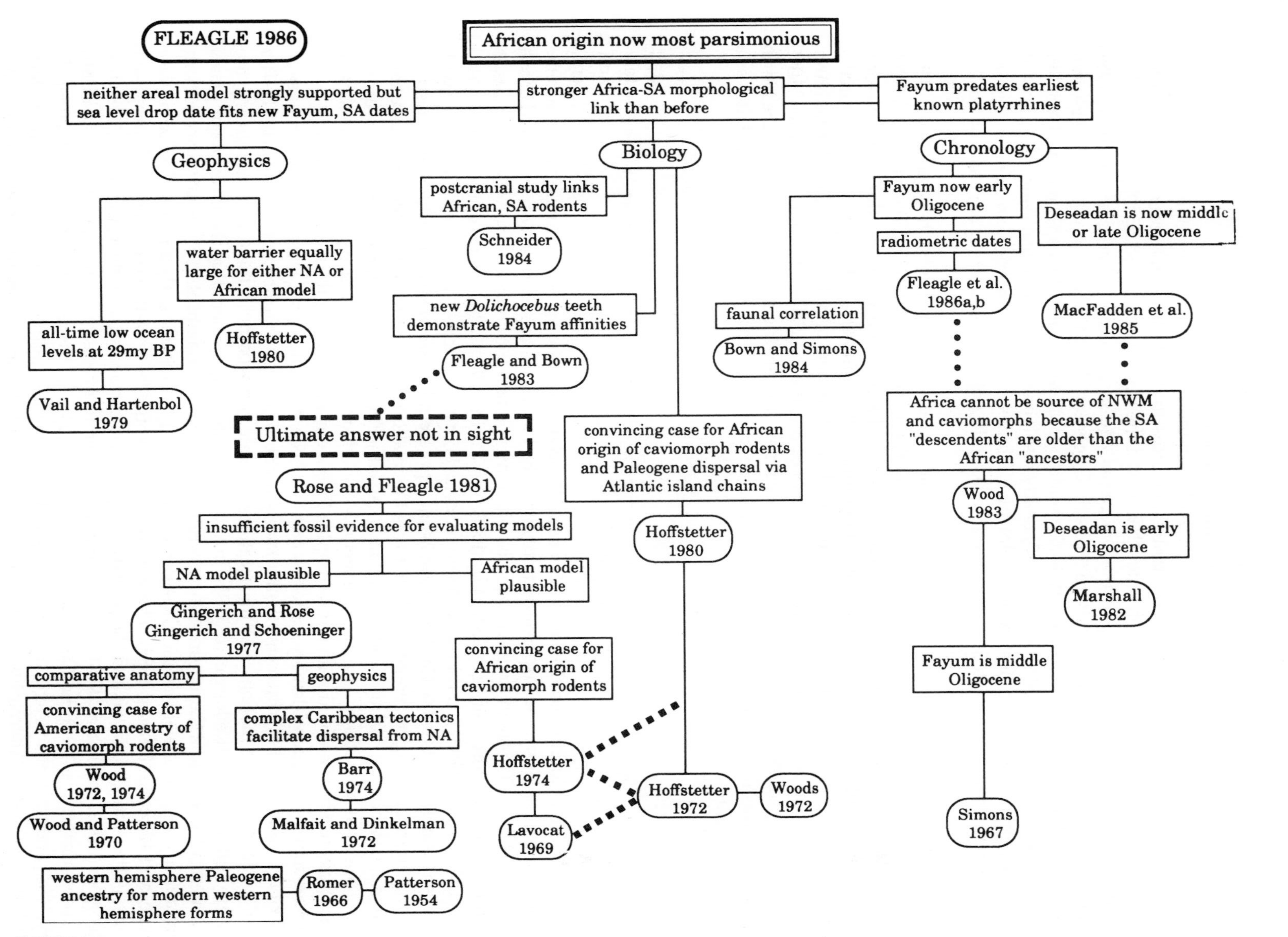
FLEAGLE 1986
African origin now most parsimonious
neither areal model strongly supported but sea level drop date fits new Fayum, SA dates
stronger Africa-SA morphological link than before
Fayum predates earliest known platyrrhines
Geophysics
Biology
Chronology
postcranial study links African, SA rodents
Schneider 1984
water barrier equally large for either NA or African model
Hoffstetter 1980
all-time low ocean levels at 29my BP
Vail and Hartenbol 1979
new Dolichocebus teeth demonstrate Fayum affinities
Fleagle and Bown 1983
Ultimate answer not in sight
Rose and Fleagle 1981
insufficient fossil evidence for evaluating models
NA model plausible
African model plausible
Gingerich and Rose
Gingerich and Schoeninger
1977
comparative anatomy
geophysics
convincing case for American ancestry of caviomorph rodents
complex Caribbean tectonics facilitate dispersal from NA
Wood 1972, 1974
Barr 1974
Wood and Patterson 1970
Malfait and Dinkelman 1972
western hemisphere Paleogene ancestry for modern western hemisphere forms
Romer 1966
Patterson 1954
convincing case for African origin of caviomorph rodents
Hoffstetter 1974
Lavocat 1969
convincing case for African origin of caviomorph rodents and Paleogene dispersal via Atlantic island chains
Hoffstetter 1980
Hoffstetter 1972
Woods 1972
faunal correlation
Bown and Simons 1984
Fayum now early Oligocene
radiometric dates
Fleagle et al. 1986a,b
Deseadan is now middle or late Oligocene
MacFadden et al. 1985
Africa cannot be source of NWM and caviomorphs because the SA "descendents" are older than the African "ancestors"
Wood 1983
Deseadan is early Oligocene
Marshall 1982
Fayum is middle Oligocene
Simons 1967

FIGURE 4-3

caviomorphs (Hoffstetter 1980; Lavocat 1980), all biological data in the Fleagle (1986) model emphasized Africa as the source of platyrrhine dispersal to South America.

An ongoing problem with an African model had been the "inverted" chronological position of the African Oligocene taxa often proposed as potential ancestors to the geologically older South American "descendants" (see Wood 1983), based on dates that placed the Fayum in the middle Oligocene (Simons 1969) and the South American Deseadan land mammal age in the early Oligocene (Marshall 1982). Subsequent redating of the Fayum locality to the early Oligocene (Bown and Simons 1984; Fleagle et al. 1986a,b) and the Deseadan *Branisella* locality to the middle-late Oligocene (MacFadden et al. 1985; later confirmed by Naeser et al. 1987; MacFadden 1990) neatly resolved this chronological issue. As long as the Fayum locality preceded the earliest evidence for platyrrhines in South America, an African origin of platyrrhines from some form related to the Fayum taxa could not be ruled out on the basis of positive fossil evidence.

The Fleagle (1986) model resigned geophysical data to a position of no confidence, but it included a reminder that a dating of all-time low ocean levels (Vail and Hartenbol 1979) at 29 million years ago (MYA) dovetailed with the revised Fayum–Deseadan chronology and as such provided the final positive (or at least uncontested) link to Africa as the most parsimonious ("least unlikely," Fleagle 1988:354) source of platyrrhine origin and dispersal. This model updated the Ciochon and Chiarelli (1980) Maximum Parsimony model and represented a sound examination of the available data. In contrast to the Cachel (1981) model, Fleagle (1986) positioned positive evidence for an African origin in such a way that the North American model was rendered less likely, rather than amounting evidence that countered the latter as a measure of the likelihood of the former.

A NEW DECADE OF DATA

Geophysical

The overwhelming evidence for the late Cretaceous–Pliocene isolation of South America renders the mechanical aspect of platyrrhine dispersal virtually irresolvable. The final separation of Africa and South America as an Albian–Turonian (mid-Cretaceous) phenomenon and the complexity of plate tectonics in the circum-Caribbean were well established by 1980 (see Tarling 1980). Flowcharts of the Ciochon and Chiarelli (1980), Cachel (1981), and Fleagle (1986) models (Figures 4-1 to 4-3) indicate the flexibility with which these research results in the Caribbean can be incorporated. The major structural geology questions relevant to platyrrhine origins remain: (1) What was the relationship of spreading rates and fluctuating sea levels in the Paleogene Atlantic, and how reliably can either be measured? (2) How extensive were the potential equatorial and South Atlantic Paleogene island chains? and (3) Can passive mechanisms of island-arc transport be identified within the complex Paleogene evolution of the Caribbean basin?

Subsequent to 1980, considerably more research has been published on Caribbean (see Bonini et al. 1984; Herrbeck and Mann 1991) than on Atlantic basin structural geology (Nurnberg and Muller 1991). The directions of relative plate motions and the locations of plate boundaries in the Caribbean during the Tertiary are still uncertain (Kellogg et al. 1985), but there is general agreement that the three major plates (North American, Cocos, and Caribbean) are converging (Guzman-Speziale et al. 1989). Studies of northern Andean orogeny indicate major tectonic phases of rapid regional uplift in the early and middle Eocene, late Oligocene, and Pliocene (Kellogg 1984). These early phases would be broadly coincident with proposed vulcanism in the Antilles (Malfait and Dinkelman 1972). Eastward translation of the Caribbean plate since the Paleogene (Erlich and Barret 1990) may be a consequence of the events precipitating the Andean uplift, as well as a principal cause of the Neogene separation of Cuba and Hispaniola (Calais and Mercier de Lepinay 1990) and the transcurrent faulting in Puerto Rico along the northern Caribbean plate boundary zone (Erikson et al. 1990). Alternatively, the westward movement of North and South America at the same time may be the ultimate agent of regional tectonic activity in the Paleogene (Ross and Scotese 1988). As information is accumulated from different databases (Guzman-Speziale et al. 1989; Rosencrantz 1990), models of Caribbean Tertiary structural geology (see Bonini et al. 1984; Ross and Scotese 1988) will be revised accordingly and our understanding should gradually improve. At this time, however, observations of plate boundary zone geology and seismicity are not adequately explained by any single current model (Herrbeck and Mann 1991), and the overall depiction of the Caribbean as a tectonically active and therefore complex region throughout the Tertiary remains the most discriminating synthesis available (Burke 1988).

Recent studies of the larger scale south Atlantic basin system also largely confirm pre-1980 syntheses. An initial ocean basin formation in the Jurassic (Nurnberg and Muller 1991) and final Africa–South America separation in the Turonian (Rand and Mabesoone 1982; Uchupi 1989) corroborate models of Mesozoic drift (e.g., Briden 1973; Newton 1976; Onstott and Hargraves 1981) and mid-Cretaceous completion of the mid-Atlantic ridge (Ramsay 1971, 1973; Rabinowitz and LaBrecque 1979). Continuing tectonic activity in Africa during the late Cretaceous (Fairhead 1988) has been invoked to explain problems of African–South American continental margin fitting and distortion (Unternehr et al. 1988), as well as dynamics of the critical Walvis Ridge–Rio Grande Rise island arc system (O'Connor and Duncan 1990). Detailed dating of hot spot activity indicates that the Rio Grande Rise connecting South America to the mid-Atlantic volcanic extrusions of Ascension and Tristan da Cunha islands ceased rising at around 70 MYA and therefore was not likely to be above sea level after that time (O'Connor and Duncan 1990). On the basis of these studies, then, the equatorial Ceara/Sierra Leone rise system bridging the western bight of Africa with northeastern Brazil, though not as active as the more southerly Walvis/Rio Grande system, would be the most likely source of emergent Paleogene island arcs (as applied by Ciochon and Chiarelli 1980, for example).

Biomolecular

Any analysis of the platyrrhine origins issue must consider the wealth of relevant data and interpretation published in the broad subject area of molecular evolution. Recent treatments of primate evolution in this context include Goodman et al. (1982), Marks (1987), Hasegawa (1990), and Miyamoto and Goodman (1990). To thoroughly represent this literature is beyond the scope of this review, but a few patterns that have emerged over the last decade factor significantly into formal models of platyrrhine origins and the larger concept of the evolutionary history of platyrrhines and catarrhines.

In 1980 three karyological/biochemical studies addressed the question of platyrrhine origins (Baba et al. 1980; Chiarelli 1980; Sarich and Cronin 1980). Chiarelli (1980) reached a minority conclusion of a polyphyletic origin of platyrrhines from a cercopithecoid early catarrhine by comparing karyotypes across several primate groups. The problems of using karyotypic banding techniques as a method for reconstructing phylogeny have long been recognized (Hershkovitz 1977) and were acknowledged by Chiarelli (1980). Given the large volume of cytogenetic literature in the last 12 years, however, it would seem that the pursuit of more refined techniques has justified the speculative nature of the results reported in the meantime. With regard to platyrrhines, virtually any possible combination of subfamily (and intergeneric) relationships has been proposed on the basis of karyotypic similarities (e.g., Dutrillaux and Couturier 1981; Ma 1981; Soares et al. 1982; Galbreath 1983; Minezawa et al. 1989; de Souza Barros et al. 1990). In a broad study across primates (using only *Cebus, Aotus,* and *Ateles* within platyrrhines), Dutrillaux (1979) concluded that platyrrhine karyotypes were more similar to lemur than to catarrhine karyotypes.

Using immunological and biochemical techniques, both Baba et al. (1980) and Sarich and Cronin (1980) supported the concept of a monophyletic Anthropoidea, though they differed on intraplatyrrhine relationships and the gradistic "level" of the platyrrhine–catarrhine last common ancestor. Subsequent to 1980, most molecular approaches to higher taxa phylogeny have used only one or two platyrrhine taxa for comparative purposes (e.g., Hayasaka et al. 1988; Koop et al. 1989) and have consistently supported anthropoid monophyly. Among these studies, however, radically different conclusions of intraplatyrrhine relationships (e.g., Goodman et al. 1982; Miyamoto and Goodman 1990; Schmitt et al. 1990) indicate that concordance among methods is as elusive in molecular biology as in comparative anatomy.

This very selective sampling of the literature on molecular evolution in platyrrhines demonstrates that this broad and active field of study has established powerful media for comparing extant groups. Within platyrrhines, biomolecular studies in large part support the integrity of natural subfamily groupings proposed on other biological grounds (Fleagle 1988). Ultimately, biomolecular approaches may provide independent means of reliably dating the divergence times of extant groups (see Miyamoto and Goodman 1990 for review), but for now differences of opinion (Hasegawa et al. 1989) and a dependence on paleontologically derived divergence times (Rosenberger 1984) render otherwise very confident molecular clock proposals (Sarich and Cronin 1980) more plausible than probable. Perhaps the most important contribution of molecular studies to the question of platyrrhine origins over the last decade has been the repeated demonstration of anthropoid monophyly. This condition has become an assumption of most working hypotheses of higher primate evolutionary history (Fleagle 1986; Fleagle and Kay 1987; Szalay et al. 1987, and many others). Coupled with cogent arguments for parsimonious paleobiogeography (Ciochon and Chiarelli 1980), this consensus supports a southern continents origin and dispersal scenario of New World monkeys.

Paleontological: The Paleogene Tropics

Deposits of Paleogene age are extremely rare in the neotropics and the importance of this well-recognized paleontological gap for the reconstruction of platyrrhine origins cannot be overstated. The major Tertiary mammalian localities pertinent to this issue are the early Neogene platyrrhine-bearing deposits in Patagonia, the La Venta locality in Colombia, and the Fayum locality in Egypt. The earliest La Venta deposits are dated to the early Miocene (Marshall 1982, 1985; Marshall et al. 1983; Takemura and Danhara 1986; Kay et al. 1987; MacFadden 1990); thus, only the Fayum deposits approach Paleogene age ranges and have yielded mammals of late Eocene/earliest Oligocene age (Bown and Simons 1984; Simons et al. 1986; Simons 1989, 1990; van Couvering and Harris 1991). Unfortunately, recent comprehensive accounts of mammalian evolutionary history (Savage and Russell 1983; Carroll 1988) are similar to earlier syntheses (Romer 1966; Cooke 1972; Maglio and Cooke 1978) that account the dearth of the evolutionary record of tropical Paleogene mammals.

The lack of a positive Paleogene fossil record for the modern tropics has been duly noted in treatments of the platyrrhine origins question: "Without the aid of a Paleocene or Eocene primate fossil record in Africa, . . . it could just as easily be argued (and indeed has been) that anthropoids originated in Africa from unknown Eocene tarsiiforms or lemuriforms and dispersed to South America via the South Atlantic route to become ancestors of the New World monkeys and to South Asia via a crossing of the Tethys to become ancestors of the presumed anthropoids of Burma's Pondaung Hills" (Ciochon and Chiarelli 1980:474). The importance of such fossils for evaluating current platyrrhine origins models is made especially obvious by the overwhelming biological and molecular evidence (see previous discussion) for the monophyly of anthropoids. Thus, despite the problems that pervade analyses of the fossil record (see later discussion), having a fossil record at a particular time and in a particular space is certainly better than not having one at all.

Paleontological: Recent Discoveries in North Africa

Early anthropoid specimens have been recovered in the last twelve years from localities in Algeria (Godinot and Mahboubi 1992), Oman/Arabia (de Bonis et al. 1988; Thomas et al. 1988) and the productive

Fayum deposits (Simons and Kay 1983, 1988; Simons 1986, 1989, 1990) and are treated in detail elsewhere in this volume (see also Fleagle and Kay 1987). These discoveries bear significantly on the platyrrhine origins question and merit a brief review.

Recent discoveries in the lower fossil-bearing strata of the Jebel Qatrani Formation in the Fayum badlands were described as the oldest known anthropoids (Simons 1989, 1990). Precise radiometric dates are unavailable, but a late Eocene/basal Oligocene time range seems likely on stratigraphic grounds. The fossils in question are certainly lower in the formation than any previously recovered primates. As with other productive fossil primate localities, these discoveries are prompting a rethinking of the anatomical and temporal divisions in the evolution of higher primate groups. Two genera, *Catopithecus* and *Proteopithecus,* have been described on the basis of craniodental remains (Simons 1989, 1990). Known from a fragmentary cranium and lower jaw, *Catopithecus* in particular offers an intriguing opportunity for post hoc analysis of previously suggested conditions of the prosimian-anthropoid transition, not to mention the platyrrhine–catarrhine dichotomy. The following discussion is based strictly on published reports (Simons 1989, 1990) and not on a direct examination of the *Catopithecus* or *Proteopithecus* fossils.

The somewhat crushed *Catopithecus* cranium is described as having a catarrhine dental formula (2-1-2-3), an ectotympanic bulla, a fused frontal bone, postorbital closure, relatively small orbits, P3 canine honing facets, and maxillary dental anatomy "almost as close to that of certain Eocene adapids as it is to that of propliopithecids" (Simons 1990:1568). Perhaps more interesting than the potential classification of *Catopithecus* is the impact of its morphological package on existing notions of early anthropoids. At the very least it suggests a much earlier platyrrhine–catarrhine split than previously postulated on the basis of the Fayum fossil record and/or a revised diagnosis of Catarrhini that excludes a 2-1-2-3 dental formula as autapomorphic. The latter seems neither practical nor justified on the basis of monotypic genera, and dental formula alone is not sufficient to define *Catopithecus* (and *Oligopithecus* for that matter) as catarrhines; however, the mere presence of these very early anthropoids and recent discoveries of other early Tertiary fossil primates in similar latitudes (Ciochon et al. 1985; Simons and Bown 1985; Simons et al. 1986; de Bonis et al. 1988; Thomas et al. 1988; Godinot and Mahboubi 1992) strongly suggest that the platyrrhine–catarrhine split was earlier than the late Eocene time projected from or implied in detailed analyses of Fayum parapithecids and propliopithecids (e.g., the Fleagle 1986 model).

Based on the known fossil record, an initial rendering of the morphological continuity among Eocene adapids, *Catopithecus* and *Proteopithecus,* and later propliopithecids leaves little room, geologically speaking, for the recovery of protoplatyrrhines in an Old World postadapid context: "The well-demonstrated relationships between certain Eocene adapids and *Oligopithecus, Proteopithecus,* and *Catopithecus,* on the one hand, and between them and the Fayum propliopithecids on the other, suggest a non-omomyid derivation for all higher primates" (Simons 1989:9960). At this time, the enormous void of information on Paleogene tropical mammals weighs more on the issue of platyrrhine origins than do the vexing problems of integrating these new Fayum taxa into existing concepts of Anthropoidea, Catarrhini, or Platyrrhini.

PALEONTOLOGICAL: FOSSIL PLATYRRHINES

No aspect of the platyrrhine origins question has changed as much over the last 12 years as the platyrrhine fossil record. Discoveries in Argentina (Fleagle and Bown 1983; Fleagle 1985; Fleagle et al. 1987; Anapol and Fleagle 1988; Fleagle 1990; Meldrum 1990; Fleagle and Rae 1992), Bolivia (Wolff 1984; MacFadden 1990), Colombia (Setoguchi and Rosenberger 1985, 1987; Luchterhand et al. 1986; Kay et al. 1987; Kay 1989, 1990a; Gebo et al. 1990; Meldrum et al. 1990; Rosenberger et al. 1991b; Meldrum and Kay 1990, 1992), and the West Indies (MacPhee and Woods 1982; Ford 1986a; Ford and Morgan 1986, 1988; Ford and Woods 1990; MacPhee and Fleagle 1991; Rivero and Arredondo 1991), as well as descriptions of new specimens or taxa (Rosenberger 1981; Hershkovitz 1981, 1984, 1988; Setoguchi 1985; Setoguchi et al. 1988, 1990; Ford 1990b; Rosenberger et al. 1991c) have substantially enhanced the primary database of platyrrhine evolution. Most of this new information, however, extends the known platyrrhine Miocene fossil record of Argentina and Colombia and as such does not alter our conception of the earliest platyrrhines as based on previous discoveries; however, the patterns evident in the platyrrhine fossil record as it has accumulated in the 1980s do support certain predictions about the as yet unknown "initial" platyrrhines and are discussed later.

The recovery of a fossil platyrrhine maxillary fragment in Bolivian deposits of Deseadan age more than 20 years ago (Hoffstetter 1969) still stands as an icon of the initial platyrrhines. New dates for this *Branisella* locality (MacFadden 1985; MacFadden et al. 1985; Naeser et al. 1987; MacFadden 1990) place it in the late rather than early-middle Oligocene but do not affect the status of the primate fossils as the earliest platyrrhines known. Descriptions (Rosenberger 1981; Rosenberger et al. 1990, 1991c) and discoveries (Wolff 1984; MacFadden 1990) of additional specimens at the same locality confirm the platyrrhine identity of the fossils, a status not always granted the type specimen in the past. The new fossil material also demonstrates upper molar variability within the taxon (Rosenberger et al. 1990) and the essential primitiveness of this early platyrrhine. Excluding the provisionally assigned Princeton mandible (Rosenberger 1981), the four other specimens could be from a single individual (Fleagle 1988), but detailed comparisons (Rosenberger et al. 1990, 1991c) suggest that two different taxa are present.

The relationships between early Old World anthropoids and early platyrrhines are far from resolved. Relatively small hypocones and lack of hypoconulids on lower molars could suggest that the Bolivian material is dentally primitive compared to *Apidium*, *Aegyptopithecus*, and *Propliopithecus* from the younger Fayum locality. This observation has been used to dismiss the importance of positive comparisons between these Fayum taxa and other fossil platyrrhines (compare Rosenberger et al. 1991c with Fleagle and Bown 1983; Fleagle 1986), unless, of course, hypoconulid loss is a derived feature of platyrrhines (Fleagle and Kay 1987; Kay 1990b). Hypoconulids are present in the *Catopithecus* mandible (Simons 1989). The differential use of *Dolichocebus* (Fleagle and Bown 1983) or *Branisella* and *Szalatavus* (Rosenberger et al. 1991c) as examples of the ancestral platyrrhine morphotype reflects the general inadequacy of the known fossil record for determining early anthropoid dental character polarities (Fleagle 1990) and/or resolving the nature and extent of convergence between fossil platyrrhine and Fayum taxa. Insofar as recent discoveries in North Africa suggest an earlier platyrrhine–catarrhine divergence, there is as yet no positive fossil evidence in South America for a pre-Oligocene platyrrhine arrival.

Paleontological: Argentina

Productive survey and collection of the Pinturas Formation in Santa Cruz Province, Argentina (Fleagle 1990) have increased the depth and breadth of the known Neogene platyrrhine radiation in southern South America. Over the last 12 years two new genera, *Soriacebus* (Fleagle et al. 1987) and *Carlocebus* (Fleagle 1990), as well as additional specimens of *Dolichocebus* and *Tremacebus* (Fleagle and Bown 1983) and numerous postcrania (Anapol and Fleagle 1988; Meldrum 1990) have been recovered. Before this work the essentially nonmodern character of the southern South America Miocene fossil platyrrhine taxa was widely recognized (e.g., Hershkovitz 1982; Gebo and Simons 1987), even in attempts to link them to extant taxa (Rosenberger 1977, 1979). The new taxa and numerous specimens subsequently recovered, while indeed quite different from most later platyrrhines (Kay 1990b), have also been proposed as near the radiation of single (Rosenberger et al. 1990) or several (Fleagle 1990) extant groups.

Detailed paleoenvironmental reconstructions of the Pinturas Formation support an interpretation of moist climate and partly if not tropically forested habitats (Bown and Larriestra 1990). Pascual and Jaureguizar (1990) propose the idea of an early Oligocene–middle Miocene Patagonian Faunistic Cycle (incorporating the traditional Deseadan, Colhuehuapian, and Satacrucian land mammal ages), characterized by steep drops in temperature and ocean levels that suggest increased opportunities for transatlantic crossing (see also Fleagle 1986, 1988). The episodic appearance of platyrrhines in the relatively well-sampled southern South America fossil record suggests that they tracked amenable but transient climates and thus are not likely to be found in earlier or later more arid deposits; conversely, their absence in those deposits may be explained by competitive exclusion via caenolestid marsupials. To date, the "impact of primates on the South American fauna is difficult to assess" (Pascual and Jaureguizar 1990:40).

Paleontological: Colombia

The number of extinct genera identified from the Miocene locality of La Venta increased twofold in the 1980s and continues to grow (Fleagle and Rosenberger 1990; Kay 1990a; Rosenberger et al. 1991b). The La Venta fauna as a whole (Hirschfeld and Marshall 1976) represents the most diverse assemblage of neotropical Miocene taxa yet known and as such has contributed greatly to our understanding of the emergence of modern South American tropical mammals. Some La Venta platyrrhines are related more directly to modern forms (Rosenberger et al. 1991a) than are others (e.g., compare Setoguchi and Rosenberger 1987; Kay 1990b; Rosenberger et al. 1990), but the general character of the assemblage is distinctly more modern than earlier platyrrhines from southern South America. While this condition is a boon to phylogenetic analyses of modern platyrrhines, the La Venta fauna does not factor significantly into proposed models for platyrrhine origins and dispersal. However, the observation that this increased fossil record has not resolved lingering debates about modern platyrrhine interrelationships (Fleagle 1990:82) is very pertinent to the role that an increased fossil record might play in analyses of platyrrhine origins (see later discussion).

In sum, then, data that have accumulated since 1980 reinforce Africa as a source of platyrrhine–catarrhine divergence and platyrrhine dispersal to South America. The timing and general pattern of Cretaceous/Tertiary continental drift as constructed in the 1970s are supported by updated studies of the Atlantic basin, and the complexity of early Tertiary interplate movement in the Caribbean has been demonstrated in numerous studies. The fossil record of Argentina, Colombia, and the Fayum has increased at a rate that exceeds the utility of comprehensive analyses, but revised dating of the localities places potential platyrrhine ancestors in Africa before the first known occurrence of primates in South America.

DISCUSSION

Looking for Patterns

Recent analyses of the platyrrhine origins problems (Ciochon and Chiarelli 1980; Fleagle 1986, 1988) clearly emphasize testable models. Is a testable model necessarily a most likely model? Conversely, is the most likely model bound by parsimony (as defined in Ciochon and Chiarelli 1980) or testability? Examining the patterns of inquiry and evidence in the larger data systems of paleontology, plate tectonics, and cladistic comparative anatomy suggests that the data of these systems are neither substantial nor scaled to the level desired in a crucible of parsimony. In other words, the data of each of these systems are observational rather than empirical and so precipitate an inductive interpretation rather than a deductive approach to hypothesis formation and testing. Therefore, an accounting of positive and negative evidence for early platyrrhine evolutionary history will be presented as a "model most likely" to complement the rigorous Maximum Parsimony model (Ciochon and Chiarelli 1980) and its modified form (Fleagle 1986) that crystallized out of the last major treatment of the issue.

Adhering to a hypothetico-deductive paradigm is not logical (is not parsimonious) when the very means with which the hypotheses are tested (i.e., the data) are in no sense equal to the interpretive power of the hypothesis. The Ciochon and Chiarelli (1980) Maximum Parsimony model is the most rigorously designed and explicitly testable model of platyrrhine origins yet proposed. While acknowledging that no such paleobiogeographic model is fully falsifiable, the very means for testing the model are described as the discovery of a single well-preserved higher primate jaw (particularly a preplatyrrhine/precatarrhine anthropoid or incipient platyrrhine) from the West African late Eocene (Ciochon and Chiarelli 1980:486). A datum such as this would, however, only become another element of the incomplete fossil record within which phylogenetic analyses necessarily operate. Based on a long history of paleontological interpretation, what that fossil jaw would be called is on the one hand independent of its "true" phylogenetic position, which, reflexively, its historically predicated classification serves to determine, at least transiently.

Aside from the difficulties of framing Popperian scientific methods around time transgressive phenomena, a model that is internally maximally parsimonious is not immune to factors external to the model but internal to the data systems the model incorporates. In the case of platyrrhine dispersal, for example, the Maximum Parsimony model proposed by Ciochon and Chiarelli (1980:483) "combines phylogenetic relationships, areal relationships and probable dispersal patterns." This model is indeed parsimonious (and thoroughly rigorous as argued throughout this analysis), relative to others presented for a North American origin. Nonetheless, it must contend with factors such as the inherent problems of long-distance ocean transport (Wood and Patterson 1970; Simons 1976; Cartmill et al. 1981), factors internal to the data systems of areal relationships and probable dispersal methods, but external to the proposed model on the grounds that such objections (Ciochon and Chiarelli 1980:477) "would appear to apply equally to Caribbean dispersal routes from North America and to the equatorial South Atlantic dispersal route from Africa." That these objections apply equally to any latest Eocene transoceanic dispersal scenario in no way nullifies them or reduces the effect they have on the strength of such dispersal scenarios. Rather, as outlined later, objections to long-distance ocean transport (anecdotal [Powers 1911; Ridley 1930] or indirect evidence [Darlington 1957; Scheltema 1971] for rafting notwithstanding) must be incorporated into an analysis of patterning within the data systems to produce a soundly predictive (albeit nonempirical) model of platyrrhine origins.

All of the many processes acting to prevent fossilization and delimit fossil discovery compel us to regard a fossil for what it is: no more indicative of an extinct natural world than a single living individual (or to be more fair, sometimes even a dozen) is of its own world. This is essentially an issue of variation and diversity through time. For a fossil specimen a new genus and species are often named (*re* platyrrhines: Luchterhand et al. 1986; Fleagle et al. 1987; Setoguchi and Rosenberger 1985, 1987; Rosenberger et al. 1991b), and this is a fair expression of the biological reality that an individual exemplifies (however fragmentary the fossil may be). At this point, however, extrapolations grounded in paleontological taxonomy become far more tenuous. Just as obvious as the finite nature of the fossil record is the observation that taxonomic diversity and abundance must increase with every series of discoveries: "It is virtually inevitable that, as the fossil record increases, we will find that the history of the group departs quite strikingly from the pathways that had been reconstructed from the neontological data alone and was an even far more diverse radiation than we had imagined" (Fleagle 1990:84). The logic of this fundamental assumption is independent of post facto taxonomic lumping and splitting and applies to classifications whether they explicitly include (Rosenberger 1981; Ford 1986b; Rosenberger et al. 1990) or exclude (Hershkovitz 1982) extinct forms in higher-order groupings based on extant taxa.

Thus, having two fossil platyrrhine taxa from a middle Miocene locality in Colombia does increase our understanding of the larger process of platyrrhine evolution, but only in the sense that having two lottery tickets doubles the chances of winning with just one ticket. Only an exponential increase in the number of lottery tickets significantly improves the chances of winning, and only an exponential increase in the number and extent (both in time and space) of fossils will significantly characterize events as time- and space-transgressive as platyrrhine origins and dispersal. Phylogenetic reconstructions based on these fossil faunas are by no means irrelevant (sensu Martin 1990; see Gingerich 1984). Indeed, the fossils are indicative of evolutionary events insofar as they exist as consequences of the same: "The intermediate forms and novel combinations found in fossil taxa provide evidence of real morphotypes and real evolutionary transformations to sort through the many theoretical possibilities and to generate new alternative hypotheses" (Fleagle 1990:84). Moreover, across groups such as platyrrhines and catarrhines, for which extensive convergence is suspected or demonstrated (Ford 1986b; Kay 1990b), fossil forms provide one of the only measures of determining character polarities (Fleagle 1990), regardless of the very different methods (compare Kay 1990b with Rosenberger et al. 1990) used to analyze them.

It is perhaps more practical to treat the paleontological data system from the bottom up to operationalize our understanding of the fossilization process. Approaching the fossil record as the recovered but by no means representative material remains of a once extinct fauna limits speculation and improves analytical efficiency.

As long as the bounds of interpretation conform to the time, place, and character of the fossils themselves, the principles of parsimony are served and contingencies minimized. When multisystem models such as those for platyrrhine origins expand the reach of the known data over the full breadth (that is, the gaps of negative evidence) of each system, they are no longer parsimoniously founded or fully falsifiable (see Ciochon and Chiarelli 1980:486). They may be proximately falsifiable, in the sense that additions to known data may invalidate a set of assumptions. Ultimately, however, the utility of such models remains compromised by virtue of that proximate falsifiability; that is, by including only known data that do not reject a model as a dictum of parsimony (Ciochon and Chiarelli 1980:481).

Rather, the point of view in this analysis is that models of platyrrhine origins are as soundly or more soundly based on the patterns of the data systems, patterns that account for known *and* unknown data within each system. There is no a priori reason for rejecting a model as unlikely simply because it is not subject to proximate falsifiability; likewise, there is no a posteriori reason for accepting an internally "most parsimonious" model as most likely on the opposite grounds. In this light, the following patterns have emerged from the data pertinent to platyrrhine origins.

Comparative Anatomy

While platyrrhines and catarrhines are almost certainly monophyletic, this in no way implies a particular morphological intimacy or distance of the living forms, as recognized in various empirical studies (Bugge 1980; Ford 1980; Maier 1980 and others). The pattern in fact seems to be that platyrrhines are primitive among living anthropoids for most character complexes (see Fleagle and Kay 1987; Ford 1988 for a thorough analysis); thus, the question of platyrrhine origins is very difficult to evaluate on the basis of comparative anatomy alone (Ford 1986b, 1990a,b; Gebo 1986, 1989; Fleagle 1988; Dagosto 1990; Gebo et al. 1990). Morphological links among platyrrhines, catarrhines, and representative early anthropoid fossils are tenuous, and similarities (particularly postcranial) either fit hypothesized primitive conditions for anthropoids (e.g., Ford 1990a) or are difficult to confirm as functionally (compare Gebo et al. 1990 with Fleagle and Meldrum 1988) or phyletically homologous (Wikander et al. 1986; Szalay et al. 1987; Beard et al. 1988; Szalay and Dagosto 1988). This is compounded by the recognized difficulty of identifying a most suitable outgroup for comparison to platyrrhines (Ford 1990a). A logical candidate for the last common ancestor of platyrrhines and catarrhines will likely be judged so only because it lacks features that disqualify it (in the sense of *Catopithecus* or *Aegyptopithecus* and their catarrhine dental formula, for example).

Paleontology

New fossil evidence rarely confirms character polarities as determined using extant and/or available fossil taxa (M1,2 hypoconulids, or wear facet X as character states in platyrrhines, catarrhines, and early Fayum anthropoids [Kay 1977, 1980, 1990b; Fleagle and Kay 1987; Simons and Kay 1988]; incisor size proportions in Eocene prosimians [Covert and Williams 1991]; and parapithecid/platyrrhine postcranial characters [Ford 1990a:141], to cite just three relevant examples), unless classification schemes are accordingly rearranged and differential diagnoses of higher taxonomic categories restructured (e.g., Fayum propliopithecids as variously pongids, hominoids, catarrhines, or basal anthropoids). Regarding recent discoveries of platyrrhine fossils, Fleagle and Rosenberger (1990:4) observed: "These new fossils demonstrate unanticipated combinations of morphological features which necessitate a reassessment of the morphological transformations involved in platyrrhine phylogeny as they have been reconstructed from extant taxa alone." The usual effect of additional fossil discoveries is to extend the absolute age, or age of origin, of lineages (Nelson 1974), and this is likely to be the effect *Catopithecus* and *Algeripithecus* (Godinot and Mahboubi 1992) have on the origin of catarrhines.

While taxonomic abundance and morphological diversity continue to expand in Holarctic Eocene primates (e.g., Gingerich 1986; Covert and Williams 1991; Beard et al. 1992), no arguments have been made for new higher taxa. This indicates that protoplatyrrhine or early anthropoid fossil primates are unlikely to be found in these relatively well-sampled North American and European regions. These two patterns are very informative, insofar as they are founded on nearly a century of collection and analysis. The Holarctic Eocene primate radiation is still classified within a facultatively prosimian grade despite the ongoing pattern of increased abundance and diversity of forms. Ironically, a pattern that extends from these is the almost instinctive unwillingness to exclude Eocene Holarctic adapids and omomyids from character polarity assessments proposed for early anthropoids (such as the primitive anthropoid anterior dental battery (Gingerich 1977, 1980, 1984; Gingerich and Schoeninger 1977; Schmid 1983; Rosenberger and Strasser 1985; Rosenberger et al. 1985; Rasmussen 1986; Szalay et al. 1987; Beard 1988; Simons 1989, 1990; Covert and Williams 1991), even though the same taxa are regularly dismissed from any potential ancestor–descendant relationships with higher primates on a variety of other grounds (see Fleagle 1988). As previously indicated, the morphological patterns of these fossils are relevant to phylogenetic reconstructions, but equal consideration must be given to the probability that Holarctic early Tertiary primates are not representative of contemporaneous prosimian and anthropoid evolution in equatorial latitudes.

Clearly, the issue of homology and what qualifies as positive proof for or against ancestor–descendant relationships remains wholly subjective and therefore not strictly empirical or objectively falsifiable (see Rosenberger and Szalay 1980:139–140 and Fleagle 1990:82 for alternative statements of this problem).

Dispersal Mechanisms

The possibility that platyrrhines and catarrhines split at a very early time appears to be missing in most scenarios (excepting Hershkovitz 1972, 1977, 1981), perhaps due to the distribution of known Paleogene fossil localities. Neither the presence of Holarctic Eocene adapids and omomyids, nor the unknown early Tertiary fossil record of the modern neotropics is sufficient reason for assuming that the platyrrhine–catarrhine split was not a very early (earliest Eocene) and/or regionally limited event. That platyrrhines are not found in the relatively abundant Paleogene fossil record of Argentina reflects as much about local and time-specific environmental conditions (but see Bown and Larriestra 1990; Pascual and Jaureguizar 1990) as it does the biogeographical distribution of early platyrrhines. Unfortunately, Paleogene deposits in Amazonia are simply unavailable. The distinctly nonmodern aspect of the early Miocene platyrrhines from Argentina (Fleagle and Bown 1983; Fleagle et al. 1987; Gebo and Simons 1987; Fleagle 1990; Kay 1990b; Meldrum 1990) likewise suggests that extant platyrrhines are a compact sample of the expanding/contracting diversity that has surely marked their adaptive radiation as a whole (see Fleagle 1990:84).

It is the perspective of this review that any late Eocene origins model must invoke a transoceanic crossing mechanism that is implausible (rafting) or suspect (waif dispersal) at best. Patterns of the plate tectonic data system are rendered with this in mind.

Continental drift is an ultimate time-transgressive phenomenon. The process itself is now canonized within the scientific community, but fine-tuning details of rates and timings of separation will always be extrapolations from static hard evidence or micromeasurements of current rates. This state of affairs is very similar to the conditions of the molecular evolution data system and its development over the last 30 years.

This is not to say that Africa and South America were conjoined at any time in the late Cretaceous/early Tertiary, but rather that the data for determining the interaction of sea levels with the extent of continental separation are not nearly as fine grained as were the actual events they depict. The identification of an all time low ocean margin level (Vail and Hartenbol 1979) at 29 MYA obviously factors into this consideration and has been cited as a fortuitous paleoclimatic event for platyrrhine dispersal (Fleagle 1986) on the condition that dispersal was relatively recent. In this context, the idiosyncrasies of the platyrrhine fossil record must also have been favorable so far given the discovery and subsequent redating (MacFadden et al. 1985; MacFadden 1990) of the isolated but timely late Oligocene platyrrhines of Bolivia. Would the discovery of a late Eocene platyrrhine in South America spur an effort to redate that low ocean margin level using more modern techniques? More indirectly, what effect does a late Eocene specimen like *Algeripithecus* (Godinot and Mahboubi 1992) have on the timing of platyrrhine separation, and can this phylogenetic separation event be considered independently of the actual platyrrhine dispersal to South America?

Conclusion: A Model Most Likely

The intuitive objections (Simons 1976) to a long oceanic crossing outweigh the present confidence of either continental separation rate extrapolations or island hopping/waif dispersal propositions. It is suggested here that colonizing platyrrhines did not raft nor were they dispersed as waifs across large segments of open ocean. No otherwise internally parsimonious latest Eocene dispersal model is exempt from this Occam's Razor of paleobiogeography. There is no convincing reason to assume that the platyrrhine–catarrhine divergence was not an early Eocene phenomenon, and that platyrrhine dispersal to South America did not occur immediately thereafter. The anatomy and stratigraphy of *Catopithecus* (Simons 1989, 1990) and *Algeripithecus* (Godinot and Mahboubi 1992) suggest that early anthropoids potentially ancestral to catarrhines emerged before the late Eocene and call for a close examination of the early splitting possibility. A most likely model of platyrrhine origins and dispersal would be one in which platyrrhines emerged in Africa in the terminal Paleocene/earliest Eocene and opportunistically dispersed to South America across still or transiently emergent insular rises connecting western Africa with northeastern Brazil.

This model predicts that platyrrhines will be found in Paleogene equatorial South American localities, and that evidence for an anthropoid grade of evolution will be recovered in equatorial African (or Asian) localities that considerably predate the Fayum. This model is openly inductive rather than deductive and does not adhere to the strict methodological guidelines recommended in other models. In aspect, however, this model does not depart from the widely accepted models of Ciochon and Chiarelli (1980) and Fleagle (1986), except in timing and mechanism as merited by patterns in the accumulation of relevant data.

SUMMARY

An examination of the patterns of recent paleontological, geophysical and biological data support Africa as a most likely source of platyrrhine origins and dispersal to South America. Previously proposed models of this platyrrhine origins hypothesis are strengthened by new fossil and substantial biomolecular evidence for anthropoid and platyrrhine/catarrhine monophyly. Revisions made here to these models relax their emphasis on a parsimonious analysis of positive data only, in favor of the openly inductive conclusion that the evolutionary history of platyrrhines and catarrhines most likely traces back to a much earlier time than the known fossil record suggests. The assumption that evolution proceeds under the mode of parsimony (literally stingy, frugal), that is, in a manner as free of contingencies as possible, stands in elegant immunity to the intractable complexities of independently demonstrating character homologies and the supreme contingencies that afflict every aspect of paleontology—the two most direct means of reconstructing evolutionary history.

LITERATURE CITED

Allard, G. O., and Hurst, V. J. (1969) Brazil-Gabon link supports continental drift. *Science* 163:528–532.

Anapol, F., and Fleagle, J. G. (1988) Fossil platyrrhine forelimb bones from the early Miocene of Argentina. *Am. J. Phys. Anthropol.* 76:417–428.

Asmus, H. E., and Ponte, F. C. (1973) The Brazilian marginal basins. In A. E. M. Nairn and F. G. Stehli (eds.): *The Ocean Basins and Margins*, vol. 1: *The South Atlantic*. New York: Plenum, pp. 87–133.

Baba, M., Darga, L., and Goodman. M. (1980) Biochemical evidence on the phylogeny of Anthropoidea. In R. L. Ciochon and A. B. Chiarelli (eds.): *Evolutionary Biology of the New World Monkeys and Continental Drift*. New York: Plenum, pp. 423–443.

Barr, K. W. (1974) The Caribbean and plate tectonics–some aspects of the problem. *Verhand. NaturForsch. Gesell.* 84:45–67.

Beard, K. C. (1988) The phylogenetic signficance of strepsirhinism in Paleogene primates. *Int. J. Primatol.* 9:83–96.

Beard, K.C., Dagosto, M., Gebo, D. L., and Godinot, M. (1988) Interrelationships among primate higher taxa. *Nature* 331:712–714.

Beard, K. C., Krishtalka, L., and Stucky, R. K. (1992) Revision of the Wind River faunas, early Eocene of central Wyoming, Part 12: New species of omomyid primates (Mammalia, Primates, Omomyidae) and omomyid taxonomic composition across the early-middle Eocene boundary. *Ann. Carnegie Mus.* 61:39–62.

Berggren, W. A., and Hollister, C. D. (1974) Paleogeography, paleobiogeography and the history of circulation in the Atlantic Ocean. In W. W. Hay (ed.): *Studies in Paleo-Oceanography*. Tulsa: Society of Economic Paleontologists and Mineralogists, pp. 126–186.

Berggren, W. A., and Hollister, C.D. (1977) Plate tectonics and paleocirculation: Commotion in the ocean. *Tectonophysics* 38:11–48.

Bond, G. (1978) Evidence for late Tertiary uplift of Africa relative to North America, South America, Australia and Europe. *J. Geol.* 86:47–65.

Bonini, W. E., Hargraves, R. B., and Shagam, R. (eds.) (1984) Caribbean-South American Plate Boundary and Regional Tectonics. *Geological Society of America Memoir* 162.

de Bonis, L., Jaeger, J. J., Coiffait, B., and Coiffait, P. E. (1988) Decouverte du plus ancien primate Catarrhinien connu dans l'Eocene superieur d'Afrique du Nord. *C. R. Acad. Sci. Ser.* 2 306:929–934.

Bown, T. M., and Larriestra, C. N. (1990) Sedimentary paleoenvironments of fossil platyrrhine localities, Miocene Pinturas Formation, Santa Cruz Province, Argentina. *J. Hum. Evol.* 19:87–119.

Bown, T. M., and Simons, E. L. (1984) First record of marsupials (Metatheria: Polyprotodonta) from the Oligocene in Africa. *Nature* 308:447–449.

Briden, J. C. (1973) Applicability of plate tectonics to pre-Mesozoic time. *Nature* 244:400–405

Bugge, J. (1980) Comparative anatomical study of the carotid circulation in New and Old World primates: Implications for their evolutionary history. In R. L. Ciochon and A. B. Chiarelli (eds.): *Evolutionary Biology of the New World Monkeys and Continental Drift*. New York: Plenum, pp. 293–317.

Burke, K. (1975) Atlantic evaporites formed by evaporation of water spilled from the Pacific, Tethyan, and Southern oceans. *Geology* 3:613–616.

Burke, K. (1988) Tectonic evolution of the Caribbean. *Ann. Rev. Earth Plan. Sci.* 16:201–230.

Burke, K., Kidd, W. S. F., and Wilson, J. T. (1973) Relative and latitudinal motion of Atlantic hot spots. *Nature* 245:133–137.

Burke, K., and Wilson, J. T. (1972) Is the African plate stationary? *Nature* 239:387–390.

Cachel, S. (1981) Plate tectonics and the problem of anthropoid origins. *Yearb. Phys. Anthropol.* 24:139–172.

Calais, E., and Mercier de Lepinay, B. (1990) Tectonique et paleogeographie de la cote Sud de 'Oriente cubain: Nouvelles constraintes pour l'evolution geodynamique de la limite de plaques decrochante nord-caraibe de l'Eocene a l'Actuel. *C. R. Acad. Sci. Ser.* 2 310:293–299.

Carroll, R. L. (1988) *Vertebrate Paleontology and Evolution*. New York: W. H. Freeman.

Cartmill, M. (1980) Morphology, function and evolution of the anthropoid postorbital septum. In R. L. Ciochon and A. B. Chiarelli (eds.): *Evolutionary Biology of the New World Monkeys and Continental Drift*. New York: Plenum, pp. 243–274.

Cartmill, M., and Kay, R. F. (1978) Craniodental morphology, *Tarsius* affinities, and primate suborders. In D. J. Chivers and K. A. Joysey (eds.): *Recent Advances in Primatology*, Vol. 3, *Evolution*. London: Academic, pp. 205–214.

Cartmill, M., MacPhee, R., and Simons, E. L. (1981) Anatomy of the temporal bone in early anthropoids, with remarks on the problem of anthropoid origins. *Am. J. Phys. Anthropol.* 56:3–22.

Chiarelli, A. B. (1980) The karyology of South American primates and their relationship to African and Asian species. In R. L. Ciochon and A. B. Chiarelli (eds.): *Evolutionary Biology of the New World Monkeys and Continental Drift*. New York: Plenum, pp. 387–398.

Ciochon, R. L. and Chiarelli, A. B. (1980) Paleobiogeographic perspectives on the origin of the Platyrrhini. In R. L. Ciochon and A. B. Chiarelli (eds.): *Evolutionary Biology of the New World Monkeys and Continental Drift*. New York: Plenum, pp. 459–494.

Ciochon, R. L., Savage, D. E., Thaw, T., and Maw, B. (1985) Anthropoid origins in Asia? New discovery of *Amphipithecus* from the Eocene of Burma. *Science* 229:756–759.

Clemens, W. A. (1979) Marsupialia. In J. A. Lillegraven, Z. Kielan-Jaworowska, and W. A. Clemens (eds.): *Mesozoic Mammals: The First Two-Thirds of Mammalian History*. Berkeley: University of California, pp. 192–220.

Coates, A. G. (1973) Cretaceous Tethyan coral-rudist biogeography related to the evolution of the Atlantic Ocean. In N.E. Hughes (ed.): *Organisms and Continents Through Time*. Special Papers in Paleontology 12:169–174.

Colbert, E. H. (1973) *Wandering Lands and Animals*. New York: Dutton.

Cooke, H. B. S. (1972) The fossil mammal fauna of Africa. In A. Keast, F. C. Erk and B. Glass (eds.): *Evolution, Mammals, and Southern Continents*. Albany: State University of New York, pp. 17–45.

Covert, H. H., and Williams, B. A. (1991) The anterior lower dentition of *Washakius insignis* and adapid-anthropoidean affinities. *J. Hum. Evol.* 21:463–467.

Cracraft, J. (1974) Continental drift and vertebrate distribution. *Ann. Rev. Ecol. Syst.* 5:215–261.

Cronin, J. E., and Sarich, V. M. (1975) Molecular systematics of the New World monkeys. *J. Hum. Evol.* 4:357–375.

Dagosto, M. (1990) Models for the origin of the anthropoid postcranium. *J. Hum. Evol.* 19:121–140.

Darlington, P. J. (1957) *Zoogeography: The Geographical Distribution of Animals*. New York: Wiley.

de Souza Barros, R., Nagamachi, C. Y., and Pieczarka, J. C. (1990) Chromosomal evolution in *Callithrix emiliae*. *Chromosoma* 99:440–447.

Delson, E., and Rosenberger, A. L. (1980) Phyletic perspectives on platyrrhine origins and anthropoid relationships. In R. L. Ciochon and A. B. Chiarelli (eds.): *Evolutionary Biology of the New World Monkeys and Continental Drift*. New York: Plenum, pp. 445–458.

Douglas, R. G., Moullade, M., and Nairn, A. E. M. (1973) Causes and consequences of drift in the South Atlantic. In D. H. Tarling and S. K. Runcorn (eds.): *Implications of Continental Drift to the Earth Sciences*, Vol. I. London: Academic, pp.517–537.

Dutrillaux, B. (1979) Chromosomal evolution in primates: Tentative phylogeny from *Microcebus murinus* (Prosimian) to man. *Hum. Genet.* 48:251–314.

Dutrillaux, B., and Couturier, J. (1981) The ancestral karyotype of platyrrhine monkeys. *Cytogenet. Cell Genet.* 30:232–242.

Emiliani, C., Gaertner, S., and Lidz, B. (1972) Neogene sedimentation on the Blake Plateau and the emergence of the Central American isthmus. *Palaeogeog. Palaeoclimatol. Palaeoecol.* 11:1–10.

Erikson, G. E. (1954) Comparative anatomy of New World primates and its bearing on the phylogeny of anthropoid apes and men. *Hum. Biol.* 26:210.

Erikson, J. P., Pindell, J. L., and Larue, D. K. (1990) Mid-Eocene – Early Oligocene sinistral transcurrent faulting in Puerto Rico associated with formation of the northern Caribbean plate boundary zone. *J. Geol.* 98:365–384.

Erlich, R. N., and Barrett, S. F. (1990) Cenozoic plate tectonic history of the northern Venezuela-Trinidad area. *Tectonics* 9:161–184.

Fairhead, J. D. (1988) Mesozoic plate tectonic reconstructions of the central South Atlantic Ocean: The role of the West and Central African rift system. *Tectonophysics* 155:181–191.

Falk, D. (1980) Comparative study of the endocranial casts of New and Old World monkeys. In R. L. Ciochon and A. B. Chiarelli (eds.): *Evolutionary Biology of the New World Monkeys and Continental Drift*. New York: Plenum, pp. 275–292.

Fleagle, J. G. (1985) New primate fossils from Colhuehuapian deposits at Gaiman and Sacanana, Chubut Province, Argentina. *Ameghiniana* 21:266–274.

Fleagle, J. G. (1986) Early anthropoid evolution in Africa and South America. In J. G. Else and P. Lee (eds.): *Primate Evolution*. Cambridge: Cambridge University Press, pp. 133–142.

Fleagle, J. G. (1988) *Primate Adaptation and Evolution*. New York: Academic.

Fleagle, J. G. (1990) New fossil platyrrhines from the Pinturas Formation, southern Argentina. *J. Hum. Evol.* 19:61–86.

Fleagle, J. G., and Bown, T. M. (1983) New primate fossils from late Oligocene localities of Chubut Province, Argentina. *Folia Primatol* 41:240–266.

Fleagle, J. G., Bown, T. M., Obradovich, J. D., and Simons, E. L. (1986a) Age of the earliest African anthropoids. *Science* 234:1247–1249.

Fleagle, J. G., Bown, T. M., Obradovich, J. D., and Simons, E. L. (1986b) How old are the Fayum primates? In J. G. Else and P. C. Lee (eds.): *Primate Evolution*. London: Cambridge University Press, pp. 3–17.

Fleagle, J. G., and Kay, R. F. (1987) The phyletic position of the Parapithecidae. *J. Hum. Evol.* 16:483–532.

Fleagle, J. G., and Meldrum, D. J. (1988) Locomotor behavior and skeletal morphology of two sympatric pitheciine monkeys, *Pithecia pithecia* and *Chiropotes satanas*. *Am. J. Primatol.* 16:227–249.

Fleagle, J. G., Powers, D. W., Conroy, G. C., and Watters, J. P. (1987) New fossil platyrrhines from Santa Cruz Province, Argentina. *Folia Primatol.* 48:65–77.

Fleagle, J. G., and Rae, T. C. (1992) Primate cranial remains from the Pinturas Formation, Argentina. *Am. J. Phys. Anthropol. Suppl.* 14:75–76.

Fleagle, J. G., and Rosenberger, A. L. (1983) Cranial morphology of the earliest anthropoids. In M. Sakka (ed.): *Morphologie Evolution Morphogenese du Crane et Origine de L'Homme*. Paris: C. N. R. S., pp. 141–155.

Fleagle, J. G., and Rosenberger, A. L. (1990) Preface: "The Platyrrhine Fossil Record". *J. Hum. Evol.* 19:1–6.

Ford, S. M. (1980) Phylogenetic relationships of the Platyrrhini: The evidence of the femur. In R. L. Ciochon and A. B. Chiarelli (eds.): *Evolutionary Biology of the New World*

Monkeys and Continental Drift. New York: Plenum, pp. 317–330.

Ford, S. M. (1986a) Subfossil platyrrhine tibia (Primates: Callitrichidae) from Hispaniola: A possible further case of island gigantism. *Am. J. Phys. Anthropol.* 70:47–62.

Ford, S. M. (1986b) Systematics of the New World monkeys. In D. Swindler (ed.): *Comparative Primate Biology*, Vol. 1: *Systematics, Evolution, and Anatomy*. New York: Alan R. Liss, pp. 73–135.

Ford, S. M. (1988) Postcranial adaptations of the earliest platyrrhine. *J. Hum. Evol.* 19:237–254.

Ford, S. M. (1990a) Locomotor adaptations of fossil platyrrhines. *J. Hum. Evol.* 19:141–174.

Ford, S. M. (1990b) Platyrrhine evolution in the West Indies. *J. Hum. Evol.* 19:237–254.

Ford, S. M., and Morgan, G. S. (1986) A new ceboid femur from the Late Pleistocene of Jamaica. *J. Vert. Paleontol.* 6:281–289.

Ford, S. M., and Morgan, G. S. (1988) Earliest primate fossil from the West Indies. *Am. J. Phys. Anthropol.* 75:209.

Ford, S. M., and Woods, C. A. (1990) New platyrrhine fossil material from Haiti. *Am. J. Phys. Anthropol.* 81:223.

Forster, R. (1978) Evidence for an open seaway between northern and southern proto-Atlantic in Albian times. *Nature* 272:158–159.

Frakes, L. A., and Kemp, E. M. (1973) Paleogene continental positions and evolution of climate. In D. H. Tarling and S. K. Runcorn (eds.): *Implications of Continental Drift to the Earth Sciences*, Vol. 1. London: Academic, pp. 539–559.

Freeland, G. L., and Dietz, R. S. (1971) Plate tectonic evolution of Caribbean-Gulf of Mexico region. *Nature* 232:20–23.

Funnell, B. M., and Smith, A. G. (1968) Opening of the Atlantic Ocean. *Nature* 219:1328–1333.

Galbreath, G. J. (1983) Karyotypic evolution in *Aotus*. *Am. J. Primatol.* 4:245–251.

Gantt, D. G. (1980) Implications of enamel prism patterns for the origin of New World monkeys. In R. L. Ciochon and A. B. Chiarelli (eds.): *Evolutionary Biology of the New World Monkeys and Continental Drift*. New York: Plenum, pp. 201–217.

Gebo, D. L. (1986) Anthropoid origins: The foot evidence. *J. Hum. Evol.* 15:421–430.

Gebo, D. L. (1989) Locomotor and phylogenetic considerations in anthropoid evolution. *J. Hum. Evol.* 18:191–223.

Gebo, D. L., Dagosto, M., Rosenberger, A. L., and Setoguchi, T. (1990) New platyrrhine tali from La Venta, Colombia. *J. Hum. Evol.* 9:737–746.

Gebo, D. L., and Simons, E. L. (1987) Morphology and locomotor adaptations of the foot in early Oligocene anthropoids. *Am. J. Phys. Anthropol.* 74:83–101.

Gingerich, P. (1977) Radiation of Eocene Adapidae in Europe. *Geobios Mem. Spec.* 1:165–185.

Gingerich, P. (1980) Eocene Adapidae, paleobiogeography, and the origin of South American Platyrrhini. In R. L. Ciochon and A. B. Chiarelli (eds.): *Evolutionary Biology of the New World Monkeys and Continental Drift*. New York: Plenum, pp. 123–138.

Gingerich, P. (1984) Primate evolution: Evidence from the fossil record, comparative morphology and molecular biology. *Yearb. Phys. Anthropol.* 27:57–72.

Gingerich, P. (1986) Early *Cantius torresi* – oldest primate of modern aspect from North America. *Nature* 319:319–321.

Gingerich, P., and Rose, K. D. (1977) Preliminary report of the American Clark Fork mammal fauna, and its correlation with similar faunas in Europe and Asia. *Geobios Mem. Spec.* 1:39–45.

Gingerich, P., and Schoeninger, M. (1977) The fossil record and primate phylogeny. *J. Hum. Evol.* 6:483–505.

Godinot, M., and Mahboubi, M. (1992) Earliest known simian primate found in Algeria. *Nature* 357:324–326.

Goodman, M., Romera-Herrera, A. E., Dene, H., Czelusniak, J., and Tashian, R. E. (1982) Amino acid sequence evidence on the phylogeny of primates and other eutherians. In M. Goodman (ed.): *Macromolecular Sequences in Systematic and Evolutionary Biology*. New York: Plenum, pp. 115–192.

Grant, N. K. (1971) South Atlantic, Benue trough, and Gulf of Guinea Cretaceous triple junction. *Geol. Soc. Amer. Bull.* 82:2295–2298.

Gregory, W. K. (1920) On the structure and relations of *Notharctus*, an American Eocene primate. *Mem. Am. Mus. Nat. Hist.* 3:51–243.

Guzman-Speziale, M., Pennington, W. D., and Matumoto, T. (1989) The triple junction of the North America, Cocos, and Caribbean plates: seismicity and tectonics. *Tectonics* 8:981–997.

Hasegawa, M. (1990) Phylogeny and molecular evolution in primates. *Jpn. J. Genet.* 65:243–266.

Hasegawa, M., Kishino, H., and Yano, T. (1989) Estimation of branching dates among primates by molecular clocks of nuclear DNA which slowed down in Hominoidea. *J. Hum. Evol.* 18:461–476.

Hayasaka, K., Gojobori, T., and Horai, S. (1988) Molecular phylogeny and evolution of primate mitochondrial DNA. *Mol. Biol. Evol.* 5:626–644.

Herrbeck, C., and Mann, P. (1991) Geologic evaluation of plate kinematic models for the North American - Caribbean plate boundary zone. *Tectonophysics* 191:1–26.

Hershkovitz, P. (1972) The recent mammals of the Neotropical region: A zoogeographic and ecological review. In A. Keast, F. C. Erk, and B. Glass (eds.): *Evolution, Mammals, and Southern Continents*. Albany: State University of New York, pp. 311–431.

Hershkovitz, P. (1977) *Living New World Monkeys (Platyrrhini), with an Introduction to Primates*. Chicago: University of Chicago Press.

Hershkovitz, P. (1981) Comparative anatomy of the platyrrhine mandibular cheek teeth dpm4, pm4, m1, with particular reference to those of *Homunculus* (Cebidae) and comments on platyrrhine origins. *Folia Primatol.* 35:179–217.

Hershkovitz, P. (1982) Supposed squirrel monkey affinities of *Dolichocebus gaimanensis*. *Nature* 298:202.

Hershkovitz, P. (1984) More on *Homunculus* dpm4 and M1 and comparisons with *Alouatta* and *Stirtonia* (Primates, Platyrrhini, Cebidae). *Am. J. Primatol.* 7:261–283.

Hershkovitz, P. (1988) The subfossil monkey femur and subfossil monkey tibia of the Antilles: A review. *Int. J. Primatol.* 9:365–384.

Hirschfeld, S. E., and Marshall, L. G. (1976) Revised faunal list of the La Venta fauna (Friasian-Miocene) of Colombia, South America. *J. Palaeont.* 50:433–436.

Hoffstetter, R. (1969) Un primate de l'Oligocene inferieur sudamericain: *Branisella boliviana* gen. et spec. nov. *C. R. Acad. Sci. Paris* 269:434–437.

Hoffstetter, R. (1972) Relationships, origins and history of the ceboid monkeys and caviomorph rodents: A modern reinterpretation. In T. Dobzhansky, M. K. Hecht, and W. C. Steere (eds.): *Evolutionary Biology*. New York: Appleton-Century-Crofts, pp. 323–347.

Hoffstetter, R. (1974) Phylogeny and geographical deployment of the primates. *J. Hum. Evol.* 3:327–350.

Hoffstetter, R. (1977) Primates: Filogenia e historia biogeographica. *Studia Geol.* 13:211–253.

Hoffstetter, R. (1980) Origin and deployment of the New World monkeys emphasizing the southern continents route. In R. L. Ciochon and A. B. Chiarelli (eds.): *Evolutionary Biology of the New World Monkeys and Continental Drift*. New York: Plenum, pp. 103–122.

Holcombe, E., and Moore, W. S. (1977) Paleocurrents in the eastern Caribbean: Geologic evidence and implications. *Mar. Geol.* 23:35–56.

Kauffman, E. G. (1973) Cretaceous Bivalvia. In A Hallam (ed.): *Atlas of Palaeobiogeography*. Amsterdam: Elsevier, pp. 353–383.

Kay, R. F. (1977) The evolution of molar occlusion in the Cercopithecidae and early catarrhines. *Am. J. Phys. Anthropol.* 46:327–352.

Kay, R. F. (1980) Platyrrhine origins: A reappraisal of the dental evidence. In R. L. Ciochon and A. B. Chiarelli (eds.): *Evolutionary Biology of the New World Monkeys and Continental Drift*. New York: Plenum, pp. 159–187.

Kay, R. F. (1989) A new small platyrrhine from the Miocene of Colombia and the phyletic position of the callitrichines. *Am. J. Phys. Anthropol.* 78:251.

Kay, R. F. (1990a) A possible "giant" tamarin from the Miocene of Colombia. *Am. J. Phys. Anthropol.* 81:248.

Kay, R. F. (1990b) The phyletic relationships of extant and fossil Pitheciinae (Platyrrhini, Anthropoidea). *J. Hum. Evol.* 19:175–208.

Kay, R. F., Madden, R. H., Plavcan, J. M., Cifelli, R. L., and Diaz, J. G. (1987) *Stirtonia victoriae,* a new species of Miocene Colombian primate. *J. Hum. Evol.* 16:173–196.

Keast, A. (1972) Continental drift and the evolution of the biota on southern continents. In A. Keast, F. C. Erk and B. Glass (eds.): *Evolution, Mammals, and Southern Continents*. Albany: State University of New York, pp. 23–87.

Kellogg, J. N. (1984) Cenozoic tectonic history of the Sierra de Perija, Venezuela-Colombia, and adjacent basins. *Mem. Geol. Soc. Am.* 162:239–261.

Kellogg J. N., Ogujiofor I. J., and Kansakar D. R. (1985) Cenozoic tectonics of the Panama and North Andes blocks. *Mem. Latin Am. Cong. Geol.* 6:40–59.

Kerr, R. A. (1978) Plate tectonics: What forces drive the plates? *Science* 200:36–38.

Koop, B. F., Tagle, D. A., Goodman, M., and Slightom, J. L. (1989) A molecular view of primate phylogeny and important systematic and evolutionary questions. *Mol. Biol. Evol.* 6:580–612.

Kumar, N., and Embley, R. W. (1977) Evolution and origin of Ceara Rise: An aseismic rise in the western equatorial Atlantic. *Geol. Soc. Amer. Bull.* 88:683–694.

Ladd, J. W. (1976) Relative motion of South America with respect to North America and Caribbean tectonics. *Geol. Soc. Amer. Bull.* 87:969–976.

Ladd, J. W., Dickson, G. O., and Pittman, W. C. (1973) The age of the South Atlantic. In A. E. M. Nairn and F. G. Stehli (eds.): *The Ocean Basins and Margins*, Vol. 1: *The South Atlantic*. New York: Plenum, pp. 555–573.

Larson, R. L., and Ladd, J. W. (1973) Evidence for the opening of the South Atlantic in the early Cretaceous. *Nature* 246:209–212.

Lavocat, R. (1969) La systematique des rongeurs histricomorphes et la derive des continents. *C.R. Acad. Sci. Paris, Ser D* 269:1496–1497.

Lavocat, R. (1974) The interrelationships between the African and South American rodents and their bearing on the problem of the origin of South American monkeys. *J. Hum. Evol.* 3:323–326.

Lavocat, R. (1978) Rodentia and Lagomorpha. In V. J. Maglio and H. B. S Cooke (eds.): *Evolution of African Mammals*. Cambridge: Harvard University Press, pp. 69–89.

Lavocat, R. (1980) The implication of rodent paleontology and biogeography to the geographical sources and origin of the platyrrhine primates. In R. L. Ciochon and A. B. Chiarelli (eds.): *Evolutionary Biology of the New World Monkeys and Continental Drift*. New York: Plenum, pp. 93–102.

Lehner P, and DeRuiter, P. A. C. (1977) Structural history of Atlantic margin of Africa. *Am. Assoc. Petrol. Geol. Bull.* 61:961–981.

Le Pichon, X. (1968) Sea floor spreading and continental drift. *J. Geophys. Res.* 73:3661–3697.

Le Pichon, X., and Fox, P. J. (1971) Marginal offsets, fracture zones, and the early opening of the South Atlantic. *J. Geophys. Res.* 76:6283–6293.

Le Pichon, X., and Hayes, D. E. (1971) Marginal offsets, fracture zones, and the early opening of the South Atlantic. *J. Geophys. Res.* 76:6294–6308.

Luchterhand, K., Kay, R. F., and Madden, R. H. (1986) *Mohanamico hershkovitzi*, gen.et sp. nov., un primate du Micoene moyen d'Amerique du Sud. *C. R. Acad. Sci. Paris* 19:1753–1758.

Luckett, W. P. (1980) Monophyletic or diphyletic origins of Anthropoidea and Hystricognathi: Evidence of the fetal membranes. In R. L. Ciochon and A. B. Chiarelli (eds.): *Evolutionary Biology of the New World Monkeys and Continental Drift*. New York: Plenum, pp. 347–368.

Luckett, W. P., and Hartenberger, J. L. (eds.) (1985) *Evolutionary Relationships Among the Rodents.* New York: Plenum.
Ma, N. S. F. (1981) Chromosome evolution in the owl monkey, *Aotus. Am. J. Phys. Anthropol.* 54:293–303.
MacDonald, W. D., and Opdyke, N. (1972) Tectonic rotations suggested by paleomagnetic results from northern Colombia. *J. Geophys.* Res. 77:5720–5730.
MacFadden, B. J. (1985) Drifting continents, mammals, and time scales: Current developments in South America. *J. Vert. Paleontol.* 5:169–174.
MacFadden B. J. (1990) Chronology of Cenozoic primate localities in South America. *J. Hum. Evol.* 19:7–22.
MacFadden, B. J., Campbell, K. E., Cifelli, R. L., Siles, O., Johnson, N., Naeser, C. W., and Zeitler, P. K. (1985) Magnetic polarity stratigraphy and mammalian biostratigraphy of the Deseadan (Late Oligocene–Early Miocene) Salla beds of northern Bolivia. *J. Geol.* 93:223–250.
MacPhee, R. D. E., and Fleagle, J. G. (1991) Postcranial remains of *Xenothrix mcgregori* (Primates, Xenotrichidae) and other late Quaternary mammals from Long Mile Cave, Jamaica. *Bull. Am. Mus. Nat. Hist.* 206:287–321.
MacPhee, R. D. E., and Woods, C. A., (1982) A new fossil cebine from Hispaniola. *Am. J. Phys. Anthropol.* 58:419–436.
Maglio, V. J., and Cooke, H. B. S. (eds.) (1978) *Evolution of African Mammals.* Cambridge: Harvard University Press.
Maier, W. (1980) Nasal structures in Old and New World primates. In R. L. Ciochon and A. B. Chiarelli (eds.): *Evolutionary Biology of the New World Monkeys and Continental Drift.* New York: Plenum, pp. 219–242.
Malfait, B. T. and Dinkelman, M. G. (1972) Circum–Caribbean tectonic and igneous activity and the evolution of the Caribbean plate. *Geol. Soc. Amer. Bull.* 83:251–272.
Maresch, W. V. (1974) Plate tectonics origin of the Caribbean mountain system of northern South America: Discussion and proposal. *Geol. Soc. Amer. Bull.* 83:251–272.
Marks, J. (1987) Social and ecological aspects of primate cytogenetics. In W. G. Kinzey (ed.): *The Evolution of Human Behavior: Primate Models.* Albany: State University of New York, pp. 139–150.
Marshall, L. G. (1982) Calibration of the age of mammals in South America. *Geobios* 6(Suppl):427–437.
Marshall, L. G. (1985) Geochronology and land-mammal biochronology of the Transamerican faunal interchange. In F. Stehli and S. D. Webb (eds.): *The Great American Biotic Interchange.* New York: Plenum, pp.49–85.
Marshall, L. G., Hoffstetter, R., and Pascual, R. (1983) Geochronology of the continental mammal–bearing Tertiary of South America. *Paleovertebrata, Mem. Extraord.* 1983:1–93.
Marshall, L. G., Berta, A., Hoffstetter, R., Pascual, R., Reig, O. A., Bombin, M., and Mones, A. (1984) Mammals and stratigraphy: Geochronology of the continental mammal-bearing Quaternary of South America. *Paleovertebrata, Mem. Extraord.* 1984:1–76.
Martin, R. D. (1990) *Primate Origins and Evolutionary History.* Princeton: Princeton University.
Martin, D. E., and Gould, K. G. (1980) Comparative study of the sperm morphology of South American primates and those of the Old World. In R. L. Ciochon and A. B. Chiarelli (eds.): *Evolutionary Biology of the New World Monkeys and Continental Drift.* New York: Plenum, pp. 369–386.
Matthew, W. D. (1915) Climate and evolution. *Ann. NY Acad. Sci.* 24:171–318.
McKenna, M. (1967) Classification, range, deployment of the prosimian primates. *Colloq. Int. CNRS* 163:603–613.
McKenna, M. (1973) Sweepstakes, filters, corridors, Noah's Arks and beached Viking funeral ships in paleogeography. In D. H. Tarling and S. K. Runcorn (eds.): *Implications of Continental Drift to the Earth Sciences.* London: Academic, pp. 295–308.
Meldrum, D. J. (1990) New fossil platyrrhine tali from the early Miocene of Argentina. *Am. J. Phys. Anthropol.* 83:403–418.
Meldrum D. J., Fleagle, J. G., and Kay, R. F. (1990) Partial humeri of two Miocene Colombian primates. *Am. J. Phys. Anthropol.* 81:413–422
Meldrum, D. J., and Kay, R. F. (1990) A new partial skeleton of *Cebupithecia sarmientoi* from the Miocene of Colombia. *Am. J. Phys. Anthropol.* 81:267.
Meldrum, D. J., and Kay, R. F. (1992) A new specimen of pitheciine primate from the Miocene of Colombia. *Am. J. Phys. Anthropol.* Suppl. 14:121.
Minezawa, M., Jordan, C. O. C., and Valdivia, B. J. (1989) Karyotypic study of titi monkeys, *Callicebus moloch brunneus. Primates* 30:81–88.
Miyamoto, M. M., and Goodman, M. (1990) DNA systematics and evolution of primates. *Ann. Rev. Ecol. Syst.* 21:197–220.
Naeser, C. W., McKee, E. H., Johnson, N. M., and MacFadden, B. J. (1987) Confirmation of a late Oligocene-early Miocene age of the Deseadan Salla Beds of Bolivia. *J. Geol.* 95:825–828.
Nairn, A. E. M., and Stehli, F. G. (1973) A model for the South Atlantic. In A. E. M. Nairn and F. G. Stehli (eds.): *The Ocean Basins and Margins,* Vol. I: *The South Atlantic.* New York: Plenum, pp. 1–24.
Nelson, G. (1974) Historical biogeography: An alternative formalization. *Syst. Zool.* 22:312–320.
Newton, A. R. (1976) Was there an Agulhas triple junction? *Nature* 260:767–768.
Nurnberg, D., and Muller, R. D. (1991) The tectonic evolution of the South Atlantic from late Jurassic to present. *Tectonophysics* 191:27–53.
O'Connor, J. M., and Duncan, R. A. (1990) Evolution of the Walvis Ridge – Rio Grande Rise hot spot system: Implications for African and South American plate motions over plumes. *J. Geophys. Res.* 95:17475–17502.
Onstott, T. C., and Hargraves, R. B. (1981) Proterozoic transcurrent tectonics: Palaeomagnetic evidence from Venezuela and Africa. *Nature* 289:131–136.
Orlosky, F. J. (1980) Dental evolutionary trends of relevance to the origin and dispersion of platyrrhine monkeys. In R. L. Ciochon and A. B. Chiarelli (eds.): *Evolutionary*

Biology of the New World Monkeys and Continental Drift. New York: Plenum, pp. 189–200.

Orlosky, F. J., and Swindler, D. R. (1975) Origins of New World monkeys. *J. Hum. Evol.* 4:77-83.

Pascual, R., and Jaureguizar, E. O. (1990) Evolving climates and mammal faunas in Cenozoic South America. *J. Hum. Evol.* 19:23–60.

Patterson, B. (1954) The geologic history of non-hominid primates in the Old World. *Hum. Biol.* 26:191–209.

Patterson, B., and Pascual, R. (1972) The fossil mammals of South America. In A. Keast, F. C. Erk, and B. Glass (eds.): *Evolution, Mammals, and Southern Continents*. Albany: State University of New York, pp. 247–309.

Perkins, E. M., and Meyer, W. C. (1980) The phylogenetic significance of the skin of primates: Implications for the origin of New World monkeys. In R. L. Ciochon and A. B. Chiarelli (eds.): *Evolutionary Biology of the New World Monkeys and Continental Drift*. New York: Plenum, pp. 331–346.

Phillips, J. D., and Forsyth, D. (1972) Plate tectonics, paleomagnetism and the opening of the Atlantic. *Bull. Geol. Soc. Amer.* 83:1579–1600.

Powers, S. (1911) Floating islands. *Pop. Sci. Monthly* 79:303–307.

Rabinowitz, P. D., and LaBrecque, J. (1979) The Mesozoic South Atlantic Ocean and the evolution of its continental margins. *J. Geophys. Res.* 84:5973–6002.

Ramsay, A. T. S. (1971) A history of the formation of the Atlantic Ocean. *Brit. Assoc. Advan. Sci.* 27:239–249.

Ramsay, A. T. S. (1973) A history of organic siliceous sediments in oceans. In N. F. Hughes (ed.): *Organisms and Continents Through Time. Special Papers in Paleontology* 12:199–234.

Rand, H. M., and Mabesoone, J. M. (1982) Northeast Brasil and the final separation of South America and Africa. *Palaeogeog. Palaeoclimatol. Palaeoecol.* 38:163–183.

Rassmussen, D. T. (1986) Anthropoid origins: A possible solution to the Adapidae–Omomyidae paradox. *J. Hum. Evol.* 15:1–12.

Raven, P. H. and Axelrod, D. I. (1975) History of the flora and fauna of Latin America. *Am. Sci.* 63:420–429.

Reyment, R. A. (1973) Cretaceous history of the South Atlantic Ocean. In D. H. Tarling and S. K. Runcorn (eds.): *Implications of Continental Drift to the Earth Sciences*, Vol. 2. London: Academic, pp. 805–814.

Ridley, H. N. (1930) *The Dispersal of Plants Throughout the World*. Kent: L. Reeve.

Rivero, M., and Arredondo, O. (1991) *Paralouatta varonai*, a new Quaternary platyrrhine from Cuba. *J. Hum. Evol.* 21:1–11.

Romer, A. S. (1966) *Vertebrate Paleontology*. Chicago: University of Chicago Press.

Rose, K. D. and Bown, T. M. (1991) Additional fossil evidence on the differentiation of the earliest euprimates. *Proc. Nat. Acad. Sci.* 88:98–101.

Rose, K. D., and Fleagle, J. G. (1981) The fossil history of non-human primates in the Americas. In A. F. Coimbra-Filho and R. A. Mittermeier (eds.): *Ecology and Behavior of Neotropical Primates*, Vol. 1. Rio de Janeiro, Academia Brasileira de Ciencias, pp. 111–167.

Rose, K. D., and Krause, D. W. (1984) Affinities of the primate *Altanius* from the early Tertiary of Mongolia. *J. Mammal.* 65:721–726.

Rosenberger, A. L. (1977) *Xenothrix* and ceboid phylogeny. *J. Hum. Evol.* 6:541–561.

Rosenberger, A. L. (1979) Cranial anatomy and implications of *Dolichocebus*, a late Oligocene ceboid primate. *Nature* 279:416–418.

Rosenberger, A. L. (1981) A mandible of *Branisella boliviana* (Platyrrhini, Primates) from the Oligocene of South America. *Int. J. Primatol.* 2:1–7.

Rosenberger, A. L. (1984) Fossil New World monkeys dispute the molecular clock. *J. Hum. Evol.* 13:737–742.

Rosenberger, A. L. (1985) In favor of the *Necrolemur*-Tarsier hypothesis. *Folia Primatol.* 45:179–194.

Rosenberger, A. L. (1986) Platyrrhines, catarrhines, and the anthropoid transition. In B. A. Wood, L. Martin, and P. Andrews (eds.): *Major Topics in Primate and Human Evolution*. Cambridge: Cambridge University Press, pp. 66–88.

Rosenberger, A. L., Hartwig, W. C., Takai, M., Setoguchi, T., and Shigehara, N. (1991a) Dental variability in *Saimiri* and the taxonomic status of *Neosaimiri fieldsi*, an early squirrel monkey from La Venta, Colombia. *Int. J. Primatol.* 12:291–301.

Rosenberger, A. L., Setoguchi, T., and Hartwig, W. C. (1991b) *Laventiana annectens*, new genus and species: Fossil evidence for the origins of callitrichine New World monkeys. *Proc. Nat. Acad. Sci.* 88:2137–2140.

Rosenberger, A. L., Hartwig, W. C., and Wolff, R. G. (1991c) *Szalatavus attricuspis*, an early platyrrhine primate. *Folia Primatol.* 56:225–233.

Rosenberger, A. L., Setoguchi, T., and Shigehara, N. (1990) The fossil record of callitrichine primates. *J. Hum. Evol.* 19:209–236.

Rosenberger, A. L., and Strasser, E. (1985) Toothcomb origins: Support for the grooming hypothesis. *Primates* 26:73–84.

Rosenberger, A. L., Strasser, E., and Delson, E. (1985) Anterior dentition of *Notharctus* and the adapid-anthropoid hypothesis. *Folia Primatol.* 44:15–39.

Rosenberger, A. L., and Szalay, F. S. (1980) On the tarsiiform origins of Anthropoidea. In RL Ciochon and AB Chiarelli (eds.): *Evolutionary Biology of the New World Monkeys and Continental Drift*. New York: Plenum, pp. 139–158.

Rosencrantz, E. (1990) Sturcture and tectonics of the Yucatan Basin, Caribbean Sea, as determined from seismic reflection studies. *Tectonics* 9:1037–1059.

Ross, M. I., and Scotese, C. R. (1988) A hierarchical tectonic model of the Gulf of Mexico and Caribbean region. *Tectonophysics* 155:139–168.

Santamaria, F., and Schubert, C. (1974) Geochemistry and geochronology of the southern Caribbean-northern Venezuela plate boundary. *Geol. Soc. Amer. Bull.* 85:1085–1098.

Sarich, V. M., and Cronin, J. (1980) South American mammal molecular systematics, evolutionary clocks, and continental drift. In R. L. Ciochon and A. B. Chiarelli (eds.): *Evolutionary Biology of the New World Monkeys and Continental Drift*. New York: Plenum, pp. 399–422.

Savage, D. E., and Russell, D. E. (1983) *Mammalian Paleofaunas of the World*. Boston: Addison–Wesley.

Scheltema, R. S. (1971) Larval dispersal as a means of genetic exchange between geographically separated populations of shallow-water benthic marine gastropods. *Biol. Bull.* 140:284–322.

Schmid, P. (1983) Front dentition of the Omomyiformes (Primates). *Folia Primatol.* 40:1–10.

Schmitt, J., Graur, D., and Tomiuk, J. (1990) Phylogenetic relationships and rates of evolution in primates: Allozymic data from catarrhine and platyrrhine species. *Primates* 31:95–108.

Schneider, E. (1984) Comparative and functional morphology of the appendicular skeleton in Oligocene Fayum Rodentia. M.S. Thesis, State University of New York, Stony Brook.

Sclater, J. G., Hellinger, S., and Tapscott, C. (1977) The paleobathymetry of the Atlantic Ocean from the Jurassic to the present. *J. Geol. 85:509–552.*

Setoguchi, T. (1985) *Kondous laventicus*, a new ceboid primate from the Miocene of La Venta, Colombia, South America. *Folia Primatol.* 44:96–101.

Setoguchi, T., and Rosenberger, A. L. (1985) Miocene marmosets: First fossil evidence. *Int. J. Primatol.* 6:615–625.

Setoguchi, T., and Rosenberger, A. L. (1987) A fossil owl monkey from La Venta, Colombia. *Nature* 326:692–694.

Setoguchi, T., Takai, M., Villarroel, A. C., Shigehara, N., and Rosenberger, A. L. (1988) New specimen of *Cebupithecia* from La Venta, Miocene of Colombia, South America. *Kyoto Univ. Spec. Pub.* 1988:7–9.

Sibuet, J. C., and Mascle, J. (1978) Plate kinematic implications of Atlantic equatorial fracture zone trends. *J. Geophys. Res.* 83:3401–3421.

Simons, E. L. (1969) The earliest apes. *Sci. Amer.* 217:28–35.

Simons, E. L. (1976) The fossil record of primate phylogeny. In M. Goodman and R. E. Tashian (eds.): *Molecular Anthropology*. New York: Plenum, pp. 35–60.

Simons, E. L. (1986) *Parapithecus grangeri* of the African Oligocene: An archaic catarrhine without lower incisors. *J. Hum. Evol.* 15:205–213.

Simons, E. L. (1989) Description of two genera and species of Late Eocene Anthropoidea from Egypt. *Proc. Nat .Acad. Sci.* 86:9956–9960.

Simons, E. L. (1990) Discovery of the oldest known anthropoidean skull from the Paleogene of Egypt. *Science* 247:1567–1569.

Simons, E. L., and Bown TM (1985) *Afrotarsius chatrathi*, first tarsiiform primate (?Tarsiidae) from Africa. *Nature* 313:475–477.

Simons, E. L., Bown, T. M., and Rasmussen, D. T. (1986) Discovery of two additional prosimian primate families (Omomyidae, Lorisidae) in the African Oligocene. *J. Hum. Evol.* 15:431–437.

Simons, E. L., and Kay, R. F. (1983) *Qatrania*, new basal anthropoid primate from the Fayum, Oligocene of Egypt. *Nature* 304:624–626.

Simons, E. L., and Kay, R. F. (1988) New material of *Qatrania* from Egypt with comments on the phylogenetic position of the Parapithecidae (Primates, Anthropoidea). *Am. J. Primatol.* 15:337–347.

Simpson, G. G. (1969) South American mammals. In E. J. Fittkau, J. Illies, H. Klinge, G. H. Schwabe, and H. Sioli (eds.): *Biogeography and Ecology in South America*, Vol. 2. The Hague: Junk, pp. 879–909.

Simpson, G. G. (1978) Early mammals in South America: Fact, controversy and mystery. *Proc. Am. Phil. Soc.* 122:318–328.

Simpson, G. G. (1980) *Splendid Isolation. The Curious History of South American Mammals*. New Haven: Yale University Press.

Smith, P. J. (1977) Origin of the Rio Grande Rise. *Nature* 269:651–652.

Smith, A. G., and Briden, J. C. (1977) *Mesozoic and Cenozoic Paleocontinental Maps*. Cambridge: Cambridge University Press.

Smith, A. G., and Hallam, A. (1970) The fit of the southern continents. *Nature* 269:651–652.

Soares, V. M .F. C., Seuanez, H. N., Pissinatti, A., Coimbra-Filho, A. F., and Alvares, J. N. (1982) Chromosome studies in Callitrichidae (Platyrrhini): A comparison between *Callithrix* and *Leontopithecus*. *J. Med. Primatol.* 11:221–234.

Stevens, G. R. (1973) Cretaceous belemnites. In A. Hallam (ed.): *Atlas of Palaeobiogeography*. Amsterdam: Elsevier, pp. 385–401.

Sykes, L., McCann, W. R., and Kafka, A. L. (1982) Motion of the Caribbean plate during the last 7 million years and implications for earlier Cenozoic movements. *J. Geophys. Res.* 87:10656–10676.

Szalay F. S. (1975) Phylogeny, adaptations and dispersal of the tarsiiform primates. In W. P. Luckett and F. S. Szalay (eds.): *Phylogeny of the Primates*. New York: Plenum, pp. 357–404.

Szalay F. S., and Dagosto, M. (1988) Evolution of hallucial grasping in the primates. *J. Hum. Evol.* 17:1–33.

Szalay F. S., and Delson, E. (1979) *Evolutionary History of the Primates*. New York: Academic.

Szalay F. S., and Li, C. K. (1986) Middle Paleocene euprimate from southern China and the distribution of primates in the Paleogene. *J. Hum. Evol.* 15:387–399.

Szalay F. S., Rosenberger, A. L., and Dagosto, M. (1987) Diagnosis and differentiation of the Order Primates. *Yearb. Phys. Anthro.* 30:75–105.

Takemura, K., and Danhara, T. (1986) Fission-track dating of the upper part of the Miocene Honda Group in La Venta badlands, Colombia. *Kyoto Univ. Overseas Res. Rep. on New World Monkeys* 5:31–37.

Tarling, D. H. (1980) The geologic evolution of South America with special reference to the last 200 million years. In R. L. Ciochon and A. B. Chiarelli (eds.): *Evolutionary Biology of the New World Monkeys and Continental Drift*. New York: Plenum, pp. 1–42.

Thiede, J. (1977) Subsidence of aseismic ridges: Evidence from sediments on Rio Grande Rise (Southwest Atlantic Ocean). *Am. Assoc. Petrol. Geol. Bull.* 61:929–940.

Thomas, H., Roger, J., Sen, S., and Al-Sulaimani, Z. (1988) Decouverte des plus anciens "Anthropoides" du continent arabo-africain et d'un Primate tarsiiforme dans l'Oligocene du Sultanat d'Oman. *C. R. Acad. Sci. Ser.* 2 306:823–829.

Thorington, R. W. (1976) The systematics of New World monkeys. In First Inter-American Conference on Conservation and Utilization of American Nonhuman Primates in Biomedical Research. *Pan African Health Organization Science Publication* 317:8–18.

Tomblin, J. F. (1975) The Lesser Antilles and Aves ridge. In A. E. M. Nairn and F. G. Stehli (eds.): *The Ocean Basins and Margins*, Vol. 3: *The Gulf of Mexico and the Caribbean*. New York: Plenum, pp. 467–500.

Uchupi, E. (1989) The tectonic style of the Atlantic Mesozoic rift system. *J. Afr. Earth Sci.* 8:143–164.

Unternehr, P., Curie, D., Olivet, J. L., Goslin, J., and Beuzart, P. (1988) South Atlantic fits and intraplate boundaries in Africa and South America. *Tectonophysics* 155:169–179.

Vail, P. R., and Hartenbol, J. (1979) Sea level changes during the Tertiary. *Oceanus* 22:71–79.

van Andel, T. H., Thiede, J., Sclater, J. G., and Hay, W. W. (1977) Depositional history of the South Atlantic Ocean during the last 125 million years. *J. Geol.* 85:651–698.

van Couvering, J. A., and Harris, J. A. (1991) Late Eocene age of Fayum mammal faunas. *J. Hum. Evol.* 21:241–260.

Washburn, S. L., and Moore, R. (1980) *Ape into Human: A Study of Human Evolution*. Boston: Little, Brown.

Wikander, R., Covert, H. H., and Deblieux, D. D. (1986) Ontogenetic, intraspecific, and interspecific variation of the prehallux in primates: Implications for its utility in the assessment of phylogeny. *Am. J. Phys. Anthropol.* 70:513–524.

Wolff, R. (1984) New specimens of the primate *Branisella boliviana*. *J. Vert. Paleont.* 4:570–574.

Wood, A. E. (1972) An Eocene hystricognathous rodent from Texas: Its significance in interpretations of continental drift. *Science* 175:1250–1251.

Wood, A. E. (1974) The evolution of the Old World and New World hystricomorphs. *Symp. Zool. Soc. Lond.* 34:21–60.

Wood, A. E. (1980a) The Oligocene rodents of North America. *Trans. Am. Phil. Soc.* 70:1–68.

Wood, A. E. (1980b) The origin of the caviomorph rodents from a source in Middle America: A clue to the area of origin of the platyrrhine primates. In R. L. Ciochon and A. B. Chiarelli (eds.): *Evolutionary Biology of the New World Monkeys and Continental Drift*. New York: Plenum, pp. 79–92.

Wood, A. E. (1983) The radiation of the Order Rodentia in the southern continents: The dates, numbers and sources of the invasions. *Scriftenr. Geol. Wiss.* 19/20S:381–394.

Wood, A. E., and Patterson, B. (1970) Relationships among hystricognathous and hystricomorphous rodents. *Mammalia* 34:628–637.

Wood, A. E., and Patterson, B. (1982) Rodents from the Deseadan Oligocene of Bolivia and the relationships of the Caviomorpha. *Bull. Mus. Comp. Zool.* (Harvard) 149:371–543.

Woods, C. A. (1972) Comparative myology of jaw, hyoid, and pectoral appendicular regions of New and Old World hystricomorph rodents. *Bull. Am. Mus. Nat. Hist.* 147:119–198.

Woods, C. A. (1982) The history and classification of South American hystricognath rodents: Reflections on the far away and long ago. In M. A. Mares and H. H. Genoways (eds.): *Mammalian Biology in South America. Special Publication Series, Pymatuning Laboratory of Ecology* 6:377–392.

CHAPTER 5

Maboko Island and the Evolutionary History of Old World Monkeys and Apes

Monte L. McCrossin
Brenda R. Benefit

INTRODUCTION

The transition from rainforest to woodland and grassland biomes profoundly influenced the evolutionary history of Old World monkeys and apes (Leakey 1967; Napier 1970; Bishop 1971; Pickford 1983). In eastern Africa this transition began toward the end of the early Miocene and was fully under way by the start of the middle Miocene (Andrews and Van Couvering 1975; Van Couvering and Van Couvering 1976; Pickford 1981). Maboko Island is a critical deposit for reconstructing the emergence of cercopithecoids and hominoids during this interval (Benefit and McCrossin 1989a). Middle Miocene deposits at Maboko preserve the greatest abundance of specimens of the primitive cercopithecoid *Victoriapithecus macinnesi* (Von Koenigswald 1969; Delson 1975a,b; Benefit 1985, 1987, 1992, 1993; Benefit and Pickford 1986; Senut 1986; Harrison 1989a; Benefit and McCrossin 1991, 1993a,b; McCrossin and Benefit 1992). Maboko is perhaps best known for *Kenyapithecus africanus*, a candidate for the earliest known ancestor of great apes and humans (Le Gros Clark and Leakey 1950, 1951; Leakey 1967; Pickford 1985; Benefit and McCrossin 1993a; McCrossin and Benefit 1993a,b). These primates represent a transitional stage between early undifferentiated catarrhines and extant Old World monkeys and apes. Their bones and teeth permit direct functional interpretation of the locomotor and dietary habits of ancestral apes and monkeys. Other noncercopithecoid catarrhines from Maboko are much less well understood. They include two "small-bodied apes" (cf. *Limnopithecus evansi* and *Simiolus leakeyorum*; Harrison 1989b; Benefit 1991), and two enigmatic forms that may be oreopithecids, *Mabokopithecus clarki* and *Nyanzapithecus pickfordi* (Von Koenigswald 1969; Harrison 1986; McCrossin 1992a). Also present at Maboko are a large bushbaby, *Komba winamensis* (McCrossin 1992b), and another, as yet undescribed, galagine species (McCrossin 1990).

Explanations of the early history and diversification of Old World monkeys and apes have largely drawn inferences from contrasting aspects of the size, diet, and locomotion of living species as well as from the fossil record (Pilbeam and Walker 1968; Napier 1970; Simons 1970; Ripley 1979; Temerin and

Cant 1983). Features unique to each superfamily are thought to have evolved as ancestral cercopithecoids and hominoids responded differently to climatic and vegetational changes.

Attempts to reconstruct the independent histories of Cercopithecoidea and Hominoidea have typically focused on changes in single adaptive complexes within each superfamily. Studies of Old World monkey evolution tend to focus on the evolution of bilophodont teeth. Jolly (1967), Napier (1970), and Simons (1970) emphasized a shift toward specialized folivory among early cercopithecoids, based on assumptions concerning the functional role of bilophodont teeth. Monkeys were thought to have been better adapted to the changing vegetational environment than apes because they could exploit leaves as an abundant food source when fruits became less available (Napier 1970). Delson (1975a; Szalay and Delson 1979) stressed the possible role of facultative folivory, especially during seasonal food shortages, on the evolution of cercopithecoid bilophodonty. Following this line of reasoning Andrews (1981; Andrews and Aiello 1984) suggested that facultative folivory in early monkeys was coupled with the biochemical evolution of enzymes enabling the earliest monkeys to tolerate plant toxins that commonly occur in unripe fruits as well as in the leaves of rainforest trees. These interpretations are based on the assumption that early Miocene apes were committed frugivores and that only cercopithecoid monkeys were eclectic in their dietary habits but that "the divergence of the cercopithecoid monkeys . . . is associated with a tendency of dietary change towards *arboreal folivory*" (Andrews 1981, our emphasis).

In contrast, reconstructions of hominoid evolution tend to concentrate on postcranial adaptations. A suspensory pattern of motion and foraging is widely inferred to have been engaged in by the last common ancestor of living hominoids (Keith 1903; Gregory 1928; Tuttle 1969; Hunt 1991). Ripley (1979) suggested that the below-branch feeding postures of apes evolved as a consequence of competition with the above-branch feeding strategies of a burgeoning adaptive radiation of Old World monkeys. The suspensory capabilities of hominoids have also been regarded as a means for apes to travel greater distances per day than quadrupedal Old World monkeys (McHenry and Corruccini 1983; Temerin and Cant 1983).

With few exceptions (e.g., Lewis 1971, 1972; Conroy and Fleagle 1972; Simons and Pilbeam 1972; Zwell and Conroy 1973), most workers have argued that African early Miocene noncercopithecoid catarrhines (creatures traditionally referred to the Superfamily Hominoidea, including *Proconsul*, *Dendropithecus*, *Afropithecus*, *Turkanapithecus*, and *Simiolus*) lack the suite of structural modifications of the forelimb and vertebral column that are characteristic of extant apes (Le Gros Clark and Leakey 1951; Le Gros Clark and Thomas 1951; Napier and Davis 1959; Morbeck 1975, 1976, 1977; O'Connor 1975, 1976; Corruccini et al. 1976; Walker and Pickford 1983; Beard et al. 1986; Gebo et al. 1988, Leakey et al. 1988a,b; Ward et al. 1991; Rose et al. 1992). Evidence from comparative anatomy and the fossil record eventually brought the combination of an "ape-like" dentition and a "monkey-like" postcranial skeleton to be regarded as representative of the ancestral condition for Old World higher primates (Washburn 1950; Sarich 1971; Corruccini et al. 1976; McHenry and Corruccini 1983).

Although manifestly lacking derived features of their molar pattern, African early Miocene noncercopithecoid catarrhines have been considered to occupy an adaptive zone like that of living Old World monkeys (Pilbeam and Walker 1968; Napier 1970; Simons 1970; Fleagle 1978; Andrews 1981). This interpretation involves the notion that aspects of the natural history of African early Miocene noncercopithecoid catarrhines emulate those of living Old World monkeys, including their inferred mode of locomotion (Le Gros Clark and Thomas 1951; Napier and Davis 1959; Morbeck 1975), their dietary adaptations (Kay 1977) and their range of body size (Fleagle 1978). An immediate consequence of this model was the idea that direct competition for food between the first cercopithecoids and African early Miocene noncercopithecoid catarrhines was resolved by a reduction in the abundance and diversity of the latter group, beginning in the middle Miocene (Andrews 1981).

Knowledge of the timing and sequence in which distinctive attributes of each superfamily were acquired is essential for understanding the emergence and diversification of cercopithecoids and hominoids. Ultimately, this knowledge must be based upon reliable assessments of the taxonomic identity and phylogenetic relationships of fossil taxa, reconstruction of lifeways from functional analysis of bones and teeth, and reconstruction of the paleoecological context of the extinct primates from geological and paleontological evidence. Such knowledge is dependent upon a rich and fairly continuous fossil record, which is seldom available.

In the past, the interpretive potential of the Maboko primate fauna was severely limited by the fragmentary nature of the fossils and by poorly understood and disputed notions of their provenience. Since 1987 we have recovered more than 1000 primate fossils at Maboko, primarily from *in situ* excavation of Beds 3 and 5 of the Maboko Formation. These fossils, which include the most complete remains of *Victoriapithecus* and *Kenyapithecus* known to date, contribute greatly to our understanding of the phylogenetic relationships, adaptations, and environmental context of early cercopithecoids and hominoids (Benefit 1991, 1993; Benefit and McCrossin 1991, 1993a, 1993b; Feibel and Brown 1991; McCrossin and Benefit 1992, 1993a,b; McCrossin 1992a). In this chapter, we review recent advances in our understanding of *V. macinnesi* and *K. africanus*. New perspectives on the evolutionary history of Old World monkeys and apes gained from this material are discussed.

HISTORY OF INVESTIGATIONS

Maboko is a small island (approximately 1 by 2 km) located on the northern side of the Winam Gulf of Lake Victoria, western Kenya. Fossils collected at Maboko by W.O. Owen and D.G. MacInnes in 1933–1934 included the remains of Old World monkeys and apes (MacInnes 1943). Important primate specimens were also quarried at Maboko Island by the British–Kenya Miocene Expedition, led by L. S. B. Leakey and D. M. S. Watson, in 1947 and 1949 (Le Gros Clark 1950, 1952). Evidence of their quarry is visible at the locality of Maboko Main today (Pickford n.d. a. b. c). In 1973 the Yale–Kenya Expedition organized by D. R. Pilbeam and P. Andrews placed small and relatively unproductive excavation units in an unspecified area in the vicinity of the trench dug by the British–Kenya Miocene Expedition (Andrews et al. 1981). From 1982 to 1984 fieldwork by M. Pickford led to the discovery of numerous primate specimens in backfill to the south of the British–Kenya Miocene Expedition trench. These discoveries helped to reestablish the importance and potential of the deposits. In an effort to recover more informative primate remains, we initiated an ongoing program of excavations on Maboko Island in 1987.

GEOLOGY AND AGE OF THE DEPOSITS

The Maboko Formation lies on a foundation of Precambrian granites and metavolcanics of Nyanzian/Kavirondan age and is capped by Ombo Phonolites (Kent 1944; Shackleton 1951; Pickford 1981; Andrews et al. 1981; Mboya 1983). The fossiliferous succession at Maboko Island represents floodplain regimes of sedimentary deposition (Shackleton 1951; Saggerson 1952; Andrews et al. 1981). Manifestations of fluviatile agents (cut-and-fill structures, conglomeratic facies, fining-upwards sequences, and graded/cross bedding) are present.

The sedimentary strata on Maboko Island are broadly divisible into two lithologic intervals, a basal fine-grained sequence and an overlying coarser interval (Feibel and Brown 1991). The lower part of the section is dominated by claystones that appear to derive in part from altered volcanic ash as indicated by the presence of nephelinitic, carbonatitic, and fenetic agglomerates. From time to time, surfaces were subaerially exposed, occasionally leading to the development of calcretes and the initiation of localized pedogenesis (Pickford 1982, 1984, 1986b; Feibel and Brown 1991). Vertebrate fossils have been found in 15 of the 20 sedimentary units recognized by Pickford (1984, 1986b) but are largely restricted to 2 major concentrations, Beds 3 and 5, in the lower portion of the 54 m aggregate section and are more widely scattered throughout the upper portion. A formal description of the Maboko sediments is being prepared by C. S. Feibel.

Radiometric age determinations by Feibel and Brown (1991) indicate that strata between the capping phonolite of Bed 20 and a tuff in Bed 8 are approximately 13.8–14.7 million years old while sediments underlying Bed 8 were deposited some time before approximately 14.7 million years ago. These determinations corroborate the long-held belief, based mainly on faunal comparisons, that assemblages from Beds 3 and 5 at Maboko Main are older than those from the c. 12.5–14.0 MYA site of Fort Ternan (Bishop et al. 1969; Shipman et al. 1981). By the same token, however, the dates on Beds 8 and 20 show that the upper part of the Maboko sequence may be coeval with Fort Ternan. Primate fossils from the upper part of the section are as of yet ill-known, including an intermediate phalanx and isolated teeth of a large-bodied hominoid most similar to *K. wickeri* from Fort Ternan. On the basis of faunal comparisons with radiometrically dated early Miocene sites west of Lake Turkana, it seems that the Maboko Main fauna is younger than 16.8 MYA, the age of the Naserte Tuff of the Lothidok Formation (Boschetto et al. 1992).

VICTORIAPITHECUS AND OLD WORLD MONKEY ORIGINS

Systematics

Cercopithecoid fossils from Maboko provide the best evidence of the morphology and adaptations of Old World monkeys during early stages of their evolutionary history. The abundance of cercopithecoid remains at Maboko (more than 1000 specimens) lies in stark contrast to the total sample (fewer than 40 specimens) of Old World monkey fossils known from all other early and middle Miocene deposits. The earliest true cercopithecoids occur at the c. 19 MYA site of Napak V in Uganda, from which a canine, M^2, and partial radius and ulna are known (Pilbeam and Walker 1968; Pickford et al. 1986). In eastern Africa, 16 craniodental specimens and a pisiform attributed to *Prohylobates* (Leakey 1985) are known from the c. 17 MYA site of Buluk (McDougall and Watkins 1985), a small number of isolated teeth have been recovered from the 16–18 MYA site of Loperot (Baker et al. 1971; Szalay and Delson 1979), an M_3 is known from Ombo; a distal humerus was discovered at Nyakach (Pickford and Senut 1988); and an unspecified but apparently small number of isolated teeth were recently recovered from Nachola (pers. comm., H. Ishida 1992). An isolated M_3 from Ongoliba, Zaire, may also be middle Miocene in age (Hooijer 1963, 1970). In addition, a total of four cercopithecoid partial mandibles with associated teeth, representing two species of *Prohylobates*, have been recovered from the North African sites of Wadi Moghara in Egypt (Fourtau 1918; Simons 1969) and Gebel Zelten in Libya (Delson 1979). Old World monkey fossils are conspicuously absent from several other well-known Miocene sites at which apes are abundant, such as Songhor, Koru, Rusinga and Fort Ternan.

The precise relationship of Maboko cercopithecoids to extant Old World monkeys is the subject of debate. Maboko cercopithecoid fossils were originally attributed to *Mesopithecus* (MacInnes 1943), a colobine monkey best known from the late Miocene site of Pikermi in Greece (Wagner 1839; Von Beyrich 1861; Gaudry 1862). Von Koenigswald (1969) erected the current genus *Victoriapithecus*. Due to the presence of features in *Victoriapithecus* that do not occur among extant Old World monkeys, Von Koenigswald (1969) referred the genus to a new subfamily, Victoriapithecinae, in order to distinguish it from Colobinae and Cercopithecinae. He suggested that the Maboko monkeys were intermediate in morphology between apes and Old World monkeys. The *Victoriapithecus* fossils were divided into two species, *V. macinnesi*, a small form with a weak or absent crista obliqua on the upper molars, and *V. leakeyi*, characterized by more quadrangular upper molars with a distinct crista obliqua (Von Koenigswald 1969).

Von Koenigswald's (1969) taxonomic and phylogenetic interpretations were challenged by Delson (1973, 1975a,b; Simons and Delson 1978; Szalay and Delson 1979). Delson emended the definitions of both species and suggested that they could be sorted into two genera, with *V. macinnesi* being at or near the ancestry of Colobinae and "*V.*" *leakeyi* being ancestral to Cercopithecinae (1973, 1975a,b; Simons and Delson 1978; Szalay and Delson 1979). *V. macinnesi* was said to have postcranial adaptations for a fully arboreal lifestyle, upper molars that lack a crista obliqua, and squarish lower molars that combine ancestrally shallow notches and bulging cingulum remnants with apparently colobine-like mesiodistally short trigonids (Delson 1973, 1975a,b). The other species, "*V.*" *leakeyi*, was considered to be more conservative dentally, with mandibular molars that are longer (relative to their width) and have mesiodistally longer trigonids and talonid basins, and upper molars (particularly M^1 and M^2) that possess a crista obliqua (Delson 1975a; Szalay and Delson, 1979). Limb bones with more terrestrial adaptations were referred to "*V.*" *leakeyi* (Delson 1973, 1975a,b; Simons and Delson, 1978; Szalay and Delson, 1979).

Early assessments of the number of species and phylogenetic affinities of fossil cercopithecoids from Maboko were based on a sample of 59 permanent teeth, of which only the P4–M3 were found associated in partial mandibles, and 11 postcranial remains (two distal humeri, two proximal ulnae, one calcaneum, five phalanges, and one caudal vertebra) (Von Koenigswald, 1969; Delson 1973, 1975a,b). Since 1982 the cercopithecoid sample from Maboko has increased 20-fold. Many new skeletal elements are now known, including the following: frontal, zygomatic, maxilla, the full set of deciduous teeth, proximal humerus, radius, lunate, pisiform, trapezium, capitate, hamate, metacarpals, ilium, ischium, femur, patella, tibia, astragalus, entocuneiform, mesocuneiform, ectocuneiform, cuboid, and metatarsals (Harrison 1989b; Benefit and McCrossin 1989a; Benefit 1992; McCrossin and Benefit 1992). The number of specimens now exceeds 1200. Many of the new permanent and deciduous teeth are associated with maxillae and mandibles, and several of the postcrania collected since 1987 are complete, adding to their significance. The new material provides a more substantive basis from which to assess the number of species, systematic position, and adaptations of the Maboko cercopithecoids than was previously possible.

Analysis of the permanent dentition of *Victoriapithecus* indicates that criteria used by Von Koenigswald (1969) and Delson (1975a,b; Simons and Delson 1978; Szalay and Delson 1979) to sort Maboko cercopithecoids into two species no longer separate specimens into two distinct morphs (Benefit and Pickford 1986; Benefit 1993). Because only isolated upper molars were known before 1973, mistakes were made in identifying such teeth to element. Differences in the shape of the *Victoriapithecus* M^2 and M^3(the latter being constricted distally while the former are not) were not understood until maxillae with associated dentition were discovered (Benefit 1987, 1993). Upper molar shape differences between the two species, as described by Von Koenigswald (1969), are attributable to the misidentification of second and third molars. Correct assignment of isolated upper molars to element helps to explain the diverse expression of crista obliqua within the *Victoriapithecus* sample. The oblique crests are differentially expressed across the molar row, being more frequent on M^1 (91 percent) than M^2 (71 percent) and on M^2 than M^3 (33 percent) (Benefit and Pickford, 1986; Benefit, 1987, 1993). One maxilla, KNM-MB 11710, was discovered with a crista obliqua on M^2 but not on M^3 (Benefit 1993). Since quadrangular upper molars are also those that usually have crista obliqua, *V. leakeyi* as defined by Von Koenigswald (1969) corresponds to the morphology of M^1 and M^2, while *V. macinnesi* corresponds to the M^3. The evidence suggests that presence or absence of the trait cannot be used to sort specimens into two species. Differences in lower molar trigonid length perceived by Delson (1973, 1975a,b) between *V. leakeyi* and *V. macinnesi* have not been substantiated by objective measurements of trigonid length (Benefit and Pickford 1986; Benefit 1987, 1993). *Victoriapithecus* lower molars are characterized by a short trigonid as is observed in colobine but not cercopithecine monkeys.

Delson (1973, 1975a,b, 1979; Delson and Andrews 1975; Szalay and Delson 1979; Strasser and Delson 1987) claimed that a semiterrestrial and a more exclusively arboreal cercopithecoid species were represented at Maboko based on differences between distal humeri KNM-MB 3 and KNM-MB 19. Distal humerus KNM-MB 3 represents a smaller individual with a more medially oriented entepicondyle, a less projecting medial trochlear keel, a higher and deeper olecranon fossa, and a more strongly developed dorsal epitrochlear fossa than KNM-MB 19. Harrison (1989a) found that the range of morphological variation between the two specimens does not exceed that observed for 7 out of 14 anthropoid species he examined. Since size differences between the two specimens correspond to those found between unpublished Bed 3 and Bed 5 cercopithecoid postcrania, the two may derive from distinct populations rather than species. Analyses of cercopithecoid postcranial material collected from Maboko before 1987 have been claimed to support the existence of either one (Harrison 1989a) or two species (Senut 1986) but not two subfamilies or even genera.

Levels of metric variation within the *Victoriapithecus* dental and postcranial samples are consistent with the existence of one rather than two species at Maboko (Benefit 1987, 1993; Harrison 1989a). Of the *Victoriapithecus* permanent and deciduous teeth, only the M_3 has a higher coefficient of variation for length and width than the average cercopithecoid species. However, *Victoriapithecus* M_3 size variation is similar to that observed for *Macaca fascicularis*, *Macaca nemestrina*, and *Macaca mulatta*. As for *M. mulatta*, high levels of *Victoriapithecus* M_3 variation may result from sexual dimorphism. The mixing of specimens from different stratigraphic levels also contributes to the high levels of M_3 variation. Lower third molars from Bed 3 are significantly longer and wider mesially than those from Bed 5 (Benefit 1993). It is concluded that the Maboko cercopithecoid sample cannot be divided into two discrete morphs on the basis of size or morphology. All Maboko cercopithecoids are therefore attributed to *V. macinnesi*, which has page priority over *V. leakeyi*.

As originally recognized by Von Koenigswald (1969), the dentition of *Victoriapithecus* differs from that of extant Old World monkeys. The Miocene monkey retains many dental features that are primitive for catarrhines. For example, the majority of upper molars and dp^4s (89 percent) possess crista obliqua and lack distal lophs. Upper dp^3s of *V. macinnesi* differ from those of extant monkeys in being neither bilophodont nor possessing a true trigon or crista obliqua (Benefit 1992). Several *V. macinnesi* M_1s (86 percent), M_2s (25 percent) and dp_4s (88 percent) possess hypoconulids, and 30 percent of dp_4s lack a distal lophid. No unequivocally derived features are shared between *Victoriapithecus* and one or the other extant subfamily. However, to the exclusion of *Victoriapithecus*, Colobinae and Cercopithecinae share the following derived features: (1) complete loss of the crista obliqua on M^{1-3} and dp^{3-4}, (2) presence of a distal loph on M^{1-3} and dp^{3-4}; (3) complete loss of the hypoconulid on M_{1-2} and dp_4; (4) invariable presence of a M_{1-2} and dp_4 distal lophid, (5) a long axis of P_4 that is aligned with the molar row rather than oriented oblique to it; and (6) enlargement of M_1 (MacInnes 1943; Von Koenigswald 1969; Benefit 1985, 1987, 1993; Leakey 1985; Benefit and Pickford 1986; Harrison 1987). It is therefore concluded that *Victoriapithecus* belongs to a clade of Old World monkeys which diverged prior to the last common ancestor of Colobinae and Cercopithecinae and represents the sister-group to Cercopithecidae (Benefit 1985, 1987, 1993; Leakey 1985; Benefit and Pickford 1986). Benefit (1987) has elevated Victoriapithecinae to family rank (Victoriapithecidae).

Leakey (1985) included *Prohylobates* within the Victoriapithecinae because it shares an oblique P_4 orientation and occasional retention of lower molar hypoconulids and upper molar crista obliqua (known only for *P.* sp. from Buluk) with *Victoriapithecus*. The paucity of cercopithecoid fossils from sites other than Maboko and the fragmentary nature of specimens from northern Africa make it difficult to precisely determine the relationship between *Victoriapithecus* and the early and middle Miocene genus *Prohylobates*.

Ancestral Cercopithecoid Morphotypes and Adaptations

Differences of opinion exist as to reconstruction of ancestral cercopithecoid morphotypes and adaptations. One school of thought suggests that the last common ancestor of extant cercopithecoids should be reconstructed on the basis of features shared by living species only (Delson and Andrews 1975). However, it is difficult to determine whether certain craniofacial and dental configurations are primitive or derived because of their distribution among extant Old World monkeys. Although both extant cercopithecoid subfamilies are bilophodont, cusp relief is much greater and shear crests longer on leaf-eating colobine than fruit-eating cercopithecine molars. As a consequence colobines have an enhanced capacity for slicing foods relative to cercopithecines (Walker and Murray 1975). Cusp tips are set buccolingually and mesiodistally closer together on cercopithecine than colobine molars, greatly reducing the size of their central trigon/talonid basins and increasing the length of the mesial shelf (Benefit 1987, 1993). As a result occluding cusp tips come into direct contact on cercopithecine molars and wear flat rapidly. They continue to efficiently grind food between wide and flat enamel rims after they are worn. Colobine molars maintain their original cusp height and ability to slice foods, especially on the buccal side of lower molars, even after substantial amounts of dentine are exposed (Walker and Murray 1975). The difference in cusp proximity between cercopithecine and colobine monkeys is also reflected in their apparent degree of flare, with cercopithecine molars seeming to bulge from tip to cervix while sides of colobine molars are vertical (Benefit 1987, 1993).

Contrasting cranial morphologies of colobines and cercopithecines are associated with diet. The face of colobine monkeys tends to be short and orthognathic, while that of cercopithecines tends to be at least moderately long and prognathic. A greater distance from the margin of the nasal aperture to orbitale occurs among the Cercopithecinae (Benefit and McCrossin 1993b). These cranial differences are related to use of the incisors and production of occlusal forces. Fruit-eating monkeys have mesiodistally broad incisors that are used to bite into fruit and strip away their outer husks. Leaf-eaters use their incisors to nip off leaves but not to chew them. The large central incisors of cercopithecine species are positioned anterior to lateral ones. Roots of the incisors form the lateral walls of the inferior portion of the nasal aperture causing it to be V-shaped inferiorly and generally teardrop-shaped as opposed to being tall and oval as in colobines (Vogel 1966, 1968). Possibly related to the difference in nasal aperture shape, the nasal bones of cercopithecine monkeys are long and narrow while those of colobines are short and broad (Vogel 1966, 1968). The interorbital septum of cercopithecines is narrow, their face is taller below the orbits and the cheek region is deep relative to facial height (Verheyen, 1962; Vogel 1966, 1968).

Given the dichotomous distribution of dental and cranial features among Old World monkeys, determination of which configurations are primitive or derived is difficult without reference to a closely related sister group (Benefit and McCrossin 1991). Among extant primates, hominoids are the most appropriate outgroup for comparison to cercopithecoids. Although the bilophodont molars of cercopithecoids are derived relative to hominoids, apes share low cusp relief and a predominantly frugivorous diet with only the Cercopithecinae. Delson (1973, 1975a, 1979; Delson and Andrews 1975; Szalay and Delson 1979; Strasser and Delson 1987) suggested that the cercopithecine dentition was primitive for cercopithecoids. Nevertheless, most workers reconstruct the earliest monkeys as having included more leaves in their annual diet than apes and cercopithecine monkeys because of their bilophodont dentition (Jolly 1967; Vogel 1968; Napier 1970; Simons 1970; Delson 1975a,b; Delson and Andrews 1975; Szalay and Delson 1979; Andrews 1981; Andrews and Aiello 1984; Temerin and Cant 1983; Harrison 1989b). The assumption that bilophodonty evolved as an adaptation to folivory is based on analogy to other mammals with lophodont teeth which happen to be folivorous.

Determination of whether the facial morphology of colobines or cercopithecines is closer to the ancestral condition is made difficult by the diversity of craniofacial configurations among extant hominoids. Extant hylobatids are similar to colobine monkeys in having a short face with a short and orthognathous snout while pongids tend to have moderately long and prognathic snouts like those of cercopithecine monkeys. Vogel (1968) tried to resolve the problem by arguing that long-faced cercopithecine baboons are derived and must have evolved from a shorter-faced form. Rather than look toward macaques or guenons that have moderately long faces as a possible baboon ancestor, Vogel selected the short-snouted Colobinae. His argument was influenced by reconstruction of the earliest cercopithecoids as having been folivorous, and thus more similar to colobine monkeys in cranial morphology. Delson (1975a,b; Delson and Andrews 1975; Szalay and Delson 1979) and Harrison (1987) accept Vogel's interpretation.

Due to its phylogenetic position as the sister group to extant Old World monkeys, Victoriapithecidae is more useful for determining morphocline polarities within the Cercopithecoidea than are extant apes. It is reasonable to assume that features shared between *Victoriapithecus* and one or the other extant subfamily are primitive for the Cercopithecoidea. The molar teeth of *Victoriapithecus* and *Prohylobates* were not perfectly bilophodont. They exhibit lower cusp relief, more closely approximated cusp tips, smaller central basins, greater flare, and a lower shearing capacity than the majority of cercopithecine monkeys

(Benefit and Pickford 1986; Benefit 1987, 1993). The last common ancestor of Old World monkeys undoubtedly had a more cercopithecine-like dentition, as has been previously suggested, but it probably did not have a more folivorous diet than extant apes. *Victoriapithecus* has broad upper central incisors like frugivorous cercopithecine monkeys. Comparison of their M^2 shear quotients with those of extant monkeys (Benefit and McCrossin 1990) indicates that the annual diet of *Victoriapithecus* from Maboko consisted of 79 percent fruits and 7 percent leaves, while that of *Prohylobates* from Buluk included 74 percent, *P. tandyi* 84 percent and *P. simonsi* 75 percent fruits. No extant ape includes fewer leaves or more fruit in its annual diet than is reconstructed for the Victoriapithecidae. These results contradict popular scenarios as to the folivorous adaptations of early cercopithecoids.

Based on the fossil evidence, the evolution of cercopithecoid bilophodont molars is better attributed to an initial adaptation for the consumption of tough fruits and possibly seeds than to facultative folivory. Kay (1975, 1978) demonstrated that the transverse lophs of cercopithecoid molars act as guides leading to the accurate interlocking of cusps and basins during occlusion, rather than as shearing crests. Both Kay (1975, 1978) and Maier (1977) have shown that two new grinding facets are acquired by Old World monkeys due to the evolution of the distal loph/lophid. In addition, the bilophodont arrangement of cusps results in well-formed central basins bordered by large occlusal faces extending from the margins of the lophs. These basins might serve to hold fruits and seeds steady between lophs and to allow higher puncturing forces to be produced during occlusion (Benefit 1987). The presence of bowl-shaped depressions on the cusp tips of *Victoriapithecus* and *Prohylobates* molars indicates that such high forces were produced during mastication. Such an adaptation would have enabled Miocene monkeys to consume tougher fruits and seeds than thin-enamelled apes such as *Proconsul*, *Rangwapithecus*, *Limnopithecus*, and *Simiolus*, some of which have greater shearing capacity than the Victoriapithecidae (Kay 1977, Benefit 1993).

Victoriapithecus cranial remains differ from the predicted ancestral cercopithecoid morphotype based on consideration of extant Old World monkeys and apes (Vogel 1966, 1968; Delson 1975a,b; Delson and Andrews 1975; Szalay and Delson 1979; Harrison 1987; Benefit and McCrossin 1989b, 1991, 1993b). *Victoriapithecus* possesses a macaque-like face with a moderately long snout, deep malar region, and narrow interorbital septum, in combination with certain colobine-like features such as a high fronto-zygomatic suture (see Figure 5–1). We therefore interpret the facial features shared by *Victoriapithecus* and moderately long–faced cercopithecines such as *Macaca* and *Cercopithecus* to be primitive for Old World monkeys (Benefit and McCrossin 1991, 1993b). Support for our new interpretation comes from the observation that at least four of the six known fossil colobine genera, *Libypithecus*, *Dolichopithecus*, *Rhinocolobus*, and *Paracolobus*, plus one extant colobine, *Nasalis*, have moderately prognathic snouts and deep cheek regions like that of the revised morphotype. Three of these genera also have cercopithecine-like narrow interorbital septa. It is more parsimonious to assume that these colobine genera have retained the

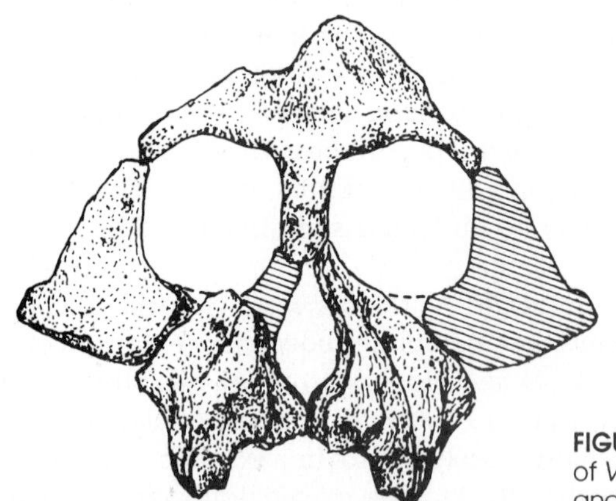

FIGURE 5-1 Facial reconstruction of *Victoriapithecus* (from Benefit and McCrossin 1991).

primitive cercopithecoid condition than that they independently evolved the supposedly derived configuration.

The revised cranial reconstruction for ancestral cercopithecoids calls into doubt current scenarios concerning the evolutionary history of the Catarrhini. According to the previous craniofacial morphotype, the gibbon-like facial morphology of *Pliopithecus* is considered to more closely represent the basal catarrhine condition than that of *Aegyptopithecus*, which is interpreted as being too long–snouted (Harrison 1987). However, the facial proportions of *Aegyptopithecus* and the Miocene hominoid *Afropithecus* closely resemble the revised ancestral cercopithecoid morphotype. It is therefore likely that those features shared by the three catarrhines (moderately long snout, greater distance from orbitale to lateral margin of the nasal aperture, anteriorly tapering premaxilla, tall facial height with relatively deep cheek regions, and long and narrow nasal bones) characterized the last common ancestor of catarrhines. These features are retained by the Cercopithecoidea. The only ancestral cercopithecoid facial feature that is derived relative to basal catarrhines is the narrow interorbital septum.

Postcranially *Victoriapithecus* is mainly semiterrestrial as first observed by Von Koenigswald (1969) and Simons (1972). The strongest indicators of its semiterrestrial habitus are its retroflected humeral entepicondyle and ulnar olecranon process as well as short phalanges (Delson 1975a). Additional material (Harrison 1989a; McCrossin and Benefit 1992) reinforces this interpretation and shows that Maboko cercopithecoids had a restricted range of hallucial abduction, ischial callosities and a tail, among other features. *Victoriapithecus* is quite small (c. 3–5 kg), in the lower part of the size range of vervet monkeys (Harrison, 1989a). Consequently, *Victoriapithecus* is one of the smallest terrestrially adapted anthropoids known.

KENYAPITHECUS

Historical Overview

Explanations of the origin of the great ape and human clade have traditionally drawn upon evidence from the comparative anatomy of living hominoids (Tuttle 1974). Keith (1903) argued that the earliest representatives of the great ape and human clade, creatures he called "giant primates" and later "troglodytians" (Keith 1923), were exclusively arboreal forms that diverged from hylobatids chiefly by means of an increase in body size. Morton (1924), however, thought that great apes arose from a group of small, gibbon-sized hominoids independent from the lineage leading to humans.

Since the discovery of *Proconsul africanus* and *Limnopithecus legetet* at Koru (Hopwood 1933a,b), theories of great ape and human origins have naturally attempted to integrate evidence from the fossil record. These explanations were often strongly influenced by interpretations of the phylogenetic relationships between Miocene and Recent genera. Until recently, Miocene apes were usually viewed as direct ancestors of extant hominoid genera. For example, *P. africanus* has been regarded as ancestral to *Pan troglodytes* (Pilbeam 1969; Conroy and Fleagle 1972; Simons and Pilbeam 1972; Walker and Teaford 1989), *Kenyapithecus wickeri* and "*Ramapithecus punjabicus*" were once viewed as precursors of *Australopithecus* and *Homo* (Lewis 1937; Leakey 1962, 1967; Simons 1961, 1964; Pilbeam 1966; Conroy and Pilbeam 1975; Simons and Pilbeam 1978; Kay and Simons 1983) and *Sivapithecus* continues to be referred to as an early representative of the lineage leading to *Pongo* (Andrews and Cronin 1982; Pilbeam 1982; Preuss 1982; Kelley and Pilbeam 1986; Martin 1986; Ward and Brown 1986).

Evolutionary scenarios deriving evidence from the fossil record have often been richly embellished by the inclusion of detailed speculation concerning the diet, locomotion and competitive interactions of extinct forms while stressing the role of adaptive shifts in the origin of new groups. For example, Simons and Pilbeam (1972:49) envisioned an ancestor–descendant transition from *P. africanus* to modern chimpanzees that involved becoming "relatively large-bodied, semi-terrestrial, predominantly frugivorous primates, possibly as a result of the radiation of the very successful omnivorous/frugivorous arboreal cercopithecines." Considering the fossil material they were based on, scenarios proposing "*Ramapithecus*" as a Miocene representative of Hominidae (Simons 1964; Pilbeam 1966; Andrews 1971; Andrews and Tekkaya 1976; Simons and Pilbeam 1978) were particularly speculative (Genet-Varcin 1969; Vogel 1975). Due to the presence of thick enamel in *Sivapithecus*, *Australopithecus*, and *Homo*, Smith and Pilbeam (1980) proposed that the thick enamel of orangutans might reflect a terrestrial stage in their ancestry. Kay (1981) demonstrated, however, that among primates there is no correlation between enamel thickness and substrate preference. According to Ward and Brown (1986:449), "enough postcranial evidence, involving features of the capitate bone, calcaneus and hallux, shows that medium to large *Sivapithecus* individuals . . . were not necessarily 'open country' or 'ground' apes, but were probably capable arborealists." Similarly, in his recent review of Miocene hominoid postcranial remains, Rose (1993) notes that "obvious morphological complexes that are indicators of the use of terrestrial locomotor modes" have not "been convincingly demonstrated" for *Sivapithecus* or any other Miocene hominoid.

In recent years, the emphasis in studies of hominoid paleobiology has largely shifted from efforts at reconstructing the natural history of extinct forms to a preoccupation with the family tree relationships of extinct taxa, primarily through the use of phylogenetic systematics. For example, in one study Martin (1986:Table 2) compiled a list of 16 derived features that *Kenyapithecus* shares with the reconstructed last common ancestor of living great apes and humans. If the transformation sequence of character states is correctly reconstructed, this renewed emphasis on systematics may allow a clearer appreciation of the relationships of Miocene noncercopithecoid catarrhines. Recent discoveries and analyses have clearly confirmed, for example, that *Proconsul* diverged prior to the last common ancestor of extant hominoids (Fleagle 1983; McHenry and Corruccini 1983; Beard et al. 1986; Ward et al. 1991). Most cladistic analyses of large-bodied hominoids fail, however, to take into account ranges of variation present in living and extinct taxa and to allow for the possibility of parallel evolution. Even more disturbing, few of these analyses propose a broad, adaptive context for why diagnostic features arose, indeed, for why the great ape and human clade originated.

The Phylogenetic Position of *Kenyapithecus*

The last few years have witnessed a greatly accelerated pace of discovery of Miocene large-bodied hominoids. Important recent finds include new genera, such as *Afropithecus* (Leakey and Leakey 1986; Leakey et al. 1988a) and *Otavipithecus* (Conroy et al. 1992). Previously known species for which substantially more complete specimens have been discovered in recent years include *P. africanus* and *P. nyanzae* (Walker and Teaford 1988), *K. africanus* (Benefit and McCrossin 1989, 1993a; McCrossin and Benefit 1993a,b), *Sivapithecus sivalensis* including all large hominoid specimens from the Chinji, Nagri, and Dhok Pathan Formations, except *Gigantopithecus giganteus*, some of which have been referred to a plethora of junior synonyms such as *S. indicus* and *S. aiyengari*; (Dehm 1983:Fig. 2A; Pilbeam et al. 1990) and *Ouranopithecus macedoniensis* (de Bonis et al. 1990). These discoveries have resulted in a greatly expanded information base for interpreting relationships between extinct and extant large-bodied apes.

In this section we focus on the phylogenetic affinities and adaptations of the Miocene ape *Kenyapithecus*, a form generally recognized as being among the earliest representatives of the great ape and human clade (Greenfield 1980; Kelley and Pilbeam 1986; Martin 1986; Ward and Brown 1986; Andrews and Martin 1987). Special attention is paid to the phylogenetic status of *K. africanus* from the Maboko Formation.

K. wickeri was described on the basis of a left maxilla with P^4–M^2 (KNM-FT 46a), a right maxilla with M^{1-2} (KNM-FT 47), a LC^1 (KNM-FT 46b) and a RM_2 (KNM-FT 48) from Fort Ternan (Leakey 1962). The type-specimen of *K. africanus*, a left maxilla with P^3-M^1 (BMNH M 16649), was originally attributed to *Proconsul africanus* (MacInnes 1943). Le Gros Clark and Leakey (1950, 1951) noted, however, that this specimen shares relatively large premolars and reduction of molar cingula with material from the Siwaliks, particularly GSI D-1, GSI D-196, and YPM 13799 (Lydekker 1879; Pilgrim 1915, 1927; Gregory et al. 1938) and erected the taxon *Sivapithecus africanus* with BMNH M 16649 as the type-specimen. After the recognition of similar features in the maxillary material from Fort Ternan and Maboko, Leakey (1967) transferred *S. africanus* to *Kenyapithecus.*

For the next two decades, the majority of researchers accepted a hominid affinity for *K. wickeri* (Leakey 1962) but regarded it as either completely synonymous (Simons 1963; Le Gros Clark 1964) or congeneric (Simons 1964; Pilbeam 1966, 1969; Andrews 1971; Walker and Andrews 1973; Andrews and Walker 1976; Simons and Pilbeam 1978) with "*Ramapithecus punjabicus*". Anticipated similarities of *Kenyapithecus wickeri* and "*Ramapithecus punjabicus*" to *Australopithecus* and *Homo* were often based on their reconstructed, as opposed to known, morphologies. Features for which little or no fossil evidence existed, but which were said to be shared by *Kenyapithecus wickeri* and "*Ramapithecus punjabicus*" included: small incisors and canines relative to cheek tooth size, reduced incisor procumbency, arched palate, arcuate tooth row and short rostrum relative to other Miocene species (Simons 1964; Pilbeam 1969; Yulish 1970; Conroy 1972; Simons and Pilbeam 1972, 1978). At the same time, some authorities preferred to retain BMNH M 16649 in *Sivapithecus,* as *S. africanus* (Madden 1980; Kay and Simons 1983) or *Dryopithecus (Sivapithecus) sivalensis* (Simons and Pilbeam 1965; Pilbeam 1969).

The discovery of more complete mandibular and craniofacial fossils (Pilbeam 1982) led to abandonment of the "*Ramapithecus*" chimera and to recognition that Asian representatives were synonymous with *Sivapithecus* (Greenfield 1980). Several craniofacial similarities between *Sivapithecus* and *Pongo,* mainly features of the frontal and subnasal regions that are unlike those of known African fossil hominoids, *Gorilla* and *Pan* (Andrews and Tekkaya 1980), suggested that *Sivapithecus* was the earliest member of the orangutan clade (Andrews and Cronin 1982; Pilbeam 1982; Preuss 1982; Ward and Pilbeam 1983).

Following discovery of craniofacial material of *Sivapithecus* and of additional large-bodied hominoid fossils from the Maboko Formation, Pickford (1982, 1985) resuscitated the genus *Kenyapithecus.* Questions quickly arose, however, as to whether *Kenyapithecus* was ancestral to both Asian and African great apes (Martin 1986), whether *K. wickeri* is specially allied to *Sivapithecus* and *Pongo* (Pickford 1985), or whether the genus is ancestral to only the *Gorilla, Pan,* and *Homo* clade (Brown and Ward 1988).

Recently, material from the middle Miocene site of Nachola has been attributed to *K. africanus* (Ishida et al. 1984). The Nachola ape is alleged to possess a subnasal pattern similar to that of *Gorilla* in which the incisive foramen is constricted and the premaxilla is obliquely oriented relative to the palatal process (Pickford 1986; Brown and Ward 1988). *Sivapithecus* and *Pongo,* in contrast, share a subnasal pattern in which the palatal process intersects the premaxilla at a shallow angle (Ward and Pilbeam 1983). Consequently, *Kenyapithecus* has been interpreted to be divergent from *Sivapithecus* and *Pongo* but to conform to the morphological pattern expected for the last common ancestor of *Gorilla, Pan,* and *Homo* (Brown and Ward 1988). However, a definitive assessment of the generic affinities of the Nachola material must await a more detailed description of the fossils, as well as knowledge of the subnasal pattern of large-bodied hominoids from Maboko and Fort Ternan. Differences in the occlusal pattern of the molar teeth (Ishida et al. 1984) and in the morphology of the mandibular symphysis may indicate that the large-bodied hominoids from Nachola belong to a different species or even genus than *K. africanus.*

Prior to this time, comparison of *Kenyapithecus* to other Miocene apes had been greatly limited by the available fossil evidence. Only 16 specimens of *K. wickeri* are known (Pickford 1985): 5 jaw fragments, 10 isolated teeth and 1 postcranial specimen (a distal humerus). The fragmentary nature of material from Fort Ternan and the absence of knowledge concerning many elements (notably, the face and subnasal regions, the deciduous dentition, crowns of P^3, M^3, and I_{1-2}, and most of the postcranial skeleton) make determination of the precise phylogenetic affinities of *K. wickeri* difficult.

The incomplete nature and disputed provenience of *K. africanus* fossils collected prior to the 1980s contributed to lingering uncertainty concerning its phylogenetic relationships, adaptations, and geographic distribution. In his original description, MacInnes (1943) did not specify the provenience of BMNH M 16649, the maxilla that was to become the type specimen of *K. africanus.* A potential for uncertainty concerning the maxilla's provenience is perfectly understandable if one acknowledges the fact that MacInnes's appendix burst while he was collecting fossils from islands in the Winam Gulf of Lake Victoria (including

Rusinga and Maboko). Nevertheless, the *K. africanus* type specimen was later stated to have been found on Rusinga Island (Le Gros Clark and Leakey 1950, 1951). Since then various lines of evidence, including elemental spectra of matrix attached to the specimen, indicate that BMNH M 16649 may have been collected from deposits on Maboko Island (Andrews and Molleson 1979). Perhaps as a result of confusion concerning the provenience of BMNH M 16649, jaws and teeth of *Proconsul nyanzae* and *P. major* from Songhor and Rusinga were attributed to *K. africanus* (Leakey 1967). Consequently, features uncharacteristic of *Kenyapithecus*, such as a slender mandibular corpus, strong superior transverse torus, and retention of beaded molar cingula, became associated with this species. In the wake of this confusion, some researchers erroneously attributed BMNH M 16649 to *P. nyanzae* (Andrews 1978; Simons et al. 1978; Szalay and Delson 1979; Conroy 1990).

Pickford (1982, 1985) revised the hypodigm of *K. africanus* based on seven new specimens from Majiwa and Kaloma and 35 additional isolated teeth collected on Maboko Island. All early Miocene material from Rusinga and Songhor was removed from the original hypodigm described by Leakey (1967), except three isolated teeth that may have been collected at Maboko but were described as coming from Rusinga (Pilbeam 1969; Simons and Pilbeam 1972; Simons et al. 1978; Pickford 1985).

While the majority of workers now attribute the Maboko and Fort Ternan material to *K. africanus* and *K. wickeri*, respectively, some would place samples from both localities into a single species (Greenfield 1979; Kay and Simons 1983; Andrews and Martin 1987), and others regard them as potentially belonging to different genera, with the Maboko species representing the more primitive of the two (Pilbeam 1969; Pickford 1985).

Suggested differences between *K. wickeri* and *K. africanus* include: (1) the development of lingual relief on maxillary central incisors, (2) the number of roots on P^3, (3) the depth of the palate, (4) the position and orientation of the anterior root of the zygomatic arch, and (5) the development of cingula on molars (Leakey 1967; Pilbeam 1969; Pickford 1985, 1986). We would add to this list the fact that the P^4 of *K. africanus* is substantially larger, relative to M^1, than in *K. wickeri*. Also, there is some indication that the mandibular molars of *K. africanus* are broader (relative to their length) than in *K. wickeri*. Because these differences are chiefly based on the maxillary type specimens, the extent of morphological variation within each species is poorly documented. Contrary to Pickford (1985, 1986a), however, P^3 appears to have possessed one lingual and two buccal roots in both *K. wickeri* (KNM-FT 46a) and *K. africanus* (BMNH M 16649) (Le Gros Clark and Leakey 1951; Leakey 1967; Pilbeam 1969; Andrews and Walker 1976). Moreover, the range of variation in cingulum development of the maxillary and mandibular molars from Maboko Island easily encompasses that seen in the much smaller sample from Fort Ternan. Pending the discovery of additional maxillary material, however, the observed differences of palate depth, zygomatic arch disposition, and premolar enlargement probably warrant treatment of the Fort Ternan and Maboko Island as two different species (Leakey 1967).

The Juvenile Mandible

Further progress toward deciphering the relationships of *Kenyapithecus* comes from the discovery of an additional 50 craniodental and 22 postcranial remains from Maboko Island since 1987. The sample of *Kenyapithecus* from Maboko Island now totals 116 specimens. The majority of the new fossils were collected from Beds 3 and 5 at Maboko Main while others derive from Bed 12 at Maboko South. One of the most significant of these finds is a nearly complete juvenile mandible (KNM-MB 20573: Figure 5–2), preserving dp_3-M_1 on both sides as well as the left permanent lateral incisor and second molar (McCrossin and Benefit 1993a). This mandible preserves the first intact and undistorted symphysis and corpus known for *K. africanus*. Comparison of *K. africanus* with *K. wickeri* and other Miocene large-bodied hominoids can now be based on a wider range of skeletal and dental elements.

As was first noted by Le Gros Clark and Leakey (1951), BMNH M 16649 shares mesiodistal elongation of the premolars and reduction of molar lingual cingula with *Sivapithecus sivalensis* and in these ways may

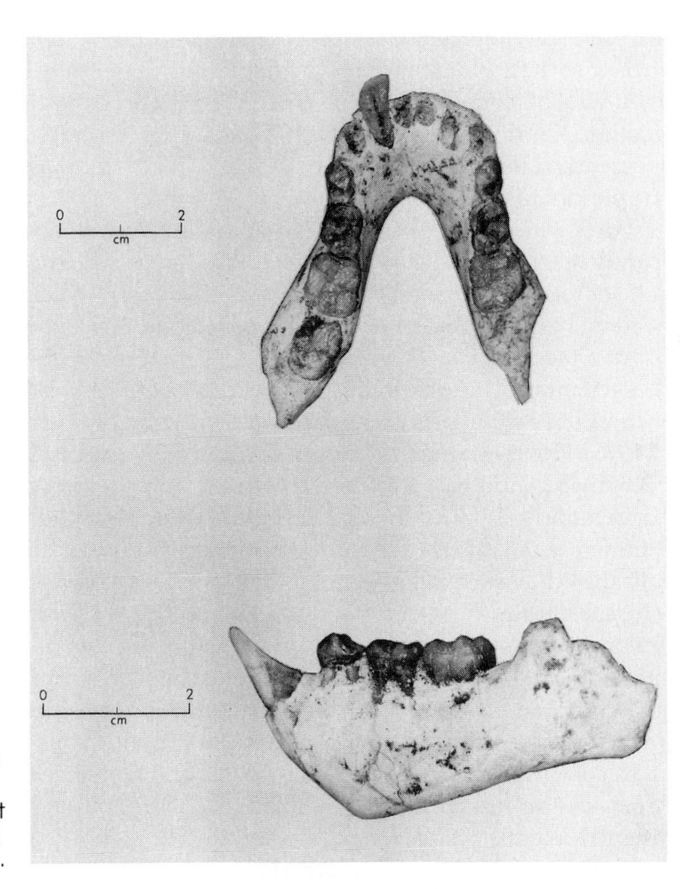

FIGURE 5-2 Juvenile mandible of *Kenyapithecus africanus* in occlusal (above) and left lateral (below) views (from McCrossin and Benefit 1993a).

be distinguished from *Proconsul*. The buccal cingula of *K. africanus* mandibular deciduous premolars and permanent molars are also markedly reduced with respect to homologues of *Griphopithecus darwini* from Candir and Pasalar, Turkey (Martin and Andrews 1991), *Dryopithecus fontani* from Saint Gaudens, France, and Lerida, Spain (Lartet 1856; Harle 1892; Smith-Woodward 1914), and *D. brancoi* from Salmendingen, Germany, and Rudabanya, Hungary (Schlosser 1901; Gregory and Hellman 1926; Kretzoi 1975; Kay and Simons 1983; Kelley and Pilbeam 1986; Andrews and Martin 1987; McCrossin and Benefit 1993a). Unfortunately, an inability to perceive differences in molar cingulum development and to distinguish *Kenyapithecus* from *Proconsul* (Andrews 1978) resulted in erroneous attribution of the Candir mandible (Andrews and Tekkaya 1976) and some of the Pasalar specimens to "*Ramapithecus wickeri*" (Andrews and Tobien 1977) and, more recently, to *K. africanus* (Alpagut et al. 1990)

The overall dp3 morphology of KNM-MB 20573 is very similar to that seen in *Pan troglodytes*. The crown is oval, dominated by a protoconid that is positioned at approximately one-third of the total mesiodistal diameter of the crown from the mesial margin, possesses a distinct metaconid that is closely apposed against the lingual face of the protoconid, and has a low and weakly developed hypoconid as well as a minute entoconid bordering the simple talonid basin (McCrossin and Benefit 1993a). *K. africanus*, *S. sivalensis* (GSP 11536) and extant great apes differ from *Proconsul* (KNM-MO 26, KNM-SO 541, KNM-ME 1) in having a mesiodistally shorter and less steeply inclined preprotocristid, as well as a reduced buccal extension of enamel onto the mesial root (McCrossin and Benefit 1993a). *Sivapithecus* and *Pongo* share a distinctive configuration of the mesial fovea, with greater development of the premetacristid. In addition, the dp3 of *S. sivalensis* is more strongly molariform than that of other hominoids (fossil and modern).

The unworn permanent lateral incisor morphology of KNM-MB 20573 similarly contributes to an understanding of the phylogenetic position of *K. africanus*. Based on their examination of the incomplete and vacant incisor sockets of a mandible fragment with P3-4 from Fort Ternan (KNM-FT 45; Andrews 1971), Andrews and Walker (1976:291) speculated that "*R. wickeri* could not have had large incisors." Based on this suggestion, Simons and Pilbeam (1978:149, 152) confidently concluded that "*Ramapithecus*", including specimens from Maboko and Fort Ternan, "had remarkably small lower incisors" that "were relatively unimportant in food preparation."

The unworn *K. africanus* I_2 is quite large, however, and differs from those of extant great apes in being significantly taller, mesiodistally narrower (relative to M_1 length), and in lacking a strongly flared distal margin (McCrossin and Benefit 1993a,b). The labiolingual thickness of the *K. africanus* I_2 (relative to M_1 length) is comparable to that of *P. nyanzae*, *Pongo* and *Pan* but substantially greater than that of *P. africanus*, *Hylobates*, and *Gorilla*. The most distinctive feature of the KNM-MB 20573 I_2 is its strongly procumbent orientation. The possibility of a procumbent lower incisor orientation was tentatively suggested for *K. wickeri* on the basis of the partial alveoli preserved on KNM-FT 45 (Andrews 1971; Walker and Andrews 1973). However, the evidence was inconclusive due to initial misidentification of the incisor sockets of KNM-FT 45 (Andrews 1971) and ignorance of whether the incisor roots were straight or strongly curved anteroposteriorly (Walker and Andrews 1973). Based on the new *K. africanus* mandible, it is now possible to confidently regard strongly procumbent orientation of the lower incisors as a derived feature that is uniquely shared by specimens from Maboko and Fort Ternan and that distinguishes *Kenyapithecus* from other Miocene hominoid genera.

Mandibular morphology, especially of the symphysis, is extremely important for differentiating Miocene hominoids (contra Simons 1972). Early Miocene apes, including *Proconsul*, differ from extant hominoids in lacking an inferior transverse torus and possessing a large superior transverse torus, inferiorly directed genioglossal pit, as well as a tall and thin corpus (MacInnes 1943; Le Gros Clark and Leakey 1951). In contrast, *Kenyapithecus*, *Sivapithecus*, and *Ouranopithecus* exhibit a pronounced inferior transverse torus and a robust corpus that is relatively short and wide (Andrews 1971; Ward and Brown 1986; de Bonis and Melentis 1980).

The mandibular morphology of *K. africanus* was unknown until the discovery of KNM-MJ 5 at Kaloma (Pickford 1982). The Kaloma mandible is fairly complete, but badly crushed and distorted, especially at the symphysis, and most of the teeth have lost a substantial amount of their enamel surfaces. The new Maboko mandible, KNM-MB 20573, is undistorted and complete with the exception of the rami but represents a juvenile individual. Comparison of the symphyseal morphology of immature and mature extant hominoid specimens indicates that the symphysis of juveniles with permanent incisors is very similar in form to that of the adult (McCrossin and Benefit 1993a). The symphyseal morphology of KNM-MB 20573 is therefore taken to closely resemble that of adult *K. africanus*. The mandible of *K. africanus* shares with *K. wickeri* and other middle Miocene apes the derived presence of a strong inferior transverse torus, a posteriorly directed genio-glossal fossa, and a robust corpus. However, the two species are unique relative to other apes in having a long axis of the symphysis that is strongly proclined, forming an angle of only 30–40 degrees to the alveolar margins of the postcanine teeth (McCrossin and Benefit 1993a,b). In contrast, symphyseal axes of *Proconsul*, *Griphopithecus*, *Sivapithecus*, and *Ouranopithecus* are more steeply oriented. Insofar as they are known to be shared only by large-bodied hominoids from Maboko and Fort Ternan, a strongly proclined symphyseal axis and procumbent orientation of the lower incisors appear to be the most diagnostic features presently known for the genus *Kenyapithecus* (McCrossin and Benefit 1993a).

The unusual procumbency of the lower incisors of *Kenyapithecus* provides an important clue as to the origin of the inferior transverse torus and other derived features that characterize middle Miocene large-bodied hominoids with thick molar enamel (Kay 1981). The strongly procumbent incisors, robust canines, and proclined symphyseal axis of *Kenyapithecus* are strikingly similar to those of pitheciines (*Pithecia*, *Chiropotes*, and *Cacajao*), and to a lesser extent orangutans, among extant primates. Pitheciines, especially *Chiropotes*, and *Pongo* use their procumbent incisors and robust, tusklike canines to open fruits and nuts with durable outer coverings that primates with a more gracile dental apparatus cannot consume (MacKinnon 1977; van Roosmalen et al. 1988)

Strain gauge tests of primate mandibles indicate that the inferior transverse torus resists anteroinferiorly directed bending moments during incisal biting (Hylander 1984). The first appearance of a massive symphysis and strong inferior transverse torus in a Miocene ape, *Kenyapithecus*, seems to be correlated with the biomechanical stresses associated with procumbent implantation of the incisors, and a shift toward the consumption of hard fruits. Strong lateral buttressing of the corpus is also first seen in *Kenyapithecus* and may be related to strong occlusal loading on the posterior teeth during the consumption of hard food objects (Kay 1981; Hylander 1984). It seems probable that unlike *Proconsul*, *Kenyapithecus* exploited hard fruits and nuts, its incisors being used to crack open these foods before they were comminuted between thick-enamelled molars (McCrossin and Benefit 1993a,b). We conclude that incisal biting played as important a role as thick molar enamel in the feeding adaptations of *Kenyapithecus*. Moreover, in light of the functional correlation of these features to seed and nut predation, it seems likely that parallel evolution may be responsible for their common presence in *Kenyapithecus* and some extant members of the great ape and human clade.

Postcrania

Because of their scarcity and fragmentary state, as well as historical difficulties in recognizing their taxonomic identity, postcranial remains have played a relatively minor role in discussions of the affinities and adaptations of middle-late Miocene large-bodied hominoids, including *Kenyapithecus*, *Dryopithecus*, and *Sivapithecus*. Le Gros Clark and Leakey (1951) provisionally attributed a clavicle (BMNH M 16335), left humerus shaft (BMNH M 16334), left proximal femur (BMNH M 16331), left femur shaft (BMNH M 16330), and right femur shaft (BMNH M 16332-3) from Maboko Island to *Proconsul* (Morbeck 1983). Subsequently, Leakey (1967) mentioned that because these remains are from Maboko they might represent *K. africanus*. Morbeck (1972) and Walker (1980), however, have suggested that BMNH M 16335 is a crocodile femur rather than a hominoid clavicle. A distal humerus from Fort Ternan (KNM-FT 2751) was also initially referred to *Proconsul* (Andrews and Walker 1976) but more probably represents *K. wickeri* (Feldesman 1982; Pickford 1985). Senut (1986) described a specimen from Maboko Island (KNM-MB 11837) as a proximal ulna of *Kenyapithecus*. Unfortunately, however, KNM-MB 11837 is a left ilium of an artiodactyl, most probably the giraffoid *Climacoceras africanus* (MacInnes 1936; Hamilton 1978). Many new postcranial remains of *Kenyapithecus* were discovered in 1992, including a proximal humerus, proximal ulna, ischium, distal femur, patella, astragalus, metatarsals and several phalanges (McCrossin in prep.). These are currently being analyzed in detail but a few preliminary observations concerning the humerus and femur can be made here.

The humerus shaft (BMNH M 16334) "shows a striking resemblance to that of the cercopithecoid monkeys rather than the modern anthropoid apes," exhibiting a strong deltopectoral crest, a distinctive forward bending of its upper portion, and a weakly developed lateral supracondylar (supinator) ridge (Le Gros Clark and Leakey 1951:97). Le Gros Clark and Leakey (1951) estimated the length of the Maboko humerus at 270 mm, slightly less than their estimated length of the femur (285 mm). It is difficult to understand, therefore, the material basis for the contention that "the humerus was probably longer than the femur" (Simons and Pilbeam 1972:51). Clearly, major limitations on phylogenetic and functional interpretation of the *Kenyapithecus* limb bones from Maboko were imposed by the absence of articular ends.

In 1992, we recovered the proximal end of the humerus, conjoining to the BMNH M 16334 shaft found by Owen in 1933. Like proximal humeral specimens of *Dendropithecus macinnesi* (or *P. africanus*; Gebo et al. 1988) and *Nyanzapithecus pickfordi* (McCrossin 1992), it lacks features shared by all living hominoids, such as a large, globose, and medially facing head, small lesser tuberosity and deep bicipital groove (McCrossin in prep.). Unlike the mobile shoulder designed for agile climbing and facultative arm-swinging indicated by the proximal humerus of *Nyanzapithecus* (McCrossin 1992a), however, the greater tuberosity of the *Kenyapithecus* humerus extends for a substantial distance above the articular surface of the head, as in semiterrestrial and terrestrial cercopithecoids (McCrossin in prep.). This arrangement is related to advantageous positioning of *m. supraspinatus* for achieving rapid and forceful protraction of the humerus and maintaining a pronograde stance in cursorial cercopithecoids (Jolly 1967).

The distal humerus from Fort Ternan (KNM-FT 2751) resembles extant great apes and humans in exhibiting a strong supinator crest, a deep olecranon fossa, and a broad trochlea with a well-developed median gutter and lateral keel delimiting a distinct *zona conoidea* between the trochlea and globular capitulum (Andrews and Walker 1976; Feldesman 1982). Compared to the long and medially directed medial epicondyle of *Hylobates* and *Pongo,* the entepicondyle of KNM-FT 2751 is abbreviated and posteromedially disposed. Because of its role as site of origin of the common tendon of the forearm and digital flexors, retroflexion and shortening of the medial epicondyle relates to a reduced emphasis on digital grasping (Jolly 1967) and may indicate an adaptation for terrestrial locomotor behaviors. The degree to which the medial epicondyle of the KNM-FT 2751 humerus is retroflected is certainly not as extreme as that seen in fully terrestrial primates such as geladas (*Theropithecus gelada*) and chacma baboons (*Papio ursinus*); its posteromedial orientation is more similar to that of semiterrestrial cercopithecins and papionins such as the patas monkey (*Erythrocebus patas*) and the pig-tailed macaque (*Macaca nemestrina*), respectively.

The Maboko femur is characterized by a small articular surface of the head, a neck that is inclined moderately, a distinct tubercle on the back of the neck, and a weakly marked superior tubercle of the trochanteric line (Le Gros Clark and Leakey 1951). According to McHenry and Corruccini (1976), morphometric comparisons indicate that the proximal end is phenetically most like that of the proboscis monkey (*Nasalis larvatus*). The *Kenyapithecus* proximal femur also closely resembles the same element of *Turkanapithecus kalakolensis* (Leakey et al. 1988b), described by Rose (1992) as "extremely similar to that of *Alouatta,* an animal whose locomotor repertoire is characterized by slow quadrupedalism, climbing and occasional suspension." The lateral portion of a distal femur discovered on Maboko in 1992 is mediolaterally broad and anteroposteriorly flat, with a subdued patellar keel like that of modern hominoids and Miocene noncercopithecoid catarrhines, including *Pliopithecus vindobonensis* (Zapfe 1960) and *Proconsul nyanzae* (Walker and Pickford 1983).

Postcranial remains of European large-bodied noncercopithecoid catarrhines are known from Eppelsheim, Germany (Pohlig 1895; McHenry and Corruccini 1976), Saint Gaudens, France (Lartet 1856; Pilbeam and Simons 1971), Klein Hadersdorf, Austria (Ehrenberg 1938; Zapfe 1960) and Rudabanya, Hungary (Kretzoi 1975; Morbeck 1983)

The Eppelsheim femur was discovered in 1820 and described as the type specimen of *Paidopithex rhenanus* by Pohlig (1895) but has been attributed to *Dryopithecus* by most investigators (Schlosser 1902; Le Gros Clark and Leakey 1951; Simons 1972; Szalay and Delson 1979). DuBois (1897), however, regarded the specimen as representing a giant gibbonlike primate, "*Pliohylobates,*" and McHenry and Corruccini (1976) found its morphology to be most similar to that of *Pliopithecus vindobonensis* and *Hylobates.*

The humerus shaft from Saint Gaudens was briefly described by Lartet (1856), who compared its gracility and its faintly marked muscle insertions to the humerus of gibbons. Le Gros Clark and Leakey (1951) discussed the Saint Gaudens humerus shaft, noting that the supracondylar crest is weakly developed. The morphology of the proximal portion of the shaft, particularly the feeble development of the deltopectoral crest, led Pilbeam and Simons (1971) to emphasize resemblances to the humerus shaft of *Pan paniscus.* Szalay and Delson (1972), however, refer to the Saint Gaudens specimen as a "juvenile humerus." Morbeck (1983) also notes that the specimen probably represents a young individual. The humerus shaft may derive from the same individual as the subadult mandible from the same locality (Lartet 1856), with incompletely erupted fourth premolars and third molar sockets that were not yet fully formed. Although normally manifesting sigmoid curvature of the humeral shaft in adults, cercopithecoid juveniles (including *Victoriapithecus*) exhibit a straight humeral shaft and faintly marked deltopectoral crest. The shaft curvature and deltopectoral crest development of the Saint Gaudens *Dryopithecus fontani* humerus might therefore be related to the immaturity of the individual it represents.

A humerus shaft and ulna from the site of Klein Hadersdorf were described by Ehrenberg (1938) and Zapfe (1960). Simons (1972:236) noted that the humerus shaft from Klein Hadersdorf exhibits "a significant feature not present in the modern apes or in hominids . . . that is that the shaft . . . is bent backward or retroflexed at a point about one third of the way down from the proximal end." According to Simons (1972:236), "flexed humeri of this sort occur principally in primates that have a quadrupedal locomotor pattern, regardless of whether they are terrestrial or arboreal." Morbeck (1983) notes that the trochlear notch of the Klein Hadersdorf ulna lacks the distinctive keel for separating medial and lateral components of the humeral trochlea seen in extant large hominoids. Szalay and Delson (1979) attribute the Klein Hadersdorf remains to *Sivapithecus darwini*, a species recently transferred to the genus *Griphopithecus* Abel 1902 by Martin and Andrews (1991).

Postcranial remains from Rudabanya (Kretzoi 1975), including a distal humerus and a fragmentary proximal ulna and radius, were described by Morbeck (1983). Although the articular surface of the humerus is extensively eroded and the ulna lacks the olecranon and anconeal processes, Morbeck (1983) noted several resemblances to the elbow joint of extant great apes and humans. Most importantly, the humerus exhibits a broad and spool-shaped trochlea with a well-developed lateral keel and a globose capitulum (Morbeck 1983). In addition, the trochlear notch of the proximal ulna fragment from Rudabanya exhibits a distinct longitudinal keel for reciprocal articulation with the median gutter of the humeral trochlea (Morbeck 1983; Begun 1992a).

Until relatively recently, there was little information concerning the postcranial anatomy of *Sivapithecus.* A wealth of information, summarized by Rose (1993), now exists concerning the arm, forearm, wrist, hand, thigh, ankle and foot morphology of *Sivapithecus* (Pilbeam et al. 1980, 1990; Rose , 1984, 1986, 1989; Spoor et al. 1992). According to Pilbeam et al. (1990: 237), *Sivapithecus* and *Pongo* "are mostly dissimilar" in terms of their postcranial anatomy. Like the humerus shafts from Klein Hadersdorf (Ehrenberg 1938; Zapfe 1960) and Maboko (Le Gros Clark and Leakey 1951), *Sivapithecus* humerus remains from Pakistan exhibit marked flexion and a strongly developed deltopectoral crest (Pilbeam et al. 1990). Because this morphology is fundamentally different from the humeral shaft anatomy of extant hominoids (Le Gros Clark and Leakey 1951), Pilbeam et al. (1990:238) proposed two competing hypotheses for the family-tree relationships of *Sivapithecus* and *Pongo*: (1) "*Sivapithecus* and *Pongo* are sister taxa, in which case a number of postcranial features shared by living hominoids must represent convergences" and (2) "*Sivapithecus* and *Pongo* are not sister taxa, in which case their palatal and facial similarities are not shared derived features but either convergent derived or shared primitive features." Pilbeam et al. (1990:239) are pessimistic about the potential for deciding which of these hypotheses is correct, stating that they are "not confident that biologically plausible procedures exist for unambiguously settling these issues."

Ultimately, this conundrum of the relationship between *Sivapithecus* and extant hominoids applies equally well to *Kenyapithecus* and *Dryopithecus* and can only be resolved by additional fossil material, especially more informative cranial and postcranial remains (Figures 5–3, 5–4). Recent discoveries from Maboko may be tipping the scales in favor of an interpretation whereby *Kenyapithecus* bears no special relationship to living great apes. Indeed, the proximal humerus anatomy of *K. africanus* is so distinctly unlike that of modern apes that we are strongly inclined to view the large-bodied hominoid from Maboko Island, once placed in the genus *Sivapithecus* (Le Gros Clark and Leakey 1951; Simons and Pilbeam 1965; Pilbeam 1969; Greenfield 1979; Madden 1980; Kay and Simons 1983) as diverging prior to the last common ancestor of living apes, including hylobatids (McCrossin in prep.). Pervasive similarities of known humeral anatomy indicate that this may be the case for *Griphopithecus* (Neudorf-Sandberg, Klein Hadersdorf, Pasalar) and *Dryopithecus* (e.g., St. Gaudens, Rudabanya) as well. Thus, *Kenyapithecus* and other large-bodied hominoids of the middle and late Miocene may merely be avatars, not ancestors, of the extant great apes. Nevertheless, some craniodental features may indicate that *Kenyapithecus* is more closely

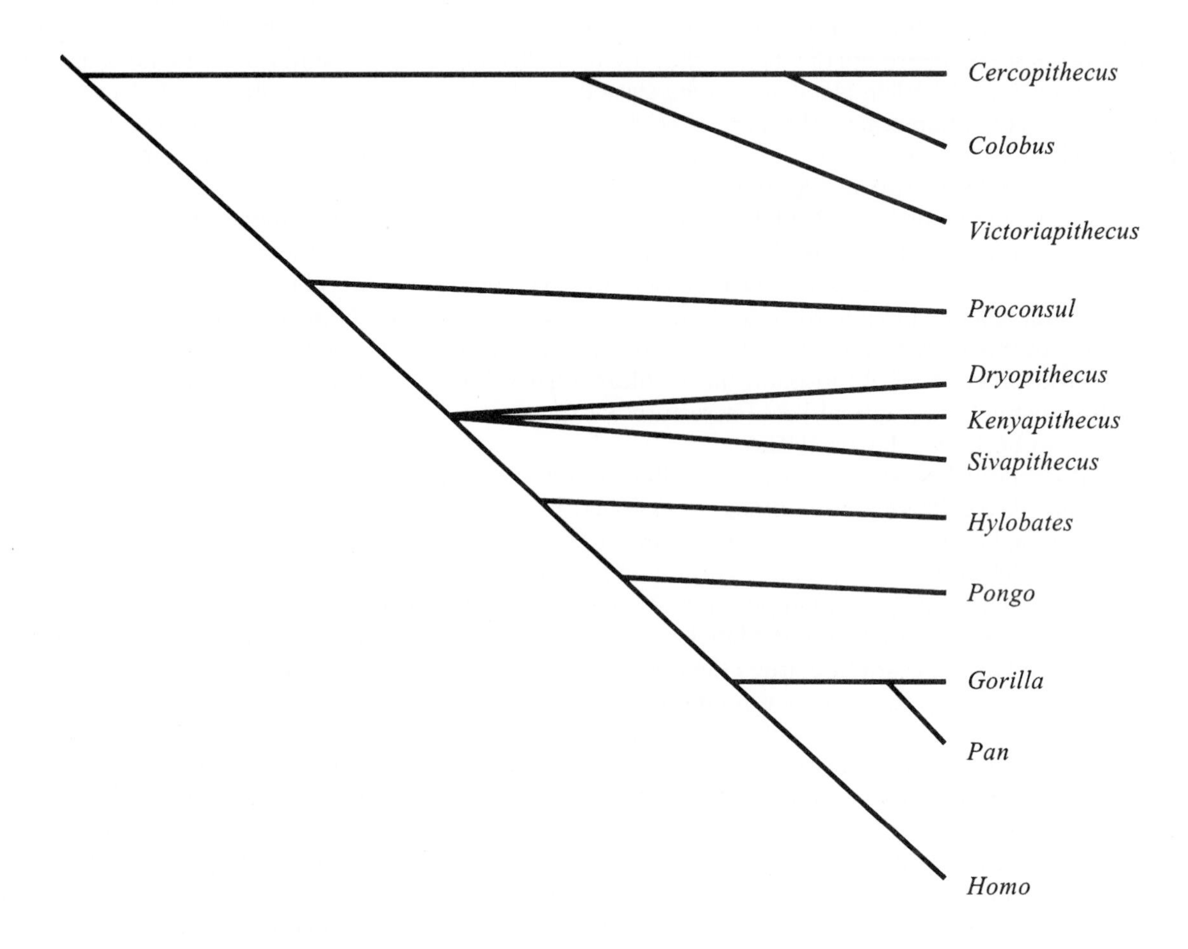

FIGURE 5-3 Relationships of *Kenyapithecus* and other large-bodied Miocene hominoids as inferred from postcranial evidence.

related to the last common ancestor of living hominoids than are *Dryopithecus* (Martin 1986) and *Griphopithecus* (McCrossin and Benefit 1993a).

Phyletic Links

Recent discoveries are also providing a clearer understanding of the distribution and polarity morphocline of craniofacial character states. Certain of the craniofacial similarities thought to specially link *Sivapithecus* with *Pongo*, including tall and narrow proportions of the orbit (Andrews and Cronin 1982; Pilbeam 1982; Ward and Brown 1986), are now known to be present also in an early Miocene hominoid, *Afropithecus turkanensis* (Leakey and Leakey 1986a, Leakey et al. 1988a). Moreover, we interpret one aspect of the craniofacial morphology shared by *S. sivalensis* (GSP 15000) and *Pongo* (Pilbeam 1982; Ward and Pilbeam 1983; Ward and Brown 1986), their distinctive supraorbital costae, as being retained from the last common ancestor of extant cercopithecoids and hominoids (Benefit and McCrossin 1989b, 1991). Supraorbital costae are raised, riblike flanges at the superolateral margin of the orbits that are formed by coalescence of the temporal line and the superciliary arch (Clarke 1977; Ward and Brown 1986). Supraorbital costae are also clearly present in *Aegyptopithecus zeuxis* (Ward and Brown 1986), *Victoriapithecus macinnesi* (Benefit and McCrossin 1991) and *Dryopithecus brancoi* from Rudabanya (RUD 44; Ward and Brown 1986). The shared presence of supraorbital costae is therefore not indicative of a close relationship between *Sivapithecus* and *Pongo* (Benefit and McCrossin 1993b). The supraorbital torus of *Gorilla, Pan,* and various

FIGURE 5-4 Relationships of *Kenyapithecus* and other large-bodied Miocene hominoids as inferred from craniodental evidence.

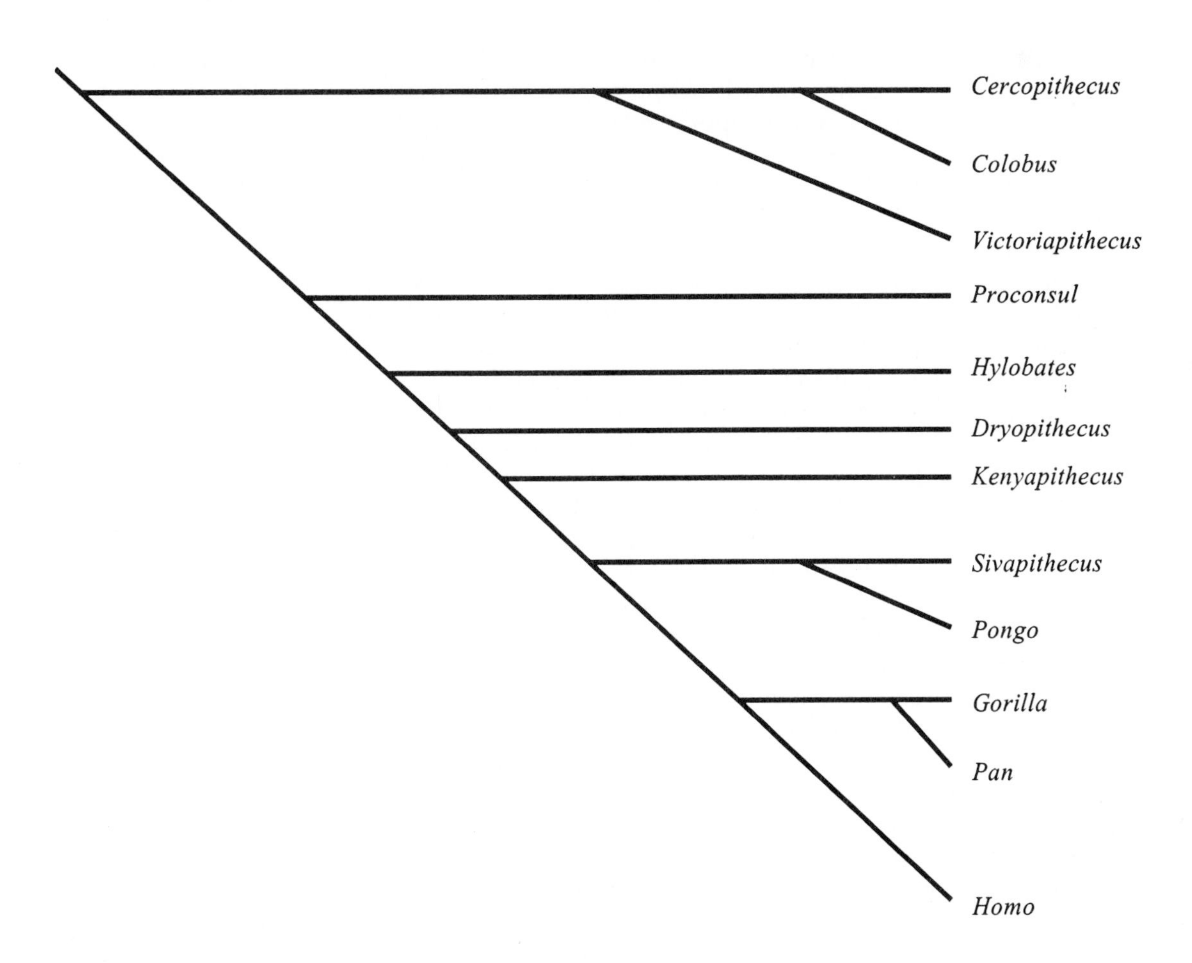

fossil hominids, in contrast, is a barlike browridge which is inflated by a frontal sinus derived from ethmoidal air cells (Cave and Haines 1940), runs continuously across the top of both orbits and a glabellar eminence and is separated from the frontal squama by a well developed post-toral sulcus. The costa supraorbitalis present on a frontal fragment of *Dryopithecus brancoi* from Rudabanya (RUD 44) should not be confused with the fundamentally different structure of a supraorbital torus (contra Begun 1992b). As noted by Ward and Brown (1986), the position and relationships of the pneumatic cavities present on RUD 44 indicate derivation from the maxillary sinus.

A reassessment is necessary of the most compelling evidence in support of a phyletic link between Miocene large-bodied hominoids and the great ape and human clade, the morphology of the subnasal region (Ward and Pilbeam 1983). Ward and Pilbeam (1983:215) described hylobatids as having a "dryomorph" subnasal pattern like that of cercopithecines, *Dendropithecus macinnesi* (KNM-SO 417), *Rangwapithecus gordoni* (KNM-SO 700), and *Afropithecus* sp. from Moroto (UMP 62-11; Leakey et al. 1988a), with a large incisive fenestrum "that opens directly into the oral cavity" and "because the anterior edge of the hard palate is retracted distally from nasospinale, no true incisive canal is present." *Sivapithecus, Pongo, Pan, Australopithecus*, and *Homo*, in contrast, exhibit a true incisive canal (Ward and Pilbeam 1983:215), due to the presence of a long and obliquely oriented nasoalveolar clivus that "overrides the anterior edge of the hard palate, producing an overlapping relationship between these two elements." Because the nasoalveolar clivus of *Gorilla* is usually more vertical, it may on occasion not be "deflected beneath nasospinale" and "the resulting configuration is reminiscent" of *Dendropithecus, Rangwapithecus*, and *Afropithecus* (Ward and Pilbeam 1983:217).

Although it is often a broad fenestrum opening directly into the oral cavity as in cercopithecoids, we have observed constricted incisive apertures in some hylobatid specimens, particularly *Hylobates agilis* and *H. syndactylus*. Indeed, oblique orientation of the nasoalveolar clivus and anterior extension of the palatal process of the maxilla in one individual of *H. agilis* (USNM 123154) precludes a direct view into the oral cavity through very small (c. 0.8 mm diam.) paired incisive canals. Unlike *Sivapithecus, Pongo, Australopithecus*, and *Homo*, the incisive opening of *D. brancoi* (RUD 12) is of fairly large caliber (Ward and Kimbel 1983:Fig.3D). Moreover, the anterior end of the palatal process of the RUD 12 maxilla is not overridden by the distal pole of the subnasal alveolar process, permitting a direct view through the incisive fossa into the oral cavity (Ward and Kimbel 1983:Fig. 3C). Consequently, a true incisive canal of the sort seen in *Sivapithecus, Pongo, Pan, Australopithecus*, and *Homo* is not present in *D. brancoi*. Thus, some extant hylobatids, the large-bodied hominoid from Nachola, *D. brancoi*, and *Gorilla* exhibit a subnasal conformation intermediate between that of *Dendropithecus, Rangwapithecus*, and *Afropithecus* (on the one hand) and that seen in *Sivapithecus, Pongo, Pan, Australopithecus*, and *Homo* (on the other hand). We interpret this intermediate pattern as typifying the last common ancestor of extant hominoids, rather than the so-called "dryomorph" pattern seen in cercopithecoids, *Dendropithecus, Rangwapithecus*, Moroto *Afropithecus*, and most hylobatid individuals (Ward and Pilbeam 1983).

The thickness and disposition of the subnasal alveolar process almost certainly relates to the breadth and procumbency of the upper incisors (Ward and Kimbel 1983). Although somewhat variable, the upper incisors of Miocene noncercopithecoid catarrhines (e.g., *Proconsul*) tend to be high-crowned and of moderate breadth. In comparison, the upper incisors of hylobatids are usually narrower and much lower crowned, while those of *Pongo* and *Pan* are substantially broader and somewhat lower crowned. As is well known, the incisors of *Gorilla* are relatively narrower than those of *Pongo* and *Pan*. Thus, there is some evidence for interpreting the upper incisor proportions of extant hylobatids and the last common ancestor of extant great apes and humans as being divergently derived from a *Proconsul*-like condition. In addition, it is reasonable to assume that the narrow incisors of *Gorilla* are a derived condition related to increased folivory. Consequently, the expansive incisive fossa usually seen in hylobatids may be a secondarily acquired feature related to reduction of the upper incisors and premaxilla from an ancestral condition for extant hominoids more like that seen in the large-bodied hominoid from Nachola and *D. brancoi* from Rudabanya. Finally, because the large incisive openings and vertical subnasal alveolar process of *Gorilla* were most probably secondarily acquired from a *Pan*-like condition of the last common ancestor of living great apes and humans, it is not phylogenetically meaningful to refer to the subnasal pattern exhibited by the Nachola and Rudabanya hominoids as an "African" pattern.

SUMMARY

Recent discoveries at the middle Miocene site of Maboko Island allow a better understanding of the evolutionary history of Old World monkeys and apes. These discoveries have overturned several aspects of conventional wisdom concerning the relationships and adaptations of Miocene catarrhines.

The evidence presented in this chapter provides a new perspective for assessing the origin and diversification of Old World monkeys and apes. Previous interpretations of ancestral Old World monkeys as having been leaf-eating (Jolly 1967; Vogel 1968; Napier 1970; Simons 1970; Delson 1975a,b; Delson and Andrews 1975; Szalay and Delson 1979; Andrews 1981; Andrews and Aiello 1984; Temerin and Cant 1983; Harrison 1989b) and according to some researchers exclusively arboreal (Jolly 1967; Napier, 1970; Simons 1970; Temerin and Cant 1983) no longer hold. The dentition of the oldest known fossil cercopithecoids *Prohylobates* and *Victoriapithecus* is similar to that of highly frugivorous *Cercocebus* and *Macaca*. Low cusp relief, short shear crests, close cusp proximity, and the presence of large bowl-shaped depressions on cusp tips characterize the molars of all early and middle Miocene cercopithecoids. These traits are correlated with hard fruit and seed consumption among extant cercopithecids (Benefit 1987; Benefit and McCrossin 1990). Upper central incisors of *Victoriapithecus* are broad, as for extant frugivores. In addition, the prognathic craniofacial morphology of Miocene cercopithecoids is more similar to that of frugivorous macaques and unlike the orthognathous faces of similarly sized colobine monkeys (Benefit and McCrossin 1991).

Folivorous dental adaptations are not found among cercopithecoids until the late Miocene. Molars of the oldest known colobine monkeys, *Microcolobus* and *Mesopithecus*, have less potential shear than African *Colobus*, which are highly specialized folivores. As for extant Asian langurs, their diets probably included equal proportions of fruits and leaves (Benefit and Pickford 1986; Benefit 1987, 1990). The craniofacial morphology of many fossil colobines (*Libypithecus, Dolichopithecus, Rhinocolobus*, and *Paracolobus*) resembles that of victoriapithecids and cercopithecines in being prognathic with deep cheek regions. Based on this evidence it seems likely that the last common ancestor of the cercopithecids (colobines and cercopithecines), as well as that of all cercopithecoids, possessed craniofacial and dental traits correlated with frugivory.

The origin of Hominoidea has been related to the advent of musculoskeletal features allowing a repertoire of orthograde, below-branch postures and modes of locomotion (Keith 1903, 1923; Gregory 1928; Schultz 1968). Suggested selective mechanisms for acquisition of suspensory adaptations primarily involve competition with cercopithecoids for ripe fruits (Ripley 1979; Temerin and Cant 1983). Aspects of craniofacial and dental morphology have been advanced as evidence that members of the great ape and human clade are known from mid-late Miocene deposits of Africa, Asia, and Europe, with *Kenyapithecus africanus* from Maboko Island as the earliest representative of this clade (Greenfield 1980; Kelley and Pilbeam 1986; Martin 1986; Andrews and Martin 1987).

New mandibular and postcranial material indicates that *K. africanus* possessed uakari-like adaptations of the anterior dentition for cracking open hard nuts and seeds and a shoulder joint designed for cursorial terrestriality. Because most of the dentognathic features suggested to ally *Kenyapithecus* with the great ape and human clade now seem to be functionally related to seed and nut predation, it seems likely that these resemblances are the product of parallel evolution. The general absence of modern ape features in the postcranial skeleton reinforces the idea that *Kenyapithecus* is an avatar, not an ancestor, of living great apes.

Our research at Maboko indicates that early cercopithecoids and thick-enamelled apes were more similar in habitus than was previously suspected. The highly frugivorous hard object diet and semiterrestrial postcranial adaptations of *Victoriapithecus* and *Kenyapithecus* probably evolved in response to changing environments in eastern Africa.

ACKNOWLEDGMENTS

We thank the Office of the President of the Republic of Kenya, as well as Richard Leakey and Mohamed Isahakia (past and present directors of the National Museums of Kenya) for permission to excavate on Maboko Island. We are grateful to Meave Leakey for her continued advice and encouragement. Nina Mudida, Richard Thorington, Linda Gordon, and Maria Rutzmoser generously allowed us access to modern primate material under their care. We would also like to thank our field crew members on Maboko Island, especially: Godfrey Wadem Dondo, Blasto Onyango, Paul Odera Abong, Joseph Onyango Miumi, Charles Omondi Agak, Cercil Odera Ajumbo, Karyluce Ogutu, Paul Asola, Monica Awasi, Prisca Akeyo and Kate Blue. Finally, our sincerest thanks go to F. Clark Howell for his untiring support of this project. Funding for various stages of this work was provided by the following: National Science Foundation, Fulbright Collaborative Grant, L. S. B. Leakey Foundation, National Geographic Society, I.T.T. International Fellowship, Wenner-Gren Foundation for Anthropological Research, Rotary International Foundation, Office of Research and Development (Southern Illinois University at Carbondale), Boise Fund (University of Oxford), and R. H. Lowie Fund (University of California at Berkeley).

LITERATURE CITED

Alpagut, B., Andrews, P., and Martin, L. (1990) New hominoid specimens from the middle Miocene site at Pasalar, Turkey. *J. Hum. Evol.* 19:397–422.

Andrews, P. (1971) *Ramapithecus wickeri* mandible from Fort Ternan, Kenya. *Nature* 228:537-540.

Andrews, P. (1978) A revision of the Miocene Hominoidea of East Africa. *Bull. Brit. Mus. Nat. Hist., Geol.* 30:85-224.

Andrews, P. (1981) Species diversity and diet in monkeys and apes during the Miocene. In C. B. Stringer (ed.): *Aspects of Human Evolution.* London: Taylor and Francis, pp. 25-61.

Andrews, P., and Aiello, L. (1984) An evolutionary model for feeding and positional behavior. In D. J. Chivers, B. A. Wood, and A. Bilsborough (eds.): *Food Acquisition and Processing in Primates.* New York: Plenum, pp. 422-460.

Andrews, P., and Cronin, J. (1982) The relationships of *Sivapithecus* and *Ramapithecus* and the evolution of the orangutan. *Nature* 297:541-546.

Andrews, P., and Martin, L. (1987) Cladistic relationships of extant and fossil hominoids. *J. Hum. Evol.* 16:101-118.

Andrews, P., Meyer, G., Pilbeam, D. R., Van Couvering, J. A., and Van Couvering, J. A. H. (1981) The Miocene fossil beds of Maboko Island, Kenya: Geology, age, taphonomy and paleontology. *J. Hum. Evol.* 10:35-48.

Andrews, P. and Molleson, T. (1979) The provenance of *Sivapithecus africanus. Bull. Brit. Mus. (Nat. Hist.) Geology Series*, 32:19-23.

Andrews, P. and Tekkaya, I. (1976) *Ramapithecus* in Kenya and Turkey. *Int. Congr. Prehist. Protohist.* 9:7-25.

Andrews, P., and Tekkaya, I. (1976) *Ramapithecus* from Kenya and Turkey. In P.V. Tobias and Y. Coppens (eds.): *Les Plus Anciens Hominides* Nice: Colloque VI, IX Congr. Union Internat. Sci. Prehist. Protohist., pp. 7-25.

Andrews, P., and Tekkaya, I. (1980) A revision of the Turkish Miocene hominoid *S. meteai. Palaeontology* 23:85-95.

Andrews, P., and Tobien, H. (1977) New Miocene locality in Turkey with evidence on the origin of *Ramapithecus* and *Sivapithecus. Nature* 268:699-701.

Andrews, P., and Van Couvering, J. A. H. (1975) Paleoenvironments in the East African Miocene. In: F. S. Szalay (ed.): *Approaches to Primate Paleobiology.* Vol. 5: *Contributions to Primatology.* Basel: Karger, pp. 62-103.

Andrews, P., and Walker, A. (1976) The primate and other fauna from Fort Ternan Kenya. In G. L. Isaac and E. R. McCown (eds.): *Human Origins: Louis Leakey and the East African Evidence.* Menlo Park, CA: W. A. Benjamin, pp. 279-304.

Baker, B. H., Williams, L. A. J., Miller, J. A., and Fitch, F. J. (1971) Sequence and geochronology of the Kenya rift volcanics. *Tectonophysics* 11:191-215.

Beard, K. C., Teaford, M. F., and Walker, A. (1986) New wrist bones of *P. africanus* and *P. nyanzae* from Rusinga Island, Kenya. *Folia Primatol.* 47:97-118.

Begun, D. R. (1992a) Phyletic diversity and locomotion in primitive European hominids. *Am. J. Phys. Anthropol.* 87:311–340.

Begun, D. R, (1992b) Miocene fossil hominoids and the chimp-human clade. *Science* 257:1929–1933.

Benefit, B. R. (1985) Dental remains of *Victoriapithecus* from the Maboko Formation. *Soc. Vert. Paleo. News Bull.* 133:21.

Benefit, B. R. (1987) The Molar Morphology, Natural History and Phylogenetic Position of the Middle Miocene Monkey *Victoriapithecus.* Ph.D. Thesis, New York University.

Benefit B. R. (1990) Fossil evidence for the dietary evolution of Old World monkeys. *Am. J. Phys. Anthropol.* 81:193.

Benefit, B. R. (1991) The taxonomic status of Maboko small apes. *Am. J. Phys. Anthropol.* 12:50-51.

Benefit, B. R. (1992) The phylogeny and paleodemography of *Victoriapithecus* – new evidence from the deciduous dentition. *Am. J. Phys. Anthropol.* Suppl. 14:48.

Benefit, B. R. (1993) The permanent dentition and phylogenetic position of *Victoriapithecus* from Maboko Island, Kenya. *J. Hum. Evol.* 25:83–172.

Benefit, B. R., and McCrossin, M. L. (1989a) New primate fossils from the middle Miocene of Maboko Island, Kenya. *J. Hum. Evol.* 18:493-497.

Benefit, B. R., and McCrossin, M. L. (1989b) The facial morphology of *Victoriapithecus. Am. J. Phys. Anthropol.* 78:191.

Benefit, B. R., and McCrossin, M. L. (1990) Diet, species diversity and distribution of African fossil baboons. *Kroeber Anthropol. Soc. Pap.* 71/72:77-93.

Benefit, B. R., and McCrossin, M. L. (1991) Ancestral facial morphology of Old World higher primates. *Proc. Nat. Acad. Sci. USA* 88:5267-5271.

Benefit, B. R., and McCrossin, M. L. (1993a) New *Kenyapithecus* postcrania and other primate fossils from Maboko Island, Kenya. *Am. J. Phys. Anthropol.* 16 (Suppl.):55–56.

Benefit, B. R., and McCrossin, M. L. (1993b) The facial anatomy of *Victoriapithecus* and its relevance to the ancestral cranial morphology of Old World monkeys and apes. *Am. J. Phys. Anthropol. 92:329-370.*

Benefit, B. R., and Pickford, M. (1986) Miocene fossil cercopithecoids from Kenya. *Am. J. Phys. Anthropol.* 69:441–464.

Bishop, W. W. (1971) The Late Cenozoic history of East Africa in relation to hominoid evolution. In K. K. Turekian (ed.): *The Late Cenozoic Glacial Ages*. New Haven: Yale University, pp. 493-527.

Bishop, W. W., Miller, J. A., and Fitch, F. J. (1969) New potassium-argon determinations relevant to the Miocene fossil mammal sequence in East Africa. *Am. J. Sci.* 267:669-699.

Boschetto, H. B., Brown, F. H., and McDougall, I. (1992) Stratigraphy of the Lothidok Range, northern Kenya, and K/Ar ages of its Miocene primates. *J. Hum. Evol.* 22: 47-71.

Brown, B., and Ward, S. (1988) Basicranial and facial topography in *Pongo* and *Sivapithecus*. In J. H. Schwartz (ed.): *Orang-utan Biology*. New York: Oxford University Press, pp. 247-260.

Cave, A. J. E., and Haines, R. W. (1940) The paranasal sinuses of the anthropoid apes. *J. Anat.* 72:493-523.

Clark, W. E. Le Gros, (1950) New paleontological evidence bearing on the evolution of the Hominoidea. *Quart. J. Geol. Soc. Lond.* 105:225-264.

Clark, W. E. Le Gros, (1952) Report on fossil hominoid material collected by the British-Kenya Miocene Expedition, 1949-51. *Proc. Zool. Soc. Lond.* 122:273-286.

Clark, W. E. Le Gros, (1964) The evolution of man. *Discovery* 25:49.

Clark, W. E. Le Gros, and Leakey, L. S. B. (1950) Diagnoses of East African Miocene Hominoidea. *Quart. J. Geol. Soc. Lond.* 105:260-262.

Clark, W. E. Le Gros, and Leakey, L. S. B. (1951) The Miocene Hominoidea of East Aftica. *Fossil Mammals of Africa*, No. 1. London: British Museum (Natural History).

Clark, W. E. Le Gros, and Thomas, D. P. (1951) Associated jaws and limb bones of *Limnopithecus macinnesi. Fossil Mammals of Africa*, No. 3. London: British Museum (Natural History).

Clarke, R. J. (1977) The Cranium of the Swartkrans Hominid, SK 847 and its Relevance to Human Origins. Ph.D. Dissertation, University of the Witwatersrand.

Conroy, G. C. (1972) Problems in the interpretation of *Ramapithecus*: With special reference to anterior tooth reduction. *Am. J. Phys. Anthropol.* 37:41-48.

Conroy, G. C. (1990) *Primate Evolution*. New York: W. W. Norton.

Conroy, G. C., and Fleagle, J. G. (1972) Locomotor behavior in living and fossil pongids. *Nature* 237:103-104.

Conroy, G. C., Pickford, M., Senut, B., Van Couvering, J., and Mein, P. (1992) *Otavipithecus namibiensis*, first Miocene hominoid from southern Africa. *Nature* 356:144-148.

Conroy, G. C., and Pilbeam, D. R. (1975) *Ramapithecus*, a review of its hominid status. In R. H. Tuttle (ed.): *Paleoanthropology, Morphology and Paleoecology*. The Hague: Mouton, pp. 59-86.

Corruccini, R. S., Ciochon, R. L., and McHenry, H. M. (1976) The postcranium of Miocene hominoids: Were dryopithecines merely "dental apes"? *Primates* 17:205–223.

de Bonis, L., Bouvrain, G., Geraads, D., and Koufos, G. (1990) New hominid skull material from the late Miocene of Macedonia in Northern Greece. *Nature* 345:712–714.

de Bonis, L., and Melentis, J. (1980) Nouvelles remarques sur l'anatomie d'un primate hominoide du Miocene: *Ouranopithecus macedoniensis. C. R. Acad. Sci. Paris* 290:755–758.

Dehm, R. (1983) Miocene hominoid primate dental remains from the Siwaliks of Pakistan. In R. L. Ciochon and R. S. Corruccini (eds.): *New Interpretations of Ape and Human Ancestry*. New York: Plenum, pp. 527–537.

Delson, E. (1973) Fossil Colobine Monkeys of the Circum–Mediterranean Region and the Evolutionary History of the Cercopithecidae (Primates, Mammalia). Ph.D. thesis, Columbia University.

Delson, E. (1975a) Evolutionary history of the Cercopithecidae. In F. S. Szalay (ed.): *Approaches to Primate Paleobiology. Contributions to Primatology*. Basel: Karger, pp. 167–217.

Delson, E. (1975b) Paleoecology and zoogeography of the Old World monkeys. In R. H. Tuttle (ed.): *Primate Functional Morphology and Evolution*. The Hague: Mouton, pp. 37–64.

Delson, E. (1979) *Prohylobates* from the early Miocene of Libya: A new species and its implications for cercopithecid origins. *Geobios* 12:725–733.

Delson, E., and Andrews, P. (1975) Evolution and interrelationships of the catarrhine primates. In W. P. Luckett and F. S. Szalay (eds.): *Phylogeny of the Primates: A Multidisciplinary Approach*. New York: Plenum, pp. 405–446.

Dubois, E. (1897) *Pithecanthropus erectus* du Pliocene de Java. Proc.–verbaux, *Bull. Soc. Belge Geol.* 9:151–160.

Ehrenberg, K. (1938) *Austriacopithecus,* ein neuer menschenaffenartiger Primate aus dem Miozan von Klein–Hadersdorf bei Poysdorf in Niederosterreich. *Stx–Ber. Akad. Wiss. Wien, mathem.–nat.* Kl. Abt. 1, 147:71–110.

Feibel, C. S., and Brown, F. H. (1991) Age of the primate–bearing deposits on Maboko Island, Kenya. *J. Hum. Evol.* 21:221–225.

Feldesman, M. R. (1982) Morphometric analysis of the distal humerus of some Cenozoic catarrhines: The late divergence hypothesis revisited. *Am. J. Phys. Anthropol.* 59:173–195.

Fleagle, J. G. (1978) Size distributions of living and fossil primate faunas. *Paleobiology* 4:67–76.

Fleagle, J. G. (1983) Locomotor adaptations of Oligocene and Miocene hominoids and their phyletic implications. In R. L. Ciochon and R. S. Corruccini (eds.): *New Interpretations of Ape and Human Ancestry.* New York: Plenum, pp. 301–324.

Fourtau, R. (1918) *Contribution a l'etude des vertebres miocenes de l'Egypte.* Survey Department, Ministry of Finance, Cairo.

Gaudry, A. (1862) *Animaux Fossiles et Geologie de l'Attique.* Paris: F. Savy.

Gebo, D. L., Beard, K. C., Teaford, M. F., Walker, A., Larson, S. G., Jungers, W. L., and Fleagle, J. G. (1988) A hominoid proximal humerus from the early Miocene of Rusinga Island, Kenya. *J. Hum. Evol.* 17:393–401.

Genet–Varcin, E. (1969) *A la Recherche du Primate Ancetre de l'Homme.* Paris: Boubee.

Greenfield, L. O. (1979) On the adaptive pattern of "*Ramapithecus*". *Am. J. Phys. Anthropol.* 50:527–548.

Greenfield, L. O. (1980) A late–divergence hypothesis. *Am. J. Phys. Anthropol.* 52:351–366.

Gregory, W. K. (1928) Were the ancestors of man primitive brachiators? *Proc. Am. Phil Soc.* 67:129–150.

Gregory, W. K., and Hellman, M. (1926) The dentition of *Dryopithecus* and the origin of man. *Amer. Mus. Anthropol. Papers* 28:1–23.

Gregory, W. K., Hellman, M., and Lewis, G. E. (1938) Fossil anthropoids of the Yale–Cambridge India Expedition of 1935. *Carnegie Inst. Wash. Publ.* 495:1–27.

Hamilton, W. R. (1978) Fossil giraffes from the Miocene of Africa and a revision of the phylogeny of the Giraffoidea. *Phil. Trans. Roy. Soc. Lond.* 283:165–229.

Harle, E. (1892) Une mandibule de singe du repaire des hyenes de Montsaunes (Haute–Garonne). *Bull. Soc. Hist. Nat. Toulouse* 26:9–11.

Harrison, T. (1986) New fossil anthropoids from the middle Miocene of East Africa and their bearing on the origin of the Oreopithecidae. *Am. J. Phys. Anthropol.* 71:265–284.

Harrison, T. (1987) The phylogenetic relationships of the early catarrhine primates: A review of the current evidence. *J. Hum. Evol.* 16:41–80.

Harrison, T. (1989a) New postcranial remains of *Victoriapithecus* from the middle Miocene of Kenya. *J. Hum. Evol.* 18:3–54.

Harrison, T. (1989b) A new species of *Micropithecus* from the middle Miocene of Kenya. *J. Hum. Evol.* 18:537–557.

Hooijer, D.A . (1963) Miocene Mammalia of Congo. *Ann. Mus. Roy. Afr. Cent. Tervuren, Belg. Sci. Geol.* 46:1–77.

Hooijer, D. A. (1970) Miocene Mammalia of Congo, correction. *Ann. Mus. Roy. Afr. Cent. Tervuren, Belg. Sci. Geol.* 67:163–167.

Hopwood, A. T. (1933 a) Miocene primates from British East Africa. *Ann. Mag. Nat. Hist. Lond.* 11:96–98.

Hopwood, A. T. (1933 b) Miocene primates from Kenya. *J. Linn. Soc. Lond. Zool.* 38:437–464.

Hunt, K. D. (1991) Positional behavior in the Hominoidea. *Int. J. Primatol.* 12:95–118.

Hylander, W. (1984) Stress and strain in the mandibular symphysis of primates: A test of competing hypotheses. *Am. J. Phys. Anthropol.* 64:1–46.

Ishida, H., Pickford, M., Nakaya, H., and Nakano, Y. (1984) Fossil anthropoids from Nachola and Samburu Hills, Samburu District, Northern Kenya. *African Study Monographs (Kyoto University)*, Suppl. 2:73–86.

Jolly, C. J. (1967) Evolution of baboons. In H. Vagtborg (ed.): *The Baboon in Medical Research,* 2. Austin: University of Texas, pp. 427–457.

Kay, R. F. (1975) The functional adaptations of primate molar teeth. *Am. J. Phys. Anthropol.* 43:195–215.

Kay, R. F. (1977) Diets of early Miocene African hominoids. *Nature* 268:628–630.

Kay, R. F. (1978) Molar structure and diet in extant Cercopithecidae. In P. M. Butler and K. Joysey (eds): *Studies in the Development, Function, and Evolution of Teeth.* London:Academic, pp. 309–339.

Kay, R. F. (1981) The nut–crackers: A new theory of the adaptation of the Ramapithecinae. *Am. J. Phys. Anthropol.* 55:141–151.

Kay, R. F. and Simons, E. L. (1983) A reassessment of the relationship between later Miocene and subsequent Hominoidea. In R. L. Ciochon and R. S. Corruccini (eds.): *New Interpretations of Ape and Human Ancestry.* New York: Plenum, pp. 577–624.

Keith, A. (1903) The extent to which the posterior segments of the body have been transmuted and suppressed in the evolution of man and allied primates. *J. Anat. Physiol.* 37:18–40.

Keith, A. (1923) Man's posture: Its evolution and disorders. *Brit. Med. J.* 1:451–454.

Kelley, J., and Pilbeam, D. R. (1986) The dryopithecines: Taxonomy, comparative anatomy, and phylogeny of Miocene large hominoids. In D. R. Swindler and J. Erwin (eds.): *Comparative Primate Biology, 1: Systematics, Evolution and Anatomy.* New York: Alan R. Liss, pp. 361–441.

Kent, P. E. (1944) The Miocene beds of Kavirondo, Kenya. *Quart. J. Geol. Soc. Lond. 10*:85–118.

Kretzoi, M. (1975) New ramapithecines and *Pliopithecus* from the Lower Pliocene of Rudabanya in north–eastern Hungary. *Nature* 257:578–581.

Lartet, E. (1856) Note sur un grand singe fossile qui se rattache au groupe des singes superieures. *C. R. Acad. Sci. Paris* 43:219–223.

Leakey, L. S. B. (1962) A new lower Pliocene fossil primate from Kenya. *Ann. Mag. Nat. Hist.* 13:689–696.

Leakey, L. S. B. (1967) An early Miocene member of the Hominidae. *Nature* 213:15–163.

Leakey M. G. (1985) Early Miocene cercopithecids from Buluk, northern Kenya. *Folia Primatol.* 44:1–14.

Leakey, R. E., and Leakey, M. G. (1986) A new Miocene hominoid from Kenya. *Nature* 324:143–146.

Leakey, R. E., and Leakey, M. G. (1988) A new Miocene small–bodied ape from Kenya. *J. Hum. Evol.* 16:369–387.

Leakey, R. E., Leakey, M. G., and Walker, A. C. (1988a) Morphology of *Afropithecus turkanensis* from Kenya. *Am. J. Phys. Anthropol.* 76:289–307.

Leakey, R. E., Leakey M. G., and Walker, A. C. (1988b) Morphology of *Turkanapithecus kalakolensis* from Kenya. *Am. J. Phys. Anthropol.* 76:277–288.

Lewis, G. E. (1937) Taxonomic syllabus of Siwalik fossil anthropoids. *Am. J. Sci.* 34:139–147.

Lewis, O. J. (1971) Brachiation and the early evolution of the Hominoidea. *Nature* 230:577–579.

Lewis, O. J. (1972) Osteological features characterizing the wrists of monkeys and apes, with a reconsideration of the region in *Dryopithecus (Proconsul) africanus. Am. J. Phys. Anthropol.* 36:45–58.

Lydekker, R. (1879) Further notices of Siwalik Mammalia. *Rec. Geol. Surv. India* 12:33–52.

MacInnes, D. G. (1936) A new genus of fossil deer from the Miocene of Africa. *J. Linn. Soc. Lond. Zool.* 39:521–530.

MacInnes, D. G. (1943) Notes on the East African primates. *J. East Afr. Uganda Nat. Hist. Soc.* 17:141–181.

MacKinnon, J. (1977) A comparative ecology of Asian apes. *Primates* 18:747–772.

Madden, C. T. (1980) East African *Sivapithecus* should not be identified as *Proconsul nyanzae. Primates* 21:133–135.

Maier, W. (1977) Die evolution der bilophodonten Molaren der Cercopithecoidea. *Z. Morph. Anthropol.* 68:26–56.

Martin, L. (1986) Relationships among extant and extinct great apes and humans. In B. Wood, L. Martin and P. Andrews (eds): *Major Topics in Primate and Human Evolution.* Cambridge: Cambridge University Press. pp. 161–187.

Martin, L., and Andrews, P. (1991) Species recognition in middle Miocene hominoids. *Am. J. Phys. Anthropol.* 12 (Suppl.):126.

Mboya, B. (1983) The genesis and tectonics of the N.E. Nyanza Rift Valley, Kenya. *J. Afr. Earth Sci.* 1:315–320.

McCrossin, M. L. (1990) Fossil galagos from the middle Miocene of Kenya. *Am. J. Phys. Anthropol.* 81:265–266.

McCrossin, M. L. (1992a) An oreopithecid proximal humerus from the middle Miocene of Maboko Island, Kenya. *Int. J. Primatol.* 13:659–677.

McCrossin, M. L. (1992b) New species of bushbaby from the middle Miocene of Maboko Island, Kenya. *Am. J. Phys. Anthropol.* 89:215–233.

McRossin, M. L., (in prep.) The Phylogenetic Relationships, Adaptations and Ecology of *Kenyapithecus* (Hominoidea, Primates). Ph.D. Dissertation, University of California at Berkeley.

McCrossin, M. L., and Benefit, B. R. (1992) Comparative assessment of the ischial morphology of *Victoriapithecus macinnesi. Am. J. Phys. Anthropol.* 87:277–290.

McCrossin, M. L., and Benefit, B. R. (1993a) Recently recovered *Kenyapithecus* mandible and its implications for great ape and human origins. *Proc. Nat. Acad. Sci. USA* 90:1962–1966.

McCrossin, M. L., and Benefit, B. R. (1993b) Clues to the relationships and adaptations of *Kenyapithecus africanus* from its mandibular and incisor morphology. *Am. J. Phys. Anthropol.* 16 (Suppl.):143.

McDougall, I., and Watkins, R., (1985) Age of hominoid–bearing sequence at Buluk, northern Kenya. *Nature* 318:175–178.

McHenry, H. M., and Corruccini, R. S. (1976) The affinities of Tertiary hominoid femora. *Folia Primatol.* 26:139–150.

McHenry, H. M., and Corruccini, R. S. (1983) The wrist of *Proconsul africanus* and the origin of hominoid postcranial adaptations. In R. L. Ciochon and R. S. Corruccini (eds.): *New Interpretations of Ape and Human Ancestry*. New York: Plenum, pp. 353–367.

Morbeck, M. E. (1972) A Re–examination of the Forelimb of the Miocene Hominoidea. Ph.D. Dissertation, University of California at Berkeley.

Morbeck, M. E. (1975) *Dryopithecus africanus* forelimb. *J. Hum. Evol.* 4:39–46.

Morbeck, M. E. (1976) Problems in reconstruction of fossil and locomotor behavior: The *Dryopithecus* elbow complex. *J. Hum. Evol.* 5:223–233.

Morbeck, M. E. (1977) The use of casts and other problems in reconstructing the *Dryopithecus (Proconsul) africanus* wrist complex. *J. Hum. Evol.* 6:65–78.

Morbeck, M. E. (1983) Miocene hominoid discoveries from Rudabanya: Implications from the postcranial skeleton. In R. L. Ciochon and R. S. Corruccini (eds.): *New Interpretations of Ape and Human Ancestry*. New York: Plenum, pp. 369–404.

Morton, D. J. (1924) Evolution of the human foot. II. *Am. J. Phys. Anthropol.* 7:1–52.

Napier, J. R. (1970) Paleoecology and catarrhine evolution. In J. R. Napier and P. H. Napier (eds.): *Old World Monkeys*. London: Academic, pp. 55–95.

Napier, J. R., and Davis, P. R. (1959) The forelimb skeleton and associated remains of *Proconsul africanus.* Fossil Mammals of Africa, No. 16. London: British Museum (Natural History).

O'Connor, B. L. (1975) The functional morphology of the cercopithecoid wrist and inferior radioulnar joints and their bearing on some problems in the evolution of the Hominoidea. *Am. J. Phys. Anthropol.* 43:113–122.

O'Connor, B. L. (1976) *Dryopithecus (Proconsul) africanus*: Quadruped or non–quadruped? *J. Hum. Evol.* 5:279–283.

Pickford, M. (n.d. a) Preliminary report on Maboko field trip–1982. Unpublished manuscript on file in Paleontology Division, National Museums of Kenya.

Pickford, M. (n.d. b) Report on fieldwork at Maboko: 27 July to 18 August 1983. Unpublished manuscript on file in Paleontology Division, National Museums of Kenya.

Pickford, M. (n.d. c) Report on fieldwork at Maboko: June–July 1984. Unpublished manuscript on file in Paleontology Division, National Museums of Kenya.

Pickford, M. (1981) Preliminary Miocene mammalian biostratigraphy for western Kenya. *J. Hum. Evol.* 10:73–97.

Pickford, M. (1982) New higher primate fossils from the middle Miocene deposits of Majiwa and Kaloma, western Kenya. *Am. J. Phys. Anthropol.* 58:1–19.

Pickford, M. (1983) Sequence and environments of the lower and middle Miocene hominoids of western Kenya. In R. L. Ciochon and R. S. Corruccini (eds): *New Interpretations of Ape and Human Ancestry*. New York:Plenum, pp. 421–440.

Pickford, M. (1984) *Kenya Palaeontology Gazetteer*, Vol. 1:- *Western Kenya*. Nairobi: National Museums of Kenya, Department of Sites and Monuments Documentation.

Pickford, M. (1985) A new look at *Kenyapithecus* based on recent discoveries in western Kenya. *J. Hum. Evol.* 14:113–143.

Pickford, M. (1986a) Hominoids from the Miocene of East Africa and the phyletic position of *Kenyapithecus*. *Z. Morph. Anthropol.* 76:117–130.

Pickford, M. (1986b) Cainozoic paleontological sites of western Kenya. *Münchner Geowiss. Abh.* (a) 8:1–151.

Pickford, M., and Senut, B (1988) Habitat and locomotion in Miocene cercopithecoids. In A. Gautier–Hion, F. Bourliere, J.–P. Gautier, and J. Kingdon (eds.): *A Primate Radiation: Evolutionary Biology of the African Guenons*. Cambridge: Cambridge University Press, pp. 35–53.

Pickford, M., Senut, B., Hadoto, D., Musisi, J., and Kariira, C. (1986) Nouvelle decouvertes dans le Miocene inferieur de Napak, Ouganda Oriental. *C. R. Acad. Sci. Paris* 302:47–52.

Pilbeam, D. R. (1966) Notes on *Ramapithecus*, the earliest known hominid, and *Dryopithecus*. *Am. J. Phys. Anthropol.* 25:1–6.

Pilbeam, D. R. (1969) Tertiary Pongidae of East Africa: Evolutionary relationships and taxonomy. *Bull. Peabody Mus. Nat. Hist.* 31:1–85.

Pilbeam, D. R. (1982) New hominoid skull material from the Miocene of Pakistan. *Nature* 295:232–234.

Pilbeam, D. R., Rose, M. D., Badgley, C., and Lipschutz, B. (1980) Miocene hominoids from Pakistan. *Postilla* 181:1–94.

Pilbeam, D. R., and Simons, E. L. (1971) Humerus of *Dryopithecus* from Saint–Gaudens, France. *Nature* 229:408–409.

Pilbeam, D. R., and Walker, A. C. (1968) Fossil monkeys from the Miocene of Napak, northeast Uganda. *Nature* 220:657–660.

Pilbeam, D. R., Rose, M. D., Barry, J. C., and Ibrahim–Shah, S. M. (1990) New *Sivapithecus* humeri from Pakistan and the relationship of *Sivapithecus* and *Pongo*. *Nature* 348:237–239.

Pilgrim, G. E. (1915) New Siwalik primates and their bearing on the evolution of man and the Anthropoidea. *Rec. Geol. Surv. India* 45:1–74.

Pilgrim G. E. (1927) A *Sivapithecus* palate and other primate fossils from India. *Mem. Geol. Surv. India (Paleontol. Indica)* 14:1–26.

Pohlig, H. (1895) *Paidopithex rhenanus* n. g. n. sp., le singe anthropomorphe du Pliocene rhenan. *Bull. Soc. Belge Geol. Pal. Hydrol.* 9:49–151.

Preuss, T. M. (1982) The face of *Sivapithecus indicus*: description of a new, relatively complete specimen from the Siwaliks of Pakistan. *Folia Primatol.* 38:141–157.

Ripley, S. (1979) Environmental grain, niche diversification, and positional behavior in Neogene primates: An evolutionary hypothesis. In M. E. Morbeck, H. Preuschoft, and N. Gomberg (eds.): *Environment, Behavior and Morphology: Dynamic Interactions in Primates*. Stuttgart: Gustav Fischer, pp. 37–74.

Rose, M. D. (1984) Hominoid postcranial specimens from the middle Miocene Chinji Formation, Pakistan. *J. Hum. Evol.* 13:503–516.

Rose, M. D. (1986) Further hominoid postcranial specimens from the Late Miocene Nagri Formation of Pakistan. *J. Hum. Evol.* 15:333–367.

Rose, M. D. (1989) New postcranial specimens of catarrhines from the middle Miocene Chinji Formation, Pakistan: Descriptions and a discussion of proximal humeral functional morphology in anthropoids. *J. Hum. Evol.* 8:131–162.

Rose, M. D. (1993) Locomotor anatomy of Miocene hominoids. In D. L. Gebo (ed.): *Postcranial Adaptation in Nonhuman Primates*. De Kalb: Northern Illinois University.

Rose, M. D., Leakey, M. G., Leakey, R. E., and Walker, A. C. (1992) Postcranial specimens of *Simiolus enjiessi* and other primitive catarrhines from the early Miocene of Lake Turkana, Kenya. *J. Hum. Evol.* 22:171–237.

Saggerson, E.P . (1952) Geology of the Kisumu District. *Rep. Geol. Surv. Kenya* 21:1–86.

Sarich, V. M. (1971) A molecular approach to the question of human origins. In P. Dolhinow and V. M. Sarich (eds.): *Background for Man*. Boston: Little, Brown, pp. 60–81.

Schlosser, M. (1901) Die menschenahnlichen Zahne aus dem Bohnerz der Schwabischen Alb. *Zool. Anz.* 24:261–271.

Schlosser, M. (1902) Beitrage zur Kenntniss der Saugethierreste aus dem suddeutschen Bohnerzen. *Geol. Palaontol. Abh.* 5:117–258.

Schultz, A. H. (1968) The recent hominoid primates. In S. L. Washburn and P. C. Jay (eds.): *Perspectives on Human Evolution*, Vol. 1. New York: Holt, Rhinehart and Winston, pp. 122–195.

Senut, B. (1986) Nouvelle decouvertes de restes post–craniens de primates Miocenes (Hominoidea et Cercopithecoidea) sur le site Maboko au Kenya occidental. *C. R. Acad. Sci. Paris* 303:1359–1362.

Shackleton, R. M. (1951) A contribution to the geology of the Kavirondo Rift Valley. Quart. *J. Geol. Soc. Lond.* 16:345–392.

Shipman, P., Walker, A., Van Couvering, J. A., Hooker, P. J., and Miller, J. A. (1981) The Fort Ternan hominoid site, Kenya: Geology, age, taphonomy and paleoecology. *J. Hum. Evol.* 10:49–72.

Simons, E. L. (1961) The phyletic position of *Ramapithecus*. *Postilla* 57:1–9.

Simons, E. L. (1963) Some fallacies in the study of hominid phylogeny. *Science* 141:879–889.
Simons, E. L. (1964) On the mandible of *Ramapithecus. Proc. Nat. Acad. Sci. USA* 51:528–535.
Simons, E. L. (1969) Miocene monkey (*Prohylobates*) from northern Egypt. *Nature* 223:687–689.
Simons, E. L. (1970) The deployment and history of Old World monkeys (Cercopithecidae, Primates). In J. R. Napier and P. H. Napier (eds.): *Old World Monkeys*. New York: Academic, pp. 97–137.
Simons, E. L. (1972) *Primate Evolution: An Introduction to Man's Place in Nature*. New York: Macmillan.
Simons, E. L., and Delson, E. (1978) Cercopithecidae and Parapithecidae. In V. J. Maglio and H. B. S. Cooke (eds.): *Evolution of African Mammals*. Cambridge: Harvard University Press, pp.100–119.
Simons, E. L., and Pilbeam, D. R. (1965) Preliminary revision of the Dryopithecinae (Pongidae, Anthropoidea). *Folia Primatol.* 3:81–152.
Simons, E. L., and Pilbeam, D. R. (1972) Hominoid paleoprimatology. In R. H. Tuttle (ed.): *The Functional and Evolutionary Biology of Primates*. Chicago: Aldine–Atherton, pp. 36–62.
Simons, E. L., and Pilbeam, D. R. (1978) *Ramapithecus* (Hominidae, Hominoidea. In V. J. Maglio and H. B. S. Cooke (eds): *Evolution of African Mammals*. Cambridge: Harvard University Press, pp. 147–153.
Smith, R. J., and Pilbeam, D. R. (1980) Evolution of the orang–utan. *Nature* 284:447–448.
Simons, E. L., Andrews P., and Pilbeam, D. R. (1978) Cenozoic apes. In V. J. Maglio and H. B. S. Cooke (eds.): *Evolution of African Mammals*. Cambridge, MA: Harvard University Press, pp. 120–146.
Smith–Woodward, A. (1914) On the lower jaw of an anthropoid ape (*Dryopithecus*) from the Upper Miocene of Lerida (Spain). *Quart. J. Geol. Soc.* 70:316–320.
Spoor, C. F., Sondaar, P. Y., and Hussain, S. T. (1992) A new hominoid hamate and first metacarpal from the Late Miocene Nagri Formation of Pakistan. *J. Hum. Evol.* 21:413–424.
Strasser, E., and Delson, E. (1987) Cladistic analysis of cercopithecid relationships. *J. Hum. Evol.* 16:81–99.
Szalay, F. S., and Delson, E. (1979) *Evolutionary History of the Primates.* New York: Academic.
Temerin, L. A., and Cant, J. G. H. (1983) The evolutionary divergence of Old World monkeys and apes. *Am. Nat.* 122:335–351.
Tuttle, R. H. (1969) Quantitative and functional studies on the hands of Anthropoidea. I. The Hominoidea. *J. Morphol.* 128:309-364.
Tuttle, R. H. (1974) Darwin's apes, dental apes, and the descent of man: Normal science in evolutionary anthropology. *Curr. Anthropol.* 15:389–398.
Van Couvering, J. A. H., and Van Couvering, J. A. (1976) Early Miocene mammal fossils from East Africa: Aspects of geology, faunistics and paleoecology. In G. L l. Isaac and E. McCown (eds.) *Human Origins: Louis Leakey and the East African Evidence*. Menlo Park, CA: W. A. Benjamin.
Van Roosmalen, M. G. M., Mittermeier, R. A., and Fleagle, J. G. (1988) Diet of the bearded saki (*Chiropotes satanas chiropotes*): A neotropical seed predator. *Am. J. Primatol.* 14:11–35.
Verheyen, W. N. (1962) Contribution à la craniologie comparee des Primates. *Ann. Mus. Roy. Af. Cent. Sci. Zool.* 105:1–247.
Vogel, C. (1966) Morphologische studien am gesichtschadel Catarrhiner primaten. *Biblio. Primatol.* 4:1–226.
Vogel, C. (1968) The phylogenetical evaluation of some characters and some morphological trends in the evolution of the skull in catarrhine primates. In B. Chiarelli (ed.): *Taxonomy and Phylogeny of Old World Primates with Reference to the Origin of Man*. Turin: Rosenberg and Sellier, pp. 21–55.
Vogel, C. (1975) Remarks on the reconstruction of the dental arcade of *Ramapithecus*. In R. H. Tuttle (ed.): *Paleoanthropology, Morphology, and Paleoecology*. The Hague: Mouton, pp. 87–98.
Von Beyrich, H. (1861) Uber *Semnopithecus pentelicus. Phys. Abh. Kon. Akad. Wiss. Berlin* 1860:1–26.
Von Koenigswald, G. H. R. (1969) Miocene Cercopithecoidea and Oreopithecoidea from the Miocene of East Africa. In L. S. B. Leakey (ed.): *Fossil Vertebrates of Africa*. 1:39–51.
Wagner, A. (1839) Fossile ueberreste von einem affenschadel und andern Saugethieren aus Griechenland. *Gelehrte Anz. Bayerisches Akad. Wiss. Munich* 38:301–312.
Walker, A. C. (1980) Functional anatomy and taphonomy. In A. K. Behrensmeyer and A. Hill (eds.): *Fossils in the Making: Vertebrate Taphonomy and Paleoecology*. Chicago: University of Chicago Press, pp. 182–196.
Walker, A., and Andrews, P. (1973) Reconstruction of the dental arcades of *Ramapithecus wickeri*. *Nature* 224:313–314.
Walker, A. C., and Pickford, M. (1983) New postcranial fossils of *Proconsul africanus* and *Proconsul nyanzae*. In R. L. Ciochon and R. S. Corruccini (eds.): *New Interpretations of Ape and Human Ancestry*. New York: Plenum, pp. 325–351.
Walker, A. C., and Teaford, M. F. (1988) The Kaswanga Primate Site: An early Miocene hominoid site on Rusinga Island, Kenya. *J. Hum. Evol.* 17:539–544.
Walker, A. C., and Teaford, M. F. (1989) The hunt for *Proconsul. Sci. Am.* 260 (1):76–84.
Walker, P., and Murray, P. (1975) An assessment of masticatory efficiency in a series of anthropoid primates with special reference to the Colobinae and Cercopithecinae. In R. H. Tuttle (ed.): *Primate Functional Morphology and Evolution*. The Hague: Mouton, pp. 203–212.
Ward, C. V., Walker, A., and Teaford, M. F. (1991) *Proconsul* did not have a tail. *J. Hum. Evol.* 21:215–220.
Ward, S. C., and Brown, B. (1986) The facial skeleton of *Sivapithecus indicus*. In D. R. Swindler and J. Erwin (eds.): *Comparative Primate Biology*, Vol. 1: *Systematics, Evolution and Anatomy*. New York: Alan R. Liss, Inc., pp. 413–452.

Ward, S. C., and Kimbel, W. H. (1983) Subnasal alveolar morphology and the systematic position of *Sivapithecus*. *Am. J. Phys. Anthropol.* 61:157–171.
Ward, S. C. and Pilbeam, D. R. (1983) Maxillofacial morphology of Miocene hominoids from Africa and Indo–Pakistan. In R. L. Ciochon and R. S. Corruccini (eds.): *New Interpretations of Ape and Human Ancestry*. New York: Plenum, pp. 211–238.
Washburn, S. L. (1950) The analysis of primate evolution with particular reference to the origin of man. *Cold Spr. Harb. Symp. Quant. Biol.* 15:67–78.
Yulish, S. (1970) Anterior tooth reduction in *Ramapithecus*. *Primates* 11:255–270.
Zapfe, H. (1960) Die Primatenfunde aus der miozanen Spaltenfullung von Neudorf an der March (Devinska Nova Ves), Tschechoslowakei. Mit anhang: Der Primatenfund aus dem Miozan von Klein Hadersdorf in Niederosterreich. *Schweiz. Palaeontol. Abh.* 78:1–293.
Zwell, M., and Conroy, G. C. (1973) Multivariate analysis of the *Dryopithecus africanus* forelimb. *Nature* 244:373–375.

CHAPTER 6

Late Miocene and Early Pliocene Hominoids from Africa

Andrew Hill

INTRODUCTION

One of the most interesting time periods in human evolution is that between about 14 MYA and 3.7 MYA. We know that during that interval bipedal hominids diverged from the lineage leading to African apes, yet we know very little about them or about the conditions that may have given rise to their characteristic and unusual postural innovation.

To know what the early hominids were like, it would help to have fossils of them, and in looking for the reasons for hominid origins, for the causes of bipedalism, it would also be useful to know what the world was like at the time, to know more about paleoenvironments. We have quite a lot of information about the diversity of the African hominoids that lived from the base of the Miocene to about 14 MYA. We also know that bipedal hominids existed at 3.7 MYA, from *Australopithecus afarensis* at Laetoli, Tanzania, and we have a comparatively decent knowledge of the subsequent hominid radiation and events that took place after that time. But for the particularly interesting 10 million years from 14 MYA onwards, we have very few specimens of hominoids, though the situation is improving. The lack of fossils comes partially from a lack of sediments of the appropriate age, but in addition hominoids seem particularly rare in such fossiliferous strata as do exist.

This review provides a brief catalog of African primate specimens claimed as hominoids and thought to be from this time span. It confines itself to large hominoids and does not deal with a number of small hominoid specimens known from the low end of the time range. It alludes to some specimens alleged to be hominoids but about which there is doubt and refers to some that were once believed to be from this time period, but which are probably not. The catalog is followed by a more general discussion of matters relevant to paleoenvironments during this time, particularly evidence for vegetation, that may have had an effect on hominid origins.

Because of ambiguities in any proposed taxonomy, I use the term "hominid" in its traditional sense, to mean the bipedal lineage that diverged from the line leading to the African apes. At the same time I realize that cladistically speaking this is incorrect, and that it does not reflect true phyletic relations.

The time period dealt with in this review has fairly artificial boundaries yet ones also justified by tradition. Historically, there has been a tendency to think that the respectable Miocene hominoid record ends with the Kenyan site of Fort Ternan at 14 MYA, and that the Pliocene to Pleistocene hominid record begins with Laetoli, Tanzania, at about 3.7 MYA. In particular, the base of the period in question is rather arbitrary. In a sense, Fort Ternan does indeed mark the end of a decent hominoid fossil record, though there are times within the earlier 25-14 MYA period that are almost as patchy as the millennia that follow. But in addition, in the context of the origin of Hominidae, Fort Ternan and 14 MYA form a good point to start because of their historical importance. Fort Ternan was for a long time the site considered to have the earliest hominid and effectively marks the greatest range in time historically postulated for the origin of the family. This remains of practical interest now only in the context of environments, as I will discuss later.

The upper boundary of the period under consideration, about 3.7 MYA, is a little more rational, as it marks the earliest time that evidence of bipedalism can be convincingly seen in the fossil record, in the form of hominid footprints at Laetoli, and thus it is the present oldest firm evidence of Hominidae. Though, as will be seen, strong hints of bipedalism can also be detected a few hundred thousand years earlier in the femur from Maka, Ethiopia, and a slim case also can be made for the even earlier existence of the taxon from dental and mandibular remains.

The treatment adopted here is roughly chronological. I describe specimens in terms of time successive sites or sets of sites in a single geological formation. However, the geographic distribution of specimens is also of interest, and consequently I begin by referring to broader regions of Africa where sites are located and specimens come from. This also saves the need for some repetition when discussing individual localities and fossils. Also in this section is a mention of some other formations and sites in the time range that have produced fossil vertebrates, but as yet no hominoids.

REGIONS

The regions that provide evidence of hominoids during this time period are very few and are restricted in geographical extent. Some can best be described as isolated sites. Others are found in larger identifiable paleogeographic units such as consistent sedimentary basins. Most, like earlier and later fossiliferous areas in Africa, are related to the Rift Valley system and occur in the eastern part of the continent, predominantly in Kenya. I will briefly discuss these general areas first, moving from north to south.

Sahabi: Libya

Sahabi is situated in northeast Libya not far south from the north African coast and the Gulf of Sirt. Early Italian explorations are recorded by Desio (1931) and Pettrochi (1934), but the most recent work there was undertaken by the International Sahabi Research Project, under the leadership of Boaz and Gaziry, from the late 1970s through the early 1980s (Boaz et al. 1979, 1982, 1987). The relevant area extends for about 400 km^2 on the west edge of the Sebkhat al Qunnayyin, and the geological succession is divided into three formations of highest Miocene to lower Pliocene age (de Heinzelin and El-Arnauti 1982, 1987). Vertebrate fossils are recorded from all formations, but the most significant, and the specimens described as hominid, come from sites in the Sahabi Formation. The age of the unit is estimated at about 5 MYA (Bernor 1982; Bernor and Pavlakis 1987).

Middle Awash: Ethiopia

The Middle Awash region is in the Rift Valley of northern Ethiopia, about 75 km south of the Hadar. Fossils were discovered there by Taieb in the 1960s and the initial paleontological exploration was conducted in the 1970s under the leadership of Kalb by the Rift Valley Research Mission in Ethiopia, which also described the geological framework (Kalb and Jolly 1982; Kalb et al. 1980, 1982a-e, 1984). Kalb (1993) provides the most recent summary of information relating to the stratigraphy. Rocks are placed in a number of formations, together forming the Awash Group. The lowest unit is the Chorora Formation, probably between 10 and 11 MYA. The overlying Adu-Asa Formation dates from younger than 6 MYA to about 4.5 MYA; the Sagantole Formation from 4.5 MYA to about 3.7 MYA. The later formations in the Awash Group, extending into the Pleistocene and including the Hadar Formation, from which large samples of *Australopithecus afarensis* derive, are not pertinent to the discussion here.

Vertebrate fossils are known from all of these units, but so far relevant hominoids have only been reported from the Sagantole Formation, discovered in the course of investigations in the region led by Desmond Clark (Clark et al. 1984). The specimens come from two sites, Belohdelie and Maka. Further investigations in the Middle Awash are currently being carried out by White and others, and it is very likely that this work will produce additional hominoid specimens in the near future.

Turkana Basin: Kenya and Ethiopia

The area around Lake Turkana, on the northern Kenya border, is particularly famous for its record of later Pliocene and Pleistocene hominids, documenting the diversity of early *Homo* and *Paranthropus*. But there are also a number of older hominoid specimens from the Turkana sedimentary basin that fall into the time period of this review. Following early explorations by Arambourg, the International Omo Research Expedition worked in Ethiopia, along the Omo River, between 1966 and 1974 (Howell and Coppens 1974). During that time Richard Leakey's Koobi Fora Research Project began on the Kenyan side of the border, along the north east shore of Lake Turkana, and continued with considerable success for a number of years (Leakey and Leakey 1978). Subsequently Leakey and his colleagues have engaged in exploration of the west side of the lake, where fossiliferous sediments extending back beyond 4.1 MYA are also found (Harris et al. 1988). The Lake Turkana sediments ranging from 4.35 MYA into the Pleistocene, are placed in a number of formations comprising the Omo Group (Feibel et al. 1989; Brown and Feibel 1991), and a few hominids, some undescribed, come from the lower parts of these units. On the east, the rock unit concerned is the Koobi Fora Formation, and a few specimens relevant to this discussion come from it. On the west, there is undescribed material that may be older than 3.7 MYA from the base of the Nachukui Formation. The lower part of the Usno Formation, along the Omo River in southern Ethiopia, falls into our time range, but none of the many isolated teeth from that formation are older than 3.05 MYA (Feibel et al. 1989). The Mursi Formation, the oldest unit of the Omo Group, is situated between the Omo and Usno rivers in southern Ethiopia. It is capped by a basalt dated at about 4 MYA (Feibel et al. 1989), but the fauna is sparse and includes no primate specimens.

Other finds are from formations either below or so far uncorrelated with the Omo Group. Fejej, for example, is an area of Ethiopia in the north east of the basin that has recently produced hominids. The unit may be comparable with at least part of the Warata Formation in northeast Kenya (Fleagle et al. 1991; Watkins and Williamson 1979), from which so far no hominoids are known. In the southwest portion of the Turkana Basin are the older sites of Lothagam and Kanapoi, both of which have produced hominoid specimens.

Nachola and Samburu Hills: Kenya

Nachola and the Samburu Hills are two fossiliferous areas in a region on the northeastern side of the Kenya Rift Valley, south of Lake Turkana. Baker (1963) originally reported fossil wood there, but the location received little attention until the 1980s, when it was investigated by a team led by Ishida in association with the National Museums of Kenya (Ishida 1984; Ishida et al. 1982, 1984b). Work began in 1980 with hominoid specimens being found in 1982, 1984, and 1986. These come from two formations, though one falls outside our time period. As was pointed out in Hill and Ward (1988), the large, and largely undescribed, assemblage of hominoids comparable to *Kenyapithecus* from the Nachola Formation is now thought to be around 15 to 16 MYA in age (Pickford 1985, 1986a-c; Ishida pers. comm.). Previously published information had suggested a date of between 10.1 to 11.8 MYA (Matsuda et al. 1984, 1986). The revision brings their age more into line with other similar hominoids from Kipsaramon in the Tugen Hills and Maboko.

Consequently, only one hominoid specimen is relevant here, that from the Namurungule Formation in the Samburu Hills. The area is about 30 km west of Baragoi. A detailed geological treatment is given by Makinouchi et al. (1984). The local sequence consists of four geological formations. The lowest, the Aka Aiteputh Formation, consists of 370 m of volcanic units, basalts, trachytes, and welded tuffs. Above this, unconformably, is the Namurungule Formation, formed by 200 m of clastic sediments, cropping out in three main areas. Above that is 120 m of clastics and lavas composing the Kongia Formation, and on top of all of these are black basaltic lavas constituting the Nagubarat Formation, lying with an angular unconformity on the Aka Aiteputh and Namurungule Formations.

Tugen Hills: Kenya

The Tugen Hills are formed by a complex set of faults within the northern Kenya Rift Valley to the west of Lake Baringo, and they extend north-south for about 75 km. The sequence is divided into a number of formations that cover the time from 16 MYA into the late Pleistocene more or less continuously with few significant breaks. Vertebrate fossils are known from many sites throughout.

Fuchs (1950) recorded the first vertebrate fossils from the region, but many more were found in connection with the work of the East African Geological Research Unit (EAGRU), which mapped the area from 1965 onward (Bishop et al. 1971; King and Chapman 1972; Chapman et al. 1978). More recently, investigations have been resumed by the Baringo Paleontological Research Unit (BPRP), now based at Yale University and working jointly with the National Museums of Kenya (Hill et al. 1985, 1986).

The lowest unit having vertebrate fossils is the Muruyur Formation, but it is now known to be outside the time period under consideration here. The Muruyur Formation is overlain by the extensive Ngorora Formation, which does fall completely in the relevant time period. This unit is the result of more or less continuous sedimentation from 13 MYA to about 8.5 MYA, and hominoids are known from three sites. Stratigraphically above the Ngorora Formation are the Mpesida Beds, a slim unit in space and time, dated about 6.5 MYA. Although there is a respectable vertebrate fauna, including a primate, as yet there are no hominoids from this formation. This is unfortunate, as at 6.5 MYA they would probably be particularly interesting. The Mpesida Beds are superseded by the Lukeino Formation, more extensive, with many sites, dated around 6 MYA, and with one hominoid specimen. Above this is the Chemeron Formation, a relatively widespread unit of sediments, and extensive in time, ranging in age from 5.6 to 1.6 MYA. Four hominoids are known from sites in the Chemeron Formation, two of them being older than 3.7 MYA and so relevant to this chapter.

Otavi Mountains: Namibia

The Otavi Mountains are in northern Namibia, where breccias have accumulated at various time periods as a filling to fissures and caves in the Precambrian dolomite. Some are now exposed at the surface by erosion; others have been excavated from beneath the surface by commercial mining operations. These breccias occasionally incorporate vertebrate fossils. Expeditions led by Conroy have discovered a number of such productive vertebrate localities, one of which, at Berg Aukas, contained a hominoid (Conroy et al. 1992; Senut et al. 1992).

Other Areas

There are formations and sites within the regions already discussed that contain vertebrate fossils, but as yet no hominoids. There are also other areas in Africa where fossiliferous sediments deposited between 14 and 3.7 MYA occur, and they are worth mentioning here for their potential importance as a source of further hominoids. In addition they should receive attention for providing information on mammalian faunas and paleoenvironments during this important interval in hominoid evolution. The following examples are limited to sub-Saharan Africa.

In the Kenya Rift Valley is the site of Nakali, found by Golden of EAGRU, subsequently worked on by Aguirre and Philip Leakey (1974) and later by Richard Leakey and Alan Walker. It is about 7 MYA in age and although the fauna is sparse, it is a very interesting one from the point of view of changes in African fauna through time.

Aterir is also in the Kenya Rift, northwest of Kapedo. First examined by Webb of EAGRU, it has subsequently been investigated on a number of occasions (Hill unpublished; see also Bishop et al. 1971). A good vertebrate fauna comes from a number of sites more or less at one level at about 4.5 MYA.

Kanam on the eastern shores of Lake Victoria in Kenya ranges from about 6 to 4 MYA and was originally explored by Louis Leakey in 1932. He discovered a hominid mandibular fragment, later referred to as *Homo kanamensis* (Leakey 1935, 1936). Serious doubts have been cast on its antiquity by Boswell (1935), Pickford (1987a), and others, and it is likely that it is a late Pleistocene intrusion. However, the Kanam sediments and fauna have clear potential and it is important that they are investigated further.

In the western Rift, approximating the border between Zaire and Uganda, are a series of formations and sites that have so far received relatively little attention. Bishop (1969) records some of the early work in the region, and they are now being investigated by expeditions led by Boaz and by Pickford and Senut. Boaz and his colleagues have been interested in the lower Semliki area, on the Semliki River near Lake Mobutu, formerly Lake Albert (de Heinzelin 1988). Relevant sites also exist on west and east sides of the lake. Pickford, Senut and collaborators have concentrated principally on sites on the Ugandan side of Lake Albert. Here the Nkondo Formation is of particular relevance, ranging in age from about 6 MYA to a little less than 5 MYA (Pickford et al. 1988, 1991). So far no ancient hominoids have been described from these localities, although a specimen labeled "Gorillinae?" from the Nkondo Formation has been alluded to by Pickford et al. (1988).

In northern Tanzania, Harrison has recently renewed investigations of an interesting sequence of sediments in the Manonga Valley (Stockley 1929; Grace and Stockley 1930; Hopwood 1931; Harrison 1992; Harrison and Verniers in press; Harrison et al. in press). From 10 sites, the Wemebere-Manonga Palaeontological Expedition has recovered an extensive and diverse series of vertebrate fossils estimated at about 4 to 6 MYA in age.

In southern Africa, the Varswater Formation, exposed at Langabaanweg, Cape Province, is of some interest. Langabaanweg 'E' quarry is dated to between 4 and 5 MYA on the basis of a very abundant and diverse fauna which as yet includes no hominoids (Hendey 1970a,b, 1972, 1973, 1974, 1976, 1984; Tankard 1975).

HOMINOID SPECIMENS AND CONTEXT

Between 14 and 3.7 MYA, there are relatively few hominoid specimens compared to the period between 25 and 14 MYA, or after 3.7 MYA. For many of these specimens a detailed anatomical appraisal is found in Hill and Ward (1988), but this review also includes a few others that have been found since that paper appeared and comments on a few mentioned there that should now be removed from consideration. The fossils are discussed stratigraphically by geological formation, in an approximately chronological sequence.

Muruyur Formation: Tugen Hills: Kenya

Hominoid specimens from the Muruyur Formation are mentioned here simply to explain why they should no longer be discussed in this time context. In our earlier review of African hominoid materials between 14 and 4 MYA (Hill and Ward 1988), Ward and I discussed a primate talus (KNM-MY 24) that we thought on stratigraphical grounds was younger than 14 MYA, as did Pickford, who originally recovered the specimen (Pickford 1988). More recently BPRP has conducted extensive work in the Muruyur Formation, resulting in nearly 50 additional specimens of a variety of large hominids that includes a large collection of teeth comparable to *Kenyapithecus* (Hill 1989; Hill et al. 1991; Brown, Hill, and Ward 1991; Hill, Brown, and Ward 1992). These hominoids come from several sites, at different levels within the formation, and we now know that none of the specimens is younger than 14 MYA. All of the relevant horizons, including the one the talus comes from, are between 16 and 15 MYA (Hill et al. 1991; Deino et al. in ms).

Berg Aukas I: Otavi Mountains: Namibia

Berg Aukas is one of a number of localities in the Otavi Mountains, Namibia, investigated by Conroy and his colleagues (Conroy et al. 1992). A number of fossiliferous breccia blocks are described there, ranging in age from the early part of the middle Miocene to the Holocene. One hominoid specimen has been reported so far. It is particularly significant in coming from a region quite distant from the majority of other specimens. It has been dated by associated rodents at about 13 MYA (Senut et al. 1992).

Berg Aukas mandible (BER I, 1'91). The first specimen discovered at Berg Aukas in the 1991 season was a mandible of a hominoid consisting of the right corpus and the symphyseal region. It extends from the mesial margin of the left canine alveolus to the base of the ascending ramus and includes crowns of P4 to M3, fragmentary crown and root of P3, partial root of the canine, and alveoli of all left and right incisors. This forms the holotype of a new genus and species *Otavipithecus namibiensis* Conroy et al. 1992. Diagnostic characters cited by Conroy et al. (1992) are the "squared off" rather than elongate appearance of the molars, which lack beaded buccal cingula. The molar cusps are described as "puffy," and the lingual slope of the hypoconid extends beyond the mid-axis of the molar, particularly in M1, restricting the size of the trigonid and talonid basins. There are protostylar ridges on M2 and M3. The molars show little differential wear and the wear pattern suggests thin enamel or high dento-enamel relief. M2 > M3 > M1 in size, and there is a large retromolar space. The incisor region is very narrow. The mandible itself does not significantly decrease in depth mesiodistally, there is only moderate development of the inferior transverse torus; and the M3 is not obscured by the anterior root of the ascending ramus in lateral view. The specimen is figured in Conroy et al. (1992).

Ngorora Formation: Tugen Hills: Kenya

The Ngorora Formation is an extensive stratigraphic unit within the Tugen Hills succession first investigated by Chapman of EAGRU (Chapman 1971). His work resulted in the discovery of a hominoid specimen and in further paleontological investigations (Bishop and Chapman 1970; Bishop et al. 1971). Aguirre conducted excavations there for one season (Aguirre and Leakey 1974) which also resulted in a hominoid tooth, not described, and perhaps not recognized, until recently (Hill et al. in ms.). More detailed stratigraphical and paleontological work was carried out by Pickford (Bishop and Pickford 1975; Pickford 1975a; 1978a). And more recently still, the unit has been worked on by BPRP. In the course of this, two other isolated large hominoid teeth have been found. Our dating program has resulted in an extremely finely calibrated and apparently precisely dated succession, which shows that the formation is temporally extensive for a terrestrial sequence, spanning about 5 MYA, from 13.2 MYA at the base to around 8.5 MYA in its most recent exposures (Hill et al. 1985, 1986; Tauxe et al. 1985; Deino et al. 1990). Sedimentation is more or less continuous throughout, and fossils are well represented. Among significant indicators of environmental conditions is one of the best fossil macrofloras in Africa, now dated at 12.6 MYA. Originally alluded to by Pickford (1978a), the horizon has been investigated in detail by BPRP (Kabuye and Jacobs 1986; Jacobs and Kabuye 1987). The three or four large hominoid specimens so far identified are all unfortunately isolated teeth: one molar, a premolar, an incisor, and possibly a canine. In addition there are a number of specimens belonging to small-bodied hominoids not discussed here.

Ngorora Molar (KNM-BN 1378). This is a specimen found in 1968 by Kiptalam Cheboi when working with Chapman. It comes from site BPRP#K060 (EAGRU 2/9) near to a track about 0.5 km south of Bartabwa (Bishop et al. 1971). The matrix indicated that it came from a black manganese layer about 4 cm in thickness, but a small excavation at the site revealed no further vertebrate material, except for fish fragments from nearby horizons. KNM-BN 1378 is a left maxillary molar crown. The specimen is figured in Hill and Ward (1988), where a detailed description is provided. The separation of the hypocone from the trigone and the large hypocone size indicate that it is an M^1 or M^2. The original reference to the specimen is by Leakey in Bishop and Chapman (1971). Leakey suggested affinities with *Kenyapithecus*, which he thought was a hominid, and with *Homo* and *Australopithecus*. It has also been alluded to in later analyses. Corruccini (1975) for example performed a crown component analysis, but without any convincing taxonomic result. Later Corruccini and McHenry (1980) saw resemblances with *Sivapithecus*, but our work (Hill and Ward 1988) failed to show any real similarities other than possibly thickened enamel, which may be primitive for all large hominoids. Neither is it particularly similar to the *Australopithecus afarensis* sample from Hadar. It does not belong to the taxon represented by the maxilla from the Samburu Hills (KNM-SH 8531; see later discussion). And although it is larger, the best phenetic comparison is with a modern chimpanzee M^2. However, at present it cannot be assigned to any known taxon, and the most suitable taxonomic allocation is to an unknown genus and species of Hominoidea.

Ngorora Premolar (KNM-BN 10489). A premolar was found from a location near Bartabwa, by Kiptalam Cheboi, working with BPRP (Hill et al. 1985; Hill and Ward 1988). The date of the site (BPRP#K065) is about 12.42 MYA; it occurs about 0.4 m below a tuff dated to that age, and assuming constant sedimentation rates in this part of the section, that thickness is likely to represent less than 0.01 MYA (Deino et al. 1990). At the same site was found a badly damaged canine tooth (KNM-BN 10556), which might also be hominoid, and if so it is unlikely to belong to the same species as the premolar. The premolar is a well-preserved left mandibular P_4 with complete crown and proximal part of the roots. It possesses features that seem to ally it with *Proconsul*, such as the contour of the buccal surface of the crown, buccal pitting and the proportions of the mesial and distal foveae. However, the shape of the crown is different, trapezoidal in this specimen rather than rectangular, and there are dissimilarities in the differential levels of trigonid and talonid. In this specimen, the trigonid is relatively high (Hill and Ward 1988). If it is in fact *Proconsul*, then it would be the last known specimen of the taxon in the record. Photographs of the tooth appear in Hill and Ward (1988).

Ngorora Incisor (KNM-BN 1461). A hominoid left central incisor tooth has recently been identified from site BPRP# K038 (EAGRU 2/1). The site, though small, has an interesting and diverse fauna, and it is situated just a few meters above the fossil plant horizon that indicates forest conditions (Jacobs and Kabuye 1987). This sedimentary thickness probably represents about 0.04 MYA. The tooth was collected by Aguirre in 1969, but not commented upon. Full details will appear in Hill et al. (in ms.), a paper that primarily describes cercopithecoids from the site. The tooth appears to resemble *Proconsul* more than other known hominoids (Ward pers. comm.). Its age is 12.49 MYA.

Ngorora Canine (KNM-TH 23144). Recently Meave Leakey (pers. comm.) has recognized that a lower right canine tooth in the BPRP 1990 collection of cercopithecoid teeth from the same site as the incisor previously discussed (BPRP# K038) appears not to be a monkey but has more features in common with hominoids. It is also 12.5 MYA old. This tooth will be described in Hill et al. (in ms.). Its taxonomic affinities are not yet exactly clear, and a designation of Catarrhini indet. is prudent. Nothing at present precludes its belonging to the same taxon as the incisor, but if so that taxon is unlikely to be *Proconsul.*

Namurungule Formation: Samburu Hills: Kenya

The Namurungule Formation in the Samburu Hills area consists of about 200 m of clastic sediments in three main areas of outcrop. Detailed geological information is given by Makinouchi et al. (1984). Only one hominoid specimen comes from the formation, from site SH-22 about 1.5 km upstream from the junction of the Asanyanait and Nakaporatelado rivers, at which point the formation is about 20 m thick. The specimen was found in a 20-cm thick unit of mud clasts set in a calcareous cement. Dating attempts are discussed in Matsuda et al. (1984, 1986) and summarized in Hill and Ward (1988). A variety of methods, K/Ar, fission track, and paleomagnetic stratigraphy, have been used, but the results are not particularly satisfactory. K/Ar determinations, if accurate, restrict the age to between 13 and 6.4 MYA. Paleontological information (Nakaya et al. 1984, Pickford et al. 1984a, b) implies an approximate age of about 9 MYA, but a date of about 8 MYA with an error of ± 2 MYA seems the best that can be suggested at present.

Samburu Hills Maxilla (KNM-SH 8531). This is one of the most interesting specimens from the whole time period. First, it is one of the most complete, consisting of a left partial maxilla with the intact crowns of P^3-M^3, an intact canine alveolus, and a partial alveolus of the lateral incisor. A preliminary description appeared in Ishida et al. (1984a), although a more definitive treatment has not yet been published and the specimen has not yet been allocated to a taxon. Ishida and colleagues remark on similarities of the specimen to the gorilla. These include its overall size, the pneumatized root of the zygomatic process, the sharp anterior margin of the nasal aperture, the degree of prognathism, the degree of arching of the palate, and the curvature of the alveolar process. The teeth, however, do not superficially resemble the gorilla, as they appear to have thick enamel rendering them bunodont in appearance. Relative molar size is $M^3 > M^2 > M^1$. They note that the specimen possesses none of the synapomorphies of the Asian great apes or of hominids. Hill and Ward (1988) indicate some similarities of the dentition to that of *Proconsul major*, but there are also significant differences. The P^3 is similar in overall anatomy to that of the gorilla, and other authors (Pilbeam 1986; Andrews and Martin 1987) have also recorded gorilla-like features. It is unlike any other known specimen, and the particular constellation of characters that it displays renders it exceptionally intriguing. Presumably it belongs to a new species and most probably a new genus.

Lukeino Formation: Tugen Hills: Kenya

The Lukeino Formation occupies a considerable geographic area along the eastern base of the Tugen Hills. The rocks were originally described by Chapman, working with EAGRU, as a member of the Kaparaina Basalts Formation (Chapman 1971), and further work was carried out by other participants in EAGRU, Martyn and McClenaghan (Martyn 1969; McClenaghan 1971). The unit was raised to the status of a formation and a type section described by Pickford in the course of later more detailed study (Pickford 1975a, b, 1978b). Since then additional investigations have been made by BPRP (Hill et al. 1985, 1986). There are a number of productive vertebrate fossil sites and the fauna emphasizes changes already detected by the top of the Ngorora Formation; new taxa of a modern aspect appear, reflected for example in elephantid proboscideans and suids, and typical middle Miocene creatures disappear from the record (Barry et al. 1985; Hill 1985a, 1987). The Lukeino Formation overlies the Kabarnet Trachytes, that date from 6.20 to 6.36 MYA. For most of its outcrop, the Lukeino Formation is overlain by the Kaparaina Basalts, and although dates on these cover a wide range (Chapman and Brook 1978), the more recent determinations by BPRP, at around 5.6 MYA are more consistent (Hill et al. 1985, 1986). Ages on materials within the formation are 6.06 and 5.62 MYA, consistent with the bracketing dates. There is only one hominoid specimen so far known.

Lukeino Molar (KNM-LU 335). This isolated tooth was found by Pickford's expedition at the site of Cheboit (BPRP# K029; EAGRU 2/219) in 1973. The locality is about 5.5 km west of Yatya. KNM-LU 335 is a relatively small first or second left lower molar crown, unerupted, with the roots missing, probably never formed. The main cusps are low and obtuse, and there are no accessory cusps. The mesial and distal marginal ridges are compressed and the hypoconulid is strongly appressed to the distal-buccal surface. It was originally described by Andrews in Pickford (1975b), where he allocated it to Hominidae but has since commented that he would now be more cautious (Andrews pers. comm.). Crown component analyses (Corruccini and McHenry 1980; McHenry and Corruccini 1980) suggested affinities with modern chimpanzees, a conclusion also reached by Kramer (1986). Hill and Ward (1988), while acknowledging the information it provides is scanty, indicated that it represents a plausible morphotype for a putative ancestor of chimpanzees and humans.

Lothagam Formation: Lothagam: Kenya

Lothagam is in the southwest of the Lake Turkana basin, in northern Kenya, between the Kerio and Lomunyenkupurat rivers, at about 2°53'N, 36°04' E. Patterson, leading a Harvard University expedition to the Turkana region, visited the site after the discovery there of fossils by Lawrence Robbins in 1965. One hominoid specimen was found (Patterson et al. 1970). Additional geological information is provided by Behrensmeyer (1976), and Smart (1976) gives a summary of the fauna. A team led by Meave Leakey has renewed work there in the last few years and has recovered much additional fauna, revised the geology, and undertaken a program of radiometric dating. It is expected that these data, along with information concerning a few additional hominoid specimens, will be published soon and will probably modify the account given here. The sedimentary succession is about 750 m thick, composed of three units labeled Lothagam 1, 2, and 3. Lothagam 1 is the principal fossiliferous unit, though some interesting fauna comes from Lothagam 3. There are some published radiometric dates, which do not closely constrain the age of the fauna, particularly not of the hominoid specimen. Plausible estimates, based upon aspects of the fauna itself, have focused on 5.5 MYA. A recent attempt to refine the age estimate suggested that the hominid specimen was older than 5.6 MYA, but probably not older than 5.8 MYA (Hill et al. 1992).

Lothagam mandible (KNM-LT 329). This specimen was found in 1967 in the upper part of Member C of Lothagam 1. It consists of a right corpus broken through the distal alveolus of P4 and through the distal part of the M3 root. Although M_1 is the only tooth with the crown preserved, the roots of M_2 are present as is the mesial root of M_3. The M_1 is in wear, with coalescing dentin islands developed on protoconid and hypoconid. Part of the lower margin of the corpus is also preserved. KNM-LT 329 was originally attributed by Patterson et al. (1970) to *Australopithecus* cf. *africanus* as it possessed a relatively gracile corpus and a small M1 crown. Since the description of *Australopithecus afarensis,* however, a number of more recent publications have attributed it tentatively to that taxon (Kramer 1986; Hill and Ward 1988; Hill et al. 1992). These authors (see also White 1986) have demonstrated that the mandible shows a number of morphological features characteristic of the *A. afarensis* sample. These include a high and narrow extramolar sulcus, a low distal extension of the inferior transverse torus, hollowing of the lingual face of the mandibular corpus beneath the mylohyoid line, and a hollowing of the lateral surface of the corpus beneath M_1. In fact the whole topographic arrangement of the lateral corpus surface as shown by stereoplot contour maps shows a great similarity with the small mandibles in the *A. afarensis* hypodigm. Dentally, morphological similarities include a serrate arrangement of the molar roots, there is a bifid mesial root to M3, and the M1 hypoconulid is mesially appressed to the crown. A variety of metric features of the molar also place it within hominids. Corpus metrics also fall within the range for A. afarensis, but while closest to that taxon, they are also consistent with some other hominoids. In sum, the morphological and metric features available on this specimen show that it can easily be accommodated within *A. afarensis*. If KNM-LT 329 is in fact a hominid, and if the age estimate of over 5.6 MYA is correct, then it would be the oldest known member of the family.

Sahabi Formation: Sahabi: Libya

The Sahabi Formation is the topmost of the three formations into which rocks at Sahabi have been divided. It lies on the lower two formations unconformably and has a thickness of from 80 to 90 m (de Heinzelin and El-Arnauti 1982, 1987). No rocks stratigraphically related to the unit are suitable for radiometric determinations, but elements of the fauna, in particular the Suidae (Cooke 1987), and the Proboscidea (Gaziry 1982, 1987) suggest a low Pliocene age. This is substantiated by reviews of the whole fauna (Bernor 1982; Bernor and Pavlakis 1987). These authors suggest a date of about 5 MYA. The Sahabi Formation appears to postdate the terminal desiccatory phase of the Messinian in the Mediterranean, implying the unit is younger than 5.3 MYA.

The paleoenvironment is described as basically semiarid, relieved by woodland along the large river that flowed through the area into the sea not far away (Boaz 1987; Dechant Boaz 1982, 1987). Fossil wood suggests that the ecotone between woodland and arid area was abrupt. Taxonomically the trees were African in character, there being no circum-Mediterranean or Eurasian components (Dechamps and Maes 1987).

Three specimens from Sahabi have been described as hominid. The taxonomic status of all of these specimens is questionable at best, and if they are in fact hominoids, as Boaz himself concedes; ". . . the taxonomic affinities, and thus the overall significance for hominid evolution, of the Sahabi remains are difficult to assess" (Boaz 1987).

Sahabi Clavicle (26P4A). Boaz has described this fossil, which comes from Member U2, as a left partial hominoid clavicle (Boaz 1980, 1987; Boaz and Meikle 1982; Boaz et al. 1979). But White and his colleagues regard it as the posterior rib of a dolphin, the taxon in question also known from the Sahabi fauna (White et al. 1983; White 1987). Whatever it is, part of the shaft is missing, including the articulating surface.

Sahabi Fibula (114P33A). This is the distal 5 cm of the shaft of a mammalian fibula described by Boaz (1987) as hominoid (also Boaz and Meikle 1982). It comes from a level below the clavicle specimen, in Member U1. There is no disagreement that it is in fact a fibula, but it is by no means clear that it has to come from a hominoid. As discussed in Hill and Ward (1988), it is unlikely to be from a Cercopithecoid simply on grounds of robusticity, nor does it show some features found in modern pongids. But it could be something other than Primate.

Sahabi Parietal (21P21A). Another fragment is a parietal, again described by Boaz (1987) as hominoid (also Boaz and Meikle 1982). It comes from a site that is probably in Member U1. The specimen possesses some features anticipated in a hominoid. Boaz refers to its flatness, its thickness, a large middle meningeal channel on the endocranial surface, and the lack of a sagittal crest. These may be necessary features, but for hominoid status they are not sufficient.

Chemeron Formation: Tugen Hills: Kenya

The Chemeron Formation is in the Tugen Hills sequence, where it crops out over quite a large area. These rocks were originally alluded to by Gregory (1896, 1921) as part of his Kamasian sediments and laterworked on by Fuchs (1934, 1950). McCall et al. (1967) divided rocks in this outcrop into two units, one of which was the Chemeron beds. This name was formalized as the Chemeron Formation in the later work of Martyn (1967, 1969). Subsequently the BPRP has conducted work on the unit (Hill 1985a; Hill et al. 1985, 1986; Hill and Ward 1988). The formation is now known to be quite extensive in time range. For most of its outcrop, it lies on the Kaparaina basalts, dated at about 5.6 MYA. But it also occurs beneath and above trachytes dated at younger than 1.6 MYA. Until recently it was thought that there would prove to be two formations in rocks formerly labeled Chemeron, in distinct mappable outcrops, and of distinct nonoverlapping ages. However, BPRP's recent work on the unit suggests that sediments extend all the way through this time period, and that the Chemeron Formation represents a continuous succession spanning some 4 MYA. Both hominoid specimens relevant to this discussion come from low in this section.

Chemeron (Tabarin) mandible (KNM-BC 13150). A mandibular fragment was found by Kiptalam Cheboi at the site of Tabarin in 1984 (Hill 1985b; Ward and Hill 1987). Tabarin (BPRP# K075) is a locality in the northern part of the Chemeron Formation. The specimen comes from a ferruginous horizon about 37 m above the Kaparaina Basalt which marks the base of the formation at this site. The Kaparaina Basalt is about 5.6 MYA in age, and other dating evidence suggests an age of roughly 4.5 MYA for the hominid. More precise details will be published by Deino and others shortly. KNM-TH 13150 is a fragment of right mandibular corpus, broken through the P_4 and M_3 sockets, and with intact and undistorted crowns of M_1 and M_2. A detailed description and analysis is provided in Ward and Hill (1987). That paper and also Hill (1985b) and Hill and Ward (1988) suggest the specimen is best compared to *Australopithecus afarensis*. The corpus is broad but shallow, small in dimensions compared with the total *A. afarensis* hypodigm, but falling within its range. There is a relatively large extramolar sulcus, broader than in the Lothagam mandible. The lateral surface of the corpus is concave anteriorly; in this respect it is similar to the Lothagam mandible, and to *A. afarensis*. It differs from the Lothagam mandible, however, in that the lateral prominence is very strongly expressed. The contours of the lingual surface, although damaged, can also be seen to be similar to *A. afarensis* and the Lothagam specimen. The occlusal surface of the teeth is weathered, but the dental crown anatomy can be seen to approximate that of *A. afarensis*. The buccal cusps show a strongly bilobate appearance, for example, and the hypoconulid is mesially appressed to the crown. Also, in most basic dental metrics, both teeth fall within the *A. afarensis* range. However, the M_2 shape index does not and is more similar to that of *Proconsul africanus*. In addition, the specimen shows a distinctive pattern of subocclusal anatomy, characteristic of *A. afarensis*. There is a serrate implantation pattern and the mesial surface of the M_3 alveolus shows that the root had a bifid apex.

Chemeron Humerus (KNM-BC 1745). A proximal humerus comes from a site in the northern exposures of the Chemeron Formation (BPRP# K037 EAGRU 2/211). The specimen comes from a level about 50 m above the Cheseton tuff, which has been dated at 5.07 MYA (Pickford et al. 1983). The associated fauna is similar to that from Tabarin and from other sites at the base of the formation, and it is likely that its age is similar to that of the Tabarin mandible. It was originally described by Pickford et al. (1983), and additional comments appear in Senut (1983) and Hill and Ward (1988). The specimen is a proximal fragment of the left humerus of a subadult. Pickford et al. (1983) allocated the specimen to Hominoidea. Senut (1983) emphasized that it could not be said to be hominid. Hill and Ward (1988), on the contrary, suggested a number of features that indicated hominid affinities and showed particular resemblance to proximal humeri of *A. afarensis,* such as AL 288-1r. The head of the humerus is elliptical rather than spherical as in living pongids, and the intertubercular groove is shallow. Usually this is shallow in hominids and deep in extant apes. There are also indications that the head is anteromedially oriented as in hominids.

Kanapoi Formation: Kanapoi: Kenya

Kanapoi is a fossiliferous region in the Lake Turkana basin area, discovered by Patterson's Harvard expedition in the 1960s. It is at the south of Lake Turkana, south of Lothagam. The rich fauna that comes from there includes one hominoid, recovered in 1965 (Patterson 1966). The Kanapoi Formation is about 70 m in thickness (Patterson et al. 1970), comprising lacustrine and riverine sediments, and overlain by a volcanic lava. This overlying basalt has been dated at 2.6 MYA and 2.78 MYA. Paleontological factors suggest that these determinations do not accurately date the fauna, including the hominoid, which is generally considered as about 4 MYA in age.

Kanapoi Humerus (KNM-KP 271). The hominoid specimen is a distal left humerus, very well preserved, and with an oblique and sharp artificially produced cut on the shaft beginning about 5 cm above the medial epicondyle. The original description is given by Patterson and Howells (1967). The humerus is a very large one, and although at various times it has been linked with most hominoid taxa, most analyses, both metric and morphological, indicate its similarities with *Homo* (McHenry and Corruccini 1975; McHenry 1976, 1984; Senut, 1980, 1981a,b, 1982, 1983; Senut and Tardieu 1985). However, it is clear that there is considerable variation in the morphology of distal hominid humeri, and Hill and Ward (1988) suggest that distal hominid humeri are very generalized, and that the characters of this specimen are quite compatible with its attribution to *A. afarensis*.

Koobi Fora Formation: Turkana Basin: Kenya

The Koobi Fora Formation is part of the Omo Group constituting many of the rocks of the Turkana basin. It ranges from 4.35 MYA into the Pleistocene. There are a number of hominoid specimens from the early part of this time scale (Walker pers. comm.), only one of which is so far described.

Allia Bay Tooth (KNM-ER 7727). A left upper M^2 crown has been described from the Allia Bay area toward the south area of eastern exposures at Lake Turkana (Leakey and Walker 1985). It occurs in the uppermost part of the Lonyumum Member of the Formation. Other teeth, so far undescribed, are also known from this region and time period. They are estimated to date at 4.10 MYA ± 0.10 MYA (Feibel et al. 1989). There may be other relevant undescribed material from the west side of the lake as well, which may date to this time.

Sagantole Formation: Middle Awash: Ethiopia

The Sagantole Formation forms part of the Awash Group of rocks, which also incorporates the underlying Adu-Asa Formation, the overlying Hadar Formation, as well as younger units. It is about 210 to 225 m thick. The original radiometric dating relevant to the Sagantole Formation (Hall et al. 1984) has recently been revised by Haileab and Brown (1992). Using tephrostratigraphy they have been able to correlate tuffs in the Sagantole Formation with dated tuffs in the Lake Turkana basin. These correlations suggest that a distinctive and very extensive horizon in the Sagantole Formation known as Cindery Tuff should be dated at about 3.8 MYA. Williams et al. (1986) provide environmental information based on the sedimentary context.

Belohdelie Frontoparietal (BEL-VP-1/1). Seven fragments of hominid fronto-parietal were found in a 1981 survey (Clark et al. 1984). They come from about 11 m below the Cindery Tuff. Hall et al. (1984) originally dated the specimen at 3.9 MYA, Clark et al. (1984) at 4.0 MYA, but Haileab and Brown (1992) suggest that it is between 3.8 and 3.9 MY in age. Three of the pieces join to form most of the frontal. It is an adult individual that has been described by Clark et al. (1984), White (1984) and most comprehensively by Asfaw (1987). The specimen shows a mix of primitive and derived characters. The individual it comes from appears to have possessed a broad upper face. There is no supraorbital sulcus, but the supraorbital margin is thick. Asfaw showed that it did not possess such a great postorbital constriction as did later *Australopithecus* crania, including estimates for *A. afarensis*. There are also other slight differences from known morphology in *A. afarensis*, but this is not inconsistent with its representing an early population of the taxon.

Maka Femur (MAK-VP-1/1). This specimen was found by White about 7 m above the projected horizon of the Cindery Tuff in the Maka drainage, about 700 m southwest of the Belohdelie specimens. Even taking a maximum date for the Cindery Tuff (see Haileab and Brown 1992) would place its age right at the recent border of our time range, if not actually younger than material from Laetoli. It has been briefly described by Clark et al. (1984) and by White (1984). The specimen is a proximal femur fragment of a subadult that shows a number of features shared with later examples of *Australopithecus*. These include low neck-shaft angle, flattened neck and proximal part of the shaft, little flare of the greater trochanter, and a relatively long neck. In these characters, it is essentially indistinguishable from the femora of *A. afarensis*.

Fejej Formation: Fejej: Ethiopia

The Fejej area is to the northeast of Lake Turkana, situated in the southern part of Ethiopia. The region was surveyed by the Paleoanthropological Inventory of the Ethiopian Ministry of Culture in 1989 (Asfaw et al. 1991), which revealed sediments ranging from Oligocene to Pleistocene in age, and including Pliocene vertebrate sites. Fossil hominoid material thought to belong to two individuals has been found and published so far (Fleagle et al. 1991). It comes from a sedimentary unit beneath a basalt attributed to the Harr Formation (Davidson 1983). The sediments may be at least partially equivalent to the Warata Formation of northern Kenya (Watkins and Williamson 1979; Asfaw et al. 1991). Davidson (1983) has published a conventional K/Ar date of 4.20 MYA on one of the basalts. Asfaw et al. (1991) have a date of 4.42 MYA on the basalt overlying one of their vertebrate sites (FJ-3), and Fleagle et al. (1991) record an Ar/Ar date from the basalt overlying the hominid site of 3.6-3.7 MYA. The vertebrate fauna is consistent with these estimates. Consequently at slightly over 3.6 MYA or slightly over 4.4 MYA, these specimens come from the very recent boundary of our time range.

Fejej Teeth (FJ-4-SB-1). This specimen number comprises remains of six mandibular teeth. All are in a comparable state of heavy wear, were found in close proximity, and are believed to come from a single aged individual. The sample consists of the following: a relatively complete right canine (FJ-4-SB-1a); a left P_3 preserving most of the crown and the cervical portion of the root (FJ-4-SB-1b); an almost complete left P_4 (FJ-4-SB-1c); a left molar, possibly M_2, with the majority of the crown and the cervical part of the roots (FJ-4-SB-1d); a fragmentary lower molar, possibly left M_1, with part of the crown and one root (FJ-4-SB-1e); and an extensively damaged fragment of the crown and one root of an unidentifiable lower molar (FJ-4-SB-1f). Taxonomically these specimens are most similar to *A. afarensis* among hominids that are already known, being almost undistinguishable in metric and nonmetric features.

Fejej Premolar (FJ-4-SB-2). A complete and unworn P_4 from an individual younger than FJ-4-SB-1.

WHAT? WHERE? WHEN?: THE BASIC DATA

The basic information needed to resolve many issues in human evolution comes from the hominoid fossil record itself. We need answers to the questions: What kinds of hominids existed? where do the fossils come from? and what is their age? For the origin of bipedal hominids, the answers to these questions are extremely inadequate, but reliable solutions are required before more sophisticated matters can be realistically addressed. Ideally we would like to know the nature of past hominoid species in this time interval, and how they are distributed in space and time. In particular, what was the first bipedal hominid like? Where and when did it evolve?

In general, more specimens are needed to establish what species existed at different times, and this taxonomic information is essential to understanding human phylogeny in more detail. Adequate representation of different skeletal parts is necessary to suggest aspects of these species' behavioral ecology.

To address these basic issues we also need a reasonable representation of specimens through time. To understand diversity we need a good geographic representation as well, and to sample hominoids throughout their range. Questions of time and geography become even more important once we tackle the origin of bipedalism, and the reasons for it.

To what extent does the existing fossil record between 14 MYA and 3.7 MYA enable us to respond to these questions? It is clear that the record is impoverished. There are very few occurrences of hominoid fossils, and the quality of individual specimens is generally poor. In particular skeletal sampling is bad. What few specimens we have consist for the most part of isolated teeth or bone fragments that give relatively little information about their possessor. Postcranial bones are few, which is clearly a problem when establishing bipedalism is one of the more interesting immediate issues.

Specimens are not evenly represented through time. Rather their distribution is polarized, and skewed toward the recent end of the time period. From the point of view of this discussion the specimens can be placed in three categories. One category is formed by some specimens at the base of our time period. They are the remains of apes essentially similar to ones we know from the earlier part of the Miocene that persist somewhat later in time. An example is the suggestion of something like *Proconsul* from the Tugen Hills Ngorora Formation.

A second large group that forms the bulk of the sample consists of the most recent fossils. Most of these occur between about 5.5 MYA and 4 MYA, and appear to be indistinguishable from specimens of *A. afarensis* known more certainly from younger strata. They include the fragmentary mandibles from Lothagam and Tabarin, the Chemeron and Kanapoi humeri, the pieces of cranium from Belhodelie, femur from Maka, and the teeth from Fejej. Some undescribed specimens from other areas in the Lake Turkana basin are also likely to fall into this category. All of these have been compared or tentatively assigned by various authors to *A. afarensis*, but with the reservation in most cases that they are provisional rather than definitive identifications. This is because, first, almost none of them preserves sufficient information to suggest bipedalism, and second, as several people have pointed out (e.g., Andrews 1992), the characters the fossils display, particularly in the teeth and jaws, are primitive and insufficient to ally them unambiguously with later bipedal hominids. If they do prove to be remains of bipedal animals, then they indicate a creature probably very much like *A. afarensis* slightly farther back in time. But it would be equally interesting if some of these could be shown not to be bipeds.

The third category consists of specimens that are neither like earlier known apes nor can be attributed even loosely to *A. afarensis*. *Otavipithecus* from Namibia falls into this category, as it appears quite clearly to be a new taxon. However, in isolation and on a larger scale, it appears not to reveal much about later hominoid evolution. The same may be said of such specimens as the Ngorora molar. Although it probably belongs to an otherwise unknown taxon, it is essentially like a Miocene ape, and its position in the greater scheme of things is largely conjectural. Only a few specimens in this category seem possibly to have links with later hominoids. One is the Lukeino molar. This appears to be like no other known taxon and suggests possible similarities between chimpanzees and later hominids. But it is an isolated tooth. A much better specimen in this category is the Samburu Hills hemimaxilla. Again this is unlike any other known species. A number of authors have alluded to its similarities to the gorilla, despite its having thick enamel, a presumed primitive character.

In summary, the present sample of hominoids suggests, first, an extension of Miocene apes essentially similar to earlier ones into a later time period, and second, something provisionally like *A. afarensis* reaching slightly farther back in time. Neither of these classes should be ignored, because they may be important for defining time ranges of known taxa. Some may refine estimates of the origination time of the hominid clade. The earliest evidence of bipedalism is best established by the footprints at Laetoli, Tanzania, dated at 3.7 MYA. But if our taxonomic inferences are correct, we have hints of hominids at about 4.5 MYA from Tabarin in the Tugen Hills, or perhaps at >5.6 MYA from Lothagam. Our present evidence weakly constrains the origin of Hominidae to probably between 6 and 9 MYA. This is a time from which as yet we have almost no hominoid specimens.

Geographic factors are also important. We speak of the African fossil record, but in fact our sample is extremely confined geographically. Most of the specimens come from Kenya, where the sites concerned can be enclosed within a rectangle about 300 by 100 km in size. There is a smaller region in Ethiopia, and one site in Namibia. As Africa is about 30,335,000 km^2 in area, this represents a sampling of not much more than 0.1 percent of the continent. This is also true of other critical intervals in human evolution. It is interesting that in earlier and later periods new regions of exploration often produce new species or even new genera, hinting at a greater diversity of hominoid taxa in Africa as a whole. This emphasizes the difficulty of appreciating diversity and delineating phylogenies on the basis of our present impoverished samples.

Did the main events of hominid evolution really take place in the area of eastern Africa where, due to geological coincidences, we happen to have our sample? Or did they occur somewhere else, where there are no fossil vertebrates of any kind preserved, and where we have no knowledge at all of certain critical species and events? Certainly this suspicion is likely to be true of the African great ape lineages, but to a significant degree it may apply to hominids as well. It is difficult to conclude much from the absence of hominoids at otherwise productive vertebrate fossil sites, but the lack of hominoids from the rich Langabaanweg assemblage is interesting. Did hominids not exist in southern Africa at about 5 MYA? And were hominids really confined to Africa during these times, or may we hope to find them elsewhere?

WHY? AND HOW?: VEGETATION AND THEORIES OF ORIGIN

The questions just discussed are fundamental to inquiring into other, perhaps more interesting issues, involving the causes of bipedalism and the process of its acquisition.

In a proximate sense, why did hominids evolve? How did bipedalism come about? Evolution is generally thought to be impelled by environmental factors, and people have traditionally thought it useful to examine paleoenvironments in order to understand the process better. In the present case, from at least Darwin onward, bipedalism has been assumed to have some connection with environmental change or with feeding behavior perhaps forced by a changing habitat. In the *Descent of Man*, Darwin writes that primates "came to be less arboreal, owing to a change in its manner of procuring subsistence, or to some change in the surrounding conditions" (Darwin 1871). In the last century, paleoanthropologists developed a scenario of relatively sudden change in these surrounding conditions, from presumed extensive forests in the Miocene of east Africa to the open savanna grasslands now known there. This in turn led to the idea of an ape descending from the trees and getting up on its back legs on the wide grassland plains.

What is the evidence for such a transition from forest to grass, and when did it take place? As mentioned in the introduction, the site of Fort Ternan in western Kenya, dated at 14 MYA, is particularly interesting in this context. It was for a long time the site that provided one of the oldest supposed examples of the Hominidae, in the form of *Kenyapithecus wickeri* for Louis Leakey (Leakey 1962, 1967) or *Ramapithecus wickeri* for most other researchers (Simons 1963; Simons and Pilbeam 1965). As *Kenyapithecus* was presumed to be a hominid, there was the implication of bipedality. As there was bipedality, so was there the assumption that the putative shift from forest to savanna grassland had already occurred by 14 MYA. Consequently considerable attention has been focused on the paleoenvironment of Fort Ternan, particularly in terms of this dichotomy between forest and savanna. The site still receives this attention, even though *Kenyapithecus* is no longer thought of as a bipedal hominid, and thus the initial motive for the question no longer exists.

The Fort Ternan literature is considerable. The evidence is involved, and it comes from diverse sources. I will not review it in detail here, but in summary it is known that a large volcano, Tinderet, existed in the vicinity of the site at the time of the site's formation. Even if there were extensive grassland around the area of the fossil site itself, it is quite possible that such a high mountain supported forest, and therefore that forest was within the site catchment. If *Kenyapithecus* lived in these forests, it could potentially have become incorporated in the Fort Ternan sediments. To what extent was there also grassland at this time?

Bonnefille (1984) has recorded grass pollen from the site but makes no claims for its extent or relative importance. Most authors, using various lines of evidence and disagreeing on details, all seem to favor an interpretation involving predominantly woodland with some grass. These include Andrews and Nesbit Evans (1979) on mammal community reconstruction, Pickford (1987b) on mollusks, Kappelman (1991) from bovid anatomy, Cerling and others (Cerling 1992; Cerling et al. 1991, 1992) using isotopes in soil carbonates, and Shipman (1982, 1986; also Shipman et al. 1980) on a variety of factors. Only Retallack and his colleagues (Retallack 1991, 1992a,b; Retallack et al. 1990; Dugas and Retallack 1993), working with fossil grasses and paleopedological inference, advocate really extensive grasslands that they compare to modern savannas. They also imply that this is a major, almost permanent change in the African environmental scene, which took place 14 MY ago.

For a number of reasons, at present I provisionally lean toward the consensus of the other authors and prefer a woodland environment with some grass. This conclusion appears to be supported by the remaining evidence for vegetation at later times in eastern Africa (Hill 1987). In the Tugen Hills sequence in the Kenya Rift Valley there is an excellent site (BPRP# K055) in the Ngorora Formation dated at 12.6 MYA (Deino et al. 1990) that provides direct evidence of the mid-Miocene flora. It occurs only a few meters below the site previously discussed (BPRP# K038), dated at about 12.5 MYA, which has produced fossil cercopithecoids and at least one hominoid tooth (KNM-BN 1461; KNM-TH 23144). The plant site preserves many tree branches with intact leaves attached, and also elements of the herbaceous component. The assemblage is autochthonous, having been preserved in an air fall tuff. It thus preserves an excellent almost instantaneous sample of the local flora (Jacobs and Winkler 1992). The fossils have been studied by Jacobs and Kabuye (1987, 1989; Kabuye and Jacobs 1986) who identify over 55 species based upon leaf morphology and epidermal micromorphology. The species composition suggests a true lowland rainforest habitat, not one simply confined to river margins.

Later, in Ethiopia, there is evidence of forest conditions at 8 MYA (Yemane et al. 1985). Then in the Tugen Hills, in the Mpesida Formation dated at about 6.5 MYA, is fossil wood that again suggests rainforest (Jacobs pers. comm.). Also at this level is an extensive area preserving intact trunks of trees embedded in place in a volcanic ash. Pickford, Senut, and colleagues report floral data in the Lake Albert basin, Uganda. Fossil fruits from the Nkondo Formation, dated at 6.5 to 5 MYA on the basis of the mammal fauna, suggest lowland tropical forest conditions, species such as *Afzelia* and *Antrocaryon* being common (Pickford et al. 1991; Dechamps et al. 1992). In the Pliocene, palynological work suggests predominant grassland at Laetoli, Tanzania, at 3.4 MYA (Bonnefille and Riollet 1987). But at Lake Turkana at about the same time, fossil fruit (Bonnefille and Letouzey 1976) and mollusks (Williamson 1985) indicate rainforest conditions.

Another approach we have applied to the Tugen Hills sequence involves stable carbon isotopes. Different plants use different metabolic pathways, resulting in different carbon isotope compositions. Two pathways are relevant here, known as C_3 and C_4. C_3 plants include all trees, nearly all shrubs, and grasses that prefer a cool growing season. C_4 plants include some shrubs and tropical grasses growing in warm areas. A C_4 signal has therefore been used to indicate savanna conditions. Various constituents of fossil soils reflect the proportion of C_3 to C_4 plants occurring at the time of soil formation, but the one most exploited so far is soil carbonates.

In the Siwaliks of Pakistan, this kind of work was carried out by Quade and colleagues (Quade et al. 1989). They found a marked shift from C_3 to C_4 dominance between 7.4 and 6 MYA. They interpreted this as a shift from trees and shrubs to extensive grassland, possibly related to the development of the monsoon system. It coincides neatly with the local disappearance of large hominoids from the region. It was thought that an analogous signal might be seen at the same time in Africa, and Cerling, an author on the Siwalik paper, carried out preliminary work in the Tugen Hills sequence. There the late Miocene is apparently also dominated by trees and shrubs, but C_4 grasses are not detected to any significant degree in the time period under consideration here. C_4 grasses only come to dominate the vegetation much later, during the middle Pleistocene (Cerling 1992).

More extensive and detailed work in the Tugen Hills is being carried out by Kingston (1992; Kingston et al. 1992). He shows considerable variation in carbon isotope composition of soil carbonates throughout the Tugen Hills section, with no clear uniform trends. There is an increase in the representation of C_4 grasses after about 8.5 MYA, but it is unlikely that either extreme closed canopy forest or open savanna grassland environments persisted in this part of the Rift Valley at any time during the last 15 MYA. Rather a much patchier vegetative structure is indicated.

A number of conclusions follow from these data. In Africa, at least in the areas for which we have information, there is no sign of a rapid decisive shift from forest to savanna grassland during the time period under consideration here. However, we have good reason to believe that bipedal hominids existed by perhaps 5.5 MYA and possibly by 8 MYA. The origin of hominids may still perhaps be explained by taking vegetative structure into account, but it is not the simple direct explanation that has generally been assumed. Notions of bipedality have to be decoupled from the origin of savanna grasslands. We might also separate ideas of bipedality from the abdication of aboreality by early hominids. There are anatomical suggestions (Susman et al. 1984; Susman and Stern 1982) that hominids, including *Australopithecus afarensis* and maybe early members of the genus *Homo*, while bipedal, may also to a certain extent have retained arboreal capabilities until quite late in time.

SUMMARY

The period between 14 and 4 MYA in Africa is critical for human evolution, for during this time the major clades of modern African hominoids diverged, including bipedal hominids. However, samples are extremely restricted in both space and time. Less than a dozen relevant specimens have been described, and none is very informative taxonomically or anatomically. Most occur near the end of the time period, from about 5.6 to 3.7 MYA, and have been compared to *Australopithecus afarensis*. Others, low in the time period, are essentially similar to Miocene apes already known. Only one or two specimens, from the middle of the time span, show unique characters. The time of origin of hominids is only very weakly constrained, to perhaps between 6 and 9 MYA, a time when there are almost no hominoid fossils. Geographic sampling is extremely restricted, emphasizing the difficulties of appreciating hominoid diversity and delineating phylogenies. The origin of bipedalism may be linked to changing ecological factors, but it is unlikely to be connected with a simple change in eastern Africa from forest to savanna grassland conditions. Savanna grasslands are not detected until hominids are well established and diverse. A variety of evidence suggests that patchy environments including a forest component persist until quite late in time.

ACKNOWLEDGMENTS

Some of the specimens and issues in this chapter were originally discussed with Steve Ward in a paper that appeared in the *Yearbook of Physical Anthropology* (Hill and Ward 1988). Joint elements of that work necessarily appear in this chapter. I am fortunate to have had the opportunity of collaborating with Steve over the last decade, and I thank him for his relentless continuing participation in the Baringo Paleontological Research Project. Our work in Africa has been financed by grants from the National Science Foundation (BNS-9208903), the LSB Leakey Foundation, the Louise Brown Foundation, and Mr. J. Clayton Stephenson. For discussion about unpublished information, I am grateful to Frank Brown, Terry Harrison, Meave Leakey, Alan Walker, and Tim White. Sally McBrearty provided essential discussion of the manuscript. I thank Robert Corruccini and Russell Ciochon for inviting me to the Clark Howell symposium, and for making it a most memorable and enjoyable occasion.

LITERATURE CITED

Aguirre, E., and Leakey, P. (1974) Nakali: Nueva fauna de *Hipparion* del Rift Valley de Kenya. *Estudios Geologicos* 30:219-227.

Andrews, P. (1992) Evolution and environment in the Hominoidea. *Nature* 360:641-646.

Andrews, P., and Martin, L. (1987) Cladistic relationships of extant and fossil hominoids. *J. Hum. Evol.* 16:101-118.

Andrews, P. and Nesbit Evans, E. (1979) The environment of *Ramapithecus* in Africa. *Paleobiol.* 5:22-30.

Asfaw, B. (1987) The Belohdelie frontal: New evidence of early hominid cranial morphology from the Afar of Ethiopia. *J. Hum. Evol.* 16:612-624.

Asfaw, B., Beyene, Y., Semaw, S., Suwa, G., White, T., and WoldeGabriel, G. (1991) Fejej: A new paleoanthropological research area in Ethiopia. *J. Hum. Evol.* 21:137-143.

Baker, B. H. (1963) The geology of the Baragoi Area. *Report of the Geological Survey of Kenya* 53:1-74.

Barry, J., Hill, A., and Flynn, L. (1985) Variation de la faune au Miocène inférieur et moyen de l'Afrique de l'est. *L'Anthropologie* (Paris) 89:271-273.

Behrensmeyer, A. K. (1976) Lothagam Hill, Kanapoi, and Ekora: A general summary of stratigraphy and faunas. In Y. Coppens, F. C. Howell, G. L. Isaac, and R. E. Leakey (eds.): *Earliest Man and Environments in the Lake Rudolf Basin: Stratigraphy, Paleoecology, and Evolution*. Chicago: University of Chicago Press, pp. 163-170.

Bernor, R. L. (1982) A preliminary assessment of the mammalian biochronology and zoogeographic relationships of Sahabi, Libya. *Garyounis Sci. Bull. Special Issue*. 4:133-142.

Bernor, R. L., and Pavlakis, P. P. (1987) Zoogeographic relationships of the Sahabi large mammal fauna (Early Pliocene, Libya). In N. T. Boaz, A. W. Gaziry, A. El-Arnauti, J. de Heinzelin, and D. D. Boaz (eds.): *Neogene Paleontology and Geology of Sahabi*. New York: Alan R. Liss; pp. 349-383.

Bishop, W. W. (1969) Pleistocene Stratigraphy in Uganda. *Geol. Survey of Uganda: Mem.* 10.

Bishop, W. W., and Chapman, G. R. (1970) Early pliocene sediments and fossils from the northern Kenya Rift Valley. *Nature* 226:914-918.

Bishop, W. W., Chapman, G. R., Hill, A., and Miller, J. A. (1971) Succession of Cainozoic vertebrate assemblages from the northern Kenya Rift Valley. *Nature* 233:389-394.

Bishop, W. W., and Pickford, M. (1975) Geology, fauna and palaeoenvironments of the Ngorora Formation, Kenya Rift Valley. *Nature* 254:185-192.

Boaz, N. T. (1980) A hominoid clavicle from the Mio-Pliocene of Sahabi, Libya. *Am. J. Phys. Anthropol.* 53:49-54.

Boaz, N. T. (1987) Sahabi and Neogene Hominoid evolution. In N. T. Boaz, A. W. Gaziry, A. El-Arnauti, J. de Heinzelin, and D. D. Boaz (eds.): *Neogene Paleontology and Geology of Sahabi*. New York: Alan R. Liss, pp. 129-134.

Boaz, N. T., and Meikle, W. E. (1982) Fossil remains of primates from the Sahabi Formation. *Garyounis Sci. Bul. Special Issue*. 4:1-48.

Boaz, N. T., Gaziry, A. W., and El-Arnauti, A. (1979) New fossil finds from the Libyan upper Neogene site of Sahabi. *Nature* 280:137-140.

Boaz, N. T., Gaziry, A. W., El-Arnauti, A., de Heinzelin, J., and Boaz, D. D. (1987) *Neogene Paleontology and Geology of Sahabi*. New York: Alan R. Liss.

Boaz, N. T., Gaziry, A. W., de Heinzelin, J., and El-Arnauti, A (1982) Results from the International Sahabi Research Project (Geology and Paleontology). *Garyounis Sci. Bull. special Issue* 4:1-142.

Bonnefille, R. (1984) Cenozoic vegetation and environments of early hominids in East Africa. In R. O. Whyte (ed.): *The Evolution of the East Asian Environment. Vol. 2: Palaeobotany, Palaeozoology and Palaeoanthropology*. University of Hong Kong: Centre of Asian Studies, pp. 579-612.

Bonnefille, R., and Letouzey, R. (1976) Fruits fossiles d' *Antrocaryon* dans la Vallée de l'Omo (Ethiopie). *Adansonia* 2:65-82.

Bonnefille, R., and Riollet, (1987) Palynological spectra from the Upper Laetolil Beds. In M. D. Leakey and J. M. Harris (eds.): *Laetoli: A Pliocene Site in Northern Tanzania*. Oxford: Oxford University Press, pp. 52-61.

Boswell, P. G. H. (1935) Human remains from Kanam and Kanjera, Kenya Colony. *Nature* 135:371.

Brown, B., Hill, A., and Ward, S. (1991) New Miocene large hominoids from the Tugen Hills, Baringo District, Kenya. *Am. J. Phys. Anthropol: Supplement* 12:55.

Brown, F., and Feibel, C. S. (1991) Stratigraphy, depositional environments and palaeogeography of the Koobi Fora Formation. In J. M. Harris (ed.): *Koobi Fora Research Project. Vol. 3: The Fossil Ungulates: Geology, Fossil Artiodactyls and Palaeoenvironments*. Oxford: Oxford University Press, pp. 1-30.

Cerling, T. E. (1992) Development of grasslands and savannas in East Africa during the Neogene. *Palaeogeog., Palaeoclimatol., Palaeoecol.* 97:241-247.

Cerling, T. E., Quade, J., Ambrose, S. H., and Sikes, N. E. (1991) Fossil soils, grasses, and carbon isotopes from Fort Ternan, Kenya: Grassland or woodland? *J. Hum. Evol.* 21:295-306.

Cerling, T. E., Kappelman, J., Quade, J., Ambrose, S. H., Sikes, N. E., and Andrews, P. (1992) Reply to comment on the paleoenvironment of *Kenyapithecus* at Fort Ternan. *J. Hum. Evol.* 23:371-377.

Chapman, G. R. (1971) The Geological Evolution of the Northern Kamasia Hills, Baringo District, Kenya. Unpublished Ph.D. thesis, University of London.

Chapman, G. R., and Brook, M. (1978) Chronostratigraphy of the Baringo Basin, Kenya Rift Valley. In W. W. Bishop (ed.): *Geological Background to Fossil Man.* London: Geological Society of London: Scottish Academic Press, pp. 207-223.

Chapman, G. R., Lippard, S. J., and Martyn, J. E. (1978) The stratigraphy and structure of the Kamasia Range, Kenya Rift Valley. *J. Geol. Soc. Lond.* 135:265-281.

Clark, J. D., Asfaw, B., Assefa, G., Harris, J. W. K., Kurashina, H., Walter, R. C., White, T. D., and Williams, M. A. J. (1984) Palaeoanthropological discoveries in the Middle Awash Valley, Ethiopia. *Nature* 307:423-428.

Conroy, G. C., Pickford, M. H. L., Senut, B., Van Couvering, J., and Mein, P. (1992) *Otavipithecus namibiensis*, first Miocene hominoid from southern Africa. *Nature* 356:144-147.

Cooke, H. B. S. (1987) Fossil Suidae from Sahabi Libya In N. T. Boaz, A. W. Gaziry, A. El-Arnauti, J. de Heinzelin, and D. D. Boaz (eds.): *Neogene Paleontology and Geology of Sahabi.* New York: Alan R. Liss, pp. 255-266.

Corruccini, R. S. (1975) A Metrical Study of Crown Component Variation in the Hominoid Dentition. Ph.D dissertation, University of California, Berkeley.

Corruccini, R. S., and McHenry, H. M. (1980) Cladometric analysis of Pliocene hominoids. *J. Hum. Evol.* 9:209-221.

Darwin, C. (1871) *The Descent of Man, and Selection in Relation to Sex.* London: John Murray.

Davidson, A. (1983) *The Omo River Project: Bulletin 2.* Ethiopian Institute of Geological Surveys: Ministry of Mines and Energy.

Dechamps, R., and Maes, F. (1987) Paleoclimatic interpretation of fossil wood from the Sahabi Formation. In N. T. Boaz, A. W. Gaziry, A. El-Arnauti, J. de Heinzelin, and D. D. Boaz (eds.): *Neogene Paleontology and Geology of Sahabi.* New York: Alan R. Liss, pp. 43-81.

Dechamps, R., Senut, B., and Pickford, M. (1992) Fruits fossiles pliocènes et pléistocènes du Rift Occidental ougandais. Signification paléoenvironnementale. *C. R. Acad. Sci.* Paris. 314:325-331.

Dechant Boaz, D. (1982) Preliminary assessment of taphonomy and paleoecology at Sahabi. *Garyounis Sci. Bull. Special Issue* 4:109-121.

Dechant Boaz, D. (1987) Taphonomy and paleoecology at the Pliocene site of Sahabi, Libya. In N. T. Boaz, A. W. Gaziry, A. El-Arnauti, J. de Heinzelin, and D. D. Boaz (eds.): *Neogene Paleontology and Geology of Sahabi.* New York: Alan R. Liss. pp. 337-348.

de Heinzelin, J. (1988) Photogeologie du Neogene de la Basse-Semliki (Zaire). *Bull. Soc. Belg. Geol.* 97:173-178.

de Heinzelin, J., and El-Arnauti, A. (1982) Stratigraphy and geological history of the Sahabi and related formations. *Garyounis Res. Bull. Special Issue* 4:5-12.

de Heinzelin, J., and El-Arnauti, A. (1987) The Sahabi Formation and related deposits. In N. T. Boaz, A. W. Gaziry, A. El-Arnauti, J. de Heinzelin, and D. D. Boaz (eds.): *Neogene Paleontology and Geology of Sahabi.* New York: Alan R. Liss, pp 1-21.

Deino, A., Tauxe, L., Monaghan, M., and Drake, R. (1990) $^{40}Ar/^{39}Ar$ age calibration of the litho- and paleomagnetic stratigraphies of the Ngorora Formation, Kenya. *J. Geol.* 98:567-587.

Desio, A. (1931) Osservationi geologiche e geografiche compiute durante un viaggio nella Sirtica. *Boll. Reale. Soc. Geogr. Ital.* Ser 6. 3:275-299.

Dugas, D. P., and Retallack, G. J. (1993) Middle Miocene fossil grasses from Fort Ternan, Kenya. *J. Paleontol.* 67:113-128.

Feibel, C. S., Brown, F. H., and McDougall, I. (1989) Stratigraphic context of fossil hominids from the Omo Group deposits: northern Turkana Basin, Kenya and Ethiopia. *Am. J. Phys. Anthropol.* 78:595-622.

Fleagle, J. G., Rasmussen, T., Yirga, S., Bown, T. M., and Grine, F. E. (1991) New hominid fossils from Fejej, southern Ethiopia. *J. Hum. Evol.* 21:145-152.

Fuchs, V. E. (1934) The geological work of the Cambridge Expedition to the East African Lakes, 1930-31. *Geol. Mag.* 71:97-112.

Fuchs, V. E. (1950) Pleistocene events in the Baringo Basin. *Geol. Mag.* 87:149-174.

Gaziry, A. W. (1982) Proboscidea from the Sahabi Formation. *Garyounis Sci. Bull. Special Issue* 4:101-108.

Gaziry, A. W. (1987) Remains of Proboscidea from the early Pliocene of Sahabi, Libya. In N. T. Boaz, A. W. Gaziry, A. El-Arnauti, J. de Heinzelin, and D. D. Boaz (eds.): *Neogene Paleontology and Geology of Sahabi.* New York: Alan R. Liss, pp. 183-203.

Grace, G., and Stockley, G. M. (1930) Geology of the Usongo area, Tanganyika Territory. *J. East Afr. Nat. His. Soc.* 37:185-192.

Gregory, J. W. (1896) *The Great Rift Valley.* London: John Murray.

Gregory, J. W. (1921) *The Rift Valleys and Geology of East Africa.* London: Seeley Service and Co.

Haileab, B., and Brown, F. H. (1992) Turkana Basin-Middle Awash valley correlations and the age of the Sagantole and Hadar Formations. *J. Hum. Evol.* 22:453-468.

Hall, C. M., Walter, R. C., Westgate, J. A., and York, D. (1984) Geochronology, stratigraphy and geochemistry of Cindery Tuff in the Pliocene hominid-bearing sediments of the Middle Awash, Ethiopia. *Nature* 308:26-31.

Harris, J. M., Brown, F. H., and Leakey, M. G. (1988) Stratigraphy and paleontology of Pliocene and Pleistocene localities west of Lake Turkana, Kenya. *Contributions in Science: Nat. His. Mus. of Los Angeles County*, 399.

Harrison, T. (1992) Paleoanthropological exploration in the Manonga Valley, northern Tanzania. *Nyame Akuma* 36:25-31.

Harrison, T., and Verniers, J. (in press) Preliminary study of the stratigraphy and mammalian palaeontology of Neogene sites in the Manonga Valley, northern Tanzania. *N. Jb. Geol. Paläont. Abh.*

Harrison, T., Verniers, J., Mbago, M. L., and Krigbaum, J. (in press) Stratigraphy and mammalian palaeontology of Neogene sites in the Manonga Valley, northern Tanzania. *Discovery and Innovation.*

Hendey, B. (1970a) A review of the geology and palaeontology of the Plio/Pleistocene deposits at Langabaanweg, Cape Province. *Ann. S. Afr. Mus.* 56:75-117.

Hendey, B. (1970b) The age of the fossiliferous deposits at Langabaanweg, Cape Province. *Ann. S. Afr. Mus.* 56:119-131.

Hendey, B. (1972) Further observations on the age of the mammalian fauna from Langabaanweg, Cape Province. *Palaeoecol. Afr.* 6:172-175.

Hendey, B. (1973) Fossil occurrences at Langabaanweg, Cape Province. *Nature* 244:13-14.

Hendey, B. (1974) The late Cenozoic Carnivora of the southwestern Cape Province. *Ann. S. Afr. Mus.* 63:1-369.

Hendey, B. (1976) The Pliocene fossil occurrences in 'E' Quarry, Langabaanweg, South Africa. *Ann. S. Afr. Mus.* 69:215-247.

Hendey, B. (1984) Southern African late Tertiary vertebrates. In: R. G. Klein (ed.): *Southern African Prehistory and Paleoenvironments.* Rotterdam: Balkema, pp. 381-395.

Hill, A. (1985a) Les variations de la faune du Miocène récent et du Pliocène d'Afrique de l'est. *L'Anthropologie* (Paris) 89:275-279.

Hill, A. (1985b) Early hominid from Baringo, Kenya. *Nature.* 315:222-224.

Hill, A. (1987) Causes of perceived faunal change in the later Neogene of East Africa. *J. Hum. Evol.* 16:583-596.

Hill, A. (1989) Kipsaramon: A Miocene hominoid site in Kenya. *Am. J. Phys. Anthropol.* 78:241.

Hill, A., Drake, R., Tauxe, L., Monaghan, M., Barry, J. C., Behrensmeyer, A. K., Curtis, G., Fine Jacobs, B., Jacobs, L., Johnson, N., and Pilbeam, D. (1985) Neogene palaeontology and geochronology of the Baringo Basin, Kenya. *J. Hum. Evol.* 14:749-773.

Hill, A., Curtis, G., and Drake, R. (1986) Sedimentary stratigraphy of the Tugen Hills, Baringo District, Kenya. In L. E. Frostick, R. W. Renaut, I. Reid, and J. J. Tiercelin (eds.): *Sedimentation in the African Rifts.* Oxford: Blackwell: Geol. Soc. Lond. Special Publication 25, 285-295.

Hill, A., Behrensmeyer, A. K., Brown, B., Deino, A., Rose, M., Saunders, J., Ward, S., and Winkler, A. (1991) Kipsaramon: A lower Miocene hominoid site in the Tugen Hills, Baringo District, Kenya. *J. Hum. Evol.* 20:67-75.

Hill, A., Brown, B., and Ward, S. (1992) Miocene large hominoids from Kipsaramon, Tugen Hills, Kenya. In J. Van Couvering (ed.): *Apes or Ancestors?* New York: American Museum of Natural History, pp. 19-20.

Hill, A. and Ward, S. (1988) Origin of the Hominidae: The record of African large hominoid evolution between 14 My and 4 My. *Yearb. Phys. Anthropol.* 31:49-83.

Hill, A., Ward, S., and Brown, B. (1992) Anatomy and age of the Lothagam mandible. *J. Hum. Evol.* 22:439-451.

Hopwood, A. T. (1931) Pleistocene mammalia from Nyasaland and Tanganyika Territory. *Geol. Mag.* 68:133-135.

Howell, C. and Coppens, Y. (1974) Inventory of remains of Hominidae from Pliocene/Pleistocene formations of the lower Omo Basin, Ethiopia (1967-1972). *Am. J. Phys. Anthropol.* 40:1-16.

Ishida, H. (1984) Outline of the 1982 survey in Samburu Hills and Nachola area, northern Kenya. *African Study Monographs, Supplementary Issue* 2:1-14.

Ishida, H., Ishida, S., Torii, M., Matsuda, T., Kawamura, Y., and Koizumi, K. (1982) Report of field survey in Kirimum, Kenya, 1980. *Study of the Tertiary Hominoids and their Palaeoenvironments in East Africa* 1:1-181. Osaka: Osaka University Press.

Ishida, H., Ishida, S., and Pickford, M. (1984a) Study of the Tertiary hominoids and their palaeoenvironments in East Africa: 2. *African Study Monographs* 2. Kyoto: Kyoto University.

Ishida, H., Pickford, M., Nakaya, H., and Nakano, Y. (1984b) Fossil anthropoids from Nachola and Samburu Hills, Samburu District, Kenya. *African Study Monographs, Supplementary Issue* 2:73-85.

Jacobs, B. F., and Kabuye, C. H. S. (1987) A middle Miocene (12.2 my old) forest in the East African Rift Valley, Kenya. *J. Hum. Evol.* 16:147-155.

Jacobs, B. F., and Kabuye, C. H. S. (1989) An extinct species of *Pollia* Thunberg (Commelinaceae) from the Miocene Ngorora Formation, Kenya. *Rev. Palaeobot. Palynol.* 59:67-76.

Jacobs, B. F., and Winkler, D. A. (1992) Taphonomy of a middle Miocene autochthonous forest assemblage, Ngorora Formation, central Kenya. *Palaeogeog. Palaeoclimatol. Palaeoecol.* 99:31-30.

Kabuye, C. H. S. and Jacobs, B. F. (1986) An interesting record of the genus *Leptaspis*, Bambusoideae from Middle Miocene flora deposits in Kenya, East Africa. *Abstracts: International Symposium on Grass Systematics and Evolution.* Washington DC: Smithsonian Institution, 32.

Kalb, J. E. (1993) Refined stratigraphy of the hominid-bearing Awash Group, Middle Awash Valley, Afar Depression, Ethiopia. *Newsletter in Stratigraphy.*

Kalb, J. E., Wood, C. B., Smart, C., Oswald, E. B., Mebrate, A., Tebedge, S., and Whitehead, P. (1980) Preliminary geology and paleontology of the Bodo d'Ar hominid site, Afar, Ethiopia. *Paleogeog. Paleoclimatol. Paleoecol.* 30:107-120.

Kalb, J. E., and Jolly, C. J. (1982) Late Miocene and early Pliocene formations in the Middle Awash Valley, Ethiopia, and their bearing on the zoogeography of Sahabi. *Garyounis Sci. Bull: Special Issue* 4:123-132.

Kalb, J. E., Oswald, E. B., Mebrate, A., Tebedge, S., and Jolly, C. J. (1982a) Stratigraphy of the Awash Group, Middle Awash Valley, Afar, Ethiopia. *Newsletter in Stratigraphy* 11:95-127.

Kalb, J. E., Jolly, C. J., Mebrate, A., Tebedge, S., Smart, C., Oswald, E. B., Cramer, D., Whitehead, P., Wood, C. B., Conroy, G. C., Adefris, T., Sperling, L., and Kana, B. (1982b) Fossil mammals and artefacts from the Middle Awash Valley, Ethiopia. *Nature* 298:17-25

Kalb, J. E., Jolly, C. J., Tebedge, S., Mebrate, A., Smart, C., Oswald, E. B., Whitehead, P., Wood, C. B., Adefris, T., and Rawn-Schatzinger, V. (1982c) Vertebrate faunas from the Middle Awash Valley, Afar, Ethiopia. *J. Vert. Paleontol.* 2:237-258.

Kalb, J. E., Oswald, E. B., Mebrate, A., Tebedge, S., and Jolly, C. J. (1982d) Stratigraphy of the Awash Group, Middle Awash Valley, Afar, Ethiopia. *Newsletter in Stratigraphy* 11:95-127.

Kalb, J. E., Oswald, E. B., Tebedge, S., Mebrate, A., Tola, E., and Peak, D. (1982e) Geology and stratigraphy of Neogene deposits, Middle Awash Valley, Afar, Ethiopia. *Nature* 298:25-29.

Kalb, J. E., Jolly, C. J., Oswald, E., and Whitehead, P. (1984) Early hominid habitation in Ethiopia. *Am. Sci.* 72:168-178.

Kappelman, J. (1991) The paleoenvironment of *Kenyapithecus* at Fort Ternan. *J. Hum. Evol.* 20:95-110.

King, B. C., and Chapman, G. R. (1972) Volcanism of the Kenya rift valley. *Phil. Trans. Royal Soc. London*. Series A. 271:185-208.

Kingston, J. (1992) Stable Isotopic Evidence for Hominid Paleoenvironments in East Africa. Ph.D. thesis, Harvard University.

Kingston, J., Hill, A., and Marino, B. (1992) Isotopic evidence of late Miocene / Pliocene vegetation in the east African Rift Valley. *Am. J. Phys. Anthropol. Suppl.* 14:100-101

Kramer, A. (1986) Hominid-pongid distinctiveness in the Miocene-Pliocene fossil record: The Lothagam mandible. *Am. J. Phys. Anthropol.* 70:457-473.

Leakey, L. S. B. (1935) *The Stone Age Races of Kenya Colony*. Oxford: Oxford University Press.

Leakey, L. S. B. (1936) Fossil human remains from Kanam and Kanjera, Kenya Colony. *Nature* 138:643.

Leakey, L. S. B. (1962) A new lower Pliocene fossil primate from Kenya. *Ann. Mag. Nat. Hist.* 4:686-696.

Leakey, L. S. B. (1967) An early Miocene member of Hominidae. *Nature* 213:155-163.

Leakey, R. E. F., and Leakey, M. (1978) *Koobi Fora Research Project*. Vol. 1: *The Fossil Hominids and an Introduction to their Context 1968-74*. Oxford: Clarendon Press.

Leakey, R. E. F., and Walker, A. (1985) Further hominids from the Plio-Pleistocene of Koobi Fora, Kenya. *Am. J. Phys. Anthropol.* 67:135-163.

Makinouchi, T., Koyaguchi, T., Matsuda, T., Mitsushio, H., and Ishida, S. (1984) Geology of the Nachola area and the Samburu Hills, west of Baragoi, northern Kenya. *Afr. Study Mono., Supplementary Issue* 2:15-44.

Martyn, J. E. (1967) Pleistocene deposits and new fossil localities in Kenya. *Nature* 215:476-477.

Martyn, J. E. (1969) The Geological History of the Country between Lake Baringo and the Kerio River, Baringo District, Kenya. Unpublished Ph.D. thesis, University of London.

Matsuda, T., Torii, M., Koyaguchi, T., Makinouchi, T., Mitsushio, H., and Ishida, S. (1984) Fission-track, K-Ar age determinations and palaeomagnetic measurements of Miocene volcanic rocks in the western area of Baragoi, northern Kenya: Ages of hominoids. *Afr. Study Mono., Supplementary Issue* 2:57-66.

Matsuda, T., Torii, M., Koyaguchi, T., Makinouchi, T., Mitsushio, H., and Ishida, S. (1986) Geochronology of Miocene hominoids east of the Kenya Rift Valley. In J. G. Else and P. C. Lee (eds.): *Primate Evolution*. Cambridge: Cambridge University Press, pp. 35-45.

McCall, G. J. H., Baker, B. H., and Walsh, J. (1967) Late Tertiary and Quaternary sediments of the Kenya Rift Valley. In W. W. Bishop and J. D. Clark (eds.) *Background to Evolution in Africa*. Chicago: University of Chicago Press, pp. 191-220.

McClenaghan, M. P. (1971) Geology of the Ribkwo Area, Baringo District, Kenya. Unpublished Ph.D. thesis, University of London.

McHenry, H. M. (1976) Multivariate analysis of early hominid humeri. In E. Giles and J. S. Friedlander (eds.): *The Measures of Man*. Cambridge: Peabody Museum Press, Harvard University, pp. 338-371.

McHenry, H. M. (1984) The common ancestor: A study of the postcranium of *Pan paniscus, Australopithecus*, and other hominoids. In R. L. Sussman (ed.) *The Pygmy Chimpanzee*. New York: Plenum, pp. 201-230.

McHenry, H. M., and Corruccini, R. S. (1975) Distal humerus in hominoid evolution. *Folia Primatol.* 23:227-244.

McHenry, H. M., and Corruccini, R. S. (1980) Late Tertiary hominoids and human origins. *Nature* 285:397-398.

Nakaya, H., Pickford, M., Nakano, Y., and Ishida, H. (1984) The late Miocene large mammal fauna from the Namurungule Formation, Samburu Hills, northern Kenya. *Afr. Study Mono.* Supplementary Issue 2:87-131.

Patterson, B. (1966) A new locality for early Pleistocene fossils in north-western Kenya. *Nature* 212:577-581.

Patterson, B., and Howells, W. W. (1967) Hominid humeral fragment from early Pleistocene of north-west Kenya. *Science* 156:64-66.

Patterson, B., Behrensmeyer, A. K., and Sill, W. D. (1970) Geology and fauna of a new Pliocene locality in north-western Kenya. *Nature* 226:918-921.

Petrocchi, C. (1934) I ritrovamenti faunistici di es-Sahabi. *Rivista delle Colonie Italiane*. 8(9):733-742.

Pickford, M. (1975a) Stratigraphy and Palaeoecology of Five Late Cainozoic Formations in the Kenya Rift Valley. Unpublished Ph.D. thesis, University of London.

Pickford, M. (1975b) Late Miocene sediments and fossils from the northern Kenya Rift Valley. *Nature* 256:279-284.

Pickford, M. (1978a) Geology, paleoenvironments and vertebrate faunas of the mid-Miocene Ngorora Formation, Kenya. In W. W. Bishop (ed.): *Geological Background to Fossil Man*. London: Geological Society of London: Scottish Academic Press, pp. 237-262.

Pickford, M. (1978b) Stratigraphy and mammalian paleontology of the late-Miocene Lukeino Formation, Kenya. In W. W. Bishop (ed.): *Geological Background to Fossil Man*. London: Geological Society of London: Scottish Academic Press, pp. 263-278.

Pickford, M. (1985) *Kenyapithecus*: A review of its status based on newly discovered fossils from Kenya. In P. V. Tobias (ed.): *Human Evolution: Past, Present and Future*. New York: Alan R. Liss, pp. 107-112.

Pickford, M. (1986a) Hominoids from the Miocene of East Africa and the phyletic position of *Kenyapithecus*. *Zeit. Morph Anthropol.* 76:115-130.

Pickford, M. (1986b) The geochronology of Miocene higher primate faunas of east Africa. In J. G. Else and P. C. Lee (eds.) *Primate Evolution*. Cambridge: Cambridge University Press, pp. 19-33.

Pickford, M. (1986c) A reappraisal of *Kenyapithecus*. In J. G. Else and P. C. Lee (eds.) *Primate Evolution*. Cambridge: Cambridge University Press, pp. 163-172.

Pickford, M. (1987a) The geology and palaeontology of the Kanam erosion gullies (Kenya). *Mainzer Geowiss. Mitt.* 16:209-226.

Pickford, M. (1987b) Fort Ternan (Kenya) paleoecology. *J. Hum. Evol.* 16:305-309.

Pickford, M. (1988) Geology and fauna of the middle Miocene hominoid site at Muruyur, Baringo District, Kenya. *Hum. Evol.* 3:381-390.

Pickford, M., Johanson, D. C., Lovejoy, C. O., White, T. D., and Aronson, J. L. (1983) A hominoid humeral fragment from the Pliocene of Kenya. *Am. J. Phys. Anthropol.* 60:337-346.

Pickford, M., Ishida, H., Nakano, Y., and Nakaya, H. (1984a) Fossiliferous localities of the Nachola-Samburu Hills area, northern Kenya. *Afr. Study Mono.* Suppl. Issue 2:45-56.

Pickford, M., Nakaya, H., Ishida, H., and Nakano, Y. (1984b) The biostratigraphic analyses of the faunas of the Nachola area and Samburu Hills, northern Kenya. *Afr. Study Mono.* Suppl. Issue 2:67-72.

Pickford, M., Senut, B., Ssemmanda, I., Elepu, D., and Obwona, P. (1988) Premiers resultats de la mission de l'Uganda Palaeontology Expedition a Nkondo (Pliocène du Bassin du Lac Albert, Ouganda). *C. R. Acad. Sci. Paris* 306:315-320.

Pickford, M., Senut, B., Vincens, A., Van Neer, W., Ssemmanda, I., Baguma, Z., and Musiime, E. (1991) Nouvelle biostratigraphie du Neogene et du Quaternaire de la région de Nkondo (Basin du lac Albert, Rift occidental ougandai). Apport a l'évolution des paléomilieux. *C. R. Acad. Sci. Paris* 312:1667-1672.

Pilbeam, D. (1986) Distinguished lecture: Hominoid evolution and hominoid orgins. *Am. Anthropol.* 88:295-312.

Quade, J., Cerling, T. E., and Bowman, J. R. (1989) Development of Asian monsoon revealed by marked ecological shift during the latest Miocene in northern Pakistan. *Nature* 342:163-166.

Retallack, G. J. (1991) *Miocene Paleosols and Ape Habitats of Pakistan and Kenya*. Oxford: Oxford University Press.

Retallack, G. J. (1992a) Comment on the paleoenvironment of *Kenyapithecus* at Fort Ternan. *J. Hum. Evol.* 23:363-369.

Retallack, G. J. (1992b) Middle Miocene fossil plants from Fort Ternan (Kenya) and evolution of African grasslands. *Paleobiol.* 18:383-400.

Retallack, G. J., Dugas, D. P., and Bestland, E. A. (1990) Fossil soils and grasses of a middle Miocene east African grassland. *Science* 247:1352-1328.

Senut, B. (1980) Nouvelles données sur l'humérus et ses articulations chez les hominidés plio-pléistocènes. *L'Anthropologie* (Paris) 84:112-118.

Senut, B. (1981a) Humeral outlines in some hominoid primates and in plio-pleistocene hominids. *Am. J. Phys. Anthropol.* 56:275-282.

Senut, B. (1981b) L'humérus et ses articulations chez les Hominidés Plio-Pléistocènes. *Cahiers Paléoanthropologie*. Paris: CNRS.

Senut, B. (1982) Reflexions sur la brachiation et l'origine des Hominidés a la lumière des Hominoides miocènes et des Hominidés plio-pléistocènes. *Geobios Mem. Special* 6:335-344.

Senut, B. (1983) Quelques remarques à propos d'un humérus d'hominoide Pliocène provenant de Chemeron (bassin du lac Baringo, Kenya). *Folia Primatol.* 41:267-276.

Senut, B., and Tardieu, C. (1985) Functional aspects of Plio-Pleistocene hominid limb bones: Implications for taxonomy and phylogeny. In E. Delson (ed.): *Ancestors: The Hard Evidence*. New York: Alan R. Liss, pp. 193-201.

Senut, B., Pickford, M., Mein, P., Conroy, G., and Van Couvering, J. (1992) Discovery of 12 new Late Cainozoic fossiliferous sites in palaeokarsts of the Otavi Mountains, Namibia. *C. R. Acad. Sci. Paris* 314:727-733.

Shipman, P. (1982) Taphonomy of *Ramapithecus wickeri* at Fort Ternan, Kenya. University of Missouri-Columbia: Museum Brief 26, Museum of Anthropology, pp. 1-37.

Shipman, P. (1986) Paleoecology of Fort Ternan reconsidered. *J. Hum. Evol.* 15:193-204.

Shipman, P., Walker, A., Van Couvering, J., Hooker, P., and Miller, J. A. (1980) The Fort Ternan hominoid site, Kenya: Geology, age, taphonomy and paleoecology. *J. Hum. Evol.* 10:49-72.

Simons, E. (1963) Some fallacies in the study of hominid phylogeny. *Science* 141:879-889.

Simons, E., and Pilbeam, D. (1965) Preliminary revision of the Dryopithecinae. *Folia Primatol.* 138:81-152.

Smart, C. (1976) The Lothagam 1 fauna: its phylogenetic, ecological, and biogeographic significance. In Y. Coppens, F. C. Howell, G. L. Isaac, and R. E. F. Leakey (eds.): *Earliest Man and Environments in the Lake Rudolf Basin: Stratigraphy, Paleoecology, and Evolution*. Chicago: Chicago University Press, pp. 361-369.

Stockley, G. M. (1929) Tinde Bone Beds and further notes on the Usongo Beds. *Ann. Rept. Geol., Survey Tanganyika* 1929:21-23.

Susman, R. and Stern, J. T. (1982) Functional morphology of *Homo habilis*. *Science* 217:931-934.

Susman, R., Stern, J. T., and Jungers, W. L. (1984) Arboreality and bipedality in the Hadar hominids. *Folia Primatol.* 43:113-156.

Tankard, A. J. (1975) Varswater Formation of the Langabaanweg-Saldanha area, Cape Province. *Trans. Geol. Soc. S. Afr.* 17:265-283.

Tauxe, L., Monaghan, M., Drake, R., Curtis, G., and Staudigel, H. (1985) Paleomagnetism of Miocene East African Rift sediments and the calibration of the Geomagnetic Reversal Timescale. *J. Geophys. Res.* 90:4639-4646.

Ward, S., and Hill, A. (1987) Pliocene hominid partial mandible from Tabarin, Baringo, Kenya. *Am. J. Phys. Anthrop.* 72:21-37.

Watkins, R. T., and Williamson, P. J. (1979) The Warata Formation, northern Kenya: its character and paleoenvironmental significance. *Proceedings of the 8th Pan-African*

Congress of Prehistory and Quaternary Studies: Nairobi 107-108.

White, T. D. (1984) Pliocene hominids from the Middle Awash, Ethiopia. In P. Andrews and J. L. Franzen (eds.): *The early evolution of man with special emphasis on southeast Asia and Africa. Courier Forschungsinstitut Senckenberg* 69:57-68.

White, T. D. (1986) *Australopithecus afarensis* and the Lothagam mandible. *Anthropos (Brno)* 23:79-90.

White, T. D. (1987) Review of Neogene Paleontology and Geology of Sahabi. *J. Hum. Evol.* 16:312-315.

White, T. D., Suwa, G., Richards, G., Watters, J. P., and Barnes, L. G. (1983) "Hominid clavicle" from Sahabi is actually a fragment of cetacean rib. *Am. J. Phys. Anthropol.* 61:239-244.

Williams, M. A. J., Getaneh, A., and Adamson, D. A. (1986) Depositional context of Plio-Pleistocene hominid-bearing formations in the Middle Awash Valley, southern Afar Rift, Ethiopia. In L. Frostick, R. W. Renaut, I. Reid, and J. J. Tiercelin (eds.): *Sedimentation in the African Rifts*. Oxford: Blackwell, Geological Society of London Special Publication 25, pp. 241-251.

Williamson, P. G. (1985) Evidence for an early Plio-Pleistocene rainforest expansion in East Africa. *Nature* 315:487-489.

Yemane, K., Bonnefille, R., and Faure, H. (1985) Palaeoclimatic and tectonic implications of Neogene microflora from the northwestern Ethiopian highlands. *Nature* 318:653-656.

CHAPTER 7

Hominid Paleobiology: Recent Achievements and Challenges

Bernard Wood

INTRODUCTION

Integration and synthesis are two activities of scientists that are persistently and unfairly undervalued. Yet, in paleontology, a historical science *par excellence*, they make an especially important contribution to our ability to interpret evolutionary history. It may not be too cynical to suggest that their relatively low status results from their being activities for which outcomes are difficult to measure and which are thus unlikely to attract that increasingly essential commodity, grant support.

A good review not only surveys evidence, attempting to identify simple patterns within it, but also evaluates competing explanatory hypotheses. Such reviews are a godsend for students and a useful exercise in intellectual stocktaking for those who write them. A good review may stimulate others to formulate research questions that were previously unrecognized, or which were at best only crudely formulated. Reliable and comprehensive reviews also assess the progress of a subject by providing an efficient summary of knowledge at the time of writing.

Most of us would count ourselves fortunate to have contributed one influential review of hominid prehistory but Clark Howell has provided his colleagues with at least two. The first, published more than 30 years ago (Howell 1960), was the catalyst for much significant research on a wide range of analytical problems relating to hominid evolution in the Middle Pleistocene of Europe and the Mahgreb. Its scope, embracing as it did geological context and paleontological and archeological evidence, combined with its diligent scholarship, ensured that it became, and has remained, a seminal influence on studies of human evolutionary history of the period. It raised fundamental questions about the morphological and behavioral distinctions between the Middle Pleistocene hominids of Europe and Africa that are still debated. Nearly two decades later the same author shifted his geographical focus, but widened his temporal scope, and undertook the immense task of surveying the African fossil evidence for hominid evolution (Howell 1978). The important features of all hominid species that were then known from African sites were described and the constituents of each hypodigm were tabulated. Despite the time that has elapsed since its publication, that review is still the most widely cited survey of the fossil evidence for African hominid evolution.

The discussion that follows is most emphatically not an attempt to update Howell's 1978 review; limitations of talent and space preclude that. However, that datum is a useful reference point from which to survey subsequent advances in our knowledge of human, and particularly African, prehistory. The plan of the chapter conforms loosely with the structure of a more conventional scientific paper. Fossils still provide the major source of information for hominid paleobiologists and the African fossil evidence that has been accumulated since 1978 is briefly reviewed. However, since that time the application of both established and novel techniques has enabled researchers to extract additional information from existing fossils, and these applications and innovations will be highlighted. The equivalent of the "Methods" section of a paper will, in this chapter, outline how advances in paleontological analysis have modified the ways in which we interpret the hominid fossil record as well as extended the scope of paleoanthropological research. The next section will assess in what ways, if any, the conjunction of fresh fossil evidence and new analytical techniques has advanced our knowledge of human evolutionary history. The concluding section will attempt to identify the research problems that are likely to preoccupy hominid paleobiologists as we move toward the turn of the century.

AFRICAN FOSSIL EVIDENCE RECOVERED SINCE 1978

The African sites that presently provide the most of the paleontological evidence for the evolutionary history of hominids are substantially the same as those specified in Howell's 1978 review (Fig. 10.1 and subsequent tables). Of the sites that have been discovered since 1978 only West Turkana has yielded remains which seriously challenge the taxonomic and phylogenetic interpretations of hominid evolution current in the mid- to late 1970s (see later discussion). This does not mean that recent discoveries at the established sites have not been controversial, for several have been, and these will be referred to. I shall follow Howell and refer to sites in three geographical regions: northern, eastern, and southern Africa.

Northern Region

Nearly half of the African land mass, namely the North African coastal plain and the Sahara, has made little or no impact on hominid paleobiology since 1978. The general lack of absolute dates may help to explain why the North African data are usually ignored, but they have posed, and continue to pose, intriguing problems that are exemplified by the Salé cranium (Jaeger 1975). It, together with other hominid remains that are linked with Tensiftien faunas and dated to around 300 KYA (Debenath et al. 1982), is very seldom included in attempts to clarify the taxonomy of Middle Pleistocene hominids. Those who have studied the Salé cranium point to its mixture of *Homo erectus* and *Homo sapiens* features, yet it is a small cranium with a cranial capacity initially estimated as 930–960 cm^3, but which Holloway (1981) subsequently assessed to be as low as 850 cm^3. Hypotheses advanced to explain the relationships between *H. erectus* and archaic *H. sapiens* can no longer ignore evidence such as that provided at Salé.

Eastern Region

The decade since 1978 coincided with a period of political upheaval within Ethiopia and thus a reduction, and for the most part a cessation, of fieldwork. It was fortunate that the major collections at the Hadar and Omo sites had already been accumulated before this disruption (Johanson et al. 1982; Coppens 1980). Prospecting has since been resumed at Hadar and additional hominids recovered (Kimbel 1992); details of the scope and significance of this new evidence are awaited with interest. Farther south in the Afar depression, but also associated with the Awash River, is the complex of localities known as "Middle Awash." The stratigraphical sequence is more than 1 km thick (Kalb et al. 1982), and hominids have been recovered from both Middle Pleistocene sediments at Bodo D'ar (Conroy et al. 1978) and Pliocene deposits at Maka (Clark et al. 1984). As at Hadar, the recent relaxation of political tension within Ethiopia has allowed paleoanthropological research at the Middle Awash to resume. Further hominid remains have been discovered (White, *in litt.*) and the results of their initial analysis are expected shortly.

A significant recent development in human origins research in Ethiopia has been the establishment, and subsequent activities, of the "Paleoanthropological Inventory of Ethiopia" organized from Addis Ababa (WoldeGabriel et al. 1992). The survey harnessed the powerful technique of remote sensing (Asfaw et al. 1990; Wood 1992b) and identified three areas, the Kesem–Kabana basin in the north and the areas of Konso–Gardula (KGA) (Asfaw et al. 1992) and Fejej (Asfaw et al. 1991; Fleagle et al. 1991) to the south. Both of the latter two areas have yielded hominid fossil remains. The right side of a mandibular corpus from KGA is likely to be older than 1.4 MYA and resembles *H. erectus* or *H. ergaster* (Asfaw et al. 1993), whereas the isolated teeth from Fejej are much older, being in excess of 3.6 MYA and perhaps as old as 4.0 MYA, and show obvious affinities with remains attributed to *Australopithecus afarensis* (Fleagle et al. 1991). Howell (1978) noted the discovery of a hominid parietal bone at the Middle Pleistocene Gomboré II locality at Melka Kunturé (Chavaillon and Coppens 1975). Subsequent discoveries of humeral (GIB) (Chavaillon et al. 1977) and frontal (Chavaillon and Coppens 1986) fragments at the same locality, and a parietal fragment at Garba III (Chavaillon et al. 1987), another locality at Melka Kunturé, strengthen the case for the presence of a hominid not unlike *H. erectus*.

Many, but not all, of the important early hominid fossils from Koobi Fora were reviewed and listed in Howell (1978). Subsequently discovered Koobi Fora hominids have in the main increased our knowledge of those species already recognized (e.g., KNM–ER 3883, 13750, 15930), but one or two specimens, particularly from the older horizons (e.g., KNM–ER 5431) may represent hominid species hitherto unrecognized at the site. Pliocene and Pleistocene sediments belonging to the same Omo group complex have been located on the west side of Lake Turkana, at the site of West Turkana. The Nachukui Formation exposed at that site apparently ranges in age from 1 to 3.5 MYA (Harris et al. 1988) and has yielded hominid remains. The Nariokotome Member samples *H. erectus/H. ergaster* (e.g., KNM–WT 15000), the Kaitio, Natoo and Lokalalei Members contain *Paranthropus boisei* (e.g., KNM–WT 17400) remains, the Lomekwi Member samples *Paranthropus* aff. *P. boisei* (e.g., KNM–WT 17000), and a skull fragment from the Kalachoro Member, KNM–WT 15001, has been allocated to *Homo* sp.

Three localities in the Baringo Basin have yielded hominid remains since 1978. These are the cranial fragments from Chesowanja (Gowlett et al. 1981), a mandible from Baringo (Wood and Van Noten 1986), and another mandibular fragment from the much earlier, Pliocene, site of Tabarin (Hill 1985).

Olduvai Gorge in Tanzania has continued to yield fossil evidence notably, in 1986, the partial skeleton, OH 62 (Johanson et al. 1987). Exploration of other Plio–Pleistocene drainage systems in Tanzania is in progress, but, as yet, no hominid remains have been recovered.

Information about dating is of no relevance to the task of identifying species or of developing hypotheses about their relationships, but it is of crucial importance for the investigation of other aspects of evolutionary history including phylogeny and scenario reconstruction (see later discussion). One of the most important recent advances in our understanding of the context of the East African fossil hominid record has come from the elegant application of geochemistry to "fingerprint" tuff layers and thus provide correlations between localities and sites that are orders of magnitude more reliable than those that depended on crude comparisons such as sediment color and particle size. The technique of tephrostratigraphy has allowed the identification of several marker tuffs in the Shungura, Koobi Fora, and Nachukui Formations (Cerling et al. 1979; Cerling and Brown 1982; Harris et al. 1988), which formed the basis of a reliable chronological framework for the fossil hominids recovered from the Turkana basin (Feibel et al. 1989). One tuff, the Tulu Bor, has proved to be particularly widely dispersed (Sarna–Wojcicki et al. 1985), and the reliable identification of its equivalent at Hadar, the Sidi Hakoma Tuff (Brown 1982), has allowed the Hadar Formation to be integrated with the Turkana Basin sequence. Likewise, other tephra have provided links between the Turkana Basin and the Middle Awash Valley and have enabled the fossil records of sites farther north in the Ethiopian Rift Valley to be matched with the Turkana basin sequence (Haileab and Brown 1992). It is difficult to overemphasize the actual and potential significance of the research program that has spawned this new dating information. Another benefit is that it will enable the excellent record of mammalian evolution from these sites to be used to examine the influence of different environments on synchronous fossil faunas.

Southern Region

The major locations of fossil hominids are, with two exceptions, those previously reported by Howell. The two exceptions are Uraha in the Karonga District in northern Malawi, the site of the discovery of a hominid mandible, and Gladysvale, a cave in the Transvaal, where two hominid teeth have recently been identified (Tobias 1992).

Since 1978 excavations in Members 4 and 5 at Sterkontein have produced a steady stream of hominid specimens (e.g., Tobias 1992). These include, in 1987, a fine partial skeleton Stw 431 (Partridge et al. 1991) and in 1984 an incomplete cranium, Stw 252; both specimens are from Member 4 (Clarke 1985). The latest phase of research at the nearby Swartkrans cave began in 1979. It concentrated on the excavation of *in situ* decalcified breccia and unconsolidated sediments within the cave and resulted in the recovery of more than 100 hominid specimens. These have been enumerated (Brain et al. 1988) and basic data for many have been published (Grine 1989; Susman 1989; Grine and Susman 1991). Of no less importance to hominid paleobiology has been the refinement of the complex stratigraphy of the Swartkrans site, which has enabled the provenance of all the Swartkrans hominids to be related, with varying degrees of confidence, to the various subdivisions and components of Members 1 to 3 (Brain 1982, 1988; Brain et al. 1988).

The cave system on the Kromdraai farm, some 1.5 km east of the Sterkfontein site, first yielded hominid remains in 1938. Broom worked at the site, which was divided into three breccia locations A–C, in 1941 and 1947, but the next significant excavation was by Brain in 1955–1956 (Brain 1975). Vrba returned to the site between 1977–1980 and concentrated on the contiguous, and possibly in the past continuous, breccia deposits known as Kromdraai C (KC) and Kromdraai B East (KBE) and West (KBW). Five members were recognized at KBE and three at KBW (Vrba 1981). Breccia recovered from Member 3 at KBE yielded five new hominid dental specimens (Vrba 1981), two of which have been described in detail (KB 5223—Grine 1982a; KB 5226—Grine 1982b). One must hope that excavations presently in progress at Kromdraai will add to the relatively small, but nonetheless significant, hominid collection from the site.

NEW DATA FROM EXISTING FOSSILS

Advances in hominid paleobiology since 1978 have not been confined to the discovery of the fossil evidence described in the preceding section. Aside from the accumulation of new fossils, one of the most exciting challenges of paleontology is that of devising ways in which fresh evidence can be extracted from the existing fossil record. The period since 1978 has seen significant advances in retrieving this evidence, some of which have utilized state–of–the–art technology and others of which owe more to the ingenuity of individual researchers who have seen the potential of well–known methods of study, such as routine radiography and standard techniques of light microscopy, when applied in novel ways. This section describes how imaging and developments in our ability to study the ontogeny of two mineralized tissues, enamel and bone, have added significantly to our knowledge of both the taxonomy and the growth and development of hominid species.

Imaging Internal Structures

The use of conventional radiography for the analysis of hominid fossils was well established long before 1978 (e.g., the excellent radiographs in Weidenreich's 1943 *Sinanthropus* monographs), but it is true to say that its potential for taxonomic and functional analysis had not been fully exploited. Most hominid isolated teeth consist of the crown and not the root and thus most emphasis had been placed on the interpretation of tooth crown morphology. But the site of weakness at the cervix of the crown and the propensity of enamel to crack when exposed to the elements resulted in many early hominid mandibles losing the crowns of the teeth, either prior to their incorporation in sediments or after their exposure to them; yet the root systems were still preserved within the mandibular corpus. Studies since 1978 have used conventional radiographic techniques to produce useful images of these retained roots (Ward et al. 1982; Wood and Uytterschaut 1987; Wood 1991a). Much useful taxonomic information has already been gained from the study of postcanine root morphology, but the potential of these data for functional analysis has yet to be exploited.

Conventional radiography has also been employed to measure the thickness of the cortical bone of both the postcranial and the cranial skeleton. Walkoff (1904) was probably the pioneer of the use of radiography for the analysis of the hominid postcranium, but it was Weidenreich (1941:69) who placed substantial reliance on radiographic evidence to substantiate his claim that among the distinguishing features of the *Sinanthropus* femora was that the medullary canal "is very narrow and the walls correspondingly thick." In a later study Kennedy (1983) used radiography to determine cortico–medullary proportions in a variety of early hominid femora, suggesting that by Weidenreich's criteria two early specimens from Koobi Fora, KNM–ER 1481 and 1472, should be allocated to *H. erectus*. Using similar logic, Kennedy (1984) went on to suggest that the relative cortical thickness of the Kabwe and Omo I femora suggested that these two femora should be allocated not to *H. erectus* but to *H. sapiens*, for their values were closer to those of a Romano–British comparative sample.

Radiography has also been successfully exploited during the analysis of the thick–walled WLH 50 cranium from the Willandra Lakes by Webb (1989), who concluded that WLH 50 is part of a subset of archaic crania with thick walls. However, by using radiography and direct observation, the author was able to demonstrate that the basis of the thick walls of the crania from the Willandra Lakes is an abundance of diploeic, or cancellous, bone in *H. erectus* crania and calottes from Java and China. Weidenreich (1943:164) had previously reported that "all three constituents of the bone take equal part in the thickening, the two tables slightly more than the diploe."

Attempts to measure enamel thickness using conventional radiography (Sperber 1985; Zilberman et al. 1992) are bedeviled by the undulating nature of the enamel–dentine junction (EDJ) and this technique is being rapidly superseded by more sophisticated imaging techniques, notably computerized tomography (see following discussion).

Computerized axial tomography (also known as CAT, or CT, scanning) was a revolutionary development in medical imaging, and it also promises to make a substantial impact on our ability to glean information from fossils. The main advantages of CT over conventional radiography are threefold. First, the arrangement of the instruments emitting and recording the radiation is such that the device minimizes the effects of superimposition, so that structures deep within an object can be visualized. Second, it enables better discrimination between materials whose density differs only slightly (e.g., fossil bone and adjacent matrix). Third, the absence of parallax distortion means that precise and accurate measurements can be made of structures visualized on the CT images. Advances are being made all the time so that since the introduction of CT there has been at least a sixfold improvement in resolution (from 3 mm to 0.5 mm). The technique has proved particularly useful for the imaging of teeth (Ward et al. 1982: Conroy and Vannier 1987, 1991), mandibular structure (Daegling 1989; Daegling and Grine 1991), and paranasal sinuses and the internal ear (Wind 1984; Zonneveld et al. 1989). Raw CT data generate two–dimensional images, but computer algorithms have been developed that allow their integration into a three–dimensional reconstruction. Such algorithms have been used to image early hominid crania whose endocranial surface is obscured by dense matrix (Wind and Zonneveld 1989), and complementary algorithms have also allowed the estimation of the volume of the matrix–filled endocranial cavities of early hominids (Conroy et al. 1990). Enamel thickness is apparently a useful taxonomic discriminator among some early hominid taxa (Beynon and Wood 1986). However, measurements can only be precisely compared if they are taken at predefined locations on carefully mechanically sectioned teeth, but such a procedure can never be extended to large samples of precious early hominid fossils. Experimental evidence and experience suggest that CT will provide useful enamel thickness data (Macho and Thackeray 1992; Spoor et al. in press). The CT technique is also applicable to the postcranial skeleton, where, for example, it can be used to provide accurate cross–sectional images of the shafts of long bones (e.g., Senut 1985; Ben–Itzhak et al. 1988). These are a promising source of evidence for more reliable estimates of early hominid body weights and also provide useful data for biomechanical analyses (Ruff and Leo 1986; Ruff 1987).

Ultrasound has yet to be used in the analysis of hominid fossil remains, but it is a much cheaper and more convenient alternative to CT for examining concealed anatomy. A recent attempt demonstrated the ability of the technique to detect the enamel/dentine interface in human teeth (Ng et al. 1989), and this modality offers the prospect of much more widely available assessments of the shape and thickness of the enamel cap of fossil hominid teeth.

Growth and Development Recorded in Hard–Tissue Microstructure

Tooth crowns are ubiquitous in the fossil hominid record and for many years their importance rested on information about their size and shape. However, since 1978 there has been a growing awareness that tooth, and particularly enamel, microstructure may provide additional information, not only about taxonomy (see previous discussion), but also about dental development and maturation (Shellis 1984). When mechanically sectioned or naturally fractured enamel surfaces are immersed in alcohol and viewed with polarizing light, exquisite details of the process of enamel formation are revealed.

A particularly important development was the ability to identify, via incremental lines and the striae of Retzius, the progress of the advancing enamel front as it made its way from the enamel–dentine junction to the surface of the mature tooth crown. The shape of this advancing wave of active enamel production was shown to be different in at least two early hominid taxa (Beynon and Wood 1987; Grine and Martin 1988). Studies of enamel development have been complemented by investigations of the timing and pattern of tooth eruption in both extant hominoids and fossil hominids and the combination of the two categories of information has the potential to be a useful taxonomic discriminator (Beynon and Dean 1988) as well as an indicator of general somatic developmental history (Smith 1991). One of the great breakthroughs of the post–1978 era of hominid paleobiological analysis has been the discovery that tooth formation provides a way in to the nexus of developmental intercorrelations. Researchers now have the opportunity to begin to study the life history of fossil hominids much in the way that the developmental histories of extant forms can be described and compared.

While many of the tissues that make up a tooth incorporate a faithful record of ontogeny, adult bone, because of its ability to respond to changes in load by means of microscopic and macroscopic remodeling, presents a confusing mixture of information about evolutionary history and immediate functional demands. However, growth studies indicate that differences in adult cranial form can be linked to distinct differences in the pattern and rate of cranial growth (Duterloo and Enlow 1970). Sites of bone formation and resorption can now be mapped on the surface of developing cortical bone using scanning electron microscopy (SEM) (Boyde and Jones 1972). Bromage (1987) has developed a high–resolution replication technique, employing a silicon–based resin, from which replicas of the cortical surface of specimens of immature hominids have been made for study with SEM. By studying the distribution of sites of bony resorption and deposition on the surface of face and mandible, the author concluded that whereas the pattern of facial growth of *Australopithecus africanus* is more apelike (Bromage 1985), that of *Paranthropus robustus* more closely resembles the modern human pattern (Bromage 1989).

RECENT DEVELOPMENTS IN ANALYTICAL METHODS

Thus far this chapter has considered post–1978 additions to the fossil evidence as well as recently introduced research strategies for the extraction of additional information from existing fossils. This section reviews and considers the no less important, and arguably the more important, task of analyzing these data. In doing so it reflects upon the general process of paleontological analysis, but it inevitably focuses on the problems of interpreting the African hominid fossil record. The framing of questions and the formulation of strategies for answering them are at the heart of the process of science. In crude terms, the simpler the question, the better the science. There is no paradox here for the best paleontologists are those who, while acknowledging the complexity of evolutionary history, are capable of dismantling a complex problem into its simpler components. What "simple" questions can we ask about early hominid evolutionary history and how do we go about the process of "intellectual" dissection? Several of my colleagues have come close to offering a sound prospectus for our science, and in presenting what follows, I shall shamelessly plunder their individual contributions to suggest a sequentially coherent analytical strategy for hominid paleobiology.

Pilbeam (1984) sensibly warned his colleagues against adopting a "progressive" or "whig–like" (Butterfield 1931) analytical approach to the reconstruction of evolutionary history. Such strategies interpret the past purely in terms of the way we can connect it to the present and this narrative tendency has recently been usefully summarized by Landau (1991). Instead, Pilbeam urges us to adopt the "historicist" strategy and to take our "analytical machine" back into the past and ask of the fossil record the sort of questions that any sensible fieldworker would ask of a newly discovered species of primate. Tooby and De Vore (1986:201) and Pilbeam (1988:92) both provide useful checklists of such questions, but both contributions sensibly duck the problem of how to identify the hominid species of whom the questions should be asked.

I will not dwell upon the species problem in paleontology, for this is a topic that could occupy all the pages of this and many other volumes. In his 1960 review Howell also wrestled with it in connection with the morphological distinctions between the Middle Pleistocene hominid remains from Asia and North Africa, on the one hand, and those from Europe, on the other. I have to confess that my own paleontological practice has not been advanced by recently proposed "solutions," such as the phylogenetic species concept (Cracraft 1987; Kimbel and Rak 1993), to the problem of defining a paleontological species, but this may say as much about my own intellectual inadequacies as it does about the logical basis and utility of the proposals themselves. My construct of a paleontological species comes closest to that of Tattersall's (1986) analysis of what species are, and how they can be recognized in the fossil record.

Species identification is ultimately a matter of judgment, but I believe that such judgments can be better informed by discriminating between the two concepts of the "degree" and "pattern" of morphological variation (Wood et al. 1991), and by the careful and sensible choice of comparative analogues, from both without the fossil record, and, if appropriate, from within it (Wood 1991b). Species cannot be diagnosed cladistically (Wood 1989—and doubtless many others); any such attempt to do so is so obviously circular that one wonders why it was ever entertained. The diagnostic, or defining, features of a fossil–based species are not necessarily novel, or autapomorphic. They may be symplesiomorphic but exist in that particular species in a unique combination.

Once fossil hominid species have been identified the remainder of the process of analysis can then proceed. I was, and still am, attracted by the classification of evolutionary hypotheses outlined by Tattersall and Eldredge (1977). They suggest that such hypotheses can be usefully divided into three levels of complexity. The simplest level contains hypotheses about the *relationships* between species. The next adds to that information more specific suggestions about *phylogeny* (i.e., ancestry and descent), and the third level attempts to offer explanations or *scenarios*, adaptive and/or ecological, for the shape of the phylogenetic tree.

Most of the post–1978 developments in the analytical armamentarium of hominid paleontologists have taken place at Tattersall and Eldredge's first and third levels, that is, those that postulate hypotheses of relationships and evolutionary scenarios. Regrettably, with one or two possible exceptions, we are little, or no, closer to understanding how to identify ancestors now than we were in 1978.

Establishing Relationships

The major development in the methods available to help establish the pattern of relationships between early hominid species has been the widespread adoption of the principles of phylogenetic analysis or cladistics. I have elsewhere attempted to set out the reasons for using cladistic methods to establish species relationships and outlined how these methods might be applied to the analysis of the hominid fossil record (Wood 1989). Attitudes toward cladistics are polarized and hardly ever neutral. While one of my colleagues viewed its introduction as "one of the most important biological revolutions of the last two decades" (Pilbeam 1986:295), others are less enthusiastic. Brauer and Mbua (1991:1) urge that more effort be devoted to examining "extant geographic variation and polymorphisms as well as the diachronic change of morphological features and patterns within the hominids" and less on "focussing on morphological discontinuities and the application of cladistic procedures," and Trinkaus (1990) counseled that "its biological limitations are too abundant for it to be more than a heuristic device for the preliminary ordering of complex human paleontological and neontological data." In some cases such skepticism about cladistic analysis is soundly based on experience or on genuine disagreement with its theoretical basis, but sometimes criticisms stem from misconceptions about the type of research problems that the cladistic method is best suited to answer. Despite its other name, phylogenetic systematics, cladistic analysis is not a method that can, unaided, generate phylogenetic trees or reconstruct evolutionary history at Tattersall and Eldredge's second, phylogenetic, level. Trinkaus, in the process of offering criticism, clearly summarized the questions that cladistic analysis is best equipped to help resolve.

Modeling—Good, Bad, and Indifferent

The jargon of management theory has intruded into discussions about the analytical techniques used by evolutionary biologists. In short, "top down" strategies are frowned upon whereas the "bottom up" approach is the subject of approbation; Tooby and De Vore (1986) refer to the two categories as, respectively, "referential" (= top–down) and "conceptual" (= bottom–up). An example of referential modeling is when researchers match the morphology of a fossil bone with a modern analogue and then make the inference that the fossil form indulged in precisely the same behavior, be it locomotion, diet, or whatever, as the extant analogue. Dunbar (1989) reminded us that whole animals have been cast in this role, for the pygmy chimpanzee has been, and by some still is, cast in the role of "faithful" analogue for the behavior of a hominid ancestor. These inferences are, of course, based on a logical fallacy that was exposed some time ago in a seminal analysis of the link between form and function (Bock and von Wahlert 1965). The latter authors advocated the distinction between levels of inference, much as Tattersall and Eldredge (1977) were to do later in a different context. Bock and von Wahlert made the point that similarities of form, function, and biological role should be distinguished. For example, a morphological and biomechanical analysis of a femur, including an expression of its ability to respond to compression, torsion and so on, is what the authors mean by *form*. Its observed use in bipedal posture and locomotion is the femur's *function*, and the involvement of bipedal running in hunting is a description of its ultimate *biological role*. Too often similar forms are accorded similar biological roles, yet Bock and von Wahlert clearly demonstrated that inferences about biological role were not justified without additional evidence of the kind that is usually not available in the fossil record. Top–down models and the use of analogues have thus developed an unfortunate reputation. However, it is illogical to damn all top down modeling because a few practitioners misuse it. Conceptual modeling (of which "strategic modeling" is a special case), or the bottom–up model, is promoted as being quite different in its approach, but I believe that the differences between properly generated top–down models and many bottom–up models are ones of degree only and I will expand on this in the following discussion.

The best sort of bottom–up models really do take the arguments back to first principles. By first principles I mean quite literally the laws of physics or chemistry. Wheeler has operated at, or close to, this level in his theoretical studies of heat and water balance (Wheeler 1984, 1985, 1991a,b, 1993). The models have predictive value and provide hitherto unexplored physiological avenues for the interpretation of the factors underlying the adoption by hominids of a habitual erect posture. But one person's first principle is another's analogy, and many other bottom–up approaches, while laudably embedded in a wide range of comparative animal models, are still based on empirical observations. Dunbar's (1988) analysis of time budgets is a case in point. It has great explanatory power, for its validation is based on observations of the social system of living primates. I would contend that these factors make it a good top–down model, and not a bottom–up model in the sense that Wheeler's models are. Ruff (1991) provides a good example of the marrying of a top–down, analogue–based, model with the use of basic physiological principles. However, this digression into pedantry detracts from my main thesis that the increasing trend to deemphasize the uniqueness of hominids (e.g., Foley 1987) is forcing hominid paleobiologists to begin the task of establishing a sound theoretical basis for developing "scenario" hypotheses to explain the events and trends of hominid evolution. Ecological principles will form an important part of this theoretical underpinning, but they are no more likely to be the panacea for scenario development than cladistics is a panacea for determining relationships between hominid species.

DEVELOPMENTS IN OUR UNDERSTANDING OF HUMAN EVOLUTIONARY HISTORY SINCE 1978—REVOLUTION OR REFINEMENT?

Taxonomy

The list of hominid species recognized by Howell (1978; Table 7-1) not only represented the consensus of his colleagues but also reflected the author's particular perspective. The latter is reflected in his espousal of a specific distinction for the "robust" australopithecine remains from Kromdraai, and in the subspecific distinction he makes between specimens attributed to *H. sapiens* from, for example, Kabwe and Djebel Irhoud. The remains from Hadar and Laetoli were referred to Hominidae gen. et. sp. indet., but Howell noted that he concurred with their assignment to *A. afarensis* which was made formally in the same year. If one were forced to choose between the two descriptions the section heading offers for the developments in our understanding of hominid taxonomy resulting from research after 1978, it would be "refinement" and not "revolution." Changes to the cast list in the human evolutionary drama that have been suggested since 1978 have been relatively minor ones.

Despite spirited resistance (Tobias 1980) the case for a specific distinction between *A. afarensis* and *A. africanus* is now widely accepted as a sound one. Doubts have been expressed about the number of species represented in the *A. afarensis* hypodigm (e.g., Olson 1981; Read 1984; Ferguson 1984), but the general consensus, to which this author subscribes (Wood 1991a), is that the existing hypodigm provides no convincing evidence for more than one species. Similar proposals about excessive intraspecific variation have also been made about the hypodigm of *A. africanus*. Just prior to the publication of Howell's review, Krantz (1977) suggested that one of the hominid specimens recovered from Member 4 at Sterkfontein, Sts 51, should be referred to *A. robustus*. More recently Clarke (1988) and Kimbel and White (1988) have also raised the possibility of taxonomic heterogeneity within *A. africanus*. My own prejudice is that despite a substantial degree of variation in the absolute size of both canine and postcanine tooth crowns, there are no significant shape dimorphisms of the type one would expect and can observe between established hominid species. Nonetheless, a substantial part of the presumed *A. africanus* hypodigm

TABLE 7-1 Hominid genera, species[a] and subspecies recognized in the African fossil record by Howell (1978)

Genus	Species	Subspecies
Australopithecus	*africanus*	
	boisei	
	crassidens	
	robustus	
Homo	*habilis*	
	erectus	
	sapiens	*rhodesiensis*
		neanderthalensis
		afer

[a]In addition, remains then known from the Laetoli Beds (Tanzania) and from several members of the Hader Formation, Afar (Ethiopia) were referred to Hominidae gen. et sp. indet.

awaits formal announcement and description, and only when that occurs can the problem of taxonomic heterogeneity in the Member 4 sample from Sterkfontein be satisfactorily resolved. Howell's separation, at the specific level, of the small Kromdraai component of the *A. robustus* hypodigm reflects a minority view, albeit a distinguished and respected one; fresh evidence from Kromdraai may help to resolve this issue.

Australopithecus boisei is one of the two hominid species recognized by Howell (1978), *Homo habilis* being the other, whose status has been affected by new discoveries and recent taxonomic interpretations. The material then referred to *A. boisei*, namely remains from Olduvai, Peninj, Chesowanja, East Turkana (now Koobi Fora), and the Omo site complex from Shungura Formation Members E to G, all date from between approximately 2.0–1.5 MYA. Substantial additions to the *A. boisei* hypodigm from the time range of the original Howell hypodigm have been made, but none of these new data lead one to seriously question the taxonomic integrity of *A. boisei*. However, what has prompted suggestions of taxonomic heterogeneity is fossil evidence from West Turkana (Walker et al. 1986) which is apparently substantially earlier in time, dated to between 2.4 and 2.5 MYA (Feibel et al. 1989). Opinions differ about the significance of this new material, which differs in several ways from the main *A. boisei* hypodigm. These differences of opinion have little to do with the detailed assessment of the fossils and are mostly a reflection of the different ways in which their proponents interpret the concept of a paleontological species. Those who equate a fossil species with a lineage, or clade, and who are thus willing to accept a significant, but undefined, amount of anagenetic change within that lineage (e.g., Bown and Rose 1987) are willing to subsume this early material into *A. boisei* (Walker et al. 1986). Others, however, see fossil species as more rigidly defined and thus doubt that the new material can be incorporated into *A. boisei* (Delson 1986; Kimbel et al. 1988). My own judgment is that while the decision is inevitably finely judged, in part because of the small sample size of the early material, the present evidence does justify its allocation to a separate species, but it is by no means clear that the species name *A. aethiopicus* remains available for any new species (Wood 1992c).

Turning to early *Homo*, Howell (1978:193) recognized a single species, *Homo habilis*, but suggested that "there appears to have been considerable phyletic evolution, especially in craniofacial and dental morphology, within the chronospecies." Howell also suggested that the *H. habilis* material recovered from the Koobi Fora succession prior to 1978 showed substantial variability. Howell's prescience was vindicated for the Koobi Fora evidence has proved to be highly variable, but researchers differ in how they interpret that variation. While some experienced observers continue to subsume *H. habilis* remains from the two main sites within one species (e.g., Tobias 1991), others have opted for a multiple species solution (Wood 1985, 1991a; Stringer 1986). My own view is that while the early *Homo* remains from Olduvai are properly included within *H. habilis sensu strictu* part of the early *Homo* hypodigm from Koobi Fora (e.g. KNM–ER 1802, 1470, 1590, and 3732) should be attributed to a separate taxon, *Homo rudolfensis* Alexeev.

In 1978 there was little sign of the controversy, much of it cladistically generated, that now surrounds *Homo erectus*. At the center of these debates is the contention that early *H. erectus* remains from Africa are morphologically distinct from the main Asian hypodigm of that taxon. It has been argued, with varying degrees of fervor (e.g., Andrews 1984; Wood 1984) that these distinctions justify erecting a separate species for specimens from Koobi Fora and West Turkana such as KNM–ER 992, 3733, 3883 and KNM–WT 15000. Given the inclusion of KNM–ER 992 in the hypodigm, the name of any new species had previously been determined by a contribution by Groves and Mazák (1975) in which they proposed that mandible as the type specimen of a new hominid species, *Homo ergaster*. While I remain (Wood 1991a) in favor of the specific distinction, I do so largely as an attempt to keep open the possibility that *H. ergaster* may yet be the sensible way to recognize the substantial regional variation among remains belonging to the *H. erectus* grade. However, others doubt that any distinctions that do exist deserve specific recognition (e.g., Turner and Chamberlain 1989; Kennedy 1991; Brauer and Mbua 1992).

In his treatment of *H. sapiens*, dividing it as he did into three subspecies, *pace* Mayr (1950), Howell anticipated some of the contemporary debates about the origins, and thus the taxonomy, of *H. sapiens*. Since 1978, proposals about the taxonomy of later *Homo* remains have polarized. On the one hand there are those who so stress the importance of gene flow in the *erectus* and post–*erectus* phases of hominid evolution that they advocate widening the definition of *H. sapiens* to include *H. erectus* (Wolpoff and Thorne 1992). In contrast, other researchers have no difficulty in discerning taxonomically significant patterns in the Upper Pleistocene hominid fossil record and consequently propose that one, or more, of these variants should be accorded specific distinction (Tattersall 1986; Stringer 1991).

Relationships and Scenarios

The nature of the relationships between hominid taxa has been the object of substantial research effort since 1978. Notwithstanding the details of the cladograms, the results of cladistic analyses of hominid species have been consistent in confirming that hominid evolution has been far from straightforward. Homoplasies are common, particularly within the masticatory system (Wood and Chamberlain 1986). The results of cladistic analyses, together with the discovery of new fossil evidence, have prompted the reexamination of the evidence for at least one established sister–taxon pairing, *P. robustus* and *P. boisei*, as well as stimulating investigations into the integrity of the *Homo* clade (Wood 1988, 1991a).

Since 1978, although there has been relatively little progress toward what Pilbeam (1984:19) has termed the "biological reconstruction" of African hominid species, what progress there has been has been sound and promising. For example, recent attempts to derive the body weights of extinct hominid taxa have resulted in estimates that are mostly at variance with ranges of body weight that we had been used to operating with (McHenry 1992). These new estimates will be essential tools in attempts to link body size with other life history variables for which data are available. Holloway's efforts over the years to provide accurate and realistically precise estimates of hominid brain volumes are available as a valuable data base, and Smith (1989) has begun setting out the ways in which these and other life history–related data can help us trace the evolution of life history variables in much the same way that we can postulate pathways for morphological evolution.

The role of climate change as a possible trigger for hominid speciation was recognized before 1978 (Coppens 1975), but it has been receiving increasing attention since that time. Hominid speciation "events," be they the appearance or the elimination of taxa, cluster at approximately 2.5 and 2.0 MYA, apparently correspond to similar events in the evolution of the African bovids (Vrba 1985), and climate change has been invoked as the common trigger mechanism. The disappearance of *A. africanus* from the southern African fossil record and the appearance in East Africa of a postcanine megadont hominid around 2.5 MYA corresponds to a shift in climate and a consequent change to a more xeric paleoenvironmental regime (Prentice and Denton 1988; Vrba 1988). It is also apparent that other large African mammals show changes in their masticatory system analogous to the postcanine megadontia seen in at least one hominid lineage (Turner and Wood 1993). The continued emphasis on intraoral food preparation in the "robust" taxa contrasts with other contemporary hominid species whose postcanine dentition did not increase in relation to body size (Wood 1991a). These are important contributions to the task of setting hominid paleontology within the broader context of mammalian evolution in Africa.

PROSPECTS AND CHALLENGES

The historicist analytical strategy urges us to accept that "real historical understanding" will not be achieved unless we make "the past our present" and attempt "to see life with the eyes of another country than our own" (Butterfield 1931:16). That "other country," even though it may have been in existence several million years before the present, has sufficient connections with the present for us to assume that the biological "first principles" that obtain today also applied to the Plio–Pleistocene. It is this very continuity of principles that allows us to apply contemporary research strategies not only to hominid evolution, but also to evolutionary history of a much greater antiquity. There are examples in related disciplines where contemporary assumptions need to be modified to take relative antiquity into account (the levels of atmospheric ^{14}C is a good example), but happily they are the exception.

The diagram shown in Figure 7-1 is a grossly oversimplified rendering of the events upon which all potential for evolution is predicated. For many years anyone studying evolutionary history was restricted to information about the shape and size of bones and teeth; I have referred to this as "classical morphology." From these data, researchers attempted to reconstruct both individual and group behavior, on the one hand, and phylogenetic history, on the other. Morphology was their only evidence for propinquity, and they relied on the proposition that the more alike two species are, then the more closely related they are likely to be. Research in genetics and developmental biology has identified many, but by no means all, of the processes and controls that translate genetic information encoded in DNA base sequences into proteins and thence into the classical phenotype. However, the epigenetic "noise" at the protein–morphology interface undoubtedly interfered with information about propinquity.

The introduction of "molecular anthropology" in the mid–1960s helped to eliminate that noise. Advances in immunology provided techniques for the comparison of the 3–D shape of proteins (Goodman 1962), and the introduction of electrophoresis enabled the discrimination of proteins and peptides by their mass and charge (Zuckerkandl 1963). Molecular anthropology effectively enabled morphological comparisons to be extended downward in scale toward the genome; the shape of a bone was replaced by the shape of a protein. This had the advantage that hypotheses of propinquity based on protein morphology were likely to be a better reflection of affinity than were those based upon classical morphology, but the comparison still stopped short of the genome. Early experiments used serum and proteins from living

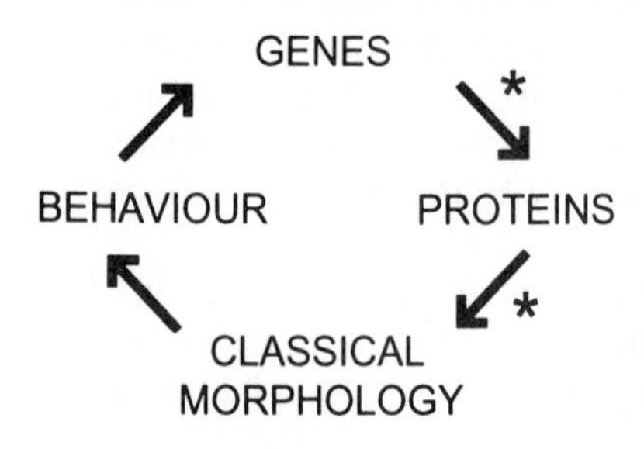

FIGURE 7-1 Simplified demonstration of the main links between genes and behavior. Developmental biology addresses the interactions at the two interfaces marked with an asterisk.

animals, but although albumin has been recovered from relatively recent fossils (Lowenstein et al. 1981), the initial promise of these methods as a means of unraveling relationships has remained unfulfilled. Now even the barrier to the genome has been broken, for DNA has been recovered and successfully amplified from ancient bone, albeit from relatively recent human remains (Hagelberg et al. 1989). These developments were made possible by the development of the polymerase chain reaction (PCR) system, which can amplify from a single copy of nuclear or mitochondrial DNA (Saiki et al. 1985). However, even if sufficiently long sequences of DNA could be extracted from Plio–Pleistocene hominid fossil bones, would it make "classical" paleontological research redundant? Emphatically not. While it is evident that DNA data are the most powerful tools we have available to determine species groupings, or for studying the relationships between those species, DNA data are a good deal less useful than classical morphology for drawing inferences about behavior. Just consider for a moment the implications of the fact that mtDNA and nuclear DNA data, from both sequencing and hybridization studies, are presently unable to resolve the relationships between *Pan, Gorilla,* and *Homo*. Yet, in terms of classical morphology and behavior, we know that the three animals are easily and obviously distinguishable.

Just as there are exciting prospects for research at the level of the genome, there are also opportunities to improve our ability to reconstruct the behavior of extinct hominid species. Archeologists interpret the archeological record in an effort to reconstruct early hominid behaviors, but discoveries of fossil footprints (Leakey and Hay 1979; are reminders that other behaviors, such as locomotion, can also be "fossilized." Details of early hominid behavior are also under the same general physiological constraints as are those of modern animals. Provided researchers can determine a more detailed record of past climates it may be possible to specify likely activity patterns, using the experimentally deduced constraints imposed by body form and the energy and water budgets of early hominids (Wheeler 1993).

Notwithstanding the developments referred to, the classical morphological evidence from the fossil record remains the main source of data for the hominid paleobiologist. But would all the unresolved problems of hominid paleobiology be solved by a better (i.e., more complete) fossil record? I have my doubts, for there is no good evidence to suggest any reliable correlation between an improvement in our comprehension of hominid evolutionary history and the size of the African hominid fossil record. More and more data are not automatically and linearly converted into greater and greater understanding. Fresh fossil evidence has the most impact when it either helps to illuminate parts of evolutionary history that are either unrepresented or poorly represented in the fossil record, or when it contributes to a sample large enough to provide reliable estimates of population parameters. At present, we sorely lack evidence about hominids between 3.5 and 4.5 MYA as well as a better fossil record between 1.75 and 2.75 MYA. Several initiatives in Ethiopia and elsewhere in Africa promise to help fill the earlier gap in our knowledge. Some African early hominid species categories now have a sufficiently good fossil record that reliable population estimates can be obtained for many dental and gnathic features (Wood 1991a).

Developments in data storage and analysis now allow the routine handling of 3–D metrical data and thus a more accurate quantitative record of classical morphology. This facility has resulted in a renewal of interest in morphometrics, and the next few years are likely to witness the increasing use of multivariate statistical methods, of both the traditional and the newer varieties, to help in the description and comparison of fossil specimens. The study of allometry and heterochrony (Gould 1977) and especially the investigation of ontogenetic series (Shea 1983, 1988) are providing researchers with the means to reconstruct evolutionary pathways and specify at which points changes to the growth trajectories of presumed ancestral forms took place.

We are also likely to see refinements and advances in the way that hard–tissue microstructure can contribute to both taxonomy and the analysis of hominid life histories. Other mineralized dental tissues, for example dentine and cementum, which until now have received much less attention than enamel, also grow incrementally and thus have the potential to record developmental history. While it does not record it as precisely as enamel, cementum has the advantage that its deposition is more susceptible to environmental factors and thus the pattern of its deposition may offer a technique for recording seasonality and for relating the time of death to any such seasonal fluctuations (Lieberman and Meadow 1992).

Decisions about the limits of the size and shape variation that it is sensible to accept within a single hominid species are also likely to become more sophisticated as we learn more about the patterns of variation within and between those species closely related to hominids. Up until now attempts to discriminate between intra– and interspecific variation have been relatively crude. The growing awareness that they need to take more cognizance of the subspecific identity of comparative material will allow researchers to resolve intraspecific variation into sexual and geographic components. This perspective is particularly important for the study of early African hominids sampled from two main, geographically distinct, regions. Cladistic analytical techniques will benefit from continued scrutiny and refinement. The extent of functional convergence in African hominids has probably been underestimated, and cladists will have to meet the challenge of how to identify convergent characters as well as acknowledge and meet the need to find ways of assessing the significance of differences between alternative cladograms. The present methods, tree lengths and computing the consistency indices, have no corresponding distribution theory to enable researchers to assess the results. Hypotheses of homology will require much closer scrutiny, with the application of ontogenetic criteria being more commonplace than is presently the case (Wood 1988; Thiranagama et al. 1991; Yoder 1992). Future studies of hominid adaptations will require information about habitats, environment and regional climates to be a good deal finer grained than is presently the case. Such data are also required if hominid paleobiologists are to capitalize on the advances in physiological and ecological modeling referred to previously.

There is no shortage of challenges to take us through to the close of the century. For example, we have been conditioned to think of increased relative brain size, improved manual dexterity, and facultative bipedalism as unique developments limited to our own genus. But what if they occurred in more than one hominid lineage? Can the characters that are shared between what we presently call *Paranthropus* and *Homo* all be explained as homoplasies, or are these two clades more closely related than we have previously assumed? Have we found in *A. afarensis* the common hominid ancestor, or is that species too specialized for that role? Can we harness biomechanical methods, isotope analysis, paleontology, and paleobotany in order to develop testable hypotheses about early hominid diets? Did *H. erectus* originate in, or subsequently migrate to, Africa? Is there evidence of a morphological "break" between *H. erectus* and *H. sapiens*, or will African remains provide crucial evidence to demonstrate species continuity since the Middle Pleistocene? All these challenges, and many more, await hominid paleobiologists as we move through this final decade of the twentieth century. There is an abundance of exciting opportunities and we are grateful to Clark Howell for helping to provide that essential, sound foundation upon which we can build the next phase of hominid paleobiological research.

SUMMARY

This chapter surveys the achievements of hominid paleobiology since 1978, the date of Howell's review of the African evidence for hominid evolution. It introduces fossil evidence found since 1978 both at well–known sites and at localities discovered subsequently. Much of this fossil evidence expands upon our knowledge of established hominid species, but some, at Olduvai and West Turkana in particular, has prompted reassessments of earlier taxonomic judgments.

The introduction of new imaging techniques and novel methods of histological examination of hard tissue microstructure have made new data accessible to paleontological analysis, and the contributions of these new data to our understanding of hominid evolutionary history are reviewed. The strengths and weaknesses of cladistic methods for the determination of species relationships are set in context as are the attempts to use somatic growth data to predict life history strategies.

Comparisons are made between taxonomic hypotheses that were current in 1978 with those of today. The case for recognizing more hominid species is outlined, and the difficulties resulting from disagreements about what constitutes an early hominid species are set out. Attention is drawn to the extent of early hominid homoplasies, and the importance for scenario reconstruction of the recent success of using trace tephra to correlate the major East African fossil sites is emphasized.

The final section looks forward to the end of the century and attempts to identify the analytical methods that are likely to contribute most to improving our future understanding of African hominid paleobiology.

ACKNOWLEDGMENTS

Research incorporated in this chapter was funded by The Leverhulme Trust, the Royal Society, and the SERC.

The assistance of Malcolm Hall, Dan Lieberman, Fred Spoor, and Alan Turner is much appreciated and Maureen Baldock's help in preparing the manuscript is gratefully acknowledged.

LITERATURE CITED

Andrews, P. (1984) An alternative interpretation of the characters used to define *Homo erectus*. *Cour. Forsch. Inst. Senckenberg* 69:167–175.

Asfaw, B., Beyene, Y., Semaw, S., Suwa, G., White, T., and Wolde Gabriel, G. (1991) Fejej: A new palaeonthropological research area in Ethiopia. *J. Hum. Evol.* 21:137–143.

Asfaw, B., Beyene, Y., Suwa, G., Walter, R. C., White, T. D., WoldeGabriel, G., and Yemane, T. (1992) The earliest Acheulian from Konso-Gardula. *Nature* 360:732–735.

Asfaw, B., Ebinger, C., Harding, D., White, T. and Wolde-Gabriel, G. (1990) Space–based imagery in paleoanthropological research: an Ethiopian example. *Nat. Geog. Res.* 6(4):418–434.

Ben–Itzhak, S., Smith, P., and Bloom, R. A. (1988) Radiographic study of the humerus in Neanderthals and *Homo sapiens sapiens*. *Am. J. Phys. Anthropol.* 77:231–241.

Beynon, A. D., and Dean, M. C. (1988) Distinct dental development patterns in early fossil hominids. *Nature* 335:509–514.

Beynon, A. D., and Wood, B. A. (1986) Variations in enamel thickness and structure in East African hominids. *Am. J. Phys. Anthropol.* 70:177–193.

Beynon, A. D., and Wood, B. A. (1987) Patterns and rates of enamel growth in the molar teeth of early hominids. *Nature* 326:493–496.

Bock, W. J., and von Wahlert, G. (1965) Adaptation and the form–function complex. *Evolution* 19:269–299.

Bown, T. M., and Rose, K. D. (1987) Patterns of dental evolution in early Eocene Anaptomorphine primates (Omomyidae) from the Bighorn Basin, Wyoming. *J. Paleont.* 61:1–162.

Boyde, A., and Jones, S. J. (1972) Scanning electron microscopic studies of the formation of mineralized tissues. In H. C. Slavikin and L. A. Bavetta (eds.): *Development Aspects of Oral Biology*. New York: Academic, pp. 243–274.

Brain, C. K. (1975) An interpretation of the bone assemblage from the Kromdraai australopithecine site, South Africa. In R. Tuttle (ed.): *Paleoanthropology, Morphology and Paleoecology*. The Hague: Mouton, pp. 225–243.

Brain, C. K. (1982) The Swartkrans site: Stratigraphy of the fossil hominids and a reconstruction of the environment of early *Homo*. *Proc. 1st Int. Congress. Hum. Palaeont.* 2:676–706.

Brain, C. K. (1988) New information from the Swartkrans cave of relevance to "robust" australopithecines. In F.E. Grine (ed.): *Evolutionary History of the "Robust" Australopithecines.* New York: Aldine de Gruyter, pp. 311–316.

Brain, C. K., Churcher, C. S., Clark, J. D., Grine, F. E., Shipman, P., Susman, R. L., Turner, A., and Watson, V. (1988) New evidence of early hominids, their culture and environment from the Swartkrans Cave, South Africa. *S. Afr. J. Sci.* 84:828–835.

Brauer, G., and Mbua, E. (1992) *Homo erectus* features used in cladistics and their variability in Asian and African hominids. *J. Hum. Evol.* 22:79–108

Bromage, T. G. (1985) Taung facial remodelling: a growth and development study. In P. V. Tobias (ed.): *Hominid Evolution: Past, Present and Future.* New York: Alan R. Liss, pp. 239–245.

Bromage, T. G. (1987) The scanning electron microscopy/replica technique and recent applications to the study of fossil bone. *Scanning Electron Microsc*: 607–613.

Bromage, T. G. (1989) Ontogeny of the early hominid face. *J. Hum. Evol.* 18:751–773.

Brown, F. H. (1982) Tulu Bor tuff at Koobi Fora correlated with the Sidi Hakoma tuff at Hadar. *Nature* 300:631–635.

Butterfield, H. (1931) *The Whig Interpretation of History*. London: George Bell.

Cerling, T. E., and Brown, F. H. (1982) Tuffaceous marker horizons in the Koobi Fora region and the Lower Omo Valley. *Nature* 299:216–221.

Cerling, T. E., Cerling, B. W., Curtis, G. H., and Drake, R. E. (1979) Preliminary correlations between the Koobi Fora and Shungura Formations, East Africa. *Nature* 279:118–121.

Chavaillon, J., Chavaillon, N., Coppens, Y., and Senut, B. (1977) Présence d'hominidé dans le site Oldowayen de Gomboré Ie Melka Kunturé, Ethiopie. *C. R. Acad. Sci. D.* 285:961–963.

Chavaillon, J., and Coppens, Y. (1975) Découverte d'Hominidé dans un site Acheuléen de Melka–Kunturé (Ethiopie). *Bull. Mém. Soc. Anthropol., Paris* 2:125–128.

Chavaillon, J., and Coppens, Y. (1986) Nouvelle découverte d'*Homo erectus* Melka Kunturé (Ethiopie). *C. R. Acad. Sci. Paris* 303:99–104.

Chavaillon, J., Hours, F., and Coppens, Y. (1987) Découverte de restes humains fossiles associés un outillage acheuléen finalé Melka Kunturé (Ethiopie). *C. R. Acad. Sci. Paris* 304:539–542.

Clark, J. D., Asfaw, B., Assefa, G., Harris, J. W. K., Kurashina, H., Walter, R. C., White, T. D., and Williams, M. A. J. (1984) Palaeoanthropological discoveries in the Middle Awash Valley, Ethiopia. *Nature* 307:423–428.

Clarke, R. J. (1985) *Australopithecus* and early *Homo* in Southern Africa. In E. Delson (ed.): *Ancestors: The Hard Evidence.* New York: Alan R. Liss, pp. 171–177.

Clarke, R. J. (1988) A new *Australopithecus* cranium from Sterkfontein and its bearing on the ancestry of *Paranthropus.* In F. Grine (ed.): *Evolutionary History of the "Robust" Australopithecines.* New York: Aldine de Gruyter, pp. 285–292.

Conroy, G. C., Jolly, C. J., Cramer, D., and Kalb, J. E. (1978) Newly discovered fossil hominid skull from the Afar depression, Ethiopia. *Nature* 276:67–70.

Conroy, G. C., and Vannier, M. W. (1987) Dental development of the Taung skull from computerized tomography. *Nature* 329:625–627.

Conroy, G. C., and Vannier, M. W. (1991) Dental development in South African australopithecines. Part I: Problems of pattern and chronology. *Am. J. Phys. Anthropol.* 86:121–136.

Conroy, G. C., Vannier, M. W., and Tobias, P. V. (1990) Endocranial features of *Australopithecus africanus* revealed by 2– and 3–D computer tomography. *Science* 247:838–841.

Coppens, Y. (1975) évolution des Hominidés et de leur environnement au cours du Plio–Pléistocéne dans la basse vallée de L'Omo en éthiopie. *C. R. Acad. Sci. Paris* 281:1693–1696.

Coppens, Y. (1980) The differences between *Australopithecus* and *Homo*; preliminary conclusions from the Omo Research Expedition's studies. In L.–K. Königsson (ed.): *Current Argument on Early Man.* Oxford: Pergamon, pp. 207–225.

Cracraft, J. (1987) Species concepts and the ontology of evolution. *Biol. Philos.* 2:329–346.

Daegling, D. J. (1989) Biomechanics of cross–sectional size and shape in the hominid mandibular corpus. *Am. J. Phys. Anthropol.* 80:91–106.

Daegling, D. J., and Grine, F. E. (1991) Compact bone distribution and biomechanics of early hominid mandibles. *Am. J. Phys. Anthropol.* 86:321–339.

Debenath, A., Raynal, J.–P., and Texier, J.–P. (1982) Position stratigraphique des restes humains paléolithiques marocains sur la base des travaux récents. *C. R. Acad. Sci. Paris* 294:1247–1250.

Delson, E. (1986) Human phylogeny revised again. *Nature* 322:496–497.

Dunbar, R. I. M. (1988) *Primate Social Systems.* London: Chapman and Hall.

Dunbar, R. I. M. (1989) Ecological modelling in an evolutionary context. *Folia Primatol.* 53:235–246.

Duterloo, H. S., and Enlow, D. H. (1970) A comparative study of cranial growth in *Homo* and *Macaca. Am. J. Anat.* 127:357–367.

Feibel, C. S., Brown, F. H., and McDougall, I. (1989) Stratigraphic context of fossil hominids from the Omo Group deposits: Northern Turkana Basin, Kenya and Ethiopia. *Am. J. Phys. Anthropol.* 78:595–622.

Ferguson, W. W. (1984) Revision of fossil hominid jaws from the Plio–Pleistocene of Hadar, in Ethiopia including a new species of the genus *Homo* (Hominoidea: Homininae). *Primates* 25:519–529.

Fleagle, J. G., Rasmussen, D. T., Yirga, S., Bown, T. H., and Grine, F. E. (1991) New hominid fossils from Fejej, Southern Ethiopia. *J. Hum. Evol.* 21:145–152.

Foley, R. (1987) *Another Unique Species.* Harlow: Longman.

Goodman, M. (1962) Immunochemistry of the primates and primate evolution. *Ann. NY Acad. Sci.* 102:219–234.

Gould, S. J. (1977) *Ontogeny and Phylogeny.* Cambridge: Belknap.

Gowlett, J. A. J., Harris, J. W. K., Walton, D., and Wood, B. A. (1981) Early archaeological sites, hominid remains and traces of fire from Chesowanja, Kenya. *Nature* 294:125–129.

Grine, F. E. (1982a) A new juvenile hominid (Mammalia: Primates) from Member 3, Kromdraai Formation, Transvaal, South Africa. *Ann. Trans. Mus. 33*:165–239.

Grine, F. E. (1982b) Note on a new hominid specimen from Member 3, Kromdraai Formation, Transvaal. *Ann. Trans. Mus.* 33:287–290.

Grine, F. E. (1989) New hominid fossils from the Swartkrans Formation (1979–1986 excavations): craniodental specimens. *Am. J. Phys. Anthropol.* 79:409–449.

Grine, F. E., and Martin, L. B. (1988) Enamel thickness and development in *Australopithecus* and *Paranthropus.* In F. E. Grine (ed.): *Evolutionary History of the "Robust" Australopithecines.* New York: Aldine de Gruyter, pp. 3–42.

Grine, F. E., and Susman, R. L. (1991) Radius of *Paranthropus robustus* from Member 1, Swartkrans Formation, South Africa. *Am. J. Phys. Anthropol.* 84:229–248.

Groves, C. P., and Mazák, V. (1975) An approach to the taxonomy of the Hominidae: Gracile Villafranchian hominids of Africa. *Cas. Miner. Geol.* 20:225–247.

Hagelberg, E., Sykes, B., and Hedges, R. (1989) Ancient bone DNA amplified. *Nature* 342:485

Haileab, B., and Brown, F. H. (1992) Turkana Basin–Middle Awash correlations and the age of the Sagantole and Hadar Formations. *J. Hum. Evol.* 22:453–468.

Harris, J. M., Brown, F. H., Leakey, M. G., Walker, A. C., and Leakey, R. E. (1988) Pliocene and Pleistocene hominid–bearing sites from west of Lake Turkana, Kenya. *Science* 23:27–33.

Hill, A. (1985) Early hominid from Baringo, Kenya. *Nature* 315:222–224.

Holloway, R. L. (1981) Volumetric and asymmetry determinations on recent endocasts: Spy I and II, Djebel Irhoud I, and the Salé *Homo erectus* specimens, with some notes on Neanderthal brain size. *Am. J. Phys. Anthropol.* 55:385–393.

Howell, F. C. (1960) European and Northwest African Middle Pleistocene hominids. *Curr. Anthropol.* 1:195–232.

Howell, F. C. (1978) Hominidae. In V. J. Maglio and H. B. S. Cooke (eds.): *Evolution of African Mammals*. Cambridge: Harvard University Press, pp. 154–248.

Jaeger, J.-J. (1975) Découverte d'un crane d'Hominidé dans le Pléistocéne moyen du Maroc. In Problemes Actuels de Paléontologie – évolution des Vertébrés. *Coll. Int. CNRS.* No. 218:897–902.

Johanson, D. C., Masao, F. T., Eck, G. G., White, T. D., Walter, R. C., Kimbel, W. H., Asfaw, B., Manega, P., Ndessokia, P., and Suwa, G. (1987) New partial skeleton of *Homo habilis* from Olduvai Gorge, Tanzania. *Nature* 327:205–209.

Johanson, D. C., Taieb, M., and Coppens, Y. (1982) Pliocene hominids from the Hadar Formation, Ethiopia (1973–1977): Stratigraphic, chronologic and paleoenvironmental contexts, with notes on hominid morphology and systematics. *Am. J. Phys. Anthropol.* 57:373–402.

Kalb, J. E., Jolly, C. J., Mebrate, A., Tebedge, S., Smart, C., Oswald, E. B., Cramer, D., Whitehead, P., Wood, C. B., Conroy, G. C., Adefris, T., Sperling, L., and Kana, B. (1982) Fossil mammals and artefacts from the Middle Awash Valley, Ethiopia. *Nature* 298:25–29.

Kennedy, G. E. (1983) Femoral morphology in *Homo erectus*. *J. Hum. Evol.* 12:587–616.

Kennedy, G. E. (1984) The emergence of *Homo sapiens*: The postcranial evidence. *Man* 19:94–110.

Kennedy, G. E. (1991) On the autapomorphic traits of *Homo erectus*. *J. Hum. Evol.* 20:375–412.

Kimbel, W. (1992) Paleoanthropology update. *I.H.O. Newsletter* 10(1):7.

Kimbel, W. H., and Rak, Y. (1993) The importance of species taxa in paleoanthropology and an argument for the phylogenetic concept of the species category. In W. H. Kimbel and L. B. Martin (eds.): *Species, Species Concepts and Primate Evolution*. New York: Plenum.

Kimbel, W. H., and White, T. D. (1988) Variation, sexual dimorphism and taxonomy of *Australopithecus*. In F. E. Grine (ed.): *Evolutionary History of the "Robust" Australopithecines*. New York: Aldine de Gruyter, pp.175–192.

Kimbel, W. H., White, T. D. and Johanson, D. C. (1988) Implications of KNM-WT 17000 for the evolution of "robust" *Australopithecus*. In F. Grine (ed.): *The Evolutionary History of the Robust Australopithecines*. New York: Aldine, pp. 259–268.

Krantz, G. S. (1977) A revision of australopithecine body sizes. *Evol. Theory* 2(2):65–94.

Landau, M. (1991) *Narratives of Human Evolution*. New Haven and London: Yale University Press.

Leakey, M. D., and Hay, R. L. (1979) Pliocene footprints in the Laetolil Beds at Laetoli, northern Tanzania. *Nature* 278:317–323.

Lieberman, D. E., and Meadow, R. H. (1992) The biology of cementum increments (with an archaeological application). *Mamm. Rev.* 22(2):57–77.

Lowenstein, J. M., Sarich, V. M., and Richardson, B. J. (1981) Albumin systematics of the extinct mammoth and Tasmanian wolf. *Nature* 291:409–411.

McHenry, H. M. (1992) How big were early hominids? *Evol. Anthropol.* 1:15–20.

Macho, G. A., and Thackeray, J. F. (1992) Computer tomography and enamel thickness of maxillary molars of Plio–Pleistocene hominids from Sterkfontein, Swartkrans, and Kromdraai (South Africa): An exploratory study. *Am. J. Phys. Anthropol.* 89:133-143.

Mayr, E. (1950) Taxonomic categories in fossil hominids. *Cold Spr. Harb. Symp. Quant. Biol.* 15:109–118.

Ng, S. Y., Payne, P. A., Cartledge, N. A., and Ferjuson, M. W. J. (1989) Determination of ultrasonic velocity in human enamel and dentine. *Arch. Oral Biol.* 34:341–345.

Olson, T. R. (1981) Basicranial morphology of the extant hominoids and Pliocene hominids: the new material from the Hadar Formation, Ethiopia and its significance in early human evolution and taxonomy. In C. B. Stringer (ed.): *Aspects of Human Evolution*. London: Taylor and Francis, pp. 99–128.

Partridge, T. C., Tobias, P. V., and Hughes, A. R. (1991) Paléoécologie et affinités entre les Australopithécinés d'Afrique du Sud: nouvelles données de Sterkfontein et Taung. *L'Anthropologie* 95:363–378.

Pilbeam, D. (1984) Reflections on early human ancestors. *J. Anthropol. Res.* 40:14–22.

Pilbeam, D. (1986) Hominid evolution and hominoid origins. *Am. Anthropol.* 88:295–312.

Pilbeam, D. R. (1988) Human origins and evolution. In A. C. Fabian (ed.): *Origins*. Cambridge: Cambridge University Press, pp. 89–114.

Prentice, M. L., and Denton, G. H. (1988) The deep-sea oxygen isotope record, the global ice sheet system and hominid evolution. In F. E. Grine (ed.): *Evolutionary History of the "Robust" Australopithecines*. New York: Aldine de Gruyter, pp. 383–403.

Read, D. W. (1984) From multivariate statistics to natural selection: a reanalysis of the Plio–Pleistocene hominid dental material. In G. N. van Vark and W. W. Howells (eds.): *Multivariate Statistical Methods in Physical Anthropology* Dordrecht: D. Reidel, pp. 377–413.

Ruff, C. (1987) Structural allometry of the femur and tibia in Hominoidea and *Macaca*. *Folia Primatol.* 48:9–49.

Ruff, C. B. (1991) Climate and body shape in hominid evolution. *J. Hum. Evol.* 21:81–105.

Ruff, C. B., and Leo, F. P. (1986) Use of computed tomography in skeletal structure research. *Yearb. Phys. Anthropol.* 29:181–196.

Saiki, R. K., Scharf, S., Faloona, F., Mullis, K. B., Horn, G. T., Erlich, H. A., and Arnheim, N. (1985) Enzymatic amplification of beta–globin genomic sequences and restriction site analysis for diagnosis of sickle cell anaemia. *Science* 230:1350–1354.

Sarna–Wojcicki, A. M., Meyer, C. E., Roth, P. H., and Brown, F. H. (1985) Ages of the tuff beds at East African early hominid sites and sediments in the Gulf of Aden. *Nature* 313:306–308.

Senut, B. (1985) Computerized tomography of a Neanderthal humerus from La Regordou (Dordogne, France): comparisons with modern man. *J. Hum. Evol.* 14:717–723.

Shea, B. T. (1983) Allometry and heterochrony in the African apes. *Am. J. Phys. Anthropol.* 56:179–201.

Shea, B. T. (1988) Heterochrony in primates. In M. L. McKinney (ed.): *Heterochrony in Evolution*. New York: Plenum, pp. 237–266.

Shellis, R. P. (1984) Variations in growth of the enamel crown in human teeth and a possible relationship between growth and enamel structure. *Arch. Oral Biol.* 29:697–705.

Smith, B. H. (1989) Dental development as a measure of life history in primates. *Evolution* 43:683–688.

Smith, B. H. (1991) Dental development and the evolution of life history strategies. *Am. J. Phys. Anthropol.* 86:157–174.

Sperber, G. H. (1985) Comparative primate dental enamel thickness: A radiodontological study. In P. V. Tobias (ed.): *Hominid Evolution: Past, Present and Future*. New York: Alan R. Liss, pp. 443–454.

Spoor, C. F., Zonneveld, F. W., and Macho, G. A. (in press) Linear measurements of cortical bone and dental enamel by computed tomography: applications and problems. *Am. J. Phys. Anthropol.*

Stringer, C. B. (1986) The credibility of *Homo habilis*. In B. Wood, L. Martin, and P. Andrews (eds.): *Major Topics in Primate and Human Evolution*. Cambridge: Cambridge University Press, pp. 266–294.

Stringer, C. B. (1991) Time for the last Neanderthals. *Nature* 351:701–702.

Susman, R. L. (1989) New hominid fossils from the Swartkrans Formation (1979–1986 excavations): postcranial specimens. *Am. J. Phys. Anthropol.* 79:451–474.

Tattersall, I. (1986) Species recognition in human paleontology. *J. Hum. Evol.* 15:165–175.

Tattersall, I., and Eldredge, N. (1977) Fact, theory and fantasy in human paleontology. *Am. Sci.* 65:204–211.

Thiranagama, R., Chamberlain, A. T., and Wood, B. A. (1991) Character phylogeny of the primate forelimb superficial venous system. *Folia Primatol.* 51:181–190.

Tobias, P. V. (1980) "*Australopithecus afarensis*" and *A. africanus*: Critique and an alternative. *Palaeont. Afr.* 23:1–17.

Tobias, P. V. (1991) *Olduvai Gorge*, Vol. 4: *The Skulls, Endocasts and Teeth of* Homo habilis. Cambridge: Cambridge University Press.

Tobias, P. V. (1992) 26th Annual Report of PARU, 25–27.

Tooby, J., and DeVore, I. (1986) The reconstruction of hominid behavioral evolution through strategic modelling. In W. Kinzey (ed.): *Evolution of Human Behavior*. Albany: State University of New York Press, pp. 183–237.

Trinkaus, E. (1990) Cladistics and the hominid fossil record. *Am. J. Phys. Anthropol.* 83:1–12

Turner, A., and Chamberlain, A. T. (1989) Speciation, morphological change and the status of African *Homo erectus*. *J. Hum. Evol.* 18:115–130.

Turner, A., and Wood, B. (1993) Comparative palaeontological context for the evolution of the early hominid masticatory system. *J. Hum. Evol.* 24:301–318.

Vrba, E. S. (1981) The Kromdraai australopithecine site revisited in 1980; recent investigations and results. *Ann. Trans. Mus.* 33:17–60.

Vrba, E. S. (1985) African bovidae: Evolutionary events since the Miocene. *S. Afr. J. Sci.* 81:263–266.

Vrba, E. S. (1988) Late Pliocene climatic events and hominid evolution. In F. E. Grine (ed.): *Evolutionary History of the "Robust" Australopithecines*. New York: Aldine de Gruyter, pp. 405–426.

Walker, A., Leakey, R. E., Harris, J. M., and Brown, F. H. (1986) 2.5–Myr *Australopithecus boisei* from west of Lake Turkana, Kenya. *Nature* 322:517–522.

Walkoff, O. (1904) Das femur des Menschen und der Anthropomorphen. Studien Entwicklungsgesch des Primatenskelett. Weisbaden.

Ward, S. C., Johanson, D. C., and Coppens, Y. (1982) Subocclusal morphology and alveolar process relationships of hominid gnathic elements from the Hadar Formation: 1974–1977 collections. *Am. J. Phys. Anthropol.* 57:605–630.

Webb, S. G. (1989) *The Willandra Lakes Hominids*. Department of Prehistory, Research School of Pacific Studies, Australian National University, Canberra, pp. 1–194.

Weidenreich, F. (1941) The extremity bones of *Sinanthropus pekinensis*. *Palaeont. Sinica. Series* D 5:1–151.

Weidenreich, F. (1943) The skull of *Sinanthropus pekinensis*: A comparative study of a hominid skull. *Palaeont. Sinica. Series* D 10:1–484.

Wheeler, P. E. (1984) The evolution of bipedality and loss of functional body hair in hominids. *J. Hum. Evol.* 13:91–98.

Wheeler, P. E. (1985) The loss of functional body hair in man: the influence of thermal environment, body form and bipedality. *J. Hum. Evol.* 14:23–28.

Wheeler, P. E. (1991a) The thermoregulatory advantages of hominid bipedalism in open equatorial environments: The contribution of increased convective heat loss and cutaneous evaporative cooling. *J. Hum. Evol.* 21:107–115.

Wheeler, P. E. (1991b) The influence of bipedalism on the energy and water budgets of early hominids. *J. Hum. Evol.* 21:117–136.

Wheeler, P. E. (1993) The influence of stature and body form on hominid energy and water budgets: a comparison of *Australopithecus* and early *Homo* physiques. *J. Hum. Evol.* 24:13–28.

Wind, J. (1984) Computerized X–ray tomography of fossil hominid skulls. *Am. J. Phys. Anthropol.* 63:265–282.

Wind, J., and Zonneveld, F. W. (1989) Computed tomography of an *Australopithecus* skull (Mrs Ples): A new technique. *Naturwiss.* 76:325–327.

Wolde Gabriel, G., White, T., Suwa, G., Semaw, S., Beyene, Y., Asfaw, B., and Walter, R. C. (1992) Kesem–Kebana: A newly discovered paleoanthropological research area in Ethiopia. *J. Field Archaeol.* 19:471–493.

Wolpoff, M. H., and Thorne, A. G. (1992) One hundred years of *Pithecanthropus* is enough. *Am. J. Phys. Anthropol. Suppl.* 14:175–176.

Wood, B. A. (1984) The origin of *Homo erectus*. *Cour. Forsch. Inst. Senckenberg* 69:99–111.

Wood, B. A. (1985) Early *Homo* in Kenya, and its systematic relationships. In E. Delson (ed.): *Ancestors: The Hard Evidence*. New York: Alan R. Liss, pp. 206–214.

Wood, B. A. (1988) Are "robust" australopithecines a monophyletic group? In F. E. Grine (ed.): *Evolutionary History of the "Robust" Australopithecines*. New York: Aldine de Gruyter, pp. 269–284.

Wood, B. A. (1989) Hominid relationships: a cladistic perspective. *Proc. Aust. Soc. Hum. Biol.* 2:83–102.

Wood, B. A. (1991a) *Koobi Fora Research Project*. Vol. 4: *Hominid Cranial Remains*. Oxford: Clarendon.

Wood, B. A. (1991b) A palaeontological model for determining the limits of early hominid taxonomic variability. *Palaeont. Afr.* 28:71–77.

Wood, B. A. (1992a) Early hominid species and speciation. *J. Hum. Evol.* 22:351–365.

Wood, B. A. (1992b) A remote sense for fossils. *Nature* 355:397–398.

Wood, B .A. (1992c) Early hominid species and speciation. *J. Hum. Evol.* 22:351–365.

Wood, B. A., and Chamberlain, A. T. (1986) *Australopithecus*: Grade or clade? In B. Wood, L. Martin, and P. Andrews (eds.): *Major Topics in Primate and Human Evolution*. Cambridge: Cambridge University Press, pp. 220–248.

Wood, B. A., and Van Noten, F. L. (1986) Preliminary observations on the BK 8518 mandible from Baringo, Kenya. *Am. J. Phys. Anthropol.* 69:117–127.

Wood, B. A., and Uytterschaut, H. (1987) Analysis of the dental morphology of Plio–Pleistocene hominids III. Mandibular premolar crowns. *J. Anat.* 154:121–156.

Wood, B. A., Yu, L., and Willoughby, C. (1991) Intraspecific variation and sexual dimorphism in cranial and dental variables among higher primates and their bearing on the hominid fossil record. *J. Anat.* 174:185–205.

Yoder, A. D. (1992) The applications and limitations of ontogenetic comparisons for phylogeny reconstruction: The case of the strepsirhine internal carotid artery. *J. Hum. Evol.* 23:183–196.

Zilberman, U., Skinner, M., and Smith, P. (1992) Tooth components of mandibular deciduous molars of *Homo sapiens sapiens* and *Homo sapiens neanderthalensis*: A radiographic study. *Am. J. Phys. Anthropol.* 87:255–262.

Zonneveld, F. W., Spoor, C. F., and Wind, J. (1989) The use of CT in the study of the internal morphology of hominid fossils. *Medicamundi* 34(3):117–128.

Zuckerkandl, E. (1963) Perspectives in molecular anthropology. In S. Washburn (ed.): *Classification and Human Evolution*. Chicago: Aldine, pp. 243–272.

CHAPTER 8

How Certain Are Hominoid Phylogenies? The Role of Confidence Intervals in Cladistics

Robert S. Corruccini

INTRODUCTION

Statistical concepts of confidence limits are now applied to phylogenetic analysis in varying manners (Felsenstein 1985; Lanyon 1985; Krajewski and Dickerman 1990; Page 1990; Faith and Cranston 1991; Faith 1991; Shaffer et al. 1991; Corruccini 1992). The value in thinking about such confidence intervals can be imagined in a theoretical situation where three or more species in a study are actually equally related in terms of phyletics—that is, they originated in an unresolvably bushlike multifurcation, rather than through orderly dichotomization. Trait selection bias, measurement error, and/or sampling error will cause variance in the shared-derived similarity and in the parsimony of one versus another dichotomy. Tree algorithms based only on the measured similarities and insisting upon pairwise dichotomies will actually create a credible dichotomous tree structure (for the same can be produced by sampling from a random number table), but a polychotomy would be the correct representation in terms of the limits of our knowledge. Hierarchical methods of pattern recognition (such as cluster analysis) are not suited to test whether there is, initially, significant structure and pattern in the data.

The pathfinder application of confidence limits to cladistics (Felsenstein 1985) utilized multiple bootstrap sampling of characters to determine how frequently a given dichotomous pairing of taxa recurs. The recommendation was for positive verification of a given dichotomy when it occurs 95 percent of the time. The 95 percent occurrence rule is a sort of significant similarity that leads to acceptance of a hypothesis of relationship. It has an unusual connotation in relation to normal inferential statistical concepts, which seek to reject incorrect hypotheses by finding them incompatible with the 95 percent confidence limits of sampling from a parent universe. While null hypotheses theoretically can never be proven true, they may eventually be judged acceptable based on repeated inability to reject them. Sanderson (1989) discusses limitations to the 95 percent occurrence rule, finding it too strenuous an expectation of cladistic repeatability.

The approach presently suggested employs inferential statistics as intended, to potentially falsify null hypotheses (in this case, polychotomies) through finding significant differences between successive branching levels compared to their confidence intervals. This follows the approach of Page (1990), who bootstrapped allozyme loci to attach a plus-or-minus interval to each branching level ("cluster height") in a majority-rule consensus tree. Another test is Faith and Cranston's (1991) random permutation of character states to see how often shorter trees than the one of maximum likelihood could happen. They found a strict consensus tree of combined data sets to be much more polychotomous than the individual sets.

Molecular (as opposed to character-based) phylogenies similarly suffer from inadequate attention to the concept of confidence intervals. With single estimates of immunological or other dissimilarity between each pair of species, one approach has been to calculate percent standard deviation of fit between tree lengths and the original immunological distance used to derive the tree (e.g., Sarich and Cronin 1980). This may underestimate true variances since the most parsimonious tree is directly related to the raw distance matrix (Corruccini 1992). Goodman et al. (1983) only depict unresolved protein sequencing multichotomies if there are exactly equally parsimonious branching arrangements. However, different proteins give different branching details, especially in the example of the relationship among *Pan, Homo,* and *Gorilla.* An approach related to that of Sarich and Cronin is suggested here that provides dispersion for immunological distance matrices and is applicable to data in the literature. This allows direct comparison of character cladistics and molecular phylogenies within the interpretational constraints of confidence intervals.

It is patently obvious after 15+ years of hominid cladistic studies (since e.g., Delson 1977) that cladistic arrangements of Hominidae and related higher primates have not reduced interpretational variation among experts and indeed have increased the complexity of opinionation in paleoanthropology (e.g., Grine 1993). While this is not necessarily bad, it certainly contradicts some of the early expectations. The amount of quantitative cladistic confidence we can have in contrasting treatments surely is worth investigating. Therefore, hominids will constitute a test case for the application of confidence limits to phylogenies.

METHODS

Morphologies

To unambiguously apply normal statistical confidence intervals to cladistic relationships, a quantity must be derived that describes the cladistic relationship and that can be resampled. This concept is undoubtedly anathema to many cladistic purists, but I shall operate as if it were valid insofar as concerns the purpose of attaching phylogenetic confidence intervals. Some support can be adduced from Springer and Krajewski (1989:374), who assert additive distance matrices reflect evolutionary trees and can be used to construct unrooted cladistic topologies because additive distances and an outgroup allow one to polarize the net amounts of shared derived change: "We should recognize that distance data and discrete character data share much in common and both have their place in phylogenetic reconstruction." Page (1990) also demonstrates that cladograms can be metrically compared by quantifying branch lengths and fitting them for congruence with each other and with the original distance matrix.

Using discrete character data, a phenetically inspired measure of phylogenetic similarity can be operationalized as the similarity contained only within shared derived states (Farris et al. 1970). Coding zero as the ancestral state and derived states in either direction from zero, the cladistic similarity here is the square root of the mean square of amount of overlap between two taxa from the value of zero. For instance, a state of 4 in one taxon and 3 in another would increment their summed squared cladistic similarity by 9; if yet another taxon has a state of 2 it would add 4 to its summed similarity to the first two taxa. This can be converted to a dissimilarity at the end by subtracting the root mean similarity from a fixed quantity. Thus shared plesiomorphous similarities do not contribute to the coefficient. Traits may be left unweighted (each character state is of equal value) or may be scaled to a maximum derived value of one. This weights each character equally rather than basing the similarity coefficient upon a count of shared derived states in common. Cluster algorithms can be applied directly to the coefficients to derive a phylogenetic tree.

But what are the sampling confidence limits of the coefficients? The bootstrap theorem allows estimation of an unknown statistical parameter by random sampling (with replacement) from an observed finite distribution (Efron and Tibshirani 1986). From a sample, cases are randomly left out (and replaced by others, thus preserving the original sample size). The central tendency of the bootstrap sample is recorded; the process is then repeated many times using a random number generator; and the resulting dispersion around the central tendency is adopted as characteristic of the infinite parent sampling universe. It is much easier to employ the bootstrap to estimate standard error (50-100 samplings) than to find true confidence intervals (1000 samplings). Thus the bootstrap is a straightforward Monte Carlo exercise in finding nonparametric estimates of parameters of an unknown distribution, using computer power to replace the lacking theory. According to Efron and Tibshirani and to Felsenstein (1985), with multiple-variable problems it is possible to bootstrap the sampling of variables (columns) rather than the cases (rows) when estimating multivariate confidence limits.

Molecules

Macromolecular measures of dissimilarity such as immunology and hybridization, hypothetically directly related to time since cladogenesis, hide the underlying genetic variables producing the measures. A method for estimating dispersion (Corruccini 1992), operating upon the assumption of unvarying molecular rates of evolution, goes through every pair of taxa and finds the disproportion of homologous branch length among the triad made up of them and each remaining taxon. Thus, each such disproportion equals $(D_{ijk'} - D_{ijk''})/[(D_{ijk'} + D_{ijk''})/2]$ where $D_{ijk'}$ is the largest and $D_{ijk''}$ the second largest of three distances among the taxa i, j, and k. With K total taxa, i will run from one to K, j from 2 to K, and k from 3 to K. The smallest of the distances among each triad is taken to represent the pair that are sister groups relative to the third, so their two distances to the third should be subequal to the extent that evolutionary rate is steady. The average of the two distances is the cluster height, and the difference relative to that height is the proportion of error in symmetry.

This sampling of branch disproportions is supplemented by the disproportion between the reciprocal asymmetric halves of the matrix, when both heterologous and homologous or driver and tracer (upper and lower triangular) halves are available: $(D_{ij} - D_{ji})/[(D_{ij} + D_{ji})/2]$ where i and j run from one to K and never equal each other. The variance between reciprocal halves of immunological matrices has been underestimated repeatedly: "Percent nonreciprocity is, on average, double what has been acknowledged in the past" (Guyer 1992:87). The total set of disproportions D can yield the mean deviation of pairwise distances and be reduced to any other statistics describing the dispersion of homologous distances and branch lengths.

Dendrograms

The 95 percent confidence limits (estimated from standard deviations) of morphological and molecular distances have been applied in a one-tailed manner toward the smaller tail, based on 50 bootstrap samplings. This essentially tests the null hypothesis, for any two branches joining, that they could randomly have been sampled from parent distributions where the distance incorporates the previous branching point. If the null hypothesis cannot be rejected, then a trichotomy can be envisioned within the confidence limits of the distances.

Upon calculation of the maximum-likelihood similarity matrix and determination of its standard deviations through bootstrapping, the taxa are clustered using weighted pair-group method. In this way the relatively well-developed numerical (not philosophical) methods of phenetics can be applied directly to phylogenetic analysis as long as characters are coded in terms of shared derived similarity (based on standard qualitative phylogenetic methods), and we can "clado-phenetically" approach problems of quantitative cladistics and confidence limits. Maximum parsimony is abandoned in view of the endless debate concerning it and the necessity of a metric approach in order to activate ordinary statistical concepts of dispersion.

DATA

The first morphological database adapted for exploring the reliability of hominoid phylogenetic dispersions is from Groves and Paterson (1991). I transcribed their data set 5 comprising 113 one-state characters for *Homo sapiens, Pan troglodytes, Gorilla,* and *Pongo,* which they employed in the course of testing the cladistic program PHYLIP to try to resolve clades among the living large-bodied hominoids.

Another cladistic analysis using many characters for few hominoids to trace phylogeny is Martin's (1986) which is also related to that of Andrews and Martin (1987). Martin lists states for 73 characters from an original list of 122 that are derived in at least one taxon among *Pongo, Sivapithecus, Gorilla, Pan, Homo,* and *Australopithecus.* Traits coded as variable by Martin were given values of 1 and derived traits were scored as 2. I took the liberty of altering one trait: His number 72, os centrale fusion with the scaphoid was coded as variable in *Pongo,* seemingly maintaining an old assertion of some occurrences of fusion in what would seem likely to have been old/arthritic specimens. Corruccini (1978) found such fusion only in African apes plus humans, and it was totally lacking in orangs, gibbons, and cercopithecoids over c. 200 normal specimens. Godinot and Beard (1991) also perpetuate the erroneous notion of such fusion in "pongids" but also mention its presence in *Indri.* Thus os centrale fusion appears to be a strongly derived trait shared among the African clade that accompanies one other trait, supraorbital torus form, in the scanty anatomical support for molecular evidence of such a clade.

Skelton et al.'s (1986) total character matrix was adopted for an analysis of within-hominid relationships and confidence intervals. They provide 69 craniodental traits with a variable number of states for *Australopithecus afarensis, A. africanus, A. robustus/boisei,* and *Homo habilis,* plus many alternative tree arrangements and a narrow decision about the most parsimonious tree.

Subsequently Skelton and McHenry (1992) have provided an expanded treatment of early hominid cladistics, entailing 77 more complexly scored traits for an outgroup, *A. afarensis, A. aethiopicus, A. africanus, A. robustus, A. boisei,* and early *Homo.*

More intricate cladistic analyses that provide full data for numerous hominids have been provided by the Wood/Chamberlain team. Wood and Chamberlain (1986) give 39 craniodental variables for the outgroups *Dryopithecus, Hylobates, Pongo, Gorilla,* and *Pan* and for the focal taxa *Ramapithecus, Australopithecus afarensis, A. africanus, A. robustus, A. boisei, Homo habilis, H. erectus,* Neandertals, and modern *H. sapiens.* Many of the traits are multistate and bidirectional, having been quantitatively derived from morphometrics. Chamberlain and Wood (1987) later expanded the treatment to 89 metrical traits transformed into more complex discrete cladistic states, for *Australopithecus afarensis, A. africanus, A. robustus, A. boisei, Homo erectus, H. sapiens,* and three different but materially overlapping definitions of *Homo habilis* and two of African *Homo sp.* The latter comprise some combinations of material now classified as *H. rudolfensis* and *H. ergaster* (Wood 1992).

None of these data sets are true Hennigian data for they admit non-uniform paths (hence homoplasy) and multistate, semiquantitative characters. This is good for it indicates that the data are realistic and not artificially tailored or pruned. However, whether expressed as continuous or as chopped-up segments of a continuum, morphometric traits are contrary to the fundamentalist spirit of cladism since Hennig forbade their use in place of binary plesiomorphous versus apomorphic scoring in which the magnitude of the difference between the two states was not supposed to matter whatsoever (for that matter any notion of statistical confidence also is likely to be considered vacuous by real cladists). The standard usage of such non-Hennigian traits in recent paleoanthropology must be seen as a justified trend.

Two large matrices of primate macromolecular data have been utilized to calculate distance variances (Corruccini 1992) to introduce the approach suggested for molecular data and provide some comparison with morphology-based cladistics. These are serum protein immunodiffusion systematics based heavily upon hemoglobin antigenic distance culled from Baba et al. (1979:Table X), and albumin immunological distance data from Cronin's (1977) overview. The goal is measuring internal inconsistency in the distance matrix as an exploratory tool, examining the stability of clustering sequence when only branch lengths are a source of variation and measurement imprecision. A truly stable molecular phylogeny would be unaffected by such sampling variations. For the results of these two distance matrices (Corruccini 1992:Figures 3, 4), a merged reconfiguration of results was estimated by taking all genera common to the two sources, scaling the cluster heights and standard deviations to a common scale, and reclustering.

RESULTS: PRIMATE MOLECULAR AFFINITIES

Figure 8-1 shows results for averaged immunodiffusion and transferrin immunological distances among primates and outgroups. A solid line represents the branches of the tree extending further than the 95 percent confidence limits in the direction of smaller distance, while the leftward extent of the "fuzzy" area to the left of every pairwise clustering represents the area of uncertainty of estimation (one-tailed). Polychotomies (with subsequent distances and standard errors averaged) are given as the solution whenever successive clusters are not at significantly different distances. That is, if higher additional forms

FIGURE 8-1 Dendogram of merged primate immunological distance from Cronin (1977) and Baba et al. (1979), combining and averaging cluster pairs with not significantly different cluster heights. Shaded boxes are superimposed over the 95 percent one-tailed confidence interval of the distances. Only the yardstick of mean deviation is applied to clusters, as this is smaller and will favor the finding of more significant dichotomies.

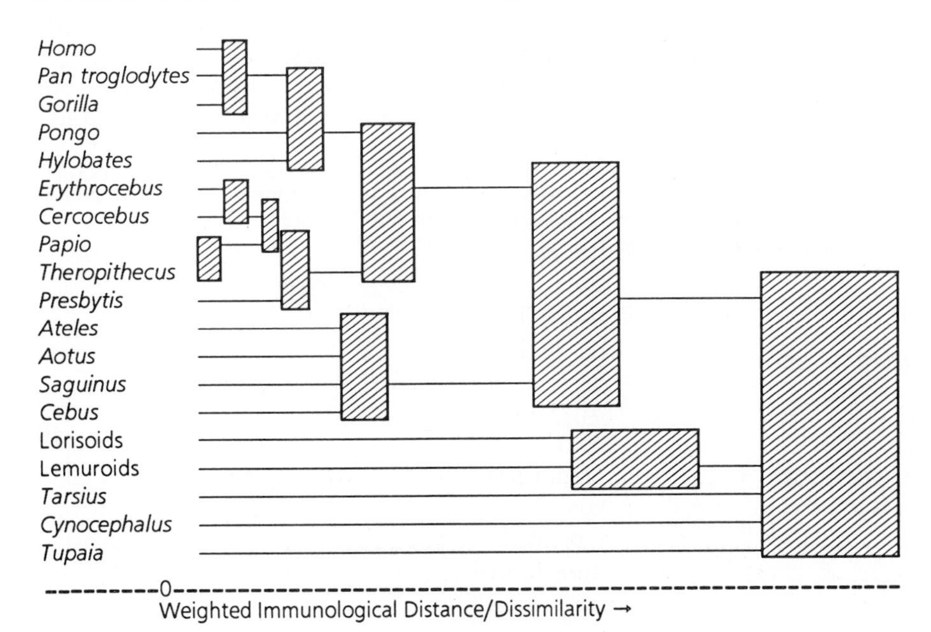

have joined a lower-level dichotomy within their minimal pairwise distance (cluster height) confidence intervals, the two nonsignificant dichotomies are combined into a smoothed-out trichotomy.

Thus, for example, although *Pongo* joins the African ape and human clade before *Hylobates* does, the bootstrap sampling of the similarities indicates that the level at which *Pongo* joins is well within the confidence limits for similarity of *Hylobates* to African apes/humans + *Pongo*; hence, they are represented as a statistically unresolvable trichotomy. On the other hand, this group of taxa has confidence intervals of similarity well below the level at which they are collectively joined next by the cercopithecoid cluster. Thus the hominoid-cercopithecoid molecular cladistic distinctness is considered reliably repeatable and statistically significant (i.e., there is a significant amount of monophyly measured within these superfamilies); this is true also of the ensuing catarrhine-platyrrhine distinction.

Similar significant clustering occurs within some other generally recognized taxonomic entities, namely colobines with cercopithecines and within strepsirhines. Other aspects of higher distance clustering, toward the bottom part of Figure 8-1, entail taxa of highly contentious relationship to primates and indeed, if only dichotomous clades are considered resolved, there is no resolution here. Much as in Cronin's (1977) results, none of *Tarsius, Cynocephalus,* nor *Tupaia* significantly more closely approach the anthropoid primates. The strepsirhines also are not *significantly* closer to anthropoid primates. Within the 95 percent confidence limits derived from molecular distance dispersions for these clades, only a five-way polychotomy can be presented at the primate (or Archontan) base.

RESULTS: HOMINID CLADISTIC ANALYSES

Groves and Paterson on Hominoids Using PHYLIP

In Figure 8-2 we see the weighted pair-group maximum likelihood dendrogram resulting from my treatment of Groves and Paterson's (1991) data. The most frequent (hence likely) initial dichotomy is between chimps and gorillas, with humans next joining and the orang decidedly an outlier. The last of these eventualities is quite "significant" by both frequency of occurrence greater than 95 percent, and by unlikelihood of the similarity of the three others to orangs falling within 95 percent their bootstrap sampling limits. On the other hand, the exclusion of humans from a trichotomy with chimps and gorillas or equally from a higher level dichotomy with an African ape (particularly chimpanzee) cannot be rejected by usual inferential statistical reasoning. As in so many other data sets (Corruccini 1992), then, the *Pan-Gorilla-Homo* trichotomy remains unresolvable into dichotomies.

Figure 8-3 attempts to superimpose the confidence limits, that is, the areas of reasonable statistical uncertainty, in a different graphical manner. Thus Figure 8-3 illustrates the same results using shaded

FIGURE 8-2 Weighted pair-group method clustering of "clado-phenetic" similarity coefficients calculated from the traits coded in Groves and Paterson's (1991) treatment. The dendogram shows one-tailed confidence intervals for the null hypothesis with horizontal dashes. Numbers at each node indicate the percent occurrence of that specific cluster membership.

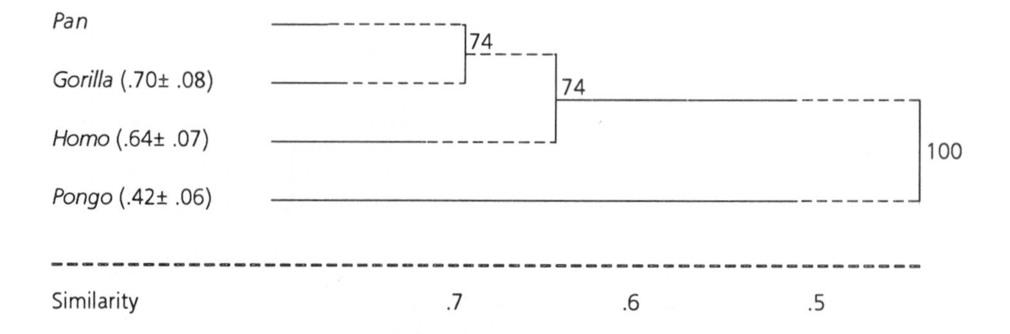

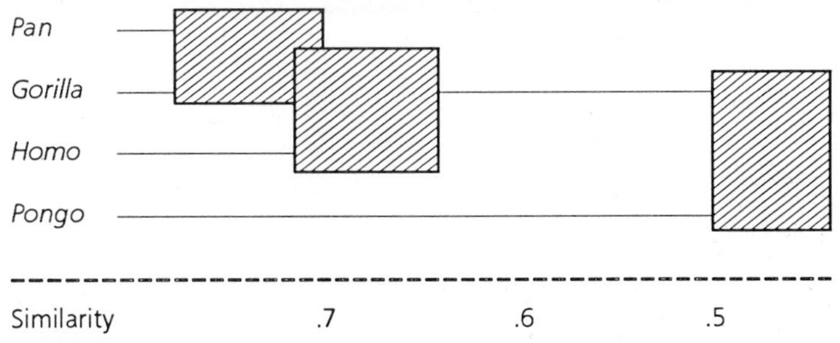

FIGURE 8-3 Alternative presentation of the results in Figure 8-2 with shaded boxes representing the confidence limits.

boxes for the amount of horizontal confidence intervals. The overlap of the *Pan-Gorilla* cluster height with the confidence interval of the subsequent clustering with *Homo* is consistent with the percent repeatabilities of specific dichotomies running rather low.

Martin on Hominoids

The results from Martin's (1986) large cladistic character base are given in Figure 8-4 using the dashed lines once again as representatives of one-tailed confidence intervals. There is tremendous reliability to three dichotomies among the six taxa: *Pongo* with *Sivapithecus*, chimps with gorillas, and humans with australopithecines. However, beyond that point there is little resolution and any further step (except hominids with the Asian clade) is about a "50-50" proposition. The confidence limits to the synapomorphic similarity coefficients agree, as the next most likely clustering after those three initial dichotomies (i.e., African apes with hominids) overlaps in its region of statistical uncertainty with the last step.

Skelton et al. on Australopithecines

The 69 cladistically coded traits of Skelton et al. (1986) yield similarity coefficients and a maximum-likelihood clustering pattern given in Figure 8-5. The top part again shows the dendrogram, the areas of statistical confidence of synapomorphic similarity (solid lines), the 95 percent confidence intervals

FIGURE 8-4 Clustering of taxa based on the cladistically defined similarity coefficients calculated over the traits used by Martin (1986). The dendrogram is repeated below after recalculating clusters by collapsing successive dichotomies that are not significantly different into polychotomies. Details as in Figure 8-2.

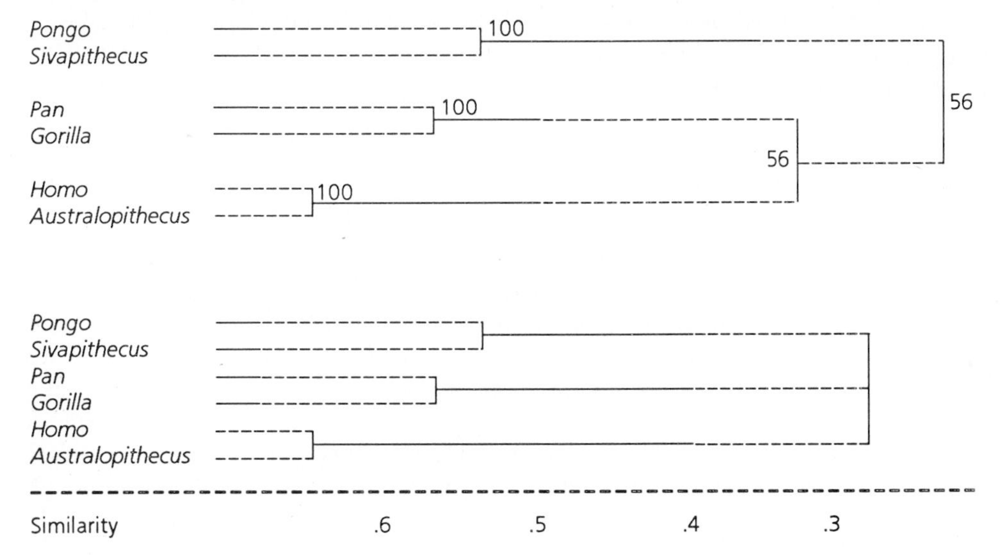

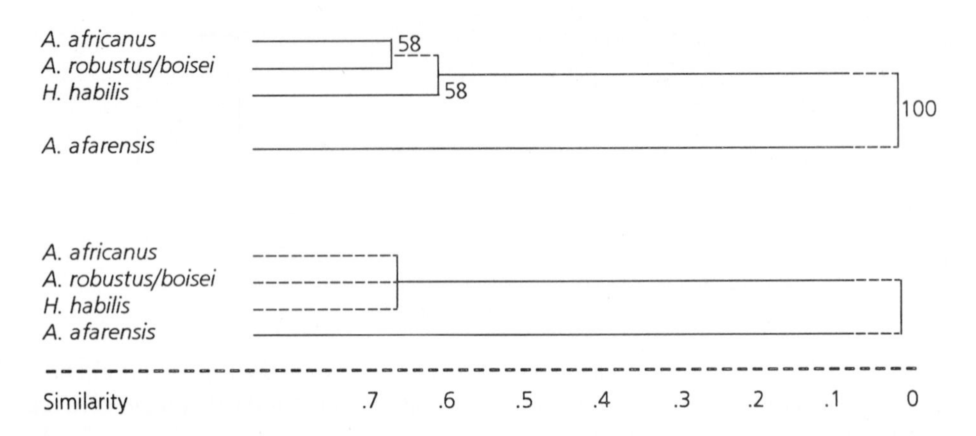

FIGURE 8-5 Phenogram based on the synapomorphic similarity coefficients from traits coded by Skelton et al. (1986), again with maximum-likelihood pair-group clustering represented above and below, the situation when nonsignificantly different sequential clusters are merged.

characterizing the similarity level at which each clustering eventuates, and repeat occurrence percentage of each branch based on 50 bootstrapped runs.

As per some recent phylogenetic interpretations, the gracile and robust australopithecine species from South Africa are linked together first as closest sister species by this treatment, which accords with Skelton et al.'s second most parsimonious arrangement supported by 45 of their 69 traits. The *africanus/robustus/boisei* link shows 58 percent frequency and a standard error of cluster height that easily overlaps with the level at which the next closest sister group, *H. habilis*, joins. The relationship of these three species in the context of these data must be considered an unresolvable trichotomy (indicated in the bottom part of Figure 8-5). *Australopithecus afarensis* is quite a significant outlier to these other species, but that is unavoidable since it was coded uniformly as having the plesiomorphous condition. This outgroup then just serves to gauge the relative shared derived similarity among the others.

This picture based upon standardized traits is altered a bit when using unweighted traits (such that a trait with two derived states has twice as much influence as one showing only states 0 and 1). The clustering sequence then changes in that, as Skelton et al. found in their most parsimonious tree (supported by 45/69 traits), there is maximum-likelihood indication (but statistically nonsignificant) of a shared derived branch among *A. robustus* and *H. habilis*. However, standard deviations of pairwise similarities are somewhat relatively smaller with standardized traits, which, in the interest of giving all benefit of the doubt to existence of dichotomies, is the reason for preferring standardized characters in this chapter.

Skelton and McHenry's Elaboration

Skelton and McHenry (1992) have refined and reorganized their characters by anatomical and functional complexes, widened the outgroup comparison and included more fossil taxa. In Figure 8-6 we see the results of their data in my system when compared for congruence within this standard framework of variance. Using the special synapomorphic similarity coefficient, the clustering sequence repeats Skelton and McHenry's exactly, except there is no difference whatever between the similarities of *Homo* or of *A. aethiopicus* to the pair of robust species and hardly any distinction between the relatedness of *A. africanus* and of *Homo* to *A. robustus/boisei*, but there is a low similarity coefficient between *A. aethiopicus* and *A. africanus*. *Homo* and *A. aethiopicus* are very dissimilar from each other because their similarities shared with *A. robustus* and with *A. boisei* are quite different. In a phenetic sense, we would say these are triangular

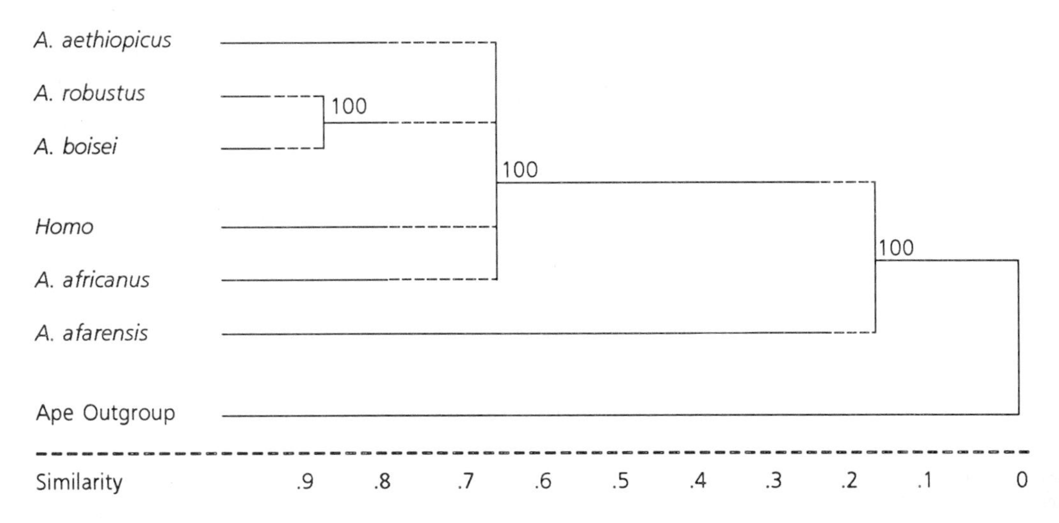

FIGURE 8-6 Phenogram of results based on Skelton and McHenry's (1992) cladistic characters.

rather than linear relationships, which automatically implies homoplasy in cladistically coded characters. Other branches of the tree are significant and hence can be represented as dichotomies in these data.

Wood and Chamberlain on Fossil Hominids

With 14 included taxa, the data of Wood and Chamberlain (1986) are quite a bit more complex than those preceding. The synapomorphic similarity matrix results in a most likely dichotomous clustering arrangement depicted in the top part of Figure 8-7. While the six outgroups are splayed out over a fair amount of cladophenetic space, the eight hominid taxa have a more compact arrangement. The percent reliability of most dichotomies is appallingly slight. The bootstrapped standard deviations also indicate no significant structure to any of the most basal relationships among recognized hominid taxa, with the exception of most of the *Homo-Australopithecus* distinction. *Dryopithecus* and *Hylobates* seem to form the most immediate sister group to hominids, then *Ramapithecus* plus *Pongo*, then African apes. Molecular and other recent information would seem to indicate a complete reversal of that order.

The bottom portion of Figure 8-7 repeats the clustering sequence when non-significant dichotomies are collapsed with preceding linkages into polychotomies. *Ramapithecus* with *Pongo* and, oddly, *Dryopithecus* with *Hylobates* are significant synapomorphic dichotomies, with the latter significantly more related to hominids. None of this would have been indicated if 95 percent repeatability had been demanded, for that way the confidence limits of this tree are generally far wider and dichotomous structure harder to establish than with the dispersions I apply to the similarity coefficient (this is especially true with lower similarity levels). Among hominids there is significant mutual similarity. *Homo* (sans *habilis*) and *Australopithecus* are significantly distinct sister groups, and the only other significant structure is *A. robustus* with *A. boisei*.

Chamberlain and Wood Later on Fossil Hominids

Figure 8-8 shows the results of my process on the more numerous traits applied to more hominid taxa given later by Chamberlain and Wood (1987). These traits, mostly divided-up craniodontometric continua, actually yield less basal hominid structure than the previous study. The newly recognized *Homo sp.* assemblage (given in two overlapping versions of included specimens to contrast with Stringer's) is a frequent outlier to all other species owing to an exclusive collection of shared traits (including such things

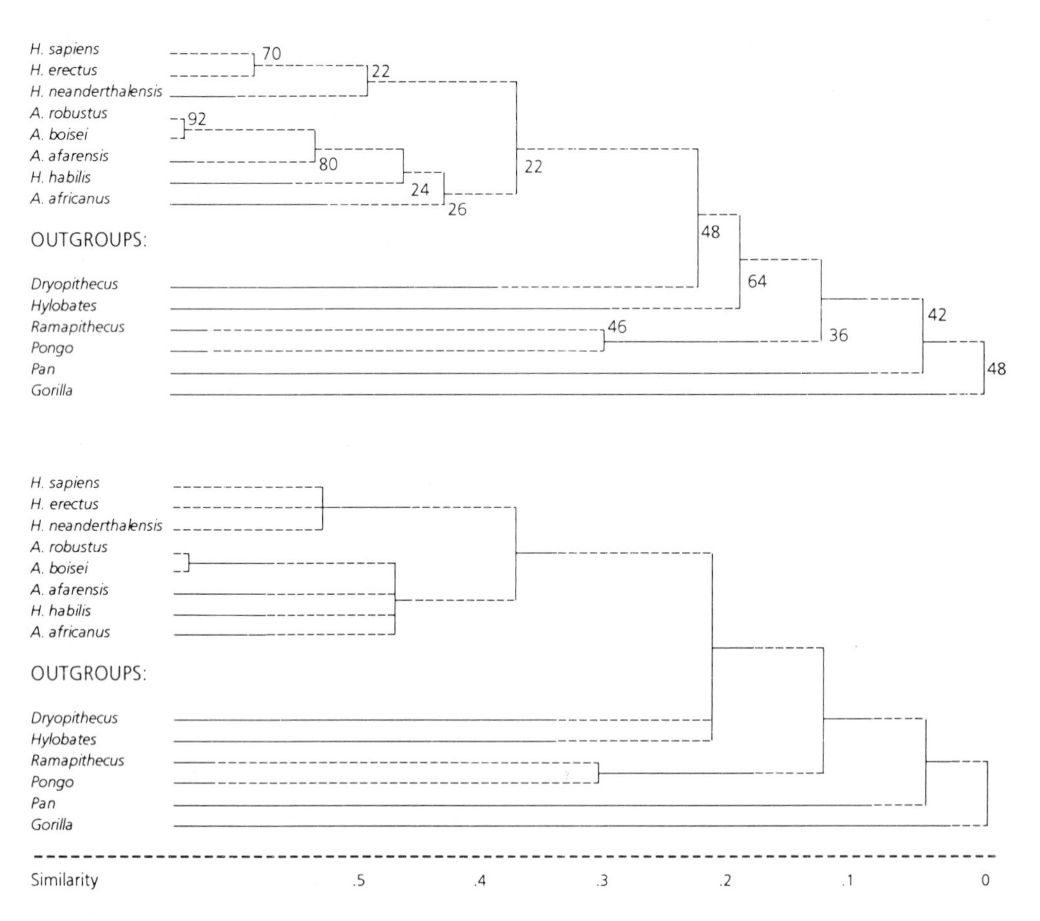

FIGURE 8-7 Phenograms of results based on Wood and Chamberlain's (1986) cladistic characters.

as nasal bone, parietal thickness and mandible dimensions). Interestingly, the 94 percent repeatability of the partially redundant *Homo sp.* assemblages would lead to their immediate relationship being considered not significant by Felsenstein's (1985) 95 percent occurrence rule. Our inferential, hypothesis-rejection approach to reliability of the tree is contrasted with Felsenstein's hypothesis-vindication approach in the top part of Figure 8-8 by indicating the other proportions of times per 100 bootstrap runs that a given dichotomy repeats.

The lack of significant lower-level groupings leaves little dichotomous structure at the base of the rest of the early hominid tree (a result that could also be produced by random variation around a single, variably measured synapomorphic similarity). The three variations of *Homo habilis* hypodigm cluster together 100 percent of the time but at most 38 percent for the specific sequence presented in the figure with an asterisk. The affinity of *A. robustus* with *A. boisei* is very strongly indicated, but all other dichotomies are very unstable. The *Homo-Australopithecus* dichotomy is more weakly expressed now and does not closely approach statistical significance. The resulting tree that expresses the statistical uncertainties as polychotomies is much simplified, but a seven-way polychotomy prevails.

IMPLICATIONS

By superimposing confidence intervals over dendrograms, polychotomies are created within grey areas of statistical "certainty" where before there had been dichotomies based on the maximum-likelihood distances, and on the strong assumption that everything should be a dichotomy. In order that this outcome not be simply automatic, the estimated dispersions of pairwise dissimilarities have been minimized by employing the more modest one-tailed confidence intervals and in other ways (Corruccini 1992). These confidence limits must tend rather more toward being conservative underestimates than exaggerations.

Clashing cladistic resolutions of basal hominid dichotomies are put in interesting perspective by this process. Both morphology and molecules fail to resolve the three-way *Pan*-hominids-*Gorilla* relationship. "An effective trichotomy arising from a polymorphic ancestor explains the conflicting data" (Rogers 1992). All the analyses agree in failing to find much significant cladistic structure within australopithecines and within *Homo,* whereas the fundamental distinction between the two taxa (when including *habilis* with *Australopithecus*) is more consistent. The method has the interesting effect of producing some graphical similarity among the different analyses from different workers, where before there had been notable differences. The increased "agreement" is admittedly curious since it lies in expanding grey areas of uncertainty to cover the areas of preexisting differences. If the more stringent 95 percent occurrence rule rather than calculated standard deviations are used to assess the morphology-based clado-phenograms, then even greater "agreement" emanates from the polychotomous structures of the trees.

It is sobering to realize that even this small number of significant dichotomies, signifying demonstrable monophyletic branch lengths, may be overestimated. Two very consistent sister group pairs were previously identified: *Pongo* with *Sivapithecus/Ramapithecus* and *A. robustus* with *A. boisei.* New fossil discoveries, as they always do, have shaken the preexisting orthodoxy. *Sivapithecus* postcrania have weakened the sister group status with *Pongo* (Pilbeam et al 1990; Larson 1992). Crossing synapomorphies have also intruded between *A. robustus* and *A. boisei* with the emergence of understanding of *A. aethiopicus,* which has some synapomorphies with *A. boisei* but not the same ones as has *A. robustus.* Troubling also is the orang-like forehead form in "*Paranthropus*" (see Clarke, this volume) as opposed to supraorbital torus synapomorphy linking African apes with the other early hominids.

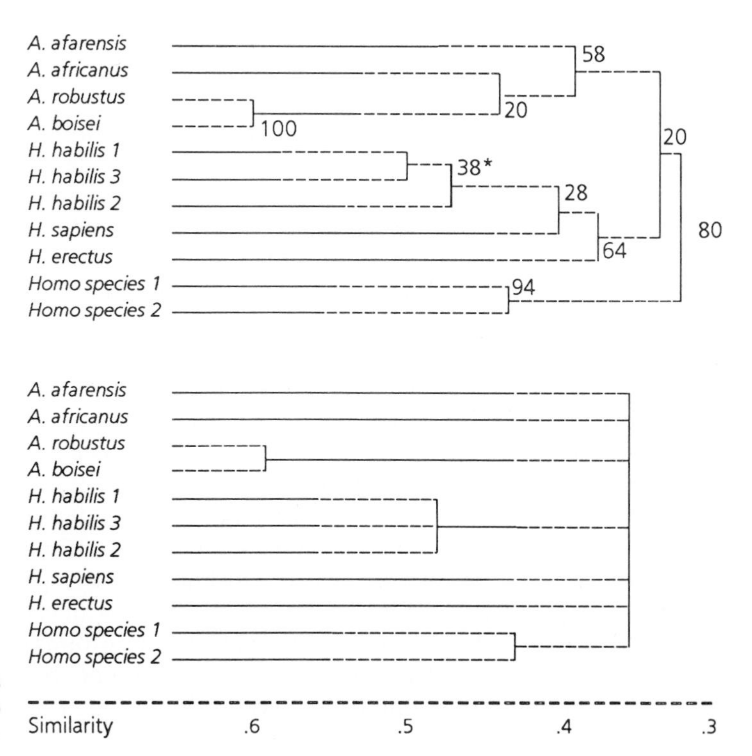

FIGURE 8-8 Phenograms of results based on Chamberlain and Wood's (1987) cladistic characters.

A Bush Rather Than a Tree?

The standard explanation for lack of dichotomous resolution of hominid cladograms would be scientific inadequacy, disagreement among different workers on traits and methodology, variable interpretative outlooks, questions about maximum parsimony, or the incomplete nature of the fossil record. Could the consistent need to resort to polychotomies, in spite of so much thoughtful work by leading experts, reflect something other than human failure to yet uncover the truly existing dichotomies? The very extent (and repeatable pattern) of cladistic grey areas might argue for the possibility of a rapid and complex bushlike radiation of early hominids, an emerging paradigm or "iconography" for the geometry of evolution (Gould 1989) that contrasts with efforts to sort out dichotomous, treelike phylogenies near the base of taxa. If species within certain hominid genera radiated suddenly from common ancestry in response to an "empty ecological barrel" (new habitat or climate), with maximum disparity evolving early rather than over time, our faulty human efforts to resolve pairwise sister group relations at the base of this radiation would logically be unrequited (although serving to generate much dissension among workers). This would be due (from our top-looking-down perspective of a few survivors) to a large noise/signal ratio. In other words, polychotomization in phylogenetic trees could be seen as reflecting biological reality rather than deficiency (Corruccini 1990).

The most influential exposition of the "life is a bush, not a tree" maxim is Gould's (1989). His book investigates the implications of the Middle Cambrian "window" afforded by the Burgess Shale. The invertebrate fossils from there had been crowbarred into extant phyla and classes by Walcott more than 50 years previously. Instead, recent work had detailed fantastic mosaic mixes of primitive and derived traits in these wonderful creatures, and indeed, the neontological criteria we employ to discriminate those derived from primitive traits are largely invalidated by a prospective (Cambrian-eyed) view rather than our prevailing retrospective viewpoint. Life from the Burgess was uniquely specialized, not primitively simple. Gould well documents analogous bottom-heavy time distributions of other, skeletonized groups that also support the basic model. Wood (this volume) discusses cladistics as having to be a "top-down" view, whereas advantages would accrue from comparison with the "bottom-up" view if only we could attain it.

"The history of life is a story of massive removal followed by differentiation within a few surviving stocks, not the conventional tale of steadily increasing excellence, complexity, and diversity" (Gould 1989). Our prejudices nurture "fatuous" concepts of life as a ladder (leading at the top to, of course, our human selves) and as a cone, increasing in diversity toward the top. Gould emphasizes "that disparity reached its peak at the outset and that life's subsequent history has been a tale of decimation, not increasing variety in design." If species relationships are indeed bushlike, not treelike, we need to rethink much of our instinctual tendencies toward the fossil record. The cone of diversity picture, with orderly branching upward toward the goal of emergence of extant forms, contrives virtually any early form as a primitive ancestor of something living later. By contrast, the Burgess revision demonstrates that early on, there are many monotypic (as far as we can see) phyla; later, a few speciose phyla. Orthodox genealogical schemes hinder treatment of the Burgess organisms and obscure the magnitude of the short-lived basal adaptive radiation. There was little hierarchy to this initial blast of disparity—the hierarchy derives from hindsight, for "history is written by the winners." Thus there are numerous mosaic trait combinations and bizarre autapomorphs; homoplasy and nonparsimony run amok in trying to make cladistic sense out of the base of adaptive radiations (Corruccini 1990).

The bushlike quality of the early phase of a major adaptive radiation makes for an undifferentiable polychotomy of many (most) of the lineages; our ability to conceptualize coherent and cohesive taxa is due only to the fact that "the removal of most lineages by extinction left only a few survivors, with big gaps between. . . . The radiation of these few surviving lineages (into a great diversity of species with restricted disparity of total form) produced the distinct groups that we know today as phyla and classes." (Gould 1989) The rule was experimentation, evolutionary lability, the juggling of characters from ensembles considered to be unmixable because of the apomorphic associations we see today. There were more varied combinations, as Gould documents, and "We now recognize clear groups, separated by great morphological gulfs, only because the majority of these experiments are no longer with us."

There was shuffling of derived traits that we today look upon in their unshuffled associations as unique signatures of discrete lineages, but that were once combined willy-nilly among multiple branches near the bush's (i.e., adaptive radiation's) base. Thus what we today refer to as homoplasy was rampant during the Cambrian explosion of complex life. "What order could possibly be found among the Burgess arthropods? Each one seemed to be built from a grab bag of characters—as though the Burgess architect owned a sack of all possible arthropod structures, and reached in at random to pick one variation upon each necessary part whenever he wanted to build a new creature. . . . Where was order, where decorum?" (Gould 1989). He adds, "Moreover, the new iconography of rapid establishment and later decimation dominates all scales, and seems to have the generality of a fractal pattern." Much the same thing does indeed occur later at lower taxonomic levels, when empty ecological barrels result from mass extinctions—for instance, mammals during the Paleocene. Thus, the same admonition must surely also hold for lower taxonomic levels approaching the species that have arisen more recently in geological time. I (Corruccini 1990) have argued that it is merely innately human to see recognizable body plans in the unknown; it is the natural thing to do. How do we first learn comparative anatomy? By organizing it and recognizing morphs, according to dissection of known (i.e., living) taxa. This breeds "neontocentrism," the bias toward interpreting past life in terms of how it is presently fitted together.

Evolution being a bush, not a tree, helps to explain why there are such intransigent complications as the primate relationship of tree shrews, microbats to megabats and to other Archonta, tarsiers to anthropoids or prosimians, the mysterious mix of indications of relationship of Cheirogaleids, the inability to resolve platyrrhine subfamilies into families, and so on. Most fossils are not going to fit comfortably into the classificatory paradigm of modern forms, especially when they are well preserved and known. Most fossils, sheerly from the statistical point of view, are not going to have any relation to later species for many more have lived than are alive today. This will be particularly true of early fossils in an adaptive radiation, near the bush's base. Most of these progenitor fossil species are not going to be minor variants on the sought-after ancestor, but quite divergent autapomorphs.

Take for example the lineage connections claimed over the years for Fayum fossils (see Fleagle, this volume; Kay and Williams 1992). *Parapithecus* for years was promoted as the ancestral or stem Cercopithecoid, notwithstanding ambiguity over the animal's dental formula. Completing the analogy to the discovery of the outrageous anatomical uniqueness of the Burgess's *Opabinia*, the recent *Parapithecus* remains show it lacked mandibular incisors—an unprecedented anomaly for any anthropoid primate. Now *Catopithecus* and *Proteopithecus* further complicate anthropoid origins. These early stem anthropoids, together with *Oligopithecus* and *Qatrania*, have anthropoid and prosimian features in different combinations (particularly the mandibular symphyseal fusion versus postorbital bone). This is predicted from the "grab bag" metaphor for early apomorphic disparity near the base of adaptive radiations.

The recognition of Burgess disparity unfolded through subsequent restudy and dissection, while the recognition of early anthropoid, early platyrrhine, hominid, and many other primate taxonomic disparities including the Fayum pattern have unfolded through the discovery of sequentially more disparate new taxa, plus anatomically more complete finds of previously poorly known taxa. The more complete the anthropoid fossil record becomes, the less explainable it is. Thus the Fayum assemblage for anthropoids and the Oligocene-Miocene platyrrhines for South American monkeys provide only peripheral indications, not direct readings of ancestry. A striking example of our "fatuous" regard for our paleoanthropological state of knowledge is suggested by de Heinzelin (this volume). He makes a fascinating case for the ecologically peripheral and taxonomically unusual state of fauna from the African rift. If this is so, all known australopithecines are atypical of what was happening in the evolutionary mainstream of events in the unfossiliferous world outside the rift, the available hominid fossil record is very nonrandom, and there is a lot we do not know. Attempts to tie the hyperdisparate fossil primate twigs into the central trunk of the bush are only exercises in self-delusion. Many fossils are overwhelmingly more irrelevant to the arising of things alive today than heretofore thought.

Confirmation of the bush paradigm can be approached by examining the historical progression of resolving fossil hominid affinities as discoveries and understanding expand. In seminars Clark Howell was forever being heard to say "The more you know the harder it is." That is, with fossil hominoids as a test case, invariably the discovery of new or more extensive fossils leads to greater discord and phylogenetic conflict rather than clearer and confirmatory connections. Say, the number of African fossil hominid specimens doubles over a period of time. Do we know twice as much or feel twice as confident in our cladogenies? Or does four times as much variance in interpretation, taxonomy and tree building result? The latter is expressly the case (see Wood 1992). If dichotomization happened regularly and left a decisive synapomorphic "trail," we would know more with ensuing discoveries; instead, such finds serve merely to highlight how little we previously understood. Howell's admonition suggests we can confirm the bushy nature of life by the time-progressive disparity in interpretations.

Related to all this is the ever thorny problem of species definition. Tattersall (1989) is correct; the correlation between morphology and speciation is sufficiently poor (among living species) that we are merely deluding ourselves about most diagnosed fossil species: "Species are hard enough to define in the living world . . . there is absolutely no necessary relationship between speciation and morphological change . . . speciation can take place with little or no morphological displacement, particularly in the hard skeletal parts that preserve in the fossil record." Therefore,

> If gradualism is the predominant pattern in human evolutionary history, then that history must at least largely be a matter of discovery: each new fossil should, as it is found, fit neatly into an emerging picture, filling in yet another gap. . . . In fact, the very reverse has occurred: with each new fossil the picture has tended to become more confused, or at least more complex, and phylogenies have regularly needed substantial readjustment to accommodate such new finds. This kind of experience is more in line with what one would expect of a human evolutionary past involving repeated speciations than with the stately progression of one or a few lineages.

I would add that the speciational complexity seems now to pertain more and more closely to the very base of the hominid bush (e.g., Grine 1993).

Accordingly, I (somewhat whimsically) offer the scenario depicted by Figure 8-9 as a possible model of the extent of reliable cladistic understanding, to date, of hominid radiations. It expresses the monophyletic nature of *Australopithecus* and of *Homo*, the dire lack of agreement over further structure within those entities, the likelihood of many other peripheral species that may or may not ever be discovered, the eternal anonymity of the true ancestral species, the fundamentally australopithecine nature of "*H.*" *habilis*, and the generally bushlike scenario that must still serve as our null hypothesis in hominid cladistics.

Do we need to revise zoological and paleontological academic training to avoid the tendency toward neontocentrism? We try to teach appreciation of cultural pluralism and diversity today in social sciences, in an attempt to head off the development of ethnocentrism and racism resulting from the traditional self-valued outlook. The analogy is a strong one to our way in comparative anatomy classes of teaching, more or less, that "our body plan is right" and therefore anything else encountered is rather wrong. But how to broaden our "tolerance of diversity" paleontologically? Students must practice comparative anatomy on known groups (by necessity) in order to recognize and learn the anatomy—but this automatically pigeonholes thinking into known and recognizable patterns of anatomical associations.

CONCLUSIONS

That life is a bush not a tree leads us to realize that we do not know nearly as much as we pretend we do. However, recognition of rapid taxonomic proliferation early in a lineage does not deny the possibility that

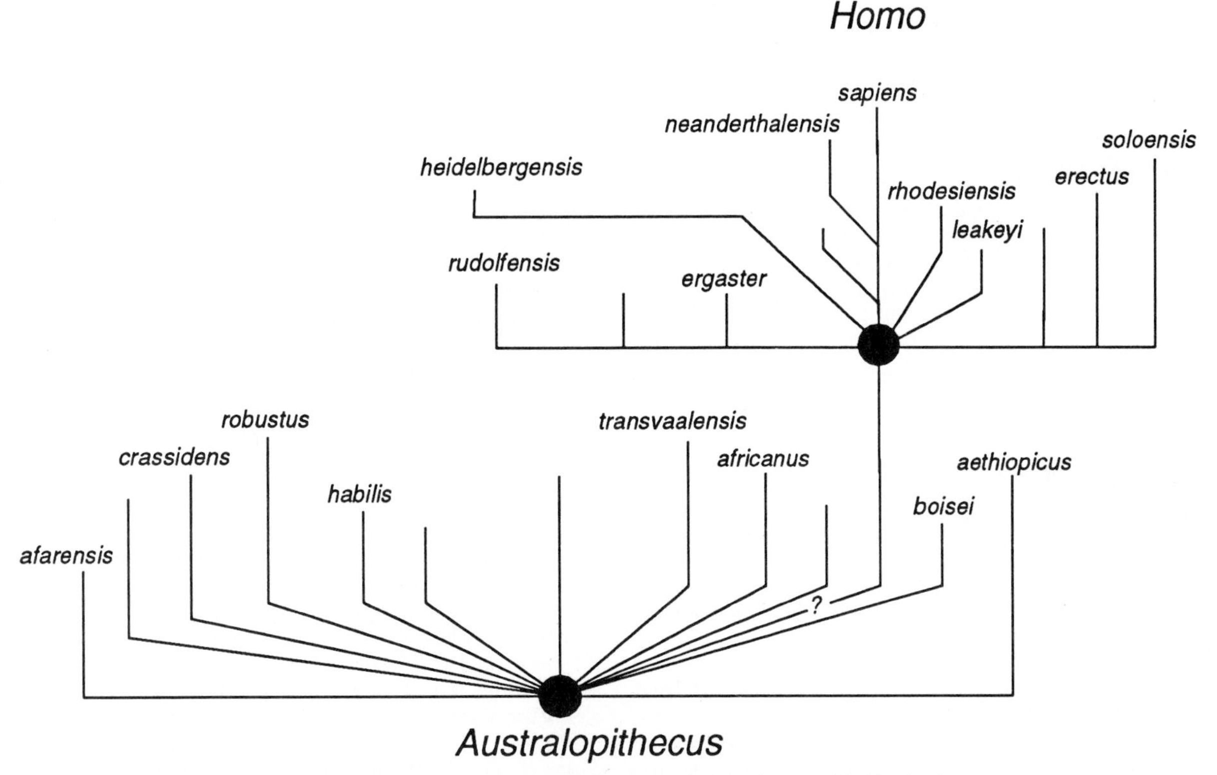

FIGURE 8-9 Hominid evolutionary scenario reflecting the results discussed in the text.

some resultant lines were more closely related to each other than to others. That there are such frequent "line crossing synapomorphies" indicates that many must represent convergent traits generally relating to very brief shared evolution with acquisition of a small number of shared derived features, followed by very long periods in which separate lineages independently changed, modifying true synapomorphies beyond recognition. If speciation is truly dichotomous then the continued search for sister groups is justifiable as long as it is pursued with reserve, retaining the possibility of larger polychotomous bushes as a working null hypothesis. Constant striving for dichotomous trees, and argumentation over contrasting dichotomies, leads to an unconscious failure to remember that most biological diversity results from sudden adaptive radiations whose bases will be tangled webs of disparity. Imposing confidence interval concepts during construction of phylogeny will assist in reminding investigators of the polychotomy option.

SUMMARY

An approach is further explored for estimating confidence intervals for phylogenetic trees based on bootstrap sampling, using morphological character and molecular data sets for hominoids and hominids. Each successive cladistic dichotomy is treated as a null hypothesis for sampling from a universe of synapomorphy-based similarity coefficients that includes the next lower similarity. Successive dichotomies that are not at significantly different similarity levels are collapsed into polychotomies. This allows comparison of robusticity of trees between different sources of data and puts a new wrinkle on comparisons of molecular and morphological systematics. Unresolvable polychotomies among fossil hominid taxa are quite common when these methods are applied. This is consistent with and perhaps indicative of the paradigm of evolution being bushlike, rather than treelike and resolvably dichotomous.

ACKNOWLEDGMENTS

Thanks to Susan Ford, Brunetto Chiarelli, and Russ Ciochon for help.

LITERATURE CITED

Andrews, P., and Martin, L. (1987) Cladistic relationships of extant and fossil hominoids. *J. Hum. Evol.* 16:101-118.

Baba, M. L., Darga, L. L., and Goodman, M. (1979) Immunodiffusion systematics of the primates. Part V. The Platyrrhini. *Folia Primatol.* 32: 207-238.

Chamberlain, A. T., and Wood, B. A. (1987) Early hominid phylogeny. *J. Hum. Evol.* 16:119-133.

Corruccini, R. S. (1978) Comparative osteometrics of the hominoid wrist joint, with special reference to knuckle-walking. *J. Hum. Evol.* 7:307-321.

Corruccini, R. S. (1990) Review of *Wonderful Life*. *Hum. Evol.* 5: 579-587.

Corruccini, R. S. (1992) Bootstrap approaches to estimating confidence intervals for molecular dissimilarities and resultant trees. *J. Hum. Evol.* 23:481-493.

Cronin, J. (1977) Anthropoid evolution: The molecular evidence. *Kroeber Anthropol. Soc. Pap.* 50:75-84.

Delson, E. (1977) Reconstruction of hominid phylogeny: A testable framework based on cladistic analyses. *J. Hum. Evol.* 6:263-278.

Efron, B., and Tibshirani, R. (1986) Bootstrap methods for standard error, confidence intervals, and other measures of statistical accuracy. *Statist. Sci.* 1:54-77.

Faith, D. P. (1991) Cladistic permutation tests for monophyly and nonmonophyly. *Syst. Zool.* 40:366-375.

Faith, D. P., and Cranston, P. S. (1991) Could a cladogram this short have arisen by chance alone? On permutation tests for cladistic structure. *Cladistics* 7:1-28.

Farris, J. S., Kluge, A. G., and Eckhart, M. J. (1970) A numerical approach to phylogenetic systematics. *Syst. Zool.* 19:172-189.

Felsenstein, J. (1985) Confidence limits on phylogenies: An approach using the bootstrap. *Evolution* 39:783-791.

Godinot, M., and Beard, K. C. (1991) Fossil primate hands: A review and an evolutionary inquiry emphasizing early forms. *Hum. Evol.* 6:307-354.

Goodman, M., Baba, M. L., and Darga, L. L. (1983) The bearing of molecular data on the cladogenesis and time of divergence of hominoid lineages. In R. L. Ciochon and R. S. Corruccini (eds.): *New Interpretations of Ape and Human Ancestry*. New York: Plenum, pp. 67-86.

Gould, S. J. (1989) *Wonderful Life*. New York: Norton.

Grine, F. (1993) Australopithecine taxonomy and phylogeny: Historical background and recent interpretation. In R. L. Ciochon and J. G. Fleagle (eds.): *The Human Evolution Source Book*. Englewood Cliffs, NJ: Prentice Hall, pp. 198-210.

Groves, C. P., and Eaglen, R.. H. (1988) Systematics of the Lemuridae (Primates, Strepsirhini). *J. Hum. Evol.* 17:513-538.

Groves, C. P., and Paterson, J. D. (1991) Testing hominoid phylogeny with the PHYLIP programs. *J. Hum. Evol.* 20:167-183.

Guyer, C. (1992) A review of estimates of nonreciprocity in immunological studies. *Syst. Biol.* 41:85-88.

Kay, R. F., and Williams, B. A. (1992) Dental evidence for anthropoid origins (abstract). *Am. J. Phys. Anthropol. Suppl.* 14:98.

Krajewski, C., and Dickerman, A. W. (1990) Bootstrap analysis of phylogenetic trees derived from DNA hybridization distances. *Syst. Zool.* 39:383-390.

Lanyon, S. M. (1985) Detecting internal inconsistencies in distance data. *Syst. Zool.* 34:397-403.

Larson, S. G. (1992) Are the similarities in forelimb morphology among the hominoid primates due to parallel evolution? (abstract) *Am. J. Phys. Anthropol. Suppl.* 14:106.

Martin, L. (1986) Relationships among extant and extinct great apes and humans. In B. Wood, L. Martin, and P. Andrews (eds.): *Major Topics in Primate and Human Evolution*. Cambridge: Cambridge University Press, pp. 161-187.

Page, R. D. M. (1990) Temporal congruence and cladistic analysis of biogeography and cospeciation. *Syst. Zool.* 39:205-226.

Pilbeam, D., Rose, M. D., Barry, J. C., and Ibrahim Shah, S. M. (1990) New *Sivapithecus* humeri from Pakistan and the relationship of *Sivapithecus* and *Pongo*. *Nature* 348:237-239.

Rogers, J. (1992) Intra-species polymorphism and the conflicting evidence for hominoid phylogeny (abstract). *Am. J. Phys. Anthropol. Suppl.* 14:140-141.

Sanderson, M. J. (1989) Confidence limits on phylogenies: The bootstrap revisited. *Cladistics* 5:113-129.

Sarich, V. M., and Cronin, J. E. (1980) South American mammal molecular systematics, evolutionary clocks, and continental drift. In R. L. Ciochon and A. B. Chiarelli (eds.): *Evolutionary Biology of the New World Monkeys and Continental Drift*. New York: Plenum, pp. 399-421.

Shaffer, H. B., Clark, J. M., and Kraus, F. (1991) When molecules and morphology clash: A phylogenetic analysis of the North American ambystomatid salamanders (Caudata: Ambystomatidae). *Syst. Zool.* 40:284-303.

Skelton, R. R., and McHenry, H. M. (1992) Evolutionary relationships among early hominids. *J. Hum. Evol.* 23:309-349.

Skelton, R. R., McHenry, H. M., and Drawhorn, G. M. (1986) Phylogenetic analysis of early hominids. *Curr. Anthropol.* 27:21-43.

Springer, M. S., and Krajewski, C. (1989) Additive distances, rate variation, and the perfect-fit theorem. *Syst. Zool.* 38:371-375.

Tattersall, I. (1989) Comment on Pilbeam. In M. K. Hecht (ed.): *Evolutionary Biology at the Crossroads*. Flushing, NY: Queen's College, pp. 139-141.

Wood, B. (1992) Origin and evolution of the genus *Homo*. *Nature* 355:783-790.

Wood, B. A., and Chamberlain, A. T. (1986) *Australopithecus*: Grade or clade? In B. Wood, L. Martin, and P. Andrews (eds.): *Major Topics in Primate and Human Evolution*. Cambridge: Cambridge University Press, pp. 220-248.

CHAPTER 9

The Craniocerebral Interface in Early Hominids

Cerebral Impressions, Cranial Thickening, Paleoneurobiology, and a New Hypothesis on Encephalization

Phillip V. Tobias

AVANT PROPOS

I salute Francis Clark Howell, my contemporary and near-twin, friend of nearly 40 years' standing and brother-seeker after the truth about fossil humans. From his first visit to Johannesburg in the early 1950s, when he was a protégé of Sherrie Washburn and I a young staff member under Raymond Dart, our paths have intersected repeatedly across the hemispheres. Always his quiet and confidence-inspiring judgment was evident, as also his broad horizons, deep thought, and balanced appraisals. In a branch of science in which flamboyance, iconoclasm, and egocentrism have all too often occupied center stage, Clark has stood out as a pillar of stoicism, poise, and dispassion. He has been a veritable Rock of Gibraltar amid the maelstrom of a vigorous and tempestuous branch of science. It is a great privilege and a very pleasant task to add my little tribute to Clark on his retirement from the University of California at Berkeley.

INTRODUCTION: THE ANATOMICAL BASIS OF PALEONEUROBIOLOGY

In the face of the exacting demands of neurobiology of the living, with its highly technical new investigative procedures, it may seem strange that one should dare to speak of the neurology of extinct creatures. Indeed, to some, paleoneurobiology may sound like a pretentious title, when to the outsider we are seen to be dealing with simply old bones. Therefore let me open the subject by summarizing the anatomical bases of paleontological and paleoneurological studies.

It is true that the raw materials for paleontological researches are fossilized bones. For any ancient group under consideration, its morphology and morphometry may be compared with those of other fossil hominoids and of extant hominoids, namely gorilla, chimpanzee, orangutan and the human species. Such comparisons may be assisted by highly refined statistical analyses and by the assessment of ontogenetic changes and of sexual dimorphism. Moreover, cognizance may be taken of size "scaling," or the structural and functional consequences of differences in size among organisms of more or less similar design, as Jungers (1985) has defined scaling in his volume on *Size and Scaling in Primate Biology*.

When one looks a little more closely, there is much more to be learned from bones than their size and form. We teachers of anatomy, in osteology classes, have long exhorted our students to "read the bones," not only with eyes but also with fingertips. So also, the paleoanatomist reads and interprets the markings and impressions of ligamentous and muscular attachments to bones, as well as the grooves, notches, foramina, and smoothings owing to the impingement on bones of other structures such as arteries, veins, nerves, ligaments, and tendons in transit. In other words, much valuable information on functionally important *soft tissue anatomy* may be garnered from the careful and observant study of even the most ancient, heavily fossilized bones. In this aspect, paleoanthropological studies form a branch of anatomy and are based preponderantly on paleoanatomical analyses.

Perhaps of no structure has this proved to be more valid than the brain, because in the calvaria or braincase we have a bony complex that in life is faithfully molded upon the contents, namely the brain, its coverings, vasculature, fluids, and the stumps of emerging cranial nerves. It is trite to point out that, under conditions of normal development, a larger brain dictates the growth of a larger calvaria. Similarly, a smaller brain is housed in a smaller calvaria. A broader brain is accommodated in a broader brain-box. A subject whose brain has a relatively larger cerebellum has a proportionately larger posterior cranial fossa. A brain with a big Broca's cap on the left has a matching deep hollow of the endocranium, compared with that on the opposite side. A specimen in which the superior sagittal sinus passes to the left, instead of to the right as is more common, usually imprints clear-cut corresponding grooves on the endocranial surface. In some brains a large part of the venous drainage, from the confluence of the sinuses to the left or right jugular bulbs, passes by way of enlarged occipital and marginal sinuses. These stamp telltale grooves on the calvaria. A cerebrum whose left occipital pole protrudes further posteriorly than the right—the phenomenon of left occipitopetalia—reflects this asymmetry faithfully in the cerebral fossae of the posterior endocranial surface.

All such features of the soft tissue contents and of the calvaria may readily be confirmed in modern cadavera. It follows that, from careful study of the interior of a calvaria, one is able to draw many conclusions about the brain and vessels that once inhabited that braincase. We can facilitate the process by filling an empty brain-case with a plastic or plaster medium and making an artificial endocranial cast. In size and form such an endocast faithfully reproduces the size and form of the endocranial cavity. The surface of the endocast takes an impression of all the markings that the brain, its coverings, and vasculature had imprinted during life on the endocranial surface. Hence one may read the sulcal, gyral, and vascular impressions directly from the surface of the endocast.

Sometimes a natural endocast forms during the process of fossilization: Once the soft tissue contents of the calvaria have disappeared, through biotic and physicochemical agencies, sand may gain access to the calvaria, especially through the foramen magnum. The filling material comes to occupy a part or the whole of the endocranial cavity, the fraction filled depending on the orientation at which the cranium has come to rest in the deposit. If appropriate chemicals such as lime are present in the surrounding rocks and cave-earth, the sandy contents of the braincase become calcified and hardened, by a process analogous with fossilization. There may result a beautifully preserved natural endocast such as that which accompanied the Taung skull (Dart 1925). Other examples stem from the Sterkfontein cave deposit and are attributable to the extinct hominid species, *Australopithecus africanus*. Details of size, form, and morphology of the brain and vessels may be read from a natural endocast, as from an artificial endocast.

The study of endocranial surfaces and of endocasts has provided the main source of raw data for paleoneurobiological research.

THE DEVELOPMENT AND RECESSION OF ENDOCRANIAL IMPRESSIONS

The surface features of the brain and its vessels are imprinted faithfully upon the endocranial surface during ontogenetic development. For example, we see reflected the meningeal vessels and the cranial venous sinuses; the major subdivisions of the brain, such as the cerebrum, the cerebellum, and, to a certain extent, the brainstem; and finer subdivisions, such as the convolutions and sulci of the cerebral cortex. Sometimes the filling material in a natural endocast extends into one or more of the foramina on the cranial base and takes a cast of the bony canal to which the foramen gives access: Thus, where cranial nerves traverse such foramina and canals, some endocasts may even preserve a cast of the stumps of one or two such cranial nerves coming off the brainstem(Figure 9-1).

The sutures between frontal, parietal, temporal, and occipital bones generally leave a clear imprint on the endocast, if, by the time of the individual's death, these sutures had not fused. The impressions of unfused sutures, however, are clearly distinguishable from those of gyri, sulci, and meningeal vessels. Then, too, certain cranial injuries, such as depressed fractures, are clearly evident on the surface of endocasts.

The clarity of the markings is inversely proportional to the age: strongest in the young, becoming progressively effaced with age (Connolly 1950). Connolly concluded that the maximum degree of fissural markings on the calvaria, in both anthropoid apes and modern humans, occurs probably in young adults.

FIGURE 9-1 Paramedian sagittal section of a juvenile human head showing the intimate relationship between cerebrum and cerebellum, on the one hand, and the endocranial surface of the calvaria, on the other hand. Note the endocranial hillocks and valleys, corresponding to grooves and protrusions of the surface of the brain.

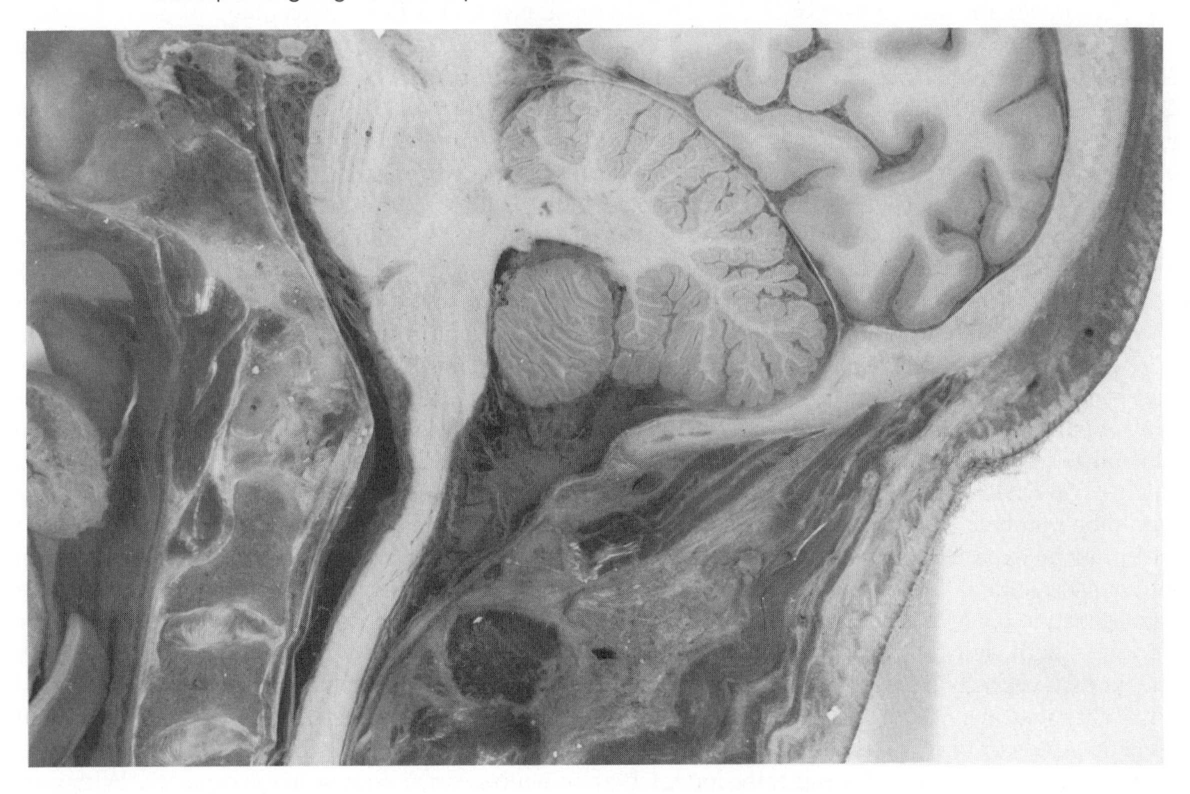

It is an interesting problem to contemplate the morphogenetic mechanisms that may be involved. Ontogenetic studies have shown that the developing neurocranial bones float apart on the surface of a sea of expanding brain (Moss 1954; Young 1957). It is the brain that, to a large extent, calls the tune in the determination of the shape and size of the calvaria—except in cases of craniosynostosis (craniostenosis), where premature closure of one or more sutures may force atypical shapes upon the enlarging brain. Presumably during these fetal, infantile, childhood, and adolescent stages, the features of the surface of the brain are most efficacious in hammering their pattern on to the endocranium. A possibly correlated variable is the changing relationship between the measurable volume of the brain and the endocranial capacity.

Many entities other than the brain are housed in the calvaria. These include the roots and intracranial trunks of no fewer than 24 cranial nerves, the dura mater and the leptomeninges, the subarachnoid space and its localized enlargements, the cisterns, numerous blood vessels including the larger meningeal and cerebral arteries and veins, the cranial venous sinuses, and blood and cerebrospinal fluid. Various studies of the nonbrain component of the endocranial capacity have yielded values ranging from 10 percent (Brandes 1927) to 33.5 percent (Mettler 1955). When data for modern humans are ranked by age, the values for the brain component range from 94 percent at birth to 80 percent at 20 years of age. In other words, the percentage of the capacity occupied by the brain is inversely correlated with age up to adulthood. Hence in the young the calvaria is closely filled. The heartbeat is faster in the young, and this factor may be important, since the brain pulsates with each beat of the heart. Are these factors—the packing density and the heart rate—causally related to the brilliantly clear etching of the surface of the brain on the young endocranium? It seems a reasonable hypothesis and one that could be tested by appropriate experiments.

Whatever the mechanism, it is a happy circumstance that a large percentage of early hominid specimens fall into the approximately 20 years and less age groups. Those in the young age groups are 35 percent of *A. africanus* (n = 65); 60.5 percent for *A. robustus* (n = 119); 73 percent for *Homo habilis* (n = 22); and 73 percent for Peking man (*H. erectus pekinensis*) (n = 22) (Tobias 1991a). When we combine these four groups, of 228 early hominid individuals whose ages at death are determinable, 127 or 56 percent fall into the putative 20 and under category. This means that a majority of early hominid fossils are of an age at which gyral and other impressions are most manifest.

On the reverse of the coin, within the adult lifetime of human individuals, the impressions become muted as the brain shrinks. There is no evidence that the endocranial capacity diminishes. Hence, it must be supposed, the nonneural component of the capacity increases with age and one's head becomes filled more and more with fluid and meninges. At the same time, the impressiones gyrorum become erased.

It follows that whereas the presence of specific endocranial markings is to be considered meaningful, the absence of a particular impression may be of no structural or functional significance.

What is the mechanism of this erasure of the impressiones gyrorum with aging? Two alternative possibilities suggest themselves: There may be an osteoclastic resorption of the hillocks (the imprint of sulci), or an osteoblastic filling in of the valleys (the imprint of gyri). Two different cellular mechanisms would be involved. The postulates could be tested: Osteoclastic resorption of the hillocks of the inner table may be expected to result in a *diminution* in the mean thickness of the inner table and possibly also in the average total thickness of the cranial vault with aging, whilst osteoblastic filling in of the valleys should lead to an *increase* in the average thickness of the inner table and possibly also of the total cranial thickness with aging. The two postulated processes may be expected to have opposite effects on endocranial capacity. Thinning of the inner table may lead to a small increase in the capacity; such a possibility was mentioned by Von Bonin (1963:41) though he adduced no measurements, either of inner table thickness or of capacity, to support this view. On the other hand, the filling in of the endocranial valleys should lead to a thickening of the inner table and hence a small *decrease* in the capacity. With good samples of young adult and old adult crania, it should be easy to test these two hypotheses. The author does not know of any data in the literature bearing on age changes in the mean endocranial capacities of human adults. However, there are some data relevant to the inferences on mean total thickness of crania.

THICKENING OF THE CRANIUM

Previous studies have shown a gradual slight increase of cranial thickness in human adults from the third decade of life onward (Todd 1924; Hartl and Burkhardt 1952; Hartl and Luther 1953; Adeloye et al. 1975). At first blush, this would appear to favor the hypothesis that the erasing of relief occurs by the filling in of the endocranial valleys, which, in turn, should increase the overall average thickness of the *inner table* of the cranium. However, a closer look is necessary before we reach a conclusion.

It has been claimed that the inner and outer tables of the cranial vault bones are functionally independent of each other (Klaauw 1946; Delattre and Fenart 1954; Simon 1955; Young 1957; Baer and Harris 1969). The *inner plate* responds primarily to the underlying brain; it appears to be the inner table that molds itself on the convoluted surface of the cerebrum in the formation of the gyral and sulcal impressions, although it must be admitted that we do not have evidence whether or not the endocranial hollowing over an especially prominent gyrus not only molds the inner table but also encroaches upon the diploë This should not be difficult to test on a series of sectioned calvariae. The variable thickness of the *outer table* may be contributed to, at least in part, by muscle action (McCown and Keith 1939; Washburn 1947; Zuckerman 1954; Mednick and Washburn 1956). Hence, an alternative possible mechanism for the thickening of cranial bones during adult life is that the outer table may thicken, irrespective of whether the inner table becomes somewhat thinner with osteoclastic erasing of the hillocks, or thicker with osteoblastic filling in of the valleys, or remains more or less unchanged.

A third possible mechanism for the thickening of the cranial bones with age may be a thickening of the diploë, irrespective of, or in addition to, changes in the outer and/or inner tables. The diploic space is known to undergo widening in certain disease conditions such as a past rachitis (Shattock 1914; Weidenreich 1943) and in some hemoglobinopathies (Adeloye et al. 1975). The subject of pathological thickening of the cranium has been reviewed in reference to the salted Piltdown I cranium, which showed marked thickening of the diploë but not of the outer or inner tables (Tobias 1992a). We know less about the variability of diploic, tabular, and total cranial thickness in healthy individuals.

In the light of this analysis, we cannot yet say what is the mechanism of the erasing of endocranial impressions with aging, nor whether it is causally related to the slight thickening of the cranial bones that occurs during adult life. Here is an area that needs further research. The cellular basis of these aging changes is wide open for investigation.

When one considers phylogenetic changes in cranial thickness, we know that most hominids, ancient and modern, have fairly thin skulls. This was true even of the so-called robust and hyper-robust australopithecine crania, whose supposed burgeoning was largely the consequence of the growth of "ectocranial superstructures" such as crests (Tobias 1967). Exceptionally, *Homo erectus* and, to a lesser degree, Neandertalers showed marked thickening of the calvarial vault-bones. What was the nature of this phylogenetic thickening? Attention has already been directed to the diploically thickened and almost certainly pathological Piltdown I calvaria (Dawson and Woodward 1913). Oakley (1960) was able to match this apparently bizarre pattern of thickening in two recent crania in the British Museum (Natural History), one of an Ona Indian from Tierra del Fuega and the other of a Bronze Age person from Sutton Courteney in Berkshire, England. In discussing the diploic thickening of the salted Piltdown I cranium, Weidenreich (1943:164) pointed out that the form of thickening contrasted sharply with that found in Peking man: In the latter hominids, "all three constituents of the bone take equal part in the thickening, the two tables slightly more than the diploë."

It seems to have been assumed for the last half century that Weidenreich's delineation of the form of thickening in *H. erectus pekinensis* typified all nonpathological cranial thickening, as in *H. erectus* in general and in the Neandertalers. However, the author does not know of any punctilious metrical study of the thickness of each of the three layers of the cranial bones in these thick-skulled hominids, that is carefully controlled as to technique of mensuration, the calvarial region, bone, position, and laterality. During a recent visit to Indonesia, the author was shown a large number of original Javanese hominid calvariae, both published and unpublished, by the generosity of Professors T. Jacob and S. Sartono and Dr. F. Aziz. One of the thickened calvarial fragments from Sangiran, presumably of *H. erectus*, presented by Aziz at the International Trinil Centennial Colloquium in November 1991, showed what to the author's naked eye appeared to be diploic, but not tabular, enlargement, that is, a form of thickening that is at variance with the pattern attributed by Weidenreich (1943) to *H. erectus pekinensis*. This fleeting, nonmetrical observation underlines the need for a careful study of patterns of cranial thickening in a phylogenetic context. When the nature of such anthropological thickening is clarified precisely, we may be a little closer to understanding this curious and enigmatic cranial thickening that beset humankind about 1.5 million years ago, lasted about 1.0 MYA, then slowly subsided.

THE ABSOLUTE SIZE OF THE ENDOCRANIAL CAVITY

The most obvious results of endocranial studies are the endocranial capacities of various hominoids. One is careful to speak of capacities and not brain sizes, for the reasons mentioned earlier, though where data for actual brain size are available, there is a very high correlation between the two sets of measures. Because of difficulties in the measuring of actual brain size and because of the availability of thousands of dried crania of recent humans and apes, far more studies have been devoted to endocranial capacities than to brain volume or weight as such.

From such studies it has been estimated that the average modern human interracial combined-sex capacity is of the order of 1350 cm^3 (Tobias 1968, 1970, 1971a, after Saller 1959:1211).

The values for the living great apes were pooled for each sex by Tobias (1971a, b) on the basis of earlier studies by Selenka, Schultz, Zuckerman, Ashton and Spence, Vallois, Oppenheim, Gaul, Hagedoorn, Bolk, Harris, Randall and Tobias. The results are summarized in Table 9-1 for separate sex samples, and in Table 9-2 for male-plus-female samples.

Since the fossil samples are most likely to include male and female specimens, we may compare their means with pooled male-plus-female data for the apes (Table 9-2).

That is, the modern human average capacity (irrespective of sex) is 2.68 times the gorilla mean, 3.33 times the orangutan value, 3.52 times the chimpanzee (*P. troglodytes*) average, and 3.93 times the bonobo (*P. paniscus*) mean.

ABSOLUTE ENDOCRANIAL CAPACITIES OF EARLY HOMINIDS

Table 9-3 lists the latest available endocranial capacities of australopithecines, classified under the four most widely recognized species. In compiling this table, the author has not followed a recent "splitting" tendency evinced by some investigators who would revive the long-sunk *Paranthropus* as a distinct genus and even Paranthropinae as a separate subfamily, and who have claimed to recognize additional species, for example *A. crassidens* and *A. aethiopicus*. The same school of thought tends to recognize *Homo ergaster* and *Homo rudolfensis* as distinct from *H. habilis*, and *Homo neanderthalensis* rather than *H. sapiens neanderthalensis*. This extreme splitting tendency, in the author's view, is not helpful in biological or phylogenetic studies, nor does it take sufficient cognizance of intraspecific variation which, in extant higher hominoids, is marked.

The values cited in Table 9-3 show that, by and large, the mean absolute capacities of *A. africanus* (440) and *A. boisei* (463) are very similar. However, Holloway (1988) has claimed "an important increment in size" from earlier to later members of the *A. boisei* lineage. To test this claim, the sample of *A. boisei* may be divided into two subsets: specimens from 2.5 to 2.0 MYA have capacities ranging from 410-?480 and a mean of 434 cm^3 (n = 3), whereas the later specimens (1.8-1.4 MYA) have a sample range of ?390-530 and a mean of 485 cm^3 (n = 4), that is, 11.75 percent greater than the early *A. boisei*. When allowance is made for the tiny samples and the probable presence of specimens of both sexes among the seven crania, and not necessarily in the same ratio in the two subsets, this early-late difference of the *A. boisei* means must

TABLE 9-1 Mean endocranial capacities of male and female pooled samples of extant great apes (cm^3)

	Male mean	Female mean	Sexual dimorphism (%)
Pan troglodytes	398.5 (n = 163)	371.1 (n = 200)	6.9
Gorilla gorilla gorilla	534.6 (n = 414)	455.6 (n = 254)	14.8
Pongo pygmaeus	434.4 (n = 203)	374.5 (n = 199)	13.8

be deemed as being of doubtful or limited importance. Within the *A. africanus* subsets, although the lowest value happens to belong to what seems to be the oldest specimen (MLD 37/38), no trend appears to be displayed within the c. 0.8 MYA lineage assumed to be represented.

The newer mean values confirm that the four australopithecine taxa show scarcely any advance in absolute terms on the mean absolute values for extant great apes (see Table 9-1).

In relation to the pooled mean values for apes, the estimated mean values for fossil hominid taxa (excluding *A. robustus* with an available sample of one) bear the following relationships (Table 9-4). When we contemplate that the modern human interracial mean is 3.52 times the chimpanzee mean, 2.68 times the gorilla value, and 3.33 times the orangutan value, the comparative interhominoid index values for *Australopithecus* species are seen to be hardly increased at all.

In other words, the earliest fossil specimens that, on dental and cranial evidence, have been assigned to the Hominidae, had mean absolute capacities of the same order of magnitude as those of great apes, but showed a small advance on the chimpanzee mean value. It is the chimpanzee that many molecular biological studies have indicated is most closely related genetically to modern *H. sapiens*, while the gorilla is somewhat less similar. This is interpreted as indicating that the human lineage departed from that of the chimpanzee somewhat more recently than the divergence of the gorilla, at a time that is provisionally dated to between 6.4 and 5.0 MYA. The divergence of the gorilla lineage from the other African hominoids is dated somewhat earlier, between 9.0 and 6.0 MYA.

To concentrate our attention on the chimpanzee, we have to use an end product of the chimpanzee lineage, one of the extant species, *Pan troglodytes*, since we have no unequivocal proto-chimpanzees of the 5-6 MYA period.

On small samples, the mean capacity values for two of the early hominid species show that the mean of the brain size of *A. afarensis* is about 8 percent greater than that of modern chimpanzee, while the *A. africanus* mean is some 15 percent greater than the chimpanzee mean (see Table 9-4).

We may infer that, by the time these two early forms of hominid had emerged as upright-walking creatures with teeth of hominid pattern, their brain size had increased only very slightly over that of the chimpanzee. The increment is slightly more than the excess of orangutan over chimpanzee but falls far short of the 31.6 percent excess by which modern gorilla exceeds modern chimpanzee (Tobias 1992b). These figures underline anew that *Australopithecus* was a genus of small-brained hominids. Keith (1915) expected that dramatic brain enlargement was a hallmark of the earliest recognizable hominids. If the various species of *Australopithecus* are correctly classified as hominids—and virtually all investigators accept this, save for Zuckerman and, perhaps, Oxnard—it follows that Keith's prognostication is con-

TABLE 9-2 Mean endocranial capacities of pooled male-plus-female samples of extant apes $(cm)^3$

Pan troglodytes	(n = 363)	383.4	Tobias, 1971a, b
Pan paniscus	(n = 11)	343.7	This study
Gorilla gorilla gorilla	(n = 668)	504.6	Tobias, 1971a, b
Pongo pygmaeus	(n = 402)	404.8	Tobias, 1971a, b

TABLE 9-3 Some classified endocranial capacities (cm^3) of australopithecines of 3.2–1.4 MYA

Species	Dating	Specimen	Capacity	Source
A. afarensis	c3.2	AL 162-28	<400	Holloway, 1988
(Hadar)	c3.2	AL 333-105	<400	Holloway, 1988
	c3.2	AL 333-45	485-500	Holloway, 1983
		Mean	c413.5	Tobias, 1987
A. africanus	c3.0	MLD 37/38	425	Conroy et al., 1990
	3.0–2.5	Sts 60	428	Holloway, 1975
		Sts 71	428	Holloway, 1975
		Sts 19	436	Holloway, 1975
		Sts 5	485	Holloway, 1975
	?c2.2	Taung 1	440	Holloway, 1975
		Mean	440.3	Conroy et al., 1990
A. robustus	1.8–1.7	SK 1585	530	Holloway, 1972
A. boisei	2.5	KNM-WT 17000	410	Walker et al., 1986
	2.1	Omo L338y-6	427	Holloway,1981a
	c2.0	KNM-ER 13750	450-480	Holloway, 1988
	1.8	KNM-WT 17400	?c390-400	Holloway, 1988
	1.8	OH 5	530	Tobias, 1963
	c1.5	KNM-ER 406	510	Holloway, 1973
	c1.5	KNM-ER 732	506	Holloway, 1973
		Mean	463.3	Tobias, 1990

founded: The earliest recognizable hominids showed only an 8-15 percent increase of absolute brain size over that of the living chimpanzee. The dramatic increase in brain-size was to come later, with the emergence of the earliest species of *Homo* to which was given the name of *Homo habilis* (Leakey et al. 1964).

Data for endocranial capacities of six specimens of *H. habilis* are given in Table 9-5.

The latest data have revealed an average endocranial capacity of *H. habilis* variously computed as 640 or 652 cm^3 (Tobias 1992b). Hence, the absolute capacity of *H. habilis* was 55-58 percent greater than in *A. afarensis*, and 45-48 percent greater than in *A. africanus*, whereas that of *A. boisei* is 5.2 percent greater

TABLE 9-4 Mean endocranial capacities of australopithecine series expressed as percentages of pooled male-plus-female means of extant great apes

		Hominid value as percentage of mean for		
Fossil taxon	**Mean**	***Pan troglodytes***	***G. gorilla***	***Pongo pygmaeus***
A. afarensis	c413.5	?108	?82	?102
A. africanus	440.3	115	87	109
A. boisei	463.3	121	92	114

than that of *A. africanus* (cf. Falk and Kasinga 1983). This means that *H. habilis* had a brain size bigger by half than the average values in *Australopithecus*. Clearly, it was with *H. habilis* that the human trend toward great cerebral expansion had begun.

Table 9-6 gives the most up-to-date values on endocranial capacity for various fossil hominid series (in several respects, these data supersede those of Tobias 1987). The data in Table 9-6 show that the mean for *H. habilis* is exceeded by the author's estimate for *H. erectus erectus* by 255.4 cm^3 or 3.1 standard deviations, and by Holloway's estimate for *H. erectus erectus* by 289.6 cm^3 or 3.5 standard deviations. In comparison with the author's estimate of the mean for a mixed Afro-Asian sample of 15 *H. erectus* specimens, the *habilis* sample falls short by 297.0 cm^3 or 3.6 standard deviations. In other words the value for the combined Asian and African sample of 15 *H. erectus* crania, 937.2 cm^3, is 46.4 percent greater than the sample mean for *H. habilis.*

RELATIVE ENDOCRANIAL CAPACITY OF EARLY HOMINIDS

Is the difference in capacity between the very early hominids and the extant great apes "real" or is it simply the consequence of differing body sizes?

Brain size in absolute terms tells us only part of the story. If one looks at the Animal Kingdom in general, one finds that smaller animals have smaller brains, while larger animals have larger brains. Even within modern humankind, it has been shown that tall people have bigger brains than short people. So the increases in brain size in ancient and modern hominids would be meaningful, only if we could estimate the body size of the ancient hominids.

Body size is sometimes expressed as stature, more commonly as body mass or weight. Cuvier first introduced the concept of relative brain weight, that is the weight of the brain expressed as a fraction of

TABLE 9-5 "Adult" values of endocranial capacity of *Homo habilis* (cm^3)

Specimen	Putative gender	Capacity	Mean
OH 7	M	674	
OH 16	M	638	688.0 (n = 3)
KNM-ER 1470	M	752	
OH 13	F	673	
OH 24	F	594	592.3 (n = 3)
KNM-ER 1813	F	510	
Total sample	M + F		640.2 (n = 6)

the weight of the body. When we are dealing with fossil taxa, body weight is generally calculated from postcranial bones. A number of methods have been devised whereby the investigator may do this. One is based on the size of the vertebrae, especially the cross-sectional area of the bodies of the vertebrae. Since the vertebrae bear the body weight, the sizes of the vertebral cross-sectional areas in *Australopithecus* and in *H. habilis* provide a means for computing the probable body weights of individuals in these species. Similarly, the lengths of the limb bones, especially those of the lower limb, provide a basis for estimating body length or stature.

It may be noted in passing that when cranial, dental, and postcranial remains are found scattered, as is commonly the case, and where more than one hominid species is known to have coexisted at the time in question, as in the entire known span of *H. habilis*, there are obvious difficulties in the ascribing of isolated postcranial bones to species. The problem is compounded by the consideration that the species are defined, exclusively or predominantly, on cranial and dental evidence. However, with the accumulation of more and more bony remains, the study of associations on living floors, morphological patterns, and paleodemographic data, the provisional assignment of postcranial bones to species has reached a degree of consensus for at least some of these skeletal elements. Of the Olduvai postcranial bones, for instance, Day (1977), Campbell (1978), and Tobias (1991b) agree in provisionally assigning OH 8 (a foot) and OH 35 (a tibia and fibula) to *H. habilis*, while the first and third authors agree similarly in respect of OH 48 (a clavicle) and OH 49 (a partial radial shaft).

TABLE 9-6 Endocranial capacity values for various fossil hominid series (cm^3)[a]

Taxon	*n*	Mean	Standard Deviation	Coefficient of Variation	95% population limits (to nearest cm^3)
A. afarensis	3	?413.5	?77.10	?18.65	352-?500[f]
A. africanus	6	440.3	22.60	5.13	383-499
A. robustus	1	530.0	-	-	-
A. boisei	7	463.3	53.66	11.58	332-595
H. habilis	6	640.2	82.23	12.85	429-852
H. erectus erectus[b]	7	895.6	93.57	10.45	667-1125
H. erectus erectus[c]	6	929.8	91.67	9.86	694-1165
H. erectus pekinensis	5	1043.0	112.51	10.79	731-1355
H. erectus (Asia and Africa)	15	937.2	135.48	14.46	647-1228
H. sapiens soloensis[d]	6	1090.8	75.39	6.91	897-1285
H. sapiens soloensis[e]	5	1151.4	99.51	8.64	896-1407

[a]In this table no attempt has been made to separate the series into presumptive male and female sub-sets.

[b]Based on Tobias's (1975) estimate, but with the incorporation of the author's new value for Trinil 2, based on Holloway's (1975) new value for Sangiran 2.

[c]Based on Holloway's (1981b) new values for six Indonesian specimens.

[d]Based on Weidenreich (1943).

[e]Based on Holloway (1980).

[f]Observed range.

Similar provisional allocations have been made for other East Aftican postcranial bones. On the basis of such provisional identifications for other hominid taxa, strengthened by those instances from Ethiopia, Kenya, Tanzania and Sterkfontein where partial skeletons, some including cranial parts, have been found, numerous attempts have been made to estimate body weight in the fossil taxa under consideration [e.g., by Genet-Varcin (1966, 1969), Lovejoy and Heiple (1970), Robinson (1972, Wolpoff (1973), McHenry (1974, 1982, 1991), Pilbeam and Gould (1974), Krantz (1977), Steudel (1980)]. Other investigators have used body height (stature) in the computation of relative brain size or relative cranial capacity (see, for example, Grüsser and Weiss 1985). We shall here use one recent set of estimates of body weight of *H. habilis* and of other taxa, namely those of Mc Henry (1975, 1976, 1982, 1984, this volume). At the same time even those critical estimates may be seen as not necessarily the last word on body weight estimates. At least they provide a provisional range of estimates of body size to which to relate the endocranial capacity values.

A second important consideration is the immense literature devoted recently to "scaling," that is, the structural and functional consequences of differences in size (or scale) among organisms of more or less similar design (Jungers 1984, 1985). Recent studies by Martin, Armstrong, and Hofman have stressed that there are metabolic constraints in brain enlargement (Martin 1980, 1981, 1982; Armstrong 1981, 1983, 1984; Hofman 1982); others have stressed the problem at which systematic level comparisons of brain scaling are most meaningful (Harvey and Mace 1982; Holloway and Post 1982).

Lande (1985) has made an important study on the quantitative genetic aspects of the problem of brain size/body size. He observes that "genetic uncoupling" of brain and body sizes in primates would have facilitated encephalization in primates, because natural selection for larger brain size would then not necessarily have favored correlated uneconomical increased body sizes: "If the genetic correlation between brain and body size within populations in the human lineage was . . . low as suggested by the data on primates, hominids would have been enabled to rapidly increase brain size in response to selection for more complex behavior without the cost of antagonistic selection to prevent the evolution of gigantism" (Lande 1985:30).

A number of different techniques have been proposed to determine the degree of encephalization, when body size is taken into account.

EQ stands for Jerison's Encephalization Quotient, which is the ratio of actual brain size to expected brain size (a kind of average for living mammals that takes body size into account) (Jerison 1970, 1973). As obtained by Jerison, expected brain size is derived from body weight by the formula 0.12 $(\text{body weight})^{0.667}$. The scaling coefficient of 0.667 has been claimed to fit the relationship between brain weight and body weight in a large sample of living mammals (Jerison 1973; Gould 1975; McHenry 1982). On a much larger sample of species, however, Martin (1982) has obtained an exponent closer to 0.75, rather than 0.67. A scaling coefficient of about 0.75 has been found to apply to primates (Bauchot and Stephan 1969), though within the primates these authors report coefficients ranging in various groups from 0.58 to 0.80 in round figures, as cited by Holloway and Post (1982).

CC is Hemmer's Constant of Cephalization and it is derived by dividing the endocranial capacity by the body weight scaled to the power of 0.23.

Table 9-7 gives results for two estimates of relative brain size. The values obtained by the two equations used here for EQ and CC confirm that (a) the various australopithecine species were slightly more encephalized than the chimpanzee; and (b) *H. habilis* was clearly more encephalized than any of the australopithecine series and represented a major step, indeed the first such, in brain aggrandizement: Its values reveal that it had attained some 50 percent of the *H. sapiens* degree of encephalization. More marked encephalization followed from *H. habilis* to *H. erectus*, the latter species reaching some 70 to 80 percent of the degree of encephalization shown by *H. sapiens*.

These coefficients, like Jerison's *Nc*, reveal that *H. habilis* is appreciably advanced in its degree of encephalization as compared with the Hadar hominids and with *A. africanus*. Since the estimated body size is built into the formulae for EQ and CC, it is clear that the larger endocranial capacity of *H. habilis* is not to be explained solely as the result of its larger estimated body mass nor, for that matter, of a higher estimated stature (Grüsser et al. cited by Grüsser and Weiss 1985): It clearly represents an advance in encephalization over the small-brained hominids, the australopithecines.

The data presented here, on relative brain size, show that, while the australopithecines were encephalized slightly more than the chimpanzee, *H. habilis* had unequivocally begun the remarkably "uncoupled" or disproportionate enlargement of the brain that is a critical hallmark of humankind.

NATURAL SELECTION AND ENCEPHALIZATION: A NEW HYPOTHESIS

What advantages were conferred by increasing encephalization during hominid phylogeny? Many selection pressures have been proposed (see summaries in Gabow 1977; Tobias 1981). The most recent and perhaps the most elegant analysis of the meaning and the evolutionary benefits of brain size is that of H. J. Jerison (1991) in his James Arthur Lecture in 1989. This work, added to his significant earlier book on *Evolution of the Brain and Intelligence* (1973), provides evidence that brain size "estimates the total information processing capacity of the brain in a species" (1991:53). In the first Andrew Abbie Memorial Lecture (1979), I listed no fewer than ten selective advantages of enlarged brain size that various workers had hypothesized. At the same time, attention was drawn to two features of hominid encephalization, namely its rapidity and its persistence (Tobias 1981).

The rapidity of hominid brain evolution had been a source of controversy between Charles Darwin and Alfred Russel Wallace (Eiseley 1956). In that oft-overlooked disputation, Darwin adhered rigidly to natural selection as an adequate explanation of the rise of humans and their brains. Wallace held that "some more rapid process of evolution than that envisaged in the Darwinian philosophy must have been at work in the production of man" (op. cit. p. 69). Seven years before Darwin published *The Descent of Man* (1871), Wallace had asserted that evolution in cultured humans was largely mental. However, he did not take the further step of postulating that this very culture had come to dominate the selective processes in humans. Instead, since he could envisage no other force to account for the rapid rise of the human brain, Wallace invoked a directive spiritual force that could not be accounted for in purely mechanistic terms.

The hypothesis I offered earlier was that, under the peculiar conditions of human social and cultural life, evolution would be expected to occur most rapidly when natural selection, and cultural or social selection drive evolutionary change in the same direction. I had been led to this view in an earlier attempt to explain the evolution of steatopygia in the Kalahari Bushmen.

It had seemed to me that, under the conditions of Bushman life, both natural and cultural selection had favored the development of steatopygia. I speculated that the same might have obtained with the Bushman's shortness of stature and with the evolution of the large brain and articulate speech of humans (Tobias 1961).

A second mechanism for rapid evolutionary change was offered, namely the genetic consequences of domestication, since the evolution of the human line resembled the domestication of animals in some respects, including the rapidity of evolution.

Closely related was Gabow's (1977) suggestion that the evolving human species—he was thinking especially of *H. erectus*—constituted a subdivided population. Such a model had been proposed by Sewall Wright (1969), and this type of population structure, it was inferred, would have had an accelerating effect on evolutionary change.

Any new explanation that may be offered on hominid encephalization must cover not only the inferred rapidity of the change that has occurred, but also the persistence of encephalization. The enlargement and reorganization of the brain were not once-off responses, but were sustained over several million years. The processes of brain change were undoubtedly involved in the evolution of the earliest hominids, the australopithecines. Further encephalization carried the lineage across a generic boundary into *Homo* and continued through consecutive chronospecies, *H. habilis*, *H. erectus*, and *H. sapiens*. Thus, the brain changes transcended systematic categories, geographical dispersal, cultural diversification,

TABLE 9-7 Mean endocranial capacity, estimated body mass, and coefficients of encephalization for a series of hominoids

	Mean endocranial capacity (cm^3)	Estimated body mass (kg)	EQ (actual value)	EQ (as percentage of *H. sapiens* value)	CC (actual value)	CC (as percentage of *H. sapiens* value)
P. troglodytes	395.0	45.0	2.6	34	33.6	31
A. afarensis	413.5	37.1	3.1	41	36.8	34
A. africanus	441.2	35.3	3.4	45	39.7	36
A. robustus	530.0	44.4	3.5	46	45.2	42
H. habilis	640.2	48.0	4.0	53	53.6	49
H. e. erectus (PVT)	895.6	53.0	5.3	70	73.4	67
H. e. erectus (RLH)	929.8	53.0	5.5	72	76.2	70
H. erectus (Asia and Africa)	937.2	53.0	5.5	72	76.8	71
H. e. pekinensis	1043.0	53.0	6.1	80	85.4	78
H. sapiens	1350.0	57.0	7.6	100	108.8	100

Note: EQ, encephalization quotient: CC, constant of cephalization; PVT, the author's estimate for *H. erectus erectus*; RLH, Holloway's (1981b) estimate for *H. erectus erectus*.

McHenry's value of 53.0 kg for *H. erectus* has been used in the above table for all of the subsets of *H. erectus*. Although mass is given in kilograms in the table, it is expressed in grams for the calculations.

Source: Modified after McHenry, 1982.

ecological radiation, and ethological variegation. Whatever selective and other causal agencies were operating, they must have continued influencing the hominid brain throughout the assumption of a bewildering array of new life-styles and environments. The suggestion offered to account for the sustained duration of encephalization was an autocatalytic positive feedback system (cf. Mayr 1963; Bielicki 1964, 1969; Tobias 1971a, 1981).

The questions one needs to answer are these: Does the occurrence of a greater relative brain size in humans connote advantages that might have been favored by natural selection? If so, what manner of advantage did the larger brain size confer?

On Lashley's (1949) and Jerison's (1963, 1970, 1973) approach, "improved adaptive capacities" provide the key to the selective advantage conferred by increased encephalization. In Jerison's (1991) newest work this view is maintained: He places hominid encephalization in the context of a specific environmental niche, namely the shrinking of the African wet forest and spread of the savanna region at the forest's edge. These conditions accompanied the emergence of the hominids about 5 or 6 MYA. In this setting, he suggests, adaptive changes occurred in the nervous systems of the ancestors of the hominids.

We may paraphrase "improved adaptive capacities" as increased adaptability or greater evolutionary flexibility. Is this the advantage of greater encephalization? The problem is more exacting than it might seem. Long ago, Mather (1943) showed that adaptedness and adaptability in evolution were inversely proportional to each other. In other words the more highly adapted an animal is to its present environment, the less evolutionary plasticity it has retained for adaptation to a new environment should conditions change. If the attainment by humankind of maximal encephalization implies that humans have attained maximal adaptability, we might expect that, on Mather's analysis, present adaptedness had been sacrificed. It is doubtful whether this deduction would stand up to close scrutiny.

Moreover, encephalization is present in a number of other mammalian lineages, for example, the cetaceans, some New World monkeys, the tree shrews or Tupaioidea, and certain rodents: The maximally encephalized end products of these evolutionary lineages certainly appear to be highly adapted in their respective econiches.

The author suggests that increased encephalization provided a means by which the organisms concerned could rise above the constraints suggested by Mather's paradox. It is hypothesized here that the relative enlargement of the hominid brain was a mechanism by which enhanced adaptability might be furnished and on which natural selection could go to work, while adaptation, that is, concurrent adaptedness, was not sacrificed and could even have been improved. In a word, it is proposed that encephalization enhanced adaptability while permitting organisms to maintain adaptation, or even to manifest more efficient adaptation.

This general proposition should have applied to all mammalian lineages characterized by progressive encephalization. In the hominid lineage it has perhaps attained its pinnacle of evolution, as reflected by modern humanity's remarkable degree of encephalization. Along this lineage, the particular property "secreted" by the expanding brain was the cognitive faculty, of such quality and degree as to generate culture. Probably no more puissant force has yet appeared on earth in its capacity to potentiate adaptation and to widen dramatically the evolutionary flexibility of its possessors. In the hominid line, particularly, culture may provide the means by which Mather's paradox has been addressed and surmounted. In this situation, adaptability and adaptedness are *not* inversely proportional.

MORPHOLOGY OF THE AUSTRALOPITHECINE BRAIN

Apart from the size, only five morphological characters distinguish australopithecine endocasts from East and South Africa from the brains of the extant apes:

1. In gross pattern, on the base of the endocranial cast the impression of the australopithecine brainstem is situated further anteriorly than in the apes, and more so in *A. robustus* and *A. boisei* than in *A. africanus*, though in all three species it shows anterior advancement toward the position that pertains in later humans.
2. The parietal lobe of the cerebrum appears well developed (Holloway 1988).
3. The cerebellar hemispheres are underslung (Tobias 1967), so that the occipital poles of the cerebrum form the most posterior part of the endocast (to which generalization, the oldest *A. boisei* endocast [that of KNM-WT 17000] and the second oldest, that of Omo L338y-6, may be exceptions—see Holloway 1981a, 1988).
4. Most of the australopithecine endocasts show the combination of right fronto-petalia (that is, the frontal pole of the right cerebral hemisphere protrudes further anteriorly than that of the left hemisphere). This combination Galaburda (1984) describes as the most common in modern man, while Holloway (1988) declares it is not found in the apes, "even the highly asymmetrical endocasts of *Gorilla*" (op. cit.:98). As in modern humans, there are exceptions to this combination among the early hominids (LeMay 1976, 1984; LeMay et al. 1982; Holloway and De LaCoste-Lareymondie 1982; Holloway 1988; Tobias 1987).
5. There is a modest bulbous protrusion in the posterior third of the inferior frontal convolution, corresponding to Broca's cap (Schepers 1946).

Some investigators, most notably Schepers (1946) and later Holloway (1974, 1975, 1985, 1988), have urged a sixth set of criteria upon us, namely the pattern of sulci, especially with regard to the position of what is taken to represent the lunate sulcus: Holloway sees this in early hominid endocasts as in a human-like posterior position, rather than in an apelike anterior position. Others, chiefly Falk (1989) and Falk et al. (1989), see only apelike sulcal patterns in the australopithecine endocasts. As far as the lunate sulcus is concerned, the author agrees with Le Gros Clark (1964) that it is really not possible to identify the lunate sulcus with certainty from the impressions on the early hominid endocasts.

We cannot be sure what cyto-architectonic alterations, if any, underlay these five gross rearrangements of the brain of various australopithecines. Suffice it to say that we have these superficial rearrangements accompanying a rather small increase in absolute and relative endocranial capacity, as the *total available information on differences between the brains of australopithecines and of present-day chimpanzees.*

MORPHOLOGY OF THE *HOMO HABILIS* BRAIN

The earliest species attributed to the genus *Homo* had endocasts testifying to major advances in the evolution of the brain. They have been described *in extenso* recently (Tobias 1987, 1991b), so a summary is all we need give here.

1. The increase in the *H. habilis* brain involves a definite broadening (mainly of the frontal and parietal lobes of the cerebrum), and a moderate heightening, but scarcely any lengthening of the cerebral hemispheres.

2. The sulcal pattern of the frontal lobes is very similar to that of modern *H. sapiens* and quite different from that of extant apes.

3. The gyral impressions on the frontal lobe include a well-marked prominence in the position of the posterior part of the inferior frontal convolution: This corresponds to the position of Broca's area and its bulbosity exceeds that in the corresponding cortical zone of *A. africanus*.

4. There is a right fronto-petalia in the few *H. habilis* endocasts in which left and right frontal poles are preserved. The posterior or caudal projection of the occipital pole is more variable: In a presumptive male of *H. habilis*, left occipito-petalia is present and, in two putative females, we find right occipito-petalia. In a modern human series reversal of the modal pattern of right fronto- and left occipito-petalia occurs more commonly in women, while the blend of right fronto-petalia with right occipito-petalia (as in one of our female *H. habilis* specimens, Olduvai hominid 24) occurs in association with *non-right-handedness* (Bear et al. 1986).

5. The impression of the superior parietal lobule is well developed and, in several endocasts of *H. habilis*, is asymmetrical with left predominance. This is a cerebral asymmetry that has not previously been reported (Tobias 1987). The anterior part of the superior parietal lobule corresponds to Brodmann area 7: Judgments of shape and visual relations involve neuronal activity in area 7 (Mountcastle et al. 1975, 1984; Eccles 1989). Roland (1985), using the injection of radio-Xenon, studied which areas of the brain showed an increase in regional cerebral blood flow when subjects were exposed to tasks requiring visuospatial judgment. *Inter alia*, they noted large increases in the posterior part of the superior parietal areas—36 percent on the left and 30 percent on the right. It would be useful to pursue the author's observation of superior parietal lobule asymmetry in *H. habilis* to see whether it is present in larger series of early hominids and in modern *H. sapiens*. If the anatomical asymmetry is confirmed, it may provide evidence of a functional asymmetry in visuospatial discrimination and judgment.

6. The parietal lobe in *H. habilis* is well expanded transversely and the inferior parietal lobule is strongly developed—in contrast with the arrangement in australopithecines and apes. The impressions of the supramarginal and angular gyri, comprising the inferior parietal lobule, are present for the first time in the hominid lineages. Since this area forms part of the larger Wernicke's area or posterior speech cortex, it has been claimed that—with the anterior speech cortex of Broca present as well—*H. habilis* is the first species in the history of the hominids to show the two most important neural bases for language abilities (Tobias 1987, 1991b). (A third area, the superior speech cortex, is part of the supplementary motor area on the superomedial surface of the cerebral hemisphere and therefore its presence cannot be detected on an endocast.)

7. One *H. habilis* endocast (Olduvai hominid 7) shows evidence of asymmetry of the sulcus lateralis (Sylvian fissure): The left sulcus ascends only slightly from anterior to posterior with a low termination,

while that on the right rises steeply. This difference tallies with the position in later hominids including modern humans (Cunningham 1892; LeMay and Culebras 1972; Le May 1976, 1977).

8. The anterosuperior part of the occipital lobe is expanded, as is the adjacent posterosuperior part of the parietal lobe. This parieto-occipital transverse expansion is more marked than the frontal transverse expansion and gives the endocast an ovoid rather than ellipsoid contour when viewed from above.

9. The pattern of the middle meningeal blood vessels is more beset with branches and anastomoses than are the patterns of the australopithecines so far described (Saban 1983; Tobias 1967, 1987).

10. Unlike the Hadar hominids and the "robust" and "hyper-robust" australopithecines, *H. habilis* endocasts show the transverse-sigmoid pattern of venous sinus drainage as in *A. africanus* and *H. sapiens*. In two out of three specimens in which the area is preserved, the superior sagittal sinus groove drains to the right, whereas in OH 24 it drains to the left. This is the same specimen that shows right fronto- and occipito-petalia, such as in modern humans is associated with non-right-handedness. In a modern human series, Henneberg and Symons (1991) found that the direction of drainage of the superior sagittal sinus was significantly contralateral to the side of occipito-petalia and to the side of the longer cerebral hemisphere. The evidence of left drainage and of right occipito-petalia in Olduvai hominid 24 accords with their results. This appears to be the first evidence of such a negative correlation in a fossil hominid, between occipito-petalia and the direction of drainage of the superior sagittal sinus.

The most important morphological traits of the *H. habilis* brain are the presence of the two main cerebral areas that in modern humans are the seat of spoken language, Broca's and Wernicke's areas. *H. habilis* was the earliest hominid to show both of these prominently enlarged.

We have therefore the revealing and provocative concurrence of two phenomena: the parts of the brain that govern spoken language became manifest at that stage when appreciable brain enlargement and marked encephalization first obtruded. These two major alterations in the structure of the brain became apparent soon after stone tools first appeared in the fossil record.

SUMMARY

In the stages by which the presumably apelike brain of the last common ancestor of humankind and chimpanzee was made over into the brain of humankind, it seems that relatively small changes accompanied the emergence of the earliest available and analyzable hominids, the australopithecines. These changes comprised a minimal increase in absolute and relative size and some limited reorganization of the overall anatomical structure of the brain. As for neurologically important changes in the brain, there is scarcely any evidence of surface alterations in the sulcal and gyral patterns. What differences have been claimed, especially in regard to the expansion of the prestriate areas of the cerebral hemispheres, are to say the least problematical.

However, major expansion of the brain and critical cortical reorganization were striking features of the change from *A. africanus* to *H. habilis*. These changes included notable augmentation of the cerebrum, strong lateral expansion of the parieto-occipital region, the appearance of a human-like sulcal pattern and the emergence for the first time of protuberances interpreted as the anterior and posterior speech cortices. That these homologs of Broca's and Wernicke's areas were used by a speaking and language-using ancestor, *H. habilis*, was proposed by Tobias (1980, 1981, 1983), and my claim has subsequently been supported by Falk (1983) and by Eccles (1989).

Thus, it is with the appearance of the *H. habilis* brain that a gigantic step was taken to a new level of organization in hominid brain evolution.

ACKNOWLEDGMENTS

My grateful thanks are owing to Peter Faugust, Val Strong, and Joel Symons.

LITERATURE CITED

Adeloye, A., Kattan, K. R., and Silverman, F. N. (1975) Thickness of the normal skull in the American blacks and whites. *Am. J. Phys. Anthropol.* 43: 23-30.

Armstrong, E. (1981) A look at relative brain size in mammals. *Neurosci. Lett.* 34: 101-104

Armstrong, E. (1983) Relative brain size and metabolism in mammals. *Science* 220:1302-1304.

Armstrong, E. (1984) Allometric considerations of the adult mammalian brain with special emphasis on primates. In W. L. Jungers (ed): *Size and Scaling in Primate Biology*. New York: Plenum, pp. 115-147.

Baer, M. J., and Harris, J. E. (1969) A commentary on the growth of the human brain and skull. *Am. J. Phys. Anthropol.* 30:39-44.

Bauchot, R., and Stephan, H. (1969) Encephalisation et niveau evolutif chez les simiens. *Mammalia* 33:225-275.

Bear, D., Schiff, D., Saver, J., Greenberg, M., and Freeman, R. (1986) Quantitative analysis of cerebral asymmetries: fronto-occipital correlation,sexual dimorphism and association with handedness. *Arch. Neurol.* 43:598-603.

Bielicki, T. (1964) Evolution of the intensity of feedbacks between physical and cultural evolution from man's emergence to present times. UNESCO Expert Meeting on Biological Aspects of Race. Moscow, August 1964, pp. 1-3.

Bielicki, T. (1969) Deviation-amplifying cybernetic systems and hominid evolution. *Mater. Pr. Anthropol.* 77:57-60.

Brandes, K. (1927) Liquorverhältnisse an der Leiche und Hirnschwellung. *Frankf. Zeit. Pathol.* 35:274-301.

Campbell, B. G. (1978) (ed.) *W. E. LeGros Clark: The Fossil Evidence for Human Evolution*, 3rd ed. Chicago, London: University of Chicago Press.

Clark, W. E. LeGros (1964) *The Fossil Evidence for Human Evolution: An Introduction to the Study of Paleoanthropology*, 2nd ed. Chicago: University of Chicago Press.

Connolly, C. J. (1950) *External Morphology of the Primate Brain*. Springfield, Ill.: Charles C. Thomas.

Conroy, G. C., Vannier, M. W., and Tobias, P. V. (1990) Endocranial features of *Australopithecus africanus* revealed by 2- and 3-D computed tomography. *Science* 247:838-841

Cunningham, D. J. (1892) Contribution to the surface anatomy of the cerebral hemispheres. *Roy. Irish Acad. Sci. Cunningham Mem.* VII. Dublin.

Dart, R. A. (1925) *Australopithecus africanus*: The man-ape of South Africa. *Nature* 115:195-199.

Darwin, C. (1871) *The Descent of Man*. London: John Murray.

Dawson, C., and Woodward, A. S. (1913) On the discovery of a palaeolithic human skull and mandible in a flint-bearing gravel overlying the Wealden (Hastings Beds) at Piltdown, Fletching (Sussex). *Quart. J. Geol. Soc.* (London) 69:117-151.

Day, M. H. (1977) *Guide to Fossil Man*, 3rd ed. London: Cassell.

Delattre, A., and Fenart, R. (1954) Le crâne acromégale, ses rapports avec la morphogenèse du crâne. *Ann. Endocrin.* 15:784-693.

Eccles, J. C. (1989) *Evolution of the Brain: Creation of the Self*. London and New York: Routledge.

Eiseley, L. C. (1956) Review of the Piltdown Forgery by J.S. Weiner. *Am. J. Phys. Anthropol.* 14:124-126.

Falk, D. (1983) Cerebral cortices of East African early hominids. *Science* 221:1072-1074.

Falk, D. (1989) Ape-like endocast of "ape-man" Taung. *Am. J. Phys. Anthropol.* 80:335-339.

Falk, D., Hildebolt, C., and Vannier, M. W. (1989) Reassessment of the Taung early hominid from a neurological perspective. *J. Hum. Evol.* 18:485-492.

Falk, D., and Kasinga, S. (1983) Cranial capacity of a female robust australopithecine (KNM-ER 407) from Kenya. *J. Hum. Evol.* 12:515-518.

Gabow, S. L. (1977) Population structure and the rate of hominid brain evolution. *J. Hum. Evol.* 6:643-645.

Galaburda, A. M. (1984) Anatomical asymmetries. In N. Geschwind and A. M. Galaburda (eds.): *Cerebral Dominance: The Biological Foundations*. Cambridge: Harvard University Press, pp. 11-25.

Genet-Varcin, E. (1966) Conjectures sur l'allure générale des Australopithèques. *Bull. Soc. Prehist. Française* 63:106-107.

Genet-Varcin, E. (1969) Structure et comportement des Australopithèques d'après certains os post-craniens. *Ann. Paleontol.* (Vertebrata) 55:137-148.

Gould, S. J. (1975) Allometry in primates, with emphasis on scaling and the evolution of the brain. In F. S. Szalay (ed.): *Approaches to Primate Paleobiology*. Basel: Karger, pp. 244-292.

Grüsser, O.-J., and Weiss, L.-R. (1985) Quantitative models on phylogenetic growth of the hominid brain. In P. V. Tobias (ed.): *Hominid Evolution: Past, Present and Future*. New York: Alan R. Liss, pp. 457-464.

Hartl, F., and Burkhardt, L. (1952)Über Strukturumbau des Skelets, besonders des Schädeldachs und Schlusselbeins, beim Erwachsenen und seine Beziehungen zur Hypophyse, nach Massgabe des spezifischen Gewichts und histologischen Befundes. *Virch. Archiv.* 322:503-528.

Hartl, F., and Luther, J. (1953) Vergleichende Messungen am Kopf und am knochernen Schädel als Beitrag zur Konstitutionsbiometrie. *Zeit. Menschl. Vererb. Konstitutionsl.* 31:381-390.

Harvey, P. H., and Mace, G. M. (1982) Comparisons between taxa and adaptive trends: Problems of methodology. In King's College Research Group (eds.): *Current Problems in Sociobiology*. Cambridge: Cambridge University Press, pp.343-361.

Henneberg, M., and Symons, J. A. (1991) Relationship of brain asymmetry to the lateralization of venous drainage from the cranial cavity. *Anat. Soc. S. Afr. Newsl.* 24:18.

Hofman, M. A. (1982) Encephalization in mammals in relation to the size of the cerebral cortex. *Brain Behav. Evol.* 20:24-96.

Holloway, R. L. (1972) New australopithecine endocast, SK 1585, from Swartkrans, South Africa. *Am. J. Phys. Anthropol.* 37:173-185.

Holloway, R. L. (1973) New endocranial values for the East African early hominids. *Nature* 243:97-99.
Holloway, R. L. (1974) The casts of fossil hominid brains. *Sci. Amer.* 231:106-115.
Holloway, R. L. (1975) Early hominid endocasts: volumes, morphology and significance for hominid evolution. In R. H. Tuttle (ed.): *Primate Functional Morphology and Evolution.* The Hague: Mouton, pp. 393-416.
Holloway, R. L. (1980) Indonesian "Solo" (Ngandong) endocranial reconstructions: some preliminary observations and comparisons with Neandertal and *Homo erectus* groups. *Am. J. Phys. Anthropol.* 53:285-295.
Holloway, R. L. (1981a) The endocast of the Omo L338y-6 juvenile hominid: Gracile or robust *Australopithecus*? *Am. J. Phys. Anthropol.* 54:109-118.
Holloway, R. L. (1981b) The Indonesian *Homo erectus* brain endocasts revisited. *Am. J. Phys. Anthropol.* 55:503-521.
Holloway, R. L. (1983) Cerebral brain endocast pattern of *Australopithecus afarensis* hominid. *Nature* 303:420-422.
Holloway, R. L. (1985) The past, present and future significance of the lunate sulcus in early hominid evolution. In P. V. Tobias (ed.): *Hominid Evolution: Past, Present and Future.* New York: Alan R. Liss, pp. 47-62.
Holloway, R. L. (1988) "Robust" australopithecine brain endocasts: some preliminary observations. In F. E. Grine (ed.): *Evolutionary History of the "Robust" Australopithecines.* New York: Aldine de Gruyter, pp. 95-105.
Holloway, R. L., and De LaCoste-Lareymondie, M. C. (1982) Brain endocast asymmetry in pongids and hominids: some preliminary findings on the paleontology of cerebral dominance. *Am. J. Phys. Anthropol.* 58:101-110.
Holloway, R. L., and Post, D. G. (1982) The relativity of relative brain measures and hominid mosaic evolution. In E. Armstrong and D. Falk (eds.): *Primate Brain Evolution: Methods and Concepts.* New York, London: Plenum, pp. 57-76.
Jerison, H. J. (1963) Interpreting the evolution of the brain. *Hum. Biol.* 35:263-291
Jerison, H. J. (1970) Gross brain indices and the analysis of fossil endocasts. *The Primate Brain* 1:225-244.
Jerison, H. J. (1973) *Evolution of the Brain and Intelligence.* New York, London: Academic.
Jerison, H. J. (1991) *Brain Size and the Evolution of Mind.* New York: American Museum of Natural History.
Jungers, W. L. (1984) Aspects of size and scaling in primate biology with special reference to the locomotor skeleton. *Yearb. Phys. Anthropol.* 27:73-97.
Jungers, W. L. (ed.) (1985) *Size and Scaling in Primate Biology.* New York: Plenum.
Keith, A. (1915) *The Antiquity of Man.* London: Williams and Norgate.
Klaauw, C. J. van der (1946) Cerebral skull and facial skull. A contribution to the knowledge of skull structure. *Archl. Neerl. Zool.* 7:16-37.
Krantz, G. S. (1977) A revision of australopithecine body sizes. *Evol. Theory* 2:65-94.
Lande, R. (1985) Genetic and evolutionary aspects of allometry. In W. L. Jungers (ed.): *Size and Scaling in Primate Biology.* New York: Plenum Press, pp. 21-32.
Lashley, K. S. (1949) Persistent problems in the evolution of mind. *Quart. Rev. Biol.* 24:28-42.
Leakey, L. S. B., Tobias, P. V., and Napier, J. R. (1964) A new species of the genus *Homo* from Olduvai Gorge. *Nature* 202:7-9.
LeMay, M. (1976) Morphological cerebral asymmetry of modern man, fossil man and nonhuman primates. *Ann. NY Acad. Sci.* 280:348-366.
LeMay, M. (1977) Asymmetries of the skull and handedness. *J. Neurol. Sci.* 32:243-253.
LeMay, M. (1984) Radiological, developmental and fossil asymmetries. In N. Geschwind and A. M. Galaburda (eds.): *Cerebral Dominance: The Biological Foundations.* Cambridge: Harvard University Press, pp. 26-42.
LeMay, M., Billig, M., and Geschwind, N. (1982) Asymmetries of the brains and skulls of nonhuman primates. In E. Armstrong and D. Falk (eds.): *Primate Brain Evolution.* New York: Plenum, pp. 263-278.
LeMay, M., and Culebras, A. (1972) Human brain: Morphologic differences in the hemispheres demonstrable by carotid arteriography. *New Eng. J. Med.* 287:168-170.
Lovejoy, C. O., and Heiple, K. F. (1970) A reconstruction of the femur of *Australopithecus africanus. Am. J. Phys. Anthropol.* 22:33-40.
Martin, R. D. (1980) Adaptation and body size in primates. *Z. Morph. Anthropol.* 71:115-124.
Martin, R. D. (1981) Relative brain size and basal metabolic rate in terrestrial vertebrates. *Nature* 293:57-60.
Martin, R. D. (1982) Allometric approaches to the evolution of the primate nervous system. In E. Armstrong and D. Falk (eds.): *Primate Brain Evolution: Methods and Concepts.* New York: Plenum, pp. 39-56.
Mather, K. (1943) Polygenic inheritance and natural selection. *Biol. Rev.* 18:32-64.
Mayr, E. (1963) *Animal Species and Evolution.* Cambridge: Harvard University Press.
McCown, T., and Keith, A. (1939) *The Stone Age of Mount Carmel II. The Fossil Human Remains from the Levalloiso-Mousterian.* Oxford: Oxford University Press.
McHenry, H. M. (1974) How large were the australopithecines? *Am. J. Phys. Anthropol.* 40:329-340.
McHenry, H. M. (1975) Fossil hominid body weight and brain size. *Nature* 254:686-688.
McHenry, H. M. (1976) Early hominid body weight and encephalization. *Am. J. Phys. Anthropol.* 45:77-83.
McHenry, H. M. (1982) The pattern of human evolution; studies on bipedalism, mastication and encephalization. *Ann. Rev. Anthropol.* 11:151-173.
McHenry, H. M. (1984) Relative size of the cheek teeth in *Australopithecus. Am. J. Phys. Anthropol.* 60:224.
McHenry, H. M. (1991) Petite bodies of the "robust" australopithecines. *Am. J. Phys. Anthropol.* 86:445-454.
Mednick, L. W., and Washburn, S. L. (1956) The role of the sutures in the growth of the braincase of the infant pig. *Am. J. Phys. Anthropol.* 14:175-186.
Mettler, F. A. (1955) Culture and the structural evolution of the neural system. James Arthur Lecture on the Evolution of the Human Brain. New York: Am. Mus. Nat. Hist.

Moss, M. L. (1954) Growth of the calvaria in the rat. The determination of osseous morphology. *Am. J. Anat.* 94:333-358.

Mountcastle, V. B., Lynch, J. C., Georgopoulos, A., Sakata, H., and Acuna, A. (1975) Posterior parietal association cortex of the monkey. *J. Neurophysiol.* 38:871-908.

Mountcastle, V. B., Motter, B. C., Steinmetz, M. A., and Duffy, C. J. (1984) Looking and seeing: The visual functions of the parietal lobe. In G. M. Edelman, W. E. Gall, and W. M. Cowan (eds.): *Dynamic Aspects of Neocortical Function*. New York: Wiley, pp. 159-193.

Oakley, K. P. (1960) Artificial thickening of bone and the Piltdown skull. *Nature* 187:174.

Pilbeam, D., and Gould, S. J. (1974) Size and scaling in human evolution. *Science* 186:892-901.

Robinson, J. T. (1972) *Early Hominid Posture and Locomotion*. Chicago: University of Chicago Press.

Roland, P. E. (1985) Cortical activity in man during discrimination of extrinsic patterns and retrieval of intrinsic patterns. In C. Chagas, R. Gattass and C. Gross (eds.): *Pattern Recognition Mechanisms*. Vatican City: Pontificiae Academiae Scientiarum Scripta Varia 54:215-246.

Saban, R. (1983) Les veines meningées moyennes des australopithèques. *Bull. Mem. Soc. Anthropol. Paris* 10:313-324.

Saller, K. (1959) *Martin's Lehrbuch der Anthropologie*, Vol. II. Stuttgart: Gustav Fischer.

Schepers, G. W. H. (1946) The endocranial casts of the South African apemen. In R. Broom and G. W. H. Schepers, *The South African Fossil Apemen: The Australopithecinae*. Transv. Mus. Mem. 2:153-272.

Shattock, S. G. (1914) Morbid thickening of the calvaria; and the reconstruction of bone once abnormal; a pathological basis for the study of the thickening observed in certain Pleistocene crania. Section VII, pp. 3-46. *XVIIth Int. Cong. Med. London*, July 1913.

Simon, E. (1955) Vordere und mittlere Schädelgrube bei Laboratoriums—und Haussaugetieren, II Mitteilung. *Acta. Anat.* 23:206-241.

Steudel, K. (1980) New estimates of early hominid body size. *Am. J. Phys. Anthropol.* 52:63-70.

Tobias, P. V. (1961) New evidence and new views on the evolution of man in Africa. *S. Afr. J. Sci.* 57:25-38.

Tobias, P. V. (1963) Cranial capacity of *Zinjanthropus* and other australopithecines. *Nature* 197:743-746

Tobias, P. V. (1967) *Olduvai Gorge*, Vol. 2: *The Cranium and Maxillary Dentition of* Australopithecus (Zinjanthropus) boisei. Cambridge: Cambridge University Press.

Tobias, P. V. (1968) Cranial capacity in anthropoid apes, *Australopithecus* and *Homo habilis*, with comments on skewed samples. *S. Afr. J. Sci.* 64:81-91.

Tobias, P. V. (1970) Brain-size, grey matter and race—fact or fiction? *Am. J. Phys. Anthropol.* 32:3-26.

Tobias, P. V. (1971a) *The Brain in Hominid Evolution*. New York, London: Columbia University Press.

Tobias, P. V. (1971b) The distribution of cranial capacity values among living hominoids. *Proceedings of the 3rd International Congress of Primatology, Zurich* 1970 1:18-35.

Tobias, P. V. (1975) Brain evolution in the Hominoidea. In R. H. Tuttle (ed.): *Primate Functional Morphology and Evolution*. The Hague: Mouton, pp. 353-392.

Tobias, P. V. (1980) L'Évolution du cerveau humain. *La Recherche* 11:282-292.

Tobias, P. V. (1981) *The Evolution of the Human Brain, Intellect and Spirit*. Adelaide: University of Adelaide Press.

Tobias, P. V. (1983) Recent advances in the evolution of the hominids with especial reference to brain and speech. *Pontificiae Academiae Scientiarum Scripta Varia* 50:85-140.

Tobias, P. V. (1987) The brain of *Homo habilis*: A new level of organization in cerebral evolution. *J. Hum. Evol.* 16: 741-761.

Tobias, P. V. (1991a) The age at death of the Olduvai *Homo habilis* population and the dependence of demographic patterns on prevailing environmental conditions. In: H. Thoen, J. Bourgeois, P. Vermeulen, Crombe, and K. Verlaeckt (eds.): *Studia Archaeologica: Liber Amicorum Jacques Nenquin*. Gent: University of Gent.

Tobias, P. V. (1991b) *Olduvai Gorge*, Vols. 4A and 4B: *The Skulls, Endocasts and Teeth of* Homo habilis. Cambridge: Cambridge University Press.

Tobias, P. V. (1992a) Piltdown: An appraisal of the case against Sir Arthur Keith. *Curr. Anthropol.* 33:243-293.

Tobias, P. V. (1992b) *Postlude on Hominid Brain Evolution*. Turin: Giulio Einaudi Editore.

Tobias, P. V. (in press) The brain of the first hominids. In *Fondation Fyssen Symposium* on "The Origins of the Human Brain: Palaeontology, Molecular Biology and Developmental Genetics," Versailles, France, 1990.

Todd, T. W. (1924) Thickness of the male white cranium. *Anat. Rec.* 27: 245-256.

Von Bonin, G. (1963) *The Evolution of the Human Brain*. Chicago: University of Chicago Press.

Walker, A., Leakey, R. E., Harris, J. M., and Brown, F. H. (1986) 2.5-Myr *Australopithecus boisei* from West Lake Turkana, Kenya. *Nature* 322:517-522.

Washburn, S. L. (1947) The relation of the temporal muscle to the form of the skull. *Anat. Rec.* 99:239-248.

Weidenreich, F. (1943) The skull of *Sinanthropus pekinensis*: A comparative study of a primitive hominid. *Palaeontol. Sin.* 10:1-298.

Wright, S. (1969) *Evolution and the Genetics of Populations*. Vol. II: *The Theory of Gene Frequencies*. Chicago: Chicago University Press.

Wolpoff, M. H. (1973) Posterior tooth size, body size and diet in South African gracile australopithecines. *Am. J. Phys. Anthropol.* 39:375-393.

Young, R. W. (1957) Postnatal growth of the frontal and parietal bones in white males. *Am. J. Phys. Anthropol.* 15:367-386.

Zuckerman, S. (1954) Correlation of change in the evolution of higher primates. In J. S. Huxley, A. C. Hardy and E. B. Ford (eds.): *Evolution as a Process*. London: George Allen and Unwin, pp. 301-352.

CHAPTER 10

Advances in Understanding the Craniofacial Anatomy of South African Early Hominids

Ron J. Clarke

INTRODUCTION

> From the South Sea, I will come back again to Africa, a country of very great extent; in which, if it were well searched, I am persuaded that all the several types of human progression might be traced and perhaps all the varieties of the species discovered. . . . James Burnet (Lord Monboddo), 1774

> Where then must we look for primaeval Man? Was the oldest *Homo sapiens* Pliocene or Miocene, or yet more ancient? In still older strata do the fossilised bones of an ape more anthropoid or a man more pithecoid await the researches of some unborn palaeontologist? . . . T.H. Huxley, 1863

Robert Broom was born in 1866, only three years after Huxley's prophecy, and Raymond Dart was born in 1893. Both of these men were to initiate the fulfillment of the yet older prophecy of Lord Monboddo in that in South Africa they made the first discoveries of the earlier types of human progression and a good many of the "varieties of the species." Broom (1925a) was to refer to Monboddo's prediction just after Dart's announcement of the Taung discovery. These discoveries and their dates are listed in Table 10-1 together with the significant discoveries made in South Africa by two other researchers, T. F. Dreyer and J. T. Robinson.

With the exception of *Australopithecus africanus*, more complete examples of all these forms of hominid were later found in East Africa beginning in 1959 with the discovery of a magnificent cranium of a *Paranthropus* (OH 5) at Olduvai, followed in 1961 by a mandible and parietals that were to become the type specimen of *Homo habilis*. Then in 1962, a splendid calvaria (OH 9) of an archaic African "*H. erectus*" was found at Olduvai. A more complete version of this—KNM-ER 3733—was found at East Lake Turkana in 1976. An archaic *Homo sapiens* calvaria similar to that from Florisbad was found at Omo in Ethiopia in 1967 and a cranium of the same form was discovered later at Laetoli, Tanzania.

TABLE 10-1 Historical sequence of hominid fossil discoveries in South Africa

Australopithecus africanus	1924	R. Dart	Taung
Archaic *Homo sapiens*	1932	T. F. Dreyer	Florisbad
Adult *Australopithecus africanus*	1936	R. Broom	Sterkfontein
Paranthropus robustus	1938	R. Broom	Kromdraai
Australopithecine sp. ancestral to *Paranthropus*	1938	R. Broom	Sterkfontein
Early African "*Homo erectus*"	1949	J. Robinson	Swartkrans
Earliest form of *Homo* (later called *H. habilis*)	1957	J. Robinson	Sterkfontein

In the following pages I will discuss each of these species of hominid and the data that have been gleaned from analysis of their craniofacial anatomy.

AUSTRALOPITHECUS AFRICANUS

When Dart (1925) described *A. africanus* on the well-preserved infant skull from Taung, he regarded it as a humanlike ape "intermediate between living anthropoids and man." He noted its humanlike characters as follows:

> [The whole cranium displays humanoid rather than anthropoid lineaments.] Dolichocephalic. Absence of supraorbital torus and forehead rises from the orbital margin in human fashion. Narrow interorbital distance without the laterally expanded ethmoids of the African anthropoids. Orbits almost circular, not subquadrate as in anthropoids. Zygomatic arches, maxillae and mandible delicate and humanoid in character. Facial prognathism slight. Nasal bones not prolonged below level of inferior margin of orbits. Dentition humanoid rather than anthropoid. Canines short. No diastema between lower canines and premolars although a diastema is present between upper canines and incisors. Incisors irregular in size, overlapping and almost vertical, unlike the symmetrical, well-spaced, markedly forward projecting incisors of apes (although the uppers do project slightly anteriorly). Parabolic dental arcade. Mandible humanoid rather than anthropoid with a short slender ramus. Vertical anterior symphyseal surface and absence of simian shelf posteriorly. Foramen magnum situated relatively far forward.

In spite of all these humanlike features, Dart had to admit that the brain, while displaying a few humanlike traits, was in size comparable to that of the apes and that, as it lacked "the distinctive localised temporal expansion which appear to be concomitant with and necessary to articulate man," it could not be that of a "true man." He therefore chose to regard *A. africanus* as a humanlike ape and was followed in this for many years by Broom when he discovered adult examples of australopithecines. Broom (1938) summed up the position when he wrote about a maxillary fragment (TM 1512) from the Lower Cave at Sterkfontein that "The face agrees fairly closely with that of the chimpanzee except for reduction of the anterior dentition and snout" (see Figure 10-1). He further wrote, "One might describe *Australopithecus* as a chimpanzee with human teeth" and "Most will agree that if it had large canines it would be near to the chimpanzee and if it had a large brain it would be near to man."

Although the Taung skull was undistorted and reasonably complete in comparison with hominid fossils since discovered, the fact that it was that of an infant meant that the adult characters of the species *A. africanus* were yet to be determined. This did not daunt Robert Broom who firmly believed that an adult would be found and who published his drawn reconstruction of how the adult would appear when found (Broom 1925a, b). In a caption to his 1925a reconstruction, he commented, "The jaw must therefore be

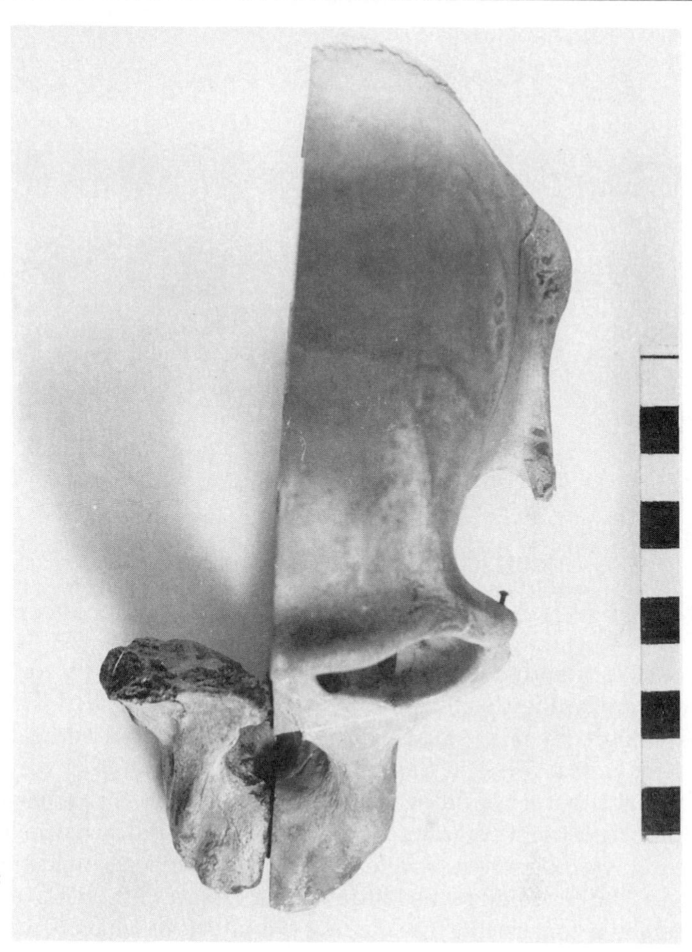

FIGURE 10-1
Superior view of left half of a chimpanzee cranium with the *Australopithecus africanus* right maxillary fragment, TM 1512.

much as indicated and the whole skull cannot differ greatly from the restoration given." His confidence proved to be well founded when in 1947 he discovered the well-preserved adult cranium Sts 5 (Mrs Ples) and it presented a remarkably similar profile to that which Broom had predicted in 1925 (Figure 10-2). Particular points of similarity are the prominent glabella, the slope of the frontal squame, the position of the malar relative to the nasal margin, the sagittal profile of the face and the sharply angled occipital profile. The only major point of difference is that from its tooth sockets Sts 5 clearly had a small canine, whereas in Broom's reconstruction of the adult Taung skull, there is a relatively large canine.

Broom's reasoning for this was sound. He observed (1925a) that the adult must have had an enlarged canine because of the presence in the infant of a diastema between the upper canine and the lateral incisor. Furthermore, Broom believed that the Piltdown skull was genuine, and as Piltdown had a large apelike canine, though smaller than that of the chimp, it was reasonable to assume that *Australopithecus* with a more chimplike cranium would have been similarly endowed with large canines. Broom wrote that "*Eoanthropus* has a human brain with still the chimpanzee jaw. In *Australopithecus* we have a being also with a chimpanzee-like jaw, but with a sub-human brain." He went on to suggest that "it seems to be the forerunner of such a type as *Eoanthropus,* which may be regarded as the earliest human variety."

Broom with Schepers (1946), with reference to the Piltdown mandible, wrote, "I must class myself with those who consider that there is no reasonable doubt that this is the mandible of the same individual as has supplied the skull. In most characters, the Piltdown mandible is remarkably chimpanzee-like, but the teeth are much more human." Broom wrote this in connection with his discussion of a Sterkfontein

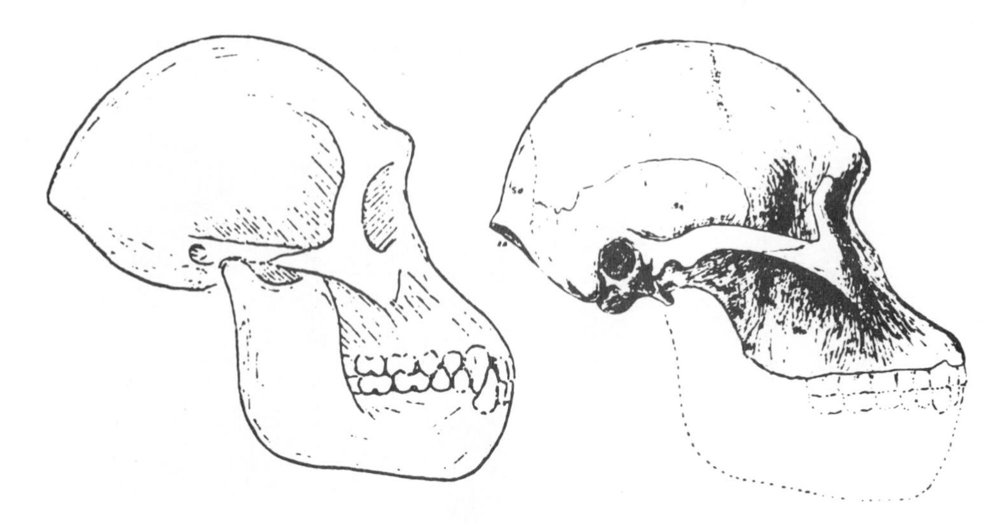

FIGURE 10-2 Broom's (1925b) prophecy of how an adult *A. africanus* would appear when found (left, courtesy of *Nature*) and his drawing of Sts 5, the adult *A. africanus* found in 1947 (courtesy of the Transvaal Museum, published in Broom and Robinson (1950)).

fossil, a mandibular symphysis of what he called a 9- or 10-year-old male (TM 1516) containing an enormous unerupted canine (Sts 50). Broom described the symphysis as almost typically anthropoid although, he said, the anterior face is a little more vertical and not so deep. He noted the very deep genioglossal fossa on the lingual surface and found it very similar to that of the chimpanzee. Broom thought that it would be useful in helping to restore the symphyseal region of *Eoanthropus*. This is an important point because whereas in 1925 the Piltdown canine had helped prompt Broom's belief that an adult *Australopithecus* would have a large canine, now in 1938, the large canine of an australopithecine was lending credit to the Piltdown mandible and Broom even thought the Sterkfontein symphysis could help in reconstructing the missing symphysis of what we now know to be a forgery!

This symphysis of TM 1516 (Figure 10-3) was important in another respect. It was found in 1938, only two years after the first adult *Australopithecus* cranium (TM 1511) was found and described as *Australopithecus transvaalensis*. It so happened that the first mandibular specimen to come into Broom's hands was this symphysis of the 9- or 10-year-old child. He wrote (Broom 1938b) that the symphysis is "so different from that of the Taungs ape that it seems advisable to place *A. transvaalensis* in a distinct genus for which the name *Plesianthropus* is proposed." Thus, although he did not designate it as the type specimen, it was this mandibular symphysis, TM 1516, and not the cranium TM 1511, that revealed the diagnostic characters that to Broom differentiated it at generic level from the Taung *Australopithecus*. Subsequent discoveries at Sterkfontein and Makapansgat have, I believe, proved that Broom was correct in separating TM 1516 taxonomically from the Taung mandible. Unfortunately, however, he placed all the Sterkfontein specimens into the same genus and species, believing that the specimens with large canines and large cheek teeth were males, while those with small canines and small cheek teeth were females. Thus his writings (e.g., Broom and Schepers, 1946) referred to "an isolated female canine," "the female upper jaw," "the much crushed left maxilla of what I regard as an old male," "the type male skull," and so on.

And so the stage was set for the general acceptance of a single species at Sterkfontein. First, it was named *A. transvaalensis*, then *Plesianthropus* and finally Robinson (1954a) sunk *Plesianthropus transvaalensis* into *Australopithecus africanus* and accorded it only a subspecific differentiation from Taung. He included in the subspecies *A. a. transvaalensis* the Makapansgat hominid fossils that Dart (1948a,b) had, I believe justifiably, classified as a distinct species *Australopithecus prometheus* and the Tanzanian Garusi (Laetoli) maxilla, which is now placed in a distinct species, *Australopithecus afarensis* (Johansen et al. 1978).

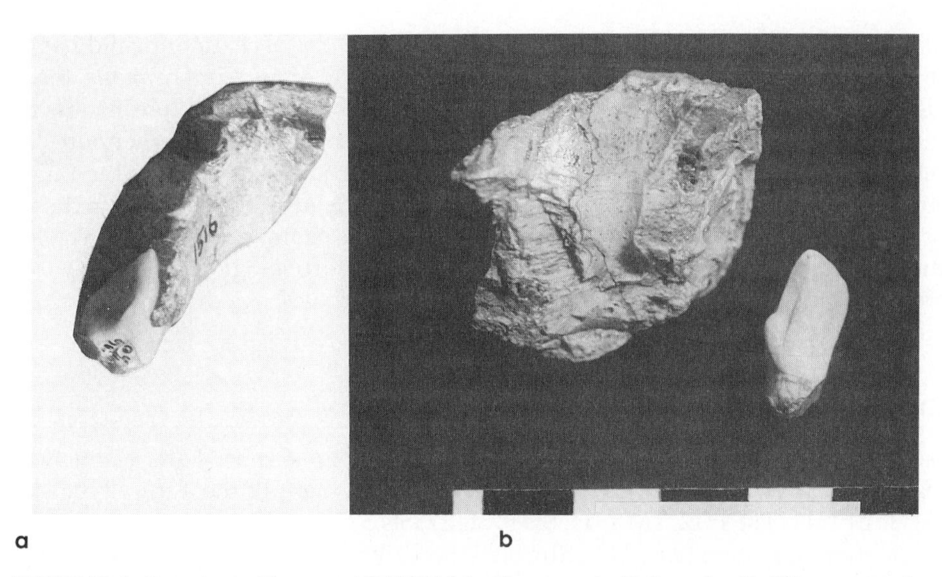

a b

FIGURE 10-3 Symphyseal fragment TM 1516 (a) with unerupted left canine Sts 50 in superior view (left) and lingual view (centre). Note deep genioglossal fossa. The canine is shown in lingual view at right (b).

The concept of the one species *A. africanus* embracing all ape-man specimens from Taung, Sterkfontein, and Makapansgat led to the species being characterized as having such a wide range of individual variation that some researchers concentrating on the smaller toothed specimens regarded it as ancestral only to *Homo* (e.g., Robinson 1967, 1972; Olson 1985), while those concentrating on the larger toothed individuals regarded it as ancestral only to *Paranthropus* (Johanson et al. 1981; Rak 1983). Those who took into account the wide range of variation saw it as ancestral to both *Homo* and *Paranthropus* (Tobias 1980; Skelton et al. 1986).

Rak (1983) produced a comprehensive analysis of facial structure in East African and South African australopithecines, providing therein a valuable work of reference to all concerned with understanding craniofacial morphology of early hominids. He accepted the Sterkfontein Member 4 and Makapansgat hominids as representing the one species *A. africanus* and concluded that "almost every aspect of the *A. africanus* face indicates that the species as a whole had already embarked on the evolutionary course leading to the robust australopithecines."

The concept of the single species was questioned by Clarke (1985a,b; 1988a,b), who suggested that there were in fact two species represented in the Member 4 breccia at Sterkfontein and in the Member 3 breccia at Makapansgat. He considered (1988a,b) that the small-toothed individuals belonged to the same species as the Taung child, (i.e., *A. africanus*) and that this species was ancestral to *Homo*. The larger toothed individuals, he said, belonged to another species that was ancestral to *Paranthropus*. This second species is discussed in the following section.

If Clarke is correct in his view about two species of australopithecine at Sterkfontein, then the question arises as to whether the infant Taung skull would have grown to resemble the smaller toothed or the larger toothed variety. That has been answered by Broom (1938b) who noted the distinction between the Taung mandible and the TM 1516 mandible with its large canine. Robinson (1954a) demonstrated that structurally the first lower deciduous molar of the Taung child resembles one from Sterkfontein in having a very small hypoconulid and a small entoconid well separated from the sharp protoconid. Both differ from *Paranthropus* of Kromdraai and Swartkrans, which has a molariform Lower dm1, while the *Australopithecus* tooth resembles that of *Homo*. Also Clarke (1977) demonstrated that the morphology of the Taung face and frontal bone was that of an immature equivalent of Sts 5 (i.e., the small toothed *Australopithecus*) and not that of *Paranthropus*, which he demonstrated (Clarke 1988a,b) to have frontal and facial similarities to the large toothed Sterkfontein *Australopithecus*. The Taung child has the anteriorly prominent glabella, the division between arcus supraciliaris and arcus supraorbitalis, the rising convex frontal squame, and the step forward from malar to lateral nasal margin that foreshadow the features of Sts 5 and not the large-toothed species represented by Sts 71 and Stw 252. Furthermore, Clarke (1990a) has shown that an infant *Paranthropus* calotte (SK 54) from Swartkrans differs from the Taung child in the frontal characters previously listed and instead foreshadows the adult *Paranthropus* in those characters.

It was shown graphically by Dart (1948c) that infant gorilla, chimp, and orangutan of the same dental age as Taung are clearly distinguishable on facial and frontal morphology and that they foreshadow the adult. There is thus no reason to believe that the Taung child would have dramatically altered its frontal and facial morphology in adulthood.

A SECOND AUSTRALOPITHECINE SPECIES AT STERKFONTEIN

Although some of the hominid cranial remains from Sterkfontein Member 4 are relatively small toothed and have facial and frontal features that are in keeping with those of the Taung type specimen of *A. africanus* (e.g., TM 1511, TM 1512, TM 1514, Sts 5, Sts 17, Sts 52 a and b), others are large toothed and do not fit the facial-frontal pattern (e.g., TM 1516, Sts 1, Sts 7, Sts 28, Sts 71). Clarke (1985a) suggested there might be two species that had been grouped as one in the Sterkfontein Member 4 breccia. After he had reconstructed a fragmented cranium (StW 252) from Sterkfontein Member 4, Clarke (1988a,b) was convinced that there were indeed two species and he presented morphological data to demonstrate this. StW 252 has enormous canines and cheek teeth that are even larger than many *Paranthropus* specimens from both East and South Africa, yet it has a thin, flattened supraorbital margin with the temporal lines curving toward the sagittal midline just behind the supraorbital margin. Nasion is situated above the fronto-maxillary suture, close to the glabella, which is not prominent. The frontal squame behind the

FIGURE 10-4 Sts 71 in right lateral view.

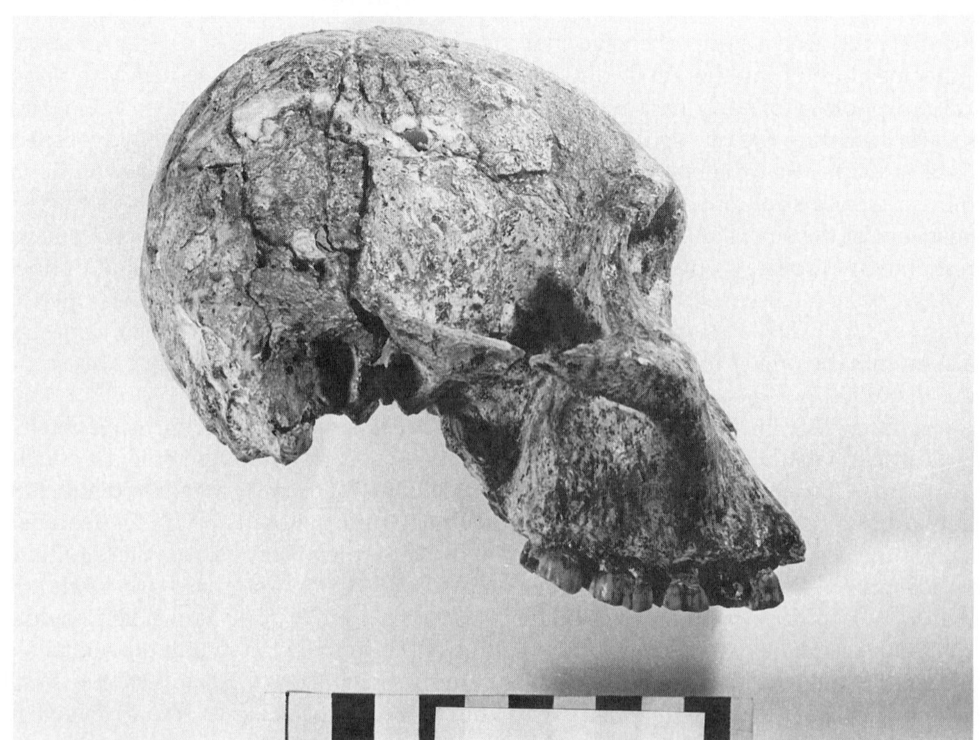

glabella is slightly concave and the occipital profile is high and flattened. In these cranial characters, StW 252 differs markedly from Sts 5 and Taung but resembles Sts 71. This latter cranium also has the cheek bones preserved and they are anteriorly situated such that in lateral profile the nasal region is hidden posterior to the cheek region (Figure 10-4). This character, together with the thin supraorbital margin, the very slight hollowing of frontal squame just behind glabella, and the large cheek teeth, is a complex like that of *Paranthropus* and unlike *A. africanus*, which has a malar set posterior to the lateral nasal margin, a thick supraorbital margin, a convex frontal squame and small cheek teeth. However, the enormous canines and large anteriorly projecting incisors with diastema between canine and lateral incisor are distinctly unlike those of *Paranthropus* but more reminiscent of pongids. Such a combination of large cheek teeth and pongidlike anterior dentition should not be unexpected in an ancestor to *Paranthropus*. If it is accepted that all hominids originally had a pongid ancestor and that *Paranthropus* specialized by developing enlarged cheek teeth, then it would be quite in order for the immediate ancestor of *Paranthropus* to have retained the pongidlike anterior dentition until it was no longer needed and became a disadvantage. Thus the ancestor of *Paranthropus* would have first developed a dietary specialization (grinding of hard foods such as dry berries and/or seeds) that ideally required large flat occlusal surfaces that formed the equivalent of upper and lower grindstones. Thus large cheek teeth would have been the first adaptation to such a dietary specialization. Only when the large canines and incisors were not only redundant but actually interfered with the side-to-side grinding action of the cheek teeth would it have been advantageous to have smaller incisors and canines (as in *Paranthropus*). Thus the dental complex seen in StW 252 does represent the expected morphology of a *Paranthropus* ancestor.

The very large lower canine Sts 50 that belongs with the adolescent mandibular symphysis TM 1516 is of the proportion that would fit with the very large upper canine and diastema of StW 252 (Figure 10-5).

Other specimens that I believe belong to this second species have been excavated out of Sterkfontein Member 4 in recent years. One such specimen is the cranium Stw 505 illustrated by Tobias (1992). Another

Figure 10-5 Cast of reconstructed StW 252 cranium with Sts 50 lower canine held in position.

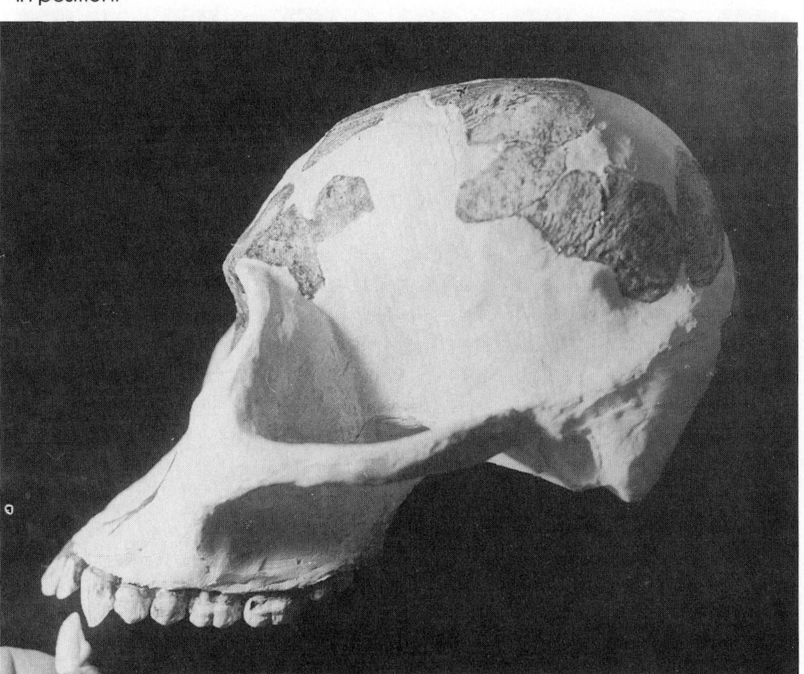

is the partial mandibular corpus Stw 384 that has three molars that are larger than those of the massive Peninj mandible. Its lower P4 approaches that of Peninj as does an isolated lower P3 (Stw 401) that fits well with this mandible, although it is not the same individual (Figure 10-6). This figure also shows an *A. africanus* mandible (Stw 404) from member 4 that in tooth size is more comparable to *H. habilis*.

It is thus clearly seen that the difference in cheek tooth size between the two Sterkfontein Member 4 hominids is as great as that between *H. habilis* and a large toothed *Paranthropus boisei*! When this size difference is considered together with the aforementioned cranial differences between Sts 5 and Stw 252, which exceed the differences between Sts 5 and *H. habilis*, then I see no logic in lumping all Sterkfontein Member 4 hominids into *A. africanus*.

Other Sterkfontein specimens that belong to this second, large-toothed species include Sts 36 and Stw 14. From Makapansgat there are also specimens of the large-toothed hominid such as the adolescent mandible MLD 2, the elderly maxilla MLD 9, and the two mandibular symphyses MLD 27 and MLD 29, both of which display deep genioglossal fossae and buccal roots of P3 lateral to the canine.

How should this second species be classified? Broom (1938a,b), on the strength of the morphology of the mandibular symphysis TM 1516 with canine Sts 50, transferred all the Sterkfontein hominids from *A. transvaalensis* to a new genus *Plesianthropus*. The type specimen of *A. transvaalensis* (the cranium TM 1511) was thereafter regarded as the type of *Plesianthropus*. As that specimen is actually an *A. africanus*, then symphysis TM 1516, which Broom and Schepers (1946) designated "the topotype," could still be regarded as belonging to the distinct genus *Plesianthropus* (which would then include all the large-toothed hominids from Member 4, if they do indeed merit generic distinction from *Australopithecus*).

Dart (1948d) classified the large-toothed MLD 2 mandible from Makapansgat as belonging to the species *Australopithecus prometheus*, which he had created to accommodate the MLD 1 occipito parietal fossil (Dart 1948a,b). I believe that MLD 1 almost certainly belongs to the second species, rather than to *A.*

Figure 10-6 From left to right. Cast of Peninj *Paranthropus* mandible, right mandible of StW 384 with StW 401 third premolar, Stw 404 right mandible, cast of OH 7 *Homo habilis* mandible.

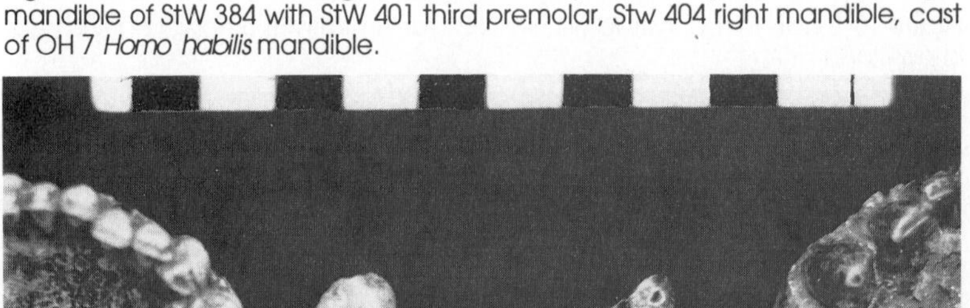

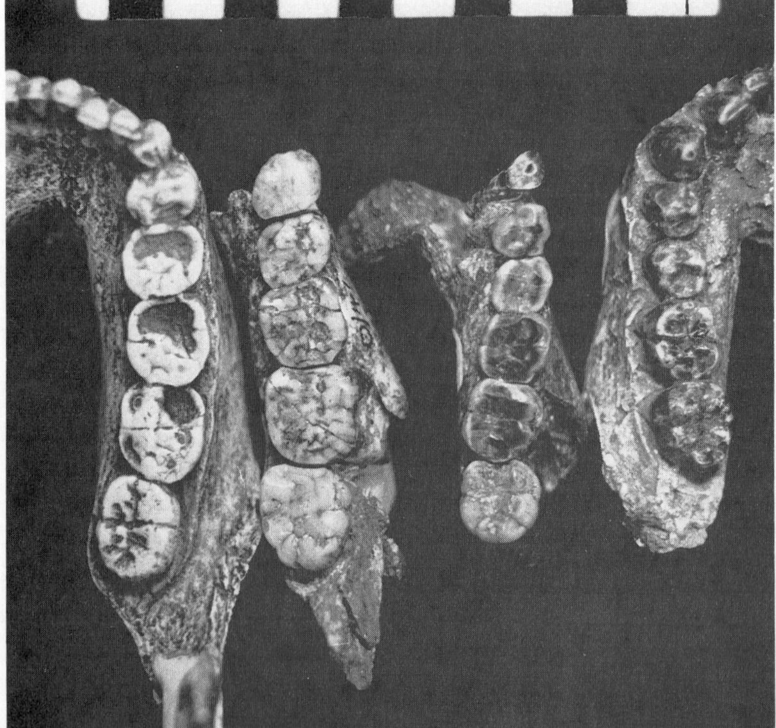

africanus, but as it is difficult to demonstrate affinity on such a fragment it is open to debate as to whether *prometheus* is a name that could be applied to MLD 2 and other specimens of the second species.

A third option is that the large-toothed Sterkfontein species should be classified as *Paranthropus* in view of the *Paranthropus* characteristics that are already well established in the species. These *Paranthropus* characters are discussed in the following section. It has indeed already been suggested by Aguirre (1970) that the MLD 2 mandible from Makapansgat belongs to the genus *Paranthropus*.

PARANTHROPUS

The genus *Paranthropus* was named by Broom on 27 August 1938 before he named *Plesianthropus* on 19 November 1938. Hence even if the name *Plesianthropus* was used to differentiate TM 1516 from *Australopithecus*, the name *Paranthropus* has priority, and I believe that the large-toothed Sterkfontein and Makapansgat hominids should be considered as belonging to an early species of that genus. This genus was created by Broom (1938a) to accommodate a craniofacial specimen and mandible (TM 1517) from Kromdraai. He noted that the face was flat, the incisors and canines were small, the molars differed in shape from those of *Plesianthropus* (i.e., *A. africanus*) and that the upper P4 was larger than that of the Sterkfontein *Australopithecus*. He concluded that "we may thus confidently place the new skull in a new genus and species" and named it *Paranthropus robustus*, meaning a strongly and stoutly built form of hominid to one side of the main line of human development. Later, Broom (1939) noted that if a ruler is placed across the cheeks of *Paranthropus*, the rest of the face is situated posterior to the ruler. He stated that "there is a remarkable degree of flattening of the lower part of the face above the incisors and canines. It was mainly on this very peculiar condition of the face that I made the Kromdraai skull the type of a new genus (*Paranthropus*)."

Gregory and Hellman (1939) agreed that there were conspicuous differences between the Sterkfontein *Australopithecus* (*Plesianthropus*) represented by TM 1511 and *Paranthropus* and noted "especially the extreme flattening of the subnasal plate in *Paranthropus*, the marked development of the crista obliqua and relatively small hypocone crest in its upper M3, its elongate-oval lower M3, the orang-like appearance of its upper premolars, its probably larger braincast."

From 1948 onward, Broom was to recover many more *Paranthropus* fossils from Swartkrans. These he placed into a new species, *Paranthropus crassidens*.

Simpson (1945) had believed *Paranthropus* to be at the most only subgenerically distinct from *Australopithecus*. Despite the new specimens from Swartkrans, Simpson's opinion was followed by Oakley (1954), Clark Howell (1955, 1968), L. S. B. Leakey et al. (1964) and Simons (1967). The generic status of *Paranthropus* was questioned by Dart (1948b,d). Washburn and Patterson (1951) recognized only the two hominid genera, *Australopithecus* and *Homo*, and were followed in this by Le Gros Clark (1955), Dobzhansky (1962), Campbell (1963), Pilbeam and Simons (1965), Tobias (1967), Wallace (1972), Brace (1973), von Koenigswald (1973), Wolpoff (1974) and Wolpoff and Lovejoy (1975).

Mayr (1950) went so far as to classify both *Australopithecus* and *Paranthropus* into the genus *Homo*, but he later (1963) accepted that generic separation was justified for *Australopithecus* and said that it was largely a matter of taste as to whether one admitted another genus, *Paranthropus*. In this context it should be noted that Dobzhansky (1962) stated, "The generic category of classification is biologically arbitrary."

The one person apart from Broom who had worked most closely with the *Paranthropus* fossils was the zoologist and paleontologist John Robinson, and he steadfastly maintained the generic distinction of *Paranthropus* (Robinson 1952, 1954a,b, 1956, 1962a, 1963, 1967, 1972) and gave strong reasons for so doing. He observed (1962a) that "the characters which distinguish *Australopithecus* and *Paranthropus* are legion since the two can be distinguished by means of almost any bits of skeleton now known in both forms." Robinson (1962a) attributed the unique architecture of the *Paranthropus* skull to specialization of the dentition and suggested that this could be due only to dietary specialization.

Clarke (1977), after a detailed study of the craniofacial anatomy of *Australopithecus, Paranthropus,* and early *Homo* remains, agreed with Robinson and has never wavered from his conviction that *Australopithecus* and *Paranthropus* should be considered as distinct genera.

It is noteworthy that of all those previously listed authors who did not recognize the generic distinction of *Paranthropus*, only two actually analyzed Broom's and Robinson's criteria for the genus. Thus Tobias (1967) agreed that the architecture of the *Paranthropus* skull is strongly related to dental specializations, but he did not consider that this warranted generic separation. Wolpoff (1974) also agreed that there are distinctions between *Australopithecus* and *Paranthropus* in cranial capacity, lower dm1 morphology, development of zygomatic arch, size of temporal fossa, size of lateral pterygoid plates, and development of a supraglenoid gutter. However, he also did not consider that this warranted generic separation.

It is astonishing that those who accept a generic separation between *Homo* and *Australopithecus* are not prepared to recognize generic separation of *Paranthropus* and *Australopithecus* when the dental and cranial distinctions between the latter two are greater than the dental and cranial distinctions between the former two, especially between *H. habilis* and *A. africanus* (see Clarke 1985a: Figures 1 and 2).

Clarke (1977, 1985a) listed the following mainly apomorphous characters that distinguish the *Paranthropus* cranium from that of other genera within the family Hominidae:

1. A brain that is on the average larger than that of *Australopithecus*, yet not as large as that of *Homo*.
2. Formation of a slightly concave, low forehead with a frontal trigone delimited laterally by posteriorly converging temporal crests.
3. Presence of a flattened "rib" of bone across each supraorbital margin.
4. A glabella that is situated at a lower level than the supraorbital margin.
5. Formation of a central facial hollow associated with a completely flat nasal skeleton, and a cheek region that is situated anterior to the plane of the piriform aperture.
6. Nasoalveolar clivus sloping smoothly into the floor of the nasal cavity.
7. Small incisive canals that open into the horizontal surface of the nasal floor.
8. Great enlargement of premolars relative to the molars and canines.
9. Great enlargement of molars and massiveness of tooth-bearing bone.
10. Anterior teeth small when compared to premolars and molars.
11. A tendency for the maxillary canine and incisor sockets to be situated in an almost straight line across the front of the palate.
12. Formation on the nasoalveolar clivus of prominent ridges marking the central incisor sockets but concavities marking the lateral incisor sockets.
13. Cusps of cheek teeth low and bulbous and situated closer to the center of the crown than in other hominid genera.
14. Formation of flat occlusal wear surfaces to the cheek teeth, accompanied by smoothly rounded borders between the occlusal surfaces and the sides of the crowns of the cheek teeth.
15. Virtually completely molarized lower dm1 with anterior fovea centrally situated and with complete margin.
16. Great increase in the size of the masticatory musculature and attachment relative to the size of the skull.
17. Temporal fossa capacious and mediolaterally expanded.
18. Formation of a broad gutter on the superior surface of the posterior root of the zygoma.
19. A tendency for the palate to be shallow anteriorly and deep posteriorly.
20. Formation of either a marked pit or a groove across the zygomaticomaxillary suture of the cheek region—at least in the South African *Paranthropus*.

It is interesting that Gregory and Hellman (1939) should have noted that the Kromdraai *Paranthropus* had oranglike premolars because *Paranthropus* has another oranglike feature of which they were unaware as the supraorbital margin was missing from the Kromdraai specimen. Clarke (1977) observed that in the morphology of the supraorbital margin *Paranthropus* did not have a supraorbital torus as defined by Schwalbe (1906). Schwalbe had given the name "torus supraorbitalis" to a single, rounded, thickened, and prominent ridge formed through fusion of the arcus superciliaris and arcus supraorbitalis. Although such a torus occurs in the gorilla, chimpanzee, and many hominid species, it does not occur in the orangutan or in *Paranthropus*. Cunningham (1909) stated that in the orang, a superciliary element (i.e., over the medial part of the orbital margin) is absent and that the supraorbital elements gradually develop in ontogeny into a projecting rim for the superior margin of the orbit. Keith (1929) also commented on the absence of a torus in the orang and figured a superimposed drawing of a chimpanzee and an orang to show this marked morphological difference between them in their frontal bones.

Clarke (1977) stated that *Paranthropus* has a similar supraorbital morphology to that of the orang in that over each orbit there is a flattened bar of bone resembling a rib. For this reason he proposed to name the bar of bone over each orbit the costa supraorbitalis or supraorbital rib.

Thus *Paranthropus*, like the orang, can be distinguished from *Australopithecus*, chimpanzee, gorilla, and *Homo* by the lack of a supraorbital torus. The prominence of the supraorbital margin in *Paranthropus* and the orang is due to the extreme postorbital constriction that leaves the supraorbital margins as noticeable but thin and flattened ribs of bone on either side of the glabella. These ribs are given even more emphasis in *Paranthropus* because of the supraglabella hollowing of the frontal squame between the medially encroaching temporal lines. The ontogenetic beginnings of this supraorbital structure can be seen in the infant *Paranthropus* frontal bone of SK 54 (Clarke, 1990a) and the phylogenetic beginnings of the structure can be seen in the *Paranthropus* ancestor as represented by Stw 252, Sts 71 and Stw 505, discussed in the previous section.

In just its supraorbital structure and large premolars, *Paranthropus* is as distinct from *Australopithecus* as the orangutan is from the chimpanzee, and nobody would argue that those two apes belong to one genus. When all the other marked cranial differences are taken into account, it is quite clear that *Paranthropus* represents a highly specialized primate that had been separate from the *Australopithecus* lineage for a very long time. *A. africanus*, however, is morphologically so close to *H. habilis* that it would seem correct to regard it as a direct ancestor of *H. habilis*. Robinson (1972) even went so far as to classify *A. africanus* within the genus *Homo*—a move that is more justifiable than the sinking of *Paranthropus* into *Australopithecus*.

HOMO HABILIS

Robinson (1957, 1958, 1962a) reported on an excavation he had conducted in the red-brown breccia of the West Pit at Sterkfontein. This breccia is now classified as Member 5 (Partridge 1978) and is known to contain quantities of early Acheulean artifacts as well as a few *H. habilis* remains (Mason 1957, 1962; Clarke 1985b, 1988a). In his 1957-1958 excavation, Robinson uncovered 286 artifacts and foreign stones associated with five hominid dental specimens that he believed to be those of *A. africanus*. These teeth are a left upper dm2 and upper M1 in maxilla fragment (Se 255), an upper right M2 (Se 1508), a buccal portion of an upper right M2 (Se 1579), and a left lower canine (Se 1937). These teeth are illustrated in Figure 10-7.

Robinson (1957, 1958, 1961, 1962b) discussed the question of who could have made the artifacts and rejected the idea that *Australopithecus* could have been responsible. *Australopithecus* remains were, he noted, common in the Sterkfontein Type Site breccia with not a single artifact present. Remains of the tool manufacturers are rare, he said, at early Stone Age sites. He therefore contended that the presence of what he believed were *Australopithecus* teeth in association with the West Pit artifacts did not mean that *Australopithecus* made the tools. Robinson suggested that *Telanthropus* (early *Homo*) was responsible for making the artifacts that were present in the red-brown breccia, and he was supported in this view by Mason (1961, 1962).

We now know that *Australopithecus* at Sterkfontein is not associated with a single artifact but that the artifact bearing Member 5 breccia of Sterkfontein yielded the skull of *H. habilis*, Stw 53, and a few other specimens (Hughes and Tobias 1977; Tobias 1978; Clarke 1985b). Thus the teeth that Robinson excavated are not those of *Australopithecus*, as he assumed, but are those of early *Homo* (Tobias 1978). The association of *Homo habilis* with Early Acheulean in Sterkfontein Member 5 was discussed by Clarke (1985b, 1988a). However, recent excavation has indicated the probability that there is a previously undetected temporal separation between the *Homo habilis* cranium StW 53 and the Early Acheulean artifacts. Unfortunately, the few isolated teeth and jaw fragments recovered from Member 5 can at present be assigned only to the genus *Homo* as they have not yet been studied and do not have obvious diagnostic characters that would differentiate between *H. habilis* and *H. erectus*. If some, however, prove to be jaws and teeth of *Homo habilis*, then John Robinson must be credited with uncovering the first specimens of that species two years before the first East Aftican specimen (Olduvai hominid 4) of *H. habilis* was found and seven years before the name and species descriptions were published by Leakey *et al.* (1964).

EARLY AFRICAN *HOMO* SPECIES

At the *Paranthropus* site of Swartkrans on 29 April 1949, John Robinson discovered a mandible (SK 15) that he recognized as being more like *Homo* than like *Paranthropus*. Broom and Robinson (1949) named it *Telanthropus capensis*.

In September of that year they found a slender mandibular fragment SK 45, which they described (Broom and Robinson 1950) as being apparently "the jaw of an early type of man" (i.e., hominine rather than australopithecine). Also in that month a maxilla fragment (SK 80) was discovered which Robinson (1953) assigned to *T. capensis* and much of the rest of the same cranium (SK 847), which Robinson did not recognize as being part of SK 80 and instead classified as a *Paranthropus*. It was not until 1969 that the real identity of SK 847 was recognized and Clarke et al. (1970) described it as the cranium of an early member of the genus *Homo*. Its diagnostic characters were the prominent nasal skeleton, the presence of a supraorbital torus, a steeply rising frontal squame and a low cranial base. Subsequently, Clarke (1977) produced a detailed anatomical study of the cranium and concluded that it should be classified as *Homo* species indet. A more complete cranium of the same kind of hominid, KNM ER 3733, was discovered in 1976 from the same geological horizon as *Paranthropus* at East Lake Turkana in Kenya and published as an early *H. erectus* (R. Leakey and Walker 1976). This cranium is virtually identical to SK 847 (Figure 10-8). Clarke (1990b) has suggested that the African crania assigned to *H. erectus* are directly ancestral to forms of archaic *H. sapiens* as represented by the Broken Hill and Ndutu crania, that they have a more sapient morphology than do the crania of Asian *H. erectus*, and that they should not be assigned to that species. He suggested that the massive-browed Olduvai Hominid 9 "*H. erectus*" calvaria is a male of the species to which SK 847 and KNM ER 3733 are females and that they correspond in ancestral terms to the male archaic *H. sapiens* represented by the Broken Hill cranium and the female archaic *H. sapiens* represented by the Ndutu cranium (Figure 10-9). If this interpretation is correct and the African *H. erectus* does not belong to that species, then it should be assigned to the species *H. leakeyi*, the name given by Heberer (1963) to the Olduvai H. 9 calvaria.

FIGURE 10-7 Sterkfontein Member 5 teeth of early *Homo* found by Robinson in 1957-1958. Top: Se 255, left dm2 and M1. Bottom, left to right: Se 1937, left lower canine, Se 2396 lingual half of P3, Se 1579 buccal portion of right M2, and Se 1508 right M2.

FIGURE 10-8 Casts of KNM ER 3733 (left) and SK 847.

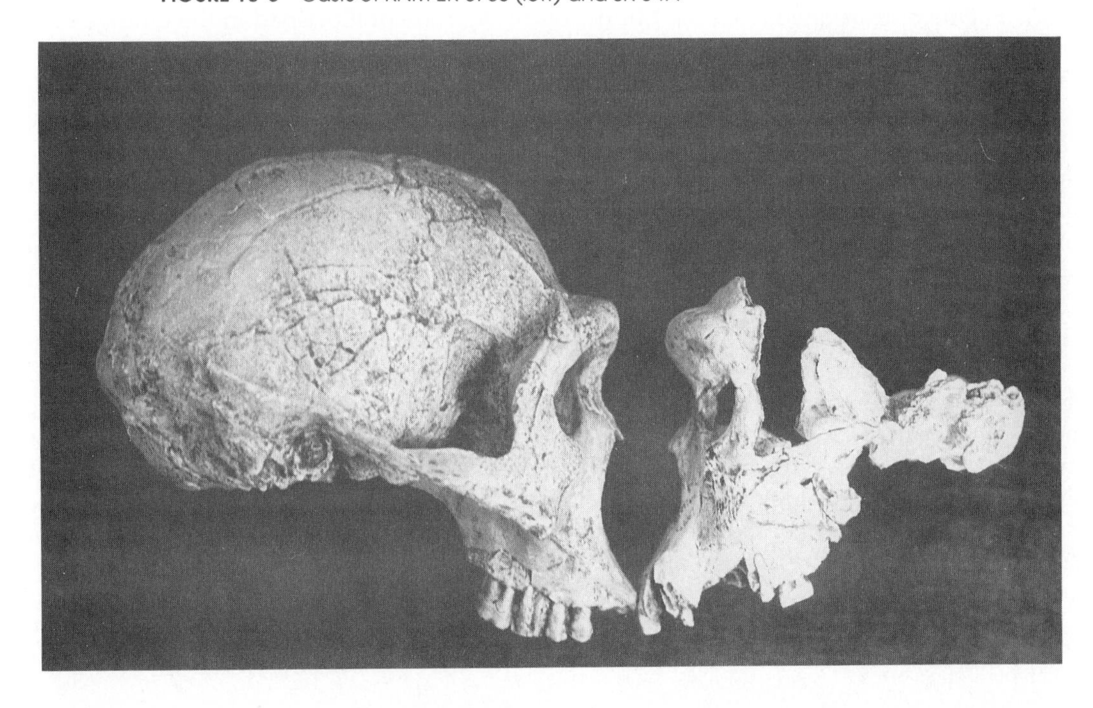

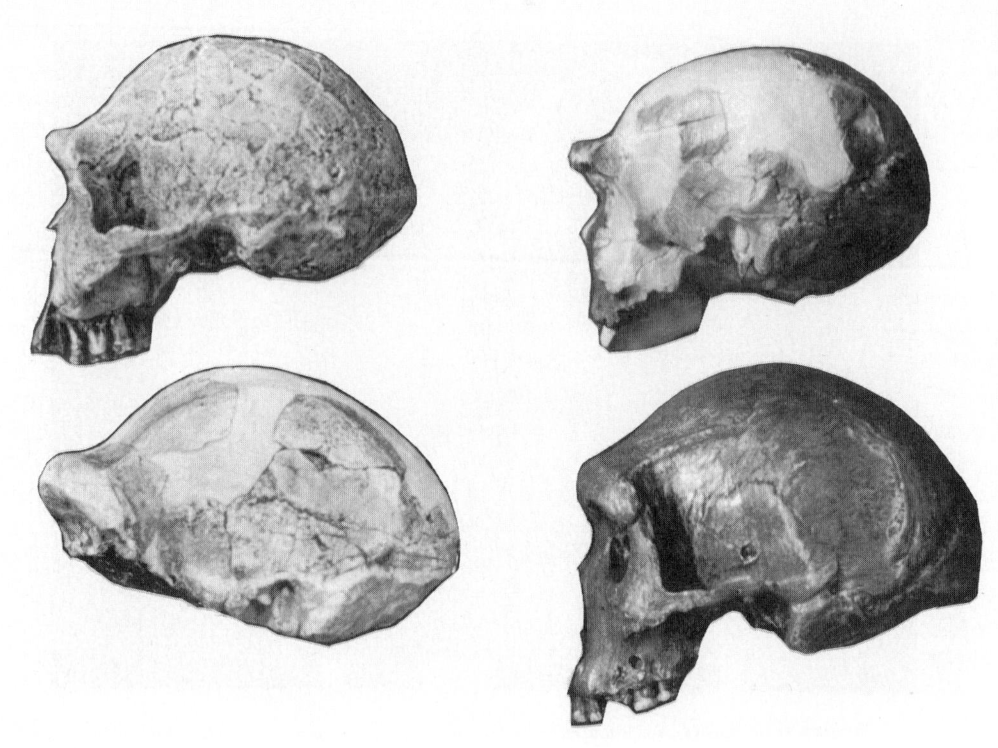

FIGURE 10-9 Casts of *Homo leakeyi*, female, KNM ER 3733 (top left), *Homo leakeyi*, male, OH 9 (bottom left), archaic *Homo sapiens* female, Ndutu (top right) and archaic *Homo sapiens* male, Broken Hill (bottom right).

EARLY *HOMO SAPIENS*

In 1932, T. F. Dreyer excavated from the Middle Stone Age site of Florisbad in the Orange Free State a craniofacial fossil that became known as the Florisbad man. Errors in his reconstruction gave the thick-skulled, low-browed hominid a rather small, modern face. A new reconstruction by Clarke (1985c) resulted in a somewhat larger and less modern-looking face. Morphological and temporal equivalents to this fossil are represented in East Africa by the Omo II calvaria (R. Leakey et al. 1969) and the Laetoli hominid 18 cranium (Magori and Day 1983). All three of these fossils have the low, receding frontal squame and rather thickened supraorbital margin that imparts to them an archaic appearance. Yet this kind of frontal bone morphology does sometimes occur in modern humans (Figure 10-10) among populations with crania that have a more vertical forehead and less thickened supraorbital margins.

Thus the archaic-looking frontal morphology of the Florisbad, Omo, and Ngaloba specimens does not in itself indicate archaic *H. sapiens* status. It must be considered together with the rest of the cranial morphology, which, taking into account the large face of Florisbad and the strongly angled occipital profile of Omo II, does speak of a form of late archaic *H. sapiens* in which archaic traits were more common than they are in more modern populations (Figure 10-10). Having said that, it should also not be thought that the kind of archaic morphology seen in the Florisbad, Ngaloba, and Omo II fossils is necessarily representative of all members of the species during that period of around 130,000 years ago. It could be expected that crania with more modern morphology might have also been present in populations of that time.

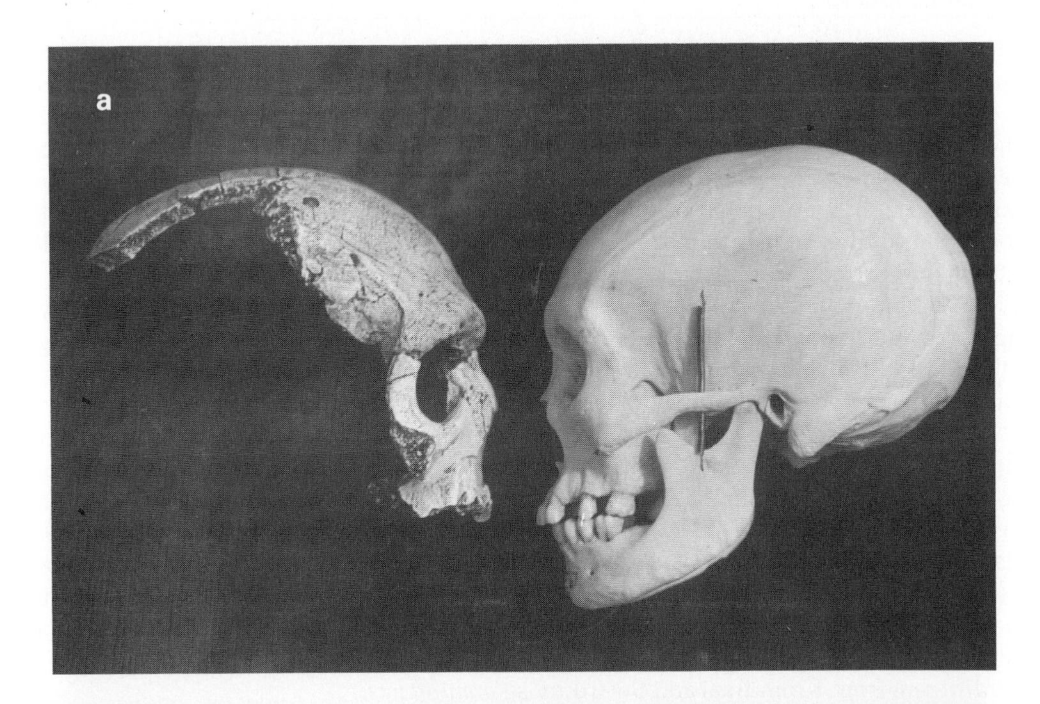

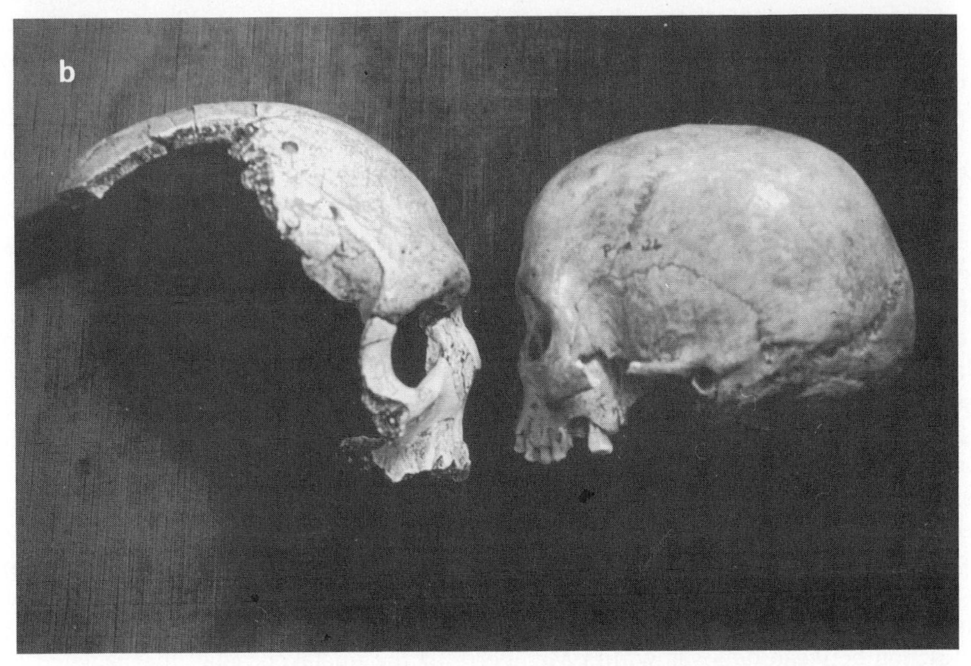

FIGURE 10-10 Top: cast of Florisbad cranium (left) compared to South African Negro cranium with unusual, sloping frontal. Bottom: cast of Florisbad cranium (left) compared to Bushman cranium.

CONCLUDING REMARKS

The task of reconstructing fossils has often been likened to a jigsaw puzzle without the complete picture as a guide. Such work is, however, comparatively easy when compared to the task of reconstructing taxonomic relationships from fragmentary fossils. This can be likened to a cryptic crossword puzzle with many of the clues missing. Those familiar with cryptic crossword puzzles will know that although there is only one correct answer to each clue, it is possible to misinterpret the clue and yet give an answer that fits the available space. It is only when one tries to mesh that answer with connecting clues that something is seen to be wrong and thus the incorrect interpretation of clues will mean that the whole puzzle cannot be satisfactorily and correctly completed.

So it is with filling in the gaps in hominid phylogeny in order to complete the evolutionary puzzle. Misreading of the fossil clues will give an answer but not necessarily the correct one and a wrong answer can make the whole puzzle more difficult to solve.

The account just given of discovery of various taxa has shown how the misreading of clues has happened frequently. The Piltdown forgery was accepted as genuine, thus giving a false view of the expected appearance of a human ancestor. Broom therefore accepted the large Piltdown-like canine of Sts 50 as being a normal variant within the *Plesianthropus transvaalensis* population at Sterkfontein and when this taxon was sunk into *A. africanus*, all the big-toothed specimens were included. When these big-toothed individuals, taken to be *A. africanus*, were compared with *Paranthropus*, many observers considered the differences not that great and accepted that *Paranthropus* should be sunk into *Australopithecus*. In hindsight, the reason for this is clear. These observers were unknowingly including early *Paranthropus* specimens from Sterkfontein in the sample of *A. africanus*, so naturally the early Sterkfontein *Paranthropus* would not appear that different from Kromdraai and Swartkrans *Paranthropus*.

The confusion was compounded when the SK 847 cranium found in 1949 was not recognized as *Homo* but was included in the Swartkrans *Paranthropus* sample, thus, for 20 years, giving the wrong impression that *Paranthropus* included individuals that were more like *Homo*.

Then in 1957, Robinson understandably assumed that the hominid teeth he found with Early Acheulean artifacts in the Sterkfontein Site belonged to *A. africanus*. This gave the false impression that *A. africanus* at least lived at the same time as an Early Acheulean tool maker even if he was not that tool maker.

It is only in hindsight and after the discovery of *H. habilis* cranial remains in East Africa as well as in the Member 5 breccia of Sterkfontein that we are able to assign correctly Robinson's teeth to the genus *Homo*, although we still cannot be sure if they are of *H. habilis* or a later species of early *Homo*.

This misreading of clues and misidentification of fossils should be adequate demonstration of the need, as Clarke (1985a) said, to rid our thoughts and our writing of the misguided concept of "evidence." This misused word is frequently taken to mean proof when in fact it only means what is seen. Thus the evidence of a large-brained human ancestor with apelike jaw at Piltdown was a forgery. The evidence of great dental size and craniofacial variation within *A. africanus* was due to two species being grouped as one, which led to the erroneous evidence of similarity between *Paranthropus* and *A. africanus*. The evidence of stone tools in association with *Australopithecus* teeth was due to erroneous taxonomic assignment of the *Homo* teeth. The evidence of great craniofacial variation within the Swartkrans *Paranthropus* sample was due to the misidentification and misclassification of SK 847, SK 27, SK 2635 *Homo* specimens as *Paranthropus*. The evidence of association between *H. habilis* and Early Acheulean may be due to there being an unclear distinction between deposits within Member 5 or the mistaken belief that isolated teeth and mandible fragments belong to *H. habilis* rather than to a later species of *Homo*. The evidence of a completely modern form of small face in the Florisbad *H. sapiens* cranium was due to faulty reconstruction.

Those last two words, "faulty reconstruction," are the key to the catalogue of errors I have just listed. Whether it be faulty reconstruction of fossils, taxonomic groups, association, stratigraphic relationships or phylogenies, it should certainly give us all pause for thought in the ever-continuing arguments about hominid evolution.

LITERATURE CITED

Aguirre, E. (1970) Identificacio de "Paranthropus" en Makapansgat. Cronica del XI Congreso Nacional de Arqueologia, Merida, 1969, pp. 98-124.

Brace, C. L. (1973) Sexual dimorphism in human evolution. *Yearb. Phys. Anthropol.* 16:31-49.

Broom, R. (1925a) On the newly discovered South African man-ape. *Nat. Hist.* 25:409-418.

Broom, R. (1925b) Some notes on the Taungs skull. *Nature* 115:569-571.

Broom, R. (1938a) The Pleistocene anthropoid apes of South Africa. *Nature* 142:377-379.

Broom, R. (1938b) Further evidence on the structure of the South African Pleistocene anthropoids. *Nature* 142:897-899.

Broom, R. (1939) A restoration of the Kromdraai skull. *Ann. Transv. Mus.* 19:327-329.

Broom, R., and Robinson, J. T. (1949) A new type of fossil man. *Nature* 164:322-323.

Broom, R., and Robinson, J. T. (1950) Man contemporaneous with the Swartkrans ape-man. *Am. J. Phys. Anthropol.* 8:151-155.

Broom, R., and Schepers, G. W. H. (1946) The South African fossil ape-men, the Australopithecinae. Transv. Mus. Mem. 2.

Burnet, J. (1774) *Origin and Progress of Language.*

Campbell, B. G. (1963) Quantitative taxonomy and human evolution. In S. L. Washburn (ed.): *Classification and Human Evolution.* Chicago: Aldine, pp. 50-74.

Clark, W. E. Le Gros (1955) *The Fossil Evidence for Human Evolution.* Chicago: Chicago University Press.

Clarke, R. J. (1977) The Cranium of the Swartkrans Hominid, SK 847 and Its Relevance to Human Origins. Unpublished Ph.D. thesis, University of the Witwatersrand (Johannesburg).

Clarke, R. J. (1985a) *Australopithecus* and early *Homo* in southern Africa. In E. Delson (ed.): *Ancestors: The Hard Evidence.* New York: Alan R. Liss, pp. 171-177.

Clarke, R. J. (1985b) Early Acheulean with *Homo habilis* at Sterkfontein. In P. V. Tobias (ed.): *Hominid Evolution: Past, Present and Future.* New York: Alan R. Liss, pp. 287-298.

Clarke, R. J. (1985c) A new reconstruction of the Florisbad cranium, with notes on the site. In E. Delson (ed.): *Ancestors: The Hard Evidence.* New York: Alan R. Liss, pp. 301-305.

Clarke, R. J. (1988a) Habiline handaxes and Paranthropine pedigree at Sterkfontein. *World Archaeol.* 20:1-12.

Clarke, R. J. (1988b) A new *Australopithecus* cranium from Sterkfontein and its bearing on the ancestry of *Paranthropus.* In F. E. Grine (ed.): *Evolutionary History of the "Robust" Australopithecines.* New York: Aldine de Gruyter, pp. 285-292.

Clarke, R. J. (1990a) Observations on some restored hominid specimens in the Transvaal Museum, Pretoria. In G. H. Sperber (ed.): *From Apes to Angels.* New York: Wiley-Liss, pp. 135-151.

Clarke, R. J. (1990b) The Ndutu cranium and the origin of *Homo sapiens. J. Hum. Evol.* 19:699-736.

Clarke, R. J., Howell, F. C., and Brain, C. K. (1970) More evidence of an advanced hominid at Swartkrans. *Nature* 225:1219-1222.

Cunningham, D. J. (1909) The evolution of the eyebrow region of the forehead, with special reference to the excessive supra-orbital development in the Neanderthal race. *Trans. Royal Soc. Edinburgh* 46:283-311.

Dart, R. A. (1925) *Australopithecus africanus*: The man-ape of South Africa. *Nature* 115:195-199.

Dart, R. A. (1948a) An *Australopithecus* from Central Transvaal. *S. Afr. J. Sci.* 1:200-201.

Dart, R. A. (1948b) The Makapansgat proto-human *Australopithecus prometheus. Am. J. Phys. Anthropol.* 6:259-284.

Dart, R. A. (1948c) The infancy of *Australopithecus.* In A. Du Toit (ed.): *Robert Broom Commemorative Volume.* Cape Town: Royal Society of South Africa, pp. 143-152.

Dart, R. A. (1948d) The adolescent mandible of *Australopithecus prometheus. Am. J. Phys. Anthropol.* 6:391-412.

Dobzhansky, T. (1962) *Mankind Evolving.* New Haven and London: Yale University.

Gregory, W. K., and Hellman, M. (1939) The dentition of the extinct South African man-ape *Australopithecus (Plesianthropus) transvaalensis* Broom. A comparative and phylogenetic study. *Ann. Transv. Mus.* 19:339-373.

Heberer, G. (1963) Uber einem neuen Archanthropinen Typus aus der Oldoway-Schlucht. *Z. Morph. Anthropol.* 53:171-177.

Howell, F. C. (1955) The age of the australopithecines of southern Africa. *Am. J. Phys. Anthropol.* 13:635-662.

Howell, F. C. (1968) Omo Research Expedition. *Nature* 219:567-572.

Hughes, A. R., and Tobias, P. V. (1977) A fossil skull probably of the genus *Homo* from Sterkfontein, Transvaal. *Nature* 265:310-312.

Huxley, T. H. (1863) *Man's Place in Nature and Other Essays.* London: J. M. Dent, Everyman's Library.

Johanson, D. C., White, T. D., and Coppens, Y. (1978) A new species of the genus *Australopithecus* (Primates: Hominidae) from the Pliocene of Eastern Africa. *Kirtlandia* 28:1-14.

Johanson, D. C., White, T. D., and Kimbel, W. H. (1981) *Australopithecus africanus*: Its phyletic position reconsidered. *S. Afr. J. Sci.* 77:445-470.

Keith, A. (1929) *The Antiquity of Man.* London: Williams and Norgate.

Leakey, L. S. B., Tobias, P. V., and Napier, J. R. (1964) A new species of the genus *Homo* from Olduvai Gorge. *Nature* 202:7-9.
Leakey, R. E. F., Butzer, K. W., and Day, M. H. (1969) Early *Homo sapiens* remains from the Omo River region of southwest Ethiopia. *Nature* 222:1132-1138.
Leakey, R. E. F., and Walker, A. C. (1976) *Australopithecus, Homo erectus* and the single species hypothesis. *Nature* 261:572-574.
Magori, C. C., and Day, M. H. (1983) Laetoli Hominid 18: An early *Homo sapiens* skull. *J. Hum. Evol.* 12:747-753.
Mason, R. J. (1957) Occurrence of stone artefacts with *Australopithecus* at Sterkfontein, Part 2. *Nature* 180:523-524.
Mason, R. J. (1961) The earliest tool-makers in South Africa. *S. Afr. J. Sci.* 57:13-16.
Mason, R. J. (1962) Australopithecines and artefacts at Sterkfontein, Part II, The Sterkfontein stone artefacts and their maker. *S. Afr. Archaeol. Bull.* 17:109-125.
Mayr, E. (1950) Taxonomic categories in fossil hominids. *Cold Spr. Harb. Symp. Quant. Biol.* 15:109-118.
Mayr, E. (1963) The taxonomic evaluation of fossil hominids. In S. L. Washburn (ed.): *Classification and Human Evolution*. Chicago: Aldine, pp. 332-346.
Oakley, K. P. (1954) Dating of the Australopithecinae of Africa. *Am. J. Phys. Anthropol.* 12:9-23.
Olson, T. R. (1985) Cranial morphology and systematics of the Hadar Formation hominids and *"Australopithecus africanus."* In E. Delson (ed.): *Ancestors: The Hard Evidence*. New York: Alan R. Liss, pp. 102-119.
Partridge, T. C. (1978) Re-appraisal of lithostratigraphy of Sterkfontein hominid site. *Nature* 275:282-287.
Pilbeam, D. R., and Simons, E. L. (1965) Some problems of hominid classification. *Am. Sci.* 53:237-259.
Rak, Y. (1983) *The Australopithecine Face*. New York: Academic.
Robinson, J. T. (1952) The Australopithecines and their evolutionary significance. *Proc. Linn. Soc. London* 1950-51, pp. 196-200.
Robinson, J. T. (1953) *Telanthropus* and its phylogenetic significance. *Am. J. Phys. Anthropol.* 11:445-501.
Robinson, J. T. (1954a) The genera and species of the Australopithecinae. *Am. J. Phys. Anthropol.* 12:181-200.
Robinson, J. T. (1954b) Phyletic lines in the Prehominids. *Z. Morph. Anthropol.* 46:269-273.
Robinson, J. T. (1956) The dentition of the australopithecinae. *Transv. Mus. Mem.* 9.
Robinson, J. T. (1957) The occurrence of stone artefacts with *Australopithecus* at Sterkfontein, Part 1. *Nature* 180:521-522.
Robinson, J. T. (1958) The Sterkfontein tool-maker. *The Leech* (University of the Witwatersrand, Johannesburg) 28:94-100.
Robinson, J. T. (1961) The australopithecines and their bearing on the origin of man and of stone tool making. *S. Afr. J. Sci.* 57:3-16.
Robinson, J. T. (1962a) The origin and adaptive radiation of the australopithecines. In G. Kurth (ed.): *Evolution und Hominisation*. Stuttgart: Gustav Fischer, pp. 120-140.
Robinson, J. T. (1962b) Australopithecines and artefacts at Sterkfontein, Part I, Sterkfontein Stratigraphy and the significance of the extension site. *S. Afr. Archaeol Bull.* 17:87-107.
Robinson, J. T. (1963) Australopithecines, culture and phylogeny. *Am. J. Phys. Anthropol.* 21:595-605.
Robinson, J. T. (1967) Variation and the taxonomy of the early hominids. In T. Dobzhansky, M. K. Hecht and W. C. Steere (eds.): *Evolutionary Biology*, I. New York: Appleton-Century-Crofts, pp. 69-100.
Robinson, J. T. (1972) *Early Hominid Posture and Locomotion*. Chicago: Chicago University Press.
Schwalbe, G. (1906) Das Schadelfragment von Brux und Verwandte Schadelformen. *Z. Morph. Anthropol.* Sonderheft. May, pp. 82-182.
Simons, E. L. (1967) The significance of primate palaeontology for anthropological studies. *Am. J. Phys. Anthropol.* 27:307-332.
Simpson, G. G. (1945) The principles of classification and a classification of mammals. *Bull. Am. Mus. Nat. Hist.* 85.
Skelton, R. R., McHenry, H. M., and Drawhorn, G. M. (1986) Phylogenetic analysis of early hominids. *Curr. Anthropol.* 27:21-43.
Tobias, P. V. (1967) *Olduvai Gorge*, Vol. 2: *The Cranium and Maxillary Dentition of* Australopithecus (Zinjanthropus) boisei. Cambridge: Cambridge University Press.
Tobias, P. V. (1978) The earliest Transvaal members of the genus *Homo* with another look at some problems of hominid taxonomy and systematics. *Z. Morph. Anthropol.* 69:225-265.
Tobias, P. V. (1980) *"Australopithecus afarensis"* and *A. africanus*: Critique and an alternative hypothesis. *Palaeontol. Africana* 23:1-17.
Tobias, P. V. (1992) New researches at Sterkfontein and Taung with a note on Piltdown and its relevance to the history of palaeo-anthropology. *Trans. Royal Soc. S. Afr.* 48, Part I:1-14.
von Koenigswald, G. H. R. (1973) *Australopithecus, Meganthropus,* and *Ramapithecus*. *J. Hum. Evol.* 2:487-491.
Wallace, J. A. (1972) The Dentition of the South African Early Hominids: A Study of Form and Function. Unpublished Ph.D. Thesis, University of the Witwatersrand, Johannesburg.
Washburn, S. L., and Patterson, B. (1951) Evolutionary importance of the South African "man-apes." *Nature* 167:650-651.
Wolpoff, M. H. (1974) The evidence for two Australopithecine lineages in South Africa. *Yearb. Phys. Anthropol.* 17:113-139.
Wolpoff, M. H., and Lovejoy, C. O. (1975) A re-diagnosis of the genus *Australopithecus*. *J. Hum. Evol..* 4:275-276.

CHAPTER 11

The Middle Ear of *Australopithecus robustus*

Does It Bear Evidence of a Specialized Masticatory System?

Yoel Rak

MASTICATORY SPECIALIZATION OF *AUSTRALOPITHECUS ROBUSTUS*

Ever since the discovery of *Australopithecus robustus*, when Broom placed a ruler across the reconstructed face of the type specimen from Kromdraai and revealed that the center of the face was sunken relative to the peripheral region, researchers have regarded many elements in the *A. robustus* skull as manifestations of a highly specialized masticatory system. These elements include huge posterior teeth with very thick enamel; a flat occlusal topography that provides the teeth with a somewhat puffy appearance; premolars that were recruited to the molar tooth row morphologically and apparently also functionally; a massive and well-buttressed mandibular corpus; a very tall, broad ramus; an unusual facial morphology and topography, which was coined "dished-face" topography; and the presence of a compound temporal/nuchal (T/N) crest and a developed, anteriorly positioned sagittal crest (see reviews in Grine 1981; White 1977; Ward and Molnar 1980; and Rak 1983). These elements express simple biomechanical strategies that were adopted by many other mammals in parallel. No other mammal, however, reached the magnitude of masticatory specialization seen in the robust australopithecines.

The profound specialization of the masticatory system in *A. robustus* has led me to think that the morphology of some robust austrolopithecine structures that are seemingly unrelated to the masticatory system might actually have been affected by the extreme specialization.

THE INCUS OF *AUSTRALOPITHECUS ROBUSTUS*

In 1979, I identified an ear ossicle of *A. robustus* (Rak and Clarke 1979). This bone—an incus—was recovered from the tympanic cavity of the temporal bone that constitutes the Swartkrans specimen SK 848, attributed to *A. robustus*. Although the temporal bone is isolated, the trumpet-shaped external auditory meatus so typical of *A. robustus* helps identify the taxon to which this specimen belongs. The

FIGURE 11-1
The incus of *A. robustus*: above, a medial view, and below, a lateral view.

incus is superbly preserved, lacking only the long process. The bone's nearly perfect state of preservation, which extends to the articular surface through which the incus connects with the malleus, permits the examination and evaluation of the bone's anatomy to be carried out with great confidence.

In the original description of the incus, attention was drawn to its unique morphology and to the great difference between it and the modern human incus. This discrepancy exceeds that between the human and the chimpanzee incudi. These results have been recently confirmed by my examination of a large sample of modern *Homo sapiens* specimens from various geographical areas: 450 incudi, probably representing 225 individuals, from India; 47 right or left incudi, representing 47 individuals, from East Africa; 24 right or left incudi, representing 24 individuals of a Middle Eastern population; 14 right or left incudi, representing 14 individuals, from the Sudan; and 36 right or left incudi, representing 36 individuals, from Thailand. These incudi were compared to one incus of a pygmy chimpanzee and 13 incudi, representing 13 individuals, of the common chimpanzee.

The body of the Swartkrans ear ossicle is inflated and bulb shaped. A deep indentation at the center of the articular surface emphasizes the protrusion of the oversized dens. The short (posterior) process is cylindrical and has a small diameter. The area of the process's attachment to the body is restricted and localized, whereas in apes and humans, the process is flat and broadens anteriorly to meet the body.

The articular surface reveals profound differences. The synovial incudomalleolar joint in humans has been described as a biaxial saddle-shaped joint, and in this respect it is identical to that of a chimpanzee. The twisting of the articular surface from the medial to the lateral sides of the bone is not found in *A. robustus*. Instead, the two parts of the articular surface (superior and inferior) face each other and form what appears to be a uniaxial joint that is shaped much like the semilunar notch of the proximal end of the ulna (Figures 11-1, 11-2, and 11-3).

I would like to discuss three interpretations of the morphological differences between the human and the *A. robustus* incudi that exceed those between the human and the chimpanzee incudi.

If the incus is considered by itself, then the first and more obvious explanation is that *A. robustus* simply precedes the split between humans and chimpanzees. Those few scientists who advocated such an early split were quick to embrace the unique anatomy of the *A. robustus* incus as evidence of such a scenario (see, for example, Oxnard 1984). This interpretation, however, is clearly not the most parsimonious one. It does not account for many other features, such as those related to bipedal walking, cranial anatomy, and dental morphologies. A second approach views this anatomy as the result of a unique specialization. This hypothesis demands a demonstrable biomechanical advantage, backed by convincing analogies in the form of parallel specializations. As a post hoc explanation, this interpretation would undoubtedly be classified (and criticized) by Gould and Lewontin (1979) as an "adaptationist program,"

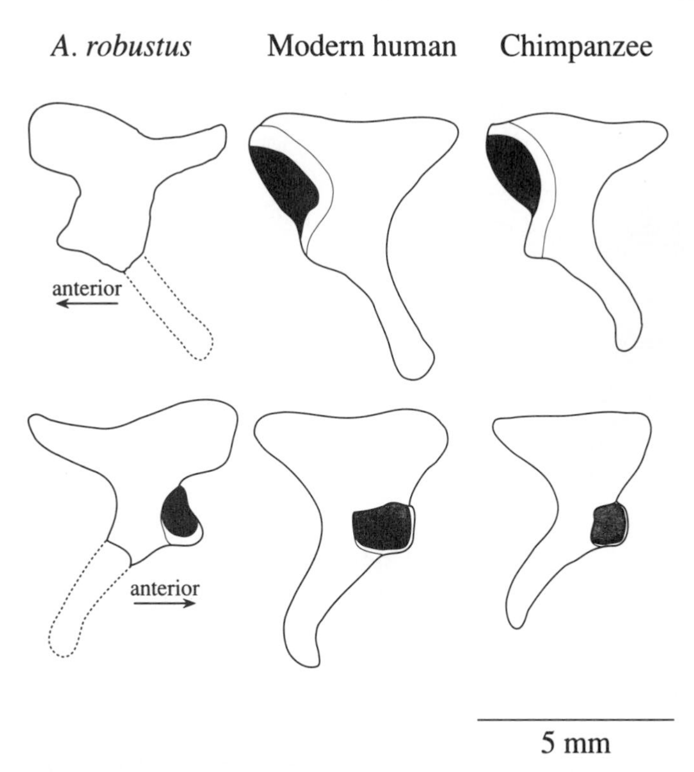

FIGURE 11-2 A comparison of the incudi of *A. robustus*, modern humans, and chimpanzees.

but it was the one favored in the original report. At the time, great similarities between a known, highly specialized ear ossicle and the incus of *A. robustus* were considered support for the claim.

Although a specialization might still be considered responsible for this unusual anatomy, I would like to suggest a third interpretation, admittedly overlooked in the original research. The morphology may be nonadaptive. I suggest that the peculiar shape of the ear ossicle of *A. robustus* is the result of a pleiotropic effect, a by-product of changes imposed by selection on other organs. I hypothesize that it was the mandible of *A. robustus* (as part of a highly specialized masticatory system) that could have been the target of the selective pressure. According to this interpretation, the morphology of the ear bone, which belongs to the same developmental system, was influenced indirectly by that pressure.

That the mandible and the two lateral ear ossicles belong to one restricted anatomic, and presumably genetic, complex can be inferred from the fact, established long ago, that the three bones have their phylogenetic and ontogenetic origin in the first gill arch. The understanding and appreciation of this anatomic relationship has been one of the more spectacular achievements of comparative anatomy, embryology, and paleontology since the beginning of the last century. As early as then, attention was drawn to the embryologic origin of the incus and the malleus in the posterior end of Meckel's cartilage (Carus 1818, Reichert 1837, and Meckel 1820, all cited by Watson 1953). This systemic relationship also seems to be reflected in the phylogenetic development of the lateral ear ossicles, as impressively documented in the fossil record (see Goodrich 1930; Watson 1953; Allin 1975). Additional support comes from gross anatomy in the form of both the peculiar, well-known innervation of the tensor tympani by the mandibular branch of the trigeminal nerve and the physical connection between the mandible and the malleus via fibers of the sphenomandibular ligament.

If this interpretation of the *A. robustus* incus is, indeed, valid, then its unique morphology clearly bears testimony to the extreme degree of specialization of the masticatory system of *A. robustus*.

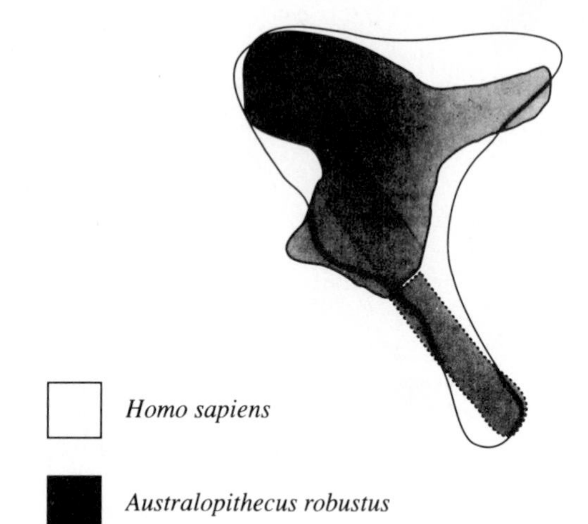

FIGURE 11-3
Superimposed outlines of *A. robustus* and *H. sapiens* ossicles.

In order to find parallel evidence, I attempted to examine ear ossicles of *Hadropithecus*, an unrelated subfossil primate whose masticatory system and that of the robust australopithecines are considered a classic case of parallelism. Unfortunately, no *Hadropithecus* ossicles have been found. The incus of *Archaeolemur*, however, another subfossil prosimian that is biomechanically less specialized than *Hadropithecus* (Rak 1983), seems not to bear any expression of the specialization of the masticatory system.

As different as the modern human and the chimpanzee mandibles are, they are nevertheless accompanied by incudi that are almost identical. The magnitude of difference between the modern human and the chimpanzee masticatory systems, including the mandibles, is apparently not sufficient to have yielded the differences that we see between the modified incus of *A. robustus* and the conservatively shaped incus of modern humans. Only *A. robustus* (and presumably *Australopithecus boisei*) had a masticatory system so specialized that it influenced even the shape of the incus.

SUMMARY

A comparison of an ear ossicle (the incus) of *Australopithecus robustus* with a modern human ossicle shows significant morphological divergence. The differences between these two bones exceed those between the human and the chimpanzee incudi. Three hypotheses to account for this unique variation are suggested: phylogenetic divergence, unique function, and developmental pleiotropy. Given the tight ontogenetic relationship between the incus and the mandible, the unusual morphology of the incus of *A. robustus* is most reasonably hypothesized as a by-product of the selective pressure that was aimed at its masticatory system.

ACKNOWLEDGMENTS

This research was conducted at the Institute of Human Origins, Berkeley, California. I thank Dr. Ian Tattersal for providing information on the incus of *Archaeolemur*, and Dr. Milford Wolpoff for offering helpful comments. Douglas Beckner skillfully drew the illustrations.

LITERATURE CITED

Allin, E. F. (1975) Evolution of the mammalian middle ear. *J. Morph.* 147:403-438.

Goodrich, E. S. (1930) *Studies on the Structure and Development of the Vertebrates*. London: Macmillan.

Gould, S. J., and Lewontin, R. C. (1979) The spandrels of San Marco and the Panglossian paradigm: A critique of the adaptationist programme. *Proc. R. Soc. Lond.* B205:581-598.

Grine, F. E. (1981) Trophic differences between "gracile" and "robust" australopithecines: A scanning electron microscope analysis of occlusal events. *S. Afr. J. Sci.* 77:203-230.

Oxnard, C. (1984) *The Order of Man*. New Haven and London: Yale University Press.

Rak, Y. (1983) *The Australopithecine Face*. New York: Academic Press.

Rak, Y., and Clarke, R. (1979). Ear ossicle of *Australopithecus robustus*. *Nature* 279 (5708):62-63.

Ward, S. C., and Molnar, S. (1980) Experimental stress analysis of topographic diversity in early hominid gnathic morphology. *Am. J. Phys. Anthropol.* 53:383-395.

Watson, D. M. S. (1953) The evolution of the mammalian ear. *Evolution* 7:159-177.

White, T. D. (1977) Anterior Mandibular Corpus of Early African Hominidae. Unpublished Ph.D. dissertation, Department of Anthropology, University of Michigan, Ann Arbor.

CHAPTER 12

Developmental Patterns of the Earliest Hominids

A Morphological Perspective

Pelaji S. Kyauka

Man and the manlike apes developed phylogenetically from one common ancestor stock. Their widely divergent evolutionary changes were due primarily to alterations in their originally identical processes of ontogenetic development. To understand the past transformations of the human body we must know first of all to what extent and in which respects the ontogenetic conditions of man have come to differ from those of apes. (Schultz 1941:59)

INTRODUCTION

The origin of anatomically modern *Homo sapiens* has been a subject of curiosity and debate for more than a hundred years. Today, there are many publications concerning the ancestors of anatomically modern *H. sapiens,* (i.e., the early hominids). The term early hominids lacks taxonomic significance, and in this chapter, the term early hominids means fossil hominids, the extinct members of the family Hominidae. Fossil hominids are divided into two genera— *Australopithecus* and *Homo* (Washburn 1951). The genus *Australopithecus* is classified into five species: *A. afarensis, A. aethiopicus, A. africanus, A. boisei,* and *A. robustus* (see Kimbel et al. 1988). The genus *Homo* comprises three species: *H. habilis, H. erectus,* and *H. neanderthalensis.* In this chapter the transitional early hominids that lived between *H. erectus* and anatomically modern *H. sapiens* (excluding *H. neanderthalensis*) are classified as archaic *H. sapiens.* The taxonomy of the early hominids is still a subject of debate; thence, the classification used in this chapter is not universal and conclusive.

During the times of Darwin (1871), Dubois (1894) and Haeckel (1868), fossil hominids were poorly known. In the past hundred years, tremendous progress in early hominid discovery, anatomy, morphology, metrics, classification, and phylogeny has been made. However, insignificant progress has been made regarding their developmental biology, that is, "the intricate and complex process that all multicellular organisms undergo during their life cycle, from fertilization, through embryogenesis, to adult stage, and thence to the differentiation of eggs and sperm to complete life cycle" (de Laat 1992:18). Very little is known about the prenatal and postnatal developmental biology of the individual fossil hominids (e.g., their age-related changes in morphology, metrics, anatomy, system complexity, and ability to function due to differentiation, morphogenesis, and growth). Data on the prenatal development of the early hominids are very scarce, although a few attempts have been made (see Frazer 1973; Hammer 1969; Leutenegger 1972, 1973, 1987). Dahlberg (1965:159) correctly summarized the situation as follows,

> In general, it must be admitted that fossils and skeletal materials have been well described, classified and studied. Some have been the subject of cephalometric techniques for limited comparative studies. Measurements have been made, recorded, and processed but very little has been gleaned regarding the specific processes of growth and development. . . . the actual mechanism that produces a specific face, tooth or part has not received sufficient attention.

This view was also echoed earlier by Abbie (1952) and Drennan (1931).

The inadequacy of knowledge about the developmental biology and the population dynamics of the early hominids is due in part to the long-held view that the developmental biology and the population dynamics of the early hominids were similar to those of anatomically modern *H. sapiens* (Dart 1925, 1948; Gould 1977; Mann 1975). This notion was a result of similarities observed between several fossil hominids and anatomically modern *H. sapiens* (see, e.g., Bronowski and Long 1953; Clark 1950; Dart 1925, 1926; Montagu 1955). Besides, current attempts to study the developmental biology and the demography of the early hominids are hampered by scarce and fragmentary fossil hominids. Also, the available early hominid specimens inadequately represent different age groups and the different parts of their skeletons. Fossil hominids representing different age groups and skeletal parts are essential for understanding the individual's developmental biology, life history, and the dynamics of the population. Moreover, reliable methods for studying fragmentary fossils are limited. In recent years, a few fragmentary bone study techniques have been developed (see Muller 1935; Steele and McKern 1969; Steele 1970). In addition, efforts to unravel the life histories and demographic attributes of the early hominids are complicated by lack of a clear indication of how well skeletal and dental developmental attributes such as size and form of individual extant hominoids correlate with their life histories or the different developmental stages such as infancy, childhood, adolescence, and adulthood.

HISTORICAL BACKGROUND FOR THE STUDY OF EARLY HOMINID DEVELOPMENTAL BIOLOGY

The developmental biology of the early hominids was not a major preoccupation of the early paleoanthropologists. But after Dart (1925) announced the discovery of the Taung child, the first *A. africanus*, interest in the developmental biology of the early hominids began to emerge. This is due in part, to a dispute started by Dart (1925), who assigned the Taung child to the family Hominidae. Dart (1925:196) argued that the Taung child was a hominid, and based on modern *H. sapiens* dental criteria, he concluded that "the specimen [the Taung child] is juvenile, for the first permanent molar tooth only has erupted in both sides of the face; i.e., it corresponds anatomically with a human child of six years of age." The major surge of interest in the studies about the developmental biology and demography of the early hominids came after Mann (1975) published the results of his research on some paleodemographic aspects of *Australopithecus* where he argued that the early hominids had a humanlike prolonged maturation period.

Dart's arguments for placing the Taung child in the taxon Hominidae, and his conclusion that the Taung child died at six years of age were highly contested (Keith 1925, 1927; Zuckerman 1928). The critics argued that the similarities Dart (1925) observed between the Taung child and *H. sapiens* were no greater than those one would observe in an immature chimpanzee at a similar dental age. Keith (1927:178) argued that,

> In certain features, as, for instance, the prominent forehead, the skull of the three-years-old Taung's ape discovered in South Africa in 1925 approaches that of a human child, but on the whole it more closely resembles the skull of young chimpanzees, . . . The adult Taung's ape, therefore, was probably not unlike an adult chimpanzee, but would have certain gorilla-like characteristics; in point of size it was half-way between them.

Although the earlier debate on the Taung child was mostly focused on the rationale of attributing the Taung child to the hominid family, a few aspects of the Taung child's developmental biology relating to dental development, brain size and facial form and size (see, e.g., Biggerstaff 1967; Dart 1925, 1926, 1929; Keith 1925, 1927; Zuckerman 1928) were discussed.

Fossil hominids, mostly neandertals, were known before the discovery of the Taung child, but attention was not focused on their developmental biology and demography. However, in subsequent years, immature *H. neanderthalensis* fossils continued to be discovered and described. Sometimes their developmental biology in relation to that of modern humans was investigated (see, e.g., Dean et al. 1986; Heim 1982; Hublin 1980; Keith 1931; Stringer et al. 1990; Tillier 1986; Thompkins and Trinkaus 1987; Trinkaus 1986; Vallois 1937; Vlcek 1970; Weidenreich 1939, 1941, 1951; Wolpoff 1979). Developmental studies on the dentition and morphology of *H. neanderthalensis* showed that *H. neanderthalensis* matured faster than *H. sapiens*. Dean et al. (1986:308) concluded that "the developing dentition of the Gibraltar child falls much closer to, or at the lower end of, the expected range of values known for modern *Homo sapiens* and as such must be considered to show an accelerated but essentially human pattern of development."

Also, developmental studies concerning archaic *H. sapiens* showed that their skeletal and dental development was unlike that of modern *H. sapiens*. For example, De Beer (1948:187) in his study of the Skhul 1 skull, an archaic *H. sapiens*, concluded that "Its dentition corresponds to the stage which in modern man is reached at age 4-4.5 years, but is more advanced in its development than a modern child of similar age."

Comprehensive and comparative dental development studies of *H. neanderthalensis* and the extant Pongidae, however, are scanty, and the few studies available (see, e.g., Dean 1985; Dean et al. 1986; Legoux 1965, 1966, 1970; Skinner 1978; Stringer et al. 1990; Thoma 1963; Wolpoff 1979) are inconclusive. Therefore, deductions regarding *H. neanderthalensis* dental and skeletal development, life cycles, and demography, cannot be made without constraints (see Stringer et al. 1990; Trinkaus and Thompkins 1990). The situation is similar for the earliest fossil hominids, but in recent years progress has been made toward improving the situation.

The most cited developmental study of the genus *Australopithecus* is that of Mann (1975). In 1975, Mann employed tooth eruption, emergence sequence, root closure, and tooth wear as measures of maturity to age *A. robustus* specimens from Swartkrans and *A. africanus* specimens from Sterkfontein. Based on similarities between *A. robustus* and *H. sapiens* in the attributes he studied, Mann (1975) concluded that the periods of dental and skeletal development in *Australopithecus* were prolonged because of a long childhood duration. Gould (1977) concurred with Mann (1975) that the life cycles of the early hominids had an extended childhood period. Similar conclusions are common in the literature (e.g., Dart 1948). In 1948, Dart concluded that MLD 2, an *A. africanus* from Makapansgat, had a prolonged dental development, although he noted the second permanent molar seemed to emerge before the fourth premolar, a condition observed in pongids. Dart (1948) justified the conclusion that MLD 2 had a prolonged dental development by arguing that Drennan (1938) observed a similar sequence in the South African San. Also, in 1951 Broom and Robinson noted that the dental emergence sequence in MLD 2 was anthropoid-like.

In 1962, Dart assumed that dental attrition in *Australopithecus* corresponded to that of modern humans, and on the basis of the assumption, he assigned ages at death to *A. africanus* fossils discovered in Makapansgat. The assumption was based on observed morphological and dental development similarities shared between modern humans and *Australopithecus*. Clark (1967:61) agreed with Dart (1962), and argued that the *Australopithecus* molars followed the same dental emergence sequence as in modern humans. He concluded that "the degree of differential wear in australopithecine permanent molars is strongly suggestive of an adolescent period similar to that of man, and considerably longer than that of the large apes of today."

The time intervals between the emergence of the first, second, and third permanent molars in modern humans are longer than in pongids. Clark (1967) assumed that the differential wear in the *Australopithecus* molars reflected the time intervals between the emergence of their permanent molars. Although widely used, dental wear is not a useful measure of maturity. In monkeys, for example, tooth emergence time intervals are shorter compared with those of humans and pongids (see Swindler 1985:Figure 5, p. 80), but dental wear is faster, probably to increase the efficiency of the molars (Bramblett 1969). Besides, dental wear depends on diet (Koski and Garn 1957; Molnar 1972), morphology, and the anatomical structure of the crown (McKee and Molnar 1988).

Biggerstaff (1967) made further investigations on the Taung child and concluded that the dental maturity of the Taung child was reached at an early age and that the generational time, growth, and development periods were shorter than in *H. sapiens*. This implied that the Taung child's dental age was likely similar to that of *P. troglodytes* at a similar stage of dental development. Biggerstaff (1967) proposed the age of the Taung child as 4 years at the time of death instead of 6 years, as suggested by Dart (1925). This conclusion must be taken with caution because of the extensive individual and geographical variations known in extant hominoids (see Trinkaus and Thompkins 1990). Besides, it is important to note that the status of the Taung child is not known, and that the Taung child may not have been a typical representative of the population.

In a study of the dental emergence sequence of *A. robustus* discovered in Swartkrans, Broom and Robinson (1952) claimed that the permanent incisors and first molars of *A. robustus* emerged at similar stages of development to those of modern humans. Also, Broom and Robinson (1952) noted that the dental emergence sequence of *A. robustus* was somehow unlike that of modern *H. sapiens* and anthropoids. Yet, when assigning age to SK 27, Broom and Robinson (1952:29) noted that, "Perhaps we may be safe in placing the *Paranthropus* child at a stage of development roughly corresponding to a modern human child of 11 years." Garn and Lewis (1963) disputed the conclusion by noting that Broom and Robinson (1952) were not aware of dental emergence sequence polymorphs.

Most of the current developmental studies about early hominids (i.e., *Australopithecus*) contradict the previous dominant view that the developmental biology and the population dynamics of the early hominids resembled those of anatomically modern *Homo sapiens* held by for example Broom (1936), Broom and Robinson (1951, 1952), Clark (1967), Dart (1925), Gould (1977), Mann (1975), and McKinley (1971). Many studies based on aspects of dental development, and a few based on skeletal remodeling (e.g., Bromage 1989), have interpreted the similarities in dental eruption and emergence sequences observed between for example, *P. troglodytes* and the early hominids as suggestive of similarities in developmental patterns. However, these similarities may not mean similar life histories, morphogenetic processes, and eruption rates. Whether the similarities between early hominids, modern pongids, or humans translate into similar individual life cycles and population demography is still in dispute.

The idea that early hominids may have had a pongidlike development pattern has a long history (see Dart 1929; Keith 1925, 1927; Zuckerman 1928). For example, Dart (1925) noted that the Taung child had a mosaic of morphological features, which by inference suggests the same for the morphogenetic processes. Again in 1929, Dart noted that in terms of measurements the Taung child was closer to the chimpanzee than *H. sapiens*. Also, the idea that the development of the early hominids was mosaic in nature is an old one (see, e.g., Dart 1925; Keith 1927; McCown and Keith 1939).

In recent years, the debate about the developmental biology of the early hominids has resurfaced. This seems to be due to the realization of the importance of developmental biology (i.e., age-related changes in size, shape, complexity and function in understanding human evolution). It is also a result of the availability of immature fossil hominids and the development of new study techniques.

Three differing interpretations regarding the development of the early hominids are apparent in the literature. A few earlier studies (e.g., De Beer 1948; Dart 1929; Hrdlicka 1927; Keith 1925, 1927; McCown and Keith 1939; Vallois 1937; Weidenreich 1939, 1941, 1951), and many current studies (Dean et al. 1986; Heim 1982; Hublin 1980; Stringer et al. 1990; Thompkins and Trinkaus 1987) concluded that most early hominids had faster rates of development. Yet, the development pattern of the early hominids continues to be a subject of debate because the studies that contend that early hominids developed like modern pongids (e.g., Beynon and Dean 1987, 1988, 1991a; Beynon and Wood 1986, 1987; Biggerstaff 1967; Bromage 1985, 1986, 1987, 1989, 1990; Bromage and Dean 1985; Conroy and Vannier 1987, 1988, 1989; Dean 1985, 1986, 1987, 1988; Keith 1927; Sacher 1975; Simpson et al. 1990; Smith 1986, 1987, 1989; Zuckerman 1928) have not provided adequate evidence to warrant rejection of conclusions that argue that early hominids developed like modern *H. sapiens* (e.g., Broom and Robinson 1951, 1952; Dart 1925; Le Gros Clark 1967; Mann 1975, 1988; Mann et al. 1987, 1990a; Wolpoff et al. 1988) or arguments put forth that the early hominids shared both pongid-like and humanlike development patterns (i.e., followed a unique development pattern, e.g., Ashton 1981; Dean and Wood 1984; Oxnard 1983; Rak and Howell 1978; Schultz 1960).

DEVELOPMENTAL STUDIES OF THE EARLY HOMINIDS: INTERPRETATIONAL PROBLEMS

The three vying interpretations concerning the developmental biology of the early hominids share similar problems associated with the study methods and the principles underlying the interpretations. Developmental studies of the early hominids use terms commonly employed in developmental studies of the extant hominids, but often, working definitions of these terms are not provided. Lack of clarification of the terms in several publications makes it impossible to assess whether different investigators are dealing with similar ideas.

The need for clarification of terms in scientific writings is explicitly expressed by Weiss (1939:1) as follows; "Ill-defined, vaguely defined, or undefined terms are a scientific curse; they make scientific language oracular instead of unequivocal, hazy instead of clear, and degrade all scientific work described in such language." The most vaguely used and ill-defined terms in developmental studies of the early hominids are tooth eruption, development pattern, and growth pattern. Tooth eruption is a continuous elevation of the tooth from the germ bud toward the aveolar ridge and occlusal plane. But, often, the application of the term tooth eruption implies emergence or piercing of the gingival tissue or the aveolar bone into the oral cavity.

In addition, the terms growth pattern and development pattern are sometimes used interchangeably, although they are not synonymous. Growth is a form of development that involves permanent increase in size and mass. On the other hand, development concerns morphogenesis, growth, and differentiation. Development concerns all individual age-related changes in form, size, mass, system complexity, and ability to function in an individual from conception to adulthood (see Liebgott 1986).

The meaning of the term growth pattern is explicitly expressed by a definition given by Weiss (1939:38): "The typical temporal order according to which growing starts in one part after another, and the typical relationship among the growth rates of different parts can be referred to in a simple terms as the growth pattern." This is also true for the term development pattern. Growth can be measured metrically, and in the literature we find many publications dealing with anatomically modern *H. sapiens* (see, e.g., Hoffman 1979) and *P. troglodytes* growth (Krogman 1969), based on measurements of size and mass.

Graphic representations of the cumulative measurements (longitudinal or cross sectional) of the different aspects of the *H. sapiens* body (e.g., total body weight, brain weight, femur length) produce specific curve pattern types. For example, measurements of the body weight with increasing age after birth produce a sigmoid curve pattern type that depicts the mode and rates of growth of the different postnatal development stages (i.e., neonatal, infant, childhood, adolescent, and adulthood). Also, when the brain size is measured with increasing age and presented graphically, the brain growth curve pattern type is different from that of the total body weight. The brain size rises sharply and levels off between the ages of 6 and 8 years (Liebgott 1986). This means that the adult brain size is reached between the ages of 8 and 10 years. These two are different growth curve pattern types in *H. sapiens*.

Developmental stages such as infancy, childhood, and adolescence, cannot be documented in fossil hominids because fossil hominids representing different age groups are lacking. These developmental stages are measures of stages of maturity in the process of development. Although it is likely that these stages can be established using morphological, anatomical, and metrical attributes of the skeleton, documentation of these attributes is lacking for both the extant hominoids and the early hominoids.

The maturity of the different immature fossil hominids (i.e., the extent to which the individual immature fossil hominid had moved toward attaining adulthood) can be predicted indirectly using observable or detectable skeletal and dental development states that can be morphological, anatomical or metrical. Currently, comparative standard states for both the extant pongids and anatomically modern *H. sapiens* are lacking, but a few attempts such as Straus's (1927) are worth noting.

Dental emergence and epiphyseal union sequences are extensively used in developmental studies of modern hominids. However, similarities in these attributes between *H. sapiens* or *P. troglodytes* and the early hominids can be misleading when applied to interpret the development of the early hominids. Although studies of this type depend on similar control factors, such as the emergence of the first permanent molar, they do not take into consideration the timing, duration, and the rate of development. Also, what these morphological, anatomical, and metrical features measure are biological ages, from which the developmental states are implied. Biological ages or developmental ages cannot always predict chronological ages. These studies, therefore, can only tell us whether the early hominids are closer to the pongids or humans of comparable stage of for example dental emergence, sequence of root formation, or skeletal epiphyseal union, but they are not informative regarding timing, duration, and rates of the events (see Simpson et al. 1990).

Many recent dental development studies have concluded that the development of the early hominids was pongidlike. This claim which is based mostly on dental development similarities seems premature because we simply do not know exactly how the similarities in dental eruption and emergence sequence or skeletal anatomy, morphology and metrics correlate with the development and growth rates, the life cycles, and the timing and duration of these events in the individual fossil hominids. For example, we now know that heterochrony (see Shea 1989) or change in rate and timing of development processes can cause major phenotypic changes with minimum genotypic changes (Bonner 1982).

Besides, the three interpretations are based on dental studies of very limited small samples (e.g., Beynon and Dean 1987; Smith 1989) that are of individuals of unknown causes of death. McKinley (1971) and Tobias (1968) noted that *A. africanus* and *A. robustus* died at young ages based on human criteria. In addition, White (1978) noted that individuals with hypoplasia, a phenomenon associated with nutritional stress and disease, died at somewhat young ages. Dental development is also affected by disease, nutrient deficiency, and environmental stress (Goodman and Rose 1990; Moller 1967; Yaeger 1976).

Furthermore, there are notable disagreements associated with the techniques and interpretations, in particular those that estimate ages from perikymata and striae of Retzius counts (e.g., Beynon and Dean 1987; Dean and Beynon 1991b) or enamel thickness (Beynon and Wood 1986). Most of these studies are still in dispute (see, e.g., Bacon 1989; Grine 1991; Mann et al. 1990a,b, 1991; Smith 1991; Stringer et al. 1990), in part due to the inaccuracy of the techniques; lack of control over age at the time of death; and unresolved questions related to timing, duration, and rates of tooth mineralization, eruption, and emergence in extant hominoids (Beynon and Dean 1991; Conroy and Mahoney 1991; Siebert and Swindler 1991; Anemone et al. 1991; Schwartz and Langdon 1991; Winkler et al. 1991).

Applications of principles derived from modern humans or pongids to interpret similar data obtained from early hominids have serious limitations. For example, it has been observed that the rate of mineralization is the same in both males and females of *H. sapiens* (Nolla 1960). Yet, the ages at which different stages of mineralization are reached are sex dependent (Burdi et al. 1970). Dental mineralization in males is delayed relative to females. Also, formation of a specific tooth depends on the timing of mineralization rather than change in growth velocity (Thompson et al. 1975; Fanning and Moorrees 1969). The canine shows the greatest delay. In addition, Nissen and Riesen (1964) noted that the emergence of the chimpanzee dentition depends on age, eruption, and loss of the deciduous teeth, length of gestation, and sex. Besides, several studies have indicated that tooth development depends on the available space (see Dean and Beynon 1991a; Aitchison 1963; Barret 1963a,b) and function (see Simpson et al. 1991).

Also, observations of several fossil hominids (LH. 2, Taung) show that the shedding patterns of the deciduous teeth of *A. afarensis* and *A. africanus* cannot be differentiated from those of *H. sapiens* or *P. troglodytes*. For example, the erupting permanent incisors are always lingual to the deciduous incisors where resorption is greatest (Kyauka in prep.). These facts are not well documented in the extant hominoids, and their application to fossil hominids requires caution.

In addition, dichotomization between "ape" and "human" is disputed because *Australopithecus* was neither an ape nor a human (e.g., Conroy and Vannier 1987). Moreover, the distinction between differences in frequency and unique specializations between pongids and modern humans is unclear. What we consider to be major differences between the apes and humans may be differences in frequency (Schultz 1952). For example, although the intrapalatal extension of the maxillary sinus in the Taung child is considered to be strong evidence for the pongidlike development pattern in *A. africanus* (Conroy and Vannier 1989), a low frequency of the same pattern is reported in *H. sapiens* (see Hajnis et al. 1967). This observation emphasizes the need for caution, particularly when dealing with data derived from limited fossil hominid specimens and features of unknown causal mechanism.

Finally, we now know that the genus *Australopithecus* is represented by roughly five species, and the genus *Homo* by at least three fossil species. It is probably unrealistic to assume that each of the *Australopithecus* or *Homo* species followed the same developmental pattern (see, e.g., Conroy and Vannier 1991a,b). Even in one *Australopithecus* species a generalized developmental pattern is difficult to comprehend because in *A. afarensis*, for example, morphological attributes suggest differential rates of evolution in different parts, which may have a profound effect on their developmental biology, life cycle, and demography. Also, most of the interpretations regarding somatological development in the early hominid are based on dental studies. However, somatological and dental development are not well correlated (see Lee et al. 1965; Simpson et al. 1990).

THE BIOLOGICAL DEVELOPMENT OF THE EXTANT HOMINOIDS

Data concerning the developmental biology of *H. sapiens*, *P. paniscus*, *P. troglodytes*, *Gorilla gorilla* and *Pongo pygmaeus* is critical for understanding the developmental biology, and the life cycles of the early hominids. Schultz (1956:888) correctly noted that,

> It is evident that a knowledge of the age changes in primates is of fundamental importance for the understanding of the evolutionary history and relationships of the primates in general and of man in particular. If man's specializations have resulted from variously changed conditions in human ontogeny, the latter has to be compared with the ontogenies of non-human primates in order to discover which of man's age changes are really unique.

Developmental specializations of an individual can be documented as morphological (gross and microscopic), anatomical (gross, microscopic, histological, radiological), and metrical attributes during prenatal and postnatal developmental periods of the individual. The relationships between skeletal development (morphological, anatomical, metrical) and the stages of development (see Liebgott 1986), in the extant pongids and modern humans are poorly known. However, it is worth noting that there are well-established developmental differences between extant pongids and modern humans both in prenatal (see Starck 1961) and postnatal stages reported in the literature (see Schultz 1949, 1956, 1962). For example, although the gestation periods of *P. troglodytes, P. paniscus, G. gorilla, P. pygmaeus,* and *H. sapiens* are not very much different (see Ardito 1978), at birth the brain capacity of *H. sapiens* is roughly 23–25 percent of the adult size. But it is 40 percent of the adult size in *P. troglodytes, P. paniscus, G. gorilla,* and *P. pygmaeus.* Also, bone ossification at birth is advanced in pongids compared with *H. sapiens* (Schultz 1941; Watts 1986). In addition, the upper limbs are always longer than the lower limbs.

During the postnal developmental period in particular, major skeletal and dental differences are evident between the pongids and modern humans. For example, at all ages, the intermembral index and robusticity are higher in *P. troglodytes, P. paniscus, G. gorilla,* and *P. pygmaeus* than in *H. sapiens.* Furthermore, in areas of the cranial region, the neurocranium reaches its adult size earlier in *P. troglodytes* and *P. pygmaeus* than in *H. sapiens.* Also, the face increases greatly in height, width, and depth with increasing age in the pongids (Moore and Lavelle 1974).

Moreover, the premaxillary suture unites late in the postnatal life in the pongids while it is lost during the prenatal life in *H. sapiens.* Besides, the pongids show posterior migration of the occipital condyles, widening of the mandibular temporal joint, straightening of the cranial base, early development of supraorbital ridges, and early loss of the bossing of the frontal with increasing age, compared with *H. sapiens.* Although the epiphyseal union and ossification in *H. sapiens, P. troglodytes, P. paniscus, G. gorilla,* and *P. pygmaeus* follow similar sequences, the timing and duration are different. In the pongids for example, earlier workers have noted that epiphyseal union ranges between 5–7 and 15–17 years, while in *H. sapiens* epiphyseal union ranges between 12 to 17 years.

Furthermore, the pongids and *H. sapiens* differ in dental development. In the pongids, the emergence sequence of the deciduous dentition is di1, di2, dm1, dm2, dc, while in humans it is di1, di2, dm1, dc, dm2. For the permanent teeth, the emergence sequence in the pongids is M1, I1, I2, P3, P4, M2, C, M3 while in *H. sapiens* it is M1, I1, I2, P3, P4, C, M2, M3. The sequences given here, especially those of the premolars, canines, and the second molar, vary extensively within individuals. Also, the resting period between M1 and I1 and P4 and C is longer in pongids than in *H. sapiens.* In addition, pongids and humans differ in bone remodeling patterns in the muzzle region, mandible, face, and cranial base. For example, in the pongids, most of the muzzle areas are depository, while in *H. sapiens* the muzzle region is mostly resorptive (see Bromage 1986).

Studies documenting anatomical, morphological and biomechanical changes with age in *H. sapiens* and the extant pongids are poorly documented in the literature. Age changes in skeletal and dental morphology and anatomy, especially the microscopic aspects, are not well studied in the extant hominoids. A few microscopic and anatomical studies of the modern bones have been conducted with the aim of establishing the validity of anatomy in estimating the age of individuals at the time of death (Enlow 1976; Ericksen 1991; Evans 1976; Ortner 1976; Simmons 1985; Singh and Gunberg 1970, 1974; Stout and Simmons 1979; Stout and Stanley 1991). Also, very little is known about biomechanical changes with increasing age (see, e.g., Currey and Pond 1989). Studies of this kind need to be pursued further to determine whether they can provide valuable information in terms of establishing the developmental stages of the fossil hominids relative to the extant hominoids. Evidently we need new baseline data based on techniques that can produce similar data from the fragmentary fossil hominids before conclusive interpretations of the life cycles or growth and development patterns of the early hominids can be provided.

THE BIOLOGICAL DEVELOPMENT OF *AUSTRALOPITHECUS*

In prenatal and postnatal development, individual organisms undergo series of dental and skeletal age-related changes that can be documented metrically, anatomically and morphologically. These series of age-related changes can be grouped into stages that can be used as measures of biological age of the individuals at different times. The series of skeletal and dental changes carry clues about the development process that can provide some insights about the life history of the individual and the demographic attributes of the taxon.

Efforts to unravel the life histories and population dynamics of *Australopithecus* have focused mostly on dental development particularly on the documentation of timing and emergence sequences of the different teeth into the oral cavity. Other aspects of the individual tooth such as timing and duration of root calcification, stages of root elongation, and crown formation have also been widely used. In addition, in recent years, the time required for the maturation (i.e., the process of attaining adult state) of the tooth crown form and structure have been estimated by studies of the enamel structures (e.g., Beynon and Dean 1987). Developmental studies of the other parts of the skeleton (e.g., Bromage 1989; Tappen 1989) are very rare, although they are of vital importance.

Although developmental studies of the early hominids have increased in recent years, most of them are based on the dentition; and the conclusions have serious methodological and interpretational limitations. The most obvious problem concerns the procedure of estimating their ages at death that uses assumed human and pongid criteria. Most of these methods (see Mann 1975; Mann et al. 1991; Smith 1986), including those that employ perikymata, striae of Retzius and cross striation counts (e.g., Beynon and Dean 1987, 1988), are not independent of assumptions that rely on modern hominid or extant pongid criteria. Although Bromage (1991) claimed that he has confirmed the enamel circadian periodicity, the problem in the techniques that use perikymata count is compounded by the fact that in certain sections of the enamel, the perikymata or cross striations cannot be counted, and only estimates are applied. For these reasons, the age estimates are very unreliable. Many of these studies are creating a vicious cycle where assumptions are made, and then the same assumptions are applied to test them (see Smith 1989).

The second major weakness concerns the deductions and extrapolations made from the data. In several instances, deductions concerning the life histories of the individual fossil hominids have been made without sound basis. Whether inferences of gestation periods, ages of weaning, birth intervals, sexual maturity, or longevity of the early hominids can be made from eruption and calcification stages, dental emergence sequences, time of root and crown formation, or brain size (see e.g., Smith 1989, 1991) without prior knowledge of the developmental rates is debatable. It is unclear whether it is possible to infer the duration and rates of the development stages (i.e., prenatal [zygote, embryo, fetus] and postnatal [neonate, infant, childhood adolescent and adulthood]) from these attributes. Certain aspects of *H. sapiens* general body growth as measured by body weight produce a sigmoid growth curve when presented graphically. The growth pattern of the early hominids cannot be represented graphically, because we lack the necessary data. Also, we have no clear knowledge on how anatomical and morphological features translate into development or growth patterns in *H. sapiens* and *P. troglodytes*.

Although morphological, anatomical, and metric similarities between early hominids and extant hominoids may not translate into the life cycles of the early hominids, they may be useful in predictions of the skeletal and dental development processes and mechanisms. In this section I will examine the status of two early hominid fossils designated as AL. 333-105 and the Taung child in relation to *H. sapiens* and *P. troglodytes* at similar stages of dental development. AL. 333-105 and the Taung child are attributed to *A. afarensis* and *A. africanus* respectively. The morphological and metric descriptions of the two immature *Australopithecus*, the Taung child and AL. 333-105 are compared with those of *P. troglodytes* and *H. sapiens* at similar stages of dental development, and their developmental biology is discussed. This study is based on the principle that age-related changes in morphology partly reflect the nature of the morphogenetic processes producing the morphology and the extrinsic forces acting on the bone.

MATERIAL AND METHODS

Taung and AL. 333-105 are immature *Australopithecus* partial crania. Previous morphological and metrical studies on them (see, e.g., Asfaw 1988; Dart 1925, 1929; Kimbel et al. 1982, 1986; Olson 1985) emphasized phylogenetic relationships. The present study emphasis is on developmental biology. Only a summary of the previous descriptions, supplemented with my own observations is given here. Also, the following metrics were estimated on Taung: the palatine length and nasion-prosthion lengths, the biorbital breadth, and postorbital constriction breadth. Due to distortion, only the postorbital constriction and biorbital breadth of AL. 333-105 estimated by Kimbel (1986) were used. Similar metrics were made on four *P. troglodytes* specimens, two (1 male and 1 female) at a similar stage of dental development to Taung, and two *P. troglodytes* (unsexed) at a similar stage of dental development to AL. 333-105. For *H. sapiens*, data published by Richards (1985) were adapted. Four indices were calculated: palatine/nasion-prosthion length; palatine length/biorbital breadth; nasion-prosthion/biorbital breadth, and postorbital breadth/biorbital breadth were calculated as percentages (see Schultz 1962).

RESULTS

Table 12-1 summarizes the data from earlier studies (Dart 1925; Kimbel et al. 1982, 1986; Asfaw 1988); and new observations on Taung and AL. 333-105 relative to *H. sapiens* and *P. troglodytes* at similar stages of dental development. Of the 19 characters, 6 are present only in *P. troglodytes*, 2 in Taung, AL. 333-105 and *P. troglodytes*, and 9 in Taung, AL. 333-105 and *H. sapiens*. Tables 12-2 and 12-3 summarize the metric data. The four craniofacial indices show that both Taung and AL. 333-105 are closer to *P. troglodytes* than they are to *H. sapiens*, contrary to the morphological data.

The data in Table 12-1 show that unlike *P. troglodytes* at a similar stage of dental development AL. 333-105 lacks the following features: a well-developed sulcus separating the frontal squama from the

TABLE 12-1 Characters compared as documented in the literature[1,2,3] including observations of this study

Character/Specimen	A	B	C	D	E	F
1. Presence of sulcus separating the frontal squama from the supraorbital ridge	Yes	No	No	No	No	Yes
2. Lack of sulcus separating the frontal squama from supraorbital ridge	No	Yes	Yes	Yes	Yes	No
3. Strong postorbital constriction	Yes	No	No	No	No	Yes
4. Weak postorbital constriction	No	Yes	Yes	Yes	Yes	No
5. Shallow palate	Yes	No	Yes	Yes	No	Yes
6. Deep palate	No	Yes	No	No	Yes	No
7. Orthognathic face	No	Yes	No	No	Yes	No
8. Flat frontal	Yes	No	No	No	No	Yes
9. Globular frontal	No	Yes	Yes	Yes	Yes	No
10. Gracile maxilla zygomatic process	Yes	No	No	No	No	Yes
11. Robust and stout maxilla zygomatic process	No	Yes	Yes	Yes	Yes	No
12. Lateroposterior inclined maxilla-zygomatic complex	Yes	No	No	No	No	Yes
13. Verticolaterally inclined maxilla-zygomatic bones	No	Yes	Yes	Yes	Yes	No
14. Presence of zygomatic tubercle	No	Yes	Yes	Yes	Yes	No
15. Presence of crest on the posterior side of the zygomatic which is partly concave	Yes	No	No	No	No	Yes
16. Zygomatic with wholly concave posterior surface	No	Yes	Yes	Yes	Yes	No
17. Lateral orbital rims thick and concave	No	Yes	Yes	Yes	Yes	No
18. Presence of *trigonum supraorbitale*	No	Yes	Yes	Yes	Yes	No
19. Presence of premaxillary suture	Yes	Yes	Yes	Yes	No	Yes

[A]*P. troglodytes* = Taung; [B]*H. sapiens* = Taung; [C]Taung; [D]AL. 333-105; [E]*H. sapiens* = AL. 333-105; [F]*P. troglodytes* = AL. 333-105; [1]Asfaw (1988); [2]Dart (1925); [3]Kimbel et al. (1982).

TABLE 12-2 Selected indices of craniofacial metrics of the Taung child

Specimen/index	Taung	Pan = Taung	*H. sapiens* (5 yrs)[a]	*H. sapiens* (3 yrs)[a]
1. PL/N-PR(100)	87.5	y = 75.76 x = 79.1	67.24	64.83
2. PL/BB (100)	54.19	y = 64.94 x = 67.1	38.29	36.20
3. N-PR/BB (100)	61.94	y = 84.8 x = 85.71	56.95	55.84
4. PB/BB (100)	85.16	y = 88.30 x = 87.34	104.63	104.49

[a] Data adapted from Richards (1985); PL = palatine length; N-PR = nasion-prosthion height; BB = biorbital breadth; PB = postorbital constriction breadth; x = female; y = male.

strong postorbital ridge; a strong postorbital constriction; a very pronounced prognathic face; a flat frontal; a gracile maxillary zygomatic process; and lateroposteriorly inclined and crests on the posterior side of the zygomatics. The zygomatic crests are extensions of the temporal lines from the lateral edges of the frontal zygomatic processes. After crossing the frontozygomatic sutures, the crests continue along the anterior side of the sphenozygomatic sutures, down to the inferior orbital fissures. Because of the crests, the frontal zygomatic processes and the upper part of the zygomatic frontal processes in *P. troglodytes* are convex (see Figure 12-3), and only the inferior part of the zygomatics shows some form of concavity. Like *P. troglodytes*, however, AL. 333-105 has a prognathic face and a shallow palate with a gentle slope starting from the anterior tip of the intrapalatal suture.

The data in Table 12-1 also show that like *H. sapiens*, AL. 333-105 retained a globular frontal with no postorbital sulcus; a stout and robust zygomaxillary process; a verticolaterally oriented zygomaxillary complex; a zygomatic with the posterior surface wholly concave; a frontal zygomatic process concavity that blends smoothly with the curvature of the temporal fossa; a thick lateral orbital rim; and a zygomatic bone with a thin edged lateral margin with a marginal tubercle. Also, the aveolar maxilla is convex compared with that of *P. troglodytes*, which is relatively straight between dm1 and the maxillary tuberosity. In addition, the frontal bone has a *trigonum supraorbitale* (see Kimbel 1986), and the temporal mandibular joint cannot be assigned into either humanlike or pongidlike.

The data in Table 12-1 and Figures 12-1 to 12-3 show that the cranial morphologies of the Taung child, like those of *H. sapiens* at a similar dental development have: a robust malar process (Figures 12-1, 12-2); a concave and robust lateral orbital rim; and a marginal tubercle on the lateral border of the zygomatic frontal process. Also, like AL. 333-105, the Taung child retained a globular frontal (Figure 12-3) until childhood, contrary to *P. troglodytes* where it ends in the infancy stage (Schultz 1940). Furthermore, unlike *P. troglodytes* at a similar dental age, the Taung child lacked a postorbital ridge (Figure 12-3)

TABLE 12-3 Selected indices of craniofacial metrics of AL. 333-105

Specimen/index	AL. 333-105	Pan = AL.333-105	*H. sapiens* (5 yrs)[a]	*H. sapiens* (3 yrs)[a]
1. PL/N-PR(100)	-	70.59	67.24	64.83
2. PL/BB (100)	-	57.4	38.29	36.20
3. N-PR/BB (100)	-	81.23	56.95	55.84
4. PB/BB (100)	93.75	93.17	104.63	104.49

[a]Data adapted from Richards (1985); PL = palatine length; N-PR = nasion-prosthion height; BB = biorbital breadth; PB = postorbital constriction breadth.

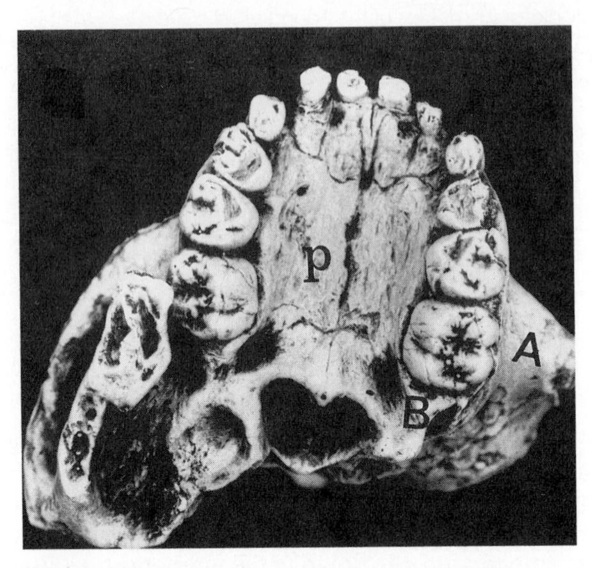

FIGURE 12-1a The inferior view of the Taung child cranium.

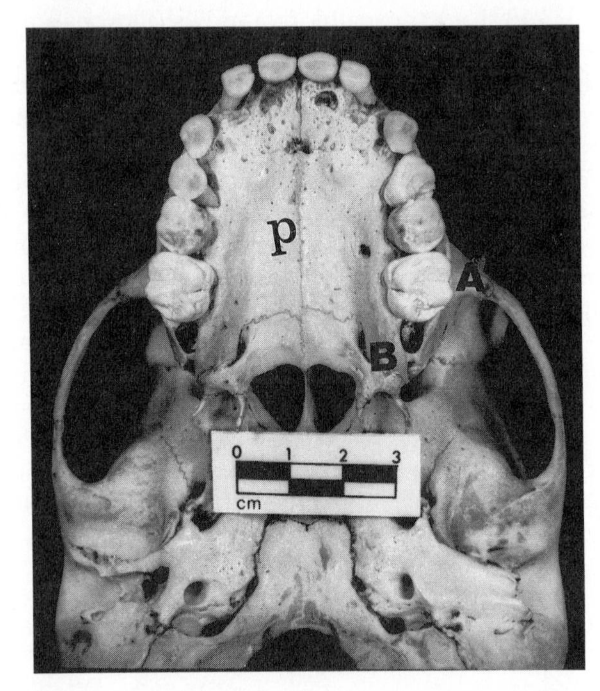

FIGURE 12-1b The inferior view of a *Pan troglodytes* cranium at a dental age similar to that of the Taung child cranium. A, malar process; B, maxillary tuberosity; P, palate.

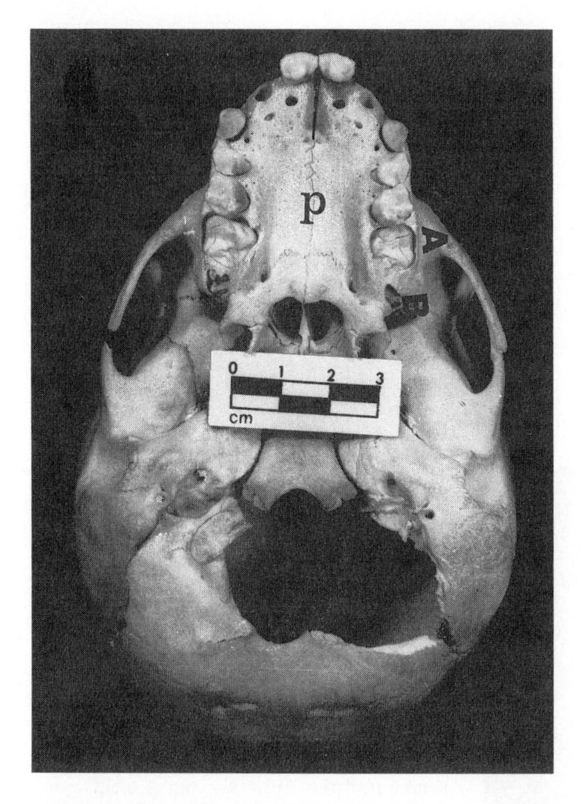

FIGURE 12-2a The inferior view of a *Pan troglodytes* cranium at a dental age similar to that of AL. 333-105. A, malar process; B, maxillary tuberosity; P, palate.

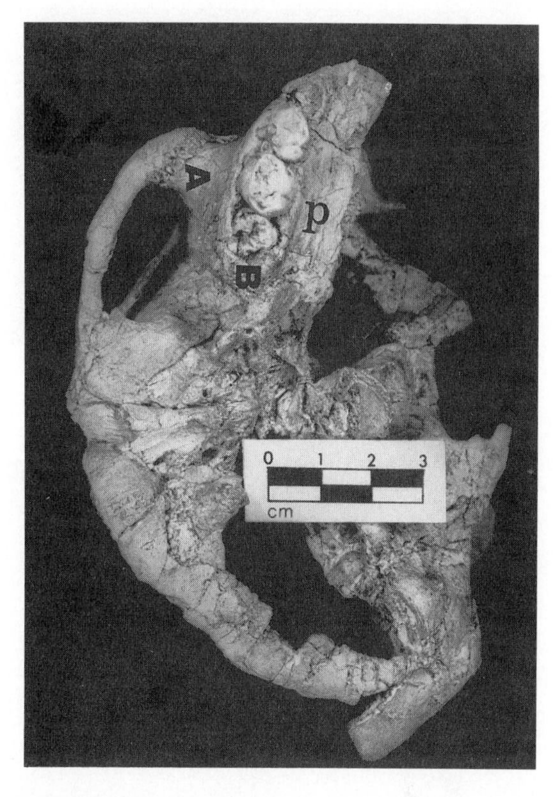

FIGURE 12-2b The inferior view of the AL. 333-105 cranium.

separated by a sulcus from the frontal squama (see Asfaw 1988), and the maxillary tuberosity appears less posteriorly positioned in Taung, *H. sapiens* and AL. 333-105 (Figures 12-1 and 12-2) than in *P. troglodytes* at similar stages of dental development respectively. The Taung child face is prognathic compared with that of *H. sapiens*.

The four selected craniofacial metric indices in Tables 12-2 and 12-3 show that both the Taung child and AL. 333-105 are closer to *P. troglodytes* than they are to *H. sapiens* at similar stages of dental development respectively. However, the prosthion-nasion/biorbital index (Table 12-2), shows that the protrusion of the Taung child's face is less than that of *P. troglodytes* but greater than that of *H. sapiens* at a similar stage of dental development. This observation supports a similar conclusion reached by Dart (1929) and Kimbel (1986) based on different analytical techniques.

DISCUSSION

The developmental morphology of an individual, and by inference of the taxon, has some bearing on the individual's life history. This view is supported by De Beer (1930), Gould (1977), and Schultz (1941, 1956) who noted that the differences in skeletal morphology between *P. troglodytes* and *H. sapiens* at adult stages are consequences of altered developmental patterns during the process of evolution.

Developmental morphology concerns stages and changes of morphology of an individual during the process of development (see, e.g., Kovacs 1971). Studies pertaining to developmental morphology of the skeletons of modern *H. sapiens*, the pongids, and the early hominids are not available. Such studies,

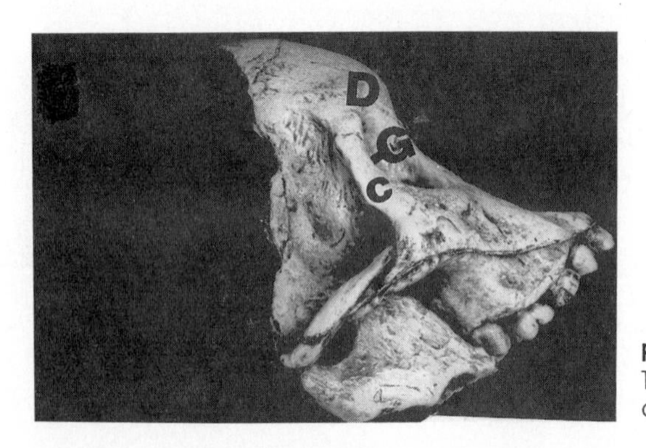

FIGURE 12-3a
The right side of the cranium of the Taung child.

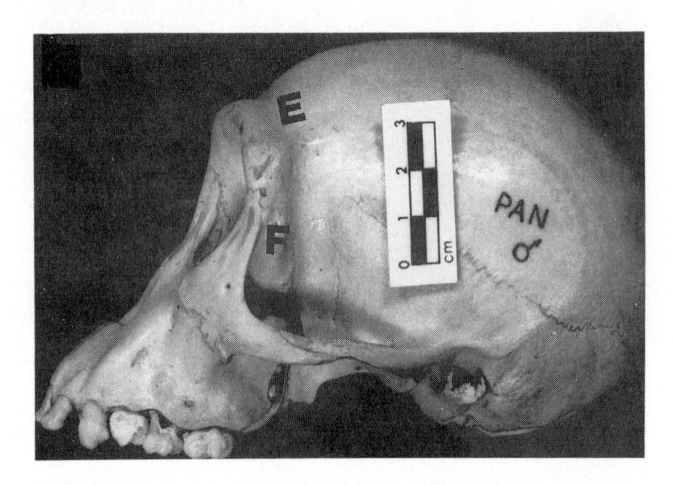

FIGURE 12-3b The left side of a *Pan troglodytes* cranium at a dental age similar to the that of the Taung child. C, marginal tubercle process; D, frontal; E, frontal sulcus; F, crest on the posterior side of the zygomatic bone; G, lateral rim of the orbit.

however, have the potential of providing data for explaining the developmental biology and evolutionary histories of the hominoids (see Straus 1927).

Morphologically, the frontal and zygomatic bones of AL. 333-105 and the Taung child have similarities to those of *H. sapiens* at similar stages of dental development (see Table 12-1). However, the adult morphology of *A. afarensis* and *A. africanus* in many aspects approaches that of adult *P. troglodytes*. The series of morphological stages and changes that occurred, for example, between the AL. 333-105 stage and the adult stage of *A. afarensis* or the Taung child stage and the adult stage of *A. africanus* are unknown and probably will be difficult to investigate because of the limited number of immature fossil hominids. Nevertheless, at a certain stage in the *A. afarensis* or *A. africanus* development process, like in *P. troglodytes*, the immature globular frontal transformed into the nearly flat adult form (see Asfaw 1988). Moreover, unlike *Pan troglodytes* at similar stages of dental development to AL. 333-105, AL. 333-105 had a globular frontal, suggesting that frontal bossing in AL. 333-105 lasted for a longer time than it does in *P. troglodytes*, that is, the flattening of the frontal bone occurs earlier in pongids (see Schultz 1940, 1941) than it did in *A. afarensis*. This condition is also true for the Taung child.

AL. 333-105 and the Taung child lack the postorbital torus and supraorbital sulcus that appears very early in *P. troglodytes*. The postorbital torus and sulcus are linked to accelerated brain growth (see Asfaw 1988; Schultz 1940) in *P. troglodytes*. Smith and Ranyard (1980), however, argued that the postorbital torus might have been a means of maintaining structural integrity due to differential growth between the splanchnocranium and neurocranium. In addition, they commented that the postorbital torus of the upper Pleistocene hominids of central Europe was partly a result of its own growth and, in part, due to remodeling of the face and the brain case.

The presence of the premaxillary suture and the shallow palate in AL. 333-105 and the Taung child suggests that developmentally, the mophogenetic processes of AL. 333-105 and the Taung child were likely similar to those of *P. troglodytes* in certain stages in these parts. Similarly, it appears that in the formation of certain cranial morphological features of AL. 333-105 and the Taung child (see Table 12-1), hominid-like morphogenetic processes were responsible. Traces of the human-like morphogenetic processes may also be inferred from the humanlike features documented in adult *A. afarensis* and *A. africanus* skeletons (e.g., in the knee joint, pelvis form, foot joint [see Latimer and Lovejoy 1990], and vertebral column structural patterns [Kyauka 1992]). The timing, duration and rates of the morphogenetic processes, however, might have been different from those of both *H. sapiens* and *P. troglodytes.*

The AL. 333-105 and the Taung child orbital rim concavity is closer to that of *H. sapiens* than to that of *P. troglodytes*. Bromage (1989) suggested that the concavity of the lateral orbital rim in the Taung child is a result of differential deposition rates (i.e., the rate of deposition is higher on the supraorbital and infraorbital part of the lateral orbital rim than on the middle section of the lateral orbital rim). The remodeling process of this region in *H. sapiens* has been described (see Enlow 1966, 1968) and it seems that resorption of the central section of the lateral orbital rim is more likely than different rates of deposition.

In addition, the lateral border of the zygomatic frontal process of the Taung child has a marginal tubercle where the temporal fascia fibers originate in *H. sapiens*. The marginal tubercle lacks in some *H. sapiens* (Hauser and De Stefano 1989) and it was not expressed in any of the *P. troglodytes* specimens used in this study.

The maxillary tuberosity in *P. troglodytes* appears more posteriorly positioned than it is in AL. 333-105, the Taung child and *H. sapiens* at similar stages of dental development. The position of the maxillary tuberosity depends on the pterygomaxillary suture that is a site of rapid bone deposition (Enlow and Bang 1965; Enlow 1968). The deposition rate at the pterygomaxillary suture is higher in *P. troglodytes* than in *H. sapiens* (Enlow 1968). According to Figure 12-1, the linear growth of the aveolar maxilla at the AL. 333-105 stage of dental development is less than in *P. troglodytes.* This observation is also true for the Taung child (Figure 12-2). This demonstrates that the posterior linear growth of the aveolar maxilla in *H. sapiens*, the Taung child and AL. 333-105 is less than in *P. troglodytes* at similar stages of dental development respectively.

In *H. sapiens* the quiescent time between the M1 and the M2 emergence is longer than in *P. troglodytes*. The M2 development in *P. troglodytes* of similar stage of dental development to AL. 333-105 or the Taung child is more advanced. It seems therefore, that the M2 development in *P. troglodytes* at a similar stage of dental development to Taung or AL. 333-105 is different from that of the Taung child and AL 333-105. Hence, the maxillary aveolar posterior linear growth in *Australopithecus* was somehow delayed compared with that of *P. troglodytes*. The difference suggests that room for the M2 was not as urgently needed at this stage in Taung or AL 333-105 as in *P. troglodytes*. The aveolar developmental pattern seems to suggest that it is possible that the maxillary aveolar development and the weaning pattern of *Australopithecus* was unlike that of *P. troglodytes*. It is unclear, however, whether the development pattern of the maxilla was responsible for the delayed development of the permanent M2 or vice versa. In addition, the shape of the maxilla of AL. 333-105 and the Taung child (Figures 12-1 and 12-2) indicates that their growth appears to have been humanlike to some degree (see Brown et al. 1983). For example, the area between the dm1 and the maxillary tuberosity is convex-like in AL. 333-105 and the Taung child while it is almost straight in *P. troglodytes*. The difference in configuration of this region between *P. troglodytes* and *Australopithecus* suggests that the remodeling pattern in this region was likely different.

Also, unlike in *P. troglodytes*, the posterior surface of the zygomatic bones of *H. sapiens*, the Taung child, and AL. 333-105 is concave with no cresting along the sphenozygomatic suture (Figure 12- 3). These features suggest humanlike morphogenetic processes in *Australopithecus* at the AL. 333-105 and the Taung child development stages.

In general, AL. 333-105 and the Taung child show a mosaic of facial morphological features that possibly imply the same for the morphogenetic processes and mechanisms that were responsible for their formation. However, as we now know, *H. sapiens* facial topography is the final product of intrinsic and extrinsic factors. It is the outcome of cranial base growth, premaxillary suture growth, remodeling, and facial bone displacements (Ashton 1981). The same processes account for the *A. afarensis*, *A. africanus* and *P. troglodytes* facial form. But the rates and duration are different. Whether orthognathy in *H. sapiens* is a result of canine size reduction (see Greenfield 1990) or partly is a function of the need to perfect bipedality is still debatable (see Brues 1966; Enlow 1968). The change from a quadruped to a biped seems to have occurred abruptly (see Helmuth 1985; Latimer and Lovejoy 1990), although the face remained basically pongidlike (i.e., prognathic in both *A. afarensis and A. africanus).*

Although the facial morphologies of the Taung child and AL. 333-105 appear prognathic, the premaxilla and the area between the nasospine and prognathion are slightly convex, while they maintain a continuous slope in *P. troglodytes*. Also in adult *A. afarensis* (see AL. 200-1) and adult *A. africanus* this region is somehow convex, and the incisor teeth are not horizontally oriented as in *P. troglodytes*, but curved downward. This subnasal region shape casts doubts on earlier observations that the permanent incisors erupt horizontally in the Taung child (Conroy and Vannier 1987, 1988). Although the morphologies of the Taung child and AL. 333-105 suggest that *A. afarensis* and *A. africanus* had both humanlike and pongidlike morphogenetic processes, the humanlike tendencies appear to exceed the pongidlike tendencies. The most likely reason for retaining both the pongidlike and humanlike morphogenetic processes could be to maintain the relationships of the different parts of the skeleton as evolution was proceeding (see Wake et al. 1983; Smith and Ranyard 1980). Wake et al. (1983) argued that ontogenetic and phylogenetic change of one element is determined by the structural and functional properties of the other elements. This argument partly explains why the evolution of *H. sapiens* ontogeny was likely mosaic in nature. Morphological studies of adult *A. afarensis* show that the various anatomical regions of the skeleton evolved differently. Retention of a mosaic developmental pattern seems to be a logical endeavor for *Australopithecus* to avoid extinction.

The available data on cranial morphology show that pongidlike morphogenetic processes can be traced in *A. afarensis, A. africanus, A. robustus, A. boisei, H. habilis,* and *H. erectus*. In addition, morphological evidence suggests a gradual shift from pongidlike in *A. afarensis* to humanlike in *H. erectus*. But it is worth noting that the mechanisms and genetics underlying the hominid morphology, including their mechanical and functional aspects, are poorly understood (see Atchley 1984; Hall 1982). Thus, the application of the Taung child and AL. 333-105 morphological characters to predicting the growth or development patterns of *A. afarensis* and *A. africanus* respectively requires caution.

Although it is reasonable to conclude that by inference, the humanlike and the pongidlike morphological features suggest the same morphogenetic processes led to the same features in *Australopithecus*, it is important to note a warning by Ghiselin (1980:181) who wrote that, "Morphology has contributed so little primarily because it has had little to contribute. It is a descriptive science of form, and only when conjoined with other disciplines does it tell us anything about causes. But once a causal mechanism has been accepted, it can provide a valuable service." Despite Ghiselin's warning, it is important to note that morphological features are widely used to differentiate the various hominoids. The rationale of using morphological characters as a source of clues for the morphogenetic processes in *Australopithecus* cannot be underestimated. This study, however, can only conclude that morphogenesis in *Australopithecus* was mosaic in nature and that the duration and rates seem to have changed with the stage of development. Also, it is argued that the globular frontal bone lasted very long in *Australopithecus* compared with *P. troglodytes*, and that we are not sure at what stage of development the globular frontal transformed into the adult *A. afarensis* or *A. africanus* frontal form.

SUMMARY

Classifying the developmental biology of the early hominids as pongidlike or humanlike based on morphological attributes has several problems. However, the presence of humanlike and pongidlike features by inference suggests the same morphogenetic processes were at work. The morphological and metric data suggest the developmental biology of the early hominids was complex. Both the morphological data of this study and most of the dental and also remodeling studies give conflicting views. This can be attributed to methodological weaknesses in revealing the actual relationships or of phylogenetic complexities (see, e.g., Anemone and Watts 1992). For example, the striae of Retzius and perikymata count techniques are still in their infancy stages of development (Foley and Cruweys 1986) and have many weaknesses.

This study has not dealt with the phylogenetic relationships of the features investigated. Yet, it has revealed several issues. First, the morphological features of AL. 333-105 and the Taung child cannot be used confidently to infer morphogenetic processes and the life cycles of either *A. afarensis* or *A. africanus*. In general, the morphological data show that the globular frontal was prolonged into childhood in *A. afarensis* and *A. africanus*. Second, it appears that the alveolar tuberosity linear growth of *A. afarensis* and *A. africanus* was delayed relative to *P. troglodytes*. This suggests that the developmental rate of the dental arcade was different compared with that of *P. troglodytes*. The developmental biology and morphogenetic processes of the early hominids, like their phylogeny, was complex, and any conclusions about their life cycles should be made with caution. Third, on the basis of the two observations, it is possible that *A. africanus* cannot be excluded from the line leading to early *Homo* in order to maintain evolutionary continuity in developmental changes, and therefore the evolution of *Homo* from *Australopithecus*. This view is supported by the adult morphology such as the presence of mandibular tori—a genetically controlled trait found in both *A. africanus* (see Sts 52b) and *H. erectus*. In short, the developmental biology of *Australopithecus*, like its phylogeny, remains a subject of debate. A clear cut "unique," "pongidlike" or "humanlike" morphogenesis or development patterns cannot confidently be assigned to *Australopithecus* based on our current knowledge.

ACKNOWLEDGMENTS

My thanks to Professors F. C. Howell and T. D. White for allowing access to their research materials. My appreciation extends to Karen Schmidt for editing my first draft of this chapter, Jean-Paul Boubli for photographic assistance and Dr. Robert Corruccini for the invitation to contribute this chapter. Financial support was partially provided by The Wenner-Gren Foundation for Anthropological Research and Lowie/Olson Fund, Department of Anthropology, University of California at Berkeley.

LITERATURE CITED

Abbie, A. A. (1952) A new approach to the problem of human Evolution. *Trans. Roy. Soc. S. Aust.* 75:70-88.

Aitchison, S. E. (1963) Some racial features of the jaws and teeth. *Brit. Dent. J.* 65:111-112.

Anemone, R. L., Watts, E. S., and Swindler D. R. (1991) Dental development of known-age chimpanzees, *Pan troglodytes* (Primate, Pongidae). *Am. J. Phys. Anthropol.* 86:229-241.

Anemone, R. L., and Watts, E. S. (1992) Dental development in apes and humans: A comment on Simpson, Lovejoy and Meindl (1990). *J. Hum. Evol.* 22:149-153.

Ardito, G. (1978) Checklist of the data on the gestation length of primates. *J. Hum. Evol.* 5:213-222.

Asfaw, B. (1988) Pliocene cranial remains from Ethiopia: New perspectives on the evolution of the early Hominid frontal bone. Ph.D. dissertation, University of California, Berkeley.

Ashton, E. H. (1981) The Australopithecinae. Their biometrical study. *Symp. Zool. Soc., Lond.* 46:67-126.

Atchley, W. R. (1984) Ontogeny, timing development and genetic variance structure. *Amer. Nat.* 123:519-540.

Bacon, A. M. (1989) Estimation de L'age a la mort des enfants actuels et fossiles a partir des stries d'accroissement de l'email dentaire. Advantages et incovenients de la methode. *Bull. Mem.Soc. Anthropol. Paris, n.s.,* 1:3-12.

Barret, M. J., Brown, T., and Luke, J. I. (1963a) Dental observations on Australian Aborigines. Mesiodistal crown diameters of deciduous teeth. *Aust. Dent. J.* 8:299-302.

Barret, M. J., Brown, T., and Macdonald, M. R. (1963b) Dental observations on Australian Aborigines. Mesiodistal crown diameters of permanent teeth. *Aust. Dent. J.* 8:150-155.

Beynon, A. D., and Dean, C. M. (1987) Crown-formation time of fossil hominid premolar tooth. *Arch. Oral Biol.* 32:773-780.

Beynon, A. D., and Dean, C. M. (1988) Distinct dental development patterns in early fossil hominids. *Nature* 335:509-514.

Beynon, A. D., and Dean, C. M. (1991a) Hominid dental development. *Nature* 351:196

Beynon, A. D., Dean, C. M., and Reid, D. J. (1991b) Histological study on the chronology of developing dentition in Gorilla and Orangutan. *Am. J. Phys. Anthropol.* 86:189-203.

Beynon, A. D., and Wood, B. A. (1986) Variation in enamel thickness and structure in East African hominids. *Am. J. Phys. Anthropol.* 70:177-193.

Beynon, A. D., and Wood, B. A. (1987) Pattern and rates of enamel growth in molar teeth of early hominids. *Nature* 326:493-496.

Biggerstaff, R. H. (1967) Time-trimmers to the Taung child, or how old is "*Australopithecus africanus*." *Am. Anthropol.* 69:217-220.

Bonner, A. R. (1982) *Evolution and Development.* Berlin: Springer-Verlag.

Bramblett, C. A. (1969) Non-metric skeletal age changes in the Darajani Baboon. *Am. J. Phys. Anthropol.* 30:161-172.

Bromage, T. G. (1985) Taung facial remodeling: A growth and development study. In P. V. Tobias (ed.): *Hominid evolution: Past, Present and Future.* New York: Liss, pp. 239-246.

Bromage, T. G. (1986). A comparative Scanning Electron Microscope study of facial growth and remodeling in early hominids. Ph.D. thesis, University of Toronto.

Bromage, T. G. (1987) The biological and chronological maturation of early hominids. *J. Hum. Evol.* 16:257-272.

Bromage, T. G. (1989) Ontogeny of the early hominid face. *J. Hum. Evol.* 18:751-773.

Bromage, T. G. (1990) Early hominid development and life history. In C.J. DeRousseau (ed.): *Primate Life History and Evolution. Monographs in Primatology* Vol. 14. New York: Wiley-Liss, pp. 105-113.

Bromage, T. G. (1991) Enamel incremental periodicity in the Pig tailed Macaque: A polychrome fluorescent labeling study of dental hard tissues. *Am. J. Phys. Anthropol.* 86:205-214.

Bromage, T. G., and Dean, M. C. (1985) Re-evaluation of the age at death of immature fossil hominids. *Nature* 317:525-527.

Bronowski, J., and Long, W. M. (1953) The australopithecine milk canines. *Nature* 172:251.

Broom, R. (1936) A new fossil anthropoid from South Africa. *Nature* 138:486-488.

Broom, R., and Robinson, J. T. (1951) Eruption of the permanent teeth in South African fossil ape-man. *Nature* 167:443.

Broom, R., and Robinson, J. T. (1952) The Swartkrans Apeman *Paranthropus crassidens. Transv. Mus. Mem.* 6:1-123.

Brown, T., Abbott, A. H., and Burgess, V. B. (1983) Age changes in dental arch in dimensions of Australian Aboriginals. *Am. J. Phys. Anthropol.* 62:291-303.

Brues, A. M. (1966) Probable mutation effect and the evolution of hominid teeth and jaws. *Am. J. Phys. Anthropol.* 25:169-170.

Burdi, A. R., Garn, S. M., and Miller, S. M. (1970) Development of the male dentition in the third trimester. *J. Dent. Res.* 49:889.

Conroy, G. C., and Mahoney, C. J. (1991) Mixed longitudinal study of dental emergence in the chimpanzee, *Pan troglodytes* (Primates, Pongidae). *Am. J. Phys. Anthropol.* 86:243-254.

Conroy, G. C., and Vannier, M. W. (1987) Dental development of the Taung skull from computerized tomography. *Nature* 329:625-627.

Conroy, G. C., and Vannier, M. W. (1988) The nature of Taung dental maturation continued. *Nature* 333:808.

Conroy, G. C., and Vannier, M. W. (1989) The Taung skull revisited. New evidence from high resolution computed tomography. *S. Afr. J. Sci.* 85:30-32.

Conroy, G. C., and Vannier, M. W. (1991a) Dental development in Southern African australopithecines. Part 1. Problems of pattern and chronology. *Am. J. Phys. Anthropol.* 86:121-136.

Conroy, G. C., and Vannier, M. W. (1991b) Dental development in Southern African australopithecines. Part 2. Dental stages assessment. *Am. J. Phys. Anthropol.* 86:137-156.

Currey, J. D., and Pond, C. M. (1989) Mechanical properties of very young bone in the axis deer (*Axis axis*) and Humans. *J. Zool. Lond.* 218:59-67.

Dahlberg, A. A. (1965) Evolutionary background of dental and facial growth. *J. Dent. Res.* 44 (suppl.): 151-160.

Dart, R. A. (1925) *Australopithecus africanus*: the ape-man of South Africa. *Nature* 115:195-199.

Dart, R. A. (1926) Taungs and its Significance. *Nat. Hist.* 26:315-327.

Dart, R. A. (1929) A note on the Taung skull. *S. Afr. J. Sci.* 26:648-658.

Dart, R. A. (1948) The infancy of *Australopithecus.* In *Robert Broom Commemorative Volume.* Special publication of the Roy. Soc. South Africa, pp. 143-152.

Dart, R. A. (1962) A cleft adult mandible and the nine other lower jaw fragments from Makapansgat. *Am. J. Phys. Anthropol.* 20:267-286.

Darwin, C. (1871) *The Descent of Man, and Selection in Relation to Sex.* London: Murray.

De Beer. G. R. (1930) *Embryology and Evolution.* Oxford: Oxford University Press.

De Beer, G. R. (1948) Embyology and evolution of man. In *Robert Broom Commemorative Volume.* Special publication of the Roy. Soc. South Africa, pp. 181-190.

Dean, M. C. (1985) The eruption pattern of the permanent incisors and permanent first molars in *Australopithecus (Paranthropus) robustus*. *Am. J. Phys. Anthropol.* 67:251-257.
Dean, M. C. (1986) *Homo and Paranthropus*: Similarities in the cranial base and developing dentition. In B. A. Wood and L. B. Martin (eds.): *Human Evolution*. Cambridge: Cambridge University Press, pp. 249-265.
Dean, M. C. (1987) The dental development status of six East African juvenile fossil hominids. *J. Hum. Evol.* 6:197-213.
Dean, M. C. (1988) Growth of teeth and development of the dentition in *Parathropus*. In F. E. Grine (ed.): *Evolutionary History of "Robust" Australopithecines*. New York: Aldine de Gruyter, pp. 43-53.
Dean, M. C., and Beynon, A. D. (1991a) Tooth crown heights, tooth wear, sexual dimorphism and jaw growth in hominoids. *Z. Morph. Anthropol.* 78:425-440.
Dean, M. C., and Beynon, A. D. (1991b) Histological reconstruction of crown formation times and initial root formation times in a modern human child. *Am. J. Phys. Anthropol.* 86:215-228.
Dean, M. C., Stringer, C. B., and Bromage, T. G. (1986) Age at death of the neanderthal child from Devil's Tower, Gilbratar and implications for studies of general growth and development. *Am. J. Phys. Anthropol.* 70:301-309.
Dean, M. C., and Wood, B. A. (1984) Phylogeny, neoteny and growth of the cranial base in hominids. *Folia Primatol.* 43:157-180.
de Laat, S. W. (1992) Developmental biology. *Communications J. ESF* 26:18-19.
Drennan, M. R. (1931) Pedomorphism in the pre-Bushman skull. *Am. J. Phys. Anthropol.* 16:203-210.
Drennan, M. R. (1938) Human growth and differentiation. *S. Afr. J. Sci.* 33:64-91
Dubois, E. (1894) *Pithecanthropus erectus*. Eine menschenahliche Ubergangsform aus Java. *Jaarb. Mijnwezen iNederl. Cost Indie* 24:5-77.
Enlow, D. J. (1976) The remodeling of bone. *Yearb. Phys. Anthropol.* 20:19-34.
Enlow D. H. (1966) A comparative study of facial growth in *Homo* and *Macaca*. *Am. J. Phys. Anthropol.* 24: 293-308.
Enlow D. H. (1968) *The Human Face*. London: Hoeber Medical Division, Harper & Row.
Enlow D. H., and Bang, S. (1965) Growth and remodeling of the human maxilla. *Am. J. Orthodont.* 51: 446-464.
Ericksen, M. F. (1991) Histologic estimation of age at death using the anterior cortex of femur. *Am. J. Phys. Anthropol.* 84:171-179.
Evans, F. G. (1976). Age changes in mechanical properties and histology of human bone. *Yearb. Phys. Anthropol.* 20:57-72.
Fanning, E. A., and Moorrees, C. F. A. (1969) A comparison of permanent mandibular molar formation in Australian Aborigines and caucasoids. *Arch. Oral Biol.* 14:999-1006.
Foley, R., and Cruweys, E. (1986) Dental anthropology: Problems and perspectives. In E. Cruweys and R. A. Foley (eds.): *Teeth and Anthropology: Problems and Perspectives*. B. A. R. International Series *291*:1-20. .
Frazer, J. F. D. (1973) Gestation period for *Australopithecus*. *Nature* 242:342.
Garn, M. S., and Lewis, A. B. (1963) Phylogenetic and intraspecific variation in tooth sequence polymorphism. In D. R. Brothwell (ed.): *Symposia of the Society for the Study of Human Biology. Dental Anthropology* Vol. 5. Oxford: Pregamon Press, pp. 53-73.
Ghiselin, M. T. (1980) The failure of morphology to assimilate Darwinism. In E. Mayr and W. B. Provine (eds.): *The Evolutionary Synthesis*. Cambridge: Harvard University, Press, pp. 180-193.
Goodman, A. H., and Rose, J. C. (1990) Assessment of systematic physiological perturbations from dental enamel hypoplasias and associated histological structures. *Yearb. Phys. Anthropol.* 33:59-110.
Gould, S. J. (1977) *Ontogeny and Phylogeny*. Cambridge, Mass. Harvard University Press.
Greenfield, L. O. (1990) Canine reduction in early man: A critique of three mechanical models. *Hum. Evol.* 5:213-226.
Grine, F. (1991) Computed tomography and the measurement of enamel thickness in extant hominids: Implications for its paleontological application. *Paleont. Africana* 28: 61-69.
Haeckel, E. (1868) *Naturliche Scopfngsgeschite: wissenschaftliche Vortrage uber die Entwicklungslehre im Allgemeinen und Diejenige von Darwin, Goethe und Lamarck im Besonderen*. Berlin.
Hajnis, K., Kustra, T., Farkas, L. G., and Fegilava, B. (1967) Sinus maxillaris. *Z. Morph. Anthropol.* 59:185.
Hall, B. K. (1982) How is mandibular growth controlled during development and evolution? *J. Craniofac. Genet. Devel. Biol.* 2:45-49.
Hauser, G. and De Stefano, G. F. (1989) *Epigenetic Variants of the Human Skull*. Stuttgart: E. Schweizerbart'sche Verlagsbuchhandlung (Nagele u. Ormilller).
Helmuth, H. (1985) Biomechanics, evolution and upright stature. *Anthropol. Anz.* 43:1-9.
Heim, J. L. (1982) *Les Enfants Neandertaliens de La Ferrassie*. Paris: Masson.
Hemmer, H. (1969) A new view of evolution of man. *Curr. Anthropol.* 10:179-180.
Hoffman, M. D. (1979) Age estimations from diaphyseal lengths. Two months to twelve years. *J. For Sci.* 24:461-469.
Hrdlicka, A. (1925) Taung's ape. *Am. J. Phys. Anthropol.* 8:379.
Hublin, J. J. (1980) La Chaise (Suard), Engis 2 et La Quina H18 : development de la morphologie occipitale externe chez l'enfant preneanderthalien et neanderthalien. *C. R. Acad. Sci. Paris* 291:669-672.
Keith, A. (1925) The fossil anthropoid from Taung. *Nature* 115:236.
Keith, A. (1927) The evolution of man. How anthropology and anatomy combine to prove man's slow ascent from lower forms of life. In J. A. Hammerton (ed.): *Universal History of the World*, Vol. 1, London: Amalgamated Press Ltd, pp. 141-185.

Keith, A. (1931) *New discoveries relating to the antiquity of man.* London: Williams and Norgate.

Kimbel, W. H. (1986) Calvarial morphology of the *Australopithecus afarensis*: A comparative phylogenetic study. Ph.D. thesis, Kent State University, Ohio.

Kimbel, W. H., White, T. D., and Johanson, D. C. (1988) Implications of KNM-WT 17000 for the evolution of "Robust" *Australopithecus*. In F. E. Grine (ed.): *Evolutionary History of the "Robust" Australopithecines*. New York: Aldine de Gruyter, pp. 259-268.

Kimbel, W. H., Johanson, D. C., and Coppens, Y. (1982) Pliocene hominid cranial remains from Hadar Formation, Ethiopia. *Am. J. Phys. Anthropol.* 57:453-499.

Koski, K., and Garn, S. M. (1957) Tooth eruption sequence in fossil and modern man. *Am. J. Phys. Anthropol.* 15:469-487.

Kovacs, I. (1971) A systematic description of dental roots. In A. A. Dahlberg (ed.): *Dental Morphology and Evolution*: Chicago and London. The University of Chicago Press pp. 211-256.

Krogman, W. M. (1969) Growth changes in skull, face, jaws, and teeth of the chimpanzee. In G. H. Bourne (ed.): *The Chimpanzee* Vol. 1. Basel/New York: Karger, pp. 104-164.

Kyauka, P. S. (1992) *Australopithecus afarensis* vertebrae structural patterns and their implications for its locomotion (abstract). *Am. J. Phys. Anthropol. Suppl.* 14:104.

Kyauka, P. S. (in preparation) A comparative study of the Laetoli Hominid Z1 skeleton and its implications for the developmental biology of *Australopithecus afarensis*. Ph. D. Dissertation, University of California, Berkeley.

Latimer, B., and Lovejoy, C. O. (1990) Metatarphalangeal joints of *Australopithecus afarensis*. *Am. J. Phys. Anthropol.* 83:13-23.

Lee, M., Chan, S. T., Law, W. D., and Chang, K. (1965) The relationship between dental and skeletal maturation in Chinese children. *Arch. Oral Biol.* 10:883

Legoux, P. (1965) Determination de l'age dentaire de l'infant Neanderthalian de Roc-de Marsal. *Rev. Franc. Ondontostom.* 12:1571-1592.

Legoux, P. (1966) *Determination de l'age dentaire des fossiles de la lignee humaine*. Paris: Maloine.

Legoux, P. (1970) Etude odontologie de infant Neanderthalien du Perch-de-l'Aze. *Arch. Inst. Paleontol. Hum. Mem.* 33:53-87.

Le Gros Clark, W. E. (1950) South African Fossil Hominoids. *Nature* 166:791

Le Gros Clark, W. E. (1967) *Man-apes or ape-man?* New York: Holt, Rinehart and Winston.

Leutenegger, W. (1972) Newborn size and pelvic dimensions of *Australopithecus*. *Nature* 240:568-569.

Leutenegger, W. (1973) Gestation period and birth weight of *Australopithecus*. *Nature* 243:548.

Leutenegger, W. (1987) Neonatal brain size and neurocranial dimensions in Pliocene hominids: Implications for obstetrics. *J. Hum. Evol.* 16:291-296.

Liebgott, B. (1986) *The Anatomical Basis of Dentistry*. Burlington, Ontario: B.C. Decker, pp. 464-482.

Mann, A. E. (1975) *Some aspects of the South African Australopithecines*. University of Pennsylvania Publications in Anthropology No. 1.

Mann, A. E. (1988) The nature of Taung dental maturation. *Nature* 333:123.

Mann, A. E., Lampl, M., and Monge, J. (1987) Maturation patterns in early hominids. *Nature* 328:673-674.

Mann, A. E., Lampl, M., and Monge, J. (1990a) Patterns of ontogeny in human evolution: Evidence from dental development. *Yearb. Phys. Anthropol.* 33:11-150.

Mann, A. E., Monge, J., and Lampl, M. (1990b) Dental caution. *Nature* 348:202.

Mann, A. E., Monge, J., and Lampl, M. (1991) Investigation into the relationship between Perikymata counts and crown formation times. *Am. J. Phys. Anthropol.* 86:175-188.

McCown, T. D., and Keith, A. (1939). *The Stone Age of Mount Carmel*, Vol. 2. *The Fossil Human Remains from the Levalloiso-Mousterian*. Oxford: Oxford University Press.

McKinley K. R. (1971) Survivorship in gracile and robust Australopithecines: A demographic comparison and a proposed birth model. *Am. J. Phys. Anthropol.* 34:417-426.

McKee, J. K., and Molnar, S. (1988) Measurement of tooth wear among Australian Aborigines: 2. Intrapopulational variation in patterns of dental attrition. *Am. J. Phys. Anthropol.* 76:125-136.

Moller, I. J. (1967) Influence of microelements on the morphology of the teeth. *J. Dent. Res.* 46:933-941.

Molnar, S. (1972) Tooth wear and culture: A survey of tooth functions among some prehistoric populations. *Curr. Anthropol.* 13:511-526.

Montagu, M. F. A. (1955) Time, morphology and neoteny in the evolution of man. *Am. Anthropol.* 57:13-27.

Moore, W. J., and Lavelle, C. L. B. (1974) *Growth of the Facial Skeleton in the Hominoidea*. New York: Academic.

Muller, G. (1935) Zur bestimmung der Lange beschadigter Extremitatenknochen. *Anthropol. Anz.* 12:70-72.

Nissen, H. W., and Riesen, A. (1964) The eruption of the permanent dentition of the chimpanzee. *Am. J. Phys. Anthropol.* 22:285-294.

Nolla, C. M. (1960) The development of the permanent teeth. *J. Dent. Child.* 27:254-266.

Olson, T. R. (1985) Taxonomic affinities of the immature hominid crania from Hadar and Taung. *Nature* 316:539-540.

Ortner, D. J. (1976). Microscopic and molecular biology of human compact bone: An anthropological perspective . *Yearb. Phys. Anthropol.* 20:39-44.

Oxnard, C. (1983) *The Order of Man: A Biomathematical Anatomy of the Primates*. Hong Kong: Hong Kong University Press.

Rak, Y., and Howell, F. C. (1978) Cranium of a juvenile *Australopithecus boisei* from the Lower Omo Basin Ethiopia. *Am. J. Phys. Anthropol.* 48:345-366.

Richards, G. D. (1985) Analysis of a microcephalic child from the late period (ca. 1100-1700 A.D.) of central California. *Am. J. Phys. Anthropol.* 68:343-357.

Sacher, G. A. (1975) Maturation and longevity in relation to cranial capacity in hominid evolution. In R. Tuttle (ed.):

Primate Functional Morphology and Evolution. The Hague: Mouton, pp. 417-441.
Schultz, A. H. (1940) Growth and development of the chimpanzee. Carnegie Institution Publication 518. *Contr. Embryol.* 170:1-63.
Schultz, A. H. (1941) Growth and development of the orangutan Carnegie. Institution of Washington Publication 525: *Contr. Embryol.* 29: 57-110.
Schultz, A. H. (1949) Ontogenetic specializations in man. *Archive Julius Klaus-Stifftung* 24:197-216.
Schultz, A. H. (1952) Vergleichende Untersuchungen an einigen menschlin Specialzation. *Bull. Schweiz. Ges. Anthropol. Ethnol.* 28:25-37..
Schultz, A. H. (1956) Postembryonic age changes. *Primatologica* 1:888-964.
Schultz, A. H. (1960) Age changes in primates and their modification in man. In J. M. Tanner (ed.): *Human Growth*. Oxford: Pergamon, pp. 1-20.
Schultz, A. H. (1962) Metric age changes and sex differences in primate skulls. *Z. Morph. Anthropol.* 52:239-255.
Schwartz, J. H. and Langdon, L. H. (1991) Innervation of the human upper primary dentition: Implications for understanding tooth initiation and rethinking growth and eruption patterns. *Am. J. Phys. Anthropol.* 86:273-286.
Shea, B. T. (1989) Heterochrony in human evolution: The case for neoteny reconsidered. *Yearb. Phys. Anthropol.* 32:69-101.
Siebert, J. S., and Swindler, D. R. (1991) Perinatal dental development in the chimpanzee (*Pan troglodytes*). *Am. J. Phys. Anthropol.* 86:287-294.
Simpson, S. W., Lovejoy, C. O., and Meidel, R. S. (1990) Hominoid dental Maturation. *J. Hum. Evol.* 19:285-299.
Simpson, S. W., Lovejoy, C. O., and Meidel, R. S. (1991) Relative dental development in hominids and its failure to predict somatic growth velocity. *Am. J. Phys. Anthropol.* 86:113-120.
Simmons, D. J. (1985) Options for bone aging with the microscope. *Yearb. Phys. Anthropol.* 28:249-263.
Singh, J. J., and Gunberg, D. L. (1970) Estimation of age at death in human males from quantitative histology of bone fragments. *Am. J. Phys. Anthropol. 33*:373-382.
Singh, I. J., Tonna, A., and Gandel, C. P. (1974) A comparative histological study of mammalian bone. *J. Morph.* 144:421-438.
Skinner, M. (1978) Dental maturation, dental attrition and growth of the skull in fossil Hominidae. Ph.D. thesis, University of Cambridge, UK.
Smith, B. H. (1986) Dental development and the evolution of life history in Hominidae. *Am. J. Phys. Anthropol.* 86:157-174.
Smith, B. H. (1987) Replies. *Nature* 328:674.
Smith, B. H. (1989) Growth and development and its significance for early hominid behavior. *Ossa* 14:63-96.
Smith, B. H. (1991a) Dental development in *Australopithecus* and early *Homo. Nature* 323:327-330.
Smith, B. H. (1991b) Dental development and the evolution of life history in Hominidae. *Am. J. Phys. Anthropol.* 86:157-174.
Smith, B. H., and Ranyard, G. (1980) Evolution of the supraorbital region in Upper Pleistocene fossil hominids from Central Europe. *Am. J. Phys. Anthropol.* 53:589-610.
Starck, D. (1961) Ontogenetic development of skull of primates. Internat. Colloq. on Evolution of Mammals, *Kon. Vlaamse Acad. Wetensch Lett. Sch. Kunsten Belgie, Brussel*, Part 1, pp. 205-214.
Steele, D. G. (1970) Estimation of stature from fragments of long limb bones. In T. D. Stewart (ed.): *Personal Identification in Mass Disasters.* Washington D.C.: Smithsonian Institution: pp. 85-98.
Steele, D. G., and McKern, T. W. (1969) A method for assessment of maximum long bone length and living stature from fragmentary long bones. *Am. J. Phys. Anthropol.* 25:319-322.
Stout, S. D., and Simmons, D. J. (1979) Use of histology in ancient bone research. *Yearb. Phys. Anthropol.* 22:228-249.
Stout, S. D., and Stanley, S. C. (1991) Percent osteonal bone versus osteon counts: The variable of choice for estimating age at death. *Am. J. Phys. Anthropol.* 86:515-519.
Straus, J. R. (1927) Growth of the Human foot and its evolutionary significance. Contrib. Embryol. *101. Carnegie Institution, Washington. Publ. 380*:93-134.
Stringer, C., Dean, C. M., and Martin, R. D. (1990) A comparative study of cranial and dental development within a recent British sample and among Neanderthals. In J. C. DeRouseau (ed.): *Primate Life History and Evolution*. Monographs in Primatology, 14. New York: Wiley-Liss, pp. 115-152.
Swindler, D. R. (1985) Nonhuman primate dental development and its relationship to human dental development. In E. S. Watts (ed.): *Nonhuman Primate Models for Human Growth and Development*. Monographs in Primatology, Vol. 6. New York: Alan R. Liss, pp. 67-94.
Tappen, N. C. (1989) Microscopic investigations of areolar brow structure in nonhuman primates and of vermiculate brow structure in hominids. *Folia Primatol.* 51:112-125.
Ten Cate, R. (1976a) Tooth eruption. In S. N. Bhaskar (ed.): *Orban's Oral Histology and Embryology*. St. Louis: Mosby, pp. 361-375.
Ten Cate, R. (1976b) Shedding of deciduous teeth. In S. N. Bhaskar (ed.): *Orban's Oral Histology and Embryology*. St. Louis: Mosby, pp. 376-394.
Thoma, A. (1963) The dentition of the Subalyuk Neandertal child. *Z. Morph. Anthropol.* 54:127-150.
Thompkins, R. L., and Trinkaus, E. (1987) La Ferrassie 6 and the development of Neanderthal pubic morphology. *Am. J. Phys. Anthropol.* 73:233-239.
Thompson, G. W., Anderson, D. L., and Popovich, F. (1975) Sexual dimorphism in dentition mineralization. *Growth* 39:289-301.
Tillier, A. M. (1986) Quelques aspects de l'ontologenese du squelette cranien des neanderthaliens. *Anthropos (Brno)* 23:207-16.
Tobias, P. V. (1968) The age of death among the Australopithecines. *The Anthropologist (Special Volume)*, Department of Anthropology, University of Delhi.

Trinkaus, E. (1986) The neanderthals and modern human origins. *Ann. Rev. Anthropol.* 15:193-218.

Trinkaus, E., and Thompkins, R. (1990) The neandertal life cycle: The possibility, probability, and perceptibility of contrasts with recent humans. In J. C. DeRouseau (ed.): *Primate Life History and Evolution. Monographs in Primatology*, Vol. 14. New York: Wiley-Liss, pp. 153-180.

Vallois, H. (1937) La duree de la vie chez l'homme fossile. *L'Anthropologie* 47:499-532.

Vlcek, E. (1970) Etude comparative onto-phylogenetique de l'infant du Pech-de-L'Aze par rapport a d'autres enfants Neandertaliens. In L' infant du Pech-de-L'Aze. *Arch. Inst. Paleontol. Hum.* 33:149-78.

Wake, D. B., Roth, G., and Wake, H. M. (1983) On the problem of stasis in organismal evolution. *J. Theor. Biol.* 101:211-224.

Washburn, S. L. (1951) Evolutionary importance of the South African "Man-apes." *Nature* 167:650.

Watts, E. S. (1986) Skeletal development. In W. R. Dukelow and J. Erwin (eds.): *Comparative Primate Biology. 3. Reproduction and Development.* New York: Alan R. Liss, pp. 415-440.

Weidenreich, F. (1939) The duration of life in fossil man in China and the pathological lesions. *Chinese Med. J.* 55:34-44.

Weidenreich, F. (1941) The brain and its role in the phylogenetic transformation of the human skull. *Trans. Am. Phil. Soc.* 31:321-442.

Weidenreich, F.. (1951). Morphology of Solo man. *Anthropol. Pap. Am. Mus. Nat. Hist.* 43:205-290.

Weiss, P. (1939) *Principles of Development.* New York: Henry Holt.

White, T. D. (1978) Early hominid enamel hypoplasia. *Am. J. Phys. Anthropol.* 49:79-84.

Winkler, L. A., Schwartz, J. H., and Swindler, D. R. (1991) Aspects of dental development in the orangutan prior to eruption of the permanent dentition. *Am. J. Phys. Anthropol.* 86:255-271.

Wolpoff, M. H. (1979) The Krapina dental remains. *Am. J. Phys. Anthropol.* 50:67-114.

Wolpoff, M. H., Monge, J. M, and Lampl, M. (1988) Was Taung human or ape? *Nature* 335:501.

Yaeger, J. A. (1976) Enamel. In S. N. Bhaskar (ed.): *Orban's Oral Histology and Embryology*. St. Louis: Mosby, pp. 45-104.

Zuckerman, S. (1928) Age-changes in the chimpanzee, with special reference to growth of brain, eruption of teeth, and estimation of age, with a note on the Taung's ape. *Proc. Zool. Soc. Lond.* 1:1-42.

CHAPTER 13

Early Hominid Postcrania

Phylogeny and Function

Henry M. McHenry

INTRODUCTION

At the time Howell's magnificent exposition of African Hominidae went to press (Howell 1978), very little was known about the postcrania of *Australopithecus* or early *Homo*. Most of the 200-plus postcranial specimens of *A. afarensis* had not been described. The *A. africanus* sample of South Africa consisted of a fragmentary partial skeleton (Sts 14), a fragmentary shoulder (Sts 7), a capitate (TM 1526), a few vertebrae, a few fragmentary pelvic fragments, and some scraps of femur. Now there are at least 100 specimens including a new partial skeleton. Almost no postcranials were known with certainty to be associated with taxonomically identifiable craniodental remains of *A. boisei*. Now there is a partial skeleton. There were only about a dozen postcranial specimens of the "robust" australopithecine of South Africa compared to the current sample of three dozen. The fragmentary partial skeleton attributed to *H. habilis* (O.H. 62) was discovered long after Howell wrote this paper as was the almost complete skeleton of *H. erectus* (KNM-WT 15000).

The purpose of this chapter is to reassess the postcranium in species of early hominids. In Howell's (1978) great treatise are taxonomic attributions, body size estimations, and descriptions that need revision based on new discoveries and interpretations. The descriptions emphasize the postcranial characteristics of species that are potentially useful to phylogenetic analysis. The focus, therefore, is on shared derived traits (synapomorphies), although the significance of primitive retentions and automorphies is not ignored.

Australopithecus afarensis

The postcranial remains of *A. afarensis* are abundant (164 specimens plus 69 specimens associated with a single individual, A.L. 288-1), beautifully described (Bush et al. 1982; Johanson et al. 1982a,b; Lovejoy et al. 1982a,b,c; Latimer et al. 1982; White 1980) and extensively interpreted (e.g. Aiello and Dean 1990; Asfaw

TABLE 13-1 Postcrania attributed to *Australopithecus afarensis*

A.L. 128-1 LEFT PROXIMAL FEMUR
A.L. 129-1A RIGHT DISTAL FEMUR
A.L. 129-1B RIGHT PROXIMAL TIBIA
A.L. 129-1C RIGHT PROXIMAL FEMUR FRAG
A.L. 129-52 LEFT ISCHIUM
A.L. 137-48A RIGHT DISTAL HUMERUS
A.L. 137-48B RIGHT DISTAL ULNA
A.L. 211-1 RIGHT PROXIMAL FRAG FEMUR
A.L. 228-1 RIGHT DISTAL FEMUR
A.L. 288-1 PARTIAL SKELETON (68 SPECIMENS)
A.L. 322-1 LEFT DISTAL HUMERUS
A.L. 332-25 INTERMEDIATE PHALANX HAND
A.L. 332-32 INTERMEDIATE PHALANX HAND
A.L. 333-3 RIGHT PROXIMAL FEMUR
A.L. 333-4 RIGHT DISTAL FEMUR
A.L. 333-5 LEFT PROXIMAL TIBIA
A.L. 333-6 LEFT DISTAL TIBIA
A.L. 333-7 LEFT DISTAL TIBIA
A.L. 333-8 RIGHT CALCANEUS FRAG
A.L. 333-9A RIGHT DISTAL FIBULA
A.L. 333-9B LEFT DISTAL FIBULA
A.L. 333-11 RIGHT PROXIMAL ULNA
A.L. 333-12 RIGHT DISTAL ULNA
A.L. 333-13 LEFT METATARSAL V PROX FRAG
A.L. 333-14 RIGHT METACARPAL V
A.L. 333-15 PROX MT II
A.L. 333-16 LEFT METACARPAL III
A.L. 333-17 RIGHT DISTAL METACARPAL V
A.L. 333-18 RIGHT DISTAL METACARPAL IV
A.L. 333-19 PROXIMAL HAND PHALANX
A.L. 333-20 PROXIMAL HAND PHALANX
A.L. 333-21 METATARSAL I DISTAL FRAG
A.L. 333-22 DISTAL PHALANX FRAG FOOT
A.L. 333-25 INTERMEDIATE HAND PHALANX
A.L. 333-26 PROXIMAL PHALANX FOOT
A.L. 333-27 LEFT DISTAL METACARPAL V
A.L. 333-28 RIGHT MEDIAL CUNEIFORM
A.L. 333-29 LEFT DISTAL HUMERUS
A.L. 333-31 PROXIMAL HAND PHALANX
A.L. 333-32 INTERMEDIATE HAND PHALANX
A.L. 333-33 PROXIMAL HAND PHALANX
A.L. 333-34 DISTAL FRAG METAPODIAL
A.L. 333-36 RIGHT NAVICULAR FOOT
A.L. 333-37 RIGHT CALCANEUS FRAG
A.L. 333-38 RIGHT PROXIMAL ULNA
A.L. 333-39 LEFT PROXIMAL TIBIA
A.L. 333-40 RIGHT CAPITATE
A.L. 333-41 RIGHT MEDIAL CONDYLE FEMUR
A.L. 333-42 LEFT PROXIMAL TIBIA
A.L. 333-46 INTERMEDIATE HAND PHALANX
A.L. 333-47 RIGHT NAVICULAR
A.L. 333-48 LEFT METACARPAL II
A.L. 333-49 PROXIMAL HAND PHAL DIST FRAG
A.L. 333-50 RIGHT HAMATE
A.L. 333-51 THORACIC CENTRUM
A.L. 333-54 LEFT MT I PROX FRAG
A.L. 333-55 LEFT CALCANEUS FRAG
A.L. 333-56 LEFT METACARPAL IV
A.L. 333-57 PROXIMAL HAND PHALANX
A.L. 333-58 RIGHT PROXIMAL FRAG METACARPAL I
A.L. 333-60 PROXIMAL PHALANX FOOT
A.L. 333-61 LEFT DISTAL SHAFT FEMUR
A.L. 333-62 PROXIMAL HAND PHALANX
A.L. 333-63 PROXIMAL HAND PHALANX
A.L. 333-64 INTERMEDIATE HAND PHALANX
A.L. 333-65 RIGHT PROXIMAL METACARPAL III
A.L. 333-69 RIGHT PROXIMAL HAND PHALANX
A.L. 333-70 METACARPAL I
A.L. 333-71 PROXIMAL PHALANX FOOT
A.L. 333-72 METATARSAL DISTAL FRAG
A.L. 333-73 BODY LUMBAR VERTEBRA
A.L. 333-75 RIGHT HEAD TALUS
A.L. 333-78 LEFT MT V PROX FRAG
A.L. 333-79 LEFT LATERAL CUNEIFORM
A.L. 333-80 RIGHT TRAPEZIUM
A.L. 333-81 THORACIC CENTRUM
A.L. 333-83 ATLAS FRAG
A.L. 333-85 LEFT DISTAL FIBULA
A.L. 333-87 LEFT PROXIMAL HUMERUS
A.L. 333-88 INTERMEDIATE HAND PHALANX
A.L. 333-89 LEFT METACARPAL V
A.L. 333-91 LEFT PISIFORM
A.L. 333-93 PROXIMAL HAND PHALANX
A.L. 333-94 LEFT FRAG CLAVICLE
A.L. 333-95 RIGHT PROXIMAL FEMUR
A.L. 333-96 LEFT DISTAL TIBIA
A.L. 333-98 RIGHT PROXIMAL RADIUS
A.L. 333-101 AXIS VERTEBRA
A.L. 333-102 PROXIMAL PHALANX FOOT
A.L. 333-106 CERVICAL VERTEBRA
A.L. 333-107 RIGHT PROXIMAL HUMERUS
A.L. 333-109 RIGHT SHAFT FRAG HUMERUS
A.L. 333-110 LEFT DISTAL FEMUR
A.L. 333-111 RIGHT DISTAL FEMUR
A.L. 333-115 LEFT PARTIAL FOOT
A.L. 333-117 NECK FRAG FEMUR
A.L. 333-118 HEAD RIB
A.L. 333-119 LEFT PROXIMAL ULNA
A.L. 333-120 DISTAL FIBULA
A.L. 333-121 PROX RADIUS
A.L. 333-122 RIGHT PROXIMAL METACARPAL IV
A.L. 333W-4 PROXIMAL HAND PHALANX
A.L. 333W-5 LEFT(?) DISTAL FRAG MC V(?)
A.L. 333W-6 RIGHT MC III PROX FRAG
A.L. 333W-7 INTERMEDIATE HAND PHALANX
A.L. 333W-8 VERTEBRAL FRAG
A.L. 333W-11 DISTAL HAND PHALANX
A.L. 333W-14 VERTEBRAL SPINE
A.L. 333W-17 FIRST RIB PROX FRAG
A.L. 333W-18 SECOND RIB PROX FRAG
A.L. 333W-19 RIB FRAG
A.L. 333W-20 PROXIMAL HAND PHALANX FRAG
A.L. 333W-21 DISTAL FRAG METATARSAL
A.L. 333W-22 RIGHT DISTAL FRAG HUMERUS

TABLE 13-1 (continued)

A.L. 333W-23 RIGHT METACARPAL II	A.L. 333X-5 RIGHT PROXIMAL ULNA
A.L. 333W-25 PROX PHALANX FOOT FRAG	A.L. 333X-6/9 RIGHT FRAG CLAVICLE
A.L. 333W-26 LEFT PROXIMAL METACARPAL V	A.L. 333X-12 THORACIC VERTEBRAE
A.L. 333W-29 PROXIMAL HAND PHALANX FRAG	A.L. 333X-13A PROXIMAL HAND PHALANX
A.L. 333W-30 RIB	A.L. 333X-13B INTERMEDIATE HAND PHALANX
A.L. 333W-31 LEFT DISTAL FRAG HUMERUS	A.L. 333X-14 PROXIMAL EPIPHYSIS RADIUS
A.L. 333W-33 RIGHT PROXIMAL RADIUS	A.L. 333X-15 PROXIMAL EPIPHYSIS RADIUS
A.L. 333W-34 INTERMEDIATE PHAL FOOT PROX FRAG	A.L. 333X-18 INTERMEDIATE HAND PHALANX
A.L. 333W-35 RIGHT MC V PROX FRAG	A.L. 333X-21A INTERMEDIATE PHALANX FOOT
A.L. 333W-36 LEFT PROXIMAL ULNA	A.L. 333X-21B INTERMEDIATE PHALANX FOOT
A.L. 333W-37 LEFT DISTAL FIBULA	A.L. 333X-26 RIGHT PROXIMAL TIBIA
A.L. 333W-38 INTERMEDIATE HAND PHALANX	KNM-BC 1745 LEFT PROXIMAL HUMERUS
A.L. 333W-39 RIGHT METACARPAL I	KNM-KP 271 LEFT DISTAL HUMERUS
A.L. 333W-40 LEFT PROXIMAL FEMUR FRAG	L.H. 21P RIGHT SHAFT CLAVICLE
A.L. 333W-41 RIB FRAG	L.H. 21Q STERNAL FRAG RIB
A.L. 333W-43 LEFT PROXIMAL TIBIA FRAG	L.H. 21R LEFT PROXIMAL FRAG ULNA
A.L. 333W-44 HEAD RIB	L.H. 21S LEFT MIDSHAFT ULNA
A.L. 333W-45 FRAG RIB	L.H. 21T DISTAL END ULNA
A.L. 333W-47 FRAG. RIB	L.H. 21V LEFT SHAFT FRAG FEMUR
A.L. 333W-50 DISTAL HAND PHALANX	L.H. 21W INTERMEDIATE HAND PHALANX
A.L. 333W-51 PROXIMAL PHALANX FRAG FOOT	L.H. 21X PROXIMAL HAND PHALANX
A.L. 333W-53 INTERMEDIATE HAND PHALANX	L.H. 21Y PROXIMAL HAND PHALANX
A.L. 333W-54 DISTAL FRAG HAND PHALANX	L.H. 21Z LEFT PROXIMAL FRAG MT II OR III
A.L. 333W-55 METACARPAL MID-SHAFT FRAG	MAK-VP 1/1 LEFT PROXIMAL FEMUR
A.L. 333W-56 RIGHT DISTAL FEMUR	

1985; Berge 1984, 1991; Berge et al. 1984; Capecchi et al. 1989; Christie 1977; Cook et al. 1983; de Arsuaga 1983; Deloison 1984, 1985, 1991; Jungers 1982, 1988a,b,c, 1991; Lamy 1983, 1986; Latimer 1988, 1991; Latimer and Lovejoy 1989, 1990a,b; Latimer et al. 1987; Lovejoy 1988; Marzke 1983, 1986; Marzke and Marzke 1987; Marzke and Shackley 1986; McHenry 1982, 1983, 1984a,b, 1985, 1986ab, 1988, 1991a,b,c; Ohman 1986; Rose 1982; Schmid 1983, 1991; Senut 1980, 1981c, 1986, 1991; Senut and Tardieu 1985; Stern and Jungers 1990; Stern and Susman 1983, 1991; Susman and Stern 1991; Susman et al. 1984; Tague and Lovejoy 1986; Tardieu 1983, 1986a,b; Tuttle 1981, 1985; White 1984; Wolpoff 1983). Table 13-1 presents this sample by element. Included here are specimens from Hadar and Laetoli as well as some of less certain taxonomic affinity from Kanapoi (Hill and Ward 1988; McHenry 1972, 1975b, 1976b; Patterson and Howells 1967; Senut 1979, 1980; Senut and Tardieu 1985), Chemeron (Hill and Ward 1988; Pickford et al. 1982), and Middle Awash (Clark et al. 1984; Hill and Ward 1988; White 1984).

By this author's latest estimates (McHenry 1991a, 1992) an average female stood about 105 cm and weighed approximately 29 kg. An average male's size is less securely known due to the fact that there are less complete specimens and also due to the problems associated with scaling body size and skeletal size in larger-bodied hominoids (see McHenry 1992, for discussion). The author's estimate for male stature is 151 cm and male body weight, 45 kg (McHenry 1991a, 1992). There are many other published estimates (e.g., Aiello 1992; Aiello and Dean 1990; Conroy 1987; Feldesman and Lundy 1988; Geissmann 1986; Hartwig-Scherer and Martin 1991; Jungers 1988b,c; McHenry 1982, 1984b, 1988).

The postcranial skeleton of *A. afarensis* has numerous synapomorphies with later hominids relative to all nonhominid species of Hominoidea. These traits are associated with bipedality either directly (i.e., reorganization of the hindlimb and lower back) or indirectly (i.e., freeing the forelimb from any role in terrestrial locomotion). These include relatively short toes (although longer than modern hominids; Stern and Susman 1983; Susman et al. 1984); proximal phalanges of the foot with dorsally oriented proximal articular surfaces; a convergent hallux; metatarsals II–V with heads expanded superiorly; a powerfully built metatarsal V with a large tuberosity; a metatarsal I with a robust and triangular diaphysis and an expanded head; a stout, anteroposteriorly expanded midtarsal region with a strong transverse and longitudinal arch; a calcaneus with a massive body, a deep dorsoplantar dimension, a tuberosity that is ovoid in cross-section, and a horizontally oriented sustentacular shelf; a distal tibia with an articular surface perpendicular to the shaft axis; a straight-shafted tibia; a distal femur with a high bicondylar angle, an elliptical lateral condyle, and a deep patellar groove with a high lateral lip; a femoral neck with a humanlike distribution of cortical and spongy bone; a pelvis with mediolaterally expanded, superointeriorly shortened, and anteriorly rotated iliac blades, robust anterior iliac spines, a distinct sciatic notch, a distinct iliopsoas groove, a rugose and large area for sacrotuberous ligament, a retroflexed auricular surface with extensive retroauricular area, a robust posterior superior iliac spine, a sigmoid curvature of iliac crest, a dorsoventrally thickened pubic symphysis, a retroflexed hamstring tuberosity, and a shortened ischial shank; lumbar lordosis and sacral retroflexion; sacral alae expanded laterally; sacroiliac and hip joints closely approximated; a univertebral articular pattern for the first rib; relatively small forelimbs; a proximal humerus with an open and shallow bicipital groove; a distal humerus with a rounded lateral well of olecranon fossa, a gracile lateral epicondyle, a moderate-sized and cranially facing medial epicondyle, and a moderate development of supracondylar ridge; a radiocarpal joint perpendicular to shaft axis; a capitate with a proximodistally shortened axis, a single and elongated facet for MCII, and a shallow excavation for MCIII articulations; metacarpals II - V are relatively short with no dorsal transverse ridge on heads; and hand phalanges relatively short (compared with apes although long for other hominids; Stern and Susman 1983).

To this list of hominid synapomorphies might be added the very humanlike characteristics of the Laetoli footprints (e.g., pronounced heel strike, adducted hallux, well-developed longitudinal arch, overall proportions). These and other humanlike qualities have been noted by several investigators (e.g., Chartieris et al. 1982; Clarke 1979; Day and Wickens 1980; Hay and Leakey 1982; M.D. Leakey 1978, 1981, 1987; M.D. Leakey and Hay 1979; Robbins 1987; Tuttle 1985, 1987, 1990, 1991; Tuttle et al. 1991a,b, 1992; White1980; White and Suwa 1987). It must be kept in mind, however, that the footprints may not be those of *A. afarensis* (a possibility explored especially by Tuttle 1985, 1987, 1988,1991; Tuttle et al. 1991a,b, 1992 and this volume; White and Suwa 1987).

The postcranium of *A. afarensis* also retains numerous primitive traits relative to *H. sapiens*. These include distal phalanges of the hand with weakly developed apical tufts, strong capsular cuffs, and well-developed tubercles for collateral ligaments; middle phalanges of the hand have pronounced ridges lateral to the insertion of flexor digitorum superficialis, and strong impressions for the insertion of this muscle; an attenuated proximal phalanx of thumb; proximal hand phalanges II–V are slender, curved, and have strong flexor sheath ridges; metacarpal I with a highly concavoconvex proximal surface, and an attenuated shaft; metacarpals II–V with large heads and bases and curved shafts; pisiform is elongated and rod-shaped; trapezium with concavoconvex articular surface for MCI; capitate with reduced area for styloid process, dorsally placed trapezoid facet, mediolaterally constricted MCIII facet, prominent palmar beak, and a waisted neck; long and narrow tuberosites ulnae and incisura trochlearis; long and narrow collum radii and tuberosites radii; radial head with a broad articular area for zona conoidea of humerus; distal humerus with a strongly developed lateral crest on the anterior surface of the trochlea, distally extended capitular surface, proximal setting of the lateral epicondyle, and a lateral shaft margin parallel with shaft; scapula with cranially oriented glenoid cavity; funnel-shaped thorax; midthoracic vertebrae with ventrally expanded centra; lumbar and sacral centra relatively small in cross section; a sacrum with only slightly developed ventral concavity, weakly developed transverse process of S1, and no upper lateral angles on superior surface of the transverse processes of S1; iliac blades face posteriorly; ischium relatively long with hamstring surface area facing mostly inferiorly; acetabulum with diminutive anterior horn; proximal femur with a poorly developed prolongation of articular surface along anteriosuperior margin of neck and a short neck relative to femoral length; relatively short femur; a knee with a rectangular shape, wide intercondylar notch, marked asymmetry of femoral condyles, and a single attachment for lateral meniscus; distal tibia with posterior tilt; distal fibula with a proximal border of distal articular surface running obliquely, articular surface facing inferomedially, less acute angle between distal articular and subcutaneous surfaces, and a broad and deep peroneal groove; a navicular with a low maximum dorsoplantar height and a large cuboid facet which faces at right angles to the lateral cuneiform; a lateral cuneiform with pongidlike plantar tuberosity; metatarsal I with a rounded head; proximal phalanges of the foot are long, curved, broad-based, narrow bodied in dorsal view, have mediolateral flare of body for flexor sheath and a more highly circumferential trochlea; and middle phalanges of the foot are relatively long.

The functional meaning of these primitive traits is the subject of much disagreement. The original describers (Johanson et al. 1982a, pp. 385-386) wrote that *A. afarensis* had an "adaptation to full and complete bipedality" and a forelimb "not primarily involved in locomotor behavior." On the other hand, several investigators called attention to their curved fingers and toes among other traits. Tuttle (1981:1989) concluded that they "probably continued to enter trees, perhaps for night rest and some foraging." Stern and Susman (1983) agreed with this assessment and went further to suggest that their bipedality "involved less extension of the hip and knee than in modern humans, and only limited transfer of weight onto the medial part of the ball of the foot" (p. 385). The argument has continued for a decade without much change in the two positions (e.g., contrast Lovejoy 1988 with Stern and Susman 1991 and Susman and Stern 1991). This author finds merit in both camps. Certainly the short thighs, laterally splayed iliac blades, and relatively long and curved toes imply a gait that is not identical to modern *H. sapiens*. Few would disagree with this general point, even those closely associated with Lovejoy (e.g., Latimer 1991). The primitive traits of the forelimb may imply a greater aboreal agility than *H. sapiens*, but pongidlike climbing is unlikely due to the fact that the hallux is adducted and the forelimbs are relatively small. The most important fact remains, however, that the postcranium of *A. afarensis* is fundamentally reorganized from the presumed common ancestor of apes and humans and shows the essential hallmarks of a hominid biped.

Australopithecus africanus

The sample of postcrania attributed to *A. africanus* has grown remarkably in the last few years thanks to the continuing efforts of Tobias, Hughes, and their staff at Sterkfontein. Table 13-2 lists 100 specimens from Makapansgat and Sterkfontein Member 4. Twenty-two of these are associated with the STS 14 partial skeleton and 16 are part of the STW 431 partial skeleton (STW 431-433, 436-439, 443, 453-463). Robinson (1972) attributes a small fragment of clavicle from Makapansgat (MLD 20) and a humeral shaft (MLD 14) from Makapansgat to *A. africanus*, but there is some doubt that these specimens are actually hominids. Howell (1978) attributes several East African specimens to this species, but more recent work indicates that *A. africanus* is confined to South Africa (Howell et al. 1987).

This larger sample provides more secure grounds for estimating body size. This author's (McHenry 1991a, 1992) recent research led to a female stature of 115 cm and a weight of 30 kg. The male may have stood 138 cm and weighed 41 kg. Perhaps these are reasonable estimates, but many assumptions are required and there are many other attempts at predicting the body size of *A. africanus* (e.g., Aiello and Dean 1990; Burns 1971; Feldesman and Lundy 1988; Geissmann 1986; Helmuth 1968; Jungers 1988b,c; McHenry 1974, 1975a, 1976a, 1988; Olivier 1976; Reed and Falk 1977; Robinson 1972; Steudel 1981; Suzman 1980; Wolpoff 1973).

In many respects the postcranium of *A. africanus* is like that of *A. afarensis* (McHenry 1986b). Especially impressive are the numerous combinations of traits in the capitate bone and pelvis that appear to be unique to these species (i.e., unlike extant apes or humans). Both species had a postcranium fundamentally reorganized from the common pattern of the Hominoidea to the bipedal pattern unique to the Hominidae. The long list of synapomorphies between *A. afarensis* and *H. sapiens*, previously given, can be applied to *A. africanus* except where elements are as yet unknown for the latter (e.g., MTI, middle phalanges of foot, and tuber calcanei).

There are a few postcranial traits in which *A. africanus* appears to be more derived (i.e., more *Homo*-like) than *A. afarensis*. Ricklan (1990) notes that the metacarpals II–V from Sterkfontein are less like those of pongids than are the metacarpals from Hadar (contra Stern and Susman 1983). Although a full

TABLE 13-2 The postcranium attributed to *Australopithecus africanus*

MLD 7 LEFT ILIUM
MLD 8 LEFT ILIUM
MLD 25 RIGHT ISCHIUM
STS 7 RIGHT DISTAL FRAG SCAPULA
STS 7 RIGHT PROX HUMERUS
STS 14 VERTEBRAL CENTRA (2)
STS 14 SACRUM FRAG
STS 14 COMPLETE COXA
STS 14 LEFT SHAFT FRAG FEMUR
STS 14 RIB COSTAE (4) FRAG
STS 14a 6TH LUMBAR VERTEBRA
STS 14b 5TH LUMBAR VERTEBRA
STS 14c 4TH LUMBAR VERTEBRA
STS 14d 3RD LUMBAR VERTEBRA
STS 14e 2ND LUMBAR VERTEBRA
STS 14f 1ST LUMBAR VERTEBRA
STS 14g 12TH THORACIC VERTEBRA
STS 14h 11TH THORACIC VERTEBRA
STS 14i 10TH THORACIC VERTEBRA
STS 14j 9TH THORACIC VERTEBRA
STS 14k 8TH THORACIC VERTEBRA
STS 14l 7TH THORACIC VERTEBRA
STS 14m 6TH THORACIC VERTEBRA
STS 14n 5TH THORACIC VERTEBRA
STS 14o 4TH THORACIC VERTEBRA
STS 34 RIGHT DISTAL FEMUR
STS 65 RIGHT FRAG COXA
STS 73 LAST THORACIC VERTEBRA
STW 8 FOUR LUMBAR VERTEBRAE
STW 25 RIGHT HEAD FEMUR
STW 26 RIGHT DISTAL FRAG MC IV
STW 28 RIGHT HAND PROX PHALANX V
STW 29 RIGHT FRAG HAND PROX PHAL I
STW 38 LEFT DISTAL SHAFT FRAG HUMERUS
STW 41 BODIES THORACIC VERTEBRAE
STW 46 LEFT DISTAL FRAG RADIUS
STW 63 LEFT MC V
STW 64 LEFT MC III
STW 65 LEFT MC IV
STW 68 RIGHT MC III
STW 99 RIGHT PROX AND SHAFT FEMUR
STW 105 RIGHT PROXIMAL RADIUS
STW 122 HAND PROX PHALANX
STW 124 LEFT SHAFT FRAG HUMERUS
STW 125 LEFT SHAFT FRAG RADIUS
STW 238 RIGHT PROX FRAG MT (IV?) V
STW 292 RIGHT DISTAL FRAG MC V
STW 293 HAND PROX PHALANX
STW 294 HAND DISTAL PHALANX I
STW 318 RIGHT DISTAL FRAG FEMUR
STW 326 RIGHT DISTAL FRAG ULNA
STW 328 RIGHT PROX FRAG HUMERUS
STW 330 LEFT FRAG MC IV
STW 331 HAND MIDDLE PHALANX
STW 339 RIGHT SHAFT FRAG HUMERUS
STW 348 RIGHT SHAFT FRAG RADIUS
STW 349 RIGHT SHAFT FRAG ULNA
STW 352 RIGHT CALCANEUS
STW 358 LEFT DISTAL FRAG TIBIA
STW 366 RIGHT DISTAL FRAG SCAPULA
STW 377 LEFT PROXIMAL FRAG MT II
STW 380 RIGHT PROXIMAL FRAG ULNA
STW 382 LEFT MC II
STW 387 LEFT PROXIMAL FRAG MT III
STW 388 RIGHT PROXIMAL FRAG MT III
STW 389 LEFT DISTAL FRAG TIBIA
STW 390 RIGHT PROXIMAL FRAG ULNA
STW 392 RIGHT HEAD FEMUR
STW 394 RIGHT DISTAL FRAG MC IV
TW 396 LEFT PROXIMAL FRAG TIBIA
STW 398 LEFT PROXIMAL FRAG ULNA
STW 399 LEFT DISTAL FRAG ULNA
STW 400 FRAG HAND PROX PHALANX
STW 403 RIGHT HEAD AND NECK FEMUR
STW 418 LEFT MC I
STW 431 RIGHT PROXIMAL FRAG RADIUS
STW 432 RIGHT PROXIMAL FRAG ULNA
STW 433 RIGHT DISTAL FRAG HUMERUS
STW 435 RIGHT PROXIMAL FRAG MT III
STW 436 RIGHT FRAG PELVIS
STW 437 RIGHT LAT ONE HALF CLAVICLE
STW 439 SACRUM
STW 441 LEFT ILIAC CREST FRAG
STW 442 RIGHT CREST FRAG ILIUM
STW 443 LEFT ACETABULAR FRAG PELVIS
STW 448 RIGHT DISTAL SHAFT FEMUR
STW 453 11TH THORACIC VERTEBRA
STW 454 10TH THORACIC VERTEBRA
STW 455 9TH THORACIC VERTEBRA
STW 457A 12TH THORACIC VERTEBRA
STW 458 1ST LUMBAR VERTEBRA
STW 459 2ND LUMBAR VERTEBRA
STW 460 3RD LUMBAR VERTEBRA
STW 461 4TH LUMBAR VERTEBRA
STW 463 5TH LUMBAR VERTEBRA
STW 470 RIGHT FOOT PROX PHALANX I
STW 477 LEFT MT III
STW 478 FRAG FOOT PROX PHALANX
TM 1513 LEFT DISTAL FEMUR
TM 1526 RIGHT CAPITATE

analysis of this comparison awaits to be published, a preliminary assessment reveals that the heads and bases of the *A. africanus* metacarpals appear to be less expanded relative to their shafts. There is some development of a styloid process of MC III in Stw 68 (*A. africanus*) but not in A.L. 333-16 and 65 (*A. afarensis*). The proximal hand phalanges of *A. africanus* are less curved than those of *A. afarensis* (although Stw 28 is strongly curved according to Susman 1988b) and have less strongly developed flexor sheath ridges (Ricklan 1987). The middle phalanx (Stw 331) has reduced ridges lateral to the insertion of flexor digitorum superficialis and a weaker impression for the insertion of the tendon of this muscle. Unfortunately there is no distal thumb phalanx preserved for *A. afarensis* to compare with the beautifully preserved one from *A. africanus* (Stw 264: Ricklan 1987, 1990). This specimen shows a very humanlike development of the ungual tuft and spines. This contrasts with the relatively narrow, chimpanzeelike apical tuffs of the distal, non-pollicial phalanges of *A. afarensis* (A.L. 333w-11 and -50). These and other traits of *A. africanus* lead Ricklan (1990) to conclude that this species had an adequate precision grip. Ricklan (1990) points out that *A. africanus* was not particularly adapted for an arboreal lifestyle, and it probably lived in an open grassland or bushland environment with few trees (Vrba 1974, 1975a,b, 1985a,b).

Australopithecus boisei

Very few postcranial specimens can be attributed with certainty to *A. boisei* (see Table 13-3) because most of the hominid postcrania deriving from strata dating to the age of *A. boisei* (i.e., 2.1 to 1.3 MYA) also contain the remains of early *Homo* and are not associated with taxonomically identifiable craniodental parts. Fortunately a fragment of mandible from the partial associated skeleton of KNM-ER 1500 has been identified as *A. boisei* (Grausz et al. 1988; Walker et al. 1989). The specimens from Area 6A of Ileret (a talus, KNM-ER 1464; a metatarsal, KNM-ER 1823; a distal humerus, KNM-ER 1824; and an atlas vertebra, KNM-ER 1825) are associated with the craniodental remains of 3 adults (KNM-ER 801, 802, and 3737) and 2 immature individuals (KNM-ER 1171 and 1816) that are attributed to *A. boisei* (Day et al. 1976; Grausz et al. 1988; R.E.F. Leakey 1972; MG Leakey and Leakey 1978; Walker et al. 1989). A calcaneus from Shungura G (Omo 33.74.895) may also be attributed to *A. boisei* on the basis of association (Deloison 1986). Howell (1978) attributes 14 additional postcranial specimens to *A. boisei* on the basis of their non-*Homo* characteristics, but subsequent discoveries of associated skeletons show that such morphologically based attributions are problematical (see McHenry 1992 for discussion).

Body size for *A. boisei* is poorly known because of this scarcity of positively identified postcranial specimens. McHenry (1992) estimates the female to have weighed 34 kg based on KNM-ER 1500. The male (represented by KNM-ER 1464) may have weighed 48.6 kg. Even more difficult is stature reconstruction, but on available evidence McHenry (1991b) estimated 124 cm for the female and 137 cm for the male.

Little is known as yet about the phylogenetic or functional meaning of the *A. boisei* postcranium. In the partial skeleton (KNM-ER 1500) the forelimbs are relatively large and the hindlimbs relatively small in comparison to modern humans (McHenry 1978) and overall proportions most resemble A.L. 288-1 (Grausz et al. 1988). Senut (1980, 1981b; Senut and Tardieu 1985) notes that the proximal radius is similar to other known members of *Australopithecus* and not like that of *Homo*. Tardieu (1981, 1983, 1986a,b) groups

TABLE 13-3 The postcranium attributed to *Australopithecus boisei*

KNM-ER 1464 RIGHT TALUS	KNM-ER 1500J RIGHT SHAFT FRAG TIBIA
KNM-ER 1500A LEFT PROXIMAL FRAG TIBIA	KNM-ER 1500K RIGHT SHAFT FRAG RADIUS
KNM-ER 1500B LEFT DISTAL FRAG FEMUR	KNM-ER 1500L SHAFT FRAG HUMERUS
KNM-ER 1500C LEFT DISTAL FRAG TIBIA	KNM-ER 1500M RIGHT PROX FRAG MT III
KNM-ER 1500D LEFT PROXIMAL FRAG FEMUR	KNM-ER 1500N SHAFT FRAG RADIUS
KNM-ER 1500E RIGHT PROX FRAG RADIUS	KNM-ER 1500O LEFT GLENOID FRAG SCAPULA
KNM-ER 1500F RIGHT PROXIMAL FRAG ULNA	KNM-ER 1500T FRAG CALCANEUS
KNM-ER 1500G RIGHT DISTAL FRAG FIBULA	KNM-ER 1823 PROX FRAG FOOT MT
KNM-ER 1500H RIGHT DISTAL FRAG TIBIA	KNM-ER 1825 LEFT FRAG ATLAS
KNM-ER 1500I RIGHT SHAFT FRAG ULNA	OMO 323-76-898 CALCANEUS

the KNM-ER 1500 distal femur and proximal tibia with other australopithecines as opposed to *Homo* (including KNM-ER 1481). The distal tibia (KNM-ER 1500L) has the fundamental adaptations for bipedality that Latimer et al. (1987) noted in the *A. afarensis* specimens.

Some very large forelimb fossils are regarded as belonging to *A. boisei* (e.g., KNM-ER 739 humerus by Aiello and Dean 1990; Day 1976b, 1978; Howell 1978; Kay 1973; R.E.F. Leakey et al. 1972; McHenry 1973; McHenry and Corruccini 1975; Senut 1978, 1980, 1981a, b, c; Senut and Tardieu 1985; Omo L 40-19 by Aiello and Dean 1990; Feldesman 1979; Howell 1978; Howell and Wood 1974; McHenry and Temerin 1979; McHenry et al. 1976; Senut 1978, 1980). It is reasonable to assume that these specimens belong to *A. boisei* because they differ in many respects from the morphology known for *Homo*. The early *H. erectus* skeleton, KNM-WT 15000 (Brown et al. 1985), has relatively small forelimbs that are quite unlike massive specimens attributed to *A. boisei*. But the recent demonstration that the partial skeleton KNM-ER 1500 belongs to *A. boisei* poses a problem: How could such a small-bodied species possess such large forelimbs? At least these three explanations are possible.

First, it is possible that male *A. boisei* were much larger than female *A. boisei* as represented by this partial skeleton (KNM-ER 1500). Certainly there is strong sexual dimorphism in craniodental size (Chamberlain and Wood 1985; Wood 1985). Body size sexual dimorphism may not be quite so strong, however (McHenry, 1991b), though body size of the male is not securely established. Second, it is possible that male forelimbs were unusually large. This possibility derives from the extension of Darwin's feed-back model of canine evolution and sexual selection. If bipedalism freed the forelimbs and forelimbs took over the role served by canines, then one might expect unusually strong sexual dimorphism in forelimb size.

A third explanation is that these large forelimbs belonged to the species with the largest hindlimbs. At Koobi Fora, the largest hindlimb fossils between 1.9 and 1.7 MY belong to *H. habilis sensu lato* or *H. rudolfensis*, following the attribution made by Wood (1991, 1992). After 1.7 MY, the largest hindlimbs belong to *H. erectus* (or *H. ergaster* (Wood 1991, 1992). The problem with this alternative is that the associated skeletons of this species (KNM-WT 15000, ER 803, ER 1808) have relatively small forelimbs.

Australopithecus robustus

Table 13-4 presents the vastly expanded sample of postcrania attributed to *A. robustus*. Most of it derives from Member 1 of Swartkrans although 11 specimens are from Member 3. Although Susman (1988a) attributed only 3 of these latter specimens to the "robust" australopithecines, all are listed here because virtually all of the taxonomically identifiable craniodental material from Member 3 belong to this species (Grine 1989).

It was this recently expanded sample of *A. robustus* that led McHenry (1991b) to characterize these "robust" hominids as having petite bodies. Most of the specimens derive from small-bodied individuals relative to modern human standards. McHenry (1992) estimates a female body weight midpoint of 32 kg and a male of 40 kg. Female stature may have been 110 cm and male, 132 cm (McHenry 1991a).

The postcranial sample of *A. robustus* is complete enough to show that this species retains many of the primitive traits of earlier hominid species (e.g., small femoral heads relative to other dimensions of the femur and pelvis, widely splayed iliac blades, long ischial shanks). Susman (1988a,b,c, 1989, 1991; Susman and Brain 1988, Susman and Stern 1991) calls attention to several derived traits that the "robust" australopithecine of Swartkrans shares with *Homo* and possibly not with *A. afarensis*. The broad apical tuft of the distal thumb phalanx (Sk 5016) is conspicuously human-like and not apelike. This is in contrast to the distal hand phalanges of *A. afarensis* but, unfortunately, no distal thumb phalanx is present in the *A. afarensis* sample. The distal thumb phalanx of *A. africanus* (Stw 294) is like *Homo* (Ricklan 1987). Two proximal hand phalanges (SKX 5018 and 22741) are less curved than those from Hadar, but a third one (SKX 27431) is as curved as those from Hadar. Susman (1988a) attributes SKX 27431 to *Homo* even though it derives from Member 3 of Swartkrans which contains only "robust" australopithecine craniodental material. The carpometacarpal joint of the thumb is broad and saddle-shaped in Susman's (1988a) description. Stern and Susman (1983) describe this joint as being concavoconvex and chimplike in *A.*

TABLE 13-4 The postcranium attributed to *Australopithecus robustus*

SK 50 RIGHT FRAG PELVIS	SKX 8761 PROX FRAG ULNA
SK 82 RIGHT PROXIMAL FEMUR	SKX 8963 DISTAL MANUAL PHALANX
SK 84 LEFT METACARPAL I	SKX 9449 MIDDLE HAND PHALANX
SK 97 RIGHT PROXIMAL FEMUR	SKX 12814 SHAFT RADIUS
SK 853 LUMBAR FRAG VERTEBRA	SKX 13476 HAND MIDDLE PHAL
SK 854 AXIS FRAG	SKX 19576 PROXIMAL HAND PHALANX
SK 3155B RIGHT ILIUM FRAG PELVIS	SKX 22511 HAND PROX PHALANX
SK 3981A T12 VERTEBRA	SKX 22741 PROXIMAL HAND PHALANX
SK 3981B L5 VERTEBRAE	SKX 22431 PROXIMAL HAND PHALANX III
SK 14147 LEFT METACARPAL V	SKX 27504 DISTAL HAND PHALANX
SK 45690 LEFT FOOT PROX PHAL I	SKX 31117 LEFT FRAG MEDIAL CUNEIFORM
SKW 3774 DISTAL HUMERUS	SKX 33355 MIDDLE HAND PHALANX
SKW 34805 DISTAL HUMERUS	SKX 33380 LEFT DISTAL 2/3 MT V
SKX 3602 RIGHT DISTAL RADIUS	SKX 35439 MIDDLE HAND PHALANX
SKX 5016 DISTAL POLLICAL PHALANX	SKX 35822 DISTAL HAND PHALANX
SKX 5017 LEFT MT I	SKX 36712 MIDDLE HAND PHALANX
SKX 5018 PROXIMAL HAND PHALANX	TM 1517 RIGHT DISTAL HUMERUS
SKX 5020 RIGHT METACARPAL I	TM 1517 RIGHT PROXIMAL ULNA
SKX 5021 MIDDLE HAND PHALANX	TM 1517 RIGHT TALUS
SKX 5022 MIDDLE HAND PHALANX	TM 1605 RIGHT ILIUM FRAG

afarensis, although Bush et al. (1982) refer to it as saddle-shaped. Susman and Brain (1988) call attention to the very humanlike morphology of the Swartkrans first metatarsal (SKX 5017) but note the apelike dorsal narrowing of the head. This bone does not appear to differ appreciably from those of *A. afarensis* as described by Latimer et al. (1982) and Latimer and Lovejoy (1990a,b).

Homo habilis sensu lato

Table 13-5 presents the postcranial sample that probably belongs to *H. habilis*, although uncertainty permeates every attribution. Uncertainty starts with the disagreement among craniodental taxonomists over the issue of whether or not there exists one or more nonrobust hominid species between 2.4 and 1.7 MYA (e.g., Chamberlain and Wood 1987a,b; R.E.F. Leakey et al. 1989; Lieberman et al. 1988; Miller 1991; Stringer 1986; Tobias 1991; Wood 1989, 1992). Uncertainty continues with the attribution of isolated postcranials that are more *Homo*-like than *Australopithecus*-like. Table 13-5 lists several hindlimb fossils (KNM-ER 1472, 1481, 3228) that are unlike known australopithecines and similar to *H. erectus*. And, finally, the potential problem exists of the choice made in Table 13-5 to attribute all Sterkfontein Member 5 specimens to *H. habilis*, since that is the only species of hominid so far identified in that deposit (Senut and Tobias 1989).

Given these many problems, one must appreciate the insecurity of any attempt to estimate body size. Assuming the attributions in Table 13-5 do represent a single species called *H. habilis*, then the small, presumably female, morph may have weighed 32 kg and the large morph, 52 kg. The statures are 125 cm and 157 cm for female and male (McHenry 1991a, 1992). If the taxonomy recommended by Wood (1991, 1992) is followed, then *H. habilis sensu stricto* weighed 32 kg and *H. rudolfensis* weighed 52 kg.

The postcranium of *H. habilis sensu stricto* retains many australopithecine traits in contrast to later species of *Homo*. Some of the most conspicuous primitive traits include large forelimbs relative to hindlimbs (Hartwig-Scherer and Martin 1991; Johanson et al. 1986; Jungers 1988b; Korey 1990; R.E.F. Leakey et al. 1989), a hand with robust and curved middle phalanges with well-marked insertions for flexor digitorum superficialis, thick and curved proximal phalanges, and a pongid-like scaphoid tubercle and trapezium articular facet (Susman and Creel 1979; Susman and Stern 1979, 1982), a foot reported to

TABLE 13-5 The postcranium attributed to *Homo habilis sensu lato*

KNM-ER 813A RIGHT FRAG TALUS	O.H. 48 LEFT SHAFT CLAVICLE
KNM-ER 813B RIGHT DISTAL SHAFT FRAG TIBIA	O.H. 62V RIGHT SHAFT HUMERUS
KNM-ER 1472 RIGHT COMPLETE FEMUR	O.H. 62W RIGHT SHAFT RADIUS
KNM-ER 1481A LEFT COMPLETE FEMUR	O.H. 62X RIGHT SHAFT ULNA
KNM-ER 1481B LEFT PROXIMAL TIBIA	O.H. 62Y LEFT NECK AND SHAFT FEMUR
KNM-ER 1481C LEFT DISTAL TIBIA	O.H. 62Z RIGHT PROXIMAL FRAG TIBIA
KNM-ER 1481D LEFT DISTAL FIBULA	STW 27 RIGHT METACARPAL III
KNM-ER 3228 RIGHT COXA	STW 88 RIGHT TALUS
KNM-ER 3735 H & I DISTAL FRAG HAND PROX PHAL (2)	STW 89 LEFT MT II
KNM-ER 3735A RIGHT DISTAL HUMERUS	STW 102 RIGHT FRAG TALUS
KNM-ER 3735B RIGHT DISTAL SHAFT FEMUR	STW 108 LEFT PROXIMAL SHAFT FRAG ULNA
KNM-ER 3735C MIDSHAFT TIBIA	STW 113 LEFT PROXIMAL ULNA
KNM-ER-3735E LEFT HEAD AND SHAFT RADIUS	STW 114 & 115 RIGHT MT V
KNM-ER 3735J SUP & CENTRAL FRAG SACRUM	STW 129 LEFT DISTAL FRAG FEMUR
O.H. 7 RIGHT PARTIAL HAND	STW 139 RIGHT HEAD RADIUS
O.H. 8 LEFT PARTIAL FOOT	STW 150 LEFT DISTAL SHAFT FRAG HUMERUS
O.H. 35 LEFT SHAFT AND DISTAL TIBIA	STW 182 RIGHT DISTAL SHAFT FRAG HUMERUS
O.H. 35 LEFT SHAFT AND DISTAL FIBULA	STW 311 RIGHT HEAD FEMUR
O.H. 43 LEFT MT III & IV	STW 355 RIGHT FOOT PROX PHAL

have an obliquely disposed subtalar joint axis, a laterally unexpanded cuboid, a squat and foreshortened talus, and many other ape-like traits (Lewis 1980, 1989), a fibula with a convex surface for the origin of the peroneus brevis muscle and a marked ridge between the two attachment areas of flexor hallucis longus muscle (Susman and Stern 1982), and a tibia with a strong crest between tibialis posterior and flexor digitorum longus muscles (Lovejoy 1978), a weakly developed interosseous ridge (Susman and Stern 1982), and marked platycnemia (Susman and Stern 1982).

There are surprisingly few traits in the postcranium of *H. habilis sensu stricto* that are shared with later *Homo* and are not present in one or more of the species of *Australopithecus*. The first carpometacarpal joint is expanded in O.H. 7 to a greater degree than is true of *A. afarensis* (Stern and Susman 1983; Susman and Creel 1979; Trinkaus 1989), but this is also true of *A. robustus* (Susman 1988a,b). The distal head phalanges of O.H. 7 have expanded apical tufts (Day 1976a; Lewis 1977; Marzke 1971; Napier 1964 Susman and Creel 1979; Susman and Stern 1979; Tuttle 1967) unlike *A. afarensis* (Stern and Susman 1983), but like *A. robustus* (Susman 1988b) and *A. africanus* (Ricklan 1987). The articular surface of the distal fibula of O.H. 35 has a proximal border that is perpendicular to the shaft, faces medially, and has a laterally facing subcutaneous surface like later *Homo* and unlike *A. afarensis* (Aiello and Dean 1990; Stern and Susman 1983), but these humanlike traits may be shared by *A. boisei* (KNM-ER 1500G). The O.H. 62 tibia is reported to lack the groove on the medial side of the shaft for the attachment of tibialis anterior and gracilis muscles like later *Homo* and unlike *A. afarensis* (Aiello and Dean 1990), but other species of *Australopithecus* may share this humanlike trait.

The postcrania that Wood (1992) refers to *H. rudolfensis* are more derived in a humanlike direction. For example, the proportions of the ilium (KNM-ER 3228) are much like those of later *Homo* and unlike all known australopithecine hips (Aiello and Dean 1990; Rose 1984). The ischial shank of this specimen is short and the ischial tuberosity closely approximates the acetabular rim like *Homo* and unlike *Australopithecus*. Most conspicuous of all is the large acetabulum relative to other pelvic or femoral dimensions (except femoral head size). The proportions of the femora (KNM-ER 1472 and 1481) are more like *H. erectus* than any species of *Australopithecus* (Corruccini and McHenry 1978; Day 1973; Kennedy 1983; McHenry and Corruccini 1976, 1978). The proportions of the talus (KNM-ER 813) are much more like *H. sapiens* than are other Plio-Pleistocene hominid tali (Wood 1974).

EARLY *Homo erectus (= Homo ergaster)*

The postcrania of the earliest *H. erectus* (= *H. ergaster*) are very well known because of the KNM-WT 15000 skeleton. Also listed in this species (Table 13- 6) are two partial skeletons from Koobi Fora (KNM-ER 803 and 1808). Less certain are the attributions of the Gombre' 1B-7594 humerus (Senut 1980, 1989). Judging from these specimens, the probable weight of the male early *H. erectus* may have been 68 kg and stature 180 cm (Ruff 1991; McHenry 1991a). Female size may have also been relatively large, but it is difficult to know with any certainty.

To the great surprise to all who study early hominid postcrania, *H. erectus* shared numerous primitive traits with *A. afarensis* and other pre-*erectus* hominids. Brown et al. (1985) note the long spinous processes, narrow lamina and small foramina of the vertebrae, anteriorly facing and concave inferior articular facets of the fifth lumbar vertebra, iliac flare, and long femoral necks. A full discussion of this species awaits the publication of the monograph on KNM-WT 15000.

The postcranium of early *H. erectus* shares some derived traits with later *Homo* that are not present in *Homo habilis sensu stricto*. Fore- and hindlimb proportions are like *H. sapiens* (Brown et al. 1985; McHenry 1978), and the ulna has a strong interosseous border, a supinator crest and a well-marked hollow for the play of the tuberosity of the radius (Aiello and Dean 1990; Day 1978; Susman 1988a). The pelvis and hindlimb of *H. erectus* is very much like the material assigned to *H. rudolfensis* by Wood (1992).

Hominidae *gen. et sp. indet.*

Table 13-7 lists hominid postcranial specimens that can be attributed to species with less confidence. All of these derive from deposits that contain craniodental evidence of at least two separate species. It is hoped that this list will shorten when more discoveries and further study reveal how isolated postcranial specimens can be sorted into species. Recent discoveries of associated partial skeletons may already contain the necessary clues, but problems remain. One problem is the species identification of post-*afarensis*–pre-*erectus*–non-"robust" specimens. For example, O.H. 62 is attributed to *H. habilis* by its discoverers (Johanson et al. 1986), but its strikingly primitive body proportions lead some to exclude it from the august company of O.H. 7 (R.E.F. Leakey et al. 1989; Hartwig-Scherer and Martin 1991; Korey 1990; Wood 1991, 1992). Even the KNM-WT 15000 *H. erectus* skeleton retains many un-*H. sapiens*–like traits such as its funnel-shaped thorax, its relatively small vertebral canal, its small cross-sectional area of the lower back and sacrum, and its long-neck femora.

SUMMARY

The sample of Plio-Pleistocene hominid postcrania has expanded enormously in the last few years. This new material permits more accurate bodysize predictions and provides a clearer picture of early hominid species. The earliest species, *A. afarensis*, had relatively strong sexual dimorphism with males weighing about 45 kg and standing about 151 cm. Females may have been 29 kg and 105 cm. The body of this species was fundamentally reorganized for bipedality, but it also retained numerous apelike features. *Australopithecus africanus* was similar in size and morphology, but in a few respects was more like later hominids, especially in features of the hand. The "robust" australopithecine of South Africa was also similar in body size but showed several more derived (*Homo*-like) features than did earlier hominids. Both the South and the East African "robust" species had relatively petite bodies (less than 50 kg with many individuals closer to 30 kg). *Homo habilis sensu stricto* was also petite and retained numerous primitive postcranial features relative to other species of *Homo*. By 1.9 MYA, some individuals (referred to *H. rudolfensis* by Wood 1992) had lost most of the primitive features of the hindlimbs seen in the australopithecines and had increased in body size to over 50 kg. The earliest *H. erectus* (or *H. ergaster*) had bodies over 65 kg in weight and 180 cm in stature but still retained some primitive features not seen in modern humans.

FIGURE 13-6 The postcranium attributed to early *Homo erectus* (= *H. ergaster*)

GOMBRE'IB-7594 LEFT DISTAL HUMERUS
KNM-ER 737 LEFT SHAFT FEMUR
KNM-ER 803A LEFT SHAFT FRAG FEMUR
KNM-ER 803B LEFT SHAFT FRAG TIBIA
KNM-ER 803C LEFT SHAFT FRAG ULNA
KNM-ER 803D LEFT MIDSHAFT RADIUS
KNM-ER 803E LEFT FRAG TALUS
KNM-ER 803F LEFT PROX FRAG MT V
KNM-ER 803G LEFT MEDIAL CONDYLE TIBIA
KNM-ER 803J LEFT FRAG MT III
KNM-ER 803K FOOT INTERM PHAL III
KNM-ER 803L FOOT INTERM PHAL II
KNM-ER 803M FOOT DIST PHAL
KNM-ER 803N LEFT SHAFT FRAG FIBULA
KNM-ER 803O RIGHT SHAFT FRAG FIBULA
KNM-ER 803P SHAFT FRAG RADIUS
KNM-ER 803Q FRAG FOOT PROX PHAL I
KNM-ER 803R FOOT PROX PHAL II OR III
KNM-ER 803S SHAFT FRAG FOOT PHAL
KNM-ER 803T LEFT PROXIMAL METACARPAL V
KNM-ER 1808M DISTAL SHAFT FRAG (DISEASED) FEMUR
KNM-ER 1808N RIGHT PROX FRAG (DISEASED) FEMUR
KNM-ER 1808O RIGHT PROX FRAG (DISEASED) FEMUR
KNM-ER 1808P LEFT SHAFT FRAG (DISEASED) FEMUR
KNM-ER 1808R LEFT FRAG (DISEASED) ILIUM
KNM-ER 1808S RIGHT FRAG (DISEASED) ILIUM
KNM-ER 1808T DIST FRAG (DISEASED) HUMERUS
KNM-ER 1808U FRAG (DISEASED) FIBULA
KNM-ER 1808V FRAG (DISEASED) TIBIA
KNM-ER 1808W SHAFT RADIUS
KNM-ER 1808X SHAFT RADIUS
KNM-ER 1808Y SHAFT FRAG (DISEASED) ULNA
KNM-ER 1808Z FRAG (DISEASED) ATLAS
KNM-ER 1808AE RIGHT SHAFT (DISEASED) HUMERUS
KNM-ER 1808AM RIGHT SHAFT (DISEASED) CLAVICLE
KNM-ER 1812D HEAD FRAG RADIUS
KNM-WT 15000C LEFT SHAFT CLAVICLE
KNM-WT 15000D RIGHT SHAFT CLAVICLE
KNM-WT 15000E RIGHT COMPLETE SCAPULA
KNM-WT 15000F RIGHT COMPLETE HUMERUS
KNM-WT 15000G LEFT FEMUR
KNM-WT 15000H RIGHT FEMUR
KNM-WT 15000I LEFT TIBIA
KNM-WT 15000K LEFT FIBULA
KNM-WT 15000L LEFT ILIUM
KNM-WT 15000L RIGHT FIBULA
KNM-WT 15000M RIGHT FEMUR
KNM-WT 15000O RIGHT ILIUM
KNM-WT 15000P RIGHT ISCHIUM
KNM-WT 15000Q LEFT ISCHIUM
KNM-WT 15000R CERVICAL #7
KNM-WT 15000S THORACIC #1
KNM-WT 15000T THORACIC #2
KNM-WT 15000U THORACIC #3
KNM-WT 15000V THORACIC #8
KNM-WT 15000W THORACIC #10
KNM-WT 15000X THORACIC #10
KNM-WT 15000Y THORACIC #11
KNM-WT 15000Z FRAG (LAMINAL/SPINE) LUMBAR #3
KNM-WT 15000AA LUMBAR #2 VERTEBRA
KNM-WT 15000AB LUMBAR #4 VERTEBRA
KNM-WT 15000AC LUMBAR #5 VERTEBRA
KNM-WT 15000AD SACRAL #1 VERTEBRA
KNM-WT 15000AE SACRAL #5 VERTEBRA
KNM-WT 15000AF COCCYGEAL #1 VERTEBRA
KNM-WT 15000AG LEFT 1ST RIB
KNM-WT 15000AH RIGHT 2ND RIB
KNM-WT 15000AI RIGHT 4TH RIB
KNM-WT 15000AJ RIGHT 10TH RIB
KNM-WT 15000AK RIGHT 7TH RIB
KNM-WT 15000AL RIGHT 6TH RIB
KNM-WT 15000AM RIGHT 8TH RIB
KNM-WT 15000AN LEFT 11TH RIB
KNM-WT 15000AO LEFT 9TH RIB
KNM-WT 15000AP LEFT 7TH RIB
KNM-WT 15000AQ LEFT 2ND RIB
KNM-WT 15000AR FRAG LUMBAR #1
KNM-WT 15000AS RIGHT 9TH RIB
KNM-WT 15000AT LEFT 3RD RIB
KNM-WT 15000AU LEFT 8TH RIB
KNM-WT 15000AV LUMBAR #2
KNM-WT 15000AW RIGHT FRAG PUBIS
KNM-WT 15000AX RIGHT FRAG PUBIS
KNM-WT 15000AY RIGHT 1ST RIB
KNM-WT 15000AZ RIGHT 1ST RIB
KNM-WT 15000BA LUMBAR #1
KNM-WT 15000BB FRAG (LAMINAL/SPINE) SACRAL #3
KNM-WT 15000BC RIGHT FRAG SACRAL #3
KNM-WT 15000BD RIGHT 4TH RIB
KNM-WT 15000BE RIGHT 4TH RIB
KNM-WT 15000BI SPINE/LAMINA THORACIC #6 0R #7
KNM-WT 15000BJ RIGHT 5TH FRAG RIB
KNM-WT 15000BK LEFT FRAG SCAPULA
KNM-WT 15000BL LEFT SPINE SCAPULA
KNM-WT 15000BM LUMBAR VERTEBRA
KNM-WT 15000BO HAND PHALANX
KNM-WT 15000BP RIGHT COMPLETE W/O ENDS ULNA
KNM-WT 15000BQ FOOT PHALANX
KNM-WT 15000BU RIGHT METACARPAL I
KNM-WT 15000BV LEFT METACARPAL I
KNM-WT 15000BW CENTRUM VERTEBRAE
KNM-WT 15000BX RIGHT MT I
KNM-WT 15000BZ LEFT COMPLETE W/O ENDS ULNA
KNM-WT 15000CA THORACIC #5 OR #6

TABLE 13-7 Postcranium attributed to Hominidae *gen. et sp. indet.*

KNM-ER 736 LEFT SHAFT FEMUR
KNM-ER 738 LEFT PROXIMAL FEMUR
KNM-ER 739 RIGHT COMPLETE W/O HEAD HUMERUS
KNM-ER 740 LEFT DISTAL FRAG HUMERUS
KNM-ER 741 LEFT PROXIMAL TIBIA
KNM-ER 815 LEFT PROXIMAL FRAG FEMUR
KNM-ER 993 LEFT DISTAL SHAFT FEMUR
KNM-ER 997 LEFT PROX FRAG MT III
KNM-ER 1463 RIGHT W/O EPIPHYSES FEMUR
KNM-ER 1465A LEFT PROXIMAL FRAG FEMUR
KNM-ER 1471 RIGHT PROXIMAL TIBIA
KNM-ER 1473 RIGHT PROXIMAL HUMERUS
KNM-ER 1475A RIGHT PROX SHAFT FRAG FEMUR
KNM-ER 1476A LEFT FRAG TALUS
KNM-ER 1476B LEFT PROXIMAL TIBIA
KNM-ER 1476C RIGHT SHAFT TIBIA
KNM-ER 1503 RIGHT PROXIMAL FEMUR
KNM-ER 1504 RIGHT DISTAL FRAG HUMERUS
KNM-ER 1505A LEFT HEAD AND NECK SHAFT FEMUR
KNM-ER 1505B LEFT DISTAL SHAFT FEMUR
KNM-ER 1591 RIGHT FRAG HUMERUS
KNM-ER 1592 RIGHT DISTAL FEMUR
KNM-ER 164B HAND PROX PHAL (2)
KNM-ER 164C C7 & T1 VERTEBRAE
KNM-ER 1807 RIGHT DISTAL SHAFT FEMUR
KNM-ER 1809 RIGHT SHAFT FEMUR
KNM-ER 1810 LEFT PROXIMAL TIBIA
KNM-ER 1822 RIGHT SHAFT (=1503) FEMUR
KNM-ER 1824 RIGHT DISTAL FRAG HUMERUS
KNM-ER 2594A LEFT PROXIMAL TIBIA
KNM-ER 2594B SHAFT FRAG TIBIA
KNM-ER 2596 LEFT DISTAL TIBIA
KNM-ER 3728 RIGHT PROXIMAL AND SHAFT FEMUR
KNM-ER 3736 LEFT PROXIMAL RADIUS
KNM-ER 3888 RIGHT PROXIMAL RADIUS
KNM-ER 3951 LEFT DISTAL FRAG FEMUR
KNM-ER 3956 LEFT MIDSHAFT HUMERUS
KNM-ER 3956A LEFT SHAFT RADIUS
KNM-ER 3956B LEFT SHAFT ULNA
KNM-ER 5428 LEFT TALUS
KNM-ER 5880 RIGHT PROXIMAL FEMUR
KNM-ER 5881 RIGHT MIDSHAFT FEMUR
KNM-ER 5882 RIGHT DISTAL SHAFT FEMUR
KNM-ER 6020 LEFT DISTAL HUMERUS
O.H. 10 RIGHT FOOT DIST PHAL
O.H. 20 LEFT PROXIMAL FRAG FEMUR
O.H. 36 RIGHT PROXIMAL AND SHAFT ULNA
O.H. 49 LEFT SHAFT, DISEASED RADIUS
O.H. 53 RIGHT SHAFT FEMUR
OMO 18-1848 HAND INTERM PHAL
OMO 28-4570 FOOT PROX PHAL
OMO 75S-1317 RIGHT PROXIMAL RADIUS
OMO 119-2718 LEFT PROXIMAL HUMERUS
OMO 141-23 LEFT PROXIMAL ULNA
OMO F511-16 LEFT MT III
OMO L 40-19 RIGHT COMPLETE ULNA
OMO L 40-19 RIGHT PROXIMAL ULNA
OMO L 754-8 LEFT SHAFT FRAG FEMUR
SK 18B LEFT PROXIMAL RADIUS
SK 85 LEFT METACARPAL IV
SKW 2954 RIGHT MC IV
SKW 3646 RIGHT MC III
SKX 247 LEFT PROX 2/3 MT III
SKX 344 FOOT MIDDLE PHAL
SKX 1084 FRAG PATELLA
SKX 1261 FOOT MIDDLE PHAL
SKX 3062 HAND MIDDLE PHAL
SKX 3342 THORACIC VERTEBRA
SKX 3498 TRIQUETRAL
SKX 3646 RIGHT PROX FRAG METACARPAL III
SKX 3699 PROX RADIUS
SKX 5019 HAND INTERM PHAL

ACKNOWLEDGEMENTS

For giving me permission to examine the original fossil material and for other kindnesses, I thank C. K. Brain, E. Vrba, and the staff of the Transvaal Museum, Pretoria; P. V. Tobias, the late A. Hughes, and the staff of the Department of Anatomy, University of Witwatersrand, Johannesburg; R. E. F. Leakey, M. D. Leakey, the late L. S. B. Leakey, and the staff of the National Museums of Kenya, Nairobi; D. C. Johanson, W. H. Kimbel, and the staffs of the Cleveland Museum of Natural History and the Institute of Human Origins; Tadesse Terfa, Mammo Tessema, Woldesenbet Abomssa, and the staff of the National Museum of Ethiopia, Addis Ababa; F. C. Howell of the Department of Anthropology, University of California, Berkeley. I am grateful to Linda McHenry, Jessica Martini, and Russell Tuttle for their help on this paper. Partial funding for this research was provided by the Committee on Research, University of California, Davis.

LITERATURE CITED

Aiello, L. (1992) Allometry and the analysis of size and shape in human evolution. *J. Hum. Evol.* 22:127-147.

Aiello, L., and Dean, C. (1990) *An Introduction to Human Evolutionary Anatomy.* London: Academic.

Asfaw, B. (1985) Proximal femur articulation in Pliocene hominids. *Am. J. Phys. Anthropol.* 68:535-538.

Berge, C. (1984) Multivariate analysis of the pelvis for hominids and other extant primates: Implications for the locomotion and systematics of the different species of australopithecines. *J. Hum. Evol.* 13:555-562.

Berge, C. (1991) Quelle est la signification fonctionnelle du pelvis tres large d'*Australopithecus afarensis* (A.L. 288-1)? In Y. Coppens and B. Senut (eds.): *Origine(s) de la bipedie chez les hominidés.* Paris: Cahiers de Paleoanthropologie, pp. 113-120.

Berge, C., Orban-Segebarth, R., and Schmid, P. (1984) Obstetrical interpretation of the australopithecine pelvic cavity. *J. Hum. Evol.* 13:573-587.

Brown, F., Harris, J., Leakey, R., and Walker, A. (1985) Early *Homo erectus* skeleton from West Lake Turkana, Kenya. *Nature* 316:788-797.

Burns, P. E. (1971) New determination of Australopithecine height. *Nature* 232:350.

Bush, M. E., Lovejoy, C. O., Johanson, D. C., and Coppens, Y. (1982) Hominid carpal, metacarpal, and phalangeal bones recovered from the Hadar Formation: 1974-1977 collections. *Am. J. Phys. Anthropol.* 57:651-678.

Capecchi, V., Panichi, A., Santarcangelo, L., and Bigazzi, R. (1989) Tentative de reconstruction de la colonne vertebrale de A.L. 288 (Lucy). In G Giacobini (ed.): *Hominidae: Proceedings of the 2nd International Congress of Human Paleontology,* Turin, September 28 - October 3, 1987. Milan: Jaka, pp. 115-118.

Chamberlain, A.T., and Wood, B. A. (1987) Early hominid phylogeny. *J. Hum. Evol.* 16:119-133.

Charteris, J., Wall, J. C., and Nottrodt, J. W. (1982) Pliocene hominid gait: New interpretations based on available footprint data from Laetoli. *Am. J. Phys. Anthropol.* 58:133-144.

Christie, P. W. (1977) Form and function of the Afar ankle (abstract). *Am. J. Phys. Anthropol.* 47:123.

Clark, J. D., Asfaw, B., Assefa, G., et al. (1984) Paleoanthropological discoveries in the Middle Awash Valley, Ethiopia. *Nature* 307:423-428.

Clarke, R. J. (1979) Early hominid footprints from Tanzania. *S. Afr. J. Sci.* 75:148-149.

Conroy, G. C. (1987) Problems of body-weight estimation in fossil primates. *Int. J. Primatol.* 8:115-137.

Cook, D. C., Buikstra, J. E., DeRousseau, C. J., and Johanson, D. C. (1983) Vertebral pathology in the Afar australopithecines. *Am. J. Phys. Anthropol.* 60:83-102.

Corruccini, R. S., and McHenry, H. M. (1978) Relative femoral head size in early hominids. *Am. J. Phys. Anthropol.* 49:145-148.

Day, M. H. (1973) Locomotor features of the lower limb in hominids. *Symp. Zool. Soc. Lond.* 33:29-51.

Day, M. H. (1976a) Hominid postcranial material from Bed I, Olduvai Gorge. In G. L. Isaac and E. R. McCown (eds.): *Human Origins: Louis Leakey and the East African Evidence.* Menlo Park, CA: W. A. Benjamin, pp. 363-374.

Day, M. H. (1976b) Hominid postcranial remains from the East Rudolf succession. In Y. Coppens, F. C. Howell, G. L. Issac and R.E.F. Leakey (eds.): *Earliest Man and Environments in the Lake Rudolf Basin.* Chicago: University of Chicago Press, pp. 507-521.

Day, M. H. (1978) Functional interpretations of the morphology of postcranial remains of early African hominids. In C. Jolly (ed.): *Early Hominids of Africa.* New York: St. Martin's, pp. 311-345.

Day, M. H., and Wickens, E. H. (1980) Laetoli Pliocene hominid footprints and bipedalism. *Nature* 286 :385-387.

Day, M. H., Leakey, R.E. F., Walker, A.C, and Wood, B. A. (1976) New hominids from East Turkana, Kenya. *Am. J. Phys. Anthropol.* 45:369-436.

de Arsuaga, J. L. (1983) Sexual variability and taxonomic variability in the innominate bone of *Australopithecus. Z. Morph. Anthropol.* 73:297-308.

Deloison, Y. (1984) Comparative study of calcaneums of primates and *Pan-Australopithecus-Homo* relationships. *C.R. Acad. Sci.* 299:115-118.

Deloison, Y. (1985) Comparative study of calcanei of primates and *Pan-Australopithecus-Homo* relationship. In P.V. Tobias (ed.): *Hominid Evolution: Past, Present and Future.* New York: Alan R. Liss, pp. 143-147.

Deloison, Y. (1986) Description d'un calcaneum fossile de Primate et sa comparaison avec des calcaneums de Pongides, d'Australopitheques et d'*Homo. C.R. Acad. Sci.* Serie III, 302:257-262.

Deloison, Y. (1991) Les Australopitheques marchaient-ils comme nous? In Y. Coppens and B. Senut (eds.): *Origine(s) de la Bipedie chez les hominidés.* Paris: Cahiers de Paleoanthropologie, pp. 177-186.

Feldesman, M. R. (1979) Further morphometric studies of the ulna from the Omo Basin, Ethiopia. *Am. J. Phys. Anthropol.* 51:409-416.

Feldesman, M. R., and Lundy, J. K. (1988) Stature estimates for some African Pliocene-Pleistocene fossil hominids. *J. Hum. Evol.* 17:583-596.

Geissmann, T. (1986) Estimation of Australopithecine stature from long bones: A.L.288-1 as a test case. *Folia Primatol.* 47:119-127.

Grausz, H. M., Leakey, R. E. F., Walker, A. C., and Ward, C. V. (1988) Associated cranial and postcranial bones of *Australopithecus boisei.* In F. E. Grine (ed.): *Evolutionary History of the "Robust" Australopithecines.* New York: Aldine de Gruyter, pp. 127-132.

Grine, F. E. (1989) New hominid fossils form the Swartkrans Formation (1979-1986 excavations): Craniodental specimens. *Am. J. Phys. Anthropol.* 79:409-450.

Hartwig-Scherer, S., and Martin, R. D. (1991) Was "Lucy" more human than her "child"? Observations on early hominid postcranial skeletons. *J. Hum. Evol.* 21:439-449.

Hay, R. L., and Leakey, M. D. (1982) The fossil footprints of Laetoli. *Sci. Amer.* 246:50-57.

Helmuth, H. (1968) Korperhohe und Gliedmasson proportionen der Australopithecinen. *Z. Morph. Anthropol.* 60:147-155.

Hill, A., and Ward, S. (1988) Origin of the hominidae: The record of African large hominoid evolution between 14 MY and 4 MY. *Yearb. Phys. Anthropol.* 31:49-84.

Howell, F. C. (1978) Hominidae. In V. J. Maglio and H. B. S. Cooke (eds.): *The Evolution of African Mammals.* Cambridge: Harvard University Press, pp. 154-248.

Howell, F. C., and Wood, B. A. (1974) Early hominid ulna from the Omo basin, Ethiopia. *Nature* 249:174-176.

Howell, F. C., Haesaerts, P., and de Heinzelin, J. (1987) Depositional environments, archaeological occurences and hominids from Members E and F of the Shungura Formation (Omo basin, Ethiopia). *J. Hum. Evol.* 16:665-700.

Johanson, D. C., Taieb, M., and Coppens, Y. (1982a) Pliocene hominids from the Hadar Formation, Ethiopia (1973-1977): Stratigraphic, chronologic, and paleoenvironmental contexts, with notes on hominid morphology and systematics. *Am. J. Phys. Anthropol.* 57:373-402.

Johanson, D. C., Lovejoy, C. O., Kimbel, W. H., White, T. D., Ward, S. C., Bush, M. E., Latimer, B. M., and Coppens, Y. (1982b) Morphology of the Pliocene partial hominid skeleton (A.L. 288-1) from the Hadar Formation, Ethiopia. *Am. J. Phys. Anthropol.* 57:403-452.

Johanson, D. C., Fidelis T.M., Eck, G. G., White, T. D., Walter, R. C., Kimbel, W. H., Asfaw, B., Manega, P., Ndessokia, P., and Suwa, G. (1986) New partial skeleton of *Homo habilis* from Olduvai Gorge, Tanzania. *Nature* 327:205-209.

Jungers, W. L. (1982) Lucy's limbs: Skeletal allometry and locomotion in *Australopithecus afarensis*. *Nature* 297:676-678.

Jungers, W. L. (1988a) Relative joint size and hominoid locomotor adaptations with implications for the evolution of hominoid bipedalism. *J. Hum. Evol.* 17:247-266.

Jungers, W. L. (1988b) Lucy's length: Stature reconstruction in *Australopithecus afarensis* (A.L.288-1) with implications for other small-bodied hominids. *Am. J. Phys. Anthropol.* 76:227-232.

Jungers, W. L. (1988c) New estimations of body size in australopithecines. In F. E. Grine (ed.): *Evolutionary History of the "Robust" Australopithecines.* New York: Aldine de Gruyter, pp. 115-126.

Jungers, W. L. (1991) A pygmy perspective on body size and shape in *Australopithecus afarensis* (A.L. 288-1, "Lucy"). In Y. Coppens and B. Senut (eds.): *Origine(s) de la Bipedie chez les Hominidés.* Paris: Cahiers de Paleoanthropologie, pp. 215-224..

Kay, R. F. (1973) Humerus of robust australopithecines. *Science* 182:396.

Kennedy, G. E. (1983) A morphometric and taxonomic assessment of a hominine femur from the Lower Member, Koobi Fora, Lake Turkana. Am. *J. Phys. Anthropol.* 61:429-436.

Korey, K. A. (1990) Deconstructing reconstruction: The OH 62 humerofemoral index. *Am. J. Phys. Anthropol.* 83:25-33.

Lamy, P. (1983) Le systeme podal de certains hominides fossilles du Plio-Pleistocene d'Afrique de l'est: Etude morphodynamique. *L'Anthropologie* 87:435-464.

Lamy, P. (1986) The settlement of the longitudinal plantar arch of some African Plio-Pleistocene hominids: a morphological study. *J. Hum. Evol.* 15:31-46.

Latimer, B. M. (1988) Functional analysis of the Pliocene hominid ankle and pedal bones recovered from the Hadar Formation, Ethiopia: 1974-1977 collections. Ann Arbor: U.M.I. Dissertation Information Service.

Latimer, B. (1991) Locomotor adaptations in *Australopithecus afarensis*: The issue of arboreality. In Y. Coppens and B. Senut (eds.): *Origine(s) de la Bipedie chez les Hominidés.* Paris: Cahiers de Paleoanthropologie, pp. 169-176.

Latimer, B. M., and Lovejoy, C. O. (1989) The calcaneus of *Australopithecus afarensis* and its implications for the evolution of bipedality. *Am. J. Phys. Anthropol.* 78:369-386.

Latimer, B., and Lovejoy, C. O. (1990a) Hallucal tarsometatarsal joint in *Australopithecus afarensis*. *Am. J. Phys. Anthropol.* 82:125-134.

Latimer, B., and Lovejoy, C. O. (1990b) Metatarsophalangeal joints of *Australopithecus afarensis*. *Am. J. Phys. Anthropol.* 83:13-23.

Latimer, B. M., Lovejoy, C. O., Johanson, D. C., and Coppens, Y. (1982) Hominid tarsal, metatarsal, and phalangeal bones recovered from the Hadar Formation: 1974-1977 collections. *Am. J. Phys. Anthropol.* 57:701-720.

Latimer, B. M., Ohman, J. C., and Lovejoy, C. O. (1987) Talocrural joint in African hominoids: Implication for *Australopithecus afarensis*. *Am. J. Phys. Anthropol.* 74:155-175.

Leakey, M. D. (1978) Pliocene footprints at Laetolil, Northern Tanzania. *Antiquity* 52:133.

Leakey, M. D. (1981) Tracks and tools. *Phil. Trans. Roy. Soc. London B* 292:95-102.

Leakey, M. D. (1987) The hominid footprints. In M. D. Leakey and J. M. Harris (eds.): *Laetoli A Pliocene Site in Northern Tanzania.* Oxford: Clarendon, pp. 490-496..

Leakey, M. D., and Hay, R. L. (1979) Pliocene footprints in the Laetoli beds at Laetoli, Northern Tanzania. *Science* 278:317-323.

Leakey, M. G., and Leakey, R. E. F. (1978) *Koobi Fora Research Project: The Fossil Hominids and an Introduction to their Context* 1968-1974, I. Oxford: Clarendon.

Leakey, R. E. F. (1972) Further evidence of Lower Pleistocene Specimens from East Rudolf, Kenya, 1971. *Nature* 237:264-269.

Leakey, R. E. F., Mungai, J. M., and Walker, A. C. (1972) New Australopithecines from East Rudolf, Kenya (II). *Am. J. Phys. Anthropol.* 36:235-252.

Leakey, R. E. F., Walker, A., Ward, C. V., and Grausz, H. M. (1989) A partial skeleton of a gracile Hominid from the Upper Burgi Member of the Koobi Fora Formation, East Lake Turkana, Kenya. In G. Giacobini (ed.): *Hominidae: Proceedings of the 2nd International Congress of Human Paleontology* Turin, September 28–October 3, 1987. Milan: Jaka, pp. 167-174.

Lewis, O. J. (1977) Joint remodelling and the evolution of the human hand. *J. Anat.* 123:157-201.

Lewis, O. J. (1980) The joints of the evolving foot, Part III. The fossil evidence. *J. Anat.* 131:275-298.

Lewis, O. J. (1989) *Functional Morphology of the Evolving Hand and Foot.* Oxford: Clarendon.

Lieberman, D. E., Pilbeam, D. R., and Wood, B. A. (1988) A probablistic approach to the problem of sexual dimorphism in *Homo habilis*: A comparison of KNM-ER 1470 and KNM-ER 1813. *J. Hum. Evol.* 17:503-512.

Lovejoy, C. O. (1978) A biomechanical view of the locomotor diversity of early hominids. In C. J. Jolly (ed.): *Early Hominids of Africa.* New York: St. Martins.

Lovejoy, C. O. (1988, November) Evolution of human walking. *Sci. Amer.* Nov.:118-126.

Lovejoy, C. O., Johanson, D. C., and Coppens, Y. (1982a) Hominid lower limb bones recovered from the Hadar Formation: 1974-1977 collections. *Am. J. Phys. Anthropol.* 57:679-700.

Lovejoy, C. O., Johanson, D. C., and Coppens, Y. (1982b) Elements of the axial skeleton recovered from the Hadar Formation: 1974-1977 collections. *Am. J. Phys. Anthropol.* 57:631-636.

Marzke, M. W. (1971) Origin of the human hand. *Am. J. Phys. Anthropol.* 34:61-84.

Marzke, M. W. (1983) Joint functions and grips of the *Australopithecus afarensis* hand, with special reference to the region of the capitate. *J. Hum. Evol.* 12:197-211.

Marzke, M. W. (1986) Tool use and the evolution of hominid hands and bipediality. In J. G. Else and P. C. Lee (eds.): *Primate Evolution.* Cambridge: Cambridge University Press, pp. 203-209.

Marzke, M. W., and Marzke, R. F. (1987) The third metacarpal styloid process in humans: Origin and function. *Am. J. Phys. Anthropol.* 73:415-431.

Marzke, M. W., and Shackley, M. S. (1986) Hominid hand use in the Pleistocene: Evidence from experimental archaeology and comparative anatomy morphology. *J. Hum. Evol.* 15:439-460.

McHenry, H. M. (1972) The postcranial anatomy of early Pleistocene hominids. Ph.D. Thesis, Harvard University; Cambridge, Mass.

McHenry, H. M. (1973) Early hominid humerus from East Rudolf, Kenya. *Science* 180:739-741.

McHenry, H. M. (1974) How large were the australopithecines? *Am. J. Phys. Anthropol.* 40:329-340.

McHenry, H. M. (1975a) Fossil hominid body weight and brain size. *Nature* 254:686-688.

McHenry, H. M. (1975b) Fossils and the mosaic nature of human evolution. *Science* 190:425-431.

McHenry, H. M. (1976a) Early hominid body weight and encephalization. *Am. J. Phys. Anthropol.* 45:77-84.

McHenry, H. M. (1976b) Multivariate analysis of early hominid humeri. In E. Giles and J. S. Friedlander (eds.): *Measures of Man.* Cambridge: Peabody Museum, pp. 338-371.

McHenry, H. M. (1978) Fore- and hindlimb proportions in Plio-Pleistocene hominids. *Am. J. Phys. Anthropol.* 49:15-22.

McHenry, H. M. (1982) The pattern of human evolution: Studies on bipedalism, mastication & encephalization. *Ann. Rev. Anthropol.* 11:151-173.

McHenry, H. M. (1983) The capitate of *Australopithecus afarensis* and *Australopithecus africanus. Am. J. Phys. Anthropol.* 62:187-198.

McHenry, H. M. (1984a) The common ancestor: A study of the post-cranium of *Pan paniscus, Australopithecus,* and other hominoids. In R. L. Susman (ed.): *The Pygmy Chimpanzee.* New York: Plenum, pp. 201-230

McHenry, H. M. (1984b) Relative cheek-tooth size in *Australopithecus. Am. J. Phys. Anthropol.* 64:297-306.

McHenry, H. M. (1985) Implications of postcanine megadontia for the origin of *Homo.* In E. Delson (ed.): *Ancestors: The Hard Evidence.* New York: Alan R. Liss, pp. 178-183.

McHenry, H. M. (1986a) Size variation in the postcranium of *Australopithecus afarensis* and extant species of Hominoidea. *J. Hum. Evol.* 15:149-156.

McHenry, H. M. (1986b) The first bipeds: A comparison of the *Australopithecus afarensis* and *Australopithecus africanus* postcranium and implications for the evolution of bipedalism. *J. Hum. Evol.* 15:177-191.

McHenry, H. M. (1988) New estimates of body weights in early hominids and their significance to encephalization and megadontia in "robust" australopithecines. In F. E. Grine (ed.): *Evolutionary History of the "Robust" Australopithecines.* New York: Aldine de Gruyter, pp. 133-148.

McHenry, H. M. (1991a) Femoral lengths and stature in Plio-Pleistocene hominids. *Am. J. Phys. Anthropol.* 85:149-158.

McHenry, H. M. (1991b) Petite bodies of the "robust" Australopithecines. *Am. J. Phys. Anthropol.* 86:445-454.

McHenry, H. M. (1991c) First Steps? Analyses of the postcranium of early hominids. In Y. Coppens and B. Senut (eds.): *Origine(s) de la Bipedie chez les Hominidés.* Paris: Cahiers de Paleoanthropologie, pp. 133-142.

McHenry, H. M. (1992) Body size and proportions in early hominids. *Am. J. Phys. Anthropol.* 87:407-431.

McHenry, H. M., and Corruccini, R. S. (1975) Distal humerus in hominoid evolution. *Folia Primatol.* 23:227-244.

McHenry, H. M., and Corruccini, R. S. (1976) Fossil hominid femora and the evolution of walking. *Nature* 259:657-658.

McHenry, H. M., and Corruccini, R. S. (1978) The femur in early human evolution. *Am. J. Phys. Anthropol.* 49:473-488.

McHenry, H. M., Corruccini, R. S., and Howell, F. C. (1976) Analysis of an early hominid ulna from the Omo Basin, Ethiopia. *Am. J. Phys. Anthropol.* 44:295-304.

McHenry, H. M., and Temerin, L. A. (1979) The evolution of hominid bipedalism: Evidence from the fossil record. *Yearb. Phys. Anthropol.* 22:105-131.

Ohman, J. C. (1986) The first rib of hominoids. *Am. J. Phys. Anthropol.* 70:209-229.

Olivier, G. (1976) The stature of australopithecines. *J. Hum. Evol.* 5:529-534.

Patterson, B., and Howells, W. W. (1967) Hominid humeral fragment from early Pleistocene of Northwestern Kenya. *Science* 156:64-66.

Pickford, M. H. L., Johanson, D. C., Lovejoy, C. O., White, T. D., and Aronson, J. L. (1983) A hominid humeral fragment from the Pliocene of Kenya. *Am. J. Phys. Anthropol.* 60:337-346.

Reed, C. A., and Falk, D. (1977) The stature and weight of Sterkfontein 14, a gracile australopithecine from Transvaal, as determined from the innominate bone. *Fieldiana (Geology)* 33:423-440.

Ricklan, D. E. (1987) Functional anatomy of the hand of *Australopithecus africanus*. *J. Hum. Evol.* 16:643-664.

Ricklan, D. E. (1990) The precision grip in *Australopithecus africanus*: Anatomical and behavioral correlates. In G. H. Sperber (ed.): *From Apes to Angels: Essays in Anthropology in Honor of Phillip V. Tobias*. New York: Wiley-Liss, pp. 177-183.

Robbins, L. M. (1987) Hominid footprints from site G. In M.D. Leakey and J. M. Harris (eds.): *Laetoli: A Pliocene Site in Northern Tanzania*. Oxford: Clarendon, pp. 497-502.

Robinson, J. T. (1972) *Early Hominid Posture and Locomotion*. Chicago: University of Chicago Press.

Rose, M. D. (1982) Food acquisition and the evolution of positional behavior: The case of bipedalism. In D. J. Chivers, B. A. Wood and A. Bilsborough (eds.): *Food Acquisition and Processing in Primates*. New York: Plenum, pp. 509-524.

Rose, M. D. (1984) A hominine hip bone, KNM-ER 3228, from east Lake Turkana, Kenya. *Am. J. Phys. Anthropol.* 63:371-378.

Ruff, C. B. (1991) Climate and body shape in hominid evolution. *J. Hum. Evol.* 20:81-105.

Schmid, P. (1983) Eine rekonstruktion des skelettes von A.L. 288-1 (Hadar) und deren konsequenzen. *Folia Primatol.* 40:283-306.

Schmid, P. (1991) The trunk of the Australopithecines. In Y. Coppens and B. Senut (eds.): *Origine(s) de la Bipedie chez les Hominides*. Paris: Cahiers de Paleoanthroplogie, pp. 225-234.

Senut, B. (1978) Revision de quelques pieces humerales Plio-Pleistocene Sud-Africaines. *Bull. Mem. Soc. Anthropol. Paris* 5:223-229.

Senut, B. (1979) Comparison des hominides de Gombore 1B et de Kanapoi: Deux pieces de genre *Homo*? *Bull. Mem. Soc. Anthropol. Paris* 1:111-117.

Senut, B. (1980) New data on the humerus and its joints in Plio-Pleistocene hominoids. *Coll. Anthropol.* 4:87-93.

Senut, B. (1981a) Humeral outlines in some hominoid primates and in Plio-Pleistocene hominids. *Am. J. Phys. Anthropol.* 275-284.

Senut, B. (1981b) *L'humerus et ses Articulations chez les Hominidés Plio-Pleistocenes*. Paris: Cahiers de Paleontologie.

Senut, B. (1981c) Outlines of the distal humerus in hominoid primates: Application to some Plio-Pleistocene hominids. In A. B. Chiarelli and R. S. Corruccini (eds.): *Primate Evolutionary Biology*. Berlin: Springer-Verlag, pp. 81-92.

Senut, B. (1986) Long bones of the primate upper limb: Monomorphic or dimorphic? *Hum. Evol.* 1:7-22.

Senut, B. (1991) Origine(s) de la bipedie humaine: Une approche paleontologique. In Y. Coppens and B. Senut (eds.): *Origine(s) de la Bipedie chez les Hominides*. Paris: Cahiers de Paleoanthropologie, pp. 245-258.

Senut, B., and Tardieu, C. (1985) Functional aspects of Plio-Pleistocene hominoid limb bones: Implications for taxonomy and phylogeny. In E. Delson (ed.): *Ancestors: The Hard Evidence*. New York: Alan R. Liss, pp. 193-201.

Senut, B., and Tobias, P. V. (1989) A preliminary examination of some new hominid upper limb remains from Sterkfontein (1974-1984). *Paleontology* 308:565-571.

Stern, J. T., and Susman, R. L. (1983) The locomotor anatomy of *Australopithecus afarensis*. *Am. J. Phys. Anthropol.* 60:279-318.

Stern, J. T., and Susman, R. L. (1991) "Total morphological pattern" versus the "magic trait": Conflicting approaches to the study of early hominid bipedalism. In Y. Coppens and B. Senut (eds.): *Origine(s) de la Bipedie chez les Hominidés*. Paris: Cahiers de Paleoanthropologie, pp. 99-112.

Steudel, K. (1981) Body size estimators in primate skeletal material. *Int. J. Primatol.* 2:81-90.

Stringer, C. B. (1986) The credibility of *Homo habilis*. In B. Wood, L. Martin and P. Andrews (eds.): *Major Topics in Primate and Human Evolution*. Cambridge: Cambridge University Press, pp. 266-294.

Susman, R. L. (1988a) New postcranial remains from Swartkrans and their bearing on the functional morphology and behavior of *Paranthropus robustus*. In F. E. Grine (ed.): *Evolutionary History of the "Robust" Australopithecines*. New York: Aldine de Gruyter, pp. 149-174.

Susman, R. L. (1988b) Hand of *Paranthropus robustus* from member I, Swartkrans: Fossil evidence for tool behavior. *Science* 240: 781-782.

Susman, R. L. (1988c) New postcranial fossils from Swartkrans Member 1: Implications for the behavior of *Paranthropus robustus*. *Am. J. Phys. Anthropol.* 75:277.

Susman, R. L. (1989) New hominid fossils from the Swartkrans Formation (1979-1986 excavations): Postcranial specimens. *Am. J. Phys. Anthropol.* 79:451-474.

Susman, R. L. (1991) Species attribution of the Swartkrans thumb metacarpals: Reply to Drs. Trinkaus and Long. *Am. J. Phys. Anthropol.* 86:549-552.

Susman, R. L., and Brain, T. M. (1988) New first metatarsal (SKX 5017) from Swartkrans and the gait of *Paranthropus robustus*. *Am. J. Phys. Anthropol.* 77:7-15.

Susman, R. L., and Creel, N. (1979) Functional and morphological affinities of the subadult hand (O.H. 7) from Olduvai Gorge. *Am. J. Phys. Anthropol.* 51:311-332.

Susman, R. L., and Stern, J. T. (1979) Telemetered electromyography of flexor digitorum profundus and flexor digitorum superficialis in *Pan troglodytes* and implications for interpretation of the O.H. 7 hand. *Am. J. Phys. Anthropol.* 50:565-574.

Susman, R. L., and Stern, J. T. (1982) Functional morphology of *Homo habilis*. *Science* 217:931-934.

Susman, R. L., and Stern, J. T. (1991) Locomotor behavior of early hominids: Epistemology and fossil evidence. In Y. Coppens and B. Senut (eds.): *Origine(s) de la Bipedie chez les Hominidés*. Paris: Cahiers De Paleoanthropologie, pp. 121-132.

Susman, R. L., Stern, J. T., and Jungers, W. L. (1984) Arboreality and bipedality in the Hadar hominids. *Folia Primatol.* 43:113-156.

Suzman, I. M. (1980) A new estimate of body weight in South African australopithecines. In R.E.F. Leakey and B. A. Ogot (eds.): *Proc. 8th Pan. Afr. Congress* Nairobi, September 1977, pp. 175-179.

Tague, R. G., and Lovejoy, C. O. (1986) The obstetric pelvis of A.L. 288-1 (Lucy). *J. Hum. Evol.* 15:237-255.

Tardieu, C. (1981) Morph-functional analysis of the articular surfaces of the knee-joint in primates. In A. B. Chiarelli and R. S. Corruccini (eds.): *Primate Evolutionary Biology*. Berlin: Springer-Verlag, pp. 68-80.

Tardieu, C. (1983) *L'articulation du Genou. Analyse Morphofonctionelle chez les Primates et les Hominides Fossiles*. Paris: Cahiers de Paleoanthropologie.

Tardieu, C. (1986a) Evolution of the knee intra-articular menisci in primates and some hominids. In J. G. Else and P. C. Lee (eds.): *Primate Evolution*. Cambridge: Cambridge University Press, pp. 183-190.

Tardieu, C. (1986b) The knee joint in three hominoid primates:Application to Plio-Pleistocene hominids and evolutionary implications. In D. M. Taub and F. A. King (eds.): *Current Perspectives in Primate Biology*. New York: Van Norstrand Reinhold, pp. 182-192.

Tobias, P. V. (1991) The environmental background of hominid emergence and the appearance of the Genus *Homo*. *Hum. Evol.* 6:129-142.

Trinkaus, E. (1989) Olduvai hominid 7 trapezial metacarpal 1 articular morphology: Contrasts with recent humans. *Am. J. Phys. Anthropol.* 80:411-416.

Tuttle, R. H. (1967) Knuckle walking and the evolution of hominoid hands. *Am. J. Phys. Anthropol.* 26:171-206.

Tuttle, R. H. (1981) Evolution of hominid bipedalism and prehensile capabilities. *Phil. Trans. Roy. Soc. London B* 292:89-94.

Tuttle, R. H. (1985) Ape footprints and Laetoli impressions: A response to SUNY claims. In P. V. Tobias (ed.): *Hominid Evolution: Past, Present and Future*. New York: Alan R. Liss, pp. 129-134.

Tuttle, R. H. (1987) Kinesiological inferences and evolutionary implications from Laetoli bipedal trails G-1,G-2/3 and A. In M. D. Leakey and J. M. Harris (eds.): *Laetoli: A Pliocene Site in Northern Tanzania.* Oxford: Clarendon, pp. 503-523.

Tuttle, R. H. (1988) What's new in African anthropology. *Ann. Rev. Anthropol.* 17:391-426.

Tuttle, R. H. (1990) The pitted pattern of Laetoli feet. *Nat. Hist.* 3:61-66, 100.

Tuttle, R. H., Webb, D. M., and Tuttle, N. I. (1991a) Laetoli footprint trails and the evolution of hominid bipedalism. In Y. Coppens and B. Senut (eds.): *Origine(s) de la Bipedie chez les Hominidés.* Paris: Cahiers de Paleoanthropologie, pp. 187-198.

Tuttle, R. H., Webb, D. M., and Baksh, M. (1991b) Laetoli toes and *Australopithecus afarensis*. *Hum. Evol.* 6:193-200.

Tuttle, R. H., Webb, D. M., Tuttle, N. I., and Baksh, M. (1992) Footprints and gaits of bipedal apes, bears, and barefoot people: Perpective on Pliocene tracks. In S. Matono, R. H. Tuttle, H. Ishida, and M. Goodman (eds.) *Topics in Primatology 3*. Tokyo: University of Tokyo Press, 221-242.

Vrba, E. S. (1974) Chronological and ecological implications of the fossil Bovidae at the Sterkfontein Australopithecine site. *Nature* 250:19-23.

Vrba, E. S. (1975a) Some evidence of chronology and paleoecology of Sterkfontein, Swartkrans and Kromdraai from the fossil Bovidae. *Nature* 254:301-304.

Vrba, E. S. (1975b) The life and times of the Transvaal apeman. *S. Afr. J. Sci.* 71:298-299.

Vrba, E. S. (1985a) Early hominids in southern Africa: Updated observations on chronological and ecological background. In P.V . Tobias (ed.): *Hominid Evolution: Past, Present, and Future*. New York: Alan R. Liss, pp. 195-200.

Vrba, E. S. (1985b) Ecological and adaptive changes associated with early hominid evolution. In E. Delson (ed.): *Ancestors: The Hard Evidence*. New York: Alan R Liss, pp. 63-71.

Walker, A., Ward, C. V., Leakey, R. E. F., and Grausz, H. M. (1989) Evolution in the *Australopithecus boisei* lineage. In G. Giacobini (ed.): *Hominidae: Proceedings of the 2nd International Congress of Human Paleontology;* Turin, September 28–October 3, 1987. Milan: Jaka, pp. 133-140.

White, T. D. (1980) Evolutionary implications of Pliocene hominid footprints. *Science* 208:175-176.

White, T. D. (1984) Pliocene homiids from the Middle Awash, Ethiopia. *Cour. Forsch. Inst. Senckenberg* 69: 57-68.

White, T. D., and Suwa, G. (1987) Hominid footprints at Laetoli: Facts and interpretations. *Am. J. Phys. Anthropol.* 72:485-514.

Wolpoff, M. H. (1973) Posterior tooth size, body size, and diet in South African gracile australopithecines. *Am. J. Phys. Anthropol.* 39:375-394.

Wolpoff, M. H. (1983) Lucy's little legs. *J. Hum. Evol.* 12:443-454.

Wood, B. A. (1974) Evidence on the locomotor pattern of *Homo* from early Pleistocene of Kenya. *Nature* 251:135-136.

Wood, B. A. (1985) Sexual dimorphism in the hominid fossil record. In J. Ghesquire, R. D. Martin and F. Newcombe (eds.): *Human Sexual Dimorphism*. London: Taylor and Francis.

Wood, B. A. (1989) Hominid relationships: A cladistic perspective. *Proc. Aust. Soc. Hum. Biol.* 2:83-102.

Wood, B. A. (1991) *Koobi Fora Research Project IV: Hominid Cranial Remains from Koobi Fora*. Oxford: Clarendon.

Wood, B. A. (1992) Origin and evolution of the genus *Homo*. *Nature* 355:783-790.

CHAPTER 14

Up from Electromyography

Primate Energetics and the Evolution of Human Bipedalism

Russell H. Tuttle

Evolution seems to have been very conservative in designing vertebrate locomotory systems for moving along the ground. What at first seemed a bewildering array of modes of locomotion—bipedal walking in humans and in birds, quadrupedal walking in mammals, trotting, galloping, and hopping—can all be reduced to two general mechanisms, a pendulum and a spring, which have been utilized either singly or in combination, to minimize the expenditure of chemical energy by the muscles for lifting and reaccelerating the center of mass within each stride. (Cavagna et al. 1977, p. R260)

A bipedal locomotor sequence, performed for a given period of time, and for a given reason, may be energetically inefficient. However, when this bipedalism is considered along with all the other positional activities the animal has to perform, the total energetic pattern may not be inefficient. (Rose 1984, p. 513)

When interpreting the relative advantages of different types of locomotion, we must be careful not to place undue emphasis on the energetic cost of locomotion simply because it can be easily quantified. (Parsons and Taylor 1977, p. 188)

INTRODUCTION

People, like other creatures, machines, marriages, friendships, and scientific societies, are subject to the law of entropy: They require regular inputs of energy if they are to forestall disintegration. Primates expend energy in quests for food and mates, to elude predation, to resist pathogens, to rebuild after cell death and injury, and to sustain basal metabolic processes, pregnancies, and their sizeable brains. Humans and nonhuman primates alike face numerous choices and compromises as they attempt to acquire and to store energy, while concurrently expending it.

In order to assess the status of primate energetics in relation to the evolution of human bipedal positional behavior, I will sketch the empirical framework under the following questions: What should we know? What do we know? and What can be done with what we know?

WHAT SHOULD WE KNOW?

Moving mammals convert chemical energy—captured through skillful and opportunistic foraging and feeding behavior and fractionally freed via digestion and cellular respiration—to mechanical energy (Goldspink 1977a).

Students of primate locomotor energetics are challenged to measure the energy that is spent by subjects as they engage in a full (or, at least, highly representative) repertoire of naturalistic positional behaviors and to make controlled interbehavioral, ontogenetic, and intertaxonal comparisons of their relative efficiency in performing them (Cavagna et al. 1977).

Ideally, the energy that is available to them, based on thorough knowledge of what they eat and what is transformed physiologically for work, should be known. Further, one must account for energy lost through air resistance and intra-articular friction and locate elastic strain energy that is stored in tendons, ligaments, and muscles during a subject's actions (Heglund 1985).

The proximate sources of power (viz. the contractions of specific muscles or parts of muscles) must be determined (Goldspink 1977b; White 1977). All factors should be precisely quantified and located temporally so that they can be expressed in equations, which also take account of speed, body size, center of mass, and segmental distributions of mass in the subjects (Taylor et al. 1982; Fedak et al. 1982; Heglund, Cavagna and Taylor 1982; Heglund, Fedak, Cavagna and Taylor 1982; Mansour et al. 1982).

This information may be used to model the evolution of hominid bipedalism and other positional behaviors. But first we must establish the reliability of inferences about body size; distribution of mass; and the kinetics, kinematics, and energetics of locomotor modes from fossil skeletal evidence and footprint trails, which are the stock-in-trade of paleoanthropologists. Following the evolutionary development of distinctive locomotor mechanisms requires reliably dated sequences of specimens, whose alpha taxonomy and phylogenetic relationships are established clearly on the basis of craniodental and postcranial evidence.

WHAT DO WE KNOW?

Studies on the energetics of primate positional behavior are still in their infancy despite a rich fare of provocative insights and theories that have been generated over the past two decades by ingenious researchers working on well-funded projects with high-tech equipment. Knowledge of anthropoid primate positional behavior and mechanics has reached the stage of robust toddlerhood, leaving much to be learned here about our closest living relatives. The evidence for emergence of hominid bipedalism and its proximate antecedents is prezygotic because we have no substantial fossils (Martin 1990) from the Late Miocene (5-11 MYA), wherein the imperative events probably occurred (Tuttle 1988; Rose 1991).

Energetics

There is virtual consensus among primate kinesiologists that apes hold the most clues to the evolution of our locomotor anatomy. However, disagreement reigns regarding which species most closely represent prehominid conditions and what they do that could represent prospective adaptations to terrestrial bipedalism. Candidate activities include nonricochetal arm-swinging, suspensory feeding, versatile quadrupedal climbing, vertical climbing, bipedal foraging, arboreal bipedal locomotion, knuckle-walking, and combinations of two or more of these positional behaviors (Tuttle 1974, 1975, 1977, 1986; Tuttle, Webb and Tuttle 1991; Senut 1989, 1991, 1992; Rose 1991; Zihlman 1992).

Given the prevailing bias toward apes as models for pre- and protohuman locomotor morphology and behavior, it is remarkable that the energetics of their locomotion is documented so rudimentarily. For instance, Pritchard's (1990) topical bibliography on ape locomotion contains only three entries (Heglund 1985; Rodman and McHenry 1980; Taylor and Rowntree 1973a) on "bioenergetic aspects". Most of these reports, and several that Pritchard missed, are focused on young chimpanzees, which walked bipedally and quadrupedally on treadmills, thereby sampling a fractional segment of their natural positional behavioral repertoire.

Although the energetics of brachiation has been explored with spider monkeys (Parsons and Taylor 1977), comparable experiments are not available for apes. Moreover, the energetics of vertical climbing and other behaviors that are currently popular as precedent to human bipedalism remain to be documented for apes. Such studies probably are feasible. Although mammalian climbing is barely explored, comparative physiologists have measured the energetics of technologically more demanding locomotor modes, including swimming, flying and burrowing, in vertebrates and invertebrates (Alexander and Goldspink 1977).

Comparative physiologists usually determine the energy consumed during steady state locomotion by measuring the rate of oxygen consumption of the subjects, which must tolerate lightweight masks (that measure gas exchange) and move on or under contraptions (treadmills, rope mills) at various speeds in more or less natural and facultative postures for prolonged periods (Goldspink 1977c; Taylor et al. 1982; Heglund 1985; Parsons and Taylor 1977; Bennett 1985). These conditions are notably disparate from naturalistic environments and varied locomotor bouts of wild apes.

A major complication in determining energy transfer for moving animals is our inability to measure elastic strain energy (Heglund et al. 1982). Because most of the energy contained in carbohydrates, lipids and proteins is lost as heat during the synthesis of ATP, only about 25 percent of their energy remains to power muscles (Cavagna et al. 1977; Taylor and Heglund 1982). Fedak et al. (1982) found that ATP utilization by muscles could not account for the work needed to supply the total kinetic energy of moving mammals and birds. Accordingly, they concluded "that elastic recoil supplies a significant fraction of the increases in kinetic energy relative to the centre of mass when these animals run at high speeds" (Fedak et al. 1982, pp. 39-40).

Tendons are especially important components because they stretch more than muscle fibers do, thereby storing greater elastic strain energy (Alexander 1984). During fast locomotion, sheep tendons, which probably closely represent hominoid primate tendons histologically and mechanically (Alexander et al. 1981), can return 93 percent of the energy used to stretch them (Ker 1981).

Only a small fraction of the total power for quadrupedal and bipedal locomotion is needed to overcome friction in the joints (because of highly efficient lubrication: Barnett et al. 1961) and air resistance, even in sprinting humans, who present broad frontal areas (Pugh 1971; Alexander 1984; Heglund 1985). Therefore, they are far less confounding than elastic strain energy is for the student of comparative locomotion.

Among the major findings of experimenters on anthropoid energetics is that, in *Ateles*, nonricochetal arm-swinging is energetically costly. Nonetheless, the advantages of moving directly between points in the canopy may balance well against costs of taking longer (perhaps predator-infested) routes via less vigorous locomotor modes (Parsons and Taylor 1977). This may be true for apes too. However, they should be studied directly, along with the energetics of ricochetal arm-swinging by gibbons.

Second, young chimpanzees (weighing 17.5 kg) expend more energy to move uphill (+15° incline) versus horizontally than is true of small mammals. However, as they move downhill, chimpanzees and much smaller mammals alike recover most of the mechanical energy that was "stored" during the ascents due to efficient acceleration of their limbs by gravity as they advance downhill (Taylor et al. 1972).

Third, young chimpanzees expend equivalent amounts of energy as they walk bipedally and quadrupedally on a treadmill at a given speed. From this, Taylor and Rowntree (1973a, p. 187) concluded that "the cost or efficiency of bipedal versus quadrupedal locomotion probably should not be used in arguments weighing the relative advantages and disadvantages that bipedal locomotion conferred on man." Likewise, Rodman and McHenry (1980) commented that once bipedalism was advantageous for other reasons, it developed uneventfully.

Previously, Taylor et al. (1970) had concluded that human running is more costly than rapid quadrupedalism in a variety of nonprimate mammals, perhaps because we use only two limbs instead of four.

Rodman and McHenry (1980) contested the presumption that human bipedalism is energetically more costly than the quadrupedalism of apes and other mammals. They showed that human walking is, in fact, somewhat less costly than that predicted for like-sized quadrupeds and calculated that wild chimpanzee travel costs are 150 percent of those predicted for like-sized pronograde quadrupeds. Accordingly, they concluded that human bipedalism is considerably more efficient than the quadrupedalism of apes. They suggested that terrestrial movement between dispersed arboreal food sources was performed bipedally by our Miocene ancestors, whose freed upper limbs remained adapted for arboreal climbing, and whose lower limbs were specialized for prolonged long-distance travel.

Rodman and McHenry's (1980) arguments on the relative efficiency of human bipedalism are generally accepted. Indeed, Heglund (1985, p. 332) stated that humans are "exceptionally efficient locomotors," at least in comparison with chipmunks, dogs, quail, and turkeys. Unfortunately, comparable quantitative data on the total mechanical power output and total metabolic power input during locomotion, which are used to calculate efficiency (Ralston 1976, p. 93), are not published for nonhuman primates.

From Heglund's (1985) limited comparative set, it appears that large size enhances locomotor efficiency because larger animals can effectively use elastic energy, their muscles have low intrinsic velocities, and they employ relatively low stride frequencies (Biewener 1990). Taylor et al. (1982) showed that in primates, as in most other mammals (Fedak and Seeherman 1979), the energetic cost of locomotion increases linearly with speed and varies in a regular, simple way with body size, independent of gait and method of movement.

Finally, Bramble and Carrier (1983; Carrier 1984) proposed that human breathing is less constrained by locomotor mechanics than that of rapidly moving quadrupeds. Whereas trotting and galloping horses, dogs, and jackrabbits synchronize their locomotor and respiratory cycles at a constant ratio of 1:1 (strides per breath), running humans employ several optional patterns (4:1, 3:1, 2:1, 1:1, 5:2, 3:2).

Unlike the thoraces of quadrupeds, human chests are freed from the stresses of direct impact loading, which is coupled with exhalation in quadrupeds. Nevertheless, practiced runners also tightly couple their breathing with the locomotor cycle, favoring a ratio of 2 strides per breath at high speeds and 4:1 at sustained slower speeds.

Bramble and Carrier (1983) suggested that because humans maintain the same gait regardless of running speed, the ability to alter one's breathing pattern might constitute an alternative means to regulate energetic cost.

Although humans are energetically efficient walkers, the cost of our running is relatively high in comparison with rapid terrestrial locomotion by most other mammals and birds (Taylor et al. 1982). Carrier (1984) sought to resolve the paradox of high energetic cost versus the fact that practiced humans are excellent long-distance runners, who rival mammals that may be faster over short distances but are unable to maintain speed for extended periods.

Endurance runners, whose muscles are highly active, must be equipped to dissipate notable amounts of metabolic heat. Otherwise, body temperatures will rise beyond tolerable levels, which may damage the brain (Taylor and Lyman 1972; Taylor and Rowntree 1973b, 1974; Baker 1982). Humans cope with this challenge by perspiring. Our abundantly vascularized, highly sensitive, atrophically haired skins are profusely endowed with sweat glands, whose voluminous secretions evaporatively cool our extensive surfaces (Montagna 1962, 1985; Newman 1970). This mechanism is particularly effective during running because the rush of air facilitates evaporation. Interestingly, we have more sweat glands on our anterior and flexor surfaces than on our posterior and extensor surfaces (Montagna 1962, p. 312).

Unlike mammals that pant to cool off, our extrarespiratory cooling system is not coupled with the locomotor cycle. Panters must stop running in order to accelerate the air flow across the evaporative surfaces in their oral and nasal cavities (Carrier 1984).

The ability to run at a variety of speeds without dangerous overheating may have enabled our ancestors to employ a variety of tactics to evade different predators and conspecific enemies. Carrier (1984) proposed that hominid locomotor stamina and ability to employ a range of speeds was advantageous for hunting smaller prey during the heat of the day. Because of indecision about the effectiveness of *Australopithecus sensu lato* as bipedal runners, he proposed that endurance hunting may not have developed until the emergence of *Homo*.

Carrier (1984) and commentators on his provocative theory did not explore the problem of how the earliest hairless hominids kept warm at night. In the absence of skinning tools, warm hides would be hard to purloin. Cuddling in nests and control of fire are options, but the former has not been detected (Tuttle 1992) and the latter is not convincingly represented in the fossil record before the heyday of *Homo erectus* (Potts 1988; Klein 1989).

While we may anticipate that sufficient skeletal materials and biomechanical knowledge may someday be available to clarify the question of running ability in Pliocene and Early-Middle Pleistocene hominids, it is probably useless to split hairs or to bristle over the relative hirsuteness of prehistoric predecessors, who are unrepresented by frozen or mummified specimens.

Laboratory Studies of Locomotion and Mechanics

Notable strides have been made to describe the mechanics of facultative bipedalism and other positional behaviors of captive apes, primarily due to pioneering experimental studies by Tuttle, Basmajian, and coworkers (1972 et seq.), who worked at Yerkes Regional Primate Research Center, in Atlanta, Georgia; Kondo (1985), Ishida, Kimura, Okada, Yamazaki, and Tuttle, working at the Primate Research Institute of Kyoto University, Inuyama, and the Department of Human Sciences of Osaka University (Ishida et al. 1978), Suita-shi; and Fleagle and coworkers (1981) at the State University of New York, Stony Brook. Among these experimentalists, only Yamazaki et al. (1979, 1983; Yamazaki 1985; Yamazaki and Ishida 1984) included energetic factors in models of the prospective hominoid biped.

From a comparison of bipedal walking by a young adult male *Homo sapiens*, and adolescent female *Pan troglodytes* and *Macaca fuscata* and computer-simulated stick figures of an apish human, a humanoid ape, and humans with an abbreviated or a hypertrophic heel, Yamazaki and coworkers (1979) concluded that, both mechanically and kinematically, chimpanzee bipedal walking is closer to that of humans than to that of Japanese monkeys. Further, they noted that large bipedal apes would have difficulty walking long distances because it would be energetically very costly and would stress their hip and knee joints notably. In order to reduce these factors, apish wanabe-bipeds would have to achieve truncal erectness and lengthen the tuber calcanei from chimpanzee to human dimensions. Yamazaki et al. (1979) consider changes in bodily proportions, particularly humanoid length of the leg, to be secondary evolutionary developments.

The sheer ingenuity of Yamazaki's computer simulation studies should not overshadow the fact that the researchers did not measure energy expenditures directly, but instead estimated them variously on the basis of electromyographic patterns and guesstimates of muscle force, derived from cross-sectional areas, mass, length, and shape and direction of muscle force, estimated from gross dissections. The inferences on humans are more convincing than those on chimpanzees and macaques because Yamazaki et al. (1979) could cite more thorough, diverse confirmatory studies on humans than on nonhuman primates.

In particular, electromyography on nonhuman primates has not been combined with strain gauge studies in the hind limb as subjects walk bipedally. Moreover, although EMG signals indicate when and relatively how active each muscle is during a particular action or behavior in a given experiment, such observations cannot be quantified precisely for unequivocal interexperimental comparisons, let alone for interspecific comparisons of actions and behaviors (Tuttle, Cortright and Buxhoeveden 1979) and reliable determinations of actual energy expenditure.

In a more comprehensive computer simulation study, Yamazaki (1985; Yamazaki et al. 1983) concluded that, on balance, hylobatid bipedal walking is closer to that of humans than are chimpanzee, spider monkey, and Japanese macaque bipedal walking. Accordingly, Yamazaki (1985, 1990a) proposed that a brachiation model best represents the predecessor of emergent hominoid bipedalism.

Kinematically, gibbon bipedal walking is closer to that of humans because they exhibit less truncal inclination and more extension of their hind limbs, especially during the propulsive toe-off, than other anthropoid primates do (Yamazaki, 1985, 1990a). And, although chimpanzee bipedal walking is closer to that of humans per the combined mechanical and kinematic indices of Yamazaki (1985), like-sized gibbons would closely rival and perhaps even surpass chimpanzees as accomplished bipeds.

Yamazaki (1985, 1990a) did not engage the vertical-climbing model (Tuttle 1975, 1981; Prost 1980; Fleagle et al. 1981), which was in full fettle at the time, though, in an earlier paper, Yamazaki and Ishida (1984) concurred that in gibbons, vertical climbing, like brachiation (Yamazaki 1990b), promotes truncal erection and knee extension; vertical climbing also develops the extensor muscles of the hip joint. All of these features facilitate terrestrial bipedalism. Moreover, Yamazaki and Ishida (1984) argued that bipedal walking on boughs strengthens the extensor muscles of the knee joint and enhances humanoid actions of the knee. In short, they supplied experimental validation for my hylobatian model (Tuttle 1974, 1975, 1981, 1987, 1988; Tuttle et al. 1991).

Yamazaki and Ishida (1984) estimated that during walking on a level terrestrial substrate, energy expenditure per unit body weight and locomotor distance is nearly the same in a 22-year-old man (62 kg) and an 8-year-old male gibbon (6.25 kg), but it increases by 30 percent when the gibbon walks on a horizontal bar. Comparisons between the energetics of hylobatid vertical climbing versus bipedal walking are confounded by the fact that the forelimbs expend an unknown proportion of energy in hylobatid vertical climbing. However, assuming that the forelimbs and the hind limbs expend equal amounts of energy, the energy load during vertical climbing is double that of level walking by the gibbon (Yamazaki and Ishida 1984).

In another study, Yamazaki (1985; Yamazaki et al. 1983) calculated that bipedal walking is energetically much more costly for gibbons than for chimpanzees and humans, which is a consequence of their smaller body size, not their pattern of walking. This is concordant with conclusions of other experimenters (Biewener 1990; Heglund 1985; Taylor et al. 1982) that in primates, as in most other mammals, the cost of locomotion simply increases with body size (and speed) regardless of locomotor mode. A chimpanzee-sized gibbon could cover more ground with fewer strides and therefore would use relatively less energy to progress bipedally.

Generally, in apes, bipedal standing and the stance phase of bipedal walking recruit notable, sustained activity in numerous muscles of their hind limbs (Tuttle et al. 1975, 1978, 1979; Tuttle and Cortright 1983; Ishida et al. 1978, 1984, 1985; Okada et al. 1976; Okada and Kondo 1982; Stern and Susman 1981). This contrasts with the relative economy of muscle activity in the lower limbs of humans as they stand and walk (Greenlaw and Basmajian 1975; Carlsöo 1972; Basmajian and De Luca 1985) and in the forelimbs of apes as they quiescently hang bimanually and even unimanually in experimental settings (Basmajian and Tuttle 1973; Tuttle and Basmajian 1974a,b,c, 1975, 1977, 1978; Tuttle and Cortright 1988; Tuttle and Watts 1985; Tuttle et al. 1983; Jungers and Stern 1981, 1984; Larson and Stern 1986; Larson et al. 1991; Stern et al. 1980).

Naturalistic Behavior

In order to design experimental studies on positional behaviors that will inform functional and evolutionary interpretations of morphological features in living and extinct primates, we must know the frequency and relative importance of various behaviors by representative subjects in their natural habitats (Tuttle, Cortright and Buxhoeveden 1979; Jouffroy 1989). Accordingly, experimentalists are heavily dependent on field primatologists for quantitative data and expert qualitative evaluations of what each species does during the daily round in all seasons, how characteristic these positional behaviors are, and the relative importance of each behavior to individual quality of life, fitness of the population, and sustenance of the species.

We fall far short on this tall order for all species of apes (Tuttle, 1986), though important chunks of data and insights have emerged during the past two decades. Periodically, I have summarized available information up to mid-1984 (Tuttle 1969a, 1970, 1972, 1975, 1977, 1986). Hunt (1991a,b, 1992) augmented these syntheses with several recent reports, including his own primary quantitative studies on the positional behavior of chimpanzees in two Tanzanian woodland and secondary forest habitats.

Hunt (1992) found that forelimb suspensory postures, which he terms "arm hanging," are characteristic of apes (except Virunga gorillas) versus Gombe baboons. Moreover, frequencies of arboreal vertical climbing distinguish chimpanzees and other apes (except Virunga gorillas) from the Gombe baboons, which represent sizeable cercopithecoid monkeys. Hunt (1992) reasonably concluded that distinctive adaptive/morphological complexes of the Hominoidea (Tuttle 1969b, 1974, 1975) are related to arboreal manually suspensory postures and vertical climbing.

Hunt's study (1991a,b, 1992) underscores the imperative for studies on the energetics of naturalistic vertical climbing and suspensory postures in forever-free apes and other primates. Likewise, we need to develop means to study *in vivo* the electromyography, elastic strain in tendons, muscles and ligaments, and the responses of bone to stress during natural spontaneous positional behaviors by these subjects. Kinesiological laboratories are limited markedly in space, cage design, and the sizes of apes that can be handled and studied productively with sophisticated gadgetry (Tuttle et al. 1991). Indeed, it is high time that we return to the forest, creatively armed, to achieve further progress with this set of problems. Thereafter, computer simulations and models would be better informed and might even replace future disruptive, invasive experiments.

Fossil Facts and Scenarios

In comparison with field observers and laboratory kinesiologists, paleoanthropologists are woefully disadvantaged in their ability to control the quality and kinds of data that will advance understanding of the evolution and effects of hominid bipedalism. At best, one may hope for reasonably complete skeletons from reliably dated sites. Such specimens are pitifully rare indeed (Tuttle 1988). This is particularly frustrating since the proportions of limbs and other locomotor organs cannot be determined unequivocally from fragmentary partial skeletons (Godinot 1990). Nonetheless, enough skeletal specimens are available from Pliocene sites in Africa and Pleistocene sites in Africa and Eurasia to fuel informed arguments on the development of bipedalism in the Hominidae during these epochs (Tuttle 1988; Trinkaus 1986, 1989; McHenry this volume).

A major insight from the wealth of hominid postcranial specimens that has accrued during the past two decades, combined with classic fossils from southern Africa, is that the several species of *Australopithecus sensu lato* and earliest *Homo* retained features for arboreal activities (Tuttle 1967, 1981; Robinson 1972, 1978; Zuckerman et al. 1973; Senut 1981; Senut and Tardieu 1985; Tardieu, 1991; Susman and Creel 1979; Susman and Stern 1979, 1991; Vrba 1979; Schmid 1983, 1991; Langdon 1985; McHenry 1982, 1986, 1991; Lewis 1989; Tobias 1991; Oxnard 1975; Oxnard and Lisowski 1980; Ashton et al. 1981; Rose 1984; Berge 1991; Deloison 1991; Jungers 1991; Preuschoft and Witte 1991), in which they probably engaged to an unknown extent, during the daily round (Tuttle 1988; Tuttle et al. 1991). They also were capable terrestrial bipeds, whose resemblance to a modern human locomotor pattern are diametrically argued to be those of a "missing link" between apes and us and virtually like ours (Lovejoy 1974, 1975, 1978, 1981, 1988; Lovejoy et al. 1973; Johanson and White 1979; White 1980; Wolpoff 1980; Latimer et al. 1987; Latimer and Lovejoy 1989, 1990a,b; Latimer 1991; Langdon et al. 1991).

Currently, we cannot resolve this moving controversy because there is no compelling consensus about the functional meaning of lower limb complexes in the Pliocene-early Pleistocene Hadar, eastern African and southern African hominids, and even for specimens in which the postcranial elements are definitely associated with craniodental remains, whose alpha taxonomy is not challenged seriously. Only if postcranial bits are virtually identical to those of modern *Homo sapiens* will no expert carp that the creatures had failed to master our brand of bipedalism.

This dilemma probably will not be resolved until we have clearly weighted the relative importance of modern human skeletal traits for standing, walking, running, squatting, rising from seated and squatting postures, ascending and descending inclined substrates, and other common terrestrial behaviors. This puzzle is reminiscent of the one that confronts students of human positional behavioral energetics: Combinations and concatenations of individual muscles, tendons, ligaments and breathing mechanisms function somewhat to very differently (to conserve and to deliver energy) during various actions, particularly vis-à-vis speed and gravity.

In the absence of definitive answers from partial skeletons and bony fragments, we might expect clarification from the Laetoli hominid footprint trails, which provide more direct evidence of positional behavior 3.5 MYA (Leakey 1981; Leakey and Harris 1987; Leakey and Hay 1971). Regrettably, they too transport us to a purgatory of problems instead of an Olympian view on how the trackmakers moved about the savanna and the extent to which they utilized arboreal resources. Even the alpha taxonomy of the printmakers is indeterminate (Tuttle 1985, 1987, 1990; Tuttle et al. 1990, 1991, 1992).

The better preserved prints are remarkably humanoid. The hallux was adducted to alignment with the lateral toes and a medial longitudinal arch is salient. The Laetoli G hominids apparently walked with a human pattern of heel strike, followed by weight transmission along the lateral sole, then medially across the ball, with toe-off falling predominantly on the hallucal pad (Tuttle 1987; Tuttle et al. 1990).

We cannot tell further how human their walking was because the features that we can measure and infer from hominoid trails—stride length, stride width, foot angles, and estimated speeds—do not distinguish the gaits of humans from those of apes and, in some features, even bears. Most critically, we do not know how fast and regularly the Laetoli bipeds walked in the moist ash (Tuttle et al. 1991, 1992).

Accordingly, there is little scientific merit in attempts to predict lengths of the lower limbs of the printmakers from the Laetoli trails (Jungers 1982; Jungers and Stern 1983; Reynolds 1983), to validate the presence of *A. afarensis* at Laetoli via the trails, or otherwise to infer the humanness of suprapedal structures and behavior in Pliocene Hominidae from them (Tuttle et al. 1991). Moreover, if Yamazaki et al. (1979) are correct that evolutionary lengthening of the calcaneus served to decrease the energy expenditure of bipedal walking, then we may allow that the Laetoli Site G hominids, whose footprints sport notable heels, were accomplished bipedal walkers even if they had lower limbs that are relatively somewhat shorter than ours.

In brief, the inferred foot structure of the Laetoli hominids evidences that they had the capacity to stand, walk, and run as we do. However, it would be foolhardy to insist that they did so, because we do not know what they were like beyond their soles, and, even more importantly, we have not quantified the extent to which features of our own feet are particularly adapted to standing, versus walking and running. If, by extending excavations at Laetoli north or south of the exposed trails, we were to find a sequence of pedal ball and toe prints (without heel prints or evidence of their taphonomic obliteration) with longer strides, I would accept that the creatures were advanced humanoid runners. Until then we must be content with the knowledge that they were walking slowly and mundanely in the moist ash, probably in order to conserve energy and perhaps out of caution, though the possibility remains that nonpedal anatomical features prevented them from walking much as we do.

Finally, apparent disconformity between the apish long, down-curved toes from Hadar (3.0 MYA) and the truly humanoid footprints at Laetoli (3.5 MYA) raises the intriguing possibilities that not only were more than one species of Hominidae in Africa contemporaneously during the Pliocene but also that they represent different levels of terrestrial bipedal adaptation and arboreal habits. Whereas toes of the Hadar hominids are longer than those of H. sapiens (McHenry, 1986; Latimer et al. 1990b), toe lengths of the Laetoli printmakers are not significantly different from those of habitually barefoot Machiguenga people (Tuttle et al. 1991). Accordingly, *pace* White and Suwa (1987), the Laetoli footprints at Site G should not be ascribed to *A. afarensis* or be used to exemplify the bipedal accomplishments of Pliocene beasts and phantoms at Hadar (Tuttle 1981, 1985, 1988, 1990; Tuttle et al. 1991).

MAKING DO WITH WHAT WE KNOW

The notable gaps in empirical knowledge and limitations of experimental approaches reasonably deter informed skeptics and philistines alike from venturing panoramic scenarios on the evolution of hominid bipedalism from treetops to globe trotting. Nonetheless, contra the ethos of Clark Howell, I will indulge in this exercise because it might stimulate some readers to investigate an aspect of the problem more creatively and productively than I have and will entertain others with a novel construct to pummel in hotel bars or honorably in print.

Research over the past two decades, particularly in Japan, has reaffirmed my confidence in a hylobatian model for the beginnings of bipedalism in our lineage. In brief, the protohominid bipeds evolved from rather long-legged, siamang-sized hominoids that were predisposed to terrestrial bipedalism by a heritage of vertical climbing on vines and tree trunks and running bipedally on horizontal and moderately inclined boughs. The ancient hylobatians differed from modern hylobatid apes in lacking excessive elongation of the upper limbs and other special features that underpin frequent brachiation, particularly ricochetal arm-swinging, and probably also in numerous craniodental (Benefit and McCrossin 1991) and epidermal characteristics. Yet, they too were versatile foragers that could adopt a variety of suspensory and bipedal postures in order to exploit virtually all regions of the canopy.

If Yamazaki and Ishida (1984) are correct that, in gibbons, the energy loads on the hind limbs during branch-walking and especially vertical climbing are greater than those of level walking on the ground, it is likely that hylobatian bipeds would be well equipped to engage in terrestrial bipedalism. A major early terrestrial challenge would be to develop a fully plantigrade pedal morphology, with concomitant shift from digital to heel contact at the beginning of each support phase of the locomotor cycle. If the Late Miocene protohominids were bipedal from the outset of their terrestrial careers (with quadrupedalism being a lesser adaptive option for them), then selection probably would be intense for a shift from prehensile to plantigrade pedal structure. Increasing body size and frequencies of sustained, erect bipedal stance on extended lower limbs would also select for a humanoid foot structure.

Standing bipedally on the ground in order to forage on overhanging branches in forest edge, mosaic, woodland, and other low-canopied and relatively open habitats (Sigmon 1971; Rose 1974, 1976, 1984; Wrangham 1980; Jolly and Plog 1986; Leutenegger 1987) could select for energy-saving osseoligamentous mechanisms in the hip and knee joints that in *H. sapiens* underpin economic, sustained stance on fully extended lower limbs (Du Brul 1962; MacConaill and Basmajian 1969; Ishida 1991; Kummer 1991). While consuming their harvests, these bipedal foragers may have squatted often (instead of sitting), thereby further selecting for sizeable heels and balance between the heel and forefoot, as well as between their feet. Frequent alternate squatting and rising from squats would enhance development of the quadriceps femoris and hamstring and gluteal muscles as knee and hip extensors, which also continued to serve them for climbing trees to forage arboreally, to evade terrestrial enemies, and to nest.

By the early Pliocene, at least some hominids began to use handy stones and sticks to extract energy-rich plant parts from protective shells and perhaps watery and nutritious tubers from beneath the ground. These extraction aids were sometimes used in the trees, as well as on the ground. Eventually, descendants of these tool-users regularly stashed stones and sticks in nests and arboreal nooks as they traveled over their ranges. When the trees produced again, tools were readily at hand for the harvest.

Further, longer sticks could be used both to draw in and to knock down plant parts and to repel competitors and predators that might try to enter a food or lodge tree (Tuttle 1992).

The Pliocene Hadar hand bones evidence that *A. afarensis* had powerful grips (Tuttle 1981). Marzke (1983, 1986) and coworkers (Marzke and Marzke 1987; Marzke and Schakley 1986) also argue cogently that their hands were capable of gripping sticks and stones firmly for vigorous pounding and throwing.

Lithic tool manufacture probably began when cutting versus smashing was required (e.g. to collect animal skins for clothing and carrying devices and to butcher large carcasses for consumption and transport away from the death spot; Potts 1984, 1988; Shipman 1986) or to sharpen hefty sticks to make digging tools and jabbing weapons. Although these activities drew the early hominids further from trees, they probably continued to return to arboreal bases for shelter until they had developed effective defenses, including control of fire, against terrestrial predators and overwhelming competitors. Accordingly, we should not be surprised that Hadar *A. afarensis* sport arboreally adapted features in their upper and lower limbs and that even Olduvai *H. habilis* evidences possible arboreal adaptation in the hand (Tuttle 1981, 1992).

Arboreality of *A. afarensis* was probably less pronounced than the SUNY claims, but somewhat more manifest than is currently admissible to Latimer and Lovejoy (1989, 1990a,b; Lovejoy 1988; Latimer et al. 1987). Even if *A. afarensis* lacked pongoid prehensile halluces, the longish, ventrally curved second-to-fifth toes would serve them well for increasing friction forces and perhaps for lowering muscle potentials during climbing (Sarmiento 1988; Rose 1984) and for holding branches and boughs as they squatted and stood to forage, to build nests, and to defend themselves in the canopy (Tuttle et al. 1991).

The last refinements of the human bipedal adaptive complex, including the heavily tendonized triceps surae muscle complex and unique epidermal and respiratory components, probably developed with regular trekking, sprinting, endurance running or some combination of these behaviors as our ancestors gained a notable foothold in open environments, at least to the extent that they could move across the landscape without undue dependence on trees for daily shelter and sustenance. For example, our remarkably versatile respiratory cycles (Bramble and Carrier 1983), the rich concentration of eccrine sweat glands in our scalps (versus few or none in those of apes), and other peculiar specializations in the development and distribution of human sweat glands probably developed relatively late in human phylogeny (Montagna 1962; Folk and Semken 1991) in response to the energetic demands for sustained and sometimes vigorous bipedal locomotion. Increased brain size, including those of developing fetuses, would require effective cooling mechanisms during and following sustained vigorous and speedy physical activities in relatively open country (Falk 1991, 1992).

Portable cutting and digging tools, carrying devices, artifactual body coverings, projectile and poisonous weapons, tool-assisted intimidation displays and evasive tactics, control of fire, and keen knowledge of and cultural accommodation to their surroundings further enabled Middle and Late Pleistocene hominids to cope with and even to flourish in a diversity of basically terrestrial habitats.

Energetic and other physiological considerations inspired Franciscus and Trinkaus (1988; Trinkaus 1989) to challenge Coon's (1962) explanation of the peculiar nasal structure in European Neandertals as an adaptation to cold climates—to warm nasally inhaled air before it reached the lungs. Instead, the large, projecting nasal apertures may represent adaptations to aridity, since the Neandertal configuration would conserve moisture during exhalation (Carey and Steegman 1981).

It would also dissipate excess body heat (Franciscus and Trinkaus 1988; Trinkaus 1989). If Neandertals necessarily engaged in vigorous activities during seasonally high temperatures, and if they wore animal furs and other insulating materials on their bodies and heads as they exercised vigorously in cold weather, it may have been necessary to cool their blood in the nasal chamber in order to keep their large brains from overheating. Perhaps more versatile use of locomotor/respiratory regimes (Trinkaus 1989), clothing, and somewhat smaller brains allowed modern *H. sapiens* to escape Neandertaloid nasal morphology, even though we are probably the nosier folk.

SUMMARY

Studies on the energetics of primate positional behavior are in their infancy. Comparative, experimental, and naturalistic studies on primate locomotion are booming. Postcranial specimens of fossil primates are receiving increased emphasis in phylogenetic constructions and functional interpretations of ancestral forms, including those in the lineage of *Homo.*

Because of the heterogeneity of studies, methods and results, and major gaps in our knowledge of naturalistic energetics and locomotor mechanics in apes, it is impossible to provide a cogent synthesis. Moreover, there are no substantial fossils from the Late Miocene that pertain to the evolution of hominid bipedalism and other hominoid positional behaviors. Accordingly, I sketch highlights of recent research and discuss the sorts of studies that are needed if we wish to more fully understand our evolutionary history, and particularly the adaptive significance of human bipedalism. The relevance of electromyographic experiments on anthropoid primates and of the 3.5 MYA Laetoli footprint trails for human evolutionary puzzles received special attention.

Finally, I reiterate the hylobatian model for the beginnings of hominid bipedalism and speculate on later developments in our evolutionary career, with special emphasis on the continued importance of arboreal behaviors, including tool-use and caching, before final development of our obligate terrestriality.

ACKNOWLEDGMENTS

I thank Susan Pfeiffer for persuading me to review the topic of primate energetics for the Human Biology Council Symposium on Physical Activity and Human Biology, which convened in Milwaukee, WI, April 3, 1991. And I am profoundly grateful for the much too brief experience of teaching the human career and for other intellectual and social interactions with Clark Howell, during my first decade at the University of Chicago, while he was mining the Omo deposits and setting a high standard for multidisciplinary paleoanthropological research and I was pursuing arboreal imperatives, resisting knuckle-walkers in our lineage, and inaugurating electromyographic research with great apes.

LITERATURE CITED

Alexander, R. McN. (1984) Walking and running. *Am. Sci.* 72:348-354.

Alexander, R. McN., and Goldspink, G. (1977) *Mechanics and Energentics of Animal Locomotion.* London: Chapman and Hall.

Alexander, R. McN., Jayes, A. S., Maloiy, M. O., and Wathuta, E. M. (1981) Allometry of leg muscles of mammals. *J. Zool.* 194:539-552.

Ashton, E. H., Flinn, R. M., Moore, W. J., Oxnard, C. E., and Spence, T. F. (1981) Further quantitative studies of form and function in the primate pelvis with special reference to *Australopithecus. Trans. Zool. Soc. Lond.* 36:1-98.

Baker, M. A. (1982) Brain cooling in endotherms in heat and exercise. *Ann. Rev. Physiol.* 44:85-96.

Barnett, C. H., Davies, D. V., and MacConaill, M. A. (1961) *Synovial Joints. Their Structure and Mechanics.* London: Longmans, Green & Co.

Basmajian, J. V., and De Luca, C. J. (1985) *Muscles Alive,* 5th ed. Baltimore: Williams & Wilkins.

Basmajian, J. V., and Tuttle, R. (1973) EMG of locomotion in gorilla and man. In R. B. Stein, K. B. Pearson, R. S. Smith, and J. B. Redford (eds.): *Control of Posture and Locomotion.* New York: Plenum, pp. 599-609.

Benefit, B. R., and McCrossin, M. L. (1991) Ancestral facial morphology of Old World higher primates. *Proc. Nat. Acad. Sci.* 88:5267-5271.

Bennett, A. F. (1985) Energetics and locomotion. In M. Hildebrand, D. M. Bramble, K. F. Liem, and D. B. Wake (eds.): *Functional Vertebrate Morphology.* Cambridge: Harvard University Press, pp. 173-184.

Berge, C. (1991) Quelle est la signification fonctionnelle du pelvis très large de *Australopithecus afarensis* (AL 288-1)? In Y. Coppens and B. Senut (eds.): *Origines de la Bipédie chez les Hominidés.* Paris: Cahiers de Paléoanthropologie, Editions du CNRS, pp. 113-119.

Biewener, A. A. (1990) Biomechanics of mammalian terrestrial locomotion. *Science* 250:1097-1103.

Bramble, D. M., and Carrier, D. R. (1983) Running and breathing in mammals. *Science* 219:251-256.

Carlsöo, S. (1972) *How Man Moves. Kinesiological Methods and Studies.* London: Wm. Heinemann.

Carey, J. W., and Steegman, A. T. (1981) Human nasal protrusion, latitude, and climate. *Am. J. Phys. Anthropol.* 56:313-319.

Carrier, D. R. (1984) The energetic paradox of human running and hominid evolution. *Curr. Anthropol.* 25:483-495.

Cavagna, G. A., Heglund, N. C., and Taylor, R. (1977) Mechanical work in terrestrial locomotion: Two basic mechanisms for minimizing energy expenditure. *Am. J. Physiol.* 233:R243-R261.

Coon, C. S. (1962) *The Origin of Races*. New York: Knopf.

Deloison, Y. (1991) Les australopitheques marchaient–ils comme nous? In Y. Coppens and B. Senut (eds.): *Origine(s) de la Bipédie chez les Hominidés*. Paris: Cahiers de Paléoanthropologie, Editions du CNRS, pp. 177-186.

Du Brul, E. L. (1962) The general phenomenon of bipedalism. *Am. Zool.* 2:205-208.

Falk, D. (1991) Breech birth of the genus *Homo*: Why bipedalism preceded the increase in brain size. In Y. Coppens and B. Senut (eds.): *Origine(s) de la Bipédie chez les Hominidés*. Paris: Cahiers de Paléoanthropologie. Editions du CNRS, pp. 259-266.

Falk, D. (1992) *Braindance*. New York: Henry Holt.

Fedak, M. A., Heglund, N. C., and Taylor, C. R. (1982) Energetics and mechanics of terrestrial locomotion. II. Kinetic energy changes of the limbs and body as a function of speed and body size in birds and mammals. *J. Exper. Biol.* 97:23-40.

Fedak, M. A., and Seeherman, H. J. (1979) Reappraisal of energetics of locomotion shows identical cost in bipeds and quadrupeds including ostrich and horse. *Nature* 282:713-716.

Fleagle, J. G., Stern, J. T., Jr, Jungers, W. L., Susman, R. L., Vangor, A. K., and Wells, J. P. (1981) Climbing: A biomechanical link with brachiation and with bipedalism. *Symp. Zool. Soc. Lond.* 48:359-375.

Folk, G. E., and Semken, A., Jr. (1991) The evolution of sweat glands. *Int. J. Biometeorol.* 35:180-186.

Franciscus, R. G., and Trinkaus, E. (1988) The Neandertal nose. *Am. J. Phys. Anthropol.* 75:209-210.

Godinot, M. (1990) An introduction to the history of primate locomotion. In F. K. Jouffroy, M. H. Stack, and C. Niemitz (eds.): *Gravity, Posture and Locomotion in Primates*. Firenze: Editrice "Il Sedicesimo," pp. 45-60.

Goldspink, G. (1977a) Muscle energetics. In R. McN. Alexander and G. Goldspink (eds.): *Mechanics and Energetics of Animal Locomotion*. London: Chapman and Hall, pp. 57-81.

Goldspink, G. (1977b) Design of muscles in relation to locomotion. In R. McN. Alexander and G. Goldspink (eds.): *Mechanics and Energetics of Animal Locomotion*. London: Chapman and Hall, pp. 1-22.

Goldspink, G. (1977c) Energy cost of locomotion. In R. McN. Alexander and G. Goldspink (eds.): *Mechanics and Energetics of Animal Locomotion*. London: Chapman and Hall, pp. 153-167.

Greenlaw, R. K., and Basmajian, J. V. (1975) Function of the gluteals in man. In R. H. Tuttle (ed.): *Primate Functional Morphology and Evolution*. The Hague: Mouton, pp. 271-279.

Heglund, N. C. (1985) Comparative energetics and mechanics of locomotion. How do primates fit in? In W. L. Jungers (ed.): *Size and Scaling in Primate Biology*. New York: Plenum Press, pp. 319-335.

Heglund, N. C., Cavagna, G. A., and Taylor, C. R. (1982) Energetics and mechanics of terrestrial locomotion. III. Energy changes of the centre of mass as a function of speed and body size in birds and mammals. *J. Exper. Biol.* 97:41-56.

Heglund, N. C., Fedak, M. A., Cavagna, G. A., and Taylor, C. R. (1982) Energetics and mechanics of terrestrial locomotion. IV. Total mechanical energy changes as a function of speed and body size in birds and mammals. *J. Exper. Biol.* 97:57-66.

Hunt, K. D. (1991a) Positional behavior in the Hominoidea. *Int. J. Primatol.* 12:95-118.

Hunt, K. D. (1991b) Mechanical implications of chimpanzee positional behavior. *Am. J. Phys. Anthropol.* 86:521-536.

Hunt, K. D. (1992) Positional behavior of *Pan troglodytes* in the Mahale Mountains and Gombe Stream National Parks, Tanzania. *Am. J. Phys. Anthropol.* 87:83-105.

Ishida, H. (1991) A strategy for long distance walking in the earliest hominids: Effect of posture on energy expenditure during bipedal walking. In Y. Coppens and B. Senut (eds.): *Origine(s) de la Bipédie chez les Hominidés*. Paris: Cahiers de Paléoanthropologie, CNRS, pp. 7-15.

Ishida, H., Kimura, T., and Yamazaki, N. (1984) Kinesiological aspects of bipedal walking in gibbons. In H. Preuschoft, D. J. Chivers, W. Y. Brockelman, and N. Creel (eds.): *The Lesser Apes*. Edinburgh: Edinburgh University Press, pp. 135-145.

Ishida, H., Kumakura, H., and Kondo, S. (1985) Primate bipedalism and quadrupedalism: Comparative electromyography. In S. Kondo (eds.): *Primate Morphophysiology, Locomotor Analyses and Human Bipedalism*. Tokyo: University of Tokyo Press, pp. 59-79.

Ishida, H., Okada, M., Tuttle, R. H., and Kimura, T. (1978) Activities of hind limb muscles in bipedal gibbons. In D. J. Chivers and K. A. Joysey (eds.): *Recent Advances in Primatology*, Vol. 3: *Evolution*. London: Academic Press, pp. 459-462.

Johanson, D. C., and White, T. D. (1979) A systematic assessment of early African hominids. *Science* 203:321-330.

Jolly, C. J., and Plog, F. (1986) *Physical Anthropology and Archeology*, 4th Ed. New York: Knopf.

Jouffroy, F. K. (1989) Quantitative and experimental approaches to primate locomotion. A review of recent advances. In P. K. Seth and S. Seth (eds.): *Perspectives in Primate Biology*, Vol. 2. New Delhi: Today & Tomorrow's Printers and Publishers, pp. 47-108.

Jungers, W. L. (1982) Lucy's limbs: Skeletal allometry and locomotion in *Australopithecus afarensis*. *Nature* 297:676-678.

Jungers, W. L. (1991) A pygmy perspective on body size and shape in *Australopithecus afarensis* (Al 288-1, "Lucy"). In Y. Coppens and B. Senut (eds.): *Origine(s) de la Bipédie chez les Hominidés*. Paris: Cahiers de Paléoanthropologie, Editions du CNRS, pp. 215-224.

Jungers, W. L., and Stern, J. T., Jr. (1981) Preliminary electromyographical analysis of brachiation in gibbon and spider monkey. *Int. J. Primatol.* 2:19-33.

Jungers, W. L., and Stern, J. T., Jr. (1983) Body proportions, skeletal allometry and locomotion in the Hadar hominids: A reply to Wolpoff. *J. Hum. Evol.* 12:673-684.

Jungers, W. L., and Stern, J. T., Jr. (1984) Kinesiological aspects of brachiation in lar gibbons. In H. Preuschoft, D. J. Chivers, W. Y. Brockelman, and N. Creel (eds.): *The Lesser Apes*, Edinburgh: Edinburgh University Press, pp. 119-134.

Ker, R. F. (1981) Dynamic tensile properties of the plantaris tendon of sheep (*Ovis aries*). *J. Exper. Biol.* 93:283-302.

Klein, R. G. (1989) *The Human Career. Human Biological and Cultural Origins.* Chicago: University of Chicago Press.

Kondo, S. (ed.) (1985) *Primate Morphophysiology, Locomotor Analyses and Human Bipedalism.* Tokyo: University of Tokyo Press.

Kummer, B. (1991) Biomechanical foundations of the development of human bipedalism. In Y. Coppens and B. Senut (eds.): *Origine(s) de la Bipédie chez les Hominidés.* Paris: Cahiers de Paléoanthropologie, Editions du CNRS, pp. 1-8.

Langdon, J. H. (1985) Fossils and the origin of bipedalism. *J. Hum. Evol.* 14:615-635.

Langdon, J. H., Bruckner, J., and Baker, H. H. (1991) Pedal mechanics and bipedalism in early hominids. In Y. Coppens and B. Senut (eds.): *Origine(s) de la Bipédie chez les Hominidés.* Paris: Cahiers de Paléoanthropologie, Editions du CNRS, pp. 159-167.

Larson, S. G., and Stern, J. T., Jr. (1986) EMG of scapulohumeral muscles in the chimpanzee during reaching and "arboreal" locomotion. *Am. J. Anat.* 176:171-190.

Larson, S. G., Stern, J. T., Jr., and Jungers, W. L. (1991) EMG of serratus anterior and trapezius in the chimpanzee: Scapular rotators revisited. *Am. J. Phys. Anthropol.* 85:71-84.

Latimer, B. (1991) Locomotor adaptations in *Australopithecus afarensis*: The issue of arboreality. In Y. Coppens and B. Senut (eds.): *Origine(s) de la Bipédie chez les Hominidés.* Paris: Cahiers de Paléoanthropologie, Editions du CNRS, pp. 169-176.

Latimer, B., and Lovejoy, C. O. (1989) The calcaneus of *Australopithecus afarensis* and its implications for the evolution of bipedality. *Am. J. Phys Anthropol.* 78:369-386.

Latimer, B., and Lovejoy, C. O. (1990a) Hallucal tarsometatarsal joint in *Australopithecus afarensis. Am. J. Phys. Anthropol.* 82:125-133.

Latimer, B., and Lovejoy, C. O. (1990b) Metatarsophalangeal joints of *Australopithecus afarensis. Am. J. Phys. Anthropol.* 83:13-23.

Latimer, B., Ohman, J. C., and Lovejoy, C. O. (1987) Talocrural joint in African hominoids: Implications for *Australopithecus afarensis. Am. J. Phys. Anthropol.* 74:155-175.

Leakey, M. D. (1981) Tracks and tools. *Phil. Trans. Roy. Soc. Lond.* B 292:95-102.

Leakey, M. D., and Harris, J. M. (1987) *Laetoli. A Pliocene Site in Northern Tanzania.* Oxford: Clarendon.

Leakey, M. D., and Hay, R. L. (1979) Pliocene footprints in the Laetolil beds at Laetoli, Northern Tanzania. *Nature* 278:317-323.

Leutenegger, W. (1987) Origin of hominid bipedalism. *Nature* 325:305.

Lewis, O. J. (1989) *Functional Morphology of the Evolving Hand and Foot.* Oxford: Clarendon.

Lovejoy, C. O. (1974) The gait of australopithecines. *Yearb. Phys. Anthropol.* 17:147-161.

Lovejoy, C. O. (1975) Biomechanical perspectives on the lower limb of early hominids. In R. H. Tuttle (ed.): *Primate Functional Morphology and Evolution.* The Hague: Mouton, pp. 291-326.

Lovejoy, C. O. (1978) A biomechanical review of the locomotor diversity of early hominids. In C. Jolly (ed.): *Early Hominids of Africa.* New York: St. Martin's, pp. 403-429.

Lovejoy, C. O. (1981) The origin of man. *Science* 211:341-350.

Lovejoy, C. O. (1988) The evolution of human walking. *Sci. Am.* 259:118-125.

Lovejoy, C. O., Heiple, K. G., and Burnstein, A. H. (1973) The gait of *Australopithecus. Am. J. Phys. Anthropol.* 38:757-780.

MacConaill, M. A., and Basmajian, J. V. (1969) *Muscles and Movements. A Basis for Human Kinesiology.* Baltimore: Williams & Wilkins.

Mansour, J. M., Lesh, M. D., Nowak, M. D., and Simon, S. R. (1982) A three dimensional multi-segmental analysis of the energetics of normal and pathological human gait. *J. Biomech.* 15:51-59.

Martin, R. D. (1990) *Primate Origins and Evolution. A Phylogenetic Reconstruction.* Princeton: Princeton University Press.

Marzke, M. W. (1983) Joint functions and grips of the *Australopithecus afarensis* hand, with special reference to the region of the capitate. *J. Hum. Evol.* 12:197-211.

Marzke, M. W. (1986) Tool use and the evolution of hominid hands and bipedality. In J. G. Else and P. C. Lee (eds.): *Primate Evolution.* Cambridge: Cambridge University Press, pp. 201-209.

Marzke, M. W., and Marzke, R. F. (1987) The third metacarpal styloid process in humans: Origin and functions. *Am. J. Phys. Anthropol.* 73:415-431.

Marzke, M. W., and Schakley, M. S. (1986) Hominid hand use in the Pliocene and Pleistocene: Evidence from experimental archaeology and comparative morphology. *J. Hum. Evol.* 15:439-460.

McHenry, H. M. (1982) The pattern of human evolution: Studies on bipedalism, mastication and encephalization. *Ann. Rev. Anthropol.* 11:151-173.

McHenry, H. M. (1986) The first bipeds: A comparison of the *A. africanus* and *A. africanus* postcranium and implications for the evolution of bipedalism. *J. Hum. Evol.* 15:177-191.

McHenry, H. M. (1991) First steps? Analyses of the postcranium of early hominids. In Y. Coppens and B. Senut (eds.): *Origine(s) de la Bipédie chez les Hominidés.* Paris: Cahiers de Paléoanthropologie, Editions du CNRS, pp. 133-141.

Montagna, W. (1962) *The Structure and Function of Skin*, 2nd ed. New York: Academic.

Montagna, W. (1985) The evolution of human skin(?). *J. Hum. Evol.* 14:3-22.

Newman, R. W. (1970) Why is man such a sweaty and thirsty naked animal: A speculative review. *Hum. Biol.* 42:12-27.

Okada, M., Ishida, H., and Kimura, T. (1976) Biomechanical features of bipedal gait in human and nonhuman primates. In P. V. Komi (ed.): *Biomechanics*. Baltimore: University Park Press, pp. 303-310.

Okada, M., and Kondo, S. (1982) Gait and EMGs during bipedal walking of a gibbon (*Hylobates agilis*) on flat surface. *J. Anthropol. Soc. Nippon* 90:325-330.

Oxnard, C. (1975) *Uniqueness and Diversity in Human Evolution*. Chicago: University of Chicago Press.

Oxnard, C. E., and Lisowski, F. P. (1980) Functional articulation of some hominoid foot bones: Implications for the Olduvai (Hominid 8) foot. *Am. J. Phys. Anthropol.* 52:107-117.

Parsons, P. E., and Taylor, C. R. (1977) Energetics of brachiation versus walking: A comparison of a suspended and an inverted pendulum mechanism. *Physiol. Zool.* 50:182-188.

Potts, R. (1984) Home bases and early hominids. *Am. Sci.* 72:338-347.

Potts, R. (1988) *Early Hominid Activities at Olduvai*. New York: Aldine de Gruyter.

Preuschoft, H., and Witte, H. (1991) Biomechanical reasons for the evolution of hominid body shape. In Y. Coppens and B. Senut (eds.): *Origine(s) de la Bipédie chez les Hominidés*. Paris: Cahiers de Paléoanthropologie, Editions du CNRS, pp. 59-77.

Pritchard, J. L. (1990) *Ape Locomotion: Anatomical, Biomechanical and Bioenergetic Aspects. A Selective Bibliography, 1972-1990*. Seattle: Primate Information Center.

Prost, J. H. (1980) Origin of bipedalism. *Am. J. Phys. Anthropol.* 52:175-189.

Pugh, L. G. C. E. (1971) The influence of wind resistance in running and walking and the mechanical efficiency of work against horizontal or vertical forces. *J. Physiol.* (Lond.) 213:255-276.

Ralston, H. J. (1976) Energetics of human walking. In R. M. Herman, S. Grillner, P. S. G. Stein, and D. G. Stuart (eds.): *Neural Control of Locomotion*. New York: Plenum, pp. 77-98.

Reynolds, T. R. (1983) Stride length of mammals, primates, humans and early hominids. *Am. J. Phys. Anthropol.* 60:244.

Robinson, J. T. (1972) *Early Hominid Posture and Locomotion*. Chicago: University of Chicago Press.

Robinson, J. T. (1978) Evidence for locomotor difference between gracile and robust early hominids from South Africa. In C. Jolly (ed.): *Early Hominids of Africa*. New York: St. Martin's, pp. 441-457.

Rodman, P. S., and McHenry, H. M. (1980) Bioenergetics and the origin of hominid bipedalism. *Am. J. Phys. Anthropol.* 52:103-106.

Rose, M. D. (1974) Postural adaptations in New and Old World monkeys. In F. A. Jenkins (ed.): *Primate Locomotion*. New York: Academic, pp. 201-22.

Rose, M. D. (1976) Bipedal behavior of olive baboons (*Papio anubis*) and its relevance to an understanding of the evolution of human bipedalism. *Am. J. Phys. Anthropol.* 44:247-261.

Rose, M. D. (1984) Food acquisition and the evolution of positional behaviour: The case of bipedalism. In D. J. Civers, B. A. Wood, and A. Bilsborough (eds.): *Food Acquisition and Processing in Primates*. New York: Plenum, pp. 509-524.

Rose, M. D. (1991) The process of bipedalization in hominids. In Y. Coppens and B. Senut (eds.): *Origine(s) de la Bipédie chez les Hominidés*. Paris: Cahiers de Paléoanthropologie, Editions du CNRS, pp. 37-48.

Sarmiento, E. (1988) Anatomy of the hominoid wrist joint: Its evolutionary and functional implications. *Int. J. Primatol.* 9:281-345.

Schmid, P. (1983) Eine Rekonstruktion des Skelettes von A.L. 288-1 (Hadar) und deren Konsequenzen. *Folia Primatol.* 40:283-306.

Schmid, P. (1991) The trunk of the australopithecines. In Y. Coppens and B. Senut (eds.): *Origine(s) de la Bipédie chez les Hominidés*. Paris: Cahiers de Paléoanthropologie, Editions du CNRS, pp. 225-234.

Senut, B. (1981) *L'Humérus et ses Articulations chez les Hominidés*. Paris: Cahiers de Paléontologie. Paléoanthropologie, Editions du CNRS.

Senut, B. (1989) *La locomotion des pré-hominidés. Hominidae*. Proceedings of the 2nd International Congress of Human Paleontology. Milan: Jaca Book, pp. 53-60.

Senut, B. (1991) Origines de la bipédie humaine: Approche paléontologique. In Y. Coppens and B. Senut (eds.): *Origine(s) de la Bipédie chez les Hominidés*. Paris: Cahiers de Paléoanthropologie, Editions du CNRS, pp. 243-257.

Senut, B. (1992) New ideas on the origins of hominid locomotion. In T. Nishida, W. C. McGrew, P. Marler, M. Pickford and F. B. M. de Waal (eds.): *Topics in Primatology*, Vol. 1: *Human Origins*. Tokyo: University of Tokyo Press, pp. 393-407.

Senut, B., and Tardieu, C. (1985) Functional aspects of Plio-Pleistocene hominid limb bones: Implications for taxonomy and phylogeny. In E. Delson (ed.): *Ancestors: The Hard Evidence*. New York: Alan R. Liss, pp. 193-201.

Shipman, P. (1986) Scavanging or hunting in early hominids: Theoretical framework and tests. *Am. Anthropol.* 88:27-43.

Sigmon, B. A. (1971) Bipedal behavior and the emergence of erect posture in man. *Am. J. Phys. Anthropol.* 34:55-60.

Stern, J. T., Jr., and Susman, R. L. (1981) Electromyography of the gluteal muscles in *Hylobates*, *Pongo*, and *Pan*: Implications for the evolution of hominid bipedality. *Am. J. Phys. Anthropol.* 55:153-166.

Stern, J. T., Jr., Wells, J. P., Jungers, W. L., Vangor, A. K., and Fleagle, J. G. (1980) An electromyographic study of the pectoralis major in atelines and *Hylobates*, with special reference to the evolution of a pars clavicularis. *Am. J. Phys. Anthropol.* 52:13-26.

Susman, R. L., and Creel, N. (1979) Functional and morphological affinities of the subadult hand (O.H. 7) from Olduvai Gorge. *Am. J. Phys. Anthropol.* 51:311-332.

Susman, R. L., and Stern, J. T., Jr. (1979) Telemetered electromyography of flexor digitorum profundus and flexor digitorum superficialis in *Pan troglodytes* and implications for interpretation of the O.H. 7 hand. *Am. J. Phys. Anthropol.* 50:565-574.

Susman, R. L., and Stern, J. T., Jr. (1991) Locomotor behavior of early hominids: Epistemology and fossil evidence. In Y. Coppens and B. Senut (eds.): *Origine(s) de la Bipédie chez les Hominidés*. Paris: Cahiers de Paléoanthropologie, Editions du CNRS, pp. 121-131.

Tardieu, C. V. (1991) Étude comparative des déplacements du centre de gravité du corps pendant la marche par une nouvelle méthode d'analyse tridimensionnelle. Mis à l'épreuve d'une hypothése évolutive. In Y. Coppens and B. Senut (eds.): *Origine(s) de la Bipédie chez les Hominidés*. Paris: Cahiers de Paléoanthropologie, Editions du CNRS, pp. 49-58.

Taylor, C. R., Caldwell, S. L., and Rowntree, V. J. (1972) Running up and down hills: Some consequences of size. *Science* 178:1096-1097.

Taylor, C. R., and Heglund, N. C. (1982) Energetics and mechanics of terrestrial locomotion. *Ann. Rev. Physiol.* 44:97-107.

Taylor, C. R., Heglund, N. C., and Maloiy, G. M. O. (1982) Energetics and mechanics of terrestrial locomotion. I. Metabolic energy consumption as a function of speed and body size in birds and mammals. *J. Exper. Biol.* 97:1-21.

Taylor, C. R., and Lyman, C. P. (1972) Heat storage in running antelopes: Independence of brain and body temperatures. *Am. J. Physiol.* 222:114-117.

Taylor, C. R., and Rowntree, V. J. (1973a) Running on two or on four legs: Which consumes more energy? *Science* 179:186-187.

Taylor, C. R., and Rowntree, V. J. (1973b) Temperature regulation and heat balance in running cheetahs: A strategy for sprinters? *Am. J. Physiol.* 224:848-851.

Taylor, C. R., and Rowntree, V. J. (1974) Panting vs. sweating: Optimal strategies for dissipating exercise and environmental heat loads. *Proceedings of the International Union of Physiological Sciences*, vol. XI, p. 348. New Delhi: XXVI International Congress.

Taylor, C. R., Schmidt-Nielsen, K., and Raab, J. L. (1970) Scaling of energetic cost of running to body size in mammals. *Am. J. Physiol.* 219:1104-1107.

Tobias, P. V. (1991) *Olduvai Gorge*, Vol. 4: *The Skulls, Endocasts and Teeth of Homo habilis*. Cambridge: University of Cambridge Press.

Trinkaus, E. (1986) The Neandertals and modern human origins. *Ann. Rev. Anthropol.* 15:193-218.

Trinkaus, E. (1989) The Upper Pleistocene transition. In E. Trinkaus (ed.): *The Emergence of Modern Humans*. Cambridge: Cambridge University Press, pp. 42-66.

Tuttle, R. H. (1967) Knuckle-walking and the evolution of hominoid hands. *Am. J. Phys. Anthropol.* 26:171-206.

Tuttle, R. H. (1969a) Quantitative and functional studies on the hands of the Anthropoidea. I. The Hominoidea. *J. Morph.* 128:309-364.

Tuttle, R. H. (1969b) Knuckle-walking and the problem of human origins. *Science* 166:953-961.

Tuttle, R. H. (1970) Postural, propulsive, and prehensile capabilities in the cheiridia of chimpanzees and other great apes. In G. H. Bourne (ed.): *The Chimpanzee*, Vol. 2. Basel: Karger, pp. 167-253.

Tuttle, R. H. (1972) Functional and evolutionary biology of hylobatid hands and feet. In D. M. Rumbaugh (ed.): *Gibbon and Siamang*, Vol. 1. Basel Karger, pp. 136-206.

Tuttle, R. H. (1974) Darwin's apes, dental apes, and the descent of man: Normal science in evolutionary anthropology. *Curr. Anthropol.* 15:389-426.

Tuttle, R. H. (1975) Parallelism, brachiation and hominoid phylogeny. In W. P. Luckett and F. S. Szalay (eds.): *Phylogeny of the Primates. A Multidisciplinary Approach.* New York: Plenum, pp. 447-480.

Tuttle, R. H. (1977) Naturalistic positional behavior of apes and models of hominid evolution, 1929-1976. In G. H. Bourne (ed.): *Progress in Ape Research*. New York: Academic, pp. 277-296.

Tuttle, R. H. (1981) Evolution of hominid bipedalism and prehensile capabilities. *Phil. Trans. Roy. Soc. London* B 292:89-94.

Tuttle, R. H. (1985) Ape footprints and Laetoli impressions: A response to the SUNY claims. In P. V. Tobias (ed.): *Hominid Evolution: Past, Present and Future*. New York: Alan R. Liss, pp. 129-133.

Tuttle, R. H. (1986) *Apes of the World. Their Social Behavior, Communication, Mentality and Ecology.* Park Ridge: Noyes.

Tuttle, R. H. (1987) Kinesiological inferences and evolutionary implications from Laetoli bipedal trails G-1, G-2/3, and A. In M. D. Leakey and J. M. Harris (eds.): *Laetoli. A Pliocene Site in Northern Tanzania*. Oxford: Clarendon, pp. 503-523.

Tuttle, R. H. (1988) What's new in African paleoanthropology? *Ann. Rev. Anthropol.* 17:391-426.

Tuttle, R. H. (1990) The pitted pattern of Laetoli feet. *Nat. Hist.* 90(3):60-65.

Tuttle, R. H. (1992) Hands from newt to Napier. In S. Matano, R. H. Tuttle, H. Ishida and M. Goodman (eds.): *Topics in Primatology*, Vol. 3: *Evolutionary Biology, Reproductive Endocrinology, and Virology*. Tokyo: University of Tokyo Press, pp. 3-20.

Tuttle, R., and Basmajian, J. V. (1974a) Electromyography of brachial muscles in *Pan gorilla* and hominoid evolution. *Am. J. Phys. Anthropol* 41:71-90.

Tuttle, R., and Basmajian, J. V. (1974b) Electromyography of forearm musculature in gorilla and problems related to knuckle-walking. In F. A. Jenkins (ed.): *Primate Locomotion*. New York: Academic, pp. 293-347.

Tuttle, R., and Basmajian, J. V. (1974c) Electromyography of the long digital flexor muscles in gorilla. In F. Barnosell (ed.): *Proceedings of the 6th Congreso Internacional de Medicina Fisica*, Vol. II, 1972. Madrid: Ministerio de Trabajo, Instituto Nacional de Prevision, pp. 311-315.

Tuttle, R., and Basmajian, J. V. (1975) Electromyography of *Pan gorilla*: An experimental approach to the problem of hominization. In S. Kondo, M. Kawai, A. Ehara, and S. Kawamura (eds.): *Proceedings from the Symposia of the 5th*

International Primatological Society, 1974. Tokyo: Japan Science, pp. 303-314.
Tuttle, R., and Basmajian, J. V. (1977) Electromyography of pongid shoulder muscles and hominoid evolution I. Retractors of the humerus and rotators of the scapula. *Yearb. Phys. Anthropol.* 20:491-497.
Tuttle, R., and Basmajian, J. V. (1978) Electromyography of pongid shoulder muscles II. Deltoid, rhomboid and "rotator cuff." *Am. J. Phys. Anthropol.* 49:47-56.
Tuttle, R. H., Basmajian, J. V., and Ishida, H. (1975) Electromyography of the gluteus max muscle in gorilla and the evolution of hominid bipedalism. In R. H. Tuttle (ed.): *Primate Functional Morphology and Evolution*. The Hague: Mouton, pp. 251-269.
Tuttle, R. H., Basmajian, J. V., and Ishida, H. (1978) Electromyography of pongid gluteal muscles and hominid evolution. In D. J. Chivers and K. A. Joysey (eds.): *Recent Advances in Primatology*, Vol. 3: *Evolution*. London: Academic, pp. 463-468.
Tuttle, R. H., Basmajian, J. V., and Ishida, H. (1979) Activities of pongid thigh muscles during bipedal behavior. *Am. J. Phys. Anthropol.* 50:123-135.
Tuttle, R., Basmajian, J. V., Regenos, E., and Shine, G. (1972) Electromyography of knuckle-walking: Results of four experiments on the forearm of *Pan gorilla*. *Am. J. Phys. Anthropol.* 37:255-266.
Tuttle, R., and Cortright, G. W. (1983) The problem of hominid bipedalism: What do we need in order to proceed? In P. K. Seth (ed.): *Perspectives in Primate Biology*. New Delhi: Today and Tomorrow's, pp. 164-174.
Tuttle, R., and Cortright, G. W. (1988) Positional behavior, adaptive complexes and evolution. In J. H. Schwartz (ed.): *Orang-utan Biology*, Oxford: Oxford University Press, pp. 311-330.
Tuttle, R., Cortright, G. W., and Buxhoeveden, D. P. (1979) Anthropology on the move: Progress in experimental studies of nonhuman primate positional behavior. *Yearb. Phys. Anthropol.* 22:187-214.
Tuttle, R., Velte, M. J., and Basmajian, J. V. (1983) Electromyography of brachial muscles in *Pan troglodytes* and *Pongo pygmaeus*. *Am. J. Phys. Anthropol.* 61:75-83.
Tuttle, R. H., and Watts, D. P. (1985) The positional behavior and adaptive complexes of *Pan gorilla*. In S. Kondo (ed.): *Primate Morphophysiology, Locomotor Analyses and Human Bipedalism*. Tokyo: University of Tokyo Press, pp. 261-288.
Tuttle, R. H., Webb, D. M., and Baksh, M. (1991) Laetoli toes and *Australopithecus afarensis*. *Hum. Evol.* 6:193-200.
Tuttle, R. H., Webb, D. M., and Tuttle, N. I. (1991) Laetoli footprint trails and the evolution of bipedalism. In Y. Coppens and B. Senut (eds.): *Origine(s) de la Bipédie chez les Hominidés*. Paris: Cahiers de Paléoanthropologie, Editions du CNRS, pp. 203-218.
Tuttle, R. H., Webb, D. M., Tuttle, N. I., and Baksh, M. (1992) Footprints and gaits of bipedal apes, bears and barefoot people: Perspectives on Pliocene tracks. In S. Matano, R. H. Tuttle, H. Ishida, and M. Goodman (eds.): *Topics in Primatology*, Vol. 3, *Evolutionary Biology, Reproductive Endocrinology and Virology*. Tokyo: University of Tokyo Press, pp. 221-242.
Tuttle, R. H., Webb, D. M., Weidl, E., and Baksh, M. (1990) Further progress on the Laetoli trails. *J. Archaeol. Sci.* 17:347-362.
Vrba, E. S. (1979) A new study of the scapula of *Australopithecus africanus* from Sterkfontein. *Am. J. Phys. Anthropol.* 51:117-129.
White, D. C. S. (1977) Muscle mechanics. In R. McN. Alexander and G. Goldspink (eds.): *Mechanics and Energetics of Animal Locomotion*. London: Chapman and Hall, pp. 23-56.
White, T. D. (1980) Evolutionary implications of Pliocene hominid footprints. *Science* 208:175-176.
White, T. D. and Suwa, G. (1987) Hominid footprints at Laetoli: Facts and interpretations. *Am. J. Phys. Anthropol.* 72:485-514.
Wolpoff, M. H. (1980) *Paleoanthropology*. New York: Knopf.
Wrangham, R. W. (1980) Bipedal locomotion as a feeding adaptation in geleda baboons, and its implications for hominid evolution. *J. Hum. Evol.* 9:329-331.
Yamazaki, N. (1985) Primate bipedal walking: Computer simulation. In S. Kondo (ed.): *Primate Morphophysiology, Locomotor Analyses and Human Bipedalism*. Tokyo: University of Tokyo Press, pp. 105-130.
Yamazaki, N. (1990a) The effects of gravity on the interrelationship between body proportions and brachiation in the gibbon. In F. K. Jouffroy, M. H. Stack, and C. Niemitz (eds.): *Gravity, Posture and Locomotion in Primates*. Firenze: Editrice "Il Sedicesimo", pp. 157-172.
Yamazaki, N. (1990b) The effects of gravity on the interrelationship between body proportions and brachiation in the gibbon. *Hum. Evol.* 5:543-558.
Yamazaki, N., and Ishida, H. (1984) A biomechanical study of vertical climbing and bipedal walking in gibbons. *J. Hum. Evol.* 13:563-571.
Yamazaki, N., Ishida, H., Kimura, T., and Okada, M. (1979) Biomechanical analysis of primate bipedal walking by computer simulation. *J. Hum. Evol.* 8:337-349.
Yamazaki, N., Ishida, H., Okada, M., Kimura, T., and Kondo, S. (1983) Biomechanical evaluation of evolutionary models for pre-habitual bipedalism. *Ann. Sci. Nat., Zool., Paris* 13e Série 5:159-168.
Zihlman, A. L. (1992) The emergence of human locomotion: the evolutionary background and environmental context. In T. Nishida, W. C. McGrew, P. Marler, M. Pickford and F. B. M. de Waal (eds.): *Topics in Primatology*, Vol. 1: *Human Origins*. Tokyo: University of Tokyo Press, pp.409-422.
Zuckerman, S., Ashton, E. H., Flinn, R. M., Oxnard, C. E., and Spence, T. F. (1973) Some locomotor features of the pelvic girdle in primates. *Symp. Zool. Soc. Lond.* 33:71-165.

CHAPTER 15

Development of Pliocene and Pleistocene Chronology of the Turkana Basin, East Africa, and Its Relation to Other Sites

Francis H. Brown

INTRODUCTION

In the fall of 1965, an anthropologist with a crew-cut hairstyle wandered into the potassium-argon laboratory at the University of California at Berkeley looking for Dr. Garniss H. Curtis. He was, of course, F. Clark Howell and was seeking a graduate student to send to the Omo Valley of Ethiopia to work on the stratigraphy of the Omo Beds that had been studied by Camille Arambourg in the 1930s. A perspicacious man, he had read of the possibility of using the potassium-argon method to date volcanic ejecta associated with early humans, and he wanted to have some idea of the stratigraphy of the beds and also some isotopic dates *before* collecting fossils from the area. Although the graduate student did not know it at the time, the latter idea was revolutionary—to examine geology and attempt to date a sedimentary sequence prior to finding a hominid was unheard of. He was also unaware that that brief encounter would lead to a lifetime of work in the region, and to a fast friendship with one of the most encyclopedic individuals ever to have worked in paleoanthropology.

The chronology of deposits in the Turkana Basin was not worked out without some difficulty, and rather acrimonious debates revolved around the dates on some units. In the beginning, the principal cause of debate was that isotopic dates on geographically widely separated volcanic units were in apparent conflict with correlations made on the basis of fossil faunas. Two laboratories were involved in the isotopic measurements to begin with—that of G. H. Curtis at Berkeley and J. M. Miller at Cambridge. A positive outcome of the conflict was that a third individual (and a third laboratory) became involved in working out the chronology of the Pliocene and Pleistocene deposits of the region—Ian McDougall at The Australian National University in Canberra.

Many pages have been written about the so-called KBS controversy, and I will not dwell on that conflict. Rather the development of the chronology is treated historically. For this, I use the time of publication of various results, although in many (perhaps most) instances the dates were available in one form or another to the anthropological community well before they were published. As presently understood, the chronology has been developed by using potassium-argon dating, $^{40}Ar/^{39}Ar$ dating, paleomagnetic polarity stratigraphy, fission track dating, tephrochronology, and classical stratigraphy. Although faunal correlations were instrumental in identifying inconsistencies in the dating schemes between various localities, I am not sufficiently versed in paleontology to cover this topic—it is better left to Basil Cooke, John Harris, or Tim White. However, the Principle of Superposition must form the basis of any coherent stratigraphic treatment, and much effort has gone into determining the sequence of beds in the Turkana Basin.

The formations of interest in the region are now five in number—the Shungura, Usno, and Mursi Formations in the Omo Valley, and the Koobi Fora and Nachukui Formations east and west of Lake Turkana, respectively. These formations have been described in summary fashion a sufficient number of times so that there is little value in repeating those descriptions here (see, e.g., de Heinzelin 1983; Brown and Feibel 1986; Harris et al. 1988). For the reader less familiar with the geographic location and stratigraphic terminology, I have included Figures 15-1 and 15-2, which show the location and stratigraphic subdivisions of these formations (except for the Mursi Formation). Brown and Feibel (1991) have reviewed the development of stratigraphic terminology at Koobi Fora, and consequently that discussion is not repeated. Numbered areas at Koobi Fora (e.g., Area 105) follow the usage in Brown and Feibel (1991).

Several considerations must be taken into account when reviewing the development of the chronology of the Pliocene and Pleistocene deposits of the Turkana Basin. Some of these are discussed later. In addition, hominid finds in the Turkana Basin since 1967 brought with them all sorts of emotional attachments and a sense of competitiveness, the effect of which is more difficult to evaluate. Clark once said, "Whether these fossils are extremely old or not is unimportant. Just tell me how old they are," or words to that effect. If all workers had been similarly detached, some problems might have been solved earlier.

The first consideration is that the potassium-argon method, though not in its infancy, was still in the early stages of development when dating began on the deposits in the Turkana Basin. The principal advances in the method were made at the University of California at Berkeley. Dr. John Reynolds invented a new type of mass spectrometer uniquely suited to analysis of noble gases, and not long after, Drs. Garniss Curtis and Jack Evernden applied their talents to date rocks and minerals of many ages. Many of the later giants of the field learned the trade at their hands. Most of the ages determined early in the development of the technique have stood the test of time (e.g., compare the results of Evernden and Curtis [1965] with those of Walter et al. [1991] for Bed I Olduvai). It is worth remembering, though, that in 1965 W. W. Bishop, an early advocate of the application of the method to geological and paleontological problems cautiously stated, "I find the attitude that all potassium-argon dates be accepted as correct unless they have been proved incorrect rather premature. Particularly is this odd as the validity of K-Ar dates for rocks of less than 30 times 10^6 years still remains unproven in the view of many geologists" (Bishop 1965).

Second is that the decay constants and abundance of ^{40}K (necessary for computing ages) has changed from the time the first dates were published to the present. The constants in use now were recommended by Steiger and Jäger (1977) and have the effect of increasing the ages computed on the basis of old constants by 2.67 percent over the time range of interest here.

It has been pointed out many times that even minute amounts of contamination of young samples by old material can lead to quite large increases in apparent potassium argon dates (e.g., Evernden and Curtis 1965; Dalrymple and Lanphere 1969; McDougall 1980). Likely contaminants in the Turkana Basin derive either from the basement, with ages of around 450 million years ago (MYA), or from Miocene volcanic rocks (with ages around 20 MYA). Contamination with very old material in amounts greater than 0.1 percent should be detectable for samples of the grain sizes commonly used. Contamination with younger material, however, may be quite difficult to detect and may account for some scatter in the data that cannot be explained analytically. Opposed to the problem of contamination is the problem of incomplete degassing, particularly of feldspar melts, which leads to apparent ages that underestimate the true age. This has been noted previously by McDougall et al. (1980), but it is seldom considered when discussing sources of error.

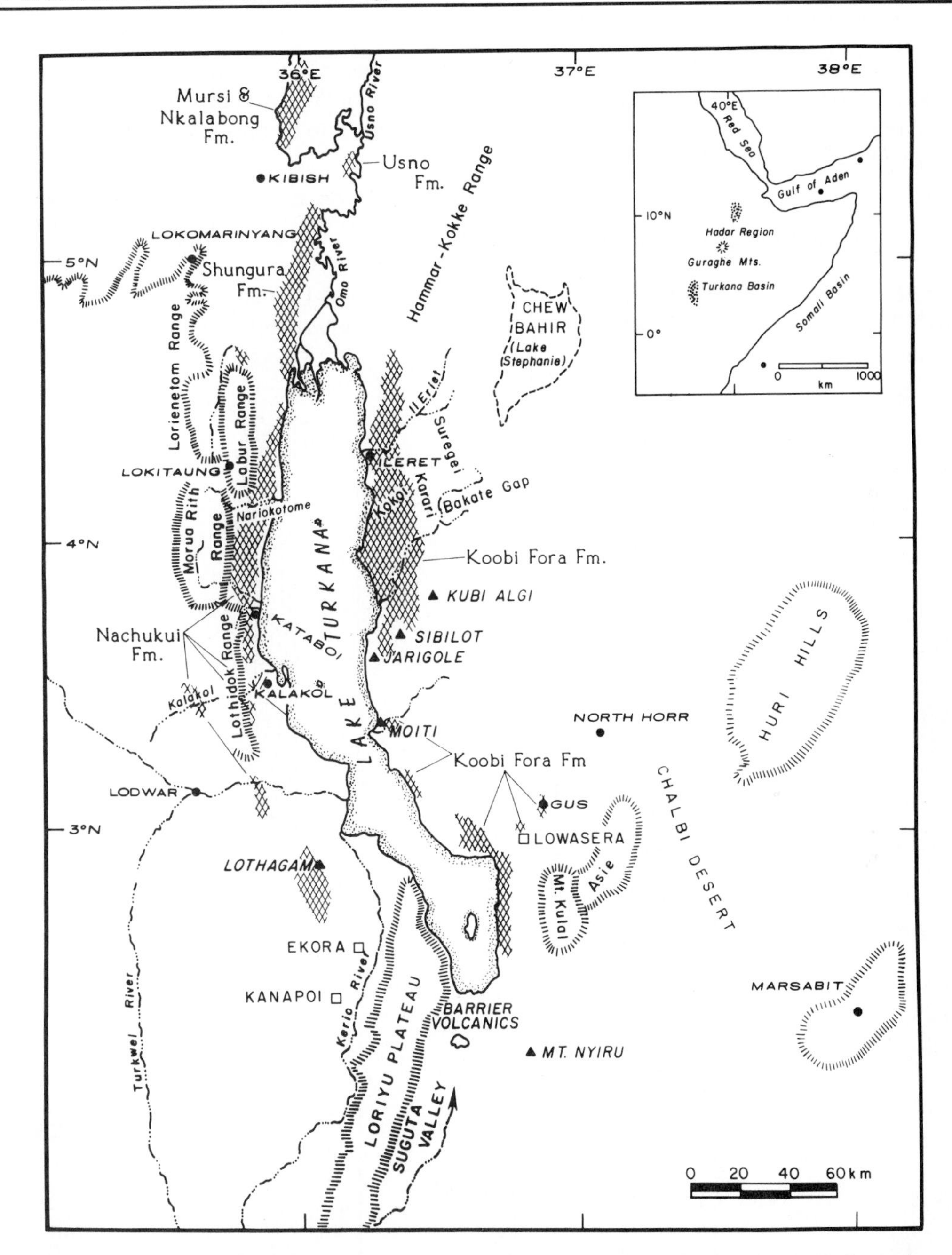

FIGURE 15-1 Map of the Turkana Basin showing most geographic names mentioned in the text.

The $^{40}Ar/^{39}Ar$ method, conceived and applied by Merrihue and Turner (1966), was even less well developed than the K/Ar method at the onset of studies in the Turkana Basin. It was applied to dating of the Turkana sequence at Koobi Fora from the earliest stages (Fitch and Miller 1970), producing results that subsequently were to lead to controversy about the age of the KBS Tuff. Some excellent total fusion results were obtained using this method quite early in the game, but some workers perhaps believed too strongly in the ability of the incremental heating (age spectrum) method to "see through" later geological events that might have affected samples, and in such "later events" themselves. This led to controversy over the interpretation of some early results, for example those of Fitch and Miller (1970) on the KBS Tuff. A decade later, understanding of the method was sufficient for it to be applied to young materials quite successfully (e.g., McDougall 1981).

FIGURE 15-2 Schematic lithologic sections of the Nachukui, Shungura, Usno, and Koobi Fora Formations with names of members given to the left of the lithologic column. A small "v" indicates a tephra layer. For simplicity, not all tephra layers are shown. Only selected correlations between sections are shown.

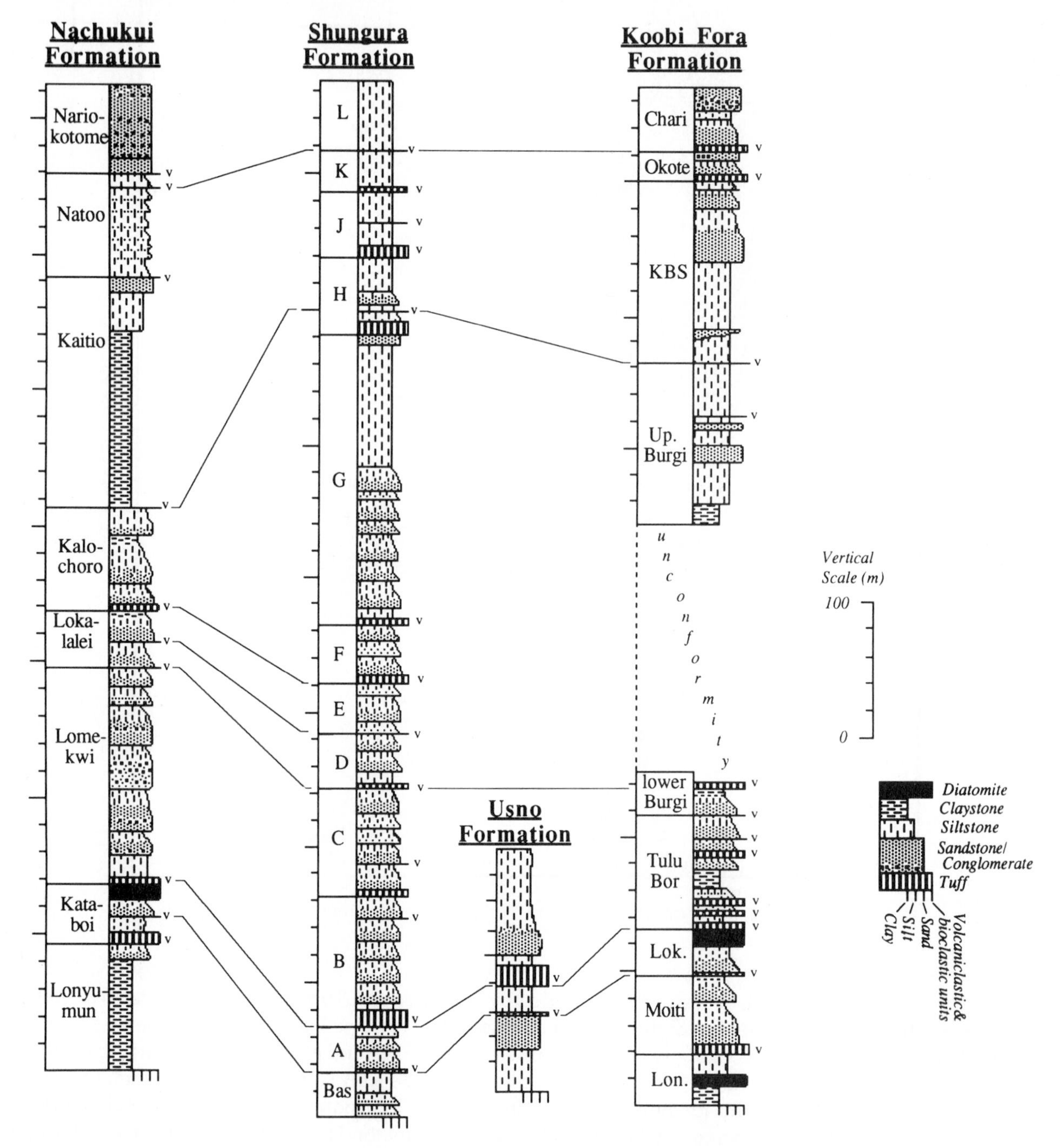

Fission track dating of young minerals, particularly those low in uranium, also had associated difficulties that were not appreciated at first. The method was vigorously applied to only one unit in the Turkana Basin—the KBS Tuff—in an attempt to find an independent method of determining its age. It could be argued that resolution of the age of the KBS Tuff was an important step in understanding just how important protocols are in applying the fission track method to young rocks. The papers by Hurford et al. (1976) and Gleadow (1980) can be profitably compared in this regard. When evaluating fission track ages, it is well to note the decay constants for spontaneous fission of ^{238}U used for computation of the ages, for two different constants are used by different workers that differ by ~20 percent; the resulting ages will also differ by this amount (see comment by Wagner 1977).

Paleomagnetic polarity zonations were also used to estimate ages of sections where no datable materials were found. This technique presupposes a reasonably precise chronology to begin with because it relies on pattern matching (normal or reversed). Therefore the method is not capable of resolving age disputes where the units in contention are assigned ages permitted by the paleomagnetic polarity. Indeed, the age of the KBS is a case in point—either an age near 2.6 MYA or near 1.9 MYA is supported by its normal polarity. If the polarity is clearly in conflict with an isotopically determined age, that alone may be sufficient reason to question the age determination, but knowing the polarity alone provides essentially no temporal information. Also, ages assigned to various polarity boundaries are themselves largely dependent on K/Ar age determinations on lava flows. As the number of dated flows increases, the ages of the boundaries are continually refined. During the period that the chronology was under development in the Turkana Basin, new estimates for several boundaries (e.g., those on the Mammoth and Kaena Subchrons, and the boundary between the Gauss and Matuyama Chrons) were made (McDougall 1979), and those estimates influenced the interpretation of the placement of various sections in time. Very recently, still newer estimates for many of these boundaries have been made by correlation of orbital perturbations of the earth with cyclical features in paleomagnetically studied deep sea sediments (Johnson 1982; Shackleton et al. 1990; Hilgen 1991a,b), and also by $^{40}Ar/^{39}Ar$ dating of critical units (Spell and McDougall 1992; Baksi et al. 1992; Tauxe et al. 1992; McDougall et al. 1992). A further difficulty, not yet fully circumvented, is that sediments may be overprinted after their deposition so that their apparent polarity does not reflect the state of the geomagnetic field when the sediments were deposited.

Some problems with dating of units from geographically disparate sections resulted because the basic stratigraphy of the deposits was poorly understood. For example at Koobi Fora several tephra layers had been incorrectly identified as being the same, whereas others that were the same had been given different names (see Cerling and Brown 1982). Although trace element analyses on some tephra had been published quite early (Brown et al. 1970), those data were essentially useless for stratigraphic correlation because no other comparative analyses existed from other areas. Even with the first few hundred analyses of the glass fraction of tephra from Koobi Fora and the lower Omo Valley, only a few correlations were achieved. It was not until many hundreds of analyses were available that it was possible to relate sections within the Koobi Fora region, and to link those with the sequence in the Shungura and Nachukui Formations. This is because of the large number (>130) of distinct tephra layers of Pliocene and Pleistocene age preserved in the Turkana Basin. It is still not known just how many tephra exist there, but if there were fewer of them, the relations would have been worked out much sooner with far fewer analyses.

For purposes of clarity, I have discussed results using the name of the unit under which it was originally published (if that designation still exists), followed by a parenthetic designation of its correlative in other formations. Thus dates on Tuff L are discussed as Tuff L (= Chari), whereas those on the Chari are discussed as Chari Tuff (= L). In a few instances it is not known with certainty just what unit was dated, and this is noted in the text along with notes where appropriate. The sequence of tephra layers and equivalences that have been established between dated units in different formations are shown in Table 15-1 (see p. 290). In addition to the correlations shown there, it is possible that the Moiti Tuff is equivalent to a tuff in the Nkalabong Formation described by Butzer (1976). In order to avoid discussing problems

TABLE 15-1 Sequence and nomenclature of dated tuffs in the Koobi Fora, Nachukui, and Shungura Formations

Nachukui and Koobi Fora Formations		Shungura Formation
Silbo Tuff		
Gele Tuff		
Nariokotome Tuff		
Chari Tuff	=	Tuff L
Lower Okote Tuff		
Malbe Tuff	=	Tuff H-4 (=Tuff I4 of some early publications)
KBS Tuff	=	Tuff H-2 (=Tuff I2 of some early publications)
		Tuff G
Kalochoro Tuff	=	Tuff F
Lokalalei Tuff	=	Tuff D
Burgi Tuff		
Hasuma Tuff	=	Tuff C
		Tuff B-10
Ninikaa Tuff		
Toroto Tuff		
Tuff B-δ		
Tulu Bor-β	=	Tuff B-β
Tulu Bor-α	=	Tuff B-α = Tuff U-10 (Usno Formation)
Topernawi Tuff		
Moiti Tuff		
Mursi Basalt		
Kataboi Basalt		no relative order implied between these
Karsa Basalt		basalt flows and/or dikes
Harr Basalt		
Kanapoi Basalt		
Lothagam Basalt		

arising from changes in decay constants for ^{40}K, I have recalculated all published ages using the constants recommended by Steiger and Jäger (1977).

As a result, the ages ascribed to various units are now all computed on the same basis, so they are comparable. A second result is that for results published before 1977, the ages used will *not* match those published in the original sources. Recalculated ages are presented in Tables 15-2 and 15-3 along with the ages originally published for the units.

The following discussion is based on what I believe to be a complete compilation of published and unpublished analytical results on volcanic rocks from the Turkana Basin and surrounding regions. Although the unpublished data are not used on diagrams, those data nonetheless have influenced my thinking in some instances. The compilation includes approximately 640 K/Ar entries, 85 ^{40}Ar/^{39}Ar entries, and 31 fission track ages. A few entries are results that were republished after discovering errors in the determinations, or after reinterpretation of results, but nonetheless the fundamental basis for the chronology of Pliocene and Pleistocene strata of the Turkana Basin is dates on volcanic feldspar measured by the K/Ar method. In order that the reader have ready access to the data, they are compiled in Tables 15-2, 15-3, and 15-4 (see pp. 302-309), but most unpublished results are not included. Because they are lengthy, these tables have been placed at the end of the chapter. All of the published K/Ar data discussed later are included in Table 15-2. $^{40}Ar/^{39}Ar$ age measurements are presented in Table 15-3, and fission track data in Table 15-4. Ages in the text are given both with and without error estimates, but errors on all dates are included in the tables.

INITIAL RESULTS

In 1969 the first K/Ar dates from the region were published on four units in the lower Omo Valley—Tuff D (Lokalalei), Tuff H-2 (KBS), Tuff B-δ, and the Mursi Basalt (Brown 1969). For Tuffs D and H-2, each sample was dated twice, and the average ages suggested for these units were 1.89 MYA and 2.50 MYA, quite close to more recent estimates. However, the individual ages on Tuff D (2.37 ± 0.12 MYA and 2.63 ± 0.12 MYA) did not agree well, and the estimated precision on Tuff H-2 was on the order of 5 percent. The age on Tuff B-δ, 3.89 ± 0.20 MYA, is now known to overestimate the age of the unit, but at the time there was no reason to doubt its validity. The date on the Mursi Basalt (4.16 ± 0.20 MYA) also appears to have been reasonable. In the same year the first total fusion $^{40}Ar/^{39}Ar$ ages from the region were published by Fitch and Miller (1969) on a feldspar separate prepared at Berkeley from a tuff in the Nkalabong Formation. If this is in fact the Moiti Tuff (now believed to be near 4 million years in age), the dates were apparently quite good—4.00 ± 0.10 and 4.10 ± 0.12 MYA.

A DECADE OF CONFUSION

From this excellent beginning, the chronology deteriorated in the following year for two reasons. First, Fitch and Miller (1970) published three believable (but erroneous) K/Ar dates on feldspar from the KBS Tuff at Koobi Fora of 2.44 ± 0.30, 2.46 ± 0.30, and 2.49 ± 1.0 MYA, and one date on a pumice lump [sic] from the same unit of 3.76 ± 2.1 MYA that was considered anomalously old. They also measured a $^{40}Ar/^{39}Ar$ incremental heating age on a feldspar concentrate for which they recommended a crystallization age of 2.68 ± 0.26 MYA. The second reason for the deterioration of the chronology was that several dates from previously undated units in the Shungura Formation were published by Brown et al. (1970) and also by Brown and Lajoie (1970) that are younger than the ages presently ascribed to them. These are the feldspar dates on Tuff D-4 (2.18 ± 0.11 MYA), Tuff F (= Kalochoro), (2.04 ± 0.10, 2.12 ± 0.11 MYA), Tuff G (1.98 ± 0.10 MYA), two whole rock dates on the Usno Basalt (3.19 ± 0.15, 3.6 ± 0.7 MYA), and a date of 2.67 ± 0.92 MYA on glass from Tuff U-10, now known to correlate with Tuff B-α (= α-Tulu Bor) exposed in the Usno Formation. Two new dates were also measured on Tuff D (= Lokalalei), 2.22 ± 0.11, and 2.37 ± 0.11 MYA, which are low compared to newer data. These latter had the effect of reducing the average age on Tuff D (= Lokalalei) to ~2.4 MYA.

The new data appeared to corroborate the temporal scale being built for the Shungura Formation and related units in the lower Omo Valley, but the security so offered was false. At almost the same time, the first ages became available on the KBS Tuff at Koobi Fora. With that, the pigs began to squeal, because the fossil faunas that had been related to the KBS Tuff were well enough known that paleontologists could see that the two dated sequences were out of kilter. This was so even though the scale for the Shungura Formation was incorrect (see, e.g., Cooke and Maglio 1972). It is worthwhile noting that Clark was a spokesman for the extinct fauna, for many of us remember him reading long lists of taxa at the 1973 Wenner-Gren conference in Nairobi. The object of his recitation was either misunderstood or unappreciated by many (perhaps most) of the participants in that meeting. The KBS Tuff (= H-2) was of special importance because the faunas above and below it in the section were reasonably well known, because an important hominid fossil KNM-ER 1470 was recovered below it, and because the tuff itself contained artifacts at the type locality. Lewin (1987) has provided a balanced view of the many lines of argument involved in the "KBS Tuff Controversy."

In an attempt to estimate the chronological scales by a more or less independent method, and also to provide additional age control on strata in the Turkana Basin, paleomagnetic work began at Koobi Fora in 1971, and in the Shungura Formation in 1973. At the same time, new K/Ar and $^{40}Ar/^{39}Ar$ ages were measured on various units in both the Shungura and Koobi Fora Formations, and fission track ages were measured on the KBS Tuff. Many of these data were in hand for the Nairobi conference of 1973 sponsored by the Wenner-Gren Foundation, but they were not published except in preliminary fashion until 1976.

In the meantime, Fitch et al. (1974) published additional dates (without supporting data) on several units from Koobi Fora. In that paper they suggested ages for several units: the Chari Tuff (= L) at Ileret (1.2, 1.31 ± 0.23 MYA), the Chari Tuff (= L) on the Karari Ridge (1.36 ± 0.10 MYA), the "Lower-Middle Tuff Complex" (1.52 ± 0.17 MYA), the Koobi Fora Tuff (1.61 ± 0.00 MYA), the "BBS Tuff Complex" (two components: 1.60 ± 0.02 and 1.75 ± 0.04 MYA), the KBS Tuff (2.68 ± 0.26 MYA), and the Ninikaa Tuff (3.27 ± 0.09 MYA) At the time of this publication the Chari Tuff on the Karari Ridge was known as the Karari Tuff. In addition, the "Lower–Middle Tuff Complex," the Koobi Fora Tuff, and the "BBS Tuff Complex" are now recognized as parts of the Okote Tuff Complex (Brown and Feibel 1985), and the Ninikaa Tuff was labeled the Tulu Bor Tuff in Figure 15-1 of Fitch et al. (1974). Two dates were suggested for tuffs in the lower part of the Koobi Fora Formation by Fitch et al. (1974), 4.0 MYA and 4.6 MYA. At the time of publication the strata from which these samples came were described as the Kubi Algi Formation (see Brown and Feibel [1986] for discussion), but the sampling localities of the dated materials are difficult to ascertain. Fitch et al. (1974) introduced the notion that feldspars in the tuffs at Koobi Fora had been affected by later thermal events and suggested ages of overprint of 1.0, 1.8, 2.06 ± 0.03, and 2.49 ± 0.02 MYA for various samples. These supposed overprints were the subject of much discussion in later papers by many workers.

The first fission track age from the region was published by Hurford (1974). This was a date of 1.8 MYA (no error estimate given) on glass from a sample that is believed to have come from the Hasuma Tuff in Area 202 at Koobi Fora. The result was used to bolster the case for thermal resetting of ages in the KBS Tuff and was not regarded as the age of cooling of the material itself. Were it not that the decay constant used (8.42×10^{-17} yr^{-1}) differs from that used later to compute fission track ages on the KBS Tuff (6.85×10^{-17}), the result hardly deserves mention. The difference in these two constants is nearly 20 percent, so that if the time of overprinting had been computed using the same constant as for the fission track ages reported on zircon, the time of overprinting would have been calculated at ~2.2 MYA (see also Wagner 1977). This demonstrates the importance of reporting all parameters used in computing ages.

Brock and Isaac (1974) used paleomagnetic polarity determinations on sections at Koobi Fora to support the age of ~2.6 MYA on the KBS Tuff. They found that the KBS Tuff was of normal paleomagnetic polarity. Beginning with the age of ~2.6 MYA on that unit, they assigned the KBS Tuff to the upper part of the Gauss Chron. From that placement, they fitted the rest of the polarity zones into a sensible framework. That framework, however, required an erosion surface of perhaps 0.7 MYA duration *above* the KBS Tuff. Not until the age of the KBS Tuff had been independently established, and the stratigraphy was under revision, was this found to be a serious error. There is indeed an unconformity in the section at Koobi Fora, but it lies *below* the KBS Tuff.

Curtis et al. (1975) published a new set of dates on what was thought to be the KBS Tuff (= H-2), but two of these dates were actually on samples of the Malbe Tuff (= H-4). Their ages on the KBS Tuff (= H-2) ranged from 1.50 to 6.90 MYA, generally with errors on the order of 1.5 percent. All but one age fell in the range 1.50–2.53 MYA, and within that range, the ages appeared to be divisible into one set at about 1.6 MYA, and another about 1.85 MYA. Those authors concluded that two different tuffs had been sampled, and that the KBS Tuff in Areas 10 and 105 did not correlate with that in Area 131. They were correct in their conclusion that two different units had been sampled. However, because their data set was marred by erroneous potassium contents on some of the samples (an error that was not corrected in the literature until 1980), the younger measured ages were incorrect. It is unfortunate that this data set was flawed, or the argument over the age of the KBS Tuff might have ceased at that time. The two tuffs that had been confuted (the KBS and Malbe Tuffs) are quite similar in age. Note also that several ages published by Curtis et al. (1975) (2.16, 2.40, and 2.53 MYA) were significantly older than their preferred ages and were, in fact, supportive of the dates obtained earlier by Fitch and Miller. These older ages were attributed, probably correctly, to detrital contamination. Curtis et al. (1975) had also dated both glass and feldspar from some samples and argued that both could not be equally affected by an "overprinting" event. They therefore believed that their ages of 1.6 MYA and 1.85 MYA were reasonably close to the true ages on a tuff that had been mistaken for the KBS, and on the KBS Tuff itself. Their dates of 1.61 and 1.68 MYA on the Malbe Tuff (= H-4) were later corrected (Drake et al. 1980).

For new published data 1976 was a banner year. Not only were new ages made available on previously dated units, but many ages were reported on previously undated ones. Brown and Nash (1976) published K/Ar dates on (1) feldspar from Tuff L (= Chari), 1.30, 1.44, 1.47, and 1.54 MYA); (2) Tuff D (= Lokalalei), 2.21, 2.48, 2.57, 2.67 MYA); (3) feldspar from Tuff B-10 (3.00, 3.04 MYA); (4) glass from Tuff B-α (3.15 MYA); and (5) feldspar from Tuff B-δ (5.12 MYA). Reported errors on these dates were again about 5 percent, and the scatter in the ages on Tuff L was particularly noticeable.

Data reported by Fitch and Miller (1976) included results reported in summary fashion earlier by Fitch et al. (1974) in addition to new results. Feldspar from the Chari Tuff yielded ages of 1.23, 1.25, and 1.26 MYA with errors of about 5 percent. Pumice and feldspar from the Koobi Fora Tuff gave a wide range of ages (1.02-6.13 MYA), none of which are near more recent estimates of the age of that unit, and with errors up to 30 percent. Ages of feldspar from the KBS Tuff were obviously too old (8.47, 8.89, 17.3, 17.7, and 19.1 MYA) and ascribed to detrital contamination. Also reported were dates on whole rock samples of the Kanapoi Basalt (3.15 ± 0.21, 3.16 ± 0.21 MYA), and the Mursi Basalt (4.3 ± 0.3, 4.5 ± 0.3 MYA). Finally, they reported three K/Ar dates (3.96 ± 0.4, 4.54 ± 0.22, and 4.78 ± 0.23 MYA) on glass from a tuff in the lower part of the Koobi Fora Formation (then called the Kubi Algi Formation). Lack of detail about the sample prevents identification of the unit from which the glass derived.

In addition to the K/Ar dates, Fitch and Miller (1976) published total fusion $^{40}Ar/^{39}Ar$ dates on many different units. Feldspar from samples of the Chari Tuff (= L) at Ileret yielded ages ranging from 1.20 ± 0.23 to 1.32 ± 0.3 MYA, whereas ages on feldspar from samples of the same tuff on the Karari Ridge had an even greater range (0.63–1.38 MYA). As with the K/Ar dates, feldspar from the Koobi Fora Tuff produced ages with such large spread (0.54–4.5 MYA) that little could be learned from them. Results on feldspar from the "BBS Tuff Complex" were not quite as scattered (0.89–1.70 MYA), but the spread was still discouragingly large. A feldspar separate from the "Lower/Middle Tuff" produced a reasonable age of 1.52 MYA (from our present perspective) but was not replicated. Feldspar from the KBS Tuff (= H-2) gave a distressingly wide range of ages (0.53–2.61MYA). The midpoint of this range is near 1.9 MYA (see Table 15-3), and samples near the midpoint contain the largest amount of radiogenic argon. This suggests that part of the problem with these dates on the KBS Tuff (= H-2) was difficulty in correcting for the atmospheric component when that component comprised nearly all of the sample. It is noteworthy that in the same paper a $^{40}Ar/^{39}Ar$ isochron date of 1.96 MYA was reported for the KBS Tuff (= H-2), a result not far from the presently estimated age of the unit. The feldspar age ascribed to the Tulu Bor Tuff (3.46 ± 0.07 MYA) by Fitch and Miller (1976) was actually measured on a sample separated from a pumice from the Ninikaa Tuff in Area 116 at Koobi Fora. Amongst the tuffs in the sections in Areas 202 and 204 previously known as the Kubi Algi Formation, to our knowledge, only the Toroto Tuff contains pumice, and it is therefore likely that the dates (3.16–4.03 MYA) measured on "sanidine concentrates from pumice samples FM7035 and FMA 246" refer to that tuff. The $^{40}Ar/^{39}Ar$ ages reported on glass from a tuff in the "Kubi Algi Formation" (4.94 ± 0.36, 5.08 ± 0.49 MYA) were most likely measured on samples of what is now called the Hasuma Tuff (= C). Fitch and Miller (1976) stated that "[t]he best age obtainable from a $^{40}Ar/^{39}Ar$ age spectrum analysis of the younger of the two tuffs is ~3.9 m.y." Findlater (1976) labels one unit the "3.9 tuff" in his discussion of tuffs in the Koobi Fora region, which is later named the Hasuma Tuff (Findlater 1978). Finally Fitch and Miller (1976) interpreted a "two-point plateau" on an age spectrum on the Kanapoi basalt as suggesting an age of 4.1 ± 1.0 MYA.

Five fission-track dates on zircons from pumice clasts in the KBS Tuff averaging 2.43 ± 0.55 MYA were reported by Hurford et al. (1976), which appeared to give strong support for the $^{40}Ar/^{39}Ar$ date of Fitch et al. (1976) of 2.49± 0.02 MYA recommended in a companion paper.

Despite all of this new data, not much headway was made on the overall chronological framework. This is because older data, now believed to be in error, continued to be used in construction of the chronology. Also, until stratigraphic problems at Koobi Fora had been rectified, it would not have been possible for any laboratory to achieve a set of ages consistent with the stratigraphy because in some cases the materials being dated were derived from different units, and the stratigraphic relations between those units were misunderstood.

The principal advances resulting from the new information were that a date somewhere around 1.2–1.4 MYA seemed appropriate for the Chari Tuff (= L), setting a minimum on the age of the KBS Tuff agreeable to all workers. Second, the new data on Tuff D (= Lokalalei) again moved the average somewhat older. Third, the marked discordance between the two dates obtained on Tuff B-δ caused Brown and Shuey (1976) to suggest that both should be disregarded in constructing a chronology. Finally, on the basis of feldspar composition, Brown and Nash (1976) suggested that the Chari Tuff (= L) might correlate with Tuff L (= Chari), and that of all the tuffs for which data were available, the KBS (= H-2) was most similar to Tuff H-2 (= KBS) of the Shungura Formation. This marked the beginning of relating the sections at Koobi Fora and in the Shungura Formation by compositional analysis of phases.

Brock and Isaac (1976) added new paleomagnetic data to their previously published work, which gave them further assurance that an age of ~2.6 MYA for the KBS Tuff was appropriate. Brown and Shuey (1976) provided a magnetochronology for the Shungura and Usno Formations in the same year, which they believed generally supported the chronology proposed for the upper part of the Shungura Formation. They suggested that the formation extended to only about 3.3 MYA at the base, and that the upper part of the formation lay above the Jaramillo Subchron but below the Brunhes Chron (that is between 0.8 and 0.9 MYA). Their magnetochronology also led them to question the date on Tuff D-4, suggesting that it might be about 0.2 million years too young. Because the ages on the Tuff F (= Kalochoro) and Tuff G were erroneously young, and also because some samples believed to be of normal polarity were in fact overprinted, the magnetochronology of the central part of the formation (Members D–G) remained confused. Also, the erroneously young age on the Usno Basalt caused Brown and Shuey (1976) to interpret the magnetochronology of the Usno Formation incorrectly.

No new dates were published between 1976 and 1980, and problems pointed out by the paleontologists persisted. Some quite imaginative solutions to these problems were proposed (e.g., Behrensmeyer 1978), as well as more mundane explanations (e.g., Brown et al. 1978). During this period though, tephrostratigraphic linkages between the Shungura and Usno Formations (Martz 1979; Martz and Brown 1981), within the Koobi Fora Formation (Cerling 1977), and between the Shungura and Koobi Fora Formations (Cerling et al. 1978, 1979) began to be established. Additional paleomagnetic work on both the Koobi Fora (Hillhouse et al. 1977) and Shungura Formation (Brown et al. 1978) was also reported during this period.

Correlation of tephra in the Turkana Basin by elemental analysis was attempted with little success by Luedtke (1975). Two years later, however, in an appendix to his thesis, Cerling (1977) showed that several tephra regarded as the same unit by the geologists at Koobi Fora were compositionally different. For example, several units identified as the Tulu Bor Tuff were not that tuff. Cerling (1977) also showed that some correlations between the KBS Tuff from one area to another at Koobi Fora were quite likely valid, but that others were almost certainly incorrect. Further, he showed that the relations between tephra layers in the interval between the KBS Tuff and the Chari Tuff (now known as the Okote Tuff Complex) were complicated, and that they could not be simply correlated except in a general way. Finally, he stated that the KBS Tuff from its type locality correlated with Tuff H-2 of the Shungura Formation, and that the Chari Tuff was equivalent to Tuff L of the Shungura Formation. Cerling et al. (1978) suggested probable correlation of an additional unit from the Koobi Fora Formation to the Shungura Formation—the Malbe Tuff with Tuff H-4.

Cerling et al. (1979) began the process of revising the stratigraphy at Koobi Fora by showing that the Chari and Karari Tuffs were identical and recommending that they be referred to by the same name. They also showed that two different units had been called the KBS Tuff. A little later, Martz and Brown (1981) noted that one of the units described as the Tulu Bor Tuff at Koobi Fora was chemically similar to Tuff B in the Shungura Formation, and that Tuff B was equivalent to Tuff U-10. No correlation was proposed because the data reported by Cerling (1977) were different enough from those reported by Martz (1979) that compositional identity could not be proven. The importance of these early efforts to correlate directly between distant outcrops cannot be overemphasized, because where linkages were secure, the dated units could no longer be considered separately. The principal importance of these early papers on tephra correlation was to demonstrate that compositional analysis of the glass phase of tuffs exposed in the sequences was not only a method of linking sections, but of uniquely identifying individual units. To workers in other regions, this came as no surprise, for the techniques had been established much earlier (e.g., Jack and Carmichael 1968; Sarna-Wojcicki 1976). What was lacking in East Africa were samples to compare from one area to another.

The paleomagnetic work of Hillhouse et al. (1977) emphasized the tenuousness of the chronological scale for Koobi Fora resulting from doubts about the identity of tuffs from one area to another, and the different dates for the KBS Tuff. Hillhouse et al. (1977) reported results on new samples, and two chronologies were contemplated, one consistent with the older date on the KBS, and the other consistent with the younger date. This was a honest attempt to reconcile all data available at the time, but in reality Hillhouse et al. (1977) could not possibly have chosen between the alternative possibilities because stratigraphic problems were too severe. For example, the "Tulu Bor Tuff" that they sampled in Area 116 is now known to be the Ninikaa Tuff and is somewhat younger than the age ascribed to it at the time; the "Tulu Bor Tuff" in Area 102 was likewise later found to be a much younger unit (the Lorenyang Tuff). Further, Hillhouse et al. (1977) used faunal correlations between the Shungura and Koobi Fora Formations to augment their arguments, but the chronology of the critical interval for them (Members F and G of the Shungura Formation) was incorrect.

On the basis of new samples from the Shungura Formation, Brown et al. (1978) revised their interpretation of its magnetochronology. The principal changes from the previous chronology were that the age of the bottom of the formation was increased to ~3.6 MYA, and that both the Mammoth and Kaena Subchrons were believed to have been identified in the lower part of Member B. Their interpretation of the lower part of the Matuyama Chron remained flawed because the dates on Tuffs F and G, used to control placement of the magnetozones, were wrong.

RESOLUTION OF CONFLICTS AND ESTABLISHMENT OF THE PRESENT SCALE

During the 1970s then, what had started out well failed because of a number of poor age determinations, for whatever reasons. The first definite progress toward the chronostratigraphic framework of the Turkana Basin as now understood came with publication of three papers in 1980, all concerned with the KBS Tuff. Even in these papers, though, there was confusion about just what was meant by the KBS Tuff. The measurements of Drake et al. (1980) and McDougall et al. (1980) were made on two different tuffs that had been recognized as compositionally distinct by Cerling et al. (1979), and which had been correlated with two stratigraphically distinct tuffs of very nearly the same age in the Shungura Formation (Tuffs H-2 and H-4). Even though the two tuffs were separated by about 16 m in the Shungura Formation, and could not be precisely the same age, analytical precision was insufficient to distinguish the age of one from another.

Gleadow (1980) redetermined the fission track age of zircon from pumice of the KBS Tuff and recommended an age of 1.87 ± 0.04 MYA. At the same time he determined that the pumices contained detrital minerals of two ages, one 23–30 MYA, the other 450–550 MYA, and suggested that K/Ar and $^{40}Ar/^{39}Ar$ ages of ~2.4 MYA that had been recommended for the KBS Tuff were likely the result of detrital contamination.

He also suggested that pumices of all compositions from KBS Tuff were of about the same age (near 1.9 MYA), and that older fission track ages that had been previously reported resulted from misidentification of acicular inclusions or dislocations as fission tracks, and perhaps bias in the grains chosen for counting.

At almost the same time, Drake et al. (1980) reported a new set of K/Ar ages measured on both feldspar and glass from the KBS Tuff and the Malbe Tuff, although all were reported as samples of the KBS Tuff. Ten feldspar ages on the KBS Tuff averaged 1.83 MYA; 9 feldspar ages on the Malbe Tuff, 1.80 MYA. One sample of the Tuff H-2 (= KBS) and two samples of Tuff H-4 (= Malbe) are included in the foregoing averages. Four ages on glass from the KBS Tuff averaged 1.83 MYA, and two glass ages on the Malbe Tuff averaged 1.88 MYA. Drake et al. (1980) also corrected the erroneous ages of ~1.6 MYA reported in Curtis et al. (1975) and reported that all KBS pumices were nearly the same age, despite compositional variation between the glass phase of the samples.

The third paper was that of McDougall et al. (1980), which reported an additional 16 K/Ar ages on anorthoclase separated from pumice clasts within the KBS Tuff (H-2), all but two of which fell in the range 1.86–1.94 MYA. The two outliers (1.97 and 1.99 MYA) were on a fine fraction of feldspar and were believed to result from detrital contamination. The 14 acceptable ages averaged 1.89 ± 0.01 MYA. In the same paper, 15 ages were reported on feldspar from the Malbe Tuff (H-4), and the 12 acceptable ages averaged 1.88 ± 0.01 MYA. McDougall et al. (1980) also computed a single isochron age on the KBS Tuff (H-2) of 1.89 ± 0.01 MYA, which was recommended as the best estimate of the unit at the time.

With those three papers, the age of the KBS Tuff (H-2) was shown to lie between 1.8 and 1.9 MYA without doubt, although the ages obtained by Drake et al. (1980) were slightly younger than those obtained by McDougall et al. (1980). McDougall et al. (1980) demonstrated that the small systematic differences resulted from the difficulty in extracting argon from alkali feldspars by measuring samples previously dated at Berkeley. Remaining argument over the best age to choose for the unit was therefore restricted to a much smaller range. The evidence was quite strong that ages near 2.4 MYA were overestimates, caused either by detrital contamination or by isotopic data of poor quality. Although not appreciated at the time, the ages reported in the two K/Ar papers were even stronger than the authors claimed because two stratigraphically proximate units had been independently determined to be of very nearly the same age by two different laboratories. These two units were identifiable by their chemical composition in sections over 100 km apart (Cerling et al. 1979). So after a decade of confusion, the age of one unit that had been problematic was determined with greater precision than any other in the sequence.

If any doubt remained about the age of the KBS, that was put to rest the following year, when $^{40}Ar/^{39}Ar$ incremental heating and total fusion ages were reported (McDougall 1981). These ages averaged 1.88 ± 0.01 MYA, in excellent agreement with the K/Ar results obtained earlier. In contrast to the $^{40}Ar/^{39}Ar$ age spectra obtained on KBS feldspars earlier, those of McDougall were essentially flat, and showed no evidence of thermal disturbance. So that line of argument was effectively stilled—the K/Ar ages (as opposed to $^{40}Ar/^{39}Ar$ ages) on the KBS Tuff did not result from overprinting of an older age.

Resolution of remaining problems with the chronology of the region was not achieved until new measurements were obtained on several units in the Shungura and Usno Formations, and on the older part of the Koobi Fora Formation. Besides supplying new information on units for which the ages were in error, age determinations were needed on additional units to more finely subdivide the sections temporally, but these were not published until the mid-1980s.

Tephrostratigraphic developments ensured that the dates obtained in one formation could be evaluated for consistency with those obtained in others, for the early success of Cerling et al. (1979), followed by that based on mineral composition (Ferguson and Gleadow 1980), demanded further efforts in this direction. An eventual result of the tephrostratigraphic correlations was revision of the stratigraphy in the Koobi Fora region. The initial set of new correlations between the Koobi Fora and Shungura Formations and their stratigraphic implications were reported in Cerling and Brown (1982), and Brown and Cerling (1982). These results were sufficient to show that the sequences at Koobi Fora and in the lower Omo Valley compared quite well overall, but that a part of the stratigraphy represented in the Shungura

Formation was apparently missing at Koobi Fora—the part ranging in age from about 2.5 to 2 MYA. Further work confirmed the correlations already made, established new ones, and extended the correlation network to the Nachukui Formation west of Lake Turkana.

Three papers appeared in 1985 that brought the chronologic framework nearly to its modern aspect. McDougall (1985) reported the results of an extensive set of K/Ar and $^{40}Ar/^{39}Ar$ measurements on seven tuffs from the Koobi Fora Formation—the Moiti (≤ 4.1 ± 0.07 MYA), Toroto (3.32 ± 0.02 MYA), Ninikaa (3.06 ± 0.03 MYA), KBS (1.88 ± 0.02 MYA), Malbe (1.86 ± 0.02 MYA), Chari (1.39 ± 0.02 MYA), and Silbo (0.74 ± 0.01 MYA), all of which were of high precision and consistent with their stratigraphic order. Dates on basalts beneath or near the bottom of the section were also obtained that set a maximum on most dated units of about 4.3 MYA. The problematic dates left from the Shungura Formation were replaced with new determinations at The Australian National University in the same year, when an age of 2.36 ± 0.05 MYA was recommended for Tuff F (= Kalochoro), an age of 2.33 ± 0.03 MYA for Tuff G, and importantly, an age of 4.10 ± 0.06 MYA for the Usno Basalt (Brown et al. 1985). Also reported in that paper were new ages for a basalt from the Mursi Formation of 3.99 ± 0.04 MYA (not the original flow dated), a basalt dike at Kataboi (4.04 ± 0.05, 4.08 ± 0.05 MYA), Tuff B-10 (2.93 ± 0.03, 2.98 ± 0.03 MYA), Tuff D (eight dates averaging 2.52 ± 0.05 MYA), Tuff H-2 (= KBS) (1.83 ± 0.02, 1.84 ± 0.02, 1.85 ± 0.02 MYA), the Lower Nariokotome Tuff (1.33 ± 0.02, 1.33 ± 0.02 MYA), and the Gele Tuff (1.27 ± 0.01, 1.23 ± 0.01 MYA). Finally, McDougall et al. (1985) reported 20 ages on pumices compositionally indistinguishable from the Morutot Tuff (= J-4) defined in the Nachukui Formation, but collected from the Lower Okote Tuff in the Koobi Fora region. All but two of these ages fell in the range 1.60–1.70 MYA, and an age of 1.65 ± 0.03 MYA was recommended for that unit. An accompanying paper by Brown and Feibel (1985) used compositional data to begin to unravel the intricacies of the Okote Tuff Complex, a task that has yet to be completed, but on which Feibel and Brown (1993) have made some headway.

Work in faraway Hadar had led to dispute about the age of another unit, the Tulu Bor Tuff. McDougall (1985) had provided sufficient information, in conjunction with its normal paleomagnetic polarity, to suggest that the Tulu Bor Tuff must lie between 3.32 and 3.42 MYA. In the Hadar sequence, the Sidi Hakoma Tuff was believed to be greater than 3.6 million years old (Walter and Aronson 1982). Based on their compositional similarity, Brown (1982) suggested that the two units correlated. Still, there was sufficient uncertainty about the chronology of strata between the Toroto Tuff (3.32 MYA) and the Moiti Tuff (≤ 4.10 MYA), that it could be argued that the age suggested by Walter and Aronson (1982) was more nearly correct. This difficulty was resolved by correlation of the Tulu Bor Tuff with a tephra layer in a deep sea core at DSDP Site 231 in the Gulf of Aden (Sarna-Wojcicki et al. 1985). Thereby, an independent estimate of the age of the Tulu Bor Tuff was obtained through calcareous nannoplankton stratigraphy, consistent with the age of the tuff derived from the data obtained in the Turkana Basin. Had I followed up on a conversation with Clark in the Omo Valley in 1973, the link to the deep sea might have been discovered substantially earlier.

Brown and Feibel (1986) revised the stratigraphy of the Koobi Fora Formation (independent of the chronological information) and showed that all ages recommended for tuffs in the Koobi Fora Formation were consistent with the new stratigraphy. In this revision, the old Kubi Algi Formation was subsumed into the expanded Koobi Fora Formation, and in conjunction with the dates on the Toroto Tuff and Moiti Tuffs, there was no longer any reason to consider seriously the ages of 3.9 and 4.5 MYA for that part of the formation.

Hillhouse et al. (1986) reinvestigated the paleomagnetic polarity stratigraphy of the Koobi Fora Formation, demonstrating that the magnetostratigraphy was consistent with the new age estimates on various units. Even so, some problems remained. For example, the recommended age of the Toroto Tuff of 3.32 MYA appeared to be somewhat older than would have been predicted from the magnetostratigraphy alone, for which an age nearer 3.2 MYA seems more appropriate. Also there was a general misfit between ages of dated units estimated on the basis of magnetostratigraphy (assuming constant sedimentation rates between boundaries), and those measured directly by the K/Ar method. This was attributed to a lag between the time of deposition and the time of magnetization of the units.

Nearly the last K/Ar ages on tephra units within the Plio-Pleistocene sequence appeared with publication of ages on the Topernawi Tuff (3.76 ± 0.04 MYA), the Burgi Tuff (2.68 ± 0.06 MYA), and further data on Tuff D (= Lokalalei) and Tuff F (= Kalochoro) (Feibel et al. 1989). These helped fill in intervals where the temporal spacing between dated units was still large. The last dates, however, were published (without supporting data) by Lewin (1987), having been determined by Hurford on splits of the original KBS Tuff used by Fitch and Miller (1970). The measured age was 1.87 ± 0.04 MYA.

Additional correlations between the Turkana Basin and the Gulf of Aden were reported by Brown et al. (1992), and the correlations have been used to refine the ages of the calcareous nannoplankton zones themselves (Rahman and Roth 1989). Correlations of other tephra below the Tulu Bor Tuff in the Turkana Basin have been made with tephra in the Sagantole Formation that underlies the Hadar Formation in the Middle Awash Valley (Haileab and Brown 1992). These, too, are consistent with the correlation of the Tulu Bor Tuff with the Sidi Hakoma Tuff. Eventually I expect that many sites in Ethiopia will be directly linked to those in the Turkana Basin through tephrostratigraphy because almost every site for which there is comparative data has revealed some tephra in common (e.g., Haileab and Brown 1992; 1994).

VOLCANIC ROCKS IN THE TURKANA BASIN

Thus far the discussion has focused on units directly related to the age of Pliocene and Pleistocene fossiliferous strata in the Turkana basin. However, a number of ages have also been measured on lavas in the basin and on its margins. These are important to our understanding of the structural and sedimentological history of the basin, and to the evolution of environments within it. Although many dates are now available for rocks of Oligocene and Miocene ages (e.g., McDougall and Watkins 1985, 1988; Boschetto et al. 1992), of particular import are Pliocene and Pleistocene volcanic rocks and volcanic centers. The interest in these lavas is twofold. First, basaltic flows often accompany structural movements, which influence the history of sedimentation in a region. Second, major volcanic edifices such as Mts. Kulal, Asie, Marsabit, and Korath may affect the history of sedimentation by blocking rivers to create lakes, or by diverting river courses.

About 4 MYA (ages range from 3.8 to 4.3 MYA), eruption of basalt flows and intrusion of basalt dikes, all of similar petrographic character, were widespread in the Turkana Basin. Included in this group of rocks are the Mursi Basalt, basalts of the Mursi Formation as mapped by Davidson (1983), the Usno Basalt, the Karsa Basalt, the Kataboi Basalt, the Chen Alia Basalt, the Harr Basalt, and basalts near Lowasera. These basalts extend from the western flank of Nkalabong in the northern part of the lower Omo Valley to the region around the southern end of Lake Turkana, some 300 km to the south. They are also apparently widespread east of Lake Turkana and may correlate with lavas of the Bulal Plateau (Charsley 1987). Shortly after eruption of these units, lacustrine sedimentation began in the Turkana Basin. Eruption continued during the initial phases of sedimentation, because some early strata are intruded by basaltic dikes at Kataboi and Namanyamanya west of Lake Turkana. It is likely that the two events—eruption and sedimentation—are related. Sedimentary deposits between 10 and 4 MYA in age are very poorly known in the Turkana Basin, with the exception of the deposits at Lothagam and still unstudied exposures between Lothidok and Lodwar. With eruption of the basalts around 4 MYA, the older drainage patterns of the region evidently changed, and sediments began to accumulate over a broad stretch of territory at least 60 x 300 km in extent.

Shortly after secure correlations became available between the Koobi Fora and Shungura Formations in 1980, Cerling and Brown (unpublished) realized that the existing model of sedimentation in the Turkana Basin was incorrect, and that the stratigraphic relations demanded a through-flowing river much of the time rather than a basin with no outlet. This idea was developed further by Feibel (1988), who refers to the through-flowing river as the "Turkana River." The major volcanic edifices of Mts. Kulal, Asie, Marsabit, and the Huri Hills are clear obstacles to drainage of the Turkana basin toward the east at the present time, but all of these are geologically young. Mts. Kulal and Asie were evidently built between about 2 and 2.5

MYA ago, and the Huri Hills and Mt. Marsabit are yet younger—2 to 0.5 million years old (Brotzu et al. 1984). Recent reconnaissance work northwest of Loiyengalani has shown that the Koobi Fora Formation extends to that region. Both the α-Tulu Bor and β-Tulu Bor Tuffs have been identified there in addition to the Kokiselei Tuff (Tuff E) and the KBS Tuff. The section appears to be condensed in the area and differs from typical sections in the Koobi Fora region in that it contains basalt flows and basaltic tuffs of local origin. One of the basaltic tuffs lies between the α-Tulu Bor and β-Tulu Bor Tuffs, showing that volcanism was active during the time that the river drained through the area. With further work it may be possible to demonstrate just how volcanism in the southeastern part of the basin affected (or even controlled) sedimentation in the northern part of the basin from which so many hominid fossils have been retrieved.

The basaltic volcanism previously discussed should not be confused with the volcanic activity responsible for the silicic tephra within the Koobi Fora, Shungura, and Nachukui Formations. Woldegabriel (1987) has demonstrated that the probable source of these tephra is in the Ethiopian highlands, or possibly the Ethiopian Rift Valley itself (see also Woldegabriel and Aronson 1987).

A SYNOPSIS OF THE PRESENT SCALE AND RELATIONS BETWEEN SITES

Within the Turkana Basin itself, little new direct information has been added to the chronology since it was last published. However, tephrostratigraphic ties from the Turkana Basin to other areas and new data at other sites allow hominids from many areas to be related temporally. The principal sites of interest are those in the Middle Awash Valley and Hadar region, Olduvai Gorge, Laetoli, and Baringo. Also of interest are ties with the deep sea record in the Gulf of Aden. Relations are shown graphically in Figure 15-3. Isotopic ages for Laetoli were taken from Drake and Curtis (1987); those for Olduvai, from Walter et al. (1991); and those for Baringo, from Chapman and Brook (1978) and Hill et al. (1992). Tephrostratigraphic correlations are summarized in Haileab and Brown (1992) for the Turkana Basin–Ethiopian Rift Valley, in Pickford et al. (1991) for the Turkana Basin–Lake Albert ties, and in Namwamba (1993) for the Turkana Basin–Baringo ties.

For most strata in the Turkana Basin, age control is now quite good, but there are still some parts of the sequence where intervals as long as 0.4 MYA lack additional control (Figure 15-3a). Feibel et al. (1989) placed hominid fossils from the Koobi Fora, Nachukui, and Shungura Formations within this temporal context. They noted that the time scale for these formations based on K/Ar and $^{40}Ar/^{39}Ar$ dating of alkali feldspars from pumices is slightly, but systematically, older than the chronology based on paleomagnetic data alone, and they advanced possible reasons for this offset. Recent publications on the age of magnetic polarity transitions in the Pliocene and Pleistocene (e.g., Baksi et al. 1992; Shackleton et al. 1990; Spell and McDougall 1992; McDougall et al. 1992) suggest that the widely adopted times of transition of the magnetic field may be in error. So far, all revisions have been toward older ages for each of the polarity transitions, bringing the time scale in the Turkana Basin based on the paleomagnetic record closer to that constructed from K/Ar and $^{40}Ar/^{39}Ar$ dating. McDougall et al. (1992) reestimated the ages of all magnetic polarity boundaries between 3.6 and 1.5 MYA using data in the Turkana Basin only and found that the fit with the astronomical calibration is exceptionally good ($r^2 = 0.999$). For the older part of the Turkana sequence, the differences amount to about 0.15 MYA, sufficiently large that the ages of hominid fossils suggested by Feibel et al. (1989) will require revision, but in most instances the ages will increase by less than 4 percent.

The temporal distribution of fossil hominids (Figure 15-3a,b) demonstrates the importance of each of the individual sites. From 3.8–4.2 MYA, there are only a handful of specimens for study, but Laetoli provides substantial material in the interval 3.5–3.7 MYA. In the Turkana Basin, the hominid record is poor between 2.9 and 3.4 MYA, but the abundant collections from Hadar largely fill this gap. Thereafter the principal record is from the Turkana Basin, first from the Shungura and Nachukui Formations (2.9–2.0 MYA), and then from the Nachukui and Koobi Fora Formations (2.0–1.4 MYA). For specimens from still later times, one must turn to Olduvai Gorge, and even there the record is sparse.

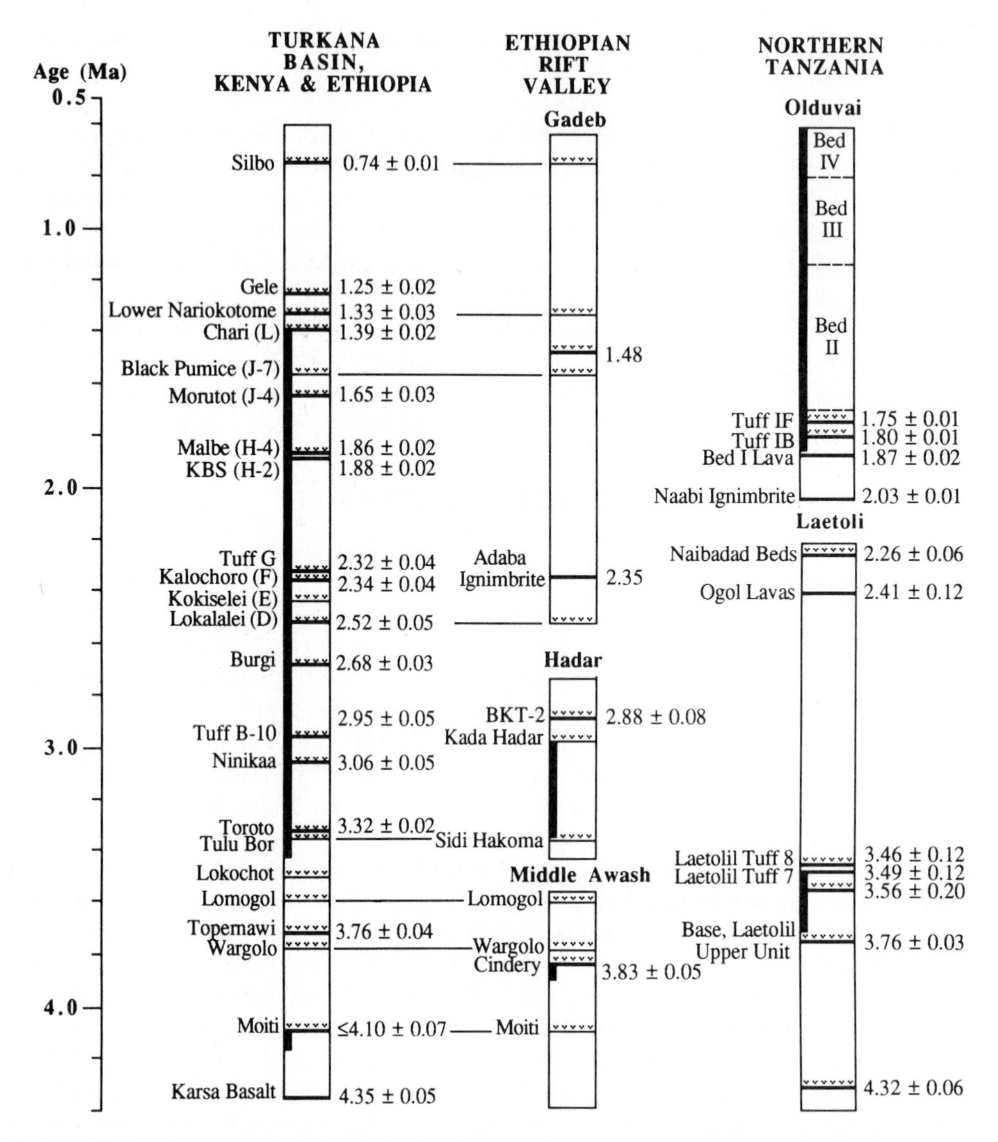

FIGURE 15-3a Ages on dated units in the Turkana Basin, and correlations of the Turkana Basin sequence with the sequence in the Ethiopian Rift Valley at the sites of Gadeb, Hadar, and Middle Awash. Only tephra layers that have yielded isotopic dates or that correlate with other sites are shown. Dated units are distinguished by a heavy line. Stratigraphic columns for Olduvai Gorge and Laetoli are included for comparison. The heavy dark line on the left edge of each column indicates an interval from which hominid fossil(s) have been collected. For this purpose, fossils were lumped into 0.1 MYA intervals.

ACKNOWLEDGMENTS

NSF grants to F. Clark Howell, first at the University of Chicago and later at the University of California, Berkeley, supported early phases of my work in the Turkana Basin. Later work was supported by grants to the author and co-principal investigators (BNS 82-10735, 84-06737, 86-05687, 88-05271). Various aspects of the work received additional funding from the National Geographic Society, the L. S. B. Leakey

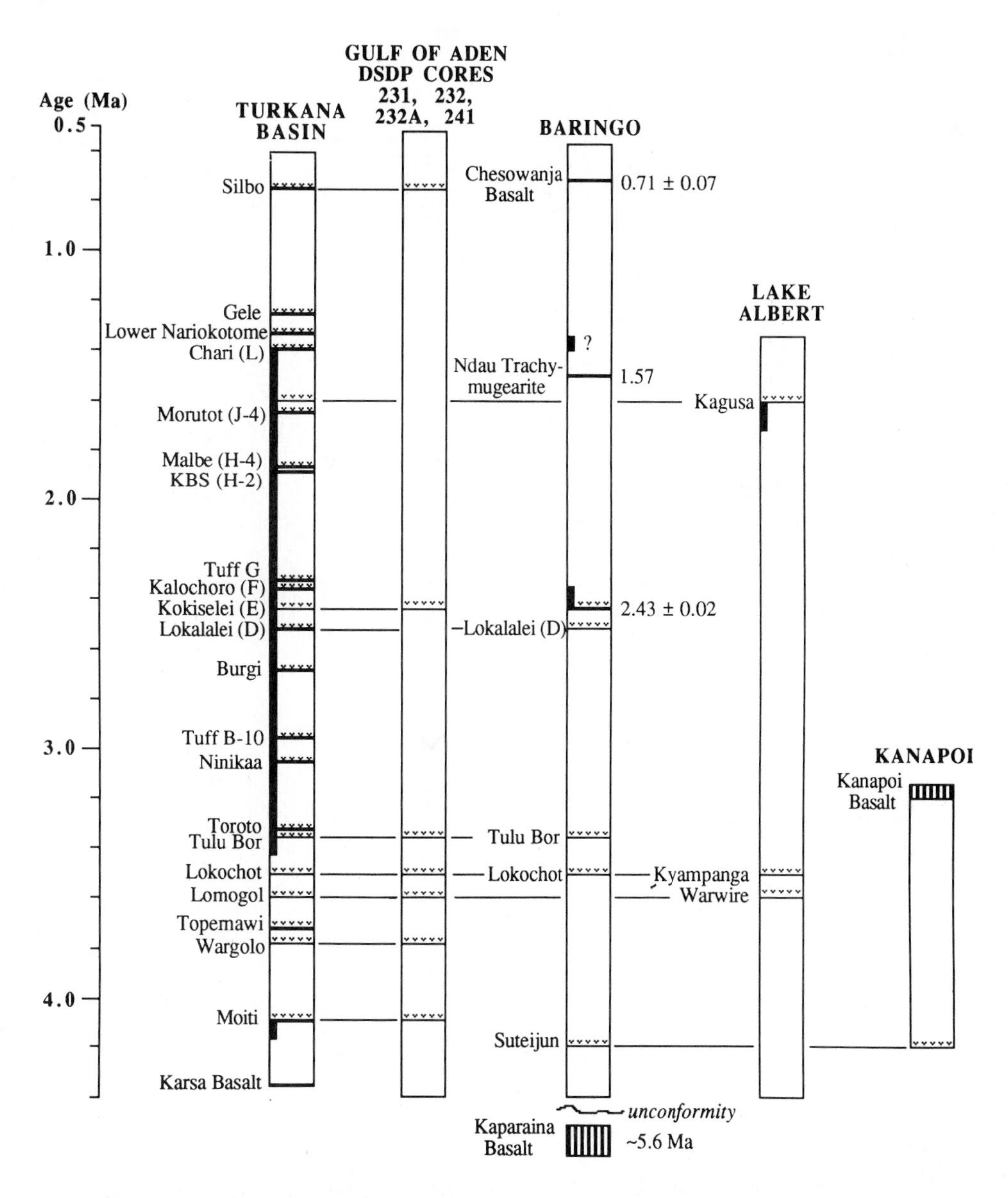

FIGURE 15-3b Correlation of the Turkana Basin sequence with deep-sea cores from the Gulf of Aden, the Baringo Basin, the Albert Basin, and Kanapoi. See text for discussion.

Foundation, and the Foundation for Research into the Origins of Man. I am deeply grateful to all of these organizations. Drs. Craig Feibel, Ian McDougall and Thure Cerling read early versions of this chapter and offered valuable suggestions for its improvement. Thanks are also due Dr. M. D. Leakey, the National Museums of Kenya, The Australian National University, the Ethiopian Institute of Geological Surveys, and the Los Angeles County Museum for help with logistics and analytical facilities. My deepest gratitude though is reserved for Clark Howell himself for introducing me to the world of paleoanthropology and to the fascinating geology of the Turkana Basin.

TABLE 15-2 Analytical data, ages, and recalculated ages of samples measured by the K/Ar method.

LABORATORY NUMBER	%K	40Ar*/g (moles)	40Ar* (%)	REPORTED AGE (MYA)	ERROR (MYA)	RECALCULATED AGE (MYA)	UNIT NAME	LOCALITY
A. Ages measured on anorthoclase feldspar								
Analytical data and calculated ages of the Silbo Tuff (from McDougall 1985)								
ANU81-116	5.475, 5.440	6.936E-12	65.3	0.733	0.008	0.73	Silbo Tuff	Area 138, K. Fora
ANU81-117(1)	5.657, 5.626	7.861E-12	74.2	0.803	0.010	0.80	Silbo Tuff	Area 138, K. Fora
ANU81-117(2)	5.657, 5.626	7.631E-12	64.3	0.780	0.010	0.78	Silbo Tuff	Area 138, K. Fora
ANU81-118	5.771, 5.776	7.290E-12	74.7	0.728	0.008	0.73	Silbo Tuff	Area 138, K. Fora
ANU81-151	5.605, 5.631	7.346E-12	39.8	0.754	0.009	0.76	Silbo Tuff	Area 138, K. Fora
ANU81-152	5.750, 5.741	7.360E-12	68.9	0.738	0.009	0.74	Silbo Tuff	Area 138, K. Fora
ANU81-153	5.624, 5.649	7.412E-12	35.4	0.758	0.010	0.76	Silbo Tuff	Area 138, K. Fora
ANU81-158	5.580, 5.069	7.236E-12	67.7	0.746	0.008	0.75	Silbo Tuff	Area 138, K. Fora
ANU81-160	5.670, 5.676	7.371E-12	63.7	0.749	0.009	0.75	Silbo Tuff	Area 138, K. Fora
Analytical data and calculated ages of the Gele Tuff (from Brown et al. 1985)								
ANU83- 3-2	4.653, 4.654	9.94E-12	73.6	1.23	0.01	1.23	Gele Tuff	Ileret, K. Fora
ANU83- 3-1	4.653, 4.658	1.023E-11	71.1	1.27	0.01	1.27	Gele Tuff	Ileret, K. Fora
Analytical data and calculated ages of the Lower Nariokotome Tuff (from Brown et al. 1985)								
ANU83- 11	5.096, 5.114	1.179E-11	74.7	1.33	0.02	1.33	L. Nariokot.	Lokapetamoi, Tk.
ANU83- 12A	4.823, 4.802	1.111E-11	49.0	1.33	0.02	1.33	L. Nariokot.	Lokapetamoi, Tk.
Analytical data and calculated ages of Tuff L (= Chari) (from Brown and Nash 1976)								
KA-2504	4.510	1.016E-11	28.8	1.27	0.10	1.30	Tuff L	Kalam Area, Omo
KA-2505	4.490	1.127E-11	27.1	1.41	0.10	1.45	Tuff L	Kalam Area, Omo
KA-2521	4.140	1.054E-11	21.0	1.43	0.10	1.47	Tuff L	Kalam Area, Omo
KA-2516	4.246	1.138E-11	28.9	1.51	0.10	1.54	Tuff L	Kalam Area, Omo
Analytical data and calculated ages of the Chari Tuff (= L) (from McDougall 1985)								
ANU78-1053(1)	4.255, 4.245	1.024E-11	66.5	1.388	0.016	1.39	Chari Tuff	Area 131, K. Fora
ANU78-1053(2)	4.255, 4.245	1.019E-11	75.1	1.382	0.016	1.38	Chari Tuff	Area 131, K. Fora
ANU78-1056(1)	4.213, 4.237	1.009E-11	75.7	1.377	0.016	1.38	Chari Tuff	Area 131, K. Fora
ANU78-1056(2)	4.213, 4.237	1.016E-11	83.5	1.386	0.016	1.39	Chari Tuff	Area 131, K. Fora
ANU78-1060(1)	4.173, 4.208	9.99E-12	75.8	1.374	0.019	1.38	Chari Tuff	Area 1, K. Fora
ANU78-1060(2)	4.173, 4.208	1.009E-11	66.5	1.388	0.019	1.39	Chari Tuff	Area 1, K. Fora
ANU78-1061(1)	4.081, 4.083	1.055E-11	75.4	1.490	0.015	1.49	Chari Tuff	Area 1, K. Fora
ANU78-1061(2)	4.081, 4.083	1.058E-11	71.1	1.494	0.015	1.49	Chari Tuff	Area 1, K. Fora
ANU78-1063(1)	4.236, 4.229	1.034E-11	79.6	1.408	0.017	1.41	Chari Tuff	Area 1, K. Fora
ANU78-1063(2)	4.236, 4.229	1.044E-11	72.2	1.422	0.015	1.42	Chari Tuff	Area 1, K. Fora
ANU78-1064(1)	4.213, 4.208	1.027E-11	79.6	1.408	0.017	1.41	Chari Tuff	Area 1, K. Fora
ANU78-1064(2)	4.213, 4.208	1.019E-11	72.2	1.395	0.015	1.39	Chari Tuff	Area 1, K. Fora
ANU78-1066	4.096, 4.051	9.82E-12	72.8	1.389	0.017	1.38	Chari Tuff	Area 1, K. Fora
ANU79- 8(1)	4.311, 4.296	1.042E-11	64.0	1.402	0.018	1.39	Chari Tuff	Area 131, K. Fora
ANU79- 8(2)	4.311, 4.296	1.116E-11	74.6	1.502	0.019	1.49	Chari Tuff	Area 131, K. Fora
ANU79- 8(3)	4.311, 4.296	1.083E-11	76.6	1.450	0.015	1.45	Chari Tuff	Area 131, K. Fora
ANU81-144(1)	4.181, 4.158	1.005E-11	77.2	1.388	0.015	1.39	Chari Tuff	Area 131, K. Fora
ANU81-144(2)	4.181, 4.158	9.8E-12	75.0	1.354	0.014	1.35	Chari Tuff	Area 131, K. Fora
ANU81-144(3)	4.181, 4.158	1.014E-11	72.9	1.402	0.018	1.40	Chari Tuff	Area 131, K. Fora
ANU81-146(1)	4.140, 4.137	9.74E-12	83.2	1.356	0.013	1.36	Chari Tuff	Area 131, K. Fora
ANU81-146(2)	4.140, 4.137	9.92E-12	82.5	1.381	0.015	1.38	Chari Tuff	Area 131, K. Fora
ANU81-146(3)	4.140, 4.137	9.82E-12	82.3	1.368	0.015	1.37	Chari Tuff	Area 131, K. Fora
ANU81-147(1)	4.163, 4.173	9.87E-12	80.8	1.365	0.012	1.37	Chari Tuff	Area 131, K. Fora
ANU81-147(2)	4.163, 4.173	1.028E-11	79.4	1.421	0.015	1.42	Chari Tuff	Area 131, K. Fora
ANU81-147(3)	4.163, 4.173	1.033E-11	76.1	1.429	0.016	1.43	Chari Tuff	Area 131, K. Fora
ANU81-148(1)	4.152, 4.149	9.87E-12	81.5	1.370	0.012	1.37	Chari Tuff	Area 131, K. Fora
ANU81-148(2)	4.152, 4.149	1.015E-11	79.7	1.409	0.015	1.41	Chari Tuff	Area 131, K. Fora
ANU81-148(3)	4.152, 4.149	9.96E-12	81.8	1.383	0.015	1.38	Chari Tuff	Area 131, K. Fora
Analytical data and calculated ages of the Chari Tuff (= L) (from Fitch and Miller 1976)								
FM7041A	4.333	9.28E-12	41.3	1.20	0.06	1.23	Chari Tuff	K. Fora region
FM7041A	4.333	9.41E-12	42.0	1.21	0.06	1.25	Chari Tuff	K. Fora region
Analytical data and calculated ages of the Chari Tuff (= L) and Tuff L (= Chari) (from Drake et al. 1980)								
KA-2863	4.15	9.3E-12	18	1.29	0.04	1.29	Chari Tuff	Area 131, K. Fora
KA-2880	4.08	9.7E-12	59	1.37	0.03	1.37	Chari Tuff	Area 131, K. Fora
KA-2865	3.92	9.4E-12	41	1.38	0.03	1.38	Chari Tuff	Area 6, K. Fora
KA-2883	3.89	9.4E-12	25	1.40	0.04	1.39	Chari Tuff	Area 6, K. Fora
KA-3254	3.94	9.6E-12	29	1.42	0.04	1.40	Tuff L	Kalam Area, Omo
KA-2815	4.05	1.04E-11	23	1.48	0.04	1.48	Chari Tuff	Area 131, K. Fora
KA-2882	4.24	1.12E-11	64	1.52	0.03	1.52	Chari Tuff	Area 131, K. Fora
Analytical data and calculated ages of the Koobi Fora Tuff (from Fitch and Miller 1976)								
FM6955	4.14	3.28E-11	50.1	4.42	0.22	4.57	K. Fora Tuff	Area 103, K. Fora
FM6955	4.14	3.30E-11	48.0	4.44	0.22	4.59	K. Fora Tuff	Area 103, K. Fora
Analytical data and calculated ages of the Morutot Tuff (= J-4) (from McDougall et al. 1985)								
ANU81-133(1)	4.757, 4.737	1.395E-11	43.3	1.694	0.020	1.690	Morutot Tuff	Area 131, K. Fora
ANU81-133(2)	4.757, 4.737	1.320E-11	38.3	1.603	0.020	1.599	Morutot Tuff	Area 131, K. Fora
ANU81-133(3)	4.757, 4.737	1.361E-11	33.2	1.652	0.022	1.649	Morutot Tuff	Area 131, K. Fora
ANU81-134	4.745, 4.741	1.600E-11	82.6	1.941	0.018	1.943	Morutot Tuff	Area 131, K. Fora
ANU83-265(1)	4.728, 4.744	1.347E-11	54.7	1.639	0.018	1.642	Morutot Tuff	Area 131, K. Fora
ANU83-265(2)	4.728, 4.744	1.368E-11	47.3	1.665	0.020	1.668	Morutot Tuff	Area 131, K. Fora
ANU83-272(1)	4.698, 4.704	1.325E-11	38.8	1.625	0.020	1.625	Morutot Tuff	Area 131, K. Fora
ANU83-272(2)	4.698, 4.704	1.351E-11	38.6	1.657	0.020	1.657	Morutot Tuff	Area 131, K. Fora
ANU83-326(1)	4.712, 4.736	1.357E-11	57.9	1.656	0.020	1.660	Morutot Tuff	Area 131, K. Fora
ANU83-326(2)	4.712, 4.736	1.338E-11	47.7	1.632	0.026	1.637	Morutot Tuff	Area 131, K. Fora
ANU83-328(1)	4.715, 4.723	1.307E-11	31.6	1.596	0.021	1.598	Morutot Tuff	Area 131, K. Fora
ANU83-328(2)	4.715, 4.723	1.338E-11	25.6	1.634	0.026	1.635	Morutot Tuff	Area 131, K. Fora
ANU83-329(1)	4.682, 4.710	1.299E-11	23.9	1.594	0.021	1.599	Morutot Tuff	Area 131, K. Fora
ANU83-329(2)	4.682, 4.710	1.313E-11	42.4	1.611	0.019	1.616	Morutot Tuff	Area 131, K. Fora

TABLE 15-2 Continued

LABORATORY NUMBER	%K	40Ar*/g (moles)	40Ar* (%)	REPORTED AGE (MYA)	ERROR (MYA)	RECALCULATED AGE (MYA)	UNIT NAME	LOCALITY
Analytical data and calculated ages of the Morutot Tuff (=J-4) (from McDougall et al. 1985) (Continued)								
ANU83-330	4.694, 4.695	2.011E-11	42.5	2.468	0.028	2.469	Morutot Tuff	Area 131, K. Fora
ANU83-331(1)	4.751, 4.726	1.320E-11	47.9	1.605	0.019	1.601	Morutot Tuff	Area 131, K. Fora
ANU83-331(2)	4.751, 4.726	1.391E-11	36.9	1.692	0.023	1.687	Morutot Tuff	Area 131, K. Fora
ANU83-331(3)	4.751, 4.726	1.345E-11	55.2	1.636	0.018	1.632	Morutot Tuff	Area 131, K. Fora
ANU83-332(1)	4.699, 4.702	1.390E-11	60.7	1.704	0.019	1.705	Morutot Tuff	Area 131, K. Fora
ANU83-332(2)	4.699, 4.702	1.386E-11	63.2	1.699	0.019	1.700	Morutot Tuff	Area 131, K. Fora
Analytical data and calculated ages of the Malbe Tuff (from Curtis et al. 1975)								
KA-2856	5.487	1.529E-11	63.3	1.61	0.02	1.61	Malbe Tuff	Area 112, K. Fora
Analytical data and calculated ages of the Malbe Tuff (= H-4) and Tuff H-4 (= Malbe Tuff) (from Drake et al. 1980)								
KA-3186	4.87	1.46E-11	41	1.73	0.03	1.73	Malbe Tuff	Area 112, K. Fora
KA-3189	5.08	1.54E-11	26	1.74	0.04	1.75	Malbe Tuff	Area 112, K. Fora
KA-3183	5.09	1.54E-11	40	1.75	0.03	1.74	Tuff H-4	Type Area, Omo
KA-3182	5.19	1.61E-11	58	1.78	0.02	1.79	Malbe Tuff	Area 112, K. Fora
KA-3187	5.12	1.59E-11	70	1.79	0.02	1.79	Malbe Tuff	Area 112, K. Fora
KA-3188	4.96	1.54E-11	67	1.79	0.02	1.79	Malbe Tuff	Area 112, K. Fora
KA-2970	5.06	1.62E-11	55	1.85	0.03	1.85	Tuff H-4	Type Area, Omo
KA-3496	5.11	1.64E-11	27	1.90	0.03	1.85	Malbe Tuff	Area 112, K. Fora
KA-2856	4.70	1.57E-11	63	1.92	0.02	1.93	Malbe Tuff	Area 112, K. Fora
Analytical data and calculated ages of the Malbe Tuff (from McDougall et al. 1980)								
ANU78-1039	5.042, 5.042	1.636E-11	68.3	1.87	0.02	1.87	Malbe Tuff	Area 112, K. Fora
ANU78-1040(1)	5.100, 5.138	1.670E-11	85.2	1.88	0.03	1.89	Malbe Tuff	Area 112, K. Fora
ANU78-1040(2)	5.100, 5.138	1.684E-11	84.5	1.90	0.03	1.90	Malbe Tuff	Area 112, K. Fora
ANU78-1041(1)	5.097, 5.093	1.655E-11	86.7	1.87	0.02	1.87	Malbe Tuff	Area 112, K. Fora
ANU78-1041(2)	5.097, 5.093	1.666E-11	83.3	1.88	0.02	1.88	Malbe Tuff	Area 112, K. Fora
ANU78-1043(1)	5.061, 5.042	6.556E-11	87.9	7.46	0.08	7.45	Malbe Tuff	Area 112, K. Fora
ANU78-1043(2)	5.061, 5.042	3.723E-11	90.6	4.11	0.04	4.24	Malbe Tuff	Area 112, K. Fora
ANU79- 17(1)	5.108, 5.112	1.651E-11	35.6	1.86	0.02	1.86	Malbe Tuff	Area 112, K. Fora
ANU79- 17(2)	5.108, 5.112	1.662E-11	37.5	1.87	0.02	1.88	Malbe Tuff	Area 112, K. Fora
ANU79- 18	4.975, 4.949	1.622E-11	41.1	1.88	0.02	1.88	Malbe Tuff	Area 112, K. Fora
ANU79- 19	5.163, 5.163	1.795E-11	64.6	2.00	0.02	2.00	Malbe Tuff	Area 112, K. Fora
ANU79-276(1)	5.107, 5.112	1.682E-11	84.4	1.90	0.03	1.90	Malbe Tuff	Area 112, K. Fora
ANU79-276(2)	5.107, 5.112	1.675E-11	84.9	1.89	0.02	1.89	Malbe Tuff	Area 112, K. Fora
ANU79-277(1)	5.100, 5.115	1.679E-11	83.0	1.89	0.02	1.90	Malbe Tuff	Area 112, K. Fora
ANU79-277(2)	5.100, 5.115	1.674E-11	84.6	1.89	0.02	1.89	Malbe Tuff	Area 112, K. Fora
Analytical data and calculated ages of the Malbe Tuff (from McDougall 1985)								
ANU81-162(1)	5.168, 5.162	1.672E-11	75.5	1.866	0.020	1.86	Malbe Tuff	Area 112, K. Fora
ANU81-162(2)	5.168, 5.162	1.654E-11	75.7	1.845	0.021	1.84	Malbe Tuff	Area 112, K. Fora
ANU81-163(1)	5.062, 5.082	1.662E-11	78.1	1.888	0.020	1.89	Malbe Tuff	Area 112, K. Fora
ANU81-163(2)	5.062, 5.082	1.660E-11	76.5	1.886	0.020	1.89	Malbe Tuff	Area 112, K. Fora
ANU81-164(1)	5.180, 5.181	1.658E-11	77.1	1.844	0.019	1.84	Malbe Tuff	Area 112, K. Fora
ANU81-164(2)	5.180, 5.181	1.656E-11	76.8	1.842	0.020	1.84	Malbe Tuff	Area 112, K. Fora
ANU81-196(1)	5.124, 5.109	1.624E-11	84.8	1.829	0.019	1.83	Malbe Tuff	Area 105, K. Fora
ANU81-196(2)	5.124, 5.109	1.631E-11	80.9	1.837	0.020	1.83	Malbe Tuff	Area 105, K. Fora
ANU81-197(1)	5.186, 5.171	1.670E-11	78.5	1.859	0.020	1.86	Malbe Tuff	Area 105, K. Fora
ANU81-197(2)	5.186, 5.171	1.677E-11	76.9	1.866	0.020	1.86	Malbe Tuff	Area 105, K. Fora
ANU81-198(1)	5.130, 5.079	1.629E-11	84.9	1.839	0.019	1.83	Malbe Tuff	Area 105, K. Fora
ANU81-198(2)	5.130, 5.079	1.620E-11	74.5	1.829	0.020	1.82	Malbe Tuff	Area 105, K. Fora
ANU81-198(3)	5.130, 5.079	1.656E-11	84.7	1.869	0.024	1.86	Malbe Tuff	Area 105, K. Fora
Analytical data and calculated ages of Tuff H-2 (= KBS) (from Brown 1969)								
KA-2187	5.182	1.668E-11	45.2	1.81	0.09	1.86	Tuff H-2	Type Area, Omo
KA-2085	5.111	n.r.	n.r.	1.87	0.09	1.92	Tuff H-2	Type Area, Omo
Analytical data and calculated ages of the KBS Tuff (= H-2) (from Fitch and Miller 1970)								
Leakey I(B2)	5.014	2.12E-11	19.5	2.36	0.3	2.44	KBS Tuff	K. Fora region
Leakey I(B2)	5.014	2.14E-11	18.8	2.38	0.3	2.46	KBS Tuff	K. Fora region
Analytical data and calculated ages of the KBS Tuff (= H-2) (from Curtis et al. 1975)								
KA-2816	5.167	6.203E-11	82.2	6.90	0.05	6.91	KBS Tuff	Area 10, K. Fora
KA-2817	5.035	1.755E-11	62.2	2.01	0.03	2.01	KBS Tuff	Area 105, K. Fora
KA-2818	5.006	2.089E-11	58.8	2.40	0.03	2.40	KBS Tuff	Area 105, K. Fora
KA-2823	4.808	1.807E-11	59.4	2.16	0.03	2.17	KBS Tuff	Area 131, K. Fora
KA-2836	4.993	1.584E-11	64.3	1.83	0.02	1.83	KBS Tuff	Area 131, K. Fora
KA-2841	4.896	1.569E-11	48.4	1.85	0.03	1.85	KBS Tuff	Area 131, K. Fora
KA-2842	5.110	1.539E-11	59.4	1.73	0.03	1.74	KBS Tuff	Area 131, K. Fora
KA-2850	5.665	1.479E-11	67.8	1.50	0.02	1.50	KBS Tuff	Area 105, K. Fora
KA-2851	~5.94	2.605E-11	69.0	2.53	0.02	2.53	KBS Tuff	Area 105, K. Fora
KA-2851R	5.936	1.698E-11	60.9	1.65	0.02	1.65	KBS Tuff	Area 105, K. Fora
KA-2852	5.785	1.545E-11	31.1	1.54	0.02	1.54	KBS Tuff	Area 105, K. Fora
KA-2855	5.468	1.481E-11	58.9	1.56	0.02	1.56	KBS Tuff	Area 105, K. Fora
KA-2857	5.543	1.558E-11	47.5	1.62	0.02	1.62	KBS Tuff	Area 105, K. Fora
KA-2859	5.643	1.613E-11	41.3	1.65	0.02	1.65	KBS Tuff	Area 105, K. Fora
KA-2861	5.359	1.485E-11	61.2	1.60	0.01	1.60	KBS Tuff	Area 10, K. Fora
KA-2869	4.598	1.441E-11	46.6	1.81	0.02	1.81	KBS Tuff	Area 131, K. Fora
Analytical data and calculated ages of Tuff H-2 (= KBS) (from Brown and Nash 1976)								
KA-2509	5.124	1.648E-11	75.5	1.81	0.09	1.85	KBS Tuff (H2)	Type Area, Omo
Analytical data and calculated ages of the KBS Tuff (= H-2) (from Fitch, Hooker and Miller 1976)								
Leakey I(B1)				2.38	0.3	2.44	KBS Tuff	K. Fora region
Leakey I(B2)				2.37	0.3	2.43	KBS Tuff	K. Fora region
FM7054	5.512	8.12E-11	47.1	8.22	0.49	8.47	KBS Tuff	K. Fora region
FM7054	5.512	8.52E-11	46.6	8.63	0.52	8.89	KBS Tuff	K. Fora region
FM7054	5.155	1.55E-10	67.9	16.8	0.8	17.29	KBS Tuff	K. Fora region

TABLE 15-2 Continued

LABORATORY NUMBER	%K	$^{40}Ar^*$/g (moles)	$^{40}Ar^*$ (%)	REPORTED AGE (MYA)	ERROR (MYA)	RECALCULATED AGE (MYA)	UNIT NAME	LOCALITY
Analytical data and calculated ages of the KBS Tuff (= H-2) (from Fitch, Hooker and Miller 1976) (Continued)								
FM7054	5.155	1.59E-10	71.3	17.2	0.9	17.68	KBS Tuff	K. Fora region
FM7054	5.155	1.71E-10	71.1	18.6	0.9	19.06	KBS Tuff	K. Fora region
Analytical data and calculated ages of the KBS Tuff (= H-2) and Tuff H-2 (= KBS) (from Drake et al. 1980)								
KA-2842	5.11	1.58E-11	59	1.78	0.03	1.78	KBS Tuff	Area 131, K. Fora
KA-3166	5.04	1.56E-11	74	1.78	0.01	1.78	KBS Tuff	Area 131, K. Fora
KA-3160	5.09	1.59E-11	70	1.80	0.02	1.80	KBS Tuff	Area 131, K. Fora
KA-2851R	5.55	1.74E-11	61	1.81	0.03	1.81	KBS Tuff	Area 105, K. Fora
KA-2836	5.10	1.62E-11	64	1.83	0.02	1.83	KBS Tuff	Area 131, K. Fora
KA-3161	5.12	1.64E-11	63	1.84	0.02	1.85	KBS Tuff	Area 131, K. Fora
KA-2869	4.60	1.47E-11	47	1.85	0.02	1.84	KBS Tuff	Area 131, K. Fora
KA-3256	4.93	1.59E-11	52	1.85	0.03	1.86	Tuff H-2	Type Area, Omo
KA-2841	4.90	1.61E-11	48	1.89	0.03	1.89	KBS Tuff	Area 131, K. Fora
KA-3167	4.87	1.61E-11	64	1.90	0.02	1.91	KBS Tuff	Area 131, K. Fora
Analytical data and calculated ages of the KBS Tuff (= H-2) (from McDougall et al. 1980)								
ANU78-1038cs(1)	5.179, 5.147	1.700E-11	48.6	1.90	0.02	1.89	KBS Tuff	Area 105, K. Fora
ANU78-1038cs(2)	5.179, 5.147	1.690E-11	85.7	1.89	0.02	1.88	KBS Tuff	Area 105, K. Fora
ANU78-1038fn(1)	5.174, 5.128	1.778E-11	77.5	1.99	0.03	1.98	KBS Tuff	Area 105, K. Fora
ANU78-1038fn(2)	5.174, 5.128	1.762E-11	40.9	1.97	0.03	1.96	KBS Tuff	Area 105, K. Fora
ANU78-1044(1)	5.136, 5.152	1.703E-11	63.5	1.91	0.02	1.91	KBS Tuff	Area 131, K. Fora
ANU78-1044(2)	5.136, 5.152	1.687E-11	82.2	1.89	0.02	1.89	KBS Tuff	Area 131, K. Fora
ANU78-1047cs(1)	5.138, 5.112	1.673E-11	78.5	1.88	0.02	1.88	KBS Tuff	Area 131, K. Fora
ANU78-1047cs(2)	5.138, 5.112	1.663E-11	46.4	1.87	0.02	1.87	KBS Tuff	Area 131, K. Fora
ANU78-1047fn(1)	5.117, 5.155	1.693E-11	53.8	1.90	0.03	1.91	KBS Tuff	Area 131, K. Fora
ANU78-1047fn(2)	5.117, 5.155	1.719E-11	63.6	1.93	0.03	1.94	KBS Tuff	Area 131, K. Fora
ANU78-1048(1)	5.127, 5.127	1.681E-11	72.9	1.89	0.02	1.89	KBS Tuff	Area 131, K. Fora
ANU78-1048(2)	5.127, 5.127	1.731E-11	75.5	1.94	0.02	1.95	KBS Tuff	Area 131, K. Fora
ANU79- 14(1)	5.146, 5.171	1.671E-11	55.5	1.87	0.02	1.87	KBS Tuff	Area 131, K. Fora
ANU79- 14(2)	5.146, 5.171	1.700E-11	55.2	1.90	0.02	1.90	KBS Tuff	Area 131, K. Fora
ANU79- 15(1)	5.129, 5.173	1.665E-11	62.5	1.86	0.03	1.87	KBS Tuff	Area 131, K. Fora
ANU79- 15(2)	5.129, 5.173	1.695E-11	77.0	1.90	0.03	1.90	KBS Tuff	Area 131, K. Fora
Analytical data and calculated ages of Tuff H-2 (= KBS) (from Brown et al. 1985)								
ANU83- 13-1	5.139, 5.115	1.629E-11	74.9	1.83	0.02	1.83	Tuff H-2	Type Area, Omo
ANU83- 13-2	5.139, 5.115	1.633E-11	75.5	1.84	0.02	1.83	Tuff H-2	Type Area, Omo
ANU83- 13-3	5.139, 5.115	1.646E-11	78.7	1.85	0.02	1.85	Tuff H-2	Type Area, Omo
Analytical data and calculated ages of the KBS Tuff (= H-2) (from McDougall 1985)								
ANU81-230(1)	5.130, 5.183	1.673E-11	79.0	1.87	0.024	1.88	KBS Tuff	Area 129, K. Fora
ANU81-230(2)	5.130, 5.183	1.676E-11	80.2	1.87	0.024	1.88	KBS Tuff	Area 129, K. Fora
ANU81-231(1)	5.217, 5.218	1.671E-11	81.2	1.85	0.020	1.85	KBS Tuff	Area 129, K. Fora
ANU81-231(2)	5.217, 5.218	1.678E-11	81.1	1.85	0.020	1.85	KBS Tuff	Area 129, K. Fora
ANU81-232(1)	5.206, 5.205	1.721E-11	71.0	1.91	0.023	1.91	KBS Tuff	Area 129, K. Fora
ANU81-232(2)	5.206, 5.205	1.709E-11	70.0	1.89	0.020	1.89	KBS Tuff	Area 129, K. Fora
ANU81-234(1)	5.160, 5.180	1.707E-11	80.4	1.90	0.020	1.91	KBS Tuff	Area 15, K. Fora
ANU81-234(2)	5.160, 5.180	1.675E-11	79.8	1.87	0.020	1.87	KBS Tuff	Area 15, K. Fora
ANU81-235	5.181, 5.147	1.674E-11	83.0	1.87	0.021	1.86	KBS Tuff	Area 15, K. Fora
ANU81-238	5.210, 5.197	1.677E-11	83.6	1.86	0.020	1.86	KBS Tuff	Area 15, K. Fora
Analytical data and calculated age of Tuff G (from Brown et al. 1970, and Brown and Nash 1976)								
LKA- 9	5.899	2.022E-11	40.7	1.93	0.10	1.98	Tuff G	Type Area, Omo
Analytical data and calculated ages of Tuff G (from Brown et al. 1985)								
ANU83- 10-1	5.846, 5.772	2.341E-11	86.5	2.32	0.04	2.31	Tuff G	Type Area, Omo
ANU83- 10-2	5.846, 5.772	2.349E-11	87.8	2.33	0.03	2.32	Tuff G	Type Area, Omo
ANU83- 31-1	5.855, 5.848	2.373E-11	81.3	2.34	0.03	2.34	Tuff G	Type Area, Omo
Analytical data and calculated ages of Tuff F (= Kalochoro) (from Brown et al. 1970, and Brown and Nash 1976)								
LKA-11	6.101	2.156E-11	45.6	1.99	0.10	2.04	Tuff F	Type Area, Omo
LKA-21	6.101	2.232E-11	43.7	2.06	0.11	2.11	Tuff F	Type Area, Omo
Analytical data and calculated ages of Tuff F (= Kalochoro) (from Brown et al. 1985, and Feibel et al. 1989)								
ANU83-364-1	6.305, 6.220	2.527E-11	87.2	2.32	0.04	2.31	Tuff F	Type Area, Omo
ANU83-364-2	6.305, 6.220	2.596E-11	86.2	2.39	0.04	2.37	Tuff F	Type Area, Omo
ANU87- 7(2)	5.606, 5.604	2.229E-11	64.3	2.29	0.03	2.29	Tuff F	Type Area, Omo
ANU87- 7(1)	5.606, 5.604	2.260E-11	69.7	2.32	0.03	2.32	Tuff F	Type Area, Omo
ANU87- 8(1)	6.278, 6.255	2.546E-11	85.7	2.34	0.02	2.34	Tuff F	Type Area, Omo
Analytical data and calculated ages of Tuff D-4 (from Brown et al. 1970, and Brown et al. 1985)								
LKA-14	5.010	1.89E-11	50.9	2.12	0.11	2.17	Tuff D-4	Type Area, Omo
ANU83- 12B	5.237, 5.235	2.547E-11	75.8	2.80	0.03	2.80	Tuff D-4	Type Area, Omo
Analytical data and calculated ages of Tuff D (= Lokalalei) (from Brown 1969)								
KA-2176	5.432	n.r.	46.4	2.37	0.12	2.43	Tuff D	Type Area, Omo
KA-2067	5.151	2.34E-11	49.0	2.56	0.12	2.62	Tuff D	Type Area, Omo
Analytical data and calculated ages of Tuff D (= Lokalalei) (from Brown et al. 1970)								
LKA-23	5.151	1.975E-11	47.6	2.16	0.11	2.21	Tuff D	Type Area, Omo
LKA-22	5.151	2.118E-11	72.7	2.31	0.11	2.37	Tuff D	Type Area, Omo
Analytical data and calculated ages of Tuff D (= Lokalalei) (from Brown and Nash 1976)								
KA-2519	5.139	1.975E-11	56.0	2.16	0.11	2.21	Tuff D	Type Area, Omo
KA-2510R	4.984	2.141E-11	53.3	2.41	0.12	2.48	Tuff D	Type Area, Omo
KA-2511	5.182	2.311E-11	69.1	2.51	0.12	2.57	Tuff D	Type Area, Omo
I1040	5.030	2.328E-11	66.0	2.60	0.12	2.67	Tuff D	Type Area, Omo
Analytical data and calculated ages of Tuff D (= Lokalalei) (from Brown et al. 1985)								
ANU83- 6-1	5.304, 5.338	2.315E-11	84.3	2.51	0.03	2.51	Tuff D	Type Area, Omo
ANU83- 6-2	5.304, 5.338	2.338E-11	80.8	2.53	0.03	2.54	Tuff D	Type Area, Omo
ANU83- 7	5.526, 5.555	2.301E-11	78.5	2.42	0.03	2.40	Lokalalei	Kangatukuseo, Tk.
ANU83- 8	5.151, 5.116	2.235E-11	83.4	2.58	0.03	2.51	Lokalalei	Kangatukuseo, Tk.

TABLE 15-2 Continued

LABORATORY NUMBER	%K	40Ar*/g (moles)	40Ar* (%)	REPORTED AGE (MYA)	ERROR (MYA)	RECALCULATED AGE (MYA)	UNIT NAME	LOCALITY
Analytical data and calculated ages of Tuff D (= Lokalalei) (from Brown et al. 1985) (Continued)								
ANU83- 14-1	5.129, 5.274	2.336E-11	40.5	2.57	0.03	2.58	Tuff D	Type Area, Omo
ANU83- 14-2	5.129, 5.274	2.279E-11	83.2	2.50	0.03	2.52	Tuff D	Type Area, Omo
ANU83- 15	5.144, 5.177	2.257E-11	90.5	2.52	0.03	2.53	Tuff D	Type Area, Omo
ANU83- 16	5.157, 5.158	2.284E-11	84.8	2.55	0.03	2.55	Tuff D	Type Area, Omo
ANU87- 14(1)	5.373, 5.350	2.242E-11	78.9	2.41	0.03	2.40	Tuff D	Type Area, Omo
Analytical data and calculated ages of the Burgi Tuff (from Feibel et al. 1989)								
ANU83-374-1	4.810, 4.823	2.265E-11	60.0	2.71	0.03	2.71	Burgi Tuff	Area 200, K. Fora
ANU83-374-2	4.810, 4.823	2.127E-11	32.6	2.54	0.03	2.55	Burgi Tuff	Area 200, K. Fora
ANU83-375-1	4.805, 4.802	2.207E-11	66.6	2.65	0.03	2.65	Burgi Tuff	Area 200, K. Fora
ANU83-375-2	4.805, 4.802	2.224E-11	79.9	2.67	0.03	2.67	Burgi Tuff	Area 200, K. Fora
Analytical data and calculated ages of the Tuff B-10 (from Brown and Nash 1976, and Brown et al. 1985)								
ANU83- 5-1	5.052, 5.046	2.570E-11	83.1	2.93	0.03	2.93	Tuff B-10	Type Area, Omo
ANU83- 5-2	5.052, 5.046	2.616E-11	82.0	2.98	0.03	2.98	Tuff B-10	Type Area, Omo
KA-2441	5.064	2.674E-11	40.0	2.96	0.1	3.04	Tuff B-10	Type Area, Omo
KA-2458	5.002	2.608E-11	64.3	2.93	0.1	3.00	Tuff B-10	Type Area, Omo
Analytical data and calculated ages of the Ninikaa Tuff (from McDougall 1985)								
ANU78-1033	4.965, 4.955	2.676E-11	92.0	3.11	0.03	3.11	Ninikaa Tuff	Area 116, K. Fora
ANU79- 21(1)	4.398, 4.447	2.442E-11	68.6	3.18	0.04	3.20	Ninikaa Tuff	Area 116, K. Fora
ANU79- 21(2)	4.398, 4.447	2.419E-11	69.6	3.15	0.04	3.17	Ninikaa Tuff	Area 116, K. Fora
ANU79-278(1)	4.510, 4.498	2.432E-11	84.6	3.11	0.03	3.11	Ninikaa Tuff	Area 116, K. Fora
ANU79-278(2)	4.510, 4.498	2.454E-11	78.6	3.14	0.03	3.13	Ninikaa Tuff	Area 116, K. Fora
ANU81-122(1)	5.024, 5.013	2.741E-11	92.4	3.15	0.04	3.14	Ninikaa Tuff	Area 116, K. Fora
ANU81-122(2)	5.024, 5.013	2.732E-11	93.0	3.14	0.03	3.13	Ninikaa Tuff	Area 116, K. Fora
ANU81-123(1)	4.545, 4.595	2.468E-11	76.8	3.11	0.04	3.13	Ninikaa Tuff	Area 116, K. Fora
ANU81-123(2)	4.545, 4.595	2.362E-11	77.1	2.98	0.04	2.99	Ninikaa Tuff	Area 116, K. Fora
ANU81-123(3)	4.545, 4.595	2.419E-11	77.3	3.05	0.04	3.07	Ninikaa Tuff	Area 116, K. Fora
ANU81-123(4)	4.545, 4.595	2.424E-11	76.7	3.06	0.04	3.07	Ninikaa Tuff	Area 116, K. Fora
ANU81-125(1)	5.031, 5.015	2.659E-11	91.6	3.05	0.03	3.04	Ninikaa Tuff	Area 116, K. Fora
ANU81-125(2)	5.031, 5.015	2.678E-11	91.7	3.07	0.04	3.07	Ninikaa Tuff	Area 116, K. Fora
ANU81-127(1)	4.156, 4.188	2.217E-11	73.4	3.06	0.04	3.07	Ninikaa Tuff	Area 116, K. Fora
ANU81-127(2)	4.156, 4.188	2.221E-11	73.4	3.07	0.04	3.08	Ninikaa Tuff	Area 116, K. Fora
ANU81-128(1)	4.804, 4.788	2.664E-11	87.7	3.21	0.03	3.19	Ninikaa Tuff	Area 116, K. Fora
ANU81-128(2)	4.804, 4.788	2.617E-11	87.5	3.14	0.03	3.14	Ninikaa Tuff	Area 116, K. Fora
ANU81-129(1)	4.501, 4.528	2.374E-11	73.7	3.03	0.03	3.04	Ninikaa Tuff	Area 116, K. Fora
ANU81-129(2)	4.501, 4.528	2.408E-11	75.4	3.07	0.03	3.08	Ninikaa Tuff	Area 116, K. Fora
Analytical data and calculated ages on pumice from a channel above the Tulu Bor Tuff (from Feibel et al. 1989)								
ANU83-362-1	4.073, 4.073	2.177E-11	55.3	3.08	0.03	3.08	Unnamed	Area 117, K. Fora
ANU83-362-2	4.073, 4.073	2.150E-11	68.5	3.04	0.03	3.04	Unnamed	Area 117, K. Fora
Analytical data and calculated ages of the Toroto Tuff (from McDougall 1985)								
ANU78-1073A(1)	4.247, 4.276	2.460E-11	80.5	3.33	0.04	3.34	Toroto Tuff	Area 204, K. Fora
ANU78-1073A(2)	4.247, 4.276	2.423E-11	88.2	3.28	0.04	3.29	Toroto Tuff	Area 204, K. Fora
ANU78-1073B(1)	4.346, 4.305	2.534E-11	86.9	3.37	0.04	3.36	Toroto Tuff	Area 204, K. Fora
ANU78-1073B(2)	4.346, 4.305	2.517E-11	48.9	3.34	0.04	3.34	Toroto Tuff	Area 204, K. Fora
ANU78-1075A	4.250, 4.294	2.478E-11	88.6	3.34	0.04	3.36	Toroto Tuff	Area 204, K. Fora
ANU78-1075B	4.347, 4.316	2.500E-11	74.0	3.32	0.04	3.31	Toroto Tuff	Area 204, K. Fora
ANU81-100	4.202, 4.269	2.440E-11	91.2	3.32	0.06	3.35	Toroto Tuff	Area 204, K. Fora
ANU81-103(1)	4.310, 4.335	2.745E-11	91.4	3.66	0.04	3.67	Toroto Tuff	Area 204, K. Fora
ANU81-103(2)	4.310, 4.335	2.662E-11	89.2	3.55	0.04	3.56	Toroto Tuff	Area 204, K. Fora
ANU81-106(1)	4.302, 4.318	2.426E-11	87.4	3.24	0.03	3.25	Toroto Tuff	Area 207, K. Fora
ANU81-106(2)	4.302, 4.318	2.509E-11	86.5	3.35	0.04	3.36	Toroto Tuff	Area 207, K. Fora
ANU81-107(1)	4.375, 4.372	2.538E-11	87.1	3.34	0.04	3.34	Toroto Tuff	Area 207, K. Fora
ANU81-107(2)	4.375, 4.372	2.549E-11	89.1	3.36	0.04	3.36	Toroto Tuff	Area 207, K. Fora
ANU81-109A	4.311, 4.301	2.488E-11	86.1	3.33	0.04	3.32	Toroto Tuff	Area 207, K. Fora
ANU81-109B	4.362, 4.365	2.507E-11	89.8	3.31	0.04	3.31	Toroto Tuff	Area 207, K. Fora
Analytical data and calculated ages of Tuff B-δ (from Brown 1969, and Brown and Nash 1976)								
KA-2096	4.625	3.120E-11	48.7	3.79	0.20	3.89	Tuff B-d	Type Area, Omo
I1029	4.474	3.979E-11	55.5	4.99	0.20	5.12	Tuff B-d	Type Area, Omo
Analytical data and calculated ages of the Topernawi Tuff (from Feibel et al. 1989)								
ANU87- 9(1)	5.360, 5.358	3.516E-11	77.2	3.78	0.04	3.78	Topernawi	Lomekwi, Tk.
ANU87- 10(1)	5.383, 5.417	3.480E-11	86.6	3.71	0.04	3.72	Topernawi	Lomekwi, Tk.
ANU87- 11(1)	5.362, 5.374	3.508E-11	77.1	3.76	0.04	3.77	Topernawi	Lomekwi, Tk.
ANU87- 12(1)	5.397, 5.403	3.726E-11	76.9	3.97	0.04	3.98	Topernawi	Lomekwi, Tk.
Analytical data and calculated ages of the Moiti Tuff (from McDougall 1985)								
ANU82-304	3.744, 3.844	2.731E-11	79.0	4.15	0.06	4.20	Moiti Tuff	Area 252, K. Fora
ANU83- 1(1)	3.971, 3.989	2.874E-11	84.7	4.16	0.04	4.17	Moiti Tuff	Area 252, K. Fora
ANU83- 1(2)	3.971, 3.989	2.837E-11	82.7	4.10	0.04	4.11	Moiti Tuff	Area 252, K. Fora
ANU83- 2(1)	4.006, 4.032	2.821E-11	91.0	4.04	0.04	4.06	Moiti Tuff	Area 252, K. Fora
ANU83- 2(2)	4.006, 4.032	2.780E-11	85.3	3.98	0.04	4.00	Moiti Tuff	Area 252, K. Fora
B. Ages measured on glass and whole rock pumice								
Analytical data and calculated ages on whole rock pumice or other material from the Koobi Fora Tuff (from Fitch and Miller 1976)								
FM6953	2.46	4.34E-12	7.2	1.05	0.13	1.02	K. Fora Tuff	K. Fora region
FM6953	2.46	2.62E-11	7.0	6.3	1.2	6.13	K. Fora Tuff	K. Fora region
FM7037A	1.577	1.25E-11	9.6	4.4	0.4	4.58	K. Fora Tuff	K. Fora, Area 103
FM7037B	2.216	8.79E-12	2.9	2.2	0.7	2.29	K. Fora Tuff	K. Fora, Area 103
FM7037B	2.216	8.97E-12	2.8	2.3	0.7	2.33	K. Fora Tuff	K. Fora, Area 103
FM7037C	1.428	9.24E-12	5.8	3.6	0.5	3.73	K. Fora Tuff	K. Fora, Area 103
FM7037D	1.685	5.31E-11	24.2	17.6	0.7	18.08	K. Fora Tuff	K. Fora, Area 103
FM7037E	2.54	2.69E-11	6.5	5.90	0.9	6.09	K. Fora Tuff	K. Fora, Area 103

TABLE 15-2 Continued

LABORATORY NUMBER	%K	40Ar*/g (moles)	40Ar* (%)	REPORTED AGE (MYA)	ERROR (MYA)	RECALCULATED AGE (MYA)	UNIT NAME	LOCALITY
Analytical data and calculated ages on glass of the Malbe Tuff (from Curtis et al. 1975, Drake et al. 1980, and Cerling et al. 1985)								
KA-2854	4.171	1.217E-11	47.3	1.68	0.01	1.68	Malbe Tuff	Area 112, K. Fora
KA-2854	3.89	1.25E-11	47	1.85	0.02	1.85	Malbe Tuff	Area 112, K. Fora
KA-3380	4.04	1.33E-11	49	1.90	0.03	1.90	Malbe Tuff	Area 112, K. Fora
ANU83- 42-1	3.924	1.26E-11	17.1	1.84	0.04	1.85	Malbe Tuff	Area 112, K. Fora
Analytical data and calculated ages on pumice from the KBS Tuff (from Fitch and Miller 1970, and Fitch et al. 1976)								
Leakey I(A)	1.49	6.11E-10	37.3	219	7	222.28	KBS Tuff	K. Fora region
Leakey I(A)	1.49	6.20E-10	38.0	221	7	225.33	KBS Tuff	K. Fora region
Leakey I(A)	1.49	6.25E-10	36.6	223	7	226.86	KBS Tuff	K. Fora region
Leakey I(B1)	0.58	3.79E-12	0.4	3.63	2.1	3.76	KBS Tuff	K. Fora region
Leakey I(B1)				3.12	1.1	3.20	KBS Tuff	K. Fora region
Leakey I(B1)	0.58	2.50E-12	8.9	2.40	1.0	2.49	KBS Tuff	K. Fora region
Analytical data and calculated ages of glass from the KBS Tuff (from Curtis et al. 1975, Drake et al. 1980, and Cerling et al. 1985)								
KA-2837	3.922	1.252E-11	7.5	1.84	0.07	1.83	KBS Tuff	Area 131, K. Fora
KA-2843	3.484	1.105E-11	17.1	1.83	0.03	1.83	KBS Tuff	Area 131, K. Fora
KA-2858	3.69	1.11E-11	19	1.73	0.04	1.73	KBS Tuff	Area 131, K. Fora
KA-2862	3.73	1.18E-11	17	1.83	0.02	1.82	KBS Tuff	Area 131, K. Fora
KA-2843	3.48	1.13E-11	17	1.87	0.03	1.87	KBS Tuff	Area 131, K. Fora
KA-2837	3.92	1.28E-11	7	1.88	0.07	1.88	KBS Tuff	Area 131, K. Fora
ANU83- 37-1	3.998	1.212E-11	5.2	1.75	0.10	1.75	KBS Tuff	Area 131, K. Fora
ANU83- 39-1	3.613	9.806E-10	2.6	1.57	0.19	1.56	KBS Tuff	Area 131, K. Fora
Analytical data and calculated ages on glass from the Ninikaa Tuff (from Cerling et al. 1985)								
ANU83- 43-1	3.489	2.038E-11	32.4	3.36	0.06	3.36	Ninikaa Tuff	Area 116, K. Fora
Analytical data and calculated ages on glass from the Allia Tuff (from Cerling et al. 1985)								
ANU83- 34-1	2.881	2.065E-11	2.5	4.13	0.45	4.13	Allia Tuff	Area 204, K. Fora
	3.001	2.15E-11	31	4.12	0.10	4.12	Allia Tuff	Area 204, K. Fora
Analytical data and calculated ages on glass from tuffs in the Tulu Bor Mb. of the Koobi Fora Fm. (from Fitch and Miller 1976)								
FM7036	2.955	2.33E-11	36	4.40	0.22	4.54	'Kubi Algi'	K. Fora region
FM7036	2.955	2.45E-11	35.8	4.63	0.23	4.78	'Kubi Algi'	K. Fora region
FM7036	2.955	2.03E-11	9.5	3.80	0.4	3.96	'Kubi Algi'	K. Fora region
Analytical data and calculated ages on glass from Tuff B-alpha (= Tulu Bor-a) (from Brown et al. 1970, and Brown and Nash 1976)								
LKA-25	2.830	1.309E-11	8.4	2.64	0.92	2.67	Tuff B-a	Shungura,Usno
I1027	2.640	1.443E-11	10.2	2.97	0.3	3.15	Tuff B-a	Shungura,Usno
C. Ages measured on whole rock basalt samples								
Analytical data and ages of basaltic rocks on Mt. Kulal (Unpublished data of McDougall and Brown)								
ANU83- 20-1	1.397	5.006E-12	65.5	2.04	0.02	2.06	Kulal Basalt	Gatab Rd,Kulal
ANU83- 21-1	0.859	3.571E-12	34	2.40	0.03	2.40	Kulal Basalt	Gatab Rd,Kulal
ANU83- 22	1.357	4.954E-12	52.9	2.10	0.02	2.10	Kulal Basalt	Gatab Rd,Kulal
Analytical data and ages of the Mursi basalt (from Brown 1969, Fitch and Miller 1976, and Brown et al. 1985)								
KA-2094	0.828	5.967E-12	24.5	4.05	0.2	4.15	Mursi Basalt	Mursi, Omo
FMA 74	0.8384	6.564E-12	7.0	4.4	0.3	4.5	Mursi Basalt	Mursi, Omo
FMA 74	0.8384	6.180E-12	7.8	4.1	0.3	4.2	Mursi Basalt	Mursi, Omo
ANU82- 24	0.926, 0.916	6.39E-12	54.0	3.99	0.04	3.97	Mursi Fm.	Nkalabong, Omo
Analytical data and ages of the Kanapoi basalt (from Fitch and Miller 1976)								
FM7004	0.7678	4.21E-12	17.5	3.06	0.21	3.16	Kanapoi Bas.	Kanapoi, Tk.
FM7004	0.7678	4.19E-12	19.0	3.05	0.21	3.15	Kanapoi Bas.	Kanapoi, Tk.
Analytical data and ages of basalt in the Chen Alia Fm. (from McDougall and Watkins 1988)								
ANU81-170	0.994	6.503E-12	33.9	3.77	0.05	3.77	Chen Alia	Area 129, K. Fora
ANU81-172	0.995	6.441E-12	27.8	3.73	0.05	3.73	Chen Alia	Area 129, K. Fora
Analytical data and ages of Lowasera basalt (unpublished data of McDougall and Brown)								
ANU87- 35	0.6562	4.466E-12	24.7	3.93	0.07	3.92	Lowasera Bas.	Lowasera
Analytical data and ages of the Usno basalt (from Brown et al. 1970, 1985)								
LKA- 2	0.7074	3.918E-12	64.6	3.11	0.15	3.19	Usno Basalt	Usno, Omo
LKA-20	0.7074	4.419E-12	9.3	3.51	0.7	3.60	Usno Basalt	Usno, Omo
ANU83- 23-1	0.691, 0.698	4.92E-12	24.8	4.11	0.06	4.10	Usno Basalt	Usno, Omo
ANU83- 23-2	0.691, 0.698	4.89E-12	23.6	4.08	0.06	4.08	Usno Basalt	Usno, Omo
Analytical data and ages of the Kataboi basalt (from Brown et al. 1985)								
ANU83- 19-1	0.679, 0.684	4.83E-12	26.8	4.08	0.05	4.10	Kataboi Bas.	Kataboi, Tk.
ANU83- 19-2	0.679, 0.684	4.78E-12	23.2	4.04	0.06	4.05	Kataboi Bas.	Kataboi, Tk.
Analytical data and ages of the Karsa basalt (from McDougall 1985)								
ANU78-1023	0.829, 0.826	6.25E-12	43.8	4.35	0.05	4.34	Karsa Basalt	Karsa, K. Fora
ANU78-1025	0.804, 0.803	5.54E-12	29.2	3.97	0.05	3.97	Karsa Basalt	Karsa, K. Fora
Analytical data and ages of basalts in the Koobi Fora region (from Fitch and Miller 1976)								
F733	0.847	3.30E-12	3.9	2.17	0.44	2.24	Basalt	Buluk Gap
F733	0.847	3.36E-12	4.0	2.22	0.44	2.29	Basalt	Buluk Gap
FM7052	0.814	5.31E-12	18.9	3.64	0.36	3.76	Kokoi Basalt	Kokoi, K. Fora
FM7052	0.814	5.22E-12	20.2	3.59	0.36	3.69	Kokoi Basalt	Kokoi, K. Fora
F7053	0.9546	6.47E-12	25.8	3.78	0.38	3.90	Basalt	K. Fora region
FM 7053	0.9546	6.51E-12	26.9	3.82	0.38	3.93	Basalt	K. Fora region
ER58W	0.7637	4.86E-12	13.5	3.5	0.4	3.7	Basalt	ENE of Derati
ER58W	0.7637	5.35E-12	14.3	3.9	0.4	4.0	Basalt	ENE of Derati
ER58W	0.7637	5.53E-12	15.3	4.0	0.4	4.2	Basalt	ENE of Derati
Analytical data and ages of Gombe Group basalts east of L. Turkana (from McDougall and Watkins 1988, and Wilkinson 1988)								
ANU81-187	0.816	5.956E-12	46.4	4.21	0.05	4.20	Gombe Group	Il Ingumwai
ANU81-189	0.764	5.984E-12	34.5	4.50	0.05	4.51	Gombe Group	Il Ingumwai
W641	1.1	3.57E-12	25.3	2.07	0.23	1.87	Gombe Group	Furaful
W641	1.1	3.57E-12	25.0	2.07	0.23	1.87	Gombe Group	Furaful
W641	1.1	4.50E-12	22.7	2.07	0.23	2.36	Gombe Group	Furaful
W688	0.66	3.12E-12	22.0	3.10	0.3	2.72	Gombe Group	S. Kokurfo Bolol
W688	0.66	3.57E-12	22.9	3.10	0.3	3.12	Gombe Group	S. Kokurfo Bolol
W699	0.68	2.23E-12	17.1	2.25	0.33	1.89	Gombe Group	Kurfufo Bolol
W699	0.69	2.68E-12	18.1	2.25	0.33	2.24	Gombe Group	Kurfufo Bolol

ABLE 15-3 Analytical data and $^{40}Ar/^{39}Ar$ ages on materials from the Koobi Fora Formation.

ABORATORY UMBER	40Ar* (%)	TOTAL FUSION AGE(MYA)	RECALC. AGE (MYA)	ERROR (MYA)	INC. TOTAL FUSION AGE (MYA)	ERROR (MYA)	PLATEAU AGE	ERROR (MYA)	ISOCHRON AGE (MYA)	ERROR (MYA)	40/36 (I)	ERROR ± 1 S.D.	UNIT NAME	LOCALITY
. 40Ar/39Ar ages measured on anorthoclase														
nalytical data and age on the Silbo Tuff (from McDougall 1985)														
NU81-152		**0.73**	0.73	0.02	0.72	0.02	0.72	0.02	0.71	0.01	298.9	4.9	Silbo Tuff	Area 128, K. Fora
nalytical data and ages on Chari Tuff (= L) (from Fitch and Miller 1976)														
M7041A	8.3	1.20	**1.23**	0.23	n.r.	n.r.	n.r.	n.r.	n.r.	n.r.	n.r.	n.r.	Chari Tuff	K. Fora region
M7041A	7.2	1.20	**1.23**	0.23	n.r.	n.r.	n.r.	n.r.	n.r.	n.r.	n.r.	n.r.	Chari Tuff	K. Fora region
MA202	9.6	0.92	**0.94**	0.16	n.r.	n.r.	n.r.	n.r.	n.r.	n.r.	n.r.	n.r.	Chari Tuff	K. Fora region
MA202	9.7	0.92	**0.94**	0.16	n.r.	n.r.	n.r.	n.r.	n.r.	n.r.	n.r.	n.r.	Chari Tuff	K. Fora region
MA213	2.2	0.61	**0.63**	0.30	n.r.	n.r.	n.r.	n.r.	n.r.	n.r.	n.r.	n.r.	Chari Tuff	K. Fora region
MA213	5.0	0.61	**0.63**	0.30	n.r.	n.r.	n.r.	n.r.	n.r.	n.r.	n.r.	n.r.	Chari Tuff	K. Fora region
MA214	6.4	0.90	**0.92**	0.32	n.r.	n.r.	n.r.	n.r.	n.r.	n.r.	n.r.	n.r.	Chari Tuff	K. Fora region
MA214	7.1	1.00	**1.03**	0.32	n.r.	n.r.	n.r.	n.r.	n.r.	n.r.	n.r.	n.r.	Chari Tuff	K. Fora region
MA215	11.7	0.89	**0.91**	0.15	n.r.	n.r.	n.r.	n.r.	n.r.	n.r.	n.r.	n.r.	Chari Tuff	K. Fora region
MA215	11.8	1.00	**1.03**	0.17	n.r.	n.r.	n.r.	n.r.	n.r.	n.r.	n.r.	n.r.	Chari Tuff	K. Fora region
MA216	23.6	1.33	**1.37**	0.11	n.r.	n.r.	n.r.	n.r.	n.r.	n.r.	n.r.	n.r.	Chari Tuff	K. Fora region
MA216	23.7	1.34	**1.38**	0.11	n.r.	n.r.	n.r.	n.r.	n.r.	n.r.	n.r.	n.r.	Chari Tuff	K. Fora region
MA219	5.1	0.80	**0.82**	0.30	n.r.	n.r.	n.r.	n.r.	n.r.	n.r.	n.r.	n.r.	Chari Tuff	K. Fora region
MA280	n.r.	1.22	**1.25**	0.01	n.r.	n.r.	n.r.	n.r.	n.r.	n.r.	n.r.	n.r.	Chari Tuff	K. Fora region
MA290	15.2	1.39	**1.43**	0.11	n.r.	n.r.	n.r.	n.r.	n.r.	n.r.	n.r.	n.r.	Chari Tuff	K. Fora region
nalytical data and ages on the Chari Tuff (= L) (from McDougall 1985)														
NU78-1060	n.r.	**1.41**	1.41	0.01	1.38	0.03	1.38	0.03	1.36	0.02	296.6	5.8	Chari Tuff	Area 1, K. Fora
NU78-1064	n.r.	**1.39**	1.39	0.01	1.40	0.04	1.40	0.04	1.39	0.02	299.8	3	Chari Tuff	Area 1, K. Fora
NU79- 8	n.r.	**1.39**	1.39	0.06	1.40	0.06	1.40	0.06	1.36	0.02	305.3	4	Chari Tuff	Area 131, K. Fora
NU81- 144	n.r.	n.r.	—	—	1.37	0.03	1.37	0.03	1.39	0.03	295.0	5.9	Chari Tuff	Area 131, K. Fora
nalytical data and ages on the Okote Tuff Complex (from Fitch and Miller 1976)														
M6955	17.4	4.32	**4.44**	0.21	n.r.	n.r.	n.r.	n.r.	n.r.	n.r.	n.r.	n.r.	K. Fora Tuff	Area 103, K. Fora
M7042A	8.7	1.69	**1.74**	0.39	n.r.	n.r.	n.r.	n.r.	n.r.	n.r.	n.r.	n.r.	K. Fora Tuff	Area 103, K. Fora
M7042B	2.2	0.53	**0.54**	0.30	n.r.	n.r.	n.r.	n.r.	n.r.	n.r.	n.r.	n.r.	K. Fora Tuff	Area 103, K. Fora
M7042B	3.8	0.53	**0.54**	0.28	n.r.	n.r.	n.r.	n.r.	n.r.	n.r.	n.r.	n.r.	K. Fora Tuff	Area 103, K. Fora
M7042C	12.7	4.40	**4.52**	1.30	n.r.	n.r.	n.r.	n.r.	n.r.	n.r.	n.r.	n.r.	K. Fora Tuff	Area 103, K. Fora
MA208	11.9	1.27	**1.30**	0.20	n.r.	n.r.	n.r.	n.r.	n.r.	n.r.	n.r.	n.r.	BBS Cmplx.	K. Fora region
MA209	16.2	1.19	**1.22**	0.12	n.r.	n.r.	n.r.	n.r.	n.r.	n.r.	n.r.	n.r.	BBS Cmplx.	K. Fora region
MA210	14.1	1.4	**1.44**	0.17	n.r.	n.r.	n.r.	n.r.	n.r.	n.r.	n.r.	n.r.	BBS Cmplx.	K. Fora region
MA223	20.1	1.12	**1.15**	0.09	n.r.	n.r.	n.r.	n.r.	n.r.	n.r.	n.r.	n.r.	BBS Cmplx.	K. Fora region
MA224	12.7	1.24	**1.27**	0.17	n.r.	n.r.	n.r.	n.r.	n.r.	n.r.	n.r.	n.r.	BBS Cmplx.	K. Fora region
MA228	7.4	0.87	**0.89**	0.22	n.r.	n.r.	n.r.	n.r.	n.r.	n.r.	n.r.	n.r.	BBS Cmplx.	K. Fora region
MA266	17.4	1.66	**1.70**	0.01	n.r.	n.r.	n.r.	n.r.	n.r.	n.r.	n.r.	n.r.	BBS Cmplx.	K. Fora region
MA278	9.28	1.48	**1.52**	0.17	n.r.	n.r.	n.r.	n.r.	n.r.	n.r.	n.r.	n.r.	BBS Cmplx.	K. Fora region
nalytical data and ages on the Malbe Tuff (= H-4) (from McDougall 1985)														
NU79-17	n.r.	**1.85**	1.85	0.02	1.84	0.03	1.84	0.03	1.86	0.02	291.7	5.9	Malbe Tuff	Area 112, K. Fora
NU81-196	n.r.	**1.85**	1.85	0.01	1.84	0.02	1.84	0.02	1.86	0.03	290.7	2.1	Malbe Tuff	Area 105, K. Fora
nalytical data and ages on the KBS Tuff (= H-2) (from Fitch and Miller 1970)														
eakey IB1	2.1	3.45	**3.54**	1.2	n.r.	n.r.	n.r.	n.r.	n.r.	n.r.	n.r.	n.r.	KBS Tuff	Area 105, K. Fora
eakey 1B2	5.1	2.64	**2.71**	0.29	n.r.	n.r.	2.61	0.26	n.r.	n.r.	n.r.	n.r.	KBS Tuff	Area 105, K. Fora
nalytical data and ages on the KBS Tuff (= H-2) (from Fitch, Hooker and Miller 1976)														
eakey IB1	n.r.	—	**—**	—	n.r.	n.r.	2.42	0.01	n.r.	n.r.	n.r.	n.r.	KBS Tuff	Area 105, K. Fora
MA517	n.r.	—	—	—	n.r.	n.r.	n.r.	n.r.	1.91	0.03	285.0	n.r.	KBS Tuff	Area 131, K. Fora
nalytical data and ages on the KBS Tuff (= H-2) (from McDougall 1981)														
NU78-1038	n.r.	**1.89**	1.89	0.03	1.890	0.026	1.890	0.026	1.860	0.01	306.1	6	KBS Tuff	Area 105, K. Fora
NU78-1038(2)	n.r.	**1.88**	1.88	0.03	1.880	0.025	1.880	0.025	1.853	0.01	303.1	2.9	KBS Tuff	Area 105, K. Fora
NU78-1038(3)	74.6	**1.900**	1.900	0.014	n.r.	n.r.	n.r.	n.r.	n.r.	n.r.	n.r.	n.r.	KBS Tuff	Area 105, K. Fora
NU78-1047	n.r.	**1.881**	1.881	0.011	1.879	0.027	1.879	0.027	1.883	0.01	296.0	2	KBS Tuff	Area 131, K. Fora
NU79-14	n.r.	**1.879**	1.879	0.023	1.925	0.037	1.925	0.037	1.859	0.02	304.1	5	KBS Tuff	Area 131, K. Fora
nalytical data and ages on the KBS Tuff (= H-2) (from Fitch and Miller 1976)														
MA203	3.2	0.52	**0.53**	0.33	n.r.	n.r.	n.r.	n.r.	n.r.	n.r.	n.r.	n.r.	KBS Tuff	K. Fora region
MA221	8.1	0.68	**0.70**	0.17	n.r.	n.r.	n.r.	n.r.	n.r.	n.r.	n.r.	n.r.	KBS Tuff	K. Fora region
MA201	4.5	0.91	**0.93**	0.54	n.r.	n.r.	n.r.	n.r.	n.r.	n.r.	n.r.	n.r.	KBS Tuff	Area 130, K. Fora
MA201	5.7	0.91	**0.93**	0.53	n.r.	n.r.	n.r.	n.r.	n.r.	n.r.	n.r.	n.r.	KBS Tuff	Area 130, K. Fora
MA220	7.4	1.06	**1.09**	0.29	n.r.	n.r.	n.r.	n.r.	n.r.	n.r.	n.r.	n.r.	KBS Tuff	K. Fora region
MA225	n.r.	1.07	**1.10**	n.r.	n.r.	n.r.	n.r.	n.r.	n.r.	n.r.	n.r.	n.r.	KBS Tuff	Area 130, K. Fora
MA226	12.8	1.36	**1.40**	0.14	n.r.	n.r.	n.r.	n.r.	n.r.	n.r.	n.r.	n.r.	KBS Tuff	K. Fora region
MA226	20.4	1.54	**1.58**	0.12	n.r.	n.r.	n.r.	n.r.	n.r.	n.r.	n.r.	n.r.	KBS Tuff	K. Fora region
MA203	16.5	1.56	**1.60**	0.19	n.r.	n.r.	n.r.	n.r.	n.r.	n.r.	n.r.	n.r.	KBS Tuff	K. Fora region
MA206	n.r.	1.75	**1.80**	n.r.	n.r.	n.r.	n.r.	n.r.	n.r.	n.r.	n.r.	n.r.	KBS Tuff	Area 130, K. Fora
MA227	n.r.	1.75	**1.80**	n.r.	n.r.	n.r.	n.r.	n.r.	n.r.	n.r.	n.r.	n.r.	KBS Tuff	Area 130, K. Fora
MA211	14.5	1.83	**1.88**	0.19	n.r.	n.r.	n.r.	n.r.	n.r.	n.r.	n.r.	n.r.	KBS Tuff	K. Fora region
MA218	17.4	2.06	**2.12**	0.19	n.r.	n.r.	n.r.	n.r.	n.r.	n.r.	n.r.	n.r.	KBS Tuff	K. Fora region
MA207	17.8	2.10	**2.16**	0.19	n.r.	n.r.	n.r.	n.r.	n.r.	n.r.	n.r.	n.r.	KBS Tuff	K. Fora region
MA294	20.9	2.12	**2.18**	0.10	n.r.	n.r.	n.r.	n.r.	n.r.	n.r.	n.r.	n.r.	KBS Tuff	Area 105, K. Fora
eakey I(B2)	3.8	2.50	**2.57**	0.50	n.r.	n.r.	n.r.	n.r.	n.r.	n.r.	n.r.	n.r.	KBS Tuff	Area 105, K. Fora
eakey I(B2)	5.1	2.64	**2.71**	0.29	n.r.	n.r.	n.r.	n.r.	n.r.	n.r.	n.r.	n.r.	KBS Tuff	Area 105, K. Fora
MA274	n.r.	2.54	**2.61**	0.23	n.r.	n.r.	n.r.	n.r.	n.r.	n.r.	n.r.	n.r.	KBS Tuff	Area 10, K. Fora
nalytical data and ages on the Ninikaa Tuff (from Fitch and Miller 1976)														
MA255	37.8	3.37	**3.46**	0.07	n.r.	n.r.	n.r.	n.r.	n.r.	n.r.	n.r.	n.r.	Ninikaa Tuff	Area 116, K. Fora

TABLE 15-3 Continued

LABORATORY NUMBER	40Ar* (%)	TOTAL FUSION AGE(MYA)	RECALC. AGE (MYA)	ERROR (MYA)	INC. TOTAL FUSION AGE (MYA)	ERROR (MYA)	PLATEAU AGE	ERROR (MYA)	ISOCHRON AGE (MYA)	ERROR (MYA)	40/36 (I)	ERROR ± 1 S.D.	UNIT NAME	LOCALITY
Analytical data and ages on the Ninikaa Tuff (from McDougall 1985)														
ANU81-123	n.r.	**3.00**	—	0.02	3.02	0.03	3.02	0.03	3.01	0.02	296.4	4.2	Ninikaa Tuff	Area 116, K. Fora
Analytical data and ages on the Toroto Tuff (from McDougall 1985)														
ANU78-1073B	n.r.	**3.32**	—	0.04	3.33	0.06	3.33	0.06	3.30	0.02	309.0	7.4	Toroto Tuff	Area 204, K. Fora
ANU78-1075A	n.r.	**3.35**	—	0.04	3.30	0.06	3.30	0.06	3.32	0.02	293.3	1.1	Toroto Tuff	Area 204, K. Fora
Analytical data and ages on the Toroto Tuff (from Fitch and Miller 1976)														
FM7035	17.1	3.08	**3.16**	0.19	n.r.	n.r.	n.r.	n.r.	n.r.	n.r.	n.r.	n.r.	Toroto Tuff	K. Fora region
FMA246	35.9	3.71	**3.81**	0.09	n.r.	n.r.	n.r.	n.r.	n.r.	n.r.	n.r.	n.r.	Toroto Tuff	K. Fora region
FM7035	29.0	3.93	**4.03**	0.20	n.r.	n.r.	n.r.	n.r.	n.r.	n.r.	n.r.	n.r.	Toroto Tuff	K. Fora region
Analytical data and ages on anorthoclase from a tuff in the Nkalabong Formation (from Fitch and Miller 1969)														
Nkalabong	32.8	3.90	**4.00**	0.10	n.r.	n.r.	n.r.	n.r.	n.r.	n.r.	n.r.	n.r.	Nkalabong	Mursi, Omo
Nkalabong	32.9	3.99	**4.10**	0.12	n.r.	n.r.	n.r.	n.r.	n.r.	n.r.	n.r.	n.r.	Nkalabong	Mursi, Omo
B. 40Ar/39Ar ages measured on materials other than alkali feldspar														
Analytical data and ages on altered tuffs or unspecified materials from the Okote Tuff Complex (from Fitch and Miller 1976)														
FMA235	8.42	6.47	**27.18**	2.86	n.r.	n.r.	n.r.	n.r.	n.r.	n.r.	n.r.	n.r.	BBS Cmplx	K. Fora region
FMA237	10.8	29.90	**30.70**	2.09	n.r.	n.r.	n.r.	n.r.	n.r.	n.r.	n.r.	n.r.	BBS Cmplx.	K. Fora region
FMA236	12.6	30.38	**31.19**	1.70	n.r.	n.r.	n.r.	n.r.	n.r.	n.r.	n.r.	n.r.	BBS Cmplx.	K. Fora region
FM6954	7.4	15.60	**16.02**	1.56	n.r.	n.r.	n.r.	n.r.	n.r.	n.r.	n.r.	n.r.	K. Fora Tuff	K. Fora region
FM 222	5.4	0.95	**0.98**	0.30	n.r.	n.r.	n.r.	n.r.	n.r.	n.r.	n.r.	n.r.	K. Fora Tuff	Area 102, K. Fora
FM 205	17.4	1.38	**1.42**	0.17	n.r.	n.r.	n.r.	n.r.	n.r.	n.r.	n.r.	n.r.	K. Fora Tuff	Area 102, K. Fora
FMA270	n.r.	1.57	**1.61**	0.00	n.r.	n.r.	n.r.	n.r.	n.r.	n.r.	n.r.	n.r.	K. Fora Tuff	Area 102, K. Fora
Analytical data and ages on whole rock pumices from the KBS Tuff (from Fitch, Hooker and Miller 1976)														
Leakey I/B1	n.r.	3.12	**3.20**	1.1	n.r.	n.r.	n.r.	n.r.	n.r.	n.r.	n.r.	n.r.		
Leakey I/B1	n.r.	2.26	**2.32**	0.5	n.r.	n.r.	n.r.	n.r.	n.r.	n.r.	n.r.	n.r.		
Analytical data and ages on clay from an altered tuff (possibly the Ninikaa Tuff) (from Fitch and Miller 1976)														
FMA233	0	27.2	**27.9**	7.7	n.r.	n.r.	n.r.	n.r.	n.r.	n.r.	n.r.	n.r.	Tulu Bor Tuff	K. Fora region
Analytical data and ages on biotite (191 MYA), and glass from tuffs of the Tulu Bor Member, Koobi Fora Formation (from Fitch and Miller 1976)														
FM7036	54.4	186.0	**191**	5	n.r.	n.r.	n.r.	n.r.	n.r.	n.r.	n.r.	n.r.	'Kubi Algi'	K. Fora region
FM7036	46.4	186.0	**191**	5	n.r.	n.r.	n.r.	n.r.	n.r.	n.r.	n.r.	n.r.	'Kubi Algi'	K. Fora region
FM7036	12.7	4.81	**4.94**	0.36	n.r.	n.r.	n.r.	n.r.	n.r.	n.r.	n.r.	n.r.	'Kubi Algi'	K. Fora region
FM7036	16.6	4.95	**5.08**	0.49	n.r.	n.r.	n.r.	n.r.	n.r.	n.r.	n.r.	n.r.	'Kubi Algi'	K. Fora region
Analytical data and ages on pumice from the Toroto Tuff (from Fitch and Miller 1976)														
FM7035	76.2	283	**291**	9	n.r.	n.r.	n.r.	n.r.	n.r.	n.r.	n.r.	n.r.	'Kubi Algi'	K. Fora region
Analytical data and ages on the Kanapoi Basalt (from Fitch and Miller 1976)														
FM7004	n.r.	4.0	**4.11**	1.0	n.r.	n.r.	n.r.	n.r.	n.r.	n.r.	n.r.	n.r.	Kanapoi Bs.	Kanapoi, Turkana

n.r. = not reported

Table 15-4. Analytical data and ages determined by the fission track method.

SAMPLE	LOCATION[a]	NO. GRAINS	R	ρ_s X 10^4	NO. OF TRACKS	ρ_i X 10^6	NO. OF TRACKS	N FLUX X 10^{15}	U (PPM)	AGE AND ERROR (MYA)	MINERAL DATED
Analytical data and ages on the KBS Tuff (from Hurford et al. 1976)											
FMA 517	131	10	—	7.05	47	4.28	1428	2.52	—	2.55 ± 0.38	Zircon
FMA 517	131	10	—	7.87	56	3.18	1129	1.59	—	2.42 ± 0.32	Zircon
FMA 517	131	6	—	7.09	36	3.32	802	1.81	—	2.37 ± 0.41	Zircon
FMA 517	131	12	—	9.55	22	2.70	311	1.11	—	2.41 ± 0.53	Zircon
ER74/131	131	8	—	13.4	21	5.26	411	1.55	—	2.43 ± 0.55	Zircon
Analytical data and ages on the KBS Tuff (from Gleadow 1980)											
FMA 559	131	14	0.666	8.78	68	3.64	1411	1.31	101	1.93 ± 0.21	Zircon
FMA 559	131	19	0.944	9.22	82	4.30	1910	1.31	119	1.71 ± 0.16	Zircon
FMA 559	131	26	0.847	6.77	115	3.00	2583	1.37	80	1.86 ± 0.15	Zircon
FMA 559	131	22	0.928	7.59	193	5.19	6597	2.05	92	1.83 ± 0.12	Zircon
FMA 560	131	12	0.934	13.10	156	5.36	3196	1.31	149	1.95 ± 0.14	Zircon
FMA 560	131	20	0.937	11.40	143	4.86	3046	1.31	135	1.87 ± 0.14	Zircon
FMA 560	131	15	0.892	8.58	91	5.73	3039	2.00	104	1.83 ± 0.17	Zircon
FMA 560	131	24	0.973	8.62	177	6.84	7019	2.00	125	1.54 ± 0.10	Zircon
FMA 517	131	17	0.822	6.26	137	2.96	3241	1.42	76	1.83 ± 0.14	Zircon
FMA 517	131	20	0.850	6.29	139	3.00	3313	1.42	77	1.82 ± 0.13	Zircon
7722-107	131	15	0.943	13.50	199	8.40	6199	1.97	155	1.93 ± 0.12	Zircon
7722-107	131	15	0.936	12.50	173	8.24	5688	1.97	152	1.82 ± 0.12	Zircon
7722-108	105	15	0.794	9.48	121	4.52	2886	1.44	114	1.84 ± 0.15	Zircon
7722-108	105	27	0.935	9.87	277	4.49	6301	1.44	114	1.93 ± 0.10	Zircon
7722-108	105	1	—	21.00	18	2.15	92	4.98	17	29.70 ± 8.00	Apatite
7722-108	105	1	—	11.30	4	1.35	24	4.98	11	25.40 ± 14.00	Apatite
7722-108	105	1	—	52.70	34	1.95	63	1.44	49	23.70 ± 5.00	Zircon
7722-108	105	1	—	46.50	24	1.51	39	1.44	38	27.00 ± 7.00	Zircon
7722-108	105	1	—	133.00	180	4.70	318	1.44	119	24.80 ± 2.00	Zircon
7722-108	105	10	0.965	76.30	504	2.78	918	4.98	22	83.00 ± 3.00	Apatite
7722-108	105	3	0.823	113.00	399	1.73	918	1.44	44	549.00 ± 3.00	Zircon
7722-108	105	5	0.993	742.00	813	5.58	611	6.47	34	504.00 ± 10.00	Sphene
7722-109B	112	6	0.887	1950.00	1078	4.14	407	1.94	78	534.00 ± 17.00	Zircon
7722-109C	112	3	0.999	2960.00	598	3.07	297	1.92	58	456.00 ± 12.00	Zircon
7722-109C	112	5	0.969	804.00	830	5.90	609	6.04	39	483.00 ± 11.00	Sphene
7722-107	131	3	0.917	1050.00	711	2.63	237	1.97	49	462.00 ± 22.00	Zircon
Analytical data and ages on African "basement rocks" (from Gleadow 1980)											
100	Isiolo	5	0.940	1000.0	2901	6.77	982	5.52	49	479.00 ± 12.00	Sphene
101	Isiolo	6	0.932	24.3	365	1.85	1386	8.75	8	70.00 ± 2.00	Apatite
102	Archers	6	0.828	25.3	334	1.57	1035	8.69	7	85.00 ± 3.00	Apatite
103	Koroli	6	0.994	842.0	1324	1.21	95	1.26	35	514.00 ± 40.00	Zircon
103	Koroli	6	0.965	33.8	446	2.20	1456	8.63	10	80.00 ± 2.00	Apatite
104	Laisamis	7	0.937	32.8	533	2.75	2237	8.56	13	62.00 ± 2.00	Apatite
136	Voi	5	0.693	63.1	1133	3.27	1470	8.44	15	99.00 ± 3.00	Apatite
141	Embu?	5	0.950	2050.0	1621	11.20	1177	5.07	88	542.00 ± 12.00	Sphene
141	Embu?	8	0.945	104.0	484	4.09	952	8.32	19	127.00 ± 3.00	Apatite

a. Koobi Fora Area Number, or other location.

LITERATURE CITED

Baksi, A. K., Hsu, V., McWilliams, M. O., and Farrar, E. (1992) $^{40}Ar/^{39}Ar$ dating of the Brunhes Matuyama Geomagnetic Field Reversal. *Science* 256:356–357.

Behrensmeyer, A. K. (1978) Correlation of Plio-Pleistocene sequences in the northern Lake Turkana Basin: A summary of evidence and issues. In W. W. Bishop (ed.): *Geological Background to Fossil Man*. Edinburgh: Scottish Academic, pp. 421–440.

Bishop, W. W. (1965) Comment on Potassium-argon dating of late Cenozoic rocks in East Africa and Italy. *Curr. Anthropol.* 6:364.

Boschetto, H. B., Brown, F. H., and McDougall, I. (1992) Stratigraphy of the Lothidok Range, northern Kenya, and K/Ar ages of its Miocene primates. *J. Hum. Evol.* 22:47-71.

Brock, A., and Isaac, G. L. (1974) Paleomagnetic stratigraphy and chronology of hominid-bearing sediments east of Lake Rudolf, Kenya. *Nature* 247:344–348.

Brock, A., and Isaac, G. L. (1976) Reversal stratigraphy and its application at East Rudolf. In W. W. Bishop (ed.): *Geological Background to Fossil Man*. Edinburgh: Scottish Academic, pp. 148–162.

Brotzu, P., Morbidelli, L., Nicoletti, M., Piccirillo, E. M., and Traversa, G. (1984) Miocene to Quaternary volcanism in eastern Kenya: Sequence and geochronology. *Tectonophysics* 101:75–86.

Brown, F. H. (1969) Observations on the stratigraphy and radiometric age of the Omo Beds, lower Omo Basin, southern Ethiopia. *Quaternaria* 11:7–14.

Brown, F. H., de Heinzelin, J., and Howell, F. C. (1970) Pliocene/Pleistocene formations in the lower Omo basin, southern Ethiopia. *Quaternaria* 13:247–268.

Brown, F. H., and Cerling, T. E. (1982) Stratigraphic significance of the Tulu Bor Tuff of the Koobi Fora Formation. *Nature* 299:212–215.

Brown, F. H., and Feibel, C. S. (1985) Stratigraphical notes on the Okote Tuff Complex at Koobi Fora, Kenya. *Nature* 316:794–797.

Brown, F. H., and Feibel, C. S. (1986) Revision of lithostratigraphic nomenclature in the Koobi Fora region, Kˇnya. *J. Geol. Soc.* 43:297–310.

Brown, F. H., Howell, F. C., and Eck, G. G. (1978) Observations on problems of correlation of Cenozoic hominid-bearing formations in the North Lake Turkana Basin. In W. W. Bishop (ed.): *Geological Background to Fossil Man*. Edinburgh: Scottish Academic, pp. 473–498.

Brown, F. H., and Lajoie, K. R. (1970) K-Ar ages of the Omo group and fossil localities of the Shungura Formation, southwest Ethiopia. *Nature* 229:483–485.

Brown, F. H., McDougall, I., Davies, T., and Maier, R. (1985) An integrated Plio-Pleistocene chronology for the Turkana Basin. In E. Delson (ed.): *Ancestors: The Hard Evidence*. New York: Liss, pp. 82–90.

Brown, F. H., and Nash, W. P. (1976) Radiometric dating and tuff mineralogy of Omo Group deposits. In Y. Coppens, F. C. Howell, G. L. Isaac, and R. E. F. Leakey (eds.): *Earliest Man and Environments in the Lake Rudolf Basin*. Chicago: University of Chicago Press, pp. 50–63.

Brown, F. H., Sarna-Wojcicki, A. M., Haileab, B., and Meyer, C. E. (1992) New correlations of Pliocene and Quaternary tephra layers between hominid-fossil bearing strata in the Turkana basin of East Africa and the Gulf of Aden, and some paleoclimatic implications. *Quat. Int.* 13/14:55–57.

Brown, F. H., and Shuey, R. T. (1976) Magnetostratigraphy of the Shungura and Usno Formations, Lower Omo Valley, Ethiopia. In Y. Coppens, F. C. Howell, G. L. Isaac, and R. E. F. Leakey (eds.): *Earliest Man and Environments in the Lake Rudolf Basin*. Chicago: University of Chicago Press, pp. 64–78.

Brown, F. H., Shuey, R. T., and Croes, M. K. (1978) Magnetostratigraphy of the Shungura and Usno Formations, southwestern Ethiopia: New data and comprehensive reanalysis. *Geophys. J. R. Astronom. Soc.* 54:519–538.

Butzer, K. W. (1976) The Mursi, Nkalabong, and Kibish Formations, Lower Omo Basin, Ethiopia. In Y. Coppens, F. C. Howell, G. L. Isaac, and R. E. F. Leakey (eds.): *Earliest Man and Environments in the Lake Rudolf Basin*. Chicago: University of Chicago Press, pp. 12–23.

Cerling, T. E. (1977) Paleochemistry of Plio-Pleistocene Lake Turkana and Diagenesis of its sediments. Ph.D. Dissertation, University of California, Berkeley.

Cerling, T. E., and Brown, F. H. (1982) Tuffaceous marker horizons in the Koobi Fora region and the Lower Omo Valley. *Nature* 299:216–221.

Cerling, T. E., Cerling, B. W., Drake, R. E., and Brown, F. H. (1978) Correlation of reworked ash deposits; the KBS Tuff, northern Kenya. Short papers of the Fourth International Conference, Geochronology, Cosmochronology, Isotope Geology (1978), *U. S. Geol. Survey Open File Report* 78-701:61–63.

Cerling, T. E., Brown, F. H., Cerling, B. W., Curtis, G. H., and Drake, R. E. (1979) Preliminary correlations between the Koobi Fora and Shungura Formations, East Africa. *Nature* 279:118–121.

Chapman, G. R., and Brook, M. (1978) Chronostratigraphy of the Baringo Basin, Kenya. In W. W. Bishop (ed.): *Geological Background to Fossil Man*. Edinburgh: Scottish Academic, pp. 207–224.

Charsley, T. J. (1987) *Geology of the North Horr area*. Report 110 (Reconnaissance) Mines and Geology Department, Ministry of Environment and Natural Resources, Nairobi. 40 pp.

Cooke, H. B. S., and Maglio, V. J. (1972) Plio-Pleistocene stratigraphy in East Africa in relation to proboscidean and suid evolution. In W. W. Bishop (ed.): *Calibration of Hominoid Evolution*. Edinburgh: Scottish Academic, pp. 303–329.

Curtis, G. H., Drake, R. E., Cerling, T. E., Cerling, B. W., and Hampel, J. H. (1975) Age of KBS Tuff in Koobi Fora Formation, East Rudolf, Kenya. *Nature* 258:395–398.

Curtis, G. H., Drake, R. E., Cerling, T. E., Cerling, B. W., and Hampel, J. H. (1978) Age of KBS Tuff in Koobi Fora Formation, East Rudolf, Kenya. In W. W. Bishop (ed.):

Geological Background to Fossil Man. Edinburgh: Scottish Academic, pp. 463–469.

Dalrymple, G. B., and Lanphere, M. A. (1969) *Potassium-Argon Dating, Principles, Techniques, and Applications to Geochronology*. San Francisco: W. H. Freeman.

de Heinzelin, J. (1983) *The Omo Group*. Musée Royal de l'Afrique Centrale, Belgique, Ann., Séries 8°, Sci. Géologiques 85 (365 pp).

Davidson, A. (1983) *The Omo River project*. Ethiopian Inst. Geol. Surveys Bull. 2.

Drake, R. E., Curtis, G. H. (1987) K–Ar geochronology of the Laetoli fossil localities. In M. D. Leakey and J. M. Harris (eds): *Laetoli, a Pliocene Site in Northern Tanzania*. Oxford: Clarendon, pp. 48–52.

Drake, R. E., Curtis, G. H., Cerling, T. E., Cerling, B. W., and Hampel, J. H. (1980) KBS Tuff dating and geochronology of tuffaceous sediments in the Koobi Fora and Shungura Formations, East Africa. *Nature* 283:368–372.

Evernden, J. F., and Curtis, G. H. (1965) Potassium-argon dating of late Cenozoic rocks in East Africa and Italy. *Curr. Anthropol.* 6:343–364.

Feibel, C. S. (1988) Paleoenvironments of the Koobi Fora Formation, Turkana Basin, northern Kenya. Unpublished Ph.D. Dissertation, University of Utah, Salt Lake City, Utah, 330 pp.

Feibel, C. S., Brown, F. H., and McDougall, I. (1989) Stratigraphic context of fossil hominids from the Omo Group deposits, northern Turkana basin, Kenya and Ethiopia. *Am. J. Phys. Anthropol.* 78:595–622.

Feibel, C. S., and Brown, F. H. (1993) Microstratigraphy and paleoenvironments. In A. Walker and R. E. F. Leakey (eds.): *Homo Erectus at Nariokotome*. Cambridge: Harvard University Press.

Ferguson, A. K., and Gleadow, A. J. W. (1980) Mineralogical characterization of some tuffs from the East Turkana Basin, Kenya, geodynamic evolution of the Afro-Arabian Rift System. *Atti dei Convegni Lincei*, 47, Accad. Naz. Lincei, Roma, pp. 165–173.

Findlater, I. C. (1976) Tuffs and the recognition of isochronous mapping units in the Rudolf Succession. In Y. Coppens, F. C. Howell, G. L. Isaac, and R. E. F. Leakey (eds.): *Earliest Man and Environments in the Lake Rudolf Basin*. Chicago: University of Chicago Press, pp. 94–104.

Findlater, I. C. (1978) Stratigraphy. In M. G. Leakey and R. E. F. Leakey (eds.): *Koobi Fora Research Project*, Vol. 1: *The Fossil Hominids and an Introduction to Their Context 1968-1974*. Oxford: Clarendon, pp. 14–31.

Fitch, F. J., Hooker, P. J., and Miller, J. A. (1976) Single whole rock K–Ar isochrons. *Geol. Mag.* 113:1–10.

Fitch, F. J., and Miller, J. A. (1969) Age determinations on feldspar from the Lower Omo Basin. *Nature* 222:1143.

Fitch, F. J., and Miller, J. A. (1970) Radioisotope age determinations of Lake Rudolf artefact site. *Nature* 226:226–228.

Fitch, F. J., and Miller, J. A. (1976) Conventional potassium-argon and argon-40/argon-39 dating of volcanic rocks from East Rudolf. In Y. Coppens, F. C. Howell, G. L. Isaac, and R. E. F. Leakey (eds.): *Earliest Man and Environments in the Lake Rudolf Basin*. Chicago: University of Chicago Press, pp. 123–147.

Fitch, F. J., Hooker, P. J., and Miller, J. A. (1976) $^{40}Ar/^{39}Ar$ dating of the KBS Tuff in Koobi Fora Formation, East Rudolf, Kenya. *Nature* 263:740–744.

Fitch, F. J., Findlater, I. C., Watkins, R. T., and Miller, J. A. (1974) Dating of the rock succession containing fossil hominids at East Rudolf, Kenya. *Nature* 251:213–215.

Gleadow, A. J. W. (1980) Fission track age of the KBS Tuff and associated hominid remains in northern Kenya. *Nature* 284:225–230.

Haileab, B., and Brown, F. H. (1992) Turkana Basin–Middle Awash Valley correlations and the age of the Sagantole and Hadar Formations. *J. Hum. Evol.* 22:453–468.

Haileab, B., and Brown, F. H. (1994) Tephra correlations between Gadeb prehistoric site, Ethiopia, and the Lake Turkana Basin. *J. Hum. Evol.* 26: (in press).

Harris, J. M., Brown, F. H., and Leakey, M. G. (1988) Stratigraphy and Paleontology of Pliocene and Pleistocene localities west of Lake Turkana, Kenya. *Contrib. Sci. Nat. Hist. Mus. Los Angeles County* 399:1–128.

Hilgen, F. J. (1991a) Astronomical calibration of Gauss to Matuyama sapropels in the Mediterranean and implications for the Geomagnetic Polarity Time Scale. *Earth Plan. Sci. Lett.* 104:226–244.

Hilgen, F. J. (1991b) Extension of the astronomically calibrated (polarity) time scale to the Miocene/Pliocene boundary. *Earth Plan. Sci. Lett.* 107:349–36.

Hill, A., Ward, S., Deino, A., Curtis, G. H., and Drake, R. E. (1992) Earliest *Homo*. *Nature* 355:719–722.

Hillhouse, J. W., Cerling, T. E., and Brown, F. H. (1986) Magnetostratigraphy of the Koobi Fora Formation, Lake Turkana, Kenya. *J. Geophys. Res.* 91:11,581–11,595.

Hillhouse, J. W., Ndombi, J. W. M., Cox, A., and Brock, A. (1977) Additional results on palaeomagnetic stratigraphy of the Koobi Fora Formation, east of Lake Turkana (Lake Rudolf), Kenya. *Nature* 265:411–415.

Hurford, A. J. (1974) Fission track dating of a vitric tuff from East Rudolf, North Kenya. *Nature 249:*236–237.

Hurford, A. J., Gleadow, A. J. W., and Naeser, C. W. (1976) Fission-track dating of pumice from the KBS Tuff, East Rudolf, Kenya. *Nature* 263:738–740.

Jack, R. N., and Carmichael, I. S. E. (1968) The chemical "fingerprinting" of acid volcanic rocks. *California Div. Mines Geol. Special Report* 100:17–32.

Johnson, R. G. (1982) Brunhes–Matuyama magnetic reversal dated at 790,000 yr B.P. by marine-astronomical correlations. *Quat. Res.* 17:135–147.

Lewin, R. (1987) *Bones of Contention*. New York: Simon & Schuster.

Luedtke, N. A. (1975) A preliminary investigation into the use of an elemental analysis in the correlation of sediments from Lake Rudolf, Kenya. M. S. Thesis, University of Rhode Island.

Martz, A. M. (1979) Petrology and chemistry of tuffs from the Shungura Formation, southwest Ethiopia. M. S. Thesis, University of Utah.

Martz, A. M., and Brown, F. H. (1981) Chemistry and mineralogy of some Plio-Pleistocene Tuffs from the Shungura Formation, Southwest Ethiopia. *Quat. Res.* 16:240–257.

McDougall, I. (1979) The present status of the geomagnetic polarity time scale. In M. W. McElhinny (ed.): *The Earth: Its Origin, Structure, and Evolution.* New York: Academic, pp. 543–566.

McDougall, I. (1981) $^{40}Ar/^{39}Ar$ age spectra from the KBS Tuff, Koobi Fora Formation. *Nature* 294:120–124.

McDougall, I. (1985) K-Ar and $^{40}Ar/^{39}Ar$ dating of the hominoid-bearing Pliocene-Pleistocene sequence at Koobi Fora, Lake Turkana, northern Kenya. *Geol. Soc. Am. Bull.* 96:159–175.

McDougall, I., Brown, F. H., Cerling, T. E., and Hillhouse, J. W. (1992) A reappraisal of the Geomagnetic Time Scale to 4 MYA using data from the Turkana Basin, East Africa. *Geophys. Res. Lett.* 19:2349–2352.

McDougall, I., Davies, T., Maier, R., and Rudowski, R. (1985) Age of the Okote Tuff Complex at Koobi Fora, Kenya. *Nature* 316:792–794.

McDougall, I., Maier, R., Sutherland-Hawkes, P., and Gleadow, A. J. W. (1980) K-Ar age estimate for the KBS Tuff, East Turkana, Kenya. *Nature* 284:230–234.

McDougall, I., and Watkins, R. T. (1985) Age of hominoid-bearing sequence at Buluk, northern Kenya. *Nature* 318:175-178.

McDougall, I., and Watkins, R. T. (1988) Potassium-argon ages of volcanic rocks from northeast of Lake Turkana, northern Kenya. *Geol. Mag.* 125:15–23.

Merrihue, C. M., and Turner, G. (1966) Potassium-argon dating by activation with fast neutrons. *J. Geophy. Res.* 71:2852-2857.

Namwamba, F. (1992) Tephrostratigraphy of the Baringo Basin, Kenya. M. S. Thesis, University of Utah.

Pickford, M., Senut, B., Poupeau, G., Brown, F. H., and Haileab, B. (1991) Correlation of tephra layers from the Turkana Basin to the Western Rift Valley. *C. R. Acad. Sci. Paris* 313:223–229.

Rahman, A., and Roth, P. H. (1989) Late Neogene calcareous nannoplankton biostratigraphy of the Gulf of Aden region. *Marine Micropaleo.* 15:1–27.

Sarna-Wojcicki, A. M. (1976) Correlation of Late Cenozoic Tuffs in the Central Coast Ranges of California by Means of Trace- and Minor-element chemistry. *U.S. Geol. Survey Prof. Paper* 972.

Sarna-Wojcicki, A. M., Meyer, C. E., Roth, P. H., and Brown, F. H. (1985) Ages of tuff beds at East African early hominid sites and sediments in the Gulf of Aden. *Nature* 313:306–308.

Shackelton, N. J., Berger, A., and Peltier, W. R. (1990) An alternative astronomical calibration of the lower Pleistocene timescale based on ODP Site 677. *Trans. Roy. Soc. Edinburgh: Earth Sci.* 81:251–261.

Shuey, R. T., Brown, F. H., and Croes, M. K. (1974) Magnetostratigraphy of the Shungura Formation, southwestern Ethiopia: Fine structure of the lower Matuyama polarity epoch. *Earth Plan. Sci. Lett.* 23:249–260.

Spell, T. L., and McDougall, I. (1992) Revisions to the age of the Brunhes–Matuyama Boundary and the Pleistocene Geomagnetic Polarity Timescale. *Geophys. Res. Lett.* 19:1181–1184.

Steiger, R. H., and Jäger, E. (1977) Subcommission on geochronology: Convention on the use of decay constants in geo- and cosmochronology. *Earth Plan. Sci. Lett.* 36:359–362.

Tauxe, L., Deino, A. D., Behrensmeyer, A. K., and Potts, R. (1992) Pinning down the Brunhes/Matuyama and upper Jaramillo boundaries: A reconciliation of orbital and isotopic time scales. *Earth Plan. Sci. Lett.* 109:561–572.

Wagner, G. A. (1977) Fission-track dating of pumice from the KBS Tuff, East Rudolf, Kenya. *Nature* 267:649.

Walter, R. C., and Aronson, J. L. (1982) Revisions of K/Ar ages for the Hadar hominid site, Ethiopia. *Nature* 296:122–127.

Walter, R. C., Manega, P. C., Hay, R. L., Drake, R. E., and Curtis, G. H. (1991) Laser-fusion $^{40}Ar/^{39}Ar$ dating of Bed I, Olduvai Gorge, Tanzania. *Nature* 354:145–149.

Wilkinson, A. F. (1988) *Geology of the Allia Bay area.* Report 109 (Reconnaissance) Mines and Geology Dept., Ministry of Environment and Natural Resources, Nairobi.

Woldegabriel, G. (1987) Volcanotectonic history of the central sector of the main Ethiopian Rift: A geochronological, geochemical and petrological approach. Ph. D. Dissertation, Case Western Reserve University.

Woldegabriel, G., and Aronson, J. (1987) The Chow Bahir rift: A "failed" rift in southern Ethiopia. *Geology* 15:430-433.

CHAPTER 16

Rifting, a Long-Term African Story, with Considerations on Early Hominid Habitats

Jean de Heinzelin

ORIGIN AND STRUCTURE OF THE AFRICAN RIFTS

"Seismic data acquisition and analysis in the petroleum industry is a multibillion dollar activity" (Blackwell Scientific Publishing, introducing J. F. Claerbout's *"Earth Sounding Analysis,"* 1992).

This seismic data acquisition activity has added immensely to our knowledge of the structure of the African continent. Out of the total database, only the tip has emerged in formal publications, but now enough information is available to frame in true proportions the spasmodic rifting phenomena that crisscrossed the face of Africa in the course of the last 230 million years (MYA).

Data acquisition occurred simultaneously along a second line of research—teleanalytical, more simply teledetection by means of both aerial mosaics and satellite imagery.

Four main stages of rifting can be discriminated:

- Karroo and its complex aftermaths, from 230 to 190 MYA.
- Lower Cretaceous, from 140 to 100 MYA.
- Paleogene, around 60-50 MYA.
- Neogene, from 30 MYA to present.

KARROO

Faulting affected the southeast part of the continent as a result of the breakdown of the former Gondwana continent, the opening of the Indian Ocean, and the eastward shift of Madagascar (Lambiase 1989; Verniers et al. 1989). Some authors postulate three successive phases: Early Permian, Late Permian, and Early Jurassic. At some places, basin sedimentation was at the onset glacigenic (Dwyka), passing to fluvio-lacustrine with coal seams (Ecca, *Glossopteris* and *Dadoxylon* flora; and Beaufort, uppermost Permian). The infilling of the typical Metangula graben started at the beginning of the Triassic (230 MYA) and ended before the Middle Jurassic transgression. About 5 km of cyclical fluvial sequences were accumulated.

In Upper Karroo times, climate evolved from temperate to warmer and seasonal, perhaps semi-arid, as a function of the migration of the poles. The record of vertebrate fossils is principally an assembly of Permian reptiles.

LOWER CRETACEOUS

After long stability, rifting returned from the other side of the African continent, as a result of the breakdown of Gondwanaland to the west and the opening of the South Atlantic Ocean (see Figure 16-1).

FIGURE 16-1 Synthetic structural map of the main early Cretaceous active troughs and faults in central/northeastern Africa and northeastern Brazil. Pp: Older cratons and deposits; MC: Post-Paleozoic sedimentary basins; Ri: Rifting. 1 to 5: faults. 6: Maximum thickness of Cretaceous and Cretaceous + Cenozoic. 7: Cross sections (not documented here). 8: Cretaceous rifts on inset map (From Guiraud and Maurin 1991).

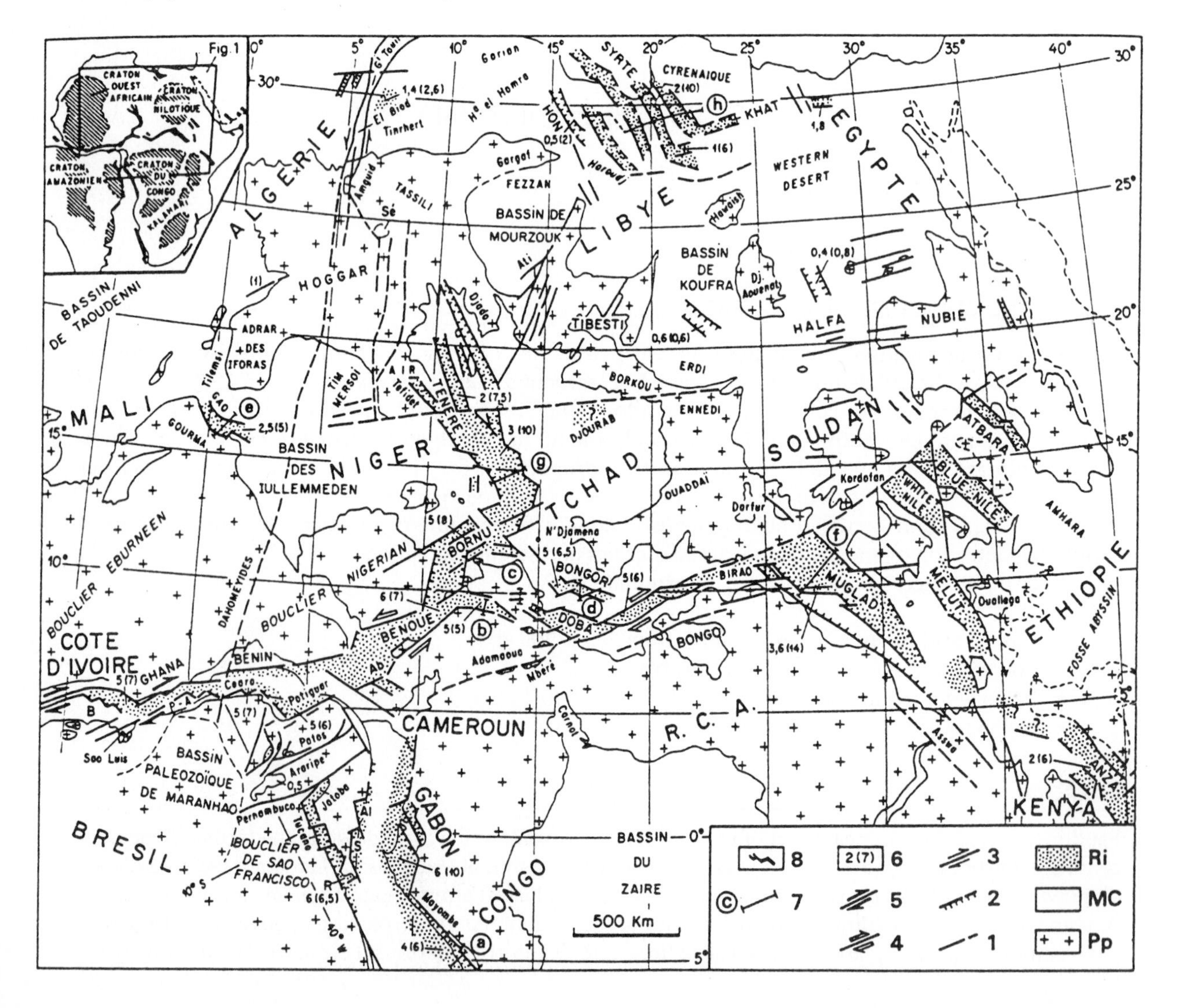

This time, the whole north equatorial interior was involved, from west to east, the Gulf of Benin margins (Ghana to Gabon and Congo), the Benue-Bornu branch, the Tenere branch and the broader array of the South-Sudanese troughs south of the Darfur-Kordofan lineament. The margin of the Tethys underwent very active faulting in the Gulf of Syrt and some lesser in the Maghreb. In the approaches to the east coast, the Anza rift became a sort of forerunner of the Turkana depression, now obliterated by later volcanism. The peculiar structure of the South Sudanese troughs deserves a comment, as they anticipate in style some later East African structures. The Muglady, Melut, White Nile, Blue Nile, and Atbara troughs consist of half-grabens, each starting with a listric fault at great depth ("thick-skin") followed by subsequent detachment faults ("thin-skin") (Mann 1989) (see Figure 16-2). At many places, the infilling of continental deposits reaches 5 to 10 km or even more. Virtually nothing is known, or published, on their composition and depositional conditions. Outcrops are scarce or covered; potentially, they are very significant regarding vertebrate paleontology. The stratigraphic evidence relies mainly on marine incursions and isotopic dating of concomitant dikes and sills (Guiraud and Maurin 1991; Mann 1989).

PALEOGENE

This episode is the least delineated and dated. Some previously established troughs became active again, as in the Gulf of Syrt, South Sudan, and the Congo and Tanzania margins. The Anza trough was rejuvenated also, including this time the Turkana area.

It is expected that many other Early Cenozoic continental sediments are still undetected; they would be of exceptional interest regarding vertebrate, specifically primate paleontology (Lambiase 1989; Mbede 1991).

NEOGENE

The last stage of faulting shaped Central and East Africa of today, forging in the same swell the highlights of higher primate paleontology as we now know it. The directing features are distributed radially from the Afar triple junction, an unusual tectonic design in which the three arms are different (see Figure 16-3).

The Gulf of Aden area is clearly an extension of the Indian Ocean oceanic ridge and its transform faults. The Red Sea area is a classic parallel sided graben of huge size, where crustal oceanization is incipient. The third arm is formed by the less evolved but more intricate African rifts of today. It became reasonably clear in recent years that the unifying structural mode is that of arcuate half-grabens 100-150 km in length and about 50 km breadth. (See Figure 16-4.)

Half-grabens are more or less (ir)regularly spaced all along, linked through smoother accommodation zones. The interference of transform, strike-slip, and oblique wrench faults has also been postulated, although the real magnitude of their displacements is still in dispute: Asswa, Tanganyika-Rukwa-Malawi

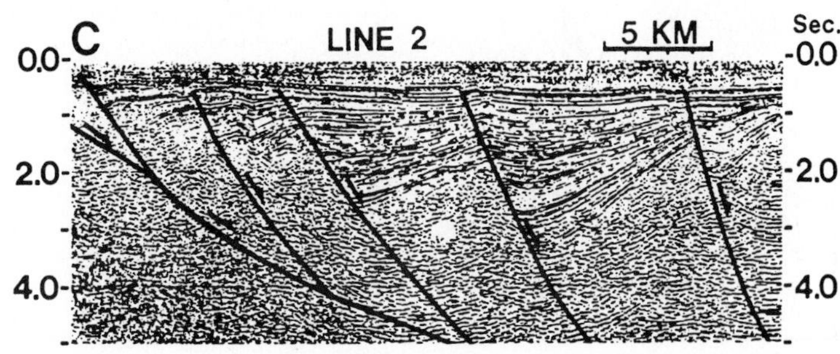

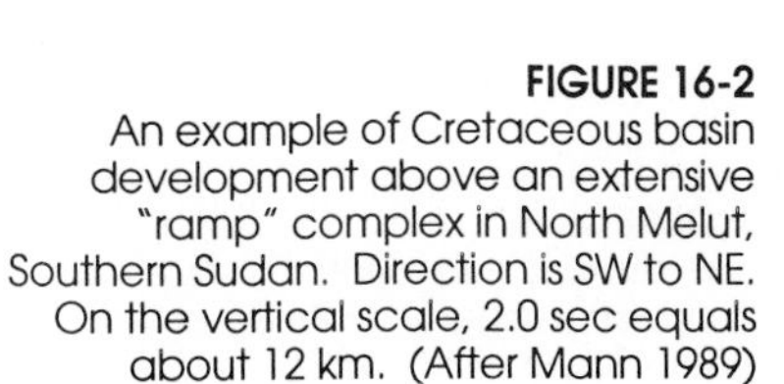

FIGURE 16-2
An example of Cretaceous basin development above an extensive "ramp" complex in North Melut, Southern Sudan. Direction is SW to NE. On the vertical scale, 2.0 sec equals about 12 km. (After Mann 1989)

and Zambezi lineaments. The transverse displacement is in any instance unimportant, a few kilometers at the best. Nowhere is a trace of oceanization detectable.

Side by side with the foregoing demonstrations, structuralists reconsidered the possible inheritance of architectures prior to the rifting. Despite repeated pleas, the evidence is not overwhelming. The cores of some Precambrian cratons seem to be avoided, indeed, but aside from that the Neogene faulting is largely autonomous, such as the Red Sea arm cutting through the Arabo-Nubian shield.

The Tanganyika-Rukwa-Malawi lineament provides another example. Only the southern half of Lake Tanganyika is imprinted on a previous Karroo trough, while the Malawi corridor cuts obliquely across the former Ruhuhu and Metangula grabens. As a result, South Tanganyika possesses at depth a maximal thickness of sediments, 5 km, of which the lower half is Karroo (Lukuga), the other half being divided in "acoustic units," Mahali and Zongwe of undefined age.

Needless to say, the age of such features remains a pure guess, Lower Miocene for some, Upper Miocene for others. This reveals enough of the uncertainties one faces in trying to ascertain the genesis of Neogene rifts (Bouroullec et al. 1991; Dunkelman et al. 1988; Rolet et al. 1991; Sander and Rosendahl 1989). A description of the geography should seem futile to everyone concerned, all the more so since Delvaux (1991) published a bibliographical synthesis on the Western Branch.

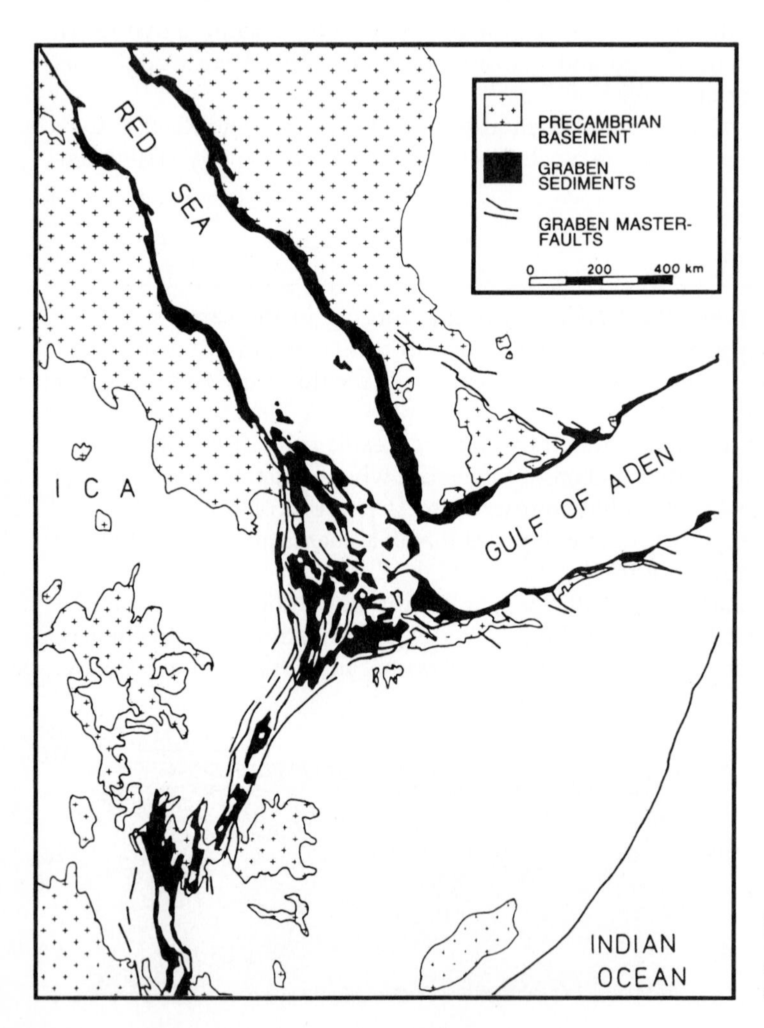

FIGURE 16-3
The Afar triple junction, a major Neogene feature. (After Bayer et al. 1989)

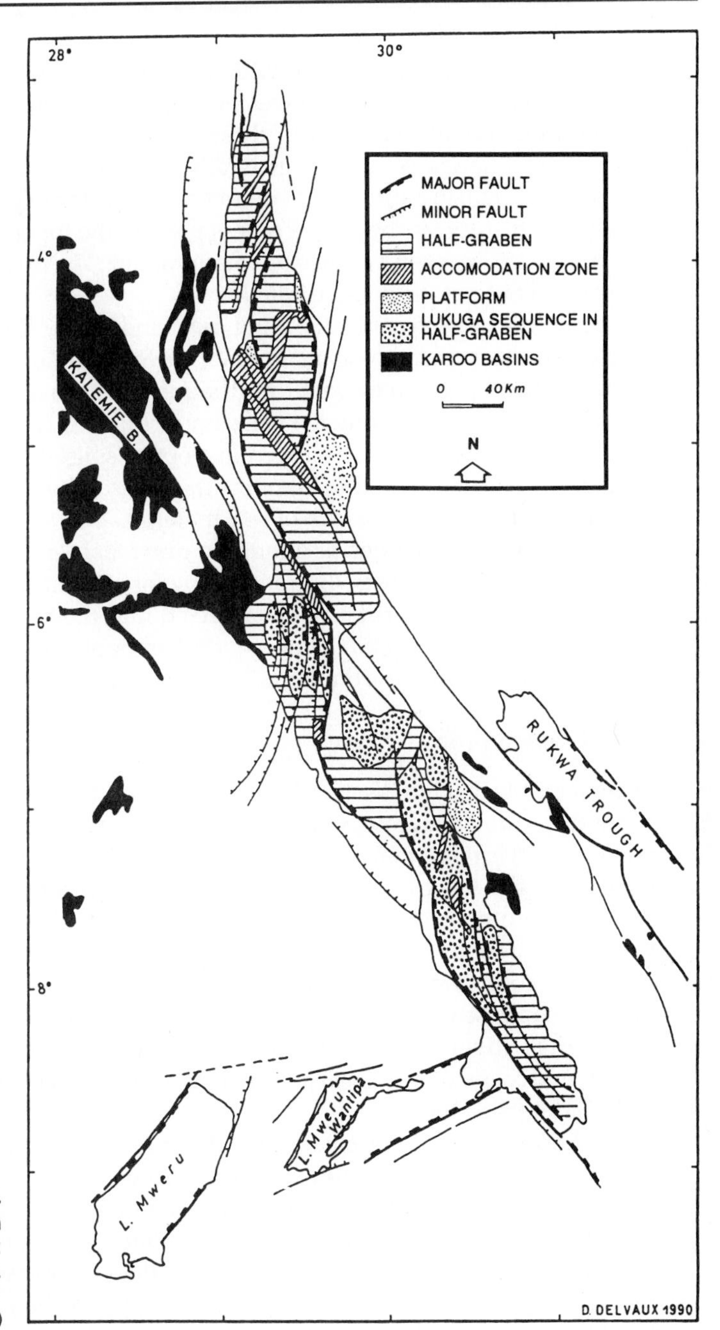

FIGURE 16-4
A typical example of the Western Branch of the East African Rift system: the Tanganyika-Mweru-Rukwa troughs. (Synthesis elaborated by Delvaux 1991 from many sources)

Foreign to Their Surroundings

This section focuses on a comparison between living conditions inside and outside the Neogene African Rifts. The bulk of information collected there from sedimentary fills, where accessible, in the course of the last decades is enormous and, for the sake of paleoanthropology, unique. This must not be taken as the unequivocal demonstration that rifts have been of special importance in the course of evolution; it means simply that, by chance, taphonomic preservation was more favorable than elsewhere.

In other words we are faced with the misleading picture of a paleontological void outside the rifts, provoked there by the balance of weathering and erosion. There is an indirect clue of the former presence of large numbers of hominids in these supposedly deserted areas, but it is seldom mentioned. The overlying soil horizons lie quite generally on a "stone line," or better said a gravel sheet that at places incorporates amazing amounts of quartz artifacts. These gravel sheets are polygenetic at large and result from a long history, as they more or less parallel the topography; the quartz artifacts are thus mixed and they indeed show several grades of deterioration. The presence of very early artifacts, of Neogene times, is suspected but not yet demonstrated.

From personal investigations in Lower and Upper Zaire (Mayumbe, Uele, Ituri), estimates of about 100 g of quartz artifacts per square m of gravel sheet are not uncommon, corresponding to 100 tons of quartz artifacts per square km. Furthermore, as quartz artifacts are the only ones preserved from weathering, the genuine content of all stone artifacts might have been much greater; cherts, obsidian, and other volcanics are indeed destroyed in tropical soils. The former figure is of regional significance only and cannot be extrapolated to all of central Africa; the Central Zaire basin is a specific case, essentially devoid of stone artifacts for lack of raw material.

With these restrictions in mind, the previously mentioned estimates widely exceed any situation encountered inside the rifts. Supposing, by fancy, that the whole formations of Shungura, Koobi Fora, and Olduvai could be artificially compacted into their coarse components: The resulting stone artifact content would vary from nil to very modest amounts per square meter. For a change, we shall question the deviations of rift environments from the norm of their surroundings.

At the Present Time

It is remarkable that the Western Branch strictly delimits, to the west, the solid, regularly spaced ecozones of intertropical Africa from, to the east, the more capricious and patchy mosaic pattern that covers East and South Africa (see map in Bonnefille, Figure 20-1). This branch at the same time forms a boundary between the ape territory to the west and the nonape territory to the east, with few minor exceptions.

A first immediate appraisal is that we find the rule to the west and deviations to the east separated by some mysterious device. Bold travelers, such as Emin Pascha and F. Stuhlmann (Stuhlmann 1894) starting by foot from Nimule (White Nile, Sudan, Asswa lineament), and going southwards to the Zambezi, supposed they would encounter the most amazing puzzle of bits of local environments, from semidesert to all sorts of savanna, marshes, permanent lakes, deltas, closed-canopy and mountain forests, prairies, and so on. On about 90 percent of the total stretch, there is no similarity between living conditions inside and outside the rift.

If there is any unifying character to detect inside, it is the low density of human population compared to the plateaus outside, with the exception of a few thresholds at high elevations such as the Kivu lake area. The Albertine-Edward trough is an exemplar situation, of which we have long personal recountings of explorations. While the Ituri and Uganda plateaus or hillsides are permanently settled in numerous villages, the rift itself is almost deserted with the exception of occasional hunters and fishermen camps. Several causes might be at work simultaneously, one of them being the high incidence of endemic diseases and of their vectors at low elevation.

Theoretically, some sort of objective mapping of the demographic pattern and of the nosologic factors in these areas might be feasible using existing records (pre-1960), but these sources are not yet integrated. Another example of barrier situations, for quite different reasons, is that of Lake Tanganyika: 600 km of stratified lake, abiotic water 125 m in depth, between rocky walls.

In Central Kenya, the Eastern Rift Branch seems somewhat less deviant from its surrounding plateaus, although north and south settlement conditions are harsher: Afar triangle, Middle Awash, Serengeti, Magadi, Eyassa are no better or worse than the Albertine-Edward trough.

Past versus Present

It might be argued that past situations inside the rifts were, to the contrary, more feasible for human settlement. We know a number of ancient situations that are in good agreement with the actual ones, and in these cases the paleoenvironmental interpretation is straightforward. Such is the upper half of the Middle Awash sequence (Plio-Pleistocene), resulting from pedialluviation, a special balance of wadi sedimentation and moderate alluviation side by side.

In the Turkana basin, the Shungura Formation (Plio-Pleistocene) of the Omo group is built out of a cyclothemic succession of alluvial plains similar to the present ones and from the same river, Omo, with sparse influxes of braided river channels, deltas, and shallow lake shores. The Koobi Fora Formation of the same basin has recently been firmly correlated with the former (Harris 1991). It differs slightly in being less regularly cyclothemic and more discontinuous; it figures a transit basin of the same Omo river on its way to the southeast and, perhaps, the Indian Ocean. Lateral drainage took importance, as well as spasmodic lake influxes, never very deep and sometimes endorrheic. In the Olduvai-Serengeti region, there is no strong discrepancy between the older deposits and the alkali-saline shallow lake basins, in the same general area at present.

There is little to say about the history of deeper basins such as Tanganyika-Malawi. We suspect that they are in existence for a long time (Lower Miocene?). If so, drilling in the deepest parts might provide a survey of many events. We know also that these lakes suffered major changes of level and that their hydraulic balance has been fluctuating on a large scale.

There are several other ancient situations that escape any simple realistic interpretation. The Hadar sequence (Upper Pliocene) is peculiar: shores of a large, shallow freshwater body surrounded by montane forests (1300 m elevation?) at less than 200 km from the oceanic base level (a dip of 6m/km?). Whatever the answer, it must also explain the peculiar taphonomy and the exceptional wealth of early hominids.

The Western Branch of the rift also affords examples of anomalous situations. In the Lower Semliki, the grits of the Mohari Formation (some Miocene) have no known counterparts elsewhere; they can be compared, in facies, to Hercynian or Alpine mollasses, strange productions from an old craton. Strange also is the long duration at the end of the Pliocene of a very large Albertine Lake with so-called "Kaiso-facies" marked by oolithic ironstone banks and endemic molluscan assemblages; the closest actual analogy is Lake Tchad, whatever this means in term of climate and vegetation.

When dealing with sedimentary sequences of this kind, vigilant afterthoughts must call their discontinuity into question. Lithologic logs at their best are cut by repeated diastemes, more often patchy and scattered in time. The least discontinuous record is that of Lake Turkana basin; otherwise time records are discontinuous between larger time gaps. We feel the practice of linking paleoclimate curves and trends across these no-information boundaries is hazardous, because scores of major fluctuations might have actually occurred (Cerling 1992).

TRACING EARLY HOMINID HABITATS

Up to now, clues have been confined to rift bottom sediments and are still relatively scarce compared to the amount of other skeletal material already collected. Artifact concentrations have been indirectly reported from Hadar but remain unstudied. In the Middle Awash, in-situ artifact occurrences have also been located, but their age and significance are still in debate (not counting the abundant Middle-Pleistocene Acheulian sites).

In the Shungura Formation of the Omo group, Turkana basin, a few direct evidences of occupation are comprised in the range of 2.4 to 2.2 MYA, defining a Shunguran culture. They pertain to two different ancient situations; on one side a dense scatter of artifacts within sandy braided river channels (unit F-1), on the other side small aggregates on flood plains approaching completion (unit F-3 and probably Lower G, unstudied).

In the Koobi Fora Formation of the Omo group, occupation sites from the Oldowan culture are centered around the KBS stuff, near 1.8 to 1.7 MYA and later cultures near 1.4 MYA. Their preferential location is at the edge of wadis, riverlets, or the heart of their deltas. In the Olduvai gorge, occupation sites of the Oldowan culture are also centered around 1.8 to 1.7 MYA but in the different environment of shores of alkaline shallow lakes.

In the Western branch of the rifts, a few scattered artifacts have been recovered from shoreline concentrations of Kaiso-facies at Kanyatsi, Lake Edward. However, it is time to point to the allegations concerning the once famous site of Senga-5, Upper Semliki. This is no more than a late Holocene slope wash in which all sorts of pieces of stone found their way, becoming a mixture of many ages. A definitive statement is still to come (Harris et al. 1987).

SUMMARY

Traces of early hominid occupation inside the rifts are sparse and diverse, which seems to denote occasional rather than preferential behavior. Probably, past environments inside rift bottoms have been rather similar to the present ones and not attractive to human occupation with few exceptions (Jaanusson 1991). If so, the Neogene rifts remained at the margins of the mainstreams of population; hominid evolution passed by in the highlands where little if any trace can be expected.

LITERATURE CITED

Bayer, H. J., El-Isa, Z., Hotzl, H., Mechie, J., Prodehl, C., and Saffarin, G. (1989) Large tectonic and lithospheric structures of the Red Sea region. *J. Afr. Earth Sci.* 8:565-587.

Bohannon, R. G., and Eittreim, S. L. (1991) Tectonic development of passive continental margins of the southern and central Red Sea with a comparison to Wilkes Land, Antarctica. *Tectonophysics* 198:129-154.

Bouroullec, J. L., Rehault, J. P., Rolet, J., Tiercelin, J. J., and Mondeguer, A. (1991) Quaternary sedimentary processes and dynamics of the northern part of the Lake Tanganyika trough, East African Rift system. *Bull. Cent. Rech. Elf-Aquitaine* 15:343-368.

Cerling, T. E. (1992) Development of grasslands and savannas in East Africa during the Neogene. *Paleogeog. Paleoclimatol. Paleoecol.* 97:241-247.

Delvaux, D. (1991) The Karroo to recent rifting in the Western Branch of the East African rift system: A bibliographical synthesis. *Mus. Roy. Afr. Centr. Tervuren Dep. Geol. Miner. Rapp.* 1989-1990, 63-83.

Dunkelman, T. J., Karson, J. A., and Rosendahl, B. R. (1988) Structural styles of the Turkana rift, Kenya. *Geology* 16:258-261.

Guiraud, R., and Maurin, J.-C. (1991) Le rifting en Afrique au Cretace inferieur. *Bull. Soc. Geol. France* 192:811-823.

Harris, J. M. (ed.) (1991) *Koobi Fora Research Project.* Vol. 3: *The Fossil Ungulates: Geology, Fossil Artiodactyls and Paleoenvironments.* Oxford: Clarendon Press.

Harris, J. W. K., Williamson, P. G., Verniers, J., Tappen, M. J., Stewart, K., Helgren, D., DeHeinzelin, J., Boaz, N. T., and Bellomo, R. V. (1987) Late Pliocene hominid occupation in Central Africa: The setting, context and character of the Senga 5A site, Zaire. *J. Hum. Evol.* 16:701-728.

Jaanusson, V. (1991) Morphological changes leading to bipedalism. *Lethaia* 24:453-457.

Lambiase, J. J. (1989) The framework of African rifting during the Phanerozoic. *J. Afr. Earth Sci.* 8:183-190.

Mann, D. C. (1989) Thick-skin and thin-skin detachment faults in continental Sudanese rift basins. *J. Afr. Earth Sci.* 8:307-322.

Mbede, E. I. (1991) The sedimentary basins of Tanzania reviewed. *J. Afr. Earth Sci.* 13:291-297.

Rolet, J., Mondeguer, A., Bouroullec, J. L., Bandora, T., Coussement, C., Rehault, J. P., and Tiercelin, J. J. (1991) Structure and different kinematic development faults along the Lake Tanganyika rift valley (East African rift system). *Bull. Cent. Rech. Elf-Aquitaine* 15:327-342.

Sander, S., and Rosendahl, B. R. (1989) The geometry of rifting in Lake Tanganyika, East Africa. *J. Afr. Earth Sci.* 8:323-354.

Specht, T. D., and Rosendahl, B. R. (1989) Architecture of the Lake Malawi rift, East Africa. *J. Afr. Earth Sci.* 8:355-382.

Stuhlmann, F. (1894) *Mit Emin Pascha ins Herz von Afrika.* Geographisches Verlagsbuchhandlung Dietrich Reimer, Berlin.

Verniers, J., Jourdan, P. P., Paulis, R. V., Frasca-Spada, L., and DeBock, F. R. (1989) The Karroo graben of Metangula, northern Mozambique. *J. Afr. Earth Sci.* 9:137-158.

CHAPTER 17

Significance of the Western Rift for Hominid Evolution

Noel T. Boaz

THE SEMLIKI RESEARCH EXPEDITION 1982-1990

Following initial discovery and investigation of Neogene fossiliferous deposits by J. de Heinzelin and co-workers (de Heinzelin 1955, 1957; Gautier 1965, 1967), work in the Semliki Valley, Zaire (Figure 17-1) was initiated by the author in June 1982 and led to the formation of the Semliki Research Expedition the following year. This multidisciplinary project has investigated the geology, paleontology, and paleoanthropology of the region in a broad natural-science approach, incorporating as well actualistic studies of the modern ecology, primatology, and ethnoarchaeology (see articles in Boaz 1990 and Boaz et al. 1992). The focus here, however, will be on the early, Miocene-to-Pliocene-aged, part of the record on the Zairean side of the Western Rift.

There are two broad regions of research in the Western Rift of Zaire: (1) an area primarily of Miocene-to-Pliocene deposits exposed along the eastern shoulder of the Western Rift southwest of Lake Mobutu, termed the "Lower Semliki," and (2) an area of Plio-Pleistocene exposures found along the northern shoreline of Lake Rutanzige and on both banks of the Semliki River, known as the "Upper Semliki" (Figure 17-1). The geology and paleontology of these two research areas will be reviewed, and their significance summarized.

Lower Semliki

Gautier (1965, 1967) published the stratigraphic framework on which field survey has been undertaken by the Semliki Research Expedition in the Lower Semliki. Fieldwork has included surveying the regions of Karugamania, Nyamavi, and Sinda-Mohari and locating, mapping, and limited sieving of fossiliferous localities in the Sinda-Mohari (cf. Hooijer 1963, 1970). De Heinzelin (1988) has provided a detailed regional stratigraphic reinterpretation of the Neogene deposits in the Lower Semliki using aerial photomosaics of

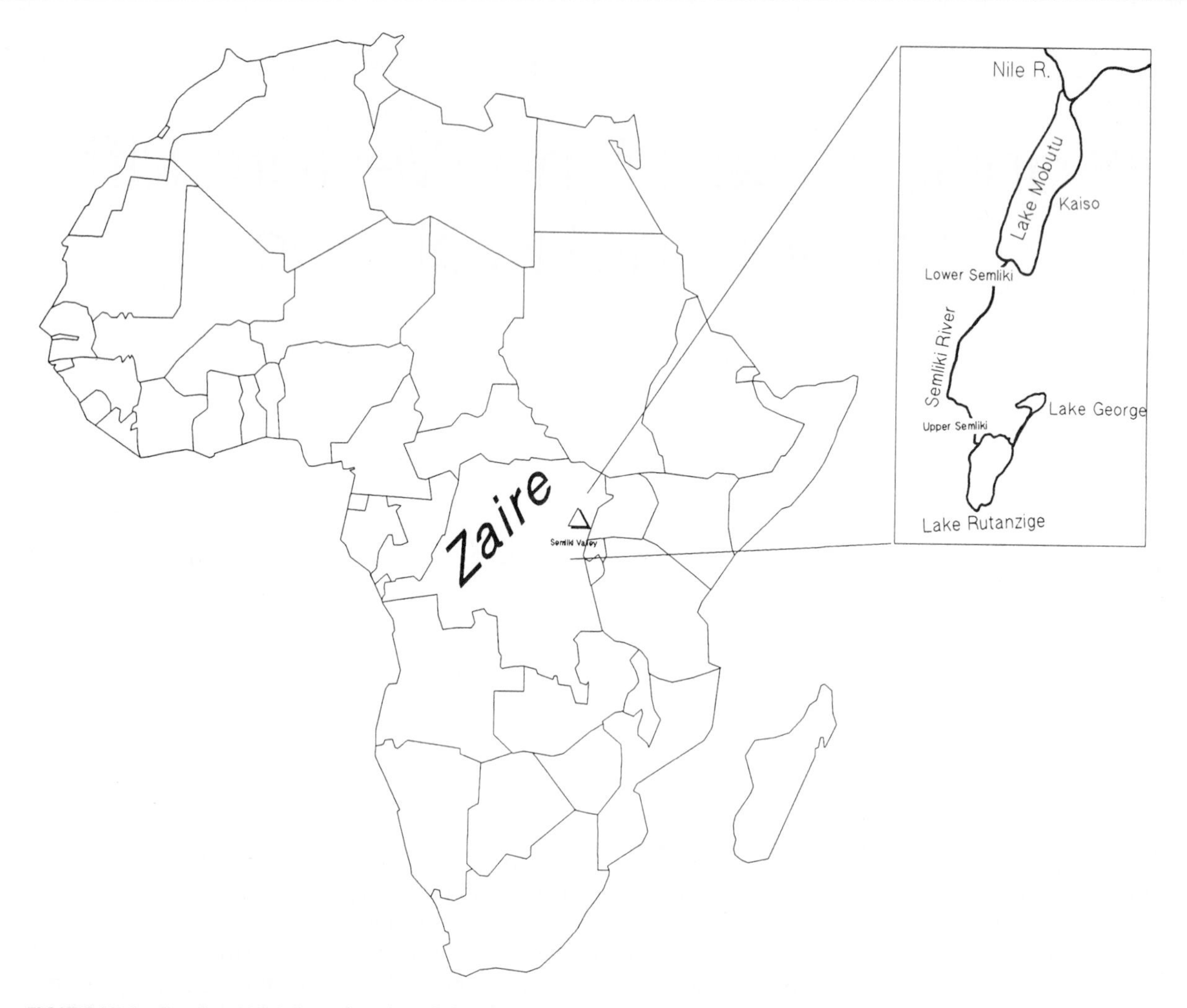

FIGURE 17-1 The Semliki Valley of eastern Zaire and western Uganda.

the area. This work builds from the now more precisely defined Upper Semliki stratigraphy and serves as the stratigraphic basis for further fieldwork in this region.

The lowermost sedimentary deposits in the Lower Semliki, of presumed Miocene age, were termed by de Heinzelin (1988) the Edo Beds, the Mohari Beds, and the Kabuga Beds, respectively. The Edo Beds are at the base of the Neogene deposits in the Lower Semliki and consist of a conglomeratic deposit of limited areal extent. They rest on pre-Cambrian basement rock and presumably record the first deposition of sediment into the then-recent Western Rift graben. The Mohari Beds are fining-upward sequences of gravels, sands, and clays that form vertical escarpments, which may be sparsely fossiliferous. The Kabuga Beds are fluvial sands with cross-bedded stratification that unconformably overlie the Mohari Beds. They have been incompletely surveyed but may be locally fossiliferous. These deposits are of great interest as they have no counterparts in the Upper Semliki. They may be in part correlative to deposits on the Ugandan side of the Western Rift, but much lithostratigraphic and biostratigraphic work needs to be done before any defensible framework can be established.

The Lower Semliki area also contains exposures of rock units at least in part equivalent to the Upper Semliki late Pliocene Lusso Beds and the middle (?) Pleistocene Semliki Beds, termed respectively, the Sinda Beds and the Ndirra/Katomba Beds (de Heinzelin 1988; Table 17-1).

TABLE 17-1 Stratigraphic Overview of the Neogene Deposits, Semliki Valley (following de Heinzelin 1988; Verniers and de Heinzelin 1990)

LOWER SEMLIKI	UPPER SEMLIKI
Unnamed	Late Pleistocene/Holocene Terraces and Katwe Ash
Katomba Beds	
	Semliki Beds (?middle Pleistocene)
Ndirra Beds	
	Lusso Beds (2.0-2.3 MYA)
Sinda Beds (c. 4.1 MYA)	
Kabuga Beds	Not present
Mohari Beds	Not present
Edo Beds	Not present

The Sinda Beds are sandy and clayey complexes interspersed with limonitic levels of "Kaiso facies," similar to those in the Lusso Beds, which have yielded abundant vertebrate and invertebrate fossils. Several fossil localities in the Lower Semliki Sinda Beds are reported in the literature (see Gautier 1965; Yasui et al. 1992) and have been verified in the field, although no fossil collections have been made by the Semliki Research Expedition pending resolution of the stratigraphic geology of the area. The known fossil faunal repertory derives for the most part from the fossiliferous Sinda Beds. The stratigraphic position of the Ongoliba Conglomerate "bone bed" lies either within the base of the Sinda Beds or in the underlying Kabuga Beds.

Makinouchi et al. (1992) have recently published a paper on the geology of the Sinda-Mohari area of the Lower Semliki, based on two field trips by a Japanese team in 1989 and 1990. This treatment, however, does not take into account much of the previous stratigraphic work in the Western Rift. There is no reference to Gautier's (1965, 1967) and de Heinzelin's (1988) observations in the same area. Makinouchi et al. (1992) ascribe all the deposits in the Sinda-Mohari area to one formation, the Sinda Beds, which they assign a latest Miocene-early Pliocene age. Whether such extension of the "Sinda Beds" is justified is beyond the scope of the current chapter, and is a question that will certainly require further geological fieldwork. Pending revisions, however, usage of the term "Sinda Beds" in this paper follows de Heinzelin (1988).

The apparent lack of volcanic or volcaniclastic rocks, datable by potassium-argon analysis, has been a major problem in assessing age relationships in Western Rift rocks (e.g. Bishop 1962; de Heinzelin and Verniers 1987). However, at least four tuffs have recently been discovered in the Kaiso region of western Uganda by Pickford (1990a). Pickford et al. (1991a) have correlated three of these to dated tuffs in the Eastern Rift (Figure 17-2). A museum sample collected by Lepersonne (1949) in the Sinda Beds of the Lower Semliki provides the first absolute dating for these beds: c. 4.1 MYA, based on a correlation with the Moiti Tuff of the Turkana Basin (F. H. Brown, pers. comm.).

The stratigraphic relationship of the Moiti Tuff horizon to the fossil collections in the Lower Semliki is problematical, but the date of this tuff does provide a much-needed benchmark. The date of 4.1 MYA probably relates to the age of the lower Sinda Beds, based on faunal comparisons with Uganda (Pickford et al. 1988, 1989). Hooijer (1963, 1970) on the other hand considered most of the Lower Semliki fauna to range in age back to the early Miocene. He termed this the "Kisegian" part of the fauna, named for Kisegi, Uganda, for outcrops of presumed correlative age. Pickford (1987) suggested dates as early as 18 MYA for

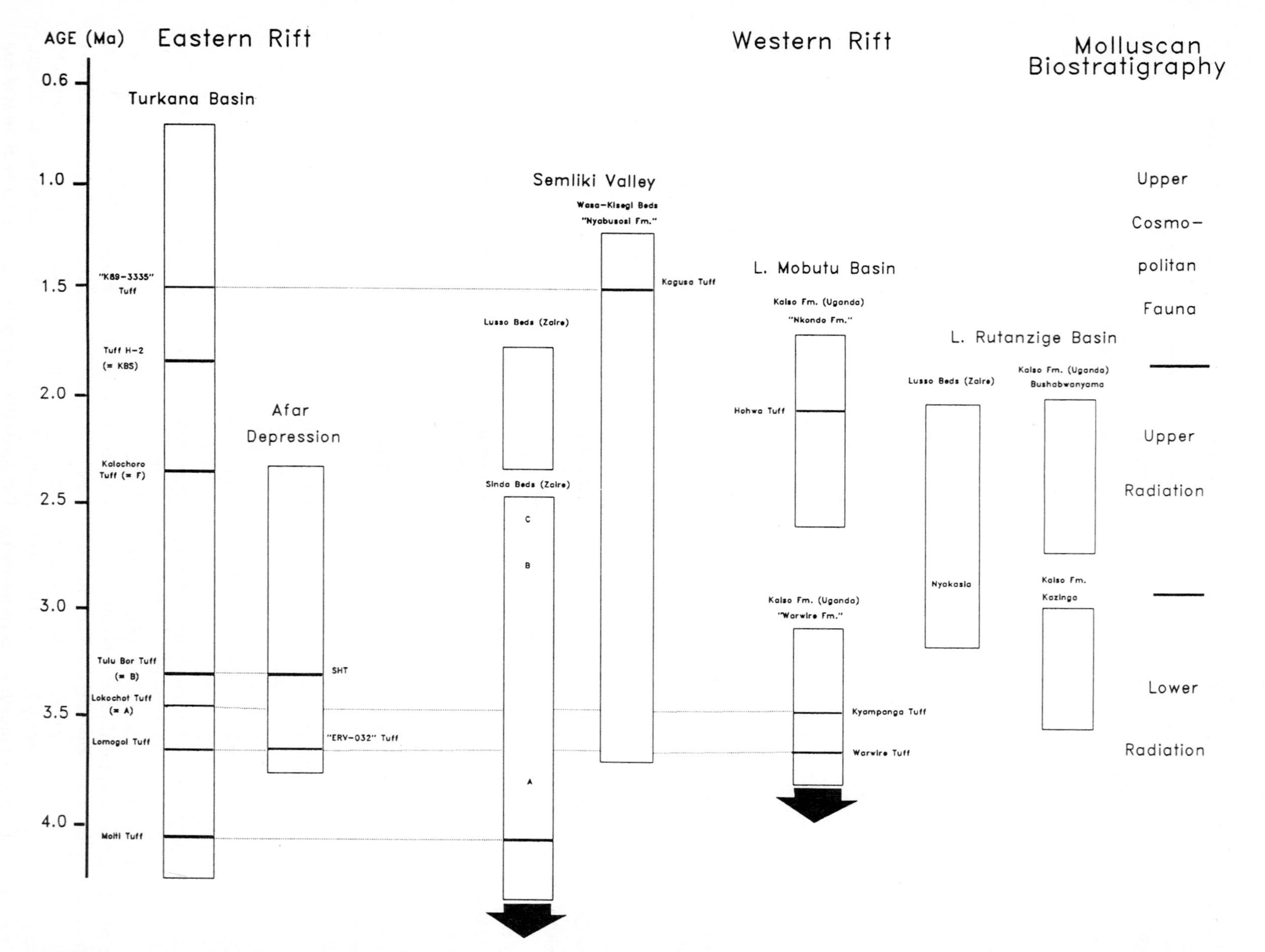

FIGURE 17-2 Relative stratigraphic positions of fossiliferous deposits of the Eastern and Western Rift Valleys.

the oldest levels in the Lower Semliki, again putatively correlative to (unfossiliferous) levels at Kisegi, Uganda. Recent reassessment of the age of the Western Rift (Ebinger 1989b), however, puts a maximum age for rift formation at 8 MYA. Pickford et al. (1989), reversing their earlier opinion, suggested that the Kisegi-Nyabusosi sequence in Uganda is no older than 7 MYA, thus indicating that the Lower Semliki would be of similar age and that the Moiti Tuff date may be a minimum age for the lower Sinda Bed fauna.

Table 17-2 presents the known invertebrate and vertebrate fauna from the Lower Semliki Neogene deposits. The invertebrate record so far known does not contribute any greater precision to the age determination of the Kabuga or Mohari Beds. Molluscan assemblages reported by Williamson (1990, 1992) from Mohari and Kabuga (Ongoliba) levels both are "lower cosmopolitan," as are lower Kaiso collections from Uganda. They are thus not reflective of age differences in these two sets of localities. Williamson's (1990, 1992) studies of the molluscs from both sides of the Western Rift show that the lower Sinda Beds ("A") share species of the "lower radiation" with the "middle" Kaiso Formation at N. Nyabrogo I-II and the Kaiso Central sites in Uganda ("Nkondo" of Pickford et al. 1988, 1989). He also found evidence in the upper Sinda Beds ("B" and "C") of the "upper radiation" of molluscs, also present at Kaiso Village Sites D and G, indicating that the Sinda Beds may span a long period of time from the early Pliocene (or earlier) to near the Plio-Pleistocene boundary.

TABLE 17-2 Vertebrate Fauna Reported from the Mio-Pliocene of the Lower Semliki (after Hooijer, 1963, 1970 (indicated by*); Yagui et al. 1992 (indicated by#); Aoki 1992 (indicated by@); Hirayama 1992 (indicated by+); Greenwood and Howes 1975 (indicated by &); Van Neer 1992 (indicated by $)).

Mammalia	
(*Brachiodus cf. aequatorialis*)*	*Trilophodon angustidens* *
Stegodon kaisensis #	?*Anancus* sp.#
Prodeinotherium sp.#	*Mammuthus subplanifrons* #
Potomochoerus (?) sp.*	?*Silvatherium sp.*#
Palaeomeryx (?) sp.*	(?*Okapia*) #
Hexaprotodon imagunculus #	*Brachypotherium heinzelini* *
Macaca sp. ("near Ongoliba") *	*Aceratherium acutirostrum* *
	Felidae sp.*
Reptilia	
Osteolaemus aff. *osborni* @	aff. *Crocodylus niloticus* @
Mecistops aff. *cataphractus* @	*Euthecodon* sp.@
Crocodylomorpha gen. et sp. indet.+	*Erymnochelys sp.*+
Carettochelyinae gen. et sp. indet.+	Trionychidae gen. et sp. indet.+
***Pisces*&**	
Sindacharax lepersonnei &	*Sindacharax* sp. "Shungura *Sindacharax*" $
Protopterus sp.&	*Gymnarchus* sp.$
Synodontis sp.&	*Hydrocynus* sp.$
Auchenoglanis sp.&	*Bagrus* sp.$
Lates rhachirinchus &	*Clarias* sp.$
Clarotes sp.&	

The vertebrate fauna from the Lower Semliki is sparse and in critical need of systematic re-collection, excavation, and revision. Hooijer's (1963) report on the fauna and his "correction" (1970) ascribed the bulk of the vertebrate fauna to the early Miocene, similar to Rusinga (Kisingiri), Kenya dated to c. 17.8 MYA (Drake et al. 1988) and Moroto at c. 17.5 MYA (Pickford et al. 1986). This interpretation is still supportable largely based on the presence in the fauna of a small anthracothere, identified by Hooijer as *Brachyodus aequatorialis*. This species is absent in East Africa after the early Miocene and no anthracotheres are known from the Ugandan Kaiso-area localities (Pickford et al. 1989). Anthracotheres occur at Moroto and at Rusinga. Similarly, results recently published by a Japanese team (Yasui et al. 1992), which visited the Sinda-Mohari area in 1989 and in 1990, can be interpreted within an early Miocene context. Although this team collected no anthracotheres, the presence of small deinotheres, ascribed to *Prodeinotherium* sp., and the absence of *Hipparion*, are characteristics of East African early Miocene faunas.

An alternative explanation of the Lower Semliki fauna is that it is a late Miocene-early Pliocene relictual fauna, preserving species in a forest refugium that recalls the East African early Miocene in both environment and species composition. If this were true then the fauna would also be expected to show evidence of more recent species as well as evidence of endemism. The presence of Pliocene taxa such as *Macaca* sp. (see later discussion), the proboscideans *Stegodon kaisensis* and *Mammuthus subplanifrons*, and the hippotamid *Hexaprotodon imagunculus* (Yasui et al. 1992) indicates advanced elements in the fauna. Other taxa that might be further supportive of a late Miocene-early Pliocene date, such as *?Agriotherium* and bovids, as reported by Yasui et al. (1992), need more complete remains before any firm conclusions can be drawn. In fact molar SN-522 (Yasui et al. 1992:107) referred to Bovidae gen. et sp. indet. seems from the photograph to be instead a small giraffid, perhaps *Okapia*, and is indicated as such in Table 17-2.

Endemism in the lower Semliki fauna may be indicated by the presence of Okapia and other elements of the mammalian fauna as they become known. But the most convincing indications come from the reptilian fauna, recently reported by Hirayama (1992) and Aoki (1992). A carettochelyid turtle (SN-074) is the first record from Africa of a family extinct in Eurasia and North America since the Eocene and which today survives only in New Guinea and northern Australia. The pelomedusid turtle *Erymnochelys*, represented by a number of specimens in the Lower Semliki, is today extinct on the African mainland and is restricted to Madagascar. Several Lower Semliki specimens also document the first known presence in the fossil record of the Zairean dwarf crocodile *Osteolaemus*, a genus limited today to the Central Forest Refuge of Zaire. Another crocodylid specimen (SN-235) indicates the presence of a previously unknown "highly longsnouted" species. The overall impression of the reptilian fauna is of a highly endemic and ecologically conservative fauna. Combined with the evidence of some fossil taxa characteristic of time levels more recent than the early Miocene, the Moiti Tuff correlation at 4.1 MYA, and the fact that the species identifications of small-bodied anthracotheres, deinotheres, and rhinocerotids considered early Miocene indicators are all based on fragmentary remains, a late Miocene-early Pliocene date for the Lower Semliki fauna seems a reasonable interpretation at the present time.

Humankind's closest living primate relatives, *Pan* and *Gorilla*, like *Okapia*, are today relictual species and denizens of the Central Forest Refuge (Kingdon 1971), which the Lower Semliki fauna seems to preserve in the fossil record. It is curious therefore that the primate fossil record from the Lower Semliki is limited to a single monkey, a molar ascribed to *Macaca* sp., found by the zoologist Xavier Misonne in 1958 "near Ongoliba." The specimen could be either from a Sinda Bed or, less likely, a Kabuga Bed context. De Heinzelin (pers. comm.) contacted Misonne following the author's first reconnaissance to the Lower Semliki in 1982 regarding the exact location of this discovery. At the time Misonne recollected that it was "near" Ongoliba, but "across the [Sinda] river." This would place the specimen within the area of outcrop of the Sinda Beds according to de Heinzelin (1988).

Further stratigraphically controlled survey and collection are needed in the Lower Semliki to clarify the fauna associated with each named and recognized geological sedimentary formation. Despite the sparseness of the fossil record, paleontological excavation and wet-sieving that have not yet been undertaken in the Lower Semliki should yield enhanced views of this important fauna. Wet-sieving in particular will provide evidence of microfauna, a critically important component in understanding the ecology of forest environments. And it is to be hoped that further paleontological investigation will yield up remains of Pliocene apes, which on biogeographic grounds should be members of the Lower Semliki fauna.

Upper Semliki

The oldest fossiliferous deposits in the Upper Semliki area are the Lusso Beds, similar in facies and age to the "Kaiso Beds" of Uganda. De Heinzelin and Verniers (1987) applied the name "Lusso Beds" to the Upper Semliki deposits earlier termed the "Kaiso Beds" by Fuchs (1934) and the "Lake Edward Beds" by Lepersonne in Hooijer (1963). Intrabasinal stratigraphic correlation between the Ugandan type sites of the Kaiso Formation (Figure 17-1) and the Zairean side of the rift is premature until the stratigraphic history and lateral facies changes are better understood for both regions. There are several regions of outcrop of fossiliferous deposits in the Upper Semliki, reviewed in Boaz (1990).

De Heinzelin and Verniers (1987) and Verniers and de Heinzelin (1990) defined and described the Lusso Beds, which are primarily lacustrine sediments typified by coarsening upwards clays, silts, and sands. The base of the Lusso Beds has not been observed. The top of the beds is an erosive contact with the coarser, mainly fluviatile sediments of the overlying Semliki Beds. There is a stratigraphic thickness of up to 50 m estimated for the Lusso Beds in the Lusso and Kanyatsi areas of the Upper Semliki, but this is likely a minimum estimate. The probable presence of older Lusso Beds exposed at Nyakasia Ravine correlative to the lower Kaiso units of Uganda, on the basis of their molluscan faunas (Williamson 1990, 1992), indicates that the stratigraphic thickness will likely prove to be significantly greater.

The fossil wood preserved in the Upper Semliki Lusso Beds provides one of the best sources of paleoecological information currently available in the African Pliocene. Dechamps and Maes's (1990) analysis has yielded a floral inventory of some 60 taxa of trees and woody plants (Table 17-3). Comparisons with modern analogues show that there was a diversity of habitats present in the Upper Semliki valley. These ranged from lowland evergreen and swamp forests to steppe. The presence of lowland forest with Central and West African affinities, documented by fossil wood, was the first such confirmation of this habitat type in the African late Pliocene fossil record.

Perhaps of greater significance than the documentation of the presence of forested environments in the Western Rift is the fact that the environment overall in Lusso times was open and even semi-arid. Indeed the overall aspect of the flora as well as the preserved vertebrate fauna (Table 17-4) is that of eastern Africa during the Plio-Pleistocene. The Lusso Beds do not preserve a forest biome, but instead an open-country biome with forest nearby. Pavlakis (1987) in a preliminary paleoecological assessment of the Lusso Beds and the Ugandan Plio-Pleistocene sites first concluded that grassland predominated in the Lusso Beds. Further research has confirmed that the Upper Semliki documents the farthest western outpost of East African-type savanna environments now known in the late Pliocene.

Williamson's findings (1990) of significant extinction among molluscan taxa in the upper Lusso Beds c. 2 MYA, indicating extreme desiccation of proto-Lake Rutanzige, argues for significant climatic change, although tectonic causation for lake draining cannot be ruled out. Williamson draws a parallel to similar extinction events in the Lake Turkana Basin in the Eastern Rift at about this same time, which supports a regional climatic causation of these extinction events. Stewart's (1990) study of the fish fauna from the Lusso Beds indicates a similar pattern of endemism and extinction. Indeed the diversity of molluscivorous fish in the Lusso Beds provides a good example of co-evolution between these two components of the lacustrine fauna. Dechamps and Maes's results for quite arid ("steppe") paleoflora at Sn5A, in the upper Lusso Beds, so far unique in its indication of aridity in the Upper Semliki Lusso Beds, may be an indication of the terrestrial aspects of this regional climatic change. Harris et al. (1987, 1990) and Boaz et al. (1992) follow other authors in noting the correlation in time of this period of climatic change, already known in

TABLE 17-3 Taxa of fossil wood from the upper Pliocene Lusso Beds, Zaire (Dechamps in Boaz 1990).

Acacia cf *abyssinica*
Acacia albida
Acacia ataxacantha
Acacia ciliolata
Acacia cf. *giraffae*
Acacia hockii
Acacia nilotica subsp. *adansoniae*
Acacia nilotica subsp. *subalata*
Acacia nilotica var. *tomentosa*
Acacia polycantha subsp. *campylacantha*
Acacia rovumae
Acacia seyal
Acacia sieberiana
Acacia sieberiana var. *woodii*
Acacia sp.
Agelaea dewevrei
Airyantha schweinfurthi
Antidesma membranacea
Anthonota macrophylla
Aphania senegalensis
Aptandra zenkeri
Baphia sp.
Baphiastrum boonei
Brachystegia cf. *laurentii*
Brachystegia cf. *microphylla*
Brachystegia cf. *utilis*
Byrosocarpus cf. *orientalis*
Canthium cf. *campylacanthum*
Cassipourea aff. *malosana*
Combretum paniculatum
Combretum sp.
Cynometra alexandrii
Dichapetulum acuminatum
Dichapetulum aff. *glandulosum*
Dichapetulum griseisepalum
Dichapetulum lokanduense
Dichapetulum lujae
Dichapetulum mombutuense
Dichapetulum mundense
Dichapetulum spp
Diospyros sp
Gramineae sp. (?"bamboo")
Grewia flavescens
Grewia mollis
Irvingia robur
cf. *Juniperus* sp.
Leptadenia cf. *hastata*
Encephalartos sp.
Loeseneriella clematiodes
Monocotyledon sp.
Magnistipula butayei
Ostryoderris gabonica
Phoenix reclinata
Phoenix sp.
Podocarpus milanjianus
Raphiostylis beninensis
Rothmannia urcelliformis
Roureopsis obliquifoliolata
Salacia sp.
Salix sp.
Sapium ellipicum

East Africa, to evolutionary changes in the hominid career, such as the appearance of the genus *Homo* and the beginnings of stone tool making (see also Stanley 1992).

Table 17-4 lists the vertebrate taxa currently known from the Lusso Beds. Because the fauna has an overall resemblance to Plio-Pleistocene East African sites, biostratigraphic comparisons of taxa present and stage of evolution in certain lineages can be made confidently. The well-controlled absolute chronologies for the East African sequences, particularly that of the lower Omo Valley/Turkana Basin, Ethiopia and Kenya (Brown et al. 1985; Feibel et al. 1989), has allowed the determination of the absolute age range for the Upper Semliki deposits. The vertebrate fauna from the Lusso Beds (Table 17-4) indicates a biostratigraphic correlation to the Eastern Rift fauna from Omo Shungura Members F and G (Cooke 1990; Gentry 1990; Boaz et al. 1992), or 2.0 to 2.3 MYA (Brown et al. 1985; Feibel et al. 1989; Brown this volume).

Cooke (1990) bases his assessment that the Lusso Beds correlate with Omo Shungura Members F and G on the three suid species *Kolpochoerus limnetes*, *Metridiochoerus jacksoni*, and *Notochoerus euilus*. This assessment agrees with earlier age estimates by Cooke and Coryndon (1970) for the "later" Kaiso fauna of Uganda. The independent analyses of bovids by Gentry (1990), of equids by Bernor and Sanders (1990), of hippopotamids by Pavlakis (1990), and of proboscideans by Sanders (1990) are consistent with this dating, even though they provide no further precision.

Biogeographic indicators in the Upper Semliki have almost exclusively pointed toward the east, with the fish and reptilian components indicating Nilotic or Sudanian affinities (Stewart 1990; Meylan 1990). This aspect of the fauna contrasts strongly with the apparent endemic Central Forest Refuge fauna of the Lower Semliki which has strong Zaire Basin affinities. Between Lower and Upper Semliki times then a major reconfiguration of the drainage of the Western Rift had occurred, with the Nile drainage capturing what had previously been the headwaters of the Zaire. This change was presumably effected by uplift of the western lip of the Western Rift and downfaulting of the rift graben.

Much of the Lusso Beds were deposited at or near the shoreline of proto-Lake Rutanzige, and it is possible with a consideration of the vertebrate fauna to reconstruct a paleoenvironmental model for the late Pliocene Upper Semliki. Although there is evidence for forested habitats from the fossil wood, the fossil vertebrates overwhelmingly indicate a more open-vegetation, savanna ecological preference. Only a single colobine monkey molar from Lu1 and remains of a pygmy hippopotamid *Hexaprotodon imagunculus* suggest the faunal component of a forest biome, the distal community of the Lusso Bed depositional environment. There are two proximal communities represented in the vertebrate assemblage of the Upper Semliki Lusso Beds, the aquatic component, which includes fish, crocodiles, and hippopotamids, and the terrestrial, which includes bovids, equids, suids, elephantids, and cercopithecine primates (*Theropithecus*). Hominids are hypothesized to have been denizens of this proximal community because of the presence of their stone tools and because of the similarity of environments here and in the known East African early hominid habitats. Pliocene apes are hypothesized to have been members of the distal forested environments and are considered less likely to be found with further fieldwork in the Upper Semliki.

A comparison of excavated faunal numbers from two sites of approximate equivalent ages in the Eastern Rift (Locality 398 at Omo, Ethiopia: Johanson et al. 1975) and in the Upper Semliki of the Western Rift (Senga 13B: Boaz et al. in press), shows that representations of taxa are proportionally quite similar (Figure 17-3). Both sites preserve a preponderance of bovids, with hippopotamids next more prevalent. Giraffids, suids, and elephantids show similar proportionate representations. Along with other paleoecological analyses presented by Boaz et al. (in press), these data indicate a general similarity in large mammal terrestrial ecologies between Eastern and Western Rift Valleys at this time in the late Pliocene. The lack of any cercopithecid fossils at Sn13B, which are known at other sites in the Upper Semliki and which would be expected at Sn13B based on the Omo L. 398 results, may be due to sampling error or may be a reflection of a relatively greater representation of wooded habitats at Omo.

THE UGANDAN SIDE OF THE WESTERN RIFT

Deposits on the eastern border of then-Lake Albert (now Lake Mobutu) near the village of Kaiso, Uganda (Figure 17-1) were the locus of the discovery, in 1919, of some of the first fossil vertebrates in Africa (Bishop 1969:21). The Kaiso Formation of western Uganda was one of the first fossil sites reported from Africa (Wayland 1926; Fuchs 1934). The deposits are typically "parallel bedded, poorly consolidated, drab, grey to greenish clays, which are very greasy when wet, interbedded with buff micaceous silts and fine sands and containing occasional rusty brown 'ironstone' bands" (Bishop 1969:23).

The "Kaiso Beds" in Uganda and the "Lake Edward Beds" (Lepersonne in Hooijer 1963) in Zaire were correlated on the basis of lithological similarity and similarity in biostratigraphic age by Bishop, Gautier, and de Heinzelin (in Gautier 1967). De Heinzelin and Verniers (1987), however, reversed this correlation and suggested that the terms "Kaiso Formation" and "Lusso Beds" be used for the Ugandan and Zairean deposits, respectively. Shortly thereafter, a number of new formational names were coined from the Nkondo area on the western shore of Lake Albert/Mobutu ("Nkondo Beds" or "Nkondo

TABLE 17-4 Invertebrate and vertebrate taxa from the upper Pliocene Lusso Beds, Zaire.

MOLLUSCA

"COSMOPOLIAN" TAXA

GASTROPODA

Bellamya unicolor
Bulinus sp.
Cleopatra bulimoides
Gabbiella humerosa
Melanoides tuberculata
Pila ovata

BIVALVIA

Aspatharia cailliaudi
Aspatharia wissmanni
Caelatura bakeri
Corbicula consobrina
Etheria elliptica
Eupera sp.
Mutela nilotica
Pleiodon ovatus
Pseudobovaria mwayana

ENDEMIC TAXA

GASTROPODA

Bellamya adami
Bellamya cylindricus
Bellamya worthingtoni
Bellamya sp. nov. A
Gastropoda gen. et sp. nov. A
Gastropoda sp. nov B
Platymelania bifidicincta
Platymelania brevissima

BIVALVIA

Caelatura sp. nov. A
Pleiodon sp. nov. A
Pseudobovaria sp. nov. A
Pseudodiplodon sengae

PISCES

Protopterus sp.
?*Hyperopisus* sp.
Gymnarchus sp.
Labeo sp.
Barbus sp.
Distichodus sp
Hydrocynus sp.
Alestes sp.
Sindacharax ?deserti
Sindacharax sp.
Characidae gen. nov. A
Characidae indet.
Auchenoglanis sp.
Bagrus sp
Clarotes sp.
Bagridae indet.
? *Clarias* sp.
Synodontis sp.
Lates niloticus
Lates cf. *rhachirhinchus*
Cichlidae indet.
Perciformes gen. et sp. indet. (A)

AMPHIBIA

Anura gen. et sp. indet.

REPTILIA

Pleurodira gen. et sp. indet.
cf. *Pelusios sinuatus*
cf. *Cycloderma*
Testudinidae indet.
Crocodylus sp.

MAMMALIA

Menelikia lyrocera
?*Kobus sigmoidalis*
Kobus ancystrocera
Kobus kob
Syncerus sp.
Tragelaphus nakuae
Alcelaphini spp.
Kolpochoerus limnetes
Notochoerus euilus
Metridiochoerus jacksoni
Hipparioninae gen. et sp. indet.
Equus sp.
Elephas recki
Giraffa sp.
cf. *Ceratotherium* sp.
Hippopotamus sp.
Hexaprotodon imagunculus
Colobinae gen. et sp. indet. (A)
Theropithecus sp.
Thryonomys sp.
Otomys sp.
Tachyoryctes sp.
Carnivora gen. et sp. indet.

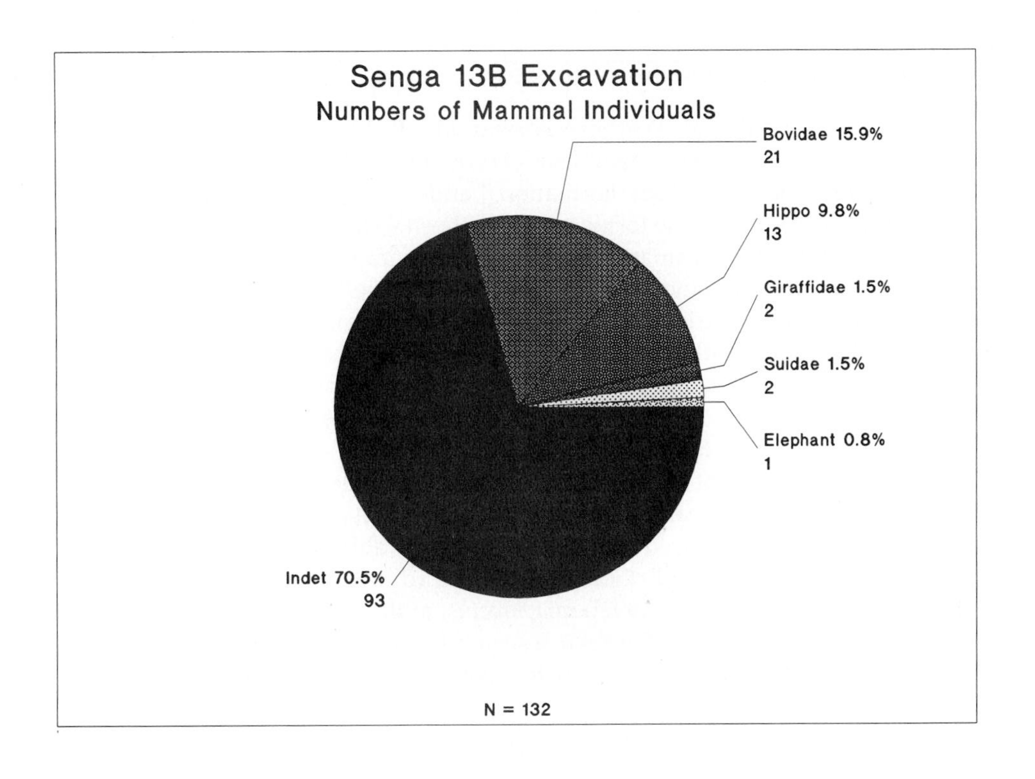

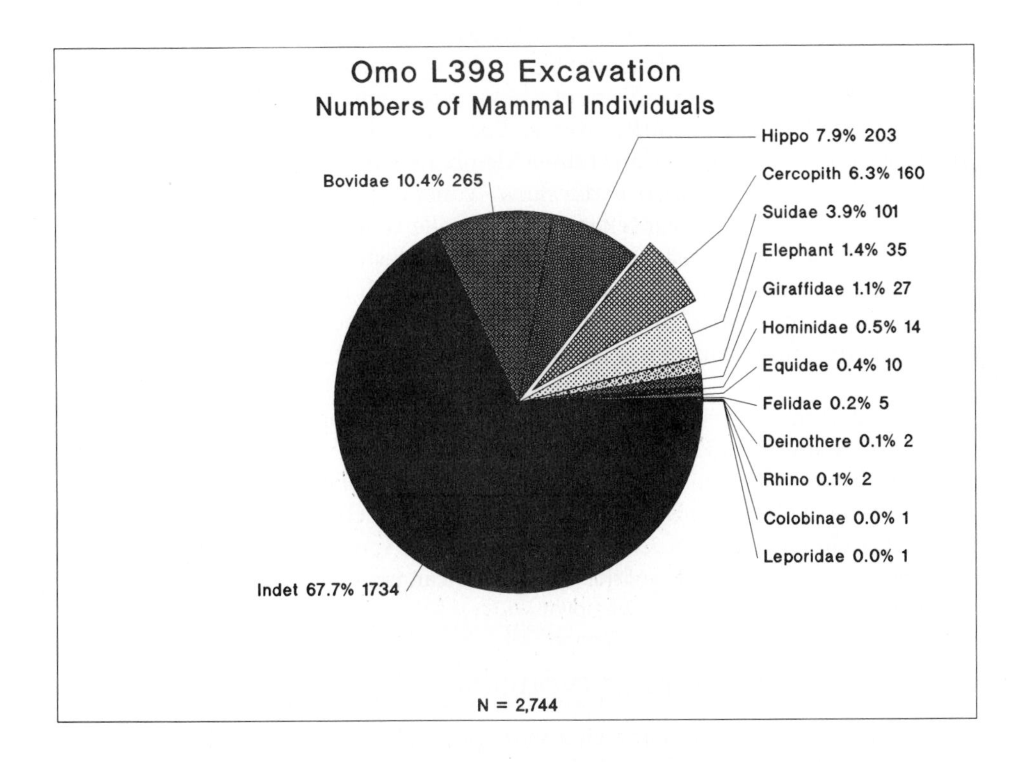

FIGURE 17-3 Numbers of individuals of fossil mammals excavated from two sites of roughly contemporary age (2.0-2.3 MYA) from the Eastern and Western Rifts, to demonstrate the similarity in relative numbers of taxa in the two areas during the Late Pliocene.

Formation" and "Warwire Beds" or "Warwire Formation"; Pickford et al. 1988, 1991) and from the Nyabusosi area in the Semliki Valley ("Nyabusosi Formation," "Katarogo Formation," "Nyakabingo Formation," "Oluka Formation," "Kakata Formation," and "Kisegi Formation"; Pickford et al. 1989). Unfortunately, descriptions of these new formations have been so far preliminary, so that the areal extent and status of the beds is difficult to evaluate. Intrabasinal stratigraphic correlation between the Ugandan type sites of the Kaiso Formation and the Zairean side of the rift is preliminary at best until the stratigraphic history and lateral facies changes are better documented for both regions. However, based on the known tuff correlations with the Eastern Rift (Figure 17-2) it would appear that the Ugandan "Warwire Beds" (c. 3.4-3.7 MYA) are somewhat later than the Zairean Lower Semliki Sinda Beds (c. 4.1 MYA), and that the Ugandan "Nyabusosi Formation" is later (c. 1.5 MYA) than the Zairean Upper Semliki Lusso Beds (biostratographically dated at 2.0-2.3 MYA).

Biogeographical connections of the western Uganda sites to the Eastern Rift seem to have become more established compared to the apparently endemic Lower Semliki fauna. Typically East African suids, bovids, and proboscideans have appeared (Tables 17-5 and 17-6), although the lower Kaiso fauna contains the Central African taxa *Okapia* sp. and ?Gorillinae. Pickford et al. (1988) reported the Zairean Basin crocodile, *Crocodylus* (or *Mecistops* [Aoki 1992]) *cataphractus*, in the Nkondo ("early Kaiso") fauna, which is absent in the fossil record of the Upper Semliki. It seems likely that the Western Rift late Neogene fossil deposits of Zaire and Uganda represent a previously unknown or poorly recognized paleobiogeographic province in Africa.

Paleoecological interpretations of the Ugandan late Miocene-early Pliocene faunas (Table 17-5) have uniformly emphasized their "humid," forested, or well-wooded characteristics (Pickford et al. 1988, 1989; Pickford 1989, 1990b). Dechamps et al. (1992) have recently reported a fossil flora from Nkondo composed of 20 taxa that indicate the presence in the late Miocene-early Pliocene of lowland forest similar to that in modern northern Kasai, Zaire. Results from the Upper Semliki Lusso Beds on the other hand show very clearly open-country, savanna paleoecological conditions with remnant forest patches, from paleobotanical, microfaunal, and large vertebrate indicators (see Boaz 1990; Boaz et al. 1992). Two specimens of the palm *Borassus aethiopicum* from the Ugandan Hohwa Member at Nkondo, approximately correlative with, or slightly older than, the Zairean Lusso Beds, show "considerably drier and more open vegetation" (Dechamps et al. 1992:325) than the lower Nkondo Beds. Pollen collected from these same beds show the preponderance of grass taxa. The Pleistocene deposits in the Western Rift preserve primarily savanna to savanna woodland faunas (Brooks and Klein in prep., Pickford et al. 1988). Finally, oxygen isotope studies on cave speleothems at Matopi Cave, Zaire, by Brook et al. (1990) indicate that at the Pleistocene Glacial Maximum the Ituri Forest, now considered a part of the Central Forest Refuge, had given way to open wooded savanna.

Because the Zairean and Ugandan sites are so close to one another major ecological differences due to geographical distance are not likely to account for the observed faunal and floral differences. The differences observed from one site to another seem most likely explicable in terms of ecological change through time. The Neogene Western Rift was a stage on which a changing mosaic of forest, woodland, and savanna moved as climatic, tectonic, and geomorphological forces came to bear on the area. Patterns of speciation, distribution, and extinction of the Central African flora and fauna were affected in profound ways by these changes.

THE AFRICAN RIFT SYSTEM AND HOMINOID EVOLUTION

Darwin (1874:171) observed that "it is somewhat more probable that our early progenitors lived on the African continent than elsewhere." Considering that there were no truly ancient fossil remains of Hominidae when Darwin expressed that opinion, it is clear that his prescient deductions were based on an assessment of close phylogenetic affinity between the human species and the African chimpanzee and gorilla, based on comparative anatomy (primarily observations of Thomas Henry Huxley). Functional

TABLE 17-5 Invertebrate and vertebrate taxa from the "early Kaiso" (molluscan "lower radiation") levels, Uganda (Cooke and Coryndon 1970; Pickford et al. 1988; and Williamson 1990).

MOLLUSCA	***MAMMALIA***
ENDEMIC TAXA	Colobinae
Bellamya turris	Gorillinae ?
Bellamya lepersonnei	Machairodontinae
Bellamya nodulosus	Hyaenidae
Bellamya Sp. Nov. B	Lutrinae
Lanistes bishopi	*Anancus kenyensis*
Pleiodon adami	*Stegodon kaisensis*
Pleiodon mohariensis	*Primelephas gomphotheroides*
Pleiodon Sp. Nov. B	*Loxodonta adaurora*
	Deinotherium bozasi
	Orycteropidae ?
	Brachypotherium sp. ?
	Rhinocerotidae
	Nyanzachoerus tulotus
	Nyanzachoerus kanamensis
	Nyanzachoerus jaegeri
	Kolpochoerus limnetes
	Hexaprotodon imagunculus
	Hippopotamus kaisensis
	Giraffidae cf. *Okapia*
	?Ugandax gautieri
	Aepyceros sp.
	Syncerus sp.
	Reduncini
	Antilopini
	Tragelaphini

considerations of human morphology (such as hairlessness) indicated to Darwin a tropical origin for humans. The biogeographic conclusion then that tropical Africa was the likely birthplace of Hominidae (rather than Asia) because of the continued presence of the chimpanzee and gorilla on that continent became unavoidable. Notwithstanding this oft-quoted opinion of Darwin's, modern paleoanthropological research was not directed to the area where the African apes live until quite recently.

Ebinger (1989a,b) using new potassium-argon dates from Zaire and Rwanda suggested a model of formation of the Western Rift that postulates initial volcanism in the Lake Mobutu section of the valley and rift propagation to the south, with volcanism beginning in the Lake Kivu area by around 7 MYA. She suggests that volcanism began in the Western Rift some 11 million years after volcanism had started in the Eastern Rift. This more recent model contrasts with earlier ideas that because of its more dissected appearance, more slumped graben walls, and fossils of presumed early Miocene age, the Western Rift was older than or as old as the Eastern Rift (Hopwood and Lepersonne 1953). If we accept Ebinger's hypothesis, dates for initiation of the major faulting are c. 8 MYA and 19 MYA for the Western and Eastern Rifts, respectively.

The Eastern Rift middle Miocene site of Fort Ternan, dated at c. 13 MYA, has been considered since its study by Gentry (1970) to indicate a record of the change from primarily forested to primarily woodland, savanna, and open-country habitats in East Africa (Shipman 1986; Kappelman 1991). The fauna preserves a preponderance of bovids, giraffids, and other taxa that are clearly ecologically more comparable to modern East African woodland than to early Miocene or modern African forest faunas. The hominoid *Kenyapithecus wickeri* discovered at Fort Ternan was thought by Leakey (1967) and others

TABLE 17-6 Invertebrate and vertebrate taxa from the "later Kaiso" (molluscan "upper radiation") levels, western Uganda (Cooke and Coryndon 1970; Pickford et al. 1987; Williamson 1990).

MOLLUSCA	***MAMMALIA***
ENDEMIC TAXA	*Theropithecus oswaldi*
Bellamya adami	*Homo* sp.
Bellamya worthingtoni	*Elephas recki*
Bellamya Sp. Nov. A	*Stegodon kaisensis*
Bellamya emerenciae	*Ancylotherium hennigi*
Bellamya cylindricus	Rhinocerotidae
Platymelania bifidicincta	*Hipparion albertense*
Playmelania brevissima	*Equus* sp.
Gastropoda Gen. Nov. Sp. A	*Notochoerus euilus*
Gastropoda Gen Nov. Sp. B	*Kolpochoerus limnetes*
Pleiodon Sp. Nov. A	*Metridiochoerus jacksoni*
Pseudobovaria Sp. Nov. A	*Sivatherium* sp.
Caelatura Sp. Nov.	*Syncerus* sp.
Pseudodiplodon sengae	Bovini indet.
	Redunca sp.
PISCES	*Kobus* sp.
Lates niloticus	*Hippotragus* sp.
Clarias sp.	cf. *Damaliscus* sp.
	cf. *Alcelaphus* sp.
REPTILIA	cf. *Parmularius* sp.
Trionyx sp.	*Aepyceros melampus*
Pelusios sp.	*Strepsiceros* cf. *maryanus*
Crocodylus sp.	cf. *Onotragus* sp.
	Hexaprotodon imagunculus
	Hippopotamus kaisensis

to represent the beginnings of the hominid lineage because of the reduced size of its canine tooth, structure of its thick-enamelled molar teeth, and the apparent orthognathy of its face. *Kenyapithecus* now seems to have been, like *Ramapithecus* in Asia, an ecological vicar of the later hominids. *Kenyapithecus* likely arose from a proconsulid source isolated in East Africa as the Eastern Rift divided the equatorial Africa forests from east to west, as the uplifted Eastern Rift walls caused a rain shadow in eastern Africa, and as global cooling caused drier conditions worldwide (Andrews and Van Couvering 1975; Brain 1981; Boaz and Burckle 1984; Boaz 1993).

Kortlandt (1972, 1974) focused attention on the Western Rift and its possible role in Neogene paleoecological change and how these changes may have consequently affected hominoid evolutionary events. He (1974:427) stated, "the tectonic formation of the western branch of the Great Rift Valley system, in combination with the Nile and Zambezi River systems, constituted a double set of barriers to creatures that could neither swim nor cross arid rain-shadow zones." Pavlakis (1987) confirmed that the Western Rift today serves to delimit on both sides taxonomic diversity in vertebrates and thus probably has acted as a significant genetic barrier in the past. Boaz and Burckle (1984) hypothesized that formation of the Western Rift escarpments set in play vegetational and biogeographic changes that became important in the African ape-hominid evolutionary divergence. Pickford (1990b) has also developed this argument.

Knowledge of fossil primates from the Western Rift is very meager. Only a single colobine and single putative gorilla canine have been reported from Nkondo, Uganda (Pickford et al. 1988). A single colobine and several *Theropithecus* molars, but no apes, have been discovered in the Zairean Lusso Beds (Boaz 1990;

Boaz et al. 1992). The well-known molar of "cf. *Macaca*" from near Ongoliba (Hooijer 1963) is the only primate record from the Lower Semliki. If the Ugandan and Lower Semliki sites were indeed relatively forested in their habitat representations then the rarity of primates is surprising (cf. Andrews and Van Couvering 1975) and may be reversed with intensive survey and wet-sieving, as has been undertaken in the Upper Semliki.

A fossil incisor from the Kaiso Beds of the Kazinga Channel, Uganda, was reported by von Bartheld et al. (1970) to be that of a hominoid, based on its enamel microstructure. This specimen has been reanalyzed by L. Martin and W. Sanders (pers. comm.) and found to be that of an ungulate. If the Nkondo lower canine of a putative hominoid (NK696'86), referred to "Gorillinae?" by Pickford et al. (1988; see also Cecchi and Pickford 1989) is confirmed, this specimen represents the only definitive fossil evidence of Pliocene nonhominid Hominoidea in the Western Rift, or in Africa for that matter. From the Eastern Rift, the left maxilla from the late Miocene of Samburu Hills, Kenya (KNM-SH 8531), dating from c. 9 MYA has been suggested as a gorilla ancestor or a common African great ape-hominid ancestor (Ishida et al. 1984).

The fossil record of Hominidae in the Western Rift is equally sparse. Senut et al. (1987) report a fragmentary hominid cranium attributed to *Homo* sp. from the Behanga Member of the Nyabusosi Formation on the eastern side of the Western Rift in Uganda. Pickford et al. (1989) estimate its age, based on its similarity to the fauna of Olduvai Bed II, at 0.7 to 1.8 MYA. Stone tools of Oldowan character are also known from this same formation (ibid.). Pickford et al. (1991) correlate this level to close to the 1.5 MYA date for the recently discovered Kagusa Tuff. The presence of Pliocene Hominidae 2.0 to 2.3 MYA on the Zairean side of the Western Rift is indicated by abundant archaeological data (Harris et al. 1987, 1990; Boaz et al. 1992).

A MODEL FOR THE AFRICAN APE-HOMINID DIVERGENCES

With the general acceptance of the "late divergence" hypothesis for the hominid-great-ape split (Andrews and Cronin 1982; Simons 1989), the discovery of Pliocene apes and pre-australopithecine hominids has become central to resolving questions of hominid origins (Boaz 1983; Hill and Ward 1988; Andrews 1992). The Western Rift divides all the known African equatorial, that is, non-South African, early hominid sites to the east, from which no known fossil apes are known, from the current ranges of *Gorilla* and *Pan* to the west (Figure 17-4).

This biogeographic pattern of apes in the west and hominids in the east likely reflects a pattern of phylogeny (see also Boaz 1993). Using the cladistic terminology of vicariant biogeography (e.g., Platnick and Nelson 1978; Endler 1982), one may hypothesize that a population of the common hominid-panid-gorillid ancestor occupying a wide area of forested Central Africa was split into two by a vicariant event. If one accepts that the African ape-hominid split is an unresolvable trichotomy (e.g., Andrews 1992), then a hypothesis that the formation of the Western Rift Valley effected this phylogenetic divergence is sufficient to explain the cladogram.

However, traditional cladistic clustering together of the African hominoids, or the increasingly well-supported notion of a chimp-human clade (Figure 17-5), causes problems for this unitary biogeographic explanation. Either nodes A1 and B1 or nodes A2 and B2 in Figure 17-5 can be explained by the vicariant event of the formation of the Western Rift, but some other mechanism needs to be invoked to explain the other.

In considering the two vicariant biogeographic cladograms in Figure 17-5, the ecological adaptations of the three extant taxa become important considerations. In Cladogram A node A1 is a split between an open-country taxon (*Homo*) and forest-living taxa (*Pan* and *Gorilla*). This is the traditional interpretation (e.g., Boaz and Burckle 1984) and has been related to the formation of the Western Rift (e.g., Kortlandt 1974). Node A2 has been left largely bereft of attention since there is a dearth of fossil evidence bearing

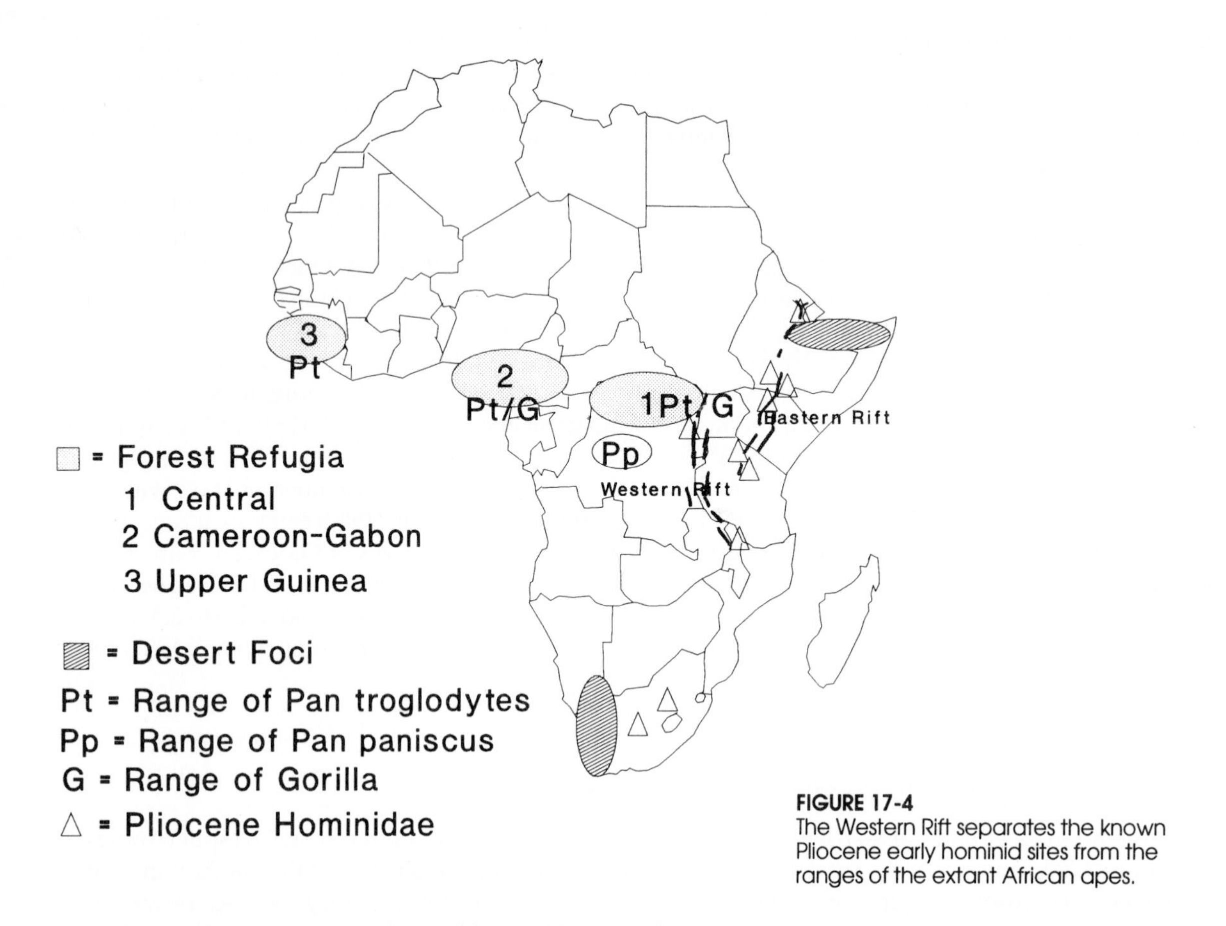

FIGURE 17-4
The Western Rift separates the known Pliocene early hominid sites from the ranges of the extant African apes.

on it, but following Boaz and Burckle (1984) it could be related to any one of several Plio-Pleistocene climatic fluctuations in which forests contracted. This explanation for Node A2 suffers from the perspective of vicariant biogeography, however, since there is an unclear pattern of separation of ranges of *Pan* and *Gorilla* (Figure 17-4). In fact *Pan* overlaps *Gorilla* entirely.

Cladogram B, however, posits a basically different biogeographic pattern (see also Boaz 1993). Node B1 represents a split between taxa that live in dense forest (*Gorilla*) and taxa that live in forest-edge/woodland savanna environments (*Homo* and *Pan*; see Moore 1992 for a recent discussion of "savanna chimp" habitats). It is suggested here that Node B1 corresponds to middle Miocene global cooling and forest contraction, dating to 12-14 MYA (Figure 17-6). The vicariant event effecting this split then was not the formation of the Western Rift, if we accept recent dating of this to about 8 MYA (Ebinger, 1989a,b), but the breakup of forests in Central Africa along an east-to-west gradient due to tropical aridity. The formation of the Eastern Rift and its attendant rain shadow effect on Central Africa probably had an additive effect (see Andrews and Van Couvering 1975).

If the Samburu Hills hominoid represents a phyletic gorillid, its putative date of c. 9 MYA accords with this model, but its geographic placement must be explained by a subsequent dispersal of lowland forest into East Africa. Since the Samburu Hills fauna seems to be characteristically East African middle-late Miocene in its open-country ecological character (Nakaya et al. 1984), there is no support from the fossil record at present for this event. That such an incursion into eastern Africa of Central African forest is possible, however, is documented during Member A of the Omo Shungura Formation at 4.1 MYA. Here are both the Central African rain forest tree *Antrocaryon* (Bonnefille and Dechamps 1983) and the Central African rain forest gastropod *Potadoma* (Williamson 1985). The putative gorillid canine from the

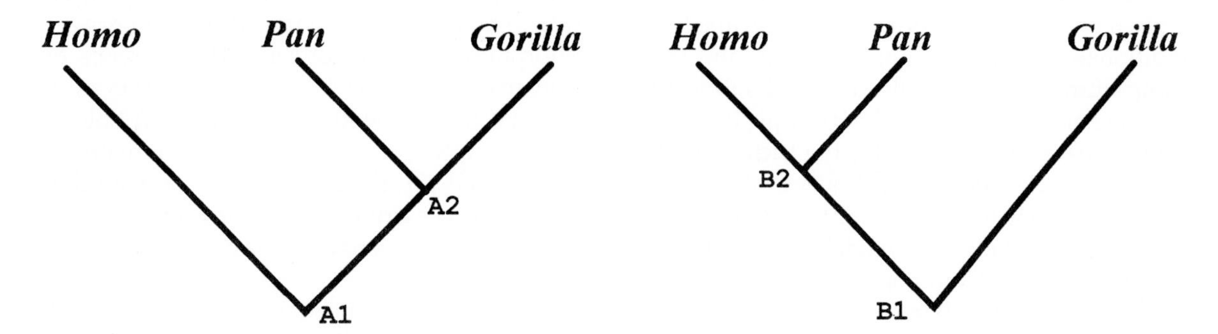

FIGURE 17-5 Biogeographic cladograms of the extant African apes and humans. Cladogram A on the left is the traditional interpretation. Cladogram B, in which humans and chimpanzees form a sister group, is the preferred cladogram in this chapter.

Nkondo Formation of Uganda is associated with a clear forest fauna (Pickford et al. 1989) that also includes the presence of *Antrocaryon* (Dechamps et al. 1992). Pending discovery of more definitive material from Samburu Hills, the lack of a clearly defined forest paleoecological context argues against the hominoid from this site representing a phyletic gorillid.

The formation of the Western Rift at c. 8 MYA is implicated in Node B2 at which point the savanna-living hominids to the east split from the open forest/woodland chimpanzees to the west. The earliest hominid currently known is the Lothagam mandible dated at 5 to 6 MYA from the Eastern Rift (Hill and Ward 1988). There is no fossil record of chimpanzees. The mosaic contraction and expansion of forests during the Pleistocene served to disperse gorilla and chimpanzee populations into overlapping ranges. Gorillas, however, did not survive in regions without mountains, such as the Zairean basin south of the Zaire River and the Upper Guinea Refuge (Figure 17-4). Montane regions such as the Central Forest Refuge in eastern Zaire and the Cameroon-Gabon Refuge (Kingdon 1971) served as refugia during periods of Pleistocene aridity.

FIGURE 17-6 · Phylogeny of the African apes and hominids in relation to the oxygen isotope curve (from Raymo and Ruddiman 1992). The vicariant event effecting the gorilla split is suggested to have been related to the precipitous global temperature drop in the middle Miocene. Major rigt activity along the Western Rift Valley is suggested to have been implicated in the split of the panid and hominid lineages in the late Miocene.

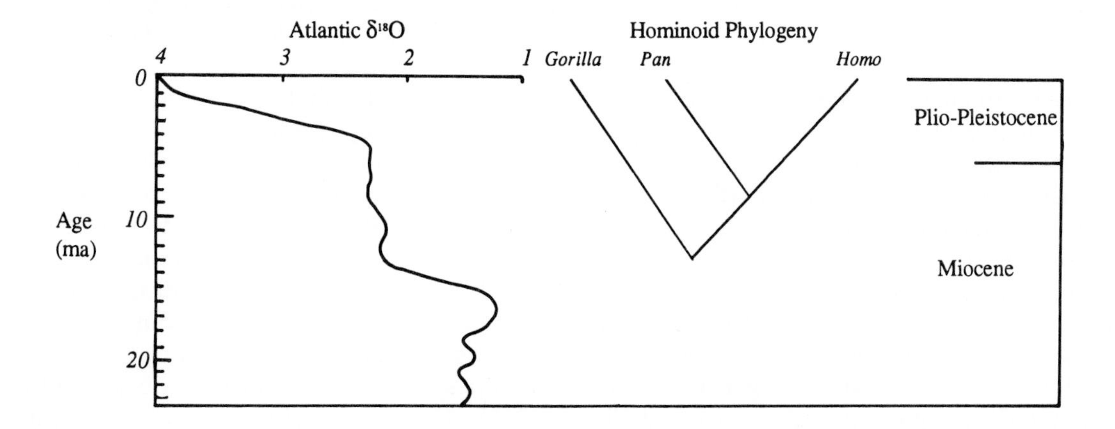

The vicariant biogeographic model for African Hominoidea presented here has important phylogenetic implications. Since it resolves the African ape-hominid trichotomy in favor of a hominid-panid lineage, the traditionally accepted synapomorphies and symplesiomorphies no longer obtain. The presence of a chimp-human clade has also received support from an increasing number of recent molecular studies (Caccone and Powell 1989; Sibley et al. 1990; Ruvulo et al. 1991; Horai et al. 1992). But morphological analyses have failed to resolve the trichotomy because of incomplete or nonexistent fossil remains and because of unclear polarity and valence of characters (see Andrews 1992). The biogeographic and molecular results allow one to model ancestral hominid and ape populations based on the presumed presence of a panid-hominid clade. This model is substantially different from past opinions on the last common ancestor of African apes and hominids.

The knuckle-walking locomotor adaptation of the African apes has been considered perhaps the primary shared derived feature of a panid-gorillid clade. If this clade is now broken up these adaptations must be considered either symplesiomorphies, shared by inheritance from a common ancestor, or homoplasies, similar adaptations developed in parallel. The apparent lack of any knuckle-walking adaptations in the proconsulids (Beard et al. 1986), *Sivapithecus* (Pilbeam et al. 1990), dryopithecids (Begun 1992), *Homo* (Susman 1979; but see Corruccini this volume), or *Australopithecus* (Susman et al. 1984), that is, in the ancestral populations or in any of the other derivative hominoid groups, indicates that knuckle-walking is not a primitive adaptation for apes. It is much more probable that it arose independently in the panid and gorillid lineages. Boaz (1993) has suggested that knuckle-walking in *Pan* and *Gorilla*, and the closely related fist-walking adaptation in *Pongo*, developed independently because of parallel increase in body size in these three lineages. Allometric increase in upper limb and shoulder size shifted the body's center of gravity in the large apes forward, making support by the forelimbs essential when on the ground. Small-bodied hominoids such as gibbons and hominids with a brachiational or "antipronograde" (Stern 1975) heritage are bipedal when on the ground.

Models for the locomotor patterns of the last common hominid-panid-gorillid ancestor then should be reconsidered. While recognizing the genetic closeness between *Pan* and Hominidae and consequently the short evolutionary distance between the two, it is reasonable to suggest that rather than using the chimpanzee as a model for understanding hominid bipedalism (e.g., Zihlman and Brunker 1979), early hominid bipedalism might provide some insights into early panid locomotion. Stern (1975) has suggested that the common ancestor was an "antipronograde" brachiator like the orangutan whose extensively elongated upper limbs might have been rendered useless for quadrupedal locomotion. Others, such as Prost (1980) have suggested that the common ancestor was a climber and that these adaptations were preadaptive for bipedalism. Tuttle (1974) suggested a small-bodied "hylobatian" model for hominid origins, following Morton (1924). Tuttle (1974:391) quotes Morton (1924:39) in describing the hypothetical prehominid in terms of "erect posture habitual, bipedal movements in the trees and on the ground, as in the modern gibbon." None of these perspectives seems antithetical to the model proposed here, which of course must await tests from the fossil record.

The vicariant biogeographic hypothesis suggested here also implies that much of the known middle Miocene hominoid record in eastern Africa, including such taxa as *Kenyapithecus*, *Afropithecus*, and *Turkanapithecus* (see Andrews 1992), are unrelated to later hominid, panid, and gorillid phylogeny. McCrossin and Benefit's conclusions (this volume) that the recently discovered humerus of *Kenyapithecus wickeri* from Maboko indicates a monkeylike quadrupedal mode of locomotion, quite unlike that of modern hominoids, supports this view. According to most biomolecular analyses and the timing of the biogeographic model presented here, when these hominoids were living in eastern Africa, the ancestors of the hominids and the modern African apes were still denizens of the Central African forests.

In a commentary on a meeting concerned with "recent advances in hominid origins" Howell (1972) hypothesized a "late divergence" of the African apes and hominids about a decade before it was generally accepted (e.g., Andrews and Cronin 1982). He stated, "[I]n my view, the origins of Hominidae lie demonstrably prior to five to six million years ago [Washburn's view], while they are not demonstrably prior to fifteen million years ago [Pilbeam's view]. So a reasonable estimate would be ~10 ± 4 million years." This estimate is still in accord with the date of hominid divergence proposed in this paper. Howell (1972:128) further noted that "it is necessary to stress that the matter of hominid origins is still largely a subject of speculation . . . and the formulation of hypotheses is challenging. However, hypotheses must be evaluated and tested in the light of empirical data, old and new, and revised or discarded accordingly. Evidently, there is a critical need for further empirical data." This chapter has proposed hypotheses that some would certainly term "speculative," and further fieldwork is needed to address the still "critical need for further empirical data." Fossiliferous localities in the Western Rift constitute the only suite of Mio-Pleistocene sites contiguous to the current ranges of both *Pan* and *Gorilla* and this is perhaps their primary importance for further work. The late Neogene of the Western Rift provides a valuable view of a Central African biogeographic zone different from that of eastern Africa during the period of time when hominids differentiated. Fossiliferous localities in eastern Zaire and western Uganda can be expected to provide tests of hypotheses on African ape and hominid emergence and to shed much light on the environmental conditions to which Central African Hominoidea responded in their evolutionary histories.

SUMMARY

The Western Rift Valley of Zaire and Uganda provides a unique natural laboratory for the investigation of problems of central importance to the evolutionary biology of hominoids—the determination of the paleoecology, biogeography, and natural history of the African ape-hominid evolutionary divergences. Subsequent to these basal splits, research on the hominid record in the Western Rift contributes important data on early hominid environments outside the well-known East and South African Plio-Pleistocene fossil sites, temporal correlation with which is now facilitated by the geochemical identification of several tuffs discovered in common between the two rifts. Late Miocene/Early Pliocene fauna and flora in Zaire and Uganda indicate the presence of dense forests. In the Plio-Pleistocene there are clear records for the first time of savanna conditions associated with archaeological documents of early hominid presence. In depositional environments with archaeological artifacts, forest taphocenoses were distal, and no ape fossils have been recovered. At Nkondo, a site dated at over 3.5 MYA in western Uganda with a forested environment, a single gorilla-like canine has been discovered. A vicariance biogeographic model for the ape-hominid divergences is proposed that posits an early split of the gorillid lineage off from a common panid-hominid lineage. Because of the reversal in polarity of characters that this model forces, the common gorillid-panid-hominid ancestor is proposed to have been a small-bodied, facultative biped. Global cooling in the Middle to Late Miocene, with associated aridity in tropical Africa, is proposed as the effector of the vicariant event for the gorillid split. The faulting and tectonic uplift of the Western Rift in the Late Miocene and the attendant paleoenvironmental changes are proposed as constituting the vicariant event that split the panid and hominid lineages. Much further work in Western Rift fossil localities will be necessary to test this hypothesis.

ACKNOWLEDGMENTS

F. Clark Howell had much to do with inspiring and facilitating the last decade of work in the Zairean Western Rift, and he is warmly thanked for his support and for providing the theoretical and methodological background on which the Semliki Research Expedition was organized. I appreciate the invitation by Russ Ciochon and Rob Corruccini to contribute to this volume. I thank my colleagues who have served as Co-Principal Investigators or project directors on the Semliki Research Expedition: Alison Brooks, J. W. K. Harris, Dorothy Dechant Boaz, Kanimba Misago, Paris Pavlakis, Horst D. Steklis, and Peter G.

Williamson, as well as the many members of the SRE teams, for the results that have led to the conclusions presented in this chapter. Meleisa McDonell, Frank Brown, John Harris, and Peter Williamson are thanked for assistance in the preparation of this chapter. Research has been supported by the National Science Foundation, Earthwatch, National Geographic Society, L.S.B. Leakey Foundation, and Wenner-Gren Foundation.

LITERATURE CITED

Andrews, P. (1992) Evolution and environment in the Hominoidea. *Nature* 360:641-646.

Andrews, P., and Cronin, J. E. (1982) The relationships of *Sivapithecus* and *Ramapithecus* and the evolution of the orangutan. *Nature* 297:541-546.

Andrews, P., and Van Couvering, J. A. (1975) Palaeoenvironments in the East African Miocene. *Contr. Primatol.* 5:62-103.

Aoki, R. (1992) Fossil crocodilians from the late Tertiary strata in the Sinda basin, eastern Zaire. *African Study Monographs, Kyoto Univ., Suppl.* 17:67-85.

Begun, D. R. (1992) Miocene fossil hominoids and the chimp-human clade. *Science* 257:1929-1933.

Beard, K. C., Teaford, M. F., and Walker, A. (1986) New wrist bones of *Proconsul africanus* and *P. nyanzae* from Rusinga Island, Kenya. *Folia Primatol.* 47:97-118.

Bernor, R. L., and Sanders, W. J. (1990) Fossil Equidae from Plio-Pleistocene strata of the Upper Semliki Zaire. In N. T. Boaz (ed.): Evolution of environments and Hominidae in the African Western Rift Valley. *Virginia Museum Nat. Hist. Mem.* 1:189-196.

Bishop, W. W. (1962) Pleistocene correlation in the Uganda section of the Albert-Edward Rift Valley. In G. Mortelmans and J. Nenguin (eds.): *Actes du IV Congres Panafricain de Prehistoire et de l'Etude du Quaternaire.* Musée Royal de l'Afrique Centrale, Annales., Sci. Hum. 40:245-253.

Bishop, W. W. (1963) The later Tertiary and Pleistocene in eastern equatorial Africa. In F. C. Howell and F. Bourliere (eds.): *African Ecology and Human Evolution.* Chicago: Aldine, pp. 246-275.

Bishop, W. W. (1969) Pleistocene stratigraphy in Uganda. *Geol. Sur. Uganda Mem.* no. 10.

Boaz, N. T. (1983) Morphological trends and phylogenetic relationships from middle Miocene hominoids to late Pliocene hominids. In R. L. Ciochon and R. S. Corruccini (eds.): *New Interpretations of Ape and Human Ancestry.* New York: Plenum, pp. 705-720.

Boaz, N. T. (ed.) (1990) Evolution of environments and Hominidae in the African Western Rift Valley. *Virginia Museum Nat. Hist. Mem.* 1.

Boaz, N. T. (1993) Origins of Hominidae. In A. Almquist and A. Manyak (eds.): *Milestones in Human Evolution.* Prospect, IL: Waveland, pp. 141-166.

Boaz, N. T., Bernor, R. L., Cooke, H. B. S., Dechamps, R., de Heinzelin, J., Harris, J. W. K., Gentry, A. W., Meylan, P., Pavlakis, P. P., Sanders, W. J., Verniers, J., Williamson, P. G., and Winkler, A. (1992) A new evaluation of the significance of the Late Neogene Lusso Beds, Upper Semliki Valley, Zaire. *J. Hum. Evol.* 22:505-517.

Boaz, N. T., and Burckle, L. H. (1984) Paleoclimatic framework for African hominid evolution. In J. C. Vogel (ed.): *Late Cainozoic Palaeoclimates of the Southern Hemisphere.* Rotterdam: Balkema, pp. 483-490.

Boaz, N. T., Pavlakis, P. P., and McDonell, M. (in press). Analysis of the Senga 13B excavation: Early hominid environments in the late Pliocene of eastern Zaire. *National Geographic Research & Exploration.*

Bonnefille, R., and Dechamps, R. (1983) Data on fossil flora. In J. de Heinzelin (ed.): *The Omo Group: Archives of the International Omo Research Expedition.* Tervuren: Musée Roy. Afr. Cent., pp. 191-207.

Brain, C. K. (1981) The evolution of man in Africa: Was it a consequence of Cainozic cooling? Alex du Toit Mem. Lecture 17. *Trans. Geol. Soc. S. Afr. Annex* 84:1-19.

Brook, G. A., Burney, D. A., and Cowart, J. B. (1990) Paleoenvironmental data for Ituri, Zaire, from sediments in Matupi Cave, Mt. Hoyo. In N. T. Boaz (ed.): Evolution of environments and Hominidae in the African Western Rift Valley. *Virginia Museum Nat. Hist. Mem.* 1:49-100.

Brown, F. H., McDougall, I., Davies, I., and Maier, R. (1985) An integrated Plio-Pleistocene chronology for the Turkana basin. In E. Delson (ed.): *Ancestors: The Hard Evidence.* New York: Alan R. Liss, pp. 82-90.

Caccone, A., and Powell, J. R. (1989) DNA divergence among hominoids. *Evolution* 43:925-942.

Cecchi, J. M., and Pickford, M. (1989) Une nouvelle technique non destructive de détermination de la structure prismatique de l'émail des dents chez les mammifères fossilies. *C. R. Acad. Sci., Ser.* 2, 308:1651-1654.

Cooke, H. B. S. (1990) Suid remains from the Upper Semliki area, Zaire. In N. T. Boaz (ed.): Evolution of environments and Hominidae in the African Western Rift Valley. *Virginia Museum Nat. Hist. Mem.* 1:197-201.

Cooke, H. B. S., and Coryndon, S. (1970) Pleistocene mammals from the Kaiso Formation and other related deposits in Uganda. In L. S. B. Leakey and R. J. G. Savage (eds.): *Fossil Vertebrates of Africa,* Vol. 2. London: Academic Press, pp. 107-224.

Darwin, C. R. (1874) *The Descent of Man and Selection in Relation to Sex,* 2nd ed. New York: Hurst and Co.

Dechamps, R., and Maes, F. (1990) Woody plant communities and climate in the Pliocene of the Semliki Valley, Zaire. In N. T. Boaz (ed.): Evolution of environments and Hominidae in the African Western Rift Valley. *Virginia Museum Nat. Hist. Mem.* 1:71-94.

Dechamps, R., Senut, B., and Pickford, M. (1992) Fruits fossiles pliocènes et pléistocènes du Rift Occidental ougandais. Signification paleoenvironmentale. *C. R. Acad. Sci. Paris, Ser. 2*, 314:325-331.

de Heinzelin, J. (1955) *Le Fossé Tectonique sous le Parallèle d'Ishango*. Inst. Parcs Natl. Congo Belge. Explor. Parc Natl. Albert, Mission J. de Heinzelin de Braucourt (1950), Fasc. 1.

de Heinzelin, J. (1957) *Les Fouilles d'Ishango*. Inst. Parcs Natl. Congo Belge. Explor. Parc Natl. Albert, Mission J. de Heinzelin de Braucourt (1950), Fasc. 2.

de Heinzelin, J. (1988) Photogéologie du Neogene de la Basse-Semliki (Zaire). *Bull. Soc. Belg. Géol.* 97:173-178.

de Heinzelin, J., and Verniers, J. (1987) Premiers resultats du Semliki Research Project (Parc National des Virunga, Zaire). I. Haute Semliki: Revision stratigraphique en cours. *Mus. Roy. Afr. Centr. Rapp. Annu. Geol.* 1985-86:141-144.

Drake, R. E., Van Couvering, J. A., Pickford, M. H., Curtis, G. H., and Harris, J. A. (1988) New Chronology for the Early Miocene mammalian faunas of Kisingiri, Western Kenya. *J. Geol. Soc. London* 145:479-491.

Ebinger, C. J. (1989a) Geometric and kinematic development of border faults and accommodation zones, Kivu-Rusizi Rift, Africa. *Tectonics* 8:117-133.

Ebinger, C. J. (1989b) Tectonic development of the western branch of the East African rift system. *Geol. Soc. Amer. Bull.* 101:885-903.

Endler, J. A. (1982) Problems in distinguishing historical from ecological factors in biogeography. *Amer. Zool.* 22:441-452.

Feibel, C. S., Brown, F. H., and McDougall, I. (1989) Stratigraphic context of fossil hominids from the Omo Group deposits: Northern Turkana Basin, Kenya, and Ethiopia. *Am. J. Phys. Anthropol.* 78:595-622.

Fuchs, V. E. (1934) The geological work of the Cambridge expedition to the East African lakes, 1930-1931. *Geol. Mag.* 71:97-112; 72:145-166.

Gautier, A. (1965) *Geological Investigation in the Sinda-Mohari (Ituri, NE-Congo): A Monograph on the Geological History of a Region in the Lake Albert Rift*. Gent: Rijksuniversiteit te Gent.

Gautier, A. (1967) New observations on the later Tertiary and early Quaternary in the Western Rift: The stratigraphic and paleontological evidence. In W. W. Bishop and J. D. Clark (eds.): *Background to Evolution in Africa*. Chicago: University of Chicago Press, pp. 73-87.

Gentry, A. W. (1970) The Bovidae (Mammalia) of the Fort Ternan fossil fauna. In L. S. B. Leakey (ed.) *Fossil Vertebrates of Africa*, Vol. 2. London: Academic Press, pp. 243-323.

Gentry, A. W. (1990) The Semliki fossil bovids. In N. T. Boaz (ed.): Evolution of environments and Hominidae in the African Western Rift Valley. *Virginia Museum Nat. Hist. Mem.* 1:225-234.

Greenwood, P. H., and Howes, G. (1975) Neogene fossil fishes from the Lake Albert-Lake Edward Rift (Zaire). *Bull. Brit. Mus. (Nat. Hist.), Geol.* 26:71-127.

Harris, J. W. K., Williamson, P. G., Morris, P. J., de Heinzelin, J., Verniers, J., Helgren, D., Bellomo, R. V., Laden, G., Spang, T. W., Stewart, K. M., and Tappen, M. J. (1990) Archaeology of the Lusso Beds. In N. T. Boaz (ed.): Evolution of environments and Hominidae in the African Western Rift Valley. *Virginia Museum Nat. Hist. Mem.* 1:237-272.

Harris, J. W. K., Williamson, P. G., Verniers, J., Tappen, M. J., Stewart, K., Helgren, D., de Heinzelin, J., Boaz, N. T., and Bellomo, R. V. (1987) Late Pliocene hominid occupation in Central Africa: The setting, context, and character of the Senga 5A site, Zaire. *J. Hum. Evol.* 16:701-728.

Hill, A., and Ward, S. (1988) Origin of Hominidae: The record of African large hominoid evolution between 14 MY and 4 MY. *Yearb. Phys. Anthropol.* 31:49-83.

Hirayama, R. (1992) Fossil turtles from the Neogene strata in the Sinda Basin, eastern Zaire. *African Study Monographs, Kyoto Univ., Suppl.* 17:49-65.

Hooijer, D. A. (1963) Miocene Mammalia of Congo. *Ann. Mus. Roy. Afr. Cent., Sci. Geol.* 46:1-77.

Hooijer, D. A. (1970) Miocene Mammalia of Congo, a correction. *Ann. Mus. Roy. Afr. Cent., Sci. Geol.* 67:163-167.

Hopwood, A. T., and Lepersonne, J. (1953) Présence de formations d'âge miocene inferieur dans le fossé tectonique du Lac Albert et de la Basse Semliki (Congo belge). *Ann. Soc. Geol. Belg.* 77:83-113.

Howell, F. C. (1972) Discussion: Man's evolutionary past. *Soc. Biol.* 19:128-135.

Horai, S., Satta, Y., and Hayasaka, K. (1992) Man's place in Hominidae revealed by mitochondrial DNA genealogy. *J. Mol. Evol.* 35:32-43.

Ishida, H., Pickford, M., Nakaya, H., and Nakano, Y. (1984) Fossil anthropoids from Nachola and Samburu Hills, Samburu District, Kenya. *African Study Monographs*, Kyoto Univ., Suppl. 2:73-85.

Johanson, D. C., Splingaer, M., and Boaz, N. T. (1975) Paleontological excavations in the lower Omo Valley, Ethiopia. In Y. Coppens, F. C. Howell, G. L. Isaac, and R. E. F. Leakey (eds.): *Earliest Man and Environments in the Lake Rudolf Basin*, Chicago: University of Chicago Press, pp. 402-420.

Kappelman, J. (1991) The paleoenvironment of *Kenyapithecus* at Fort Ternan. *J. Hum. Evol.* 20:95-129.

Kingdon, J. (1971) *East African Mammals*, Vol. 1. London: Academic.

Kortlandt, A. (1972) *New Perspectives on Ape and Human Evolution*. Amsterdam: Stichtung voor Psychobiologie.

Kortlandt, A. (1974) New perspectives on ape and human evolution (Book review with comments). *Curr. Anthropol.* 15:427-448.

Leakey, L. S. B. (1967) An early Miocene member of the Hominidae. *Nature* 213:155-163.

Lepersonne, J. (1949) Le fossé tectonique du Lac Albert-Semliki-Lac Édouard. Résumé des observations geologiques effectuées en 1938, 1939, 1940. *Ann. Soc. Geol. Belg.* 72:1-92.

Makinouchi, T., Ishida, S., Sawada, Y., Kuga, N., Kimura, N., Orihashi, Y., Bajope, B., Yemba, M., and Ishida, H. (1992) Geology of the Sinda-Mohari region, Haut-Zaire Province, eastern Zaire. *African Study Monographs,* Kyoto Univ., Suppl. 17:3-18.

Meylan, P. (1990) Fossil turtles from the Upper Semliki, Zaire. In N. T. Boaz (ed.): Evolution of environments and Hominidae in the African Western Rift Valley. *Virginia Museum Nat. Hist. Mem.* 1:163-170.

Moore, J. (1992) "Savanna" chimpanzees. In T. Nishida, W. C. McGrew, P. Marler, M. Pickford, F. B. M. de Waal (eds.): *Topics in Primatology,* Vol. 1: *Human Origins.* Tokyo: Univ. Tokyo Press, pp. 99-118.

Morton, D. J. (1924) Evolution of the human foot. II. *Am. J. Phys. Anthropol.* 7:1-52.

Nakaya, H., Pickford, M., Nakano, Y., and Ishida, H. (1984) The late Miocene large mammal fauna from the Namurungule Formation, Samburu Hills, northern Kenya. *African Study Monographs,* Kyoto Univ. Suppl. 2:87-131.

Pavlakis, P. P. (1987) Biochronology, Paleoecology and Biogeography of the Plio-Pleistocene Fossil Mammal Faunas of the Western Rift (East-Central Africa) and Their Implication for Hominid Evolution. Ph.D. Dissertation, Department of Anthropology, New York University.

Pavlakis, P. P. (1990) Plio-Pleistocene Hippopotamidae from the Upper Semliki. In N. T. Boaz (ed.): Evolution of environments and Hominidae in the African Western Rift Valley. *Virginia Museum Nat. Hist. Mem.* 1:203-223.

Pickford, M. (1987) Implications de la succession des mollusques fossiles du Bassin du lac Albert (Ouganda). *C. R. Acad. Sci. Ser.* 2, 305(4):317-322.

Pickford, M. (1989) Evidence for climatic changes near the Miocene-Pliocene boundary in tropical Africa. *Boll. Soc. Paleontol. Ital.* 28(2-3):317-320.

Pickford, M. (1990a) Tempo and mode of molluscan evolution in the Pliocene of the Albertine Rift, Uganda-Zaire. *C. R. Acad. Sci. Ser.* 2, 311(9):1103-1109.

Pickford, M. (1990b) Uplift of the roof of Africa and its bearing on the evolution of mankind. *Hum. Evol.* 5:1-20.

Pickford, M., Senut, B., Hadoto, D., Musisi, J., and Kariira, C. (1986) Découvertes récentes dans les sites miocènes de Moroto (Ouganda oriental): Aspects biostratigraphiques et paléoécologiques. *C. R. Acad. Sci. Ser.* 2, 302:681-686.

Pickford, M., Senut, B., Puppeau, G., Brown, F. H., and Haileab, B. (1991) Correlation of tephra layers from the Western Rift Valley (Uganda) to the Turkana Basin (Ethiopia/Kenya) and the Gulf of Aden. *C. R. Acad. Sci. Ser.* 2, 313:223-229.

Pickford, M., Senut, B., Ssemmanda, I., Elepu, D., and Obwona, P. (1988) Premiers resultats de la mission de l'Uganda Palaeontology Expedition à Nkondo (Pliocène du Bassin du Lac Albert, Ouganda). *C. R. Acad. Sci. Ser.* 2, 306:315-120.

Pickford, M., Senut, B., Roche, H., Mein, P., Ndaati, G., Obwona, P., and Tuhumwire, J. (1989) Uganda Palaeontology Expedition: Resultats de la deuxième mission (1987) dans la region de Kisegi-Nyabusosi (bassin du lac Albert, Ouganda). *C. R. Acad. Sci. Paris, Ser 2,* 308:1751-1758.

Pilbeam, D. R., Rose, M. D., Barry, J. C., and Shah, I. (1990) New *Sivapithecus* humeri from Pakistan and the relationship of *Sivapithecus* and *Homo. Nature* 348:237-239.

Platnick, N. I., and Nelson, G. (1978) A method of analysis for historical biogeography. *Syst. Zool.* 27:1-16.

Prost, J. H. (1980) Origin of bipedalism. *Am. J. Phys. Anthropol.* 52:175-189.

Raymo, M. E., and Ruddiman, W. F. (1992) Tectonic forcing of late Cenozoic climate. *Nature* 359:117-122.

Ruvulo, M., Disotell, T. R., Allard, M. W., Brown, W. M., and Honeycutt, R. L. (1991) Resolution of the African hominoid trichotomy by use of a mitochondrial gene sequence. *Proc. Nat. Acad. Sci.* 88:1570-1574.

Sanders, W. J. (1990) Fossil Proboscidea from the Pliocene Lusso Beds of the Western Rift, Zaire. In N. T. Boaz (ed.): Evolution of environments and Hominidae in the African Western Rift Valley. *Virginia Museum Nat. Hist. Mem.* 1:171-187.

Senut, B., Pickford, M., Ssemmanda, I., Elepu, D., and Obwona, P. (1987) Découverte du premier Homininae (*Homo* sp.) dans le Pléistocène de Nyabusosi (Ouganda Occidental). *C. R. Acad. Sci. Ser.* 2, 305:819-822.

Sibley, C. G., Comstock, J. A., and Ahlquist, J. E. (1990) DNA hybridization evidence of hominoid phylogeny: A reanalysis of the data. *J. Mol. Evol.* 30:202-236.

Shipman, P. (1986) Paleoecology of Fort Ternan reconsidered. *J. Hum. Evol.* 15:193-204.

Simons, E. L. (1989) Human origins. *Science* 245:1343-1350.

Stanley, S. M. (1992) An ecological history of the genus *Homo. Paleobiology* 18:237-257.

Stern, J. (1975) Before bipedality. *Yearb. Phys. Anthropol.* 19:59-68.

Stewart, K. M. (1990) Fossil fish from the Upper Semliki. In N. T. Boaz (ed.): Evolution of environments and Hominidae in the African Western Rift Valley. *Virginia Museum Nat. Hist. Mem.* 1:141-162.

Susman, R. L. (1979) Comparative and functional morphology of hominoid fingers. *Am. J. Phys. Anthropol.* 50:215-236.

Susman, R. L., Stern, J. T., and Jungers, W. L. (1984) Arboreality and bipedality in the Hadar Ethiopia hominids. *Folia Primatol.* 43:113-156.

Tuttle, R. (1974) Darwin's apes, dental apes, and the descent of man: Normal science in evolutionary anthropology. *Curr. Anthropol.* 15:389-426.

Van Neer, W. (1992) New late Tertiary fish fossils from the Sinda region, eastern Zaire. *African Study Monographs,* Kyoto Univ., Suppl. 17:27-47.

Verniers, J., and de Heinzelin, J. (1990) Stratigraphy and geological history of the Upper Semliki: A preliminary report. In N. T. Boaz (ed.): Evolution of environments and Hominidae in the African Western Rift Valley. *Virginia Museum Nat. Hist. Mem.* 1:17-39.

von Bartheld, F., Erdbrink, D. P., and Krommenhoek, W. (1970) A fossil incisor from Uganda and a method for its determination. *Konink. Nederl. Akad. Wetensch., Ser. B,* 73:426-431.

Wayland, E. J. (1926) The geology and palaeontology of the Kaiso Bone-beds. *Geol. Surv. Uganda, Occ. Paper* 2:5-12.

Williamson, P. G. (1985) Evidence for an early Plio-Pleistocene rainforest expansion in East Africa. *Nature* 315:487-489.

Williamson, P. G. (1990) Late Cenozoic mollusc faunas from the North Western African Rift (Uganda-Zaire). In N. T. Boaz (ed.): Evolution of environments and Hominidae in the African Western Rift Valley. *Virginia Museum Nat. Hist. Mem.* 1:125-139.

Williamson, P. G. (1992) Tempo and mode of molluscan evolution in the Plio-Pleistocene of the Edward-Albert Rift, Uganda-Zaire: A reply to Pickford. *C. R. Acad. Sci., Ser. 2,* 315:1139-1145.

Yasui, K., Kunimatsu, Y., Kuga, N., Bajope, B., and Ishida, H. (1992) Fossil mammals from the Neogene strata in the Sinda basin, eastern Zaire. *African Study Monographs,* Kyoto Univ., Suppl. 17:87-107.

Zihlman, A., and Brunker, L. (1979) Hominid bipedalism: Then and now. *Yearb. Phys. Anthropol.* 22:132-162.

CHAPTER 18

An Hypothesis of Heterochrony in Response to Climatic Cooling and Its Relevance to Early Hominid Evolution

Elisabeth S. Vrba

INTRODUCTION

It can hardly be said that paleoanthropologists "are unable to see the forest for the trees"—the hominid tree is so puny that it would be inconspicuous within the luxuriant genealogical forest were it not for our partisan focus. But we do face the challenge of gaining a broad view of that forest—particularly of the numerous genealogical trees of Neogene large mammals in their larger environmental and temporal context—in order to see the single hominid tree more clearly. Clark Howell is preeminent among those who took up this challenge. As a tribute to his work, I discuss the changes in hominid evolution as an integral part of wider biotic and climatic changes. I specifically explore a hypothesis of how ontogenies evolved in relation to climate. We are currently testing this hypothesis by an analysis of hominoid brain development (Vrba and Vaisnys in prep.). Although the analysis is not yet completed, I shall refer to our preliminary results.

Hominine Encephalization in the Latest Pliocene

Hominine encephalization in the latest Pliocene started a new trend of higher evolutionary rates than before (Figure 18-1). Disagreements remain on how to estimate hominid body weights (McHenry 1991, 1992), encephalization quotients (McHenry 1988; Martin 1983), fossil taxonomic and sexual identities, and taxon branching sequence and chronology (Howell 1978; Wood 1992). Also, there is a gap in the record of hominid crania about 2.6–1.9 million years ago (MYA). As a result, the precise Late Pliocene timing of onset of this encephalization trend (Figure 18-1) remains uncertain. Nevertheless, all agree that this change is fundamental to human evolution, that the magnitude of encephalization increase even to earliest known *Homo* was considerable (Figure 18-1), and that the setting was the African late Pliocene. Figure 18-1 is not a genealogical statement. It leaves open whether early *Homo* is most closely related to *Australopithecus*

Editor's Note: A set of examples and glossary of terms pertaining to heterochrony appear at the end of this chapter.

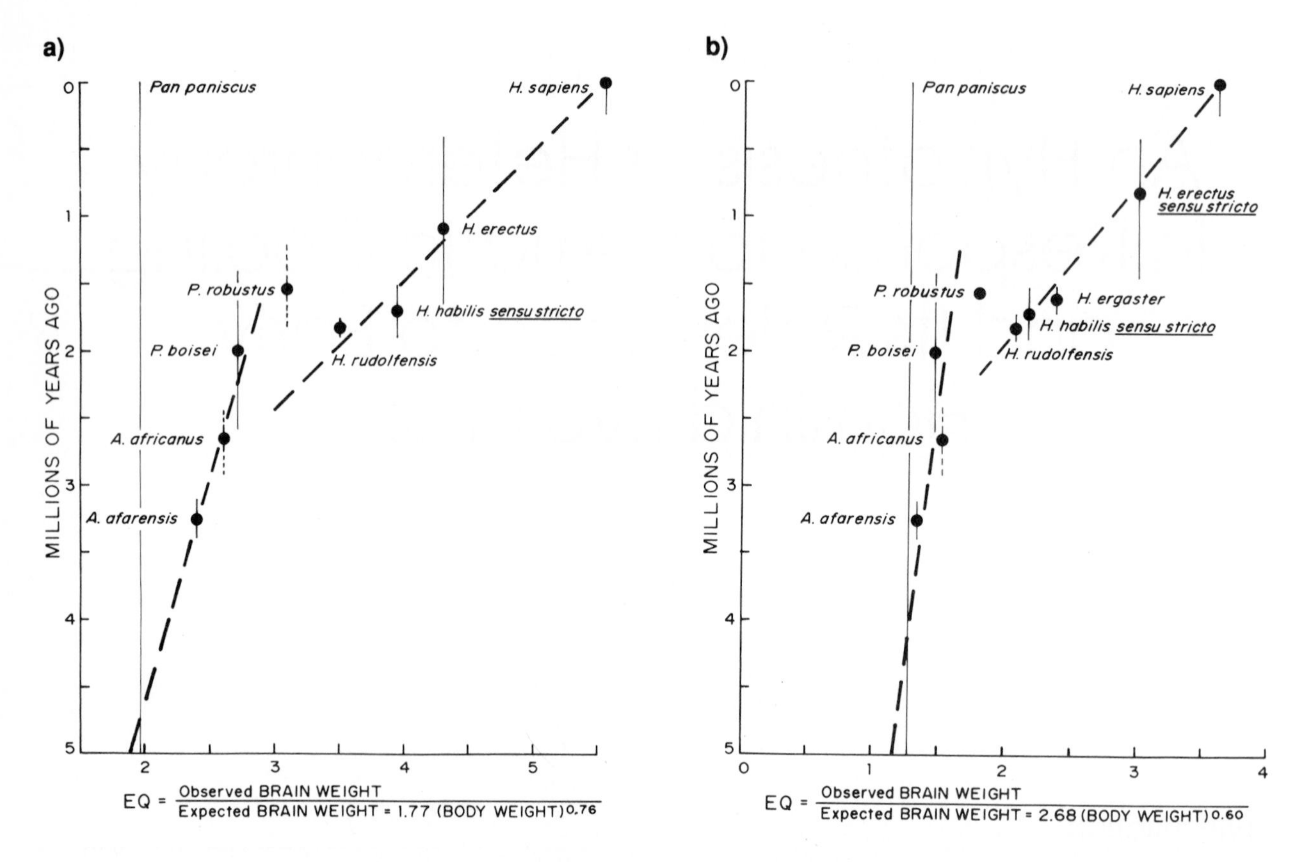

FIGURE 18-1 Evolutionary trends in hominoid encephalization quotients (EQs) obtained using Martin's (1983) formulae for (a) placental mammals and (b) Cercopithecoidea. The graphs do not represent genealogy. Estimates of hominid body weights from McHenry (1991, 1992) and of cranial capacities from Holloway (1970, 1972, 1978), Tobias (1971), Johanson and Edey (1981), and McHenry (1988), and of values for *Pan paniscus* from Martin (1983). Wood (1992) and others he cites view the traditional *Homo habilis* as comprising at least two taxa, *H. rudolfensis* and *H. habilis*, the EQs for which were calculated based on individual specimens as classified in Wood (1992) and McHenry (1992). In (b) a similar procedure was used to subdivide *H. erectus* by separation of *H. ergaster* (Wood 1992). Relative to the EQ axis, all solid dots are placed at mean EQ values. Relative to the time axis, solid dots for *A. africanus* and *P. robustus* are placed at age estimates after Vrba (1982, 1985a, dashed line = error estimate), and for other taxa at mean ages for crania within age ranges (solid lines) as reviewed in Wood (1992).

afarensis, *A. africanus*, or to another perhaps as yet undiscovered taxon (Wood 1992). But it does suggest strongly that the crucial time period for onset of this trend was 2.3–3.0 MYA. As in all evolution, interactions between environment, ontogeny, and population genetic processes must have played a causal role. I propose a particular combination of these factors to explain hominine brain expansion.

Hominine Evolution and Climatic Cooling

I argued previously that stone tools near 2.5 MYA, encephalized hominid fossils some time later, and robust australopithecines appeared strikingly close in time to widespread—indeed global—pulses of **turnover** (boldface terms are discussed in the glossary at the end of the chapter) in other organismal groups; and that the particular patterns of this coincidence suggest a common **initiating cause** of global cooling at this time (Vrba 1974, 1975, 1982, 1985a, 1988). For instance, the mammalian taxa involved in the massive African pulse of first appearances at the end of the Pliocene (Figure 18-2, representing both Eurasian immigration and speciation) on their own without reference to the climatic record bear the unmistakable stamp of aridification and at least seasonal cooling (e.g., Coppens 1975; Wesselman 1984; Vrba 1985a,c). Most date

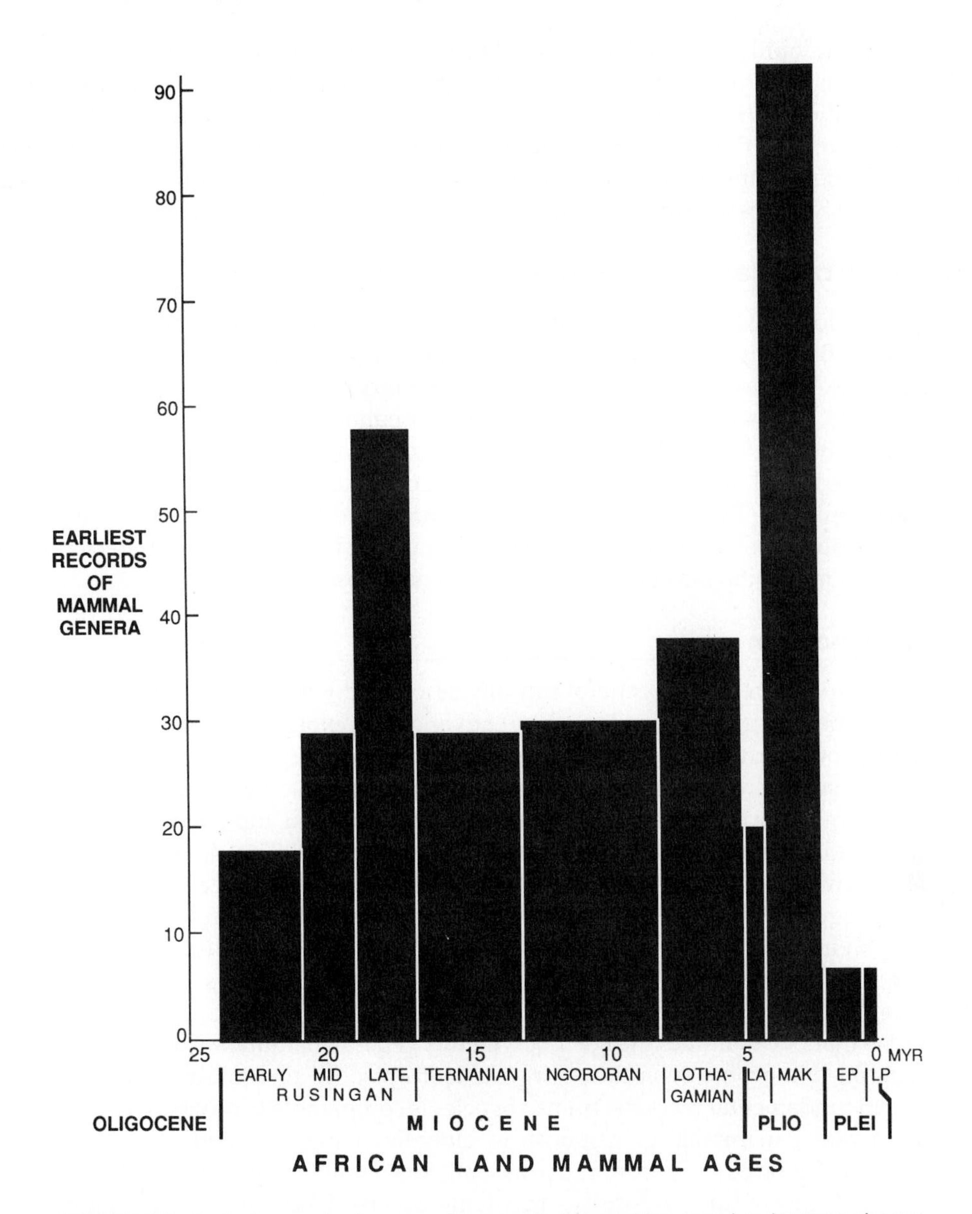

FIGURE 18-2 Earliest records of mammal genera in African Neogene land mammal ages (based on Savage and Russell, 1983.) LA = Langebaanian, MAK = Makapanian, EP, LP = Early, Late Pleistocene, PLIO = Pliocene, PLEI = Pleistocene, MYR = millions of years.

close to 2.5 MYA (Vrba 1985c; additional personal observations), as did turnover on other continents (Azzaroli 1983; Lindsay et al. 1980).

During the mid-1970s, when 2.5-2.0 MYA was first suggested as a period of extraordinary hominid evolution in response to climatic change, there was no consensus on a major global climatic event at this time. Chronological estimates either placed securely identified earliest *Homo* and robusts at later or at disputed earlier dates (Howell 1978). The hypothesis of climatic forcing of hominid evolution has since then gained support both from well-dated evidence for global refrigeration at this time (reviewed in Vrba et al. 1985; Partridge et al. 1986; Burckle et al. 1986; Vrba et al. 1989), and from documented associations of earliest robust australopithecine (Walker et al. 1986) and *Homo* (Hill et al. 1992) fossils with radiometric dates of 2.6–2.4 MYA ago. Furthermore, from this first appearance during a phase of exceptional cooling, the subsequent evolution of *Homo* was accompanied by a progressive intensification of the cold extremes of the astronomical cycles (Vrba et al. 1989). Thus, if we are looking for unusual outside factors to aid our explanation of the unique human brain, the unprecedented cooling trend (unprecedented at least during the Cenozoic) since 2.5 MYA is an excellent candidate.

Hominine Evolution by Heterochrony: Previous Arguments

Heterochrony has a long history among postulated causes of human evolution. Kollman (1905), Bolk (1926, 1929) and De Beer (1958) claimed that humans bear a predominant stamp of **paedomorphosis** (specifically **neoteny**: Figures 18-3, 18-4, 18-5) with **peramorphic** characters like our long legs in the minority (Table 18-1). These claims were subsequently expanded (Montagu 1962, 1981; Lorenz 1970; Gould 1977; Manley-Buser 1986). Others disagreed based on particular aspects of morphology (Moss et al. 1982; Bromage 1985) and on evidence of the very opposite kind of human heterochrony: a massive imprint of **hypermorphosis** (Table 18-1; Shea 1989; McKinney and McNamara 1991). The problem is that hypermorphosis is traditionally expected to result in **pera**morphic hypertrophy of characters (Figure 18-3). How, then, are we to model its cooccurrence with **paedo**morphosis?

A Proposal for a Resolution

I argue that cold temperatures are in general causally associated with *both* hypermorphosis and paedomorphosis in the same species. Reasons have been advanced to expect this both in transient **ecophenotypes** and during long-term evolution. Evidence from many organismal groups supports both expectations. I discuss different models of how ontogeny might evolve by prolongation of ancestral growth rate phases toward a descendant that is both hypermorphosed and paedomorphosed.

I first outline old and new concepts of heterochrony (Figures 18-3, 18-4, 18-5). Next, I focus on the different modes in which hypermorphosis may be accompanied by paedomorphosis (I subsume all such modes, such as in Figures 18-6, 18-7, 18-8a, under the term hyperpaedomorphosis) in response to cold temperature, and on implications for sexual dimorphism (Figure 18-9). Finally, I explore the application to Hominidae, including a preliminary report of our results (Vrba and Vaisnys in prep.) on hominid brain evolution (Figure 18-10).

I hypothesize that one hominid lineage followed this general mammalian tendency and underwent progressive hyperpaedomorphic evolution during major episodes of cooling. Darwin recognized (in Hodge 1983) that in fundamental ways the human species is comparable to others. The general evolutionary patterns of savanna mammals in response to climatic changes provide the logical context for understanding the evolution of early Hominidae as well. It is possible that major physical environmental changes initiated not simply **turnover-pulses** involving diverse kinds of taxa (Vrba 1985b, 1993), but pulses of particular kinds of heterochrony.

HETEROCHRONY: GENERAL COMMENTS

All evolution is first expressed as developmental novelty. The rules governing morphogenesis can constrain and facilitate the directions of evolution (Maynard Smith et al. 1985; Wake and Roth 1989; Arnold et al. 1989). I refer to two ways in which environmental stimuli are known to affect ontogenies:

Short-Term Ecophenotypic Responses

It is well known that development can produce a range of ecophenotypes in response to differing environmental conditions (review in Stearns 1982). This range is known as the norm of reaction. While

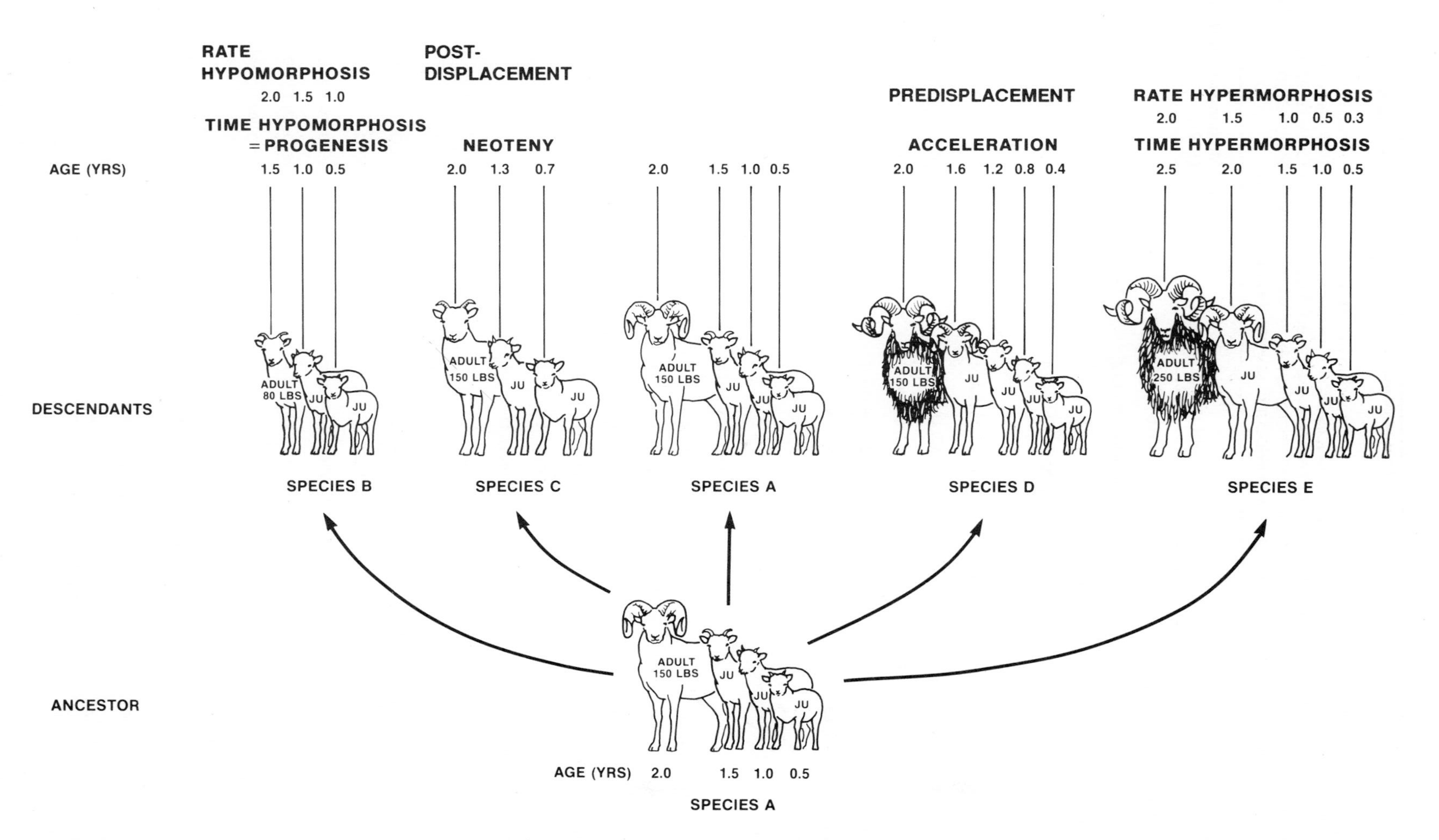

FIGURE 18-3 Illustration of some basic, pure modes of heterochrony (directly comparable with Figures 18-4 and 18-5). The ancestor is based on the Bighorn Sheep after Geist (1971). The descendants are hypothetical.

TABLE 18-1 Human characters listed as paedomorphic (= paed. in the table) by neoteny (= neot.), or as peramorphic (= pera.) by hypermorphosis (= hyper.) or acceleration (= accel.) by Gould (1977) and McKinney and McNamara (1991). Blank spaces indicate that no opinion was expressed. Gould (1977) regarded peramorphic characters as minor departures from a dominant paedomorphic theme, while McKinney and McNamara (1991) represent the opposite view. Both sources agree with respect to several characters (—->).

		Gould (1977)		**McKinney & McNamara (1991)**
1.	orthognathy	paedo. by neot.	——->	
2.	reduction of body hair	paedo. by neot.	——->	
3.	loss of pigmentation (skin, eyes, hair)	paedo. by neot.		
4.	external ear shape	paedo. by neot.		
5.	epicanthic eyefold	paedo. by neot.		
6.	central position of foramen magnum	paedo. by neot.	——->	
7.	high brain weight	paedo. by neot.		pera. by hyper.
8.	late persistance of cranial sutures	paedo. by neot.		pera. by hyper.
9.	labia majora	paedo. by neot.		
10.	hand, foot structure	paedo. by neot.		hyper. with neot.
11.	pelvic structure	paedo. by neot.		pera. by accel.
12.	ventral direction of female sexual canal	paedo. by neot.		
13.	dental variations	paedo. by neot.		
14.	absent brow ridges	paedo. by neot.		
15.	absent cranial crests	paedo. by neot.		
16.	thinness of skull bones	paedo. by neot.		
17.	position of orbits	paedo. by neot.		
18.	brachycephaly	paedo. by neot.		
19.	relatively small teeth	paedo. by neot.		
20.	late tooth eruption	paedo. by neot.		pera. by hyper.
21.	unrotated big toe	paedo. by neot.		
22.	prolonged infantile dependency	paedo. by neot.		pera. by hyper.
23.	prolonged growth period	paedo. by neot.		pera. by hyper.
24.	long life span	paedo. by neot.		pera. by hyper.
25.	large body size	paedo. by neot.		pera. by hyper.
26.	skull base and upper respiratory tract			pera. by accel. or hyper.
27.	pronounced leg length	pera. by hyper.	——->	
28.	vertebral column at center of chest cavity	pera. by hyper.		
29.	early fusion of sternum	pera. by hyper.		
30.	fusion of centrale with naviculare	pera. by hyper.		

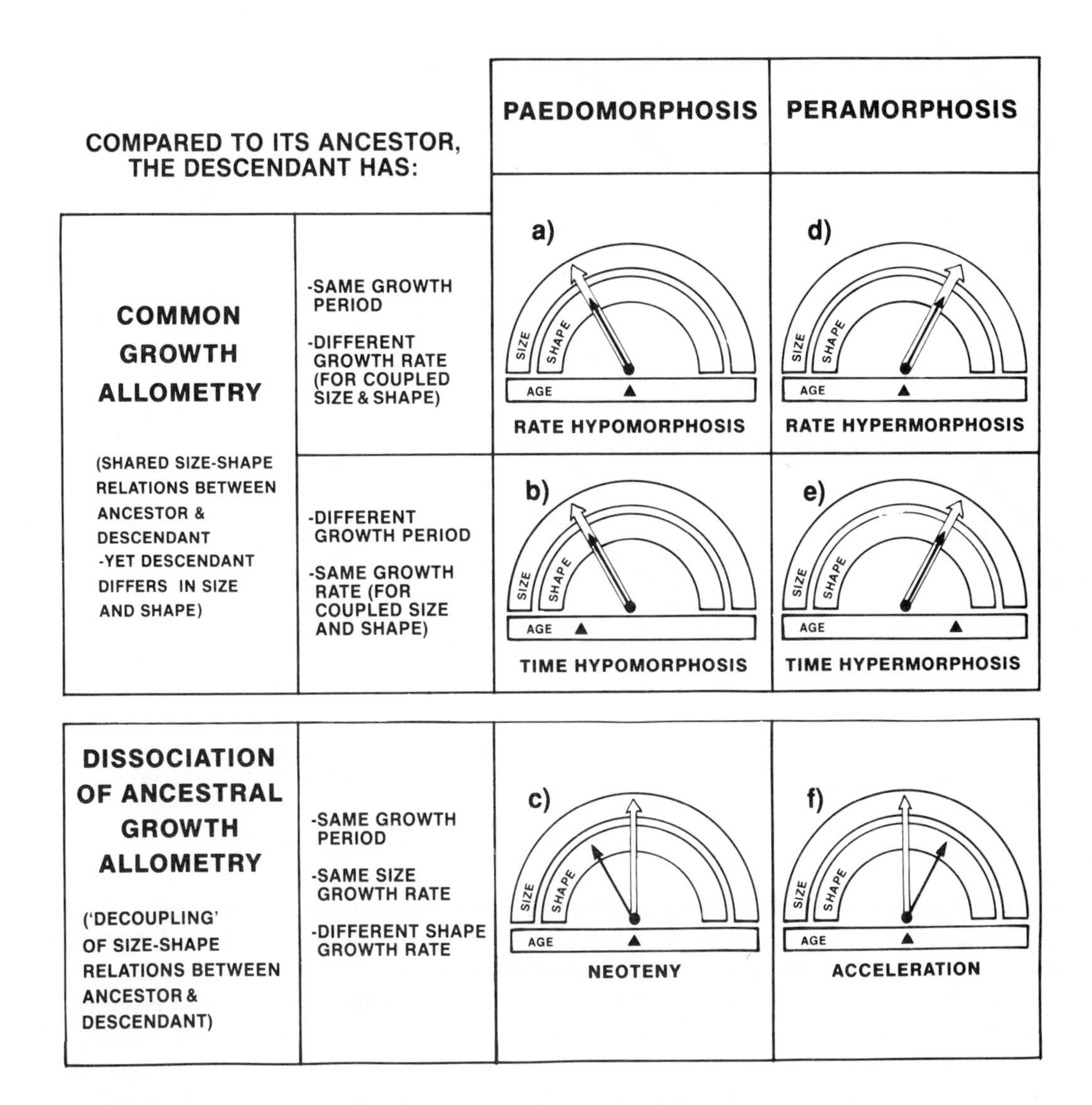

FIGURE 18-4 The "Clock Model" of changes in size, shape, and timing between ancestors and descendants (adapted from Gould 1977 with additions by Shea 1988).

heterochrony and related terms are generally applied to changes in the rates and timing of development during long-term evolution, they can also serve in comparisons among individual ontogenies. Such ecophenotypes are not inheritable. The same kinds of ecophenotypes tend to occur consistently in response to the same environmental conditions in related species (e.g., Gould 1977; Stearns 1982).

Long-term Evolutionary Responses

Long term evolutionary responses could occur either when ecophenotypes are selectively assimilated (contributions in Wake and Roth 1989; reviewed by Smallwood pers. comm.) or when novel genes and ontogenetic phenotypes are directly fixed by selection.

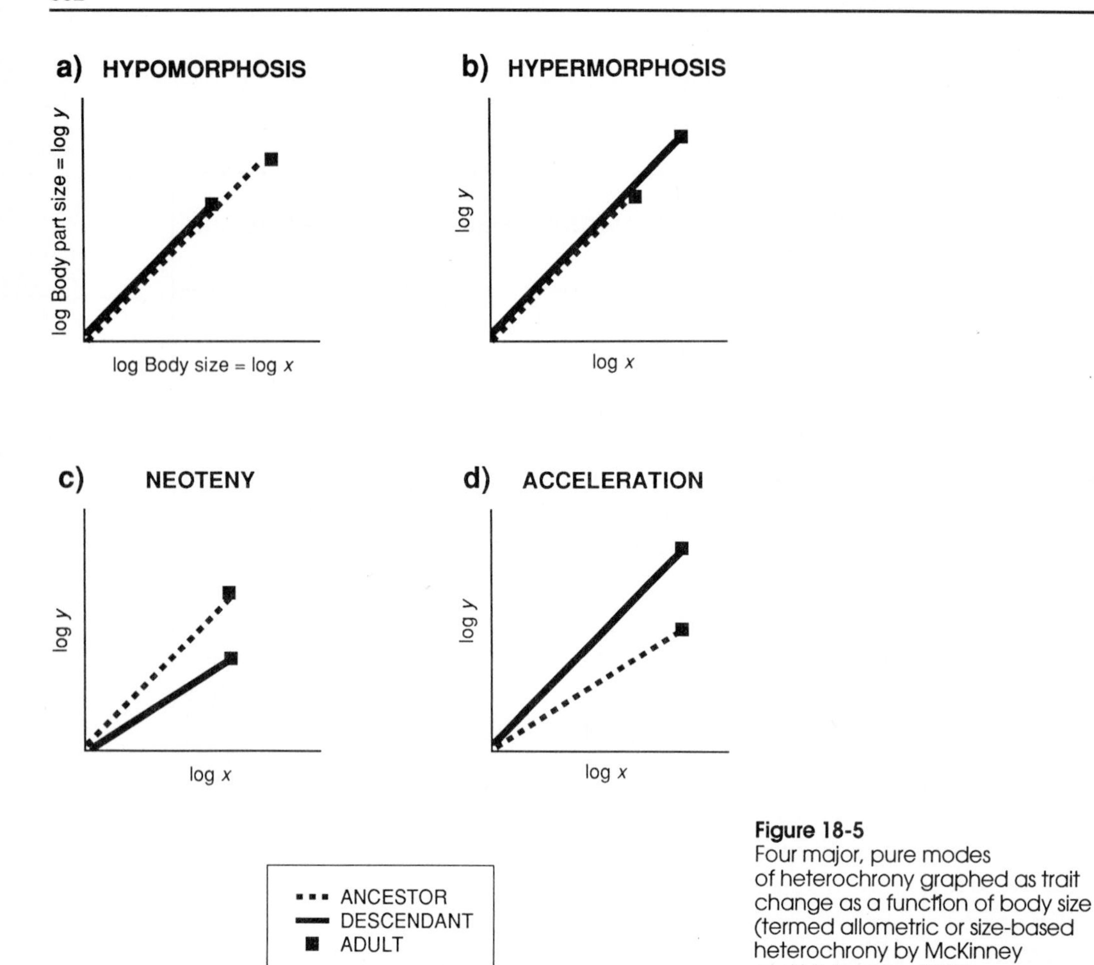

Figure 18-5
Four major, pure modes of heterochrony graphed as trait change as a function of body size (termed allometric or size-based heterochrony by McKinney and McNamara 1991).

Let us first recall traditional concepts of heterochronic modes in pure form (Figures 18-3, 18-4, 18-5), bearing in mind that more than one mode usually co-occur (Gould 1977; Alberch et al. 1979; Shea 1988; McKinney and McNamara 1991 give subtly different versions).

Paedomorphosis refers to retention of ancestral juvenile characters in descendant adults (Figure 18-3b,c). It can result from three basic kinds of evolution: (1) **Hypomorphosis** (Figure 18-3b, which subsumes progenesis = time **hypomorphosis**, Shea 1988) produces descendants of reduced size but of common growth allometry; (2) neoteny and (3) postdisplacement are similar in producing paedomorphosis at large size (in Figure 18-3c, the juvenilized descendant is as large as the ancestor), and in that the ancestral size–shape relationship has become dissociated. In both, dissociation is caused by relative reduction of *net* rate of shape growth. But in neoteny this results from rate reduction throughout ontogeny, while in postdisplacement it is due to delayed onset of shape development.

Peramorphosis refers to descendant adults that have derived shapes (Figure 18-3d,e). (1) Hypermorphosis is the inverse of hypomorphosis (a descendant of increased size but of common allometry), (2) **acceleration**, the inverse of neoteny (Figure 18-3d: a descendant of increased shape growth rate relative to size growth rate), and (3) pre-displacement of post-displacement (also represented by Figure 18-3d: a descendant in which onset of shape growth is relatively earlier).

The classification I use (Figures 18-3 to 18-9) is based on Gould (1977) and Shea (1988). The proposed revisions of McKinney and McNamara (1991) focus on decrease versus increase in developmental rates as the primary classificatory device. Thus, for instance, they lump under neoteny both classical neoteny (Figure

(Figure 18-3c) and ***rate hypomorphosis*** (one route to result B in Figure 18-3). I prefer to retain a primary separation between dissociative heterochrony (Figure 18-3c,d), and purely allometric changes such as hypo- and hypermorphosis (Figure 3b,e), because I suspect that dissociative heterochronies have played a fundamental role in evolution.

Heterochrony is often conceived of as *age-based;* that is, shape and body size change relative to ontogenetic age (as in Figures 18-7, 18-8, and 18-9, which will be discussed later). But one can validly study *size-based* heterochrony by allometric analysis (McKinney and McNamara 1991): change in "shape" *sensu lato* (meaning parts of the phenotype) relative to body size (Figures 18-5, 18-6). One needs to translate with caution between these two concepts. The rate profiles of body weight growth in ancestor and descendant need to be similar for direct linkage.

Recent detailed analyses of particular ontogenies have resulted in a more sophisticated appreciation of the broad categories in Figures 18-3, 18-4, and 18-5. Rates for particular characters that can be described by a single equation and single, pure modes are seldom found (Gould 1977; Shea 1988; McKinney and McNamara 1991). I illustrate this below as I focus on paedomorphosis at large size; namely, not produced by hypomorphosis, but by neoteny and especially by processes that result in descendant adults that are both larger than ancestors and relatively juvenilized.

FIGURE 18-6 Three examples of combinatorial modes of heterochrony that result both in hypermorphosis of body size with paedomorphic trait shape when expressed as a function of body size. (Note the distinction between such allometric or size-based heterochrony, McKinney and McNamara 1991, and age-based heterochrony as in Figures 18-7 through 18-9.) Descendants I: hypermorphosis combined with neoteny; II: hypermorphosis with postdisplacement; III: hypermorphosis with early offset of trait growth.

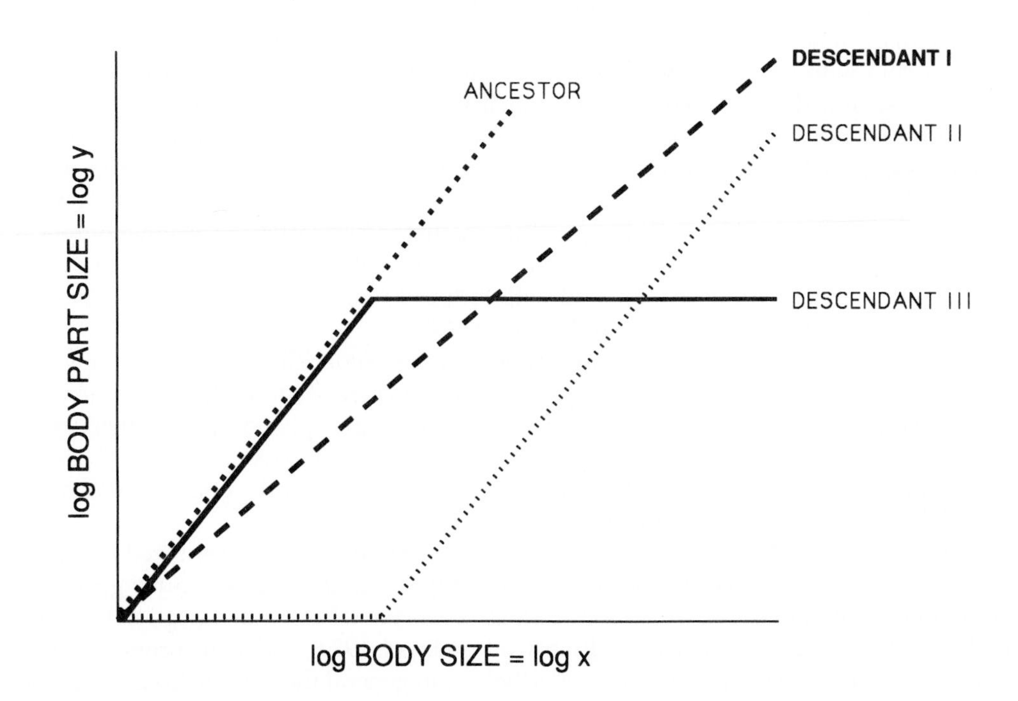

HYPERPAEDOMORPHOSIS AND COLD CLIMATE

The Genetic and Cellular basis

Evidence is accumulating, although still piecemeal from different taxa, on how external stimuli like temperature influence rates of gene expression and levels of gene products to mediate onset and offset of growth phases. For instance, in domestic mammals, reception by the hypothalamus of information on cooling (or warming) from thermoreceptors in the skin and in the hypothalamus itself elicits hypothalamic production of hormones that influence the metabolism (Hammel et al. 1963; Jenkinson 1972; see also Iggo 1972 for discussion of thermoreception in primates). McKinney and McNamara (1991) review hypothalamic regulation of postnatal growth in mammals: The hypothalamus produces neurotransmitter growth hormone releasing factor (GHRF), which stimulates pituitary production of somatotropin or growth hormone (GH). This in turn stimulates liver and other cells to produce insulin-like growth factor I (IGF I), which acts as a mitogen and stimulates growth in various tissues. While GH and IGF I affect postnatal growth, IGF II influenced by thyroid hormone acts earlier during fetal development to increase mitosis in both brain and body tissues (Tanner 1988, on human ontogeny). In a few studies the precise cellular and genetic basis of heterochrony is known to implicate single genes (e.g., Ambros 1988).

Evidence

I know of no report of a direct link from temperature to gene to gene product that affects heterochronic growth in primates. But one can reasonably postulate such linkages based on direct evidence from other organisms and on circumstantial evidence for many groups including primates. In salamanders low temperature culminates in neotenic phenotypes as follows (reviewed in Gould 1977; Raff and Kauffman 1983): Cold retards production of thyrotropin releasing factor by the hypothalamus, which regulates production of thyrotropin by the pituitary. Thyrotropin reduces thyroid hormone levels to result in delayed onset and offset of ontogenetic growth phases. Development in cold lakes of giant adult neotenes in the axolotl, *Ambystoma tigrinum*, and of normal metamorphosis in warmer waters, are evident not only today but also in successive glacial and interglacial strata in the Pleistocene record (Tihen 1955). This exemplifies the dissociation, in cold temperature, of body size growth from a relatively delayed shape growth. It links cold with levels of thyroid hormones, and these levels in turn with onset and offset of ontogenetic phases, a link that also seems to be present in mammals. For instance, Khamsi and Eayrs (1966) showed that thyroid hormone increase or decrease in rats relatively accelerates or decelerates shape growth (but not size growth except at very extreme levels of thyroid hormones). Scow and Simpson (1945) reported retardation of growth and maturation with strong paedomorphosis in thyroidectomized rats.

A great deal of indirect evidence of temperature-mediated heterochronic ecophenotypes is known in many groups. For instance, colder temperature is associated with larger body size by more prolonged growth phases—namely hypermorphosis—and with reduced reproductive rates in diverse vertebrates including mammals (Stearns 1982; Khamsi and Eayrs 1966).

During long-term evolution, a strong association between colder climates and heritable hyper- and paedomorphic phenotypes is also well documented. In general, correlations with environmental variables of such obligate heterochronies are similar to those observed among ecophenotypes of other closely related forms. For instance in salamanders some entire clades are characterized by parallel paedomorphic evolution to small body sizes correlated with high temperatures, and others by obligate neoteny associated with cold (Gould 1977; Wake and Larson 1987).

Bergman's (1847) and Allen's (1877) Rules imply association of large juvenilized bodies with cold temperature. Both have been investigated for ecophenotypes within species and for genetically based differences among related lineages. Exceptions to these rules have been noted (e.g., to Bergman's Rule: Ralls and Harvey 1985; Geist 1987; to Allen's Rule: Clutton Brock 1979), and some proposed explanations (see later discussion) have been questioned. Nevertheless, in general their predictions are upheld (Koch

1986; Davis 1981; Kurten 1959; Heintz and Garutt 1965; Geist and Bayer 1988). The Bergman association of large bodies with cold is realized by **rate** and **time hypermorphosis.** For instance, numerous African tropical ungulate species have shorter periods of growth to smaller body sizes in warm lowlands, while their close relatives at higher altitudes grow for longer periods and become larger. Examples include species of Reduncini antelopes (Vrba et al. in press) and the subspecies of the African buffalo, *Syncerus caffer caffer* and *S. c. nanus* (Sinclair 1977). Allen's Rule, of relative decrease in extremities of the body (e.g., horns, antlers, limbs, ears, muzzles) as temperature declines, probably usually represents paedomorphosis. Thus, one can visualize a marriage of modes c and e in Figure 18-3. I posit that this combination again and again proved to be the best evolutionary route to the simultaneous attainment of large body size *and* avoidance of the massive hypertrophy of skeletal appendages usually associated with hypermorphosis (Figure 18-3e; Gould 1977) *and* to reduction of sexual dimorphism (see later discussion).

Adaptive Scenarios

For my current arguments it is unimportant which combination of selective causes might have most often promoted hyperpaedomorphosis. Rather I want to point out that considerable support already exists for the causal association of large juvenilized bodies with cold, vegetatively open environments. Still, it is interesting to note in passing some adaptive reasons that have been proposed: Many followed Bergman's (1847) and Allen's (1877) original hypotheses that cooler temperature promotes larger body size with reduced extremities in homeotherms for reasons of thermoregulation. Others suggested that the primary selective causes are nutritional (e.g., Newman 1960; review in Baker 1988). Schoener (1983; see also Janis 1982) posited a selective advantage of large size in ungulates evolving to seasonally cooler open grasslands as follows: The ability to digest fibrous foods such as grasses is improved by the longer digestive tracts that arise at large body sizes from allometric scaling (Demment and Van Soest 1985). Martin (1986) argued for increased competitiveness in rodent taxa through larger body sizes and increased aggression in open environments. Additional selective hypotheses relate to predator escape, sexual selection, and other factors (reviews in Estes 1991; McKinney and McNamara 1991). Proposals that reduced sexual dimorphism should be selectively favored in open environments that are at least seasonally cold will be reviewed later.

In addition, consider the following. Ecophenotypes themselves cannot be selected. But they are the flexible responses of norms of reaction that were selected to produce viable adults in spite of variable environments (Stearns 1982). We observe in nature that temperature decrease elicits increased ecophenotypic expression of hypermorphosis (often known to result from delayed growth phases) and paedomorphosis (e.g., Koch 1986; Davis 1981; Harrison et al. 1988). These changes in growth physiology can be inferred to function more efficiently in producing a viable adult under cold conditions than the ancestral ontogenetic growth schedule would have done. If this inference is valid, the following is a reasonable conclusion: If the cold extreme were to intensify beyond the temperature range that a given norm of reaction encountered previously, then mutations that reinforce the existing facultative tendencies (for example, that prolong growth even more than was possible within the ancestral norm of reaction) should be favored both by ontogeny and by selection.

Hypo- and Peramorphosis in Warm Climates

Conversely, I suggest that acceleration and hypomorphosis often evolve in warmer environments. In fact, the celebrated correlation of dwarfing of mammals on islands may well have less to do with absence of predators and resource depletion (e.g., Lomolino 1985) than with the fact that island refugia for large mammals come into being at times of global warming and sea level rise—namely, maximal warming periods—and islands at all times enjoy a more mesic climate than the mainland uplands of the ancestors. Prothero and Sereno's (1982) results for North American fossil rhinoceroses is relevant: Dwarfed species were associated with mesic forest-swamp "climatic islands" on the Miocene landmass, surrounded by savanna uplands on which larger rhinoceros taxa lived.

DIFFERENT ONTOGENETIC ROUTES TO HYPERPAEDOMORPHOSIS

Figure 18-6 shows simple logarithmic graphs for three descendants that all grow to larger body weights than the ancestor. That is, the descendants are hypermorphosed in the body size–based sense. In pure hypermorphosis (Figure 18-5b) the descendant line is simply an extension of the ancestral one. In contrast, in Figure 18-6 each descendant adult value lies below the ancestral log-log line, that is, part size is also paedomorphosed. Descendant I in Figure 18-6 illustrates that the descendant adult value could be absolutely larger than, yet relatively paedomorphic with respect to that of the ancestor. (Note that the three descendant trajectories in Figure 18-6 could also be visualized as *stopping short of* the ancestral body weight, rather than beyond as here, which would be hypopaedomorphosis. If the trajectories of descendants I, II, and III *matured at* the ancestral body weight, they would represent classical paedomorphic cases of I: neoteny, II: postdisplacement of onset, and III: hypomorphosis by early offset of character change.)

It will be useful for the subsequent discussion to make two broad distinctions among the many possible modes of evolutionary juvenilization with size increase. First, can the actual growth of the body part be described as a single phase, or as multiple phases? In the practical sense "single phase" means that the growth phase can be described by a single (albeit possibly complex) equation. All four cases in Figure 18-5 and all three in Figure 18-6 belong in the single-phase category. (Although descendant II technically has a phase of zero growth preceding one of part size increase, the actual growth is describable as a single phase from onset to maturation.) In contrast, each case in Figures 18-7 and 18-8 has more than one ancestral growth phase, and each phase is proportionally extended in the descendant (a process termed *sequential hypermorphosis* by McKinney and McNamara 1991; Figure 18-7, Figure 18-8a,b) or proportionally truncated (sequential hypomorphosis; Figure 18-8c,d).

A second distinction is this: The paedomorphic shape of a descendant adult may result from relative *reduction*, or from relative *increase* of the body part measure. Paedomorphoses of all kinds are usually conceived of as occurring by reduction relative to age or body size. Paedomorphosis under Allen's Rule involves reduction relative to body size. The seemingly counterintuitive paedomorphic modes by relative increase of part size have hardly been considered previously. In the next section I explore in more detail how net *paedomorphic increase* in a feature with multiple growth phases might occur by sequential hypermorphosis.

SEQUENTIAL HYPERMORPHOSIS CAN INCLUDE PAEDOMORPHIC RESULTS

Let us focus particularly on sequential "stretching" of all growth phases of a character that in the ancestral organisms has higher exponential growth rates in the earlier phase or phases than in the later ones. And, rather than think of three successive growth phases as in Figure 18-7, let us simplify the problem to two phases as in Figure 18-8a: an earlier and a later juvenile phase before adulthood.

Many characters are relatively larger during juvenile than adult stages. For instance, the orbits in bovids and the hindfeet of rodents complete a larger proportion of their total size growth during earlier rather than later ontogeny. That is, the early juveniles are larger in these characters relative to body or skull size than are the adults; and, in an age-based sense, they are larger in these characters relative to developmental time elapsed, although the adults have absolutely larger values.

Hafner and Hafner (1988) analyzed ontogeny and heterochrony of hindfeet and other characters in rodents of the Family Heteromyidae: Kangaroo rats, *Dipodomys*, are bipedal saltatory inhabitants of semiarid to arid regions in North America. They are hypermorphosed with longer growth periods—as well as paedomorphic in some features like the hindlimbs—relative to their ancestor, which the authors regard as well represented by the ontogeny of pocket mice (e.g., *Perognathus*). The early juveniles of all these taxa have relatively larger hindfeet. Hindfoot growth in the pocket mouse later slows down at the expense of relatively faster growth in other body features such that adult hindfeet look smaller. But the kangaroo rat, in spite of growth to larger adult body size than the pocket mouse, maintains juvenile proportions.

I suggest that this case fits hyperpaedomorphosis as in Figure 18-8a (or Figure 18-7) very well: By equal extension of all growth phases for hindfeet during evolution, the earliest phase with the fastest exponential growth rate came to dominate the adult *Dipodomys* proportions so that even on a hypermorphosed body the hindfeet still appear large and, thus, juvenile in shape. Another way to put it is that the hindfeet of adult *Dipodomys* owe their large size disproportionately to "shape units" that accrued during the earliest juvenile growth phase.

Let us state this more formally, in the sense of age-based heterochrony, for sequential heterochrony in general and for the particular case of sequential hyperpaedomorphosis, Case (a) following, with earlier

FIGURE 18-7 Generalized representation of sequential hypermorphosis in a character relative to age assuming three growth phases of positive allometry relative to age, of decreasing exponential rates from earliest to latest, n = 2 (the factor by which all ancestral growth phases are multiplied to yield the descendant ontogeny), and earlier offset of female shape growth.

PAEDOMORPHOSIS FROM SEQUENTIAL HYPERMORPHOSIS

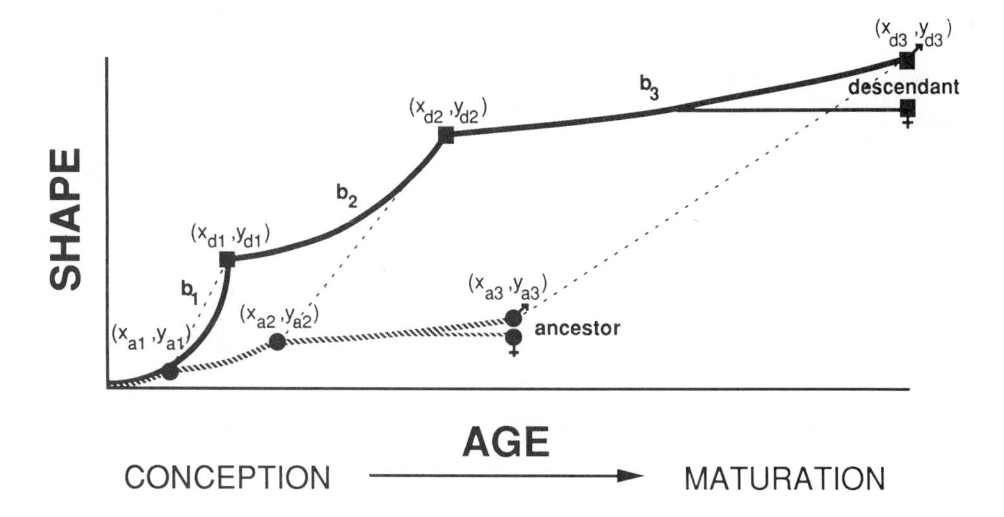

Given growth equations of the general form $y=kx^b$, then for $x_{di}-x_{d(i-1)}=n[x_{ai}-x_{a(i-1)}]$, $n>1$; and rate $b_i>b_{i+1}$,

$$\frac{y_{di}-y_{d(i-1)}}{y_{d(i+1)}-y_{d(i-1)}} > \frac{y_{ai}-y_{a(i-1)}}{y_{a(i+1)}-y_{a(i-1)}}$$

faster exponential growth rate for a feature. Other cases of sequential heterochrony, (b) hyperperamorphosis, (c) hypoperamorphosis, (d) hypopaedomorphosis, are treated similarly in the APPENDIX and illustrated in Figure 18-8b–d. The following argument can be read with reference to Figure 18-8a.

Sequential heterochrony occurs when (1) and (2) are true:

1. The ancestral ontogeny is multiphasic, such that the character develops through at least two phases of distinct exponential growth rate. The following arguments concern two juvenile phases that culminate in adulthood (Figure 18-8):

   0 to x_1: ancestral onset to offset ages of early juvenile (e.g., foetal) phase;

   0 to x_2: descendant onset to offset ages of early juvenile phase;

FIGURE 18-8 Four kinds of sequential age-based heterochrony assuming two growth phases of differing exponential rates. Depending on the size of n (the factor by which all ancestral growth phases are multiplied to yield the descendant ontogeny) and on whether the exponent a of the earlier growth phase equation is larger or smaller than exponent b for the later phase, different kinds of paedo- and peramorphosis result. (See discussion in the text and APPENDIX. Note that these examples show cases in which a>1 and b>1. The principles also apply to cases in which 0<a, b<1.)

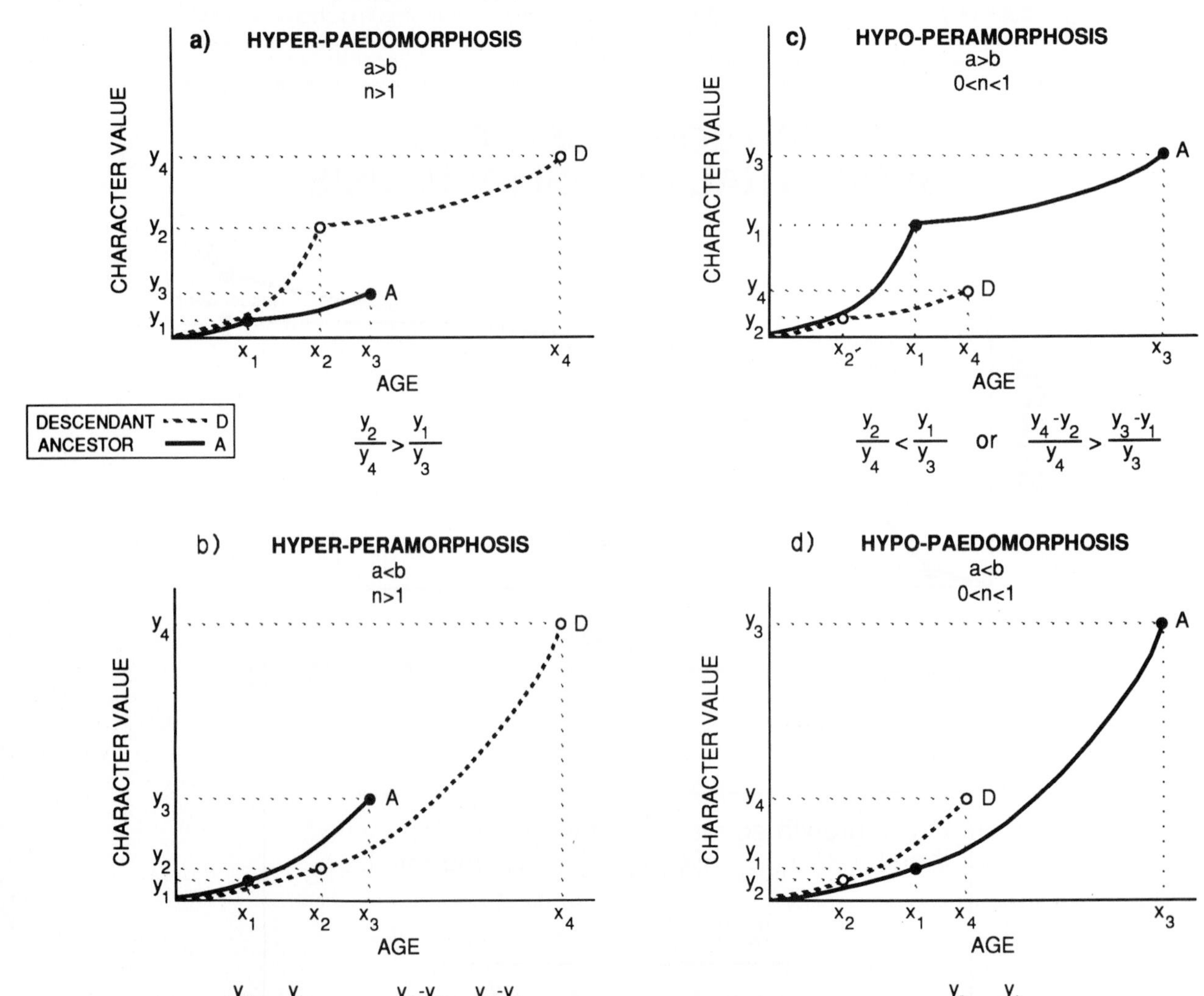

x_1 to x_3: ancestral onset to offset ages of later juvenile phase;
x_2 to x_4: descendant onset to offset ages of later juvenile phase;
y: shape units accumulated at respective ages;
a: growth rate of earlier phase in ancestor and descendant ($y = x^a$);
b: growth rate of later phase in ancestor ($y = k_{anc}x^b$) and descendant ($y = k_{desc}x^b$).

2. Evolution occurs by proportional time extensions or contractions of each of these ontogenetic phases: while the ancestral phase-specific growth rates are maintained in the descendant, the phases are lengthened or shortened by a constant proportion, n, with $n > 0$.

Case (a) Sequential HyperPaedomorphosis, n > 1, with a > b

The earlier phase has a higher exponential growth rate for the character than the later phase.

Given: $a > b$; $x_2 = nx_1$; $(x_4 - x_2) = n(x_3 - x_1)$; $n > 1$ (In Figure 18-8a, $n = 2$).

In this case evolution by *hypermorphosis* results in a descendant adult in which the proportion of shape units derived from the earlier juvenile phase (or *paedomorphic* shape) is increased;

namely, $$\frac{y_2}{y_4} > \frac{y_1}{y_3}$$

Demonstration: $y_1 = x_1^{\,a}$; and $y_2 = (nx_1)^a$

$$y_3 = x_1^{\,a} + (x_3 - x_1)^b \text{; and } y_4 = (nx_1)^a + (x_4 - x_2)^b \quad (1)$$

Therefore $$\frac{y_2}{y_4} = \frac{n^a x_1^{\,a}}{n^a x_1^{\,a} + n^b (x_3 - x_1)^b} = \frac{x_1^{\,a}}{x_1^{\,a} + n^{b-a}(x_3 - x_1)^b}$$

and $$\frac{y_1}{y_3} = \frac{x_1^{\,a}}{x_1^{\,a} + (x_3 - x_1)^b} \quad (2)$$

For a > b and $n > 1$, $b - a < 0$, *and* $n^{b-a} < 1$; therefore

$$\frac{y_2}{y_4} > \frac{y_1}{y_3}.$$

That is, the proportion of shape units derived from the earlier juvenile phase (or paedomorphic shape) increases in the descendant.

In sum, a particular kind of hypermorphosis can indeed result in paedomorphic proportions. The implication at first seems counterintuitive: *More extreme* expression of a character to *larger* proportion relative to the body can yield *paedo*morphosis. Each of the graphs in Figures 18-5 through 18-8 refer to to a single character at a time, while a neighboring morphological feature in the same organism could in principle represent a different kind of heterochrony. Thus, a single episode of sequential heterochrony can affect characters differently depending on whether the ancestral character has faster earlier or faster later growth rates. For instance, sequential hypermorphosis is a single coherent kind of heterochrony that can in one stroke produce paedomorphic characters accompanied in mosaic fashion by other characters that are peramorphic.

Case (b) in Figure 18-8 is labelled hyper*pera*morphosis because the proportion of shape units derived from extension of the last growth phase dominates in the descendant. In practice, if the presence of sequential phases is not noticed, this might be taken as a case of postdisplacement. (If the X-axis in Figure 18-8b represented body weight, a comparison of logarithmic plots of only the last phases in ancestor and descendant would look similar to Descendant II in Figure 18-6: hypermorphosis combined with *paedo*morphosis as the descendant adult value lies below the extended ancestral log line.) Similarly in case (a) (previously discussed and Figure 18-8a), if the sequential phases are not noticed, this might be taken as a case of predisplacement. Perhaps many reported cases of simple heterochronies are in reality sequential heterochronies that have remained unrecognized due to inadequate data.

The possible pathways to a hyperpaedomorphic result here discussed (Figures 18-6, 18-7, 18-8a) by no means exhaust the possibilities. Note also that all arguments for heterochrony require a genealogical context. Concepts like paedomorphosis require reference to an outgroup or ancestral morphotype in cladistic terms (Miyazaki and Mickevich 1984; Nelson 1978; Fink 1982; Kluge 1989; Vrba et al. in press).

HYPERPAEDOMORPHOSIS AND SEXUAL DIMORPHISM

Most hyperpaedomorphic evolutionary modes (Figures 18-6, 18-7) imply reduced sexual dimorphism. Consider that dimorphism commonly results from bimaturism (difference in male and female offset times of body growth) which may or may not be accompanied by differential growth rates (Shea 1986). Traditional hypermorphosis (see Figures 18-3, 18-4, 18-5) is expected to *increase* character dimorphism: The greater the bimaturism, the more shape difference rides along allometrically, as demonstrated for gorillas (Shea 1982, 1986). But the situation is different for paedomorphosis.

Pure neoteny entails growth offset at equivalent body weights in ancestral and descendant males, and also in ancestral and descendant females. Inspection of, for instance, Figure 18-5c shows that this will decrease dimorphism in the descendant.

Under sequential hypermorphosis with higher earlier growth rate (Figure 18-7), provided the descendant offset of female weight growth is postponed in proportion to other phase postponements, less sexual dimorphism per body weight results. Most of the increased size of the descendant character is accounted for by earlier growth phases that are shared by males and females.

Given postdisplacement (Figure 18-6: II, the descendant log-log line from shape growth onset onward is parallel to the ancestral one), if descendant offset of female growth is postponed in proportion to the postdisplacement (along the body weight axis), and if the last part of ancestral shape growth remains incomplete (as exemplified in Figure 18-6 by the lower adult body part measure for Descendant II than for the ancestor), then the part representing male–female differentiation is likely to be curtailed or even eliminated.

Under early growth offset (Figure 18-6: III), male–female differentiation is likely to be at least reduced and possibly eliminated. Thus, Leutenegger and Larson's (1985) conclusion that sexual dimorphism always increases in favor of males with body size increase (see also Cheverud et al. 1985), holds for classical hypermorphosis (Figure 18-3e; Figure 18-4d,e, Figure 18-5b), and for sequential hypermorphosis in characters with higher *late* growth rates in ontogeny (Figure 18-8b), but it does not hold for all heterochronies that involve body size increase.

Jarman (1983), based on studies of placental and marsupial herbivores, distinguished between dimorphism in body weight, in structure and weapons (such as teeth in primates, and horns in ungulates), and in appearance. Under appearance he included many late-developing epigamic characters, such as hair growth, skin flaps and color patterns. Under sequential hypermorphosis, such *late*-appearing sex differences are expected to evolve toward stronger dimorphism because of their hyper-*pera*morphic trajectory (Figure 18-8b), with a faster later developmental phase. In contrast, many characters like the brain and cranium, and the orbital prominence in herbivores and hindfeet in rodents that were already mentioned, have fast early growth. Sequential hypermorphosis of the kind in Figure 18-7 is predicted to reduce sexual difference in such characters: An even greater proportion of growth comes to be shared by males and females because it falls in the early phases.

Estes (1991) argued that selection for social interactions peculiar to nomadic existence in open grasslands reduces sexual dimorphism. He scored sexual dimorphism in gregarious bovids and showed its positive correlation with absolute dominance mating systems and segregation of sexes, and conversely, between low degree of dimorphism and social organization in which sexes are either usually associated or are together in bisexual groups outside the breeding season or during mass movements.

I compared Estes's (1991) sexual dimorphism scores with male body weights and with closed-to-open vegetation cover. The result (Figure 18-9) at first glance is not encouraging for the hypothesis of reduced sexual dimorphism in large-bodied savanna inhabitants. While there is a strong association in the 30–500 kg range between low degrees of sexual dimorphism and open grasslands, large-bodied species clearly occur in a variety of environments. I suggest that this plot would make a more convincing case for the hypothesis if the *genealogical* relationships among taxa were shown. The arrows in Figure 18-9 indicate information from the few available cladistic analyses: Each arrow points from earlier to later branches on the cladogram for a given group. In these cases the plesiomorphic-apomorphic polarity has indeed been simultaneously toward larger body sizes, toward accompanying reduced sexual dimorphism, and toward more open vegetational habitats.

Other support for this hypothesis is reviewed in Estes (1991; also Jarman 1983; Geist and Bayer 1988 for American ungulates). For instance, male horn size and sex differences in body weight relate to species body weight similarly and curvilinearly, both peaking at intermediate body weight and diminishing at both very small and very large extremes of body weight (Jarman 1983).

The precise nature of the heterochronic processes that produced these correlations remain to be investigated. But three points can be illustrated here by relative body and horn size in bovids (compare Mentis 1972 and Geist 1971 with cladistic hypotheses in Vrba 1979, 1987a,b, in prep. for Tragelaphini; Gatesy et al. 1992): (1) In most taxa living in cold (or seasonally cool) steppe or semidesert both males and females grow for longer than did their ancestors (one form of hypermorphosis). (2) At the same time, in most cases the males evolved toward relatively shorter and lighter horns (Geist 1988) suggesting

FIGURE 18-9 Sexual dimorphism scores (Estes 1991), adult male body weights (Macdonald 1984; Smithers 1983; Spinage 1986), and vegetation habitats in species of African savanna Bovidae and Equidae. Four cladistic analyses available for these taxa all suggest evolutionary trends towards larger body size with reduced sexual dimorphism in more open habitat: arrow 1: Tragelaphini (Vrba in prep.); 2: Reduncini (Vrba et al. in press); 3: Hippotragini (Gatesy et al 1992); 4: Alcelaphini (Vrba 1979).

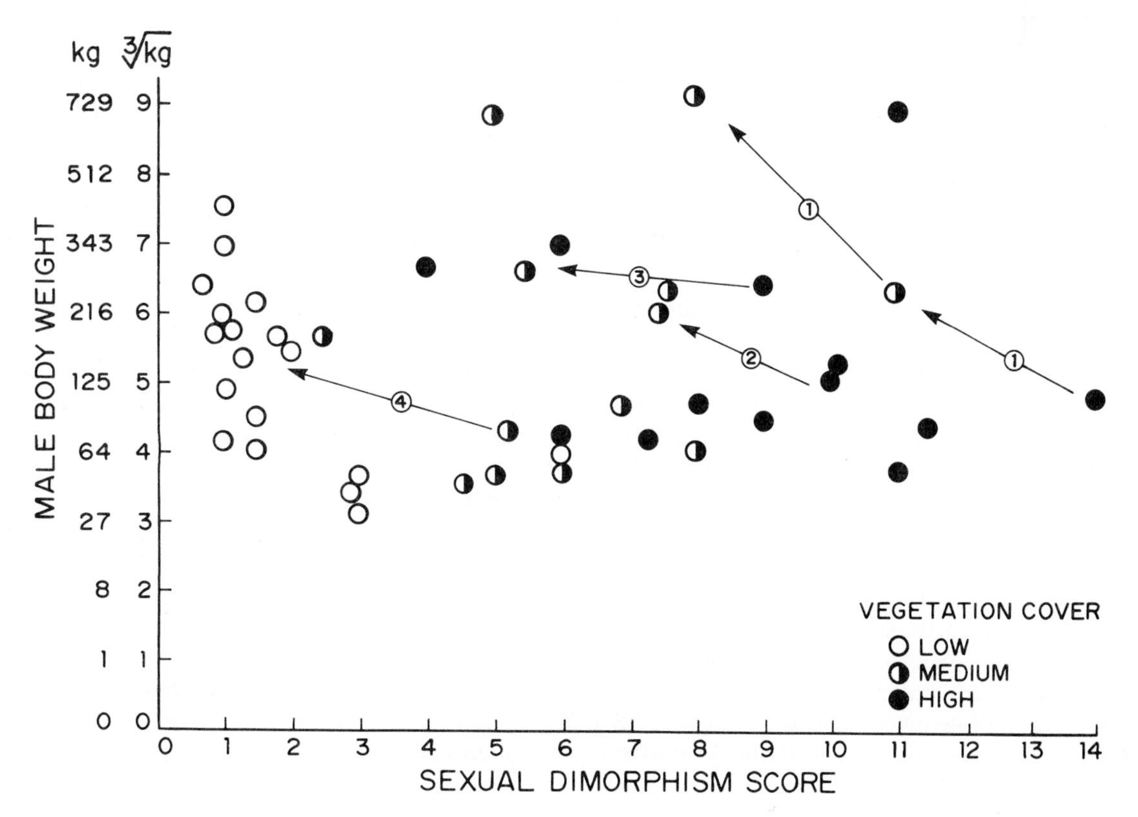

paedomorphosis. (3) The females evolved from a primitive absence of horns (Janis 1982) toward horns closely similar in size to those of males. Together (2) and (3) account for a major component of decreasing sexual dimorphism as bovids evolved toward larger body size (Figure 18-9). Note that the evolution of female bovid horns accords very well with hyperpaedomorphosis. Horn growth onset occurs near the beginning of the final juvenile growth phase (the approximate equivalent of the last phases in Figures 18-7 and 18-8). If onset of horn growth is a function of body weight, a weight rubicon in males that is usually not reached by female bovids, then hypermorphosed lengthening of all growth phases is expected to move this rubicon to an earlier stage (relative to the growth cycle) in females and in males and therefore result in female hornedness. It is a fact that species with horned females also have relatively earlier horn growth in males (Estes 1991).

HYPER-PAEDOMORPHOSIS IN HOMINIDAE

Taxonomic Context: Hypermorphosis in Primates

Primate radiation was accompanied by the massive Cainozoic cooling trend (Vrba et al. 1989) during which many mammal lineages became hypermorphosed by either increased growth rates (rate hypermorphosis) or prolongation of growth (time hypermorphosis) or both. The mammalian brain completes a large proportion of its total growth rapidly early in ontogeny (Count 1947; Holt et al. 1975; Sacher 1982). Sacher and Staffeldt (1974) showed that fetal growth rates of the brain are nearly constant between mammalian species, while those of body size vary widely. They proposed the "minimax theory" to explain this: Because the growth rate of neural tissue is slower than that of other organs, the maximum rate of fetal growth is limited to that of neural tissue. The growth rates of mammalian fetal brains are so similar because all are growing at this common maximum rate.

This implies, I suggest, that major advances in encephalization quotient (EQ) were accomplished by time hypermorphosis and not by rate hypermorphosis: Rate hypermorphosis cannot increase growth rates that are already at a physiological upper limit. Thus, it can only augment the slower postnatal rates of brain growth along with growth rates of the body as a whole. In contrast, prolongation of the maximum fetal brain growth rates by time hypermorphosis is possible, with a potentially large advance in EQ (Figures 18-7, 18-8a).

Sequential time hypermorphosis involves more fundamental and comprehensive evolutionary changes than does rate hypermorphosis of growth phases: The former affects size increase, shape reorganization, *and* reduced birthrates in a prolonged life cycle (with fundamental demographic and selective implications, Lovejoy 1981; Ross 1988), while the latter alters only size and shape. Thus, I propose that time hypermorphosis of sequential growth phases has occurred less frequently in the evolution of larger bodies than its rate counterpart, more likely to be eliminated by selection upon first appearance. Primates are remarkable among mammals for their generally prolonged life histories (Sacher 1959; Gould 1977). This suggests that time hypermorphosis played an unusually important role in primate evolution, although it was probably still rare relative to rate hypermorphosis. (Time hypermorphosis is expected also to involve rate increase as the body grows larger simply as a consequence of the prolongation of ancestral phases of exponentially increasing rates. But rate hypermorphosis can in principle—and is known to [e.g., Shea 1982]—occur without growth phase extensions.) A higher incidence of time hypermorphosis in primates is consistent with Peters's (1983) result that growth rates in primates, especially in anthropoids, increase more slowly with body weight than they do in mammals as a whole. Sacher (1982) gave evidence for two major events in primate evolution: At the beginning of the strepsirrhine radiation, some 55 MYA, a growth pattern evolved that resulted in a 12 percent ratio of neonatal brain to body weight (shared by all extant primates in contrast to the 6 percent ratio in other mammals), and in increases in precociality at birth and in adult EQ. A second episode of extraordinary primate evolution was associated with the origin of Anthropoidea some 35 MYA when EQs increased dramatically. It is interesting that for both these time periods there is evidence of global cooling (contributions in Vrba et al. 1985; Burckle et al. 1986). Concerning the event 55 MYA, Sacher (1982:106) wrote "this shift was presumably accomplished early in fetal development . . . . by retarding non-neural cell proliferation." I propose that time hypermorphosis was an additional and major component of both evolutionary events.

The evidence suggests that most trends to larger bodies in mammals evolved by rate hypermorphosis. But rare events of sequential time hypermorphosis (although less rare in primates) left major cladistic imprints of encephalization. This combination may explain the "taxon-level effect" in the scaling of brain on body weight (Pagel and Harvey 1989): Brain weight differs more for a given difference in body weight as the mammal species compared are more distantly related. That is, among closely related species brain weight evolves more slowly than body weight. Pagel and Harvey (1989:1590) proposed that "higher taxa . . . may differ in the length of the early growth period, whereas the length of the [later] growth period may be more characteristic of differences among closely related species." I suggest that, while large clades do indeed differ in the length of the early brain growth phase, due to rare phylogenetic events of time hypermorphosis, differences among closely related species are mostly due to changes in growth rate.

Taxonomic Context: Heterochrony and Sexual Dimorphism in Hominoids

In several instances heterochrony has been scrupulously documented largely due to Shea (1982, 1983, 1986, 1988; Shea and Bailey 1989). He demonstrated pervasive peramorphosis, by rate hypermorphosis with increased sexual dimorphism, in *Gorilla* (Shea 1982), neoteny with hypomorphosis and reduced sexual dimorphism in the Pygmy Chimpanzee (Shea 1983; presuming in each case an ancestral ontogeny like that of the Common Chimpanzee), and rate hypomorphosis in West African Pygmies relative to other humans (Shea and Bailey 1989). Thus, it is eminently reasonable to investigate possible heterochrony in the evolution of *Homo* in general (as already discussed by Gould 1977; Lovejoy 1981; and others) and of the hominine brain in particular.

Estes (1991) argued that the correlation of low sexual dimorphism with large bisexual groups also holds in primates as a whole. But tests of the above hypotheses across primates, that compare characters in vegetatively open relative to closed habitats, are difficult. The problem is that this order is known for its remarkably low representation in colder, more open areas (e.g., Macdonald 1984). Thus, in the medium to large-bodied cercopithecoids only a small number of taxa evolved to be able to live in open savanna and deserts, as *Homo* and *Papio* did. Both these taxa did evolve to larger body sizes (e.g., Wood 1992 for hominids; Delson 1984 for cercopithecids). *Papio* itself certainly does not support the prediction of decreased sexual dimorphism (Wood 1976).

Climatic Context

In the case of Africa, major global cooling produced a large-scale extension of open habitats and of arid areas during Pleistocene glacials and during earlier cold periods like the 2.5 MYA cold onset (reviewed in Vrba et al. 1985; Burckle et al. 1986; Partridge et al. 1986; Vrba et al. 1989). This does not imply that all environments changed. For instance, some of the hominid sites almost certainly were riverine and lacustrine, moist, wooded refugia during some Plio-Pleistocene periods of major climatic changes (Vrba 1988). Also, it would be ridiculous to imply that early hominids faced extremes of cold as did some populations of Neanderthals and *Homo sapiens*. The early hominids probably had mesic conditions for much of the year. Yet, it is documented (Brain 1985; Greenacre and Vrba 1984) that cold winter temperatures can result in open vegetation within tropical and subtropical areas in spite of mesic, even hot, temperatures during other seasons and reasonably high annual rainfall. African savannas can be bitterly cold in winters, especially in areas of high elevation, which is precisely the context in which most hominid fossils have been found. In evaluating modern relative to ancient African climates, we should also recall that we are today situated near the peak of a brief warm interglacial. Without implying that all Pleistocene periods were cold, one can note that 2.5 MYA marked the onset of the modern Ice Age during which the world was gripped by cold most of the time, with relatively brief interglacials, and that glacial extremes intensified throughout the Pleistocene (Vrba et al. 1989: Figure 3).

To adapt the late Evelyn Hutchinson's phrase, the ecological theater in which the hominine evolutionary play was performed was beset by progressive seasonal cooling. It is especially significant that the onset of global glaciations followed upon a long mesic middle Pliocene period of exceptional warmth (Vrba et al. 1989). It is such *relative change* in temperature and in its seasonality that is important to evolution in general. I argue that it was also important to the late Pliocene evolution of hominids.

Ecophenotypes, Allen's and Bergman's Rules

Humans show ecophenotypic variation in response to local developmental environments as clearly as do other taxa (Harrison et al. 1988). For instance in relation to temperature, Tanner (1988) noted a well-marked seasonal effect on growth velocity in modern humans from temperate industrialized countries: Average growth in height (an aspect of shape growth) is faster during spring warming and slowed down relative to body weight during autumn cooling. Baker (1988:505), citing Roberts (1978), wrote: "Roberts' regressions [of body weight on mean annual temperature in modern humans] clearly establish that Bergman's rule applies as validly to *Homo sapiens* as to any other polytypic species." He added that the only deviations involve populations that have not yet been in their current environments for very long. Baker (1988) also reviewed evidence that relative sitting height and relative arm span in modern humans support Allen's Rule. He noted (p. 507) that "the Allen rule also applies to a proximal-distal trend in arms and legs. Thus, the lower leg-to-thigh ratio is greater in hotter populations, and even the foot length relative to breadth is greatest in the populations native to hot climate regions."

Some studies of modern populations suggest a genetic basis and others an ecophenotypic cause for conformity with Bergman's and Allen's Rules. Consistent patterns within and among lineages through time almost certainly indicate genetic differences. Given the strong cooling trend since the Late Pliocene, Bergman's Rule predicts that hominids should have become progressively hypermorphosed—a hypermorphocline (McKinney and McNamara 1991). There is ample evidence of hypermorphoclines in both the hominine and robust australopithecine lineages (e.g., McHenry 1992). Modern humans are clearly time hypermorphosed relative to other living hominoids. There seems to be universal agreement that humans have an exceptionally long growth period, in fact that our entire life cycle is enormously prolonged (Gould 1977; Shea 1989; Smith 1991; McKinney and McNamara 1991). It is interesting that hypermorphosis in robust australopithecines seems to have occurred in a different way to that in hominines: by increased growth *rates* rather than prolongation of growth (Beynon and Dean 1988).

As mentioned earlier, species in a particular taxonomic group often share particular kinds of heterochrony, and the details of growth patterns and environmental correlations of these heterochronies tend to be closely comparable between such species and between ecophenotypes and genetically based phenotypes (e.g., Gould 1977; Wake and Larson 1987). Thus, the modern human evidence may provide clues to what happened in hominine evolution. For instance, modern cold-adapted (relative to warm-adapted) humans show delayed onset and offset of growth phases (e.g., of tooth eruption, Tanner 1988; of leg elongation, Schultz 1926) as predicted by classical sequential hypermorphosis (Figures 18-7, 18-8a,b). Smith (1991) examined dental developmental sequences of early hominids. She concluded that the pattern of growth and aging becomes more "stretched out" like that of modern humans only at the appearance of the genus *Homo*, and that this pattern evolved substantially during the past 2 MY. Living cold-adapted humans also show net reduction of some shape measures within their hypermorphosed trajectories (e.g., facial height, Schultz 1926; and the characters consistent with Allen paedomorphosis). This suggests a neotenic component or postdisplacement within the overall hypermorphosis, or a truncation of the last hypermorphosed phase. In each case, the origin of the paedomorphosis is different from that of the relatively enlarged paedomorphic characters represented in Figures 18-7 and 18-8a and exemplified earlier by the hindfeet of kangaroo rats. A single evolutionary episode of pure sequential hypermorphosis could simultaneously account for *enlarged* paedomorphic and enlarged peramorphic features in a descendant. However, a single evolutionary episode of pure sequential hypermorphosis could not account for absolutely *reduced* paedomorphic features together with enlarged peramorphic features in the same animal. In this case, an additional kind of heterochrony needs to be invoked.

Brain Growth and Evolution

It is widely acknowledged that the delayed offset of fetal brain growth rates in humans relative to other hominoids is in some sense paedomorphic (Gould 1977), even by the most severe critics of human neoteny (Shea 1988; McKinney and McNamara 1991). But precisely why this should happen needs additional exploration.

Vaisnys and I (in prep.) are currently analyzing a large cross-sectional data set on brain weights, body weights, and in many cases corresponding ages of chimpanzees and modern humans (Ashton 1950; Bischoff 1880; Blinkov and Glezer 1968; Burn et al. 1975; Biological Tables 1941; Bukhschtab 1884; Count 1947; Dobbing and Sands 1973; Falkner and Tanner 1986; Fischer 1983; Gavan 1971; Grether and Yerkes 1940; Hagedoorn 1924; Haug 1958; Hdrlicka 1925; Holloway 1980; Jordaan 1976a,b; Keeling and Riddle 1975; Keith 1895; Krogman 1969; Larroche 1967; Lestrel and Read 1972, 1975; Martin 1983; McClure pers. comm., 1992; Mikhailets 1952; Muhlman 1957; Oppenheim 1911; Passingham 1975; Schultz 1930, 1940, 1941, 1965, 1969; Schulz, et al. 1962; Selanka 1899; Shantha and Manocha 1969; Spence and Yerkes 1937; Tanner 1988; Tanner 1990; Tanner and Whitehouse 1982; Zuckerman 1928). We are comparing different models of how growth phases should be subdivided, and different equations for best fit to the phases. We are testing hypotheses of sequential hypermorphosis, assuming that the hominid ancestor's ontogeny shared some basic characteristics of growth phases with the Common Chimpanzee. Some preliminary results are as follows.

The human data fit quite well to the equations for three growth phases in the chimpanzee provided each chimpanzee phase length is doubled (Figure 18-10). This result would conform to the classical case of sequential hyperpaedomorphosis of a character with faster exponential growth rates in earlier growth phases (Figure 18-7): Under this model, human encephalization (Figure 18-1) resulted from simple proportional extension of all ancestral brain growth phases, in which process the growth accrued during the fast earlier phases automatically came to dominate. This model is consistent with the fact that humans alone of all primates have high fetal rates of brain growth for nearly 21 months after conception (e.g., Martin 1983). The corresponding size-based allometric plots, that depict growth of human brain volume (or weight) relative to body weight, are familiar to everyone (e.g., Gould 1977; Martin 1983): isometric growth (with a log slope of 1) until about 1 year after human birth, followed by a major reduction in rate (slope near 0.01, Holt et al. 1975) until offset.

However, while there seems to be little doubt that sequential hypermorphosis of some kind occurred, we suspect that other phase subdivisions than those in Figure 18-10 and, particularly, subtle rate modifications within the hypermorphosed human phases, may result in equations that fit even better. For instance, we are testing the assumption that human postnatal growth is subject to a consistent small rate reduction. This would imply a neotenic component. We are also testing whether the last phase may be cut off short of its full proportionally hypermorphic extent in the human relative to a chimpanzee-like ancestor. Various features of the two ontogenies (Tanner 1988) suggest the latter: For instance, the steep combined curve of the earlier two phases remains steep until near birth in chimpanzees (7.60 months) and until at least 15 months in humans in our data set (Vrba and Vaisnys in prep.) and 20.75 months in humans according to Tanner (1988). Thus, this part of brain ontogeny is lengthened in humans relative to chimpanzees by at least a factor of 2. Yet, the intervals in the two taxa between this marker and the onset of the adolescent growth spurt are closer to each other: only 1.7 times longer in humans (Tanner 1988; see

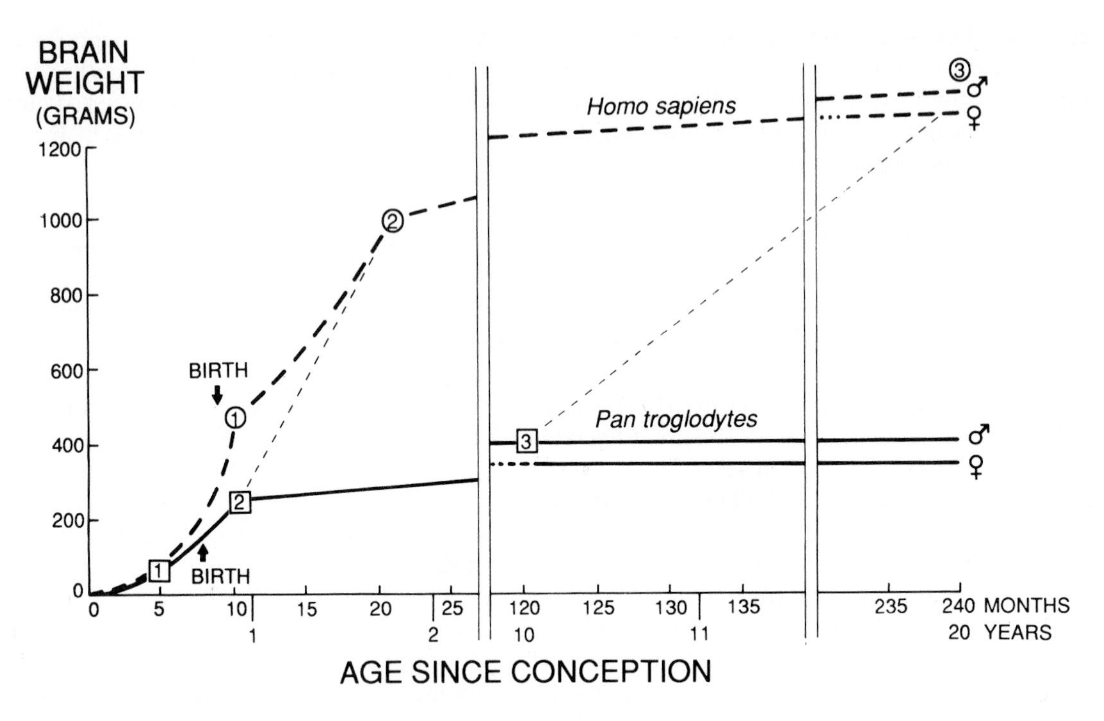

FIGURE 18-10 Change in brain capacity with age in modern *Homo sapiens* and in *Pan troglodytes*. (Compiled from numerous sources cited in the text.)

also Weitz 1979). Each of these possibilities implies a strong paedomorphic component in the evolution of human encephalization (Figure 18-1).

Sexual Dimorphism

Humans are less dimorphic than chimpanzees, the least dimorphic great apes. Wood (1985, 1976; Wood et al. 1991) found *Homo* less dimorphic both overall and in the shape component in his analyses of modern primates. Humans have relatively reduced dimorphism in cranial capacity and other cranial features (even more reduced than in Pygmy Chimpanzees, Zihlman 1979), and in canines. But humans show more dimorphism in the hip girdle, and in other characters that participate in the late adolescent growth spurt, like hair growth and leg dimensions (Tanner 1988).

Hypermorphosis in Robust Australopithecines, Genus *Paranthropus*

Hypermorphosis in robust australopithecines seems to have occurred in a different way to that in hominines: by *rate* hypermorphosis, that is, by increased growth rates rather than prolongation of growth. This conclusion requires that the increased rates of dental development in *Paranthropus* relative to *Australopithecus africanus* observed by Beynon and Dean (1988) reflect increased rates for the cranium as a whole. As mentioned earlier, Sacher and Staffeldt (1974) proposed that the maximum rate of fetal growth should be limited to the maximum growth rate of neural tissue. I suggest that this is why robust australopithecines (or, for that matter, gorillas: Shea 1982) did not become as encephalized as *Homo* in spite of hypermorphosis. While rate hypermorphosis in *Paranthropus* did hypertrophy other features (such as of the face and dentition), it could not increase the rate of the ancestral earliest *brain* growth phase, which was already at a maximum. Let us assume for the moment that this is correct, and also that an overwhelming proportion of brain increase in robusts occurred early in ontogeny (as in other hominoids,

Figure 18-10). This might help to explain why the larger *P. boisei* does not have a higher brain capacity than the smaller *P. robustus* (Holloway 1972; McHenry 1988) and, thus, why *P. boisei* ends up with a lower EQ (Figure 18-1). In contrast, progressive prolongation by *time* hypermorphosis of fast early brain growth and EQ increase with body weight are possible as shown in hominines.

While the trend to larger bodies in *Homo* occurred by a switch to the rarer evolutionary mode, sequential time hypermorphosis, the analogous trend in *Paranthropus* continued to follow the mode of size increase that is more general in Hominoidea, rate hypermorphosis. Hence *Homo* diverged into a new EQ trend while *Paranthropus* continued within the ancestral one (Figure 18-1).

SUMMARY: CONCLUSIONS AND QUESTIONS

Hyper-paedomorphosis (hypermorphosis or enlargement, with paedomorphosis or juvenilization) may have been a general evolutionary response to climatic cooling and may have occurred in the evolution of *Homo*. Paedomorphosis in general, including hyper-paedomorphosis, most often entails reduced sexual dimorphism. In anthropomorphic terms: Hyper-paedomorphosis seems an efficient way to achieve in one stroke larger body size without the grotesque proliferation of bodily protruberances and extremities *and* therefore larger body size with reduced sexual dimorphism in structure. The observation that this mode of heterochrony is common suggests that the onset of such phenotypic novelty is relatively readily brought about by ontogenies—evolutionarily "easy" in the sense of not needing to be pieced together by a myriad of small independent genetic changes. (Of course gene mutations are random in the sense that they occur independently from what ontogeny and selection might "prefer". This does not contradict a proposal that certain ontogenetic expressions of those mutations are more likely than others: Arnold et al. 1989.) An important concern in evolutionary biology is the study of ontogenetic mechanisms involved in the novel production of phenotypic variants, among which selection can then sort.

I suggest that the brain of one Late Pliocene hominid lineage started on a sequential hyperpaedomorphic trajectory in response to Late Pliocene cooling. This trend was amplified by subsequent Pleistocene glaciations to culminate in the human brain. Sequential hypermorphosis occurs by proportional prolongation of multiple ancestral ontogenetic phases of different growth rates. Such evolution should be most successful in lineages already predisposed to hypermorphic extension of life cycles (as primates are: Sacher and Staffeldt 1974; Martin 1983).

I report preliminary results of an analysis (Vrba and Vaisnys in prep.) of a large data set on brain weights, ages since conception, and body weights for Common Chimpanzees and modern humans. These results support a basic pattern of sequential hypermorphosis in the evolution of the human brain, but studies of possible modifications to this basic model are still in progress. In its simplest form this model suggests that linear extension of growth phases was the chief mechanism that led to dramatic exponential increases in encephalization, and to the even more dramatic acceleration in intellectual capabilities in our lineage. It has been claimed (e.g., McKinney and McNamara 1991) that the massive and undisputed overprint of human hypermorphosis necessitates the demise of human paedomorphosis. Our results suggest that these claims were premature.

Additional implications, as well as some intriguing thoughts for future testing, include the following.

1. A single evolutionary episode of sequential heterochrony can in principle affect characters differently, depending on the ancestral distribution of growth rates among growth phases. For instance, sequential hypermorphosis is a single coherent kind of heterochrony that can potentially in one stroke produce *paedomorphic* characters (those that grew faster early in the ancestral ontogeny) accompanied in mosaic fashion by other characters that are *peramorphic* (or hypertrophied, those that grew faster late in the ancestral ontogeny). This is a counterintuitive combination. Arguments over human heterochrony have been fueled largely by the confusing mosaic of different kinds of features (Table 18-1): Why should the braincase (Bolk 1926; Gould 1977), foot (Manley-Buser 1986), and so on, be paedomorphic, while our

leg length (Tanner 1988), sternebrae (Schultz 1969), and other characters are peramorphic? Sequential hypermorphosis may provide a resolution for at least some of these arguments.

2. Sequential hypermorphosis has involved a *reorganization* in terms of *proportional dominance among* areas of the brain. Proportional dominance among brain parts is important to brain function (Edelman 1989). This occurred because not only does the whole brain show a sequential hypermorphic extension of ancestral growth phases, but so do its subparts. In fact, parts of the brain, notably the neocortex and the cerebellum do grow at disproportionately high rates during particular growth phases (Passingham 1975; Dobbing and Sands 1973). If these ideas are valid, our understanding of human brain evolution will be greatly illuminated by learning about the relative ontogenies of *subparts* of hominoid brains.

3. The global cooling event near 2.5 MYA may have precipitated a change in the "rules"—both the rules imposed by environmental selection, and those of development. Within Hominidae, this was possibly the beginning of hypermorphosis by prolongation of growth rather than by only increasing growth rates, at least in one branch: Hominine encephalization in the latest Pliocene started a new trend of higher evolutionary rates than before (Figure 18-1). This raises questions: If the specific distinction between *H. habilis sensu stricto* and *H. rudolfensis* (Figure 18-1) that is claimed by some (Wood 1992) is valid, why did one, *H. rudolfensis*, evolve toward larger body size accompanied by increased encephalization (in agreement with the hyperpaedomorphosis hypothesis) while *H. habilis* also became encephalized but at significantly smaller body size (McHenry 1991, 1992; Wood 1992)? Is Beynon and Dean's (1988) result, of a slower eruption rate of the permanent teeth in *A. africanus* and possibly early *Homo*, than in *Paranthropus*, indicative that the first two extinct taxa had already embarked on the route to hyperpaedomorphosis by extended growth phases? What happened during the 1.9–1.5 MYA period during which huge brain capacities appeared (Rightmire 1985)? Did *H. erectus* also originate during a particularly severe cooling phase (as previously suggested, Vrba 1989) that remains as yet unresearched by climatologists? Why did no significant brain expansion occur between 1.5-.4 MYA in the *H. erectus* lineage (Rightmire 1985)?

4. Hypermorphosis in robust australopithecines seems to have occurred in a different way to that in hominines: by *rate* hypermorphosis, that is, by increased growth rates rather than prolongation of growth (Beynon and Dean 1988). I suggest that this is why robust australopithecines (or, for that matter, gorillas, Shea 1982) did not become encephalized in spite of hypermorphosis: *Rate hypermorphosis* could not increase the rate of the ancestral earliest brain growth phase, which was already at a maximum (Sacher and Staffeldt 1974). In contrast, prolongation of this phase was possible, as in hominines. Of the two Late Pliocene hominid lineages that responded to cooling by hypermorphosis, why did one do so by rate increase while the other emphasised growth phase extension?

5. I suspect that, in general, particular forms of heterochrony are causally associated across different taxa with particular kinds of climatic change. Were there not only turnover-pulses of migrations, speciations and extinctions, but also "heterochrony pulses" across biotic groups in response to common climatic causes? Hominid evolutionary events may have formed parts of both.

There are many unanswered questions, but the potential for testing these ideas is there.

ACKNOWLEDGMENTS

I am grateful to Russ Ciochon and Robert Corruccini for inviting me to participate in this conference to honor Clark Howell. I thank Rimas Vaisnys, Brian Shea, and Bernard Wood for very useful comments. Clark Howell has made enormous contributions to our field right across the very broad spectrum of his interests. His many remarkable characteristics include that he has been a major stimulus to the research of others. He constantly inspires his colleagues and students to try out new research directions and to formulate new ideas. ("Write something provocative" Clark replied when I asked him what he would like to see in his Festschrift.) He always welcomes the appearance of his colleagues' and students' results with both enthusiasm and tough critique and—if by chance they are any good—he selflessly promotes them. Thank you, Clark—you have made and continue to make an enormous difference!

APPENDIX:

Sequential Heterochrony occurs when (1) and (2) are true:

1. The ancestral ontogeny is multiphasic, such that the character develops through at least two phases of distinct exponential growth rate. The following arguments concern two juvenile phases before growth offset (Figure 18-8):
   0 to x_1: ancestral onset to offset ages of early juvenile phase;
   0 to x_2: descendant onset to offset ages of early juvenile phase;
   x_1 to x_3: ancestral onset to offset ages of later juvenile phase;
   x_2 to x_4: descendant onset to offset ages of later juvenile phase;
   y: shape units accumulated at respective ages;
   a: growth rate of the earlier phase in ancestor and descendant ($y = x^a$);
   b: growth rate of the later phase in ancestor ($y = k_{anc}x^b$) and descendant ($y = k_{desc}x^b$).
2. Evolution occurs by proportional time extensions or contractions of each of these ontogenetic phases. That is, while the ancestral phase-specific growth rates are maintained in the descendant, the phases are lengthened or shortened by a constant proportion, n, with $n > 0$. (See text for Case (a).)

Case (b) sequential Hyperperamorphosis, n > 1, a < b: The earlier phase has a lower exponential growth rate for the character than the later phase.

Given: $a < b$; $x_2 = nx_1$; $(x_4 - x_2) = n(x_3 - x_1)$; $n > 1$ (In Figure 18-8b, $n = 2$).

In this case *hypermorphosis* results in a descendant adult in which the proportion of shape units derived from the later growth phase (or *peramorphic* shape) is increased;

namely, $$\frac{y_2}{y_4} < \frac{y_1}{y_3}$$

Demonstration: $y_1 = x_1{}^a$; and $y_2 = (nx_1)^a$

$$y_3 = x_1{}^a + (x_3 - x_1)^b \text{; and } y_4 = (nx_1)^a + (x_4 - x_2)^b \quad (1)$$

Therefore $$\frac{y_2}{y_4} = \frac{n^a x_1{}^a}{n^a x_1{}^a + n^b (x_3 - x_1)^b} = \frac{x_1{}^a}{x_1{}^a + n^{b-a} (x_3 - x_1)^b}$$

and $$\frac{y_1}{y_3} = \frac{x_1{}^a}{x_1{}^a + (x_3 - x_1)^b} \quad (2)$$

For $a < b$ and $n > 1$, $b - a > 0$, and $n^{b-a} > 1$;

Therefore $$\frac{y_2}{y_4} < \frac{y_1}{y_3}, \; or \; \frac{y_4 - y_2}{y_4} > \frac{y_3 - y_1}{y_3}.$$

That is, the proportion of shape units derived from the later juvenile phase (or peramorphic shape) increases in the descendant.

Case (c) sequential hypoperamorphosis, 0 < n < 1, a > b (Figure 18-8c: The earlier phase has a higher exponential growth rate than the later phase.

Given: $a > b$; $x_2 = nx_1$; $(x_4 - x_2) = n(x_3 - x_1)$; $0 > n > 1$ (In Figure 18-8c, $n = 0.5$).

In this case *hypomorphosis* results in a descendant adult in which the proportion of shape units derived from the later growth phase (or *peramorphic* shape) is increased;

namely, $\frac{y2}{y4} < \frac{y1}{y3}$

Demonstration: Steps (1) and (2) are as for Case (b).
For $a > b$ and $0 < n < 1$, $b - a < 0$, and $n^{b-a} > 1$:

Therefore $\frac{y2}{y4} < \frac{y1}{y3}$, *or* $\frac{y4 - y2}{y4} > \frac{y3 - y1}{y3}$.

That is, the proportion of shape units derived from the later juvenile phase (or peramorphic shape) increases in the descendant.

Case (d) sequential hypo-paedomorphosis, 0 > n > 1, a < b (Figure 18-8d: The earlier phase has a lower exponential growth rate than the later phase.

Given: $a < b$; $x_2 = nx_1$; $(x_4 - x_2) = n(x_3 - x_1)$; $0 < n < 1$ (In Figure 18-d, $n = 0.5$).

In this case *hypomorphosis* results in a descendant adult in which the proportion of shape units derived from the earlier growth phase (or *paedomorphic* shape) is increased;

namely, $\frac{y2}{y4} < \frac{y1}{y3}$

Demonstration: Steps (1) and (2) are as for Case (b).
For $a < b$ and $0 < n < 1$, $b - a > 0$, and $n^{b-a} < 1$:

Therefore $\frac{y2}{y4} > \frac{y1}{y3}$

That is, the proportion of shape units derived from the earlier juvenile phase (or paedoamorphic shape) increases in the descendant.

GLOSSARY

ACCELERATION: In classical peramorphosis by acceleration the descendant has growth offset at the same time and is of similar size and weight as the ancestral adult, but of more advanced shape. Thus, the ancestral and descendant growth allometries are dissociated because the descendant rate of shape growth is relatively accelerated.

ECOPHENOTYPES: Different ontogenetic expressions of a feature, in response to differing local environments, that are not genetically based.

GLOBAL CLIMATIC CHANGE: Alterations of climatic signals in many parts of the world during a particular time period, such as those that can result from astronomical forcing. My use of the term recognizes that not all localities are affected, and that particular climatic results may be out of phase.

HETEROCHRONY: Includes all evolutionary changes in the timing of appearance of characters during ontogeny, and in the rates of shape and size development. From one to most descendant characters may be affected.

HYPERMORPHOSIS: Classical hypermorphosis is a form of peramorphosis in which the descendant adult is more highly developed in both shape and size than the ancestral adult. It occurs by extension (either by prolonged growth, or by increased growth rate) of a common growth allometry to larger size in the descendant by ontogenetic scaling. There is no dissociation of shape-size relations.

HYPOMORPHOSIS: Classical hypomorphosis is a form of paedomorphosis in which the descendant adult is more juvenilized in both shape and size than the ancestral adult. It occurs by truncation of a common growth allometry to smaller size in the descendant by ontogenetic scaling. There is no dissociation of shape-size relations.

INITIATING CAUSES OF SPECIATION AND EXTINCTION: Physical and biotic events on earth that initiate the causal chain that results in speciation and extinction. The term does not imply that extraterrestrial or subsequent terrestrial causes are not operating as well in the larger causal chain.

NEOTENY: In classical paedomorphosis by neoteny, the descendant adult retains shape characters that appeared in ancestral juvenile stages. Growth offset occurs at the same time, and size growth rates are similar, in ancestor and descendant; but their growth allometries are dissociated because the descendant rate of shape growth is relatively retarded. The result is a descendant adult of similar size and weight as the ancestral adult, but of juvenilized shape relative to the ancestor.

PAEDOMORPHOSIS [from Greek *paid*, child, and *morph*, shape]: The retention of ancestral juvenile characters (including proportions or shapes) by later ontogenetic stages of descendants.

PERAMORPHOSIS [from Greek *pera*, beyond, and *morph*, shape]: Signifies that the descendant adults have more derived characters (including proportions or shapes) than the ancestral adult.

RATE HYPERMORPHOSIS: A kind of hypermorphosis in which growth offset [cessation] in the descendant occurs at the same time as in the ancestor, but descendant growth rates are increased relative to the ancestor. Thus, in the same developmental time, the descendant adult becomes more advanced in both shape and size than the ancestral adult.

RATE HYPOMORPHOSIS: A kind of hypomorphosis in which the growth offset in the descendant occurs at the same time as in the ancestor, but descendant growth rates are reduced relative to the ancestor. Thus, in the same developmental time, the descendant adult becomes more juvenilized in both shape and size than the ancestral adult.

SEQUENTIAL HYPERMORPHOSIS: The ancestral ontogeny is multiphasic, such that the character develops through at least two phases of distinct exponential growth rate. Evolution occurs by proportional time extension of each of these ontogenetic phases. It differs from classical hypermorphosis in that ancestral and descendant growth allometries are dissociated.

SEQUENTIAL HYPOMORPHOSIS: The ancestral ontogeny is multiphasic, such that the character develops through at least two phases of distinct exponential growth rate. Evolution occurs by proportional time truncation of each of these ontogenetic phases. It differs from classical hypomorphosis in that ancestral and descendant growth allometries are dissociated.

TIME HYPERMORPHOSIS: A kind of hypermorphosis in which growth offset in the descendant occurs later than in the ancestor, while they share the same size and shape growth rates. Descendant adults are larger, with peramorphic morphology as in the case of rate hypermorphosis, but for a different reason.

TIME HYPOMORPHOSIS [also called PROGENESIS]: A kind of hypomorphosis in which growth offset in the descendant occurs earlier than in the ancestor, while they share the same size and shape growth rates. Descendant adults are smaller, with juvenilized morphology, just as in the case of rate hypomorphosis, but for a different reason.

TURNOVER OF LINEAGES: Includes speciation, extinction and migration, all of which change the composition of species in particular areas.

TURNOVER-PULSE: Concentration of turnover events against the time scale. For example, if a high number of first and last records of species in different lineages occur together within a time interval of 100,000 years or less, preceded and postdated by a million years of predominant stasis in the same monophyletic groups, I would regard this as evidence of a turnover-pulse. I expect that turnover-pulses also occur in more refined time intervals, and that at least some stratigraphic sequences will be of sufficient quality to show this.

LITERATURE CITED

Alberch, P., Gould, S. J., Oster, G. F., and Wake, D. B. (1979) Size and shape in ontogeny and phylogeny. *Paleobiology* 5:296-317.

Allen, J. A. (1877) The influence of physical conditions in the genesis of species. *Radical Review* 1:108-140.

Ambros, V. (1988) Genetic basis for heterochronic variation. In M. L. McKinney (ed.): *Heterochrony in Evolution*. New York: Plenum, pp. 269-284.

Arnold, S. J., Alberch, P., Csanyi, V., Dawkins, R. C., Emerson, S. B., Fritzsch, B., Horder, T. J., Maynard Smith, J., Starck, M., Vrba, E. S., Wagner, G. P., and Wake, D. B. (1989) How do complex organisms evolve? In D. B. Wake and G. Roth (eds.) *Complex Organismal Functions: Integration and Evolution in Vertebrates*. Dahlem Konferenzen Chichester: John Wiley & Sons, pp. 403-433.

Ashton, E. H. (1950) The endocranial capacities of the Australopithecinae. *Proc. Zool. Soc. Lond.* 120:715-721.

Azzaroli, A. (1983) Quaternary mammals and the "End-Villafranchian" dispersal event—a turning point in the history of Eurasia. *Palaeogeog. Palaeoclimatol. Palaeoecol.* 44:117-139.

Baker, P. T. (1988) Human adaptability. In G. A. Harrison, J. M. Tanner, D. R. Pilbeam and P. T. Baker (eds.): *Human Biology*. Oxford: Oxford University Press, pp. 439-544.

Bergmann, C. (1847) Uber die Verhaltnisse der Wermekonomie der Thiere zu ihrer Grosse. *Gottinger Studien* 3. 1:595-708.

Beynon, A. D., and Dean, M. C. (1988) Distinct dental development patterns in early fossil hominids. *Nature* 335:509-514.

Biological Tables (Tabulae biologicae) (1941) Growth of Man. 20. London.

Bischoff (1880) *Hirngewicht des Menschen, anatomische, physiologische und physikalische Tabellen*. Bonn: Bonn Verlag.

Blinkov, S. M., and Glezer, I. I. (1968) *The Human Brain in Figures and Tables. A Quantitative Handbook*. New York: Plenum.

Bolk, L. (1926) *Das Problem der Menschwerdung*. Jena: Fischer Verlag.

Bolk, L. (1929) Origin of racial characteristics in man. *Am. J. Phys. Anthropol.* 13:1-28.

Brain, C. K. (1985) Temperature-induced environmental changes in Africa as evolutionary stimuli. In E. S. Vrba (ed.): *Species and Speciation*. Pretoria: Transvaal Museum. *Transvaal Museum Monograph* No. 4:45-52.

Bromage, T. G. (1985) Taung facial remodeling: A growth and development study. In P. V. Tobias (ed.): *Hominid Evolution: Past, Present, and Future*. New York: Alan R. Liss, pp. 239-246.

Bukhshtab, I. (1884) *Facts Relating to the Weight, Volume, and Specific Gravity of the Brain*. St. Petersburg: S. Petersburg.

Burckle, L. H., Denton, G. H., Partridge, T. C., and Vrba, E. S. (1986) Palaeoclimate and evolution III. *S. Afr. J. Sci.* 82:493-522.

Burn, J., Birkbeck, J. A., and Roberts, D. F. (1975) Early fetal brain growth. *Hum. Biol.* (Wayne State University) 47:511-522.

Cheverud, J., Leutenegger, W., and Dow, M. (1985) Phylogenetic autocorrelation and the correlates of sexual dimorphism in primates. *Am. J. Phys. Anthropol.* 63:145.

Clutton-Brock, T. H. (1979) The functions of antlers. *Behavior* (1- 2):108-125.

Coppens, Y. (1975) Evolution des hominides et de leur environment au cours du Plio-Pleistocene dans la basse vallee de l'Omo en Ethiopie. *C. R. Acad. Sc. Paris* 281:1693-1696.

Count, E. W. (1947) Brain and body weight in man. *Ann. NY Acad. Sci.* 46:993-1122.

Davis, S. J. (1981) The effects of temperature change and domestication on the body size of Late Pleistocene to Holocene mammals of Israel. *Paleobiology* 7:101-114.

De Beer G. (1958) *Embryos and Ancestors*. Oxford: Oxford Unversity.

Delson, E. (1984) Cercopithecid biochronology of the African PlioPleistocene: Correlation among eastern and southern hominid-bearing localities. *Cour. Forsch. Inst. Senckenberg* 69:199-218.

Demment, M. W., and Van Soest, P. J. (1985) A nutritional explanation for body-size patterns of ruminant and nonruminant herbivores. *Am. Nat.* 125:641-672.

Dobbing, J., and Sands, J. (1973) Quantitative growth and development of human brain. *Arch. Disease Childhood* 48:757.

Edelman, G. M. (1989) *The Remembered Present*. New York: Basic Books.

Estes, R. D. (1991) The significance of horns and other male sexual characters in female bovids. *App. Anim. Behav. Sc.* 29:403-451.

Falkner, F. and Tanner, J. M. (Eds.) 1986 *Human Growth: A Comprehensive Treatise*. Vols. 1 and 2. New York: Plenum.

Fink, W. L. (1982) The conceptual relationship between ontogeny and phylogeny. *Paleobiology* 8:254-264.

Fischer, K. W. (Ed.) 1983 *Levels and Transitions in Children's Development*. San Francisco: Jossey Bass.

Gatesy, J., Yelon, D., DeSalle R, and Vrba, E. S. (1992) Phylogeny of the Bovidae (Artiodactyla, Mammalia), based on mitochondrial ribosomal DNA sequences. *Mol. Biol. Evol.* 9:433-446.

Gavan, J. A. (1971) Longitudinal, postnatal growth in chimpanzee. In G. H. Bourne (ed.): *Behavior, Growth, and Pathology of Chimpanzees*. Baltimore: Unversity Park Press, pp. 47-102.

Geist, V. (1971) *Mountain Sheep: A Study in Behavior and Evolution*. Chicago: University of Chicago Press.

Geist, V. (1987) Bergmann's rule is invalid. *Canad. J. Zool.* 65:1035-1038.

Geist, V., and Bayer, M. (1988) Sexual dimorphism in the Cervidae and its relation to habitat. *J. Zool. Lond.* 214:45-53.

Gould, S. J. (1977) *Ontogeny and Phylogeny*. Cambridge: Harvard Unversity Press, 501 pp.

Greenacre, M. J., and Vrba, ES (1984) A correspondence analysis of biological census data. *Ecology* 65:984-997.

Grether, W. F., and Yerkes, R. M. (1940) Weight norms and relations for chimpanzee. *Am. J. Phys. Anthropol.* 27:181-190.

Hafner, J. C., and Hafner, M. S. (1988) Heterochrony in rodents. In M. L. McKinney (ed.): *Heterochrony in Evolution: A Multidisciplinary Approach.* New York: Plenum, pp. 217-235.

Hagedoorn, A. (1924) *Schädelkapazität von Anthropomorphen.* Amsterdam: Nederlandsche Tydschriften van Geneeskunde.

Hammel, H. T., Jackson, D. C., Stolwijk, J. A. J., Hardy, J. D., and Stromme, S. B. (1963) Temperature regulation by hypothalamic proportional control with an adjustable set point. *J. Appl. Physiol.* 18:1146-1154.

Harrison, G. A., Tanner, J. M., Pilbeam, D. R. and Baker, P. T. (eds.) (1988) *Human Biology.* Oxford: Oxford Unversity Press.

Haug, H. (1958) *Quantitative Untersuchungen an der Sehrinde.* Stuttgart: Stuttgart Verlag.

Heintz, A., and Garutt, V. E. (1965) Determination of the absolute age of the fossil remains of mammoth and wooly rhinoceros from the permafrost in Siberia by the help of radiocarbon (C_{14}). *Nor. Geol. Tidsskr.* 45:73-79.

Hill, A., Ward, S., Deino, A., Curtis, G., and Drake, R. (1992) Earliest *Homo. Nature* 355:719.

Hodge, M. J. S. (1983) The development of Darwin's general biological theorizing. In D. S. Bendall (ed.): *Evolution from Molecules to Men.* Cambridge: Cambridge Unversity Press, pp. 43-62.

Holloway, R. L. (1970) New endocranial values for the australopithecines. *Nature* 227:199-200.

Holloway, R. L. (1972) New australopithecine endocast SK 1585, from Swartkrans, South Africa. *Am. J. Phys. Anthropol.* 37:173-186.

Holloway, R. L. (1978) Problems of brain endocast interpretation and African hominid evolution. In C. J. Jolly (ed.): *Early Hominids of Africa.* London: Duckworth, pp.379-401.

Holloway, R. L. (1980) Within-species brain-body weight variability: A reexamination of the Danish data and other primate species. *Am. J. Phys. Anthropol.* 53:109-121.

Holt, A. B., Cheek, D. B., Mellits, E. D., and Hill, D. E. (1975) Brain size and the relation of the primate to the nonprimate. In D. B. Cheek (ed.): *Fetal and Postnatal Cellular Growth: Hormones and Nutrition.* New York: John Wiley, pp. 23-44.

Howell, F. C. (1978) Hominidae. In V. J. Maglio and H. B. S. Cooke (eds.): *Evolution of African Mammals.* Cambridge, Massachusetts: Harvard Unversity Press, pp. 154-248.

Hrdlicka, A. (1925) Weight of brain and of internal organs in American monkeys. *Am. J. Phys. Anthropol.* 8:201-211.

Iggo, A. (1972) Cutaneous thermoreceptors. In G. M. O. Maloiy (ed.): *Comparative Physiology of Desert Animals.* London: Academic, pp. 327-343.

Janis, C. (1982) Evolution of horns in ungulates: ecology and paleoecology. *Biol. Rev.* 57:261-318.

Jarman, P. J. (1983) Mating system and sexual dimorphism in large, terrestrial, mammalian herbivores. *Biol. Rev.* 59:485-520.

Jenkinson, D. M. (1972) Evaporative temperature regulation in domestic animals. In G. M. O. Maloiy (ed.): *Comparative Physiology of Desert Animals.* London: Academic, pp. 345-354.

Johanson, D. C. and Edey, M. (1981) *Lucy: The Beginnings of Humankind.* New York: Simon & Schuster.

Jordaan, H. V. F. (1976a) Newborn: Adult brain ratios in hominid evolution. *Am. J. Phys. Anthropol.* 44:271-178.

Jordaan, H. V. F. (1976b) Newborn brain: Body weight ratios. *Am. J. Phys.Anthropol.* 44:279-284.

Keeling, M. E., and Riddle, K. E. (1975) Reproductive, gestational, and newborn physiology of the chimpanzee. *Lab. Anim. Sci.* 25:822.

Keith, A. (1895) Growth of brain in men and monkeys. *J. Anat. Physiol.* 29:282-303.

Khamsi, F., and Eayrs, J. I. (1966) A study of the effects of thyroid hormones in growth and development. *Growth* 30:143-156.

Kluge, A. G. (1989) A concern for evidence and phylogenetic hypothesis of relationships among *Epicrates* (Boidae, Serpentes). *Syst. Zool.* 38:7-25.

Koch, P. L. (1986) Clinal variation in mammals: Implications for the study of chronoclines. *Paleobiology* 12:269-281.

Kollman, J. (1905) Neue Gedanken uber das alte Problem von der Abstammung des Menschen. *Corresp. Bl. Dtsch. Ges. Anthropol. Ethnol. Urges* 36:9-20.

Krogman, W. M. (1969) Growth changes in skull, face, jaws, and teeth of the chimpanzee. In G. H. Bourne (ed.): *Anatomy, Behavior, and Diseases of Chimpanzees.* Baltimore: University Park, pp. 104-164.

Kurtén, B. (1959) On the bears of the Holsteinian Interglacial. *Stockholm Contrib. Geol.* 2:73-102.

Larroche, J. C. (1967) Maturation morphologique du systeme nerveux central: Ses rapports avec le developpement ponderal du foetus et son age gestationnel. In A. Minkowski (ed.): *Regional Development of the Brain in Early Life.* Philadelphia: FA Davis Co., pp. 247-267.

Lestrel, P. E. (1975) Hominid brain size versus time: revised regression estimates. *J. Hum. Evol.* 5:207-212.

Lestrel, P. E., and Read, D. W. (1972) Hominid cranial capacity versus time: A regression approach. *J. Hum. Evol.* 2:405-411.

Leutenegger, W., and Larson, S. (1985) Sexual dimorphism in the postcranial skeleton of New World primates. *Folia Primatol.* 44:82-95.

Lindsay, E. H., Opdyke, N. D., and Johnson, N. M. (1980) Pliocene dispersal of the horse *Equus* and late Cenozoic mammalian dispersal events. *Nature* 287:135.

Lomolino, M. (1985) Body size of mammals on islands. *Am. Nat.* 125:310-316.

Lorenz, K. Z. (1970) *Studies on Animal and Human Behavior.* Cambridge: Harvard Unversity Press.

Lovejoy, C. O. (1981) The origin of man. *Science* 211:341-350.

Macdonald, D. (1984) *The Encyclopaedia of Mammals*: 1. London: George Allen and Unwin.

Manley-Buser, K. A. (1986) A heterochronic study of the human foot. *Am. J. Phys. Anthropol.* 69:235-255.

Martin, R. A. (1986) Energy, ecology and cotton rat evolution. *Paleobiology* 12:370-382.

Martin, R. D. (1983) *Human Brain Evolution in an Ecological Context*. Second James Arthur Lecture on the Evolution of the Human Brain. New York: American Museum of Natural History Publications.

Maynard Smith, J., Burian, J., Kauffman, S., Alberch, P., Campbell, J., Goodwin, B., Lande, R., Raup, D., and Wolpert, L. (1985) Developmental constraints and evolution. *Quart. Rev. Biol.* 60:265-287.

McClure, W. (1992) Records of the Yerkes Region Primate Research Center. (pers. comm.) Atlanta, GA: Emory University.

McHenry, H. M. (1988) New estimates of body weight in early hominids and their significance to encephalization and megadontia in "robust australopithecines." In F. E. Grine (ed.): *The Evolutionary History of the Robust Australopithecines*. New York: Aldine, pp. 133-148.

McHenry, H. M. (1991) Petite bodies of the "robust" australopithecines. *Am. J. Phys. Anthropol.* 86:445-454.

McHenry, H. M. (1992) Body size and proportions in early hominids. *Am. J. Phys. Anthropol.* 87:407-431.

McKinney, M. L., and McNamara, K. J. (Eds.) (1991) *Heterochrony. The Evolution of Ontogeny*. New York: Plenum.

Mentis, M. T. (1972) A review of some life history features of the large herbivores of Africa. *Lammergeyer* 16:1-89.

Mikhailets, V. Y. (1952) Increase in weight of the brain during the intrauterine development of man. *Nauch. Zap. Uzhgorodsk. Gos. Univ. 5. Sb. Stud. Rabot Uzhgorod* 2:47-50.

Miyazaki, J. M., and Mickevitch, M. F. (1984) Evolution of Chesapecten (Mollusca: Bivalvia, Miocene-Pliocene) and the Biogenetic Law. *Evol. Biol.* 40:369-409.

Montagu, M. F. A. (1962) Time, morphology, and neoteny in the evolution of man. In M. F. A. Montagu (ed.): *Culture and the Evolution of Man*. New York: Oxford University Press, pp. 324-342.

Montagu, M. F. A. (1981) *Growing Young*. New York: McGraw-Hill.

Moss, M. L., Moss-Salentijn, L., Vilmann, H., and Newell-Morris, L. (1982) Neuro-skeletal topology of the primate basicranium: Its implications for the "fetalization hypothesis." *Gegenbaurs Morph. Jahrb., Leipzig* 128:58-67.

Muhlman, L. (1957) *Die Abhangigkeit des Hirngewichtes*. Munich: Munchen Verlag.

Nelson, G. (1978) Ontogeny, phylogeny, paleontology, and the biogenetic law. *Syst. Zool.* 27:324-345.

Newman, R. W. (1975) Human adaptation to heat. In A. Damon (ed.): *Physiological Anthropology*. New York: Oxford Unversity Press.

Oppenheim, S. (1911) Zur Typologie des Primatenkraniums. *Z. Morph. Anthrop.* 14:1-203.

Pagel, M. D., and Harvey, P. H. (1989) Taxonomic differences in the scaling of brain on body weight among mammals. *Science* 244:1589-1593.

Partridge, T. C., Vrba, E. S., Burckle, L. H., and Denton, G. H. (eds.) (1986): Palaeoclimate and Evolution II. *S. Afr. J. Sci.* 82:62-96.

Passingham, R. E. (1975) The brain and intelligence. *Brain Behav. Evol.* 11:1-15.

Peters, R. H. (1983) *The Ecological Implications of Body Size*. Cambridge: Cambridge Unversity Press.

Prothero, D. R., and Sereno, P. C. (1982) Allometry and paleoecology of medial Miocene dwarf rhinoceroses from the Texas Gulf Coastal Plain. *Paleobiology* 8:16-30.

Raff, R. A., and Kaufman, T. C. (1983) *Embryos, Genes and Evolution*. New York: Macmillan.

Ralls, K., and Harvey, P. H. (1985) Geographic variation in size and sexual dimorphism of North American weasels. *Biol. J. Linnean Soc.* 25:119-167.

Rightmire, G. P. (1985) The tempo of change in the evolution of mid- Pleistocene *Homo*. In E. Delson (ed.): *Ancestors: The Hard Evidence*. New York: Alan R. Liss, pp. 255-264.

Roberts, D. F. (1978) *Climate and Human Variability*. Menlo Park, CA: Cumming.

Ross, C. (1988) The intrinsic rate of natural increase and reproductive effort in primates. *J. Zool. Lond.* 214:199-219.

Sacher, G. A. (1959) Relation of life span to brain weight and body weight in mammals. In G. E. W. Wolstenholme and M. O'Connor (eds.): *The Lifespan of Animals*. London: Churchill, pp. 115-141.

Sacher, G. A. (1982) The role of brain maturation in the evolution of primates. In E. Armstrong and D. Falk (eds.): *Primate Brain Evolution: Methods and Concepts*. New York: Plenum, pp. 97-112.

Sacher, G. A., and Staffeldt, E. F. (1974) Relation of gestation time and brain weight of placental mammals: Implications for the theory of vertebrate growth. *Am. Nat.* 105:593-615.

Savage, D. E., and Russell, D. E. (1983) *Mammalian Paleofaunas of the World*. London: Addison-Wesley.

Schoener, T. W. (1983) Field experiments on interspecific competition. *Am. Nat.* 122:240-285.

Schultz, A. H. (1926) Fetal growth of man and other primates. *Quart. Rev. Biol.* 1:465-521.

Schultz, A. H. (1930) Notes on the growth of anthropoid apes, with especial reference to deciduous dentition. Report of the Laboratory and Museum of Comparative Pathology of the Zoological Society of Philadelphia. 58:34-45.

Schultz, A. H. (1940) Growth and development of the chimpanzee. Carnegie Inst. Washington, Publ. 518. *Contrib. to Embryol.* 28:1-63.

Schultz, A. H. (1941) The relative size of the cranial capacity in primates. *Am. J. Phys. Anthropol.* 28:273-287.

Schultz, A. H. (1965) The cranial capacity and the orbital volume of hominoids according to age and sex. In A. H. Schultz (ed.): *Homenaje a Juan Comas en su 65 aniversario*, Volumen II. Mexico City: Editorial Libros de Mexico, pp. 337-357.

Schultz, A. H. (1969) The skeleton of the chimpanzee. In G. H. Bourne (ed.): *Anatomy, Behavior, and Diseases of Chimpanzees*. Baltimore: University Park Press, pp. 50-101.

Schulz, D. M., Giordano, D. A., and Schulz, D. H. (1962) Weights of organs of fetuses and infants. *Arch. Pathology* 74: 244-250.

Scow, R. O., and Simpson, M. E. (1945) Thyroidectomy in the newborn rat. *Anat. Rec.* 91:209-226.

Selenka, E. (1899) Schadel des Gorilla und Schimpanse. Menschenaffen. *Studien Entwicklungsgesch.* 7:93-160.

Shantha, T. R., and Manocha, S. L. (1969) The brain of chimpanzee. In G. H. Bourne (ed.): *Anatomy, Behavior, and Diseases of Chimpanzees.* Baltimore: University Park Press, pp. 188-234.

Shea, B. T. (1982) Ontogenetic allometry and scaling. A discussion based on the growth and form of the skull in African apes. In W. L. Jungers (ed.): *Size and Scaling in Primate Biology.* New York: Plenum, pp. 175-206.

Shea, B. T. (1983) Paedomorphosis and neoteny in the Pygmy Chimpanzee. *Science* 222:521-522.

Shea, B. T. (1986) Ontogenetic approaches to sexual dimorphism in anthropoids. *Hum. Evol.* 1:97-110.

Shea, B. T. (1988) Heterochrony in primates. In M. McKinney (ed.): *Heterochrony in Evolution.* New York: Plenum, pp. 237-266.

Shea, B. T. (1989) Heterochrony in human evolution: The case for neoteny reconsidered. *Yearb. Phys. Anthropol.* 32:69-101.

Shea, B. T., and Bailey R. C. (1989) Allometric growth of body proportions in Efe pygmies of Zaire. *Am. J. Phys. Anthropol.* 78:300-301.

Sinclair, A. R. E. (1977) *The African Buffalo.* Chicago: University of Chicago Press.

Smith, B. H. (1991) Dental development and the evolution of life history in Hominidae. *Am. J. Phys. Anthropol.* 86:157-174.

Smithers, R. H. N. (1983) *The Mammals of the Southern African Subregion.* Pretoria: University of Pretoria.

Spence, K. W., and Yerkes, R. M. (1937) Weight, growth and age in chimpanzee. *Am. J. Phys. Anthropol.* 22:229-250.

Spinage, C. A. (1986) *The Natural History of Antelopes.* New York: Facts on File Publications.

Stearns, S. C. (1982) The role of development in the evolution of life histories. In J. T. Bonner (ed.): *Evolution and Development.* Berlin: Springer Verlag, pp. 237-258.

Tanner, J. M. (1988) Human growth and constitution. In J. M. Tanner et al. (eds.): *Human Biology.* New York: Oxford Unversity Press, pp. 339-432.

Tanner, J. M. (1990) *Foetus into Man.* Cambridge, MA: Harvard University Press.

Tanner, J. M., and Whitehouse, R. H. (1982) *Atlas of Children's Growth: Normal Variation and Growth Disorders.* London: Academic.

Tihen, S. A. (1955) A new Pliocene species of *Ambystoma* with remarks on other fossil ambystomatids. *Contr. Mus. Paleontol. Univ. Michigan* 12:229-244.

Tobias, P. V. (1971) *The Brain in Hominid Evolution.* New York: Columbia Unversity Press.

Vrba, E. S. (1974) Chronological and ecological implications of the fossil Bovidae at the Sterkfontein Australopithecine Site. *Nature* 256:19-23.

Vrba, E. S. (1975) Some evidence of chronology and palaeoecology of Sterkfontein, Swartkrans and Kromdraai from the fossil Bovidae. *Nature* 254:301-304.

Vrba, E. S. (1979) Phylogenetic analysis and classification of fossil and recent Alcelaphini (Family Bovidae, Mammalia). *J. Linn. Soc. (Zool.)* 11:207-228.

Vrba, E. S. (1982) Biostratigraphy and chronology, based particularly on Bovidae, of southern African hominid-associated assemblages: Makapansgat, Sterkfontein, Taung, Kromdraai, Swartkrans; also Elandsfontein (Saldanha), Broken Hill (now Kabwe) and Cave of Hearths. In H. de Lumley and M. A. de Lumley (eds.): *Proceedings of Congrés International de Paléontologie Humaine.* Nice: Union Internationale des Sciences Prehistoriques et Protohistoriques 2:707-752.

Vrba, E. S. (1985a) Ecological and adaptive changes associated with early hominid evolution. In E. Delson (ed.): *Ancestors: The Hard Evidence.* New York: Alan R. Liss, pp. 63-71.

Vrba, E. S. (1985b) Environment and evolution: Alternative causes of the temporal distribution of evolutionary events. *S. Afr. J. Sci.* 81:229-236.

Vrba, E. S. (1985c) African Bovidae: Evolutionary events since the Miocene. *S. Afr. J. Sci.* 81:263-266.

Vrba, E. S. (1987a) New species and a new genus of Hippotragini (Bovidae) from Makapansgat Limeworks. *Palaeont. Africana* 26:47-58.

Vrba, E. S. (1987b) A revision of the Bovini (Bovidae) and a preliminary revised checklist of Bovidae from Makapansgat. *Palaeont. Africana* 26:33-46.

Vrba, E. S. (1988) Late Pliocene climatic events and hominid evolution. In F. E. Grine (ed.): *The Evolutionary History of the Robust Australopithecines.* New York: Aldine Publishing Co., pp. 405-426.

Vrba, E. S. (1989) The environmental context of the evolution of early hominids and their culture. In R. Bonnichsen and M. H. Sorg (eds.): *Bone Modification.* Orono, Maine: Center for the Study of the First Americans, pp. 27-42.

Vrba, E. S. (1993) Turnover-pulses, the Red Queen, and related topics. *Am. J. Sci.* 293-A:418-452.

Vrba, E. S., Burckle, L. H., Denton, G. H., and Partridge, T. C. (eds.) (1985): Palaeoclimate and Evolution I. *S. Afr. J. Sci.* 85:224-275.

Vrba, E. S., Denton, G. H., and Prentice, M. L. (1989) In N. G. Gejvall (ed.): *Early Human Behaviour.* Ossa (International Journal of Skeletal Research) 14:127-156.

Vrba, E. S., Vaisnys, J. R., Gatesy, J. E., DeSalle, R., and Wei, K.-Y. (1994) Tests of paedomorphosis using allometric characters: The example of extant Reduncini (Bovidae, Mammalia). *Syst. Biol.* (In press).

Wake, D. B., and Larson, A. (1987) Multidimensional analysis of an evolutionary lineage. *Science* 238:42-48.

Wake, D. B., and Roth, G. (eds.) (1989): *Complex Organismal Functions: Integration and Evolution in Vertebrates.* New York: John Wiley & Sons.

Walker, A., Leakey, R. E., Harris, J. N., and Brown, F. H. (1986) 2.5-MYA *Australopithecus boisei* from west of Lake Turkana, Kenya. *Nature* 322:517-522.

Weitz, C. (1979) *Introduction to Physical Anthropology and Archaeology.* Englewood Cliffs, NJ: Prentice Hall.

Wesselman, H. B. (1984) *The Omo Micromammals. Systematics and Paleoecology of Early Man Sites from Ethiopia*. New York: Karger.

Wood, B. A. (1976) The nature and basis of sexual dimorphism in the primate skeleton. *J. Zool. Lond.* 180:15-34.

Wood, B. (1985) Sexual dimorphism in the hominid fossil record. In J. Ghesquiere, R. D. Martin and F. Newcombe (eds.): *Human Sexual Dimorphism*. Symposia of the Society for the Study of Human Biology. Philadelphia: Taylor and Francis, pp. 105-123.

Wood, B. (1992) Origin and evolution of the genus *Homo*. *Nature* 355:783-790.

Wood, B. A., Li, Y., and Willoughby, C. (1991) Intraspecific variation and sexual dimorphism in cranial and dental variables among higher primates and their bearing on the hominid fossil record. *J. Anat.* 174:185-205.

Zihlman, A. L. (1979) Pygmy chimpanzee morphology and the interpretation of early hominids. *S. Afr. J. Sci.* 75:165-168.

Zuckerman, S. (1928) Age-changes in the chimpanzee, with special references to growth of brain, eruption of teeth, and estimation of age; with a note on the Taungs Ape. *Proc. Zool. Soc. Lond.* I:1-42.

CHAPTER 19

Taphonomy and the Fluvial Environment

Examples from Pliocene Deposits of the Shungura Formation, Omo Basin, Ethiopia

Dorothy Dechant Boaz

INTRODUCTION

Over 400 papers have been published on data collected from the Plio-Pleistocene deposits of the Omo Group Formations found in the Lower Omo Basin of southern Ethiopia. A thorough description of Western discovery and exploration of the region is provided by Howell and Coppens (1983), along with the names of many of the scientists who have contributed to our knowledge of its past.

Among the Omo Group Formations, the Shungura Formation has been recognized as unique in being one of a few relatively continuous sequences of continental Plio-Pleistocene sedimentary deposits, intercalated with readily datable tuffs and yielding abundant fossiliferous remains (de Heinzelin 1983). During the time period extending from around 4.1 to 2.0 MYA, deposition of the Shungura Formation was dominated by sedimentation by a large perennial river, the ancient Omo River (Brown and Feibel 1988). The overall fluviatile setting afforded a variety of depositional environments conducive to fossilization of vertebrate remains. Varying degrees of transport, from relatively minimal on floodplains to significant in channel environments, ultimately influenced the characteristics of resultant fossil assemblages.

A number of excavations were undertaken at localities established in sediments of the Shungura Formation from 1969 to 1974. Several articles describe the paleontological assemblages from these excavations and their depositional settings (de Heinzelin and Haesaerts 1983; Howell et al. 1987; Johanson et al. 1976; Boaz 1977a,b, 1979, 1985), but publication of a detailed study of their taphonomic histories has been lacking.

The sedimentary environment and faunal composition of four of the excavated samples were described by Johanson et al. (1976). Three of these four assemblages, including Localities 345, 338y, and 398, are analyzed here. I attempt to extract taphonomically relevant evidence from mammalian remains of each assemblage and their sedimentary contexts, my goal being to approach a reasonable understanding of their taphonomic histories, including reconstruction of events leading to accumulation and preservation of the bones. The contrasts and similarities among the three assemblages and their depositional environments are examined.

Localities 345, 338y, and 398, dating from 2.51 to 2.34 MYA, lie within Members C, E, and F of the Shungura Formation (de Heinzelin and Haesaerts 1983). Although artifacts have been found and described from several localities in Members E and F (see Howell et al. 1987, and references therein), they do not occur at Localities 345, 338y, and 398. Because associated flakes or stone tools are lacking, hominids are not considered to have influenced the bone assemblages from these localities. Reptiles, fish, and birds making up some component of these three samples will not be discussed here.

EXCAVATION OF LOCALITY 345

Locality 345 was established in 1969 when G. Eck recovered 43 specimens, including some associated remains of a cercopithecid (Eck and Jablonski 1987), and a hominid parietal fragment (Howell, pers. comm.), from the surface of the site. The excavation took place on a ridge between two erosion gullies (Johanson et al. 1976). An area of 65m^2 was excavated during 1970 (by the late D. L. Cramer and D. C. Johanson) and 1973 (by N. T. Boaz) yielding 500 specimens *in situ* (see Figure 19-1, after ibid.:403, Figure 19-1). All excavated sediment was fine screened.

Dating

At L 345 the fossils occur 10 to 15 meters below Tuff D in Submember C-9 (de Heinzelin and Haesaerts 1983). The depositional age of Tuff D has been established as 2.50 ± 0.05 MYA and that of Submember C-9 as 2.51 MYA (Feibel et al. 1989).

Sedimentary Environment

The sediments at Locality 345 are illustrated in Figure 19-2. De Heinzelin and Haesaerts (1983:59) hypothesized accumulation of a "partly articulated, but autochthonous" skeletal sample in a "temporary shallow pond with lateral seepage." They described the fossils as lying .6 m above the base of Submember C-9 in a layer of heterogenous sand slightly solidified by limonitic crusts. This layer of sand, deposited between finer grained sediments, may have accumulated at the site during a time when a lateral connection existed between the pond (or inlet) and a nearby channel. After plotting longbone orientations on a rose diagram (see Figure 19-7) Johanson et al. (1976) concluded the assemblage had been subjected to flowing water prior to burial. They described the setting as a backswamp or near-channel "slough" with recurrent influxes of water to account for orientation yet minimal movement of the bones. A rootcast layer, found 5 to 15 cm above the fossiliferous horizon, suggested subaerial exposure after burial of the bone layer (Johanson et al. 1976). The upper 4.5 m of the 6-m thick section of C-9 found at L 345 consisted of a fining upward sequence of silty sand, silt, and a reduced silty clay (see Figure 19-2, after de Heinzelin and Haesaerts 1983:329, profile of L345 in Sector 9) indicative of overbank sedimentation subsequent to accumulation of the bone layer.

Faunal Composition

Taxonomic Representation. From the total sample of 543 fossils, 237 pieces are assigned to 6 mammalian taxa and 19 to mammalia indeterminate, as shown in Table 1. Reptile and fish remains numbered 165, 90 fragments could be identified only as animalia, and 12 specimens were unavailable for identification. (Twenty rootcast samples were excavated and catalogued.)

The six mammalian taxa identified from the L 345 sample include: *Hippopotamus protamphibius*, *Theropithecus brumpti*, *Australopithecus* sp. indet., *Tragelaphus nakuae*, Elephantidae indet., and Suidae indet.

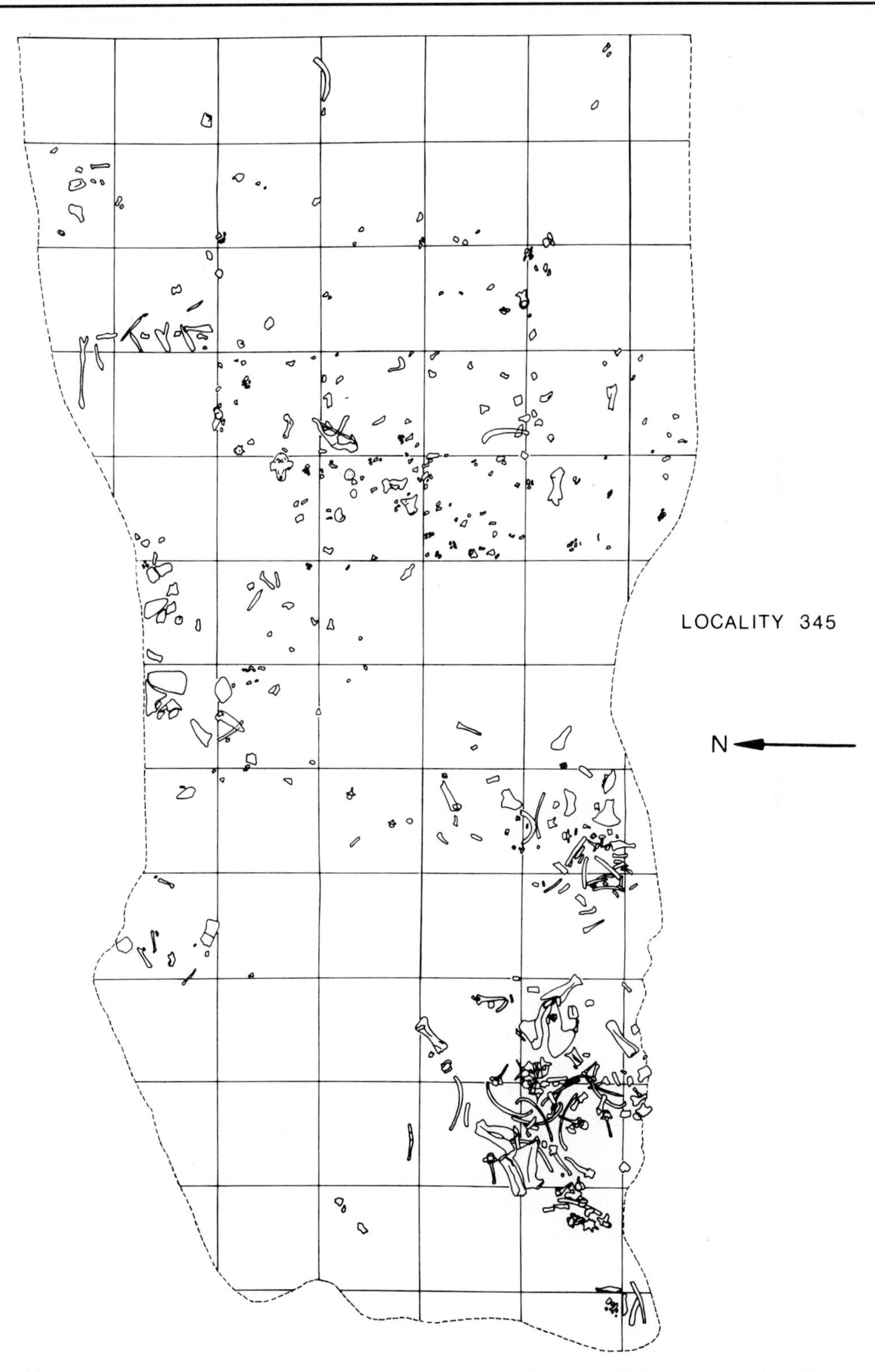

FIGURE 19-1 Horizontal distribution of fossils from the excavation at L 345 (from Johanson et al. 1976:403, Figure 1).

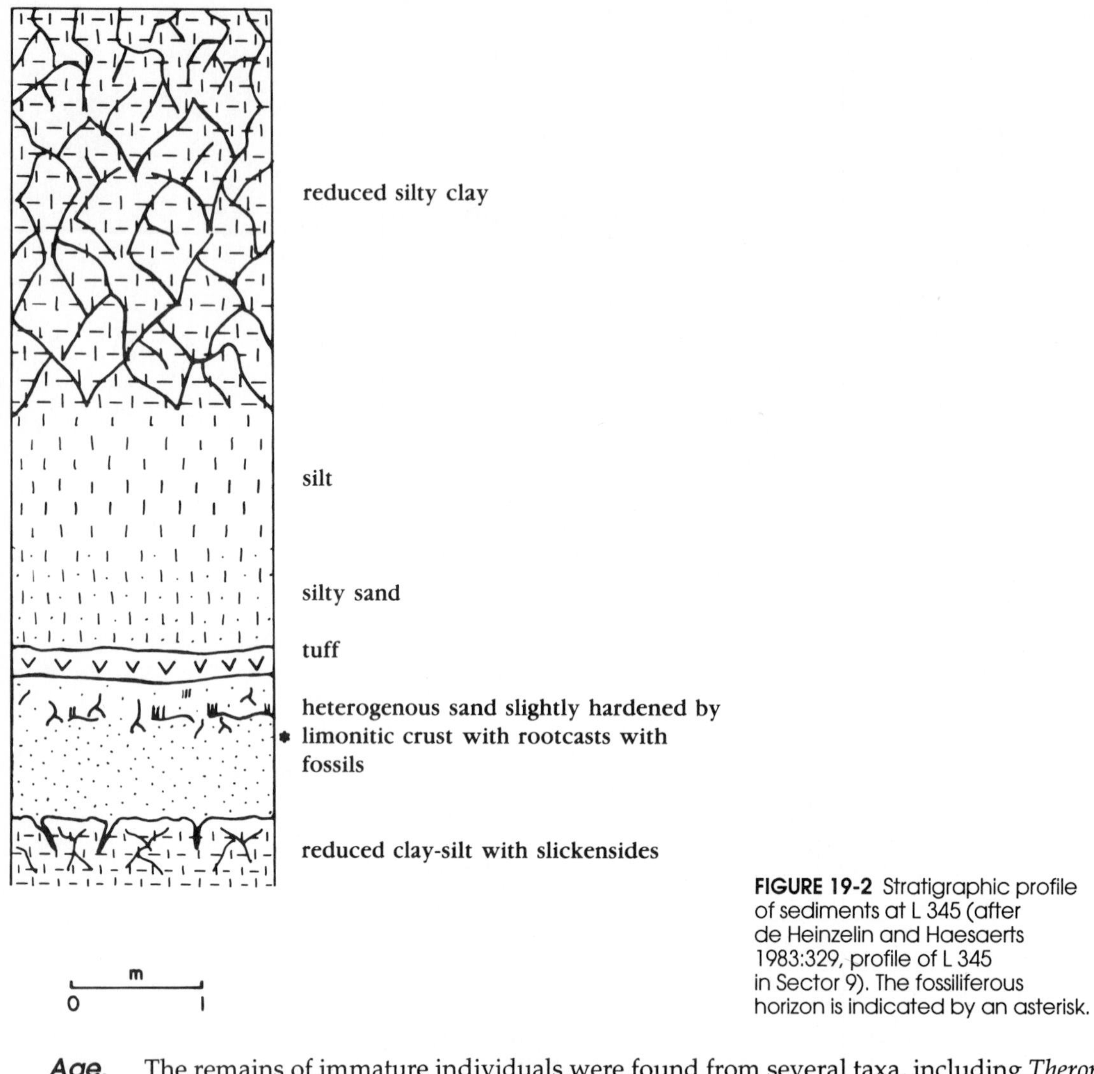

FIGURE 19-2 Stratigraphic profile of sediments at L 345 (after de Heinzelin and Haesaerts 1983:329, profile of L 345 in Sector 9). The fossiliferous horizon is indicated by an asterisk.

Age. The remains of immature individuals were found from several taxa, including *Theropithecus* sp., Cercopithecidae indet., *Hippopotamus* sp., and Suidae indet.

MNI. Using Shotwell's (1955, 1958) method for determining minimum number of individuals from a bone sample, and based on some observed variation in robusticity among the skeletons, the sample is calculated to consist of the remains of one *Tragelaphus nakuae* (mature); five *Theropithecus* (two mature males, one mature gender indeterminate, one immature male, and one immature indeterminate); four *Hippopotamus* (three mature and one immature); one elephantid; two suids (one mature and one immature); and one hominid (mature), as shown in Table 19-1. These computations were based mainly on crania, mandibulae, and teeth.

Assemblage Modification

Skeletal Articulation and Association. No records of articulations between excavated skeletal elements are available for L 345. However, partial associations did occur among some of the hippopotamid parts (Johanson et al. 1976) and perhaps among cercopithecid remains (see Figures 19-3 and 19-4).

patron's name:WYNNE, STEPHEN ANDREW

title:SOCIAL DARWINISM IN AMERI
author:HOFSTADTER, RICHARD,
item id:1040829-1001
due:5/24/2006,23:59

title:AN INTRODUCTION TO THE PR
author:BENTHAM, JEREMY,
item id:1040825-1001
due:5/24/2006,23:59

title:IN THE MINDS OF MEN : DAR
author:TAYLOR, IAN T.,
item id:1040832-1001
due:5/25/2006,23:59

title:DID DARWIN GET IT RIGHT?
author:JOHNSTON, GEORGE SIM.
item id:1040826-1001
due:5/25/2006,23:59

title:INTEGRATIVE PATHS TO THE
item id:1040811-1001
due:6/1/2006,23:59

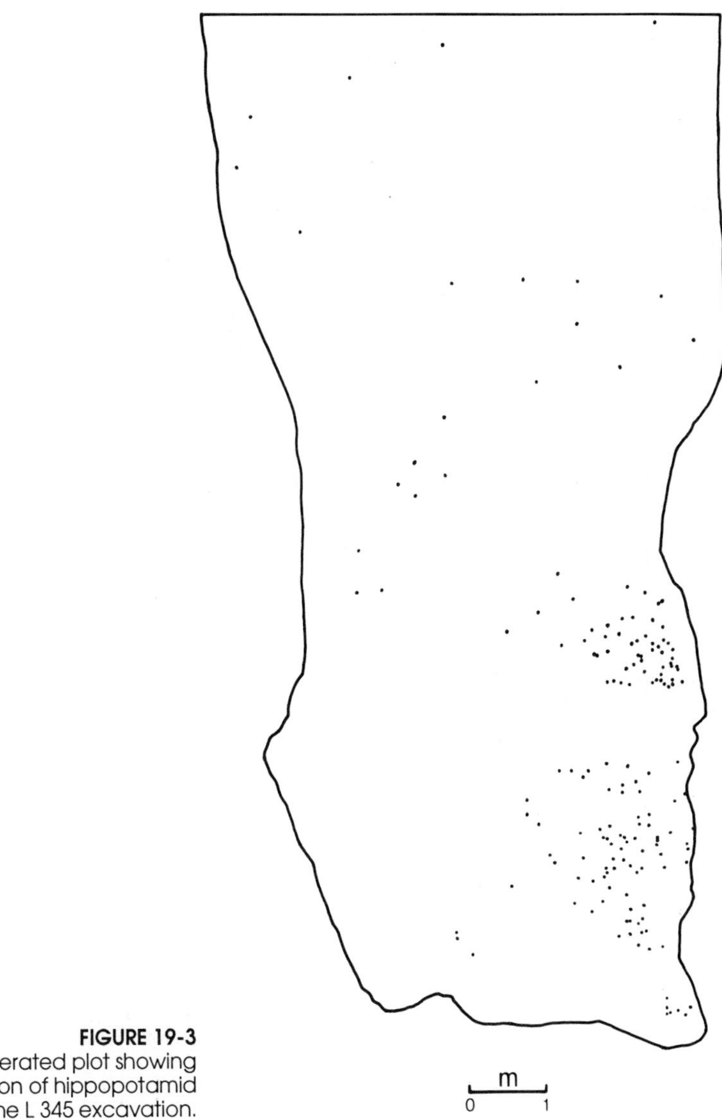

FIGURE 19-3
Computer-generated plot showing the distribution of hippopotamid fossils from the L 345 excavation.

Specimens determined to be from one hippopotamid individual, probably of the genus *Hippopotamus protamphibius* (ibid.) include most of the axial skeleton, parts of both forelimbs (including some carpals), metacarpals and phalanges, a distal right femur fragment, left astragalus, and some phalanges of the pes. Their horizontal distribution on the excavation floor is shown in Figure 19-5. A second hippopotamid individual may be represented by the cluster of elements lying just east of that previously mentioned (see Figure 19-3).

Skeletal-Part Representation. Skeletal-part representation is a measure of the completeness of individual skeletons. Fossil samples are characterized by skeletal loss, resulting from biological and/or geological agents rendering initially whole carcasses incomplete. Different skeletal elements exhibit variable degrees of resistance to destruction by carnivores and scavengers (Binford and Bertram 1977;

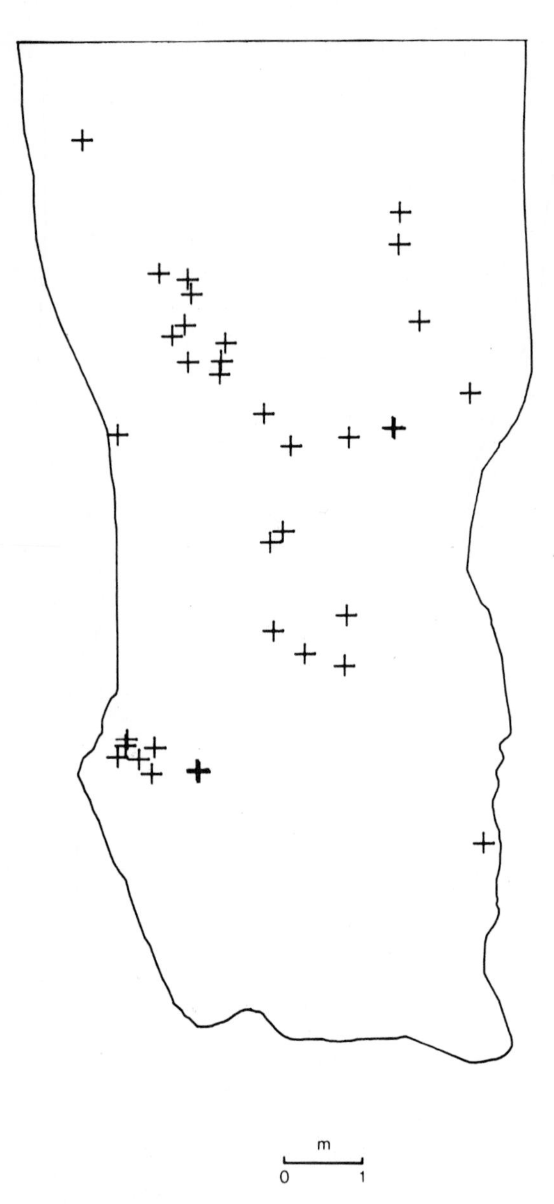

FIGURE 19-4
Computer-generated plot showing the distribution of cercopithecine fossils from the L 345 excavation.

Brain 1981; Lyman 1984, 1985) and to transport by flowing water (Voorhies 1969; Dodson 1973; Behrensmeyer 1975; Boaz and Behrensmeyer 1976; Gifford and Behrensmeyer 1977; Korth 1979; Shipman 1981; Schick 1984, 1986).

Skeletal completeness can be determined for *Hippopotamus*, *Theropithecus*, and *Tragelaphus* individuals from the data provided in Table 19-1. The few cranial and dental pieces of hippopotamids are grouped with closely associated postcrania to form a sample derived from three mature and one immature *Hippopotamus*. The cercopithecid cranial and dental remains have been assessed as *Theropithecus* (and some as *T. brumpti*) (Eck and Jablonski 1984, 1987; Meikle 1982), and a number of the postcrania have been assigned to *T. brumpti* on the basis of multivariate analyses (Ciochon 1986), and nonmetric traits (H. Krentz, pers. comm.). The bovid postcranial remains are combined with the *T. nakuae* crania and teeth (identified by A. Gentry, pers. comm.) on the basis of size. There is no evidence suggesting the presence of another bovid.

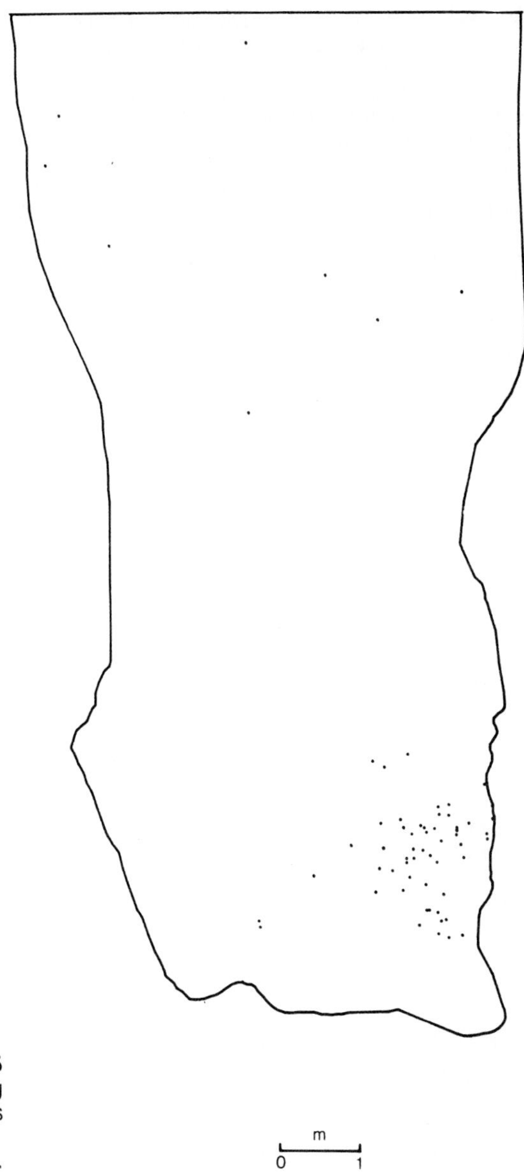

FIGURE 19-5
Computer-generated plot showing the distribution of skeletal elements thought to be from one hippopotamid individual at L 345.

As shown in Table 19-1, the *Hippopotamus* individuals are represented by the greatest diversity and highest numbers of skeletal elements. When compared with the total number of different elements found in three whole modern *Hippopotamus* individuals, scapulae (83 percent), sacra (67 percent), ribs (59 percent) and tibiae (50 percent) occur in the highest frequencies for the mature sample. The single immature *Hippopotamus* is poorly represented by one radius.

When compared with total elements found in three modern *Theropithecus* individuals, crania (including maxillae) (100 percent), teeth (72 percent), hemimandibles (67 percent), femora (67 percent), innominates (50 percent), ulnae (50 percent), and tibiae (50 percent) occur in the highest frequencies in the mature *Theropithecus* sample. Innominates (50 percent) dominate the sample of two immature *Theropithecus* individuals. For the *T. nakuae* sample, hemimandibles (50 percent) and astragali (50 percent) are found in frequencies most similar to those found in the skeleton of a single modern *Tragelaphus* individual.

TABLE 19-1 Taxonomic (by family) distribution of skeletal elements from L 345, grouped according to degree of susceptibility to transport as defined by Voorhies (1969)/Behrensmeyer (1975) Transport Groups. Hip = Hippopotamidae, Bov = Bovidae, Cer = Cercopithecidae, Sui = Suidae, Ele = Elephantidae, Hom = Hominidae, Mam Indet = Indeterminate Mammal element, N = total numbers of each element for all taxa; % of Sample indicates the relative frequencies of different skeletal elements. Percentages for elements making up less than 0.5% of the sample are not shown. Transport gp % indicates the relative frequencies of "easily transportable" (Group I), variably transportable (Groups I/II, II, and II/III), and "lag" (Group III) elements for the whole assemblage. NISP = number of identifiable skeletal parts, MNI = minimum number of individuals, RTA = relative taxonomic abundance, at the family level, based on MNI. AST/CALC = astragalus/calcaneum, OTH PODIAL = other podial, METAPOD = metapodial, INNOM = innominate, HEMIMAND = hemimandible, SKULL/HC = skull and/or horn core, WH MAND = whole mandible, ISO TOOTH = isolated tooth, IND T FR = indeterminate tooth fragment, LB SH FR = long-bone shaft fragment.

Voorhies/Behrensmeyer Transport Groups										
	Hip	Bov	Cer	Sui	Ele	Hom	Mam Indet	N	% of Sample	Transport gp %
GROUP I										
RIB	53						1	54	21.4	
VERTEBRA	41	2	5				1	49	19.4	
SACRUM	2							2	.8	44.2
STERNUM										
SESAMOID	1							1		
PATELLA										
GROUP I/II										
SCAPULA	5		1					6	2.4	
ULNA		1	4					4	1.6	
PHALANX	5	1		1				7	2,8	12.1
AST/CALC	3	1						4	1.6	
OTH PODIAL	6		1					8	3,2	
GROUP II										
FEMUR	2		4					6	2.4	
TIBIA	3		3					6	2.4	
HUMERUS	2		1					3	1,2	15.4
RADIUS	3		2					5	2.0	
METAPOD	10							10	4.0	
INNOM	2		5					7	2.8	
GROUP II/III										
HEMIMAND	2	1	6				1	10	4.0	4.2
GROUP III										
SKULL/HC			2			1	3	6	2.4	
WH MAND										24.2
ISO TOOTH	23	5	21	2	1			51	20.2	
IND T FR								1		
TRANSPORT GRP UNTESTED										
EPIPHYSIS							1	1		
FIBULA			2					2	.8	
MAXILLA			1					1		
LB SH FR							12	12	4.8	
								256		
NISP	163	11	58	3		1	19			
MNI	4	1	5	2	1	1				
RTA	28.6	7.1	35.7	14.3	7.1	7.1				

Overall, the mature *Hippopotamus* skeletons are most complete, followed by mature and immature *Theropithecus*, and the mature *Tragelaphus*.

Damage. For several reasons predepositional damage was difficult to distinguish from postdepositional and excavation damage: (1) A number of the pieces had chalky, friable textures due to postdepositional diagenetic changes and some were crushed by sediment, making excavation difficult; (2) Many of the elements, whole at the time of burial, were broken and scratched during excavation, and transport from the field to the laboratory caused additional degradation.

By careful examination of the available sample, the probable predepositional condition of most of the skeletal assemblage was assessed. Many of the *Hippopotamus* elements were whole at the time of burial, although breakage patterns on some vertebral processes and ribs indicate chewing by mammalian carnivores. Three bones had puncture marks from carnivore teeth. The outer table of bone was absent in a number of cases signifying abrasion, although the sediments suggest low energy deposition. This apparent "abrasion" possibly resulted from carnivore or herbivore gnawing (Brain 1981; Binford 1981; Haynes 1991), from trampling (Behrensmeyer and Dechant Boaz 1980; Fiorillo 1984, 1989; Myers et al. 1980; Gifford and Behrensmeyer 1977; Behrensmeyer et al. 1989; Oliver 1989), or from diagenetic changes during fossilization (Wells 1967; Hare 1980; Behrensmeyer et al. 1989) combined with corrosion during or after excavation. Many of the bones of chalky or powdery appearance had a pitted surface texture. Pitting may be caused by prolonged gnawing (Binford 1981), by trampling (Behrensmeyer et al. 1989) or by the activities of microorganisms (?) while a bone is submerged (pers. obs.). Pitting was found on *Hippopotamus* ribs, but also on other elements showing no other types of damage from carnivores.

The available cercopithecid sample showed a high incidence of carnivore chewing, particularly on the postcrania. Some spiral fractures and striation, puncture, and gnawing marks were found. Most of the longbones were fragmentary, missing distal or proximal ends. Pitting was observed on the chewed pieces. "Abraded" areas were also present on a number of the postcrania. One cranium was crushed, probably postdepositionally by sediment (G. Eck, pers. comm.). A number of isolated teeth had broken roots and chipped enamel from unknown causes.

Many of the remains of the other four mammalian species were fragmentary and damage was undiagnostic. The hominid parietal fragment showed no obvious features of carnivore chewing.

Weathering. Most of the 108 specimens (76 percent) examined for degree of surface weathering showed none (Stage 0), while 93 percent of the sample fell within Stage 0 and 1 weathering categories (Behrensmeyer 1978). A few pieces were more heavily weathered at Stages 2, 3, and 4, indicating some variation in exposure duration and depositional microenvironment prior to or during burial of the assemblage (Lyman and Fox 1989).

Transport Groups. The sedimentary environment of L 345 has been described as "a temporary shallow pond with lateral seepage" (de Heinzelin and Haesaerts 1983:59). It is possible that the carcasses accumulated at L 345 were subjected periodically to running water. As noted by Johanson et al. (1976), however, it is unlikely that transport of many elements to or from the site occurred. Grain sizes and sedimentary structures indicate minimal current velocities during burial (see Figure 19-2).

As a means of corroborating sedimentary evidence, fossil assemblages can be analyzed for evidence of fluvial transport. In Table 19-1, the different elements are arranged by appropriate transport group based on flume studies by Voorhies (1969) and settling velocity studies by Behrensmeyer (1975). Highly transportable (Group I) elements dominate the overall sample (44 percent), with lag (Group III) elements following in abundance (24 percent). When coupled with stratigraphic evidence, the abundance of Group I "highly transportable elements" indicates no "winnowing" of the sample by flowing water and argues for an autochthonous (or biological) accumulation.

As the assemblage was not "winnowed" before burial, predators and scavengers must have played major roles in influencing the composition and distribution of the skeletal sample. That considerable damage was inflicted, particularly on the cercopithecid sample, is evident from discussion in previous sections. The transport groupings in Table 19-1 provide some evidence of differential damage to the various taxa.

Light, transportable elements, such as vertebrae and ribs, are among those most susceptible, while dense, untransportable elements, such as hemimandibles and teeth, are among those most resistant to destruction by carnivores and scavengers (Brain 1969, 1981). The Group I sample is dominated by mature *Hippopotamus* remains, suggesting the adult hippopotamid individuals died at the site just prior to burial. The *Theropithecus* sample, dominated by durable Group III elements, likely suffered greater damage and dispersal before burial than the *Hippopotamus*. The other taxa also suffered greater damage and dispersal before burial than the *Hippopotamus*, but probably died close to the site of deposition.

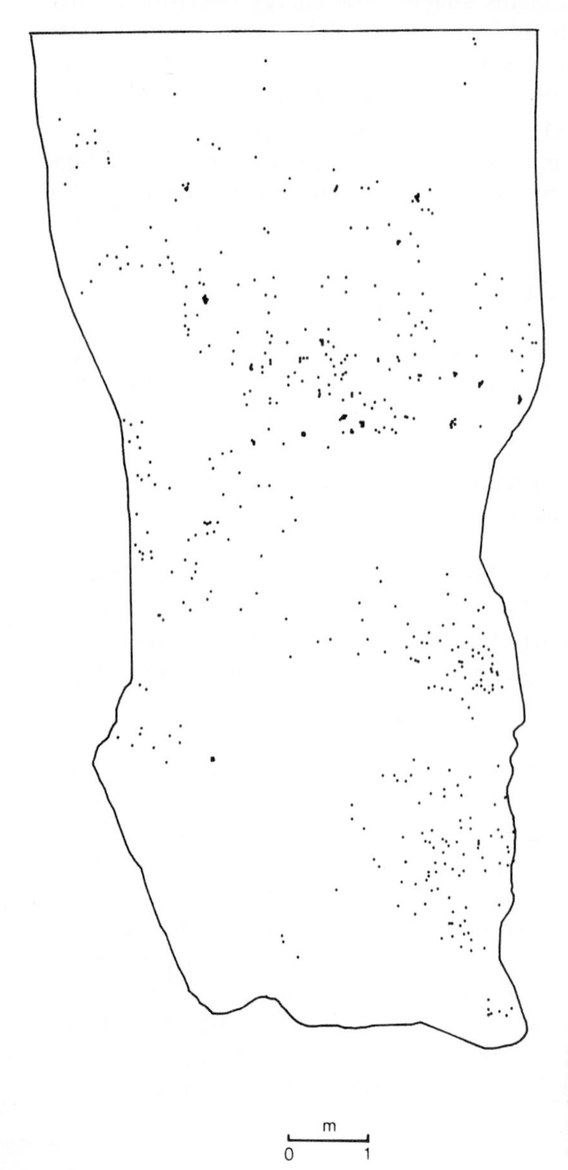

FIGURE 19-6
Computer plot showing the horizontal distribution of all bones found on the floor of L 345 during excavation.

Dispersal and Orientation. As shown in Figure 19-6, some horizontal movement of the originally articulated carcasses (hippopotamid) and possible skeletal units (cercopithecid, bovids and suid) occurred before their burial. Disarticulation and dispersal of the elements likely was initiated by predation and scavenging.

By plotting the longbones on a rose diagram, Johanson et al. (1976) determined that an "east-west orientation was preferred" (see Figure 19-7, after ibid:404, Figure 2). They reasoned that although the sediments indicated quiet-water deposition, some slight but steady current was able to orient the bones, probably parallel to the direction of flow. They describe some pieces showing a preferred polarity with the heavy ends dipping upstream in the easterly direction, suggesting that the current flowed from east to west. This author's interpretation of the rose diagram is that it does not provide strong evidence for orientation, but rather for a random distribution of the longbones, which could have resulted from carnivore/scavenger activity, or possibly from trampling (Fiorillo 1989).

Reconstruction of Burial Events

Sediments and characteristics of the fossil assemblage at L 345 argue for an autochthonous accumulation. The stratigraphic profile indicates "quiet-water" deposition (silts and clays) before and after burial of the bones, with periodic lateral connection between the site and a nearby channel contributing coarser sands during accumulation of the assemblage (de Heinzelin and Haesaerts 1983). The discovery of a root cast layer signifies periodic subaerial exposure of the site.

FIGURE 19-7 Rose diagram of elongated elements from L 345, with the circle indicating the 10-bone interval (from Johanson et al. 1976:404, Figure 2).

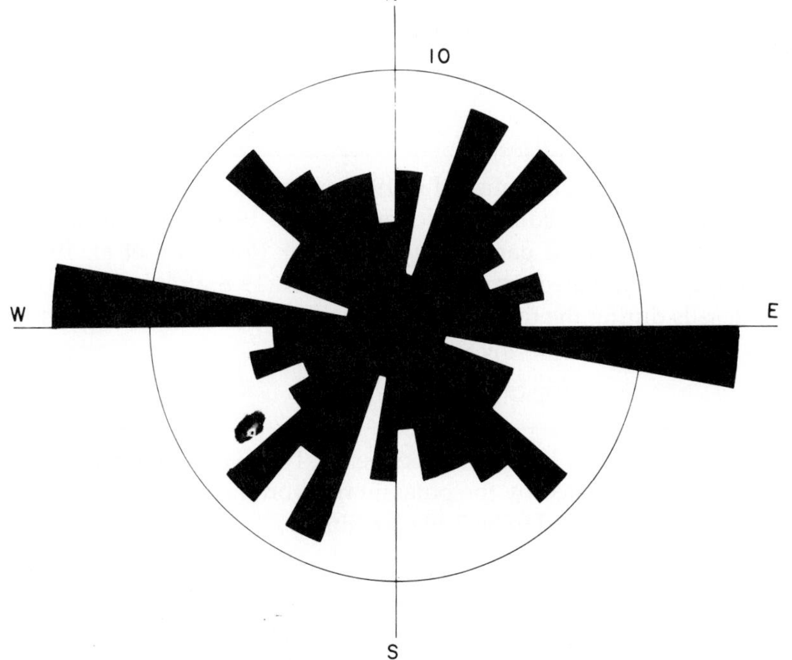

Several features in combination argue for a biologically accumulated assemblage, as opposed to a fluvially concentrated one (Badgley 1986a). These include the clusters of associated hippopotamid remains containing an abundance of "easily transportable" skeletal elements. The high incidence of carnivore chewing on cercopithecid postcrania coupled with a prevalence among cercopithecid and other taxa of durable isolated teeth, hemimandibles and crania indicate significant damage from predators and/or scavengers prior to burial of the assemblage. Such an assemblage could result from fluvial concentration if the river sampled from a landscape characterized by a few bones heavily scavenged by carnivores. However, the taxonomic homogeneity found at L 345 coupled with the high relative abundance of several taxa in what is likely their preferred habitat (i.e., *Theropithecus brumpti* preferring riverine forest-woodland or backswamp habitats [Eck and Jablonski 1987] and *Hippopotamus protamphibius* preferring backswamp or near-channel habitats, based on inference from the preferences of modern *Hippopotamus amphibius*) supports the argument for biological accumulation of the assemblage.

Defleshing and disarticulation of the skeletons were probably initiated by the activities of large scavengers (mammalian and avian?), insects and microorganisms. Lightly dispersed skeletons, such as those of the hippopotamids, indicate slow defleshing and disarticulation. Pitting found on many of the remains may have resulted from prolonged submersion. Periodic gentle currents, during times of overbank flow, dispersed isolated elements. Overbank alluviation eventually buried the assemblage.

Contrast in skeletal-part representation among the taxa suggests that hippopotamid individuals suffered less damage and may have been buried more rapidly than the other taxa. Although footprints were not observed during excavation, it is conceivable the site was a watering hole near the main channel of the ancient Omo River. Hippopotamids probably resided in the immediate environment and died there. Monkeys and bovids may have frequented the area to drink and were killed by predators. The elephantid, hominid, and suid bones may represent animals killed nearby whose bones were dispersed and destroyed by intense scavenging.

This reconstruction of burial events suggests the skeletons accumulated over a relatively short period of time and the taxa could have been sympatric inhabitants of the local animal community (Badgley 1986a; Behrensmeyer 1988).

EXCAVATION OF LOCALITY 338y

Locality 338y (previously 338ss) was established in 1969 when J. de Heinzelin found a juvenile hominid occipital fragment eroding from the sediments (Johanson et al. 1976). An 8 m^2 test excavation followed with the discovery of both associated parietals by G. G. Eck (Johanson et al. 1976; de Heinzelin and Haesaerts 1983). M. Splingaer excavated some 158 m^2 (see Figure 19-8, after Johanson et al. 1976:408, Figure 5) that yielded 2,291 fossils during the three following field seasons. Six specimens were collected from the surface. All excavated sediment was fine screened.

Dating

At L 338y, the fossiliferous layer lies 12 to 15 meters below Tuff F in Submember E-3 (de Heinzelin and Haesaerts 1983). Tuff F has been dated by the potassium-argon method to 2.34 MYA (Feibel et al. 1989). The age of Submember E-3 is estimated to be 2.39 MYA (ibid.).

Sedimentary Environment

The sediments at Locality 338y are illustrated in Figure 19-9 (after de Heinzelin and Haesaerts 1983:335, profiles). The stratigraphy at L 338y, consisting of four subunits (1-4) of Submember E-3, has been described by de Heinzelin and Haesaerts (1983:80,81). Subunits E-3-1 and E-3-2 contain the *lower* and *main* fossiliferous occurrences, respectively (Howell et al. 1987). These subunits are composed of fining-upwards sequences of interfingering silts, silty sands, sandy silts and sands, occasionally with concretions

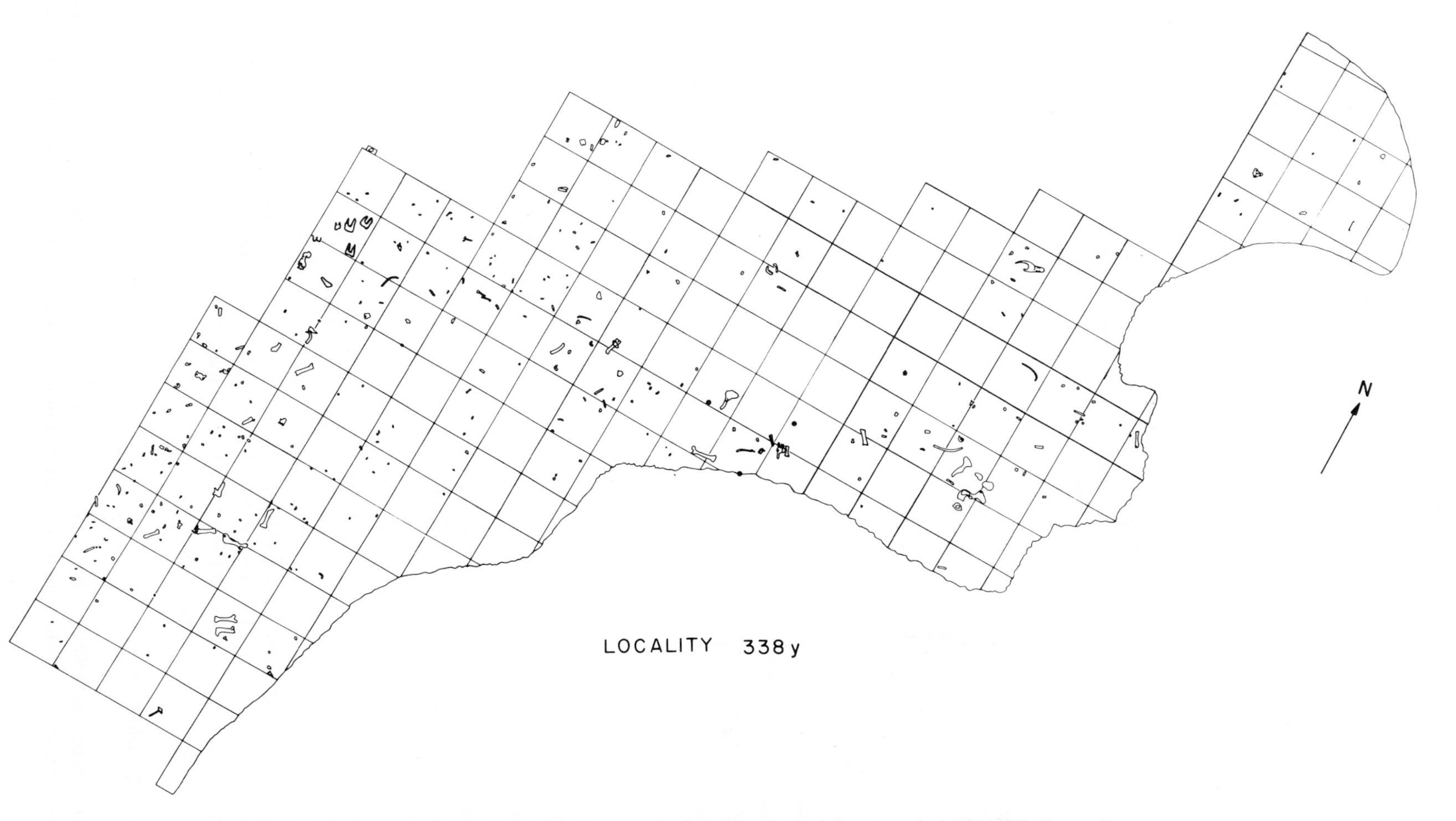

FIGURE 19-8 Horizontal distribution of the fossils excavated from L 338y (from Johanson et al. 1976:408, Figure 5).

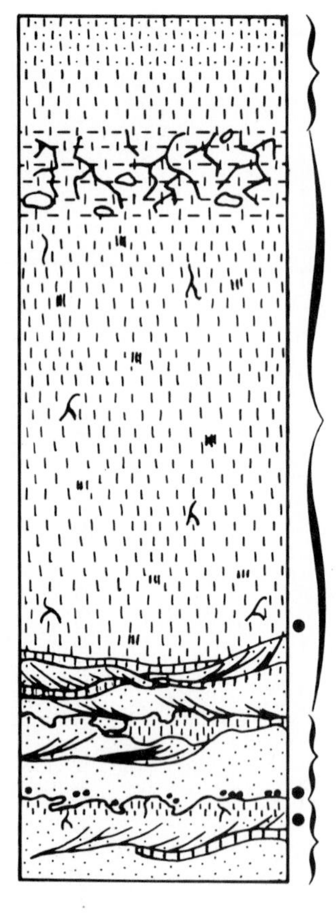

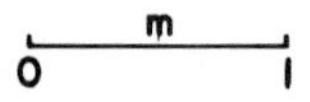

FIGURE 19-9 Stratigraphic profile of sediments at L 338y (after de Heinzelin and Haesaerts 1983:335 profiles, and from description of the stratigraphy in Howell et al. 1987:689-690). The *upper, main,* and *lower* fossiliferous horizons are indicated by dots.

or other coarser materials. The *main* fossiliferous bed lies encased in a medium sand, covered by sandy silt, and overlain by obliquely stratified medium sands with lenses of small gravels and silt pellets (de Heinzelin and Haesaerts 1983:81). The sedimentary composition of these two subunits, coupled with indications of trampling (irregular pockets of sand in finer sediments), other bioturbations (tubelike structures in sediments), and subaerial exposure (reduced rootcasts) at several interfaces, suggests deposition in a braided stream situation (de Heinzelin and Haesaerts 1983; Howell et al. 1987:Figure 4).

Another sparsely fossiliferous layer, termed the *upper* fossiliferous occurrence (Howell et al. 1987:689), was found in subunit E-3-3 at the abrupt interface between an obliquely stratified medium sand with lenses of silt and a silt. Overlying reduced clays with slickensides at the top of subunit E-3-3 and the subsequent silt deposits of subunit E-3-4 indicate a change from braided conditions to overbank alluviation with periodic subaerial exposure.

Faunal Composition

Taxonomic Representation. From the total sample of 2,297 fossils, 445 specimens have been assigned to mammalian taxa and 259 to Mammalia indeterminate, as shown in Table 19-2. A number of fragments

(341) were identifiable only as animalia, and 16 bones were unavailable for identification. (Four rootcast specimens were excavated and catalogued.)

Fifteen unique mammalian taxa were identifiable from the L 338y sample: *Australopithecus aethiopicus/boisei* (Rak and Howell 1978; Howell et al. 1987), *Theropithecus brumpti, Papio* sp., *Elephas recki shungurensis, Giraffa gracilis, Kolpochoerus limnetes, Metridiochoerus andrewsi, Tragelaphus nakuae, Lepus* sp., *Xerus* sp., Equidae indet. cf. *Hipparion* sp., Mellivorinae sp. nov., Felidae indet., Viverridae indet., and Hippopotamidae indet.

Other fragmentary specimens, possibly but not definitely subsumable under the taxonomic categories previously represented, are categorized as *Theropithecus* sp., Cercopithecinae indet., *Giraffa* sp., Giraffidae indet., *Kolpochoerus* sp., Suidae indet., Tragelaphini indet., Bovidae indet., Elephantidae indet., and Hominidae indet.

Age. The remains of immature individuals from a number of taxa were found, including *Australopithecus aethiopicus/boisei, Theropithecus* sp., Cercopithecinae indet., *Elephus recki shungurensis, Kolpochoerus* sp., Suidae indet., Bovidae indet., and Hippopotamidae indet.

Relative Abundance of Taxa. There are basically two methods for determining the relative abundance of taxa from a skeletal assemblage. Sample sizes are assessed either by calculating the minimum number of individuals (MNI) from the total number of bones or by simply using the total number of identifiable bones (NISP). Badgley (1986a,b) has proposed that the former method be used when the sedimentary context indicates "quiet-water" deposition and when features of the assemblage (such as articulation/association of skeletal parts) suggest that initially "whole" carcasses (or smaller units) have been disarticulated at the site. Such an assemblage accumulates through the actions of biological agents (ibid.). The latter method is used when the bones are obviously unassociated, having derived from a number of disarticulated, widely dispersed skeletons and occur in sediments deposited from flowing water (ibid.).

Three fossiliferous horizons have been described at L 338y: an *upper* fossil layer with sparse but well preserved and fairly complete specimens; a *main* fossil layer including heavily mineralized, rolled teeth and bone along with other more delicate remains (i.e., juvenile hominid cranial pieces); and a *lower* fossil layer with several rolled specimens (de Heinzelin and Haesaerts 1983, Howell et al. 1987).

In his field notes, de Heinzelin (1971b) described the bones from the main fossiliferous layer as "scattered in silt," while "more concentrated in a sand." Some remains were likely autochthonous, lying *in situ* perhaps where they fell with some dispersal of the skeletons in a north/south direction; others showed signs of having rolled long distances. When examined by this author, the teeth, making up the bulk of the sample, were found to be extensively fractured and undiagnostic, while some of the cranial and postcranial bones were cracked but recognizable. Clearly the L 338y sample has a complicated taphocoenosis.

Because of the apparent mixing of autochthonous and allochthonous components, I decided it would be more accurate to calculate the relative abundance of taxa by considering each element to be from a different skeleton (except where associations between bones are apparent). The skeletal part totals (NISP) per family, from a sample of 445, are shown in Table 19-2, along with percentages indicating relative abundance (RTA). Hippopotamids dominate the sample, with bovids, cercopithecids and suids being fairly numerous.

Assemblage Modification

Skeletal Articulation and Association. No articulated skeletal parts were found among the sample of excavated fossils from L 338y. Some association is apparent, however, in the recovery of three nearly associated cranial fragments from a single juvenile *Australopithecus aethiopicus/boisei* individual (Rak and

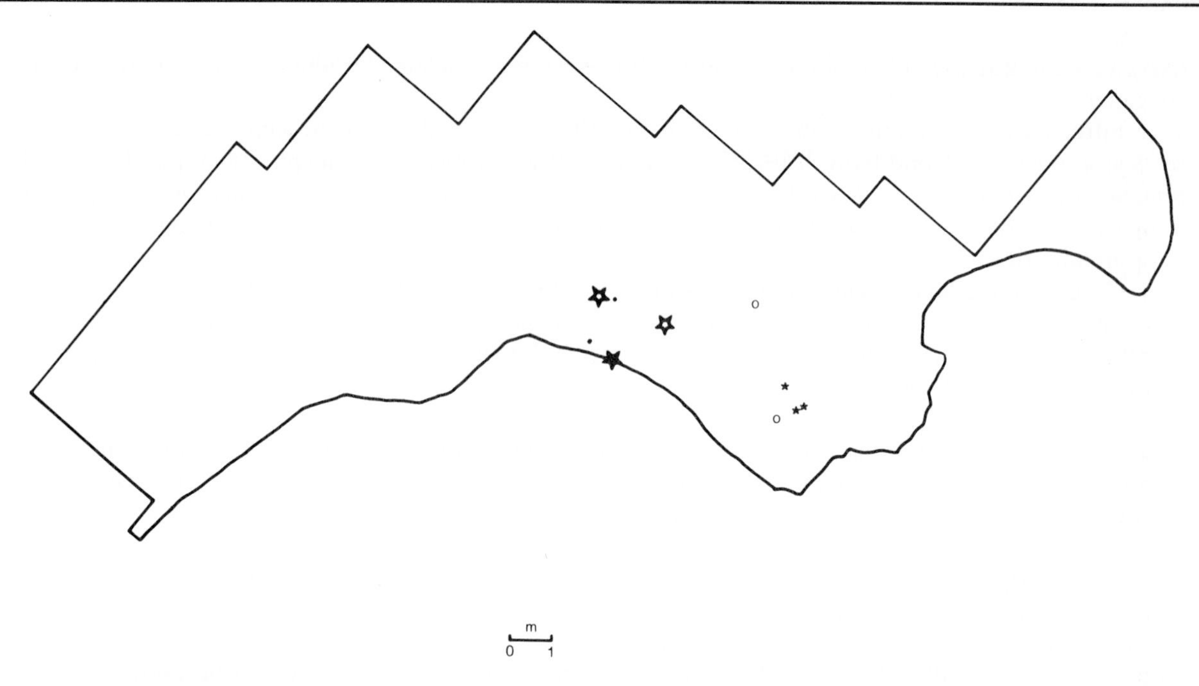

FIGURE 19-10 Computer-generated plot showing three sets of associated skeletal elements of hippopotamids (dots, circles, and small stars), and three associated cranial fragments of a juvenile *Australopithecus aethiopicus/boisei* (large stars) from L 338y.

Howell 1978; Rak 1983; Howell et al. 1987). These pieces were found approximately 1.5 m apart as illustrated in Figure 19-10. Other remains possibly derived from several hippopotamid individuals, based on appearance and proximity, are (1) a femur and innominate, (2) two innominates and a lumbar vertebra, and (3) an atlas and other cervical vertebra, also shown in Figure 19-10.

The specimens previously mentioned are among the most complete for the L 338y sample, a fact that enhances their identifiability and suggests they suffered less transport and damage prior to burial than the greater portion of the assemblage (as noted by de Heinzelin and Haesaerts 1971, 1983).

Skeletal-Part Representation. As each specimen is considered to be from a different individual (except where skeletal associations were obvious), skeletal-part representation, as a measure of the completeness of individual carcasses, is poor for all taxa.

As indicated in Table 19-2, most of the mammalian individuals are represented by teeth. Within the relatively small mammalian postcranial sample (17 percent of all the remains), hippopotamid and bovid elements are most numerous, along with many indeterminate mammalian pieces.

Damage. The L 338y fossil assemblage consisted primarily of mammalian and reptilian tooth fragments. A few postcranials of hippopotamids and bovids were well preserved, but, in general, the elements were broken and fragile. Some abrasion was apparent, as well as pitting. Spirally fractured longbone pieces were present but rare as compared to the L 398 sample, to be discussed later. Trampling (as noted by de Heinzelin and Haesaerts 1983) probably contributed to breakage (Behrensmeyer and Dechant Boaz 1980; Fiorillo 1984, 1989; Myers et al. 1980; Behrensmeyer et al. 1986, 1989).

Distinctively large holes were found in a crocodilian scute and a chelonian plastron fragment. These likely resulted from predepositional events and may be attributable to predation or scavenging by crocodilians or large mammalian carnivores. A smaller hole was found in the well-preserved right parietal of a juvenile *Australopithecus aethiopicus/boisei* from the site. The damage appears to be predepositional and

TABLE 19-2 Taxonomic (by family) distribution of skeletal elements from L 338y, grouped according to degree of susceptibility to transport as defined by Voorhies (1969)/Behrensmeyer (1975) Transport Groups. Hip = Hippopotamidae, Bov = Bovidae, Cer = Cercopithecidae, Sui = Suidae, Ele = Elephantidae, Hom = Hominidae, Gir = Giraffidae, Equ = Equidae,Mus = Mustelidae, Fel = Felidae, Viv = Viverridae, Lep = Leporidae, Sci = Sciuridae, Mam Indet = indeterminate mammal element, N = total numbers of each element for all taxa; % of Sample indicates the relative frequencies of different skeletal elements. Percentages for elements making up less than 0.5% of the sample are not shown. Transport gp % indicates the relative frequencies of "easily transportable" (Group I), "variably transportable" (Groups I/II, II, and II/III), and "lag" (Group III) elements for the whole assemblage. NISP = number of identifiable skeletal parts, RTA = relative taxonomic abundance, based on NISP numbers being equal to the number of individuals per taxon. AST/CALC = astragalus/calcaneum, OTH PODIAL = other podial, METAPOD = metapodial, INNOM = innominate, HEMIMAND = hemimandible, SKULL/HC = skull and/or horn core, WH MAND = whole mandible, ISO TOOTH = isolated tooth, IND T FR = indeterminate tooth fragment, LB SH FR/BF = longbone shaft fragment/bone fragment.

	Voorhies/Behrensmeyer Transport Groups																
	Hip	Bov	Cer	Sui	Ele	Hom	Gir	Equ	Mus	Fel	Viv	Lep	Sci	Mam Indet	N	% of Sample	Transport gp %
GROUP I																	
RIB	2	2												28	32	4.5	
VERTEBRA	5	6												9	20	2.8	
SACRUM		1													1		7.9
STERNUM																	
SESAMOID		1												1	2		
PATELLA																	
GROUP I/II																	
SCAPULA							1							1	2		
ULNA		2													2		
PHALANX	1	4	3	1										2	11	1.6	3.9
AST/CALC		5	1										1	1	8	1.1	
OTH PODIAL	1							1						2	4	.6	
GROUP II																	
FEMUR	4		1											3	8	1.1	
TIBIA	3													1	4	.6	
HUMERUS	2					1	1							4	8	1.1	5.2
RADIUS		4												1	5	.7	
METAPOD		4													5	.7	
INNOM	4		1											1	6	.9	
GROUP II/III																	
HEMIMAND	4	3		1											8	1.1	1.2
GROUP III																	
SKULL/HC	1	2	1		1	1				1				7	14	2.0	
WH MAND																	81.8
ISO TOOTH	62	42	69	41	6	1		1	2		1	1			229	32.5	
IND T FR	61	39	12	16	7		3							190	325	46.2	
TRANSPORT GRP UNTESTED																	
EPIPHYSIS																	
FIBULA																	
MAXILLA	1		1												2		
LB SH FR/BF														8	8	1.1	
															704		
NISP	152	115	89	59	14	3	5	2	2	1	1	1	1	259			
RTA	34.2	25.8	20.0	13.3	3.1	.7	1.1	.4	.4	.2	.2	.2	.2				

might have been caused by the canines of a large predator, such as *Panthera pardus* (Rak and Howell 1978; Brain 1970).

Weathering. The bones were generally in a friable state, and weathering was obscured, in some cases, by surface corrosion or pitting. Of a sample of 155 unobscured fossils, 80 percent showed no weathering (Stage 0), while 94 percent fell within Stages 0 and 1 (Behrensmeyer 1978). A few pieces were more heavily weathered at Stages 2, 3, and 4, indicating some variation in exposure duration and depositional microenvironment prior to and/or during burial of the assemblage (Lyman and Fox 1989).

Transport Groups. The sedimentary environment of the main fossil occurrence at L 338y has been described as a "braided stream situation . . . with temporary floods producing sandbanks and silt lenses" (Howell et al. 1987:690).

From a sample of 694 mammalian elements sorted into Voorhies/Behrensmeyer transport groups, 82 percent (primarily isolated teeth) fell within Group III (Table 19-2). The abundance of these "untransportable" teeth and bones suggests a residual sample of remains from heavily "winnowed" skeletons. Trampling is suspected to have been a factor contributing to degradation of the more delicate elements of the skeletal sample. The concentration of tooth fragments probably reflects their endurance as well as their resistance to transport.

Dispersal and Orientation. The horizontal distribution of fossils from Locality 338y is shown in Figure 19-11. Many of the catalogued specimens were not mapped, and it is likely they were found during screening of individual meter squares by M. Splingaer and his excavation crew. As a consequence, the illustrations in Figures 19-8 and 19-11 show horizontal distribution of less than half the sample, although the better preserved specimens were mapped (de Heinzelin and Haesaerts 1971).

FIGURE 19-11 Computer-generated plot showing the horizontal distribution of all bones mapped on the floor of L 338y during excavation.

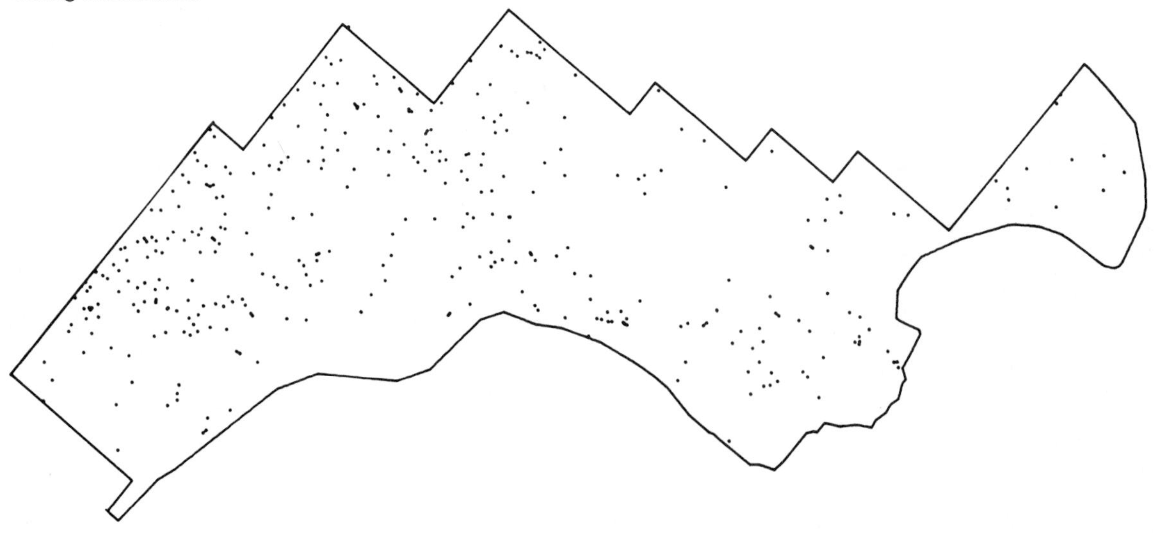

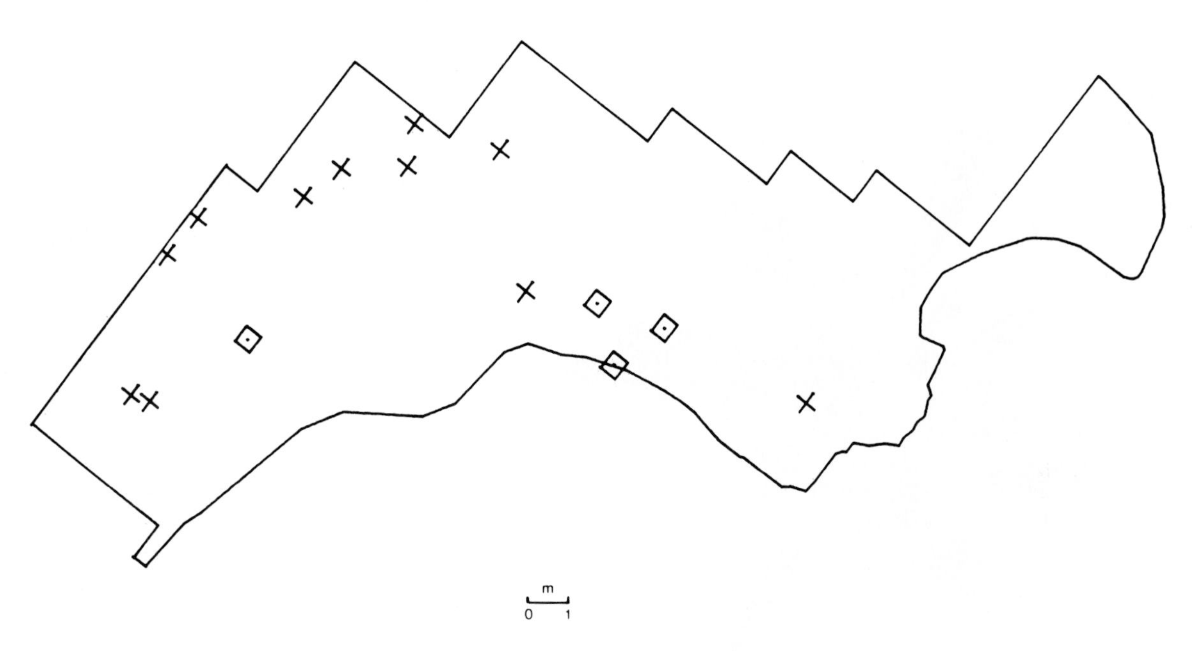

FIGURE 19-12 Computer-generated plot showing the distribution of cercopithecid (pluses) and hominid (squares) fossils from L 338y.

In general, the pieces are not clustered on the excavation floor. Possible associations among hippopotamid bones are noted in Figure 19-10. The distributions of mapped cercopithecine and hominid fossils are shown in Figure 19-12.

Horizontal distribution was likely influenced by trampling and kicking, predation and scavenging, and water current.

In a preliminary report, de Heinzelin (1971c) plotted orientations of bones mapped by M. Splingaer. His sample included fossils from both the *main* and *lower* fossiliferous horizons (see Figure 19-9). The findings indicated a preferred orientation which, when coupled with an observable dip in obliquely stratified sands, suggested a north/south flow direction. Plotting a smaller sample of longbones only, Johanson et al. (1976) were unable to distinguish a preferred orientation (see Figure 19-13, after ibid:411, Figure 7). Original current orientation may have been randomized by trampling (Fiorillo 1989).

Vertical distribution of skeletal parts along the W/E (20 m) and S/N (24 m) axes are shown in Figures 19-14 and 19-15. The depths used to make these plots were taken from M. Splingaer's field notes. The bone bed is inclined due to a dip in the bedding of approximately 9°W; the *in situ* sample had a vertical scatter of 120 to 150 cm (see Figure 19-14). This spread corresponds to the combined thicknesses of subunits E-3-1 and E-3-2, as indicated by de Heinzelin and Haesaerts (1983) and the stratigraphic profile (see Figure 19-9).

Both vertical plots show that some of the fossils (those lying in the horizontal plane at shallow depths) had probably eroded from the elevated (due to dipping of the beds) areas of the site. Although they were originally buried and fossilized with the other bones, these pieces were probably found on the surface as excavation began and were not *in situ*.

That two separate fossiliferous horizons, a *main* and a *lower*, existed (de Heinzelin and Haesaerts 1983; Howell et al. 1987) is not clear from the vertical profiles. What may have been obvious in the field is less so when plotted at the scale of these figures and out of sedimentary context. The apparent continuous vertical distribution of bones, particularly as shown in Figure 19-14, supports an interpretation of continuous, rather than episodic, deposition of the bone bed.

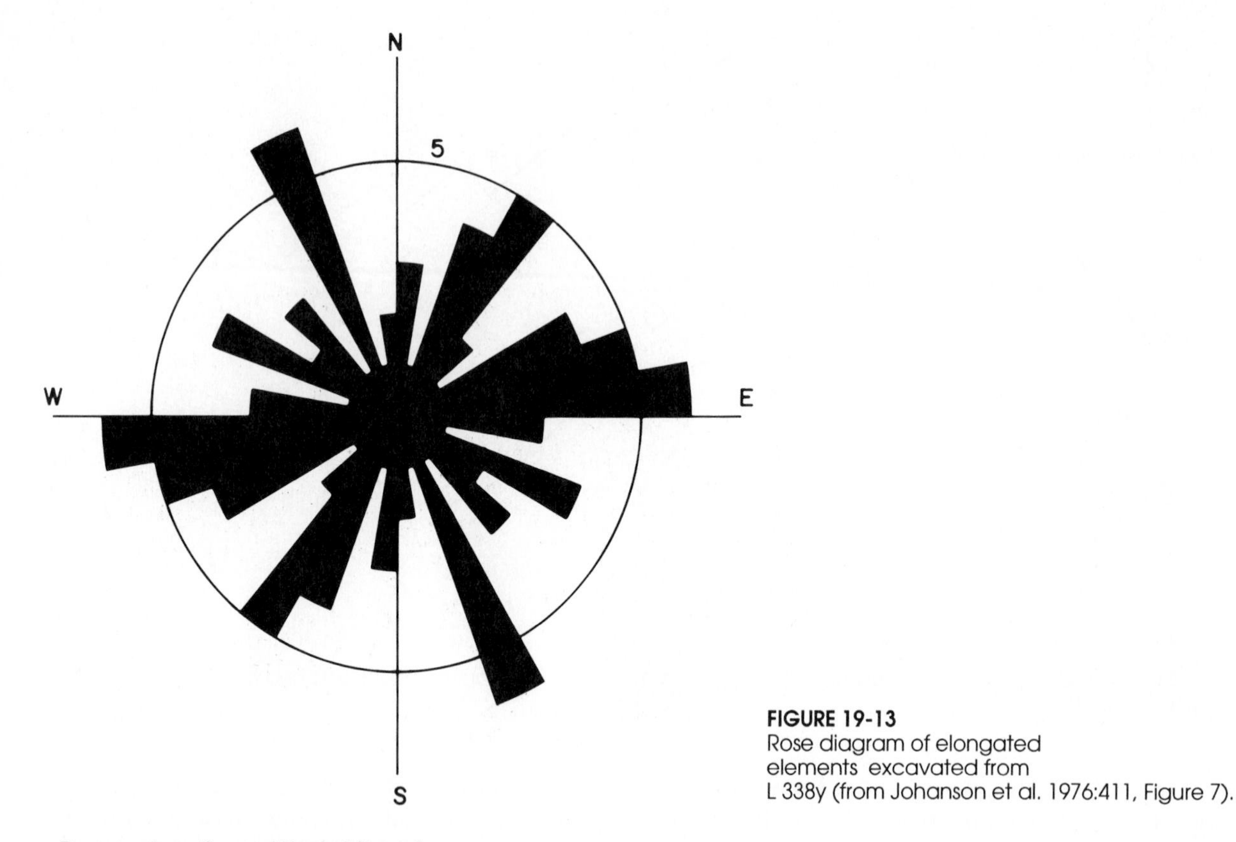

FIGURE 19-13
Rose diagram of elongated elements excavated from L 338y (from Johanson et al. 1976:411, Figure 7).

Reconstruction of Burial Events

Sediments and features of the fossil assemblage at L 338y suggest a complicated taphocoenosis with mixing of autochthonous and allochthonous components (de Heinzelin and Haesaerts 1971, 1983). The stratigraphic profile illustrates several episodes of recurrent channeling, characteristic of braided stream deposition, during accumulation of the *lower, main* and *upper* fossiliferous horizons. Periods of subaerial exposure are indicated by rootcasts, tubulations, and evidence of trampling (irregular pockets of coarse sand).

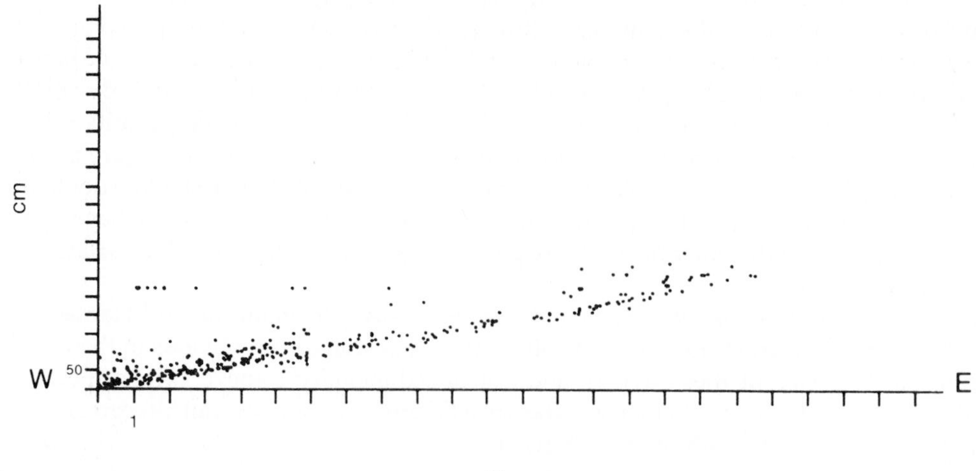

FIGURE 19-14
Computer-generated plot showing the vertical distribution, along the W/E axis, of all skeletal elements mapped during the excavation of L 338y.

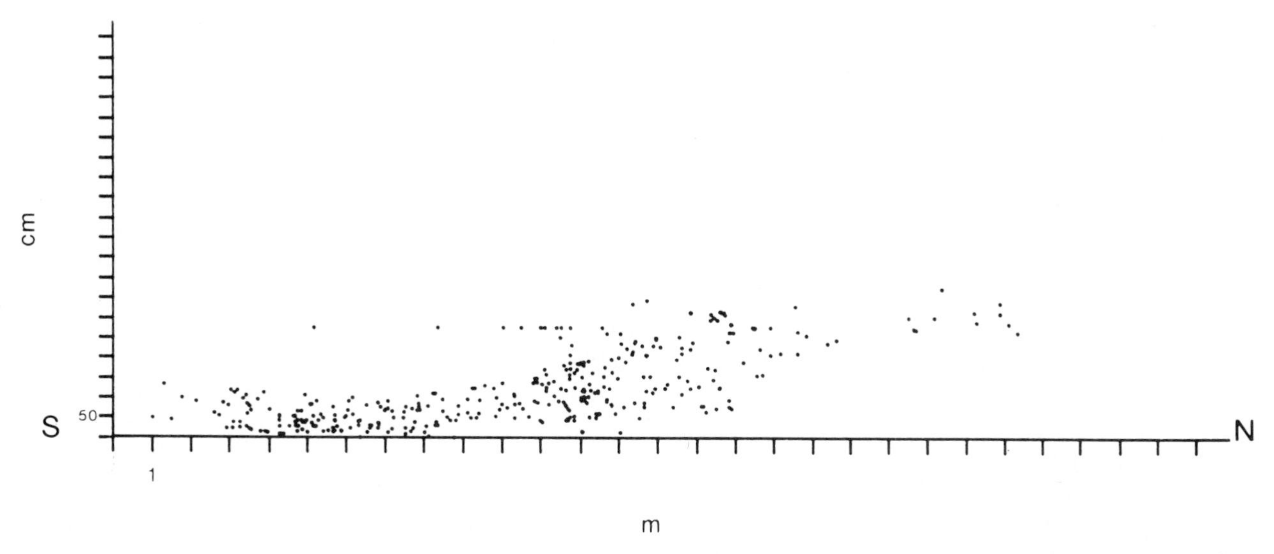

FIGURE 19-15 Computer-generated plot showing the vertical distribution, along the S/N axis, of all skeletal elements mapped during the excavation of L 338y.

The *main* fossiliferous sample contains heavily mineralized, rolled pieces along with some "fresh," well-preserved specimens (including the hominid finds) (de Heinzelin and Haesaerts 1983:81). Numerous fragmented "lag" elements (mostly isolated teeth) indicate concentration was influenced primarily by fluvial action. Hypothetically, channel bars and alluvial islands formed from accretion of bed load sediments carrying bones from upstream during times of high current velocity (Leopold and Wolman 1957). The association of some "easily transportable" elements (mostly vertebrae and some postcranials of hippopotamids and bovids, see Figure 19-10) suggests that, during times of periodic emersion, individuals dying close to the site were incorporated in the sample. Disarticulation and dispersal of these individuals probably resulted from the activities of predators and scavengers, as well as from trampling. Domination of the sample by hippopotamid remains (34 percent) likely reflects their proximity, during life, to the depositional environment, as modern hippos spend most of their time in the water. The bone assemblage probably resulted from a number of small-scale depositional and erosional episodes caused by fluctuating water levels.

Carnivore damage is apparent on a few of the more complete specimens, such as the juvenile *Australopithecus aethiopicus/boisei* partial cranium (Rak and Howell 1978). This individual presumably died locally, after which time most of its skeleton was eaten by predators/scavengers or otherwise removed (i.e., it was transported downstream; it was trampled and disintegrated).

Johanson et al. (1976:407) hypothesized bones of the *main* fossiliferous occurrence, including the hominid cranial remains, "washed" in during a time of alluviation. However, the cranium could only have floated in as part of a buoyant carcass (Dechant Boaz 1982; Fiorillo 1991) or as a defleshed but undamaged element (Boaz and Behrensmeyer 1976). The former scenario is feasible. The latter is improbable considering the hole found in the right parietal, attributed to the bite of a large felid (Rak and Howell 1978). With such a hole the defleshed skull would sink. A whole, defleshed cranium deposited at the site would be unattractive to a large felid. The specimen is too well preserved to have been transported as a "lag" element. No one has tested the transport behavior of a fresh human head.

Dispersal of the surviving pieces of this cranium (two parietals and an occipital) and the postcranials of several hippopotamids conceivably was caused by *in situ* scavenging or by the river's flow. As exemplified by the findings made during excavation of modern wildebeest skeletons on the Mara River, Kenya (Dechant Boaz 1982), whole carcasses can easily become disarticulated and dispersed under the influences of scavenging and of repeated cycles of deposition and erosion during flooding.

This reconstruction of events leading to burial of the bones suggests a large portion of the sample (isolated teeth) took longer to accumulate than the remaining few (postcranials and some cranial elements). This phenomenon was recognized by de Heinzelin and Haesaerts (1971) when excavation of the site was in progress. The sample of isolated teeth and broken bone debris was deposited with bed load sands when stream velocity was high. The bovids, hippopotamids, and the hominid individual, represented by well-preserved specimens, may have been contemporaneous members of the local animal community. The hippopotamid conceivably died at the site in its preferred habitat, the braided stream. During periods of emersion, a water hole situation may have developed with animals from the surrounding environment coming to drink, being killed by predators or dying from other causes, and their skeletons being disarticulated and trampled at the site (de Heinzelin 1971a,b). Fine-grained deposits, accumulated from overbank alluviation, eventually buried the assemblage (Howell et al. 1987).

In their preliminary report on Locality 338y, de Heinzelin and Haesaerts (1971) hinted at the possibility that hominid "occupation sites" were located near the bone accumulation at L 338y. A "suspected" hominid footprint was observed, although poorly defined, and a putative piece of charcoal was found *in situ*. "Artificial perforations" in a crocodile scute, chelonian carapace fragment, and hominid right parietal were reported. These are attributed, here and elsewhere (Rak and Howell 1978), to damage from carnivore teeth. The absence of stone tools was noted.

EXCAVATION OF LOCALITY 398

Locality 398 was established in 1970 after 13 specimens were recovered from the surface of the site. R. L. Ciochon and D. C. Johanson began excavation of the site in 1971. In 1972 and 1973, some 5.5 m of unfossiliferous sediment overlying the bone bed were removed by jackhammer (Atlas Copco) to expedite the work of D. C. Johanson (1972) and N. T. Boaz (1973)(Johanson et al. 1976). During the three field seasons of excavation, 178 m^2 were exposed (see Figure 19-16, after Ibid.:412, Figure 8), yielding 2,717 fossils. Measurements of depth were taken in all seasons; measurements of trend/plunge were taken on most of the bones. All excavated sediment was fine screened. (The screen material, sorted and identified by M. McCrossin in 1981, yielded 10,527 vertebrate tooth and bone fragments. Four hundred and seventy of these, including two hominid tooth fragments, could be identified at the mammalian family level only.)

Dating

At Locality 398, the fossiliferous layer occurs within Tuff F, in Tuff F′ sediments (de Heinzelin and Haesaerts 1983). Tuff F has been dated by the potassium-argon method to approximately 2.34 MYA (Feibel et al. 1989).

Sedimentary Environment

The sediments at Locality 398 are illustrated in Figure 19-17. The stratigraphy at L 398, consisting of three subunits (F′-α, F′-β, and F′–γ) of Tuff F′, has been described by de Heinzelin and Haesaerts (1983:84). Tuff F′ is the name given to those sections of the vitric Tuff F where channels were cut and eventually filled in with reworked tuff, or tuffitic sediments. The channel deposits are divided into three subunits grading upward from coarse grained, cross-bedded tuffite (F′-α), through medium grained to silty tuffite with poor or even stratification (F′-β), to fine-grained tuffite with some concretions, mottling, and incipient soil formation (F′-γ) (ibid.). The fossils excavated from L 398 were found at two levels within subunit F′-a: the *main* fossiliferous horizon at 0 to 10 cm above the basal contact, and the *upper* fossiliferous horizon approximately 1 m above the main bed (ibid.).

During excavation, the fossils were found buried in cross-bedded sands, occasionally associated with clay balls, and sometimes concentrated in shallow depressions within the basal sterile silt (Johanson et al. 1976). The authors (ibid.:408) hypothesized a turbulent, rapidly flowing stream or river to account for

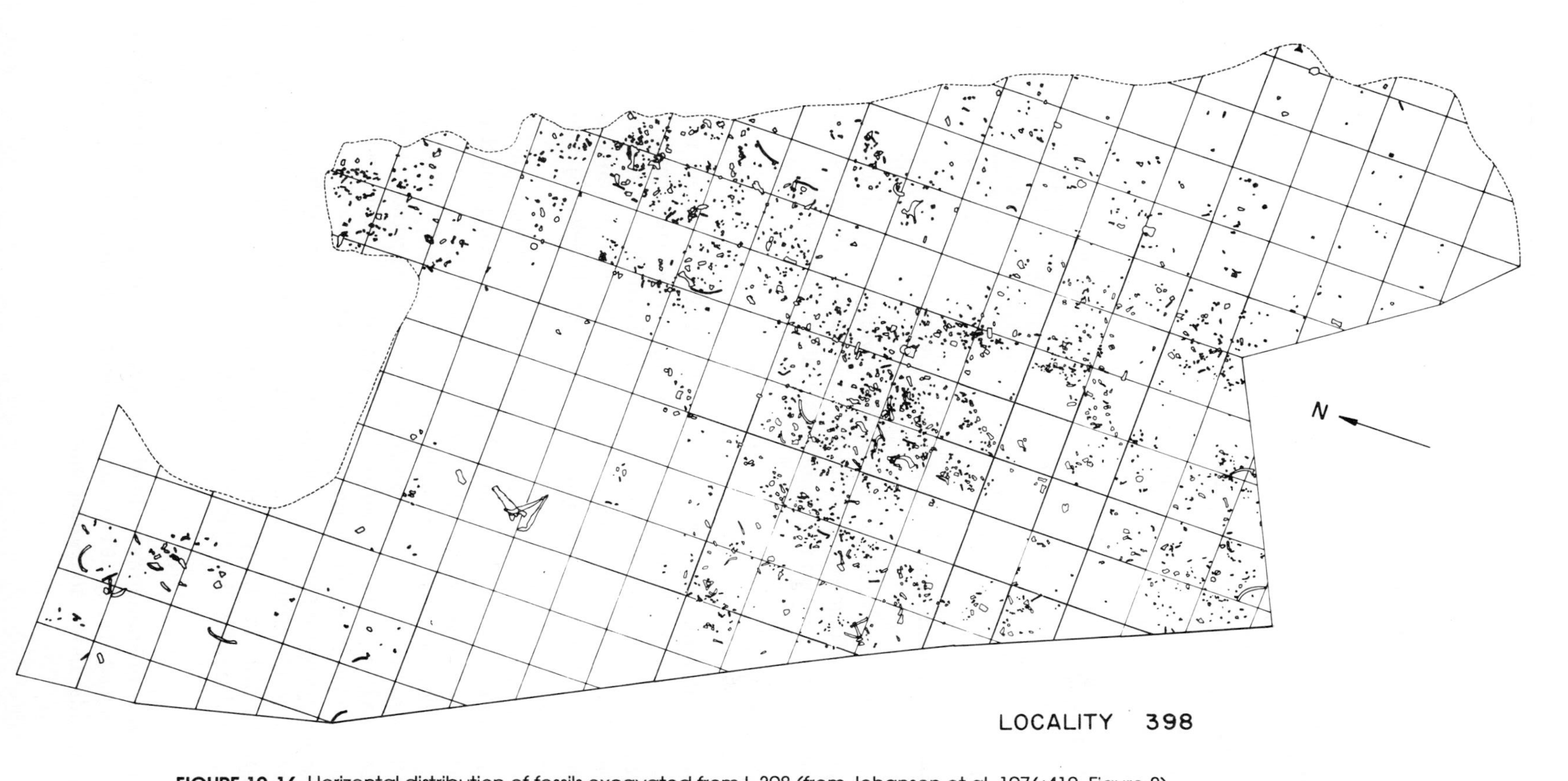

FIGURE 19-16 Horizontal distribution of fossils excavated from L 398 (from Johanson et al. 1976:412, Figure 8).

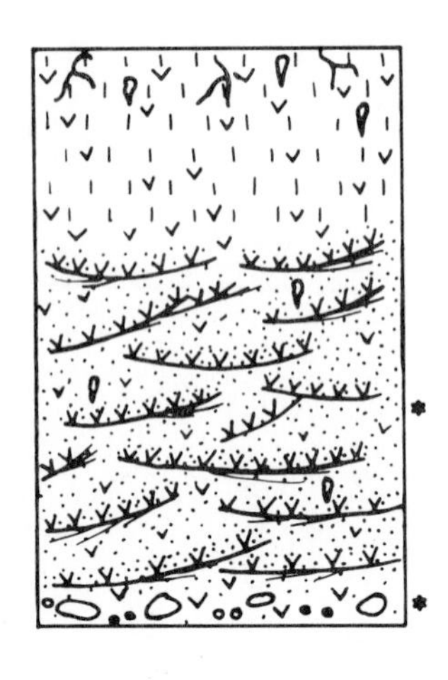

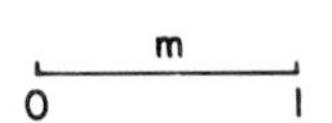

FIGURE 19-17
Stratigraphic profile of sediments of F′-α at L 398 (after de Heinzelin and Haesaerts 1983:85, Figure 32, and from description of the stratigraphy ibid.:84). The *main* and *upper* fossiliferous horizons are indicated by asterisks.

bones lying at steep inclinations, and the presence of dense, immovable elements (i.e., a large *Hippopotamus* ulna and *Crocodylus* cranium). Deposition of several fossils a meter above the *main* fossiliferous bed was attributed to transportation in a channel heavily laden with suspended sediment. They found an abundance of small, rolled bone fragments indicative of rapid current. The recovery of some heavily abraded, "polished" fossils suggested reworking of previously buried assemblages.

Faunal Composition

Taxonomic Representation. From the total sample of 13,244 specimens recovered, 2,648 were excavated, 66 surface collected, and 10,530 found in screening. Of these remains, 1,344 have been assigned to mammalian taxa, and 696 to Mammalia indeterminate, as shown in Table 19-3. Fish and reptile specimens numbered 1,494; 3 ?*Eutheria* sp. shell fragments found in the screens were catalogued. The bulk of the assemblage (9,688) was identifiable only as animalia fragments. Nineteen specimens were unavailable for study.

Thirty unique mammalian taxa were identifiable from the L 398 sample: *Australopithecus aethiopicus*, aff. *Homo* sp., *Theropithecus* sp., *Papio* sp., small Papionini sp., Colobinae indet. (either *Paracolobus mutiwa* or *Rhinocolobus turkanaensis* [Leakey 1987]), *Diceros bicornis, Giraffa gracilis, Giraffa jumae, Giraffa pygmaeus, Sivatherium* sp., *Tragelaphus gaudryi, Tragelaphus nakuae, Aepyceros* sp., Alcelaphini indet., Bovini indet., Reduncini indet., *Hipparion ethiopicum, Hipparion* aff. *sitifense, Kolpochoerus limnetes, Metridiochoerus andrewsi, Notochoerus scotti, Deinotherium bozasi*, Elephantidae indet., *Felis caracal, Dinofelis* sp., Viverridae indet., *Lepus* sp., Muridae indet., and Hippopotamidae indet.

Other fragmentary specimens, possibly but not definitely subsumable under the taxonomic categories represented, were identified as cf. *A. aethiopicus* (or Hominidae indet.), Cercopithecinae indet., *Kolpochoerus* sp., *Metriodiochoerus* sp., *Notochoerus* sp., Suidae indet., *Giraffa* sp., Giraffidae indet., *Tragelaphus* sp., Bovidae indet., Equidae indet., Felidae indet., Rhinocerotidae indet., and Proboscidea indet.

Age. The remains of immature individuals were found for the following taxa, including *A. aethiopicus/boisei, Theropithecus* sp., Cercopithecinae indet., *Diceros bicornis*, Elephantidae indet., Felidae indet., Hippopotamidae indet., and Suidae indet.

TABLE 19-3 Taxonomic (by family) distribution of skeletal elements from L 398, grouped according to degree of susceptibility to transport as defined by Voorhies (1969)/Behrensmeyer (1975) Transport Groups. Hip = Hippopotamidae, Bov = Bovidae, Cer = Cercopithecidae, Sui = Suidae, Ele/Dei = Elephantidae/Deinotheriidae, Hom = Hominidae, Gir = Giraffidae, Equ = Equidae, Rhi = Rhinocerotidae, Fel = Felidae, Viv = Viverridae, Lep = Leporidae, Mur = Muridae, Mam Indet = indeterminate mammal element, N = total numbers of each element for all taxa; % of Sample indicates the relative frequencies of different skeletal elements. Percentages for elements making up less than 0.5% of the sample are not shown. Transport gp % indicates the relative frequencies of "easily transportable" (Group I), "variably transportable" (Group I/II, II, and II/III), and "lag" (Group III) elements for the whole assemblage. NISP = number of identifiable skeletal parts, RTA = relative taxonomic abundance, based on NISP numbers being equal to the number of individuals per taxon. AST/CALC = astragalus/calcaneum, OTH PODIAL = other podial, METAPOD = metapodial, INNOM = innominate, HEMIMAND = hemimandible, SKULL/HC = skull and/or horn core, WH MAND = whole mandible, ISO TOOTH = isolated tooth, IND T FR = indeterminate tooth fragment, LB SH FR = longbone shaft fragment.

	Voorhies/Behrensmeyer Transport Groups																
	Hip	Bov	Cer	Sui	Ele/ Dei	Hom	Gir	Equ	Rhi	Fel	Viv	Lep	Mur	Mam Indet	N	% of Sample	Transport gp%
GROUP I																	
RIB	9													83	92	4.6	
VERTEBRA	10	7			3									13	33	1.6	
SACRUM																	6.7
STERNUM																	
SESAMOID	1														1		
PATELLA	2	1													3		
GROUP I/II																	
SCAPULA	2	2	1		2									4	11	.6	
ULNA	2	1				1								3	7		
PHALANX	7	6		2	1		1			1				1	19	1.0	4.3
AST/CALC	5	11		2			1	1		1				2	23	1.1	
OTH PODIAL	5	13		1	1		2							1	23	1.1	
GROUP II																	
FEMUR	2	4	3	2										1	12	.6	
TIBIA	2	5		1											8		
HUMERUS	6	4	1	1										4	16	.8	3.9
RADIUS		2													2		
METAPOD	2	21					2	2	2						29	1.5	
INNOM		2												7	9		
GROUP II/III																	
HEMIMAND	2	13	4	4										2	25	1.2	1.3
GROUP III																	
SKULL/HC	1	22	1	1										17	42	2.1	
WH MAND			1												1		83.8
ISO TOOTH	248	168	124	158	144	15	29	8	5	4		1	1		905	45.2	
IND T FR	42	103	32	12	21		3	1	1					453	668	33.4	
TRANSPORT GRP UNTESTED																	
EPIPHYSIS																	
FIBULA	1	1													2		
MAXILLA	1	1		1							1			1	5		
LB SH FR														104	104	5.2	
															2040		
NISP	350	387	167	185	172	16	38	12	8	6	1	1	1	696			
RTA	26.0	28.8	12.4	13.8	12.8	1.2	2.8	.9	.6	.4	.1	.1	.1				

Relative Abundance of Taxa. Various aspects of the L 398 fossil assemblage and stratigraphy signify its derived nature. Only one articulated individual, a *Crocodylus* cranium and mandible, was found during excavation (Johanson et al. 1976). Most of the pieces were abraded, and many have been rounded into bone "pebbles." The sediments indicate high-energy transport and deposition in a channel environment. Most of the sample appears to have been carried, as bed load, a considerable distance, with reworked, previously buried bones a significant component of the assemblage.

Because of the allocthonous nature of the sample, the relative abundance of mammalian taxa was calculated by considering each element to be from a different skeleton.

The skeletal part totals (NISP) per family, from a sample of 1,344, are shown in Table 19-3, along with percentages indicating relative abundance. Bovids and hippopotamids dominate the sample, with suids, cercopithecids, and elephantids being fairly abundant.

Assemblage Modification

Skeletal Articulation and Association. As mentioned previously, only one incidence of skeletal articulation/association, between a *Crocodylus* cranium and mandible, was discovered (see Figure 19-16). During excavation no features of the sample were found to indicate derivation of different skeletal elements from the same individual (Boaz 1979).

Skeletal-Part Representation. With each specimen defining a different individual, skeletal representation, as a measure of the completeness of individual skeletons, is poor for all taxa.

Most of the mammalian individuals are represented by crania, mandibulae, and teeth, as shown in Table 19-3. Postcranial remains of bovids and hippopotamids are most numerous, followed by those of suids and elephantids/deinotheriids.

Damage. Although fragmented, the fossils excavated from L 398 were better preserved than those from either L 345 or L 338y. Bone surfaces were less friable and predepositional damage could be assessed more easily.

Nearly all of the bones were fragmentary, and over 50 percent of the postcrania showed some degree of pitting of the bone surface. Several elongated bones were pitted only on one side, suggesting they had been lying for some time on a damp but subaerially exposed surface.

Although a few carnivore gnaw marks and tooth punctures were apparent, most of the bones were abraded ("polished") and pitted to such an extent that any primary damage was undetectable. An abundance of spirally fractured shaft fragments was present, possibly produced by scavenging (Hill 1989; Brain 1981; Haynes 1983) or trampling (Agenbroad 1989; Myers et al. 1980; Fiorillo 1989) prior to entrainment in the channel environment or by postdepositional processes (Klein and Cruz-Uribe 1984).

Several elements showed no damage, including some delicate bovid postcranials. The occurrence of these pieces along with numerous abraded longbone fragments indicates an autochthonous component to this basically allochthonous assemblage.

Weathering. Of the 150 bones analyzed for degree of surface weathering, 61 percent showed none (Stage 0), while 27 percent showed a few small cracks (Stage 1) (Behrensmeyer 1978). Twelve percent of the sample fell within weathering Stages 2, 3, and 4. Although repeated cycles of subaerial exposure, deposition and erosion (which seem to characterize the history of the L 398 assemblage) would create a sample of more weathered bones, the prevalence of bones weathered to Stages 0 and 1 could be a measure of the instability of bones reaching Stages 2, 3, and 4, as these likely would disintegrate before becoming buried and fossilized in a high-energy transport situation (Behrensmeyer et al. 1989).

Transport Groups. From a sample of 1,934 mammalian skeletal elements sorted into Voorhies/Behrensmeyer transport groups, Group III lag elements (mostly isolated teeth) dominate, making up 84 percent of the sample for all taxonomic categories (Table 19-3). Teeth are among the most durable elements of the mammalian skeleton. They are resistant to movement in currents due to their high density (Behrensmeyer 1975).

Johanson et al. (1976:410) described L 398 as an "obviously derived taphonomic situation" where features of "transport, sorting, and differential loss of bones of small animals" are apparent. It seems likely the fossils in this sample are from individual carcasses subjected to extensive disarticulation, dispersal, and degradation. The gritty(?) tuffitic silts and sands transported along with the bones may have accelerated the normal rate of abrasion.

Dispersal and Orientation. The horizontal distribution of fossils excavated from L 398 is shown in Figure 19-18. As discussed previously, the sample consists of presumably unassociated teeth and bones from a variety of mammals, reptiles, and fish.

As illustrated in Figure 19-23, the distribution of hippopotamid remains from L 398 is wide and random. Hominid and particularly cercopithecid remains (Figure 19-19) appear more clustered than the hippopotamids from L 398, but they are probably unrelated considering the conditions under which deposition occurred.

Figure 19-20 shows the rose diagram (after Johansen et al. 1976:Figure 9) made by plotting longbones excavated in 1972 and 1973. These authors concluded a northeast to southwest direction of flow.

Vertical distribution of skeletal parts along the E/W (10 m) and N/S (19 m) axes are shown in Figures 19-22 and 19-23. The depths used to make these plots were measured during the 1971 and 1972 field seasons. A few depths were taken in 1973, but these could not be correlated with those of the previous years and are absent from the figures. (The different areas excavated in 1971, 1972, and 1973 are shown in Johanson et al. 1976:412, Figure 8).

The bone bed is inclined due to dip in bedding of approximately 10°S, and the *in situ* sample had a vertical scatter of approximately 120 cm (see Figure 19-22). Johanson et al. (1976) noted dip in a westerly direction also, although this is not particularly evident in Figure 19-21.

FIGURE 19-18 Computer-generated plot of all bones found during excavation of L 398.

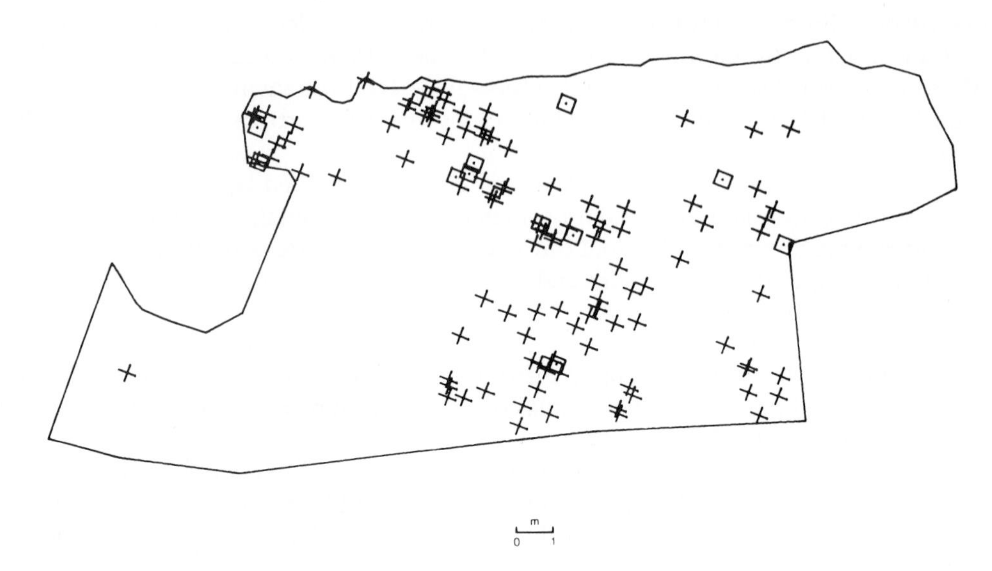

FIGURE 19-19 Computer-generated plot showing the distribution of cercopithecid (pluses) and hominid (squares) fossils from L 398.

The fossils are vertically distributed through 120 cm (see Figure 19-22), corresponding to the site description and stratigraphic profile provided by de Heinzelin and Haesaerts (1983) (see Figure 19-17). Their definition (ibid.:84) of two fossiliferous horizons, a *main* and an *upper*, is possibly confirmed by Figure 19-22, although the *main* layer appears to be greater than the 10 cm thickness described by them and the two layers tend to merge, particularly at the northern end of the excavation. (It would be useful, if possible, to add the 1973 depth measurements with those shown, to see if a separation of two bone beds is more distinct with a larger sample.)

Johanson et al. (1976) explained vertical displacement of the sample by suggesting that some bones were part of the suspended load and became buried at higher levels than those elements transported as bed load (Voorhies 1969).

Reconstruction of Burial Events

The sediments and features of the fossil assemblage at L 398 point to an essentially allocthonous accumulation. The stratigraphic profile (Figure 19-17) indicates a fining-upward sequence, typical of channel deposition, from coarse to medium-grained, cross-bedded tuffite containing pumices and bones (the *main* fossiliferous horizon), through silty, poorly stratified tuffite containing rootcasts, to tuffitic clay or silt with rootcasts and sparse bones (the *upper* fossiliferous horizon) (de Heinzelin and Haesaerts 1983:84). Tuff F′-α, containing the *main* fossiliferous horizon, was deposited as the channel became "choked" with vitric tuff (de Heinzelin et al. 1976).

The discovery of only one incidence of association among the fossil remains (a *Crocodylus* mandible and cranium) coupled with a predominance of "lag" remains (84 percent of the sample) suggests most individuals died upstream and were transported considerable distances as bed load before being buried. Extensive "polishing" and the large number of fragments attest to long-distance transport in the channel environment. (Abrasion rate may have been accelerated by association with gritty tuffitic silts and sands.) Probably much of the sample was subjected to repeated cycles of deposition, erosion, and subaerial exposure as it progressed downstream (Dechant Boaz 1982; Hanson 1980). Extensive pitting of the bone may represent submersion in shallow water. Some pieces showed pitting on one side only; these may have rested for some time on a damp surface prior to burial. As they were transported downstream, the

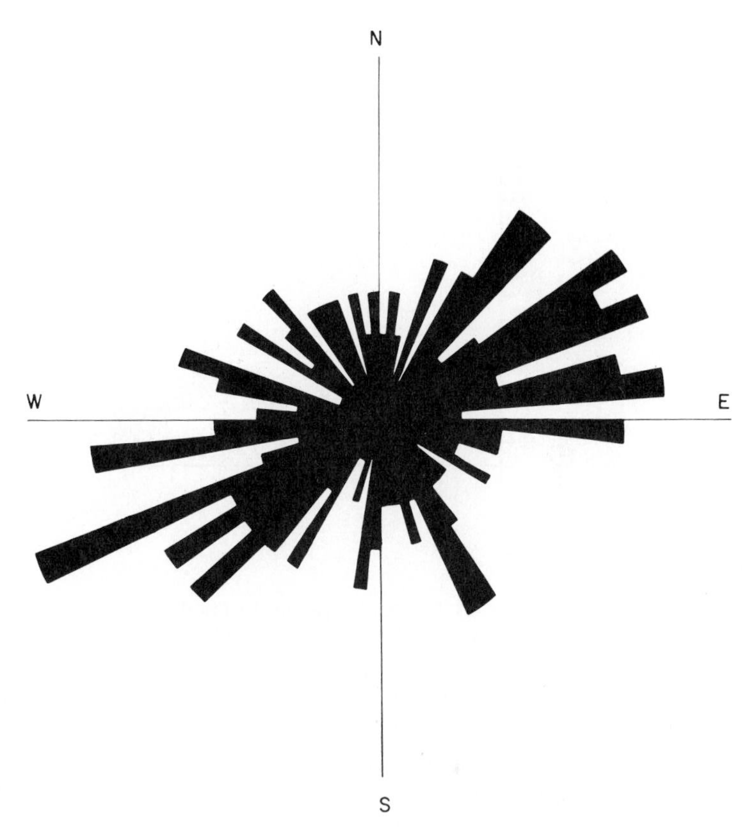

FIGURE 19-20
Rose diagram of elongated elements from L 398, showing their preferred direction and polarity (from Johanson et al. 1976:413, Figure 9).

elements (elongated ones in particular) adapted orientations in response to the direction of flow. Eventually, the bones were buried on the channel floor by vitric sands and silts in a fining-upward sequence (de Heinzelin et al. 1976) typical of a channel undergoing avulsion.

This reconstruction of events leading to burial of the bones suggests that the sample took years to accumulate, although the channel environment was unique in some respects and may require special consideration. The deposition of Tuff F could have been relatively rapid, depending on catchment and the conditions of volcanic eruption. Since the channel was eroding into Tuff F, it would incorporate older sediments and the bones buried in them. Fresh carcasses and bones entrained in the channel would be mixed with more ancient remains, creating a temporally mixed sample.

COMPARISON OF THE THREE EXCAVATED LOCALITIES

Sedimentary Environment

In the fluvial environment, skeletons are influenced by current to varying degrees (Hanson 1980). The three samples described here can be categorized according to how extensively they have been affected by flowing water. Skeletal composition and sedimentary context suggest that the L 345 assemblage experienced little alteration due to flow, the L 338y assemblage was somewhat affected by flow, and the L 398 sample was strongly influenced by it.

The sedimentary environment at L 345 was a pond or backswamp receiving sediment mainly from overbank flow. Bones were not transported to or from the site by rapid currents. The buried assemblage consisted of animals dying proximally, either from disease, starvation, senility, or predation.

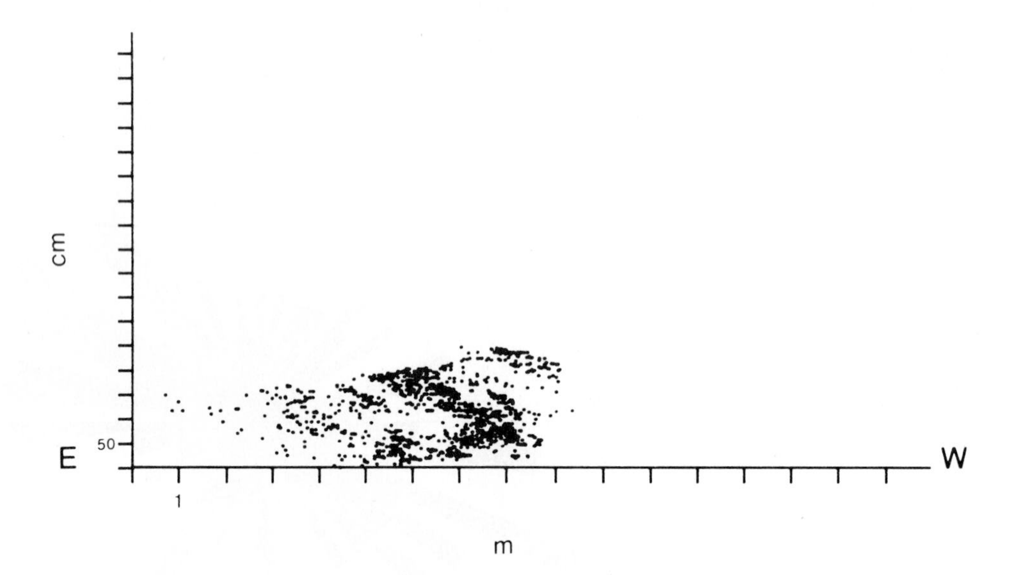

FIGURE 19-21 Computer-generated plot showing the vertical distribution, along the E/W axis, of skeletal elements mapped during the 1971 and 1972 excavations of L 398.

The sedimentary environment at L 338y was a braided stream where, during periods of emersion, animals may have come to drink. The skeletons of individuals dying on or near alluvial islands were mixed with those transported from upstream.

L 398 was formed by the filling of a channel eroded in to Tuff F. The channel incorporated bones of a variety of animals probably because it was cutting through different sedimentary environments that, potentially, preserved the bones of mammals dying in a diversity of habitats.

FIGURE 19-22 Computer-generated plot showing the vertical distribution, along the N/S axis, of skeletal elements mapped during the 1971 and 1972 excavations of L 398.

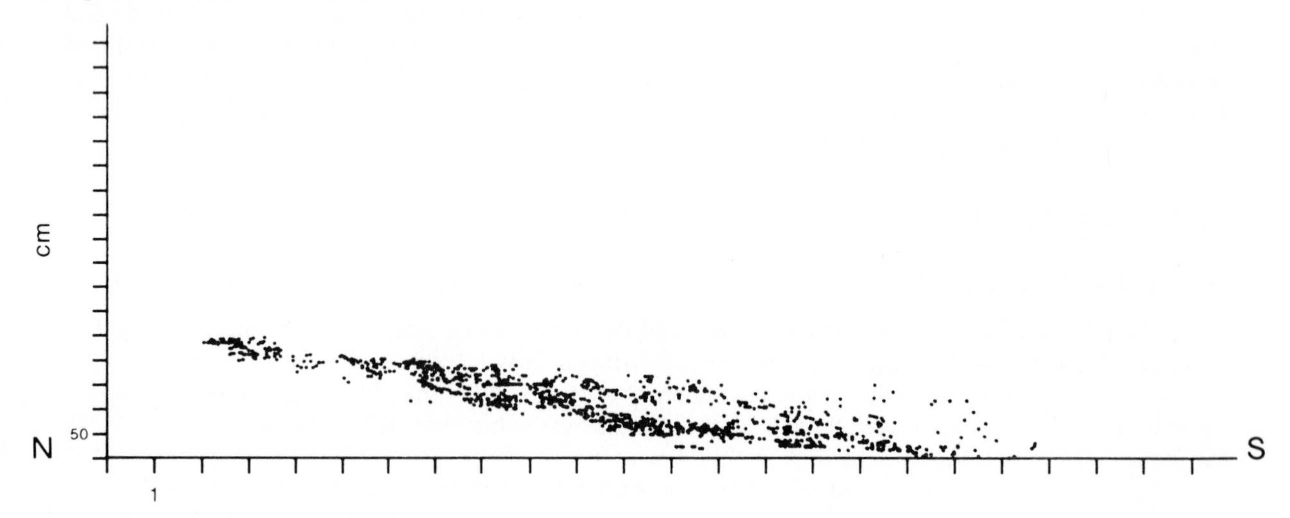

TABLE 19-4 A comparison of taxonomic composition at the family level of assemblages from the three localities. NISP = Number of Identifiable Skeletal Parts, # Ind = Number of Individuals determined to be represented by each sample, based on appropriate quantification method (i.e., MNI for L 345, NISP for L 388y and L398), as explained in the text and as shown in Tables 19-1, 19-2, and 19-3. RTA = Relative Taxonomic Abundance at the family level within each assemblage, based on MNI calculations for L 345, and NISP calculations for L 388y and L 398.

	L 345			L 338y			L 398		
	NISP	# Ind	RTA	NISP	# Ind	RTA	NISP	# Ind	RTA
Hippopotamidae	163	4	28.6	152	152	34.2	350	350	26.0
Bovidae	11	1	7.1	115	115	25.8	387	387	28.8
Cercopithecidae	58	5	35.7	89	89	20.0	167	167	12.4
Suidae	3	2	14.3	59	59	13.3	185	185	13.8
Elephantidae/ Deinotheriidae	1	1	7.1	14	14	3.1	172	172	12.8
Hominidae	1	1	7.1	3	3	.7	16	16	1.2
Giraffidae				5	5	1.1	38	38	2.8
Equidae				2	2	.4	12	12	.9
Rhinocerotidae							8	8	.6
Mustelidae				2	2	.4			
Felidae				1	1	.2	6	6	.4
Viverridae				1	1	.2	1	1	.1
Leporidae				1	1	.2	1	1	.1
Sciuridae				1	1	.2			
Muridae							1	1	.1
Totals	237	14		445	445		1344	1344	

Faunal Diversity

The lists of unique mammalian taxa given for each locality (see Taxonomic Representation sections of text) show an increase in diversity, L 345 being composed of six unique mammalian taxa, L 338y of fifteen, and L 398 of thirty. These contrasts in faunal diversity may be due in part to the different sample sizes, L 345 being the smallest (237), L 338y next in size (445), and L 398 the largest (1,344). It is more likely, however, that the contrasts in faunal diversity are due to circumstances under which the taphocoenoses were formed. In her taphonomic analysis of mammalian fossil remains from Siwalik, Badgley (1986a) recorded the largest sample size from her composite "Taphonomic Assemblage I," which derived from active channel sedimentary environments.

A comparison of faunal diversity is made in Table 19-4, using NISP values for the braided stream (L 338y) and channel (L 398) assemblages, and MNI values for the backswamp (L 345) assemblage. A marked difference in faunal diversity at the family level is apparent between the L 345 assemblage and the other two. Similar patterns of faunal diversity were found in comparisons between "lag" assemblages and assemblages composed of associated skeletal remains analyzed from the Siwalik (Badgley 1986a) and from Careless Creek Quarry (Fiorillo 1991).

TABLE 19-5 A comparison of skeletal remains of immature individuals found from the three localities. N = the number of elements per assemblage; % of Assemblage = the percentage that immature (vs. adult) remains make up of each of the three samples.

	L345 N	L338y N	L398 N
Crania		1	
Maxillae		2	
Hemimandibles	1	2	
Deciduous Teeth	3	28	18
Unfused Long Bones	3	3	1
Unfused Axial Bones	6	1	
Epiphyses	1		
Total	14	37	19
% of Assemblage	5.5	5.3	.9

Age. In Table 19-5, the remains of immature individuals are compared among the three samples. In all three cases, the immature remains make up a small percentage of the total, although the L 398 assemblage has a lower percentage than the other two. The skeletal elements of adult mammals are more resistant to

FIGURE 19-23 Computer-generated plot comparing the distribution of hippopotamid bones from L 398 and L 345 at the same scale.

TABLE 19-6 A comparison of skeletal-part frequencies among the three assemblages, calculated from all remains attributable to six taxonomic families shared by the three samples. These common families include Hippopotamidae, Bovidae, Cercopithecidae, Suidae, Elephantidae (with Deinotheriidae at L 398), and Hominidae. Frequencies from the three fossiliferous assemblages are compared with skeletal-part frequencies found in the "Average Mam" skeleton, calculated from an averaging of elements found in six modern representatives of the families previously mentioned , including *Hippopotamus amphibius*, a composite African bovid, *Theropithecus gelada*, *Phacochoerus africanus*, *Loxodonta africana*, and *Homo sapiens*.

	L 345		L 338y		L 398		Average Mam	
	N	%	N	%	N	%	N	%
Crania	4	1.7	9	2.1	29	2.2	1	.5
Hemimandibles	9	3.8	8	1.9	25	1.9	2	.9
Teeth	52	21.9	356	82.4	1065	82.4	32	14.5
Vertebrae	48	20.3	11	2.5	20	1.5	43	19.5
Ribs	53	22.4	4	.9	9	.7	28	12.6
Limb/Girdle	41	17.3	23	5.3	67	5.2	20	9.0
Metapodials	10	4.2	4	.9	23	1.8	17	7.7
Podials	13	5.5	8	1.9	39	3.0	27	12.2
Phalanges	7	3.0	9	2.1	16	1.2	51	23.1
Total	237		432		1293		221	

destruction than those of the young because they are relatively larger and denser and have reached more advanced stages of suture and epiphyseal fusion (Scott 1989; Gifford-Gonzalez 1989). Some of the least durable elements (i.e., unfused axial elements, epiphyses) are most abundant in the L 345 sample.

Skeletal-Part Representation. In Table 19-6, skeletal-part frequencies (derived from six taxa common to the three localities) are compared among the three samples, and to those frequencies found in a composite mammal skeleton. As skeletal-part composition measures degree of skeletal loss, it is clear that individuals represented in the L 345 sample, matching most closely, as they do, the skeletal-part frequencies from the average mammal, have suffered the least alteration of skeletal integrity. Overall, the L 338y and L 398 assemblages show a marked similarity in skeletal-part frequencies, undoubtedly due to the effects of fluvial sorting.

Damage. All three samples showed signs of carnivore damage including punctures and gnawing marks, and some evidence of pitting on bone surfaces. The L 345 sample was poorly preserved with friable, abraded surfaces found on many of the elements. "Abrasion" was attributed to gnawing, trampling, postfossilization processes or to excavation technique.

A gradient of fragmentation was notable, with the L 345 sample being the least fragmentary, the L 338y sample more fragmentary, and the L 398 sample most fragmentary, with many spirally fractured shaft pieces. Trampling undoubtedly influenced degree of fragmentation of the L 338y sample, while transport in a channel environment, coupled with carnivore chewing, weathering, trampling and postdepositional damage, led to the accumulation of "polished" bone pebbles and fragments common in the L 398 assemblage.

Weathering. Of the samples analyzed for surface weathering from L 345 and L 338y, few had weathered beyond Stages 0 or 1 (Behrensmeyer 1978). This similarity in weathering pattern may indicate rapid burial at the two sites, as other variables affecting weathering rate (i.e., taxon, depositional microenvironment, accumulation history, and spatial distribution [Lyman and Fox 1989]) are not particularly similar between the two sites. Repeated subaerial exposure prior to "permanent" burial could have destroyed more weathered, and thus more fragile, postcranial elements of the L 338y sample, leaving mostly the "unweathered" allocthonous component for analysis.

"Weathered" specimens (Stages 2, 3, and 4) make up a slightly higher percentage of the L 398 sample (12 percent) than is the case for L 345 (7 percent) and L 338y (6 percent). It is difficult to determine the variables affecting stages of weathering at these three sites as some are constant between them (i.e. particularly taxa of L 338y and L 398), while others are not (i.e., depositional microenvironment, accumulation history, spatial distribution [Lyman and Fox 1989]).

Transport Groups. Concentrations of durable, dense elements are clues to the degree of transport undergone by assemblages buried in fluvial contexts. Damage from transport and compaction may become cumulative as skeletal elements experience repeated cycles of deposition, subaerial exposure, burial, erosion, and entrainment, resulting in highly fragmented samples. These residual features of highly transported assemblages are least evident in the L 345 sample and become increasingly obvious in the L 338y and L 398 samples (see Tables 19-1, 19-2, and 19-3). The L 345 assemblage, from a backswamp environment, is composed of 44 percent "highly transportable" elements, and 24 percent "lag" elements. The L 338y and L 398 samples are composed of 82 percent and 84 percent "lag" elements, respectively, mostly isolated teeth.

Skeletal Disarticulation and Dispersal. The pattern of element distribution, particularly of associated hippopotamid remains from L 345 and of possibly associated segments of hippos from L 338y, seems to indicate that carnivores and/or scavengers were influential in disarticulating and dispersing carcasses at these two sites (Hill 1975, 1979). In contrast, the hippo remains from L 398 seem widely scattered and randomly distributed. When plotted at the same scale, clusters of associated hippopotamid elements are apparent in the L 345 sample, while a dispersed pattern of unrelated elements is observed at L 398 (Figure 19-23).

CONCLUSIONS

Three fossil assemblages from fluviatile sediments of Members C, E, and F of the Shungura Formation, Ethiopia, show evidence of modification from biological and geological agents prior to ultimate burial and preservation. Two general types of taphonomic histories could be reconstructed based on sedimentary and faunal evidence.

The L 345 sample, formed in a channel-fill environment, probably accumulated under the influence of predation and other attritional circumstances and was not subjected to transport by fluvial action. Associated remains were recognized and, in contrast to the L 338y and L 398 samples, skeletal integrity was relatively high. Taxonomic diversity was low for the L 345 sample and, because the assemblage probably accumulated by selective forces (i.e., predator prey-preference), its taxonomic composition does not provide an accurate reflection of the actual ecological diversity of that time (Badgley 1986a). Habitat-specific information might be gleaned from such an assemblage, as the amount of time and space represented by its accumulation would be less than for fluvially concentrated assemblages (Behrensmeyer 1988; Badgley 1986a).

The L 338y and L 398 assemblages, formed predominantly as channel-lag deposits, accumulated under the influence of fluvial transport. Associated remains were a possibility in the L 338y sample, but essentially nonexistent in the large L 398 sample, and skeletal integrity was low as both assemblages were dominated by isolated teeth. Taxonomic diversity was high for both these samples and, at the family level, they had similar compositions (see Table 19-4). Accumulated as "lag" deposits, they likely derived from extensively winnowed skeletons. The assemblages could include remains excavated by flowing water from previously deposited channel and floodplain sediments, consequently sampling from a diversity of habitats and a range of time periods (Dechant Boaz 1982; Badgley 1986a; Behrensmeyer 1988).

SUMMARY

Three large fossil samples from Members C, E, and F of the Shungura Formation are examined for taphonomic and sedimentary evidence pertinent to the histories of their formation. These assemblages, dating from 2.3 to 2.5 MYA, accumulated in fluviatile sediments deposited by the ancient Omo River system. Each assemblage is described in terms of its depositional environment, taxonomic composition, immature elements present, and minimum number of individuals (Locality 345) or relative taxonomic abundance (Localities 338y and 398). Taphonomic parameters discussed for each assemblage include skeletal-part representation, damage and weathering, transport groups, and horizontal and vertical distribution of specimens. Analysis of these variables, within sedimentary context, suggests that the L 345 remains accumulated as a result of predation (or other forms of attrition) in a localized channel-margin setting (i.e., a pond or waterhole). The L 338y sample, accumulated primarily as a braided stream channel lag deposit, contains a minor but significant autocthonous component (including the parieto-occipital portion of a juvenile *Australopithecus aethiopicus/boisei* cranium). The L 398 assemblage consists of a concentration of lag elements, including 15 isolated hominid teeth, in a river channel environment. A gradational change towards decrease in skeletal integrity and increase in taxonomic diversity is apparent when the assemblages are arranged in sequence from least to most transported (L 345-L 338y-L 398). Habitat-specific information may be gleaned for taxa represented by the L 345 (and some of the L 338y) specimens. The L 398 fossils, derived from a diversity of habitats and a range of time periods, would not provide habitat-specific information for those taxa represented in the sample.

ACKNOWLEDGEMENTS

The ideas and data on which this chapter are based were presented originally in a section of my Ph.D. dissertation. My acknowledgements, therefore, cover the field and laboratory research and writing efforts of then and now. The governments of Kenya and Ethiopia are thanked for access to modern and fossil samples. For research permission in Kenya, I am grateful to E. K. Ruchiami, S. W. Taiti and S. Tipis. For sponsorship in Kenya I thank L. Aggundey and R. E. F. Leakey. Funding for the field research was given generously by the L. S. B. Leakey and Wenner-Gren Foundations. B. Benefit, N. Boaz, M. McCrossin, and members of the MaraSara Camp provided invaluable field and laboratory assistance. For unselfishly giving of their time and computer ability I thank B. Howell and G. Richards. F. C. Howell and T. White read earlier drafts of this work; they also provided refuge and use of their facilities in the Laboratory for Human Evolutionary Studies. E. Kroll deserves special appreciation for generously sharing her ideas and expertise on site analysis and computer plotting of excavated samples. I am indebted to K. Behrensmeyer for introducing me to the study of taphonomy, and for the guidance and encouragement she and A. Hill have given. For support in many forms I thank my parents; for their appreciation I thank my children. Above all, I am grateful to Professor F. C. Howell for giving me the opportunity to become part of the Omo Research Expedition, for encouraging my undertaking of this work, and especially for his generosity and friendship.

LITERATURE CITED

Agenbroad, L. D. (1989) Spiral fractured mammoth bone from nonhuman taphonomic processes at Hot Springs Mammoth Site. In R. Bonnichsen and M. H. Sorg (eds.): *Bone Modification*. Orono, Maine: Center for the Study of the First Americans, pp. 121-137.

Badgley, C. (1986a) Taphonomy of mammalian fossil remains from Siwalik rocks of Pakistan. *Paleobiology* 12:119-142.

Badgley, C. (1986b) Counting individuals in mammalian fossil assemblages from fluvial environments. *Palaios* 1:328-338.

Behrensmeyer, A. K. (1974) Taphonomy and the fossil record. *Am. Sci.* 72:558-566.

Behrensmeyer, A. K. (1975) The taphonomy and paleoecology of Plio-Pleistocene vertebrate assemblages east of Lake Rudolf, Kenya. *Mus. Comp. Zool. Bull.* 146:473-578.

Behrensmeyer, A. K. (1978) Taphonomic and ecologic information from bone weathering. *Paleobiology* 4:150-162.

Behrensmeyer, A. K. (1988) Vertebrate preservation in fluvial channels. *Palaeogeog. Palaeoclimatol. Palaeoecol.* 63:183-199.

Behrensmeyer, A. K., and Dechant Boaz, D. (1980) The recent bones of Amboseli National Park, Kenya, in relation to East African paleoecology. In A. K. Behrensmeyer and A. P. Hill (eds.): *Fossils in the Making: Vertebrate Taphonomy and Paleoecology.* Chicago: University of Chicago Press, pp. 72-92.

Behrensmeyer, A. K., Gordon, K. D., and Yanagi, G. T. (1986) Trampling as a cause of bone surface damage and pseudo-cutmarks. *Nature* 319:768-771.

Behrensmeyer, A. K., Gordon, K. D., and Yanagi, G. T. (1989) Nonhuman bone modification in Miocene fossils from Pakistan. In R. Bonnichsen and M. H. Sorg (eds.): *Bone Modification.* Orono, Maine: Center for the Study of the First Americans, pp. 99-120.

Binford, L. R. (1981) *Bones: Ancient Men and Modern Myths.* New York: Academic.

Binford, L. R., and Bertram, J. B. (1977) Bone frequencies and attritional processes. In L. R. Binford (ed.): *For Theory Building in Archaeology.* New York: Academic, pp. 77-153.

Boaz, N. T. (1977a) Paleoecology of early Hominidae in Africa. *Kroeber Anthropol. Soc. Papers* 50:37-61.

Boaz, N. T. (1977b) Paleoecology of Plio-Pleistocene Hominidae in the Lower Omo basin, Ethiopia. Ph.D. dissertation, University of California, Berkeley.

Boaz, N. T. (1979) Early hominid population densities: New estimates. *Science* 206:592-595.

Boaz, N. T. (1985) Early hominid paleoecology in the Omo Basin, Ethiopia. In Y. Coppens (ed.): *L'Environnement des Hominidés au Plio-Pléistocène.* Paris: Fondation Singer-Polignac, Masson, pp. 279-308.

Boaz, N. T., and Behrensmeyer, A. K. (1976) Hominid taphonomy: Transport of human skeletal parts in an artificial fluviatile environment. *Am. J. Phys. Anthropol.* 45:53-60.

Brain, C. K. (1969) The contribution of Namib desert Hottentots to an understanding of australopithecine bone accumulations. *Scientific Papers of the Namib Desert Research Station* 39:13-22.

Brain, C. K. (1970) New finds at the Swartkrans australopithecine site. *Nature* 225:1112-1119.

Brain, C. K. (1981) *The Hunters or the Hunted? An Introduction to African Cave Taphonomy.* Chicago: University of Chicago Press.

Brown, F. H., and Feibel, C. S. (1988) "Robust" hominids and Plio-Pleistocene paleogeography of the Turkana Basin, Kenya and Ethiopia. In F. E. Grine (ed.): *Evolutionary History of the "Robust" Australopithecines.* New York: Aldine de Gruyter, pp. 325-342.

Ciochon, R. L. (1986) The Cercopithecid Forelimb: Anatomical Implications for the Evolution of African Plio-Pleistocene Species. Ph.D. dissertation, University of California, Berkeley.

Dechant Boaz, D. (1982) Modern Riverine Taphonomy: Its Relevance to the Interpretation of Plio-Pleistocene Hominid Paleoecology in the Omo Basin, Ethiopia. Ph.D. Dissertation, University of California, Berkeley.

de Heinzelin, J. (1971a) Omo Research Expedition—L338ss-P386. Excav. Marcel Splingaer—General Statement. Unpublished report.

de Heinzelin, J. (1971b) Omo Research Expedition—L338ss-P386. Excav. Marcel Splingaer—General Statement. Unpublished report.

de Heinzelin, J. (1971c) Omo Research Expedition—L338ss-P386. Excav. Marcel Splingaer—General Statement. Unpublished report.

de Heinzelin, J. (ed.) (1983) *The Omo Group. Stratigraphic and Related Earth Sciences Studies in the Lower Omo Basin, Southern Ethiopia.* (Contributions by R. Bonnefille, F. H. Brown, T. E. Cerling, Y. Coppens, R. Dechamps, G. G. Eck, A. Gautier, P. Haesaerts, J. de Heinzelin, F. C. Howell, R. T. Shuey, R. Stoops and B. Van Vliet) With a geological map of the Shungura Formation (color, 2 sheets, scale 1:10,000). *Musée Royal de l'Afrique Centrale, Tervuren, Belgique, Annales Sciences Géologiques,* no. 85.

de Heinzelin, J., and Haesaerts, P. (1971) Omo Research—L338ss-P386. Excav. Marcel Splingaer—Factual Description. Unpublished report.

de Heinzelin, J., and Haesaerts, P. (1983) The Shungura Formation. In J. de Heinzelin (ed.): *The Omo Group. Stratigraphic and Related Earth Sciences Studies in the Lower Omo Basin, Southern Ethiopia.* Musée Royal de l'Afrique Centrale, Tervuren, Belgique, Annales Sciences Géologiques, no. 85, pp. 25-127 (and appendices).

de Heinzelin, J., Haesaerts, P., and Howell, F. C. (1976) Plio-Pleistocene formations of the lower Omo basin with particular reference to the Shungura formation. In Y. Coppens, F. C. Howell, G. L. Isaac, and R. E. F. Leakey (eds.): *Earliest Man and Environments in the Lake Rudolf Basin.* Chicago: University of Chicago Press, pp. 24-49.

Dodson, P. (1973) The significance of small bones in paleoecological interpretation. *Contrib. Geol.* 12:15-19.

Eck, G. G., and Jablonski, N. G. (1984) A reassessment of the taxonomic status and phyletic relationships of *Papio baringensis* and *Papio quadratirostris* (Primates: Cercopithecidae). *Am. J. Phys. Anthropol.* 65:109-134.

Eck, G. G., and Jablonski, N. G. (1987) The skull of *Theropithecus brumpti* compared with those of other species of the genus *Theropithecus.* In Y. Coppens and F. C. Howell (eds.): *Les Faunes Plio-Pléistocène de la Vallée de l'Omo (Éthiopie).* Cercopithecidae de la Formation de Shungura.

Tome 3. Paris: Cahiers de Paléontologie, CNRS, pp. 18-122.

Feibel, C. S., Brown, F. C., and McDougall, I. (1989) Stratigraphic context of fossil hominids from the Omo Group Deposits: Northern Turkana Basin, Kenya and Ethiopia. *Am. J. Phys. Anthropol.* 78:595-622.

Fiorillo, A. R. (1984) An introduction to the identification of trample marks. *Curr. Research* 1:47-48.

Fiorillo, A. R. (1989) An experimental study of trampling: Implications for the fossil record. In R. Bonnichsen and M. H. Sorg (eds.): *Bone Modification.* Orono, Maine: Center for the Study of the First Americans, pp. 61-72.

Fiorillo, A. R. (1991) Taphonomy and depositional setting of Careless Creek Quarry (Judith River Formation), Wheatland County, Montana, U.S.A. *Palaeogeog. Palaeoclimatol. Palaeoecol.* 81:281-311.

Gifford-Gonzalez, D. (1989) Ethnographic analogues for interpreting modified bones: Some cases from East Africa. In R. Bonnichsen and M. H. Sorg (eds.): *Bone Modification.* Orono, Maine: Center for the Study of the First Americans, pp. 179-246.

Gifford, D. P., and Behrensmeyer, A. K. (1977) Observed formation and burial of a recent human occupation site in Kenya. *Quat. Res.* 8:245-266.

Hanson, C. B. (1980) Fluvial taphonomic processes: Models and experiments. In A. K. Behrensmeyer and A. P. Hill (eds.): *Fossils in the Making.* Chicago: University of Chicago Press, pp. 156-181.

Hare, P. E. (1980) Organic geochemistry of bone and its relation to the survival of bone in the natural environment. In A. K. Behrensmeyer and A. P. Hill (eds.): *Fossils in the Making.* Chicago: University of Chicago Press, pp. 208-219.

Haynes, G. (1983) Frequencies of spiral and green-bone fractures on ungulate limb bones in modern surface assemblages. *Am. Antiq.* 48:102-114.

Haynes, G. (1991) *Mammoths, Mastodonts, and Elephants: Biology, Behavior, and the Fossil Record.* Cambridge: Cambridge University Press.

Hill, A. P. (1975) Taphonomy of Contemporary and Late Cenozoic East African Vertebrates. Ph.D. dissertation, University of London, London.

Hill, A. P. (1979) Butchery and natural disarticulation: an investigatory technique. *Am. Antiq.* 44:739-744.

Hill, A. P. (1989) Bone modification by modern Spotted Hyenas. In R. Bonnichsen and M. H. Sorg (eds.): *Bone Modification.* Orono, Maine: Center for the Study of the First Americans, pp. 169-178.

Howell, F. C., and Coppens, Y. (1983) Introduction. In J. de Heinzelin (ed.): *The Omo Group. Stratigraphic and Related Earth Sciences Studies in the Lower Omo Basin, Southern Ethiopia.* Musée Royal de l'Afrique Centrale, Tervuren, Belgique, Annales Sciences Géologiques, no. 85, pp. 1-5.

Howell, F. C., Haesaerts, P., and de Heinzelin, J. (1987) Depositional environments, archeological occurrences and hominids from Members E and F of the Shungura Formation (Omo basin, Ethiopia). *J. Hum. Evol.* 16:665-700.

Johanson, D. C., Splingaer, M., and Boaz, N. T. (1976) Paleontological excavations in the Shungura formation, lower Omo basin, 1969-1973. In Y. Coppens, F. C. Howell, G. L. Isaac, and R. E. F. Leakey (eds.): *Earliest Man and Environments in the Lake Rudolf Basin.* Chicago: University of Chicago Press, pp. 402-420.

Klein, R. G., and Cruz-Uribe, K. (1984) *The Analysis of Animal Bones from Archaeological Sites.* Chicago: University of Chicago Press.

Korth, W. W. (1979) Taphonomy of microvertebrate fossil assemblages. *Ann. Carnegie Mus.* 48:235-285.

Leakey, M. (1987) Colobinae (Mammalia, Primates) from the Omo Valley, Ethiopia. In Y. Coppens and F. C. Howell (eds.): *Les Faunes Plio-Pléistocène de la Vallée de l'Omo (Éthiopie). Cercopithecidae de la Formation de Shungura.* Tome 3. Paris: Cahiers de Paléontologie, CNRS, pp. 149-169.

Leopold, L. B., and Wolman, M. G. (1957) River channel patterns: Braided, meandering and straight. *U. S. Geol. Surv. Prof.* Papers 282B:39-85.

Lyman, R. L. (1984) Bone density and differential survivorship of fossil classes. *J. Anthropol. Archaeol.* 3:259-299.

Lyman, R. L. (1985) Bone frequencies: Differential transport *in situ* destruction, and the MGUI. *J. Archaeol. Sci.* 12:221-236.

Lyman, R. L., and Fox, G. L. (1989) A critical evaluation of bone weathering as an indication of bone assemblage formation. *J. Archaeol. Sci.* 16:293-317.

Meikle, W. E. (1982) Population Studies of Plio-Pleistocene *Theropithecus.* Ph.D. dissertation, University of California, Berkeley.

Myers, T. P., Voorhies, M. R., and Corner, R. G. (1980) Spiral fractures and bone pseudotools at paleontological sites. *Am. Antiq.* 45:483-490.

Oliver, J. S. (1989) Analogues and site context: bone damages from Shield Trap Cave (24CB91), Carbon County, Montana, U.S.A. In R. Bonnichsen and M. H. Sorg (eds.): *Bone Modification.* Orono, Maine: Center for the Study of the First Americans, pp. 347-379.

Rak, Y. (1983) *The Australopithecine Face.* New York: Academic.

Rak, Y., and Howell, F. C. (1978) Cranium of a juvenile *Australopithecus boisei* from the lower Omo basin, Ethiopia. *Am. J. Phys. Anthropol.* 48:345-365.

Schick, K. (1984) Processes of Palaeolithic Site Formation: An Experimental Study. Ph.D. dissertation, University of California, Berkeley.

Schick, K. (1986) *Stone Age Sites in the Making: Experiment in the Formation and Transformation of Palaeolithic Occurrences.* Oxford: BAR, International Series No. 319.

Scott, K. (1989) La Cotte de St. Brelade, Jersey, Channel Islands. In R. Bonnichsen and M. H. Sorg (eds.): *Bone Modification.* Orono, Maine: Center for the Study of the First Americans, pp. 335-346.

Shipman, P. (1981) Applications of scanning electron microscopy to taphonomic problems. *Ann. NY Acad. Sci.* 376:357-386.

Shotwell, J. A. (1955) An approach to the paleoecology of mammals. *Ecology* 36:327-337.
Shotwell, J. A. (1958) Inter-community relationships in Hemphillian (mid-Pliocene) mammals. *Ecology* 39:271-282.
Voorhies, M. R. (1969) *Taphonomy and Population Dynamics of an Early Pliocene Vertebrate Fauna, Knox County, Nebraska.* Contrib. Geol. Special Paper #1. Laramie: University of Wyoming Press.
Wells, C. (1967) Pseudopathology. In D. Brothwell and A. T. Sandison (eds.): *Diseases in Antiquity.* Springfield, Ill.: Charles C. Thomas, pp. 5-19.

CHAPTER 20

Palynology and Paleoenvironment of East African Hominid Sites

Raymonde Bonnefille

AVANT-PROPOS

Although I have not been involved in the field of paleoanthropology during the last 10 years, it is a great honor and a great pleasure for me to contribute to this celebration honoring Dr. F. Clark Howell. Clark asked me to join the International Omo Expedition in 1972. I had just begun research on "Palynology and Paleoecology of the Plio-Pleistocene in Ethiopia" and I had already participated in the two previous field seasons of "La mission paléontologique française dans la Basse vallée de l' Omo." Working on pollen studies in the area was exciting because it was completely new, but finding fossil pollen preserved in the sediments from the outcrops was a very tedious task. Clark was convinced that a multidisciplinary approach would bring new insight in understanding the environment of the early hominids. His enthusiasm and his strength induced me to continue looking for direct evidence of past vegetation. The several weeks of fieldwork during these years and the stimulating workshops that followed them were among my greatest scientific experiences. Clark's encouragement and advice certainly induced and guided my further research. I will always remember Clark working under a tent, in the Omo, despite the great heat, carefully identifying and labeling any single faunal remain that had been found by the collecting team during the day. At dusk, he would say; "time for a drink, now," and I appreciated the brief but always stimulating talks around the dinner table when each team member reported about the news and discoveries of the day. What mattered, to him, was to bring new evidence, new data. "If you have to do it, do it now" was his greatest lesson and one that I will never forget.

For showing us the truth, for the luck of those days when it was such hard work to discover any piece of knowledge, but also for the excitement and great pleasure that you gave us, Clark, thank you.

INTRODUCTION

The timing of appearance and extension of grasslands in East Africa is important to understand how and when early hominids evolved from hominoids in the middle to the late Miocene (Kennedy 1978; Nesbit-Evans et al. 1981; Pickford 1983; Hill and Ward 1988). It is critical also to understand if the emergence of the genus *Homo,* around 2.5 million years ago would have coincided with an arid period and the spread of the savanna (Tobias 1991).

The paleoecological interpretation of fossil Miocene sites in East Africa has long been a subject of debate. One point of view holds that forest and woodland dominated the landscape (Andrews and Nesbit-Evans 1979; Pickford 1987), while another favors savanna or wooded grassland (Shipman 1986; Retallack et al. 1990). The paleoecological reconstitutions are based on three different approaches. The first has been to study the fauna and to interpret the past vegetation from assumptions made about the relationship of the faunal assemblages (Shipman 1986; Pickford 1987) and morphological characters of the fossils such as Bovids to their present day habitats (Kappelman 1991). A second approach has used the direct evidence from macrobotanical remains, fossil wood (Dechamps et al. 1992), and pollen (Bonnefille 1984, 1985). A third approach, developed more recently, is based on stable carbon isotopic composition of pedogenic carbonate and organic matter from soils and paleosols (Cerling 1984; Cerling et al. 1991). The conclusions drawn on isotopic data (Cerling 1992) are in fairly good agreement with pollen results obtained some years ago at the various East African hominid sites (Bonnefille and Vincens 1985). New results have come out (Yemane et al. 1987), which therefore now deserve additional comments.

PRELIMINARY REMARKS

Botanical Information from Pollen Data

Pollen data provide three kinds of information. The first concerns the floral taxonomical composition of the past vegetation, pollen identification being generally confident at the generic level. Second, the distribution of arboreal pollen (AP) against nonarboreal pollen (NAP) in a pollen spectrum leads to some estimate of the total tree cover, under the condition that differential pollen production is taken into account. Third, local contributors can be discriminated concerning the sedimentological context at each site.

It is a fact that we now have a much better knowledge of the present pollen rain than in the course of earlier studies (Bonnefille 1984, 1985; Bonnefille and Vincens 1985). The latter information has been obtained from the analysis of several hundred sites. Such an extensive modern pollen data set documents the great diversity of the plant communities existing in the intertropical region (Bonnefille et al. 1992). Furthermore, a new well-documented fossil pollen diagram has been provided for upper Miocene lignite in Ethiopia (Yemane et al. 1985; Yemane et al. 1987). A mid-Miocene leaf flora from Kenya was also carefully identified (Jacobs and Kabuye 1987) and abundant fossil wood and fruits collected from neogene deposits in Kenya (Coppens et al. 1988) and in Uganda (Dechamps and Maes 1990; Dechamps et al. 1992). But the investigation for fossil pollen, recently pursued in the Western Rift Valley (Senut et al. 1990; Pickford et al. 1991) had very little success (Nakimera-Ssemmanda 1991).

Vegetation Types and Botanical Terminology

Vegetation may be classified in several ways. It may be classified by the direct reference to the physiognomic aspect (Keay 1959); the ecological relationship between vegetation and environment or climate (Walter 1971); the geographical distribution of species composition (White 1983); or the relationship between plant form, vegetation structure and climatic conditions (Box 1981). Therefore, the diversity in nomenclature of the vegetation types in Africa is notoriously great. As a result of an international collaboration, the first version of the vegetation map of Africa (Keay 1959) coordinates French and English terms, according to the international recommendations of the Yangambi meeting (1956).

The Use of the Term "Savanna"

The term "savanna" (savane) is used for a vegetation in which perennial grasses, at least 80 cm high, are usually burnt. The term "steppe" (steppe) is used for vegetation in which annual plants are often abundant, and perennial grasses less than 80 cm high. Steppe vegetation predominates in the drier areas. But none of these criteria can be recognized in the fossil data. Moreover, while considering the use of the term savanna, a confusion may arise from the fact that savanna is considered as a derived vegetation, mainly produced by human activities, in which grazers play an important role. Despite these controversial point of views, the term savanna is still used for designating, in spatial terms, the broad transition zone between the closed tropical forests and the open desert steppes (Bourlière 1983).

In the temperate zone, we find either pure grassland (prairie) or woodlands (forest), but in the tropical regions, the term savanna, used in its broad sense, includes a continuum of physiognomic types of vegetation in which the relative density of trees and shrubs versus grass cover can show great variations, from woodland to grassland (Menaut 1983). In the tropical region, the savanna occupies areas under summer rain. The distribution of grasses and woody plants depends upon climate, and in a lesser proportion upon soil.

Broadly speaking, it is correct to summarize the information given by the fossil pollen data in saying that "early hominids were living in a savanna environment." But this is a very approximate term to describe the great varieties of habitats and the wide range of climatic conditions that it can include.

The fossil pollen record provides more precision that can be kept by using the appropriate descriptive terminology using the relative density of trees versus grasses or herbaceous plants (Pratt et al. 1966; White 1983). This classification has the advantage of being statistically evaluated by pollen counts. It is also directly comparable to the relative proportion of C_3 (trees) versus C_4 (tropical grasses) plants determined by isotopic analysis (Cerling 1984).

The Vegetation Map of Africa

From the ecological point of view, the savanna is found under a continuum gradient of climatic regimes. Mean annual rainfall can vary from 500 to 1200 mm, and the dry season can last from 3 to 9 months. Steppes occur in arid or semiarid climate, under rainfall less than 500 mm and a dry season of more than 7 months (Trochain 1980). From the floral point of view, the transition region between the equatorial rainforest and subdesert steppe is very rich in species. It has been subdivided into five distinct units (II-III-IV-XI-XII), as illustrated in Figure 20-1. These subdivisions are mainly based on the taxonomical composition of the flora, and on the distribution of the plants. Each defined unit has a total of more than 1000 endemic species, and more than 50 percent of its species are confined to each unit (White 1983). The Guineo-Congolian region (I) extends as a broad band north and south of the equator. The Zambezian region (II) has the richest flora; it includes forest, deciduous, and evergreen woodland, mosaic of woodland and grassland and so on. In the Sudanian region (III), there is no true forest, apart of swamp and riparian forest. The Somali-Masai region (IV) is occupied by deciduous Bushland and steppe, in which *Acacia* and *Commiphora* are frequent. The area around the lake Victoria (XII) contains a mixed flora with elements of the units I, II, III, VIII; it is called a mosaic (White 1983). The regions X, XI are transitional. A special unit concerns the homogenous vegetation found everywhere in Africa at altitudes above 1300 m that consequently is called Afromontane (VIII). It can occur in any of the other units, even on an isolated mountain.

Looking at Figure 20-1, it is clear that East African hominid sites, located in the Rift valleys, extend from the eastern margin of the equatorial rainforest in the south, near lake Albert, to the subdesert steppe in northern Kenya and Ethiopia. Although the landscape at Laetoli, Olduvai, Omo, Turkana, and Hadar is dominated by the typical thorntree (*Acacia*) and may exhibit physiognomic resemblances, the plant associations are florally different. It is noteworthy that these sites, located at various altitudes, do not

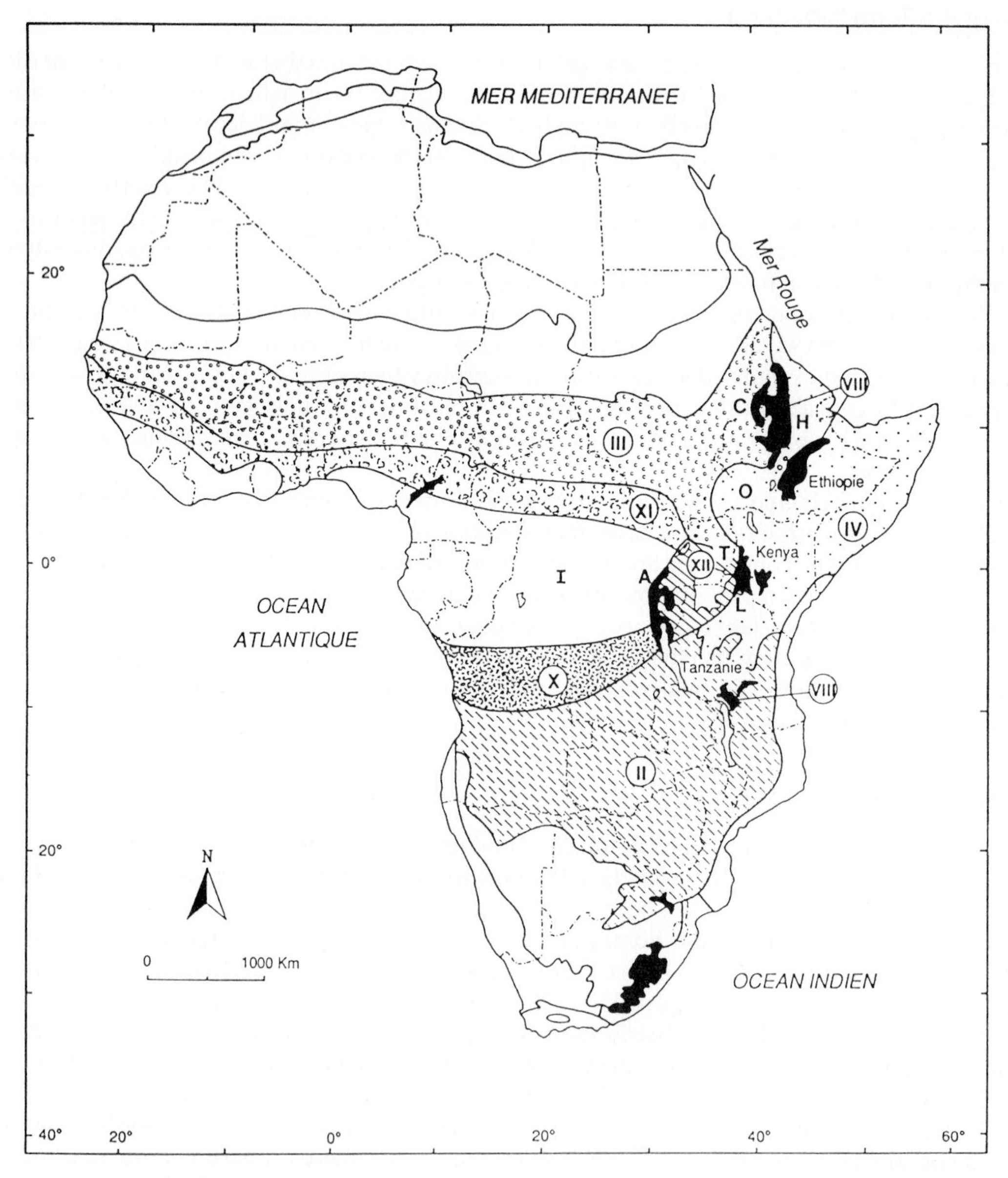

I	Guineo-Congolian Region	II	Zambezian Region
III	Sudanian Region	IV	Somalia - Masai
VIII	Afromontane Region (in black)	X et XI	Transition zone
XII	Lake Victoria regional mosaix		

(each unit is defined on richness of endemic species)

Palynological sites : C: Chilga ; H : Hadar ; O : Omo/Turkana ; T : Fort Tenan ; L : Laetoli ; A : Lake Albert area.

FIGURE 20-1 Mapping units, simplified vegetation map (after White 1983). Each unit is defined by richness of endemic species. I: Guineo-Congolian Region. II: Zambezian Region. III: Sudanian Region. IV: Somalia- Masai. VIII: Afromotane Region (in black). X and XI: Transition zone. XII: Lake Victoria Regional Mosaic.

experience the same climatic conditions, although they yield habitats for similar fauna. Variations in vegetation types as well as in climatic conditions and therefore in available habitats should also have existed in the past even in the Miocene (Axelrod and Raven 1978). Moreover in each of the broad scale phytogeographical units (II, III, IV, and XII) defined in the Savanna biome (Figure 20-1), several ecosystems, ranging from forest to woodland, wooded grassland, and grasslands can exist under the same global climatic conditions. For example, in the Omo Turkana basin, an *Acacia* steppe occupies the plains, whereas a riverine forest with tall trees is found in the vicinity of the Omo river. Short grasslands are widespread on the flat lands surrounding the lake, and an impoverished montane forest occurs on Mont Kulal. The diversity in local vegetation type provides the diversity in habitats and this has to be taken into account when the past environment is reconstructed.

Paleoenvironmental Reconstruction

While interpreting the pollen data in terms of environment, we should be aware that plant taxa could have been, in the past, associated differently from what is observed now. Another limitation to such interpretation, and not the least, comes from the fact that botanical knowledge on plant distribution concerns the species, which we do not know by looking at pollen grains. Paleodata found at one single site do not provide information on the geographical diversity. In extrapolating this interpretation to a larger geographical area, there is a great risk of minimizing the real diversity. We have to reconstruct past vegetation considering geographical patterns through time, and this needs a lot of paleobotanical data, carefully identified and precisely located in the stratigraphy. Taking into account all the well-known limitations of interpreting the pollen data, the fossil pollen assemblages or spectra, provide significant relative or absolute frequencies of several tens of identified pollen taxa for each sample. These are statistical data thoroughly more meaningful than single occurrences. Despite the effort that was made, the Plio-Pleistocene pollen record remains too discontinuous to provide by itself an unequivocal reconstruction of the paleoenvironment of early hominid sites. It becomes evident that any reconstruction of the past environment would have to include other information from sedimentations, fauna, geochemical analysis and so on.

In following such a multidisciplinary approach, we would like to draw attention to the fact that different data do not necessarily provide directly equivalent information. There are two main reasons for that. One is the differential time span included in the sampling of the different fossil indicators, for example, the information given by pollen vis à vis micro or macro fossil faunal assemblages. Another bias of the interpretation may arise from the differential responses of the various biological markers to the climatic variations. Changes in the vegetation can be short-term changes, such as documented on the few thousand or the hundred thousand years cycle. Faunal changes are long-term changes that could only be compared to a general trend over a long period of time, if we had enough information on the past vegetation. It is the aim of this chapter to reexamine the available pollen data with this awareness in mind.

FORT TERNAN POLLEN SPECTRUM

Some years ago, sediment samples from Miocene sites in East Africa were investigated in order to find preserved fossil pollen. These attempts made at Songhor, Fort Ternan, Ngorora, Baringo, and more recently in the lake Albert area (Pickford et al. 1991) were not successful. Although more than 100 samples were processed, a single pollen spectra was obtained from a sample taken at the base the upper Chogo paleosol that is exposed in the 8-m–thick sequence of the excavation wall, at Fort Ternan National Monument where *Kenyapithecus wickeri* was discovered. The pollen spectra has been completed from Bonnefille (1984). Information on the life form and on the photosynthetic pathway of the plants corresponding to the identified pollen taxa have been added (Table 20-1).

TABLE 20-1 Pollen spectra from the Upper Chogo Paleosol at Fort Ternan.

Pollen Taxa	N	Counts	%	AP	NAP	AFROM	U	C_3	C_4
Acalypha (Euphorbiaceae)		3			x	x		x	x
Amaranthaceae/Chenopodiaceae	8		2.5		x		x		
Anthospermum (Rubiaceae)	1				x	x			
Artemisia (Compositae)	1			x	x	x			
Brucea (Simarubaceae)		2			x		x		x
Celtis (Combretaceae)	4		1.2	x		—		x	
Combretaceae	1			x		—			
Compositae tubuliflorae	7		2		x		x		
Croton (Euphorbiaceae)		1			x	x		x	
Cyperaceae	92		27		x		x	x	x
Gramineae	176		52		x		x	(x)	x
Juniperus (Cupressaceae)	6		1.7	x		x		x	
Lannea (Anacardiaceae)	1			x		—		x	
Olea africana	2			x		x		x	
Plantago coronopus (Plantaginaceae)	1				x	x			
Podocarpus (Podocarpaceae)	9		2.6	x		x		x	
Potamogeton pectinatus (Potamogetonaceae)	1					x		x	
Spores monoletes (Pteridophyta)	4		1.2		x		x		
Typha (Typhaceae)	2				x		x	x	
Urticaceae	9		2.6		x	x			
Unidentified	4		1.7						
TOTAL	335								

SOURCE: C3/C4 attributions of plants after Cerling et al. (1991) and Aucour et al. (in press); AP (arboreal pollen) includes trees and shrubs; NAP = herbaceous plants; AFROM = afromontane forest element (White 1983); U = ubiquitous.

Information on the Local Vegetation

The identification of the aquatic plant *Potamogeton*, and the record of algae attributed to *Botryoccocus* are clear and coherent indications that a swamp with permanent water was existing at the sampling site, contemporaneously with Ramapithecine remains.

The open character of the vegetation is clearly indicated by the significant 52 percent grass pollen. The occurrence of grasses is now well confirmed at the site by the identification of abundant macroremains (Retallack et al. 1990). Cuticular morphology suggests that the grasses belong to the Panicoideae and

Chlorideae, both subfamilies including C_4 grasses of savanna grassland (Smith 1982). *Typha*, a tall C_3 reed of wet bottom lands at various altitudes has also been identified, but its pollen percentage, much lower than that for the other grasses, indicates that it would not be attributed to herbaceous plants; the Cyperaceae (sedges) represent 27 percent of the total pollen count. They include C_4 genera, such as *Cyperus papyrus* and C_3 species as well (Aucour et al. in press).

The grasses could have existed around the swamp, but unless they did not contribute to the production of the organic matter of the paleosol, the most abundant ones cannot be C_4 plants. The stable carbon isotopic composition of paleosol gives strongly negative values which indicate local ecosystems dominated by C_3 plants (Cerling 1992). This excludes *Leersia, Loudetia, Oryza* that are C_4 and can make savanna grass swamp today in Africa (Thompson and Hamilton 1983). Among the aquatic grass *Miscanthidium* (or *Miscanthus*, Clayton and Renvoise 1982) are a widespread component of African highland bogs that can have extensive cover on swamps in Africa. They belong to the Andropogoneae subfamily and have a C_4 photosynthetic pathway (Aucour et al. in press).

In conclusion, the great pollen percentages of grasses and sedges indicate their local occurrence near the site. The plants dominating the grassland that locally surrounded the swamp had to be C_3 and therefore had more taxonomical affinities with the modern component of the highland meadows than with the tropical savanna (Livingstone and Clayton 1980).

The Miocene Pollen Flora

Another remarkable feature of the fossil pollen assemblage is the attribution of most of the fossil arboreal pollen taxa to the Afromontane phytogeographical element of the African flora (White 1983). Among the conifers, *Podocarpus, Juniperus* (cedar tree), and *Olea* (olive tree) are presently common components of the driest montane forest in East Africa. Together with *Anthospermum* and *Plantago, Artemisia,* such an association characterizes pollen assemblages of highlands above 2000 meters altitude in East Africa (Bonnefille and Buchet 1987). *Celtis* (1.2%), and Combretaceae (single occurrence) are more frequent in deciduous woodlands found between 1500 and 1800 m altitude (Friis 1991). *Lannea* a widespread genus in the tropics, has 17 species distributed in East Africa, in deciduous woodland, wooded grassland, river valleys (Kokwaro 1986), and in dry evergreen and riverine forests, but also in lowland and upland rain-forests, from 300 to 2400 m (Polhill 1966). *Acalypha,* which can be herb, shrub, or trees (Livingstone 1967), are found in forests edges (Kendall 1969), humid forests from 1300 m to 2600 m in Ethiopia (Friis 1991), on Kilimanjaro (Livingstone 1967), and in riparian forest from lowlands. Their pollen are common in surface samples from mid-altitude semideciduous forest on the eastern escarpment of Lake Albert, at 1200 m (Nakimera-Ssemanda 1991). Urticaceae include 15 genera of nettles reported in East Africa, from 100 to 4000 m altitudes. Some species grow along streams in savanna, montane sclerophylle and humid forests, the bamboo zone, and the ericaceous belt. They are conspicuously abundant in the forests of East African mountains (Friis 1991). Most plants of the Urticaceae family have well-dispersed small pollen grains (Hamilton 1972). The plants would grow in any forest that provided the proper moist, shady environment (Livingstone 1967). They are particularly common between 2000 to 3000 m, especially in moister areas (Hamilton 1982).

In conclusion, the arboreal pollen spectra of the Fort Ternan sample indicates that a woodland or a forest was existing in the vicinity of the site. But the environment at the site itself could not have been a forest. Modern samples taken from lakes or swamp surrounded by a forest show 60 to 70 percent of AP (Bonnefille and Buchet 1987), whereas the total proportion of arboreal pollen at Fort Ternan is less than 10 percent. This forest had to be open in order to allow grasses to grow. Its floral composition includes trees and the herbaceous plant taxa that are exclusively distributed in the montane vegetation of the Afromontane phytogeographical unit. This indicates that the Afromontane flora is of ancient origin, at least from the Miocene.

Comparison with Other Paleoenvironmental Indicators

Pollen indicators of the equatorial rainforest have now been identified in Cameroon (Brenac 1988), in Gabon (Jolly 1987) as well as in the Congo (Elenga 1992). None of them is present in the Fort Ternan pollen spectra. Among the 500 pollen spectra that constitute the modern pollen data set compiled for the main ecosystems presently recorded in East Africa (Bonnefille et al. 1992), the closest analogues are to the Fort Ternan paleosols, compared to those of modern soils (Cerling et al. 1991). Any other deduction based on past climate would be highly hypothetical, because we have no information on the Miocene topography. By extrapolating the present situation, we may come to an erroneous interpretation. The occurrence of a montane forest could be understood in view of the assumption that a volcano existed nearby (Shipman et al. 1981). This could explain the low value of the isotopic composition of soil organic matter that averages minus 27 percent and indicate a C_3 plant ecosystem (Cerling et al. 1991).

A mosaic distribution of patches of montane forest and grassland is a typical feature of the landscape on the high plateau of Malawi (Chapman and White 1970). It is also found in Southwest Congo, on the Bateke plateaux, at 800 m altitude. In this region, depressions of a few kilometers wide, occupied by a mosaic of wooded grassland, are surrounded by Loudetia grassland. Pollen and isotopic studies, undertaken in those depressions, have shown that the organic matter deposited during the last thousand years registers ^{13}C values that average 28 percent, although a proportion of 40 percent grass pollen versus 35 percent arboreal was recorded in the same sediments (Elenga 1992). Perhaps such a situation offers a modern example of the controversial results between botanical and isotopic data found at Fort Ternan (Cerling 1991; Retallack et al. 1990). Presently, the Bateke plateaux of Congo receives 1600 mm mean annual rainfall and has a three-month dry season. But the trees that are found there today, belong to the equatorial Guineo-Congolese flora, which has no representative in the paleobotanical data at Fort Ternan.

In conclusion, pollen, fossil leaves, paleosols, and isotopic studies carry the evidence that grasses were present and locally abundant at Fort Ternan, 12 MYA. But they had to be C_3 plants. From the faunal and the pollen data it is clear that both forest trees and open vegetation were available in the vicinity of the site, where Ramapithecines were buried. Such a vegetation pattern would fit the attribution of a mosaic-like vegetation inferred by the study of paleosoils (Retallack et al. 1990). However, the C_4 grasses that constitute extensive grassy plains in East Africa today had not yet appeared. During the mid-Miocene time, grasslands may have existed. We do not know how extensive they could have been. They were not similar to the modern "savanna biome."

CHILGA POLLEN DIAGRAM, ETHIOPIA

A very abundant and well-diversified pollen and spore flora was extracted from the lignite deposited at Chilga, a lacustrine basin in the northwestern Ethiopian highlands, 12° N, 37′ E, to the north of Lake Tana (Yemane et al. 1985). The basalt layer, at the bottom of the basin was dated at 8 MYA (Yemane et al. 1987). In the well-preserved and abundant assemblages of fossil pollen, *Afrocrania, Brachystegia, Isoberlinia*—distinctly recognizable pollen—were referred to as exotic, to the modern Ethiopian flora. It is true that their present geographical distribution does not reach Ethiopia. Among the Legume family, with microphylle leaves, *Brachystegia* and *Isoberlinia* have common species represented by big trees that are widely spread in the deciduous woodland of the Zambezian region (unit II of Figure 20-1). This vegetation, sometimes called by its vernacular name "Miombo," is characteristic of the tropical region, under strongly seasonal climate, south of the Equator (White 1983). *Isoberlinia* and *Brachystegia* have also been identified by fossil wood or leaves in a Miocene flora from the Lake Albert area (Lakhampal and Prakash 1970; Chaney 1933). These well-documented occurrences attest to the establishment of a seasonal climate in East Africa for 20 million years. The Chilga pollen diagram indicates that a climate with a strongly seasonal distribution of rainfall was established in Ethiopia 8 million years ago.

In our present understanding of the climate mechanism in East Africa, the dry season is being reinforced by the northeastern monsoonal wind originated in the high pressure center of the high plateau of Tibet (Rudiman and Kutzbach 1991). The occurrence of a strongly seasonal upper Miocene climate, inferred from the Chilga pollen record, would be in agreement with a maximum elevation of the Tibetan plateau, already achieved 8 MYA (Harrison et al. 1992).

The percentage of grass pollen in the Chilga diagram never exceeded 20 percent . This implies more than 80 percent of arboreal pollen, an indication of a forest environment. The great abundance of fern favors the interpretation of a humid climate. But the trees are represented by genera that can be found now, both in the humid montane forests and in deciduous woodland. But again, none of the lowland rainforest indicators from West Africa (Elenga 1992; Jolly 1987; Brenac 1988) has been identified, although the total pollen count on the whole diagram is really significant. The great abundance of fern spores, together with the occurrence of indicators of humid forest such as *Alchornea, Mitragyna,* Sapotaceae that have not been found in any of the Plio-Pleistocene pollen assemblages (Bonnefille and Vincens 1985) points to climatic conditions certainly more humid during the Upper Miocene. Pollen mixing could have arisen, either by sampling several vegetation types present in the basin—a plausible fact if relief and subsident basin were both present in the catchment area. Nothing is known about the Miocene paleogeography, except that the paleoaltitude of the Ethiopian plateau could have been much lower than 2000 m (Yemane et al. 1985). It has also been established that both pollen evidence and changes in the sediments, notably clay mineral composition, indicate possible climatic changes through the Miocene time as well. The strong fluctuations lead to the conclusion of rapid climatic fluctuations in the upper part of the Chilga sequence (Yemane et al. 1987). To what extent these variations reflect tectonic changes or climatic variations or both would need a better control of the time span covered by the deposits and a closer sampling resolution for the pollen sequence. On the other hand, alternatively drier and humid conditions could well be related to the climate changes known to have occurred at 7 to 8 MYA.

In conclusion, the Chilga fossil pollen data indicates a climate much more humid during the Upper Miocene than later in the Plio-Pleistocene. But there is no evidence of an equatorial primary rainforest as far as 12° North, 8 MYA. The evidence for a seasonal deciduous woodland is coherent with older paleobotanical identifications from Miocene sites in Kenya and Uganda. It confirms that a seasonal climate was already established in East Africa. In order to explain this, the duration of the dry season should have been much shorter during the Upper Miocene than in the present day. The abundance of fern spores in the fossil Chilga record has no equivalent in the modern pollen analyses from the montane forest in Ethiopia today (Bonnefille and Buchet 1987; Bonnefille 1985). We note that grass pollen is present but their low percentages exclude the existence of extensive grassland on the Ethiopian Plateau, although this does not mean that they could not have existed elsewhere.

THE PLIO-PLEISTOCENE RECORD

Regarding the stratigraphical location of the Plio-Pleistocene pollen spectra, additional and more precise work has been added since the pollen studies were carried out (Feibel et al. 1989; Harris 1991; Walter et al. 1991). However, there is, up to now, little change (de Heinzelin in press) from the previously published data at Omo (Bonnefille and Dechamps 1983), Omo/Turkana (Bonnefille 1984, 1985; Bonnefille and Vincens 1985), at Olduvai, Laetoli (Bonnefille and Riollet 1987), and at Hadar (Bonnefille and Buchet 1987).

Regarding the interpretation of the gathered data, the attribution of the results to different vegetation types can be summarized as shown in Figure 20-2.

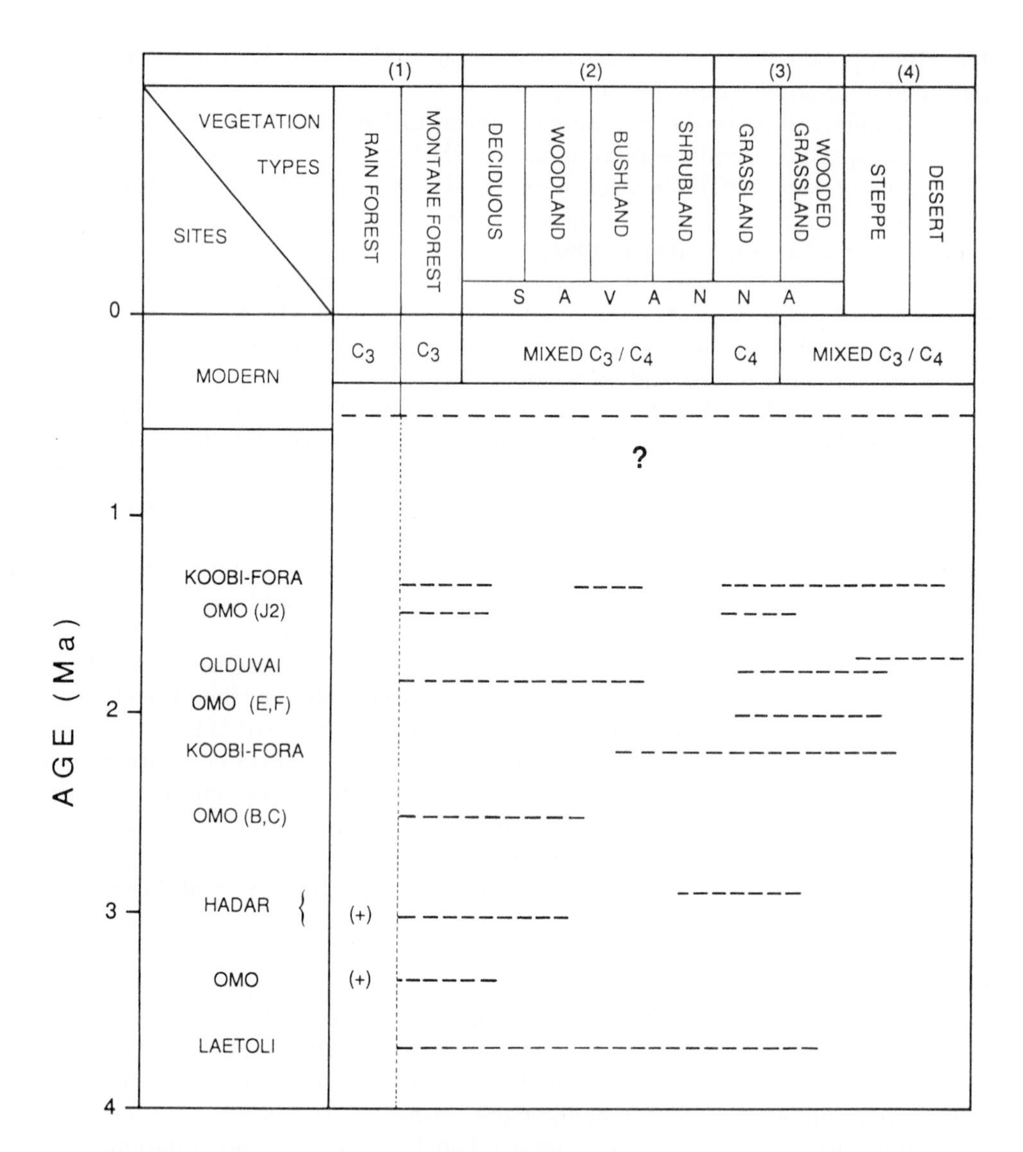

Figure 20-2 Vegetation types inferred from pollen studies at East African fossil localities. Isotopic composition of modern soils after Cerling (1992).
The main vegetation types based on physiognomy (after Pratt et al. 1966; White 1983):
1. Forest: a closed stand of trees of several strata, large trees 40 to 60 m high, with an interlaced canopy. Montane forest is an evergreen type of forest with a smaller height of trees and a homogenous composition everywhere above 1300 m.
2. Woodland: a stand of trees up to 18 m height with an open or continuous canopy cover of more than 20 percent (Pratt et al. 1966); at least 40 percent (White 1983) grasses and herbs dominate the ground cover. Subtypes are defined by reference to the genera of the dominant trees. Forest and woodland can either be evergreen or deciduous. In Bushland, 40 percent of the land is covered by bushes. Shrubs that vary in height from 10 cm to 2 m or more.
3. Grassland: land dominated by grasses and occasionally other herbs, sometimes with widely scattered or grouped trees and shrubs; canopy cover not exceeding 2 percent , usually subject to burning.
4. Steppe: vegetation with short grass, abundant annual plants between widely spaced perennial herbs, and scattered trees (*Acacia* or *Commiphora*) being the most common trees in East Africa.
N.B. In a broad sense, the term savanna includes all categories between woodland and desert.

CONCLUSION

We would like to focus on a few selective points that have not been emphasized yet.

The information that has been gathered up to now covers the period between less than 4 to about 1.3 MYA. But nothing is known for the last million years. We know that climatic changes have occurred during that period. They should have modified the distribution of vegetation types and also the distribution of species within the broad-scale vegetation unit. Most probably, long glacial periods favor the expansion of grassland, of arid plants, and of C_4 plants. The paleobotanical record is still discontinuous, and the missing gaps may be strongly misleading our interpretation in evaluating the impact of climatic changes on the environment. One important conclusion is the good agreement between the pollen and the isotopic results, which both reflect direct changes in the vegetation cover. An overall aridity trend from the mid Miocene documented from the isotopic record is not contradictory with paleobotanical data.

In the Plio-Pleistocene, there is no proof for an existence of an equatorial rainforest at the sites themselves. The occurrence of isolated elements from this forest such as fossil fruits or fossil leaves or fossil wood have to be taken as indicating floral affinities with the Guineo-Congolian vegetation unit. But they cannot attest to an existing rainforest ecosystem proper. The most plausible explanation is that Guineo-Congolese species may have persisted in a well diversified riverine or fringing forest that could have occurred, in the past, near the great river Omo, or close to the lake, at Hadar.

Aridity events are noticed several times, when vegetation varies from woodland to subdesert steppe. This means an increasing length of the dry season, important for survival of living individuals. They have been shown in all the hominid sites. Their greatest frequencies are around 1.8 MYA and they can be synchronous. But how environmental changes have affected the evolution of hominids will stand as a matter of assumption until we gather more complete paleobotanical data.

This overview emphasizes that two or three distinct vegetation units—the montane forest, the savanna *sensu lato*, and the riverine forest—all occur in investigated sites where fossil hominid remains have been collected. These offer several ecological niches and indicate that more than one habitat could have been available for early hominids. This possibility should certainly receive more attention from the paleoanthropologists.

ACKNOWLEDGMENTS

We thank J. de Heinzelin, E. Vincent, T. Cerling, and L. Beaufort for their critical comments on the manuscript.

LITERATURE CITED

Andrews, P., and Nesbit-Evans, E. (1979) The environment of *Ramapithecus* in Africa. *Paleobiology* 5:22-30.

Aucour, A. M., Bonnefille, R., and Hillaire-Marcel, C. (in press) Late Quaternary biomass change, a record from a highland peatbog (Equatorial Africa). *Quat. Res.*

Axelrod, D. I., and Raven, P. H. (1978) Cretaceous and Tertiary vegetation history in Africa. In M. J. A. Werger, J. Balesdent, A. Mariotti, and B. Guillet (eds.): *Biogeography and Ecology of Southern Africa*. The Hague: W. Junk, pp. 77-130.

Bonnefille, R. (1984) Cenozoic vegetation and environments of early hominids in East Africa. In R. O. Whyte (ed.): *The Evolution of the East Asian Environment. II Palaeobotany, Palaeozoology, and Palaeoanthropology*. Hong Kong: University of Hong Kong; Centre for Asian Studies, pp. 579-612.

Bonnefille, R. (1985) Evolution of the continental vegetation: The paleobotanical record from East Africa. *S. Af. J. Sci.* 81: 267-270.

Bonnefille, R., and Buchet, G. (1987) Contribution palynologique à l'histoire récente de la forêt de Wenchi (Ethiopie). *Mém. Trav. E.P.H.E.* 17:143-158.

Bonnefille, R., Chalié, F., Guiot, J., and Vincens, A. (1992) Quantitative estimates of full glacial temperatures in Equatorial Africa from palynological data. *Climate Dynamics* 6:251-257.

Bonnefille, R., and Dechamps, R. (1983) Data on fossil flora. In J. de Heinzelin (ed.): *The Omo Group, Archives of the International Omo Research Expedition*, Vol. I. Mus. R. Afr. Centr. Tervuren, pp. 191-207.

Bonnefille, R., and Riollet, G. (1987) Palynological spectra from the upper Laetoli beds. In M. D. Leakey and J. M.

Harris (eds.): *The Pliocene sites of Laetoli, Northern Tanzania*. Oxford: Oxford University Press, pp. 52-60.

Bonnefille, R., and Vincens, A. (1985) Apport de la palynologie à l'environnement des Hominidés d'Afrique orientale. *Coll. Int. Fond. Singer Polignac*, Paris: Masson, pp. 238-277.

Bourlière, F. (1983) *Ecosystems of the World 13, Tropical Savannas*. New York: Elsevier.

Box, E. O. (1981) *Macroclimate and Plant Forms: An Introduction to Predictive Modeling in Phytogeography*. London: Junk.

Brenac, P. (1988) Evolution de la végétation et du climat dans l'Ouest-Cameroun entre 25,000 et 11,000 B.P. Acte X° Symposium A.P.L.F., Bordeaux. *Inst. Fr. Pondichéry Trav. Sec. Sci. Tech.* 25:91-103.

Cerling, T. E. (1984) The stable isotopic composition of soil carbonate and its relationship to climate. *Earth Planet. Sci. Lett.* 71: 229-240.

Cerling, T. E., Quade, J., Ambrose, S. H., and Sikes, N. E. (1991) Current events. Fossils soils, grasses, and carbon isotopes from Fort Ternan, Kenya: grassland or woodland? *J. Hum. Evol.* 21:295-306.

Cerling, T. E. (1992) Development of grasslands and savannas in East Africa during the Neogene. *Paleogeog. Paleoclimatol. Paleoecol.* 97:241-247.

Chaney, R. W. (1933) A tertiary flora from Uganda. *J. Geol.* 41:703-709.

Chapman, J. D., and White, F. (1970) *The Evergreen Forests of Malawi*. Oxford: Oxford University Press.

Clayton, W. D., and Renvoize, S. A. (1982) *Gramineae. Flora of Tropical East Africa*. Rotterdam: A. A. Balkema.

Coppens, Y., Dechamps, R., Delteil-Desneux, F., and Koeniger, J. C. (1988) Les bois fossiles du Néogène des confins éthiopico-kenyens (Turkana, basse vallée de l'Omo), leur signification paléoclimatique. *C. R. Acad. Sci. Paris* 306:1409-1412.

Dechamps, R., and Maes, R. (1990) Woody plant communities and climate in the Pliocene of the Semliki Valley, Zaire. *Virginia Mus. Nat. Hist. Mem.* 1:71-94.

Dechamps, R., Senut, B., and Pickford, M. (1992) Fruits fossiles pliocènes et pléistocènes du Rift Occidental ougandais. Signification paléoenvironnementale. *C. R. Acad. Sci. Paris* 314:325-331.

de Heinzelin, J. (in press) *Bull. Soc. Belg. de Géologie*.

Elenga, H. (1992) Végétation et climat du Congo depuis 24 000 ans B. P. Analyse palynologique de séquences sédimentaires du Pays Bateke et du littoral. Thése de Doctorat, Université d'Aix-Marseille III.

Feibel, C. S., Brown, F., and Dougall, M. (1989) Stratigraphic context of fossil hominids from the Omo group deposits, northern Turkana Basin, Kenya and Ethiopia. *Am. J. Phys. Anthropol.* 78:595-622.

Friis, I. (1991) *The Forest Tree Flora of Northeast Tropical Africa (Ethiopia, Djibouti and Somalia), a Conspectus of Previous Studies*. Copenhagen: Botanisk Museum.

Hamilton, A. (1972) The interpretation of pollen diagrams from Highland Uganda. *Paleoecol. Africa* 7:45-149.

Hamilton, A. (1982) *Environmental History of East Africa. A Study of the Quaternary*. London: Academic Press.

Harris, J. M. (ed.) (1991) *Koobi Fora Research Project*. Vol. 3: *The Fossil Ungulates: Geology, Fossil Artiodactyls and Paleoenvironments*. Oxford: Clarendon.

Harrison, T. M., Copeland, P., Kidd, W. S. F., and Yin A. (1992) Raising Tibet. *Science* 255:1663-70.

Hill, A., and Ward S. (1988) Origin of the Hominidae: The record of African large hominoid evolution between 14 My and 4 My. *Yearb. Phys. Anthropol.* 31:49-83.

Jacobs, B. F., and Kabuye, C. H. A. (1987) A Middle Miocene (12.2 my old) forest in East African rift valley, Kenya. *J. Hum. Evol.* 16:147-155.

Jolly, D. (1987) Représentation pollinique des forêts sempervirentes du Nord-Est Gabon. D. E. A., Univ. Montpellier.

Kappelman, J. (1991) The paleoenvironment of *Kenyapithecus* at Fort Ternan. *J. Hum. Evol.* 20:95-129.

Keay, R. W. (1959) *Vegetation Map of Africa south of the Tropic of Cancer, Explanatory notes*. Oxford: Oxford University Press.

Kendall, R. L. (1969) An ecological history of the Lake Victoria basin. *Ecological Monographs* 39 (2):121-176.

Kennedy, G. E. (1978) Hominoid habitat shifts in the Miocene. *Nature* 271:11-12.

Kokwaro, J. O. (1986) *Anacardiaceae. Flora of Tropical East Africa*. Rotterdam: A. A. Balkema.

Lakhampal, R. N., and Prahash V. (1970) Cenozoic plants from Congo. I. Fossil woods from the Miocene of Lake Albert. *Ann. Sci. Geol. Mus. R. Afr. Centr.* (Tervuren) 64:1-20.

Livingstone, D. A. (1967) Postglacial vagetation of the Rawenzori Mountains in equatorial Africa. *Ecol. Monog.* 37:25-52.

Livingstone, D. A., and Clayton, W. M. D. (1980) An altitudinal cline in tropical African grass floras and its paleoecological significance. *Quat. Res.* 13:392-402.

Maley, J. (1991) The African rain forest vegetation and paleoenvironments during Late Quaternary. *Climatic Change* 19:79-98.

Menaut, J. C. (1983) The vegetation of African savannas in ecosystems of the world. 13. In F. Bourliere (ed.): *Tropical Savannas*. New York: Elsevier, pp. 109-149.

Nakimera-Ssemmanda, I. (1991) Histoire des végétations et du climat dans le Rift occidental depuis 13,000 an B. P. Etude palynologique de séquences sédimentaires des lacs Albert et Edouard. Diplôme E. P. H. E., Bordeaux.

Nesbit-Evans, J. M., Van Couvering, J. A. H. and Andrews, P. (1981) Paleoecology of Miocene sites in western Kenya. *J. Hum. Evol.* 10: 99-116.

Pickford, M. (1983) Sequence and environments of the Lower and Middle Miocene hominoids of Western Kenya. In R. L. Ciochon and R. S. Corruccini (eds.): *New Interpretation of Ape and Human Ancestry*. New York: Plenum, pp. 421-439.

Pickford, M. (1987) Fort Ternan (Kenya) paleoecology. *J. Hum. Evol.* 16:305-309.

Pickford, M., Senut B., Vincens, A., Van Neer, W., Ssemmanda, I., Baguma, Z. and Musiim, E. (1991) Nouvelle biostratigraphie du Néogène et du Quaternaire de Nkondo (Bassin du lac Albert, Rift occidental ougandais). *C. R. Acad. Sci. Paris* 312:1667-1672.

Polhill, R. M. (1966) Ulmaceae. In C. E. Hubbard and E. Milne-Redhead (eds.): *Flora of Tropical East Africa*. p. 14.

Pratt, D. J., Greenway, D. J., and Gwynne M. D. (1966) A classification of East African rangeland with an appendix on terminology. *J. Appl. Ecol.* 3:369-382.

Retallack, G. J., Dugas, D. P., and Bestland, E. A. (1990) Fossil soils and grasses of a Middle Miocene East African grassland. *Science* 247:1325-1328.

Rudiman, W. F., and Kutzbach, J. E. (1991) Plateau uplift and climatic change. *Sci. Amer.* 264:66-75.

Senut, B., Pickford, M., Bonnefille, R., Gayet, M., Roche, E., Kasande, R., and Obwena, P.(1990) Nouvelles découvertes paléontologiques, archéologiques dans le bassin du lac Edouard. *C. R. Acad. Sci. Paris* 311:1011-1016.

Shipman, P. (1986) Paleoecology of Fort Ternan reconsidered. *J. Hum. Evol.* 15:193-204.

Shipman, P., Walker, A.,Van Couvering, J. A., Hooker, P. J., and Miller, J. A. (1981) The Fort Ternan hominoid site, Kenya: Geology, age, taphonomy and paleoecology. *J. Hum. Evol.* 10:49-72.

Smith, B. N. (1982) General characteristics of terrestrial plants (agronomic and forests). C_3, C_4 and brassulacean acid metabolism plants. In A. Mitsui and C. C. Black (eds.): *Handbook of Biosolar Resources*. Boca Raton: CRC Press, pp. 99-118.

Thompson, K., and Hamilton, A. C. (1983) Peatlands and swamps of the African continent. In A. J. P. Gore (ed.): *Ecosystem of the World. 4B*. Amsterdam: Elsevier, pp. 331-373.

Tobias, P. V. (1991) Nouvelles hypothèses sur l'apparition de l'Hominidé dans un environment africain instable. *L'Anthropologie* 95:379-390.

Trochain J.L. (1980) *Ecologie Végétale de la Zone Intertropicale non Désertique*. Université Toulouse.

Walter, H. (1971) *Ecology of Tropical and Subtropical Vegetation*. Edinburgh: Oliver and Boyd.

Walter, R. C., et al. (1991) Laser fusion $^{40}Ar/^{39}Ar$ dating of Bed I, Olduvai Gorge, Tanzania. *Nature* 354:145-149.

White, F. (1983) *The Vegetation of Africa*. Paris: UNESCO.

Yemane K, Bonnefille, R.,and Faure, H. (1985) Paleoclimatic and tectonic implications of Neogene microflora from the northwestern Ethiopian highlands. *Nature* 318:653-656.

Yemane, K., Robert, C. and Bonnefille, R. (1987) Pollen and clay assemblages of a late Miocene lacustrine sequence from the northwestern Ethiopian highlands. *Paleogeog. Paleoclimatol. Paleoecol.* 60:123-141.

CHAPTER 21

Early Stone Age Technology in Africa

A Review and Case Study into the Nature and Function of Spheroids and Subspheroids

Kathy D. Schick
and
Nicholas Toth

INTRODUCTION

F. Clark Howell's contribution to Early Stone Age archeological research in Africa is perhaps exemplified in his codirection of the Omo project in Ethiopia (Howell 1976; Howell and Coppens 1976; Howell et al. 1987) and in his investigations at Isimila (Howell et al. 1962) in Tanzania. This research, along with his archeological fieldwork at Ambrona and Torralba in Spain (Howell 1966, 1989) and Yarimburgaz in Turkey has spanned three decades and has provided crucial information regarding the origins and development of paleolithic technology and behavior over time.

The interdisciplinary Omo project became a model for large–scale collaborative paleoanthropological research in Africa, incorporating physical anthropologists, archeologists, vertebrate and invertebrate paleontologists, paleobotanists, geologists, and geochronologists. This project discovered the earliest known archeological sites in the world, establishing a chronological benchmark for the origins of flaked stone technologies (and, by definition, the earliest recognizable archeological record) between about 2.3–2.4 MYA and allowing investigations of the relationship between the emergence of the paleolithic and other hominid evolutionary events.

This chapter begins with a review of the state of knowledge regarding the Early Stone Age of Africa and note some of the major issues that are still unresolved. We then focus on a particular archeological problem within this technological stage: the nature and function of the artifact classes including "subspheroids" and "spheroids" ("bolas," "stone balls"). We use an experimental archeological approach to address this problem. Finally, the prehistoric patterns manifested in Beds I and IF of Olduvai Gorge, Tanzania, are examined.

The African Early Stone Age

At our present state of knowledge, the earliest archeological sites are dated to approximately 2.4–1.5 MYA and consist of simple flaked cores (choppers, discoids, polyhedrons, heavy–duty scrapers), debitage (flakes and fragments), occasional retouched pieces (light–duty scrapers), and battered pieces (hammerstones, subspheroids and spheroids). These simple technologies have been assigned to the Oldowan Industrial Complex (including the Oldowan and Developed Oldowan Industries), named after the site locality of Olduvai Gorge in northern Tanzania.

Between 1.5 and 1.7 MYA, new technological elements begin to emerge in the prehistoric record: larger, often bifacially flaked picks, handaxes, and cleavers. These forms, usually produced from large flakes struck from boulder cores, have been assigned to an early phase of the Acheulean Industrial Complex. At present, the earliest documented Stone Age occurrences are found within ancient lake margin and alluvial deposits in the African Rift Valley and within cave infillings in karstic regions of South Africa. These localities include:

- *The Omo Valley, Ethiopia*: Archeological sites from stream deposits in Members E and F of the Shungura Formation have been radiometrically dated to about 2.3–2.4 MYA. These occurrences consist mainly of quartz flakes and fragments, although some larger core forms have been recovered (Chavaillon 1970, 1976; Merrick 1976; Merrick and Merrick 1976; Howell et al. 1987).
- *Gona (Hadar), Ethiopia*: Several occurrences on the east and west side of the Gona River have been excavated from channel and floodplain deposits and consist primarily of lava cores and debitage (Corvinus 1975, 1976; Corvinus and Roche 1976, 1980; Harris 1983; Roche 1980). Although dating of the site is still problematic, a date of approximately 2.4 MYA has been suggested by geological researchers (R. Walter, pers. comm.).
- *Gadeb, Ethiopia*: Several archeological occurrences have been excavated in channel deposits at the Gadeb locality in the high Ethiopian Plain. Artifacts consist primarily of lava cores and debitage, but some ignimbrites and obsidian were also used (Clark and Kurashina 1976, 1979). The researchers have suggested the earlier Developed Oldowan assemblages here date to around 1.5 MYA, with later Acheulean sites also present.
- *Melka Kunture, Ethiopia*: Oldowan or Developed Oldowan sites containing artifacts of lava and obsidian have been excavated from from riverine deposits, the oldest of which may be approximately 1.5 million years old (Chavaillon et al. 1979).
- *West Turkana, Kenya*: The oldest archeological sites have been identified spanning the time period from approximately 2.3 to 1.3 MYA. Most of these Oldowan occurrences contain artifacts made from lava and include cores, debitage, and occasional retouched pieces (Leakey and Lewin 1992; Kibunjia et al. 1992). Surface Acheulean materials have also been found that appear to be eroding out of sediments dated to approximately 1.7 MYA; if confirmed, this would be the earliest Acheulean documented anywhere in the world.
- *East Turkana (Koobi Fora), Kenya*: A large number of Oldowan and Developed Oldowan archeological sites have been excavated from floodplain, channel, deltaic, and lake margin deposits in this region, spanning a time period from approximately 1.9 to 1.3 MYA. The predominant raw material is lava, with smaller quantities of ignimbrite, chert, and quartz. Larger cores and more numerous retouched forms become more common over time in this sequence (Bunn et al. 1980; Harris and Isaac 1976; Isaac and Harris 1978; Toth 1985). An early Acheulean site here may date to c. 1.3 MYA.
- *Chesowanja, Kenya*: Two archeological sites dated to approximately 1.5 MYA have been excavated from within the Chemoigut Formation in the Lake Baringo Basin. Artifacts consist primarily of lava debitage with some simple core forms (Gowlett et al. 1981; Harris and Gowlett 1980).
- *Olduvai Gorge, Tanzania*: Numerous archeological occurrences assigned to Oldowan, Developed Oldowan, and Acheulean Industries have been excavated from lake margin, floodplain and

channel deposits in Beds I and II, spanning a time range from approximately 1.86 to 1.35 MYA. The principal raw materials that were used included quartz, lava, and chert. There is a tendency toward higher uses of quartz over time, more retouched forms and subspheroids/spheroids, and beginning in upper Bed II times (c. 1.5 MYA) large unifacial and bifacial forms assigned to the Acheulean Industrial Complex (Leakey 1971, 1975; Potts 1988).

- *Peninj, Tanzania*: Two early Acheulean sites dated to about 1.5 MYA were excavated in river channel deposits. Artifacts include crude lava picks, handaxes, and cleavers as well as a range of smaller lava and quartz artifacts (Isaac 1967).
- *Senga 5A, Zaire:* At this site, located in the Western Rift Valley and provisionally dated to between 2.0 to 2.3 MYA based upon faunal correlations (but see chapters by de Heinzelin and by Boaz, this volume), several hundred artifacts and several thousand faunal specimens have been excavated from low–energy lake margin deposits. The artifacts consist primarily of quartz cores, flakes, and fragments (Harris et al. 1987).
- *Mwimbi, Chiwondo Beds, Malawi*: A small sample of artifacts consisting of quartz and quartzite cores, debitage, and retouched pieces have been excavated at this locality within the Rift Valley in northern Malawi (Kaufulu and Stern 1987). The artifacts were found in sandy sediments infilling a paleogully; Plio–Pleistocene fauna from these Beds suggest an age of greater than 1.6 million years.
- *Sterkfontein, South Africa*: In the Member 5 cave breccia here, quartz and quartzite artifacts have been recovered, including simple core forms, debitage, retouched pieces, and a few early Acheulean cleavers and handaxes. *Homo habilis* (STW 53) was also recovered from this Member, along with a rich fauna dominated by artiodactyls. Age estimates for this deposit are approximately 1.5 to 2.0 MYA (Mason 1962; Robinson 1962).
- *Swartkrans, South Africa*: Artifacts have been found in Members 1, 2, and 3 in the brecciated deposits within this karstic cave in the Transvaal region. Simple cores, debitage, and retouched pieces were made in quartz, chert, and quartzite. Clark (1991) suggests that the archeological sample was washed into the cave from the hillslope outside. A rich faunal sample has also been recovered, including *Australopithecus* (or *Paranthropus*) *robustus* and *Homo* sp. from Members 1, 2, and 3. Age estimates for these deposits range from 1.8 to 1.5 for Member 1, 1.5 to 1.0 MYA for Member 2, and less than 1.0 MYA for Member 3 (Brain 1981; Clark 1991; Leakey 1970).

Other sites that may be in the time range considered here include an Oldowan industry at Ain Hanech in Algeria (Arambourg and Balout 1952; Balout 1955; Sahnouni 1987), the lower part of the Casablanca sequence on the Moroccan coast (Biberson 1961, 1966), and lake margin deposits at el 'Ubeidiya in the Jordan Valley of Israel (Bar–Yosef and Tchernov 1972; Goren 1981; Stekelis 1966).

For discussion of the nature of the evidence at these sites and the diverse questions and problems surrounding their interpretation, a number of reviews of the African early Stone Age are available (e.g., Clark 1980; Harris 1983; Isaac 1982, 1984; Klein 1989; Potts 1992; Schick and Toth 1993; Toth and Schick 1986). Within the past 15 years, major controversies regarding these early archeological occurrences have centered on such questions as what patterns of hominid transport stone and animal parts produced these stone–bone concentrations (e.g., home bases, scavenging stations, stone caches), the means by which hominids obtained parts of animal carcasses (hunting vs. scavenging), and which hominid taxon or taxa might be responsible for the archeological residues. Although these topics are beyond the scope of this chapter, the preceding reviews may be referred to for more information on these subjects.

Possible Hominid Tool-Makers

Between 2.4 and 1.3 MYA, a number of fossil hominid forms existed (Bilsborough 1992; Foley 1987; Howell 1978, 1982; Johanson et al. 1987; Susman 1988, 1992; Tobias 1991; Walker 1981; Wood 1991, 1992). While there is still much debate as to precisely how many species can be identified and what are the phylogenetic relationships between them, taxa currently identified by various researchers include:

- *Australopithecus (Paranthropus) aethiopicus*
  Time range: ~2.6–2.2 MYA
  Key sites: West Turkana, Omo
  Key fossil: KNM-WT 17000, Omo Shungura C mandible
  Cranial capacity: ~410 cc (one specimen)
- *Australopithecus (Paranthropus) boisei*
  Time range: ~2.3–1.2 MYA
  Key sites: Omo, East Turkana, Olduvai, Peninj
  Key fossils: OH 5, Peninj, KNM-ER 406, KNM-ER 732
  Mean cranial capacity: ~520 cc
- *Australopithecus (Paranthropus) robustus*
  Time range: ~2.0–1.0 MYA
  Key sites: Swartkrans, Kromdraai
  Key fossils: SK 48, SK 46, SK 23, SK 12, SK 23, SK 876, TM 1517
  Mean cranial capacity: ~520 cc
- *Homo habilis*
  Time range: ~1.9–1.5 MYA (a recent claim for early *Homo* at 2.4 MYA has also been made by Hill et al. [1992])
  Key sites: Olduvai, Koobi Fora
  Key fossils: OH 7, OH 13, OH 24, OH 62, KNM-ER 1813, KNM-ER 3735, KNM-ER 1805, Stw 53, SK 847(?)
  Mean cranial capacity: ~610 cc
- *Homo rudolfensis*
  Time range: ~1.9–1.7 MYA
  Key sites: East Turkana
  Key fossils: KNM-ER 1470, KNM-ER 1590, KNM-ER 3732, KNM-ER 1802, KNM-ER 1482, KNM-ER 1483, KNM-ER 1801
  Mean cranial capacity: ~750 cc
- *Homo erectus* (although some anthropologists have come to prefer the designation *Homo ergaster* to differentiate the earlier African forms from later African and Asian forms)
  Time range: ~1.8–<1.0 MYA
  Key sites: East Turkana, West Turkana, Olduvai
  Key fossils: KNM-ER 3733, KNM-ER 3883, KNM-ER 992, KNM-WT 15000, OH 9
  Mean cranial capacity: ~875 cc

At present it is not possible to conclusively demonstrate which of these forms were responsible for the paleolithic occurrences found during this time period. Although many anthropologists have argued that the larger brained, less dentally specialized fossils assigned to *Homo* are more likely candidates for more habitual tool–makers and tool–users, others, notably Susman (1988, 1992), have suggested that the robust australopithecines were at least as likely to have been early stone tool–makers.

Typology of African Early Stone Age Technological Patterns

The early Stone Age sites dating from approximately 2.4 to 1.4 MYA exhibit a range of technological patterns. Leakey (1971) has grouped sites at Olduvai Gorge into several different industries with some trends evident over time:

- *Oldowan*: At these sites, lava tends to be the predominant raw material, and characteristic core forms include choppers, polyhedrons, discoids, and heavy duty scrapers, with some spheroids/subspheroids and light duty scrapers. This industry characterizes the earliest sites at Olduvai Gorge, with examples spanning from the lowest sites in Bed I to the lower part of Middle Bed II.
- *Developed Oldowan A*: Found in Lower Bed II and lower part of Middle Bed II (characterized by smaller proportions of choppers and other large core–tool forms and greater proportions of spheroids, subspheroids, and modified battered nodules and blocks as well as more light–duty tools [scrapers, awls, etc.] than seen in the earlier Oldowan industry).
- *Developed Oldowan B*: Found in Middle and Upper Bed II, and perhaps continuing into Bed III (characterized by greater quantities of light–duty scrapers and other small tool forms as well as low numbers of bifaces).
- *Early Acheulean*: Found in Middle Bed II, and continuing through Beds III, IV, and the Masek Bed (characterized by higher proportions of bifaces that also tend to be longer in absolute terms and relative to their breadth and thickness in comparison with the Developed Oldowan bifaces, and with relatively small proportions of spheroids, subspheroids, and light–duty tools).

Whether these industrial designations can be easily applied to other early African localities is not yet clear. Some researchers (such as Isaac 1984) have suggested that two major groupings, an Oldowan Industrial Complex (G. Clark's Mode 1 industry, Clark 1971) and an early Acheulean Industrial Complex (Clark's Mode 2 industry), might be a more appropriate classification scheme to encompass all of the African Early Stone Age phenomena, at least at our present state of knowledge.

There appear to be important changes in stone technologies at Olduvai Gorge sites, including the dramatic rise in numbers and proportions of spheroids and subspheroids from lower Bed I through upper Bed II times. These forms have long perplexed prehistorians with regard to their mode of manufacture and possible uses.

Interestingly, these spherical and subspherical artifact forms tend for the most part to be made out of quartz and quartzite. One important question here, then, concerns the effects of lithic raw material type upon artifact form. As Clark (1980), Jones (1979), and Stiles (1979) have pointed out, certain types of raw materials may possess unusual characteristics of size, shape, or fracture that, during manufacture or use, tends to make them assume certain morphologies. The next section of this chapter will address the question of the role of raw material in producing this peculiar class of Early Stone Age artifacts.

EXPERIMENTS INTO THE NATURE OF SPHEROIDS AND SUBSPHEROIDS

Problem and Experimental Design

Spherical, often flaked and battered pieces of stone commonly classified as "spheroids" and "subspheroids" (Leakey 1971) represent one of the most enigmatic and ubiquitous types of artifacts in the Early and Middle Stone Age of Africa. The more symmetrical examples of these have sometimes been called "stone balls" or "bolas stones." In East Africa the majority of these forms are made from quartz, although a minority are also made from lava. Elsewhere in the Old World (e.g. North Africa, the Middle East, China, Western Europe) these forms tend to be made in either quartz or limestone. (For an excellent review of the nature and distribution of this artifact class see Willoughby 1985, 1987.)

For a century prehistorians have speculated upon the functional significance of these artifact forms. Suggestions have included their use as tethered bolas stones in hunting, hand-thrown missiles, club heads, bone processing tools, vegetable processing tools, or even as some sort of symbolic ("nonutilitarian") objects.

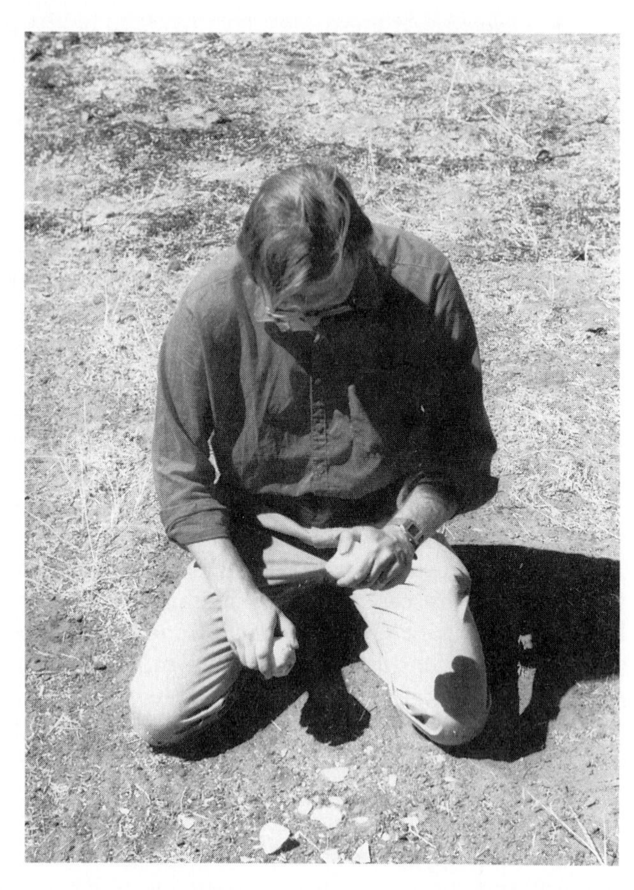

FIGURE 21-1
Reducing a quartz core using a quartz hammer during the experimental program.

Although experimental archeological approaches have addressed a range of questions pertaining to the early stone age of Africa (Jones 1979; Schick and Toth 1993; Toth 1985, 1987, 1991), little has been published regarding experimental archeological investigations relating to the manufacture and potential uses of these forms. In 1988, while conducting archeological research in central Africa, we had the opportunity to study this problem through an experimental approach.

Our experiments were conducted in Zambia and focused upon Precambrian quartz similar to that found at Olduvai Gorge as the principal raw material. The interpretation of the results of this study primarily concerns quartz spheroids/subspheroids; further studies are planned for other raw materials such as limestone and lava.

The chipping, rounding, and heavy battering that are exhibited on many quartz specimens from East Africa, and their often highly spherical morphology suggested to us that there had been appreciable stone material (small flakes and fragments, rock dust) removed from the initial clast in the process of their formation.

At the outset of this experimental work, hundreds of casual experiments in flaking quartz cores with quartz hammers were conducted for extended periods of time to gain an in–depth appreciation of the mechanics of fracture and attrition through battering of this raw material during the course of flaking (Figure 21-1). We were particularly interested to see if there might be a simpler explanation for the forms typed as spheroids and subspheroids: Might they simply be byproducts of using angular chunks or cores of quartz as hammers to flake other pieces of stone? For several weeks we concentrated on flaking quartz, producing a wide range of the cores and retouched forms that are found in the Oldowan and Developed Oldowan using angular clasts of quartz as hammerstones for extended periods of time.

To exemplify the patterns of flaking in the production of quartz flakes from cores, one round of experimentals in flaking 25 chunks of quartz (ranging from 8 to 12 cm in maximum dimension) produced 30 "exhausted" cores (some of the chunks broke into more than one piece during flaking and made two cores). Using Mary Leakey's (1971) classification system to type these cores, the following breakdown resulted:

Choppers:	1(3%)
Discoids:	3 (10%)
Polyhedrons:	6 (20%)
Heavy duty scrapers:	8 (27%)
Subspheroids:	12 (40%)

It should be noted that many of these core forms assumed a subsperical shape simply by serving as cores for flake removal. Many of the cores produced, especially the spheroids and polyhedrons, could have been (and were) subsequently used as hammerstones in the absence of smooth river cobbles.

Although some of the subspheroid cores exhibited an amount of battering from prolonged impacts from the quartz hammer, others exhibited little or no battering but were still assigned to this artifact class based on their overall morphology. This is consistent with Mary Leakey's classification, which also assigns large, largely unbattered specimens (almost certainly cores for flake production) into this category (see, for example Leakey 1971: plates 21, 25).

The primary focus of this experimentation was to observe the effect of long–term battering on pieces of quartz used as hammerstones to flake other cores. It became apparent that subangular pieces of quartz would naturally assume a subspherical and finally spherical morphology if the hammer did not accidentally break apart during use.

To exemplify the pattern observed from the more casual experimental study, two pieces of quartz (one angular chunk, one polyhedron core) were selected as hammers. Changes in the weight, dimensions, and morphology of these hammers were carefully recorded at 30-minute intervals for up to 4 hours of

FIGURE 21-2 Stages in spheroid formation. From left: (1) an unmodified chunk of quartz; (2) a quartz polyhedral core with flakes removed; (3) a quartz hammer (subspheroid) that has been used for one hour of continuous flaking; (4) a quartz hammer (spheroid) that has been used for two hours; (5) a quartz hammer (spheroid/"Stone ball"/"bolas stone") that has been used for four hours.

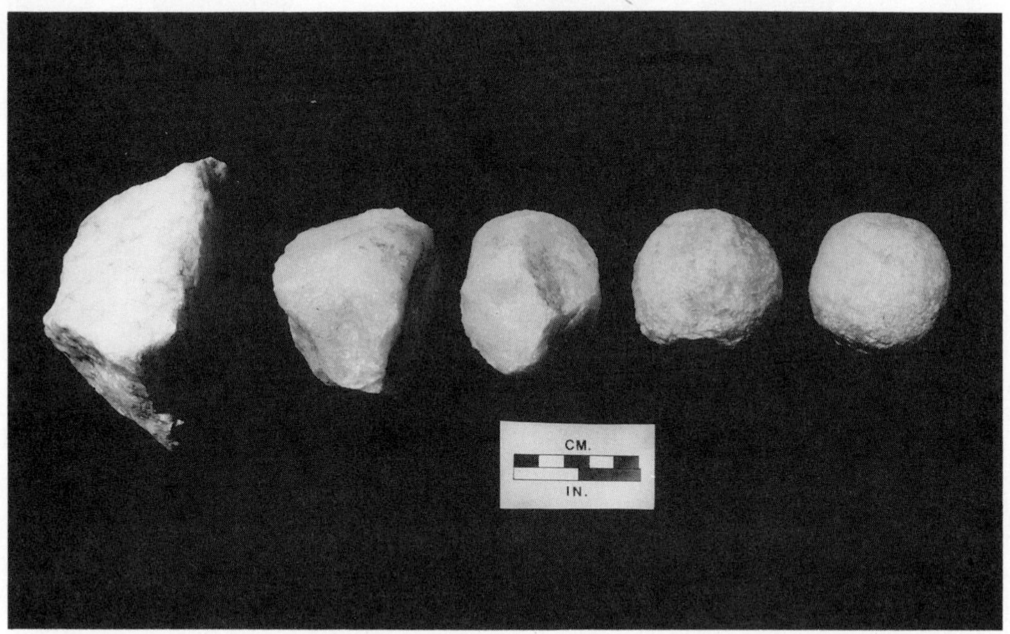

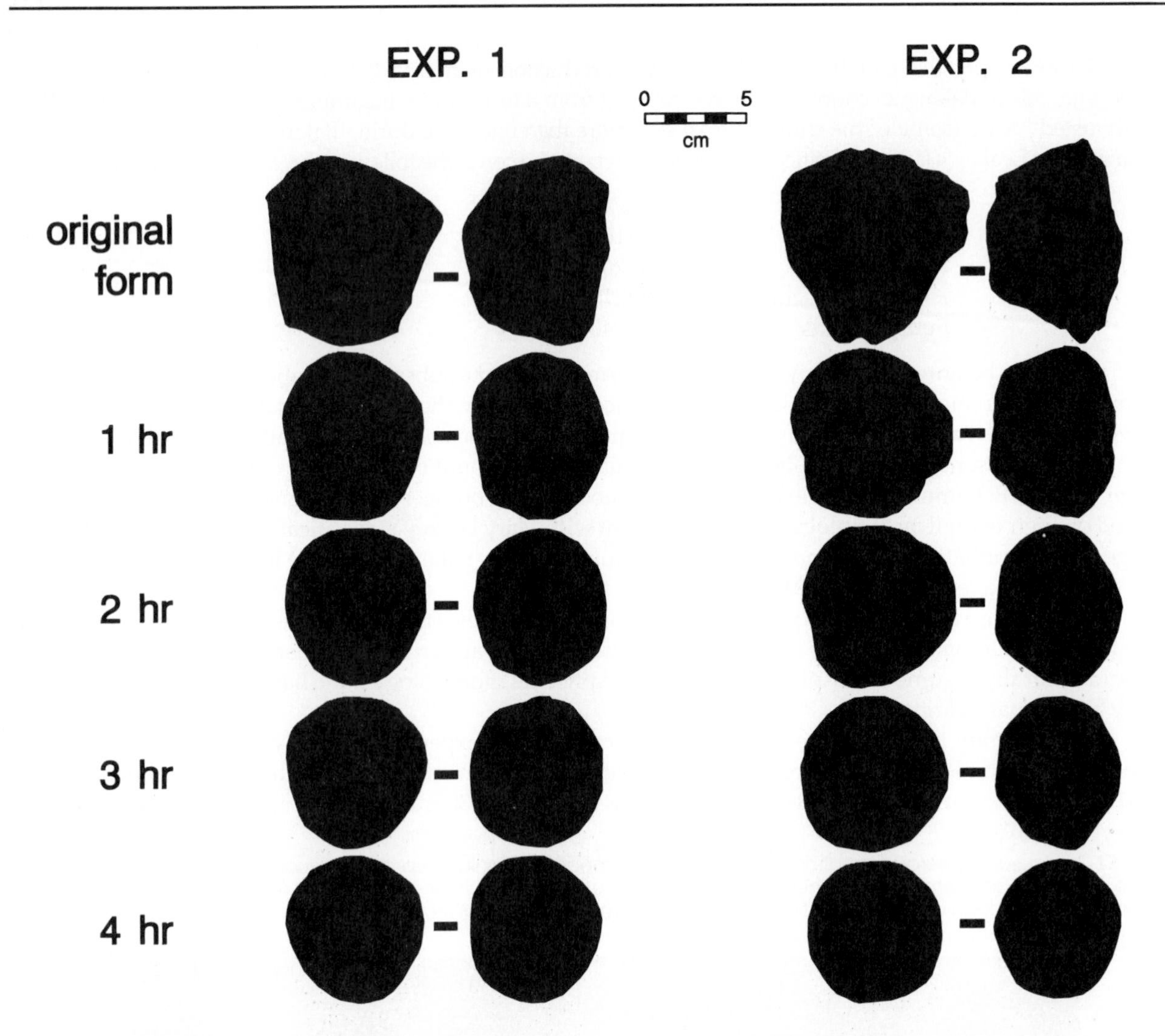

FIGURE 21-3 Changes in the morphology of the quartz percussors from Experiment 1 and 2, shown at 30-minute intervals.

continuous flaking with the quartz hammer (averaging 25 blows per minute or 1500 blows per hour). (See Figures 21-2 through 21-5 and Table 21-1.)

Results

Our experiments demonstrated that use of an angular quartz hammer to flake other cores naturally tended to produce battered, rounded, and very spherical artifacts in a relatively short period of time (2 to 4 hours). Although no deliberate attempt was made to arrive at these spherical end products, quartz clasts used as hammers became progressively more rounded and spherical over time through the incidental shatter of more angular or protruding edges. In addition, the tendency to strike the core with the end of the longest axis of the hammer (for greater ease of use and reduction of shock to the flaker's hand) tended to equalize the length–breadth ratio of the hammer over time through such deterioration from impact.

An average of 168 grams of material were removed from these quartz hammers over 4 hours of use, leaving on average 63 percent of the original hammerstone weight. This worked out to a mean of 0.028 grams of material removed by each blow over a 4-hour period. However, this loss of mass in the hammer

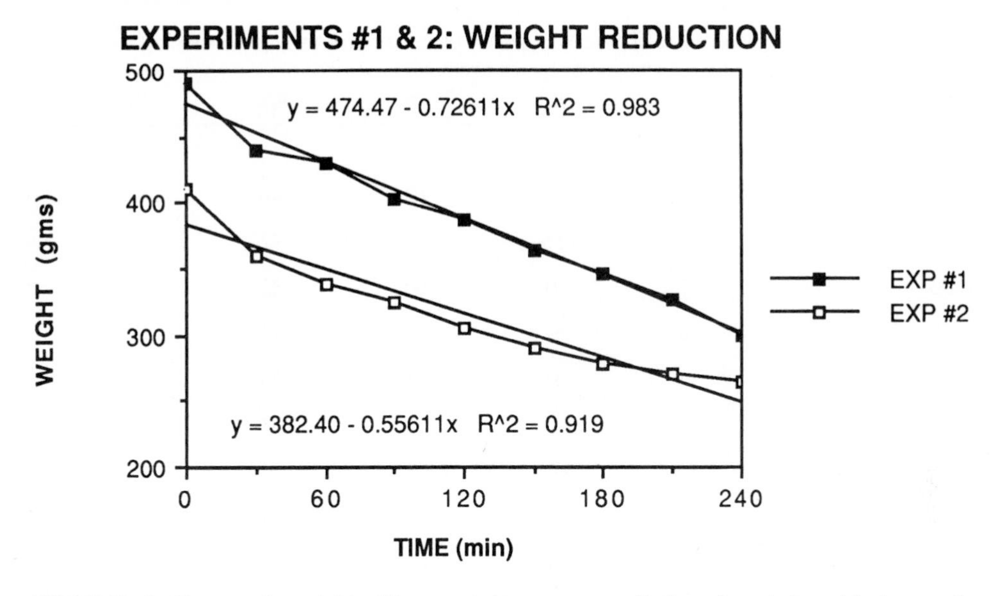

FIGURE 21-4 Changes in weight of the quartz hammers used in Experiments 1 and 2, shown at 30-minute intervals.

FIGURE 21-5 Changes in the shape of the quartz hammers (as seen in breadth/length and thickness/breadth ratios). A perfect sphere (L = B = Th) would be in the upper right corner of the graph. Note that the hammers tend to become more spherical over time, although occasionally a spall along a plane of weakness in the rock will remove substantial material and change the shape.

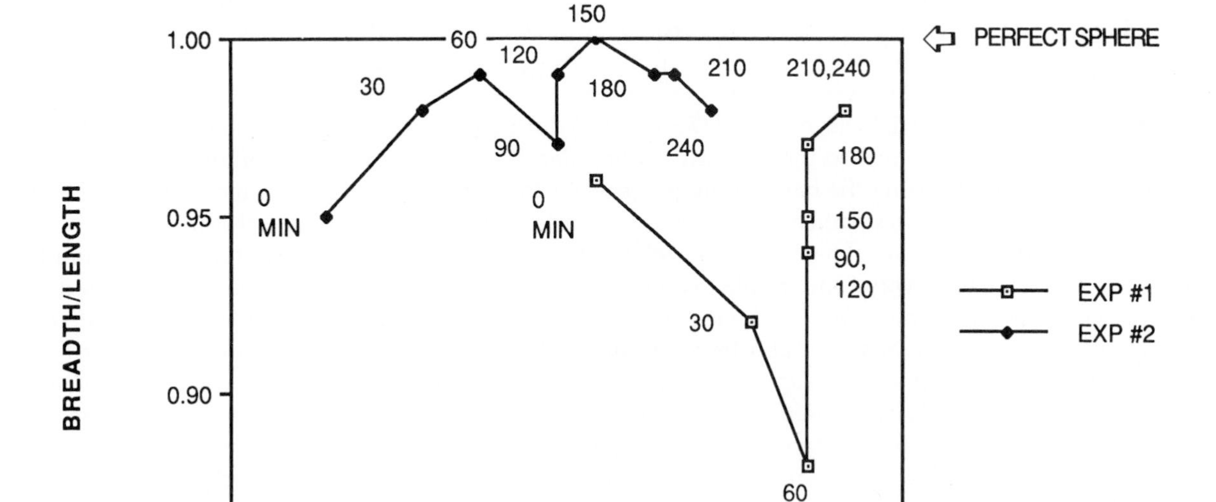

TABLE 21-1 Changes in weight and dimensions of pieces of quartz used as hammerstones in flaking experiments

Time	Blows	Wt.	L.	B.	Th.	B/L	Th./B	%left
Experiment 1								
0	0	410	78	75	63	.96	.84	100
30	750	359	72	66	61	.92	.92	83
60	1500	339	69	61	58	.88	.95	83
90	2250	324	65	61	58	.94	.95	79
120	3000	306	64	61	57	.95	.93	75
150	3750	289	64	60	57	.94	.95	70
180	4500	278	62	60	57	.97	.95	68
210	5250	271	60	59	57	.98	.97	66
240	6000	265	60	59	57	.98	.97	65
Experiment 2								
0	0	490	93	88	62	.95	.70	100
30	750	440	81	79	59	.98	.75	90
60	1500	430	77	76	59	.99	.78	88
90	2250	402	74	72	59	.97	.82	82
120	3000	387	73	72	59	.99	.82	79
150	3750	364	70	70	59	1.00	.84	74
180	4500	347	69	68	59	.99	.87	71
210	5250	327	67	66	58	.99	.88	67
240	6000	299	64	63	57	.98	.90	61

was not linear. More material was removed (70.5 grams) during the first hour of hammering than during the last hour (30.5 grams) due to the fact that the angular nature of the initial hammer allowed larger fragments to spall off during the hammering process. As the percussor assumed a more spherical and rounded shape, it tended to stabilize in form with less material being removed over time.

A symmetrical, spheroidal hammer may well have come to be favored by Early Stone Age hominids since it would have been very comfortable to hold and it had proven its reliability over a long period of time (no fatal flaws to shatter in the knapper's hand). An analogous pattern can be seen in similar battered and spherical stones from Neolithic and Iron Age sites in Africa, produced through prolonged use of a stone hammer to pound the upper surfaces of grinding stones used in processing cereals and other foods. Although these stone balls were used to pound and roughen another stone surface rather than to strike off flakes, they also achieved a spherical form as a byproduct of this stone–on–stone impact.

On a microscopic level, another analogous pattern can be observed: Angular sand grains of quartz that are subjected to innumerable random impacts in deserts during high winds assume a characteristic spherical, rounded, and battered shape that can resemble subspheroids and spheroids to a remarkable degree (see, for example, Reineck and Singh 1980: Fig. 222, a and c).

It was observed that many of the fragments that were removed from quartz hammerstones as they became more spherical could potentially be identified in the archeological record as well. These flakes, usually less than 2 cm in size, were characterized by having a shattered striking platform, a flat bulbar surface, and evidence of battering on their dorsal surface. Sometimes a quartz hammer would split in half along a fault plane, or break into several fragments. Again, battering was often observable on the outer surface of such fragments.

Conclusions from the Experiments

While these results do not *prove* that battered quartz subspheroids and spheroids were produced from their use as percussors in flaking stone, it does give strong evidence that there is a simpler, least–effort explanation for these forms. And, if the majority of these forms are indeed heavily used hammerstones, as we believe, they still could have been used for other activities once they had arrived at their rounded, spherical, and battered morphology.

If our interpretation is correct, a prediction can be made: In a prehistoric area where a range of raw materials was available to Early Stone Age hominids, those sites having quartz as their predominant raw material should have relatively high proportions of subspheroids and spheroids in their assemblage composition, since this would be the principal rock type to serve as hammers in flaking other pieces of quartz. Sites having lava as the predominant raw material should have high proportions of hammerstones, as this type of raw material does not tend to "snowball" into a spheroidal shape through battering as readily as quartz (Toth 1982). This prediction is testable at archeological sites in Beds I and II at Olduvai Gorge using the detailed data reported by Leakey (1971, 1975). The next part of this chapter will focus upon this question.

AN ARCHEOLOGICAL CASE STUDY: OLDUVAI GORGE, TANZANIA

The Sample

The excavated archeological sites at Olduvai Gorge provide an excellent sample to look at patterning of quartz use and spheroid manufacture by early tool–making hominids. Twenty–six archeological sites assigned to the Oldowan, Developed Oldowan A, Developed Oldowan B, and early Acheulean were analyzed in this study (see Table 21-2). These sites span a time period from approximately 1.86 to 1.35 MYA (Hay 1976; Manega 1993; Walter et al. 1991, 1992). Our age estimates for the Olduvai sites are approximations based on radiometric dates and stratigraphic positioning. Since quartz and quartzite grade into one another in the Olduvai region (the great majority being quartz), which can make it difficult to neatly identify certain rock types (R. Hay, pers. comm.), we will deal here with materials designated by Leakey as "quartz/quartzite" under the general category of "quartz."

Hay (1971, 1976) has conducted extensive studies of the petrology of the artifacts at Olduvai sites and has identified a number of the potential sources that hominids may have frequented to get rocks for their tool–making and tool–using activities. Isaac (1980, 1981) has emphasized the importance of "landscape archeology" in understanding behavioral patterns in prehistoric times. Both of these approaches will be considered later. Although higher energy site formation processes may have greatly affected the proportion of lighter artifacts relative to heavier artifacts (Potts 1988; Schick 1986, 1987a, 1987b, 1991), it is unlikely that these forces would have greatly affected the relative proportions of the various types of heavy–duty tools and hammerstones, which were the focus of this study.

Of special interest to this study were (1) changes in types of raw materials used over time from Bed I through Bed II times; (2) relationships between artifact type (particularly hammerstones and spheroids or subspheroids) and raw materials used; and (3) relationships between frequency of use of different raw materials at each site and its distance from proposed sources of those raw materials.

In particular, relationships were examined among the following variables:

1. Change over time in the use of quartz (relative to other raw materials) for heavy duty tools;
2. Change over time in the frequency of spheroids and subspheroids (relative to other heavy duty tools);
3. Change over time in the frequency of hammerstones (relative to heavy–duty tools);
4. Frequency of quartz relative to the frequency of spheroids and subspheroids (among heavy duty tools in an assemblage);
5. Frequency of lava and frequency of hammerstones in an assemblage;
6. Frequency of spheroids and subspheroids and the distance of the site from the Naibor Soit quartz source;
7. Frequency of quartz in an assemblage and the distance of the site from the Naibor Soit quartz source;
8. Frequency of lava in an assemblage and the approximate distance of the site from alluvial fans, major sources of the Sadiman lava raw materials.

TABLE 21-2 (PART 1). Data for the 18 Olduvai Gorge sites in Beds I and II (derived from Mary D. Leakey, 1971). For ASS. TYPE (Assemblage type): O = Oldowan, DOA = Developed Oldowan B, A = Acheulean. For Bed: L, M, U, LM, and UM refer to subdivisions of Beds I and II (lower, middle, upper, lower middle, and upper middle). Ages are approximations based upon radiometric dates derived for Olduvai volcanics (Hay1976; Walter et al. 1991, 1992; Manega pers. comm.) and interpolations based upon relative stratigraphic position of sites. N. Art. = total number of artifacts; N. SPH = total number of spheriods; %QTZ SPH = % of spheroids and subspheroids in quartz; N HS = number of hammerstones; %LAVA HS = hammerstones in lava. "—" denotes data not available.

SITE	ASS. TYPE	BED	AGE	N. ART.	N SPH.	%QTZ SPH	N HS	%LAVA HS
DK	O	I:L	1.86	1198	7	0.0	48	81.3
FLK ZINJ	O	I:M	1.78	2470	0	0.0	13	100.0
FLK N L5	O	I:U	1.76	151	1	0.0	10	100.0
FLK N L1-2	O	I:U	1.75	1205	12	33.3	62	93.5
HWK E L1	O	II:L	1.70	154	1	100.0	21	100.0
HWK E L2	DOA	II:L	1.60	313	11	100.0	6	100.0
HWK E L3	DOA	II:LM	1.50	1283	83	88	11	100.0
HWK E L4	DOA	II:LM	1.50	601	23	100.0	10	80.0
FLK N SANDY	DOA	II:LM	1.50	234	35	91.4	11	81.8
MNK SKULL	O	II:LM	1.45	689	6	66.7	15	93.3
EF-HR	ACH	II:UM	1.45	522	9	77.8	4	50.0
MNK MAIN	DOB	II:UM	1.40	4399	159	93.1	64	76.6
FC WEST OCC	DOB	II:UM	1.40	1184	48	87.5	23	82.6
FC TUFF	DOB	II:UM	1.40	673	22	90.9	8	100.0
SHK ALL	DOB	II:UM	1.40	1807	318	95.3	—	—
TK LOWER	DOB	II:U	1.35	2153	31	87.1	0	0.0
TK UPPER	DOB	II:U	1.35	5180	76	92.1	15	73.3
BK	DOB	II:U	1.35	6801	199	86.9	18	77.8

TABLE 21-2 (PART 2) DIST RM = approximate distance in km from Naibor Soit quartz source; DIST AF = approximate distance from alluvial fan; N HDT = number of heavy-duty tools; %QTZ HDT and %LAVA HDT = %of quartz or lava among heavy-duty tools; %SPH HDT = % of spheroids and subspheroids among all heavy-duty tools; %HS/HS + HDT= % of hammerstones among hammerstones and heavy-duty tools combined.

SITE	DIST RM	DIST AF	N HDT	%QTZ HDT	%LAVA HDT	%SPH HDT	%HS/ HS + HDT
DK	2.00	0.50	123	5.7	94.3	5.3	28.1
FLK ZINJ	2.00	2.00	38	50.0	47.4	0.0	25.5
FLK N L5	2.00	2.00	30	16.7	83.3	3.3	25.0
FLK N L1-2	2.00	2.00	134	17.2	82.1	8.9	31.6
HWK E L1	2.25	1.25	50	20.0	80.0	2.0	29.6
HWK E L2	2.25	1.25	67	67.2	31.3	16.4	8.2
HWK E L3	2.25	1.25	269	58.0	35.3	30.8	3.9
HWK E L4	2.25	1.25	105	59.0	27.6	21.9	8.7
FLK N SANDY	2.00	2.00	87	74.7	21.8	40.2	11.2
MNK SKULL	3.00	2.00	41	24.4	73.2	14.6	26.8
EF-HR	2.25	0.75	88	26.1	72.7	10.2	4.3
MNK MAIN	3.00	2.00	379	77.0	20.3	41.9	14.4
FC WEST OCC	3.00	2.50	174	67.8	29.9	27.5	11.7
FC TUFF	3.00	2.50	71	77.5	16.9	30.9	10.1
SHK ALL	3.50	2.50	798	80.0	19.0	39.8	—
TK LOWER	1.25	0.50	76	84.2	18.9	40.8	0.0
TK UPPER	1.25	0.50	190	80.0	18.9	40.0	7.3
BK	3.75	3.50	490	78.1	17.5	40.6	3.5

NOTE: There are no data on hammerstones at SHK as they were not collected. The raw material breakdowns for artifacts at BK are estimates, since Leakey's detailed artifact breakdown includes only the 1963 season (n = 6801), while her raw material percentages are based on the results of 5 different field seasons (sample size not reported). We are assuming that the overall raw material breakdown would be representative of the 1963 sample.

Results and Conclusions

The major results of this analysis are shown in Figures 21-6 through 21-10 and are ranked below in groups of very high, moderately high, and relatively low correlation between the variables (full explanation of the variables is given in the figure captions).

Very high:
- Quartz/Spheroids ($r^2 = 0.765$)
- Spheroids/Time ($r^2 = 0.716$)

High:
- Hammerstones/Time ($r^2 = 0.616$)
- Lava/Hammerstones ($r^2 = 0.598$)
- Quartz/Time ($r^2 = 0.594$)
- Lava/Time ($r^2 = 0.589$)

Very low:	Lava/Distance from alluvial fan ($r^2 = 0.073$)
	Spheroids/Distance to Naibor Soit ($r^2 = 0.058$)
	Quartz/Distance to Naibor Soit ($r^2 = 0.039$)

Our primary prediction based upon the experimental results appears to be validated among these early sites at Olduvai Gorge: As the frequency of quartz rises at archeological sites, the frequency of spheroids and subspheroids does as well. And most important here, there is an inverse relationship between the use of lava and the use of quartz over time, and, in parallel, an inverse relationship between the proportion of hammerstones and the proportion of spheroids and subspheroids over time. As the use of lava and the proportion of hammerstones decline from Bed I through Bed II times, there is a corresponding rise in the use of quartz and the proportion of spheroids and subspheroids.

Simply put, as lava is used less, quartz is used more; and as hammerstones become less common, spheroids and subspheroids become more common. This pattern strongly suggests that quartz is gradually replacing lava over time, not only among the cores (heavy duty-tools) but also among the percussors.

Thus, it would appear that the increasing frequency of subspheroids and spheroids that are found in Developed Oldowan sites from Upper Bed I through Bed II can largely be explained by Early Stone Age hominids' concentrating increasingly upon quartz over time as a predominant raw material in their stone tool–making, for both their cores and their hammerstones. The repeated use of quartz chunks or exhausted cores as percussors would naturally produce battered artifacts that would formally be classed as subspheroids and spheroids without any necessary intent or premeditation on the part of the hominids to produce these forms.

FIGURE 21-6 Changes in the percentages of quartz and lava in the heavy-duty tool category over time at Olduvai Gorge. Note that between approximately 1.86 (far right) and 1.35 MYA, the percentage of lava falls at most sites while the percentage of quartz rises. There appears to be an inverse relationship between these two raw materials over time.

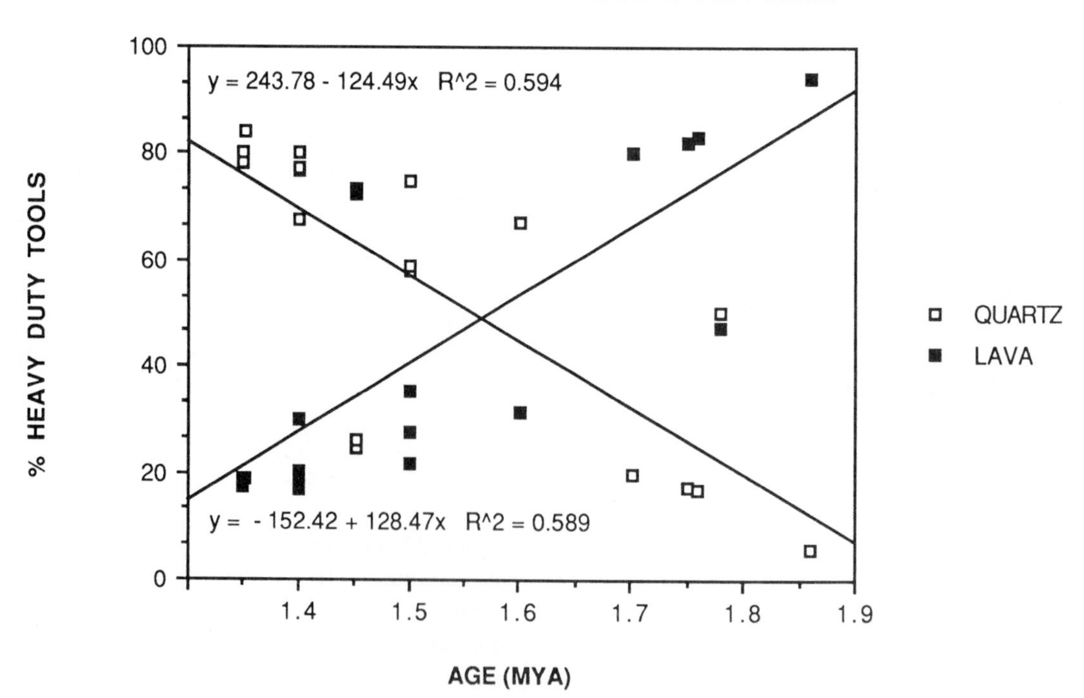

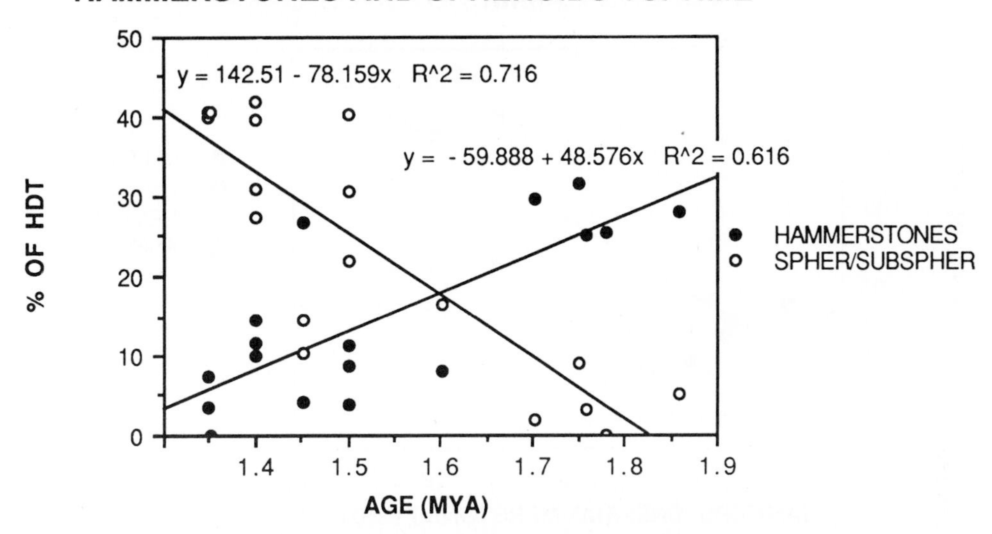

FIGURE 21-7 Changes in the percentages of hammerstones and spheroids/subspheroids over time (all raw material types). As hammerstones were not included in the heavy-duty tool category by Leakey, their percentages were calculated by dividing the number of hammerstones by the sum of hammerstones and all heavy-duty tools, and then multiplying times 100. Note that between approximately 1.86 MYA (far right) and 1.35 MYA, the percentage of hammerstones falls at most sites while the percentage of spheroids/subspheroids rises. There appears to be an inverse relationship beween the two classes of artifacts over time.

The inverse correlation observed between lava/hammerstones and quartz/spheroids over time is corroborative evidence that in the later, quartz-dominated Developed Oldowan assemblages, subspheroids and spheroids were serving the same function that lava hammerstones had in the earlier, Oldowan occurrences.

FIGURE 21-8 Relationships of the percentage of quartz in the heavy-duty tool category versus the percentage of spheroids and subspheroids in the heavy-duty tool category. Note that sites with higher percentages of quartz also have higher percentages of spheroids and subspheroids, notably in the Developed Oldowan.

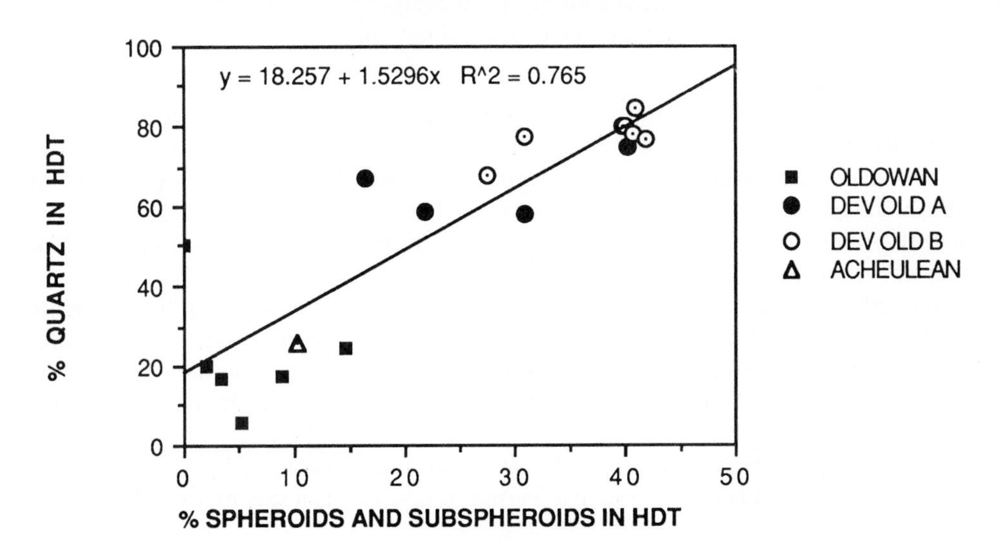

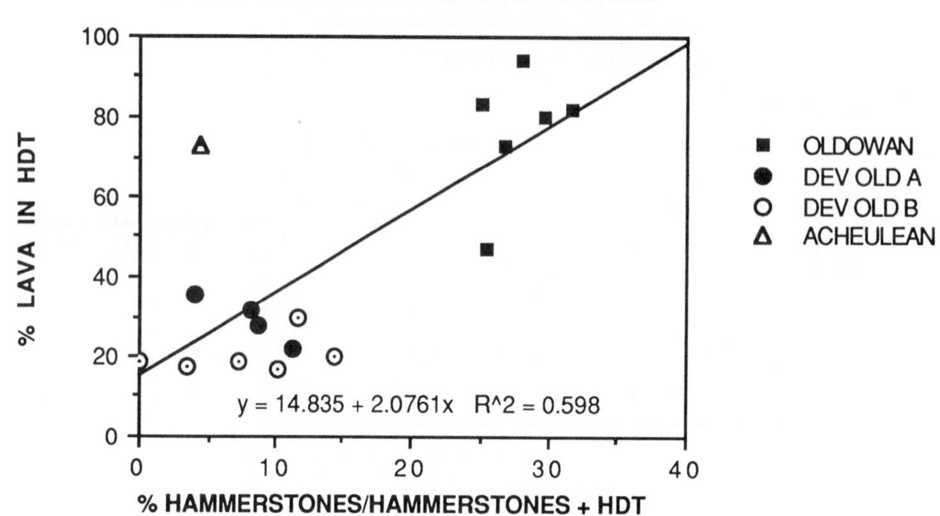

Figure 21-9 Relationships of the percentage of lava in the heavy-duty tool category versus the percentage of hammerstones relative to hammerstones + heavy-duty tools (hammerstones were not assigned to the heavy-duty tool class by Leakey). Note that sites with high percentages of lava in the heavy-duty tool class tend to have relatively high percentages of hammerstones, notably the earlier Oldowan sites.

Also of extreme interest in this study are the variable sets with low correlation values. A seemingly plausible hypothesis that frequency of quartz in a site assemblage might decline with increased distance from the Naibor Soit source (due to transport costs) was not confirmed. There appears to be a very low correlation between the proximity of a site to Naibor Soit and frequency of quartz among heavy duty tools ($r^2 = 0.024$) or among debitage ($r^2 = 0.008$). Within the "window" of exposure and excavation at Olduvai Gorge, then, there does not appear to be any significant variation in the frequency of use of quartz over a distance stretching from 1 km to 4 km away from Naibor Soit. Similarly, landscape archeological research by Blumenschine and Masao (1991) detected no significant relationship between distance from Naibor Soit and the proportion or size of quartz artifacts in a study area extending between approximately 2 km to 3 km from the quartz inselberg. The study here indicates this lack of relationship is maintained among the excavated sites for a distance of at least 4 km from the Naibor Soit source.

This overall lack of correlation may be due to several factors:

1. Developed Oldowan hominids intentionally selecting quartz in preference to other raw materials and habitually transporting it relatively long distances (several kilometers) on the landscape.
2. The presence of other sources of quartz that were available in the Olduvai basin in Bed I and II times but that now lie buried below the present surface of the Serengeti Plain and therefore unknown to present–day prehistorians (the only reason that the inselberg Nabor Soit is known is that it still protrudes above the Serengetti today).
3. The Developed Oldowan hominids' ranging patterns during foraging took them preferentially to the northeastern side of the Olduvai lake basin, where quartz would have been the predominant raw material available (ranging in the southeastern side of the basin would have taken them nearer the lava sources in alluvial fans and streams flowing from the volcanic highlands).

To us, the lingering question is not so much "Why did Bed II hominids make so many spheroids?" as it is "Why did Bed II hominids develop such a passion for quartz as a raw material?" Although quartz

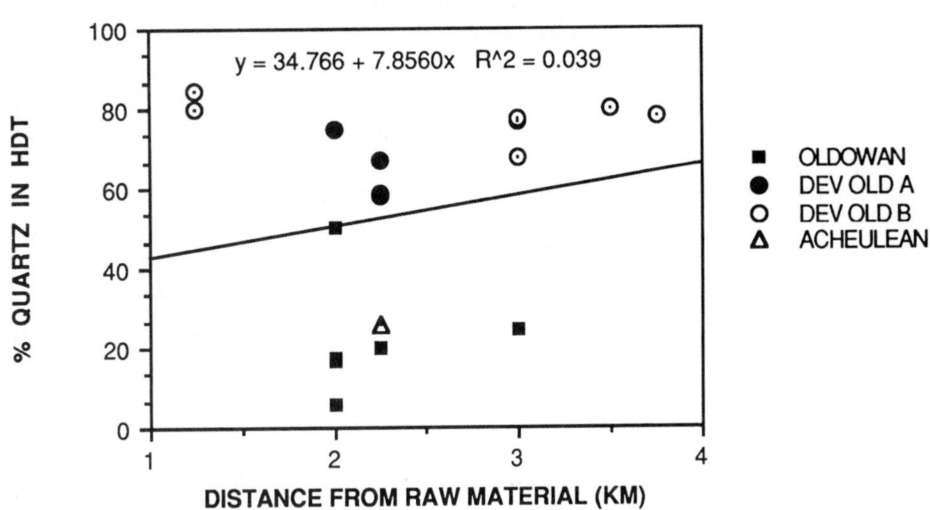

FIGURE 21-10 Relationships between the percentage of quartz in heavy-duty tools and the distance from the presumed quartz source, Nabor Soit. Note the low correlation between the distance from Nabor Soit and the percentage of quartz used by the hominids for heavy-duty tools.

is notoriously difficult for the archeologist to analyze due to its tendency to fracture not only conchoidally but also along flaws and crystal lattices, often with irregular release surfaces, functional experiments have shown that flaked quartz produces edges that are very sharp for cutting activities, sharper than most lavas that were used by early hominids. It may have been for superior cutting edges that Bed II hominids started concentrating on this material at many sites.

In any case, the rise in spheroids and subspheroids at Olduvai appears to be well correlated with this marked increase in the use of quartz as a raw material. This is almost certainly due to the use of pieces of quartz as percussors in lieu of lava hammerstones. The friable nature of quartz makes it amenable to progressive battering and rounding in the course of its use as a percussor. Thus, the appearance and development of the Developed Oldowan may be explained by the rise in the use of quartz by hominids from Bed I through Bed II times.

Although this functional interpretation of subspheroids and spheroids as percussors is perhaps less romantic than their use as bolas stones or missiles, it does highlight important behavioral information that these artifacts might yield. In particular, as more of these relatively rounded and spherical quartz artifacts appear to be products of repeated battering through their use as percussors, their rise in frequency throughout the Developed Oldowan may indicate greater selectivity for superior raw materials, longer-term curation of certain artifact forms by early hominids, or greater duration or frequency of occupation of specific sites.

SUMMARY AND CONCLUSIONS

The earliest flaked stone technologies appear in the prehistoric record, at the present state of knowledge, about 2.4 MYA. At Olduvai Gorge, archeological sites are first found near the base of Bed I at about 1.86 MYA. During the next 500,000 years, several temporal trends can be seen in the Early Stone Age archeological record at Olduvai. These include: (1) a trend toward more light–duty retouched pieces such as scrapers and awls; (2) a trend toward an emphasis on the use of quartz as a raw material; (3) a rise in the number of subspheroids and spheroids in many assemblages; and (4) the emergence of large Acheulean forms such as picks, handaxes, and cleavers.

Experimentation suggests that the simplest explanation for the artifact classes of subspheroids and spheroids is that these forms are hammerstones that have been used for an extended length of time for flaking cores. As time progresses, a semi–angular chunk of quartz (or a reused core) becomes more spherical and rounded as battering continues. After about 4 hours of continuous battering, these quartz hammers are almost perfectly symmetrical in shape (the classic "spheroid," "bolas," "stone ball" morphology).

The rise in the incidence of spheroids from Bed I to Bed II times can best be explained by early hominids shifting their preference of raw materials from lava to quartz over time. This shift in raw material selectivity may be primarily due to the superior cutting edges produced by quartz and the ease of removing flakes, but it could also reflect changes in foraging patterns of hominids that favored the northern side of the Olduvai lake basin, where the quartz inselbergs were exposed. The rise in spheroids may also indicate either more prolonged curation of stones as hammers through time or more habitual reoccupation of sites where these quartz percussors have been left. These technological changes appear to be happening at the same time as the emergence of early Acheulean forms, often made on large flakes from special lava quarry sources (Hay 1976).

It is possible that we may be observing a shift during Bed I to Bed II times from more casual, opportunistic tool–makers to more habitual, committed tool–makers through time. Interestingly, in Bed I and Lower and Middle Bed II times both *Homo habilis* and *Australopithecus boisei* are known; it is during Upper Bed II times at Olduvai that the first fossils definitely attributed to *Homo erectus* are found, which may herald a major evolutionary change as well as significant technological and behavioral changes through time. It is also in Upper Bed II that *A. boisei* makes its last appearance at Olduvai. Marean (1989) has suggested that stone tool–using hominids at Olduvai favored more closed habitats in lava–dominated parts of the Olduvai basin prior to 1.6–1.7 MYA, but afterward shifted to a broader use of both open and closed habitats as they adapted to changing, drier environmental conditions.

Surprisingly, little geographic patterning was seen in the proportion of quartz, the proportion of subspheroids and spheroids, or the size of quartz subspheroids, spheroids or cores in relation to the proximity of Olduvai sites to the Naibor Soit quartz inselberg at Olduvai. This suggests that the 1–4 km transport distances of quartz seen at Olduvai were not high enough to show profound geographic patterning (or, alternatively, there were other sources of quartz in the basin margin that were used by early hominids but that cannot be found today). In any case, it would appear that Early Stone Age hominids were habitually transporting stones at least several kilometers from their sources and accumulating them in significant numbers at discrete concentrations on the ancient landscapes, sometimes with large quantities of fragmented mammalian remains as well.

We hope that this case study shows how a combined use of archeological fieldwork, experimental archeology, lithic assemblage analysis, and spatial analysis can help shed light on a long–standing problem in the paleolithic record. It is hoped that further research will hopefully clarify relationships between raw material types, flaking techniques and strategies, tool functions, stone artifact morphologies, paleolithic site assemblage compositions, and land–use patterns throughout the human evolutionary record.

ACKNOWLEDGMENTS

Aspects of this research were made possible by grants from the L. S. B. Leakey Foundation, Indiana University, Council for International Exchange of Scholars (Fulbright Research Fellowship), and the Social Science Research Council. We would like to extend our appreciation to the staff of the Livingstone Museum in Zambia for their assistance. Special thanks to Frank Brown, Desmond Clark, Richard Hay, Paul Menaga, Bob Walter, and Tim White for feedback on various aspects of this chapter. Finally, we would like to acknowledge the invaluable set of data that was generated by Mary Leakey in her dilligent excavations and detailed publication of the Bed I and II archeological occurrences at Olduvai Gorge (1971).

LITERATURE CITED

Arambourg, C., and Balout, L. (1952) Du nouveau a l'Ain Hanech. *Bull. Soc. Hist. Afr. Nord* 43:152–169.

Bar–Yosef, O., and Tchernov, E. (1972) *On the Paleo–ecological History of the Site of 'Ubeidiya*. Jerusalem: Israel Academy of Science and Humanities.

Balout, L. (1955) *Prehistoire de l'Afrique du Nord*. Paris: Arts et Metiers Graphiques.

Biberson, P. (1961) *Le Paleolithique Inferieur du Maroc Atlantique*. Morocco: Service des Antiquites, Publication No. 17.

Biberson, P. (1966) Galets amenages du Maghreb et du Sahara. Paris: Musee Nationale d'Histoire Naturelle, *Fiches Typologiques Africaines* 2:33–64.

Bilsborough, A. (1992) *Human Evolution*. London: Blackie Academic & Professional.

Blumenschine, R. J., and Masao, F. T. (1991) Living sites at Olduvai Gorge, Tanzania? Preliminary landscape archeology results in the basal Bed II lake margin zone. *J. Hum. Evol.* 21:451–462.

Brain, C. K. (1981) *The Hunters or the Hunted? An Introduction to African Cave Taphonomy*. Chicago: University of Chicago Press.

Bunn, H. T., Harris, J. W. K., Isaac, G., Kaufulu, Z., Kroll, E., Schick, K., Toth, N., and Behrensmeyer, A. K. (1980) FxJj 50: An early Pleistocene site in northern Kenya. *World Archaeol.* 12:109–136.

Chavaillon, J. (1970) Decouverte d'un niveau Oldowayen dans la basse vallee de l'Omo (Ethiopie). *Bull. Soc. Prehist. Francaise* 67:7–11.

Chavaillon, J. (1976) Evidence for the technical practices of early Pleistocene hominids. In Y. Coppens, F. C. Howell, G. L. Isaac, and R. E. F. Leakey (eds): *Earliest Man and Environments in the Lake Rudolf Basin*. Chicago: University of Chicago Press, pp. 565–573.

Chavaillon, J. (1979) From the Oldowan to the Middle Stone Age at Melka Kunture (Ethiopia): Understanding cultural changes. *Quaternaria* 21:87–114.

Clark, G. (1971) *World Prehistory: A New Outline*. Cambridge: Cambridge University Press.

Clark, J. D. (1980) Raw material and African lithic technology. *Man Env.* 4:44–55.

Clark, J. D. (ed.) (1982) *The Cambridge History of Africa*, Vol. I: *From the Earliest Times to 500 BC*. Cambridge: Cambridge University Press.

Clark, J. D. (1991) Stone artifact assemblages from Swartkrans, Transvaal, South Africa. In J. D. Clark (ed.) *Cultural Beginnings: Approaches to Understanding Early Hominid Life—Ways in the African Savanna*. Bonn: Monographien Band 19. Dr. Rudolf Halbelt GMBH.

Clark, J. D., and Kurashina, H. (1976) New Plio–Pleistocene archaeological occurrences from the plain of Gadeb, Upper Webi basin, Ethiopia, and a statistical comparison of the Gadeb sites with other Early Stone Age assemblages. In J. D. Clark and G. L. Isaac (eds.): *Pretirage du Colloque V du IXe Congres de l'Union Internationale des Sciences Prehistoriques et Protohistoriques*. Nice, pp. 158–216.

Clark, J. D., and Kurashina, H. (1979) Hominid occupation of the east–central highlands of Ethiopia in the Plio–Pleistocene. *Nature* 282:33–39.

Corvinus, G. (1975) Paleolithic remains at the Hadar in the Afar region. *Nature* 256:468–471.

Corvinus, G. (1976) Prehistoric exploration at Hadar, Ethiopia. *Nature* 261:571–572.

Corvinus, G. and Roche, H. (1976) La prehistoire dans la region d'Hadar (basin de l'Awash, Afar, Ethiopie): premiers resultats. *L'Anthropologie* 80:315–324.

Corvinus, G. and Roche, H. (1980) Prehistoric exploration at Hadar in the Afar (Ethiopia) in 1973, 1974, and 1976. In R. E. F. Leakey and B. A. Ogot (eds.): *Proceedings of the VIIIth Panafrican Congress of Prehistory and Quaternary Studies, Nairobi 1977*. Nairobi: The International Louis Leakey Memorial Institute for African Prehistory, pp. 186–188.

Foley, R. (1987) *Another Unique Species: Patterns in Human Evolutionary Ecology*. Harlow, Essex: Longman.

Goren, N. (1981) The Lithic Assemblages of the Site of 'Ubeidiya, Jordan Valley. Jerasulem. Ph.D. Dissertation, Institute of Archeology, Hebrew University.

Gowlett, J. A., Harris, J. W. K., Walton, D., and Wood, B. A. (1981) Earliest archaeological sites, hominid remains and traces of fire from Chesowanja, Kenya. *Nature* 292:125–129.

Harris, J. W. K. (1983) Cultural beginnings: Plio–Pleistocene archaeological occurrences from the Afar, Ethiopia. *Afr. Archeol. Rev.* 1:3–31.

Harris, J. W. K., and Gowlett, J. A. (1980) Evidence of early stone industries at Chesowanja, Kenya. In R. E. F. Leakey and B. A. Ogot (eds.): *Proceedings of the VIIIth Panafrican Congress of Prehistory and Quaternary Studies, Nairobi 1977*. Nairobi: The International Louis Leakey Memorial Institute for African Prehistory, pp. 208–212.

Harris, J. W. K. and Isaac, G. L. (1976) The Karari Industry: Early Pleistocene archaeological evidence from the terrain east of Lake Turkana, Kenya. *Nature* 263:738–740.

Harris, J. W. K., Williamson, P., Verniers, J., Tappen, M., Stewart, K., Helgren, D., de Heinzelin, J., Boaz, N., and Bellomo, R. (1987) Late Pliocene hominid occupation in central Africa: The setting, context, and character of the Senga 5 site, Zaire. *J. Hum. Evol.* 16:701–728.

Hay, R. L. (1971) Geological background of Beds I and II: stratigraphic summary. In M. D. Leakey: *Olduvai Gorge*, Vol. 3: *Excavations in Beds I and II 1960–1963*. Cambridge: Cambridge University Press, pp. 9–18.

Hay, R. L. (1976) *Geology of the Olduvai Gorge*. Berkeley: University of California Press.

Hill, A., Ward, S., Deino, A., Curtis, G., and Drake, G. (1992) Earliest *Homo*. *Nature* 355:719–722.

Howell, F. C., Cole, C., and Kleindienst, M. (1962) Isimila, an Acheulian occupation site in the Iringa highlands. *Actes du IVe Congres Panafricaine de Prehistoire et de l'Etude du Quaternaire*. Tervuren: Musee Royale de l'Afrique Centrale, pp. 43–80.

Howell, F. C. (1966) Observations on the earlier phases of the European Lower Paleolithic. *Am. Anthropol.* 68 (2/2):88–201.
Howell, F. C. (1976) Overview of the Pliocene and earlier Pleistocene of the lower Omo basin, southern Ethiopia. In G. L. Isaac and E. R. McCown (eds): *Human Origins: Louis Leakey and the East African Evidence*. Menlo Park, CA: W. A. Benjamin, pp. 227–268.
Howell, F. C. (1978) Hominidae. In V. J. Maglio and H. B. S. Cooke (eds.): *Evolution of African Mammals*. Cambridge Harvard University Press, pp. 154–248.
Howell, F. C. (1982) Origins and evolution of African Hominidae. In J. D. Clark (ed.): *The Cambridge History of Africa*, Vol. 1: *From the Earliest Times to 500 BC*. Cambridge: Cambridge University Press, pp. 70–156.
Howell, F. C. (1989) The evolution of human hunting (Book review). *J. Hum. Evol.* 18:583–594.
Howell, F. C., and Coppens, Y. (1976) An overview of Hominidae from the Omo succession, Ethiopia. In Y. Coppens, F. C. Howell, G. L. Isaac, and R. E. F. Leakey (eds.): *Earliest Man and Environments in the Lake Rudolf Basin: Stratigraphy, Paleoecology, and Evolution*. Chicago: University of Chicago Press, pp. 522–532.
Howell, F. C., Haesaerts, P., and de Heinzelin, J. (1987) Depositional environments, archeological occurrences and hominids from Member E and F of the Shungura Formation (Omo Basin, Ethiopia). *J. Hum. Evol.* 16:665–700.
Isaac, G. L. (1967) The stratigraphy of the Peninj group—early Middle Plestocene formations west of Lake Natron, Tanzania. In W. W. Bishop and J. D. Clark (eds.): *Background to Evolution in Africa*. Chicago: University of Chicago Press, pp. 229–257.
Isaac, G. L. (1980) Casting the net wide: A review of archaeological evidence for early hominid land–use and ecological relations. In L. K. Konigsson (ed.): *Current Arguments on Early Man*. Oxford: Pergamon, pp. 226–251.
Isaac, G. L. (1981) Stone Age visiting cards: Approaches to the study of early land use patterns. In I. Hodder, G. L. Isaac, and N. Hammond (eds.): *Patterns of the Past*. Cambridge: Cambridge University Press, pp. 131–155.
Isaac, G. L. (1982) The earliest archaeological traces. In J.D . Clark (ed.): *The Cambridge History of Africa*, Vol. 1: *From the Earliest Times to 500 BC*. Cambridge: Cambridge University Press, pp. 157–247.
Isaac, G. L. (1984) The archaeology of human origins: Studies of the Lower Pleistocene in East Africa 1971–1981. *Advances in World Archaeology* 3:1–87.
Isaac, G. L., and Harris, J. W. K. (1978) Archaeology. In M. G. Leakey and R. E. F. Leakey (eds.): *Koobi Fora Research Project*, Vol. 1: *Fossil Hominids and an Introduction to Their Context 1968–1974*. Oxford: Clarendon Press, pp. 64–85.
Johanson, D. C., Masao, F. T., Eck, G. G., White, T. D., Walter, R. C., Kimbel, W. H., Asfaw, B., Manega, P., Ndessokia, P., and Suwa, G. (1987) New partial skeleton of *Homo habilis* from Olduvai Gorge, Tanzania. *Nature* 327:205–209.
Jones, P. (1979) Effects of raw materials on biface manufacture. *Science* 204:835–836.
Kaufulu, Z. M., and Stern, N. (1987) The first stone artifacts to be found *in situ* within the Plio–Pleistocene Chiwondo Beds in Northern Malawi. *J. Hum. Evol.* 16:729–740.
Kibunjia, M., Roche, H., Brown, F., and Leakey, R. (1992) Pliocene and Pleistocene archeological sites west of Lake Turkana. *J. Hum. Evol.* 23:431–438.
Klein, R. G. (1989) *The Human Career: Human Biological and Cultural Origins*. Chicago: University of Chicago Press.
Leakey, M. D. (1970) Stone artefacts from Swartkrans. *Nature* 225:1222–1225.
Leakey, M. D. (1971) *Olduvai Gorge*, Vol. 3: *Excavations in Beds I and II 1960–1963*. Cambridge: Cambridge University Press.
Leakey, M. D. (1975) Cultural patterns in the Olduvai sequence. In K. W. Butzer and G. L. Isaac (eds.): *After the Australopithecines*. The Hague: Mouton, pp. 477–493.
Leakey, R., and Lewin, R. (1992) *Origins Reconsidered: In Search of What Makes Us Human*. New York: Doubleday.
Manega, P. C. (1993) Geochronology, Geochemistry and Isotopic Study of the Plio-Pleistocene Hominid Sites and the Ngorongoro Volcanic Highlands in Northern Tanzania. Ph.D. Dissertation, University of Colorado.
Marean, C. W. (1989) Sabretooth cats and their relevance for early hominid diet and evolution. *J. Hum. Evol.* 18(6):559–582.
Mason, R. (1962) Australopithecines and artefacts at Sterkfontein. Part 2. The Sterkfontein artefacts and their makers. *S. Afr. Archaeol. Bull.* 17(66):109–125.
Merrick, H. V. (1976) Recent archaeological research in the Plio–Pleistocene deposits of the lower Omo Valley, southwestern Ethiopia. In G. L. Isaac and E. R. McCown (eds.): *Human Origins: Louis Leakey and the East African Evidence*. Menlo Park, CA: W. A. Benjamin, pp. 461–482.
Merrick, H. V. and Merrick, J. P. S. (1976) Archaeological occurrences of earlier Pleistocene age from the Shungura Formation. In Y. Coppens, F. C. Howell, G. L. Isaac, and R. E. F. Leakey (eds.): *Earliest Man and Environments in the Lake Rudolf Basin*. Chicago: University of Chicago Press, pp. 574–584.
Potts, R. (1988) *Early Hominid Activities at Olduvai*. New York: Aldine de Gruyter.
Potts, R. (1992) Why the Oldowan? Plio–Pleistocene tool–making and the transport of resources. *J. Anthropol. Res.* 47:153–176.
Reineck, H. and Singh, I. (1980) *Depositional Sedimentary Environments*. Berlin: Springer-Verlag.
Robinson, J. T. (1962) Australopithecines and artefacts at Sterkfontein. Part 1. Sterkfontein stratigraphy and the significance of the Extension Site. *S. Afr. Archaeol. Bull.* 27(66):87–107.
Roche, H. (1980) *Premiers Outils Tailles d'Afrique*. Paris: Societe d'Ethnographie.
Sahnouni M (1987) *L'Industrie sur Galets du Gisement Villafranchian Superieur de Ain Hanech*. Algers: Office des Publications Universitaires.

Schick, K. D. (1986) *Stone Age Sites in the Making: Experiments in the Formation and Transformation of Archaeological Occurrences.* Oxford: British Archaeological Reports International Series 319.

Schick, K. D. (1987a) Experimentally derived criteria for assessing hydrologic disturbance of archeological sites. In D. T. Nash and M. D. Petraglia (eds.): *Natural Formation Processes and the Archaeological Record.* Oxford: British Archaeological Reports International Series 352, pp. 86–107.

Schick, K. D. (1987b) Modeling the formation of Early Stone Age artifact concentrations. *J. Hum. Evol.* 16:789–808.

Schick, K. D. (1991) On making behavioral inferences from early archaeological sites. In J. D. Clark (ed.): *Cultural Beginnings: Approaches to Understanding Early Hominid Life–ways in the African Savanna.* Bonn: Dr. Rudolf Habelt GMBH Monographien, Band 19, pp. 79–107.

Schick, K. D., and Toth, N. (1993) *Making Silent Stones Speak: Human Evolution and the Dawn of Technology.* New York: Simon & Schuster.

Stekelis, M. (1966) *Archeological excavations at 'Ubeidiya 1960–1963.* Jerusalem: The Israel Academy of Sciences and Humanities.

Stiles, D. N. (1979) Early Acheulean and Developed Oldowan. *Curr. Anthropol.* 20:126–129.

Susman, R. L. (1988) Hand of *Paranthropus robustus* from Member I, Swartkrans: Fossil evidence for tool behavior. *Science* 240:781–784.

Susman, R. L. (1992) Who made the Oldowan tools? Fossil evidence for tool behavior in Plio–Pleistocene hominids. *J. Anthropol. Res.* 47 (2):129–151.

Tobias, P. V. (1991) *Olduvai Gorge,* Vol. 4: *The Skulls, Endocasts, and Teeth of Homo habilis.* Cambridge: Cambridge University Press.

Toth, N. (1982) The Stone Technologies of Early Hominids at Koobi Fora: An Experimental Approach. Ph.D. Dissertation, University of California, Berkeley.

Toth, N. (1985) The Oldowan reassessed: A close look at early stone artifacts. *J. Archaeol. Sci.* 12:101–120.

Toth, N. (1987) Behavioral inferences from early stone artifact assemblages: An experimental model. *J. Hum. Evol.* 16:763–787.

Toth, N. (1991) The importance of experimental replicative and functional studies in Paleolithic archaeology. In J. D. Clark (ed.): *Cultural Beginnings: Approaches to Understanding Early Hominid Life–ways in the African Savanna.* Bonn: Dr. Rudolf Habelt GMBH Monographien, Band 19, pp. 109–124.

Toth, N., and Schick, K. D. (1986) The first million years: The archaeology of protohuman culture. *Advances in Archaeological Method and Theory* 9:1–96.

Walker, A. (1981) The Koobi Fora hominids and their bearing on the origins of the genus *Homo.* In B. A. Sigmon and J. S. Cybulski (eds.): *Homo erectus: Papers in Honor of Davidson Black.* Toronto: University of Toronto Press, pp. 193–215.

Walter, R. C., Manega, P. C., Hay, R. L., Drake, R. E., and Curtis, G. H. (1991) Laser–fusion $^{40}Ar/^{39}Ar$ Dating of Bed I, Olduvai Gorge, Tanzania. *Nature* 354:145–149.

Walter, R. C., Manega, P. C., and Hay, R. L. (1992) Tephrochronology of Bed I, Olduvai Gorge: An Application of Laser–fusion $^{40}Ar/^{39}Ar$ dating to calibrating biological and climatic change. *Quat. Int.* 13/14:37–46.

Willoughby, P. R. (1985) Spheroids and battered stones in the African Early Stone Age. *World Archaeol.* 17:44–60.

Willoughby, P. R. (1987) Spheroids and battered stones in the African Early and Middle Stone Age. *Cambridge Monographs in African Archaeology* 17.

Wood, B. (1991) *Koobi Fora Research Project 4: Hominid Cranial Remains from Koobi Fora.* Oxford: Clarendon.

Wood, B. (1992) Origin and evolution of the genus *Homo.* *Nature* 355:783–790.

CHAPTER 22

The Acheulian Industrial Complex in Africa and Elsewhere

J. Desmond Clark

INTRODUCTION

It gives me much pleasure to discuss the Acheulian Industrial Complex, as it is called, because it was the Acheulian that gave Clark Howell his introduction to fieldwork in Africa and it was the first time that he and I met in central Tanganyika in 1956 when he, Maxine Kleindienst, and Glen Cole were excavating at the Isimila Karonga near Iringa. This was one of the first well-excavated open-air sites with a rich series of stratified and alternating Acheulian and core and flake assemblages. It has also yielded some of the largest and heaviest bifaces I have ever seen. It is one of those sites where the assemblages show varying proportions of biface, core, and flake components, and where, sometimes, the bifaces are absent. This variable composition is universal throughout the whole extent of the Old World where this complex is found (Howell et al. 1962, 1972). (See Figure 22-1.)

The Acheulian, first recognized from the Somme terrace gravels near Amiens in northern France, was characterized initially by large, bifacially flaked handaxes (Bordes 1984). Later, however, at the turn of the century when relatively undisturbed activity areas at localities around St. Acheul were discovered and excavated (e.g., Comment's Workshop and other sites: Howell 1966:167-169; Bordes and Fitte 1953), the full range of bifaces, cores, flakes and retouched flakes became apparent. It is important that this should be realized as subsequently, in East Africa, the discovery was made of the oldest yet known Industrial Complex, the Oldowan, and it was found in stratified sediments prior to the appearance of the Acheulian at Olduvai Gorge. The Oldowan in its later and "developed" stages and the Acheulian (Leakey 1971:124-222, 1975) have tended to be seen as two separate and distinct entities even though they existed contemporaneously for some 1.5 million years. When the possibility existed of two contemporaneous hominid species each responsible for a separate technological tradition, the dual phyla hypothesis received some support. The greatly increased understanding of hominid biological evolution now available shows that both these complexes were, from 1.5 million years ago, made first by *Homo erectus* and later, from

FIGURE 22-1 Comparisons of Acheulian assemblages to show component variability through time at two stratified sites: Kalambo Falls and Isimila. At Kalambo Falls, Zambia, six stratigraphically related assemblages (earliest at the bottom) from the Mkamba Member of the Kalambo Falls Formation are presented while at Isimilia, Tanzania, six assemblages from Sandstones 1a, 1b and 3 from Isimilia korongo are illustrated. (After Clark 1975)

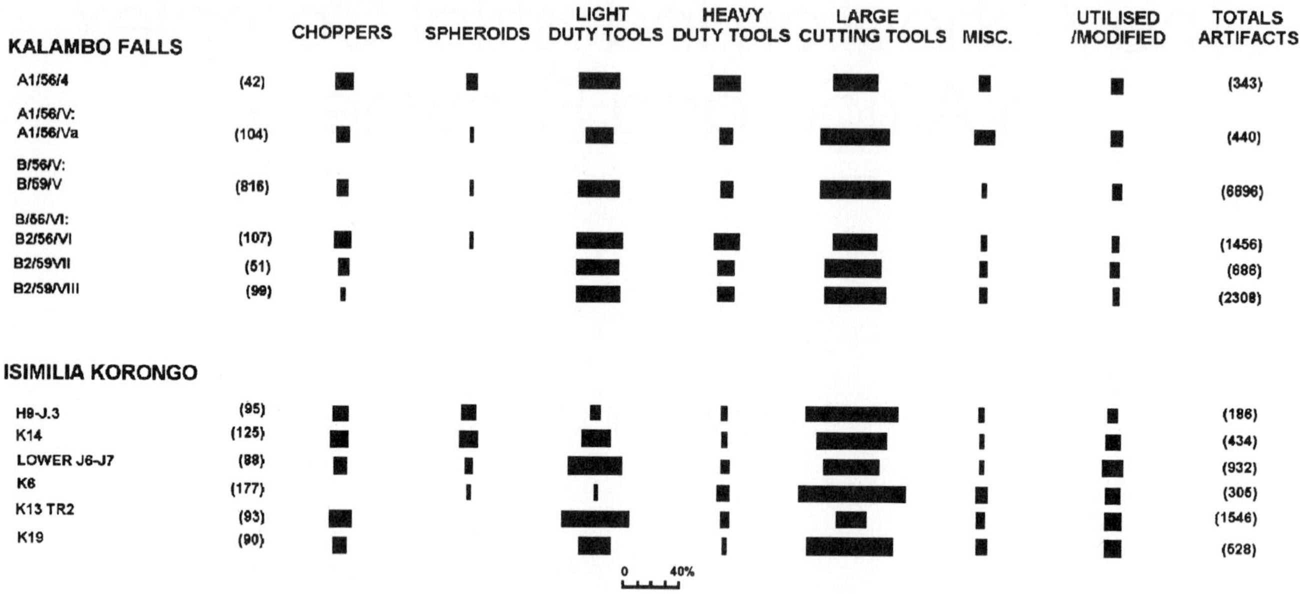

KALAMBO FALLS		CHOPPERS	SPHEROIDS	LIGHT DUTY TOOLS	HEAVY DUTY TOOLS	LARGE CUTTING TOOLS	MISC.	UTILISED /MODIFIED	TOTALS ARTIFACTS
A1/56/4	(42)								(343)
A1/56/V: A1/56/Va	(104)								(440)
B/56/V: B/59/V	(816)								(6696)
B/56/VI: B2/56/VI	(107)								(1456)
B2/59VII	(51)								(686)
B2/59/VIII	(99)								(2308)
ISIMILIA KORONGO									
H9-J.3	(95)								(186)
K14	(125)								(434)
LOWER J6-J7	(88)								(932)
K6	(177)								(305)
K13 TR2	(93)								(1546)
K19	(90)								(628)

0 40%

about 0.5 million or more years ago, by Archaic *Homo sapiens*. It must now be accepted that the stone tool assemblages resulted from the activities of these two grades of hominid and that the variability the assemblages show is a product of a number of interacting factors, some ecological, some cultural and some, possibly, psychological (Figure 22-2).

Throughout the range of time and space in which the biface component is found, a high percentage of the assemblages show that the core/flake component is invariably present. This is well seen in numbers of assemblages from Africa, Europe, and western and southern Asia. There are also relatively rare

FIGURE 22-2 Comparison of assemblage variablity at Acheulian and Developed Oldowan sites in East and South Africa. Middle Pleistocene industrial assemblages in the "Open Savanna" and "Closed Savanna"/Forest Culture Areas of Sub-Saharan Africa are used in this comparison. (After Clark 1975)

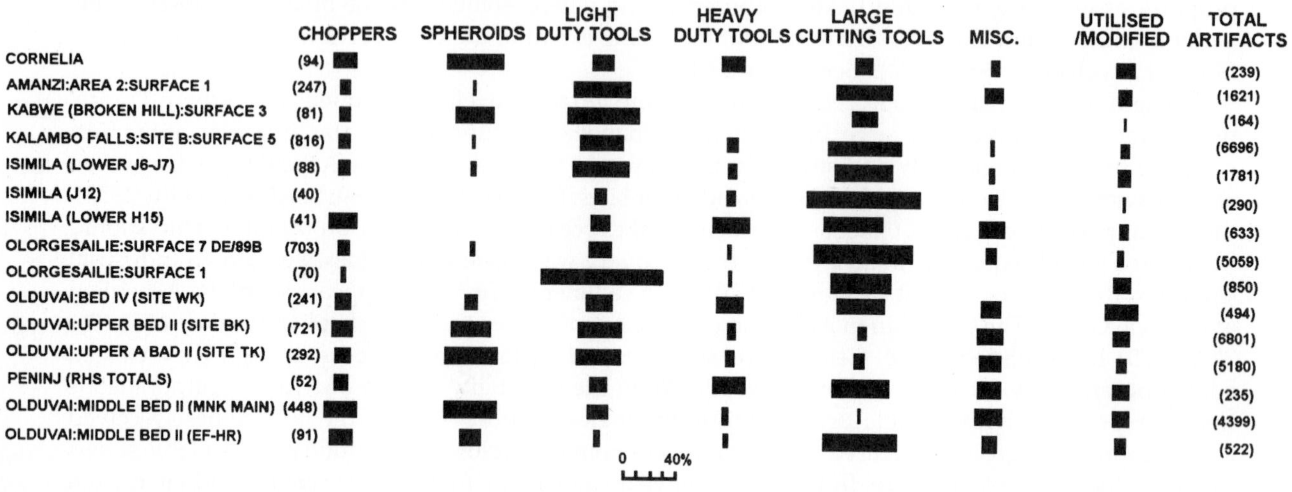

		CHOPPERS	SPHEROIDS	LIGHT DUTY TOOLS	HEAVY DUTY TOOLS	LARGE CUTTING TOOLS	MISC.	UTILISED /MODIFIED	TOTAL ARTIFACTS
CORNELIA	(94)								(239)
AMANZI:AREA 2:SURFACE 1	(247)								(1621)
KABWE (BROKEN HILL):SURFACE 3	(81)								(164)
KALAMBO FALLS:SITE B:SURFACE 5	(816)								(6696)
ISIMILA (LOWER J6-J7)	(88)								(1781)
ISIMILA (J12)	(40)								(290)
ISIMILA (LOWER H15)	(41)								(633)
OLORGESAILIE:SURFACE 7 DE/89B	(703)								(5059)
OLORGESAILIE:SURFACE 1	(70)								(850)
OLDUVAI:BED IV (SITE WK)	(241)								(494)
OLDUVAI:UPPER BED II (SITE BK)	(721)								(6801)
OLDUVAI:UPPER A BAD II (SITE TK)	(292)								(5180)
PENINJ (RHS TOTALS)	(52)								(235)
OLDUVAI:MIDDLE BED II (MNK MAIN)	(448)								(4399)
OLDUVAI:MIDDLE BED II (EF-HR)	(91)								(522)

0 40%

occurrences where only bifaces and unmodified flakes are present though these might represent only a part of the total assemblage if more extensive areas of the occupation horizon had been excavated. Again, other assemblages are totally lacking in bifaces and this is the case in all the regions of the Old World where the biface component was present. Moreover, in considerable areas of Europe and Asia, bifaces were not made and the core/flake component alone represents the tool kit of Middle Pleistocene hominids (Klein 1989) (Figure 22-3). This is the case in eastern Europe and in eastern and central Asia, though some have claimed that biface assemblages are found there, and also in southeast Asia and the Far East. Possible explanations for this regional division will be discussed later. Where bifaces are present with varying percentages of the core/flake component, the assemblage is usually recognized as Acheulian. Where they are absent, the assemblages are known as Developed Oldowan (in Africa) or chopper/chopping tool assemblages in Asia. The latter term is now becoming obsolete and Core/Flake Complex is preferred by this author. We are concerned here with the Acheulian and with the cause and effect of the regional cultural dichotomy.

An Acheulian Industry is usually defined as having varying proportions of large bifacial tools—handaxes, cleavers, picks and other such large tools—associated with cores of different kinds from which flakes were removed by hard hammer technique and sometimes reworked. This retouching of both bifaces and flake tools is the result often of flaking with a soft hammer. In many regions, due to the way in which the raw material occurred, the bifaces were made from large flakes and sometimes this required special preparation of the nucleus prior to removal of the flake; this will be discussed later. The Acheulian shows in its bifaces an attempted standardization and symmetry of form and is the first Industrial Complex to do this. The overall impression given by the Acheulian is of conservative conformity within the parameters

FIGURE 22-3 Distribution of Acheulian and Core/Flake Techno-Complexes in the Old World.

of the technology and the "mental templates" persisting throughout the long range of time that the complex is known to have existed (Figure 22-4). Other than the variability necessitated by the raw material used, no regional specialization such as becomes visible in the Middle Paleolithic/Middle Stone Age is identifiable other than, of course, the biface and core/flake dichotomy (Villa 1981; Jones 1981).

Temporal changes, however, *are* apparent and imply very slow evolutionary development without any periods of rapid modification, except in the terminal stages. The first appearance of bifaces by which the Acheulian is recognized denotes a sudden change, an innovation, with the addition of large tools to the old core/flake tool kit. It has been suggested that the Acheulian biface may be derived from the proto-biface of the Oldowan but this is nowhere demonstrated, and early Acheulian bifaces are very different from proto-bifaces. Similar changes cannot be recognized within the stone artifact assemblages of the Middle Pleistocene hominids and, although changes do occur, they appear to be gradual, not speedy. Of course, it is possible that when dating methods have been further refined, this view will need to be modified.

The Acheulian makes its first appearance 1.6-1.5 MYA in dated contexts in East Africa and Ethiopia and it persisted until about 200 KYA, perhaps in some regions into less than 100 KYA but, as yet, its termination is not well dated. It lasted, therefore, about 1.5 million years. This long-lasting technocomplex can be divided into three chronologically separate stages: an earlier, a later, and a terminal stage. Each is characterized by the relative frequency of certain tool forms and technology. All three stages are present in Africa, but apart from the Jordan Rift site of 'Ubeidiya, the earlier stage is not clearly seen in Eurasia.

EARLIER ACHEULIAN

Earlier Acheulian assemblages have been recognized in East, South, and North Africa on the basis of chronostratigraphy. Until recently the biface assemblage from the EF-HR site in Middle Bed II at Olduvai Gorge, occurring above the faunal break and dating to c. 1.4 MYA, was the oldest known appearance of Acheulian handaxes and cleavers. Although no hominid remains were associated with this assemblage the *H. erectus* calvarium from the LLK II locality in the upper part of Bed II suggests that the Acheulian was introduced by this earliest large hominid. In 1991 the Paleoanthropological Inventory Project of Ethiopia found, in the Konso-Gardula area of the Ethiopian Rift, rich Acheulian assemblages in 10 stratified, thin horizons with associated fauna (Asfaw et al. 1992). Interstratified tephra samples, dated by laser-fusion 40Ar/39Ar method, show that the age of the sediments with Acheulian lies at 1.3 MYA at the top and 1.6-1.8 MYA in the lower part of the sequence. This sequence equates, therefore, with the Chari and possibly Okote Tuffs at Koobi Fora. The assemblages are dominated by handaxes and trihedral picks (some large:—more than 250 mm in length). Flakes are mostly unworked and retouched small tools appear to be absent. A half mandible from one locality is attributed to *H. erectus*. I am indebted to my colleague, Dr. T. D. White, for this information. In the Middle Awash area of the Afar Rift, the locality of Dakanihyalo (Figure 22-5) has yielded similar bifaces with trihedral picks and a fauna of comparable age to Konso-Gardula but, as yet, there is no radiometric age (Clark et al. 1984). In the Lake Natron Basin two artifact horizons have, similarly, produced early forms of handaxe, cleaver, and pick with the core/flake component at Peninj in the upper part of the Humbu Formation and dated between 1.6 and 1.4 MYA (Isaac 1967).

In South Africa, rare handaxes and cleavers occur with the core/flake component in Member 5 of the Sterkfontein Cave sequence, associated with a hominid calvaria ascribed to *H. habilis* and thought to date to c. 1.6 MYA (Clarke 1985).

In northwest Africa at the Sidi Abderrahman quarries of the Atlantic coast of Morocco at Casablanca, the early Acheulian occurs in sediments at the end of an early transgression beach (Maarifian). The fresh assemblage from the STIC quarry with handaxes, trihedral picks, and cleavers is typologically early and equated with deposits in another quarry (Thomas I) with remains of *H. erectus*. A minimum age of 0.7 MYA has been obtained but the true age is regarded as being considerably older (Raynal and Texier 1989; Biberson 1961). On the Algerian plateau, the lake site of Ternifine (Tighenif) yielded similar early

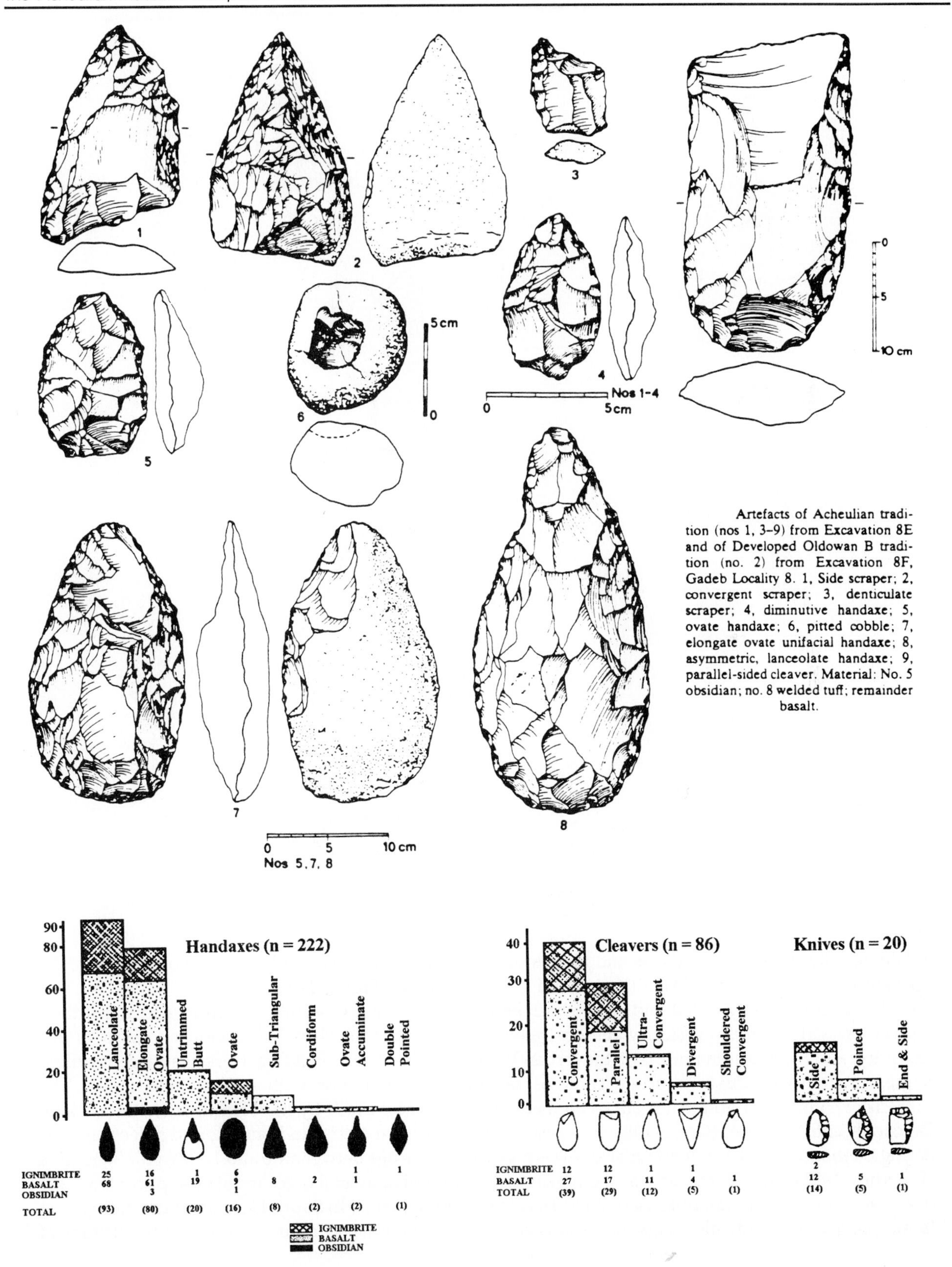

FIGURE 22-4 Acheulian artifact forms and raw materials from Gadeb, Ethiopia. (After Clark et al. 1984)

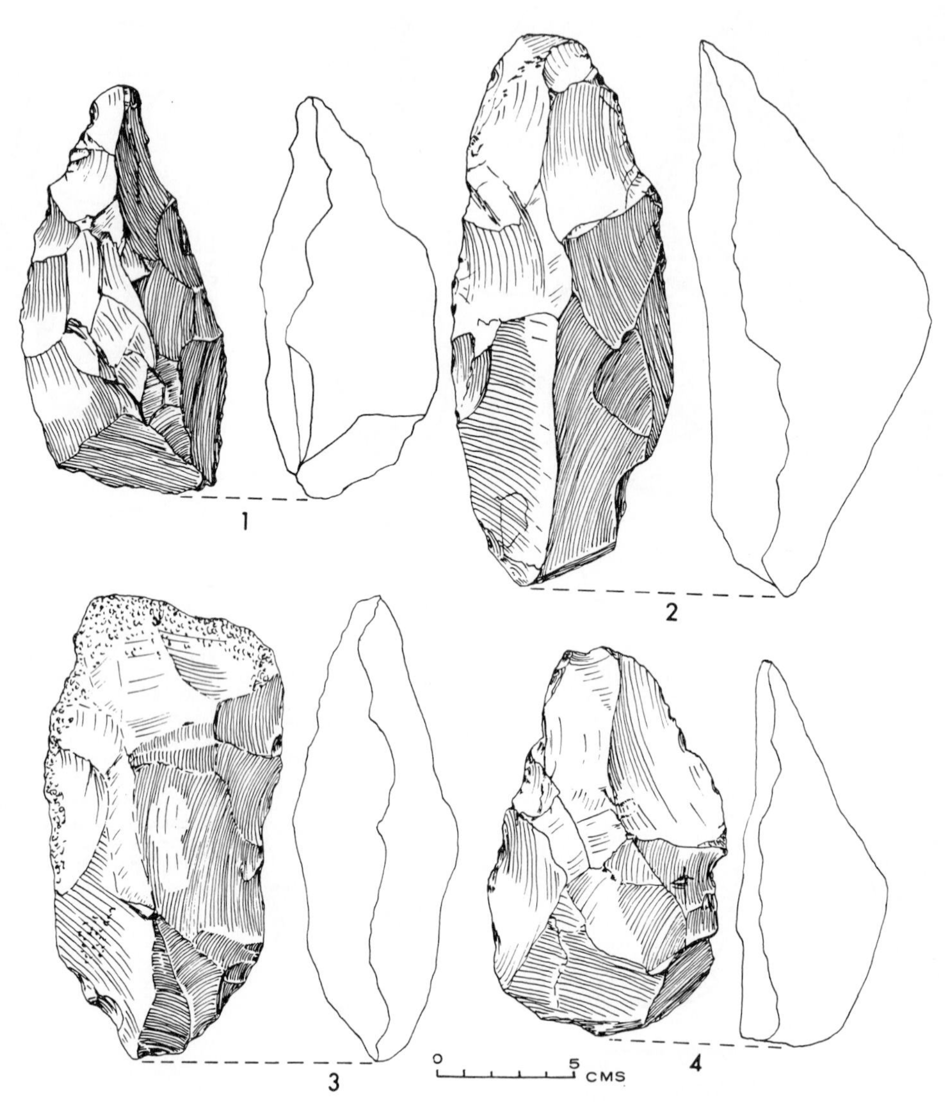

FIGURE 22-5 Earlier Acheulian trihedral picks (1 and 2), cleaver (3), and handaxe (4) in quartzite from Dakanihyalo, Middle Awash Valley, Ethiopia.

Acheulian bifaces, as well as flakes and flake tools with three mandibles and a parietal of *H. erectus;* the age is thought to be about 0.7 MYA (Balout and Tixier 1957). (See Figure 22-6.)

These are the only dated and well-researched early Acheulian assemblages from Africa but their association with *H. erectus* is clearly demonstrated. The only other locality that is unquestionably of this antiquity, and the oldest in Eurasia, is the Jordan Valley, Israel, site of 'Ubeidiya. Here a faulted, stratified sequence of fluvio-lacustrine origin contains several beach horizons with artifacts and fauna. The core/flake component dominates although there are rare bifaces made by hard hammer technique with flake tools and spheroids (Goren 1981; Bar-Yosef and Goren-Inbar 1993). Radiometric and paleomagnetic dating methods as well as microfaunal analysis give a probable age between 1.0 and 1.4 MYA for 'Ubeidiya (Tchernov 1987). That *H. erectus* was the maker of the 'Ubeidiya assemblages receives support from the discovery in 1991 of a mandible ascribed to *H. erectus* from Dmanisi in South Georgia. If the dating—on faunal associations and paleomagnetism—is correct, this is of immense importance for dating the time of

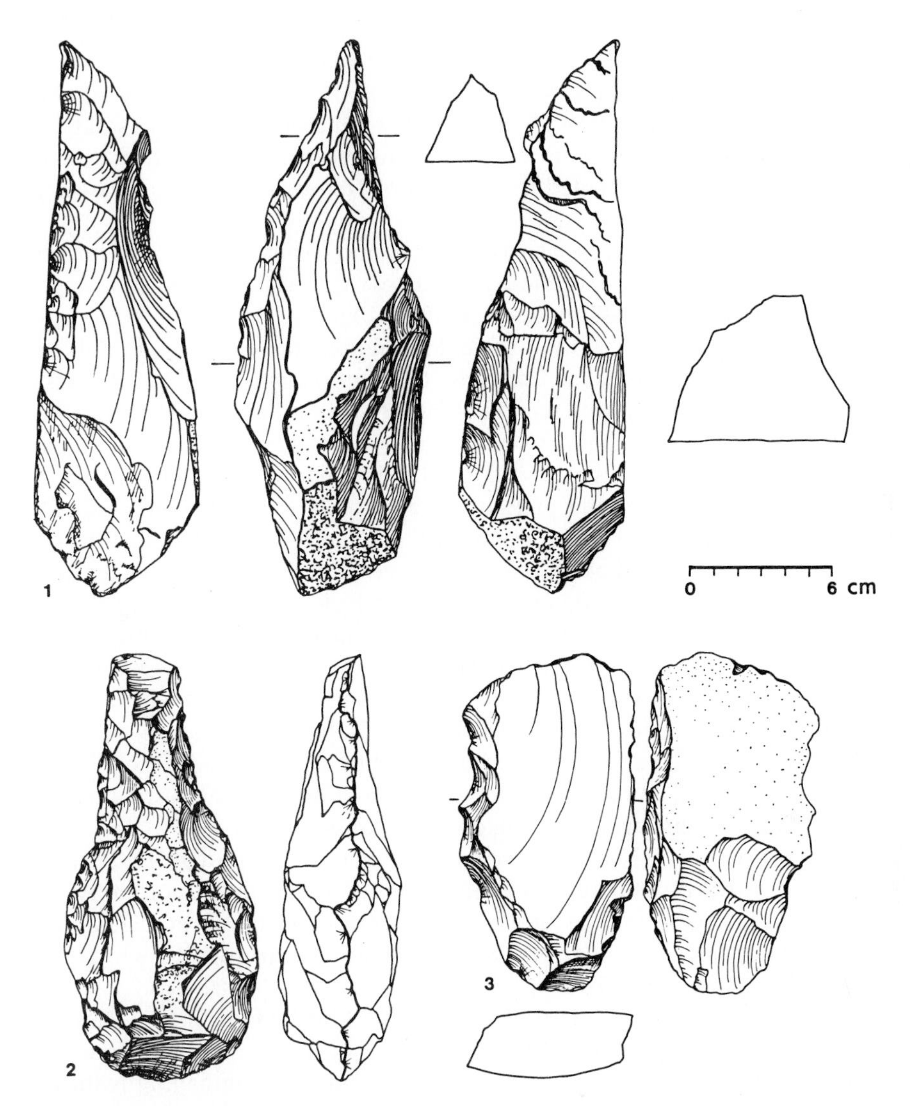

FIGURE 22-6 (1) Earlier Acheulian trihedral picks from STIC Quarry, Sidi Abderrahman, Casablanca, Morocco, (2) chisel-ended handaxe, and (3) proto-cleaver, both from Ternifine, Algeria (after Clark 1992a).

the migration of *H. erectus* into Asia or a possible challenge to the African origin of this hominid (Dzaparidze et al. 1989).

In these early assemblages the core/flake (Oldowan) component dominates with varying percentages of bifaces. These latter are all technologically simple, nearly always exhibiting deep, hard hammer flake scars. The trihedral picks are especially representative and they all point to an African origin for the Acheulian.

LATER AND TERMINAL ACHEULIAN

Large numbers of Acheulian sites that are assumed, or shown, to be of later age are found in almost all regions of the continent and in different ecological habitats, and there is a suggestion that demographic increase and improved technology may have been partly responsible. Most of these sites are undated, other than by fauna, and those that have been dated by radiometric or other means show a significant range of dates from c. 1.4 to less than 0.2 MYA. There is now a noticeable increase in soft hammer technique, but it is not possible to use this to demonstrate any evolving refinement since the nature of raw materials that were brought into use often controlled the amount of "retouch" that was necessary to produce an effective artifact.

Many of these stratified localities, precisely excavated and researched, are well known. In East Africa there are Olorgesailie (c. 1.0-0.7 MYA), Kariandusi (0.9-0.8 MYA), Isimila, and Kapthurin between 200 and 300 KYA but probably older. Other assemblages, such as Kilombe, come from strata showing reversed magnetism and so date to less than 78,000 years ago. In Ethiopia, the sequences at Melka Kunture date to 1.3-1.1 MYA and contain *H. erectus* remains; Gadeb on the high plateau dates to 1.4 to less than 0.78 MYA and Bodo and others in the Middle Awash section of the Afar Rift dated to c. 0.5 MYA (Klein 1989:209-212, 224-233; Clark et al. in prep). Bodo belongs with the Archaic *H. sapiens* population that evolved from *H. erectus*. In central Africa *H. rhodesiensis* is morphologically similar and, although undated radiometrically, is considered contemporary with a later Acheulian; as also is Elandsfontein with the "rhodesioid" calvarium and Lake Eyasi with its cranial fragments. All are associated with biface and core/flake components and could be 0.5 MYA old or somewhat younger. Even later assemblages, probably terminal Acheulian, come from many sites in the continent. Noted here are the mound spring site of Amanzi (Deacon 1970) and the fluvial context site at Kalambo Falls (Clark 1969), both with wood and other plant remains. Kalambo Falls also has good evidence for fire as does the Cave of Hearths (as its name implies) in the Transvaal in South Africa (Mason 1962:158-169). Also to be mentioned are the Central Sahara lakeside site of Erg Tihodaine (Arambourg and Balout 1955), the mound springs at Bir Sahara (with a minimum age of 350 KYA); Kharga and Dakhleh Oases in the eastern Sahara (Wendorf and Schild 1980: 21-24, 225-228, 243-247) and Sidi Zin in Tunisia, the last showing very interesting standardization and symmetry of form in succeeding stratified spring deposits (Gobert 1950).

A fluvial context site in the Gilf Kebir, located by radar imagery has an age of c. 270 KYA. Also terminal Acheulian, but undated, is the assemblage assigned to the refined Stage VIII of the Acheulian sequence in coastal Morocco. In addition to handaxes and cleavers, there is a range of retouched flake tools, a blade component and one tanged flake. It is older than the 6-8 m marine transgression which dates on Uranium series to 140-120 KYA (Biberson 1961:331-98).

I have previously attempted to show that the regional Middle Paleolithic/Middle Stone Age is a refined continuum of the late Acheulian, all the technological elements of which were already present in the earlier complex. The presence or absence of the bifaces is the distinguishing feature. At the eastern Sahara spring sites a form of backed knife (bifacial and unifacial) makes its appearance and can be compared with those found in the East African Acheulian and the Prondnik forms found in early Last Glacial contexts in Poland and Germany (Clark 1992a:30-31). (See Figure 22-7.) Another characteristic of these spring sites is the abundance of handaxes and the scarcity of flakes and small flake tools. The variability in tool form and component is well seen at some of these sites (Figure 22-8 on page 460) and is included here (on the evidence of two biface tips) as a terminal African Acheulian variant the Pre-Aurignacian, the equivalent of the Amudian in the Levant "Mugharan Tradition."

A significant feature of the later Acheulian is the variability in primary technology. This all shows a deep understanding of the properties of the raw materials which ranged from coarse (dolerite, quartzite, diabase, and silicified granite) to medium (various lavas, limestone, quartz, and fine quartzite) to fine (flint, chert, obsidian, and hornfels). Flaking treatment was varied according to texture, hardness, shape, and size. The technique used varied from boulder on boulder (throwing), to direct hammerstone percussion and special preparation of cores for obtaining large flakes from which to make bifaces. One such is the Kombewa technique that produces a flake with a flake surface on both faces. Another is the proto-Levallois method where dorsal face and platform are prepared to produce either side- or end-struck flakes (Figure 22-9 on page 461). The most elaborate and seemingly the most ingenious is the Tabal-

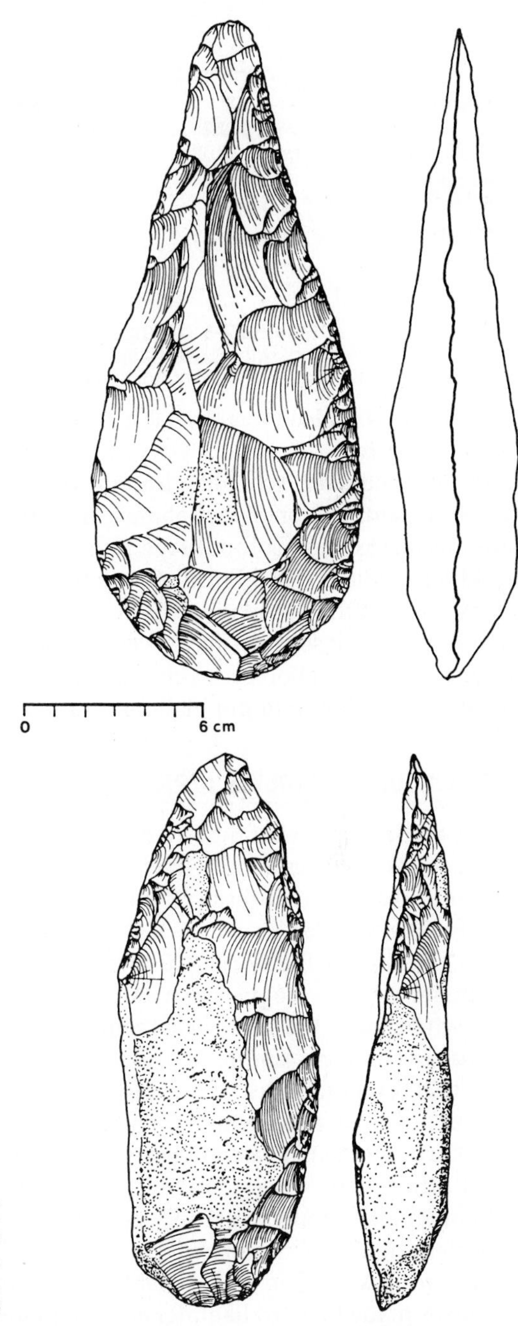

FIGURE 22-7
(1) Ovate acuminate handaxe and (2) Prondnik-type knife, both from Mound Spring K010, Kharga Oasis, Egypt. (after Clark 1992a)

balat/Tachengit method of preforming the biface (usually a cleaver) on the core before it is struck (Figure 22-10 on page 462). In the terminal Acheulian smaller Levallois cores are present, as in the Middle Paleolithic.

The determining factor controlling the distribution and dispersal of the Acheulian technocomplex can be seen as climate and environment, and it seems likely that ecology controlled where and when the Acheulian populations dispersed (Figure 22-11 on page 463). Nowhere is this better seen than in the Sahara from Penultimate Glacial (Riss [Saalian]) times onwards. The earlier migration into Eurasia around 1.0

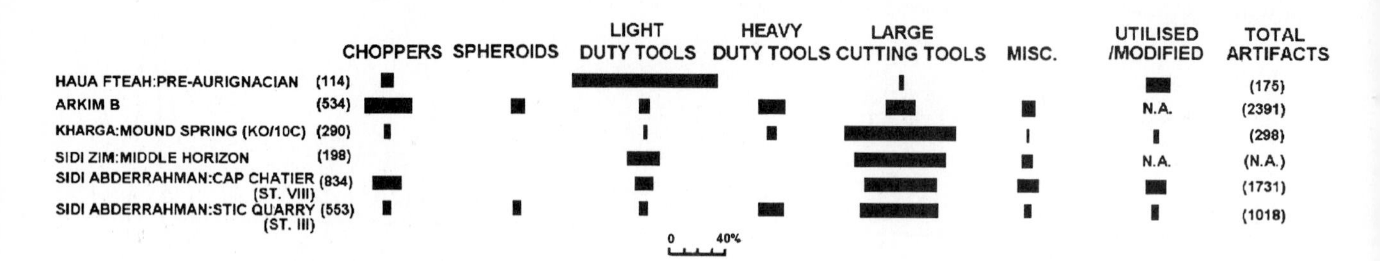

FIGURE 22-8 Comparison of component variability in the Acheulian and other industrial assemblages of Middle and early Upper Pleistocene age sites from North Africa. (After Clark 1975)

MYA or less was also most probably determined by climatic events in the Sahara, but the evidence is not yet forthcoming. A favorable climate in the Sahara, as was the case some 300 KYA, brought the Ethiopian fauna and the later Acheulian into the desert (oxygen-isotope Stages 7-9). The glaciation (190-130 KYA) and the desertification of the Sahara caused migration to the peripheries. With the amelioration of the Last Interglacial (oxygen-isotope Stage 5) the Sahara was again occupied by the makers of the various regional Middle Paleolithic/Middle Stone Age Industries that replaced the Acheulian. If the desertification of the Sahara caused the populations to move into peripheral regions, in particular the coastal areas along the Mediterranean, it seems to have been the interglacials that made further movements possible and perhaps necessary if population pressure was involved. It is suggested that this was the mechanism that triggered movements of hominid populations within and out of the continent.

THE ACHEULIAN OUTSIDE AFRICA

Before considering what the Acheulian may imply concerning technical skills, intellectual and communicative abilities, and general behavior, it is necessary to review the distribution of this large technocomplex in time and space in Eurasia. Of necessity this has to be very brief, though it is very well documented at least for Europe (Bordes 1984:4-124) and western Asia (Bar-Yosef 1975). After the initial diaspora 1.0 MYA or less, the Middle Pleistocene cultural assemblages show the same combinations of biface and core/flake components that were present in Africa and, as we have seen, there are extensive regions of Europe and Asia where only the core/flake component is present. This earlier migration put *H. erectus* into the subtropical and tropical parts of Asia. Archaic *H. sapiens* fossils in Europe, West, South, and East Asia show that this grade may have evolved in the respective regions or perhaps, as in the case of Europe, were the result of a later movement carrying the Acheulian complex into that continent.

Earlier Acheulian assemblages are known from Israel and Syria, although none is very satisfactorily dated other than by geomorphology and fauna. Perhaps the earliest after 'Ubeidiya, on context and technology, is Evron-Quarry in western Galilee, Israel, where simply modified handaxes occur with an important small tool component (Ronen 1991). The Middle Orontes Valley in northern Syria has produced a number of Acheulian sites, notably Latamne, a single horizon activity area with early Acheulian-type handaxes made by hard hammer and an interesting core/flake tool component (Clark 1967; Besançon et al. 1978; Sanlaville et al. 1993). On the Syrian coast at Lattakiya, the assemblages from Jebel Berzine localities represent a core/flake industry and are considered to be contemporary (Besançon et al. 1978). These sites and others may be as old as, or considerably older than, 0.5 MYA and appear to be in part contemporary with a cold climate.

Possibly somewhat later in age is the site of Gesher Benot Ya'aqov on the Jordan River, Israel (Goren-Inbar et al. 1992). Several occupation levels, the lower ones waterlogged, contain Acheulian artifacts and Middle Pleistocene fauna. The age, on radiometric and other dating methods, lies between 240 and 730 KYA. The site is potentially important for the rich plant and other organic remains (including

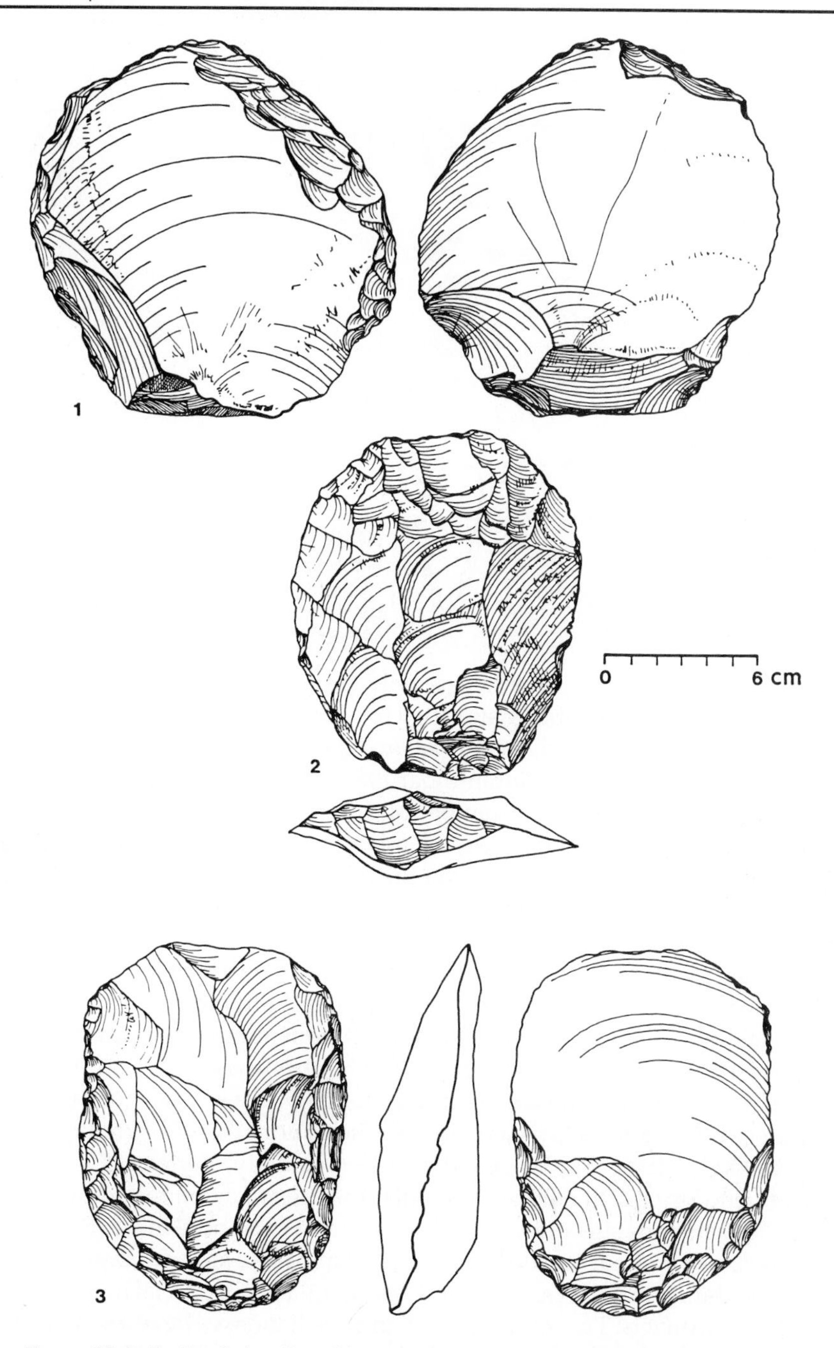

Figure 22-9 Earlier Acheulian side-scraper on a Kombewa flake, (1) Ternifine Later Acheulian Levallois flake, (2) Tachengit; and cleaver, (3) Tachengit. (After Clark 1992a)

modified wood), a rich fauna and Acheulian bifaces in basalt, flint, and limestone. This is a rare example of a large, stratified Acheulian assemblage in the Levant. The high values for cleavers in basalt, often made by the Kombewa method, mirror others from Africa, in particular those from the Afar Rift in Ethiopia. Although the number of Lower Paleolithic sites with Acheulian bifaces is not small, very few of them have, as yet, been dated except relatively (e.g., Ohel 1986). Concerning the Lower Paleolithic of the Levant

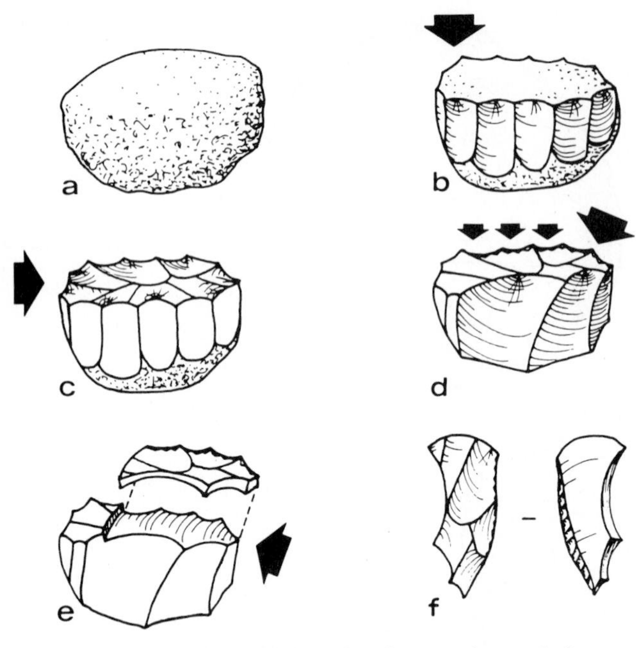

FIGURE 22-10
Stages in the production of a preformed cleaver by the Tabalbala-Tachengit method. (After Clark 1992a)

(Israel, Lebanon, and Syria) reference is made here only to one later Acheulian site, Berekat Ram on the Golan Heights, as this is a sealed assemblage stratified between two basalt flows dated respectively to 233 (upper) and 800 (lower) KYA. It is a sealed, well-excavated, Lower Paleolithic open site which is rare in the Levant and is associated with a paleosol (Goren-Inbar 1985). Artifacts are made from local flint, and a full range of flaking waste is associated with a number of different flake scraper forms, including what are known as "Upper Paleolithic" forms with the Levallois technique and six handaxes. Typologically this assemblage looks to be nearer the terminal stage of the Acheulian though this could be deceptive. There are significant differences between Berekat Ram and the more coastal sites of Tabun and Mount Carmel. Figure 22-12 (on page 464) compares some of these late or terminal Acheulian variants that represent the transition to the Middle Paleolithic seen in the "Mugharan tradition" with its several variants such as the Jabrudian and the blade-dominated Amudian. This represents both innovation and continuity.

The Lower Paleolithic of Europe is even better known from numerous excavated and published sites (Howell 1966; Bordes 1984). Assemblages with and without bifaces are, as in Africa, a feature of the European Lower Paleolithic. In eastern Europe the Acheulian biface is absent or rare and known mostly, if not entirely, from open sites such as, in Germany, Karlich (more than 78 KYA: Bosinski et al. 1980) and Bilzingsleben (425-200 KYA: Mania and Vlcek 1981) and in Hungary, Verteszöllös (perhaps 210-160 KYA: Vertes 1965). In southern Europe, in the south of Italy, bifaces are again rare or absent from the earlier sites, such as the fluvio-lacustrine sites of Isernia (Coltorti et al. 1982) and Venosa (Piperno and Segre 1982), the former dating to 700 KYA or earlier. Bifaces, usually handaxes, are well represented in central Italy (e.g., Fontana Ranuccio dating to 450 KYA: Segre and Ascenzi 1984). Of special note here is a bone handaxe and other modified bone artifacts. Two other Acheulian bone handaxes have also been found at Castel di Guido, near Rome, and are probably of comparable age.

The sequence is well seen at St. Acheul and other localities on the Somme (Bordes 1984:13-46); at l'Arago in the French Pyrenees where the cave yielded many Tayacian flake tools with some Acheulian bifaces and the Archaic *H. sapiens* cranium and mandible dated to c. 450 KYA (de Lumley 1975:774-776); also in Spain (Santonja and Villa 1990), notably the elephant butchery sites at Aridos and Torralba/Ambrona (Klein 1987) and in England (Roe 1981; Singer and Wymer 1976). Of concern also are Swanscombe with the Archaic *H. sapiens* fossils dated to 460-250 KYA (Waechter 1973), Boxgrove (Roberts 1986), an extensive Acheulian workshop locality with fauna, associated with an interglacial raised beach, and the cave site of Pontnewydd in Wales dating to 245-215 KYA (Green 1984).

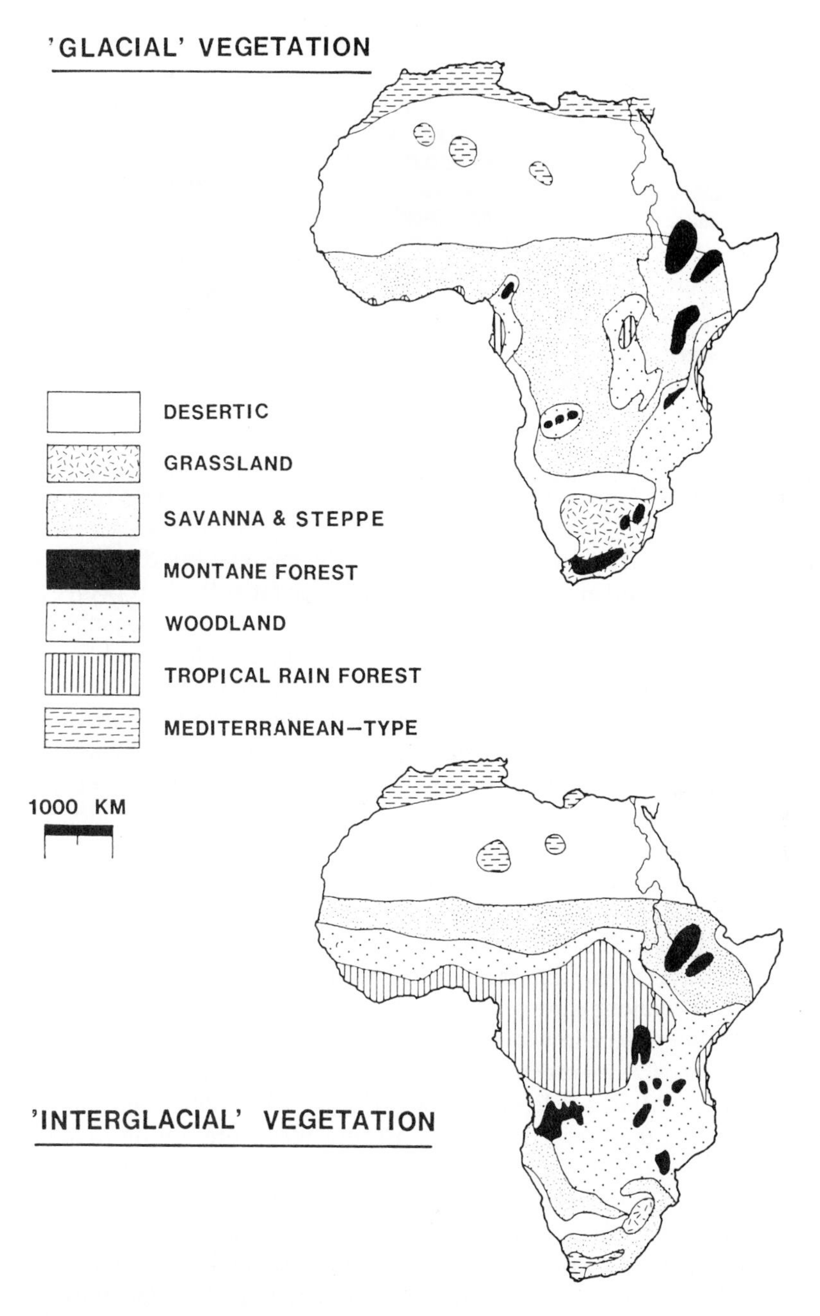

FIGURE 22-11 Hypothetical vegetation patterns in Africa under glacial and interglacial climates. (After Clark 1992b)

The late Acheulian is often found in caves as well as in open sites in France and Spain such as Biache (but only one biface [Sommé et al. 1978] and dated to >196-159 KYA). The Acheulian lies at the base of the El Castillo sequence (Cabrera 1984:109-140) and occurs in several French caves (e.g., Combe Grenal, Peche de l'Aze, Grainfollet: Bordes 1955, 1972). Again, as in Africa, the evolved nature of these industries of the pre- and early Last Interglacial is comparable to the terminal Acheulian industries of the "Mugharan

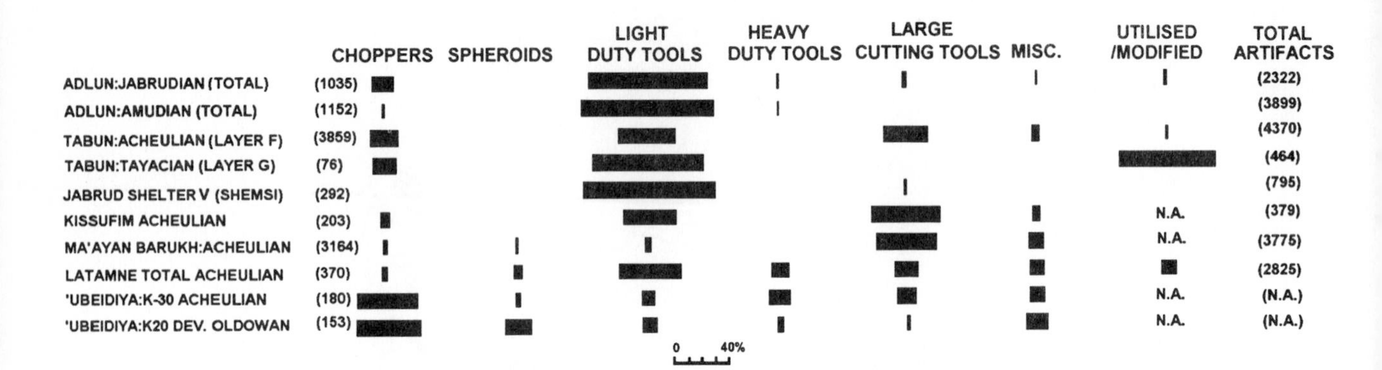

FIGURE 22-12 Comparison of component variability in the Acheulian and related industrial assemblages of Middle and early Upper Pleistocene age sites in the Levant. (After Clark 1975)

tradition" in the Levant and can be seen as evolving into the regional Middle Paleolithic from which the biface has disappeared (Copeland and Hours 1981). Notable so far as the European evidence is concerned is the effect of raw material on the composition of the Acheulian. Most European Acheulian assemblages are made from flint and, where this material occurs, it is more often used for the small flake tool component. Where quartzite is used, however, as in Spain (Freeman 1975), the Garonne (Tavoso 1978) and southern Italy (Alimen 1975), large flakes are a common primary form for bifaces and these assemblages resemble those from northern Africa where similar raw material was used. Variability in the European Acheulian can be seen to be a product of the material used and not of separate divergence (Villa 1981). Another trait in common with the African Acheulian is the use of the Levallois technique from what have been called Middle Acheulian times.

Proceeding eastwards, an unknown quantity in the spread of the Acheulian Complex is the Arabian Peninsula. Survey and preliminary excavations suggest that the Acheulian is well represented in the north and south (Whalen and Pease 1991; Whalen and Rowse 1992). Quartzite seems to be the material used and the handaxes and cleavers resemble those from the Horn of Africa. Whether there was a migration across the Straits of Bab-el-Mandeb during a time of low sea level is, as yet, unproven but cannot be ruled out, and this is a major region to be investigated.

In the Crimea, Caucasus, Armenia, and Azerbaijan, the Acheulian probably makes a late appearance, immediately preceeding the Middle Paleolithic (Klein 1966; Clark 1992b). Claims for Acheulian in Central Asia are not convincing and the Lower Paleolithic assemblages there belong to the core/flake complex (Davis et al. 1980). It is present but not well represented in northern Iraq and Iran. It is, however, well represented in Pakistan and peninsular India where the two major components of Middle Pleistocene technology—the Acheulian and Core/Flake traditions—combine and separate as in Africa and Europe. This is clearly demonstrated by several well-stratified and excavated assemblages (Misra 1989; Misra et al. 1982, 1988; Jayaswal 1982; Pant and Jayaswal 1991; Paddayya 1982, 1985; Sharma and Clark 1983:23-115). Probably the furthest north an Acheulian occurs is in Nepal (Corvinus 1990). Quartzite is the predominant raw material used in India for handaxes and cleavers, made often from large flakes sometimes obtained by the proto-Levallois and Kombewa methods. Recent radiometric and isotopic dates for the Acheulian in India show it to be beyond the lower limit of the Uranium-series method and to be more than 350 KYA old though the exact upper and lower limits still have to be determined (Mishra 1992).

The Acheulian biface and its associated flake tool component are not found east of the western boundary of the tropical evergreen forest in Assam. To the east and south of this only the core/flake tradition is present. There are, as yet, no distinct quantitative or qualitative stone artifact assemblages from South East Asia, but it is likely that they follow the pattern seen in China at Zhoukoudian (Pei and Zhang

1985) dated between 0.5 and 0.24 MYA and to more than 0.78-1.0 MYA in the Nihewan Basin in northern China (Schick et al. 1991). These are assemblages of cores and flakes, sometimes modified as "choppers" or scraper forms. There are also some assemblages in central and southern China, and also in Korea, that have, in addition to the small tool component, larger tools both unifacial and bifacial, which bear some superficial resemblance to earlier Acheulian handaxes. These come mostly from secondary contexts and there are few, if any, fully representative assemblages from sealed horizons. One pointed biface from Lantien, dating between 1.2 and 0.8 MYA, is the most likely representative of the Acheulian handaxe, but it is unique. These assemblages are mostly undated, but if they are largely comparable in age to Dingcun in the Fen Valley where hornfels and quartzite were used, then they are of late Middle to early Upper Pleistocene age (Huang 1987; Qui Zhonglang 1985; Bae 1988).

The standardization of form and the symmetry of the Acheulian biface, the absence of soft hammer flaking and of the specialized techniques for the production of large "pre-forms," and the scarcity of regularly retouched flake tools set these assemblages apart from the Acheulian; in technique and time they resemble more closely the African Sangoan Industrial Complex. Another feature of the Far Eastern and, perhaps, of the Central Asian Paleolithic from the Lower to the Upper stages is its unspecialized nature and lack of formal types of retouched tools. This is not the place to discuss the reason for this conservative continuity, but it seems likely that the stone equipment represents only the basic requirements for cutting, splitting, chopping, and scraping to make the finished equipment in other, but perishable, materials. It was possible to demonstrate recently that the different processes needed to work bamboo with a metal tool could equally well be carried out using stone artifacts comparable to those at Zhoukoudian (Clark 1992b: 210).

DISCUSSION

This streamlined review of the great Acheulian technocomplex shows its early and first appearance in Africa at 1.6-1.5 MYA, its early movement into tropical and subtropical Asia, and later movement into temperate Europe. This complex combined in various ways the biface and core/flake components of Middle Pleistocene cultural manifestations, sometimes to the exclusion of one or the other—most clearly seen in the dominance of the core/flake component in south and eastern Asia. It is the Acheulian biface component that can tell us something about hominid behavior and the significance of this stone tool kit for its makers. While throughout the Lower Paleolithic there is a wide range of variability in form and proportions, this is all within the parameters of a very conservative pattern, and there are no regionally adapted entities except at the very end. The Acheulian represents a very long and slowly evolving technology, suggesting that behavioral changes were equally slow and gradual. Within this mode, however, some interesting points emerge. The bifaces and the small flake tools show a high degree of standardization in form, proportions, and symmetry that conceals the shape and form of the primary piece. Quantitative analyses show this very clearly and one or two dominant shapes are easily identifiable. These bifaces are, moreover, usually shapely and well made with controlled soft hammer flaking often over both faces, and the symmetry, scar pattern, and thinness of the finished piece go beyond the flaking required to make an efficient tool. This is especially well seen in the later Acheulian in the selection of raw materials for large and small tools and in the primary methods that evolved to produce, where necessary, the large flakes. This shows some considerable ingenuity as well as skills that are the most visibly evolved in this time range. The bifaces are some of the best evidence for planning, recognition of the objective, and execution of the pattern envisaged in the "mental template" in the minds of *H. erectus* and Archaic *H. sapiens*. As Wynn (1979) has said, this is indeed a "window" into the mind of the biface makers. It does not in any way mean that they were ahead of those hominids who did not make and use bifaces and whose ingenuity was surely directed toward other materials and in other directions. It does, however, provide some means of understanding the capabilities of Middle Pleistocene hominids and a means for assessing less subjectively some of the other evidence put forward to reconstruct Lower Paleolithic life-ways.

That they were highly mobile is seen by the sometimes considerable distances over which they transported raw material. Site size varies considerably from small scatters to large concentrations of bifaces and other artifacts. Some, though not all, of these concentrations can be seen as the result of fluvial resorting due to the propensity of Acheulian groups to be associated with stream channel contexts. The reasons for the larger activity areas, greater numbers of retouched tools and greater range of artifact types remain unclear. Do they represent larger group size, regular re-use of favorable sites, or anticipation of future requirements in the form of retouched artifacts that can be cached and/or carried to other locations? These are possibilities that need to be tested with new methods and new data on changing evidence for plant and animal resources, on sourcing of materials, and on application of weathering criteria.

It is to be expected that seasonal movements and resource use were now more structured and that territorial boundaries began to be defined. Occupation of some activity areas may have been more prolonged than that of others. In Europe, sites such as Terra Amata, Lazaret (de Lumley 1975), and Bilzingsleben, as well as Latamne in Syria (Clark 1967, 1968), have evidence of what are claimed as the bases of temporary structures. Alternative explanations are possible, but the likelihood remains, and since caves are known to have been used at this time, it might be expected that some kind of protective structure would have been built at open, unprotected sites. When fire first began to be used is, as yet, unknown. Evidence for use in Africa 1.0 MYA is equivocal but burnt bone at Verteszöllos; charcoal, burnt bone, ash and burned stones at Zhoukoudian; charcoals and other burnt material at Torralba and Ambrona, Terra Amata, l'Arago and Gesher Benot Ya'aqov all speak of more regular use around some 0.5 MYA and, therefore, by the time of the Riss (Saale) glaciation, fire-using and, no doubt, regular conserving were more general, as at La Cotte de St. Brelade (Callow and Cornford 1986), Kalambo Falls, and the Cave of Hearths. Besides enhancing opportunities for social communication, fire-using could have had a dramatic effect on expanding the possibilities for human diet.

The standard of the Middle Pleistocene hominids' hunting skills has been challenged. The wooden spear from Clacton-on-Sea is suggestive of some hunting but scavenging appears to have been the main source of animal protein. The association of Acheulian tool kits with elephant and hippopotamus carcases is, however, significant, and although the collective hunting of elephant at Torralba and Ambrona has been challenged, that from other meat-processing sites such as Aridos (Santonja and Villa 1990), Mwanganda (Clark and Haynes 1969), and the even earlier *Deinotherium* and *Elephas reckii* skeletons with tools at Olduvai Gorge are certain indications of the association of stone tools with single large animal carcases. The cause of death is unknown, but the evidence for butchery and meat-processing is compelling. These large animal carcases with very tough hides and great quantities of meat (and, where hippos are concerned, of fat also) would, when fresh, have been accessible to hominids with knives and other stone tools but not to most other large predators. Butchery and also, perhaps, intentional driving of mammoth and rhino over a cliff are dramatically demonstrated at La Cotte de St. Brelade dated, on TL samples, to c. 240 KYA, where disarticulated skulls, mandibles and postcranial bones had been piled along a cliff wall and may have been kept there in "cold storage" (Scott 1986).

This raises the often posed question as to what the bifaces may have been used for. The answer must still be that we do not know. Experiment shows that they are very efficient for butchering large carcases, cutting through ligaments and joints, cutting branches from trees, cutting toe holds for tree-climbing and for working wood. Two handaxes from Hoxne have polish identified as meat polish, and more microwear studies on flint tools are needed as well as biochemical studies. In many regions, however, the raw materials used do not preserve evidence of edge wear or polish and association is the best approach here. Most likely the bifaces were used for a range of different purposes besides forming a source of material when flakes for use as knives or further modification were required. Paleoclimatic evidence does point to the Acheulian biface component's having been associated with woodland/savanna and the newly evolved, Middle Pleistocene tropical grasslands in sub-Saharan Africa. The same might be the case in India also and Acheulian is well represented in association with temperate forest and grassland in Europe.

The Acheulian provides virtually no evidence for any symbolic behavior except, perhaps, the apparent use of red ochre (Wreschner 1980; Marshack 1981) and the believed engraved bones from Bilzingsleben and Peche de l'Aze. Finally, a small grooved scoria pebble from the later Acheulian site of Berekat Ram in Israel exhibits what is, it is suggested, intentionally made grooves so that this specimen might even be an Acheulian figurine (Goren-Inbar 1986). If this scant and debatable evidence for the emergence of symbolic behavior on the part of later Acheulian populations is in no way definitive, it does point to the need to look for possible further traces of this kind.

As to the meaning of the dichotomy between bifaces and core/flake traditions, many possible explanations must be cited. The most compelling, however, would seem to be that they were the result of the relationship between climate, ecology, and raw materials and the different uses to which these latter were put. In the Middle Awash region of the Afar Rift in Ethiopia, the change within a single Formation from Core/Flake assemblages in the earlier sediments, associated with a fluvio-lacustrine environment, to Acheulian biface assemblages in the overlying beds, is related to increasing shifts in sedimentation within a pedi-alluvial context, and this wadi-fan environment became increasingly dominant (Clark et al. in prep.). This cultural switch within the time range of 0.5 MYA or more may reflect this environmental shift and may be the explanation for the technological shift as different resources became available. Clearly this landscape approach needs to be intensified.

In closing, it is suggested that the Acheulian in Africa is the ancestral complex from which that continent's Middle Paleolithic/Middle Stone Age regional variants evolved. From one or more of these came the later biological and technological "breakthrough" with the appearance of anatomically modern humans and their eruption into Eurasia 100,000 years ago.

LITERATURE CITED

Alimen, M. H. (1975) Les isthmes hispano-marocain et si-cielo-tunisien aux temps acheuléens. *L'Anthropologie* 79:399-436.

Arambourg, C., and Balout, L. (1955) L'Ancien lac de Tihodaïne et ses gisements préhistoriques. In L. Balout (ed.): *Congrès Panafricain de Préhistoire, Actes de la IIe Session, 1952.* Algiers, pp. 281-292.

Asfaw, B., Beyene, Y., Suwa, G., Walter, R. C., White, T. D., Wolde-Gabriel, G., and Yemane, T. (1992) Konso-Gardula: The earliest Acheulian. *Nature.* 360:732-735.

Bae, Kidong (1988) The significance of the Chongokni stone industry in the tradition of the Paleolithic cultures of East Asia. Ph.D.Dissertation in Anthropology, University of California, Berkeley.

Balout, L., and Tixier, J. (1957) L'Acheuléen de Ternifine. *C. R. Congrès de préhistoire de France,* 15e session, 1956, pp. 214-218.

Bar-Yosef, O. (1975) Archaeological occurrences in the Middle Pleistocene of Israel. In K. W. Butzer and G. L. Isaac (eds.): *After the Australopithecines.* The Hague: Mouton, pp. 571-604.

Bar-Yosef, O., and Goren-Inbar, N. (1993) The lithic assemblages of 'Ubeidiya: A Lower Palaeolithic site in the Jordan Valley. *QEDEM: Monographs of the Institute of Archaeology,* Hebrew University of Jerusalem.

Besançon, J., Copeland, L., Hours, F., and Sanlaville, P. (1978) The Paleolithic sequence in Quaternary formations of the Orontes river valley, Northern Syria: A preliminary report. *Institute of Archaeology, Bull.* (London) 15: 149-170.

Biberson, P. (1961) *Le Paléolithique inférieur du Maroc atlantique.* Pub. du Service des Antiquités du Maroc, Rabat. Fasc. 17.

Bordes, F. (1955) Les Gisements du Peche-de-l'Aze (Dordogne). *L'Anthropologie* 59 (1-2): 1-38.

Bordes, F. (1972) *A Tale of Two Caves.* New York: Harper and Rowe.

Bordes, F. (1984) *Leçons sur le Paléolithique,* Vol. 2: *Le Paléolithique en Europe.* Cahiers du Quaternaire No. 7. CNRS Paris.

Bordes, F., and Fitte, P. (1953) L'Atelier Comment. *L'Anthropologie* 57:1-45.

Bosinski, G., Brannacker, K., Lanser, K. P., Stephan, S., Urban, B., and Würges, K. (1980) Altpaläolithische Funde von Kärlich, Kr Mayen-Koblenz (Neuwieder Becken). *Archäol. Korrespond.* 10:295-314.

Cabrera Valdes, V. (1984) *El Yachtsmen de la Cueva de "El Castillo."* Bibliotheca Praehistorica Hispana, Madrid.

Callow, P., and Cornford, J. M. (1986) *La Cotte de St. Brelade, 1961-1978: Excavations by C. B. M. McBurney.* Norwich: Geo Books.

Clark, J. D. (1967, 1968) The Middle Acheulian occupation site at Latamne, Northern Syria. *Quaternaria* 9:1-68; 10:1-71.

Clark, J. D. (1969) *Kalambo Falls Prehistoric Site,* Vol. I. Cambridge: Cambridge University Press.

Clark, J. D. (1975) A comparison of the Late Acheulian Industries of Africa and the Middle East. In K. W. Butzer and G. L. Isaac (eds.): *After the Australopithecines.* The Hague: Mouton, pp. 605-659.

Clark, J. D. (1992a) The Earlier Stone Age/Lower Paleolithic in North Africa and the Sahara. In R. Kuper (ed.): *New Light on the Northeast African Past.* Köln: Heinrich Barth Institut, pp. 17-37.

Clark, J. D. (1992b) African and Asian perspectives on the origins of modern humans. In M. J. Aitken, C. B. Stringer and P. A. Mellars (eds.): *The Origins of Modern Humans and the Impact of Chronometric Dating. Phil. Trans. Roy. Soc. (Biol. Sci.)* 337 (1280):201-215.

Clark, J. D., Asfaw, B., Assefa, G., Harris, J. W. K., Kurashina, H., Walter, R. C., White, T. D., and Williams, M. A. J. (1984) Paleoanthropological discoveries in the Middle Awash Valley, Ethiopia. *Nature* 307 (5950):423-428.

Clark, J. D., and Haynes C. V. (1969) An elephant butchery site at Mwanganda's village, Karonga, Malawi and its relevance for Paleolithic archaeology. *World Archaeol.* 1(3):390-411.

Clark, J. D., de Heinzelin, J., Schick, K., Hart, W. K., White, T. D., Wolde Gabriel, G., Walter, R. C., Suwa, G., Asfaw, B., and Vrba, E. (In prep.) Middle Pleistocene Discoveries in the Middle Awash Valley, Ethiopia.

Clarke, R. J. (1985) Early Acheulian with *Homo habilis* at Sterkfontein. In P. V. Tobias (ed.): *Hominid Evolution: Past, Present, and Future.* New York: Liss, pp. 287-298.

Coltorti, M., Cremaschi, M., and 10 others. (1982) Reversed magnetic polarity at an early Lower Paleolithic site in central Italy. *Nature* 300:173-176.

Copeland, L., and Hours, F. (1981) La fin de l'Acheuléen et l'avênement du Paléolithique Moyen en Syrie. *Colloques Internationaux du CNRS* No. 598 (Préhistoire du Levant, 1980), pp. 225-238.

Corvinus, G. (1990) A note on the discovery of handaxes in Nepal. *Man Env.* 15(2):9-11.

Davis, R. S., Ranov, V. A., and Dodonov, A. E. (1980) Early Man in Soviet Central Asia. *Sci. Amer.* 243(6):130-137.

Deacon, H. J. (1970) The Acheulian occupation at Amanzi Springs, Uitenhage District, Cape Province. *Ann. Cape Prov. Mus. Nat. Hist.* 8:89-189.

Dzaparidze, V., Bosinski, G., and 15 others. (1989) Der altpaläolithische Fundplatz Dmanisi in Georgien (Kaukasus). *Sonderausdruck aus Jahrbuch des Römisch-Germanischen Zentralmuseums,* Mainz.

Freeman, L. G. (1975) Acheulian sites and stratigraphy in Iberia and the Maghreb. In K. W. Butzer and G. L. Isaac (eds.): *After the Australopithecines.* The Hague: Mouton, pp. 661-743.

Gobert, E. G. (1950) Le gisement paléolithique de Sidi Zin. *Karthago* 1:1-51.

Goren, N. (1981) The Lithic Assemblages of the Site of 'Ubeidiya, Jordan Valley. Doctoral dissertation, Hebrew University, Jerusalem.

Goren-Inbar, N. (1985) The Lithic Assemblage of the Berekhat Ram Acheulian site, Golan Heights. *Paléorient* 11(1):7-28.

Goren-Inbar, N. (1986) A figurine from the Acheulian site of Berekhat Ram, 1986. *Mitekufat Haeven* (Jerusalem) 19 (n. s.): 7-12.

Goren-Inbar, N., Belitzky, S., Verosub, K., Werker, E., Kisley, M., Heimann, A., Carmi, I., and Rosenfeld, A. (1992) New Discoveries at the Middle Pleistocene Acheulian site of Gesher Benot Ya'aqov, Israel. *Quat. Res.* 38:117-128.

Green, H. S. (1984) *Pontnewydd Cave.* Cardiff: National Museum of Wales.

Howell, F. C. (1966) Observations on the earlier phases of the European Lower Paleolithic. *Am. Anthropol.* 68(2) part 2:88-201.

Howell, F. C., Cole, G. H., and Kleindienst, M. R. (1962) *Isimila, an Acheulian occupation site in the Iringa Highlands.* Actes du IVe Congrès Panafricain Préhistoire, pp. 43-80.

Howell, F. C., Cole, G. H., Kleindienst, M. R., Szabo, B. J., and Oakley, K. P. (1972) Uranium-series dating of bone from the Isimila Prehistoric Site, Tanzania. *Nature* 237:51-52.

Huang, W. (1987) Bifaces in China. *Acta Anthropol. Sin.* 6 (1):61-68.

Isaac, G. L. (1967) The stratigraphy of the Peninj Group—Early Middle Pleistocene Formations West of Lake Natron, Tanzania. In W. W. Bishop and J. D. Clark (eds.): *Background to Evolution in Africa.* Chicago: Chicago University Press, pp. 229-257.

Jayaswal, V. (1982) *Chopper-Chopping Component of the Paleolithic in India.* Delhi: Agam Kala Prakashan.

Jones, P. R. (1981) Experimental implement manufacture and use: A case study from Olduvai Gorge. *Phil. Trans. Roy. Soc.* (London) B292:189-195.

Klein, R. G. (1966) Chellean and Acheulean on the territory of the Soviet Union: A critical review of the evidence as present in the literature. *Am. Anthropol.* 18(2) part 2:1-45.

Klein, R. G. (1987) Reconstructing how early people exploited animals: Problems and prospects. In M. H. Nitecki and D. V. Nitecki (eds.): *The Evolution of Human Hunting.* Nw York: Plenum, pp. 11-45.

Klein, R. G. (1989) *The Human Career: Human Biological and Cultural Origins.* Chicago: Chicago University Press.

Leakey, M. D. (1971) *Olduvai Gorge,* Vol. 3: *Excavations in Beds I and II, 1960-1963.* Cambridge: Cambridge University Press.

Leakey, M. D. (1975) Cultural patterns in the Olduvai Sequence. In K. W. Butzer and G. L. Isaac (eds.): *After the Australopithecines.* The Hague: Mouton, pp. 477-490.

deLumley, H. (1975) Cultural evolution in France in its paleoecological setting during the Middle Pleistocene. In K. W. Butzer and G. L. Isaac (eds.): *After the Australopithecines.* The Hague: Mouton, pp. 475-508.

Mania, D., and Vlcek, E. (1981) *Homo erectus* in middle Europe: The discovery from Bilzingsleben. In B. A. Sigmon and S. Cybulski (eds.): *Homo erectus.* Papers in honor of Davidson Black. Toronto: University of Toronto Press, pp. 133-151.

Marshack, A. (1981) On Paleolithic ochre and the early uses of color and symbol. *Curr. Anthropol.* 22(2):188-191.

Mason, R. (1962) *Prehistory of the Transvaal.* Johannesburg: Witwatersrand University.

Mishra, S. (1992) The age of the Acheulian in India: New evidence. *Curr. Anthropol.* 33(3):325-328.

Misra, V. N. (1989) Stone Age India: An ecological perspective. *Man Env.* 14:17-64.

Misra, V. N., Rajaguru, S. N., and Ragavan, H. (1988) Late Middle Pleistocene environment and Acheulian culture around Didwana, Rajasthan. *Proc. Ind. Nat. Sci. Acad.* 54A 3:425-438.

Misra, V. N., Rajaguru, S. N., Raju, D. R., and Ragavan, H. (1982) Acheulian occupation and evolving landscape around Didwana in the Thar Desert. *Man Env.* 7:112-131.

Ohel, M. (1986) *The Acheulian of the Yiron Plateau, Israel.* Oxford: BAR International Series, 307.

Paddayya, K. (1982) *Acheulian culture of Hunsgi Valley.* Pune: Deccan College.

Paddayya, K. (1985) Acheulian occupation sites and associated fauna from Hunsgi-Baichbal valleys, peninsular India. *Anthropos* 80:653-658.

Pant, J. C., and Jayaswal, V. (1991) *The Stone Age settlement of Bihar.* Delhi: Agam Kala Prakashan.

Pei, W., and Zhang, S. (1985) A study of the lithic artifacts of *Sinanthropus. Paleontol. Sin.* 12:1-277.

Piperno, M., and Segre, A. G. (1982) *Pleistocene e Paleolitico inferiore di Venosa: Nuove Ricerche.* Firenze: Estratto degli atti della XXIII Missione Scientifici della Instituto Italiano di Preistoria e Protostoria 1980, pp. 549-596.

Qui Zhonglang (1985) The Middle Paleolithic of China. In Wu Rukang and J. W. Olsen (eds.): *Paleoanthropology and Paleolithic Archaeology in the People's Republic of China.* Orlando: Academic, pp.190-210.

Raynal, J.-P., and Texier, J.-P. (1989) Découverte d'Acheuléen ancien dans la carrière Thomas 1 à Casablanca et problème de l'ancienneté de la présence humaine au Maroc. *C. R. Acad. Sci. Paris* 308 (Série II):1743-1749.

Roberts, M. B. (1986) Excavation of the Lower Paleolithic site at Amey's Eartham Pit, Boxgrove, West Sussex: A preliminary report. *Proc. Prehist. Soc.* 52:215-245.

Roe, D. A. (1981) *The Lower and Middle Paleolithic Periods in Britain.* London: Routledge and Kegan Paul.

Ronen, A. (1991) The Lower Paleolithic site Evron Quarry in western Galilee, Israel. Köln: *Sonderv. Geol. Inst. Univ. Köln* 82 (Festschrift Karl Brunnacker):187-212.

Sanlaville, P., Besançon, J., Copeland, L. and Muhesen, Sultan (1993) Le Paléolithique de la Vallée moyenne de l'Oronte (Syrie). BAR International series 587.

Santonja, M., and Villa, P. (1990) The Lower Paleolithic of Spain and Portugal. *J. World Prehist.* 4(1):45-94.

Schick, K. D., Toth, N., Wei, Qi, Clark, J. D., and Etler, D. (1991) Archaeological perspectives in the Nihewan Basin, China. *J. Hum. Evol.* 21:13-26.

Scott, K. (1986) The large mammal fauna. In P. Callow and J. M. Cornford (eds.): *La Cotte de St. Brelade, 1961-1978: Excavations by CBM McBurney.* Norwich: Geo Books, pp. 109-137.

Segre, A., and Ascenzi, A. (1984) Fontana Rannucio: Italy's earliest Middle Pleistocene Hominid site. *Curr. Anthropol.* 25:230-233.

Sharma, G. R., and Clark, J. D. (eds.) (1983) *Paleoenvironment and Prehistory in the Middle Son Valley (Madhya Pradesh, North Central India).* Allahabad: Abinash Prakashan.

Singer, R., and Wymer, J. J. (1976) The sequence of Acheulian industries at Hoxne, Suffolk. In J. Combier (ed.): *L'Evolution de l'Acheuléen en Europe.* Nice: UISPP IXe Congrès, Colloque X, pp. 14-30.

Sommé, J., Munaut, A. V., Puissegur, J.-J., Chaline, J., Tuffreau, A., Piningre, J. F., Poplin, F., and Vandermeersch, B. (1978) Le gisement Paléolithique de Biache-Saint Vaast (Pas-de-Calais). *Bull. Assoc. Franc. Quat.* 1-3:27-67.

Tavoso, A. (1978) Le Paléolithique inférieur dans le bassin du Tarn. In *La Préhistoire Française sous la direction de H. de Lumley,* 1^{a}2° Editions. Paris: CNRS, pp. 893-908.

Tchernov, E. (1987) The age of the 'Ubeidiya Formation: An early Pleistocene hominid site in the Jordan Valley, Israel. *Israel J. Earth Sci.* 36:3-30.

Vertes, L. (1965) Typology of the Buda Industry, a pebble tool Industry from the Hungarian Lower Paleolithic. *Quaternaria* 7:185-193.

Villa, P. (1981)Matières premières et formations culturelles dans l'Acheuléen français. *Quaternaria* 23:19-35.

Waechter, J. d'A. (1973) The later Middle Acheulian industries of the Swanscombe area. In D. E. Strong (ed.): *Archaeological Theory and Practice.* New York: Seminar, pp. 67-86.

Wendorf, F., and Schild, R. (1980) *Prehistory of the Eastern Sahara.* New York: Academic.

Whalen, N., and Pease, D. W. (1991) Archaeological survey in south-east Yemen, 1990. *Paléorient* 17(2):127-131.

Whalen, N., and Rowse, D. A. (1992) Early mankind in Arabia. *Aramco World, Houston* 43(4):16-23.

Wreschner, E. (1980) Red ochre and human evolution: A case for discussion. *Curr. Anthropol.* 21:631-644.

Wynn, T. (1979) The intelligence of later Acheulian hominids. *Man* 14:371-391.

CHAPTER 23

Southern Africa Before the Iron Age

Richard G. Klein

Paleoanthropology began in southern Africa during the second half of the last century, almost as early as in Europe (J. Deacon 1990a). Stone age archeology was particularly successful early on, and by the 1920s, Goodwin and van Riet Lowe (1929) were able to define three stages—the Earlier, Middle, and Later Stone Ages—that were later recognized throughout sub-Saharan Africa. In the 1920s and 1930s, southern Africa also produced the first australopithecines (Clark 1967), and it was only in 1959, with the recovery of an australopithecine at Olduvai Gorge, that southern Africa began to lose paleoanthropological primacy. This was partly due to highly successful follow-up investigations at Olduvai and other east African sites and partly due to the decided advantages that east Africa presented, including: (1) a much larger number of internal drainage basins to trap sediments, fossils, and artifacts; (2) sedimentary fills that are more commonly antacid (and thus preserve fossils more readily); and (3) sedimentary sequences that often include volcanic extrusives that can be dated by the potassium/argon ($^{40}K/^{40}Ar$) method. There has been no active vulcanism in southern Africa throughout the entire later Cenozoic, and in contrast to east Africa, in southern Africa, only the prehistoric record that falls within the 40,000–30,000 year range of conventional radiocarbon (^{14}C) dating is independently well dated.

However, paleoanthropological and related paleoecological research has proceeded in southern Africa, and important, sometimes unique discoveries continue to be made. For example, fieldwork near Cape Town has revealed an extraordinarily prolific early Pliocene fossil site that is paleoecologically far more instructive than any comparable east African locality; the South African australopithecine sample, which is still accumulating, remains roughly twice as large as its east African counterpart; and sites in various parts of South Africa have produced what are arguably some of the earliest modern or near-modern human fossils found anywhere in the world. These are accompanied by abundant archeological evidence suggesting that modern behavior evolved after modern or near-modern morphology. My purpose here is to summarize the prehistory of southern Africa, with special emphasis on its singular contributions. Following H. J. Deacon and Thackeray (1984), I define southern Africa as the region south of the Zambezi River, including the modern political units of South Africa, Lesotho, Swaziland, southern Angola, Namibia, Botswana, Zimbabwe, and southern Mozambique (Figure 23-1). This definition is founded mainly on a fundamental and arguably long-standing zonal difference, reflected historically in the contrast between the deciduous miombo woodlands north of the Zambezi and the bushveld, thorn-tree savannas, grasslands, and shrublands to the south. The time interval to be considered is the entire known span of hominoid evolution, from the early Miocene, roughly 19–20 million years ago (MYA), to the late Holocene, between 1800 and 500 years ago, when Iron Age farmers (first) and European colonists (later) progressively truncated the southern African stone age.

THE MIOCENE

Following Harland et al. (1990), the time span of the Miocene is taken here as 23.3 to 5.2 MYA. Early and middle Miocene deposits in east Africa have provided numerous apelike (hominoid) taxa, including some

FIGURE 23-1 Approximate locations of the sites mentioned in the text.

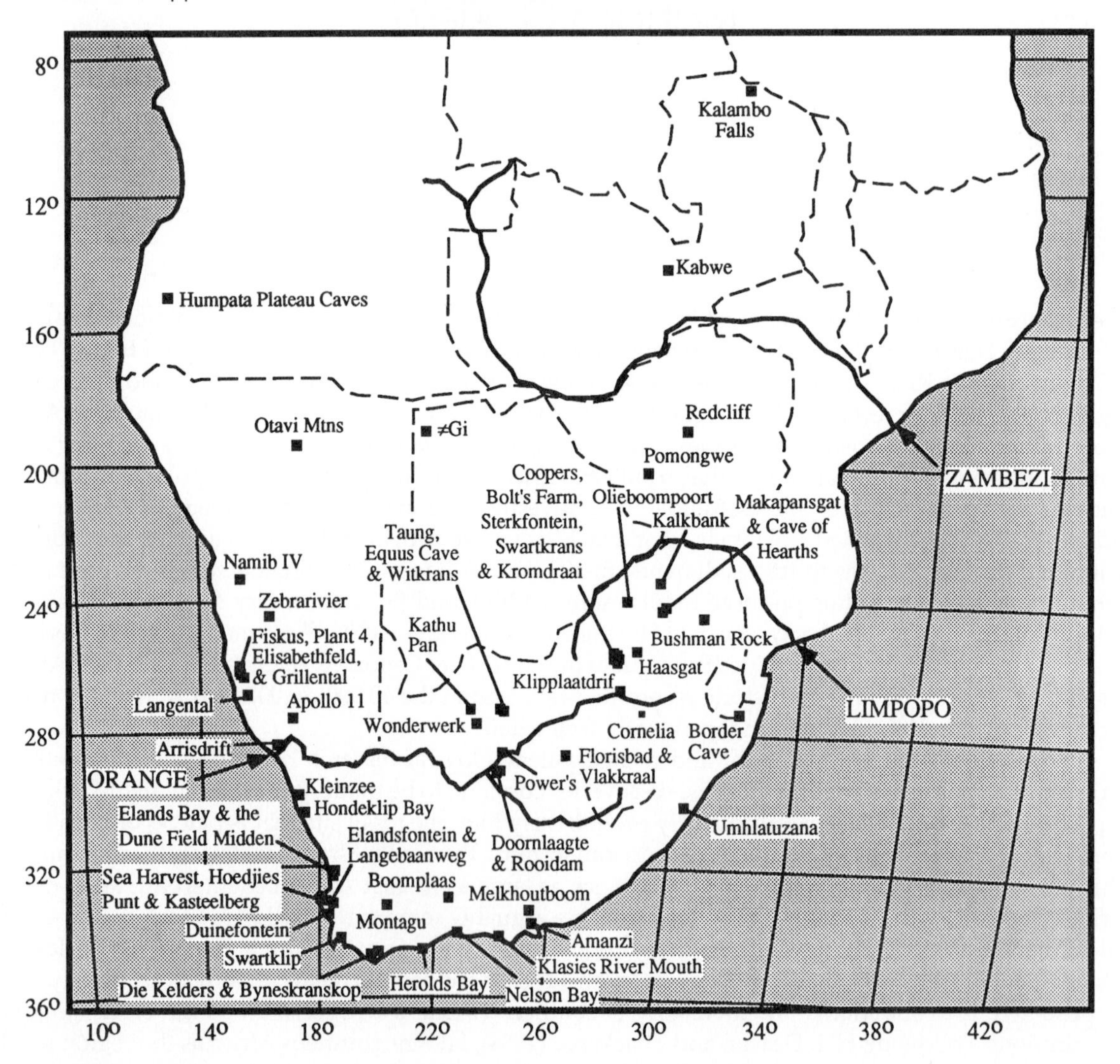

like *Kenyapithecus* that could lie near the ancestry of both people (hominids) and the living African apes. A combination of genetic and fossil evidence indicates that people probably emerged in east Africa from an apelike ancestor during the late Miocene, some time between 10 and 5 MYA. Unfortunately, southern Africa has added little to our understanding of Miocene primate evolution, at least in part because it is relatively poor in fossiliferous, terrestrial Miocene localities. Until 1976, there were only five well-documented sites (Langental, Plant 4, Grillental, Fiskus, and Elisabethfeld), concentrated near the Namibian coast between Lüderitz and Bogenfels (Hamilton and Van Couvering 1977; Hendey 1984). The context of the fossils at each site has not been well studied, but the sediments are variably fluviatile or lacustrine. The individual faunal samples are small and highly fragmented, but they appear to be broadly similar in age. Overall, they are heavily dominated by small mammals (rodents, lagomorphs, and insectivores), but they also contain a creodont carnivore, a gomphothere, a hyrax, a rhinoceros, and a handful of artiodactyls. Primates are absent, perhaps because the total sample size is small. The identified taxa suggest broad contemporaneity with the "Rusinga-like" fauna of east Africa, which probably dates to between 19 and 17 MYA (Pickford 1981).

The Miocene faunal history of southern Africa was significantly enriched in 1976, when diamond prospectors found a new and far more productive locality at Arrisdrift in far southern Namibia, on the north bank of the Orange River, approximately 30 km from its mouth. The fossils occur in a highly consolidated gravel lag at the base of a shallow channel incised into a high terrace of the Orange. Only 180 sq m of deposit, comprising a small part of the total, has been removed, but it produced several thousand identifiable bones from at least 37 vertebrate taxa (Table 23-1; Hendey 1984). The full taxonomic composition of the fauna remains to be established, but the presence of relatively advanced ruminant artiodactyls and the absence of creodont carnivores indicate that it postdates the "Rusinga-like" fauna from the coastal sites to the north, and it may be only slightly older than the fauna from the famous Fort Ternan locality in Kenya. This has been dated to about 14 MYA by $^{40}K/^{40}Ar$ and $^{40}Ar/^{39}Ar$ (Shipman et al. 1981; Shipman 1986a; Pickford 1986). In sum, it seems likely that the Arrisdrift bones accumulated some time between 17 and 15 MYA.

Like the early Miocene coastal samples, the Arrisdrift assemblage contains no primates, and since it is relatively large, the implication may be that primates did not exist nearby. Primates would not be expected in the modern Arrisdrift setting, which is hyperarid and barren, but the chevrotain, the deer-like ruminant *Climacoceras*, and other taxa in the fossil assemblage imply a relatively moist and wooded environment, perhaps broadly similar to the one in which *Kenyapithecus* prospered near Fort Ternan (Kappelman 1991; Cerling et al. 1991). In addition, the possibility that renewed excavation might uncover primates has clearly been enhanced by the discovery of a hominoid mandible in the Otavi Mountains of northern Namibia (Andrews 1992; Conroy et al. 1992a, 1992b). The specimen includes the right corpus with the crowns of P_4-M_3, the partial crown and root of P_3, and the partial root of the canine, together with all four incisor alveoli and the partial alveolus of the left canine. It came from a brecciated karst fill that also contains rodent taxa that imply an age of 14-12 MYA and a relatively mesic environment, perhaps broadly like the one suggested for Arrisdrift. The Otavi mandible represents the first Miocene hominoid to be found south of equatorial east Africa, and it has been assigned to a previously unknown genus and species, *Otavipithecus namibiensis*. Its phylogenetic relationships aside, it substantially enlarges the known distribution of Miocene hominoids, and it further underlines their remarkable evolutionary success, before a trend toward cooler, drier, and more seasonally variable climatic conditions after 12-10 MYA surely eliminated them from Namibia and substantially reduced their distribution, diversity, and abundance elsewhere.

THE PLIOCENE

So far, later Miocene fossiliferous localities, dating to between 10 and 5.2 MYA, are unknown in southern Africa. This is especially unfortunate, because the late Miocene witnessed dramatic faunal change (Brain

TABLE 23-1 The mid-Miocene vertebrate fauna of Arrisdrift

HIGHER TAXA	LOWER TAXA
OSTEICHTHYES (fishes)	at least one species
AMPHIBIA (frogs)	at least one species
REPTILIA	
Squamata (snakes, lizards)	at least two species
Crocodilia (crocodiles)	*Crocodylus* sp.
Chelonia (tortoises, turtles)	at least one species
AVES (birds)	at least three species
MAMMALIA	
Isectivora	
Chrysochloridae (golden moles)	gen. & sp. indet.
Macroscelididae (elephant shrews, myohyracines)	*Myohyrax* cf. *oswaldi*
Soricidae (shrews)	gen. et sp. not det.
Insectivora not det.	at least two species
Tubulidentata (aardvarks)	*'Orycteropus'* cf. *minutus*
Carnivora	
Amphicyonidae (bear-dogs)	cf. *Amphicyon major*
	Amphicyon cf. *steinheimensis*
?Ursidae (bears)	?Hemicyoninae indet.
Mustelidae (weasels, etc.)	*?Ischyrictis* sp.
Viverridae mongooses, etc.)	gen. & sp. indet.
Carnivora not det.	at least two species
Hyracoidea (hydraxes or dassies)	*Prohyrax* sp.
Deinotheriodea (deinotheres)	*Prodeinotherium hobleyi*
Proboscidea (elephants, mastodons, etc.)	
Gomphotheriidae	gen. & sp. not det.
Perissodactyla	
Rhinocerotidae (rhinoceroses)	*Dicerorhinus* sp.
Artiodactyla	
Suidae (pigs)	gen. & sp. not det.
	Lopholistriodon moruoroti
Tragulidae (chevrotains)	*Dorcatherium* cf. *pigotti*
Climacoceridae (deer-like ruminants)	*Climacoceras* sp.
Bovidae (antelopes, etc.)	gen. & sp. not det.
Pecora (deer, giraffes, antelopes, etc.)	gen. & sp. not det.
Lagomorpha	
Ochotonidae (pikas)	*Kenyalagomys* sp.
Rodentia	
Pedetidae (springhares)	*Parapedetes* sp.
Rodentia not det.	at least four species

Source: After Hendey 1984:91

1984; Vrba 1985a), probably stimulated mostly by the trend toward a cooler, drier climate that was mentioned at the end of the last section. This trend apparently culminated in a "Terminal Miocene Event" between about 6.5 and 5 MYA, when a rapid increase in the Antarctic ice sheet may have been accompanied by a sharp decline in global temperatures. A concomitant drop in sea level led to total or near total desiccation of the Mediterranean, and climates probably became significantly drier at lower and middle latitudes. Grasslands and savannas therefore tended to expand at the expense of forests and woodlands, and previously prominent browsing species, including the highly successful hominoids, experienced an evolutionary crisis. Many browsers became extinct, while others evolved hypsodonty and/or other adaptations that permitted them to survive, even thrive, in the newly created or expanding grasslands. The earliest stages in the development of characteristically African grazing species are not well understood, because potentially informative late Miocene sites are so far rare even in east Africa. However, the intermediate stages, dating to the early Pliocene are well established, thanks in no small part to discoveries in southern Africa. Again following Harland et al. (1990), the Pliocene is defined here as the interval between 5.2 and 1.64 MYA. In east Africa, hominids were almost certainly present during this entire interval, but this was not clearly so in southern Africa, and for present purposes, it is useful to divide the

Pliocene between an early part, antedating 3 MYA, when hominids may have been locally absent, and a later part after 3 MYA, when they were certainly present.

The Early Pliocene

Small samples of certain or probable early Pliocene mammalian fossils have been found at five localities in the northwestern Cape Province of South Africa, including especially Kleinzee and Hondeklip Bay (Hendey 1984). However, the sites in the northwestern Cape are of little significance compared to the extraordinary early Pliocene occurrence at Langebaanweg in the southwestern Cape, roughly 110 km NNW of Cape Town. Thanks primarily to field and laboratory work by Q. B. Hendey (1981a, 1982, 1984), the Langebaanweg fossil collection comprises hundreds of thousands of taxonomically identifiable invertebrate and vertebrate fossils, including what may be the largest samples of early Pliocene bird and mammal bones recovered anywhere in the world.

The overwhelming majority of Langebaanweg vertebrate fossils come from two lithostratigraphic units of the Varswater Formation—the Quartzose Sand Member (QSM) and the Pelletal Phosphorite Member (PPM), both of which are extensively exposed in the open-cast phosphate mine known as "E" Quarry. The QSM consists primarily of fine-grained, white quartz sands laid down on the estuarine floodplain of a river that probably entered the sea immediately southwest of "E" Quarry. The PPM consists of relatively coarse, highly phosphatic, well-sorted sands laid down primarily in the channel of the same river as its bed migrated progressively northwards. Within the PPM there are two distinct, especially fossiliferous channel fills, known as bed 3aS (farther south) and bed 3aN (farther north). The PPM partly truncates the QSM and must therefore be younger, but the two units share the same vertebrate fauna, and they are probably separated by very little time. There is nothing in the deposits that can be dated radiometrically, but the very extensive mammalian fauna shares taxa with east African faunas that are dated to roughly 5 MYA. An age of about 5 MYA is further supported by Hendey's (1981b) analysis of the full Langebaanweg sedimentary succession in relation to later Cenozoic global sea-level and climatic changes.

The Langebaanweg mammalian fauna is striking above all for its richness. It comprises at least 83 species (Table 23-2), representing all 14 extant orders of African mammals but the Sirenia (dugongs and manatees). Unlike the early and middle Miocene faunas of Namibia, in which all taxa below the family level are extinct, the Langebaanweg fauna contains at least 30 extant genera, and its archaic character is overwhelmingly conspicuous only at the species level. Although it includes a handful of distinctly "non-African" elements—most notably, a bear, a wolverine, a monk seal, a peccary, a boselaphine (nilgai-like) antelope, and two species of ovibovine (musk-ox–like) antelopes—its soul is unmistakably African, and it has provided two species of alcelaphine (hartebeest-wildebeest-like) antelopes that are among the oldest to exhibit the extreme hypsodonty that largely underlies the subsequent success of their tribe.

The abundance of most Langebaanweg taxa has not been systematically estimated, but the enormous volume of the sample may be roughly gauged from the numbers of measurable teeth used by Klein (1982) to construct mortality profiles for the three-toed horse (2 dP^2s and 10 P^2s), the rhinoceros (13 dP^4s and 47 P^4s), the "giant bushpig" (20 M_3s), the sivathere (340 dP_4s and 224 M_3s), the giraffe (52 dP_4s and 47 M_3s), the boselaphine (11 dP_4s and 38 M_3s), the primitive buffalo (2 dP_4s and 10 M_3s), and the alcelaphines (33 dP_4s and 215 M_3s). Even the carnivores, which are rare in most fossil faunas, tend to be represented by many individuals per taxon (Hendey 1974, 1978a, 1978b, 1980). The total sample is in fact many times larger, richer, and more diverse than any Plio/Pleistocene east African sample, and it seems improbable that any widespread taxon is absent simply by chance. In this light, the most conspicuous absentee is surely an australopithecine.

The oldest well-documented australopithecine fossils in east Africa are only slightly older than 4 MYA (White 1984), but most specialists believe that the existence of australopithecines in east Africa by 5

TABLE 23-2 The early Pliocene mammalian fauna of Langebaanweg

TAXA	STRATIGRAPHIC UNITS QSM	PPM 3aS	PPM 3aN
INSECTIVORA			
Chrysochloridae (golden moles)			
Chrysochloris sp.	x	x	x
Soricidae (shrews)			
Myosorex sp.	x		
Suncus sp.	x		
Soricidae gen. et sp(p.) not det.	x	x	x
Macroscelididae (elephant shrews)			
Elephantulus sp.	x	x	x
CHIROPTERA (bats)			
Vespertilionidae			
Eptesicus sp.	x		
PRIMATES (monkeys, etc.)			
Cercopithecidae gen. & sp. indet.	x		
PHOLIDOTA (pangolins)			
Phataginus sp.	x		
TUBULIDENTA (aardvarks)			
Gen. & sp. not. det.	x		x
CARNIVORA			
Canidae			
Gen. & sp. not det. (?aff. *'Canis' brevirostris*) (fox)		x	?
Vulpes sp. (fox)		x	x
Ursidae			
**Agriotherium africanum* ("giant" bear)		x	x
Mustelidae			
**Plesiogulo monspessulanus* (wolverine)	x	?	
Mellivora benfieldi (honey badger)		x	x
**Enhydriodon africanus* (otter)		x	x
Phocidae ("true" seals)			
**Homiphoca capensis*	x	x	x
Viverridae			
'Viverra' leakeyi (?aff. *Civettictis*) (civet)	x		x
*Viverrinae gen. & sp. not det. (?aff. *Pseudocivetta*) (civet)	x	x	x
Genetta sp. (genet)	x		
Herpestes spp. A, B (mongooses)	x		
Herpestinae spp. C, D, E (mongooses)	x		
Herpestinae not det.		x	x
Hyaenidae			
*"*Adcrocuta" australis*	x	?	?
**Ictitherium preforfex*		x	x
Hyaena abronia (ancestral striped hyena)	x	x	x
**Euryboas* sp. ("hunting hyena")	x	x	x
Hyaenidae sp. E		x	
Hyaenidae not det.		x	x
Felidae			
*"*Machairodus"* sp. (leopard-size saber-tooth)		x	
**Homotherium* sp. (lion-size saber-tooth)	x	x	?
Felis sp. (wildcat)	x		
Felis aff. *issiodorensis* (lynx)	x	x	x
Felis obscura		x	
**Dinofelis diastemata* (false saber-tooth)	x	x	x
Felidae not det.		x	
Carnivora not det.			
Gen. & sp. not det. (Canidae or Viverridae)	x		
Gen. & sp. not det. (?Procyonidae)	x		
Gen. & sp. not det. (?Lutrinae)	x		
PROBOSCIDEA			
Gomphotheriidae			
**Anancus* sp. (gomphothere)	x	x	
Elephantidae			
**Mammuthus subplanifrons* ("mammoth")	x	?	x

TABLE 23-2 Continued

TAXA	STRATIGRAPHIC UNITS QSM	PPM 3aS	PPM 3aN
HYRACOIDEA			
Procaviidae			
Procavia cf. *antiqua* (hyrax or dassie)	x		?
PERISSODACTYLA			
Equidae			
**Hipparion* cf. *baardi* (three-toed horse)	x	x	x
Rhinocerotidae			
Ceratotherium praecox (white rhinoceros)	x	x	
ARTIODACTYLA			
Tayassuidae			
**Pecarichoerus* (or *Barberahyus*) *africanus* (peccary)		x	x
Suidae			
**Nyanzachoerus* cf. *pattersoni* (or *kanamensis*) ("giant bushpig")	x		
**Nyanzachoerus* cf. *jaegeri* ("giant bushpig")		x	
Hippopotamidae (hippopotamuses)			
Gen. & sp. not det.		x	x
Giraffidae			
**Sivatherium hendeyi* (ox-bodied, short-necked giraffe)	x	x	x
**Palaeotragus* cf. *germaini* (okapi-like)			x
Giraffa sp. (true giraffe)	x	x	x
Bovidae			
Tragelaphus sp. A (nyala-like)	x	x	x
Tragelaphus sp. B (nyala-like)			x
**Mesembriportax* (or *Miotragerocera*s) *acrae* (kudu-like relative of the nilgai)	x	x	x
**Simatherium demissum* (buffalo)	x	x	x
Kobus subdolus (kob-like)		x	x
Kobus sp. B (kob-like)			x
**Damalacra neanica* (hartebeest-like)		x	x
**Damalacra acalla* (hartebeest-like)	x	x	x
Raphicerus paralius (steenbok)	x	x	x
Gazella sp. (gazelle)	x	x	x
*Ovibovini gen. & at least 2 spp. not det. (musk-ox-like)	x	x	x
LAGOMORPHA			
Leporidae			
Pronolagus sp. (rock hare)	x	x	x
RODENTIA			
Bathyergidae (rodent moles)			
Bathyergus sp.	x	x	x
Cryptomys sp.	x		?
Hystricidae (porcupines)			
Gen. & sp. not det. A	x		
Gen. & sp. not det. B		x	x
Cricetidae (rats, mice, gerbils, etc.)			
Mystromys sp. A	x		
Mystromys cf. *darti*	x		
Mystromys cf. *hausleitneri*	x		
Gerbillus or *Desmodillus* sp.	x		
Dendromus sp.	x		
Steatomys or *Malacothrix* sp.	x		
Muridae (rats, mice)			
Aethomys spp. A, B	x		
Mus spp. A, B	x		
Rhabdomys sp.	x		
**Euryotomys pelomyoides*	x		
Muscardinidae (dormice)			
Graphiurus sp.	x		
Rodentia not. det.	x	x	x
CETACEA (whales and dolphins)			
Gen. & spp. not det.	x	x	x

Asterisks mark extinct genera. Source: After Hendey 1984:97-99).

MYA will be demonstrated when suitably rich fossiliferous sites are investigated. If Langebaanweg were in east Africa, it would surely qualify, and its failure to provide australopithecines therefore suggests they were not present nearby. A priori, the most plausible reason is that they were unable to extend their range so far south and west (32°58′S, 18°09′E) at this remote time, probably because they were still essentially tropical creatures who could not adapt to the temperate conditions that had replaced tropical ones in far southern Africa during the late Miocene (Hendey 1981a). In this connection, it is noteworthy that no australopithecines, even ones that postdate Langebaanweg by 2–3 million years, are known further north or south than 27°32′, the latitude of the famous Taung site. Surface and in situ artifact discoveries in fact suggest that the first hominids to colonize Africa south of Taung may have been relatively advanced hand-axe makers of the genus *Homo*, who perhaps arrived no more than a million years ago.

Given the probability that hominids were absent or at least very rare in the ancient Langebaanweg environment, three features of the fauna merit special attention. The first and simplest is the presence of numerous burnt bones (Hendey 1982). The burning almost certainly resulted from natural (?seasonal) veldt fires, and it serves as a useful reminder that evidence of fire is not by itself evidence of people. The second and third features are more complex. They involve the ungulate age profiles (Klein 1982) and the abundance of carnivores respectively.

The Langebaanweg ungulate age profiles divide clearly between the two mathematically and demographically related types that characterize all more or less stable populations of large mammals: (1) profiles that comprise various age classes in rough proportion to their live abundance and that are therefore relatively rich in prime-age (reproductively active) adults; and (2) profiles that are much poorer in prime-age adults relative to very young and old individuals (Figure 23-2). In paleobiology, a profile comprising various age classes in rough proportion to their live abundance is commonly called "catastrophic," because its fossil occurrence is most likely to reflect a great flood, volcanic eruption, epidemic disease, or other catastrophe that affected all individuals equally, regardless of their age. A profile dominated by the very young and the old is usually called "attritional," because its fossil occurrence is most likely to reflect everyday, routine mortality factors like predation, accidents, and endemic disease that disproportionately affect the young and the old.

In keeping with advance expectations, at Langebaanweg, catastrophic profiles characterize those species that dominate the channel fills of the PPM, where the sedimentary matrix and the very dense packing of bone suggest death by flash (?seasonal) floods. Attritional profiles characterize those species that are rare in the channel fills and much more common in the floodplain deposits of the QSM, where the sparser distribution of bones, often as partial skeletons, sometimes chewed or associated with carnivore coprolites, suggests death from predation or other more or less routine mortality factors. The difference in mortality profiles between the most abundant channel fill (PPM) species (the sivathere) and the most common floodplain (QSM) species (the rhinoceros) is illustrated in Figure 23-2. The difference between the sedimentary facies in relative species abundance probably reflects differences in where the species normally were when flash floods occurred.

The Langebaanweg mortality profiles might have no particular paleoanthropological relevance, except that the attritional profiles depart from theoretical expectations in one very conspicuous respect—they contain many fewer young individuals than must have died for attritional reasons. Since bone preservation is superb, the most likely explanation for the absence of very young individuals is selective pre- (rather than post-) depositional destruction of very young bones, and in the context of the Lange-

FIGURE 23-2 Top left: Schematic catastrophic-age (mortality) profile for a population of large mammals that is basically stable in size and structure. The grey bars represent the number of individuals that survive in each successive age cohort, the hatched bars the number that die between successive cohorts. Top right: Separate plot of the hatched bars, showing the corresponding schematic attritional age profile. The basic form of corresponding catastrophic and attritional profiles is the same for all large mammals, but the precise form will differ from population to population, depending on species biology and specific mortality factors. Middle and Bottom: Large ungulate age (mortality) profiles from Langebaanweg and Klasies River Mouth Cave 1. Individual ages were estimated from dental crown heights as explained in Klein (1982).

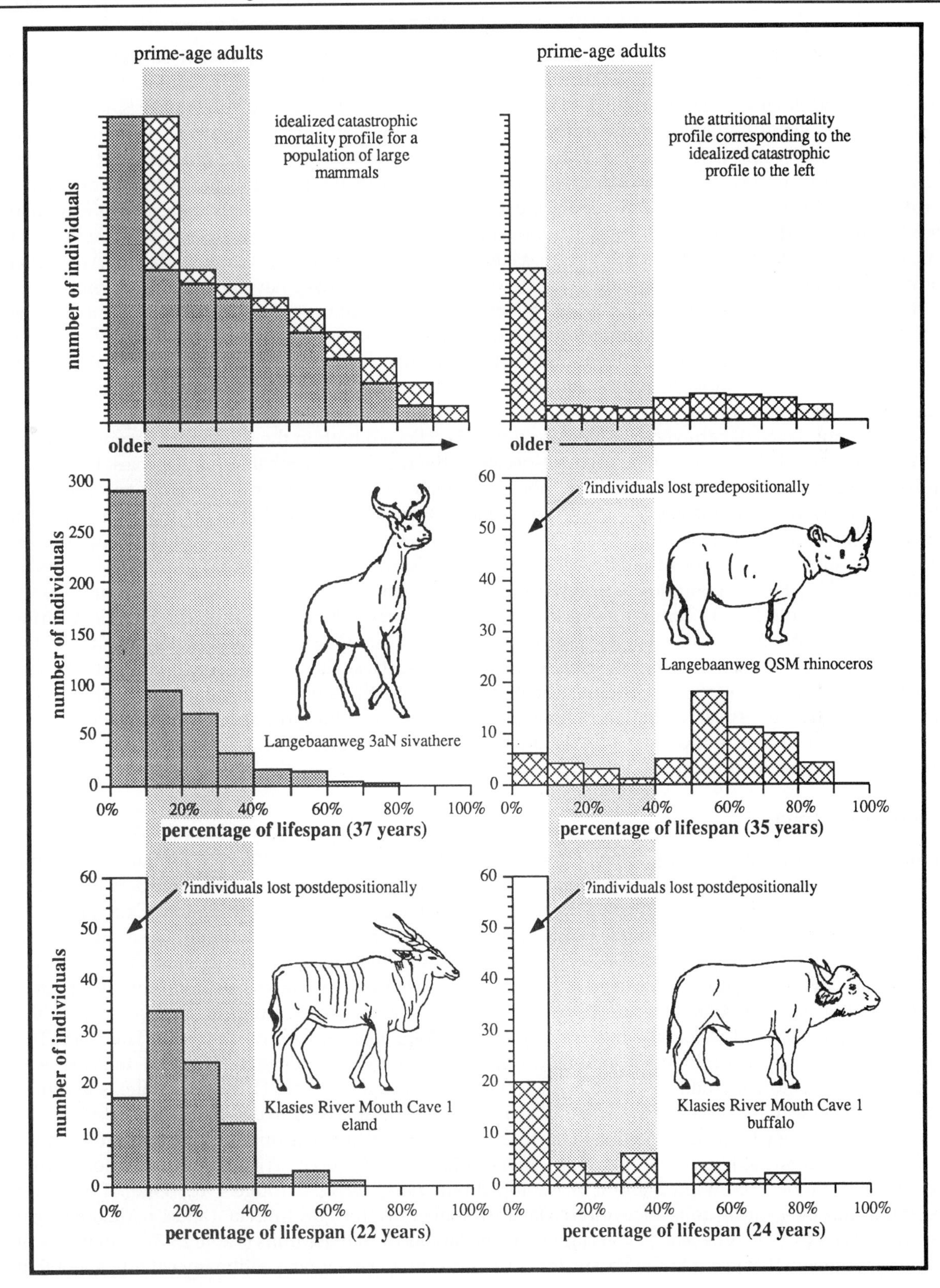
prime-age adults
prime-age adults
idealized catastrophic mortality profile for a population of large mammals
the attritional mortality profile corresponding to the idealized catastrophic profile to the left
number of individuals
older
older
300
250
200
150
100
50
0
60
50
40
30
20
10
0
?individuals lost predepositionally
Langebaanweg 3aN sivathere
Langebaanweg QSM rhinoceros
0% 20% 40% 60% 80% 100%
percentage of lifespan (37 years)
percentage of lifespan (35 years)
?individuals lost postdepositionally
?individuals lost postdepositionally
Klasies River Mouth Cave 1 eland
Klasies River Mouth Cave 1 buffalo
percentage of lifespan (22 years)
percentage of lifespan (24 years)

baanweg fauna, the most likely destructive agents are the numerous hyenas, whose modern representatives rapidly and completely consume most very young carcasses, even of buffalos, rhinoceros, elephants, and other very large ungulates (Blumenschine 1987; Goddard 1970; Laws 1966; Sinclair 1977). The relatively small number of very young individuals in the Langebaanweg attritional profiles thus represents what would be available to scavenging hominids, assuming that like modern wildlife biologists, they could not locate most young carcasses before hyenas did. It follows that archeological attritional profiles in which very young individuals are very well represented probably reflect active human predation rather than scavenging. Such attritional profiles in fact characterize archeological samples from late Pleistocene sites like Klasies River Mouth, as illustrated in Figure 23-2 and discussed later. Attritional profiles that contain as few young individuals as those at Langebaanweg have not yet been found at any archeological site that preserves bone as well as Langebaanweg, but very few such sites that antedate the late Pleistocene have been found, and those that do exist are mainly places where people may not have been the main bone accumulators. Ultimately, when suitable, very early archeological sites are found, attritional profiles like those at Langebaanweg might constitute the firmest evidence for early human scavenging (vs. active hunting).

The abundance of carnivores at Langebaanweg is reflected both in the number of carnivore bones and in the ratio of the number of carnivore species to the number of probable prey species. Counts of individual carnivores and prey animals are not available, but the frequencies of carnivore and prey taxa are reasonably well established. Thus, excluding the smaller taxa that probably fed mainly on insects, micromammals, or plants, there are 20 carnivore species in the Langebaanweg fauna. Similarly, apart from small prey such as small rodents and other micromammals, there are 27 potential prey species. Similar counts based on the large middle Pleistocene faunal sample from Elandsfontein, only 10 km southeast of Langebaanweg, disclose 11 carnivore species and 28 prey species (Hendey 1974; Klein and Cruz-Uribe 1991; also see later discussion). For the extensively sampled late Pleistocene and Holocene sites of the region, the counts reveal 12 carnivore species and 32 prey species (Klein 1983). These data indicate that sometime between the early Pliocene and the mid-Pleistocene there was a substantial reduction in the diversity of carnivore species and, if anything, an increase in potential prey species. It seems likely that the decrease in carnivore taxa was at least in part caused by the arrival and subsequent evolutionary success of meat-eating hominids.

The Late Pliocene

In southern Africa, faunal samples that possibly or probably date to the late Pliocene (defined here as the interval between 3 and 1.64 MYA) have been recovered in brecciated cave fills on the Humpata plateau of southern Angola (Pickford et al. 1990), at Baard's Quarry (Langebaanweg, southwestern Cape Province, South Africa) (Hendey 1978c), and at Bolt's Farm, Coopers, and Haasgat (southern Transvaal, South Africa) (Brain 1958; H. B. S. Cooke 1978; Keyser and Martini 1990). However, from a paleoanthropological perspective, by far the most important late Pliocene sites are the five known australopithecine caves—Taung in the northern Cape; Sterkfontein, Kromdraai (B), and Swartkrans near Krugersdorp in the southern Transvaal; and Makapansgat (Limeworks) near Potgietersrus in the north-central Transvaal. The context of the single australopithecine known from Taung is poorly understood, because the cave was quarried away before it could be studied by specialists (Partridge et al. 1991), but the contexts in which australopithecine fossils occur in the Transvaal caves have been analyzed in detail.

At each Transvaal cave the sedimentary fill has been assigned to a formation named after the cave, and each formation has been divided among 3 to 6 stratigraphically successive members, numbered from one at the bottom to between 3 and 6 at the top (Brain 1981, 1988; Partridge 1982, 1986). The oldest australopithecine-bearing units uncovered so far are Members 3 ("the grey breccia") and 4 at Makapansgat and Member 4 (= "the Type Site") at Sterkfontein. The faunas from these units (Table 23-3) contain few of the archaic genera that occur in the Varswater Formation at Langebaanweg and many extant genera that are not found there. They are therefore clearly younger than the Langebaanweg fauna, but for more precise age assignment, they are equally dependent on comparisons to faunas from radiometrically dated sites in east Africa. Faunal dating is complicated by the possibility that elements from different units within the cave fills were inadvertently mixed during removal, particularly in the early stages of investigation at each site. Nonetheless, the most recent attempts at dating (Delson 1988; Vrba 1982, 1985b) show fairly clearly that Makapansgat Member 3 probably accumulated around 3 MYA and Member 4 only slightly later, while Sterkfontein Member 4 probably formed between 2.8 and 2.3 MYA, perhaps mainly around 2.5 MYA. These age estimates are supported, or at least not contradicted, by paleomagnetic determinations at Sterkfontein (D. L. Jones et al. 1986) and especially at Makapansgat (McFadden and Brock 1984).

The oldest sediments in the remaining three caves are all probably or certainly younger. An age near 2 MYA has been suggested for Taung (Delson 1984, 1988; Vrba 1982) and for the only unit at Kromdraai B (Member 3 East) in which australopithecine remains have been found *in situ* (Vrba 1982), but the associated faunas are very sparse, and any age estimates are necessarily tentative. The much more abundant fauna from Member 1 at Swartkrans (Table 23-3) suggests an age between 1.8 and 1.5 MYA (Vrba 1982), straddling the Plio/Pleistocene boundary as defined here. Member 1 also contains artifacts and fossils of archaic *Homo*. A broadly similar Plio/Pleistocene age (between roughly 1.8 and 1.3 MYA) has been suggested for Member 5 (= roughly "the Extension Site") at Sterkfontein (Vrba 1982). Unlike Swartkrans 1, Sterkfontein 5 has not provided any australopithecines, but it does contain artifacts and fossils of archaic *Homo*. Together with the like-aged finds from Swartkrans 1, they comprise the oldest artifacts and fossils of *Homo* so far found in southern Africa. The stratigraphically younger units at both Swartkrans (Members 2-5) and Sterkfontein (Member 6) clearly date from the Pleistocene (in the narrow sense) (Brain et al. 1988; Partridge 1982, 1986; Vrba 1982). However, Swartkrans Members 2 and 3 have provided australopithecine fossils that may be among the youngest found anywhere, as discussed later.

The South African australopithecine sample bears on the same basic issues of early hominid taxonomy, phylogeny, and ecology as its east African counterpart, and it also raises a unique question: How did australopithecine bones become incorporated in the various cave fills? Taxonomically, the South African australopithecines are now usually divided between two species—*Australopithecus africanus* and *A. robustus*. In the vernacular, these are often called the "gracile" and "robust" australopithecines respectively, but this terminology is misleading, insofar as it implies a substantial difference in body mass or robusticity. In fact, the species did not differ markedly in size: individual *A. africanus* probably weighed between 33 and 67 kg and averaged roughly 115 cm tall in females and 138 cm in males, while individual *A. robustus* probably weighed between 28 and 62 kg and averaged perhaps 110 cm tall in females and 132 cm in males (Jungers 1988; McHenry 1988, 1991a, 1991b). Both species had relatively small endocranial capacities, ranging from 430 to 520 cc in *A. africanus* and from 500 to 530 cc in *A. robustus* (Holloway 1975, 1983). Insofar as the species differed in size or robusticity, it was in the postcanine dentition and in the bony craniofacial structures to which the masticatory muscles attached. In *A. robustus*, the molars and especially the premolars were greatly expanded, and the craniofacial architecture was clearly specialized to allow the application of substantial vertical force between the upper and lower cheek-tooth rows during mastication. Based on this distinctive craniofacial morphology, Broom (1938, 1949) argued that (what is here called) *A. robustus* merited a separate genus, *Paranthropus*. He further suggested that *Paranthropus* comprised two distinct species, *P. robustus* at Kromdraai and *P. crassidens* at Swartkrans. Some prominent authorities (Clarke 1985a; Grine 1985, 1988; Howell 1978, 1988) have revived one or both of Broom's views, but only the argument in favor of separate generic status is possibly gaining favor.

A. africanus is represented at Makapansgat (primarily in Member 3, but also probably in 4), at Sterkfontein (Member 4), and at Taung. *A. robustus* occurs at Kromdraai B (Member 3 East and possibly other units) and at Swartkrans (Members 1-3). Based on the faunal dates that were previously presented, it follows that *A. africanus* was present in southern Africa between roughly 3 MYA and 2 MYA, while *A. robustus* was present between roughly 2 MYA and perhaps 1 MYA (the terminal date is less certain, as discussed in the next section). The same faunal dates imply that *Homo* first appeared locally at approxi-

TABLE 23-3 Late Pliocene large mammals from Makapansgat (MAK).

TAXA	MAK Mb3	STK Mb4	STK Mb5	SWK Mb1
PRIMATES				
Hominidae				
**Australopithecus africanus*	x	x		
**Australopithecus robustus*				x
Homo sp.			x	x
Cercopithecidae				
**Parapapio broomi* (extinct baboon)	x	x		
**Parapapio jonesi* (extinct baboon)	cf.	x		x
**Parapapio whitei* (extinct baboon)	x	x		
Cercocebus sp. (mangabey)	x			cf.
Papio hamadryas robinsoni (hamadryas baboon)		x		x
Papio izodi (extinct baboon)		x		
Papio sp. (?chacma baboon)			x	
**Dinopithecus ingens* (extinct baboon)				x
Theropithecus oswaldi ("giant" gelada baboon)	cf.			x
**Cercopithecoides williamsi* (extinct large colobine monkey)	x	x		
TUBULIDENTATA				
Orycteropodidae				
Orycteropus afer (aardvark)	x			
CARNIVORA				
Canidae				
Canis mesomelas (black-backed jackal)		x	x	x
Canis adustus (side-striped jackal)	cf.			
Vulpes pulcher (extinct fox)	cf.			x
Mustelidae				
Aonyx cf. *capensis* (clawless otter)				x
Hyaenidae				
**Chasmoporthetes nitidula* ("hunting hyena")		x	x	x
**Chasmoporthetes silverbergi* ("hunting hyena")		x	x	
Hyaena hyaena (striped hyena)	x			x
Hyaena brunnea (brown hyena)		x		x
Crocuta crocuta (spotted hyena)		x	x	x
**Pachycrocuta brevirostris* (extinct hyena)	x	x	x	
Proteles sp. (aardwolf)				x
Felidae				
Felis caracal (caracal)	x			
Felis serval (serval)	x			
Panthera pardus (leopard)		x		x
Panthera leo (lion)		x		x
Acinonyx jubatus (cheetah)	x			x
**Dinofelis barlowi* (false saber-tooth)	x	x		
**Megantereon cultridens* (saber-tooth)		x	x	cf.
**Homotherium crenatidens* (saber tooth)	cf.	x	x	
PROBOSCIDEA				
Elephantidae				
Elephas recki (Reck's elephant)	cf.	x	x	
HYRACOIDEA				
Procaviidae				
**Gigantohyrax maguirei* ("giant" hyrax)	x			
Procavia transvaalensis (extinct rock hyrax)	x			x
Procavia antiqua (extinct rock hyrax)	x			x
PERISSODACTYLA				
Rhinocerotidae				
Ceratotherium simum (white rhinoceros)	x			
Diceros bicornis (black rhinoceros)	x			
Chalicotheriidae				
Ancylotherium hennigi (chalicothere)	x			

TABLE 23-3 Continued

	MAK	STK		SWK
	Mb3	Mb4	Mb5	Mb1
Equidae				
Hipparion libycum (three toed horse)	x			x
Equus capensis ("giant" Cape zebra)		x		x
Equus burchelli (Burchell's zebra)			x	
ARTIODACTYLA				
Hippopotamidae				
Hippopotamus amphibius (hippopotamus)	x			
Suidae				
**Notochoerus capensis* (extinct pig)	x			
**Potamochoeroides shawi* (extinct pig)	x			
**Tapinochoerus meadowsi* (extinct "warthog")				x
**Metridiochoerus* sp. (extinct "warthog")		x		
Suidae gen. & sp. indet.			x	
Giraffidae				
**Sivatherium maurusium* (sivathere)	x			
Giraffa camelopardalis (giraffe)	x			
Giraffidae gen. & sp. indet.				x
Bovidae				
Tragelaphus cf. *strepsiceros* (greater kudu)	?			x
Tragelaphus aff. *angasi* (nyala)	x	x		
Tragelaphus pricei (bushbuck-like)	x			
Taurotragus cf. *oryx* (eland)	?		x	
**Simatherium* cf. *kohllarseni* (ancestral "giant" buffalo)	x			
Syncerus sp. (Cape buffalo)		x		x
Hippotragus cf. *equinus* (roan antelope)		x		
Hippotragus cookei (roan-sable like)		cf.		
*?Hippotragini gen. & sp. nov. (roan-sable like)	x			
**Wellsiana torticornuta* (roan-sable like)	x			
Redunca darti (extinct reedbuck)	x	x		
Redunca cf. *arundinum* (southern reedbuck)				x
Beatragus sp. or *Alcelaphus* sp. (hartebeest-like)				x
**Rabaticeras porrocornutus* (extinct hartebeest)				x
**Parmularius braini* (bastard hartebeest)	x			
**Parmularius* sp. nov. (bastard hartebeest)	x			
*?*Damalops* sp. (hartebeest-like)	x			
Damaliscus sp. or **Parmularius* sp.(bastard hartebeest)		x	x	x
Damaliscus sp. (bastard hartebeest)			x	
Connochaetes sp. (wildebeest)		x	cf.	x
**Megalotragus* sp. ("giant hartebeest")	x	cf.	cf.	x
Cephalolphus sp. (small duiker)	x			
Pelea cf. *capreolus* (vaalribbok)				x
Oreotragus major (extinct klipspringer)	cf.		x	x
**Makapania broomi* (musk-ox-like)	x	cf.		cf.
Antidorcas recki (Reck's springbok)		cf.	x	x
Antidorcas bondi (Bond's springbok)		cf.		
Gazella vanhoepeni (extinct gazelle)	x			
?Gazella sp. (?gazelle)		x		x
Antilopini gen. et sp. indet. (2 spp.) (gazelles or springboks)	x			
LAGOMORPHA				
Leporidae				
Pronolagus sp. (rock hare)	x			
RODENTIA				
Hystricidae				
**Xenohystrix crassidens* ("giant" porcupine)	x			
Hystrix makapanensis (extinct porcupine)	x			
Hystrix africaeaustralis (porcupine)	x			x

Asterisks mark extinct genera.

Source: After Cooke 1978; Maguire et al. 1980; Delson 1984, 1988; Turner 1984; Vrba 1987); Sterkfontein (STK) after McKee 1991, and Swartkrans (SWK) after Brain et al. 1989.

mately 2 MYA, at perhaps the same time as *A. robustus*. These chronological relationships and the craniofacial morphology of *A. africanus* can be used to argue that *A. africanus* was ancestral to *A. robustus* (Johanson and White 1979) or perhaps to both *A. robustus* and early *Homo* (McHenry and Skelton 1985; Tobias 1978). However, the derivation of *A. robustus* from *A. africanus* seems increasingly unlikely, now that robust australopithecine fossils dating to 2.5 MYA have been found in east Africa (J. M. Harris et al. 1988; Walker et al. 1986). The phylogenetic status of *A. africanus* is problematic, in part because it remains unknown or at least unconfirmed in east Africa, where so far, only robust australopithecines are well documented between 3 and 2 MYA. However, pending further discoveries, the developing consensus (Figure 23-3) is probably that (1) *A. africanus* was ancestral to *Homo* and (2) *A. robustus* (including a coeval east African variant, *A. boisei*) evolved from a more primitive robust australopithecine that co-existed with *A. africanus* between 3 and 2 MYA. This more primitive robust australopithecine and *A. africanus* may both in turn derive from the most primitive known australopithecine, *A. afarensis*. *A. afarensis* has been bracketed in east Africa between greater than 4 MYA and roughly 3 MYA (Howell 1988), and either it or its immediate ancestor is the species that might be expected at Langebaanweg 5 MYA, if australopithecines were present in the vicinity. As previously argued, their absence in the Langebaanweg fauna (and perhaps throughout southern Africa before 3 MYA) suggests that the earliest hominids were physiologically and behaviorally restricted to equatorial latitudes.

The long period of sympatry between the robust australopithecines on the one hand and the lineage probably including *A. africanus* and *Homo* on the other obviously implies ecological separation. Following Robinson (1954, 1963), most authorities have emphasized a likely dietary difference, and it has often been suggested that *A. robustus* used its massive postcanine dentition and functionally related chewing musculature mainly to grind vegetal foods. In fact, both macroscopic and microscopic differences in molar wear between *A. robustus* and *A. africanus* suggest that *A. robustus* focused more on hard, fibrous plant tissues (Grine 1981, 1986; Kay and Grine 1988), while the same evidence and observations on later (including living) *Homo* may be used to argue that the nonrobust lineage was adapted to a more eclectic diet, including meat. If this interpretation is correct, then the apparent absence of robust australopithecines in southern Africa before about 2 MYA might reflect a relative lack of suitable plant foods, particularly edible bulbs and roots. These probably became much more abundant following a shift between 2.3 and 1.8 MYA from generally more wooded conditions (reflected in the faunas from Makapansgat 3 and Sterkfontein 4) to much grassier ones (reflected in the faunas from Swartkrans 1 and Sterkfontein 5) (Vrba 1985b, 1988). The faunal change probably reflects a marked decline in precipitation that is implied by sedimentological differences between Sterkfontein 4 (below) and Sterkfontein 5 (above) (Partridge 1985a, 1986). The decline in precipitation in turn probably followed on major global cooling around 2.4 MYA (Prentice and Denton 1988) that may also have prompted significant vegetational and faunal change in eastern Africa (Vrba 1985b, 1988).

It has often been assumed that only *Homo* or its immediate ancestor produced the earliest known stone artifacts, such as those that were deposited in Swartkrans Member 1 and Sterkfontein Member 5 between roughly 1.8 and 1.3 MYA and at contemporaneous and even older east African sites antedating 2 MYA (J. W. K. Harris 1983). This follows from (1) the certainty that *A. robustus* was not ancestral to stone–tool making *Homo*; (2) the frequent assumption that stone tool making originated largely to facilitate the acquisition of meat, while *A. robustus* was mainly vegetarian; and (3) the presence of only one readily identifiable initial stone artifact tradition and the way in which this apparently anticipates the further development of artifact making after 1–0.8 MYA, when only *Homo* survived. However, morphological studies by Susman (1988, 1991) show that the hands of *A. robustus* were eminently suited for stone-tool manufacture, and captive orangutans and chimpanzees, whose hands are much less suitable, will flake stones, given sufficient incentive (Wright 1972; Toth et al. 1993). It therefore seems possible that *A. robustus* was responsible for at least some of the crude stone artifacts found in Swartkrans 1, Sterkfontein 5, and coeval or somewhat older sites in east Africa. In addition, both Swartkrans 1 and Sterkfontein 5 contain pointed, polished, and striated bone fragments whose form and features have been replicated on modern, experimental fragments used to dig bulbs and edible roots from rocky soil (Brain 1988; Brain et al. 1988). A priori, there is no reason to suppose that *A. robustus* would not have used such tools.

The question of how bones accumulated in the South African australopithecine caves is not fully resolved, but very plausible suggestions are available, thanks mainly to systematic observations by Brain (1981) at Swartkrans, Sterkfontein, and Kromdraai, and by Maguire (1985; Maguire et al. 1980) at Makapansgat. Taung is not considered here, because the early destruction of the cave and the small size of the surviving faunal sample preclude compelling inferences.

At the time(s) that bones of australopithecines and other creatures accumulated at Swartkrans and Sterkfontein, each cave was probably a subterranean receptacle linked to a relatively flat or gently undulating surface by a strongly inclined or nearly vertical shaft (Brain 1958). Since most terrestrial vertebrates could not have easily negotiated the shaft, their bones probably represent talus debris that accumulated in cones at its base. The problem then is to identify the agent or agents that concentrated bones near the shaft entrance, perhaps on ledges within a small shelter that overhung it or in trees that clustered around it. Based primarily on patterns of bone damage and skeletal part representation, Brain has concluded that the agent varied through time, beginning at each site (in Swartkrans Member 1 and Sterkfontein Member 4) with leopards, other large cats, and perhaps hyenas that utilized the overhang or trees above the shaft entrance. He believes that a combination of people (early *Homo*), leopards, and brown hyenas, sheltering near the cave mouth, may have produced the bone assemblage in Swartkrans Member 2, while people (early *Homo*) alone were responsible for the bones in Swartkrans 3 and Sterkfontein 5.

Brain's studies therefore suggest that both kinds of australopithecines were entirely passive participants in the bone accumulation process, representing the hunted, not the hunters (nor even scavengers). However, he thinks they may have taken refuge in or near the cave mouths, particularly at night, and that it was their presence that may have attracted large predators to the sites. This is suggested above all by the abundance of australopithecines in both Sterkfontein 4 (where they comprise roughly 14 percent of the individual animals in the sample) and Swartkrans 1 (where they comprise about 27 percent) (Brain 1981). It is perhaps further implied by the abundance of fossil baboons, which, based on modern analogy, may likewise have favored cave mouths as night-resting spots, and which are the only species at either site whose numbers approach or exceed those of the australopithecines.

If we accept that the australopithecines were taken mainly as they slept in the cave mouths, it may be noteworthy that remains of early *Homo* are much less abundant at each site. This might imply that predators found sleeping bands of early *Homo* far harder to approach, perhaps in part because they had incipient control over fire. This possibility is suggested at Swartkrans, where Members 1 and 2, which contain the overwhelming majority of australopithecine bones and the best evidence for carnivore activity, provided no burnt bones in a total excavated sample of approximately 100,000 pieces, while overlying Member 3, which contains cut-marked bones and little indication of carnivore presence, provided 270 burnt bones in an excavated sample of about 60,000 pieces (Brain 1988; Brain and Sillen 1988; Brain et al. 1988). If the burnt bones in Member 3 were genuinely burnt by people, their putative (faunal) age of roughly 1 MYA would make them the oldest evidence for human use of fire in southern Africa and nearly as old as patches of baked earth that constitute equally intriguing (and equivocal) evidence for fire in east Africa (J. D. Clark and Harris 1985; Barbetti 1986). Control over fire by 1 MYA would clearly help explain how and why people were able to extend their range not only to the extreme south of Africa by this time, but also to the extreme north and from there to southern Eurasia.

In form, the Kromdraai Cave was probably broadly similar to Swartkrans and Sterkfontein, and Brain (1981) has suggested that large cats, hyenas, or both introduced most of the bones of australopithecines and other large vertebrates. This remains plausible, though excavations supervised by Vrba (1981; also Vrba and Panagos 1982) suggest that some bones probably derive from animals that fell in and were unable to climb out.

A priori, the explanation for bone accumulation at the Makapansgat (Limeworks) Cave may be different from that at Sterkfontein, Swartkrans, and Kromdraai, since it was probably a more tunnel-like

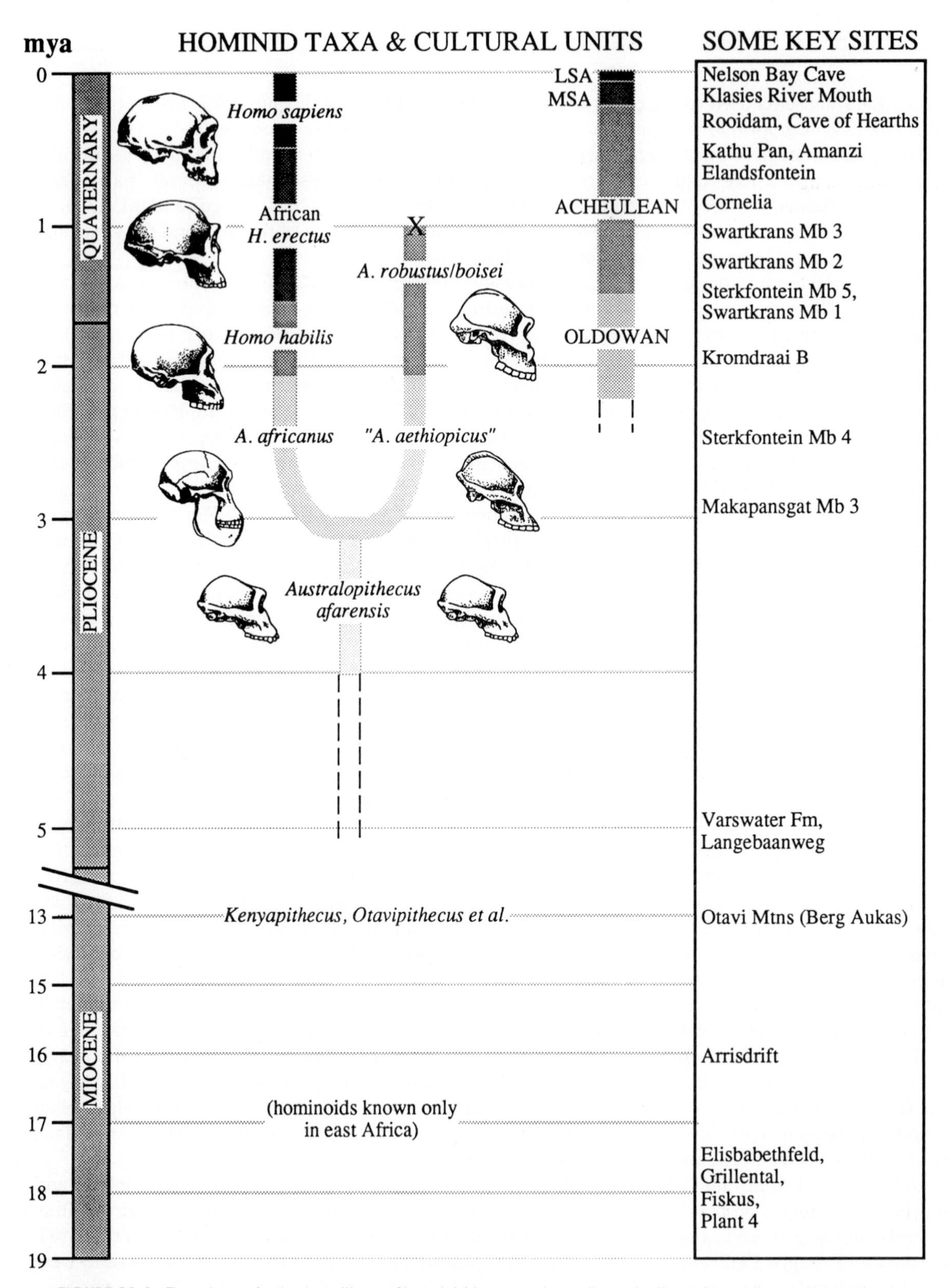

FIGURE 23-3 The chronological positions of hominid taxa, major culture-stratigraphic units, and some key Southern African sites mentioned in the text. Note that the absolute time scale is more condensed for the Miocene than it is for the Pliocene and Quaternary.

cave, with a nearly horizontal entrance leading in from the flank of a steep-sided valley. Either australopithecines or carnivores could probably have walked in and out at will, and either could have lived inside the cave. In addition, the fauna from the principal australopithecine unit, Member 3, itself suggests a different mode or agent of deposition, since it is relatively much poorer in primate bones and much richer in bovid remains than the comparable units at Sterkfontein, Swartkrans, and Kromdraai. The unique features of Makapansgat are especially important to consider in light of Dart's (1957) contention that the animal bones from Member 3 were australopithecine tools and weapons. His argument turned particularly on the large sample of antelope skeletal parts, which he noted were not represented in their anatomically expectable frequencies. Some, like mandibles and distal humeri, were disproportionately abundant, and Dart believed this was because the australopithecines found them especially useful. Since stone artifacts were absent, he concluded that Makapansgat reflected an "osteodontokeratic" (bone–tooth–horn) cultural stage that antedated the stone age.

Dart's hypothesis suffered greatly when Brain (1967, 1969) showed that skeletal part discrepancies like those in Makapansgat 3 could be a result of differences in bone density. Such differences almost inevitably lead to differential pre- or postdepositional loss of softer bones in fossil assemblages, whatever the agent of accumulation. More recently, Maguire (1985; Maguire et al. 1980) reported the presence of hyena coprolites in Makapansgat 3 and showed also that the bones are commonly damaged like those from recently observed hyena breeding dens. She concluded that hyenas were the principal bone accumulators in Makapansgat 3, and this is supported by resemblances between the Makapansgat assemblage and much younger (late Pleistocene) South African assemblages where bone damage, the presence of coprolites, and the absence of stone tools strongly implicates hyenas (Cruz-Uribe 1991; Klein 1975; Klein et al. 1991). Features that these later assemblages share with Makapansgat include a high frequency of carnivores (including hyenas themselves) and a tendency for the ratio of postcranial to cranial remains to increase with species size, while the cranial bones of very large animals come almost entirely from juveniles and the postcranial bones come from adults.

Neither the high frequency of carnivores nor the peculiar cranial/postcranial pattern has been observed in archeological assemblages (meaning ones associated with abundant stone artifacts and usually containing numerous cut-marked bones). The difference is readily explained by (1) the higher frequency of interaction between hyenas and other carnivores; and (2) the lesser ability of hyenas to transport very large bones and their greater tendency to destroy smaller and softer bones altogether. In addition, the overall pattern of antelope skeletal part representation in the late Pleistocene hyena assemblage from Swartklip 1 near Cape Town, where bone preservation closely resembles that at Makapansgat, is remarkably similar to the pattern calculated by Dart for Makapansgat (Figure 23-4). Such differences as do exist between the two assemblages could be attributed to somewhat different mixes of antelope size classes and to somewhat different counting methods. In sum, like Swartkrans, Sterkfontein, and Kromdraai, Makapansgat offers little on australopithecine behavior or ecology, except to show that the australopithecines commonly fell prey to large carnivores.

THE QUATERNARY

The Quaternary is taken here to span the past 1.64 million years, comprising the Pleistocene before 12,000 years ago (KYA) and the Holocene thereafter. The earliest Quaternary has already been introduced, since deposits that straddle the Plio/Pleistocene boundary occur in the Swartkrans and Sterkfontein australopithecine caves, in Members 1 and 5 respectively. As already noted, the overwhelming majority of hominid fossils from Swartkrans 1 represent *A. robustus*, but early *Homo* is also present. Both *A. robustus* and early *Homo* further occur in Swartkrans 2, and *A. robustus* is present in Member 3. Early *Homo* is the only hominid so far documented in Sterkfontein 5. The robust australopithecine fossils from Swartkrans 3 are especially noteworthy, because they could be among the youngest known. The associated mammal taxa and tentative thermoluminescence dates suggest that Member 3 probably formed sometime between 1.5 and

0.8 MYA (Brain 1988; Brain et al. 1988), while the youngest robust australopithecines from east Africa antedate 1.4–1.2 MYA (F. H. Brown and Feibel 1988; Leakey and Hay 1982). (The robust form in east Africa is *A. boisei*, which is regarded here as a variant of *A. robustus*). The Swartkrans specimens may therefore bear directly on robust australopithecine extinction, which has been variously attributed to climatic change, including, for example, a change in the nature of glacial/interglacial cycles about 0.85 MYA (Williams et al. 1988), or to biotic change, perhaps above all to the expanding ecological niche of evolving *Homo* (Klein 1988a). Unfortunately, however, the true cause remains elusive, if only because the last appearance of the robust australopithecines is so poorly fixed. In both southern and eastern Africa, they could have disappeared anytime between 1.4 and 0.8 MYA. The problem is partly imprecise dating (especially in southern Africa) and partly the lack of large fossil samples dating between 1.4–1.2 and 0.8 MYA (especially in east Africa).

Early *Homo* and the Oldest Southern African Artifacts

The fossils of early *Homo* from Swartkrans 1–2 have sometimes been assigned to *H. erectus* (or to *H.* cf. *erectus*), while those from Sterkfontein 5 have sometimes been placed in *H. habilis* (or in *H.* aff. *habilis*) (Day 1986). This may appear contradictory, since the relevant deposits at both sites are believed to have formed at about the same time (between roughly 1.8 and 1.5–1.3 MYA), while the much more numerous specimens of early *Homo* from east Africa are usually taken to imply that *H. habilis* evolved into *H. erectus* about 1.8–1.7 MYA (Klein 1989a with references). However, the Swartkrans and Sterkfontein specimens are arguably too fragmentary for species diagnosis, and they are perhaps best referred to simply as *Homo* sp. (Rightmire 1984). This is especially true, given continuing uncertainty about the meaning or validity of *H. habilis* in east Africa. The east African sample is relatively large, but it is also remarkably heterogeneous, and it could reflect the co-existence of two contemporaneous species of very early *Homo* between roughly 2.2–2.1 and 1.8–1.7 MYA (Stringer 1986; Wood 1985, 1992).

As already indicated, besides important hominid fossils, Swartkrans 1–3 and Sterkfontein 5 also contain the oldest artifacts yet found in southern Africa. At both sites, besides the simple, informal bone implements that have already been discussed, the artifacts include simple flaked stone pieces: choppers, flakes, cores, chunks, and other mainly nondescript objects at Swartkrans (Brain et al. 1988) and essentially similar pieces accompanied by at least three crude bifaces (one hand-axe and two cleavers) at Sterkfontein (Clarke 1985b). The Swartkrans assemblage has been assigned to the "Developed Oldowan" (Brain et al. 1988), while the Sterkfontein one has been placed in the "early Acheulean" (Clarke 1985b, 1988). The terms reflect an appreciation of the much fuller record of early stone-tool industries in east Africa, where the oldest known stone artifacts are assigned to the Oldowan Tradition, which was present from 2.3–2.2 MYA or before, and assemblages including bifaces are assigned to the Acheulean Tradition, which supplanted the Oldowan about 1.5–1.4 MYA (Isaac 1975, 1984; Isaac and Harris 1978; Potts 1991; Toth and Schick 1986).

Oldowan assemblages are dominated by simple flakes, choppers, and a variety of other crude, flaked pieces that are probably better described as points along a graded continuum of flaking than as discrete types (Potts 1991; Toth 1985, 1987; Volman 1984). The Acheulean is distinguished from the Oldowan primarily by the addition of bifaces (hand-axes, picks, and/or cleavers; Figure 23-5) that are generally regarded as a logical outgrowth of Oldowan bifacial choppers. Oldowan choppers and other tools typically continue alongside bifaces at Acheulean sites, and assemblages in which Oldowan types heavily outnumber bifaces have sometimes been assigned to a separate "Developed Oldowan B" tradition (Leakey 1971, 1975). This is presumed to have evolved from the "Developed Oldowan A," which is a late, relatively refined variant of the Oldowan in the narrow sense (without bifaces). However, most authorities now believe that variation in biface abundance need imply only differences in the activity mix at different sites, or perhaps in some cases, differences in local raw material availability. The "Developed Oldowan B" is therefore now commonly subsumed within the Acheulean (Gowlett 1988; Stiles 1991).

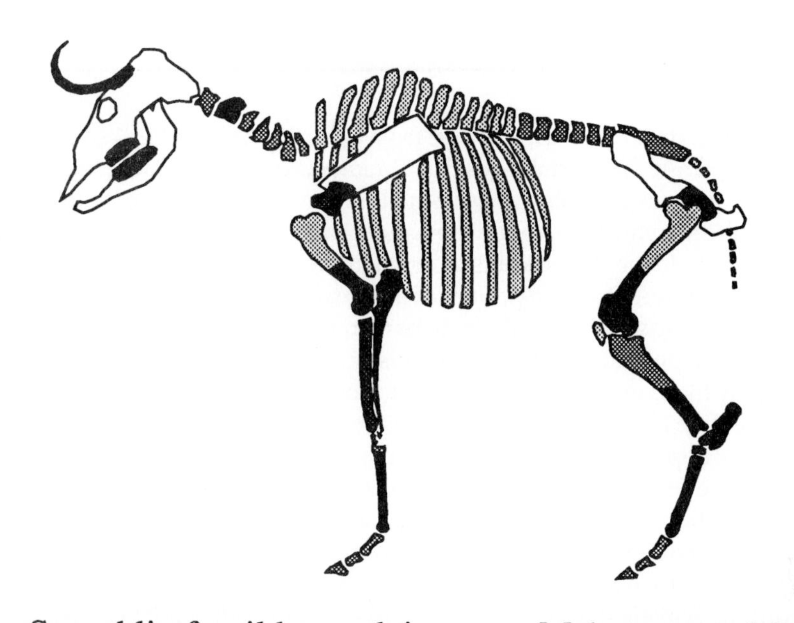

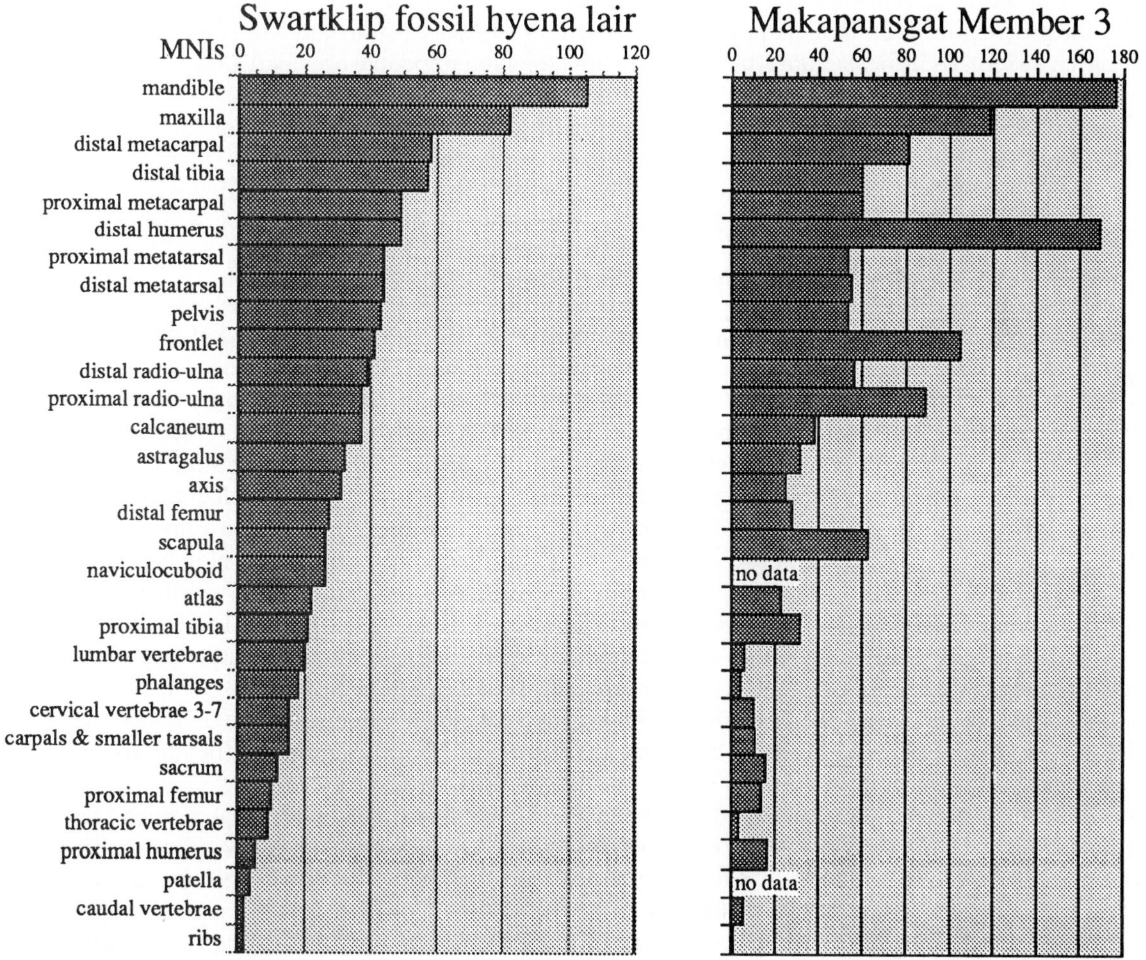

FIGURE 23-4 Bottom: The Minimum Number of Individuals (MNI) represented by various bovid skeletal parts in the Swartklip late Pleistocene hyena lair (Klein 1975) and in Makapansgat Member 3 (Dart 1957).
Top: A schematic black wildebeest skeleton on which various parts are shaded in rough proportion to their abundance at Swartklip.

On dating alone, the artifact assemblage from Swartkrans Member 1 might belong to the Oldowan in the narrow sense, while the assemblages from Members 2 and 3 are more likely to derive from the early Acheulean. In fact, they are probably all too small for any firm assignment, and small sample size could also explain why the assemblages from Swartkrans 2 and 3 lack the bifaces present in Sterkfontein 5. In sum, on present knowledge, Swartkrans 1 could represent the Oldowan, while Swartkrans 2–3 and Sterkfontein 5 could represent an early phase of the ensuing Acheulean. The more basic point is that small sample size and dating uncertainty limit the behavioral information that Swartkrans and Sterkfontein can provide. However, the possible bone digging tools in Swartkrans 1-3 and Sterkfontein 5 complement and supplement slightly older or like-aged evidence for occasional, informal artifactual utilization of bone at Olduvai Gorge (Leakey 1971; Shipman 1989); and the animal bones in both Swartkrans 3 and Sterkfontein 5, where early *Homo* appears to have been the prime accumulator, supplement coeval or somewhat older east African evidence for early human interest in animals far larger than those commonly obtained by chimpanzees. In addition, like broadly comparable east African sites (Bunn 1981; Potts and Shipman 1981), Sterkfontein 5 has provided at least one cut-marked bone that demonstrates a link between stone tools and butchery. The South African sites do not clarify how early *Homo* obtained bones (or animals)—whether by hunting, scavenging, or a mix of the two—but this has not been persuasively resolved even with the much more abundant materials available from east Africa (Bunn and Kroll 1986; Potts 1988, 1991; Shipman 1986b). As indicated later, determining how (or even if) people obtained animals is very difficult even at much later archeological sites, dated to the mid-Pleistocene, after 0.8 MYA.

With the arrival of stone-tool makers in southern Africa in the late Pliocene or early Quaternary, it becomes possible to outline the local prehistory in more strictly archeological (vs. chronological) terms. The remainder of this chapter is therefore organized according to the widely accepted subdivision of the Stone Age among Earlier, Middle and Later units, abbreviated ESA, MSA, and LSA respectively.

The Earlier Stone Age (ESA): The Acheulean Industrial Tradition

The ESA comprises both the Oldowan and Acheulean Industrial Traditions that have already been introduced. The Oldowan is not certainly represented in southern Africa, but a very early Acheulean is probably present at both Swartkrans and Sterkfontein. An assemblage of relatively crude Acheulean artifacts from the 6 m terrace of the Klip River (a tributary of the Vaal) (Mason 1962, 1985; Partridge 1985b) may be just as old, but confirming geologic/paleontologic evidence is lacking. Most other local Acheulean sites are surface occurrences that are impossible to date, and even important sealed sites, like those at Amanzi Springs (southern Cape) (H. J. Deacon 1970) and Doornlaagte (northern Cape) (Butzer 1974; Mason 1967), are very difficult to place in time, except on potentially circular typological grounds. The problem is that even sealed sites cannot generally be linked to any dated external stratigraphy and few contain faunal remains or materials suitable for conventional radiometric dating. Excepting Swartkrans and Sterkfontein, faunal remains that could be useful for age determination are known from only five localities (see Figure 23-1): the Vaal "Younger Gravels" (= the Rietputs Formation) (H. B. S. Cooke 1949; Helgren 1977, 1978, 1979; Wells 1964); Elandsfontein (= Saldanha = Hopefield) (Singer and Wymer 1968; Klein 1978a; Klein and Cruz-Uribe 1991); Cornelia (Butzer et al. 1974); Kathu Pan (Beaumont et al. 1984; Klein 1988b); and Namib IV (Shackley 1980; Klein 1988b).

Like the older southern African faunas that have already been considered, the Acheulean faunas can be dated only by reference to east Africa, particularly to Olduvai Gorge Upper Bed II through Bed IV, where the fauna is bracketed between roughly 1.6 and 0.6 MYA (Leakey and Hay 1982). Taking Upper Bed II/IV as a standard, aside from Sterkfontein and Swartkrans, the oldest fossiliferous southern African Acheulean site may be Cornelia, where an Upper Bed II/IV-like fauna overlies the Acheulean horizon. The composite fauna and broadly associated Acheulean from the Vaal "Younger Gravels" may span the entire Upper Bed II/IV interval, but evolved forms of *Elephas recki* and *Metridiochoerus andrewsi* from the only well-documented association, at Power's Site (Klein 1988b; Power 1955; Table 23-4 here), imply an

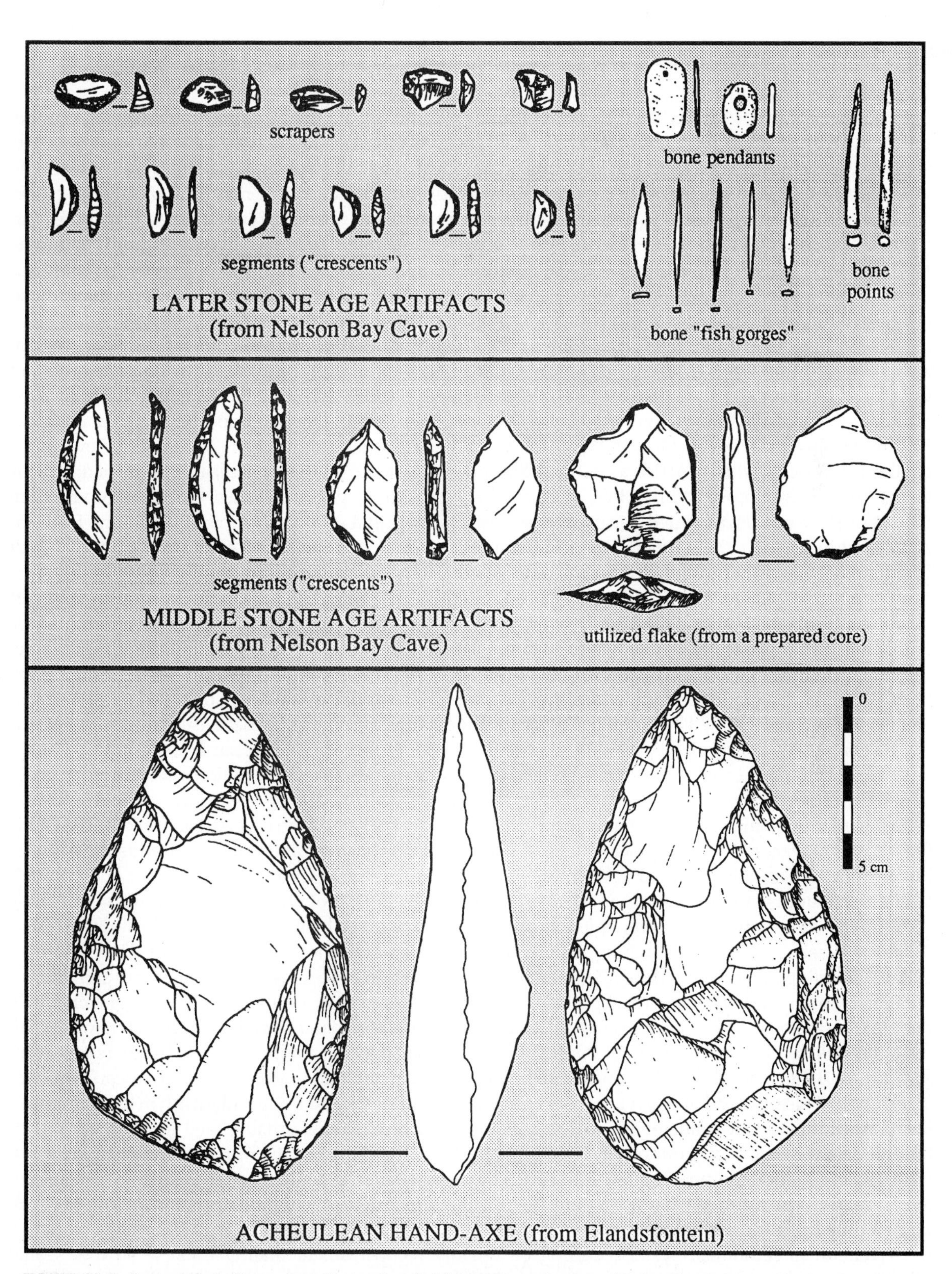

FIGURE 23-5 Typical Early Stone Age (Acheulean), Middle Stone Age, and Later Stone Age artifacts. (Acheulean hand-axe redrawn from original by Volman unpub.; Middle and Later Stone Age artifacts redrawn from originals by J. Deacon. All pieces to the same scale.)

age roughly equivalent to Bed IV. The Acheulean at Elandsfontein may be somewhat younger, perhaps partly overlapping and partly postdating Bed IV. The Elandsfontein fauna is by far the largest from any southern African Acheulean site, but its dating is complicated by two problems: (1) The faunal assemblage was mainly surface collected rather than systematically excavated, and it may thus comprise taxa of somewhat different ages; and (2) both the local historic fauna and the occurrence of fossil bovid taxa that have never been found anywhere else suggest that the Elandsfontein region enjoyed a unique Quaternary faunal history, reflecting its location at the far southwestern tip of the Africa. At present, the least equivocal faunal evidence for a post–Bed IV Acheulean comes from Kathu Pan and Namib IV, where the artifacts are associated with a very advanced form of *Elephas recki* (Klein 1988b).

Including Swartkrans and Sterkfontein, the faunal data thus imply a long time span for the Acheulean in southern Africa, certainly from before 1 MYA until after 0.6 MYA. How long after remains uncertain. At Montagu Cave (Keller 1973), Wonderwerk Cave (Beaumont 1982; Beaumont et al. 1984; Malan and Wells 1943), the Cave of Hearths (Mason 1962; 1988a), Olieboompoort Cave (Mason 1962), Kathu Pan (Beaumont et al. 1984; Butzer 1984; Klein 1988b), and Kalambo Falls (in Zambia, beyond the northern limit of southern Africa as defined here) (J. D.Clark 1974), the Acheulean is stratified directly below the succeeding MSA, but this does not help in dating, partly because the MSA occupations themselves are not firmly dated and partly because a substantial gap could separate each one from the Acheulean below. At the moment, the least ambiguous dates for the latest Acheulean come (1) from Rooidam in the northern Cape Province (South Africa), where lacustrine limestones overlying typologically "very late" Acheulean artifacts have been dated by uranium disequilibrium to roughly 174 KYA (Szabo and Butzer 1979), and (2) from Duinefontein 2 in the southwestern Cape, where a combination of faunal and geomorphic data show that the MSA was probably underway by the beginning of global-isotope stage 6, roughly 195 KYA (Butzer unpub.; Volman 1984). These southern African dates are consistent with $^{40}K/^{40}Ar$ dates from Ethiopia indicating that the MSA was underway there by c. 200 KYA (Wendorf et al. 1975; Clark 1988). However, in both southern Africa and east Africa, secure dating of the Acheulean/MSA interface will probably require the broad application of still partly experimental methods like thermoluminescence, electron-spin resonance, and amino-acid racemization (on ostrich eggshell.)

The dating uncertainties make it difficult to perceive any time trends within the southern African Acheulean, though the overall pattern was probably broadly the same as in east Africa (Isaac 1975). There, the pace of artifactual change appears to have been agonizingly slow, and only two trends are roughly discernible: (1) Later bifaces tend to be thinner and more extensively trimmed than earlier ones, and (2) later assemblages tend to include a greater variety of recognizable tool types. At least in southern Africa, these advances were further accompanied during the late Acheulean by the invention of the Levallois-like, "Victoria West" prepared core technique for producing flakes of predetermined size and shape (Volman 1984). Assuming that greater artifactual refinement is a reflection of younger age, it can be argued that the southern (cooler) and western (drier) portions of southern Africa were occupied only by later Acheulean people (H. J. Deacon 1975), perhaps only after 1 MYA. Population expansion during the later Acheulean was perhaps facilitated by incipient control of fire, which may be implied by the burnt bones from Swartkrans 3 that have already been discussed, by burnt bat guano that underlies the Acheulean sequence at the Cave of Hearths and by ash that occurs within it (Mason 1962), and by abundant ash in the Acheulean layers of Montagu Cave (Butzer 1973). The burning in each case is reasonably well established, but its human origin is not. A natural origin is perhaps particularly likely for the burnt guano at the Cave of Hearths, and natural ignition of plant food debris or vegetal bedding could have produced the ash at both the Cave of Hearths and Montagu. To date, hearths that unequivocally demonstrate control over fire are known only from succeeding MSA sites.

Acheulean settlement systems cannot be reliably reconstructed, mainly because so few sealed occurrences are known and these probably represent a biased sample of those that originally existed. Thus, the relatively small number of Acheulean cave sites might imply that Acheulean people occupied caves less often than their MSA and LSA successors (H. J. Deacon 1989), but caves and cave fills themselves have

TABLE 23-4 The large mammals associated with Acheulean artifacts at Elandsfontein (EFT), Kathu Pan (KP), and Power's Site (PS) (Vaal "Younger Gravels").

	EFT	KP	PS
PRIMATES			
Hominidae			
Homo sapiens ('archaic' human)	x		
Cercopithecidae			
Theropithecus oswaldi (giant gelada baboon)	x		
PHOLIDOTA			
Manidae			
Phataginus sp. (pangolin)	x		
CARNIVORA			
Canidae			
Canis mesomelas (black-backed jackal)	x		
Lycaon pictus (hunting dog)	x		
Mustelidae			
Ictonyx striatus (striped polecat)	x		
Mellivora capensis (honey badger)	x		
Aonyx capensis (clawless otter)	x		
Viverridae			
Viverra civetta (civet cat)	x		
Herpestes ichneumon (Egyptian mongoose)	x		
Suricata suricatta (slender-tailed mongoose)	x		
Hyaenidae			
Hyaena brunnea (brown hyena)	x		
Crocuta crocuta (spotted hyena)	x		
Felidae			
Felis caracal (caracal)	x		
Panthera leo (lion)	x		
**Megantereon gracile* (sabertooth cat)	x		
PROBOSCIDEA			
Elephantidae			
Loxodonta atlantica (Atlantic elephant)	x		
Elephas recki (Reck's elephant)		x	x
PERISSODACTYLA			
Equidae			
Equus capensis ('giant' Cape zebra)	x	x	x
Equus quagga (plains zebra)	?	x	cf.
Rhinocerotidae			
Diceros bicornis (black rhinoceros)	x		
Ceratotherium simum (white rhinoceros)	x	x	x
ARTIODACTYLA			
Hippopotamidae			
Hippopotamus amphibius (hippopotamus)	x	x	x
Suidae			
**Kolpochoerus paiceae* (extinct bushpig)	x		
**Metridiochoerus andrewsi* (extinct warthog)	x		x
Phacochoerus aethiopicus (warthog)		x	x
Giraffidae			
**Sivatherium maurusium* (sivathere)	x		
Giraffa cf. *camelopardalis* (giraffe)		x	
Bovidae			
Tragelaphus strepsiceros (greater kudu)	x	cf.	
Taurotragus oryx (eland)	x	cf.	x
**Pelorovis antiquus* ('giant' buffalo)	x	x	x
Syncerus cf. *caffer* (Cape buffalo)		x	x
Redunca arundinum (southern reedbuck)	x	x	x
Hippotragus gigas ('giant' hippotragine)	x		
Hippotragus leucophaeus (blue antelope)	x		
Hippotragus sp. (roan/sable)		x	
Damaliscus aff. *lunatus* (bastard hartebeest)	x		

TABLE 23-4 Continued

	EFT	KP	PS
?Damaliscus niro (bastard hartebeest)	X	?	?
?Damaliscus sp. nov. (bastard hartebeest)	X		
?Parmularius sp. nov. (bastard hartebeest)	X		
**Rabaticeras arambourgi* (extinct hartebeest)	X	?	?
Alcelaphus buselaphus (Cape hartebeest)	?	?	
**Megalotragus priscus* ('giant hartebeest')	X	X	X
Connochaetes gnou (black wildebeest)	X	?	?
Raphicerus melanotis (Cape grysbok)	X		
Gazella sp. (gazelle)	X	?	
Antidorcas recki (Reck's springbok)	X	?	
Antidorcas australis (southern springbok)	X		
Antidorcas bondi (Bond's springbok)		cf.	X
*Tribe indet. gen. & sp. nov. (strange antelope)	X		
LAGOMORPHA			
Lepus sp.	X		
RODENTIA			
Bathyergidae			
Bathyergus suillus (Cape dune molerat)	X		
Hystricidae			
Hystrix africaeaustralis (porcupine)	X	X	
Pedetidae			
Pedetes capensis (springhare)		X	

* Asterisks mark extinct genera

The list for Elandsfontein is from Hendey (1974) and Klein and Cruz-Uribe (1991); the lists for Kathu Pan and Power's Site are from Klein (1988b). Klein (1988b) and Klein and Cruz-Uribe (1991) also provide abundance estimates for various taxa. The Elandsfontein sample is much larger and better preserved than the other two.

a limited lifetime, and many Acheulean caves probably collapsed long ago or were flushed of their deposits. In all the excavated caves that lack Acheulean deposits, MSA or LSA layers lie directly on bedrock, and there is no sterile layer that might imply that Acheuleans avoided caves. Insofar as present evidence suggests a difference between Acheulean and later settlement patterns, it is in the degree of water dependence. Both the spatial distribution of surface and sealed Acheulean sites (H. J. Deacon 1975) and faunal or sedimentological indications for relatively moist conditions near each sealed site (Klein 1988b) imply that Acheuleans may have been more closely tied to standing water. This may mean that unlike later (especially LSA) people, Acheuleans lacked impermeable water containers (H. J. Deacon 1975).

Data to reconstruct Acheulean subsistence are equally sparse. It seems likely that Acheulean populations focused heavily on plant foods, but in southern Africa, plant remains are preserved only at Amanzi Springs (H. J. Deacon 1970) and Kalambo Falls (J. D. Clark 1969), and only at Kalambo Falls are there items (seeds and fruits) that could have been eaten. Bones are more commonly preserved, and they often come from elephants, rhinoceros, buffalos, and other large ungulates that could imply formidable hunting prowess. Added to this, experiments have shown that hand-axes are very effective for heavy-duty butchering (P. R. Jones 1980). However, inferences about hunting and butchery are complicated by an interpretive dilemma that fossiliferous southern African sites share with most other Acheulean sites in Africa and in Europe. In each case, the artifacts and bones accumulated in the open air near an ancient pond or water course. Hand-axes, cleavers, and associated flake tools demonstrate that Acheulean people were present, and cut-marked or hacked bones often show that people must have been involved in the bone accumulation. At the same time, however, gnawed or chewed bones usually show that carnivores also played a role. In southern Africa, the problem is particularly acute at the spectacular Elandsfontein site, where carnivore-chewed bones and hyena coprolites substantially outnumber cut bones and stone artifacts. In addition, like the Langebaanweg large ungulate attritional mortality profiles previously discussed, those from Elandsfontein are remarkably deficient in very young animals. A reasonable conclusion is that the mortality profiles reflect carnivore scavenging much more than human hunting. In sum, it seems likely that many, if not most, of the animals found at Elandsfontein and probably also at localities like Kathu Pan or Power's Site were killed by carnivores or died naturally near ancient water sources. The artifacts that occur at each site may simply show that people were commonly drawn to the same watering points. By themselves, they do not prove that the people hunted or even scavenged most of the animals represented.

In order to evaluate Acheulean hunting proficiency, the minimum requirement is a site where it can be assumed that people accumulated most of the bones. A priori, caves are obviously better candidates than open-air sites, and in southern Africa, Wonderwerk Cave and the Cave of Hearths might qualify. However, neither has provided a large enough bone sample for meaningful analysis, and the prominence of carnivores in the Cave of Hearths sample suggests that hyenas may have been the main bone accumulators (Klein 1988b). Fossiliferous Acheulean cave sites are known in northwest Africa (Howell 1960; Jaeger 1975) and in Europe (for example, in southern France [Villa 1991]), but the faunal remains have not been analyzed for information on Acheulean hunting ability. For the moment, perhaps the most secure basis for assessing this comes from faunas accumulated by the MSA successors to the Acheuleans, as discussed later. If it is fair to project backwards from the MSA, Acheulean people were probably very ineffective hunters.

As previously indicated, human remains accompany Acheulean (or "Developed Oldowan") artifacts at Swartkrans and Sterkfontein. Remains from later sites comprise a fragmentary mandible from the Cave of Hearths (Tobias 1971) and a mandibular ramus fragment and skullcap from Elandsfontein (Rightmire 1984; Singer 1954). To this meager list should perhaps be added the famous "Rhodesian Man" skull from a cave at Kabwe (formerly Broken Hill), Zambia. Its artifactual associations are unknown, but faunal elements that were possibly associated suggest it may date from broadly the same interval (between ?0.7 and ?0.4 MYA) as the Elandsfontein skull, which it closely resembles (Rightmire 1984).

Together with a small number of additional human fossils from Acheulean sites in east and north Africa (references in Klein 1989a), the southern African specimens are usually said to show that earlier Acheulean artifacts, antedating roughly 0.7–0.5 MYA, were made by *H. erectus*, while later ones were made by early or "archaic" *H. sapiens*. This interpretation reflects the view that natural selective factors and gene flow drove widespread human populations along a common evolutionary track, even after 1 MYA, when populations had become widely dispersed in both Africa and Eurasia. However, the sparse record of mainly fragmentary and poorly dated fossils allows at least one other interpretation. This follows from (1) the high degree of morphological variability in the composite African and Eurasian sample of "archaic" *H. sapiens* (Kennedy 1991; also references in Klein 1989a); and (2) the possibility or even likelihood that classic (or type) *H. erectus* in the Far East was largely contemporaneous with "archaic" *H. sapiens* in Africa and Europe (Brooks and Wood 1990; references in Klein 1989a). Combined with the suggestion that the most ancient specimens of African *H. erectus* lack derived features that distinguish the (probably younger) classic Far Eastern form (Andrews 1984; Wood 1984), these observations could mean that early African *H. erectus* was ancestral to more than one later human lineage, including not only Far Eastern *H. erectus*, but perhaps also a separate lineage in Europe that culminated in the Neanderthals and another in Africa that produced the modern or near-modern makers of the MSA discussed later. If this interpretation is correct, African *H. erectus* would require a new name, for which *H. ergaster* is available. Sorting out the alternatives is clearly critical, but will require a much denser and better dated fossil record, not just in Africa but also in Eurasia.

The Middle Stone Age (MSA)

The MSA is better known in southern Africa than anywhere else, thanks to the persistent interest of southern African archeologists and to the occurrence of many rich sites. Because the MSA is more recent than the Acheulean, it is much more commonly found in caves, and it is also much better dated. Absolute dates and geomorphic inferences show that it was underway throughout southern Africa by 128 KYA and that it terminated before 40–30 KYA (Brooks et al. 1990; Grün et al. 1990a, 1990b; Vogel and Beaumont 1972; Volman 1984). The terminal date is imprecise, partly because it may have varied from place to place, but more because it is everywhere at or beyond the limit of conventional radiocarbon dating. In addition, there is the problem that most known sites were abandoned during the MSA/LSA transition, probably because adverse climatic conditions had dramatically reduced human populations in most parts of southern Africa (H. J. Deacon and Thackeray 1984).

Artifactually, the MSA is distinguished from the preceding Acheulean mainly by the absence of large bifaces (hand-axes, picks, and cleavers). MSA assemblages are dominated instead by flakes or flake-blades that were often removed from well-prepared cores. Retouched tools tend to be relatively rare and comprise mostly scrapers, points, and denticulates (Sampson 1974; Volman 1984). Arguably, the number of distinct stone-tool types or subtypes is larger than in the Acheulean, and MSA assemblages may have varied more through time and space. However, assemblages separated by thousands of years or thousands of kilometers still tend to differ more in the frequencies of types than in their presence or absence, and most differences need reflect only differences in the activities undertaken at different sites or perhaps, in some cases, differences in raw material availability. The only clear exception is the intriguing phenomenon known as the Howiesons Poort Industry or Variation, which is marked by a blossoming of well-made backed pieces, mainly segments (or "crescents") and trapezoids, in otherwise typical MSA assemblages at many sites south of the Limpopo River. At one time, the Howiesons Poort appeared to postdate the MSA in the narrow sense and was regarded as a possible transitional industry to the LSA. Above all, the segments resemble LSA pieces of the same name, though they are typically much larger (Figure 23-5). However, it is now clear that the Howiesons Poort was supplanted by stereotypic MSA industries lacking numerous well-made backed elements and that it antedates segment-rich LSA industries by tens of thousands of years.

The age of the Howiesons Poort is controversial (Parkington 1990a), and the possibility exists that various occurrences differ significantly in age. However, their tendency to occupy the same position within deeply stratified MSA sequences implies broad contemporaneity, and sedimentologic observations at Boomplaas Cave and Klasies River Mouth in the southern Cape Province suggest they span the transition from the Last Interglaciation (= global isotope stage 5) to the Last Glaciation, roughly 80–70 KYA (H. J. Deacon 1989). This estimate is tentatively supported by amino-acid racemization dates on ostrich egg shell from several key sites (Brooks pers. comm.). If it is correct, it could mean that characteristic Howiesons Poort artifacts formed part of an MSA response to climatic stress (Deacon 1989). Future finds may show that they also relate in some way to the origins of biologically modern humans, but the Howiesons Poort substantially postdates the oldest modern or near-modern human fossils from Klasies River Mouth and Border Cave discussed later.

Like the Acheuleans, MSA people, including the makers of Howiesons Poort artifacts, do not seem to have realized that bone, ivory, and shell can be carved, polished, or ground into "points," "awls," "hide-burnishers," and other formal artifact types. They used bone and related materials very casually and informally, and this may partly explain why, like the Acheuleans, they left no evidence for art. In this respect and all others that have been noted, the MSA recalls the broadly contemporaneous Mousterian of Europe and western Asia, from which it is in fact distinguished more by historical tradition than by content.

Like contemporaneous Mousterian sites, no MSA site has provided any indisputable structural remains ("ruins"), perhaps because MSA shelters or windbreaks were usually too flimsy to leave archeological traces. However, again like Mousterian sites (and unlike Acheulean ones), MSA sites commonly contain lenses of ash and charcoal that mark fireplaces or hearths. The abundance of such lenses suggests that MSA (and Mousterian) people could make fire at will. No indisputable fire-making artifacts have yet been identified, and none may ever be found, if they were usually made of wood. Wood and other macrobotanical remains (excepting charcoal) are rarely preserved in MSA sites, but wood working is implied by a possible throwing-stick fragment from the MSA spring deposits at Florisbad in the Orange Free State (J. D. Clark 1955).

In Europe and western Asia, Mousterian people appear to have been the first to bury the dead, and it is reasonable to suppose that MSA people did likewise. However, except for an infant's burial at Border Cave that may have been intrusive into MSA layers, no MSA graves have yet been found, and the known MSA human remains tend to be fragmented and dispersed, like the animal bones with which they are associated. This might mean that MSA people were cannibals (White 1987; H. J. Deacon 1989), or it might mean that MSA graves were often exhumed by hyenas or other carnivores. Because frost-fracturing of cave ceilings and walls was less common in Africa than in Europe, MSA cave deposits tend to be much thinner than those in like-aged Mousterian sites, and, in general, MSA people may have been forced to dig shallower graves that carnivores were very likely to disturb. Equally important, even Mousterian sites have produced relatively few burials, and far fewer MSA sites have been excavated. Thus, MSA graves may yet be found, perhaps in on-going excavations at Die Kelders Cave 1 (southwestern Cape Province), where the MSA deposits are unusually thick and the bone preservation is excellent.

MSA Faunal Remains and Ecology

At most southern African MSA sites, bones were long ago removed by acid ground waters, and even where they survive, leaching has often reduced both their number and their quality. However, analyzed bone samples have been recovered from a wide variety of cave and open-air sites (Table 23- 5). Geomorphic/sedimentologic observations indicate that bone samples from two of the open-air sites — Florisbad and Duinefontein 2—date to the late middle Pleistocene, before 128 KYA, and the faunas retain some of the archaic ungulates that occur in preceding Acheulean sites. The remaining samples probably all date from the earlier part of the late Pleistocene, between 128 and 40 KYA, and they are dominated by extant taxa. The only extinct species they contain are an equid and four bovids, all of which survived to the end of the Pleistocene, 12–10 KYA, or even more recently (Klein 1984). Thus, the MSA faunas fully confirm site stratigraphies and absolute dates showing the MSA succeeded the Acheulean.

From a human behavioral perspective, by far the most important MSA faunas are the large samples from the Klasies River Mouth Caves (KRM) and a somewhat smaller sample, presently being enlarged, from Die Kelders Cave 1 (DK1) (Table 23- 6). Radiometric dates, faunal remains, and geochemical/sedimentologic observations indicate that MSA people initially occupied the KRM caves during the regression from the maximum sea level of the Last Interglaciation (= global oxygen-isotope stage 5), roughly 120 KYA, and they returned intermittently until the early part of the Last Glaciation (= isotope stage 4), perhaps 60 KYA (H. J. Deacon 1989; Grün et al. 1990b). More tentatively, sedimentologic and faunal evidence imply that MSA people sporadically occupied DK1 during the early part of the Last Glaciation, between perhaps 75 and 60 KYA (Grine et al. 1991; Tankard and Schweitzer 1976). Following the latest MSA occupation, both sites were abandoned for tens of thousands of years and were reoccupied only by relatively late LSA people during the Holocene.

Like many other fossil faunas, for analytic purposes, those from KRM and DK1 may be usefully divided between "micro" and "macro" components. The microfauna, comprising mostly small rodents and insectivores, comes largely from artifact-poor layers, where it was probably accumulated by owls (D. M. Avery 1982, 1987). It thus informs on past environment, but not on human behavior. In contrast, the macrofauna comes mainly from layers where abundant artifacts and cut-marked bones indicate that MSA people were the main bone accumulators. It can therefore be used to infer MSA subsistence ecology, particularly when it is compared to macrofaunas from local LSA sites and from late Pleistocene hyena lairs, like the one at Swartklip discussed previously. For comparative purposes, the most useful LSA samples are from Nelson Bay Cave (NBC) (J. Deacon 1978; Inskeep 1987; Klein 1983) near KRM and from Byneskranskop 1 (BNK1) (Schweitzer and Wilson 1981) near DK1 (see Figure 23-1). The comparison between KRM and NBC is especially apt, because it can be restricted to the large interglacial samples from both sites (the Last Interglaciation at KRM and the Present Interglaciation or Holocene at NBC). The lack of any apparent paleoenvironmental difference increases the likelihood that any differences between the faunas reflect differences in human behavior. The comparison between the DK1 and BNK1 macrofaunas is more poorly controlled in this sense, because the DK1 (MSA) faunal sample accumulated during a "glacial" interval, while the BNK1 (LSA) sample accumulated mainly during the Present Interglaciation. Differences between the faunas could therefore partly reflect differences in past environment. There is also the problem that the DK1 and BNK1 faunal samples are smaller than their KRM and NBC equivalents. Nonetheless, to the extent that DK1 and BNK1 can be meaningfully compared, they complement and supplement the inferences that KRM and NBC suggest.

The comparisons between the MSA and LSA faunas reveal some obvious similarities and some even more striking differences. The most conspicuous similarities are in basic taxonomic composition, in skeletal part representation, and in ungulate mortality profiles. The caves from which the samples come are all located in broadly similar surroundings on or near the south coast of South Africa, and their faunas comprise essentially the same set of terrestrial and marine species. The terrestrial species are in fact dominated by basically the same ungulates that dominate local fossil hyena assemblages, which probably means that both MSA and LSA people competed with hyenas to some extent and that like hyenas, they were both at or near the top of the trophic pyramid. However, the MSA and LSA faunas can both be readily distinguished from the fossil hyena samples in skeletal part representation, in mortality profiles (which are uniformly attritional in the carnivore samples, but sometimes catastrophic in the stone age ones), and perhaps most significantly, in the relative abundance of carnivores (previously discussed and in Cruz-Uribe 1991 and Klein et al. 1991). Carnivores are considerably more common in the hyena samples, suggesting that the fossil hyenas resembled living hyenas and differed from stone age people in their greater tendency to encounter (and feed on) other carnivores.

Together with marine food debris from Herolds Bay Cave (Brink and Deacon 1982) and from two open-air MSA shell middens (Sea Harvest and Hoedjies Punt) at Saldanha Bay (Volman 1978), the marine debris from KRM and DK1 comprise the oldest known evidence for systematic human use of coastal resources anywhere. Marine fossils (mainly intertidal shells and bones of fur seals and penguins) are particularly abundant in the Last Interglacial levels of KRM (Klein 1976a; Thackeray 1988; Voigt 1973). They are much less common in the presumed early Last Glacial layers at both KRM and DK1, probably because the coastline had been displaced seawards by a global drop in sea level. To compensate for reduced access to marine resources, the early Last Glacial occupants of KRM may have intensified their use of plants, perhaps especially of the geophytes whose subterranean corms were intensively exploited by local Holocene (LSA) people (H. J. Deacon 1976). Increased corm use may be implied by the heavily carbonized bands and lenses that are conspicuous in the early Last Glacial deposits at KRM and at Boomplaas Cave, located about 200 km to the northwest (H. J. Deacon 1989). However, no corm residues are preserved at KRM or Boomplaas, and aside from a few charred seeds or nuts from the MSA layers of Bushman Rock Shelter (Transvaal) (Plug 1981) and Border Cave (Natal) (Beaumont 1973), no likely food plant remains have been identified at any MSA site.

With regard to skeletal part representation, every MSA and LSA faunal sample tends to be somewhat unique, perhaps minimally because each was exposed to a different amount of postdepositional leaching and profile compaction. However, every reasonably large sample, including those from KRM, NBC, DK1, and BNK1, also exhibits a consistent difference in skeletal part representation between (1) smaller ungulates, which tend to be well represented by a wide variety of parts including proximal limb bones, and (2) larger ungulates, which tend to be represented disproportionately by bones of the feet and skull. The basic pattern is in fact extremely widespread throughout the world and extends beyond faunas in which all species are wild to ones in which the smaller and larger ungulates are both domestic. Since the foot and skull bones that dominate the large ungulate sample at KRM have relatively little meat, marrow, or grease value, Binford (1984) argued that MSA people did not hunt large ungulates, but simply scavenged carcasses from which the choicer parts had already been removed. However, scavenging vs. hunting is obviously unlikely to explain the near ubiquity of the basic pattern, and it is inconsistent with the KRM large ungulate mortality profiles, which show that the most abundant large ungulates were

TABLE 23-5 Southern African Middle Stone Age Sites with Faunal Remains

CAVES	REFERENCES
Zebrarivier Shelter (Namibia)	Cruz-Uribe and Klein (1982); Wendt (1972)
Pomongwe (Zimbabwe)	Brain (1981); C.K. Cooke (1963)
Redcliff (Zimbabwe)	C.K. Cooke (1978); Cruz-Uribe (1983); Klein (1978b)
Bushman Rock (Transvaal)	Brain (1981); Plug (1981)
Cave of Hearths (Transvaal)	Mason (1962, 1988a); H.B.S. Cooke (1962)
Border Cave (Natal)	D.M. Avery (1992); Beaumont (1973, 1980); Beaumont et al. (1978); Klein (1977)
Umhlatuzana Shelter (Natal)	Kaplan (1990)
Witkrans Cave (Cape)	J. D. Clark (1971)
Wonderwerk Cave (Cape)	Beaumont (1982); Beaumont et al. (1984); Klein (1988b)
Die Kelders Cave 1 (Cape)	Grine et al. (1991); Schweitzer (1973)
Boomplaas Cave A (Cape)	H.J. Deacon (1979, 1989); H.J. Deacon et al. (1984); Klein (1983)
Herolds Bay Cave (Cape)	Brink and Deacon (1982); H.J. Deacon (1989)
Klasies River Mouth Caves (Cape)	H.J. Deacon (1989); H.J. Deacon and Gelejnse (1988); Klein (1976a, 1983, 1989b); Singer and Wymer (1982)

OPEN-AIR SITES	REFERENCES
Gi (Botswana)	Brooks et al. (1980, 1990); Helgren and Brooks (1983); Klein (unpub.)
Kalkbank (Transvaal)	A.J.V. Brown (1988); Mason (1988b); H.B.S. Cooke (1962)
Vlakkraal (Orange Free State)	H.B.S. Cooke (1963)
Florisbad (Orange Free State)	Brink (1987); H.B.S. Cooke (1963); Kuman and Clarke (1986)
Kathu Pan (Cape)	Beaumont et al. (1984); Klein (1988b)
Duinefontein 2 (Cape)	J. Deacon (1976); Klein (1976b)

almost certainly hunted. More likely, at KRM and most other sites, the pattern reflects differences in carcass size as these determine (1) the likelihood that particular skeletal parts will be transported from a carcass to a base camp, and (2) the likelihood that they will survive there in identifiable condition (Klein 1989b).

Mortality profile assessment at KRM, NBC, DK1, BNK1, and most other archeological sites must take into account the probability that postdepositional leaching and profile compaction selectively removed many relatively fragile teeth and bones of very young animals. Fortunately, this is not an insuperable obstacle to interpretation, since very young individuals dominate both attritional and catastrophic profiles, and the two types are therefore most readily distinguished by the ratio of prime to postprime adults. With this in mind, both catastrophic (prime-age rich) and attritional (prime-age poor) profiles occur in the known MSA and LSA samples, but probable or certain attritional profiles dominate heavily. Among the clearest attritional profiles are those for buffalo at both KRM (MSA) and NBC (LSA), while the only unequivocal catastrophic profile is for eland at KRM (MSA) (see Figure 23-2) (the NBC [LSA] eland profile is arguably also catastrophic, but comprises too few animals for secure diagnosis.) The dominance of attritional profiles shows that even LSA people were like lions, hyenas, and other large predators in their inability to obtain large numbers of prime-age adults relative to very young and old animals. Depending on the prey species, prime adults are variably protected by especially acute special senses, agility, or gregariousness, and in the special case of buffalo, by large size and aggressiveness. A priori, attritional profiles are probably to be expected in stone age sites, since prime adults are essential for reproduction, and no predator can acquire very many without threatening its own long-term survival.

Seen in this light, only the KRM eland profile requires special explanation, and the KRM and NBC buffalo profiles might seem unremarkable. However, they are both notably richer in very young individuals than the large ungulate attritional profiles from Langebaanweg and Elandsfontein that were previously discussed. The contrast is even more striking than it appears at first glance, since bone preservation is much poorer at KRM and NBC, and many more very young individuals were probably lost to selective postdepositional destruction than at Langebaanweg and Elandsfontein. Assuming that carnivore consumption explains the relative paucity of very young individuals in the Langebaanweg and Elandsfontein attritional profiles and that stone age people generally could not locate carcasses before other carnivores, the relative abundance of very young buffalo at KRM and NBC implies active hunting, at least of very young buffalo.

The catastrophic shape of the KRM eland profile is special, not simply because it is unique but also because it would seem to be paradoxical. It is based on eland accumulated over thousands or tens of thousands of years, during which repeated catastrophic mortality should have sapped the vitality of the species and perhaps led to its extinction. Yet the eland clearly survived, and it does not appear to have declined through time near KRM. It is the most abundant large ungulate from top to bottom, both in "interglacial" and "glacial" layers. The apparent paradox can be resolved by two historic observations: (1) Among all indigenous African ungulates, eland are uniquely amenable to driving, but (2) eland herds tend to be widely dispersed and hard to locate. The KRM eland mortality profile implies that the people probably drove eland herds into traps or over cliffs where age-related differences in vulnerability lost their meaning. Suitable cliffs exist very near KRM. However, the eland's survival implies that MSA people did not drive eland very often, probably because they rarely found herds in a suitable position. In short, despite their special knowledge of eland behavior, the KRM people probably obtained very few eland overall.

The behaviorally significant differences between the MSA and LSA faunas can be summarized as follows (Klein 1989a; Grine et al. 1991):

1. Both KRM (MSA) and NBC (LSA) contain intertidal shells and bones of fur seals and sea birds, but unlike NBC, KRM contains virtually no fish bones. In addition, at KRM, the sea birds are mainly penguins, while those at NBC are mainly cormorants, gulls, and other airborne species (G. Avery 1990). DK1 and BNK1 contrast in the same way, and the most plausible explanation is that MSA people fished and fowled much less regularly than LSA people did. It is notable in this regard that only LSA sites have provided artifacts that are similar to ethnographically recorded fishing and fowling gear. The likely fishing artifacts include grooved or ringed stones that were probably net- or line-sinkers and carefully fashioned bone slivers that were probably "gorges," baited and tied to a line (see Figure 23-5). The possible or likely fowling implements include bone projectile points and perhaps some of the same "gorges" that served for fishing.

2. At NBC (LSA), ungulate species occur in rough proportion to their historical live abundance in the region, but at KRM (MSA), they tend to be represented in rough proportion to the danger involved in hunting them. In particular, compared to the NBC fauna, the KRM fauna is much richer in relatively docile eland and much poorer in more dangerous Cape buffalo and bushpig (Figure 23-6). Historically, the reverse was true near both sites—eland were very rare and buffalo and bushpig were abundant. The same contrast distinguishes DK1 (MSA) and BNK1 (LSA), though it is numerically less striking, because both samples are much smaller. The observed difference in ungulate species abundance may again be tied to differences between MSA and LSA artifact assemblages. Probably most significant, only LSA assemblages contain stone and bone artifacts that closely recall parts of historically observed, composite arrows. With the bow and arrow LSA people could have attacked truly dangerous game at greatly reduced personal risk, and even if most attacks were unsuccessful, the higher number of attempts would have yielded more animals.

   The abundance of eland at KRM is particularly notable in light of the inference from the eland mortality profile that the KRM people did not obtain eland very often. The implication is that they even more rarely acquired other species that were much more abundant nearby but are much rarer than eland in the site. Conversely, the rarity of eland at NBC need not imply that LSA people obtained eland less often than their predecessors, only that they obtained other, locally more common species much more often. In short, the composite evidence suggests that local LSA people probably obtained a significantly larger number of animals overall.

TABLE 23-6 The large mammal taxa associated with Middle Stone Age artifacts at Klasies River Mouth Main Site (KRM), Die Kelders Cave 1 (DK1) and Boomplaas Cave A (BPA). Klein (1983) and Grine et al. (1991) provide abundance estimates for each taxon in the separate layers of each site. Asterisks designate extinct genera.

TAXA	KRM	DK1	BPA
INSECTIVORA			
Erinaceus frontalis (hedgehog)		x	
PRIMATES			
Homo sapiens (people)	x	x	
Papio ursinus (chacma baboon)	x		x
CARNIVORA			
Canis mesomelas (black-backed jackal)	x		x
Lycaon pictus (hunting dog)			x
Aonyx capensis (clawless otter)	x		
Herpestes ichneumon (Egyptian mongoose)	x		
Galerella pulverulenta (Cape grey mongoose)	x	x	x
Atilax paludinosus (water mongoose)	x		
Genetta sp. (genet)	x		x
Hyaena brunnea (brown hyena)	x	x	cf.
Felis libyca (wildcat)	x	x	x
Felis caracal (caracal)	x	x	x
Panthera pardus (leopard)	x		x
Arctocephalus pusillus (Cape fur seal)	x	x	
Mirounga leonina (southern elephant seal)	x		
PROBOSCIDEA			
Loxodonta africana (elephant)	x		
HYRACOIDEA			
Procavia capensis (rock hyrax)	x	x	x
PERISSODACTYLA			
Equus quagga (plains zebra)	x	x	x
Diceros bicornis (black rhinoceros)	x	x	
ARTIODACTYLA			
Potamochoerus porcus (bushpig)	x		
Phacochoerus aethiopicus (warthog)	x		
Hippopotamus amphibius (hippo)	x	x	
Tragelaphus scriptus (bushbuck)	x		
T. strepsiceros (greater kudu)	x	x	
Taurotragus oryx (eland)	x	x	x
Syncerus caffer (Cape buffalo)	x	x	x
**Pelorovis antiquus* ("giant" buffalo)	x	x	
Hippotragus leucophaeus (blue antelope)	x	x	x
Redunca arundinum (southern reedbuck)	x	x	
R. fulvorufula (mountain reedbuck)	x	x	x
Damaliscus dorcas (bontebok/blesbok)	x	x	x
Connochaetes gnou (black wildebeest)	x	x	x
Alcelaphus buselaphus (Cape hartebeest)	x		?
Pelea capreolus (vaalribbok)	x	x	x
Raphicerus campestris (steenbok)			cf.
Raphicerus melanotis (Cape grysbok)	x	x	
Oreotragus oreotragus (klipspringer)		x	x
Antidorcas cf. *australis* (southern springbok)	x	x	x
LAGOMORPHA			
Lepus capensis (Cape hare)	x	x	x
Pronolagus crassicaudatus (red rock hare)			cf.
RODENTIA			
Hystrix africaeaustralis (porcupine)	x	x	x
Bathyergus suillus (Cape dune molerat)		x	
CETACEA			
Gen. & spp. not det.	x	x	

3. Cape fur seals occur in all the sites, but the fur seal mortality profile at KRM differs sharply from the profile at NBC and other coastal LSA sites. Whereas the KRM seal bones come from a wide range of juveniles and adults, the bones at NBC and other LSA sites come overwhelmingly from individuals that were about 9 months old (Figure 23-7). This is the approximate age when fur seals are weaned, and the LSA pattern suggests that LSA people timed their coastal visits to coincide with the weaning season, when newly weaned seals often die or become exhausted at sea and can be literally harvested on the beach. The broader age distribution at KRM may indicate that MSA people were unaware of the seasonal peak in seal availability and that they simply killed or scavenged occasional individuals they encountered during seasonally unfocused visits. The DK1 (MSA) fur seal sample is too small to determine whether it resembles the one from KRM, but a larger sample that could be used for this purpose is being excavated now.

4. At both DK1 (MSA) and BNK1 (LSA), the inhabitants collected large numbers of angulate tortoises (*Chersina angulata*), but on average, the tortoises are much larger at DK1 (Figure 23-8). Since DK1 was occupied under relatively cool, moist conditions that are unlikely to have accelerated tortoise growth, the larger tortoise size is more likely to reflect relatively light MSA predation (collecting) pressure. Greater LSA pressure would be a natural consequence of the larger, denser human populations that more effective hunting-foraging probably made possible. The contrast in tortoise size cannot be checked at KRM and NBC, because tortoises are rare in both faunal samples, just as they were near both sites historically. Theoretically, a comparable size difference might characterize the shellfish species that DK1, BNK1, KRM, and NBC share, but DK1 has provided virtually no shells, the NBC shells have yet to be measured, and the majority from KRM were biased toward larger and more complete specimens during excavation (Singer and Wymer 1982; Voigt 1973). However, shells of brown mussel (*Perna perna*) that were systematically recovered during recent excavation of the MSA deposits at KRM are clearly much larger on average than brown mussel shells from the overlying (mid-to-late Holocene) LSA layers (Thackeray 1988); and shells of the limpet *Patella granatina* in the Sea Harvest open-air MSA midden are much larger than those found in a variety of nearby LSA sites. The Sea Harvest (MSA) limpets in fact approximate the largest ones found on totally unexploited local intertidal rocks today (Buchanan et al. 1978), and they therefore suggest that MSA collecting pressure was exceptionally light.

MSA Human Remains and Modern Human Origins

In southern Africa, associations between human fossils and MSA artifacts have been reported at Florisbad (a partial cranium comprising facial, frontal, and parietal fragments) (Rightmire 1984; Clarke 1985c); Border Cave (an infant's skeleton, an adult skull, two partial adult mandibles, and some postcranial bones) (Beaumont et al. 1978; Beaumont 1980; Rightmire 1979, 1984), Die Kelders Cave 1 (13 isolated teeth) (Grine et al. 1991), and Klasies River Mouth (five partial mandibles, two fragmentary maxillae, three small cranial fragments, isolated teeth, and six postcranial bones) (Singer and Wymer 1982; Rightmire and Deacon 1991). To these can probably be added human remains from the fossil hyena lairs at Equus Cave (a mandible and 12 isolated teeth) (Grine and Klein 1985; Klein et al. 1991) and Sea Harvest (Saldanha) (a phalange and a tooth) (Hendey 1974; Klein 1983; Grine and Klein 1993.; the Sea Harvest hyena lair is spatially and stratigraphically distinct from the MSA midden, just discussed). Geomorphic context suggests an MSA age for both the Equus Cave and Sea Harvest fossils, but radiometric confirmation is available only for Sea Harvest. This is a ^{14}C date of greater than 40,000 years (UW–292) on associated ostrich eggshell.

Most of the fossils are fragmentary and difficult to diagnose, and the relatively complete specimens from Border Cave may come from post-MSA graves that intruded into the MSA layers. They are generally much better preserved than the MSA animal bones with which they are supposed to be associated, and none was excavated under tightly controlled conditions. Still, with or without the Border Cave specimens, the MSA fossils collectively imply that MSA inhabitants of southern Africa looked far more modern than their Neanderthal contemporaries in Europe. Especially noteworthy are two specimens from Klasies River Mouth—a frontal-nasal fragment that lacks any sign of a supraorbital torus and a relatively complete mandible with a strongly developed mental eminence and no retromolar space. Only the Florisbad skull is obviously archaic in any way, and it is almost certainly significantly older than the other fossils. It probably dates from the late mid-Pleistocene, before 128 KYA (Butzer 1984; Kuman and Clarke 1986), while the others are all of earlier late Pleistocene age, between roughly 128 and 40 KYA. In its relatively massive face and thick supraorbital margin, the Florisbad skull recalls the earlier-middle Pleistocene Kabwe skull, but it is clearly advanced over Kabwe in its more domed frontal and lack of a continuous supraorbital torus (Clarke 1985c; Rightmire 1984, 1987). Together with broadly comparable and possibly like-aged skulls from Ngaloba (Laetoli) and Omo (Omo II) in east Africa (Clarke 1985c), it may represent an evolutionary link between Kabwe (and allies) on the one hand and late Pleistocene MSA people on the other, but a much more complete and firmly dated mid-Pleistocene fossil record will be necessary to confirm this.

By themselves, the southern African MSA fossils are arguably insufficient for phylogenetic analysis, but when they are combined with like-aged fossils from east and north Africa (references in Klein 1989a), they clearly suggest that modern people were evolving in Africa at a time when the Neanderthals and other kinds of archaic people were the sole occupants of Eurasia. The implication might be that the Neanderthals and other archaic Eurasians represent evolutionary dead ends that were extinguished by modern African emigrants. This hypothesis was recently weakened by the elimination of some genetic data that appeared to demand an African origin for all living humans (Hedges et al. 1992; Templeton

FIGURE 23-6 The minimum numbers of individuals by which eland, Cape buffalo, and wild pig (bushpig and/or warthog) are represented in the MSA layers of Klasies River Mouth Cave 1 and Die Kelders Cave 1 and in the pre-pastoralist LSA layers of Nelson Bay Cave and Byneskranskop Cave 1.

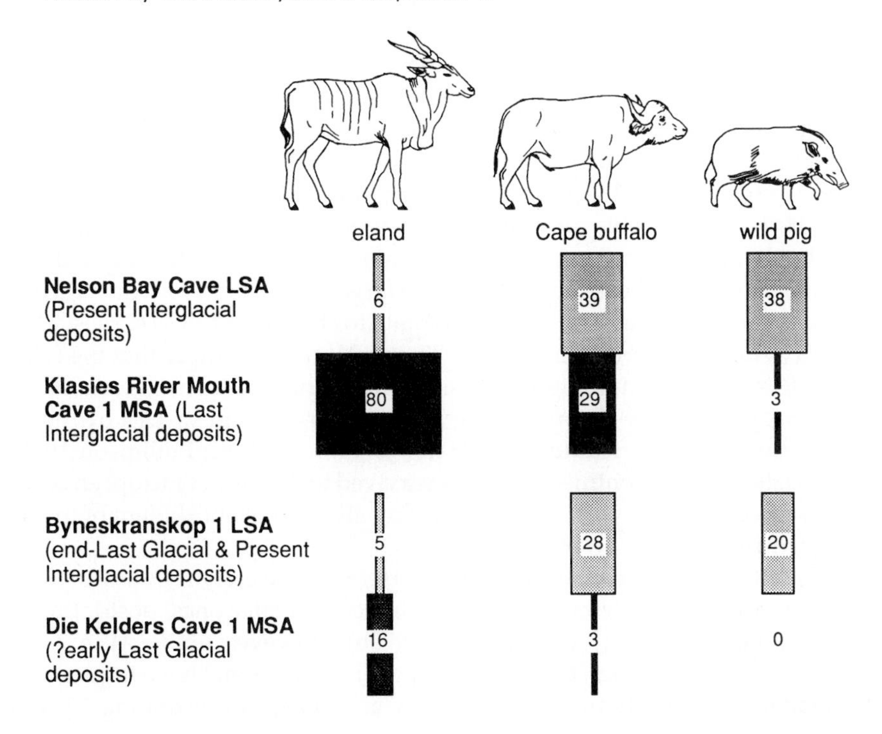

1992), but it remains highly plausible on the fossil record alone. If there is a paleontological objection to it, it is perhaps the apparent disjunction between modern morphology and modern behavior. While MSA people were anatomically much more modern than their Neanderthal contemporaries, they were equally primitive in every detectable behavioral (archeological) respect.

The answer to this problem may be that the disjunction is more apparent than real and that African MSA people would be better regarded as "near modern" or simply as the most probable immediate ancestors of fully modern people. The Klasies River Mouth fossils, for example, are remarkably variable in their expression of fully modern morphology, especially in the chin region (Caspari and Wolpoff 1990; Wolpoff and Caspari 1990a,b), and it may be that there was still significant anatomical change to come. Conceivably this change included a neurological advance that made possible fully modern behavior (or more precisely, the fully modern ability to wield culture in adaptation), and it was this that allowed fully modern people to displace the Neanderthals and other archaic Eurasians. The composite archeological and fossil records suggest the displacement occurred 50–40 KYA, with the advent of the LSA in Africa and of the Upper Paleolithic in western Eurasia.

The hypothesis that neurological change in near-modern Africans underlies the emergence of fully modern behavior may prove difficult to test, even with a much more complete fossil record. However, the basic notion that modern people originated in Africa and that their spread depended on the development of fully modern behavior would clearly be strengthened if it could be shown that fully modern behavior appeared first in Africa. Demonstrating this (or its converse) will require the recovery of large artifact and faunal samples from African sites that were occupied more or less continuously between 60 and 30 KYA. In southern Africa, such sites are very rare, probably because very dry climatic conditions depressed human populations during this interval. Sites of this age are also rare in northern Africa, probably for the same reason, but a limited number of excavations suggest that they are much more common in east Africa (references in Klein 1989a), and it is probably there that on-going searches for pivotal evidence should concentrate.

The Later Stone Age (LSA)

As already indicated, the MSA/LSA interface is very difficult to date, partly because it occurred near the 40–30 KYA limit of conventional radiocarbon dating and partly because it is recorded at very few sites. Apparently inconsistent ^{14}C dates, such as one of c. 38 KYA for the early LSA at Border Cave in Natal (Beaumont et al. 1978) and one of c. 32 KYA for the very late MSA at Boomplaas Cave in the southern Cape (H. J. Deacon 1979; H. J. Deacon et al. 1984), may indicate that the LSA appeared at different times in different places, but only a minute amount of contamination can make a ^{14}C sample that is much older than 30 KYA seem much younger. Therefore, until a much larger suite of relevant dates becomes available, it is probably best to regard 38 KYA as a minimum estimate for the latest MSA/earliest LSA. Preliminary data from east Africa (J. D. Clark 1988; van Noten 1982; van Neer 1989) imply that the LSA may actually have begun before 40 KYA. The terminal date is far easier to fix. In the moister northern and eastern portions of southern Africa, LSA people were absorbed or replaced between 2 and 1.8 KYA by the rapid spread of Bantu-speaking Iron Age mixed agriculturalists (Hall 1987; Maggs 1984; Phillipson 1985). In the drier western and central parts of the subcontinent, the LSA survived to the time of European contact, beginning at the very end of the fifteenth century A.D. and accelerating after the establishment of the first permanent European settlement in 1652.

The LSA is difficult to characterize artifactually, partly because early LSA assemblages, antedating 22 KYA, are so far poorly known or described, and partly because later ones, postdating 22 KYA, varied significantly through time and space (J. Deacon 1984, 1990b). However, unlike MSA assemblages, LSA assemblages, even the very earliest ones, tend to include well-made formal bone implements and bone or shell beads and pendants. In addition, most later LSA assemblages (postdating 22 KYA) are readily

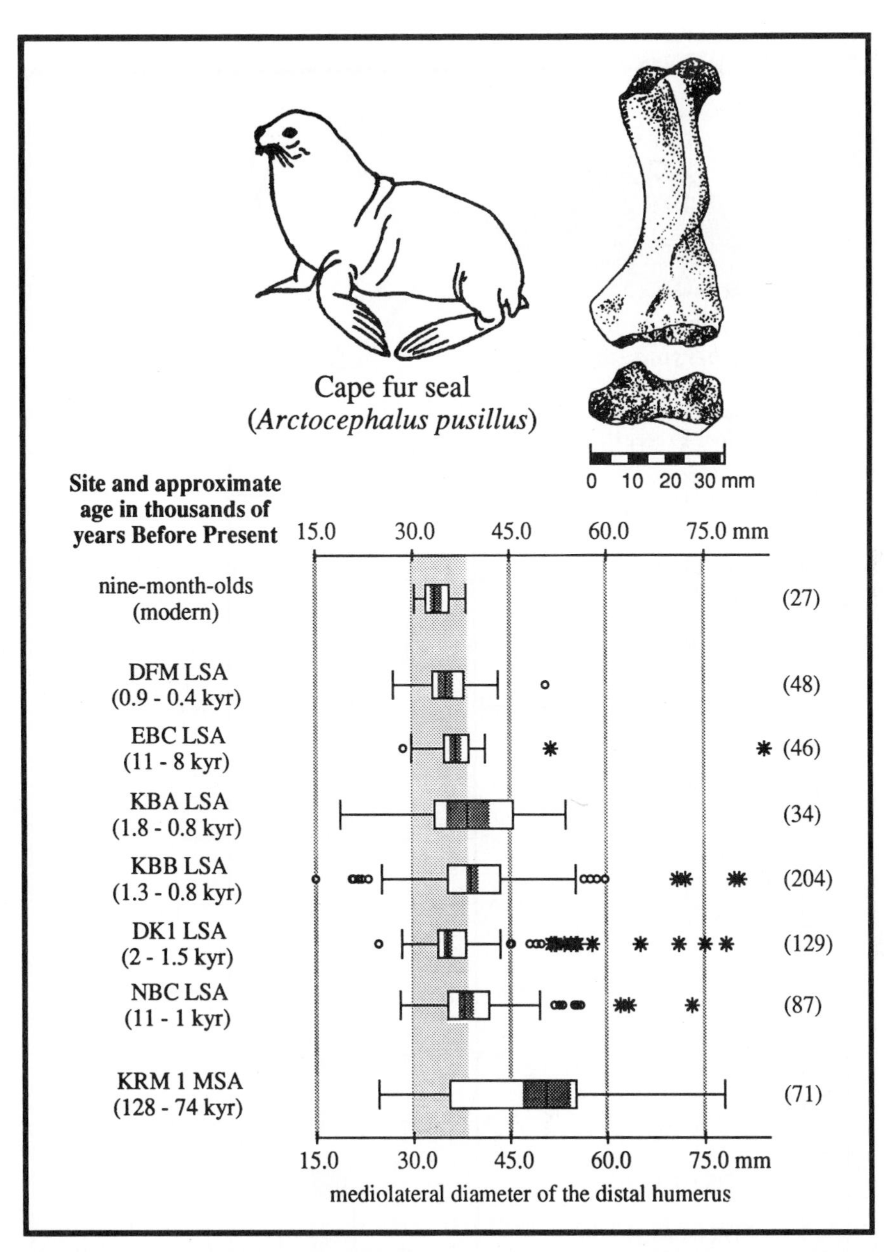

FIGURE 23-7 Boxplots summarizing the mediolateral diameters of Cape fur seal distal humeri in the LSA layers of the Dune Field Midden (DFM), Elands Bay Cave (EBC), Kasteelberg A (KBA), Kasteelberg B (KBB), Die Kelders Cave 1 (DK1), and Nelson Bay Cave (NBC), and in the MSA layers of Klasies River Mouth Cave 1 (KRM1). In each boxplot (as described by Velleman 1989), the vertical line near the center is the median, the vertical lines at the ends mark the range of more or less continuous data, the hachured rectangle is the 95 percent confidence interval for the median, and the open box encloses the middle half of the data between the 25th and 75th percentiles. Circles and starbursts indicate extreme values (points that are far removed from the main body of data). The number of measured humeri in each sample is in parentheses after each boxplot. The LSA samples differ from the MSA one in two related ways: the measurements in each LSA sample tend to be much more closely clustered around the median and the median of each LSA sample tends to more closely approximate the median of modern 9-month-olds. (The broad vertical bar marks the range for 9-month-olds.) The LSA pattern probably reflects the tendency of LSA people to concentrate their coastal visits on the seal weaning season when 9-month-old seals become especially abundant on the beach. Differences among the LSA medians probably reflect relatively small differences in the season of site occupation.

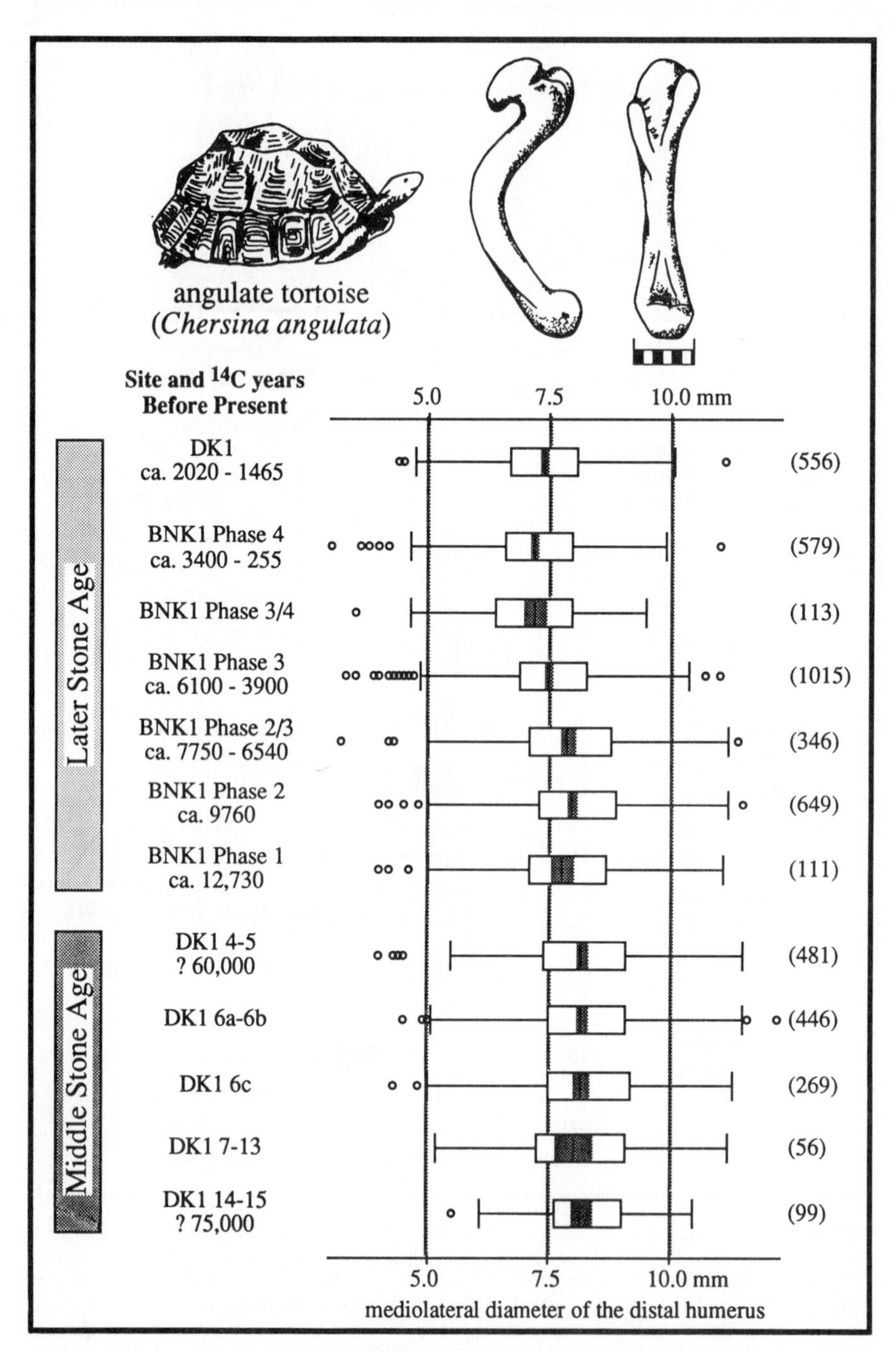

FIGURE 23-8 Boxplots summarizing the mediolateral diameters of angulate tortoise distal humeri from the MSA and LSA deposits of Die Kelders Cave 1(DK1) and the LSA deposits of nearby Byneskranskop Cave 1 (BNK1). In each boxplot (as described by Velleman 1989), the vertical line near the center is the median, the vertical lines at the ends mark the range of more or less continuous data, the hachured rectangle is the 95 percent confidence interval for the median, and the open box encloses the middle half of the data between the 25th and 75th percentiles. The number of measured humeri in each sample is in parentheses after each boxplot. Circles indicate extreme values (points that are far removed from the main body of data). The plots show that MSA tortoise humeri tend to be significantly larger than LSA humeri, suggesting that LSA people collected tortoises more intensively.

distinguished from MSA assemblages by the occurrence of microlithic artifacts, especially microblades before about 12 KYA and small convex scrapers and backed elements after about 8 KYA (see Figure 23-5). It is not yet known whether microliths were also a common feature of the early LSA, and they are not always present even in some later industries. They are especially rare or absent in the Albany/Oakhurst/Smithfield A Industrial Complex, which was widespread in southern Africa between about 12 and 8 KYA and which could be confused with the MSA, if artifact size were the only criterion. However, compared to MSA assemblages, Albany/Oakhurst/Smithfield A assemblages tend to include many fewer flakes or flake-blades removed from well-prepared cores, many more large convex scrapers, and perhaps most important of all, numerous well-made formal bone and shell artifacts.

The LSA inhabitants of southern Africa often engraved artistic designs in shell and bone, and they also appear to have been the first to paint or engrave on rock. They certainly produced most of the paintings and engravings that adorn exposed rock surfaces throughout the subcontinent, since these would have weathered away after no more than a few thousand years. Deep caves like those that preserve much older (Upper Paleolithic) art in southwest France and northern Spain are rare in southern Africa, but at Apollo 11 rockshelter in Namibia buried rock slabs with painted animals have been bracketed between 27.5 and 19 KYA by radiocarbon (Wendt 1976; J. Deacon 1990b). The Apollo 11 paintings are probably as old or older than any known in Europe. Later engraved or painted stones, all postdating 10 KYA, have been found at several southern African sites, including Wonderwerk Cave, Boomplaas Cave, and Klasies River Mouth Cave 5 in the Cape Province (H. J. Deacon et al. 1974; Thackeray et al. 1981; Singer and Wymer 1969).

In all essential features, including the degree of internal variability, the manufacture of numerous formal bone artifacts, and the production of art and ornaments, the LSA clearly resembles the broadly contemporaneous Upper Paleolithic complex of Eurasia. To the extent that the LSA and the Upper Paleolithic differ, especially in stone artifact typology and technology, this simply reinforces the conclusion that both LSA and Upper Paleolithic people were far more inventive and innovative than their predecessors and that they made far greater use of culture in adapting to natural and social environmental change. Together, LSA and Upper Paleolithic people are in fact the first for whom it is possible to infer both "Culture" and "cultures" (or ethnicity) in the classic anthropological sense.

The LSA scraper/backed element industries that existed after 8 KYA are especially well studied and are commonly lumped in the Wilton Industrial Complex or Tradition. The forms and the numbers of both scrapers and backed elements varied spatially and temporally within the Wilton, probably because different Wilton groups had different stylistic norms and different functional needs (J. Deacon 1984). Ethnohistoric observations and occasional archeological finds indicate that the scrapers were usually mounted or inserted on wooden or bone handles, and both ethnohistoric observations and microwear studies show that they were commonly used to prepare skins. This may explain why they tend to be most abundant in regions where aboriginal people traditionally made clothing mainly from leather and least abundant in regions where they made it mainly out of bark cloth (H. J. Deacon and J. Deacon 1980). Ethnohistoric observations and archeological finds suggest that backed elements were commonly used to arm arrows or other light projectiles, and they appear to be especially common in regions where ethnographic or archeological evidence suggests that people favored barbed projectiles, each requiring several backed pieces (J. Deacon 1984).

Early LSA (and yet earlier) people undoubtedly made artifacts from relatively perishable plant and animal tissues, but it is only in some Wilton sites that such artifacts are abundantly preserved. The remarkable Wilton horizons of Melkhoutboom Cave, for example, have provided pieces of cordage, netting, and matting made from sinews and plant fibers; small pieces of leather garments; cut or broken sections of reeds that probably represent couplings or parts of arrow shafts; and numerous wooden objects—both shavings and finished products that seem to have been fire sticks, fire drills, points, pegs, and probably holders for stone bits (H. J. Deacon 1976). No stone artifacts were actually found in mounts at Melkhoutboom, but many pieces retain traces of the vegetal mastic that was used to mount them.

Ethnographic observations and use-wear analyses suggest that LSA people commonly worked wood with small, oblong, or rectangular, steeply retouched pieces that archeologists call "adzes" (Binneman and Deacon 1986). These were themselves usually mounted on wooden or bone handles, and they tend to be especially common at sites (or layers within sites) that contain pollens, wood, or charcoal, implying that wood was readily available nearby.

Bones are preserved at many LSA sites, and large, informative samples have been recovered from deeply stratified caves, including especially Nelson Bay Cave and Byneskranskop Cave 1, which have already been discussed, and Boomplaas Cave (H. J. Deacon 1979; Klein 1983) and Elandsbay Cave (Klein and Cruz-Uribe 1987; Parkington 1981, 1987, 1990b). As already noted, LSA faunal assemblages tend to be taxonomically very similar to preceding MSA assemblages. The species are overwhelmingly modern, and there are only five—an equid and four bovids—that were absent historically (A. J. V. Brown and Verhagen 1985; Klein 1984; Thackeray 1984). All five apparently disappeared between roughly 12 and 8 KYA in the transition from the Last Glaciation to the Present Interglaciation. This was a time of profound environmental change that may have substantially reduced the numbers and ranges of the extinct species, but all five survived the last interval of similar change centered on 12–8 KYA. The only conspicuous difference 12–8 KYA was the presence of more sophisticated (LSA) hunter-foragers (as previously discussed and summarized again later), who perhaps precipitated the extinctions in their attempt to maintain a doomed lifestyle.

It was already noted that MSA and LSA faunal assemblages accumulated under similar environmental conditions suggest (1) that LSA people obtained dangerous prey much more often, (2) that they fished and fowled far more regularly, and (3) that they were perhaps the first to recognize seasonal variability in the availability of key resources. Where tortoises and limpets can be compared between MSA and LSA sites, LSA specimens tend to be much smaller, suggesting that LSA people exerted greater collecting pressure, probably because LSA populations were larger and denser. The faunal differences are mirrored in artifactual differences implying that LSA populations possessed a far more effective hunting-fishing-fowling technology. The difference may even have extended to plant exploitation, since only LSA assemblages contain bored stones identical to historically observed digging stick weights. They are also much richer in well-formed grinding stones, at least some of which were surely used to process plants.

In sum, both faunal and artifactual evidence indicates that LSA people were more effective hunter-foragers than their predecessors. However, it is possible to exaggerate the degree of LSA sophistication, and as already indicated, LSA people were unable to obtain a higher proportion of prime-age adult ungulates, even if they were able to obtain a larger number of ungulates overall. Additionally, most variation in taxonomic abundance among LSA sites is far more readily explained by environmental differences than by cultural ones, and LSA people were obviously still highly vulnerable to the vagaries of nature. This is most dramatically shown by their virtual disappearance from the west coast and much of the interior of southern Africa when conditions became extremely dry between roughly 8 and 4 KYA (H. J.Deacon and Thackeray 1984; J. Deacon 1984, 1990b; Parkington 1987).

So far, human remains have not been found at any southern African LSA site antedating 12 MYA, but it is assumed that all LSA people were anatomically modern. Those who lived in southern Africa after 12 MYA closely resembled the stone age Khoisan people encountered by Europeans beginning about 500 years ago (Rightmire 1984). From a strictly economic perspective, the Khoisan are said to have comprised two distinct groups (Parkington 1984; Smith 1990; Smith et al. 1991)—sheep and cattle pastoralists (or Khoi) who occupied the western and southern coastal forelands, and hunter-foragers (or San), who were found mainly in the adjacent mountains or in the drier interior regions where herding was impractical. Together with pottery, pastoralism was introduced about 1800 years ago (H. J. Deacon et al. 1978; J. Deacon 1984; Klein 1986; Schweitzer 1979), but the mechanism remains unclear. Its spread was almost certainly linked to the nearly simultaneous expansion of Iron Age mixed farmers through the moister, eastern third of the subcontinent, but unlike the Iron Age farmers, the pastoralists physically resembled the hunter-foragers they replaced. Together with continuity in basic stone and bone artifacts (J. Deacon 1984) and a tendency for the pastoralists to rely heavily on hunting and foraging (Klein and Cruz-Uribe 1989; Smith 1987), this suggests that pastoralism may have dispersed as much by diffusion as by population movement.

Almost certainly, herding, or more precisely herding-foraging, supplanted hunting-foraging because it allowed larger populations. This is circumstantially suggested by a reduction in average tortoise size in pastoralist (vs. prepastoralist) LSA sites (Klein and Cruz-Uribe 1989). LSA pastoralists were thriving when Europeans first rounded the Cape of Good Hope at the end of the fifteenth century, and it was partly their willingness to barter stock for European goods that encouraged the Dutch East India Company to establish a permanent refreshment station at (what is now) Cape Town in 1652. Unfortunately for the pastoralists, European traders were soon followed by European farmers and European diseases, and the result was the near total disintegration of pastoralist societies by 1750 (Elphick 1977; Smith 1990). Hunter-forager groups in the more remote interior suffered a similar fate in the century that followed. A few descendants of the historic Khoisan survive in southern Africa today, but they are playing almost no role in determining the future of the subcontinent in which they so recently prospered.

SUMMARY

The prehistory of southern Africa from the early Miocene through the Stone Age may be summarized as follows:

1. Southern Africa contains very few fossiliferous, Miocene sites, particularly by comparison to east Africa. The principal ones are in Namibia, where they have been dated by faunal comparisons of east Africa to the early and middle Miocene, between roughly 19 and 14–12 million years ago (MYA). In all cases, the taxa represented show that Namibia was far better watered than it is at present. Much more mesic conditions are especially implied in the far south, where the relatively rich mid–Miocene fauna from Arrisdrift indicates woods or bush in a region that is presently hyperarid. So far, only one site—in the Otavi Mountains of northern Namibia—has provided a hominoid fossil, but others may eventually be found, perhaps especially in an enlarged sample from Arrisdrift. Late Miocene sites, dating between 12 and 5.2 MYA, are totally unknown in southern Africa, but a late Miocene trend toward much cooler and drier conditions probably extinguished any hominoids that had existed in southern Africa previously.

2. Southern Africa also contains relatively few early Pliocene sites, dating between roughly 5.2 and 3 MYA, but this lack is largely offset by the extraordinarily prolific early Pliocene occurrence at Langebaanweg near the southwestern tip of the subcontinent. The Langebaanweg sample comprises hundreds of thousands of invertebrate and vertebrate specimens, recovered from the floodplain and channel deposits of a river that entered the sea nearby. More than 80 mammalian species are represented, including at least one from every extant African mammalian order but the Sirenia (dugongs). Comparisons to dated mammalian faunas in east Africa show that the Langebaanweg bones accumulated about 5 MYA, but despite the extraordinary size and richness of the sample, hominids are conspicuously absent. This suggests that they were still essentially tropical creatures who could not adapt to relatively cool, temperate conditions like those that characterized extreme southern Africa by the early Pliocene. Archeological evidence in fact suggests that people did not colonize the temperate parts of southern Africa until the early Pleistocene, perhaps only about 1 MYA.

In the absence of hominids, it is especially noteworthy that the Langebaanweg fauna contains numerous burnt bones, that very young ungulates are remarkably rare in otherwise typical attritional mortality samples from the floodplain deposits, and that large carnivore species are much more varied and numerous (vs. ungulate species) than in the mid-Pleistocene–to–historic fauna of the region. The burnt bones surely derive from (?seasonal) veldt fires, and they are a potent reminder that burning need not imply human presence. Complete carnivore consumption of very young carcasses probably explains the rarity of very young ungulates in the attritional samples, and these samples thus illustrate what would be available to scavenging humans. It follows that active hunting of very young animals is probably implied by archeological samples in which young animals are significantly more abundant. Such samples are known from local late Pleistocene archeological sites like Klasies River Mouth and Nelson Bay Cave. The arrival and subsequent success of carnivorous hominids sometime between 5 and 1 MYA perhaps partly explains the reduction in carnivore taxonomic diversity that occurred between the early Pliocene and the mid-Pleistocene.

3. The late Pliocene, between roughly 3 and 1.64 MYA, is known primarily from the Transvaal australopithecine caves of Makapansgat, Sterkfontein, Swartkrans, and Kromdraai. Makapansgat and Sterkfontein have provided abundant fossils of *Australopithecus africanus*, dated (by faunal comparisons to east Africa) between roughly 3 and 2.3 MYA. Swartkrans and Kromdraai have provided fossils of *Australopithecus robustus*, dated between roughly 2 and 1 MYA. Sterkfontein and Swartkrans also contain occasional bones of early *Homo*, dated between perhaps 1.8 and 1.3 MYA. The phylogeny of the australopithecines and early *Homo* remains controversial, but the combined south and east African samples currently suggest that *A. africanus* and *A. robustus* represent distinct lineages that split from a common ancestor (*A. afarensis)* at roughly 3 MYA. *A. africanus* was probably directly ancestral to *Homo*, while *A. robustus* became extinct sometime between 1.2 and 0.8 MYA.

At both Swartkrans and Sterkfontein, the same deposits that contain *Homo* have also provided the oldest known stone artifacts in southern Africa. At Swartkrans, the artifacts could represent the Oldowan Tradition, the Acheulean Tradition, or both (in different levels). At Sterkfontein, they include bifaces and thus clearly represent the early Acheulean. Besides stone artifacts, Swartkrans has also provided informally shaped bone implements that may have been used to dig roots or bulbs from stony soil. Early *Homo* probably manufactured most of the artifacts at both sites, but *A. robustus* could have made some, particularly at Swartkrans.

At Sterkfontein, Swartkrans, and Makapansgat, bone damage, the pattern of skeletal part representation, and other features of the faunal assemblages imply that the bones in the main australopithecine units were introduced mainly by large cats (especially at Swartkrans and Sterkfontein) or hyenas (especially at Makapansgat). Early *Homo* may have introduced most of the bones to the immediately younger deposits that have provided no australopithecine fossils at Sterkfontein and relatively few at Swartkrans. At Swartkrans, burnt bones that appear only in the deposits postdating the main australopithecine unit may indicate incipient human use of fire, at or even somewhat before 1 MYA. The infrequent robust australopithecine fossils from the same deposits are among the youngest known anywhere and may therefore bear on robust australopithecine extinction. This has been speculatively attributed to climatic change, especially to a marked shift in the nature of glacial/interglacial cycles about 0.85 MYA, or to niche absorption by evolving *Homo*.

4. The Acheulean artifacts recovered at Sterkfontein may be as old as any found in east Africa, where the transition from the Oldowan to the Acheulean occurred roughly 1.5–1.4 MYA. Also as in east Africa, the Acheulean in southern Africa probably terminated during the late middle Pleistocene, perhaps around 0.2 MYA. On typological grounds, sometimes supported by associated fauna, the vast majority of southern African Acheulean sites probably date between 1 and 0.2 MYA, and the drier, western and cooler, southern regions may not have been inhabited until roughly 1 MYA. Conceivably, they were colonized only when people had achieved the control over fire that may be implied at Swartkrans. Acheulean people were undoubtedly hunter-foragers, relying especially on plants, but food plant remains have not been preserved at any site. Animal bones are sometimes preserved, above all at the spectacular Elandsfontein site near Cape Town, but their utility for elucidating Acheulean hunting or scavenging success is limited, since the samples come almost exclusively from ancient stream or pond settings, where sedimentary context, bone damage, numerous carnivore coprolites, mortality profiles, and so forth imply that people may have been only incidentally involved in the bone accumulation.

The sparse human fossil record associated with the Acheulean in Africa and western Eurasia is usually said to indicate that *H. erectus* made earlier Acheulean artifacts, antedating roughly 0.5 MYA, while "archaic" *H. sapiens* made later ones. In southern Africa, *H. erectus* is arguably represented at Swartkrans, while fossils assigned to "archaic" *H. sapiens* are associated with Acheulean artifacts at Elandsfontein and the Cave of Hearths. However, a much fuller and better dated fossil record may one day show that African and European Acheulean people were on different evolutionary tracks, culminating after 0.2 MYA in near-modern or modern people (in Africa) and in the Neanderthals (in Europe). In this event, the makers of African Acheulean artifacts will probably be assigned to a single evolving species.

5. The Acheulean was succeeded in southern Africa by the Middle Stone Age (MSA), which differed from the late Acheulean mainly in the absence of large bifacial tools. Like the Acheuleans, MSA people did not make formal bone artifacts, and they left no unequivocal evidence for art. Their artifact assemblages also varied remarkably little through time and space, and the faunal remains they left behind suggest they were relatively ineffective hunter-foragers.

In virtually every detectable archeological respect, the MSA closely resembles the broadly coeval Mousterian of Europe. However, while the makers of European Mousterian artifacts were Neanderthals who were morphologically distinguishable from all living people, the makers of MSA artifacts appear to have been anatomically modern or near modern. This suggests that modern people evolved in Africa and subsequently spread to replace the Neanderthals and other equally archaic Eurasians. The failure of early modern or nearmodern humans to supplant their non-modern contemporaries early on probably reflects the lack of any behavioral difference. Replacement became possible only when modern anatomy was supplemented by fully modern behavior, or perhaps more precisely by the fully modern ability to wield culture in adaptation. The emergence of this ability is probably signaled by the appearance of the Later Stone Age (LSA), between perhaps 50,000 and 40,000 years ago. Unfortunately, this was an interval when adverse climatic conditions had dramatically reduced human populations throughout southern Africa, and the early LSA is therefore very poorly known. The reasons for its emergence also remain unclear, but the most economic hypothesis may be one that reflects the last in a long series of highly adaptive neurological advances. Searches for evidence to illuminate the appearance of the LSA should perhaps concentrate on east Africa, where climatic conditions apparently permitted denser human populations in the crucial interval.

6. Only the later LSA, postdating 22,000 years, is well documented in southern Africa. Microlithic artifacts are prominent in most but not all later LSA assemblages, which are unified more by the absence of typical MSA pieces (especially flakes and flake-blades from carefully prepared cores) and by the presence of well-made formal bone artifacts and items of personal adornment. Faunal remains imply that later LSA people hunted and foraged much more effectively than their MSA predecessors, probably in large part because they possessed a much more sophisticated hunting-foraging technology, including especially the bow and arrow. In virtually every detectable respect, LSA people were advanced over MSA people in the same way that all historically observed stone-tool–using people were. However, LSA people were still highly vulnerable to the vagaries of nature, and their populations dropped dramatically when climatic conditions deteriorated. The LSA persisted until the dramatic expansion of Bantu-speaking Iron Age people into southern Africa beginning about 2000 years ago. The Iron Age colonists essentially replaced LSA populations in the moister northern and eastern portions of the subcontinent where Iron Age mixed farming was practical. Elsewhere, at about the same time, the LSA was altered by the addition of pottery and in some regions by the introduction of domestic stock. However, it survived and was terminated only by European contact, especially after the establishment of the first permanent European settlement at Cape Town in 1652.

ACKNOWLEDGMENTS

I thank the National Science Foundation and the South African Museum for supporting my research on southern African prehistory. K. Cruz-Uribe, R. G. Milo, and T. P. Volman kindly commented on a draft of the manuscript.

LITERATURE CITED

Andrews, P. A. (1984) On the characters that define *Homo erectus*. *Cour. Forsch. Inst. Senckenberg* 69:167-178.

Andrews, P. A. (1992) An ape from the south. *Nature* 356:106.

Avery, D. M. (1982) Micromammals as palaeoenvironmental indicators and an interpretation of the Late Quaternary in the southern Cape Province, South Africa. *Ann. S. Afr. Mus.* 85:183-374.

Avery, D. M. (1987) Late Pleistocene coastal environment of the southern Cape Province of South Africa: Micromammals from Klasies River Mouth. *J. Archaeol. Sci.* 14:405-421.

Avery, D. M. (1992) The environment of early modern humans at Border Cave, South Africa: Micromammalian evidence. *Palaeogeog., Palaeoclimatol., Palaeoecol.* 91:71-87.

Avery, G. (1990) Avian fauna, palaeoenvironments and palaeoecology in the Late Quaternary of the western and southern Cape, South Africa. University of Cape Town, unpublished Ph.D. Dissertation.

Barbetti, M. (1986) Traces of fire in the archaeological record, before one million years ago? *J. Hum. Evol.* 15:771-781.

Beaumont, P. B. (1973) Border Cave—a progress report. *S. Afr. J. Sci.* 69:41-46.

Beaumont, P. B. (1980) On the age of Border Cave hominids 1-5. *Palaeont. Afr.* 23:21-33.

Beaumont, P. B. (1982) Aspects of the northern Cape Pleistocene Project. *Palaeoecol. Afr.* 15:41-44.

Beaumont, P. B., de Villiers, H., and Vogel, J. C. (1978) Modern man in sub-Saharan Africa prior to 49,000 B. P.: a review and evaluation with particular reference to Border Cave. *S. Afr. J. Sci.* 74:409-419.

Beaumont, P. B., van Zinderen Bakker, E. M., and Vogel, J. C. (1984) Environmental changes since 32 000 BP at Kathu Pan, northern Cape. In JC Vogel (ed.): *Late Cainozoic Palaeoclimates of the Southern Hemisphere*. Rotterdam: Balkema, pp. 329-338.

Binford, L. R. (1984) *Faunal Remains from Klasies River Mouth*. New York: Academic Press.

Binneman, J., and Deacon, J. (1986) Experimental determination of use wear on stone adzes from Boomplaas Cave, South Africa. *J. Archaeol. Sci.* 13:219-228.

Blumenschine, R. J. (1987) Characteristics of an early hominid scavenging niche. *Curr. Anthropol.* 28:383-407.

Brain, C. K. (1958) The Transvaal Ape–Man-Bearing Cave deposits. *Transv. Mus. Mem.* 11:1-131.

Brain, C. K. (1967) Hottentot food remains and their bearing on the interpretation of fossil bone assemblages. *Sci. Papers Namib Desert Res. Station* 32:1-11.

Brain, C. K. (1969) The contribution of Namib Desert Hottentots to an understanding of australopithecine bone accumulations. *Sci. Papers Namib Desert Res. Station* 39:13-22.

Brain, C. K. (1981) *The Hunters or the Hunted? An Introduction to African Cave Taphonomy*. Chicago: University of Chicago Press.

Brain, C. K. (1984) The Terminal Miocene Event: A critical environmental and evolutionary episode? In J. C. Vogel (ed.): *Late Cainozoic Palaeoclimates of the Southern Hemisphere*. Rotterdam: A. A. Balkema, pp. 491-498.

Brain, C. K. (1988) New information from the Swartkrans cave of relevance to "robust" australopithecines. In F. E. Grine (ed.): *Evolutionary History of the "Robust" Australopithecines*. New York: Aldine de Gruyter, pp. 311-316.

Brain, C. K., Churcher, C. S., Clark, J. D., Grine, F. E., Shipman, P., Susman, R. L., Turner, A., and Watson, V. (1988) New evidence of early hominids, their culture and environment from the Swartkrans cave, South Africa. *S. Afr. J. Sci.* 84:828-835.

Brain, C. K., and Sillen, A. (1988). Evidence from the Swartkrans cave for the earliest use of fire. *Nature* 336:464-466.

Brink, J. S. (1987) The archaeozoology of Florisbad, Orange Free State. *Mem. Nas. Mus. (Bloemfontein)* 24:1-151.

Brink, J. S., and Deacon, H. J. (1982) A study of a last interglacial shell midden and bone accumulation at Herolds Bay, Cape Province, South Africa. *Palaeoecol. Afr.* 15:31-40.

Brooks, A. S., Crowell, A. L., and Yellen, J. E. (1980) Gi: a Stone Age archaeological site in the northern Kalahari Desert, Botswana. In R. E. F. Leakey and B. A. Ogot (eds.): *Proceedings of the Eighth Panafrican Congress of Prehistory and Quaternary Studies* (Nairobi, 1977). Nairobi: The International Louis Leakey Memorial Institute for African Prehistory, pp. 304-309.

Brooks A. S., Hare, P. E., Kokis, J. E., Miller, G. H., Ernst, R. D., and Wendorf, F. (1990) Dating Pleistocene archeological sites by protein diagenesis in ostrich eggshell. *Science* 248:60-64.

Brooks, A. S., and Wood, B. A. (1990) The Chinese side of the story. *Nature* 344:288-289.

Broom, R. (1938) The Pleistocene anthropoid apes of South Africa. *Nature* 142:377-379.

Broom, R. (1949) Another new type of fossil ape-man. *Nature* 163:57.

Brown, A. J. V. (1988) The faunal remains from Kalkbank, northern Transvaal. *Palaeoecol. Afr.* 19:205-212.

Brown, A. J. V., and Verhagen, B. T. (1985). Two *Antidorcas bondi* individuals from the Late Stone Age site of Kruger

Cave 35/83, Olifantsnek, Rustenburg District, South Africa. *S. Afr. J. Sci.* 81:102.
Brown, F. H., and Feibel, C. S. (1988) "Robust" hominids and Plio-Pleistocene paleogeography of the Turkana Basin, Kenya and Ethiopia. In F. E. Grine (ed.): *Evolutionary History of the "Robust" Australopithecines.* New York: Aldine de Gruyter, pp. 325-341.
Buchanan, W. F., Hall, S. L., Henderson, J., Olivier, A., Pettigrew, J. M., Parkington, J. E., and Robertshaw, P. T. (1978) Coastal shell middens in the Paternoster area, southwestern Cape. *S. Afr. Archaeol. Bull.* 33:89-93.
Bunn, H. T. (1981) Archaeological evidence for meat-eating by Plio-Pleistocene hominids from Koobi Fora and Olduvai Gorge. *Nature* 291:574-577.
Bunn, H. T., and Kroll, E. M. (1986) Systematic butchery by Plio/Pleistocene hominids at Olduvai Gorge, Tanzania. *Curr. Anthropol.* 5:431-452.
Butzer, K. W. (1973) A provisional interpretation of the sedimentary sequence from Montagu Cave (Cape Province), South Africa. *Univ. Calif. Anthropol. Rec.* 28:89-92.
Butzer, K. W. (1974) Geo-archaeological interpretation of Acheulian calc-pan sites at Doornlaagte and Rooidam (Kimberley, South Africa). *J. Archaeol. Sci.* 1:1-125.
Butzer, K. W. (1984) Archeogeology and Quaternary environment in the interior of southern Africa. In R. G. Klein (ed.): *Southern African Prehistory and Paleoenvironments.* Rotterdam: A. A. Balkema, pp. 1-64.
Butzer, K. W., Clark, J. D., and Cooke, H. B. S. (1974) The geology, archaeology, and fossil mammals of the Cornelia Beds, *O. F. S. Mem. Nas. Mus. (Bloemfontein)* 9:1-84.
Caspari, R., and Wolpoff, M. H. (1990) The morphological affinities of the Klasies River Mouth skeletal remains. *Am. J. Phys. Anthropol.* 81:203.
Cerling, T. E., Quade, J., Ambrose, S. H., and Sikes, N. E. (1991) Fossil soils, grasses, and carbon isotopes from Fort Ternan, Kenya: a grassland or a woodland. *J. Hum. Evol.* 21:295-306.
Clark, J. D. (1955) A note on a wooden implement from the level of Peat 1 at Florisbad, Orange Free State. *Nav. Nas. Mus. (Bloemfontein)* 1:135-140.
Clark, J. D. (1969) *Kalambo Falls Prehistoric Site*, Vol. 1. Cambridge: Cambridge University Press.
Clark, J. D. (1971) Human behavioral differences in southern Africa during the later Pleistocene. *Am. Anthropol.* 73:1211-1236.
Clark, J. D. (1974) *Kalambo Falls Prehistoric Site*, Vol. 2. Cambridge: Cambridge University Press.
Clark, J. D. (1988) The Middle Stone Age of East Africa and the beginnings of regional identity. *J. World Prehist.* 2:235-305.
Clark, J. D., and Harris, J. W. K. (1985) Fire and its roles in early hominid lifeways. *Afr. Archaeol. Rev.* 3:3-27.
Clark, W. E. Le Gros (1967) *Man Apes or Ape Men?* New York: Holt, Rinehart and Winston.
Clarke, R. J. (1985a) *Australopithecus* and early *Homo* in southern Africa. In E. Delson (ed.): *Ancestors: the Hard Evidence.* New York: Alan R. Liss, pp. 171-177.
Clarke, R. J. (1985b) Early Acheulean with *Homo habilis* at Sterkfontein. In P. V. Tobias (ed.): *Hominid Evolution: Past, Present and Future.* New York: Alan R. Liss, pp. 287-298.
Clarke, R. J. (1985c) A new reconstruction of the Florisbad cranium with notes on the site. In E. Delson (ed.): *Ancestors: The Hard Evidence.* New York: Alan R. Liss, pp. 301-305.
Clarke, R. J. (1988) Habiline handaxes and paranthropine pedigree at Sterkfontein. *World Archaeol.* 20:1-12.
Conroy, G. C., Pickford, M., Senut, B., and Van Couvering, J. (1992a) *Otavipithecus namibiensis*, first Miocene hominoid discovered from southern Africa (Berg Aukas, Namibia). *Amer. J. Phys. Anthropol.* Suppl. 14:62.
Conroy, G. C., Pickford, M., Senut, B., Van Couvering, J., and Mein, P. (1992b). *Otavipithecus namibiensis*, first Miocene hominoid from southern Africa. *Nature* 356:144-148.
Cooke, C. K. (1963) Report on excavations at Pomongwe and Tshangula Caves, Matopos Hills, Southern Rhodesia. *S. Afr. Archaeol. Bull.* 18:73-151.
Cooke C. K. (1978) The Redcliff Stone Age site, Rhodesia. *Occ. Pap. Nat. Mus. Mon. Rhodesia A* 1:45-73
Cooke, H. B. S. (1949) Fossil mammals of the Vaal River deposits. *Geol. Surv. S. Afr. Mem.* 35(3):1-109.
Cooke, H. B. S. (1962). Notes on the faunal material from the Cave of Hearths and Kalkbank. In R. J. Mason: *Prehistory of the Transvaal.* Johannesburg: Witwatersrand University Press, pp. 447-453.
Cooke, H. B. S. (1963). Pleistocene mammal faunas of Africa, with particular reference to southern Africa. In F. C. Howell, and F. Bourliére (eds.): *African Ecology and Human Evolution.* Chicago: Aldine, pp. 65-116.
Cooke, H. B. S. (1978) Faunal evidence for the biotic setting of early African hominids. In C. Jolly (ed.): *Early Hominids of Africa.* New York: St. Martin's Press, pp. 267-281.
Cruz-Uribe, K. (1983) The mammalian fauna from Redcliff Cave, Zimbabwe. *S. Afr. Archaeol. Bull.* 38:7-16.
Cruz-Uribe, K. (1991) Distinguishing hyena from hominid bone accumulations. *J. Field Archaeol.* 18:467-486.
Cruz-Uribe, K., and Klein, R. G. (1982) Faunal remains from some Middle and Later Stone Age archaeological sites in South West Africa. *J. SWA Wissenschaftl. Ges.* 36-37:91-114.
Dart, R. (1957) The osteodontokeratic culture of *Australopithecus africanus. Mem. Transv. Mus.* 10 :1-105.
Day, M. H. (1986) *Guide to Fossil Man* (4th ed.). Chicago: University of Chicago Press.
Deacon, H. J. (1970) The Acheulian occupation at Amanzi Springs, Uitenhage District, Cape Province. *Ann. Cape Prov. Mus.* 8:89-189.
Deacon, H. J. (1975) Demography, subsistence and culture during the Acheulean in southern Africa. In K. W. Butzer, and G. L l. Isaac (eds.): *After the Australopithecines.* The Hague: Mouton, pp. 543-569.
Deacon, H. J. (1976) Where hunters gathered: A study of Holocene stone age people in the eastern Cape. *South Afr. Archaeol. Soc. Monogr. Ser.* 1:1-231.

Deacon, H. J. (1979) Excavations at Boomplaas Cave—a sequence through the Upper Pleistocene and Holocene in South Africa. *World Archaeol.* 10:241-257.

Deacon, H. J. (1989). Late Pleistocene palaeoecology and archaeology in the southern Cape, South Africa. In P. Mellars, and C. Stringer (eds.): *The Human Revolution: Behavioural and Biological Perspectives in the Origins of Modern Humans.* Edinburgh: Edinburgh University Press, pp. 547-564.

Deacon, H. J., Deacon, J., and Brooker, M. (1976) Four painted stones from Boomplaas Cave, Oudtshoorn District. *S. Afr. Archaeol. Bull.* 31:141-145.

Deacon, H. J., Deacon, J., Brooker, M., and Wilson, M. L. (1978) The evidence for herding at Boomplaas Cave in the southern Cape, South Africa. *S. Afr. Archaeol. Bull.* 33:39-65.

Deacon, H. J., Deacon, J., Scholtz, A., Thackeray, J. F., and Brink, J. S. (1984) Correlation of palaeoenvironmental data from the Late Pleistocene and Holocene deposits at Boomplaas Cave, southern Cape. In J. C. Vogel (ed.): *Late Cainozoic Palaeoclimates of the Southern Hemisphere.* Rotterdam: A. A. Balkema, pp. 339-351.

Deacon, H. J., and Geleijnse, V. B. (1988) The stratigraphy and sedimentology of the Main Site sequence, Klasies River, South Africa. *S. Afr. Archaeol. Bull.* 43:5-14.

Deacon, H. J., and Deacon, J. (1980) The hafting, function and distribution of small convex scrapers with an example from Boomplaas Cave. *S. Afr. Archaeol. Bull.* 35:31-37.

Deacon, H. J., and Thackeray, J. F. (1984) Late Pleistocene environmental changes and implications for the archaeological record in southern Africa. In J. C. Vogel (ed.): *Late Cainozoic Palaeoclimates of the Southern Hemisphere.* Rotterdam: A. A. Balkema, pp. 375-390.

Deacon, J. (1976) Report on stone artefacts from Duinefontein 2, Melkbosstrand. *S. Afr. Archaeol. Bull.* 31:21-25.

Deacon, J. (1978) Changing patterns in the late Pleistocene/early Holocene prehistory of southern Africa as seen from the Nelson Bay Cave stone artifact sequence. *Quat. Res.* 10:84-111.

Deacon, J. (1984) Later Stone Age people and their descendants in southern Africa. In R. G. Klein (ed.): *Southern African Prehistory and Paleoenvironments.* Rotterdam: A. A. Balkema, pp. 221-328.

Deacon, J. (1990a) Weaving the fabric of Stone Age research in southern Africa. In P. Robertshaw (ed.): *A History of African Archaeology.* London: James Currey, pp. 39-58.

Deacon, J. (1990b) Changes in the archaeological record in South Africa at 18 000 BP. In C. Gamble and O. Soffer (eds.): *The World at 18 000 BP.* London: Unwin Hyman, pp. 170-188.

Delson, E. (1984) Cercopithecid biochronology of the African Plio-Pleistocene: Correlation among eastern and southern hominid-bearing localities. *Cour. Forsch. Inst. Senckenberg* 69:199-218.

Delson, E. (1988) Chronology of South African australopithecine site units. In F. E. Grine (ed.): *Evolutionary History of the "Robust" Australopithecines.* New York: Aldine de Gruyter, pp. 317-324.

Elphick, R. (1977) *Kraal and castle: Khoikhoi and the founding of white South Africa.* New Haven: Yale University Press.

Goddard J (1970) Age criteria and vital statistics of a black rhinoceros population. *E. Afr. Wildl. J.* 8:105-121.

Goodwin, A. J. H., and van Riet Lowe, C. (1929) The Stone Age Cultures of South Africa. *Ann. S. Afr. Mus.* 27:1-289.

Gowlett, J. A. J. (1988) A case of Developed Oldowan in the Acheulean? *World Archaeol.* 20:13-26.

Grine, F. E. (1981) Trophic differences between 'gracile' and 'robust' australopithecines: A scanning electron microscope analysis of occlusal events. *S. Afr. J. Sci.* 77:203-220.

Grine, F. E. (1985) Australopithecine evolution: The deciduous dental evidence. In E. Delson (ed.): *Ancestors: The Hard Evidence.* New York: Alan R. Liss, pp. 153-167.

Grine, F. E. (1986) Dental evidence for dietary differences in *Australopithecus* and *Paranthropus*: A quantitative analysis of permanent molar microwear. *J. Hum. Evol.* 15:783-822.

Grine, F. E. (1988) Evolutionary history of the "robust" australopithecines: A summary and historical perspective. In F. E. Grine (ed.): *Evolutionary History of the "Robust" Australopithecines.* New York: Aldine de Gruyter, pp. 509-520.

Grine, F. E., and Klein, R. G. (1985) Pleistocene and Holocene human remains from Equus Cave, South Africa. *Anthropology* 8:55-98.

Grine, F. E., and Klein, R. G. (1993) Late Pleistocene human remains from the Sea Harvest site, Saldanha Bay, South Africa. *South. Afr. J. Sci.* 89:145-152.

Grine, F. E., Klein, R. G. , and Volman, T. P. (1991). Dating, archaeology and human fossils from the Middle Stone Age levels of Die Kelders, South Africa. *J. Hum. Evol.* 21:363-395.

Grün, R., Beaumont, P. B., and Stringer, C. B. (1990a). ESR dating evidence for early modern humans at Border Cave in South Africa. *Nature* 344:537-539.

Grün, R., Shackleton, N. J., and Deacon, H. J. (1990b). Electron-spin-resonance dating of tooth enamel from Klasies River Mouth Cave. *Curr. Anthropol.* 31:427-432.

Hall, M. (1987) *The Changing Past: Farmers, Kings and Traders in Southern Africa, 200-1860.* Cape Town: David Philip.

Hamilton, W. R., and Van Couvering, J. A. (1977) Lower Miocene mammals of South West Africa. *Namib. Bull. (Suppl. 2, Transv. Mus. Bull.)*:9-11.

Harland, W. B., Armstrong, R. L., Cox, A. V., Craig, L. E., Smith, A. G., and Smith, D. G. (1990) *A Geologic Time Scale 1989.* Cambridge: Cambridge University Press.

Harris, J. M., Brown, F. H., Leakey, M. G., Walker, A. C., and Leakey, R. E. (1988). Pliocene and Pleistocene hominid-bearing sites from west of Lake Turkana, Kenya. *Science* 239:27-33.

Harris, J. W. K. (1983) Cultural beginnings: Plio-Pleistocene archaeological occurrences from the Afar, Ethiopia. *Afr. Archaeol. Rev.* 1:3-31.

Hedges, S. B., Kumar, S., Tamura, K., and Stoneking, M. (1992) Human origins and analysis of mitochondrial DNA sequences. *Science* 255:737-739.

Helgren, D. M. (1977) Geological context of the Vaal River faunas. *S. Afr. J. Sci.* 73:303-307.
Helgren, D. M. (1978) Acheulian settlement along the Lower Vaal River, South Africa. *J. Archaeol. Sci.* 5:39-60.
Helgren, D. M. (1979) River of diamonds: An alluvial history of the Lower Vaal Basin, South Africa. *Univ. Chicago Dept. Geog. Res. Pap.* 185:1-389.
Helgren, D. M., and Brooks, A. S. (1983) Geoarchaeology at Gi, a Middle Stone Age and Later Stone Age site in the Northwest Kalahari. *J. Archaeol. Sci.* 10:181-197.
Hendey, Q. B. (1974) The Late Cenozoic Carnivora of the South-Western Cape Province. *Ann. S. Afr. Mus.* 63:1-369.
Hendey, Q. B. (1978a) Late Tertiary Hyaenidae from Langebaanweg, South Africa, and their relevance to the phylogeny of the family. *Ann. S. Afr. Mus.* 76 265-297.
Hendey, Q. B. (1978b) Late Tertiary Mustelidae (Mammalia, Carnivora) from Langebaanweg, South Africa. *Ann. S. Afr. Mus.* 76:329-357.
Hendey, Q. B. (1978c) The age of the fossils from Baard's Quarry, Langebaanweg, South Africa. *Ann. S. Afr. Mus.* 75:1-24.
Hendey, Q. B. (1980) *Agriotherium* (Mammalia, Ursidae) from Langebaanweg, South Africa, and relationships of the genus. *Ann. S. Afr. Mus.* 81:1-109.
Hendey, Q. B. (1981a) Palaeoecology of the late Tertiary fossil occurrences in "E" Quarry, Langebaanweg, South Africa, and a reinterpretation of their geological context. *Ann. S. Afr. Mus.* 84:1-104.
Hendey, Q. B. (1981b) Geological succession at Langebaanweg, Cape Province, and global events of the late Tertiary. *S. Afr. J. Sci.* 77:33-38.
Hendey, Q. B. (1982) *Langebaanweg: A Record of Past Life.* Cape Town: The South African Museum.
Hendey, Q. B. (1984) Southern African late Tertiary vertebrates. In R. G. Klein (ed.): *Southern African Prehistory and Paleoenvironments.* Rotterdam: A. A. Balkema, pp. 81-106.
Holloway, R. L. (1975) Early hominid endocasts: Volumes, morphology, and significance for hominid evolution. In R. H. Tuttle (ed.): *Primate Functional Morphology and Evolution.* The Hague: Mouton, pp. 393-415.
Holloway, R. L. (1983) Human brain evolution: A search for units, models and synthesis. *Canad. J. Anthropol.* 3:215-230.
Howell, F. C. (1960) European and Northwest African Middle Pleistocene hominids. *Curr. Anthropol.* 1:195-231.
Howell, F. C. (1978) Hominidae. In V. J. Maglio and H. B. S. Cooke (eds.): *Evolution in African Mammals.* Cambridge, Mass.: Harvard University Press, pp. 154-248.
Howell, F. C. (1988). Foreword. In F. E. Grine (ed.): *Evolutionary History of the "Robust" Australopithecines.* New York: Aldine de Gruyter, pp. xi-xv.
Inskeep, R. R. (1987) *Nelson Bay Cave, Cape Province, South Africa. The Holocene Levels.* Oxford: British Archaeological Reports, International Series 357: Vols. 1 and 2.
Isaac, G. L l. (1975) Stratigraphy and cultural patterns in East Africa during the middle ranges of Pleistocene time. In K. W. Butzer, and G. L l. Isaac (eds.): *After the Australopithecines.* The Hague: Mouton, pp. 495-542.
Isaac, G. L l. (1984) The archaeology of human origins: Studies of the Lower Pleistocene in East Africa 1971-1981. *Advances in World Archaeology* 3:1-87.
Isaac, G. L l., and Harris, J. W. K. (1978) Archaeology. In M. G. Leakey and R. E. F. Leakey (eds.): *Koobi Fora Research Project,* Vol. 1. Oxford: Clarendon, pp. 64-85.
Jaeger, J. J. (1975) The mammalian faunas and hominid fossils of the Middle Pleistocene in the Maghreb. In K. W. Butzer, and G. L l. Isaac (eds.): *After the Australopithecines.* The Hague: Mouton, pp. 399-410.
Johanson, D. C., and White, T. D. (1979). A systematic assessment of early African hominids. *Science* 202:321-330.
Jones, D. L., Brock, A., and McFadden, P. L. (1986) Palaeomagnetic results from the Kromdraai and Sterkfontein hominid sites. *S. Afr. J. Sci.* 82:160-163.
Jones, P. R. (1980) Experimental butchery with modern stone tools and its relevance for Palaeolithic archaeology. *World Archaeol.* 12:153-175.
Jungers, W. L. (1988) New estimates of body size in australopithecines. In F. E. Grine (ed.): *Evolutionary History of the "Robust" Australopithecines.* New York: Aldine de Gruyter, pp. 115-125.
Kaplan, J. (1990) The Umhlatuzana Rock Shelter sequence: 100 000 years of Stone Age history. *Natal Mus. J. Hum.* 2:1-94.
Kappelman, J. (1991) The paleoenvironment of *Kenyapithecus* at Fort Ternan. *J. Hum. Evol.* 20:95-129.
Kay, R. F., and Grine, F. E. (1988) Tooth morphology, wear and diet in *Australopithecus* and *Paranthropus* from southern Africa. In F. E. Grine (ed.): *Evolutionary History of the "Robust" Australopithecines.* New York: Aldine de Gruyter, pp. 427-447.
Keller, C. M. (1973) Montagu Cave in prehistory. *Univ. Calif. Anthropol. Rec. 28*:1-150.
Kennedy, G. E. (1991) On the autapomorphic traits of *Homo erectus. J. Hum. Evol.* 20:375-412.
Keyser, A. W., and Martini, J. E. J. (1990). Haasgat: A new Plio-Pleistocene fossil occurrence. *Palaeoecol. Afr.* 21:119-129.
Klein, R. G. (1975) Paleoanthropological implications of the non-archeological bone assemblage from Swartklip 1, south-western Cape Province, South Africa. *Quat. Res.* 5:275-288.
Klein, R. G. (1976a) The mammalian fauna of the Klasies River Mouth sites, southern Cape Province, South Africa. *S. Afr. Archaeol. Bull.* 31:75-98.
Klein, R. G. (1976b) A preliminary report on the Duinefontein 2 "Middle Stone Age" open-air site (Melkbosstrand, South-Western Cape Province, South Africa). *S. Afr. Archaeol. Bull.* 31:12-20.
Klein, R. G. (1977) The mammalian fauna from the Middle and Later Stone Age (later Pleistocene) levels of Border Cave, Natal Province, South Africa. *S. Afr. Archaeol. Bull.* 32:14-27
Klein, R. G. (1978a) The fauna and overall interpretation of the "Cutting 10" Acheulean site at Elandsfontein (Hopefield), southwestern Cape Province, South Africa. *Quat. Res.* 10:69-83.

Klein, R. G. (1978b) Preliminary results of the analysis of the mammalian fauna from the Redcliff Stone Age cave site, Rhodesia. *Occ. Pap. Nat. Mus. Mon. Rhodesia* 4(2):74-80.

Klein, R. G. (1982) Patterns of ungulate mortality and ungulate mortality profiles from Langebaanweg (early Pliocene) and Elandsfontein (middle Pleistocene), south-western Cape Province, South Africa. *Ann. S. Afr. Mus.* 90:49-94.

Klein, R. G. (1983) Palaeoenvironmental implications of Quaternary large mammals in the Fynbos Biome. *S. Afr. Natl. Sci. Progr. Rep.* 75:116-138.

Klein, R. G. (1984) Mammalian extinctions and stone age people in Africa. In P. S. Martin, and R. G. Klein (eds.): *Quaternary Extinctions: A Prehistoric Revolution.* Tucson: University of Arizona Press, pp. 553-573.

Klein, R. G. (1986) The prehistory of Stone Age herders in the Cape Province of South Africa. *S. Afr. Archaeol. Soc. Goodwin Ser.* 5:5-12.

Klein, R. G. (1988a) The causes of "robust" australopithecine extinction. In F. E. Grine (ed.): *Evolutionary History of the "Robust" Australopithecines.* New York: Aldine de Gruyter, pp. 499-505.

Klein, R. G. (1988b) The archaeological significance of animal bones from Acheulean sites in southern Africa. *Afr. Archaeol. Rev.* 6:3-26.

Klein, R. G. (1989a) *The Human Career.* Chicago: University of Chicago Press.

Klein, R. G. (1989b) Why does skeletal part representation differ between smaller and larger bovids at Klasies River Mouth and other archeological sites? *J. Archaeol. Sci.* 16:363-381

Klein, R. G., and Cruz-Uribe, K. (1987) Large mammal and tortoise bones from Elands Bay Cave and nearby sites, Western Cape Province, South Africa. *British Archaeological Reports International Series* 332:132-163

Klein, R. G., and Cruz-Uribe, K. (1989) Faunal evidence for prehistoric herder-forager activities at Kasteelberg, Vredenburg Peninsula, western Cape Province, South Africa. *S. Afr. Archaeol. Bull.* 44:82-97.

Klein, R. G., and Cruz-Uribe, K. (1991) The bovids from Elandsfontein, South Africa, and their implications for the age, paleoenvironment, and origins of the site. *Afr. Archaeol. Rev.* 9:21-79.

Klein, R. G., Cruz-Uribe, K., and Beaumont, P. B. (1991) Environmental, ecological, and paleoanthropological implications of the Late Pleistocene mammalian fauna from Equus Cave, northern Cape Province, South Africa. *Quat. Res.* 36:94-119.

Kuman, K., and Clarke, R. J. (1986) Florisbad—new investigations at a Middle Stone Age hominid site in South Africa. *Geoarchaeology* 1:103-125.

Laws, R. M. (1966) Age criteria for the African elephant, *Loxodonta africana africana. E. Afr. Wildl. J.* 4:1-37.

Leakey, M. D. (1971) *Olduvai Gorge: Excavations in Beds I and II, 1960-1963.* Cambridge: Cambridge University Press.

Leakey, M. D. (1975) Cultural patterns in the Olduvai sequence. In K. W. Butzer, and G. L l. Isaac (eds.): *After the Australopithecines.* The Hague: Mouton, pp. 476-493.

Leakey, M. D., and Hay, R. L. (1982) The chronological position of the fossil hominids of Tanzania. In M. A. de Lumley (ed.): *L'Homo erectus et la place de l'homme de Tautavel parmi les hominidés fossiles.* Nice: 1er Congrès International de Paléontologie Humaine, pp. 753-765.

Maggs, T. (1984) The Iron Age south of the Zambezi. In R. G. Klein (ed.): *Southern African Prehistory and Paleoenvironments.* Rotterdam: A. A. Balkema, pp. 329-360

Maguire, J. M. (1985) Recent geological, stratigraphic and palaeontological studies at Makapansgat Limeworks. In P. V. Tobias (ed.): *Hominid Evolution: Past, Present and Future.* New York: Alan R. Liss, pp. 151-164.

Maguire, J. M., Pemberton, D., and Collett, M. H. (1980). The Makapansgat Limeworks grey breccia: Hominids, hyaenas, hystricids or hillwash? *Palaeont. Afr.* 23:75-98.

Malan, B. D., and Wells, L. H. (1943). A further report on the Wonderwerk Cave, Kuruman. *S. Afr. J. Sci.* 40:258-270.

Mason, R. J. (1962) *Prehistory of the Transvaal.* Johannesburg: Witwatersrand University Press.

Mason, R. J. (1967) Prehistory as a science of change: new research in the South African interior. *Occ. Pap. Archaeol. Res. Unit. Univ. Witwatersrand* 1:1-19.

Mason, R. J. (1985) Sterkfontein and the Klipplaatdrif Gravels. In P. V. Tobias (ed.): *Hominid Evolution: Past, Present and Future.* New York: Alan R. Liss, pp. 299-301.

Mason, R. J. (1988a) Cave of Hearths, Makapansgat, Transvaal. *Occ. Pap. Archaeol. Res. Unit. Univ. Witwatersrand* 21:1-711.

Mason R. J. (1988b) A Middle Stone Age faunal site at Kalkbank, northern Transvaal: Archaeology and interpretation. *Palaeoecol. Afr.* 19:201-203.

McFadden, P. L., and Brock, A. (1984). Magnetostratigraphy at Makapansgat. *S. Afr. J. Sci.* 80:482-483.

McHenry, H. M. (1988) New estimates of body weight in early hominids and their significance to encephalization and megadontia in "robust" australopithecines. In F. E. Grine (ed.): *Evolutionary History of the "Robust" Australopithecines.* New York: Aldine de Gruyter, pp. 133-148.

McHenry, H. M. (1991a) Femoral lengths and stature in Plio-Pleistocene hominids. *Am. J. Phys. Anthropol.* 85:149-158.

McHenry, H. M. (1991b). Petite bodies of the "Robust" australopithecines. *Am. J. Phys. Anthropol.* 86:445-454.

McHenry, H. M., and Skelton, R. R. (1985) Is *Australopithecus africanus* ancestral to *Homo*? In P. V. Tobias (ed.): *Hominid Evolution: Past, Present and Future.* New York: Alan R. Liss, pp. 221-226.

McKee, J. K. (1991). Palaeo-ecology of the Sterkfontein hominids: a review and synthesis. *Palaeont. Afr.* 28:41-51.

Parkington, J. E. (1981) The effects of environmental change on the scheduling of visits to the Eland's Bay Cave, Cape Province, S. A. In I. Hodder, G. Isaac, and N. Hammond (eds.): *Patterns of the past.* Cambridge: University of Cambridge Press, pp. 341-349.

Parkington, J. E. (1984) Soaqua and Bushmen: Hunters and robbers. In C. Schrire (ed.): *Past and Present in Hunter-Gatherer Studies.* Orlando, FL: Academic Press, pp. 151-174.

Parkington, J. E. (1987) Changing views of prehistoric settlement in the western Cape. *British Archaeological Reports International Series* 332:4-23.

Parkington, J. E. (1990a) A critique of the consensus view on the age of Howieson's Poort assemblages in South Africa. In P. Mellars (ed.): *The Emergence of Modern Humans: An Archaeological Perspective.* Edinburgh: Edinburgh University Press, pp. 34-55.

Parkington, J. E. (1990b) A view from the south: Southern Africa before, during, and after the Last Glacial Maximum. In C. Gamble and O. Soffer (eds.): *The World at 18 000 BP.* London: Unwin Hyman, pp. 214-228.

Partridge, T. C. (1982) The chronological positions of the fossil hominids of southern Africa. In M. A. de Lumley (ed.): *L'Homo erectus et la place de l'homme de Tautavel parmi les hominidés fossiles.* Nice: 1er Congrès International de Paléontologie Humaine, pp. 617-675.

Partridge, T. C. (1985a) The palaeoclimatic significance of Cainozoic terrestrial stratigraphic and tectonic evidence from southern Africa: A review. *S. Afr. J. Sci.* 81: 245-247.

Partridge, T. C. (1985b) The Klipplaatdrif gravels: Morphology, age and depositional environment with special reference to comparative evidence from the Sterkfontein Extension Site. In P. V. Tobias (ed.): *Hominid Evolution: Past, Present and Future.* New York: Alan R. Liss, pp. 303-309.

Partridge, T. C. (1986) Palaeoecology of the Pliocene and Lower Pleistocene hominids of Southern Africa: How good is the chronological and palaeoenvironmental evidence? *S. Afr. J. Sci.* 82:80-83.

Partridge, T. C., Bollen, J. F., Tobias, P. V., and McKee, J. K. (1991) New light on the provenance of the Taung skull. *S. Afr. J. Sci.* 87:340-341.

Phillipson D. W. (1985) *African Archaeology.* Cambridge: Cambridge University Press.

Pickford, M. (1981) Preliminary Miocene biostratigraphy of western Kenya. *J. Hum. Evol.* 10:73-97.

Pickford, M. (1986) The geochronology of Miocene higher primate faunas of East Africa. In J. G. Else, and P. C. Lee (eds.): *Primate Evolution.* Cambridge: Cambridge University Press, pp. 19-33.

Pickford, M., Fernandes, T., and Aco, S. (1990) Nouvelles découvertes de remplissages de fissures à primates dans le "Planalto da Humpata", Huíla, Sud de l'Angola *C. R. Acad. Sci., Série II,* 310:843-848.

Plug, I. (1981) Some research results on the late Pleistocene and early Holocene deposits of Bushman Rock Shelter, eastern Transvaal. *S. Afr. Archaeol. Bull.* 36:14-21.

Potts, R. B. (1988) *Early Hominid Activities at Olduvai.* New York: Aldine de Gruyter.

Potts, R. B. (1991) Why the Oldowan? Plio-Pleistocene toolmaking and the transport of resources. *J. Anthropol. Res.* 47:153-176.

Potts, R. B., and Shipman, P. (1981) Cutmarks made by stone tools on bones from Olduvai Gorge, Tanzania. *Nature* 291:577-580.

Power, J. H. (1955) Power's site, Vaal River. *S. Afr. Archaeol. Bull.* 10:96-101.

Prentice, M. L., and Denton, G. H. (1988) The deep-sea oxygen isotope record, the global ice sheet system and hominid evolution. In F. E. Grine (ed.): *Evolutionary History of the "Robust" Australopithecines.* New York: Aldine de Gruyter, pp. 383-403.

Rightmire, G. P. (1979) Implications of Border Cave skeletal remains for later Pleistocene human evolution. *Curr. Anthropol.* 20:23-35.

Rightmire, G. P. (1984) The fossil evidence for hominid evolution in southern Africa. In R. G. Klein (ed.) *Southern African Prehistory and Paleoenvironments.* Rotterdam: A. A. Balkema, pp. 147-168.

Rightmire, G. P., and Deacon, H. J. (1991) Comparative studies of late Pleistocene human remains from Klasies River Mouth, South Africa. *J. Hum. Evol.* 20:131-156.

Robinson, J. T. (1954) Prehominid dentition and hominid evolution. *Evolution* 8:324-334.

Robinson, J. T. (1963) Adaptive radiation in the australopithecines and the origin of man. In F. C. Howell, and F. Bourlière (eds.): *African Ecology and Human Evolution.* Chicago: Aldine, pp. 385-416.

Sampson, C. G. (1974) *The Stone Age Archaeology of Southern Africa.* New York: Academic Press.

Schweitzer, F. R. (1979) Excavations at Die Kelders, Cape Province, South Africa: The Holocene deposits. *Ann. S. Afr. Mus.* 78: 101-233.

Schweitzer, F. R., and Wilson, M. L. (1981) Byneskranskop 1, a late Quaternary living site in the southern Cape Province, South Africa. *Ann. S. Afr. Mus.* 88:1-203.

Shackley, M. (1980) An Acheulean industry with *Elephas recki* fauna from Namib IV, South West Africa (Namibia). *Nature* 284:340-341.

Shipman, P. (1986a) Paleoecology of Fort Ternan reconsidered. *J. Hum. Evol.* 15:193-204.

Shipman, P. (1986b) Studies of hominid-faunal interactions at Olduvai Gorge. *J. Hum. Evol.* 15:691-706.

Shipman, P. (1989) Altered bones from Olduvai Gorge, Tanzania: Techniques, problems, and implications of their recognition. In R. Bonnichsen, and M. H. Sorg (eds.): *Bone Modification.* Orono, Maine: Center for the Study of the First Americans, pp. 317-334.

Shipman, P., Walker, A., Van Couvering, J. A., Hooker, P. J., and Miller, J. A. (1981) The Fort Ternan hominoid site, Kenya: Geology, age, taphonomy and paleoecology. *J. Hum. Evol.* 10:49-72.

Sinclair, A. R. E. (1977) *The African Buffalo.* Chicago: University of Chicago Press.

Singer, R. (1954) The Saldanha Skull from Hopefield, South Africa. *Am. J. Phys. Anthropol.* 12:345-362.

Singer R, and Wymer, J. J. (1968) Archaeological investigations at the Saldanha skull site in South Africa. *S. Afr. Archaeol. Bull.* 25:63-74.

Singer R, and Wymer, J. J. (1969) Radiocarbon date for two painted stones from a coastal cave in South Africa. *Nature* 244:508-510.

Singer R, and Wymer, J. J. (1982) *The Middle Stone Age at Klasies River Mouth in South Africa.* Chicago: University of Chicago Press.

Smith, A. B. (1987) Seasonal exploitation of resources on the Vredenburg Peninsula after 2000 B. P. *British Archaeological Reports International Series* 332:393-402.

Smith, A. B. (1990) The origins and demise of the Khoikhoi: The debate. *S. Afr. Hist. J.* 23:3-14.

Smith, A. B., Sadr, K., Gribble, J., and Yates, R. (1991). Excavations in the south-western Cape, South Africa, and the archaeological identity of prehistoric hunter gatherers within the last 2000 years. *S. Afr. Archaeol. Bull.* 46:71-91.

Stiles, D. (1991) Early hominid behaviour and culture tradition: Raw material studies in Bed II, Olduvai Gorge. *Afr. Archaeol. Rev.* 9:1-19.

Stringer, C. B. (1986) The credibility of *Homo habilis*. In B. A. Wood, L. Martin, and P. Andrews (eds.): *Major Topics in Primate and Human Evolution.* Cambridge: Cambridge University Press, pp. 266-294.

Susman, R. L. (1988) New postcranial remains from Swartkrans and their bearing on the functional morphology and behavior of *Paranthropus robustus*. In F. E. Grine (ed.): *Evolutionary History of the "Robust" Australopithecines.* New York: Aldine de Gruyter, pp. 149-172.

Susman, R. L. (1991). Who made the Oldowan tools? Fossil evidence for tool behavior in Plio-Pleistocene hominids. *J. Anthropol. Res.* 47:129-151.

Szabo, B. J., and Butzer, K. W. (1979) Uranium-series dating of lacustrine limestones from pan deposits with Final Acheulian assemblages at Rooidam, Kimberley District, South Africa. *Quat. Res.* 11:257-260.

Tankard, A. J., and Schweitzer, F. R. (1976) Textural analysis of cave sediments: Die Kelders, Cape Province, South Africa. In D. A. Davidson, and M. L. Shackley (eds.): *Geoarchaeology.* London: Duckworth, pp. 289-316.

Templeton, A. R. (1992). Human origins and analysis of mitochondrial DNA sequences. *Science* 255:737.

Thackeray, J. F. (1984) Climatic change and mammalian fauna from Holocene deposits in Wonderwerk Cave, northern Cape. In J. C. Vogel (ed.): *Late Cainozoic Palaeoclimates of the Southern Hemisphere.* Rotterdam: Balkema, pp. 371-374.

Thackeray, J. F. (1988). Molluscan fauna from Klasies River, South Africa. *S. Afr. Archaeol. Bull.* 43:27-32.

Thackeray, A. I., Thackeray, J. F., and Vogel, J. C. (1981) Dated rock engravings from Wonderwerk Cave, South Africa. *Science* 214:64-67.

Tobias, P. V. (1971) Human skeletal remains from the Cave of Hearths, Makapansgat, Northern Transvaal. *Am. J. Phys. Anthropol.* 34:335-368.

Tobias, P. V. (1978) The South African australopithecines in time and hominid phylogeny, with special reference to the dating and affinities of the Taung skull. In C. Jolly (ed.): *Early Hominids of Africa.* New York: St. Martin's Press, pp. 45-84.

Toth, N. (1985) The Oldowan reassessed: A close look at early stone artifacts. *J. Archaeol. Sci.* 12:101-120.

Toth, N. (1987) The first technology. *Sci. Amer.* 255(4):112-121.

Toth, N. and Schick, K. (1986) The first million years: The archaeology of protohuman culture. *Advances in Archaeological Method and Theory* 9:1-96.

Toth, N., Schick, K., Savage-Rumbaugh, E. S., Sevcik, R. A., and Rumbaugh, D. M. (1993) Pan the tool-maker: Investigations into the stone tool-making and tool-using capabilities of a bonobo *(Pan paniscus). J. Archaeol. Sci.* 19:81-91

Turner, A. (1984) Biogeography of Miocene—Recent larger carnivores in Africa. In J. C. Vogel (ed.): *Late Cainozoic Palaeoclimates of the Southern Hemisphere.* Rotterdam: Balkema, pp. 499-506.

van Neer, W. (1989) Contribution to the archaeozoology of Central Africa. *Annales (Sciences Zoologiques) du Musée Royal de l'Afrique Centrale* (Tervuren, Belgium) 259:1-140.

van Noten, F. (1982) *The Archaeology of Central Africa.* Graz: Akademische Druck-u. Verlagsanstalt.

Velleman, P. F. (1988) *Data Desk Handbook.* Vol. 1. Northbrook, Ill.: Odesta Corporation.

Villa, P. (1991) Middle Pleistocene prehistory of southwestern Europe: The state of our knowledge and ignorance. *J. Anthropol. Res.* 47:193-217.

Vogel, J. C., and Beaumont, P. B. (1972). Revised radiocarbon chronology for the Stone Age in South Africa. *Nature* 237:50-51.

Voigt, E. A. (1973) Stone Age molluscan utilisation at Klasies River Mouth Caves, South Africa. *S. Afr. J. Sci.* 69:306-309.

Volman, T. P. (1978) Early evidence for shellfish collecting. *Science* 201:911-913.

Volman, T. P. (1984) Early prehistory of southern Africa. In R. G. Klein (ed.): *Southern African Prehistory and Paleoenvironments.* Rotterdam: A. A. Balkema, pp. 169-220.

Vrba, E. S. (1981) The Kromdraai Australopithecine Site revisited in 1980: Recent investigations and results. *Ann. Transv. Mus.* 33(3):18-60.

Vrba, E. S. (1982) Biostratigraphy and chronology, based particularly on Bovidae, of southern hominid-associated assemblages: Makapansgat, Sterkfontein, Taung, Kromdraai, Swartkrans; also Elandsfontein (Saldanha), Broken Hill (now Kabwe) and Cave of Hearths. In M. A. de Lumley (ed.): *L'Homo erectus et la Place de l'Homme de Tautavel parmi les Hominidés Fossiles.* Nice: 1er Congrès International de Paléontologie Humaine, pp. 707-752.

Vrba, E. S. (1985a) African Bovidae: Evolutionary events since the Miocene. *S. Afr. J. Sci.* 81:263-266.

Vrba, E. S. (1985b) Early hominids in southern Africa: Updated observations on chronological and ecological background. In P. V. Tobias (ed.): *Hominid evolution: Past, present and future.* New York: Alan R. Liss, pp. 195-200.

Vrba, E. S. (1987). A revision of the Bovini (Bovidae) and a preliminary revised checklist of Bovidae from Makapansgat. *Palaeont. Afr.* 26:33-46.

Vrba, E. S. (1988) Late Pliocene climatic events and hominid evolution. In F. E. Grine (ed.): *Evolutionary History of the "Robust" Australopithecines.* New York: Aldine de Gruyter, pp. 405-426.

Vrba, E. S. and Panagos, D. C. (1982) New perspectives on taphonomy, palaeoecology and chronology of the Kromdraai apeman. *Palaeoecol. Afr.* 15:13-26.

Walker, A. C., Leakey, R. E. F., Harris, J. M., and Brown, F. H. (1986) 2.5-Myr *Australopithecus boisei* from west of Lake Turkana, Kenya. *Nature* 322:517-522.

Wells, L. H. (1964) The Vaal River Younger Gravels faunal assemblage: A revised list. *S. Afr. J. Sci.* 60:91-93.

Wendorf, F., Laury, E. L., Albritton, C. C., Schild, R., Haynes, C. V., Damon, P. E., Shafiqullah, M., and Scarborough, R. (1975) Dates for the Middle Stone Age of East Africa. *Science* 187:740-742.

Wendt, E. (1972) Preliminary report on an archaeological research programme in South West Africa. *Cimbebasia* B, 2:1-61.

Wendt, E. (1976) "Art mobilier" from Apollo 11 Cave, South West Africa: Africa's oldest dated works of art. *S. Afr. Archaeol. Bull.* 31:5-11.

White, T. D. (1984) Pliocene hominids from the Middle Awash, Ethiopia. *Cour. Forsch. Inst. Senckenberg* 69:57-68.

White, T. D. (1987). Cannibals at Klasies? *Sagittarius* 2(1):6-9.

Williams, D. F., Thunell, R. C., Tappa, E., Rio, D., and Raffi, I. (1988) Chronology of the Pleistocene oxygen isotope record: 0-1.88 m.y. B. P. *Paleogeog. Paleoclimatol. Paleoecol.* 64:221-240.

Wolpoff, M. H., and Caspari, R. (1990a) Metric analysis of the skeletal material from Klasies River Mouth, Republic of South Africa. *Am. J. Phys. Anthropol.* 81:319.

Wolpoff, M. H., and Caspari, R. (1990b) On Middle Paleolithic/Middle Stone Age hominid taxonomy. *Curr. Anthropol.* 31:394-395.

Wood, B. A. (1984) The origins of *Homo erectus*. *Cour. Forsch. Inst. Senckenberg* 69:99-112.

Wood, B. A. (1985) Early *Homo* in Kenya, and its systematic relationships. In E. Delson (ed.): *Ancestors: the Hard Evidence.* New York: Alan R. Liss, pp. 206-214.

Wood, B. A. (1992) Origin and evolution of the genus *Homo*. *Nature* 355:783-790.

Wright, R. (1972) Imitative learning of a flaked-stone technology: The case of an orangutan. *Mankind* 8:296-306.

CHAPTER 24

ESR Dating of Tooth Enamel

A Universal Growth Curve

Naomi Porat
and
Henry P. Schwarcz

INTRODUCTION

In constructing a time scale for human evolution, it has been necessary to invent a number of new dating methods that can span the time beyond the range of ^{14}C (less than 50,000 yr) and be more widely applicable than Ar/K (or $^{40}Ar/^{39}Ar$) dating, whose use is limited to volcanic terrains. Two such methods, based on similar physical principles, are electron spin resonance (ESR) and thermoluminescence dating (TL). They are both applicable to a wide range of archaeological sites over the time range from 1,000 yr (1 KY) to 2 million yr (2 MY). ESR dating is applicable to tooth enamel (Grün et al. 1987), which is ubiquitous at prehistoric sites, while both TL and ESR dating can be used to date burnt flint (Valladas et al. 1988; Porat and Schwarcz 1991), which is found at many sites younger than about 500 KY.

In this chapter we discuss some aspects of the dating of tooth enamel by ESR. This method is based on the observation that materials that have been buried in archaeological sites are bombarded by environmental radiation. As a result of this bombardment electrons are liberated and some of these can be trapped at defect sites in the material. The existence of these trapped electrons can be detected by the presence of a characteristic signal on an ESR spectrometer (Figure 24-1). The electrons trapped at a particular site are distinguished by the shape and position of the signal in the ESR spectrum. From the height of the signal we can determine the radiation dose received by the sample. To do this, we generally must subject the sample to a known artificial dose in order to determine its sensitivity. This is the additive dose method (ADM) originally developed for TL dating. A typical growth curve is shown in Figure 24-2.

When the accumulated dose (or AD) in the material has been determined, we can calculate the age from the ratio of the AD to the sum of the internal + external dose rates. In the case of tooth enamel, we assume that these rates have changed with time for a variety of reasons. Therefore we must use an integral equation to solve for the elapsed time that would have been necessary to produce the observed dose, assuming a knowledge of how the dose rate has changed with time (Schwarcz and Grün 1992).

The range of ESR dating of enamel is limited by the fact that there are a finite number of sites in a sample where electrons can be trapped. As these fill up, the sample saturates, and further doses do not increase its intensity. For many materials the age limit is reached before saturation, when the rate of electron trapping equals the rate of loss of electrons by detrapping; this condition is called the steady state. The detrapping process determines the lifetime of the signal in a tooth and should be at least 10 times longer than the age being determined. For tooth enamel, the lifetime of the trapped electrons has been estimated to be 10^9 yr at room temperature (Schwarcz 1985), based on heating experiments. Therefore, it is unlikely that teeth ever approach a steady state condition, and the age should be limited only by saturation.

The external dose rate for teeth (or other ESR-datable materials) is determinable by a number of methods: TL dosimetry; gamma ray spectrometry or gamma scintillometry; chemical analysis of the sediment for U, Th, and K; or by the isochron method of Blackwell and Schwarcz (1992). Tooth enamel as well as attached dentine and cementum absorb uranium (U) from the enclosing sediment, and the history of this U-uptake process must be known and included in the calculation of the dose received by the tooth (Grün et al. 1987). Uranium series analyses of tooth enamel can be used to refine the estimate of the U-uptake history, since the ratio of U to its daughter isotopes (e.g., thorium-230) is also determined by the uptake history (Grün et al. 1988; McDermott et al. 1993). Specifically, two models of U-uptake are usually considered: early-uptake (EU), in which the present-day content of U was assumed to have been acquired soon after burial, and linear uptake (LU) in which the present-day content of U was assumed to have been acquired at a constant rate since burial. The EU model generates the lowest possible ESR dates for a given set of data, while the LU model gives dates that are greater than or equal to the EU dates for the same data set.

APPLICATIONS OF ESR DATING TO HOMINID EVOLUTION

ESR dating of tooth enamel has now been applied to a large number of sites in the Old World, ranging in age from about 40 KY to over 2 MY. At several of these sites the ESR dates could be cross-checked with other independent dating methods, principally U-series dates on calcite speleothems and TL dates on burnt flint. In general the best agreement is reached between ESR dates and other methods when we assume the LU model. However, recently McDermott et al. (1993) have used U-series dating to suggest that the EU is more appropriate for some samples.

Over the past few years both TL and ESR dating have been used to revise the time scale for the last stages of hominid evolution, that is, the emergence of *Homo sapiens sapiens*. These studies were initially mostly focused in Israel where a number of critical sites exist in the Upper Galilee and in caves of the Mt. Carmel range. In 1988, Valladas et al. used TL of burned flint to obtain a date of 92 ± 6 KY for the site of Qafzeh in the Galilee. ESR dates were carried out on teeth from the same site by Schwarcz et al. (1988). The LU age (120 ± 8 KY; revised ages as reported by Grün and Stringer 1991) was somewhat greater than the TL result, while the EU estimate (100 ± 10 KY; revised estimate) agreed with the TL date. Both the TL and ESR data show that the anatomically modern hominids at Qafzeh must date from the last interglacial, whereas the oldest modern hominids from Europe are dated to about 45 KY. An interglacial age was subsequently obtained by ESR for the *Homo sapiens sapiens* site of Skhul in Mt. Carmel (Stringer et al. 1989), while a similar age was obtained for the Neanderthal-bearing layer in the adjacent site of Tabun (Grün et al. 1991). Another Neanderthal site, Kebara, also found in Mt. Carmel, a few km away from Tabun and Skhul, gave concordant TL and ESR dates of approximately 60 KY (Valladas et al. 1988; Schwarcz et al. 1989).

Not only did these dates prove the greater antiquity of our species, but they also lent support to the model of an African origin for *Homo sapiens sapiens*. This was further supported by ESR data on the South African site of Border Cave where an age of about 70 KY was obtained for the stratigraphic level from which the modern hominid cranium may have been recovered (Grün et al. 1990). Also, at the South African site of Klasies River Mouth, a date of about 90 KY was obtained for the level from which a hominid mandible of modern aspect was recovered (Grün and Stringer 1991).

RELIABILITY OF ESR DATING

The application of ESR dating of tooth enamel as outlined has in some cases resulted in ages that are in disagreement with other evidence. In general a first requirement of accurate ESR ages is that they should be in proper stratigraphic order; this requirement is generally satisfied at all sites that have been studied (see, e.g., Grün et al. 1991). Even where this requirement is met, however, it is possible that ESR ages may be in error for a number of possible reasons. Chief amongst these are:

1. A U-uptake history that does not follow either the LU or EU model;
2. Significant, unrecognized changes in the external dose rate through the burial history;
3. Diagenetic changes in the tooth enamel causing it to alter its physical state and ESR signal;
4. For very old teeth, saturation of the ESR signal;
5. For young teeth, changes in sensitivity of the enamel soon after burial.

It is possible to detect some of these effects, to correct for them, or to avoid samples that display them. For (1), the U-series disequilibrium of the tooth enamel gives an independent estimate of the U-uptake history, as does fission-track mapping. Changes in dose rate can be dealt with using the isochron approach (Blackwell and Schwarcz 1992), as long as internal dose provides a significant fraction of the total dose rate and is variable between subsamples.

It is much more difficult to detect the effects of diagenesis leading to loss of signal, since the additive dose method (ADM) apparently produces growth curves that look quite similar for altered and unaltered samples and even gives quite comparable errors in the estimates of AD. Therefore it was thought desirable to develop some alternate methods for recognizing samples that were diagenetically altered. We have begun to develop a battery of such methods using X-ray diffraction, infra-red spectroscopy, and trace element analysis (Porat et al. in prep.).

Of the various effects leading to errors in AD, it would be particularly difficult to recognize changes through time in the sensitivity of tooth enamel, because the present-day sensitivity of the enamel would seem "normal," as inferred from the shape of the growth curve. Therefore we decided to test whether teeth of varying age and AD would reproduce the growth curves that we get when we artificially irradiate teeth today. In effect, this would test an important assumption behind the ADM, namely that artificial irradiation reproduces the growth of the signal induced by natural irradiation.

THE UNIVERSAL GROWTH CURVE

All samples of fossil tooth enamel, regardless of their source, display an essentially identical ESR spectrum (see Figure 24-1), with a characteristic g value of 2.0018. As a result of local differences in their dose history, ESR signals from enamel samples differ greatly in intensity due to differences in their ages and the dose rates (internal and external) to which they have been exposed, but their peak shapes are essentially identical. We proposed to compare these signals from various samples that we have studied in a manner that would disregard the differences in signal intensity due to local dose effects.

A simple approach to this was to plot the observed *natural intensity*, I_n, of the ESR signal against the AD as obtained by the ADM. I_n is the intensity of the ESR signal displayed by the sample before it has been given added doses of gamma radiation as required by the ADM. By making such a comparison we hoped to be able to identify anomalous samples because they would depart from the normal relationship between signal intensity and acquired dose (AD). In addition, we hoped to learn whether there were intrinsic differences in the behavior of tooth enamel from different sources (e.g., different species of animals).

CONSTRUCTION OF THE UGC

In order to test this proposition, we assembled data from the McMaster ESR Laboratory (MESRL), including published data as well as a large number of analyses of samples for which complete data have not yet been published. All ESR analyses were carried out on about 50 mg samples of powdered enamel, sieved to a standard size range (250 to 75μ). The intensities, weight-normalized, are given in arbitrary units that represent scale divisions on the chart-recorder of the ESR spectrometer. We have found that the response of the spectrometer remains fairly constant over long periods of time, so we were able to use data collected over three years of analyses in constructing the UGC. Ideally, all intensity data should be normalized by dividing by the intensity of the $g = 2.0018$ peak in a standard sample of enamel. All samples come from archaeological sites, the names of which are given in Figure 24-3.

Average I_n values were plotted against the average AD values obtained for each sample. The data can be fitted by a quadratic curve, which we call the Universal Growth Curve (UGC), shown in Figure 24-3. If all tooth enamel (regardless of its source) behaved uniformly with respect to its ESR response to radiation, then we would expect all analyses to fall on a single locus that could be fitted by a quadratic or cubic polynomial and would flatten at high doses due to saturation. The curve of Figure 24-3 agrees well with this model for AD values less than 50 kilorad (krad), but there is obviously a great deal of scatter around the curve at higher doses, where the curve begins to flatten.

The UGC of Figure 24-3 was obtained using average values of the AD for each site. Many subsamples of enamel are normally analyzed for each tooth at a site and these can also be plotted on the same coordinates, producing the distribution shown in Figure 24-4. These data have been plotted on log-log coordinates in order to even the distribution of points. However, the best-fit cubic curve shown on Figure 24-4 is essentially identical to the UGC of Figure 24-3. The scatter of points about the curve on Figure 24-4 is considerably greater than in the UGC and shows the benefit of using an average value for a site to estimate the AD.

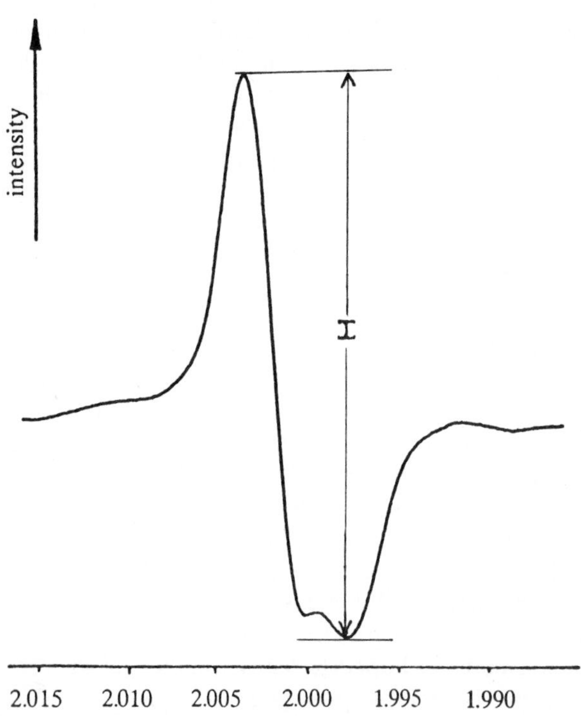

FIGURE 24-1
ESR spectrum of fossil tooth enamel. The intensity, I, of the signal is indicated.

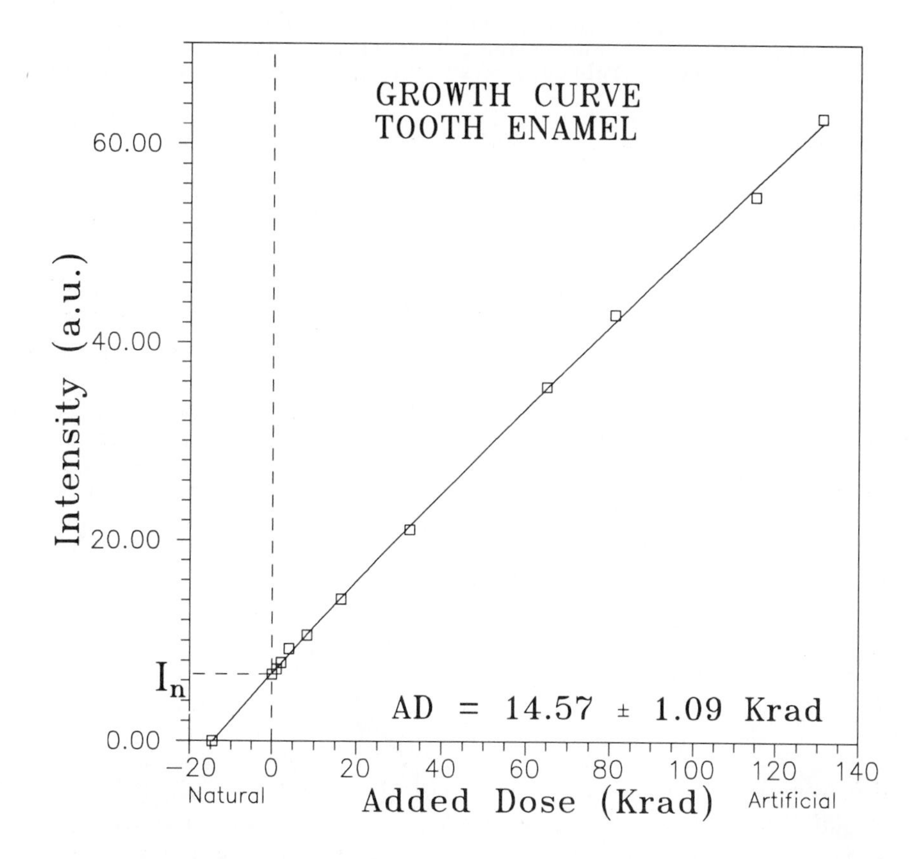

FIGURE 24-2 Typical growth curve obtained using the additive dose method for a sample of tooth enamel. Each point on the curve indicates the intensity of the g = 2.0018 signal for an aliquot of powdered enamel that has received an added gamma dose as shown. The intensity of the natural signal is I_n. The extrapolated intercept at I = 0 gives the accumulated dose AD. The horizontal line shows the response expected for a sample in saturation.

PROPERTIES OF THE UGC

Universality

As proposed in the model, the great majority of tooth enamel samples lie on a single curve of I_n vs. AD, indicating that the ESR response of tooth enamel to radiation does not depend significantly on the source of the enamel. This diagram includes data for teeth of horse, cow, elephant, rhinoceros, deer, bear, and pig. There is no tendency for teeth from a particular species to define a separate curve. This is encouraging news for future applications of ESR dating to tooth enamel, as it appears to be possible to compare data for different species from the same site.

ADs without Irradiation?

The fit of the data is sufficiently good to a single curve that we could actually make a good estimate of the AD of the sample simply from the I_n value. Indeed, the error in this estimate is not much more than the error in the AD measurements themselves, as estimated by the "jackknifing" procedure of Grün and

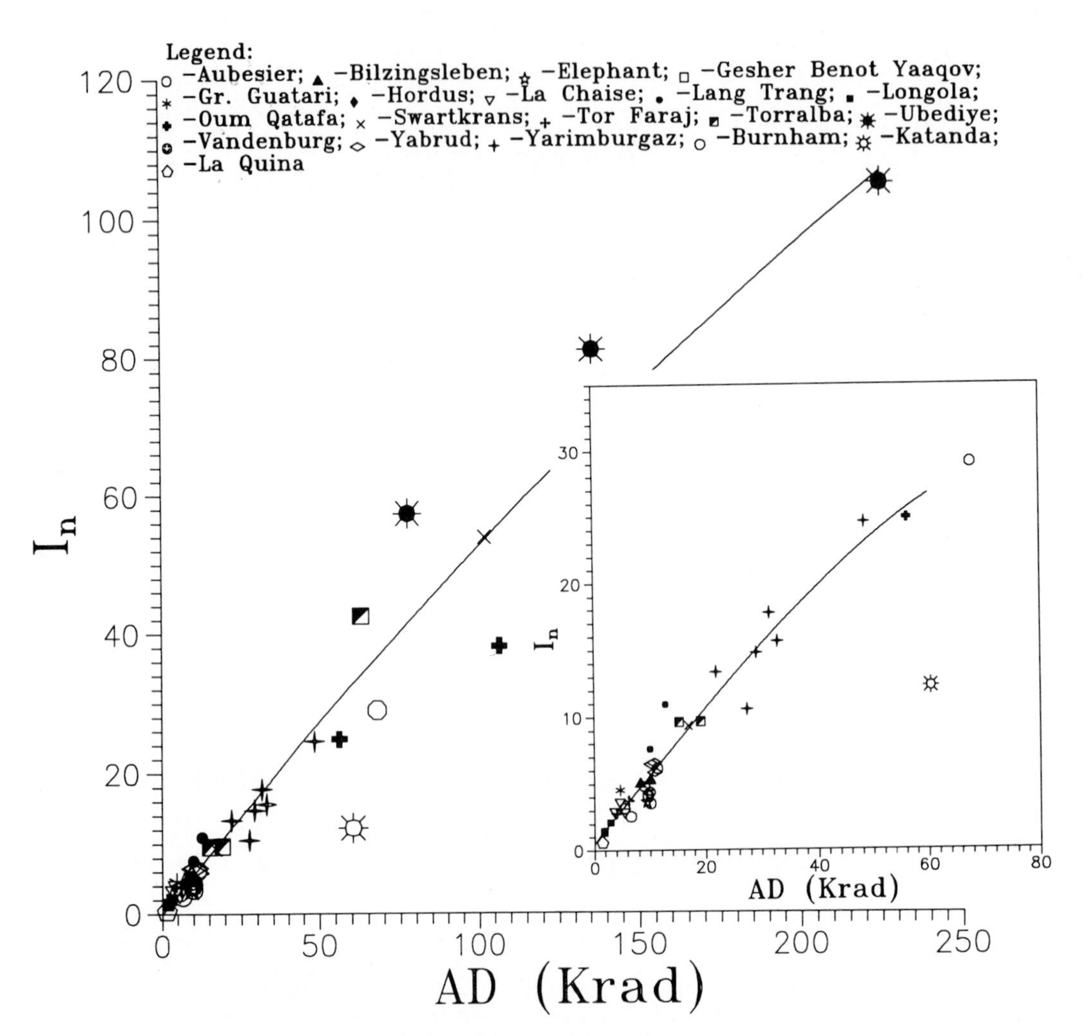

FIGURE 24-3 Universal Growth Curve (UGC) for enamel samples from archaeological sites: Plot of normalized natural intensity I_n vs. AD as determined by ADM: (1) lower dose range; (2) total dose range. A quadratic curve has been fitted to the points.

MacDonald (1989). At low I_n values, much of the scatter may be due to day-to-day variation in the sensitivity of the spectrometer, which could be corrected by normalizing to the intensity of a standard sample. For samples with large AD values, the scatter of points around the best-fit curve may be due in part to errors in AD that inevitably arise when the intercept of the growth curve (at $I = 0$) is very far from I_n.

In a sense, the curves of Figure 24-3 and 24-4 should be thought of as approximations to a plot of I_n vs. the "true" AD acquired by the sample, where this true AD has been determined by some independent physical method. It would be possible in principle to obtain such a curve for at least the low end of this curve, where the effects of U-uptake by the tooth and growth of daughter isotopes of uranium are small. For high doses, we can only assume that the AD obtained by the ADM is a good estimate of the "true" dose absorbed by the tooth during its burial history.

Note that if we can obtain a value of AD from I_n alone, then we could determine the age of the sample directly, using the usual data for dose rate. Thus, it would be possible to determine ESR ages without use of gamma-irradiation facilities.

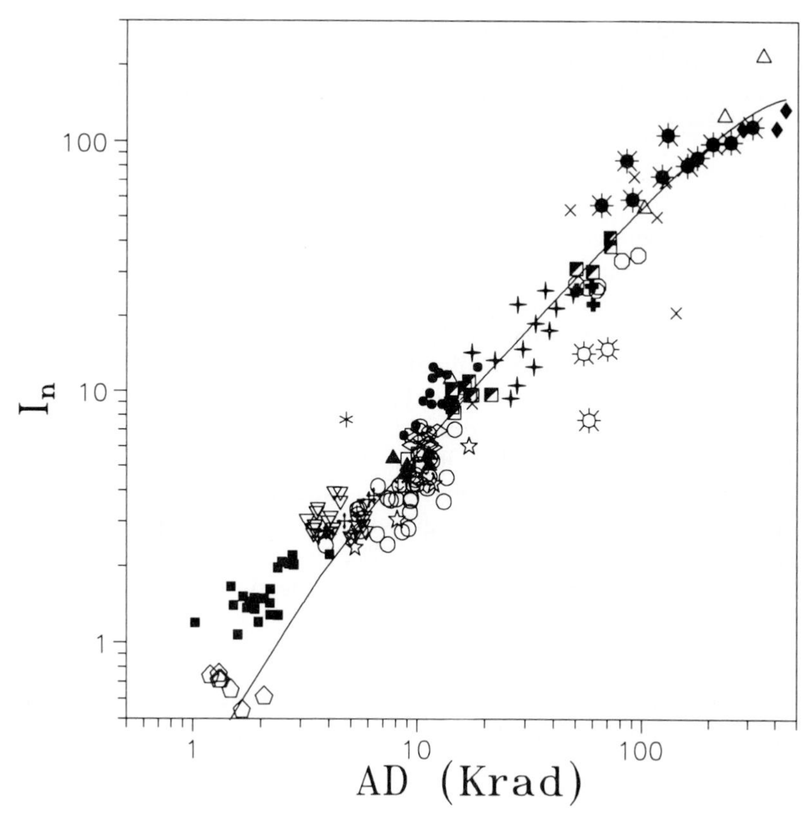

FIGURE 24-4 Universal growth curve for subsamples of tooth enamel from archeological sites. The cubic curve fitted to these data agrees closely with the curve in Figure 24-3.

Saturation

The UGC deviates only slightly from linearity between 5 and 200 krad. In the high-dose region (greater than 200 krad) the curve begins to flatten as expected if saturation had begun to occur. It is unlikely however that the flattening is a result of saturation, for the following reason: The AD values shown have all been obtained by the ADM. In using this method, we assume that when a sample is in saturation there will be no further increase in signal intensity when the sample is irradiated by gamma rays. Thus, a sample in saturation would display an essentially flat growth curve, and an apparently infinite AD. But in fact even the teeth from the Miocene Hordus formation showed an increase in intensity with dose. This allowed us to determine a finite AD value for this sample even though its true dose was certainly much greater (thousands of krads).

We assume therefore that these teeth are in the condition referred to as steady state, that is, where the rate of loss of the ESR signal is matched by the rate of growth due to irradiation. The ESR intensity of a material at steady state can still increase if we expose it in the laboratory to gamma rays at higher dose rates than it experienced in nature. Normally steady state occurs because thermal detrapping of electrons is occurring on a time scale comparable to the lifetime of the trapped electrons. But the evidence from isothermal heating experiments on enamel (Schwarcz 1985) suggests that the lifetime at ambient temperature is greater than 10^9 yr, suggesting that the samples could not be in steady state. Other comparisons between experimentally determined lifetimes of trapped electrons and estimates based on measurements of old samples suggest that these experimental estimates generally are an *underestimate* of the lifetime (Porat and Schwarcz 1993).

Processes other than thermal detrapping can, however, contribute to loss of signal. In the case of tooth enamel, it is likely that slow recrystallization of the tooth would also contribute, since old bones and teeth are generally found to have recrystallized to some extent. Further study of the physical characteristics of these high-AD samples is needed to better define the processes leading to the limiting dose values.

Supralinearity

Likewise, at very low doses there is a tendency for some samples to lie above the UGC, and a smooth curve through the data points would have a significantly lower slope at AD less than 5 krad. This is an effect that has not been noted before in any studies of ESR dating, that is, an increase in the radiation sensitivity of the $g = 2.0018$ signal with increasing dose. If this were generally true, then AD estimates using the additive dose method might underestimate the dose and therefore the age of a sample. This is because the method consists of exposing portions of the sample to added doses of gamma rays, determining the sensitivity of the ESR signal to radiation, and then back-extrapolating to determine the AD. If the sensitivity of the sample today is higher than it was in the past (Figure 24-5), then this will give an erroneous estimate of the AD. Such an effect is referred to as supralinearity (Aitken 1985) and is commonly observed in TL responses of minerals, where the growth curve can be regenerated after the sample has been heated and evaluated at low doses. It is not possible to zero the ESR signal in enamel by heating, however, because a strong organic radical signal grows in the teeth when they are heated.

Outliers

On Figure 24-3, a few teeth seem to lie far off the curve (>2 standard deviations away from the least-squares fit). These may be "problem samples" for which ages would be suspect. In each case, however, only one

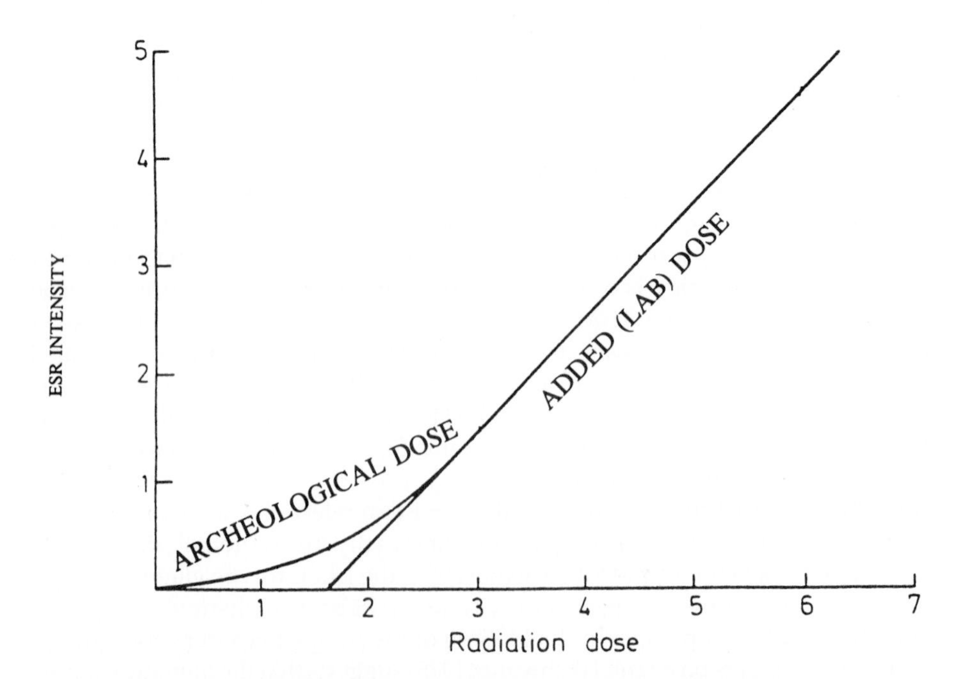

FIGURE 24-5 Effect of supralinearity in the growth curve: The additive dose method assumes that the high-dose side of the growth curve has the same form as the low-dose end. If, as shown, the slope of the historical growth curve was lower, then the additive dose method will underestimate the AD and, therefore, the age.

of the subsamples for a site (out of the ten or more analyzed) lies off the curve, showing that most enamel samples are "well-behaved." In each case only one sample from a set of subsamples lies far off the UGC (Figure 24-4), and as a result the averaged data do not deviate far from the UGC (Figure 24-3), with the exception of one tooth from Torralba which lies above the UGC, and one sample from Oumm Qatafa.

The presence of supralinearity will, as noted, tend to make low ADs appear even lower. This effect will mainly affect dates on very young sites (less than 40 KY) or on sites with very low dose rates such as the cave of La Chaise-de-Vouthon studied by Blackwell et al. (1992). ESR dates for such sites should be tested against ^{14}C or U-series dates (as at La Chaise). The conflict between ^{14}C and ESR dates for the youngest samples from Border Cave (Grün et al. 1991; Miller and Beaumont 1992) may be partly attributable to this effect.

One possible limitation of the UGC as a method for identifying bad samples lies in the observation that teeth of many species all lie on the same UGC. This suggests that any sample of carbonate hydroxyapatite might lie on this curve as well, as long as it possessed the same g = 2.0018 signal. It is possible that partly recrystallized teeth might also lie on the curve (although with lower AD values corresponding to the time of recrystallization). It is, however, unlikely that inorganically precipitated apatite would exhibit the same relation between I_n and AD as is observed for tooth enamel, since the composition of such a diagenetic phase would presumably differ somewhat from biogenic apatite.

CONCLUSIONS

The ESR method as applied to tooth enamel has tremendous potential for the dating of prehistoric sites. The range of the method is from a few thousand years up to perhaps 3 million years. The material required for dating, tooth enamel of large animals, is virtually ubiquitous at archaeological sites and is generally sufficiently well preserved for ESR dating. The principal limitation of the method lies in its low precision. This arises from uncertainties in the two quantities on which the age calculation is based, the acquired dose (AD) and the dose rate. In this chapter we have focused on the evaluation of AD. We have shown that enamel is a well-behaved material from the standpoint of the long-term growth of the ESR signal and the relation between normalized intensity and AD. We have attempted to develop a "universal growth curve" for enamel samples. Samples whose ADs and intensities have been determined can be compared with this curve; outliers would then be excluded as unsuitable for ESR dating.

The UGC may also be usable as a dating tool itself, as it could allow us to determine AD directly from normalized intensity. This would require the use of a standard sample material to be used by any lab wishing to compare the intensities recorded on their ESR spectrometer with those on the UGC. Such an approach would however eliminate the need for subjecting subsamples to artificial irradiation in order to build up a growth curve for the determination of AD. This would have a number of advantages. First, it would make ESR dating available to those labs that have an ESR spectrometer but lack a well-calibrated and intense source of gamma rays. Second, it would permit the analysis of much smaller samples, which would in turn permit the use of many more subsamples from a single tooth. The use of numerous subsamples is particularly important in the application of the isochron method (Blackwell and Schwarcz 1992). Finally, the use of smaller samples would allow us to date smaller teeth such as those of rodents, from which growth curves cannot be built up. Indeed, the possibility exists of using the UGC to date human teeth, which could be analyzed in a nondestructive fashion.

SUMMARY

The electron spin resonance (ESR) dating method as applied to tooth enamel has now been shown to be capable of producing dates over a time range from a few thousand up to about 3 million years. Since large mammalian teeth are essentially ubiquitous at archaeological sites, this method is widely applicable to solving problems of chronology in both paleoanthropology and archeology. Several technical problems in the method remain unsolved, however, and the precision of ESR dates is still no better than ±10 percent of the age. In an attempt to improve this precision, we have compared ESR data from a large number of sites that we have studied and have shown that the sensitivity of ESR signals in tooth enamel is a very well-behaved natural phenomenon. Teeth of widely ranging age, animal species, and environment of deposition, all produce ESR signals that lie on a single Universal Growth Curve (UGC) of signal height versus dose. From the analysis of this relationship we learn that: (1) there may be a previously unrecognized increase in ESR sensitivity with dose, at low doses, which could lead to underestimates of age; (2) the limit of the ESR dating method occurs beyond doses of about 200 kilorads, at which point the UGC flattens out due to saturation or steady state; and (3) by further refinement of the UGC it may be possible to eliminate the need to subject most samples to added doses of gamma radiation in order to obtain their AD values.

ACKNOWLEDGMENTS

The studies described here, carried out at McMaster University, were made possible largely through the assistance of Professor F. Clark Howell, who has supported our research in many ways over the past few years. This chapter represents a small token of our appreciation to him for this assistance.

This research was supported by grants to HPS from the Social Sciences and Humanities Research Council (Canada) and from the National Science Foundation, USA to the University of California (BNS 8801699, to F. C. Howell).

LITERATURE CITED

Aitken, M. J. (1985) *Thermoluminescence Dating*. New York: Academic Press.

Blackwell, B., Porat, N., Schwarcz, H. P., and Debénath, A. (1992) ESR dating of tooth enamel: Comparison with U-series dates at La Chaise-de-Vouthon (Charente), France. *Quat. Sci. Rev.* 11:231-244.

Blackwell, B., and Schwarcz, H. P. (1992) Electron spin resonance (ESR) isochron dating: solving the external gamma problem. *App. Radiation and Isotopes* 44:243-252.

Grün, R., Beaumont, P., and Stringer, C. B. (1990) ESR dating evidence for early modern humans at Border Cave in South Africa. *Nature* 344:537-539.

Grün, R., Chadam, J., and Schwarcz, H. P. (1988) ESR dating of tooth enamel: Coupled correction for U-uptake and U-series disequilibrium. *Nuclear Tracks Rad. Meas.* 14:237-241.

Grün, R., and Macdonald, P. D. M. (1989) Non-linear fitting of TL/ESR dose response curves. *App. Radiation and Isotopes* 40:1077-1080.

Grün, R., Schwarcz, H. P., and Zymela, S. (1987) ESR dating of tooth enamel. *Can. J. Earth Sci.* 24:1022-1037.

Grün, R., and Stringer, C. B. (1991) Electron spin resonance dating and the evolution of modern humans. *Archaeometry*, 33:153-199.

Grün, R., Stringer, C. B., and Schwarcz, H. P. (1991) ESR dating of teeth from Garrod's Tabun Cave collection. *J. Hum. Evol.* 20:231-248

McDermott, F., and Grün, R. (1993) Mass-spectrometric U-series dates for Israeli Neanderthal/early hominid sites. *Nature*, in press.

Miller, G. H., and Beaumont, P. (1992) Dating the Middle Stone Age at Border Cave, South Africa, by the epimerization of isoleucine in ostrich shells. *Proc. Roy. Soc. Lond.* B337:149-158.

Porat, N., and Schwarcz, H. P. (1991) Use of signal subtraction methods in ESR dating of burned flint. *Nuclear Tracks Rad. Meas.* 18:203-212

Porat, N., and Schwarcz, H. P. (1993) Problems in the determination of lifetimes of ESR signals in flint and chert by isothermal annealing. Submitted to *Nuclear Tracks Rad. Meas.*

Schwarcz, H. P. (1985) ESR studies of tooth enamel. Proc. Fourth Specialist Seminar on TL and ESR Dating, Worms, 1984. *Nuclear Tracks Rad. Meas.* 10:865-867.

Schwarcz, H. P., Buhay, W. M., Grün, R., Valladas, H., Tchernov, E., Bar-Yosef, O., and Vandermeersch, B. (1989) ESR dates for the Neanderthal site of Kebara, Israel. *J. Archaeol. Sci.* 16:653-661.

Schwarcz, H. P., and Grün, R. (1992) ESR dating and the origin of modern man. *Proc. Roy. Soc. Lond.* B337:145-148.

Schwarcz, H. P., Grün, R., Vandermeersch, B., Bar-Yosef, O., Valladas, H., and Tchernov, E. (1988) ESR dates for the hominid burial site of Qafzeh in Israel. *J. Hum. Evol.* 17:733-737.

Stringer, C. B., Grün, R., Schwarcz, H. P., and Goldberg, P. (1989) ESR dates for the hominid burial site of Es Skhul in Israel. *Nature* 338:756-758.

Valladas, H., Reyss, J. L., Valladas, G., Bar-Yosef, O., and Vandermeersch, B. (1988) Thermoluminescence dates of Mousterian "Proto-Cro-Magnon" remains from Israel and the origin of modern man. *Nature* 331:614-616.

CHAPTER 25

The Evolution of Human Cognition and Cultural Capacity

A View from the Far East

Geoffrey G. Pope
and
Susan G. Keates

INTRODUCTION

In contrast with the formative period of Western paleoanthropological studies that identified Central Asia as the "Cradle of Mankind" (Osborn 1927; Weidenreich 1939), the majority of modern overviews of the behavioral evolution of the Hominidae have been written largely on the basis of data deriving almost exclusively from Europe, Africa, and the Near East. Such reviews have neglected the vast area of Eurasia that encompasses the Far East (East and Southeast Asia). This is especially true of recently renewed attempts to describe and analyze the archeological evidence documenting the evolution of human cultural capacity, cognition, and sapient behavior (Mellars 1989; Clarke 1990; Klein 1992). Just as the controversy over *Australopithecus afarensis* eventually resulted in a concrete anatomical definition of the Hominidae (habitually striding bipeds), the Replacement–Regional Continuity debate is rapidly forcing paleoanthropologists to define both the term "human" and criteria for recognizing sapient behavior in the paleobehavioral (archeological) record of the genus *Homo*. Although no consensus has yet arisen as to the characteristic and definitive archeological residues of modern human behavior, the time depth that one ascribes to demonstrably human behaviors has a direct impact on attempts by anthropologists and other behavioral scientists to understand the diverse sociocultural complexity of modern humans (see Pope this volume). Nowhere is this more apparent than in the Far East where, as this chapter shows, it is possible to document a long technological continuity that is not only without precedent in other parts of the Old World, but also indicative of recognizably Early Pleistocene sapient behavior.

In the few instances where the Paleolithic archeology of the Far East has been dealt with in some detail, it has been variously considered as indicative of cultural isolation (Aigner 1978a, 1981), cultural "backwardness" (Movius 1948; and see Clarke 1990) or even "acultural" hominid adaptations (Binford and Ho 1985; Binford and Stone 1986; Bowdler 1988, 1990). Alternatively, some studies have postulated that the unformalized nature of the Far Eastern Paleolithic results from an increased emphasis on nonlithic resources (Ikawa-Smith 1978; van Heekeren 1957, 1972; Harrisson 1970; Pope 1983, 1988, 1989a). Yet other studies have maintained that the totality of the evidence is currently insufficient to reach any meaningful conclusion about the significance of the Far Eastern Paleolithic evidence (Klein 1992; Jones 1989; Mellars 1989; and see Hutterer 1985). Particularly inaccurate is any contention that insufficient research accounts for the currently recognized distinctness of the Far Eastern Paleolithic (cf. Klein 1992). At the very least, it is now obvious that the lithic paleocultural record of the Far Eastern hominids contrasts markedly with Africa and Europe (Howell 1978).

This chapter reviews and analyzes the current evidence for behavioral evolution in relation to recent theories supporting an African origin of modern humans (Cann et al. 1987; Stringer and Andrews 1988). Specifically, it is now possible to show that a long continuity in lithic technology argues against the behavioral "superiority" of invading geologically recent African-derived modern humans as indicated by the archeological record. Here we attempt to correct the impression that the Far Eastern archeological record is supportive of a Replacement Model or that it is too poorly understood to allow any important conclusions about the evolution of human behavior. Although investigations of some of the sites and assemblages are still at a preliminary stage, it is already clear that the Far Eastern evidence is wholly inconsistent with the so-called Replacement Model for modern human origins. The Far Eastern assemblages strongly suggest that models of paleocultural evolution developed for Europe, Africa, and the Near East are inappropriate to developing an understanding of the evolution of human behavior in this part of the Pleistocene world.

ARCHEOLOGICAL IMPLICATIONS OF THE EVE HYPOTHESIS

The "Eve Hypothesis," also known as "Noah's Ark," "Out of Africa," or the "Replacement Model," was originally based on mitochondrial DNA studies (Cann et al. 1987, Stringer 1989a,b) and maintains that all human races derive from a geologically very recent African population containing a single female whose mitochondrial descendants replaced all non-African fossil populations throughout the entire Old World. Though some of the proponents of replacement deny it (Stringer and Andrews 1988; Stringer 1989a,b, 1990), archeological and paleobehavioral implications of this interpretation inherently suggest that invading populations of African origin possessed some technological, biological, or behavioral advantage that facilitated the total replacement of all the Eurasian descendants of *Homo erectus*. A number of researchers have cited the stratigraphically abrupt introduction of more formalized tool kits (especially in Europe), the advent of art, and the introduction of other complex social behaviors (intentional burials, language, organized big game hunting, etc.) as indicative of new biologically based behaviors that would have conferred a decisive adaptive advantage (Pfeiffer 1969, 1985; Marshack 1972, 1975; Mellars 1989; Binford 1989). This view has been supported by an increasing tendency to view fossil hominids, especially Neanderthals and *Homo erectus* as essentially nonhuman in their behavior (Binford and Ho 1985; Binford and Stone 1986; Binford 1989).

In all these interpretations the question of the existence of truly human behavior (including language, foresight, planning, storage, social stratification, ethnicity, long distance social networks, and trading) is usually operationalized in the form of evidence for repetitive standardized complexity and diversity of artifact assemblages and sites. Numerous recent discussions of the qualitative and quantitative contrasts between the Lower, Middle, and Upper Paleolithic (again especially in Europe, the Near East and southern Africa) repeatedly underscore this approach (Klein 1992; Mellars 1989; Shea 1989). If consistent, standardized complexity and diversity is indeed the principal archeological indicator of recognizably human behavior, the assemblages we describe later show that such behavior is documented, though not ubiquitous, at an extremely early date in China. Furthermore, on the basis of the totality of the Far Eastern evidence, it is very difficult to support the interpretation that postulated modern human immigrants employed a superior lithic technology that facilitated the complete replacement of indigenous East Asian and Southeast Asian populations. Indeed, our research over the years at two sites in northern China, excavations in Thailand and Indonesia, strongly suggests an ancient regional continuity in Asia that contrasts markedly with the record of marked technological change recognized in Europe, the Near East, and Africa.

Nowhere has the tendency toward dehumanization of hominid ancestors been more apparent than in Southeast Asia, where recent debate has centered around the contention that *Homo erectus* may simply have not made stone artifacts at all in this part of the world (Bowdler 1988; Hutterer 1985; Jones 1989). Some workers (Bowdler 1988; Bartstra 1985, 1989; Bartstra et al. 1988; Groves 1989) have gone so far as to suggest that the advent of standardized lithic technology was coincident with the replacement event and the arrival of anatomically and behaviorally modern *Homo sapiens* in Southeast Asia and Australia. We believe that on the basis of radiometric and archeological data (see later discussion), it is already possible to show that there is no support for this interpretation and that it is the geological contexts of the Southeast Asian archeological finds that have been misconstrued in an attempt to reify what is essentially a specific application of the culturally minimalistic Binfordian perspective and a restatement of Movius's original view of Asia as an area of "cultural retardation" (Movius 1944, 1948; see also Sieveking 1960).

In contrast to traditional attempts to understand hominid-generated artifact assemblages in terms of the sequential appearance of phases, temporal shifts in typological frequencies or implied/explicit stages of complexity or idiosyncratic cultural characteristics, the Far Eastern Paleolithic is best understood and most compatible with the recognition of *H. erectus* as a widely distributed manufacturer of geographically variable tool kits. The geographic flexibility in the technological repertoire of this species strongly supports a relatively early origin c. 1 MYA and long persistence of complex behaviors associated with the highly malleable behavior characteristic of modern humans.

As part of our treatment of the Paleolithic evidence from the Far East, it has been necessary to develop a new approach to an understanding of the probable "meanings" of lithic assemblages that in the past have usually been studied by traditional typological classifications originally developed on the bases of the European and African data. One of us (S. G. K.) introduces a tool classification system based on the location of use-wear. These tool categories are not categorical statements of the type of function and are without reference to implied conjectural sociocultural templates (Keates in press in prep.). The much more cryptic and less numerous Paleolithic assemblages from Southeast Asia are dealt with (G. G. P.) in terms of the interrelated questions on the basis of their regional distinctness, which illustrates contrasts in raw material selection, artifact production, and variability of associated geological-paleoenvironmental contexts.

Taken as a whole, our several years of research in the Far East favor the interpretation that regional adaptations have a great temporal depth and continuity. This supports neither the notion of "invading" technologically superior cultures, nor the view of this edge of Eurasia as an isolated cultural backwater inhabited by a side branch of the Hominidae, somewhat removed from the mainstream of human evolution (Rightmire 1984, 1990).

THE ARCHEOLOGICAL EVIDENCE FROM THE FAR EAST

The Far East as used here encompasses mainland and island Southeast Asia and East Asia (Figure 25-1). Two environmental aspects unite these regions. Both are strongly influenced by the seasonal monsoon (Liu 1991a,b) and both are characterized by extant environments that are extremely similar to Pleistocene environments (Pope 1982, 1983, 1985a, 1988; Whyte 1972, 1974, 1984; Whitmore 1975, 1981). The marked climatic oscillations that occurred in glacial Europe never developed in the Far East, though there has been a long history of attempts to understand the Far East in European frameworks (see especially Sun 1991; Aigner 1978b, 1981; Pope 1985, 1992).

Within the Far East, present and past environments can be divided into the tropical and subtropical monsoon forests of Southeast Asia, the highly karstic areas of subtropical and tropical South China, the temperate and open woodlands of North China and the (previously glacial and periglacial) environments

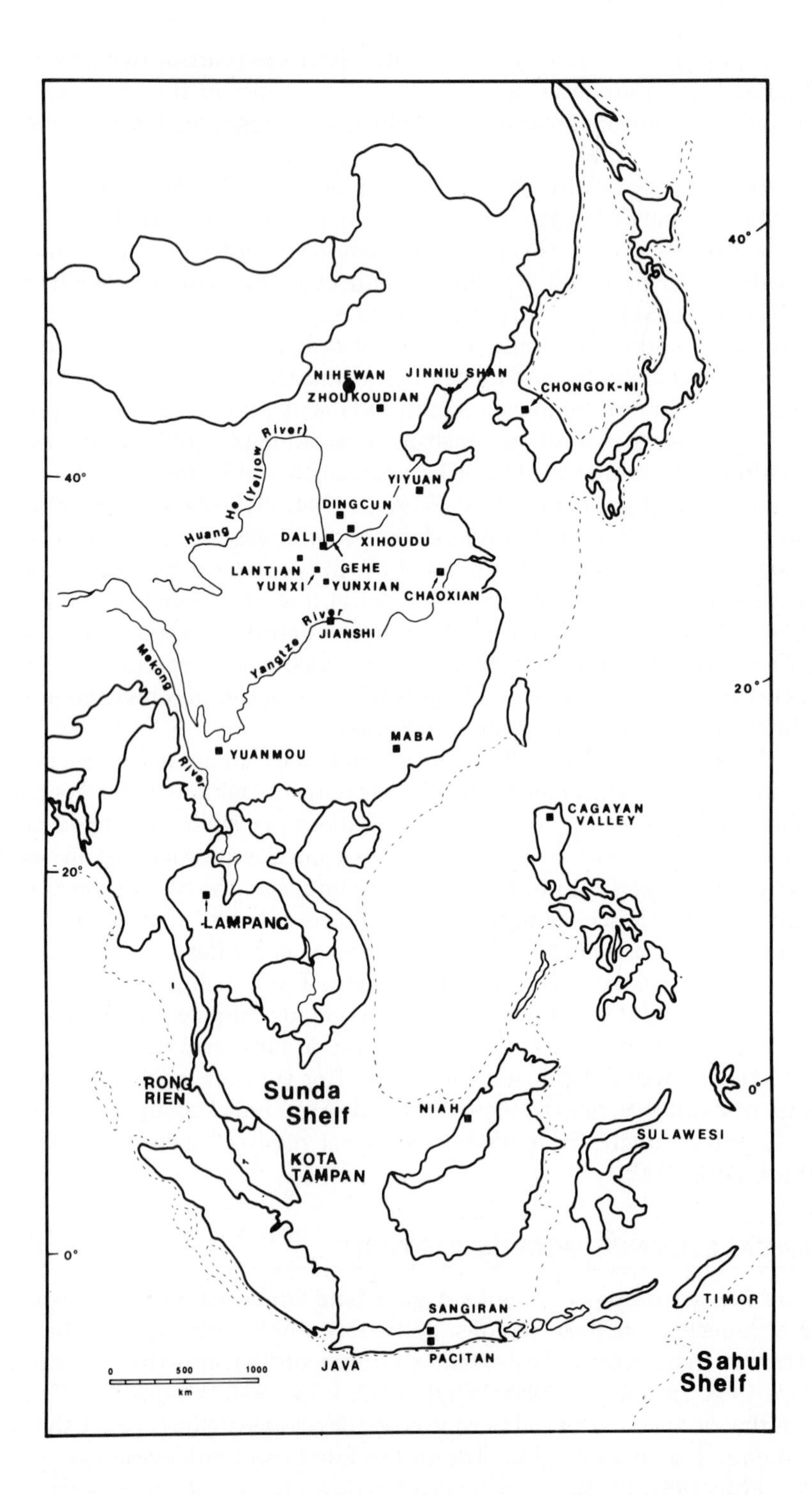

FIGURE 25-1 Important Paleolithic and hominid localities in the Far East. Lampang localities include Mae Tha South, Ban Don Mun, and Kao Pah Nam. Nihewan localities include Xiaochangliang, Donggutuo, and Xujiayao.

of Siberia. Although there is a general consensus that the paleoclimates underwent some fluctuation during the nearly 1 million years of hominid presence, there is substantial disagreement about the severity and geographic extent of such fluctuations. As suggested previously, we think the evidence for marked paleoclimatic change as it has been deduced from paleontological, geological, and palynological fluctuations (Lee 1939; Aigner 1981) is minimal at best.

Among paleoanthropologists, there is substantially more, if not unanimous, agreement that hominids first reached China and Indonesia (Java) approximately 1 MYA via Southeast Asia and the now submerged Sunda Shelf (Pope 1983; Shutler and Braches 1988) before occupying the higher latitudes of the East Asian mainland. The rapidity of this dispersal as indicated by both northern and southern dates of equal antiquity (c. 1 MYA) has, until recently, gone unappreciated (Pope 1992). The tropical and subtropical Southeast Asian environments first encountered by the early Asian immigrants represented a substantial faunal filter that probably, because of its tropical forest character, excluded a number of otherwise ubiquitous Plio-Pleistocene Eurasian mammals (i.e., equids, giraffoids, and camelids: Pope 1983, 1988). The complete absence of these species in Southeast Asia is a strong indication of the novel adaptive challenges that Southeast Asian habitats posed to early hominid immigrants. On the basis of currently known paleoanthropological evidence, hominids were previously confined to the more open areas of the African and South Asian savannas and woodlands. Whether or not *H. erectus* ever colonized glacial or interglacial Europe is still uncertain (Howell 1986; Cook et al. 1982; Stringer 1984).

From a purely biogeographical perspective, the rapidity of dispersal and long duration of *H. erectus* in a diverse number of paleoenvironments surely has important behavioral implications that cannot be ignored in attempts to assess the degree of development of sapient behavior in this species. No other species of primate has occupied such a diversity of environments and habitats for so long. Even if one accepts the highly improbable possibility that *H. erectus* did not make stone tools in Southeast Asia, the ubiquitous presence of artifacts (some of them standardized) in neighboring China and India demands the recognition of this taxon as a culturally dependent species. The behavioral meaning of the geographical variability of the archeological residues from such a diverse and widespread number of habitats must certainly be approached with this firmly in mind.

SOUTHEAST ASIA

Island Southeast Asia

Few Island Southeast Asian lithic assemblages are amenable to detailed statistical analyses. Even in the few cases where large assemblages have been recovered, their artifactual nature and chronostratigraphic positions have often been questioned. This situation is not due solely to the application of inadequate techniques, but also stems from the fact that the informality, low raw numbers, and unique geological contexts contrast markedly with presumably penecontemporaneous industries in Europe and Africa. Nowhere is this situation better exemplified than in our understanding of the Pacitanian (Patjitanian), which despite over one-half century of continuing research remains poorly understood (von Koenigswald 1936a,b, 1956; Hutterer 1977, 1985; Ikawa-Smith 1978; van Heekeren 1957, 1972; Movius 1978; Bartstra 1984, 1985, 1989). Von Koenigswald subsequently modified his original concept (1936) of the Pacitanian and came to regard it as not being associated with *H. erectus* (von Koenigswald 1975).

The Pleistocene record of colonization and occupation is inadequately understood, but it is already apparent that a number of recent interpretations should be given little credence on the basis of what we do know. Potentially early artifacts have been recovered from paleomagnetically normal sediments in one instance (Jacob et al. 1978) and numerous surface finds have at times been provisionally associated with *H. erectus* (e.g., von Koenigswald 1936a,b), but *in situ* excavations have as yet failed to reliably establish artifacts in early geological and chronometric contexts (Bartstra 1985). This is a far different conclusion than that put forward by workers who believe that early Southeast Asians may not have made stone tools (Hutterer 1985; Bellwood 1985; Bowdler 1988, 1990; Jones 1989). Such contentions obscure the difficulties of identifying use and manufacture sites in the high-energy fluvial and extrusive volcanic depositional environments of the Sangiran Dome and other artifact "localities," which have nonetheless yielded both numerous fossil hominids and artifacts (von Koenigswald 1939; van Heekeren 1972; von Koenigswald and Ghosh 1972; Bartstra 1985; Sémah 1982, 1984). Seasonally intense precipitation and catastrophic burials resulting from massive lahar movements and deposition are known to be primary geotaphonomic agents responsible for the failure to discover intact archeological sites.

Work on other islands of Southeast Asia has yielded *in situ* evidence from middle Late Pleistocene contexts (Leang Burung 2, Sulawesi: Glover 1981; Niah Great Cave, Borneo: Harrisson 1970; Tabon, Philippines: Fox 1970, 1978), but no early hominid specimens have been recovered from these localities despite decades of investigations (Figure 25-1). The Cagayan Valley has not yielded convincing evidence of pre-Holocene hominid occupation. Few workers now support a direct association of any of these assemblages with *H. erectus* (but see von Koenigswald 1975). Previously suggested reliance on nonlithic technology (Hutterer 1977, but see Hutterer 1985; Pope 1983, 1989) in conjunction with the complex depositional environments are much more likely explanations of the lack of indication of *H. erectus*–age artifacts. This has been virtually ignored by proponents of a noncultural interpretation of *H. erectus* behavior (cf. Puech 1983).

Mainland Southeast Asia

In mainland Southeast Asia, only northern Thailand has yielded unambiguous evidence for early artifact manufacture in the form of assemblages from Mae Tha, Mae Tha South, and Ban Don Mun localities (Pope et al. 1981, 1986, 1987) from Lampang Province and Phrae Province (cf. Sørensen 1976, 1988). Despite the clear artifactual nature of the Lampang finds (Figure 25-2 a,b,c) and isotopic (^{40}K-^{40}Ar 0.8±0.3 and 0.6 ± 0.2 MYA, Sasada et al. 1987) and paleomagnetic studies (Barr et al. 1976; MacDonald and MacDonald 1978; Barr and MacDonald 1981) placing an overlying basalt at approximately 0.73 MYA, these data have been dismissed as unreliable (Bowdler 1988, 1990; Jones 1989; Anderson 1988) and questioned without the benefit of either stratigraphic or artifactual investigation. A number of the artifacts recovered by Sørensen from Mae Tha and localities in the Phrae Basin (n = 2512) are very similar in both morphology and lithic material composition to the artifacts reported from Lampang Province (Sørensen 1976, 1988; Pope et al. 1986,1987). Sørensen (1976, 1988) has reported an age of c. 0.73 MYA for the artifacts based on paleomagnetic and geomorphological evidence.

One important characteristic of the northern Thai localities and most other Southeast Asian mainland localities is the low density of artifacts at any one locality or level. The northern Thai artifacts also share in common their almost exclusive manufacture from quartzite river cobbles and the confinement of flaking to one side or end of a cobble, with the occasional presence of an isolated "back-nicking scar" (Pope et al. 1986: see Figure 25-2a,b). Sørensen (1976) has speculated (and we agree) that these isolated back-nick scars may represent material testing of sediment coated cobbles. Heider (1960) may have described the same phenomenon on unifacial pebble tools deriving from surface collections in southern Thailand (Keates in prep.). It has been pointed out that despite their simplicity, these cores and large flakes exhibit a distinct patterning (Pope et al. 1986). This patterning takes the form of a few flakes (one to four) struck from one end of quartzite cobbles (see later discussion). It is also apparent to those who have worked in both northern Thailand and southern China (see Qian and Zhou et al. 1991) that there is a decided preference for quartzite despite the presence of more tractable raw materials such as chert.

Despite the use of quartzite, a simple and repetitive patterning is evident in early northern Thai artifacts, and the widespread presence of back-nicking seems to attest to a consistent resource procurement strategy. The characteristically small numbers of artifacts preclude meaningful statistical analyses of the Lampang artifacts (see later discussion of the "Lannathaian"—Sørensen, 1988). The fact that the low

FIGURE 25-2a Quartzite artifacts from Lampang northern Thailand (**a**); a quartzite core from Kao Pah Nam rockshelter; (**b**) a "nosed scraper" from Mae Tha South; (**c**) the same artifact category from Mae Tha South. See pages 537, 538, and 539.

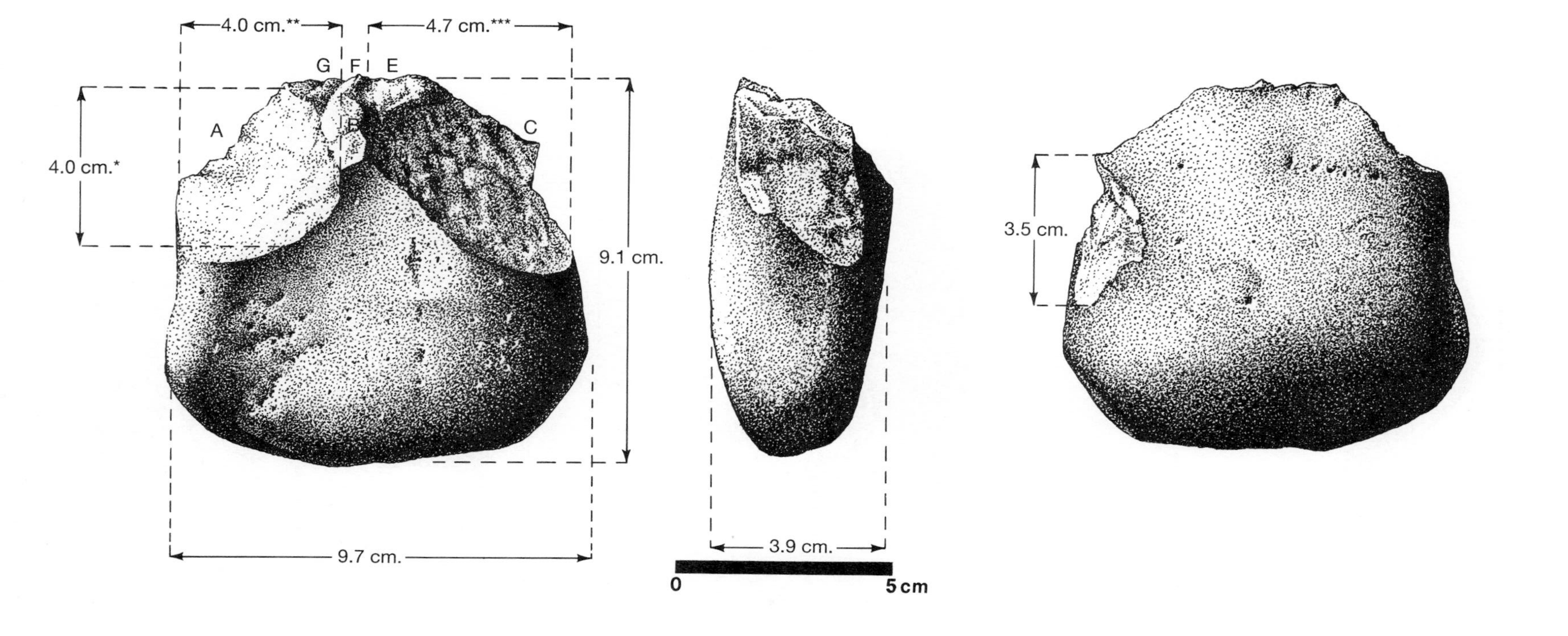

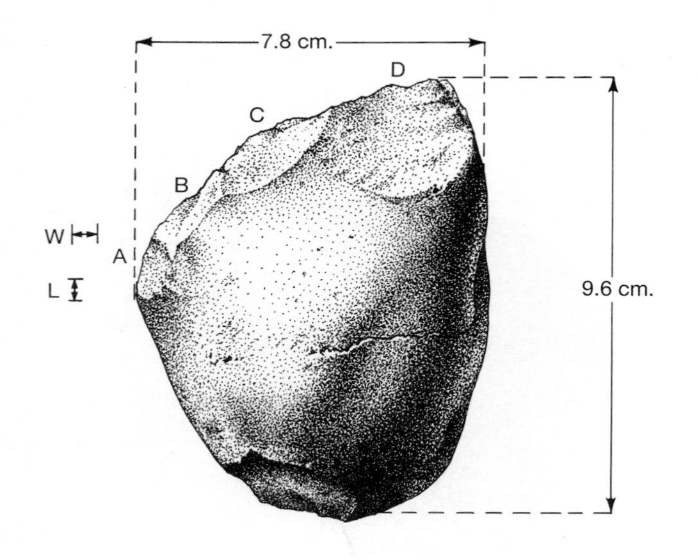

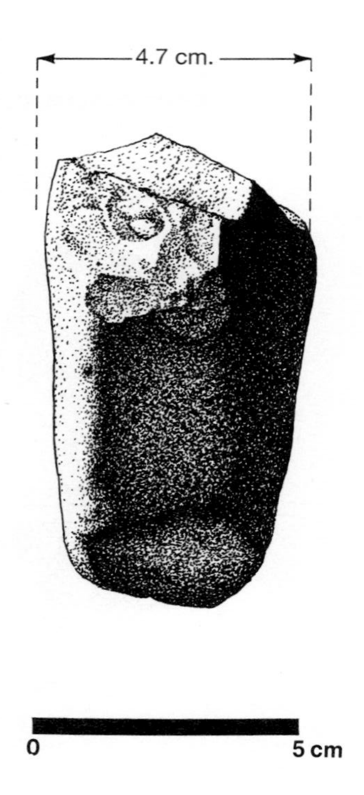

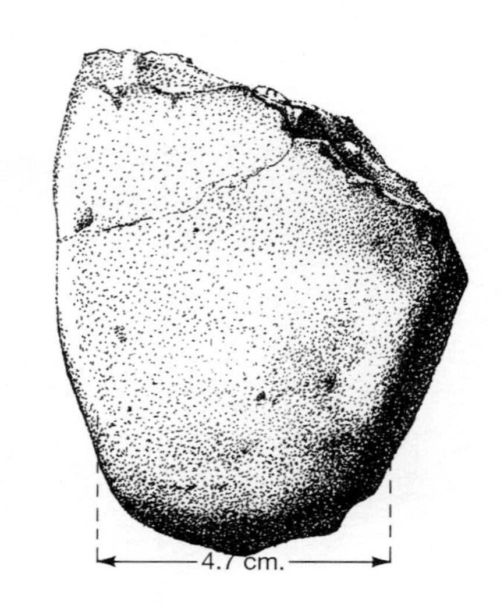

FIGURE 25-2b

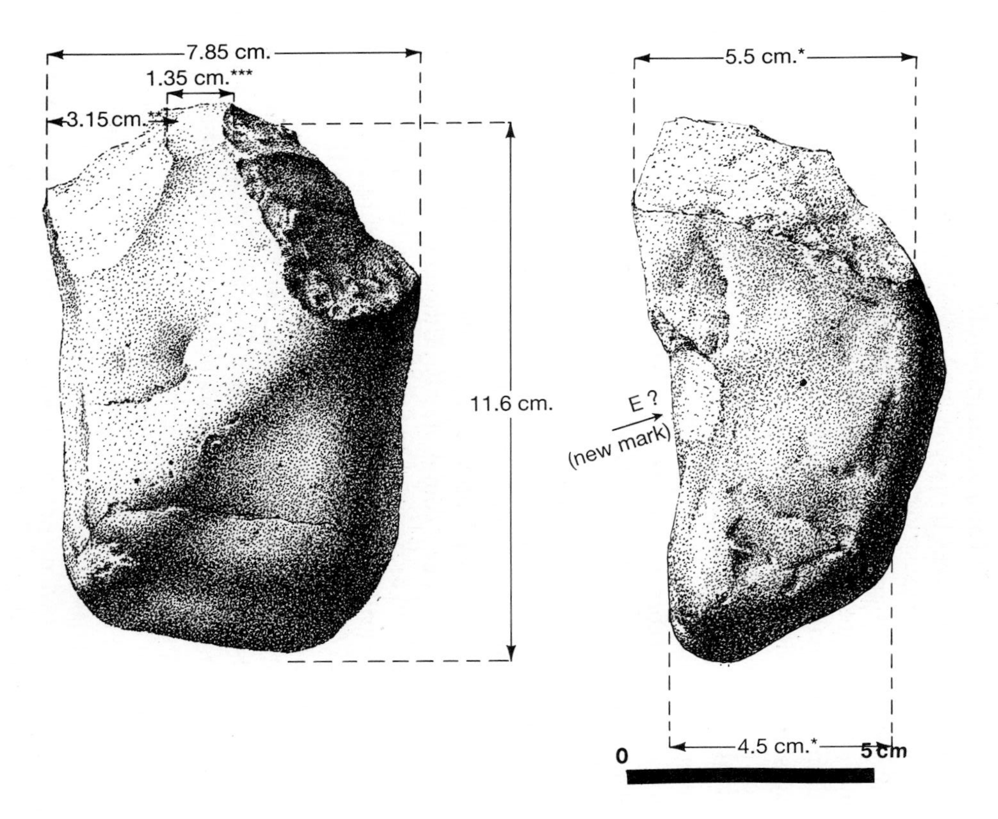

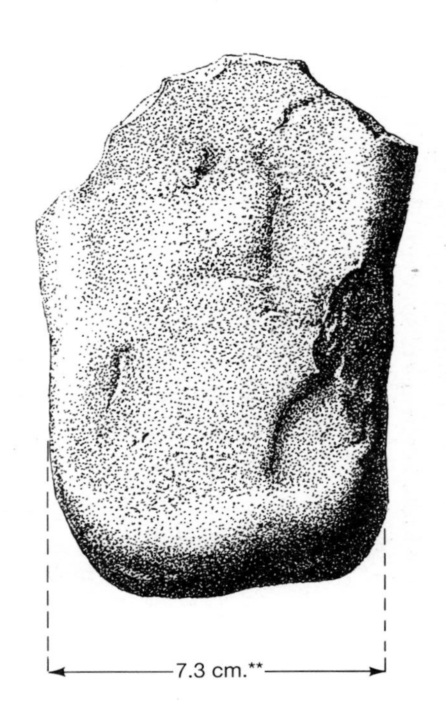

FIGURE 25-2c

densities are not just an artifact of sampling strategies is also supported by the densities of artifact scatters from both the Yuanmou Basin of Yunnan to the north (Qian and Zhou et al. 1991) and the Late Pleistocene locality of Lang Rongrien on the Krabi Peninsula of southern Thailand (Anderson 1988, 1990). Similarly, pebble artifacts from exposed terrace deposits from eight localities in the Kwae Noi Valley, southern Thailand represent a total of only 104 specimens which include four flakes (Heider 1960).

In mainland Southeast Asia, the problem of the definition of assemblages has been especially persistent and difficult due to inadequate provenience and chronometric data. There is absolutely no consensus about the chronological or artifactual definition of the Hoabinhian. The same is also true of other subregional "industries" such as the Anyathian (De Terra and Movius 1943), Fingnoian (van Heekeren 1948), and the Tampanian (Collings 1938; Sieveking 1960). There is, however, a growing willingness to recognize the Hoabinhian (Colani 1927) as a very late chronological development. Kamminga (1990) has gone so far (somewhat following Gorman 1970, 1971) as to suggest that the Hoabinhian is associated with agricultural subsistence. Despite the inability of recent Southeast Asian researchers (Olsen and Ciochon 1990) to date assemblages, a Middle Pleistocene antiquity of some Southeast Asian assemblages is supported by their obvious formal resemblances to the excavated *in situ* artifacts from the Yuanmou Basin, southern China (Qian and Zhou et al. 1991), which lends strong credibility to the possibility that some of the mainland Southeast Asian assemblages date to the early Middle or late Early Pleistocene.

Other reported assemblages from Lampang, northern Thailand also support a late Early Pleistocene date for Southeast Asian lithic technology (Pope et al. 1981, 1986, 1987; Barr et al. 1976; Barr and MacDonald 1981; see later discussion). A new industry, the "Lannathaian," has been proposed on the basis of several finds deriving from northern Thailand (Sørensen 1988). This regional industry is based on a number of surface finds (some of which are known to derive from deflated "blow outs") occurring throughout northern Thailand. In his recognition of the artifacts as a distinct industry Sørensen (1988; see also 1976) suggested a perhaps unprecedented 68 formal lithic categories for Southeast Asian lithic assemblages. Our own inspection of a portion of the specimens attributed to the Lannathaian convinces us that while some artifacts are present, a great number of them cannot be convincingly considered as anything but natural. We further suggest that this industry can possibly be differentiated from the Hoabinhian on the basis of the absence of discoidal cores (van Heekeren 1948). We also propose "Lampangian" as a tentative name for assemblages that are older than the Hoabinhian and contain far fewer artifact categories than recognized for the "Lannathaian."

Subsequent work at Mae Tha South (Pope et al. 1986, 1987) confirms Sørensen's (1976) original stratigraphic placement of artifacts below a radiometrically and paleomagnetically dated basalt flow yielding minimum values of 0.73 MYA. The *in situ* excavation of artifacts from below this same basalt (Pope et al. 1986) at the nearby site of Ban Don Mun has yielded the same results (contra Chareonwongsa 1988:21; Anderson 1990; Bowdler pers. comm. 1990). Although we take strong exception with the multitude of typological categories recognized by Sørensen, we agree that there is a distinct, repetitive, and formalized pattern to such artifacts as the "nosed scrapers" (Figure 25-2b,c) of Sørensen's classification. Specifically, a number of cores from such sites as Kao Pah Nam, Mae Tha, and Ban Don Mun exhibit a pattern consisting of the removal of three to four flakes from one end of quartzite cobbles. Furthermore, as previously noted , the same pattern and selection of raw material are evident at sites in the Yuanmou Basin, southern China. At this point it is possible to say not only that there is a formal, though simple aspect to the Paleolithic assemblages from this region, but also that these assemblages extend back into the late Early Pleistocene. Pope (1982, 1983, 1988, 1989a) has repeatedly suggested that these types of lithic assemblages result from a consistent reliance on nonlithic tool technology and resources (e.g., bamboo) resulting in the production of widely scattered low-density accumulations. The absence of identifiable debitage and small flake artifacts may be a function of the geological contexts or the product of a pattern of landscape use that we do not yet understand. However the excavation of the rockshelter site of Kao Pah Nam has produced the same pattern of low density and absence of small flakes discerned at the open sites. Only future researchers can resolve this problem. Once again it seems unlikely that such research will substantially alter the current picture.

The best-known and most geographically circumscribed of the mainland Southeast Asian assemblages, the Tampanian from Kota Tampan (Figure 25-1), an open-air site in Malaysia has also been the object of several critical evaluations that have questioned purported primary contexts and the artifactual nature of the specimens (Harrisson 1978; Shutler and Forgie 1990; Hutterer 1985; Anderson 1990; Reynolds 1990). The geomorphological dating of the terrace by Walker (Walker and Sieveking 1962) has also been questioned (Haile 1971; Verstappen 1975; Harrisson 1975). This problematic "workshop" locality (Majid 1990; Majid and Tjia 1988) has yielded more artifacts than any other ostensible single archeological locus in Southeast Asia (but see later discussion of Nui Do). This locality is probably not the site originally reported by previous workers (Collings 1938; Sieveking 1960; Walker and Sieveking 1962), but it appears to be close to the previous excavations (Majid 1990).

The majority of the artifacts from new excavations in 1987 are reported as concentrated in one square meter in a single 30–40 cm thick gravel layer of a 4 × 4 m excavation. The artifacts were dated at 31± 3 KYA by the fission track method applied to an overlying volcanic ash (Stauffer et al 1980; Majid 1990; Majid and Tjia 1988). However, the relationship between the ash layer and the artifacts is imprecisely established (Majid 1990). The number of artifacts is high (n = 1759) compared to numbers reported by previous workers (n = 208, Collings 1938; n = 254, Sieveking, 1960). Debitage from this excavation makes up 95.8 percent of the specimens and the artifacts are primarily on quartzite and a few on quartz and tuff (Majid 1990). Tools are mainly on unstandardized flakes (n = 31) and core tools (n = 10) are also present. One of the "partly intuitive" categories, "blanks" are classified as a further flake tool category and these are defined as "*intended* for trimming or use" (Majid 1990:83; italics added). A number of the flake blanks may, according to Majid (1990:80, 86-87), represent waste flakes, and similarly a number of waste flakes may "very well have been intended to be blanks." A definite attribution of some flake artifacts to either "category" was characterized as impossible, because the quartzite does not produce "distinct flaking marks" (Majid 1990:80, 87). The assemblage remains unamenable to statistical analysis due to the use of loosely defined categories [e.g., "Anvils" (n = 15), cores (n = 20), "debitage" (chips, waste flakes, chunks; n = 1632), "flake blanks" (n = 20), and "hammerstones" (n = 18; Majid 1990:84, 88, Figure 11)]. The site has been interpreted as a "workshop" because of the range of artifact types and because some cores and flakes (frequency unspecified) could be refitted (Majid 1990).

The artifactual nature of some or all of the specimens collected by Collings (1938) and Sieveking (1960) has been questioned (Harrisson 1975; Shutler 1984; Shutler and Forgie 1990) and Collings's (1938) specimens exhibit pronounced natural wear and patination (Movius 1948:403). Those illustrated in Majid (1990:Figures 4, 6, 8-10) may or may not be artifacts (cf. Shutler and Forgie 1990), and Shutler (pers. comm.) has pointed out that due to the raw material (quartzite) it is difficult to positively ascertain the artifactual nature of the specimens.

The recovery of the presumed workshop material from a gravel-bearing strata casts doubt on the artificial origin of this body of evidence. Thus even the raw numbers of artifacts of the most dense site in Southeast Asia may have been greatly overestimated. It is difficult to incorporate this evidence in any coherent argument about Paleolithic technology that is ostensibly based on reliable data. Whatever the true total number of artifacts is ultimately decided to be, the fact remains that the largest known paleolithic assemblage in Southeast Asia has yielded fewer than 50 tools. We do not doubt that artifacts are present at Kota Tampan, though we have some doubt that the locality can be uncritically accepted as a workshop. However, the site does seem to conform to the general observation that Southeast Asian Paleolithic assemblages are minimally patterned.

A similar claim for the presence of a large number of diverse artifacts (n = c. 2500) has been made for Nui Do in Vietnam (Chi et al. 1961; Boriskovskii 1966:40-71, 1969; Kinh and Tieu 1976, cited in Olsen and Ciochon 1990:768). As Olsen and Ciochon (1990:768) point out, the "proposed" artifacts consist of 90 percent utilized flakes including "seven bifaces, 89 chopper-chopping tools, and 37 cleavers." These authors conclude that not only is this and other proposed early "sites" in Vietnam undatable, but that it probably represents a "palimpsest" (Olsen and Ciochon 1990). As with Kota Tampan, Nui Do represents yet another example of a locality that can be shown to be neither non–time-transgressive nor representative of true Acheulean assemblages. One very interesting observation of Olsen and Ciochon (1990:768) is that the alleged "cleavers" and "handaxes" represent surface finds. As previously pointed out, there is good evidence for suggesting that such a pattern for relatively few bifaces may be universal throughout most of the Far East.

If the dates for Kota Tampan, and much more uncertainly, Nui Do, are correct, this is further evidence for the long persistence of artifact manufacture techniques. At the same time it is important to point out that the great number of artifacts present at Kota Tampan undoubtedly partially reflect the relatively protected nature of the site. Conversely the lack of any debitage at pebble and cobble tool (core tools) sites in Southeast Asia must logically be due at least in part to the effects of postdepositional geological forces and differential sorting. Additionally, there can also be little doubt that at least some previous workers may have failed to record smaller specimens of debitage and amorphous flakes.

The Late Pleistocene rockshelter site of Lang Rongrien has yielded an assemblage of 45 stone (chert and shale) and 3 dubious osteo-keratic (bone and antler) artifacts (Anderson 1988, 1990:54) deriving from the lowermost layers (0.4-0.8 m in thickness) that have been dated by the ^{14}C method to a maximum age of approximately 37 KYA or older (Anderson 1988, 1990). In spite of the tendency of rockshelters to facilitate the concentration and preservation of artifacts, the actual numbers are low. Although 47.9 percent of the artifacts are flake tools and 20.8 percent of the artifacts are described as core tools, the lack of formalized morphology is apparent. However, Anderson (1988, 1990) also notes that with regard to the shape of worked edges, the artifacts are more uniform. No refit matches were found and it has been tentatively suggested that the artifacts were not manufactured (but may have been retouched) at the site. Although this predominantly flake tool industry dates to Upper Paleolithic equivalent times, it is obvious that the assemblage bears little in the way of formalized tool types. There can also be little doubt that the manufacturers of the artifacts were anatomically modern *Homo sapiens*. A pattern of low density of recognizable artifacts is also indicated by Majid's (1982) study of Harrisson's (1970) Niah Cave lithic sample in which less than half of the specimens were identified as artifacts (48.6 percent, n = 1872 of total n = c. 3848), including the preceramic sample (n = c. 1209).

It is very difficult if not impossible to seriously suggest that the nature of site assemblages like Mae Tha, Ban Don Mun, Kao Pah Nam, Lang Rongrien, Niah, Kota Tampan, and Nui Do can be attributed to minimal cultural capacity, especially since the manufacturers of the Late Pleistocene assemblages were undoubtedly modern humans. This is a conclusion also supported by numerous ethnographic studies repeatedly documenting the nature of lithic adaptations in forests (see especially Forde 1934; Golson 1977; White 1977; White et al. 1977; Blackwood 1964). Those workers who dismiss the early northern Thai assemblages as nonexistent would still be faced with the problem of explaining why the assemblages of invading moderns derived from Africa produced assemblages very similar to the undoubted Middle and perhaps Early Pleistocene assemblages of southern China. A far more parsimonious interpretation recognizes that "simple" formality has a very long history in Southeast Asia and southern China.

EAST ASIA

The vast majority of Paleolithic evidence in East Asia has been recovered from china where artifacts and hominid fossils date back to approximately 1 million years. The degree to which paleoanthropological research has continued to be influenced by methodologies originally developed in Europe and Africa has been substantial (Pope in press). Normative archeological research has concentrated on the reification of the existence of Asian Acheulean and Mousterian assemblages (Huang 1987; von Koenigswald 1936a,b,

1956; and see Teilhard de Chardin 1937). The correlation of Asian geological and biostratigraphic sequences with European glacial stages (Aigner 1978b, 1981), the recognition of African savanna-like paleoenvironments (Hutterer 1985), the search for ancient ethnographic entities (Jia et al. 1972) and a general willingness to demonstrate a cultural complexity in the archeological record similar to that perceived in Europe (Huang 1990) have all impeded our understanding of the archeological record in East Asia.

Modern Chinese research has also generated a few highly influential "indigenous" research questions. The most important of these has been the dichotomous "Large Tool" (core assemblages) and "Small Tool" (flake assemblages) "Traditions" (Jia et al. 1980; Huang 1987). The question of the reality and meaning of this dichotomy between the Large Tool and Small Tool Tradition persists as a widely accepted framework for understanding the Chinese paleolithic record (Table 25-1). In its initial development (Jia et al. 1972), this interpretation regarded low-density core tool assemblages (containing trihedral picks, bifacial choppers, and large flakes) from the sites of Kehe, Lantian, Dingcun, and Sanmenxia as precursors of the Inner Mongolian E'maokou assemblages. The small flake tool assemblages from Zhoukoudian Locality 1 and 15, Hutouliang, Shiyu, Salawusu, and Xiaonanhai were interpreted as precursors of microlithic assemblages. The discovery of the Dingcun (Movius 1956) and the Xujiayao assemblages (Jia and Wei 1976; Jia et al. 1979) with their respective large tool and small flake tool assemblages from Shanxi Province in northern China precipitated the recognition of the current Small Tool-Large Tool framework. The subsequent discovery of Small Tool assemblages from the Nihewan Basin sites (Donggutuo and Xiaochangliang), northern China resulted in the extension of the dichotomy back into the Early Pleistocene (Jia and Huang 1985) and forward in time into the Holocene. However, problems with this scheme still exist in the form of assemblages from Shuidonggou, Xiachuan, and Hutouliang, which document assemblages containing both large and small tools. These sites are now regarded as hybridized industries reflecting a "mixed influence" of the Small and Large Tool Traditions (see Jia and Huang, 1985b:262 for a graphic presentation of this framework). Interesting, but not surprising to those familiar with paleolithic studies in China, is the fact that assemblage variability resulting from different activity facies or environmental or geographical contexts has rarely been discussed in relation to the observed assemblage categories. Recently, however, Jia and Huang (1985b) argued that paleoenvironmental factors may also account for the differences between Chinese assemblages, an approach to Chinese prehistory very similar to that of Aigner (1981). Furthermore, southern Chinese assemblages were and continue to be treated as too poorly understood to be assessed in this framework with the implicit and sometimes explicit view that the South was a marginal area of cultural development.

The perception of Chinese geographic and cultural isolation during the Pleistocene (Aigner 1978b, 1981) has frequently been used to bolster the notion of a record of stage by stage, *in situ* cultural "progress" only slightly influenced by developments in the West (Jia 1980:58-60; Jia and Huang 1990:201-202). However, even these current research concerns have their roots in observations and opinions once current among previous European and American workers (Boule 1927; Breuil 1927; Bordes 1950). Methodological approaches developed in the West after World War II such as taphonomy, refit studies, use-wear analyses, and even interdisciplinary excavations are still in their infancy in China. The constellation of debates raised by "processual archeology" in the West never influenced the study of paleolithic archeology in China where the nature and course of sociocultural evolution was subsumed partially by Marxist theory and more importantly by a deluge of new data resulting from massive salvage excavations.

The parameters of discussion and research in China are somewhat similar to those for Southeast Asia in that a distinction (however improbable) has been recognized between core tool and flake tool assemblages, but differs from the Southeast Asian parameters in regarding indigenous cultural development of separate cultures in China as paradigmatic. The extent to which clear-cut classification into the two traditions has been influenced by collecting techniques and taphonomic agencies is probably extensive, but largely unrecognized. This cannot be overemphasized. The absence of small flakes from the early Large Tool Assemblages reflects the fact that these localities are geological aggregates which have yielded isolated specimens where geological agents are the most likely source of transport and deposition. Xihoudu does preserve both flakes and core artifacts and has therefore been considered to be "not directly comparable to

TABLE 25-1 Important Paleolithic and hominid localities in the Far East (from Keates in prep.)

Locality	Deposition context	Age (MYA)	Dating tech-nique	Artefacts ("Tradi-tions")	Total n	Raw material	Other features
North China							
Xiaochangliang nw Hebei	Open-air lacustrine *in situ*	0.73–0.97	P,B	C, Wfrg, Wf,Ftl (S)	c. 1,884	Chert, basalt, quartz, sandstone	Possible cut-mark
Donggutuo nw Hebei	Open-air fluvial ? *in situ*	0.73–0.97	P,B	C, Wfrg, Wf,Ftl (S)	c. 10,000	Chert, basalt, quartz, limestone	Cut-mark
Zhoukoudian Loc. 13	Cave/rockshelter	0.73 MP (e/m)	P B	Ctl, ?Fl	n†1 = 1	Chert	Burnt bones
Zhoukoudian Loc. 1 Beijing Shi	Cave *in situ*	0.46–0.24 MP(1)	P, B, U-s ESR, F, T, A	C, Wf, Flt, Ctl (S)	c. 10,000 (S)	Quartz, chert (? flint), sandstone, etc.	*H. erectus* cranial and post-cranial n† = 45, juvenile and adult; substantial ash deposits; burnt bones and antlers, gnaw marks; ? cut marks; ? bone tools
Xihoudu s Shanxi	Open-air fluvial	0.73/1.67	P, B	C, Fl, Ctl (S/L)	c. 31	Quartzite	Burnt bones
Kehe, s Shanxi (11 Sites, pooled)	Open-air fluvial	MP (? 1)	B	C, Fl, Ctl, Ftl (L)	138	Quartzite, vein quartz	
Lantian, s Shaanxi (Gongwangling, Chenjiawo)		EP/MP	P, B	C, Fl, Ctl, Ftl (L)	c. 25	Quartzite, quartz	*H. erectus* calotte; *H. erectus mandible*
Yunxian n Hubei	Open-air fluvial river terrace	MP (? l or t) 0.40	B U-s	C, Fl, Ftl, Ctl (S/L) (surface and *in situ*)	68	Quartz, quartzite	*H. erectus* crania n = 2
Dingcun, s Shanxi (10 sites, pooled)	Open-air fluvial river terrace	0.16–0.21	U-s	C, Fl, Ctl, Ftl Sp and/or Hst (L)	2,005	Hornfels, quartzite, chert, basalt, limestone, etc.	*H sapiens* parietal n = 1 and teeth n = 3
Xujiayao n Shanxi	Open-air lacustrine *in situ*	> 0.1 0.104–0.125 > 0.04 0.169 ± 0.002	U-s U-s ^{14}C	C, Wf, Ftl, Ctl (S)	14,039; (32 bone 'artefacts')	Vein quartz, chert, volcanic stone	*H. sapiens* n†1 = 11, juvenile and adult; Sph n = 1,073; burnt bones; ash; predominance of *E. przewalskyi* n† = 91 and *Coelodonta antiquitatis* n†= 11
Dali n Shaanxi	Open-air fluvial ? *in situ*	0.209 ± 0.23 MP; ?LP(e)	U-s B	C, Wf, Flt (S)	481	Chert, quartzite, quartz	*H. sapiens* cranium
Shiyu n Shanxi	Open-air fluvial	28,945 ± 1370 B.P. 28,135 ± 1330 B.P.	^{14}C ^{14}C	C, Wf, Ftl (S)	15,000	Vein quartz, quartzite, siliceous limestone, agate, etc.	*H. sapiens sapiens* n = 1; charred stones; ash deposits; predominance of *E. przewalskyi* n†1= 120 and *E. hemionus* n†1 = 88

South China							
Yuanmou, n Yunnan	Open-air lacustrine	0.73–0.97	P	Ctl	3	Quartzite, quartz	
Guanyindong c Guizhou	Cave	? MP (m or l) 0.057 ± 0.03 (upper layers) 0.076 and 0.070–0.119 (lower layers)	B U-s U-s	C, Fl, Fltl, Ctl,	2,323 (layers 3-8) (of total n = 3,000)	flint, quartz, Sandstone etc.	
Baise w Guangxi (series of river terrace localities)	Open-air fluvial	0.73–1.0	G	C, Fl, Ctl Fl, Ctl	ca. 3,069 (mainly surface) 100 (excavated specimens)	Quartzite, sandstone Quartz, quartzite, metamorphic rock, sandstone, flint	
Mainland Southeast Asia							
Kao Pah Nam	Open rock-shelter	MP	B	Fl, C	5	Quartzite, indurated sandstone, chert	back-nicking; burnt cobbles (basalt); bone artefact; hearth; possibly burnt mammalian bones;
Ban Don Mun	Open air	0.73	P	Fl, C	3	Quartz, quartzite	
Mae Tha Lampang, n Thailand	Open air	0.73		C	6	Quartzite, indurated sandstone	back-nicking
Lang Rongrien s Thailand	Cave	38,110 B.P. (layer 9) 28,171 B.P. (layer 8)	^{14}C	C, Wf, Ftl, Ctl	45	Chert, shale	hearth features (oxidized soil, charcoal, fire-fractured rocks, bone fragments)
Kota Tampan nw Malaysia	Open-air river terrace	31,000 ± 3,000 bp	F	C, An, Hst, Wf, Ftl, Ctl	1,759	Quartzite, quartz, tuff	
Island Southeast Asia							
Sambungmachan c. Java	Open-air fluvial *?in situ*	0.73–0.97	P, B	Ctl, Ftl	2	Basaltic andesite	
Pacitan, sc Java (series of 'sites')	Open-air fluvial	? LP	G	C, Fl, Ftl, Ctl	2,419	Silicified limestone	
Sangiran c Java	Open-air fluvial	?MP	G	C, Fl, Ctl, Ftl	797	Chalcedony, agate, silicified wood, limestone, etc.	
Leang Burung 2 sw Sulawesi	Cave	0.031–0.019	^{14}C ^{14}C	C, Fl, Ftl	c. 6,800 *in situ:* 5,485	Chert, quartz, limestone	burnt bones, ash, shells, charcoal
Niah n Sarawak	Cave	LP (1) ?0,04–0.006		C, Fl, Ctl, Ftl	c. 1,209 (preceramic)	Sandstone, limestone, quartzite, chert	

Key to abbreviations: n = north; nw = north-west; e = east; s = south; w = west; sw = south-west; c = central; sc = south-central; A = Amino acid; B = Biostratigraphic; Ep = Early Pleistocene; MP = Middle Pleistocene; LP = Late Pleistocene; e = early; m = middle; l = late; t = terminal; ESR = Electron Spin Resonance; F = Fission-Track; G = Geomorphological; ^{14}C = Radiocarbon; P = Paleomagnetic; K/Ar = K; P-A = Potassium-Argon; T = Thermoluminescence; U-s = Uranium-series; n† = number of individuals; n†1 = minimum number of artefacts or individuals; C = Cores; An = Anvils; Sph = Spheroids; Hst = Hammerstones; Wfrg = Waste fragments; Wf = Waste flakes; Tl = Tools; Ctl = Core tools; Ftl = Flake tools; "Traditions" of Chinese authors: L = Large tool traditions; S = small tool tradition; S/L small and/or large tool tradition.

any other Paleolithic materials yet reported from China" (Jia 1985:137). For this reason and because of the early date attributed to the assemblage, it has been considered to be the progenitor of all subsequent assemblages. The artifactual nature of many of the 30 artifacts and the Plio-Pleistocene date of the assemblage is highly questionable (Pope 1982, 1985a, 1988). Far more convincing as evidence of well-preserved early assemblages is the evidence from the two demonstrably Early Pleistocene Nihewan Basin assemblages of Donggutuo and Xiaochangliang that preserve numerous artifacts in association with fauna derived from low-energy depositional contexts.

The Nihewan Basin

The Nihewan Formation is located in northwestern Hebei Province (c. 114° E by 40° N) approximately 150 km northwest of Beijing and Zhoukoudian (Figures 25-1, 25-3, 25-4, 25-5). This basin has long been recognized as one of the richest and most extensive fossiliferous Plio-Pleistocene sequences in China (Licent 1924; Barbour 1924, 1925; Barbour et al. 1927; Teilhard de Chardin and Piveteau 1930; Teilhard de Chardin 1941). More recent research efforts have resulted in the recovery of *in situ* Lower Paleolithic artifact assemblages dated by stratigraphic, paleontological, and paleomagnetic studies to at least c. 0.73 – 0.97 MYA (Huang 1985; Jia and Wei 1987; You et al. 1978; Wei 1985; Wei et al. 1985; Pope et al. 1990). Because the Nihewan Paleolithic sites will remain pivotal in any arguments about the archeological evidence for paleocultural behavior in the Far East, they deserve a somewhat detailed explication of their geological and chronological contexts.

Lithostratigraphy. The Nihewan Formation derives its name from a small farming village located approximately 3 km (Figures 25-3, 25-4, 25-5) from the type section (at Xiasha Gou) first quarried for mammalian fossils by Licent (1924) and Barbour (1924; Barbour et al. 1927). The deposits are exposed along both sides of the east-west flowing Sanggan River and are the result of fluvio-lacustrine deposits that began accumulating at least as early as the Late-Middle Pliocene. The sediments dip northwest at an angle of about 8-10° and become thicker toward the center of the basin and thinner toward its margins. The exposures have been divided geographically into three separate sections: (a) the Nihewan village section on the north side of the Sanggan River Valley, (b) Shangsha Gou to the northwest, and (c) the Donggutuo Platform on the south flank of the Sanggan River (Pope unpubl.).

The lithostratigraphy (Figure 25-6) of the Nihewan Basin has been investigated for over 60 years (Licent 1924; Barbour 1924; Barbour et al. 1927; Teilhard de Chardin and Piveteau 1930) and is now generally understood as an accumulation of subaqueously deposited Plio-Pleistocene loess and fluviolacustrine

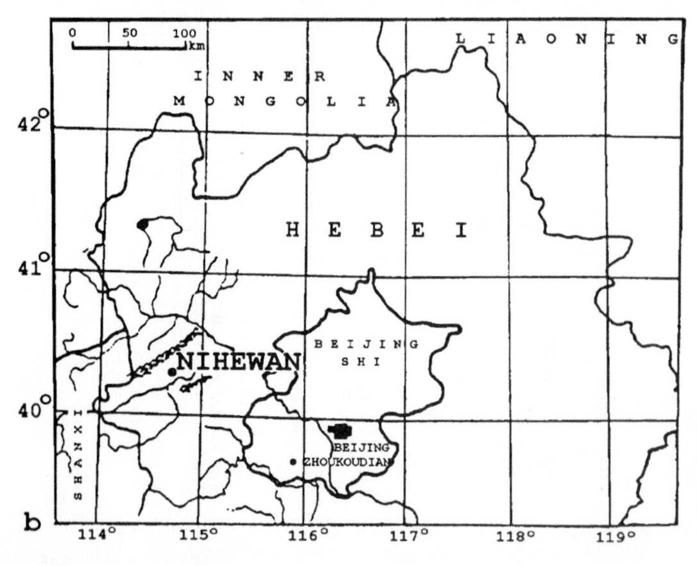

FIGURE 25-3
The location of the Nihewan Basin in Hebei Province (a) and its geographic location relative to Zhoukoudian and Beijing (b).

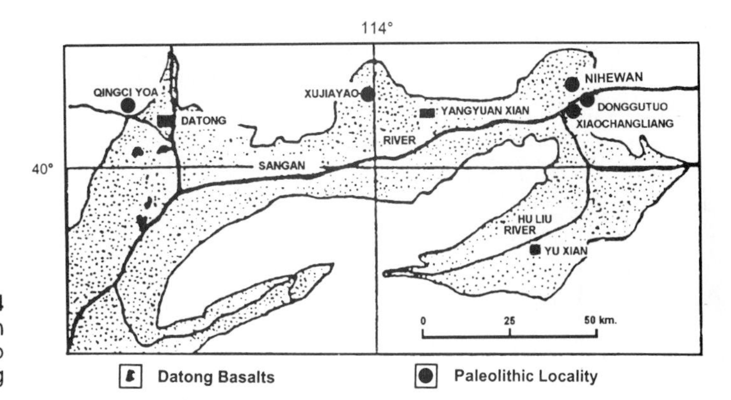

FIGURE 25-4
The Nihewan (Sanggan River) Basin and the locations of Donggutuo and Xiaochangliang

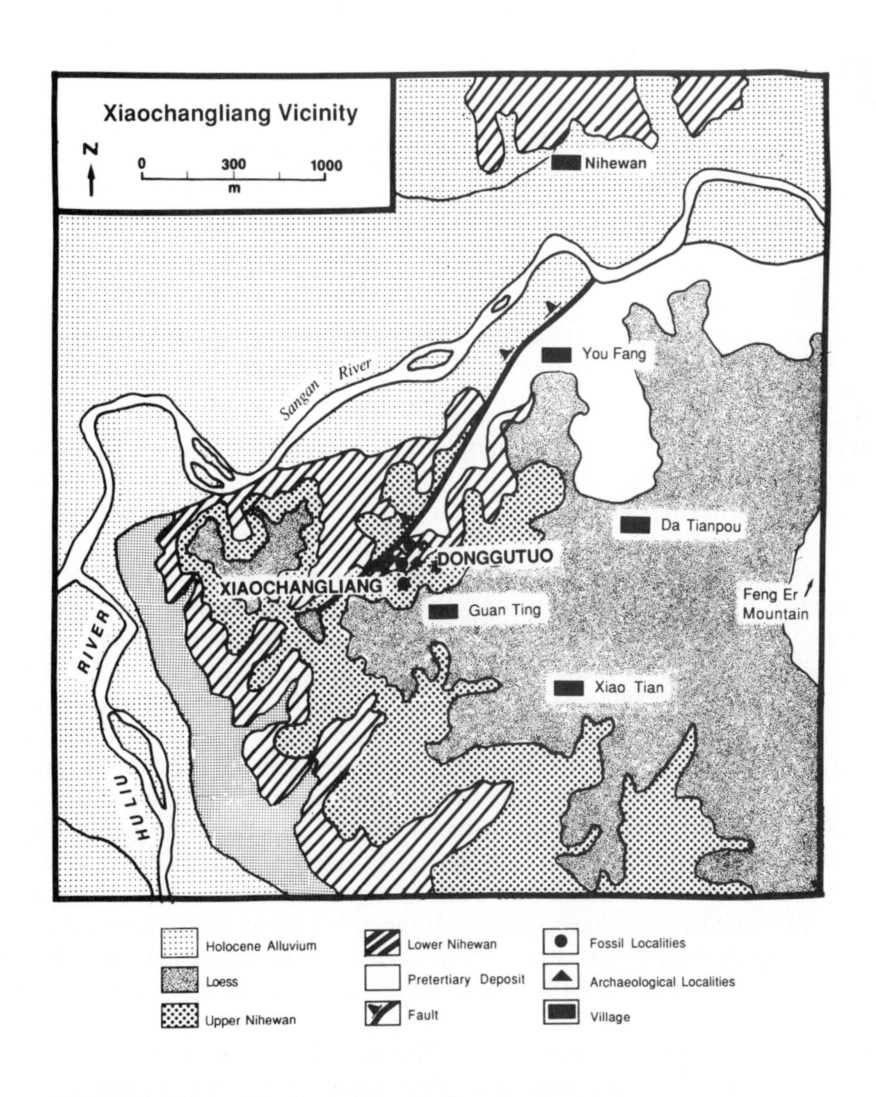

FIGURE 25-5 Map of the Donggutuo Platform Area.

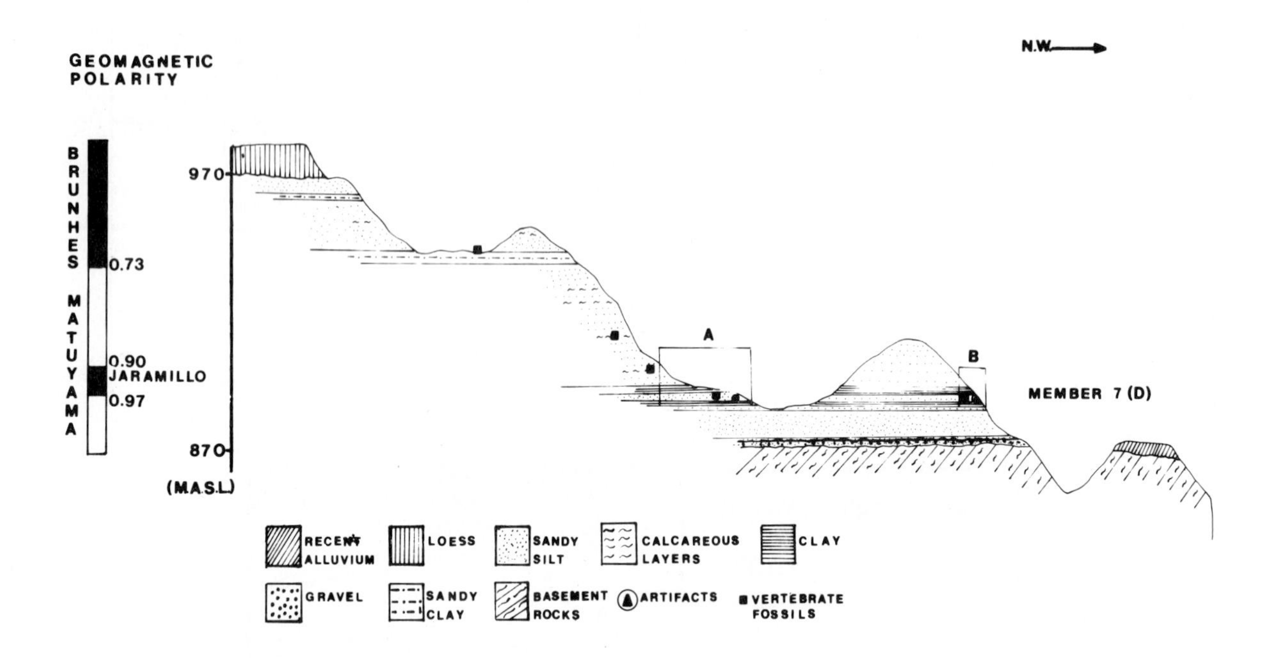

FIGURE 25-6 Schematic cross section of the deposits (approximately 72 m in thickness) of the Xiaochangliang promontory and their relationship to the paleomagnetic stratigraphy for the area. The top of the contact between the Malan Loess and the fluviolacustrine Nihewan deposits is situated approximately 970 M.A.S.L. (Meters Above Sea Level). (After Li and Wang 1982, with modifications.)

sediments overlying and surrounding basement outcrops of cryptocrystalline pre-Tertiary basement rock (Nihewan Cenozoic Research Team 1974; Wei 1978, 1988; Pope et al. 1990). The aeolian deposits of Malan Loess cap the fluviolacustrine deposits. The base of these loess deposits date to approximately 0.125 MYA (Liu et al. 1985).

The Nihewan sequence has been subdivided in various ways (see Wei 1978 for a review; also Liu and Xia 1983), but there is an emerging though perhaps oversimplistic consensus that the entire succession can be divided into 14 or 15 members ("beds" or "layers" of authors) beginning from bottom to top with number 1, the so-called Late Pliocene "Red Hipparion Clays" (Nihewan Cenozoic Research Team 1974; You et al. 1978, 1980; Wei 1978, 1988; Pope 1982). Unfortunately, biostratigraphy has played a major role in previous lithostratigraphic divisions. The "Red Hipparion Clays" vary locally in their development and rest unconformably on pre-Tertiary metamorphic and igneous rocks. The red clays are separated from the sediments of the Nihewan Formation proper by an erosional unconformity and a basal conglomerate that underlies the fluvio-lacustrine sediments of the Nihewan Formation. The genus *Hipparion*, however (generally regarded as a taxon restricted to the Pliocene of Eurasia), may in fact not be restricted to the "Red Hipparion Clays" since it has also been reported from the uppermost Lower Nihewan Formation (see later discussion) and Xiaochangliang. Although the zonation of this species as well as the age of *Hipparion* were previously thought to date to about 2.5 MYA (Eisenmann 1975; Eisenmann and Brunet 1973) its Last Appearance Datum (L.A.D) now appears to date to about 1 MYA in Eurasia (Pope 1988).

Biostratigraphy. The fossiliferous areas of the Nihewan Formation are a complex succession of fluviolacustrine clays, silts, sands, and gravels. Local faulting occurs in various locations with blocks being down-thrusted toward the center of the Sanggan River Basin (Liu and Xia 1983, 1989). The Nihewan Formation has been divided into an Upper and Lower Nihewan Formation with Member 7 representing the uppermost portion of the Lower Nihewan Formation and members 8-14 representing the Upper Nihewan Formation (You et al. 1978, 1980; Pope unpubl.). The Nihewan Formation as used here thus includes members 2-14 and excludes the pre-Tertiary basement rocks, the "Red Hipparion Clays," and the Upper Pleistocene aeolian Malan Loess "cap." All of the fossil-bearing strata can be attributed to the depositional phases of a large lake and its tributaries that began forming either in the Late Pliocene Fen He Erosion Cycle (Pope 1982) or in the Early Pleistocene. This lake apparently fluctuated in size throughout the Pleistocene (Nihewan Cenozoic Research Team 1974), at times becoming markedly saline (M. Palacios pers. comm.). The Formation reaches a maximum thickness of more than 620 m at Nihewan (based on core samples) and becomes thinner at its outer margins on either side of the Sanggan River. Most of the natural outcrops are between 100-150 m thick (Nihewan Cenozoic Research Team 1974).

The "Nihewan Fauna" has generally been regarded as representative of the "Chinese Villafranchian" (Barbour 1924; Teilhard de Chardin and Piveteau 1930). *Elephas* (*Palaeoloxodon*), *Equus*, *Hipparion (s.l.), and Paracamelus* have all been reported from the Nihewan Basin (Wei 1978, 1983; Xu and You 1982; You et al. 1978). *Leptobos* has also been reported, but its presence is based on very few specimens. However, despite previous research, it is still uncertain as to whether or not these taxa all derive from a single synchronic stratigraphic level of the Nihewan Formation. The documentation of the sympatric and synchronic occurrence of the so-called ELE (*Equus*, *Leptobos*, *Elephas*) group at Nihewan would be extremely useful for the development of long-range correlations with other areas of Eurasia.

Recently Chinese scientists have shown that there are "stratigraphic" and even direct localized associations of *Equus* and *Hipparion* in at least four localities at Nihewan (Wei 1978; Xu and You 1982). If this proves to be true it would be the first such confirmed stratigraphic overlap in Eurasia outside of Pakistan (Hussain and Bernor 1984; see also Eisenmann 1975; Eisenmann and Brunet 1973). This co-occurrence is usually dated in Eurasia to about 2.5-3.5 MYA. Such an association also has a direct bearing on the relatively early age (c. 2 MYA) claimed for the archeological locality of Xihoudu, in southern Shanxi (Jia 1980, 1985). Whether such an association (if actual) is the result of the relatively late survival of *Hipparion* (c. 1 MYA) or the relatively early appearance of hominids in China is of paramount importance for determining chronometric frameworks for other Eurasian hominid localities. The discovery of *Hipparion* (outside of Africa) in association with artifacts would represent an important advance in attempts to establish a reliable Plio-Pleistocene framework for all of Eurasia.

Similarly, there is some indication that *Elephas namadicus* may appear as early as the Early Pleistocene (0.73-1.8 MYA) in China (Li 1984; Zhou and Yu 1974). Alternatively, early species of this genus may represent *Elephas (Archidiskodon) planifrons* (Maglio 1973). This possibility may also be testable at Nihewan. The existence of *Leptobos* and *Bison* in East Asia has also been the focal point of much debate (Aigner 1972, 1981). The first appearance of these ungulates represents an important datum for the establishment of long-distance correlations with Europe and North America.

Scattered but often dense concentrations of artifacts and mammalian fossils (Figures 25-7 and 25-8) have been recovered from both the Upper and Lower Nihewan Formation but seem to be most dense in Member 7 (Pope unpubl.). These gray-yellow, sandy silts contain numerous fossil localities and at least six archeological localities including Xiaochangliang and Donggutuo (Member D), which on the basis of paleomagnetic studies have been accorded an age of between 1.52–3.0 MYA (You et al. 1978) or older than 0.96 MYA (Cheng et al. 1978; Liu and Xia 1983; Li and Wang 1982; Jia and Wei 1987). This stratum has been traced for several kilometers and has been interpreted as the result of perilacustrine floodplain and tributary deposition.

As a result of a long and continuing research effort detailing the microfaunal stratigraphy of the nearby Luochuan Loess Plateau, overlap zones, F.A.D.(First Appearance Datum) and L.A.D.(Last Appearance Datum) points of ochotonids, lagomorphs, and myospalacids are becoming much clearer for northern China (Liu et al. 1985). This work provides a firm basis for tying the Xiaochangliang and Donggutuo exposures into a regional sequence for which a well developed paleomagnetic stratigraphy also exists (Liu et al. 1985; Figure 25-6). Several species of myospalacids are now recognized spanning the Upper Wucheng and Lower and Middle Lishi Loess. These have been divided into reasonably well-defined subzones:

FIGURE 25-7 Excavations at the Xiaochangliang promontory.

Myospalax chaoyatseni and *M. arvicolinus* between 0.9-0.6 MYA (Liu et al. 1985; and others). Since the artifact-bearing portions of the Xiaochangliang and Donggutuo levels appear to correspond to the lower part of the Lower Lishi Loess, this work provides a valuable source for the assessment of both the biostratigraphic and paleoenvironmental contexts of these sites.

No isotopic studies of igneous rocks have yet been carried out in the Donggutuo Platform area due to an absence of suitable Tertiary and Quaternary rocks. However, both kinds of dates taken in conjunction with biostratigraphic data and TL and uranium series dates produced by studies of the Central Loess Plateau near Luochuan, Shanxi (Liu et al. 1985), currently suggest that the Jaramillo event occurs about 70–80 m below the base of the Malan Loess throughout northern China. At Xiaochangliang the Jaramillo appears to occur between 50–60 m below the Malan Loess (Pope pers. observ.). Future work needs to concentrate on correlating the datable Datong deposits with the fossiliferous and artifact-bearing deposits of the Yang Yuan and Yu Xian portions of the Nihewan Formation. These deposits have so far supplied a minimum age for the top of the Nihewan Basin sequence.

Paleomagnetic Dating. One of the most persistent questions at Nihewan and indeed at almost all northern Chinese sites has been the question of chronology. Age estimates for the appearance of the earliest Chinese hominids range from c. 3 MYA to c. 1 MYA. The paleomagnetic studies from Nihewan present a similar divergence of opinions (see Liu et al. 1977; Jia and Wei 1987; You et al. 1978; Li and Wang 1982; Pope 1988).

Past studies are in agreement that the earliest archeological sites (Member 7 at Xiaochangliang and Member D at Donggutuo) at Nihewan antedate the beginning of the Jaramillo event at 0.97 MYA, but some studies (You et al. 1978, Cheng et al. 1978) suggest these early sites are far earlier. The paleomagnetic study of Li and Wang (1982) was based on 614 oriented samples from three 100 m thick sections near Nihewan

FIGURE 25-8 The white sandy silt of Xiaochangliang which has yielded artifacts in association with mammalian fauna.

village each about 100 m thick. They concluded that the Jaramillo subchron occurs at about 70 m below the base of the Malan Loess (Figure 25-6). In contrast, You et al. (1978) assumed the highest reported occurrence of a relatively generalized *Hipparion* (not the more specialized *Proboscidipparion*) at about 2.0 MYA and therefore correlated the base of normally magnetized sediments just above Member 7 with the base of the Olduvai subchron at c. 1.85 MYA. Ultimately, the question of the actual age of Member 7 can only be settled with new studies. Our conclusion based on 275 samples collected at 0.5 m intervals and analyzed by John Hillhouse (U.S. Geological Survey) indicates that artifact-bearing levels at Xiaochangliang occur within or just below the Jaramillo Subchron.

Xiaochangliang and Donggutuo

The Xiaochangliang–Donggutuo sites are significant in that they are two of the few Early Pleistocene localities in Asia where mammalian fossils occur in direct association with numerous stone artifacts. Fossil mammals are especially well preserved at Xiaochangliang, and preliminary investigations show that the Xiaochangliang site represents a tool manufacture and activity site produced by hominids along an Early Pleistocene lake margin (Jia and Wei 1987). Donggutuo, however, appears to represent a geological concentration that preserves no demonstrable evidence of hominid utilization or occupation. No discrete stratigraphically concentrated horizons of artifacts indicate a paleolandscape surface. Instead artifacts appear to be distributed randomly in natural layers of silt that overlay a gravel-boulder bed indicative of a high energy-fluvial environment. At Xiaochangliang, artifacts are confined to a circumscribed strata of white sandy silts and orange sands (Figs. 25-7 and 25-8) which on the basis of current observations do not exceed 0.25 m in thickness.

However, in both sites a wide range of similar artifact categories is present, including cores, waste fragments ("chunks"), waste flakes, micro-debitage, and utilized and/or retouched flakes (Keates in prep.). In comparison with other open-air Asian Early Pleistocene localities, the range in artifacts is unprecedented, since few other Early and Middle Pleistocene archeological localities in Asia include this range of artifacts. In comparison with the other earliest Paleolithic assemblages, these localities present a different set of problems that do not easily lend themselves to the previously accepted parameters of the Large-Small tool scheme previously mentioned. Following the Large Tool-Small Tool scheme, both localities can be classified as representative of the Small Tool industries. Both assemblages are also similar in the selection of raw materials from nearby outcrops.

Donggutuo

Donggutuo (40°13′ 13" N x 114° 39′ 34" E) refers to a series of excavations on the east and west flanks of a gully conducted from the years 1981 to 1992. This locality was discovered and first investigated in 1981 (Wei 1985, 1989, pers. comm.; Jia and Wei 1987; Schick et al. 1991). Lithic artifacts and fossilized vertebrate materials were excavated from five trenches [T1–T5] divided into five archaeological layers (Wei 1988, pers. comm.).

Our inspection of these trenches and subsequent study of the recovered specimens has convinced us that although the Donggutuo specimens are artifacts, the trenches probably do not represent hominid utilization or occupation sites. Like the other four trenches, the largest and most productive of these, T 1 appears to represent thick colluvial deposits without distinct artifact-bearing horizons. The entire 4-m thickness is underlain by a large bed of rounded gravels ranging up to several centimeters in diameter. The extremely fragmented condition of the faunal remains is consistent with the interpretation that the deposits represent a colluvial aggregation. Despite this possibility, the artifact frequencies and technology of the artifacts show obvious and strong similarities to the Xiaochangliang assemblage.

Artifact frequencies are higher than those known from Xiaochangliang (n = c. 10,000; Jia and Wei 1987). The majority of Donggutuo artifacts are manufactured from fine and coarse grained chert (Table 25-2). Of the total number of Donggutuo artifacts published (n = 1443, Wei 1985, 1989) about 8.1 percent (n = 117) were conservatively judged to be tools and 91.9 percent debitage (n = 1326) (Table 25-3) (Keates in prep.).

Xiaochangliang

The Xiaochangliang locality is approximately 1 km from the site of Donggutuo (Jia and Wei 1987). As the result of previous Chinese work and our own work in 1989 and 1990, the area is now recognized as a complex of four contemporaneous excavation localities. The original archeological localities are situated on a promontory (Figure 25-7) that is part of the Donggutuo (or Datienhua of authors) platform area of

TABLE 25-2 Comparative raw material frequencies of Donggutuo tools

Category	T1	T2	T3	T4	T5	Total	
	n	n	n	n	n	n	%
Chert							
fine	42	12	10	10	5	79	67.5
fine-coarse	16	1	3	6	5	31	26.5
Basalt	2	1	0	0	2	5	4.3
Quartz	1	0	0	1	0	2	1.7

TABLE 25-3 Comparative frequencies of Dongguto tool categories

Category	T1 n	T2 n	T3 n	T4 n	T5 n	Total n	Total %
Side utilized	25	7	7	8	7	54	46.2
End utilized	4	0	0	1	0	5	4.3
Side denticulates	15	2	4	5	2	28	23.9
End denticulates	3	0	0	1	0	4	3.4
Denticulate points	2	2	0	0	0	4	3.4
Point utilized	4	1	1	2	0	8	6.8
Notches	8	2	1	0	3	14	12.0
Total n	61	14	13	17	12	117	
Total %	52.1	12.0	11.1	14.5	10.3		100

the Nihewan Formation in the same stratum as the "gray silts" (Figure 25-8) of Member 7, which can be traced for several kilometers throughout the local Guan Ting village area.

Artifacts occur throughout 0.2-1.0 m of fine gray silt in association with mammalian fossils. This locality was the object of preliminary study and excavation in 1978 (You et al. 1978) and was surveyed again in 1988, 1990, and 1991. An area approximately 10 x 100 m along part of the promontory was investigated and cataloged as IVPP sites 78005 A and B (You et al. 1978, 1980). Certain portions of the excavation squares have been characterized as containing primarily flakes and/or cores, but no further documentation in the form of detailed grid plots was generated (see You et al. 1978, 1980). From our own inspections in 1988, and excavations in 1990, the greatest density of artifacts was concentrated in an approximately 0.5-m thick stratum in direct association with vertebrate fossils, some of which show unambiguous evidence of cut marks. At the time of their excavation, Chinese workers did not think it important to plot the provenience of artifacts and fauna. Again from our previous inspections, it is clear that cores, waste flakes, and waste fragments occur in association with tools. It is significant to note that these cultural remains were recovered *in situ*, as indicated by current exposures of lithics in the walls and floors of previous test pits.

At Xiaochangliang 804 stone artifacts were reported by the original excavators (You et al. 1978, 1980; You 1983) from the two site areas A and B. The vast majority of the artifacts, 790 (98 percent) were manufactured on locally obtained fine-grained, vesicular chert. Two basalt artifacts and two sandstone artifacts were also recovered in 1978, as were six specimens of modified bone (You et al. 1978, 1980; You 1983). Our sample comprises 8.7 percent (n = 70) of the You et al. assemblage and also includes specimens recovered *in situ* during the 1988 and 1990 field seasons (n = 80; and see Pope et al. 1990). The lithic categories include cores, waste fragments ("chunks"), waste flakes, including micro-debitage, and utilized and/or retouched flakes (Table 1, 2 ; You et al. 1978, 1980; Keates in prep.). The comparative frequencies of waste materials and tools are 98.1% and 1.9% (You et al. 1978, 1980; You 1983) and 68.7% and 31.3% (Keates in prep.). (*Note*: In many cases the standards of Chinese lithic analysis all but precludes a meaningful and useful assessment of the make-up of published Paleolithic assemblages. However, our own analyses generally concur with the general conclusions that Chinese researchers have reached. Currently the major problem does not lie with the overall characterization of the assemblages but with the insufficient body of statistical data that document the assemblages themselves).

Comparison of the Donggutuo and Xiaochangliang Assemblages

The majority of tools at Donggutuo (47.9 percent) and Xiaochangliang (31.9 percent) are side utilized flakes manufactured from chert (Tables 25-4 and 25-5). Other categories are end utilized flakes, side denticulates, end denticulates, point utilized, and notch flakes. The Xiaochangliang tool categories and to some extent their frequencies are comparable to Donggutuo with the following exceptions: Donggutuo includes no end and side utilized flakes, but includes denticulate points (Table 25-2). The point utilized flakes show evidence of use-wear on one or both edges of the point. Both tool assemblages include mainly irregular flakes. The majority of tools were neither shaped by secondary retouch, nor do they exhibit evidence of rejuvenation. A common feature of the few retouched specimens is their pointed shape, although some of these do not exhibit use-damage on their points (Keates in prep.). At least two tools, each showing evidence of use, resemble "multifaceted core-tools" with graver tips. Each of these artifacts shows a distinct "beak" or rounded and abraded knoblike attribute that evinces clear evidence of use wear.

The majority of artifact forms were produced through the simple reduction of nodules by direct hard-hammer percussion. Bipolar specimens occur in very low frequencies at both sites (anvils used for bipolar percussion were not identified in the assemblages, and none were found during our work at Xiaochangliang). The cores are generally of irregular morphology, and a few relatively regular shaped cores are of cylindrical shape. Cores frequently exhibit acute platform angles. A characteristic feature of the cores are platform tables that were worked to a flat or relatively flat surface. These have been identified on the majority of cores and are interpreted as a core preparation technique. The presence of this technique may be understood with reference to the fracturing characteristics of the raw material as deduced from the results of our core reduction experiments. These experiments showed that both the chert and basalt have mechanical weaknesses that "make it disintegrate." This technique may have been carried out to exert some control over these materials to prevent disintegration and perhaps to produce "more regular" tool blanks. This modification indicates foresight, purpose, an understanding of material characteristics, and a mental template (Keates in prep.).

The large frequency of apparently unexhausted cores at the sites, the small frequency of tools with evidence of shape retouch and/or rejuvenation, and their predominantly unilateral utilization indicate that tools were in general not modified for reuse. These aspects of the technology may be directly related to the availability, accessibility, and proximity to the massive local raw material outcrops. These variables may have "fostered" what appears to be an uneconomic use of raw material. The generally small size of

TABLE 25-4 Comparative raw material frequencies of Xiaochangliang tools

Category	A	B	A/B	TOTAL	
	n	n	n	n	%
Chert					
fine	2	1	30	33	70.2
fine-coarse	0	0	11	11	23.4
Quartz	0	0	3	3	6.4
Total	2	1	17	47	100

Note: Tools under A/B cannot be attributed to either section A or section B, since the excavators (You et al. 1978, 1980) did not record their provenience according to section.

TABLE 25-5 Comparative frequencies of Xiaochangliang tool categories

Categories	A n	B n	A/B n	TOTAL n	TOTAL %
Side utilized	1	0	14	15	31.9
End utilized	0	0	11	11	23.4
Side and end utilized	0	0	1	1	2.1
Side denticulates	0	0	5	5	10.6
End denticulates	0	0	3	3	6.4
Point utilized	1	1	4	6	12.8
Notches	0	0	6	6	12.8
Total	2	1	44	47	100

artifacts is not a distance-related phenomenon, but appears to be related to the fracturing characteristics of the local materials (Keates in prep.).

Primary flakes are very uncommon in the Donggutuo assemblage, and a similar conclusion may be true for Xiaochangliang based on the study of a small part of the original Xiaochangliang collection and our own work at the site. The primary lithic technology at Xiaochangliang and Donggutuo can be characterized as one that relied almost exclusively on local chert outcrops as the source of raw materials. The almost exclusive use of such a fine-grained material at an early Lower Paleolithic site is without precedent in Early Pleistocene Chinese assemblages. The proximity of the outcrops to the sites and the abundance of chert indicates that this may have been a significant factor in both site and raw material selection (Keates in prep.).

Based on the museum study of previously excavated artifacts and of artifacts recovered during the summers of 1988 and 1990, it has become apparent that the Xiaochangliang lithic industry also includes flake-blades not previously encountered in Lower Paleolithic Chinese (and East Asian) assemblages (Figures 25-10 and 25-11). The significance and extent of this technology must be assessed.

The Xiaochangliang assemblage is similar to other presumably Early Pleistocene assemblages in northern China such as Kehe, Xihoudu, and Lantian only in the presence of some trihedral "picks," which though similar in form are much smaller in size. The Nihewan sites differ from these other northern sites in the co-occurrence of flakes and cores and the high density and large total numbers of artifacts. These "Large Tool Tradition" (Jia 1985) sites may not represent true occupation or utilization sites. Debitage and small flakes have been noticeably absent from most of these and other early assemblages, and it has been suggested that many of these sites in fact represent secondary contexts in which artifacts have been transported and/or redeposited (Jia 1985). The significance of these characteristics is poorly understood, but it may be the result of a number of factors including an emphasis by past workers on the collection of "core tools" at the expense of flakes. However, it is already clear that the Xiaochangliang and Donggutuo sites preserve *in situ* assemblages of artifacts and bones documenting the manufacture and utilization of stone tools. It is also clear that Xiaochangliang does represent a hominid activity site.

The archeological assemblages from both Xiaochangliang and Donggutuo are of great interest for a number of reasons. Firstly, open-air (nonspeleological) paleolithic assemblages, preserving a number of formalized stone tools associated with both debitage and cores in direct and proximate (horizontal and vertical) association with faunal remains have never been verified in Asia, let alone China (Binford and Stone 1986; Jia 1980; Jia and Wei 1980). The raw material of the artifacts consists almost wholly of a local fine- to coarse-grained chert that outcrops at numerous locations in the Nihewan Basin and the immediate vicinity of Xiaochangliang. Previously known Lower and Middle Paleolithic assemblages in China (and the Far East in general) have been manufactured primarily from coarse-grained quartzite and vein quartz (Pope 1989a). This aspect of the Nihewan assemblages is very important to long-standing debates about the relationship between raw material availability (Pope 1983, 1984, 1988, 1989a), standardization of lithic technology (Hutterer 1977, 1985), and the cultural capacity of early hominids (Binford 1985, 1987, Binford and Ho 1985; Binford and Stone 1986; Jia 1980; Jia and Wei 1980).

Furthermore, a very important aspect of the Nihewan Basin sites is their geographical location between both Zhoukoudian Locality 1 (to the southwest) and the poorly understood Siberian and Mongolian complexes to the north (Larichev et al. 1987, 1988), and subtropical China, where assemblages are characterized almost exclusively by the artifacts of the "chopper-chopping tool complex" (Aigner 1981; Pope 1989a). In addition to the Nihewan Basin's geographic location, the occurrence of Paleolithic artifacts in well-stratified, fine-grained, well-sorted perilacustrine and lacustrine silts, indicative of an open-air depositional environment, is unprecedented among Asian Lower Paleolithic sites. The biostratigraphy, regional lithostratigraphy, and paleomagnetic data are also in general agreement in according a "mid-Villafranchian," Lower Lishi Loess age to the site (Cheng et al. 1978; Liu et al. 1985; Jia and Wei 1987; Li and Wang 1982).

An understanding of the interrelationships between the distribution and availability of lithic raw materials, the geographic variability of paleolithic assemblages and site formation processes can contribute greatly to interpretations of early hominid cultural capacities and adaptational strategies.

THE AFFINITIES OF FAR EASTERN INDUSTRIES

"Acheulean"

Since the first description of the Pacitanian artifacts, there has been a continuing search for the Acheulean in Asia (Chung 1984; Huang 1985). Although handaxes have been recovered from China and Korea, their frequencies and formalization do not recall the Acheulean. Although large symmetrical bifacial artifacts do occur, their frequencies are always minute compared with the total numbers of the assemblages. Once again the clear pattern in the Far Eastern Paleolithic record is one of continuity of largely informal tool assemblages although formality in the location of flake scars. All of the handaxes in China are surface finds and may conceivably date to the early Upper Paleolithic. Where controlled excavations have been conducted in association with the surface finds, no handaxes have been recovered (Huang 1987, 1990, pers. comm.; Huang et al. 1988, 1990). At Baise two handaxes have been recovered as surface finds. In Korea the site of Chongok-Ri (Chongok-Ni) has yielded artifacts that can be arguably described as handaxes, but they bear a much stronger and more striking similarity to the distinctive Late Pleistocene Dingcun "pick" assemblage and in some ways recall the artifacts from Lantian and Yunxian (Dai and Chi 1964; Aigner 1981; Li 1991). In the case of Chongok-Ri there is now a virtual consensus that the artifacts are redeposited and much later than originally thought.

Despite more than one hundred years of research, definitive chronometric evidence of Early Pleistocene lithic tool manufacture has eluded researchers in island Southeast Asia. Bartstra suggests a Late Pleistocene and Holocene dating of the Pacitanian based on excavations and geomorphological analysis of the fluvial deposits from which he (1976, 1984, 1985) has recovered Pacitanian artifacts. In fact, the redeposition of fossils due to geological agencies such as lahar flows, violent tropical flash floods, and especially mud volcanoes is such a problem that one must beware of the difficulty of establishing the original provenience of any artifact assemblages even if they are recovered *in situ* (Pope 1982, 1983). Despite the contention of De Vos et al. (1982) that there is clear biostratigraphic differentiation between so-called Jetis (Djetis) and Trinil faunas, other researchers have been unable to confirm distinct biostratigraphic zonations in Java (Watanabe and Kadar 1985; Itihara et al. 1985).

FIGURE 25-9 A unifacial fine-grained worked chert artifact from Lampang (Wang River Valley), northern Thailand; (a) Surface with several flakes removed; (b) surface showing extensive battering scars resulting from uses as an anvil or hammerstone.

The "Middle/ Upper Paleolithic Transition" in Asia

Although the age of handaxes in the Far East is uncertain, it is impossible to agree with Klein's (1992:9) contention that our understanding of the transition to the Upper Paleolithic and the "impression of remarkable continuity and conservatism . . . is . . . based on only a small number of excavated sites, often poorly dated and unevenly described, and it may not be sustained as research expands." In fact, however "poorly dated" or "unevenly described," the fact remains that after more than half a century of archeological research based on scores of sites, true Acheulean assemblages remain absent in the Far East. It is highly unlikely that future research, no matter how carefully conducted, will change this picture. Once again the fact that all of the handaxes from China are surface finds strongly suggests that handaxes may be an Upper Paleolithic development in this geographic region. On the basis of current evidence, the notion that handaxes are a late development, while heretical in Europe, Africa, and the Near East is well founded in the Far East. We further take strong exception to Klein's (1992:9) contention that we should look to Sri Lanka for "what may have actually happened." Whatever actually happened in China and other regions of the Far East can only be determined by what is actually found there.

Based on an awareness of both the northern and southern mainland Southeast Asian materials, an informal approach to lithic manufacture seems to have a great time depth in Southeast Asia in which minimally formalized assemblages of low density were manufactured for hundreds of thousands of years with very little in the way of lithic innovation. Furthermore, there is no cogent case for relegating particular artifact types to even broad ranges of time (see especially Harrisson 1970; Olsen and Ciochon 1990; but see Bartstra 1984, 1985 for a different view).

In Praise of Anecdotal Information

The observation that the Asian Paleolithic assemblages do not readily lend themselves to statistical analysis is not tantamount to viewing Pleistocene archeology of this region as unknown or unknowable. Two other pieces of admittedly anecdotal evidence from Thailand not only pose a serious question about the validity of using lack of standardization and complexity as an archeological indicator of the cultural capacity of forest adapted people, but also offer an opportunity to deduce early hominid capacities for such abstractions as foresight.

During the course of several field seasons of the Thai-American expedition, an undatable surface find of a unifacial core artifact manufactured on locally available very fine-grained chert (Figure 25-9a,b) was recovered from the Wang river valley near the site of Kao Pah Nam (see later discussion). On one side of the artifact (Figure 25-9a) numerous facets had been removed to produce an artifact that is very similar to handaxes as they are known from Europe and Africa. The other side of the artifact (Figure 25-9b) exhibits numerous pits resulting from extensive battering. There can be no doubt that the raw material is highly suitable for the manufacture of any form of standardized artifact. The overall shape and multifaceted surface attest to the manufacturer's ability to produce a symmetry characteristic of a handaxe simply by repeating the same technique on the other side. Regardless of whether the manufacturer of the tool was a *Homo erectus* or a modern human, the manufacturer's ability is obvious. In other words if the artifact is of recent origin, then the preference for unifacial artifacts undeniably persists into recent prehistory. If the artifact is considerably older, then the ability to make symmetrical artifacts persists well into the past in Southeast Asia.

Another anecdotal observation also sheds light on the question of foresight in early Southeast Asian hominids. Excavations from the Middle Pleistocene (based on faunal evidence) rockshelter of Kao Pah Nam in the Wang river valley, northern Thailand (Pope et al. 1981) revealed apparently burnt bones in association with fire cracked basalt cobbles, which have been interpreted as a hearth (Pope 1985b; Pope et al. 1986). The fact that these were the only basalt materials (the rest being limestone and quartzite) found at the site and the probable location of the source near a basalt flow at least a kilometer from the site led to the conclusion that the basalt was deliberately brought in. Later it was learned that repeated exposure of the limestone to alternate heating resulted in the manufacture of caustic quicklime (Pope 1989a). We suggest that the occupants of Kao Pah Nam were aware of the unpleasant effects of combining water, limestone, and heat and took precautions to avoid the unpleasant results.

The use of anecdotal information, while generally eschewed, should not be ignored in the study and understanding of artifact assemblages that currently cannot and perhaps never can be described on the basis of statistical techniques requiring large samples of the kind that have come to be the *sine qua non* of modern archeological investigations. However, we suggest that this kind of deduction not only forces the conclusion that these hominids had an intimate and culturally complex idea of the cultural landscape, but they also possessed foresight based on the details of memory.

Binford (1987) has also addressed the question of evidence for foresight at Ambrona and Torralba in a similarly anecdotal way. Binford has reasoned that the co-occurrence of artifacts manufactured from native stone at Torralba and faunal remains does not argue for planned hunting strategies because nonnative rock would have been transported to the meat procurement site if the activity had been preplanned. The logic here is that transporting material to a potential ambush site shows more foresight than picking a site where both suitable lithic materials and animals are known to occur. The transport of exotic stone to activity sites may represent more conclusive evidence of foresight and planning, but it can be alternatively interpreted as representing a precautionary strategy across an unfamiliar landscape. Choosing a locality where all the resources are concentrated at a single locus may represent evidence for true complex planning. By this reasoning the evidence from Kao Pah Nam must represent both planning and a retained familiarity (detailed knowledge) of both the landscape and in this case the physiochemical properties of at least one kind of rock.

As noted previously archeological conclusions about subsistence in the Asian tropics will always be further complicated by both the nature of the environment and the prevalence of extensive nonlithic resources. Some time ago it was observed that on the basis of archeological standards of lithic evidence, some living groups cannot be shown to actually exist: "Indeed it is doubtful whether an archaeologist finding the remains of a Semang hearth after the wooden tools had rotted away in the hot, humid climate would in the few rough stones recognize a human industry at all" (Forde 1934:17; see also De Morgan 1924:28).

Particularly relevant to the question of the cultural complexity and capacity of Pleistocene hominids are logical deductions that can be made about the Pacitanian (Patjitanian) and some unreported surface finds from Thailand. The Pacitanian "Industry" presents a similar example of what can be termed "unutilized cultural capacity." A number of the bifacial Pacitanian artifacts are conformable with the "Abbevillian" (the Acheulean of current usage) as it has been recognized in Europe. Bartstra believes that these more standardized Javanese artifacts should be associated with anatomically modern humans in Java. Although this point is impossible to argue, what is significant is that the Pacitanian is an outstanding example of an exception that proves a rule since its geographic context is in the midst of an area that has usually produced unformalized forms. We suggest that the most parsimonious way of interpreting the occasional occurrence of standardized forms is the interpretation that for whatever reason the manufacturers of both the unifacial artifacts from Thailand and the highly standardized forms from Pacitan were capable of, but not always interested in, making particular forms. Rather than suggesting an area of cultural retardation or isolation, this observation suggests a highly malleable repertoire of artifact manufacture. This may suggest that standardized stone tools were not adaptive. However, this does not necessarily indicate that the local hominid technologies did not produce standardized nonlithic assemblages. We think that whatever the age of the Baise, Dingcun, Kehe, Chongok-Ri, and Pacitanian assemblages and whoever the manufacturers were, the conclusion that these hominids had the ability, but not the desire to manufacture large numbers of handaxes cannot be avoided.

PALEONEUROLOGICAL AND LINGUISTIC EVIDENCE

Such an interpretation is also much more in accord with the paleoneurological evidence for the evolution of the hominid brain (Falk 1987; Holloway and de la Coste-Lareymondie 1982). Although studies of hominid endocasts disagree about the ultimate antiquity of an anatomically modern cerebral cortex, there is agreement that by about 1.5-2.0 MYA cerebrally asymmetrical brains were present in early members of the genus *Homo* (Falk 1987; Holloway and de la Coste-Lareymondie 1982). Furthermore there is absolutely no evidence for the synchronization of neuroanatomical change in the brain with the Middle/Upper Paleolithic transition (Tattersall 1992). The archeological work of Toth (1982, 1985) also supports this conclusion and indicates that early members of *Homo* exhibited preferential handedness. The connection between cerebrally asymmetric brains, handedness and language though not understood in detail has received almost paradigmatic acceptance by neurologists and psychologists (Kimura 1992). This makes arguments for prolonged minimal cultural capacity in *Homo erectus*, Archaic *Homo sapiens*, and Neandertals highly unlikely.

We believe that a greater awareness on the part of archeologists of the paleoneurological evidence would deter a willingness to postulate minimal cultural capacities for early hominids not only in Asia, but for all parts of the Old World. Other anatomical arguments that point to the evolution of osseous elements associated with the soft tissues of the vocal tract (Laitman and Reidenberg 1987; Lieberman 1989) have suggested that regardless of the neurological structure of the brain, even as late as Neanderthal times the genus *Homo* lacked the physical capacity for the production of the full range of sounds of modern speech. Despite the obvious difficulty (impossibility in our view) of reconstructing vocal tracts from fossil remains, such arguments ignore a very important aspect of many languages, tonality. The reconstruction of minimal linguistic capabilities in Asia is especially ironic in the Far East where the vast majority of Asian language speakers employ a system in which a single phoneme conveys a variety of totally discrete meanings. There is of course absolutely no credible evidence for the antiquity of tonal languages.

There is however, a large body of clinical, anatomical, and behavioral evidence for the connection of handedness, cerebral asymmetry, language, and manual manipulation in modern humans. It is also possible to point out that both human speech and the manufacture of artifacts involve the imposition of arbitrary forms on the environment. We are not arguing that early hominid speech was equal in complexity to modern speech or that early tool-making capacity was equal to that of modern humans. The archeological record indicates that early artifact production was probably much influenced by the shape and nature of the raw material (see also Jones 1979, 1981). We are arguing that the continuing geographic variability and the paleoneurological evidence strongly supports the interpretation that by the time early hominids reached the Far East recognizably sapient human behavior had already evolved. We believe that this is one of the important conclusions that can be drawn from the Pleistocene archeological record of Asia. One of us (Pope 1989a) has postulated that an emphasis on nonlithic technology in the Far East was a further expression of highly flexible sapient behavior.

CONCLUSIONS: PATTERNS IN PLEISTOCENE BEHAVIOR

Outside of the Far East, the search for the causes, processes, and origins of the explosion in cultural complexity of the European and Near Eastern (and to some extent the African) Upper Paleolithic has been the subject of intense debate and discussion (Rigaud 1989; Shea 1989; Mellars 1989; Klein 1992). The geologically rapid increase in the complexity, formalization, and diversity of artifacts in the last forty thousand years has been "explained" on the bases of both biological and socioecological models involving causative explanation ranging from neuroanatomical change, to inter-populational dynamics, to climatological change (Cavalli-Sforza et al. 1988; Wilson and Cann 1992). The "Eve" hypothesis as applied to the European Paleolithic record combines elements of all these models (cf. Mellars 1989). Most recently an increased resolution of the Paleolithic evidence and its chronometric contexts as they are known in the West has underscored the fact that there is no consistent correlation between lithic technologies and hominid taxa (Grün et al. 1990; Schwarcz et al. 1989; Stringer and Andrews 1988). This greatly diminishes the likelihood that biological-anatomical change was a "prime mover" in cultural evolution of the last fifty thousand years. Again this is highly consistent with our current understanding of both the antiquity of humanlike cerebral organization and the evolution of cranial capacity.

What has been long overlooked in attempts to discern the origins of modern cultural complexity is the fact the Euro-African models and the interpretive problems that they seek to solve are largely irrelevant to the Far East. As we have shown in this chapter, the questions in the Far East are entirely different. In this area of the world, continuity, not rapid change, early, but sporadic standardization and geographic variability are the outstanding problems that are evident in the Asian paleolithic record. The voluminous literature on the meaning and origins of paleolithic art is a prime example of a question that is currently without relevance in the Far East. Its absence in the Far East is conspicuous (Keates in press).

We are not suggesting that the concept of a transition to the Upper Paleolithic is irrelevant in places like China and Southeast Asia (see also Ikawa-Smith 1978), only that the main issues generated by a study of the available data already indicate that an entirely new perspective must be brought to our understanding of this part of Eurasia. On the other hand, there does seem to be a period of marked change in northern Asia in the last forty to thirty thousand years, but from China to Indonesia there is no "Middle Paleolithic" base for such change since the concept of a Middle Paleolithic is without relevance in most of the Far East.

New lithic forms do seem to appear in the Far East at a time penecontemporaneous with the Upper Paleolithic in Europe, but once again it is the origins of geographic rather than temporal variability that suggests itself as a primary research focus. Currently the most useful and accurate way of understanding the Far Eastern Paleolithic is by geographical regions rather than by chronological periods. Specifically although a formalized blade technology does emerge or is introduced in northern China, its geographic variability and distribution remain poorly understood.

Based on the occurrence of occasional formalization as evidenced at both early sites such as the Nihewan localities and the existence of overall unformalization at late sites such as those from Southeast Asia, continuity is overwhelmingly the best way of understanding the Far Eastern Paleolithic. Attempts to demonstrate the introduction of new technologies into this region of Pleistocene Asia have absolutely no support from the data. Contrary to frequently expressed doubts about the chronology and resolution of the Asian evidence, the Paleolithic of the Far East is sufficiently well known to argue that future research will not alter this picture. Although much remains to be done, much has already been done. This fact remains essentially unappreciated by the majority of archeologists who do not have direct linguistic access to the voluminous Chinese literature. This body of evidence incorporates hundreds of Chinese site reports that despite their widely varying recency or degree of sophistication continue to produce a very consistent picture of the evidence for paleocultural behavior at the edge of Pleistocene Eurasia. We believe that the main "surprises" of future research will not be the confirmation of results predicted on the basis of fieldwork in the West. Instead, we feel that new discoveries will increasingly underscore the reality of the substantial differences to be found in the Far Eastern archeological record.

ACKNOWLEDGMENTS

We are very grateful to our numerous Asian colleagues all of whom have struggled with us to understand the immensity of the Far Eastern Paleolithic Record. Special gratitude is due to Wei Qi for his willingness to share his pioneering work at the Nihewan. Other members of the Institute of Vertebrate Paleontology and Paleoanthropology to whom we are indebted include Huang Wanpo, Huang Weiwen, Jia Lanpo, Gai Pei, and Li Yi. In Thailand we have benefited greatly from our long-standing collaboration with Vadhana Subhavan and Somsak Pramankij (Museum of Prehistory, Mahidol University) and the director of the Museum, Professor Sood Sangvichien. Finally, G. G. P. wishes to acknowledge the years of support, friendship, and guidance that F. Clark Howell has so steadfastly provided.

LITERATURE CITED

Aigner, J. S. (1972) Relative dating of north Chinese faunal and cultural complexes. *Arctic Anthropol.* 9:36-79.

Aigner, J. S. (1978a) Important archaeological remains from North China. In F. Ikawa-Smith (ed.): *Early Paleolithic in South and East Asia.* The Hague: Mouton, pp. 163-232.

Aigner, J. S. (1978b) Pleistocene faunal and cultural stations in south China. In F. Ikawa-Smith (ed.): *Early Paleolithic in South and East Asia.* The Hague: Mouton, pp. 129-162.

Aigner, J. S. (1981) *Archaeological Remains in Pleistocene China.* München: C. H. Beck.

Anderson, D. D. (1988) Excavations of a Pleistocene rockshelter in Krabi and the prehistory of southern Thailand. In P. Chareonwongsa and B. Bronson (eds.): *Prehistoric Studies: The Stone and Metal Ages in Thailand.* Bangkok: Papers in Thai Antiquity, pp. 43-59.

Anderson, D. D. (1990) Lang Rongrien rockshelter: A Pleistocene-Early Holocene archaeological site from Krabi, southwestern Thailand. University Monograph 71:1-86. University of Pennsylvania: The University Museum.

Barbour, G. B. (1924) Preliminary observation in Kalgan area. *Bull. Geol. Soc. China* 3(3):167-168.

Barbour, G. B. (1925) *Physiographic History of the Yangtze. National Geological Survey of China.* Peking: Institute of Geology National Academy.

Barbour, G. B., Licent, E., and Teilhard de Chardin, P. (1927) Geological study of the deposits of the Sangkanho Basin. *Bull. Geol. Soc. China* 6:263-278.

Barr, S. M., and MacDonald, A. S. (1981) Geochemistry and geochronology of late Cenozoic basalts of Southeast Asia. *Geol. Soc. Am. Bull.* Part I, 92:508-512 and Part II, 92:1069-1142.

Barr, S. M., MacDonald, A. S., Haile, N. S., and Reynolds, P. H. (1976) Paleomagnetism and age of the Lampang basalt (northern Thailand) and age of the underlying pebble tools. *J. Geol. Soc. Thailand* 2:1-10.

Bartstra, G.-J. (1976) Contributions to the study of the Palaeolithic Patjitan Culture, Java, Indonesia, Part I. In J. E. van Lohuizen-De Leeuw (ed.): *Studies in South Asian Culture*, Vol. 6. Leiden: E. J. Brill, pp. 121.
Bartstra, G.-J. (1984) Dating the Pacitanian: Some thoughts. In P. Andrews and J. L. Franzen (eds.): *The Early Evolution of Man with Special Emphasis on South East Asia and Africa. Cour. Forsch. Senckenberg* 69:253-258.
Bartstra, G.-J. (1985) Sangiran, the stone implements of Ngebung, and the Paleolithic of Java. *Mod. Quatern. Res. Southeast Asia* 9:99-113.
Bartstra, G.-J. (1989) Late *Homo erectus* or Ngandong man of Java. *Palaeohistoria* 29:1-7.
Bartstra, G.-J., Soegondho, S., and van der Wijk, A. (1988) Ngandong man: Age and artifacts. *J. Hum. Evol.* 17:325-337.
Bellwood, P. (1985) *Prehistory of the Indo-Malaysian Archipelago*. New York: Academic.
Binford, L. R. (1981) *Bones: Ancient Men and Modern Myths*. New York: Academic.
Binford, L. R. (1983) *In Pursuit of the Past: Decoding the Archaeological Record*. New York: Thames and Hudson.
Binford, L. R. (1985) Human ancestors: Changing views of their behavior. *J. Anthropol. Archaeol.* 4:327-492.
Binford, L. R. (1987) Were there elephant hunters at Torralba? In M. H. Nitecki and D. V. Nitecki (eds.): *The Evolution of Human Hunting*. New York: Plenum, pp. 47-106.
Binford, L. R. (1989) Isolating the transition to cultural adaptations: An organizational approach. In E. Trinkaus (ed.): *The Emergence of Modern Humans*. Cambridge: Cambridge University Press. pp. 18-41.
Binford, L. R., and Ho, C. K. (1985) Taphonomy at a distance: Zhoukoudian, "The cave home of Beijing man"? *Current Anthropol.* 26:413-442.
Binford, L. R., and Stone, N. M. (1986) Zhoukoudian: A closer look. *Cur. Anthropol.* 27:453-475.
Black, D. (1927) On a lower molar hominid tooth from the Chou Kou Tien deposit. In Sun, C. Y., and Ting, K. V. (eds): *Palaeontol. Sin.*, Series D, Vol. 7.
Black, D., Teilhard de Chardin, P., Young, C. C., and Pei, W. C. (1933) Fossil man in China: The Chou Kou Tien cave deposits with a synopsis of our present knowledge of the late Cenozoic in China. *Mem. Geol. Surv. China*, Series A, No. 11, Peking.
Blackwood, B. (1964) *The Technology of a Modern Stone Age People in New Guinea*. In T. K. Penniman and B. M. Blackwood (eds.): *Occas. Pap. Technol.* 3:1-60. Oxford: Oxford University Press.
Bordes, F. (1950) Review of P. Teilhard de Chardin, "Early Man in China." *L'Anthropologie* 54(1-2):82-91.
Boriskovskii, P. I. (1966) Vietnam in primeval times [Part II]. *Sov. Anthropol. Archaeol.* 7:3-19.
Boriskovskii, P. I. (1969) Vietnam in pimeval times, Part III. *Sov. Anthropol. Archaeol.* 8:70-95.
Boule, M. (1927) Introduction. In M. Boule, H. Breuil, E. Licent, and P. Teilhard de Chardin (eds.): *Le Palaeolithique de la Chine*. Archives de l'Institut de Paléntologie Humaine 4:I-VIII. Masson: Paris.
Bowdler, S. (1988) Early Southeast Asian prehistory: A view from Down Under (Abstract). Ass. Southeast Asian Archaeol. Western Europe, Sec. Internat. Conf., Paris, Musée Guimet, 19th-23rd Sept. 1988.
Bowdler, S. (1990) The earliest Australian stone tools and implications for Southeast Asia, Conference Paper. 14th Congress Indo-Pacific Prehistory Association, Yogyakarta, 25th Aug.-2nd Sept. 1990.
Bowdler, S. (1991) The evolution of modern humans, *Homo sapiens sapiens*, in East Asia: Implications of archeological evidence from Australia and Southeast Asia. INQUA Abstracts. Beijing: INQUA, p. 33.
Breuil, F. (1927) Archeologie. In M. Boule, H. Breuil, E. Licent, and P. Teilhard de Chardin (eds.): *Le Paléolithique de la Chine*. Archives de l'Institut de Paleontologie Humaine 4:103-136.
Cann, R. L., Stoneking, M., and Wilson, A. C. (1987) Mitochondrial DNA and human evolution. *Nature* 325:31-36.
Cavalli-Sforza, L. L., Piazza, A., Menozzi, P., and Mountain, J. (1988) Reconstruction of human evolution: Bringing together genetic, archaeological, and linguistic data. *Proc. Nat. Acad. Sci. U.S.A.* 85:6002-6006.
Chareonwongsa, P. (1988) The current status of prehistoric research in Thailand. In P. Chareonwongsa and B. Bronson (eds.): *Prehistoric Studies: The Stone and Metal Ages in Thailand*. Bangkok: Papers in Thai Antiquity, pp. 17-41.
Cheng, G., Lin, J., Li, S., and Liang, Q. (1978) A preliminary paleomagnetic survey of the Nihewan bed. *Sci. Geol. Sin.* 7:247-252.
Chi, N. D., Lan, V. L., and 10 others (1961) *Traces of Primitive Man on the Territory of Vietnam*. Hanoi: Su Hoc.
Chung, Y. W. (1984) Acheulean handaxe culture of Chongok-Ni in Korea. In R. O. Whyte (ed.): The Evolution of the East Asian Environment, Vol. 2. Hong Kong: University of Hong Kong Press, 894-914.
Clarke, R. J. (1990) The Ndutu cranium and the origin of *Homo sapiens*. *J. Hum. Evol.* 19:699-736.
Collings, H. D. (1938) A Pleistocene site in the Malay Peninsula. *Nature* 142:575-576.
Colani, M. (1927) *L'Age de la pierre dans la province de Hoa-Binh.* Hanoi: Tonkin.
Cook, J., Stringer, C. B., Currant, A. P., Schwarcz, H., and Wintle, A. G. (1982) A review of the chronology of the European Middle Pleistocene hominid record. *Yearb. Phys. Anthropol.* 25:19-65.
Dai, E., and Chi, H. (1964) Palaeoliths from Lantian, Shanxi. *Vert. PalAs.* 8(2):152-161.
De Morgan, J. (1924) *Prehistoric Man*. London: Kegan Paul, Trench, Trubner & Co., Ltd.
De Terra, H., and Movius, H. L. (1943) Research on early man in Burma. *Trans. Am. Phil. Soc.* 32: 271-393.
De Vos, J., Sartono, S., Hardja-Sasmita, S., and Sondaar, P. Y. (1982) The fauna from Trinil, type locality of *Homo erectus*: A reinterpretation. *Geol. Mijnbouw* 61:207-211.

Eisenmann, V. (1975) Nouvelles interpretations des restes d'Equides (Mammalia, Perissodactyla) de Nihowan (Pleistocene inferieur de la Chine du Nord): Equus teilhardi nov. sp. Extrait de *Geobois* 8:125-134.

Eisenmann, V., and Brunet, M. J. (1973) Presence simultanee de cheval et d'Hipparion dans le Villafranchien moyen de France a Roccaneyra (Puy-de-Dome); etude critique de ces semblabes (Europe et Proche-Orient). Int'l Colloquium on the problem "The boundary between the Neogene and Quaternary." Collection of Paper IV. Moscow.

Falk, D. (1987) Hominid paleoneurology. *Ann. Rev. Anthropol.* 16: 13-30.

Forde, C. D. (1934) *Habitat, Economy & Society*. London: Methuen.

Fox, R. B. (1970) *The Tabon Caves: Archaeological explorations and excavations on Palawan Island, Philippines.* National Museum of the Philippines, Monograph 1. Quezon City: New Mecury Printing Press.

Fox, R. B. (1978) The Philippine Paleolithic. In F. Ikawa-Smith (ed.): *Early Paleolithic in South and East Asia*. The Hague: Mouton, pp. 60-83.

Glover, I. C. (1981) Leang Burung 2: An Upper Palaeolithic rock shelter in South Sulawesi, Indonesia. *Mod. Quatern. Res. Southeast Asia* 6:1-38.

Golson, J. (1977) Simple tools and complex technology. In R. V. S. Wright (ed.): *Stone Tools as Cultural Markers*. pp. 154-161.

Gorman, C. F. (1970) Excavations at Spirit Cave, North Thailand: Some interim interpretations. *Asian Perspectives* 13:79-107.

Gorman, C. F. (1971) The Hoabinhian and after: Subsistence patterns in Southeast Asia during the late Pleistocene and early recent periods. *World Archaeol.* 2(3):300-320.

Groves, C. P. (1989) A regional approach to the problem of the origin of modern humans in Australasia. In P. Mellars and C. B. Stringer (eds.): *The Human Revolution: Behavioural and Biological Perspectives on the Origins of Modern Humans.* Edinburgh: Edinburgh University Press, pp. 274-285.

Grün, R., Beaumont, P. B., and Stringer, C. B. (1990) ESR dating evidence for early modern humans at Border Cave in South Africa. *Nature* 344:537-539.

Haile, N. S. (1971) Quaternary shore lines in West Malaysia and adjacent parts of the Sunda shelf. *Quaternaria* 15:333-343.

Harrisson, T. (1970) The prehistory of Borneo. *Asian Perspectives* 13:17-45.

Harrisson, T. (1975) Tampan: Malaysia's Palaeolithic reconsidered. *Mod. Quatern. Res. Southeast Asia* 1:53-69.

Harrisson, T. (1978) Present status and problems for Palaeolithic studies in Borneo and elsewhere. In F. Ikawa-Smith (ed.): *Early Paleolithic in South and East Asia*. The Hague: Mouton, pp. 37-57.

Heider, K. G. (1960) A pebble-tool complex in Thailand. *Asian Perspectives* 2:63-67.

Holloway, R. L., and de la Coste-Lareymondie, M. C. (1982) Brain endocast asymmetry in pongids and hominids: Some preliminary findings on the paleontology of cerebral dominance. *Am. J. Phys. Anthropol.* 58: 101-110.

Howell, F. C. (1978) Hominidae. In V. J. Maglio and H. B. S. Cooke (eds.): *Evolution of African Mammals*. Cambridge: Harvard University Press, pp. 154-248.

Howell, F. C. (1986) Variabilité chez *Homo erectus*, et probleme de la presence de cette espece en europe. *L'Anthropologie* 90(3):447-481.

Huang, W. (1985) On the stone industry of Xiaochangliang. *Acta Anthropol. Sin.* 4:301-307.

Huang, W. (1987) Bifaces in China. *Acta Anthropol. Sin.* 6:61-68.

Huang, W. (1990) Bifaces in China. *J. Hum. Evol.* 4:87-92.

Huang, W. (1991) Evidence for early man's activities from the lateritic beds of South China. *Quatern. Sci.* 12(4):373-379.

Huang, W., Liu, Y., Li, C., and Yuan, X. (1988) Tentative opinions on the age of Baise stone industry. *Treatises in Commemoration of the 30th Anniversary of the Discovery of Maba Human Cranium*. Guandong Provincial Museum and the Museum of the Qujiang County. Beijing: Cultural Relics Publ. House, pp. 95-101.

Huang, W., Leng, J., Yuan, X., and Xie, G. (1990) Advanced opinions on the stratigraphy and chronology of Baise stone industry. *Acta Anthropol. Sin.* 9:105-112.

Hussain, T., and Bernor, R. L. (1984) Evolutionary history of Siwalik Hipparions. *Cour. Forsch. Senckenberg* 69:181-189.

Hutterer, K. L. (1977) Reinterpreting the Southeast Asian Paleolithic. In J. Allen, J. Golson, and R. Jones (eds.): *Sunda and Sahul*. London: Academic, pp. 31-72.

Hutterer, K. L. (1985) The Pleistocene archaeology of Southeast Asia in regional context. *Mod. Quatern. Res. Southeast Asia* 9:1-25.

Ikawa-Smith, F. (1978) Introduction: The Early Paleolithic tradition of East Asia. In F. Ikawa-Smith (ed.): *Early Paleolithic in South and East Asia*. The Hague: Mouton, pp. 1-10.

Itihara, M., Kadar, D., and Watanabe, N. (1985) Concluding remarks. In N. Watanabe and D. Kadar (eds.): Quaternary geology of the hominid fossil bearing formations in Java. *Geol. Res. Dev. Cent. Spec. Publ.* 4:367-378.

Jacob, T., Soejono, R. P., Freeman, L., and Brown, F. H. (1978) Stone tools from Mid-Pleistocene sediments in Java. *Science* 202:885-887.

Jia, L. (1980) *Early Man in China*. Beijing: Foreign Languages.

Jia, L. (1985) China's earliest Paleolithic assemblages. In R. Wu and J. W. Olsen (eds.): *Palaeoanthropology and Palaeolithic Archaeology in the People's Republic of China*. Orlando: Academic, pp. 135-145.

Jia, L., Gai, P., and You, Y. (1972) Report to excavation in Shi Yu, Shanxi—a Palaeolithic site. *Acta Archaeologia Sinica* 1:39-60.

Jia, L., and Huang, W. (1985a) The Late Palaeolithic of China. In R. Wu and J. W. Olsen (eds.): *Palaeoanthropology and Palaeolithic Archaeology in the People's Republic of China.* Orlando: Academic, pp. 211-223.

Jia, L., and Huang, W. (1985b) On the recognition of China's Palaeolithic cultural traditions. In R. Wu and J. W. Olsen (eds.): *Palaeoanthropology and Palaeolithic Archaeology in*

the People's Republic of China. Orlando: Academic, pp. 259-265.
Jia, L., and Huang, W. (1990) *The Story of Peking Man*. Beijing: Foreign Language and Hong Kong: Oxford University Press.
Jia, L., and Wang, J. (1957) Earlier cradle of man—the Nihewan strata. *Science Bulletin* 1.
Jia, L., and Wei, Q. (1980) Some animal fossils from the Holocene of north China. *Vert. PalAs.* 18(4):327-333.
Jia, L., and Wei, Q. (1976) A Palaeolithic site at Hsue-chia-yao in Yangkao County, Shansi Province. *K'ao Ku Hsueh Pao* 2:97-114.
Jia, L., and Wei, Q. (1987) Stone artifacts from lower Pleistocene at Donggutuo site near Nihewan (Nihowan), Hebei province, China. *L'Anthropologie* 91:727-732.
Jia, L., Wei, Q., and Li, C. (1979) Report on the excavation of Hsuchiayao man site in 1976. *Vert. PalAs.* 17(4):277-293.
Jones, P. R. (1979) Effects of raw materials on biface manufacture. *Science* 204:835-836.
Jones, P. R. (1981) Experimental implement manufacture and use; a case study from Olduvai Gorge, Tanzania. *Phil. Trans. Roy. Soc. Lond.* B 292:189-195.
Jones, R. (1989) East of Wallace's Line: Issues and problems in the colonization of the Australian Continent. In P. Mellars and C. B. Stringer (eds.): *The Human Revolution: Behavioural and Biological Perspectives on the Origins of Modern Humans*. Edinburgh: Edinburgh University Press, pp. 743-782.
Kamminga, J. (1990) Lecture presented at 14th Congress Indo-Pacific Prehistory Association, Yogyakarta, 25th Aug.-2nd Sept. 1990.
Keates, S. G. (in press) Archaeological evidence of hominid behaviour in Pleistocene China. *Cour. Forsch. Senckenberg.*
Keates, S. G. (in prep.) Hominid behaviour in the Nihewan Basin, North China. D. Phil. Dissertation, University of Oxford.
Kimura, D. (1992) Sex differences in the brain. *Sci. Am.* 267:118-125.
Kinh, P., and Tieu, L. T. (1976) The Lower Palaeolithic site of Nui Do. *Viet. Stud.* 46:50-106.
Klein, R. G. (1989) *The Human Career: Human Biological and Cultural Origins*. Chicago: University of Chicago Press.
Klein, R. G. (1992) The archeology of modern human origins. *Evol. Anthropol.* 1(1):5-14.
Laitman, J. T., and Reidenberg, J. S. (1987) Advances in understanding the relationship between the skull base and larynx with comments on the origin of speech. *Hum. Evol.* 3(1):99-109.
Larichev, V., Khol'ushkin, U., and Laricheva, I. (1987) The Lower and Middle Paleolithic of Northern Asia: Achievements, problems, and perspectives. *J. World Prehist.* 1:1-27.
Larichev, V., Khol'ushkin, U., and Laricheva, I. (1988) The Upper Paleolithic of Northern Asia: Achievements, problems, and perspectives. *J. World Prehist.* 2:359-396.
Lee, J. S. (1939) *The Geology of China*. New York: Nordeman Publishing.
Li, T. (1991) Unearthing of Chaoxian man's fossil skull, *China Cultural Report* 5:1.
Li, Y. (1984) The early Pleistocene mammalian fossils of Danangou, Yuxian, Hebei. *Vert. PalAs.* 22(1):60-68.
Li, H., and Wang, J. (1982) Magnetostratigraphic study of several typical geologic sections in north China. In *Quaternary Geology and Environment of China*. Quaternary Research Association of China. Beijing: China Ocean.
Licent, E. (1924) Voyage aux terrasses du Sangkan Ho. Publications du Musée Hoang Ho Pai Ho (Tien Tsinn) 4:1-14.
Lieberman, P. (1989) The origins of some aspects of human language and cognition. In P. Mellars and C. B. Stringer (eds.): *The Human Revolution: Behavioural and Biological Perspectives in the Origins of Modern Humans*. Edinburgh: Edingurgh University Press, pp. 391-414.
Liu, C., Zhu, X., and Ye, S. (1977) A paleomagnetic study on the cave deposits of Zhoukoudian (Choukoutien), the locality of Sinanthropus. *Sci. Geol. Sin.* 1:25-30.
Liu, T. (1991a) *Loess, Environment and Global Change*. Beijing: Science Press.
Liu, T. (1991b) *Quaternary Geology and Environment*. Beijing: Science Press.
Liu, T., Lu, Y., Zheng, H., Wu, Z., and Yuan, B. (1985) *Loess and the Environment*. Beijing: China Ocean.
Liu, X., and Xia, Z. (1983) A suggestion on the division and correlation of the Nihewan formation. *Mar. Geol. and Quatern. Geol.* 3(1):75-85.
Liu, X., and Xia, Z. (1989) A suggestion on the division and correlation of the Nihewan formation. In Q. Wei and F. Xie (eds.): *Selected Treatises on Nihewan*. Beijing: Cultural Relics Publishing House, pp. 421-432.
MacDonald, S. B., and MacDonald, A. S. (1978) Age of the Lampang basalt and underlying pebble tools. Dept. of Geological Sciences, Chiangmai University, Special Publication 2:1-10.
Maglio, V. J. (1973) Origin and evolution of the Elephantidae, *Trans. Am. Phil. Soc.* 63:1-149.
Majid, Z. (1982) The West Mouth, Niah, in the prehistory of Southeast Asia. *The Sarawak Mus. J. Spec. Monograph* 31(52) NS. No. 3, p. 200.
Majid, Z. (1990) The Tampanian problem resolved: Archaeological evidence of a late Pleistocene lithic workshop. *Mod. Quatern. Res. Southeast Asia* 11:71-96.
Majid, Z., and Tjia, H. D. (1988) Kota Tampan, Perak: The geological and archaeological evidence for a late Pleistocene site. *J. Malay. Branch R. Asiatic Soc.* 61(2):123-134.
Marschack, A. (1972) Cognitive aspects of Upper Paleolithic engraving. *Curr. Anthropol.* 13:445-477.
Marshack, A. (1975) Exploring the mind of ice age man. *Nat. Geog.* 147: 64-89.
Mellars, P. (1989) Major issues in the emergence of modern humans. *Curr. Anthropol.* 30:349-385.
Movius, H. L. (1944) *Early Man and Pleistocene stratigraphy in southern and eastern Asia*. Papers of the Peabody Museum, Harvard University 19:1-125.
Movius, H. L. (1948) The Lower Paleolithic cultures of southern and eastern Asia. *Trans. Am. Phil. Soc.* 38(4):329-420.

Movius, H. L. (1956) New palaeolithic sites, near Ting-Ts'un in the Fen River, Shansi province, North China. *Quaternaria* 3:13-26.

Movius, H. L. (1978) Southern and Eastern Asia: Conclusions. In F. Ikawa-Smith (ed.): *Early Paleolithic in South and East Asia*. The Hague: Mouton, pp. 351-355.

Nihewan Cenozoic Research Team (1974) Observation on the late Cenozoic of Nihewan basin. Vert. PalAs. 12:99-110.

Olsen, J. W., and Ciochon, R. L. (1990) A review of evidence for postulated Middle Pleistocene occupations in Viet Nam. *J. Hum. Evol.* 19:761-788.

Osborn, H. F. (1927) Recent discoveries relating to the origin and antiquity of man. *Science* 65:481-488.

Pfeiffer, J. E. (1969) *The Emergence of Man*. New York: Harper & Row.

Pfeiffer, J. E. (1985) *The Emergence of Humankind*, 4th ed. New York: Harper & Row.

Pope, G. G. (1982) Hominid evolution in East and Southeast Asia. Ph.D. Thesis, University of California, Berkeley.

Pope, G. G. (1983) Evidence on the age of the Asian hominidae. *Proc. Nat. Acad. Sci. USA* 80:4988-4992.

Pope, G. G. (1984) The antiquity and paleoenvironment of the Asian hominidae. In R. O. Whyte (ed.): *The Evolution of the East Asian Environment*, Vol. II. Hong Kong: University of Hong Kong Press. pp. 822-847.

Pope, G. G. (1985a) Taxonomy, dating, and paleoenvironment: The paleoecology of the early Far Eastern hominids. Mod.Quatern. *Res. Southeast Asia* 9:65-80.

Pope, G. G. (1985b) Evidence of Early Pleistocene hominid activity from Lampang, northern Thailand. *Indo-Pacific Prehistory Association Bulletin* 6:2-9.

Pope, G. G. (1988) Recent advances in Far Eastern paleoanthropology. *Ann. Rev. Anthropol.* 17:43-77.

Pope, G. G. (1989a, October) Bamboo and human evolution. *Nat. Hist.* pp. 48-57.

Pope, G. G. (1989b) Proposal for archaeological research in the Nihewan to the National Science Foundation, Washington DC, USA Unpubl. MS, pp. 1-36.

Pope, G. G. (1992a) The craniofacial evidence for the origin of modern humans in China. *Yearb. Phys. Anthropol.* 35:243-298.

Pope, G. G. (in press b) An historical and scientific perspective on paleoanthropological research in the Far East. *Cour. Forsch. Senckenberg.*

Pope, G. G., Frayer, D. W., Liangchareon, M., Kulasing, P., and Nakabanlang, S. (1981) Paleoanthropological investigations of the Thai-American Expedition in Northern Thailand (1978-1980): An interim report. *Asian Perspectives* 21:148-163.

Pope, G. G., Barr, S., MacDonald, A., and Nakabanlang, S. (1986) Earliest radiometrically dated artifacts from mainland Southeast Asia. *Curr. Anthropol.* 27: 275-279.

Pope, G. G., Nakabanlang, S., and Pitragool, S. (1987) Le Paléolithique du nord de la Thaïlande découvertes et perspectives nouvelles. *L'Anthropologie* 91:749-754.

Pope, G., An, Z., Keates, S., and Bakken, D. (1990) New discoveries in the Nihewan Basin, northern China. *The East Asian Tertiary/Quaternary Newsletter* 11:68-73.

Puech, P.-F. (1983) Tooth wear, diet, and the artefacts of Java man. *Curr. Anthropol.* 24(3):381-382.

Qian, F., Zhou, G. et al. (1991) *Quaternary Geology and Palaeoanthropology of Yuanmou, Yunnan, China*. Beijing: Science.

Reynolds, T. (1990) The Hoabinhian: A review. In G. L. Barnes (ed.): *Hoabinhian Jomon, Yagoi Early Korean States*. Oxford: Oxbow Books, pp. 1-21.

Rigaud, J.-P. (1989) From the Middle to the Upper Paleolithic: Transition or convergence. In E. Trinkaus (ed.): *The Emergence of Modern Humans*. Cambridge: Cambridge University Press, pp. 142-153.

Rightmire, G. P. (1984) Comparisons of *Homo erectus* from Africa and southeast Asia. *Cour. Forsch. Senckenberg* 69:83-99.

Rightmire, G. P. (1990) *The Evolution of* Homo erectus. Cambridge: Cambridge University Press.

Sasada, M., Ratanasthien, B., and Soponpongpipat, P. (1987) New K-Ar ages from the Lampang basalt, northern Thailand. *Bull. Geol. Surv. Japan* 38(1):13-20.

Schick, K., Toth, N., Wei, Q., Clark, J. D., and Etler, D. (1991) Archaeological perspectives in the Nihewan Basin, China. *J. Hum. Evol.* 21:13-26.

Schwarcz, H. P., Buhay, W. M., Grün, R., Valladas, H., Tchernov, E., Bar-Yosef, O., and Vandermeersch, B. (1989) ESR dating of the Neanderthal Site, Kebara Cave, Israel. *J. Archaeol. Sci.* 16:653-659.

Sémah, F. (1982) Pliocene and Pleistocene geomagnetic reversals recorded in the Geomolong and Sangiran domes (Central Java). Mod. Quatern. *Res. Southeast Asia* 7:151-164.

Sémah, F. (1984) The Sangiran dome in the Javanese Plio-Pleistocene chronology. *Cour. Forsch. Senckenberg* 69:245-252.

Shea, J. J. (1989) A functional study of the lithic industries associated with hominid fossils from Kebara and Quafzeh Caves, Israel. In P. Mellars and C. B. Stringer (eds.): *The Human Revolution: Behavioural and Biological Perspectives on the Origins of Modern Humans*. Edinburgh: Edinburgh University Press, pp. 611-625.

Shutler, R., Jr. (1984) Kota Tampan, once again. *Heritage* 6:91-101.

Shutler, R., Jr. and Braches, F. (1988) The origin, dating and migration routes of hominids in Pleistocene East and Southeast Asia. In P. Whyte, J. S. Aigner, N. G. Jablonski, G. Taylor, D. Walker, and P. Wang (eds.): *The Palaeoenvironment of East Asia from the Mid-Tertiary*. Hong Kong: Centre of Asian Studies, 2:1084-1089.

Shutler, R., Jr., and Forgie, M. (1990) Kota Tampan, Malaysia: A reinterpretation. *Proc. Twelfth In. Symp. Asian Studies* Hong Kong: Asian Research Service, pp. 569-576.

Sieveking, A. de G. (1960) The Palaeolithic history of Kota Tampan, Perak. *Asian Perspectives*, 2:91-102.

Sørensen, P. (1976) Preliminary note on the relative and absolute chronology of two early Palaeolithic sites from North Thailand. In A. K. Ghosh (ed.): *Le Paléolithique Inferieur et Moyen en Inde, en Asie Centrale, en Chine et dans le sud-est Asiatique*. Union Internationale des Sciences

Prehistoriques et Protohistoriques, IX Congres, Nice, pp. 237-251.
Sørensen, P. (1988) The Lannathain culture: The Early Palaeolithic culture of northern Thailand (Abstract). Ass. Southeast Asian Archaeol. Western Europe, Sec. Internat. Conf., Paris, Musée Guimet, 19th-23rd September 1988.
Stauffer, P. H., Nishimura, S., and Batchelor, B. C. (1980) Volcanic ash in Malaya from a catastrophic eruption of Toba, Sumatra, 30,000 years ago. *Phys. Geol. Indonesian Island Arcs*, pp. 156-164.
Stringer, C. B. (1984) The definition of *Homo erectus* and the existence of the species in Africa and Europe. *Cour. Forsch. Senckenberg* 69:131-145.
Stringer, C. B. (1989a) The origin of early modern humans: A comparison of the European and non-European evidence. In P. Mellars and C. B. Stringer (eds.): *The Human Revolution: Behavioural and Biological Perspectives on the Origins of Modern Humans*. Edinburgh: Edinburgh University Press, pp. 232-244.
Stringer, C. B. (1989b) Documenting the origin of modern humans. In E. Trinkaus (ed.): *The Emergence of Modern Humans*. Cambridge: Cambridge University Press, pp. 97-141.
Stringer, C. B. (1990) The Asian connection. *New Scientist* 178:33-37.
Stringer, C. B., and Andrews, P. (1988) Genetic and fossil evidence for the origin of modern humans. *Science* 239:1263-1268.
Sun, D. (1991) The hunting for the Quaternary glaciers in China. In D. Liu (ed.): *Quaternary Geology and Evironment in China*. Beijing: Science, pp.1-15.
Tattersall, I. (1992) Biological and cultural innovation in human evolution. *Evol. Anthropol.* 1:112.
Teilhard de Chardin, P. (1937) The Pleistocene of China stratigraphy and correlations. In G. G. McCurdy (ed.): *Early Man*. Philadelphia. pp. 211-220.
Teilhard de Chardin, P. (1941) *Early man in China*. Institut de Geo-Biologie Pekin, No. 7:1-99.
Teilhard de Chardin, P., and Piveteau, J. (1930) Les mammiferes fossiles de Nihowan (Chine). *Annales de Paléontologie* 19:1-134.
Thorne, A., and Wolpoff, M. (1981) Regional continuity in Australasian Pleistocene hominid evolution. *Am. J. Phys. Anthropol.* 55:337-349.
Toth, N. (1982) The stone technologies of early hominids from Koobi Fora, Kenya. Ph.D. Diss. Berkeley: University of California.
Toth, N. (1985) Archeological evidence for preferential right-handedness in the lower and middle Pleistocene, its possible implications. *J. Hum. Evol.* 14:607-614.
Trigger, B. G. (1989). *A History of Archaeological Thought*. Cambridge: Cambridge University Press.
van Heekeren, H. R. (1948) Prehistoric discoveries in Siam, 1943-44. *Proc. Preh. Soc.* 14:24-32.
van Heekeren, H. R. (1957) *The Stone Age of Indonesia*. The Hague: Martinus-Nijhoff.
van Heekeren, H. R. (1972) *The Stone Age of Indonesia*. Verhandelingen 61. The Hague: Martinus Nijhoff (2nd ed.).
Verstappen, H. T. (1975) On palaeo climates and landform development in Malesia. *Mod. Quatern. Res. Southeast Asia* 1:3-35.
von Koenigswald, G. H. R. (1936a) Early Palaeolithic stone implements from Java. *Bull. Raffles Museum Ser.* B 1: 52-60.
von Koenigswald, G. H. R. (1936b) Über altpalaeolithische Artefakte von Java. *T. K. Nederl. Aardrijkskd. Genoot.* 53(2):41-44.
von Koenigswald, G. H. R. (1939) Das Pleistocän Javas. *Quartär* 2:28-53.
von Koenigswald, G. H. R. (1956) *Meeting Prehistoric Man*. London: Thames and Hudson.
von Koenigswald, G. H. R. (1975) Bemerkungen zu 'Notes about Sangiran (Java, Indonesia)' by G.-J. Bartstra. *Quartär* 26:167.
von Koenigswald, G. H. R., and Ghosh, A. K. (1972) Stone implements from the Trinil beds of Sangiran, central Java. *Proc. K. Nederl. Akad. Wettensch.* Series B, 76(1):1-34.
Walker, D., and Sieveking, A. de G. (1962) The Palaeolithic history of Kota Tampan, Perak. *Proc. Prehist. Soc.* 28:103-139.
Watanabe, N., and Kadar, D. (1985) Quaternary geology of the hominid fossil bearing formations in Java. *Geol. Res. Dev. Cent. Spec. Publ.* 4.
Wei, Q. (1978) New discoveries from the Nihewan formation and other aspects regarding its study. In *Treatises on Paleoanthropology*. Beijing: Science Publishing Society, pp. 136-150.
Wei, Q. (1983) A new Megaloceros from Nihewan beds. *Vert. PalAs.* 21(1):87-95.
Wei, Q. (1985) Paleoliths from the lower Pleistocene of the Nihewan beds in the Donggutuo site. *Acta Anthropol. Sin.* 4(4):289-300.
Wei, Q. (1988) The stratigraphic, geochronological, and biostratigraphic background of the earliest known Paleolithic sites in China. *L'Anthropologie* 92(3):931-938.
Wei, Q. (1989) Paleoliths from the lower Pleistocene of the Nihewan beds in the Donggutuo site. In Q. Wei and F. Xie (eds.): *Selected Treatises on Nihewan*. Beijing: Cultural Relics Publishing House, pp. 115-128.
Wei, Q., Meng, H., and Cheng, S. (1985) New Palaeolithic site from the Nihewan (Nihowan) beds. *Acta Anthropol. Sin.* 4(3):223-232.
Wei, Q., and Xie, F. (1989) *Selected Treatises on Nihewan*. Beijing: Cultural Relics Publishing House.
Weidenreich, F. (1939) On the earliest representatives of modern mankind recovered on the soil of East Asia. *Peking Nat. Hist. Bull.* 13:161-174.
White, J. P. (1977) Crude, colourless and unenterprising: Prehistorians and their views on the stone age of Sunda and Sahul. In J. Allen, J. Golson, and R. Jones (eds.): *Sunda and Sahul*. London: Academic, pp. 13-30.
White, J. P., Modjeska, N., and Hipuya, I. (1977) Group definitions and mental templates. In R. V. S. Wright (ed.): *Stone Tools as Cultural Markers*. Australian Institute of Aboriginal Studies, Canberra. New Jersey: Humanities, pp. 380-390.

Whitmore, T. C. (1975) *Tropical Rainforests of the Far East*. Oxford: Clarendon.
Whitmore, T. C. (1981) Paleoclimate and vegetation history. In T. C. Whitmore (ed.): *Wallace's Line and Plate Tectonics*. Oxford: Clarendon, pp. 36-43.
Whyte, R. O. (1972) The Graminae, wild and cultivated, of monsoonal and tropical Asia. *Asian Perspectives* 15:127-151.
Whyte, R. O. (1974) Grasses and grasslands. In *Natural Resources of Humid Tropical Asia*. Paris: UNESCO, pp. 239-262.
Whyte, R. O. (1984) The Gramineae in the palaeoenvironment of east Asia. In R. O. Whyte (ed.): *The Evolution of the East Asian Environment*, Vol. II. Hong Kong: University of Hong Kong Press, pp. 622-650.
Wilson, A. C., and Cann, R. L. (1992) The recent African genesis of humans. *Sci. Amer.* 266(4):66-73.
Wolpoff, M. H., Spuhler, J. N., Smith, F. H., Radovcic, J., Pope, G., Frayer, D. W., Eckhardt, R., and Clark, G. (1988) Modern human origins. *Science* 241:772-773.
Wu, J., and Wu, X. (1984) Hominid fossils from China and their relation to those of neighboring regions. In R. O. Whyte (ed.): *The Evolution of the East Asian Environment*, Vol. 2. Hong Kong: University of Hong Kong Press, pp. 787-795.
Wu, R., and Dong, X. (1985) *Homo erectus* in China. In R. Wu and J. W. Olsen (eds.): *Palaeoanthropology and Palaeolithic Archaeology in the People's Republic of China*. Orlando: Academic, pp. 79-88.
Wu, X., Yuan, Z., Han, D., Qi, T., and Lu, Q. (1966) Report of the excavation at Lantian Man Locality of Gongwangling in 1965. *Vert. PalAs.* 10:23-29.
Xu, Q., and You, Y. (1982) Four post-Nihewanian Pleistocene mammalian faunas of North China: Correlation with deep-sea sediments. *Acta Anthropol. Sin.* 1:180-187.
Yin, T. (1933) Les Volcans Quaternaires de Tatung, Shansi. *Bull. Geol. Soc. China* 12:355-375.
You, Y. (1983) New data from the Xiaochangliang Paleolithic site in Hebei and their chronology. *Prehistory* 1:46-50.
You, Y., Tang, Y., and Li, Y. (1978) Paleolithic discoveries in the Nihewan formation. *Chinese Quat. Res.* 1(5):1-13.
You, Y., Tang, Y., and Li, Y. (1980) New discovery of palaeoliths in the Nihewan Formation. *Quat. Sin.* 5(1):1-13.
Zhang, S. (1985) The Early Palaeolithic of China. In R. Wu and J. W. Olsen (eds.): *Palaeoanthropology and Palaeolithic Archaeology in the People's Republic of China*. Orlando: Academic, pp. 147-186.
Zhou, M., and Yu, P. (1974) *Chinese Fossil Proboscidea*. Peking: Academia Sinica.

CHAPTER 26

The Movius Line Reconsidered

Perspectives on the Earlier Paleolithic of Eastern Asia

Kathy D. Schick

INTRODUCTION

Clark Howell's integrative approach to paleoanthropology has stressed the vital importance of both hominid paleontology and archeology to human evolutionary studies. Studies of either human biological evolutionary change or the nature of human technological adaptation cannot be fully understood in isolation: Study of the hominids should appreciate the behavioral and cultural patterns indicated by archeological evidence, and archeological interpretations and inferences should ultimately be understood within the overall pattern of human biological change and population migrations and radiations. In the final analysis, understanding of overall trends of human evolutionary change must integrate the evidence from both subfields of study.

This dictate to be interdisciplinary and integrative is directly relevant to studies of the biology and archeology of early hominid populations in eastern Asia. In 1977, a paleoanthropological delegation from the United States to the People's Republic of China was sponsored by the National Academy of Sciences (Howells and Tsuchitani 1977). Chaired by F. Clark Howell and involving scientists from diverse disciplines including physical anthropology, archeology, and geology, this delegation was able to visit sites and scientific institutions, to examine artifact and fossil collections, and to confer with Chinese colleagues. This laid an important foundation for a new burst of international cooperation and exchange of information between the East and West during the past decade.

More recently, this opening of the eastern Asian "door" has culminated in larger scale collaboration in the form of joint Chinese–United States excavations and studies of some of the earliest archeological sites in China (Clark and Schick 1988; Schick et al. 1991; Schick and Dong 1993) as well as collaborative studies of hominid fossils (Li and Etler 1992). The renewed flow of information provides the opportunity to take a fresh look at long-recognized questions regarding the nature and potential meaning of differences between eastern and western archeological patterns.

The nature of hominid biological evidence from the Early and Middle Pleistocene in eastern Asia and its evolutionary affinities and relationships with other hominid populations through the Pleistocene has been a matter of much concern and debate in recent years (e.g., Thorne and Wolpoff 1981; Rightmire 1992; Groves 1989; Wood 1991). Equally, the archeological evidence during this period of time shows a number of discontinuities with that seen in many places in western Eurasia and Africa. Clearly, the questions and problems posed by the archeological evidence are directly relevant to many larger scale questions regarding human evolutionary change, in particular questions regarding the amount, timing, and direction of the flow of genes and culture among hominid populations during pre-*Homo sapiens* times.

UNDERLYING QUESTIONS: LINES OF BIOLOGICAL AND CULTURAL CONNECTIONS DURING HUMAN EVOLUTION?

There are a number of important, long-standing questions regarding the variability seen in the Stone Age, even in the early stages of the Paleolithic. Why do stone technologies made by hominids in one region or one part of the world sometimes differ, at times very markedly, from those found elsewhere at more or less the same period of time? The technological variation seen over the geographical range of early hominids is especially difficult to explain when we consider that during the earlier Paleolithic there is sometimes very strong continuity seen in technological traditions over very long periods of time, even over the course of hundreds of thousands of years.

Discussions of technological similarities and differences between tool traditions are commonly loaded with myriad assumptions regarding what biological ties may indicate about cultural ties and vice versa. For instance, a common expectation is that the stronger or more direct the biological links (for instance, if there is considerable gene flow or close ancestor–descendant relationships among populations), the larger the degree of "cultural" and technological similarity (the "culture/tribe/people" hypothesis; Coles and Higgs 1969).

Conversely, less sharing of common culture or technology has often been considered an indication of less contact between populations, either in cultural interaction and diffusions or biological transmission (gene flow). While we should realize that such assumptions are not always borne out in the real world (e.g., closely related biological populations sometimes have drastic differences in culture and even in language, and culture can be transmitted quickly between biologically distant populations), there are some grounds for argument that strong cultural continuities commonly follow biological linkages.

Conservative maintenance of technological traditions appears, in fact, to be almost a rule throughout much of our Stone Age past, particularly in the earlier phases of our prehistory. To explain the continuity in technological traditions over time or space, we generally implicitly infer the maintenance of some learned, cultural traditions, of rules and ideas regarding how things, including tool–making, are to be done, that have been passed on from one individual to another and from one generation to the next (hence the common reference in many prehistoric studies to stone tool "cultures," although such terminology is widely discouraged (Clark et al. 1966)). In a sense, for much of earlier prehistory, "sameness" appears to be the norm and measure, while difference or change is the anomaly. If this technological continuity is broken, then, and distinct change or variability is observed over time or space among the tool traditions of related populations, this invites explanation and discussion.

This problem is presented by the archeological patterns observed in the early stages of the Paleolithic, during the spread of hominids out of Africa and into the southern and temperate latitudes of Eurasia. A strong, even dramatic difference has long been recognized between the stone technologies of the earliest East Asian hominids and those produced by many hominid populations in the rest of the Old World (e.g., Africa, Europe, and western Asia, during Early and Middle Pleistocene times) (Movius 1948, 1949; Teilhard de Chardin 1941). In recent years, particularly within the past decade, it has become possible to investigate more directly and critically many archeological sites and artifacts in eastern Asia, to ask whether this pattern indeed holds up under closer scrutiny and, if it does, why?

I will review here the major arguments and hypotheses regarding the nature of the early Paleolithic in eastern Asia and examine general technological patterns in this part of the world in view of current evidence available from Early and Middle Pleistocene archeological sites there. Finally, I will consider this east Asian evidence in an integrative, comparative framework, considering the larger gamut of hominid migrations out of Africa and into Europe and Asia during Early Palaeolithic times.

BACKGROUND: THE "MOVIUS LINE"

Movius: An East Asian Technological "Flavor"

In Movius's perception, the eastern Asian stone tool cultures showed profound differences from the stone industries elsewhere in the Old World: "Perhaps the most salient feature characterizing the Lower Paleolithic culture complex of Southeastern Asia, Northern India and China as a whole is the *absence* of certain characteristic types of Lower Paleolithic implements (hand-axes and Levallois flakes), as much as it is the *presence* of others (choppers and chopping–tools)" (Movius 1969:72).

FIGURE 26-1 Map showing general regions in Early or Middle Pleistocene times in which Acheulean biface cultures (and also Mode 1 industries) are found, and regions exhibiting principally non-Acheulean, Mode 1 industries. Movius's line of demarcation is shown separating his "hand-axe cultures" to the west and his "chopper-chopping tool" cultures in eastern and southeastern Asia (horizontal pattern). Since Movius's work, other non-hand-axe industries have been identified in localities in eastern Europe and central Asia (vertical pattern outside the eastern Asian "chopper-chopping tool area"). (Drawn by David Johnson)

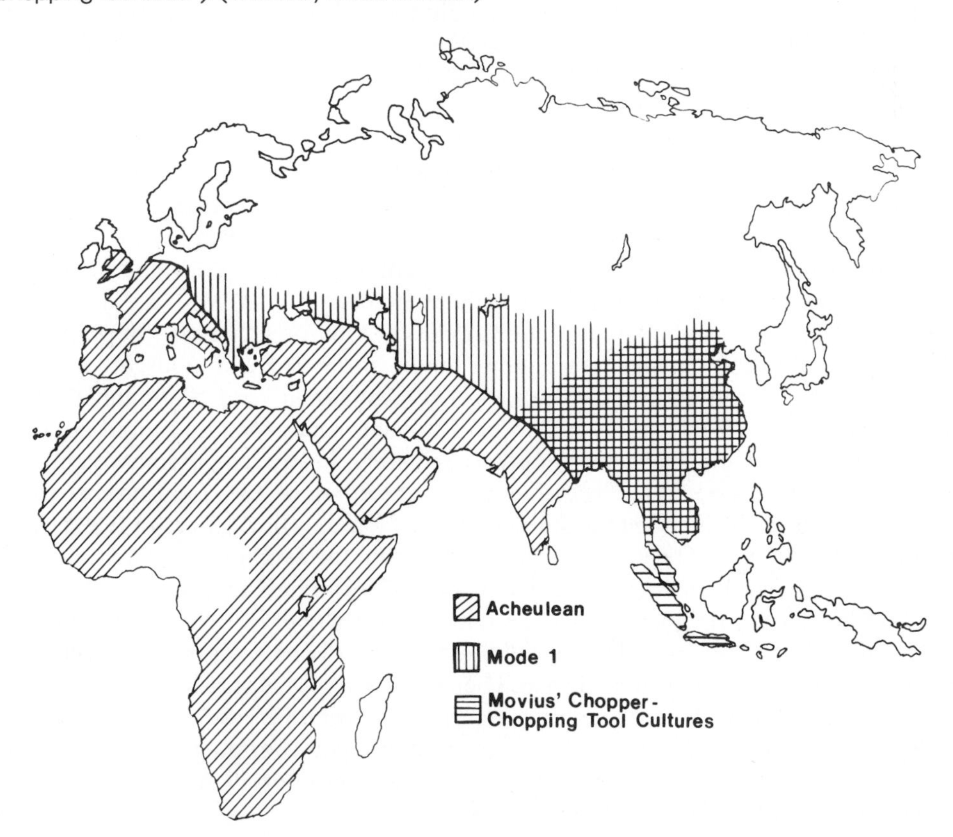

Movius recognized two major technological traditions in the earlier Stone Age (Figure 26-1):

1. A western tradition of "hand-axe cultures," stretching from Africa northward into much of Europe and eastward into western Asia (through the Middle East to central and southern India), thought to represent a large "culture area" sharing common styles of tool–making, especially including the consistent manufacture of large forms of bifacial tools such as hand-axes and cleavers. This technological phenomenon, currently referred to as the "Acheulean Industrial Complex" (Clark et al. 1966), is a long–lived technological tradition that persisted for several hundreds of thousands of years within Africa (starting at least 1.5–1.7 MYA and lasting until at 200,000 to 100,000 years ago), and eventually extended into much of Europe and western Asia.
2. A tradition east of this boundary, which has come to be called the "Movius Line," that he termed the "chopper–chopping tool culture," with artifact assemblages dominated by simple cores or core tools made on pebbles, especially forms he called "choppers" (in Movius's terminology, unifacially flaked or worked on one side of a cobble) and "chopping tools" (bifacially flaked or worked on two sides of a cobble).

Movius identified several centers of the East Asian chopper and chopping tool culture during the Middle Pleistocene: in China, a culture called the "Choukoutienian" (made primarily out of quartz, found at the "Peking Man" cave site of Choukoutien—now called Zhoukoudian—just southwest of Beijing) (Figure 26-2); in northwest India, a culture he called the "Soanian" (mainly made of quartzite cobbles and boulders, found in ancient river gravel terraces of the Soan and Indus River systems) that coexisted with a bifacial hand-axe culture in northwest India; and in Burma, the "Anyathian" (made mainly on fossil wood and silicified tuff, found within terraces of the Irrawaddy River Valley). (He also viewed rolled, quartzite pebble tools found in northern Malaya, called the "Tampanian," as another likely variant of these chopper and chopping–tool cultures, although their age was not certain.) Sites assigned to all of these "cultures" were characterized by simply flaked cobbles and chunks (commonly known as "pebble tools") as well as numbers of simple flakes, very few of which show modification or "retouch" of their edges (Movius 1948, 1949).

In addition, Movius recognized another tool culture in Java, which he called the Patjitanian (now "Pacitanian"), that, in addition to the pebble tools found in these other eastern Asian tool industries, also contained some larger flaked tools that he called "proto–hand-axes." He considered these to be substantively different, however, from Acheulean biface industries of the west (being more unifacially worked and plano–convex in morphology), and to be an independent or convergent development in Java toward technologies with some superficial resemblances to the Acheulean.

More recent investigations of many of these varieties of Movius's chopper-chopping-tool complex indicate that they are of uncertain age. Found either on the surface or within largely undatable terrace gravels of river systems in southeast Asia, the so–called "Anyathian" and "Patjitanian" cultures could represent admixtures of material from diverse sources and ages and probably should not be considered as integral "industries" from any particular period of prehistory. In any case, their assignment to Early or Middle Pleistocene times has not been verifiable, and more recent reports indicate that the Patjitanian tools could in fact stem from fairly recent Pleistocene or even Holocene times (Barstra 1982). Unfortunately, no indisputable artifacts have yet been recovered directly from the early hominid fossil localities in Java at Modjokerto, Sangiran, Trinil, or Ngandong (Barstra 1982). (Although a report has recently been made of artifacts of Middle Pleistocene age at the Sangiran locality [Semah et al. 1992], better documentation of the stratified context of the artifacts and the age of the deposit will be necessary to establish this.)

Zhoukoudian, of course, still stands firm as a bona fide early Asian site, and there are artifacts now known from a number of other Early Paleolithic archeological occurrences in eastern Asia, which I will briefly describe here. So, although Movius's "centers" of his chopper–chopping tool culture have not stood the test of time, finds at various sites have confirmed the presence of tool–making hominids in eastern Asia during later Early and Middle Pleistocene times.

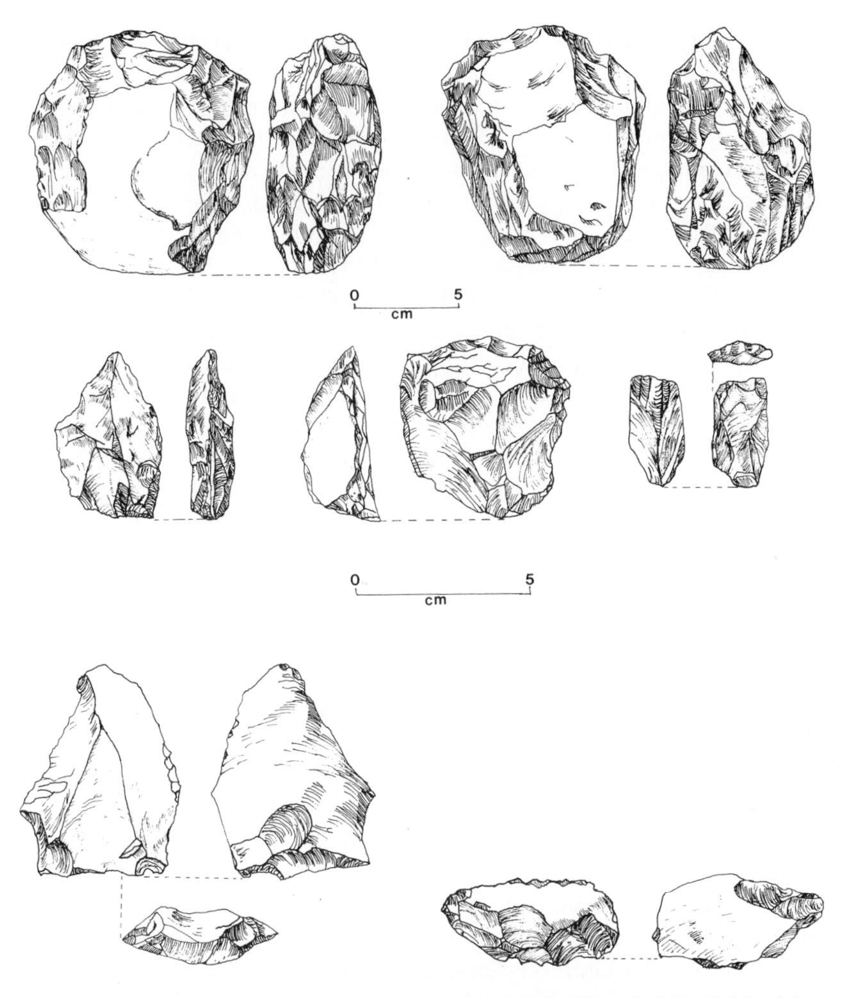

FIGURE 26-2 Simple cores, flakes, and retouched flakes found at the Peking Man site (Zhoukoudian), dated to between 500,000 and 250,000 years ago. Such artifacts are characteristic of the early stone industries Movius described and which he considered part of a sphere of "chopper and chopping tool cultures" in eastern and southeastern Asia. (Drawn by David Johnson after Movius 1949)

Movius's Inference: Eastern Asia as a "Cultural Backwater"

In Movius's view, there were two major stages or phases of the Early Stone Age in the Old World. First to appear was the chopper and chopping-tool culture that "comprised a sort of basic substratrum" (1969:73) of tool–making cultures that he thought had spread at some very early time throughout most of the Old World. Sometime afterwards, in his construct, these industries were changed dramatically by the hand-axe complex throughout most of the western part of the region, described as a "huge triangle, embracing the whole of Africa, the Middle East, and Peninsular India, as well as the southern, central and western portions of Europe."

On the other hand, in the eastern reaches of the earlier, chopper and chopping-tool culture, "this ancient tradition persisted and continued to develop uninfluenced by contemporary innovations found elsewhere" (1969:73). In some islands off the shore of mainland southeast Asia, such as in Malaysia and Java, Movius detected a separate development toward bifacial technologies—the "Patjitanian," discussed further later—that he thought had developed independently of and uninfluenced by the strictly "Acheulean" bifacial technologies of the west.

An explicit corollary to Movius's argument was that the eastern Asian sphere was somewhat off the center stage or main path of human evolution. In other words, populations there had branched away from the mainstream of human biological and cultural evolution at some fairly early date and then maintained a separate biological trajectory and a large degree of cultural isolation for a long period of evolutionary time (i.e., most of the Pleistocene). Moreover, Movius hypothesized that the eastern tool traditions represented a "marginal region of cultural retardation" (Movius 1969:75).

How does this paradigm stand up against current evidence regarding patterns observable in the archeological record? Movius did much of this research and interpretation about 50 years ago, before many details of the Stone Age sequence, including more finely tuned absolute dating and concerted search for sites in Africa and the rest of the world had been carried out. It is now time, perhaps, that this characterization of an east–west dichotomy in stone industries be reevaluated in light of what is now known about the nature and sequence of technological change in the early Paleolithic of Africa, Europe, and Asia.

BRIEF OVERVIEW OF WORLDWIDE EARLY PALEOLITHIC TRENDS

In order to explore possible differences between "east" and "west" during the Stone Age, a brief overview will be given here of the major characteristics of archeological patterns in different major areas of the Old World during the early stages of the Paleolithic (Clark 1970; Wymer 1982).

Africa

At the present time and after decades of research and search for hominid fossils, the earliest record of bipedal hominids and indisputable archeological evidence is to be found in Africa. The earliest hominid fossils appear in East Africa approximately 3.5–4.0 MYA. The fossil record continues in both eastern and southern Africa between 3 and 1 MYA, and expands more extensively throughout eastern, southern and then northern Africa within the past 1 million years. The earliest archeological remains (i.e., stone artifacts) also make their first appearance in East Africa at approximately 2.4 MYA, become common in East and South Africa beginning approximately 2 MYA, and spread to northern Africa by at least approximately 1 MYA (Isaac 1984; Klein 1989; Leakey 1971; Toth and Schick 1986).

We now can outline the basic time progression in technological traditions in the Early Stone Age of Africa (largely equivalent to the Lower Paleolithic of Europe and the Mediterranean region of western Asia and northern Africa). These phases of the Early Stone Age are

1. *The Oldowan tradition.* Starting about 2.4 MYA with the earliest stone tool industries, artifacts found are characterized by simply flaked cores on pebbles and chunks, associated flaking debris, and usually some flakes that had been retouched or modified. This conforms to Clark's "Mode 1" technology, signifying the simple reduction of cores and resultant flakes (Figure 26-3) (Clark 1961).
2. *The Acheulean tradition.* By approximately 1.5–1.7 MYA, another stone technological tradition also appears on the scene, the Acheulean, characterized by relatively large, bifacially flaked forms such as hand-axes and cleavers, that sometimes constitute very high proportions of a site's artifacts (Figure 26-4). This fits Clark's (1961) "Mode 2" technology, signifying the systematic production of relatively large, bifacially flaked core tool forms. This biface tradition continued over a long period of time, in some places in parallel or contemporaneous with the Oldowan–style artifacts. In its later phases (by approximately 200,000 to 300,000 years ago), it is often characterized by extremely well–made, symmetrical bifacially flaked tools (Figure 26-5).

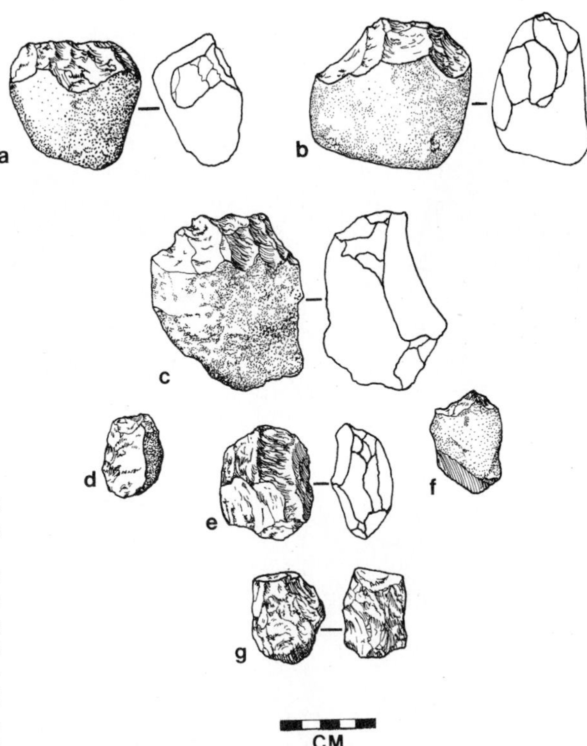

FIGURE 26-3
Africa, Mode 1 technology. These artifacts represent the simple core forms (choppers, etc.) and flakes typical of the earliest stone industries (Oldowan) and also characteristic of later sites in Africa and Europe that are contemporaneous with Acheulean sites. Artifacts shown are from Olduvai Gorge. Identified types are (a) unifacial chopper, (b) bifacial chopper, (c) heavy-duty scraper, (d) light-duty scraper, (e) bifacial discoid, (f) awl, (g) polyhedron. (Drawn by David Johnson after Leakey 1971)

A major advantage of classifying sites according to such technological "modes" is that it allows us to make initial characterizations of sites separated in time and/or space on a relatively neutral basis, in terms of basic technological practices of the hominid tool–makers, rather than through the loaded terminology of "cultural designations." Thus, we can avoid implying specific cultural connections among hominids responsible for different sites simply because their artifacts show some degree of similarity. Arguments among archeologists can then center on questions about whether technological similarities might support a hypothesis of cultural connection, or whether they are superficial and could well have developed independently in different regions.

Significantly and perplexingly here, Mode 1, Oldowan–type sites are not always completely replaced by the Acheulean, but co–exist in many regions over long periods of time, even hundreds of thousands of years. The reasons for this are still controversial. It could result from any of a variety of factors, for example, the functional and/or seasonal or environmental needs, cultural divisions between groups or lineages of hominids, raw material effects on artifact forms, and/or the effects of geological site formation agencies that may accentuate the proportion of large bifaces at some sites by the action of flood waters winnowing away smaller artifacts (especially smaller "pebble tools" and flakes) (Leakey 1971; Isaac 1977; Schick 1986).

Nevertheless, we must recognize that there is variability among the earlier Stone Age industries of Africa: Mode 1 simple core technologies continue through time along with the Mode 2 biface technologies. The proportion of Mode 1 to Mode 2 artifact forms influences, in fact, whether a particular site is classified as an Acheulean site (with more, better–made bifaces) or a more evolved development of the Oldowan Tradition (with fewer and cruder bifaces) (Leakey 1971).

Europe and Western Asia

Hominids appear to have migrated into Europe and Asia sometime after their first appearance in Africa and after the emergence of the relatively larger bodied and larger brained hominid form often referred to

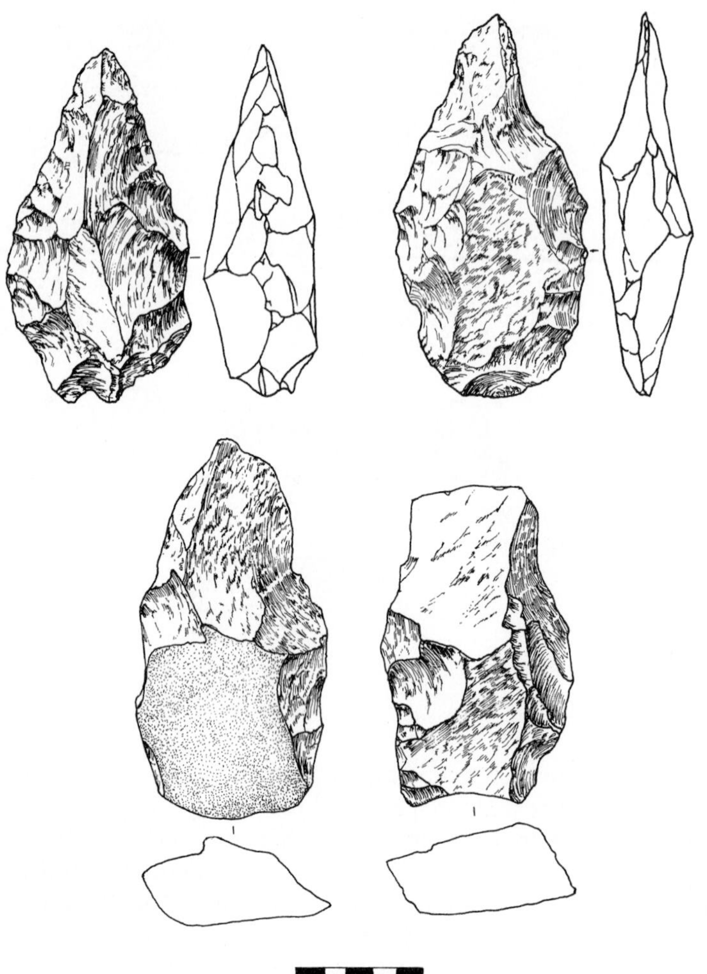

FIGURE 26-4 Africa, earlier Mode 2 technology. Bifacial tools from Bed II of Olduvai Gorge, dated to approximately 1.5 MYA. Top: hand-axes; bottom: cleaver. Such large bifacial forms are typical of many earlier Acheulean industries in Africa. (Drawn by David Johnson, after Leakey 1971)

as *Homo erectus*. For a number of years, the evidence has pointed to an initial emergence of hominids from Africa approximately 1 MYA, although there is now some evidence that the first hominid incursions into Eurasia may have occurred somewhat earlier than this.

The site of el 'Ubeidiya in Israel has long been cited as the oldest known definite evidence of hominid occupation outside of Africa. The site has been dated to about 1 MYA, although there is now evidence that it might extend somewhat further back in time to between 1 and 1.4 MYA (Tchernov 1987). Further to the north, the recently discovered site of Dmanissi in Georgia, where a hominid mandible reported to be early *Homo erectus* has been found, suggests hominid presence in this region by at least 1 MYA and perhaps as long ago as 1.6 MYA. It may be, however, since such early hominid sites in Eurasia are presently so rare, that the earliest hominid migrations may have been tentative and may not have involved large–scale population incursions for several hundred thousand years. But such patterns do raise the question of exactly when and at what stage(s) of technological development hominid migration waves extended from Africa into the rest of the Old World.

Some technological differences among the early European archeological sites have long been recognized (Howell 1966). Some contain essentially simple core and flake industries. For instance, sites such as Clacton–on–Sea (Singer et al. 1973), the lower levels of Swanscombe in England (Wymer 1964), and Verteszöllös in Hungary (Vertes 1965) contain basically Oldowan–style, Mode 1 technology.

At other sites, such as deposits at St. Acheul and Abbeville within ancient terrace deposits along the Somme River in France (Howell 1966), Terra Amata in France (de Lumley 1969), Hoxne (Singer and Wymer 1976) and the upper levels of Swanscombe in England (Waechter 1973), and Torralba and Ambrona in Spain (Howell 1966), to name only a few better–known sites, there are numbers of bifacial tools, especially hand-axes, that are unmistakably Acheulean in that they share a large, complex number of technological features with similar tool industries in Africa. In Africa, of course, Acheulean sites are contemporary with these as well as much older, indicating an African origin for this technological tradition and its probable subsequent spread into western Eurasia.

Now with the addition of more sites to the early archeological record of Europe and western Asia, and with concerted attempts to date these sites via paleomagnetic, faunal, or any available radiometric means, it seems clear that there is some element of a *time sequence* in the different technological traditions observed (see Klein 1989:212–216):

1. *Earlier sites* include Isernia La Pineta in Italy (Coltorti et al. 1982), Le Soleihac (Bonifay et al. 1976) and Vallonnet Cave (de Lumley 1975) in France, Prezletice and perhaps Stranska Skala (Valoch 1986) in the former Czechoslovakia, and Karlich (Kulemeyer 1986) in Germany (Figure 26-6). On paleomagnetic and other evidence these sites appear to predate the last major magnetic reversal that began the Brunhes Chron approximately 780,000 YA. Isernia, Prezletice, and Stranska Skala are contained within reversed sediments, presumably late Matuyama Chron. Based on recalibration of the paleomagnetic sequence (Spell and McDougall 1992; Tauxe et al. 1992), these sites would date to before

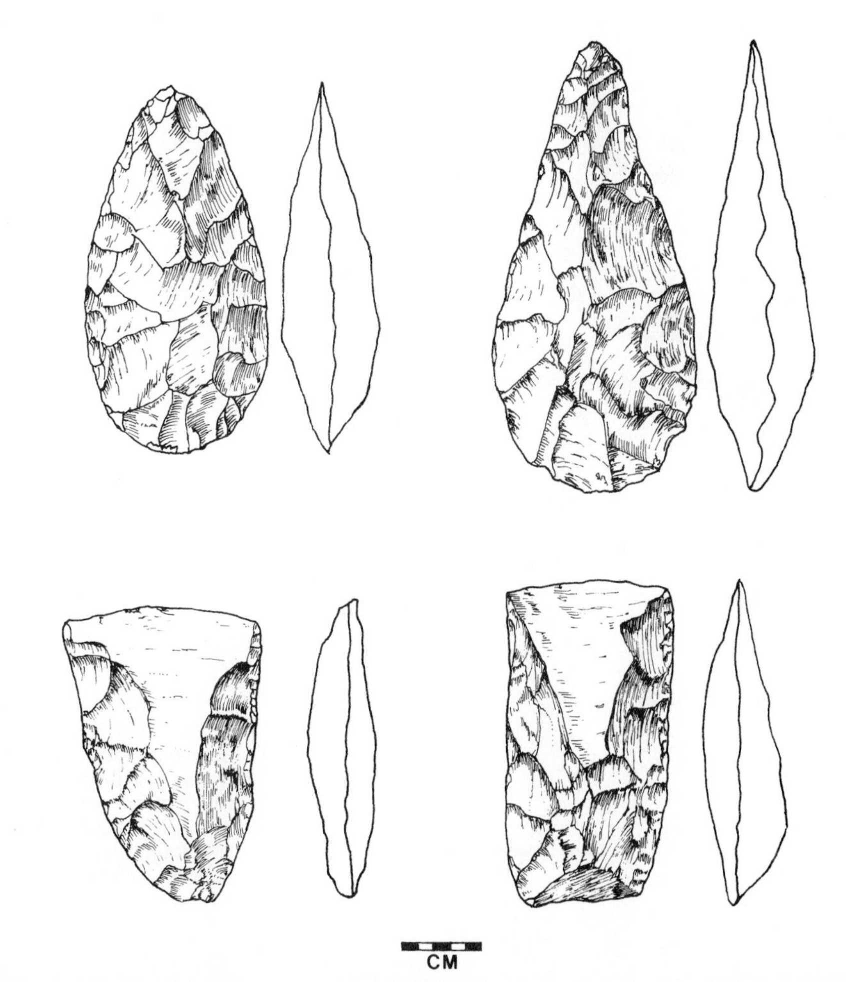

FIGURE 26-5 Africa, later Mode 2 technology. Acheulean hand-axes (above) and cleavers (below) from a later Middle Pleistocene site at Kalambo Falls, Zambia (dated to approximately 300,000 YA). Such later Acheulean forms often show a high degree of symmetry and finesse in their manufacture. (Drawn by David Johnson after Clark 1969)

780,000 YA. Vallonnet and Solheilac are in normal sediments inferred to be within the Jaramillo Subchron (between 930,000 and about 1 MYA) (Klein 1989). This would date them to 1 million or so years ago. Significantly, all of these sites bear principally simple, *Mode 1 technologies*, dominated by choppers, other simple cores, and flakes, although a few of them are said to contain one (Soleihac) or very few (Prezletice) roughly bifacial forms.

2. *Many later sites*, particularly starting about 500,000 to 600,000 years ago, begin to show definite *Acheulean technologies* that continue until 100,000 to 200,000 years ago (Figures 26-7 and 26-8). As is the case with Acheulean sites in Africa, sites containing Acheulean, Mode 2, industries in Europe also contain quantities of artifacts of Mode 1 technology.

Just as in Africa, where Developed Oldowan sites are contemporaneous with Acheulean occurrences, some Middle Pleistocene sites in Europe with Mode 1 technology (for example, Arago in France) are roughly contemporaneous with Acheulean ones (Villa 1991). Such Mode 1 occurrences are often characterized by designations such as "Clactonian," "Tayacian," or a miniaturized Mode 1 industry, the "Taubachian." These two technological modes overlap in western Europe. In eastern Europe and western Asia, however, Acheulean technologies are largely absent, and essentially Mode 1 technologies are present, as at Verteszöllös in Hungary and Bilzingsleben in Germany (Mania and Vlcek 1981).

In Europe and western Asia, then, during the Early Pleistocene and the earlier Middle Pleistocene, Mode 1 rather than Acheulean technology appears to be found at the earliest sites (excepting Ubeidiya).

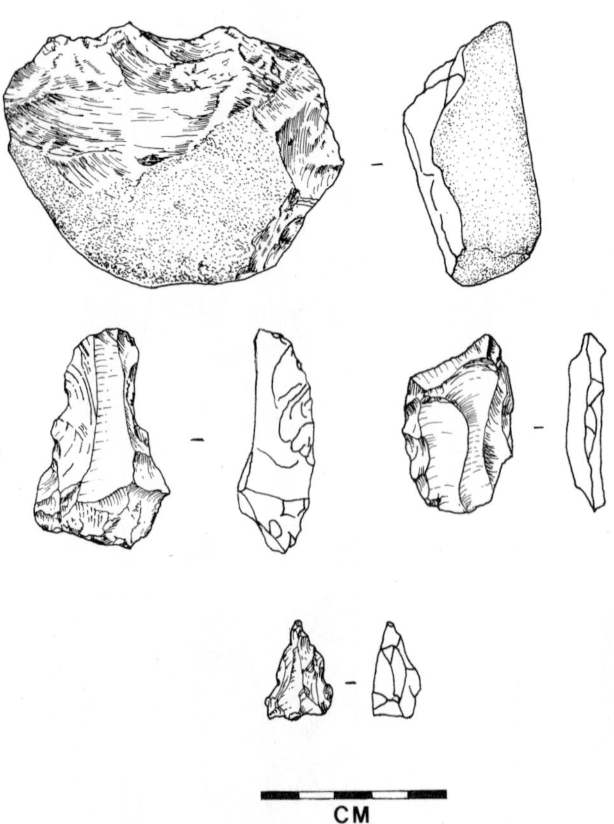

FIGURE 26-6 Europe, early Mode 1 technology. Simple core and retouched flakes from the site of Isernia in Italy, dated to before the last major magnetic reversal, or to at least 800,000 years ago. Top: limestone chopper; middle and bottom: flint pieces with denticulated ("toothed") retouch. (Drawn by David Johnson after Coltorti et al. 1982)

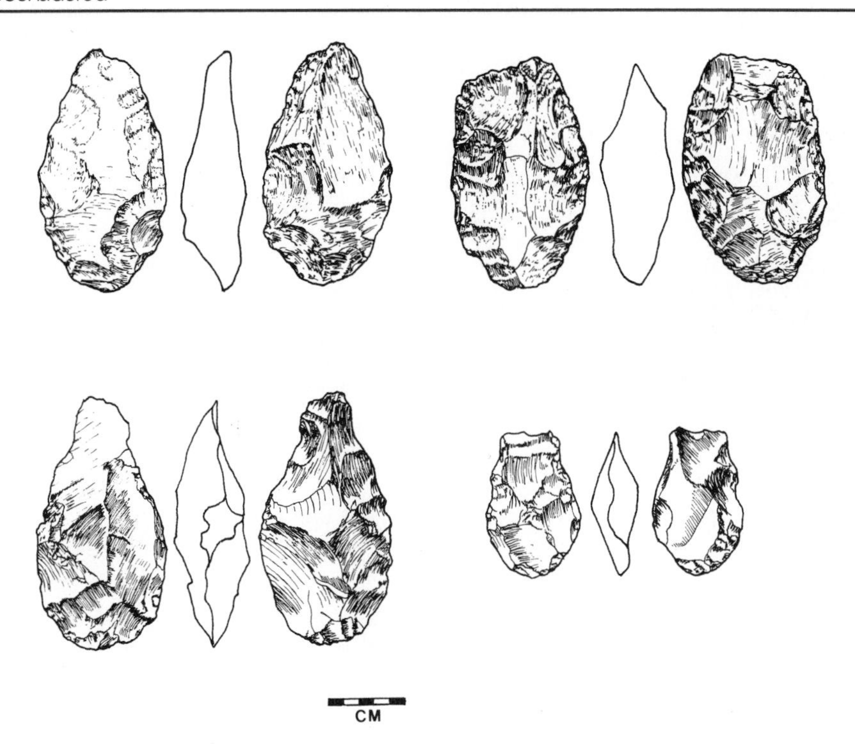

FIGURE 26-7 Europe, Mode 2 technology. Acheulean bifacial tools from the site of Torralba in Spain, dated to approximately 400,000 years ago. (Drawn by David Johnson, after Klein 1989, redrawn from Howell 1966).

After the Acheulean penetrated beyond the Levant (apparently not until the Middle Pleistocene), Mode 1 industries continued alongside the Acheulean in western Europe but appear to be carried on in eastern Europe and also northern central Asia to the exclusion of hand-axe industries (see Figure 26-9).

Eastern Asia

As previously summarized, Movius considered the absence of Acheulean bifacial industries to be a distinguishing characteristic of the early Stone Age of eastern Asia. This lack of Acheulean technology was considered a profound differentiating feature separating tool–making cultures in eastern Asia from counterparts in western Asia, Europe, and Africa. Although the various explanations offered by Movius for this pattern are largely conjectural and do not necessarily conform to all of the evidence available, his basic characterization of the major characteristics of early stone technologies in eastern Asia still holds up.

Even with many new surveys and excavations, new site discoveries, and new dating methods (see Wu and Olsen 1985), the Early Paleolithic of eastern Asia can still be characterized largely as a simple core and flake technological tradition (Hutterer 1977). Aigner has also reiterated "the conservative nature of Chinese Pleistocene stone tool technology and the absence of outside cultural influences as indicated by lithic similarities" (1978:223).

Although some sites currently have problems about whether their finds are veritable artifacts or about the precise stratigraphic provenience and age of their finds (Rendell and Dennell 1985; Pope et al. 1986), many verifiable new sites can now be dated to the Early or Middle Pleistocene, especially in China. Some of the important sites to add to the pattern since Movius's synthesis include

1. *Early Pleistocene archeological sites* within paleomagnetically reversed sediments deposited more than 780,000 years ago. These include Lantian in Shanxi Province (Zhang 1985) and sites in the Nihewan

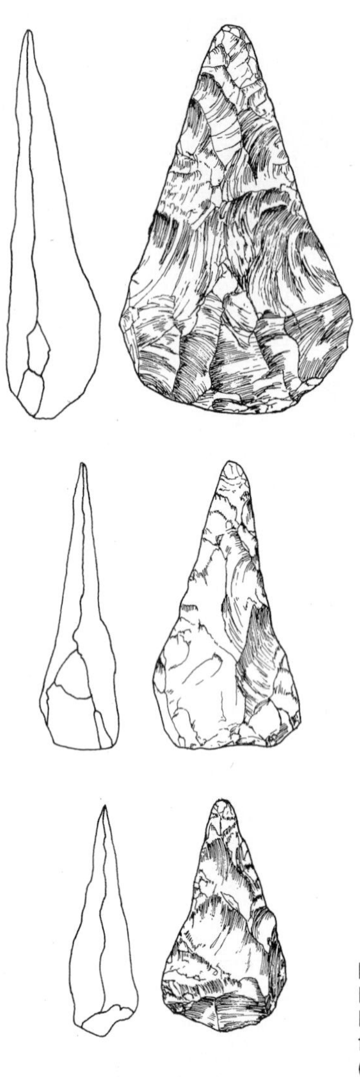

FIGURE 26-8
Europe, later Mode 2 technology.
Finely made pointed Acheulean hand-axes from later Middle Pleistocene times (c. 200,000 YA) from the site of Swanscombe, England. (Drawn by David Johnson after Wymer 1982)

basin of northern Hebei Province (Donggutuo and Xiaochangliang) (Huang 1985; Jia and Wei 1987; Schick et al. 1991; Tang et al. 1981; Wei 1988; Wei and Xie 1988) (Figure 26-10).

2. *Middle Pleistocene sites* between 780,000 years ago and approximately 100,000 to 200,000 years ago. Best known is Zhoukoudian Locality 1, whose sequence of deposits is now dated from approximately 450,000 years ago up to about 250,000 years ago (see Figure 26-2) (Black et al. 1933; Jia and Huang 1990; Wu and Lin 1973). More recent discoveries include other sites in the Nihewan basin such as Maliang and Cenjiawan (Schick et al. 1991; Wei 1988); Kehe in Shanxi Province (Jia 1980; Zhang 1985), Jinniushan (Huang et al. 1987) and Miaohoushan in northeast China in Liaoning Province and in southern China the caves of Daye or Shilongtou in Hubei Province and Guanyindong in Guizhou Province (Zhang 1985, 1989).

At all of these sites, the stone artifact industries can only be characterized as simple *Mode 1 core and flake technologies*. The size and particular morphologies of cores and flakes do vary somewhat from site to site, but this is very likely due to the initial size, shape, and particular flaking qualities of the raw materials

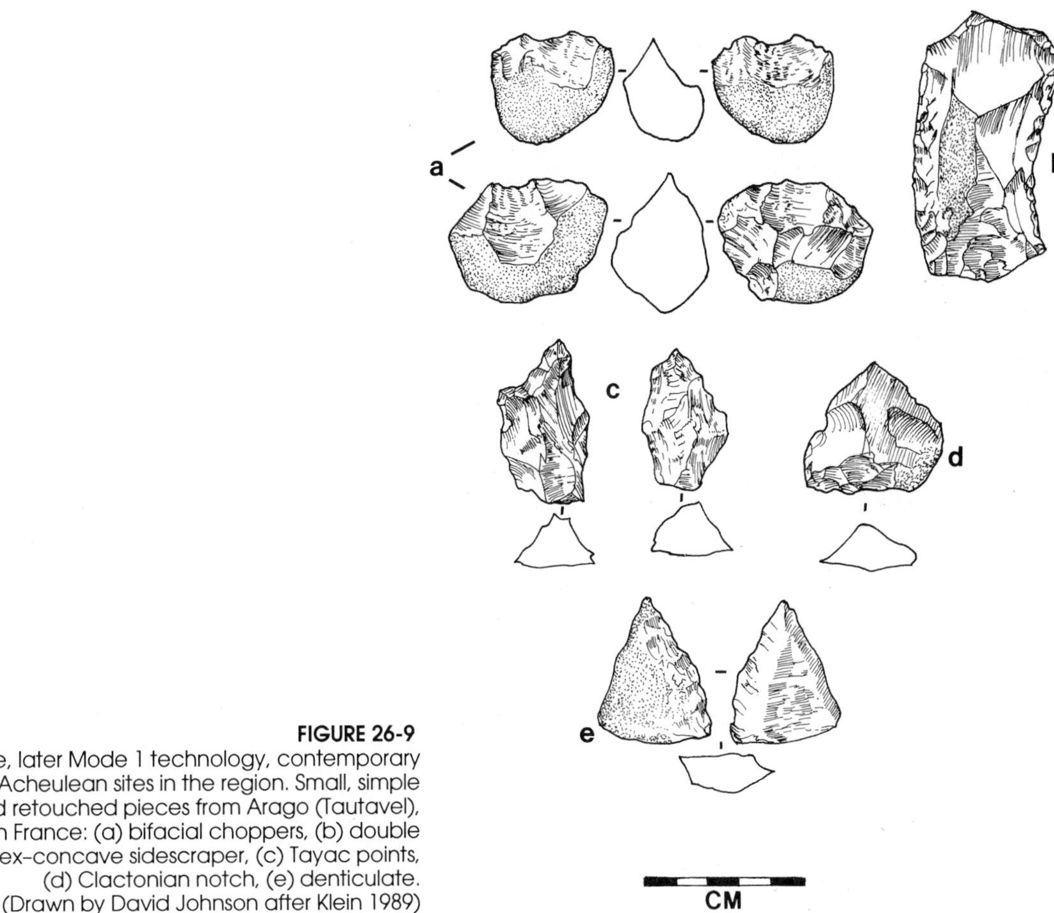

FIGURE 26-9
Europe, later Mode 1 technology, contemporary with Acheulean sites in the region. Small, simple cores and retouched pieces from Arago (Tautavel), southern France: (a) bifacial choppers, (b) double convex-concave sidescraper, (c) Tayac points, (d) Clactonian notch, (e) denticulate. (Drawn by David Johnson after Klein 1989)

worked (often smaller chunks of quartz were used, but at some sites other rock types such as blocks of chert, nodules of silicified limestone, or large quartzite cobbles, were employed).

While many Early Paleolithic sites elsewhere in southeast Asia have been difficult to verify (usually due to questions regarding the stratigraphic provenience, absolute age, or even artifactuality of the purported "tools"), other sites in Tadzhikistan on the central Asian Plain have been documented in later Middle Pleistocene times. The technologies at these sites, Karatau and Lakhuti I, are similarly dominated by simple core and flake assemblages (Davis et al. 1980; Ranov and Davis 1979).

CURRENT EVIDENCE: DOES THE "MOVIUS LINE" STAND?

Does an East–West Technological Dichotomy Really Exist?

Yi and Clark (1983) marshalled a strong argument against many implicit and explicit features of Movius's model of technological differentiation between the east and west. In sum, they argued that the schema of technological classification employed was outdated, overlooked the technological variability within eastern Asia, placed Eurocentric stress upon western technological developments, and ignored in particular the smaller tool and flake aspects of the eastern Asian industries. Moreover, as a clincher argument, they pointed to bifacial traditions evident at some sites in eastern Asia (e.g., Chon–Gok–Ni in Korea and Dingcun in northern China) (Bae 1988). These, they argued, should dispel notions of real east–west

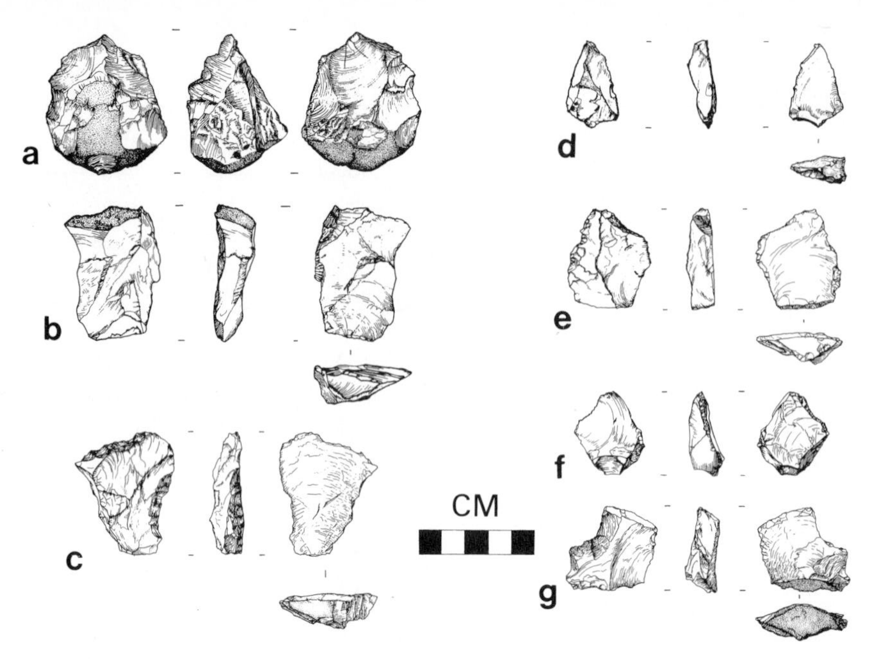

FIGURE 26-10 Asia, Mode 1 technology. Stone artifacts from the site of Donggutuo in the Nihewan basin, dated to approximately 1 MYA: (a) core, (b) flake, (c) modified piece (denticulate and concave scraper edge), (d) small flake, (e) modified piece (denticulate), (f) retouched piece (concave scraper), (g) modified piece (notch). (Drawn by Rachael Freyman)

technological differentiation. In other words, in their argument, east–west technological differences are actually quite complicated, and, moreover, they are not really as profound or "real" as they might seem.

The latter argument—that basically Acheulean technologies are found at various sites in the east and that, therefore, this paradigm of a technological dichotomy between east and west is basically a misconception—does not really ring true. The eastern biface technologies appear to represent a real phenomenon in certain places in eastern Asia, possibly by later Middle or early Late Pleistocene times. Unfortunately, reasonable radiometric approximate dates have been obtained for only a few sites, for example, Dingcun (c. 200,000 YA) (Chen and Yuan 1988) and Chon–gok–ni (<200,000 YA) (Bae 1988), but it is hoped that more will be available soon for more of these occurrences (e.g., Baise and other localities in southern China) (Figure 26-11).

Despite their bifacial nature, however, these technologies differ in very significant ways from the Acheulean tradition of western Eurasia and Africa. For one, the bifacial core forms do not show the characteristic biface thinning strategies evident in the west, and the resultant tools tend to be much thicker, sometimes even trihedral. They are in many ways reminiscent of the Sangoan industries in central Africa that followed the Acheulean and preceded flake tool–dominated Middle Stone Age technologies (J. D. Clark, pers. comm.). They do not show the same technological sets of procedures evident in Acheulean technologies throughout their identified area of spread.

Despite their bifacial or Mode 2 nature, then, these eastern biface technologies should really not be considered part of the overall *Acheulean* tradition, which maintained a high degree of technological coherence and continuity throughout a vast region over very long periods of time. They appear to represent some bona fide technological development that appears relatively late in the Early Paleolithic period in eastern Asia, possibly as an independent invention although all such "biface sites" in this part of the world are not necessarily contemporaneous (corroboration and extension of this pattern will require additional well–dated sites). It is important to recognize that this biface technology appears to be regionally, and perhaps temporally, quite sporadic in eastern Asia, in strong contrast to the strong regional and temporal continuities of the Acheulean. In this sense, there is a magnitude of difference between Acheulean biface technology and this eastern Asian phenomenon. This technology, if it can be demonstrated, at least in part, to represent a chronologically confined development, may perhaps have arisen through a combination of specific functional requirements in those regions in this time period and of newly developed cultural rules for producing stone technologies.

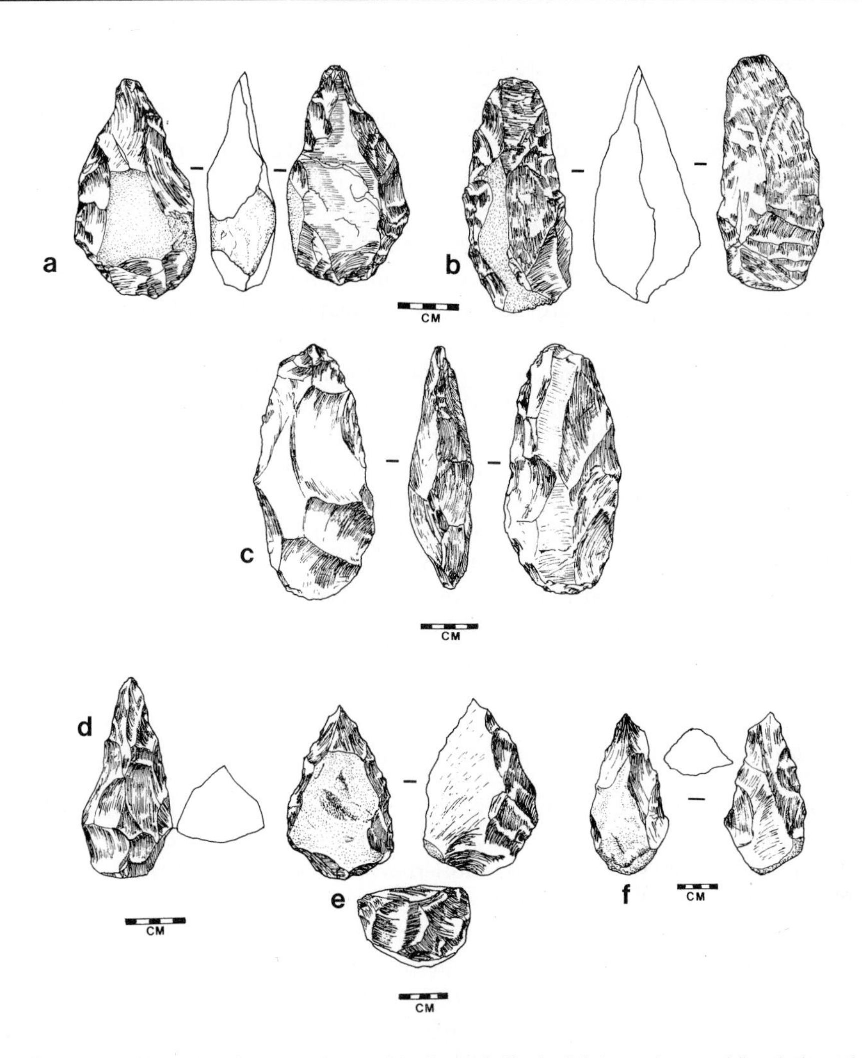

FIGURE 26-11 Asia, Mode 2 technology. Such tools tend to be relatively thick and even trihedral and are not typical of contemporary Acheulean industries in Europe and western Asia (although early Acheulean tools in Africa in the Early Pleistocene sometimes have similar morphologies). Most are not well dated, but two sites (Chon-gok-ni in Korea and Dingcun in China) may be approximately 200,000 years old. Bifaces from (a) and (b) Chon-gok-ni, Korea; (c) and (d) Dingcun, northern China; (e) Sanmenxia, northern China; and (f) Pingliang, a locality in Lantian County, northern China. (Drawn by David Johnson)

Some other points in Yi and Clark's case ring true, however, and some have also been raised by others. In particular, the characterization of eastern technologies as "chopper-chopping tool traditions" distorts the issue and is not currently used by most researchers. It is preferable to refer to these industries as Mode 1 technologies—made of simple flaked cores and especially flakes, along with a small proportion of retouched or modified flakes.

It appears appropriate to stress the smaller sized component of the artifact sets found at eastern Asian sites: It is generally the flakes, flake fragments, and, to a lesser extent, modified or retouched flakes (scrapers, notches, awls, etc.) that actually dominate these industries, rather than the core forms that have often received most of the attention. Furthermore, as shown by experimental research into Oldowan technologies, purported core "tools" may often be no more than cores for producing flakes. Also, the varieties of early core forms may reflect constraints laid down by the initial raw material form and extensiveness of flaking, rather than by any predetermined goals or target forms in the eyes of the tool-makers (Toth 1985).

Mode 1 Technology Does Not a "Culture" Make

Thus, we cannot regard Mode 1–style technologies as diagnostic "traditions" in terms of a strong, coherent set of rules of tool–making. Such simple core and flake technologies are in a sense least–effort strategies for producing sharp flakes and edges from blocks or cobbles of stone (Toth 1985). All of the simple core–and–flake technologies of Movius's chopper–chopping tool cultures as well as the Oldowan tradition of Africa can be produced merely through the application of the very basic principle of hard–hammer percussion to chunks and cobbles of stone. Such technologies do not require elaborate, defined sets of rules and strategies for tool manufacture that would indicate shared cultural rules (in contrast to the Acheulean, as explained later).

Thus, the practice of Mode 1 technology does not indicate some "stage" of human cultural, intellectual, and/or biological development. Mode 1 or essentially "Oldowan–like" technologies are found at all periods of time throughout much of the world during the Stone Age and do not indicate necessary levels of cultural or evolutionary development. Stone technologies are adaptive features of hominids within a much larger sphere of tools and culture: They serve as a minimal mirror or window into the past but do not describe fully all features of the cultures they represent.

Despite all of these caveats, however, the Paleolithic of eastern Asia still shows some very distinctive discontinuities with what is seen in much of the Old World. The large biface technologies of the Acheulean eventually stretched from Africa throughout many parts of Europe, the Near East, and peninsular India during Middle Pleistocene times. They are, however, still conspicuously absent in the East. Why is this?

The Acheulean: A Complex Technological Tradition

The Acheulean, unlike the Oldowan, is a tool tradition representing a complex set of norms, rules, and procedures for the making of stone tools. Novices in stone tool manufacture (who commonly become infatuated with bifacial technology and impassioned with producing either Acheulean–style bifaces or bifacial points) soon learn that biface production is not as easy as it may look in the hands of a master.

Common problems encountered by beginning knappers include removing too much width before the piece is adequately thinned (producing thick, narrow, even quaduhedral products), failure to maintain a good plane (e.g., producing bowed or extremely sinuous bifaces or just a lot of flake waste and an amorphous core), poor control over the outline shape (producing very asymmetrical products or whittling the biface down through a series of overcorrections), failure to extend the bifacial edge through more obtuse areas of the blank, removing the tip end through uncontrolled flaking, or breaking the biface in half with too strong a blow after it has been substantially thinned.

A number of procedures are involved in the manufacture of Acheulean bifaces. These are quite complex and fairly rigidly adhered to by early hominids, particularly by later Early Pleistocene and Middle Pleistocene times. Rules tend to get more elaborate and more rigidly adhered to later in the Middle Pleistocene. These procedures include:

1. The initial selection of stone finding a large, flat stone nodule (Figure 26-12) to work down into a biface (Figure 26-13), or striking large flakes from a boulder core (Figure 26-14) to be flaked down into the biface products (Figure 26-15). The production of large flakes from boulder cores in and of itself demands a certain amount of skill and expertise and is not something at which most novices are competent. There is also some geographic patterning in this—bifaces generally are made on large flakes in many parts of Africa and some other areas such as Spain, France, Israel, India, and Syria—and on nodules in most other parts of the world—at least partially according to raw materials prevailing in different regions, especially lavas or quartzites, but also probably influenced by cultural rules or traditions.

2. General shaping of the flake or nodule through bifacial flaking. This involves defining the major plane of the piece and its two faces and then flaking the blank, either alternating back–and–forth from one face to the other or working preferentially one side and then concentrating on the opposite face.
3. Directing blows into the main mass of the piece (to thin it) by setting up striking platforms nearly perpendicular to the faces of the biface through simple or more elaborate platform preparation on the opposite face (often producing typical "biface trimming flakes," sometimes even with faceted platforms from this preparation).
4. Developing and maintaining a twofold bilateral symmetry around the edge of the biface (in plan view and through the cross-sectional thickness of the artifact), this becoming more developed and pronounced in the later Acheulean (see Wynn 1989).
5. Often, final shaping of the tip of the piece, either through finer flaking and thinning into a point or curved, convex edge, or the truncation of the tip end with a well–placed blow to produce a cleaver bit (if the cleaver end was not already defined, in the case of large flake blanks, by the end of the initial flake struck from a boulder core). This set of technological procedures in the Acheulean involves a much more elaborate set of rules and sequence of procedures than anything observable in Mode 1 industries.

Moreover, sites with Acheulean bifaces very often contain large numbers of them (both hand-axes and cleavers) and they also commonly make up a very large percentage of the total assemblage of tools

FIGURE 26-12
A large nodule of flint can be shaped into an Acheulean biface following the techniques and procedures outlined in the text. Many of the Acheulean bifaces of Europe were manufactured from such nodular pieces of rock.

FIGURE 26-13 The finished product: a finely made hand-axe manufactured from the flint nodule in Figure 26-12.

and cores. And in any one time horizon at an Acheulean site, we often observe that hominids followed a very standardized way of making their bifaces (in terms of initial blank forms—large flakes versus nodules—as previously mentioned, flaking techniques, and size and shape characteristics of the final bifacial form). It seems that Acheulean tool–makers generally adhered to a very strong set of rules regarding the manufacture of their tools.

OTHER EXPLANATIONS FOR TECHNOLOGICAL DIFFERENCES

So, if these Acheulean bifaces are absent in eastern Asia, we can ask a legitimate, anthropological question: Why did the Acheulean, with this very strong, powerful tradition of tool–making, maintained by hominids through Early and Middle Pleistocene times across vast reaches of time, tremendous distances, and presumably drastically different environmental zones, never really penetrate the eastern reaches of Asia?

Raw Material Differences

Some of the suggested explanations for the distinctively different "flavor" of stone technologies in eastern Asia have centered upon possible differences in that part of the world in the availability of raw materials that could be fashioned into tools. These can be grouped into two major categories.

Lack of suitable stone for biface production in eastern Asia, and subsequent loss of the Acheulean tool–making tradition. A more–or–less casual hypothesis sometimes offered to explain the lack of Acheulean technologies in eastern Asia centers on a purported lack of suitable raw material for large bifacial tools in eastern Asia (i.e., good quality cherts, lavas, quartzites, etc. available in nodules or large blocks). While this may in some regions be the case, this precept cannot be applied unilaterally to the whole of eastern Asia. There are many regional or local deposits of volcanic deposits, quartzites, cherts or siliceous limestones that are perfectly suitable for biface manufacture (Figure 26-16). Later Middle

Pleistocene/early Later Pleistocene sites previously mentioned also testify to the presence of raw materials suitable for the production of Mode 2 technologies.

Availability of better, nonlithic raw materials for tools in eastern Asia (e.g., the "bamboo hypothesis"). The idea that bamboo may have replaced stone as a favored raw material for many tool functions has been promoted by various researchers for a few decades (see Ciochon et al. 1990, for an overview). This idea was suggested quite early by a Russian researcher, Boriskovskii (1968), who explained the limitations of the stone tool kit of Vietnam as probably due to the widespread use of both bamboo and shells as materials for tools.

In a 1978 compilation by Ikawa–Smith of current views on southeast Asian archeology, Harrisson developed similar themes on this "raw material replacement" explanation for the distinctive paucity of the southeast Asian stone tool kit:

> I prefer the concept that hardwood, bone, and other materials were extensively used to supplement and elaborate simple stone–cutting tools right through the Paleolithic. . . . In forested lands, especially, stone was only the primary coarse tool. Wood, bone, and shell did the finer work, especially where stone was so scarce that it clearly had enduring value and was persistently reused. . . . Looked at this way, the very slow, painful evolution of Early Paleolithic stone typologies is not necessarily indicative of "cultural levels" as a whole. (Harrisson 1978:43–44)

FIGURE 26-14 An alternative method for producing Acheulean tool forms: striking a large flake from a boulder (and then shaping this flake into the final hand-axe or cleaver). This method was commonly followed at sites in various regions, such as southern, eastern, and northern Africa, Spain, and the Levant.

FIGURE 26-15 Typical finished Acheulean tool forms produced from large flakes (top row, from left): ovate hand-axe, pointed hand-axe, cleaver, pick. In the bottom row are other common components of Acheulean industries: a hammerstone (left) and typical flakes and flake tools.

In the introduction to the same volume, François Bordes echoed and expanded on Harrisson's ideas, pointing especially to why in eastern Asia the stone tools found may be predictably simple and more elaborate tools may be archeologically "invisible":

> But the lack of flint or other suitable stone, and probably as Tom Harrisson, we believe, hypothesized, the use of vegetal resources which were not available to Western man (like bamboo) oriented their technology in quite a different way. Bamboo is a very versatile material which can be used for knives, bows and arrows, and spears (none of which need any stone blade), containers, construction material, etc. To work bamboo man needed only the simplest of tools, such as choppers and chopping–tools to cut them, and ordinary flakes to shape them. Unhappily, under tropical conditions bamboo does not keep, so it is quite possible that we shall never see, in these regions, more than the most basic tools. (Bordes 1978:ix–x)

These ideas have been reasserted recently by Pope (1989), who noted the overlap between the projected natural vegetation zone for bamboo and the distribution of Movius's simple stone tool cultures in eastern Asia. Clark (1992) has suggested a possible environmental basis for the technological differentiation, with Mode 2 Acheulean tools being replaced by simpler Mode 1 core–and–flake industries in forest and steppe zones where bamboo and wood replaced stone as the dominant material for tools (discussed further later). Further, he links the absence of hand-axe industries to environmental differences not only in eastern Asia but also in central Asia, where bifaces tend to be found in the more tropical south (peninsular India and northeast Pakistan) but absent in the Early Paleolithic of the more northern forest–steppe.

Recent ethnoarcheological research conducted in New Guinea (West Irian Jaya) and southern China (Schick and Toth 1993) was able to investigate directly the myriad ways bamboo can be used for simple tasks of day–to–day life (as Boriskovskii observed earlier in Vietnam), and also to carry out direct

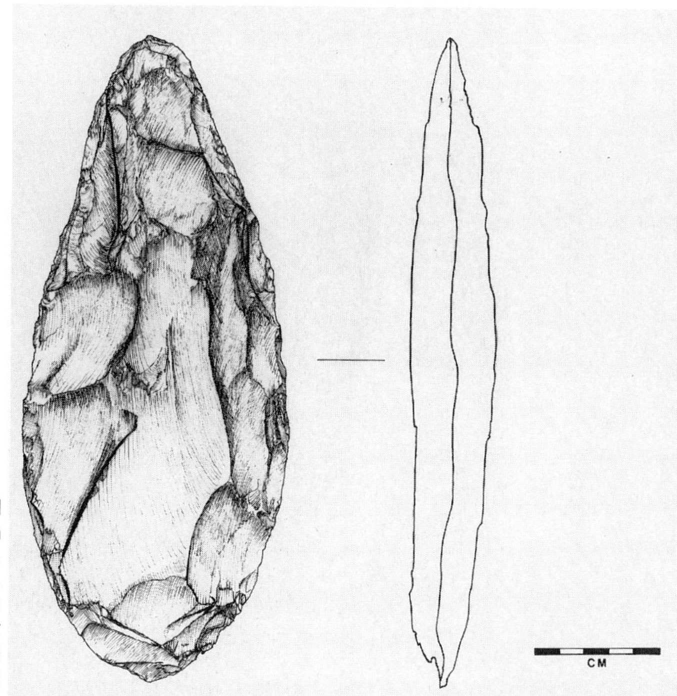

FIGURE 26-16 A hand-axe manufactured by the author near the Zhoukoudian ("Peking Man") site from a piece of tabular, local limestone. In many areas of eastern Asia, raw materials suitable for Acheulean-type technology were available but often were not used extensively and, if used, were not shaped into Acheulean forms. (Drawn by David Johnson)

experiments into how simple stone tools can be used to fashion bamboo implements. Interestingly, even the remaining expert stone adze–makers of Irian Jaya do not make use of stone regularly in any other facets of their lives (Toth et al. 1992). Many metal imports have, of course, replaced tools in indigenous materials. But significantly, despite the influx of metal tools, their preferred tool in cutting up dead animals is a simple piece of split bamboo, which, like hand-axes, is remarkably useful in butchering activities (Figures 26-17 and 26-18).

Barriers to Migrations and Cultural Diffusion

There have been some suggestions made in recent years that the eastern and western Asian hominid populations were more–or–less cut off from one another for much of the earlier Pleistocene due to profound barriers, especially geographic ones such as the mountains and tropical rain forest of central Asia, that effectively cut off flow of biological or cultural information between populations in eastern and western Asia (Aigner 1978:223; Clark 1992; Walker and Sieveking 1962). (This might be seen as a modified version of Movius's earlier idea, with each biogeographical area cut off from the other but without any loaded implications of one being "retarded" or "stagnant").

On the basis of physical traits among fossils, a number of researchers have suggested that eastern Asian hominids formed a lineage that had become separated from western hominid populations, perhaps sometime in the Early Pleistocene, and contributed little if any to subsequent hominid evolution (Andrews 1984; Groves 1989; Rightmire 1992; Wood 1991), although there are strong opponents to this hypothesis as well (e.g., Thorne and Wolpoff 1981; Wolpoff 1989).

Different Environments, Different Functions, Different Tools

Other arguments have centered on different requirements of different ecological environments in eastern and, especially, southeastern Asia to explain Movius's eastern culture area. Similar lines of argument have

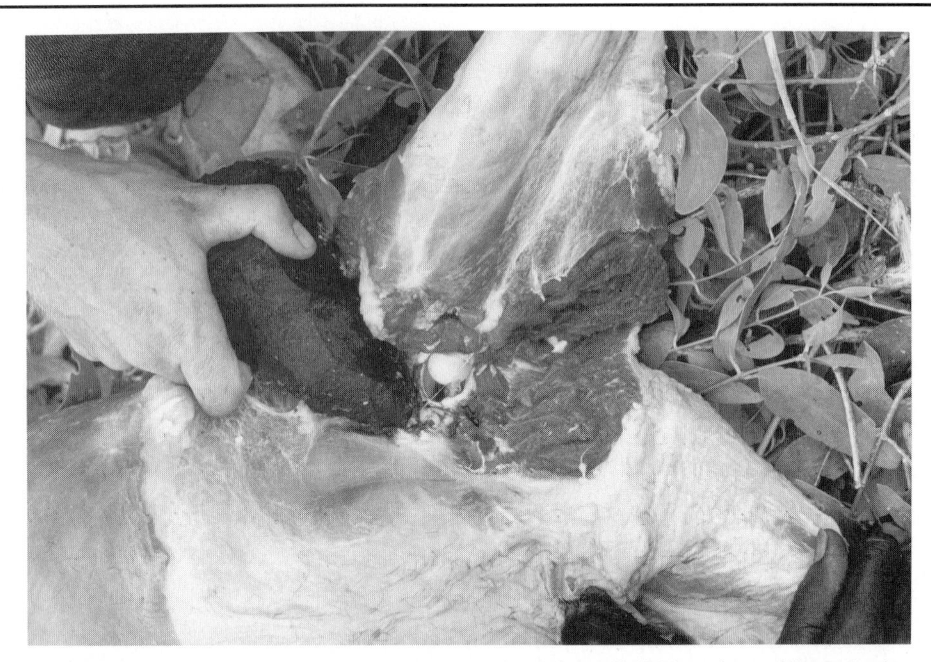

FIGURE 26-17 Use of a hand-axe in heavy-duty butchery. The side edges of these tools (usually the upper one-half/two-thirds of the length) tend to be quite effective for cutting and sawing, and the pointed or rounded tip is useful for hacking at tough connective tissue.

also been advanced to explain the differences between Acheulean and Mode 1 technologies observed in central and western Asia (Clark 1992).

One hypothesis has proposed that southeast Asian lithic industries are descendant from the Acheulean but highly transformed or "readapted" within tropical forest environments where plant collecting and some small animal hunting replaced subsistence based upon large game. Thus, large Acheulean cutting tools were replaced by flake tools more suited to the procurement and processing of plants and smaller animals (Watanabe 1985). In this construct, the higher latitude sites in more temperate environments in eastern Asia, such as Zhoukoudian, continued in the "chopper-chopping tool" vein because they had already passed through the "filter" of the southeast Asian forests (Watanabe 1985:13–14).

Another ecologically based hypothesis has suggested that dispersed food resources in tropical forests miligated against transporting an elaborate stone tool kit from place to place and favored the use of ubiquitous materials such as bamboo and wood. Furthermore, the various plant materials favored for tools can be easily worked with relatively simple stone tools (Hutterer 1977).

Cultural Interruptions during Hominid Migrations

Another factor revolves around possible problems these early hominids might have had in maintaining rather involved, complex cultural traditions such as those involved in Acheulean tool–making during migrations into new areas (Toth and Schick 1993). As hominids migrated into new areas, one major requirement they would have faced is locating raw material sources for their stone tools. This is especially crucial with regard to Acheulean technologies, as in most environments the majority of stone sources are not suitable (clasts are too small, textures are too coarse or too friable etc.) for the demands of Acheulean bifacial flaking.

It could have taken some time before the hominids could locate suitable raw materials in many regions to which they had spread, and in the meantime they would likely have made other, simpler stone tools. This break in Acheulean–style flaking, especially if it lasted for many years or for a generation, could have periodically led to the demise of this tradition during the foray of hominid groups out of areas with established Acheulean traditions. (This may be particularly likely if these earlier hominid forms had some limitations in linguistic and cognitive abilities relative to modern *Homo sapiens.*) Thus, in a sense, the Acheulean tradition may have been somewhat "fragile" in that a one–generation break in its maintenance due to lack of, or problems in locating, suitable raw materials could have led to its local extinction (at least until subsequent migrations reintroduced the tradition to local hominid populations with sufficient knowledge of regional raw materials).

CONCLUSION: THE MOVIUS LINE IS STILL A PARADOX

The Acheulean: A Late Entrant into Eurasia?

It is necessary that we should consider the Acheulean's absence in eastern Asia in a worldwide, comparative framework. As was noted in the Paleolithic overview presented earlier, the Acheulean does not seem to have spread out of Africa with all of the earliest migrants to Eurasia. In fact, it appears that, aside from a fairly early Acheulean presence in the Levant area of the Near East, the earliest migrant populations deeper into Eurasia did not carry with them a distinctly Acheulean tool–making tradition. The earliest sites so far documented in Europe and Western Asia show primarily Mode 1 industries, with rare Mode 2 technologies only occasionally present, and these are not demonstrably Acheulean but consist of relatively crude bifaces or "protobifaces." This pattern persists for some time, probably for several hundred thousand years.

It would seem, then, that over most of the geographic extent of early hominid migrations into Europe and Western Asia from Early Pleistocene times through the earlier Middle Pleistocene, there is some sort of "delayed transmission" of the Acheulean. The Acheulean tradition did not spread in a monolithic wave with the earliest hominids into Eurasia but was preceded by essentially Mode 1 technologies. The relatively late Mode 1 industries in Europe could also attest to later migrations in which the Acheulean

FIGURE 26-18 Use of a simple split piece of bamboo in animal butchery by people in the highlands of Irian Jaya (western New Guinea). Bamboo is very easily worked with relatively simple stone tools and will efficiently perform many operations, especially cutting softer materials.

tradition was similarly lost, or could attest to functional differences. The fact that the Acheulean never seems to have reached eastern Asia should be considered in this context.

The Levallois Method: Corroboration and a Possible Key?

There is another special, stylized technology, the Levallois method, that may serve as corroboration of important technological differentiation between east and west and also as a possible key to potential reasons behind these differences. The overall extent of distribution of the final Acheulean shows extensive overlap with the Levallois method of flaking (Figure 26-19). This method essentially involves special shaping and "preparation" of a core—especially its platforms—in the flaking process. In the most common type (the Levallois "tortoise core"), a discoidal core is reduced around its circumference through special platform preparation and with the production of a relatively large, final flake when the core is getting rather small. In another variant, the Levallois point core, the core is shaped for the final removal of a sharp, pointed flake. Movius clearly noted that the Levallois is "almost completely lacking throughout the Far East, as far as Southeastern Asia and Northern China are concerned," that the Levallois technique is commonly found with specialized (Acheulean) hand-axe cultures, and furthermore that "the known distribution of the two is very nearly coincidental" (Movius 1969:71).

In view of this parallel distribution of Acheulean and Levallois technologies, both involving strong sets of prescriptive rules regarding flaking operations, we may have to reconsider the involvement here of some sort of long–term impedance or bottlenecks in cultural transmission between east and west during the Early and Middle Pleistocene.

The transmission of technological rules and standards of the Acheulean was apparently relatively slow to move from Africa into western Eurasia. By the Middle Pleistocene, however, this tool–making tradition was fairly well implanted there (though not universal), and by the later Middle Pleistocene, the Levallois technique is also a recurrent feature of many later Acheulean assemblages. But *neither* the Acheulean tradition nor, perhaps even more enigmatically, the Levallois technique seems to have breached the mid–Asian barrier. In view of the continuation of simple Mode 1 flake industries in eastern Asia, we must consider why the Levallois never spread to eastern Asia, especially since the Levallois technique cannot be argued to have as stringent raw material requirements or the commonly inferred specialized

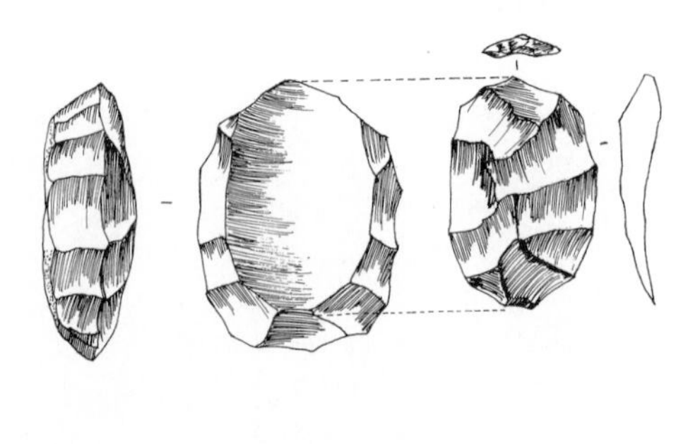

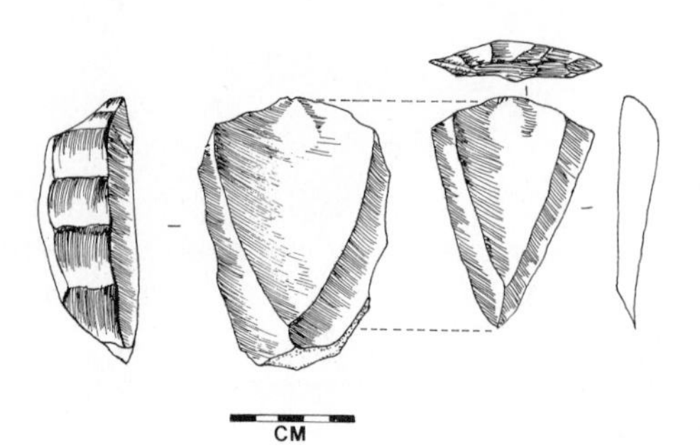

FIGURE 26-19 Levallois cores and flakes. Top: Levallois "tortoise core" (left)—a specially prepared core with radial flaking and platform preparation around its circumference—and the "Levallois flake" (right), the relatively large, final flake removed from such a core. Bottom: Levallois point core (left), prepared so that a sharp pointed flake (right) can be produced. Enigmatically, eastern Asia also does not show evidence of Levallois technique during Middle Pleistocene times, a pattern not easily explained in terms of different functional requirements of tools in the region. (Drawn by David Johnson, after Bordes 1970)

functions of the Acheulean. The absence of *both* technologies in eastern Asia is not easily explained on grounds of lack of suitable raw material, preeminent use of nonlithic raw materials for tools, or different functional requirements in the eastern Asian environments.

SUMMARY

Thus, in some ways, the "Movius Line" is still an enigma. The problem posed more than 40 years ago by Movius, that there is a real difference between Paleolithic technological patterns in eastern Asia and the rest of the world, still stands to a large extent, as does the question, why? We have many alternative hypotheses for the simple stone technologies in eastern Asia—different tool requirements, absence of proper stone raw materials, presence of nonlithic materials for tools, cultural interruptions during migrations, and so on—and any could partially be responsible for this pattern. At the present, we cannot resolve this question entirely. It is perhaps harder to explain the failure of the Levallois flaking method to spread to eastern Asia during the later Acheulean (and on into the Middle Paleolithic) in terms of some of these hypotheses, and it may indicate some break in cultural transmission.

It will be interesting in the future to explore whether these technological differences may be mirroring similar bottlenecks in gene flow between eastern Asia and western regions (Europe and Africa) (as per the strong differences between Asian forms of *Homo erectus* and related hominid forms in Africa and Europe now recognized by many researchers). Foley has suggested that culture tends to change relatively slowly, changing and diversifying much as biological organisms do over time, and thus stone tool assemblages can potentially serve as a valuable sign or marker for genetic lineages (Foley 1987). However, as Olsen and Miller-Antonio (1992) point out, the relatively low rate of technological change throughout the Paleolithic in China should make us realize the problems in "equating hominid type with typology." Based upon the archeological evidence, however, we cannot rule out that the rate or "fluidity" of transmission or sharing of cultural norms or rules, and perhaps of gene flow, may have been impeded or hindered between eastern and western Asia, perhaps over long periods during Early and Middle Pleistocene times.

Importantly, if such bottlenecks in cultural and gene flow were the case, this does not say that improved population contacts could not have been renewed by Late Pleistocene times and each area contributed to later hominid evolution. There have been strong arguments on both sides of this issue (see Rightmire 1992). Perhaps refined DNA studies of modern populations and (if techniques are improved to extract, amplify, and identify DNA from hominid bones themselves) fossil DNA comparisons may ultimately help us resolve these issues.

It is obvious, however, that understanding the overall evolutionary patterns of hominid populations during the Pleistocene will be aided by, and perhaps even require, consideration of both the biological and the archeological evidence. Both sets of evidence can, in their own way, reflect important connections and separations among hominid populations throughout human evolution and help us interpret the growing body of evidence for our evolutionary course throughout the Pleistocene.

ACKNOWLEDGMENTS

A number of people over the years have been instrumental in helping develop and assemble the ideas and the information contained in this study. First of all, I would like to thank F. Clark Howell for giving me the opportunity to study Acheulean technologies firsthand at the Spanish site of Ambrona, and for his work in laying the foundation for dialogue and research collaboration between Western and Chinese archeologists. Next, I thank many Chinese colleagues, especially Professors Jia Lanpo, Wei Qi, and Xie Fei, for making it possible to explore with them the nature of the Chinese Early Paleolithic evidence. The Institute of Vertebrate Paleontology and Paleoanthropology (IVPP) in Beijing and the Institute of Cultural Relics of Hebei Province have provided important logistical and institutional support to carry out joint surveys and excavations involving Chinese and U.S. researchers. Many discussions over the years with colleagues, especially J. Desmond Clark, F. Clark Howell, and Dong Zhuan have helped form the basis for many of the ideas presented here.

LITERATURE CITED

Aigner, J. (1978) Important archaeological remains from North China. In F. Ikawa–Smith (ed.): *Early Paleolithic in South and East Asia.* The Hague: Mouton, pp. 153–232.

Aigner, J. S. (1981) *Archaeological Remains in Pleistocene China.* Munich: Verlag C.H. Beck.

Andrews, P. (1984) An alternative interpretation of the characters used to define *Homo erectus. Courier Forsch. Inst. Senckenberg* 69:167–175.

Bae, K. (1988) The Significance of the Chongokni Stone Industry in the Tradition of the Palaeolithic Cultures of East Asia. Ph.D. Dissertation, Anthropology Department Library, U.C. Berkeley.

Baksi, A. K., Hsu, V., McWilliams, M. O., and Farrar, E. (1992) $^{40}Ar/^{39}Ar$ dating of the Brunhes Matuyama geomagnetic field reversal. *Science* 256:356–357.

Barstra, G. (1982) *Homo erectus*: The search for his artifacts. *Curr. Anthropol.* 23:318–320.

Black, D., Teilhard de Chardin, P., Yang, Z., and Pei, W. (1933) Fossil man in China. *Mem. Geol. Surv. of China, Series A* 11:1–166.

Bonifay, E., Bonifay, M.–F., Panattoni, R., and Tiercelin, J.–J. (1976) Soleihac (Blanzac, Haute–Loire): Nouveau site prehistorique du debut du Pleistocene moyen. *Bull. Soc. Prehist.Française* 73:293–304.

Bordes, F. (1970) *Tools of the Old and New Stone Age.* New York: Natural History Press.

Bordes, F. (1978) Foreword. In F. Ikawa–Smith (ed.): *Early Paleolithic in South and East Asia.* The Hague: Mouton, pp. ix–x.

Boriskovskii, P. I. (1968) Vietnam in primeval times. *Soviet Anthropol. Archaeol., Part 1* 7(2):14–32.

Chen, T., and Yuan, S. (1988) Uranium–series dating of bones and teeth from Chinese Paleolithic sites. *Archaeometry* 30:59–76.

Ciochon, R., Olsen, J., and James, J. (1990) *Other Origins: The Search for the Giant Ape in Human Prehistory.* New York: Bantam.

Clark, G. (1961) *World Prehistory: In New Perspective.* Cambridge: Cambridge University Press.

Clark, J. D. (1969) *Kalambo Falls Prehistoric Site,* Vol. 1. Cambridge: Cambridge University Press.

Clark, J. D. (1970) *The Prehistory of Africa.* London: Thames and Hudson.

Clark, J. D. (1992) African and Asian perspectives on the origins of modern humans. *Trans. Roy. Soc. Lond.* B 337:201–215.

Clark, J. D., Cole, G. H., Isaac, G. L., and Kleindienst, M. R. (1966) Precision and definition in African archaeology: Implications for archaeology of the recommendations of the 29th Wenner–Gren Symposium—"Systematic Investigation of the African Later Tertiary and Quaternary, 1965.' *S. Af. Archaeol. Bull.* 11(83)(Part III):114–121.

Clark, J. D., and Schick, K. D. (1988) Context and content: Impressions of Palaeolithic sites and assemblages in the People's Republic of China. *J. Hum. Evol.* 17:438–448.

Coles, J. M., and Higgs, E. S. (1969) *The Archaeology of Early Man.* Middlesex, England: Penguin.

Coltorti, M., Cremaschi, M., Delitala, M. C., Esu, D., Fornaseri, M., McPherron, A., Nicolletti, M., van Otterloo, R., Peretto, C., Sala, B., Schmidt, V., and Sevink, J. (1982) Reversed magnetic polarity at an early Lower Palaeolithic site in central Italy. *Nature* 300:173–176.

Davis, R. S., Ranov, V. A., and Dodonov, A. E. (1980) Early man in Soviet Central Asia. *Sci. Am.* 6:130–137.

Foley, R. (1987) Hominid species and stone–tool assemblages: How are they related? *Antiquity* 61:380–392.

Groves, C. P. (1989) *A Theory of Human and Primate Evolution.* Oxford: Oxford University Press.

Harrisson, T. (1978) Present status and problems for Paleolithic studies in Borneo and Admacent Islands. In F. Ikawa–Smith (ed.): *Early Paleolithic in South and East Asia.* The Hague: Mouton, pp. 37–57.

Howell, F. C. (1966) Observations on the earlier phases of the European Palaeolithic. *Am. Anthropol.* 68(2/2):88–201.

Howells, W. W., and Tsuchitani, P. J. (eds.) (1977) *Paleoanthropology in the People's Republic of China.* Washington, D.C.: National Academy of Sciences.

Huang, W. (1985) On the stone industry of Xiaochangliang. *Acta Anthropol. Sin.* 4:301–307.

Huang, W., You, Y., Gao, S., and Wei, H. (1987) On the problems of the karst cave and the deposits of the site of Jinniushan man. *Carsologica Sinica* 6(1):61–67.

Hutterer, K. (1977) Reinterpreting the southeast Asian Palaeolithic. In J. Allen, J. Golson, and R. Jones (eds.): *Sunda and Sahul: Prehistoric Studies in Southeast Asia, Melanesia and Australia.* New York: Academic Press, pp. 31–71.

Ikawa–Smith, F. (ed.) (1978) *Early Paleolithic in South and East Asia.* The Hague: Mouton.

Ikawa–Smith, F. (1978) Introduction: The Early Paleolithic tradition of East Asia. In F. Ikawa–Smith (ed.): *Early Paleolithic in South and East Asia.* The Hague: Mouton, pp. 1–10.

Isaac, G. L. (1977) *Olorgesailie: Archaeological Studies of a Middle Pleistocene Lake Basin in Kenya.* Chicago: University of Chicago Press.

Isaac, G. L. (1984) The archaeology of human origins: Studies of the Lower Pleistocene in East Africa 1971–1981. In F. Wendorf and A. Close (eds.): *Advances in World Archaeology* 3. New York: Academic Press, pp. 1–87.

Jia, L. (1980) *Early Man in China.* Beijing: Foreign Languages Press.

Jia, L. (1985) China's earliest Palaeolithic industries. In R. Wu and J. W. Olsen (eds.): *Palaeoanthropology and Palaeolithic Archaeology in the People's Republic of China*. New York: Academic Press, pp. 135–145.
Jia, L., and Wei, Q. (1987) Artefacts lithiques provenant du site Pleistocene ancien de Donggutuo pres de Nihewan (Nihowan), Province d'Hebei, Chine. *L'Anthropologie* 91:727–732.
Jia, L., and Wang, J., (1978) *Hsihoutu—A Culture of Early Pleistocene in Shansi Province*. Beijing: Cultural Relics Press.
Jia, L., and Huang, W. (1990) *The Story of Peking Man*. New York: Oxford University Press.
Jinniushan Combined Excavation Team. (1978). A preliminary study of Palaeolithic artifacts of Jinniushan, Yingkou, Liaoning. *Vert. PalAs*. 14(2):120–127.
Klein, R. G. (1989) *The Human Career: Human Biological and Cultural Origins*. Chicago: University of Chicago Press.
Leakey, M. D. (1971) *Olduvai Gorge. Volume 3*. London: Cambridge University Press.
Li, T., and Etler, D. A. (1992) New Middle Pleistocene hominid crania from Yunxian in China. *Nature* 357:404–407.
Lumley, H. de (1969) A Palaeolithic camp site at Nice. *Sci. Am*. 220(5):42–50.
Lumley, H. de (1975) Cultural evolution in France in its paleoecological setting during the Middle Pleistocene. In K. W. Butzer and G. L. Isaac (eds.): *After the Australopithecines*. The Hague: Mouton, pp. 745–808.
Mania, D., and Vlcek, E. (1981) *Homo erectus* in middle Europe: The discovery from Bilzingsleben. In B. A. Sigmon and S. Cybulski (eds.): *Homo erectus—Papers in Honor of Davidson Black*. Toronto: University of Toronto Press, pp. 131-151.
Movius, H. (1944) *Early man and Pleistocene stratigraphy in southern and eastern Asia*. Papers of the Peabody Museum of Archaeology and Ethnology, Harvard University 19(3).
Movius, H. (1948) The Lower Palaeolithic cultures of southern and eastern Asia. *Trans. Am. Phil. Soc*. 38, pt. 4:329–420.
Movius, H. (1969) Lower Paleolithic archaeology in southern Asia and the Far East. Reprinted in W. W. Howells (ed.): *Early Man in the Far East, Studies in Physical Anthropology No. 1*. New York: Humanities Press, pp. 17–82.
Olsen, J. W., and Miller–Antonio, S. (1992) The Palaeolithic in Southern China. *Asian Perspectives* 31(2):129–160.
Pope, G. (1989) Bamboo and human evolution. *Nat. Hist*. 10:48–57.
Pope, G., Barr, S., Macdonald, A., and Nakabanlang, S. (1986) Earliest radiometrically dated artifacts from Southeast Asia. *Curr. Anthropol*. 27:275–279.
Ranov, V. A., and Davis, R. S. (1979). Toward a new outline of the Soviet Central Asian Paleolithic. *Curr. Anthropol*. 20(2):249–270.
Rendell, H., and Dennell, R. W. (1985) Dated Lower Palaeolithic artifacts from northern Pakistan. *Curr. Anthropol*. 26:393.
Rightmire, G. P. (1992) *Homo erectus*: Ancestor or evolutionary side branch? *Evol. Anthropol*. 1(2):43–55.
Schick, K. (1986) *Stone Age Sites in the Making: Experiments in the Formation and Transformation of Archaeological Occurrences*. Oxford: British Archaeological Reports International Series 319.
Schick, K., and Dong, Z. (1993) Early Paleolithic of China and Eastern Asia. *Evol. Anthropol*. 2(1):22-35.
Schick, K. and Toth, N. (1993) *Making Silent Stones Speak: Human Evolution and the Dawn of Technology*. New York: Simon & Schuster.
Schick, K., Toth, N., Wei, Q., Clark, J. D., and Etler, D. (1991). Archaeological perspectives in the Nihewan Basin, China. *J. Hum. Evol*. 21:13–26.
Semah, F., Semah, A.–M., Djubiantono, T., and Simanjuntak, H. T. (1992) Did they also make stone tools? *J. Hum. Evol*. 23(5):439–446.
Singer, R., and Wymer, J. J. (1976) The sequence of Acheulian industries at Hoxne, Suffolk. In J. Combier (ed.): *L'Evolution de l'Acheuleen en Europe, 14-30*. Nice: Union International des Sciences Prehistoriques et Protohistoriques (Colloque X).
Singer, R., Wymer, J. J., Gladfelter, B. G., Wolff, R. G. (1973) Excavation of the Clactonian Industry at the golf course, Clacton–on–Sea, Essex. *Proc. Prehist. Soc*. 39:6–74.
Spell, T. L., and McDougall, I. (1992) Revisions to the age of the Brunhes–Matuyama Boundary and the Pleistocene geomagnetic polarity timescale. *Geophy. Res. Lett*. 19:1181–1184.
Tang, Y., You, Y., and Li, Y. (1981) Some new fossil localities of early Pleistocene from Yangyuan and Yuxian basins, northern Hebei. *Vert. PalAs*. 12(3):256–268.
Tauxe, L., Deino, A. D., Behrensmeyer, A. K., and Potts, R. (1992) Pinning down the Brunhes/Matuyama and upper Jaramillo boundaries: A reconciliation of orbital and isotopic time scales. *Earth Planet. Sci. Lett*. 109:561–572.
Tchernov, E. (1987) The age of the 'Ubeidiya Formation, an Early Pleistocene hominid site in the Jordan Valley, Israel. *Israel J. Earth Sci*. 36:3–30.
Teilhard de Chardin, P. (1941). Early man in China. *Institut de Geo-Biologie Pub*. 7:1–99.
Thorne, A., and Wolpoff, M. H. (1981) Regional continuity in Australasian Pleistocene hominid evolution. *Am. J. Phys. Anthropol. 55*:337–349.
Toth, N. (1985) The Oldowan reassessed: A close look at early stone artifacts. *J. Archaeol. Sci*. 12:101–120.
Toth, N., Clark, J. D., and Ligabue, G. (1992). The last stone ax makers. *Sci. Am*. July:88–93.
Toth, N., and Schick, K. (1986) The first million years: The archaeology of proto–human culture. In M. B. Schiffer (ed.): *Advances in Archaeological Method and Theory*. New York: Academic Press, pp. 1-96.
Toth, N., and Schick, K. (1993) Early stone industries and inferences regarding language and cognition. In K. R. Gibson and T. Ingold (eds.): *Tools, Language and Cognition in Human Evolution*. Cambridge: Cambridge University Press, pp. 346–362.

Valoch, K. (1986) The central European early Palaeolithic. *Anthropos* (Brno) 23:189–206.

Vertes, L. (1965) Typology of the Buda industry: A pebble–tool industry from the Hungarian Lower Paleolithic. *Quateraria* 7:185–195.

Villa, P. (1991) Middle Pleistocene prehistory in southwestern Europe: The state of our knowledge and ignorance. *J. Anthropol. Res.* 47(2):193–217.

Waechter, J. D.'A. (1973) The late Middle Acheulian industries of the Swanscombe area. In D. E. Strong (ed.): *Archaeological Theory and Practice.* New York: Seminar, pp. 67–86.

Walker, D., and Sieveking, A. D. (1962) The Palaeolithic industry of Kota Tampan, Perak, Malaya. *Proc. Prehist. Soc.* 28:103–139.

Watanabe, H. (1985) The chopper–chopping tool complex of eastern Asia: an ethnoarchaeological–ecological reexamination. *J. Anthropol. Archaeol.* 4:1–18.

Wei, Q. (1985) Palaeolithis from the Lower Pleistocene of the Nihewan Beds in the Donggutuo site. *Acta Anthropol. Sin.* 4(4):289–300.

Wei, Q. (1988) Le Cadre Stratigraphique, Geochronologique et Biostratigraphique des sites les plus anciens connus en China. *L'Anthropologie* 92:931–938.

Wei, X., and Xie, F. (1990). *Selected Treatises on Nihewan.* Beijing: Cultural Relics Publishing House.

Wolpoff, M. H. (1989) Multiregional evolution: The fossil alternative to Eden. In P. Mellars and C. B. Stringer (eds.): *The Human Revolution.* Edinburgh: Edinburgh University Press, pp. 62–108.

Wood, B. A. (1991) *Koobi Fora Research Project IV: Hominid Cranial Remains from Koobi Fora.* Oxford: Clarendon.

Wu, R., and Olsen, J. W. (eds.) (1985) *Palaeoanthropology and Palaeolithic Archaeology in the People's Republic of China.* New York: Academic Press.

Wu, R., and Lin, S. (1973) Peking Man. *Sci. Am.* 248:78–86.

Wymer, J. J. (1964) Excavations at Barnfield Pit, 1955–1960. In C. D. Pveu (ed.): *The Swanscombe Skull.* London: Royal Anthropological Institute, pp. 19–61.

Wymer, J. J. (1982). *The Palaeolithic Age.* London: Croom Helm Ltd.

Wynn, T. (1989) *The Evolution of Spatial Competence.* Illinois Studies in Anthropology No. 17. Chicago: University of Chicago Press.

Yi, S., and Clark, G. A. (1983) Observations on the lower Palaeolithic of Northeast Asia. *Curr. Anthropol.* 24:181–202.

Zhang, S. (1985) The Early Palaeolithic of China. In R. Wu and J. W. Olsen (eds.): *Palaeoanthropology and Palaeolithic Archaeology in the People's Republic of China.* New York: Academic Press, pp. 147–186.

Zhang, S. (1989) The Early Palaeolithic of North China. In R. Wu, X. Wu, and S. Zhang (eds.): *Early Humankind in China.* Beijing: pp. 97–158.

CHAPTER 27

Torralba and Ambrona

A Review of Discoveries

Leslie G. Freeman

INTRODUCTION

In the 1960s, F. Clark Howell began a program of multidisciplinary investigations at the Spanish Mesetan sites of Torralba and Ambrona that quickly became classic. Torralba and Ambrona remain among the best-preserved, most carefully excavated, and informative mid-Pleistocene localities known from Western Europe to the present day. It is my belief that in the future these excavations will be increasingly recognized as among Howell's foremost contributions.

This chapter reviews the work of the team that excavated and analyzed Acheulian residues and bones at Torralba and Ambrona under Howell's supervision and outlines the implications of the analysis of those residues.

The conclusions reached by Howell's team in the 1960s seemed interesting but unexceptionable at the time. Careful attention to microstratigraphy revealed several stratified levels of paleontological and archeological materials. Intimate spatial associations of tools and faunal materials and some otherwise seemingly inexplicable marks on the bones suggested that humans had visited the site to hunt or at least to butcher large game animals, although it was always recognized that some of the animals could have died natural deaths without human intervention and might have had nothing to do with hominid scavenging or butchering at all. We stated that elephants were neither the exclusive nor the principal object of human attention: that other animals, especially horses, were as abundant or more so in some levels. We detected and recognized geologically caused rearrangements of residues, some due to faulting and more impressive ones due to freeze-thaw cycles in a harsh climate, and we were attentive to the possibility of winnowing and realignment due to flow in sheets and channels, though we could not detect edges of channels in any part of the Torralba Acheulian deposits. We carried out extensive analyses for paleoenvironmental reconstruction, including the contour-mapping of old temporary surfaces. We knew that carnivores had been present at least occasionally at both sites and had occasionally gnawed at a bone.

Worked wood—cut and hacked, and sometimes charred—was recovered, as was charcoal in abundance, but nothing we could definitely identify as a hearth.

Statistical analyses indicated that the visible spatial associations we could see were part of larger patterns of consistent and repeated frequency relationships. In the 1970s, T. P. Volman showed that the frequency relationships detected when whole levels were compared had a spatial component in individual levels, that different sets of stone tools and body parts were consistently found in different parts of the ancient landscape: marshy waterlogged low-lying areas were the loci of death and discovery of carcasses, and the loci of preliminary disjointing of body parts. Higher areas were the setting of intermediate stages of butchering and bone breaking. Still higher and drier were the few situations where final processing of carcasses took place, with some amount of stone flaking or tool repair (Freeman 1978).

Throughout the 1960s and 1970s there was little in the way of challenge or contradiction of those interpretations, although new studies of site formation processes and taphonomy suggested by the late 1970s that some revision of interpretations was necessary. Beginning in 1981, however, those conclusions were disputed, often with little or no justification and less regard for the facts. Certainly conclusions reached 30 years ago can stand a deal of revision in light of new information and criticism. But the conclusions Howell's team reached about Torralba—conclusions for which I take a major share of responsibility—were not simply evaluated and evenhandedly criticized; the interpretations were distorted by the critics to become unrecognizable caricatures, and the caricature then savaged. Now some say that results from the Torralba/Ambrona excavations, where "faunal assemblages are in disturbed context" (Villa 1991:206), are unreliable or at least suspect. To advance science, those critics advocate dismissing our results, to rely instead on other sites, whose deposits are in fact no more intact, whose stratigraphy is no less complex, whose age is no less uncertain, whose samples are smaller, whose excavations were if anything less carefully controlled, and whose excavators have proposed interpretations no less "simplistic" or "anecdotal" and "unsystematic" (Villa 1991:202, 204) than those we proffered.

The best answer to criticism comes from the sites themselves; were the data they provide better known, much of the debate about them would evaporate. A chapter of this length unhappily cannot do justice to excavations whose results require substantial monographic publication. The final monograph on Torralba, finished in the early 1980s, has been ready for press for some time, and at one time was even accepted for publication. Its appearance has paradoxically been delayed several years due to just such misconceptions about the site and its residues as a prompter publication might have dispelled.

Despite the deplorable impression that very little about the sites has seen print, part of the information to be reviewed here has long been available. For Torralba alone, there are more than a dozen largely nonrepetitive articles in English, based specifically on the analysis of recovered materials and distribution patterns from the site; together they total more than 300 pages. Though there are fewer sources about Ambrona, some quite extensive preliminary treatments of our work there have appeared (for example; Howell, Butzer, and Aguirre 1963; Howell and Freeman 1982; Howell, Freeman, Butzer, and Klein 1992). This chapter reviews aspects of the research conducted at Torralba and Ambrona during the 1960s and 1980s, in light of the most salient questions that have been raised.

The chapter is intended as a clarification of the record, not a debate with critics: truth is usually not well served by rhetoric. In the few passages where irritation mars the presentation, I beg the reader's indulgence with my loss of patience. I intend simply to state the facts about Howell's (and later, our) excavations as I understand them and to make it clear that for any understanding of mid-Pleistocene adaptations in mid-latitude Europe the data they offer must be taken into consideration.

Dismissing the results of research at Torralba and Ambrona is unwise—it would mean casting aside a great deal of important information about the nature of environmental change, site-formation processes, and hominid adaptations. Even the most vocal critic of our work cannot help admitting, in the midst of a slighting comment about my procedural inadequacies, that "the interaction between hominids and faunal remains seems clear. In fact, the results are not in conflict with the results that Freeman obtained" (Binford

1987:95). Encouraged by so forceful an advocate, even an analyst as short-sighted as I cannot fail to be hopeful that a new overview will sharpen our vision of the significance of these Mesetan sites.

THE EXCAVATIONS

History of Research

On June 17, 1909, the Marqués de Cerralbo visited the hamlet of Torralba del Moral near Medinaceli in Soria, Spain.There, in 1888, trenches cut for the Madrid-Zaragoza railway had revealed bones of extinct Pleistocene mammals, including huge elephants. To his surprise, Cerralbo found Acheulian hand-axes and other stone tools in association with these remains (Cerralbo 1909, 1913a,b). This high-altitude site (1113 meters above m.s.l.) was soon famous as one of the earliest human hunting stations known from Europe, though Cerralbo did not live to see it published in extenso. After Cerralbo's death in 1922, despite the site's recognized importance, no one returned to explore it until the 1960s.

It was of course Howell who initiated new fieldwork. In 1961 he also rediscovered the Ambrona site, an analogue to Torralba, situated at a slightly higher elevation (1140 m above m.s.l.) about 3 km away. Though Cerralbo located and tested Ambrona some time prior to 1916, it was only known from the briefest published references (Obermaier 1916:190, 1925:180), before Howell's work.

Beginning in 1961, Howell directed three seasons' excavations at Torralba, removing most of the site sediments left intact after Cerralbo's extensive excavations. The seven-week 1961 season proved that a portion of the site was still undisturbed, yielding hundreds of animal fossils and scores of stone artifacts. However, the site stratigraphy proved much more complex than suspected in 1961, when most finds seemed to come from a single archeological level. As a third-year graduate student at Chicago, I joined Howell's team as an assistant in 1962, and after an initial three-week period excavating at Ambrona with Howell and Dr. Pierre Biberson, I spent six weeks working at the Torralba site. Again in 1963, I spent a month digging at Ambrona—Thomas Lynch supervised work there until his departure in June—before undertaking ten weeks' work at Torralba as site supervisor. Emiliano Aguirre undertook limited excavations at Ambrona in 1973 to improve the on-site museum. As co-director with Howell and the late Dr. Martín Almagro, I returned for full-scale excavations at Ambrona in 1980-1981. Howell alone directed one last season there (1983), in which other excavations at El Juyo in Cantabrian Spain kept me from participating. In total, the 1980s excavations lasted 203 days.

Size of Sites and Exposures

The Torralba site was much smaller than its sister, Ambrona. When Cerralbo began work, it may have extended over as much as 3000 sq m or perhaps slightly more. Although he gave a much lower estimate of his exposures at Torralba, we learned that he had in fact opened at least 1000 sq m. For a careful, modern excavation, Howell's fieldwork was also undertaken on a very large scale, as its duration would suggest. In 1961, he exposed approximately 450 sq m over the site surface, and during 1962 and 1963, we dug another 576 sq m all told, evidently in a richer and stratigraphically more complex part of the site. (While we left some intact sediments at Torralba as a witness, they are neither contiguous, nor easily accessible.) At Ambrona, the largest European Acheulian site known at present (more than 6000 sq m were "intact" in 1962), Howell's exposures through 1983 attained the truly impressive extent of some 2800 sq m.

Excavation Techniques

Methods of excavation employed at Torralba and Ambrona from the 1960s on were as close to state-of-the-art as Howell or I could make them under the circumstances. Obviously, appropriate tools and techniques must vary with the nature of deposits, the availability of water, and other factors. At both Torralba and Ambrona, some levels are fine clays that when dry come away in chunks, separating cleanly from the finds they encase, while others are fluviatile/colluvial deposits that are sometimes sandy and friable, sometimes indurated to a near rock-hard consistency. At both sites, excavators used small pick-hammers on the indurated and clayey sediments, as well as knives and trowels, "crochets," and brushes of several kinds as digging tools.

The excavation was mostly done by workers—farmers from the surrounding hamlets—but they were as well-trained and capable as most students I have since had on field crews of my own; some became as technically virtuous excavators as any I have ever known. They were adequately overseen. One trained student assistant supervised the four workers excavating in two contiguous squares, advising them as needed, measuring, drawing, excavating in particularly delicate situations, and so on.

In any excavation (if the excavator is honest), some materials are inevitably recovered "out of context," and Torralba/Ambrona are no exception: some finds were made whose level was known but whose exact horizontal position, orientation, and so on, were indeterminate. Others were recovered in screening. These pieces were not plotted, but bagged by square, sector, and level.

At all times, our excavations were conducted with careful attention to provenience and microstratigraphy. The procedures used were not perfect. They never are. But I do not see how anyone could have done a much better job of excavating Torralba and Ambrona than Howell and his crews. Speaking as one who has at least as much experience directing meticulous excavations as any other Old World prehistorian, I find no contradiction of that evaluation in the work of others since then. The excavations were visited and inspected by a large number of first-class excavators and sedimentologists. The methods used were praised at the time—in the 1980s no less than in the 1960s. Only one person has seen fit to challenge our procedures; I can only characterize her comments as both uninformed and unreasonable (Villa 1990).

Sediments and Stratigraphy

At both sites natural levels of deposition are primarily differentiated by texture, and we were as scrupulous as possible in detecting minor changes in sediments as we proceeded. At Torralba, there are 10 "major" archeological horizons that seem to have accumulated in colluvial screes, or in and on dry channel deposits, or in fans along a pond-margin, or in the marshy shallows of a pond. Many recovered bones show localized—rarely complete—polish or abrasion, perhaps by waterborne sand flowing over partly buried pieces; some elongated pieces have polish or abrasion restricted to one or both ends, like the wear on expedient butchering tools of bone from some U.S. buffalo jumps described by Frison (1991:302–308). Most archeological residues were found atop former temporarily stable surfaces, at the contact between layers of sediment that differed texturally. Very occasionally, the presence of a continuous sheetlike horizon of bones and artifacts within an otherwise uniform level was the only indication of a former temporary surface.

Field designations of levels differ from the final Occupation designations (these are "final" only at Torralba). As excavation progressed, and some levels were subdivided (or in some cases where relationships between different spreads of material were temporarily unclear), field designations became cumbersome ("B4aa"; "Occupation X"; etc.). Microfaulting required that final correlation of Torralba levels be done in the laboratory, using all maps and sections as well as three-dimensional stereoscopic plots and models of the site. Publications that appeared at different times reflect these changes, which has produced some confusion on the part of readers.

The earlier excavations of Cerralbo, in the form of wide trenches, cut through the upper site deposits, producing an interruption in distributions in all the levels affected. Though his trench did not always remove the basal levels, where it was present it appears as a blank in the distributions, and that gap reappears in the same area in maps of all the levels affected. It appears, and is clearly labelled, on the partial map of Occupation 7 published by Freeman and Butzer (1966). Binford, in the course of an *ad hominem* attack, suggested that the distribution gap, and the resulting apparent alignment of materials along its edges in several levels, may be "the structured result of differential erosion," adding that "Freeman never considered this possibility since he already assumed the hominid behavioral cause of his structured results" (Binford 1987:58). On the contrary, I have never suggested that the gap means anything other than Cerralbo's trench, or that its structure is due to prehistoric cultural behavior. It is characteristically careless of Binford to suggest both that I have done so, and to offer the innovative "reinterpretation" that the disturbance is instead really some sort of stream channel.

There were several cases of detached islets or spreads of archeological material that occurred atop a single temporary surface but were separated from each other horizontally by interruptions or large gaps in item distributions, sometimes caused by removal of the intervening surface by Cerralbo's excavations, or happening to coincide with a zone of microfaulting. Since the Torralba stratigraphy is so complex, with fluviatile/colluvial levels pinching out laterally or merging to produce a single surface where there were two before, I still believe that the only safe practice, in the absence of some obvious proof that the islets are contemporaneous (such as finding, in different islets, conjoinable bone or stone fragments, or bones from the same identifiable individual) was to keep their contents separate for analytical purposes, even when it seemed likely that they had accumulated at "approximately the same time." Where solid evidence of contemporaneity was lacking, separation was consistently our practice.

There were nine such "sublevels" all told: four (designated 1a-1d) on the Level 1 surface, apart from the major contiguous expanse of Level 1 itself; two in each of Levels 2 and 4; and one in Level 3. Four of the individual spreads involved were quite small, but five were large enough to provide considerable material. Even without counting these segregated islands of material that occurred in the same deposits, the natural archeological strata distinguished in the field were finer and more numerous than the geological units of deposition recognized by Karl Butzer, who analyzed the sedimentology at both sites.

Vertical and Horizontal Control

We excavated by natural archeological strata, also recording absolute depths to the nearest cm below an arbitrary horizontal datum, whose position was marked on stakes in each square. Levelling (within a square) was sometimes done with line levels; at other times "parallax triangles" were used, following the practice of the late François Bordes. In the 1980s excavations at Ambrona, an optical level was used for vertical control.

Horizontal control was provided by a grid of 3 meter squares, and all visible finds in each square were located with tapes and plumb bob and piece-plotted at a scale of 1:20. When two pieces were found in direct and intimate contact, they were often given the same feature number. This was explained in notes on the plans, and in the site log or inventory such finds were differentiated as necessary by adding letters to the feature number. Unless the pieces were themselves very similar, the letters were not always placed on the piece labels themselves (there was little reason to do so, since the inventory was expected to resolve any possible confusion).

Orientations and inclinations of pieces were generally visible from or noted on the plans, but where recovered pieces were markedly disconformable to the lie of the stratum that contained them, special measurements, photographs, and notations were made. Naturally, we drew continuous sections showing both geological and archeological levels following all square walls, and abundant photography documents our procedures and finds and the stratigraphic distinctions we made.

Screening

At Torralba there was no available water for wet-sieving or washing finds (and in the 1960s, the needs of the Ambrona farmers for garden irrigation kept us from using the trickle of water seasonally available in the Río Ambrona below that site). Contrary to some of the critics (surprisingly, these include Klein: see

1987:22-23), we did dry-screen samples of sediments at Torralba. In the 1962 excavations at Torralba, small amounts of sediment were sporadically passed through round screens with a mesh of about 5 mm. In 1963, we more systematically screened 15-20 percent samples of sediment (by square and level) from the archeological horizons, and 100 percent of the sediment from three selected squares designated as controls (screening did not include any of the later culturally sterile deposits overlying the archeological levels, though that procedure might also have been informative). In that year, the screens used were specially constructed large rectangular ones, still with a 5 mm mesh. The requirements of backdirt disposal dictated the technique employed: we unbolted the screens from their stands, lay them over wheelbarrows, and shoveled excavated sediment through them directly into the wheelbarrows. (Figure 27-1, taken to document the appearance of one edge of Cerralbo's trench through the area we excavated, shows a screen and its stand.) Screening at Torralba yielded a disappointingly small amount of material.

At Ambrona, in the 1980s, in addition to dry-screening, it became possible to wash sediments in bulk through fine-mesh screens in the stream below the site. This, of course, permitted more complete retrieval of small finds, including microfaunal remains. The richest source of finds was the clayey pond/marsh sediment, much better represented at Ambrona than at Torralba. It is, however, noteworthy that washing did not yield appreciable quantities of small flaking debris.

RESULTS AND INTERPRETATION

Paleoenvironments and Site Formation Processes

The archeological deposits at Torralba and Ambrona were studied by K. W. Butzer in 1962 and 1963, and he returned to Ambrona during the 1980-1981 field seasons. What follows is a brief summary of his results, focused particularly on the Torralba site, digested from his most recent treatment to appear in the forthcoming monograph. His interpretation of the nature of the sedimentary column at that site is based both on his field examination of morphology, sediment sizes, and particle or item orientations as well as on macroscopic and microscopic analysis of 77 sediment samples taken during the course of excavation and later processed by Dr. Réné Tavernier in Gent. I have intercalated the results of the pollen analysis, based on identifications by the late F. Florschütz and J. Menéndez-Amor, to add relevant vegetational detail to the paleoclimatic picture. They analyzed 161 samples (of which many were sterile) taken in two partially overlapping series at 10-cm intervals through the Torralba column.

The archeological horizons are found in cold-indicative Pleistocene sediments lying above Triassic Keuper clays. Later lubrication and deformation of the plastic Keuper resulted in a series of microfaults,

FIGURE 27-1 Screens at Torralba (1963).

with thrusts of a few centimeters to as much as a meter, that affected the site sediments.

Following a series of sterile units formed under cold conditions, Member IIb of the Torralba Formation, up to 30-cm of coarse, subangular to subrounded gravel, incorporating fine lenses of clay, was formed ("A-Gravel"). The deposit suggests a frost-weathered detritus transported over some distance. Cobbles and larger rocks have been rearranged into stone rings of 25- to 40-cm diameter on slopes of 2-5 degrees, elongated into ellipsoidal "garlands" of rock on slopes of 5-10 degrees, and on even steeper slopes torn apart into stone stripes, perpendicular to the contours, or scatters in which individual pebbles either point downhill or lie parallel to the contours. These are typical "patterned ground" phenomena of periglacial upland environments, attributed to seasonal or diurnal freeze-thaw cycles. The stone rings and garlands are contemporary with the accumulation of the A-Gravel or with the human occupation directly on top of them. Rare artifacts are found reworked in the gravels and clay lenses of this unit. However, the earliest archeological level coincides with the immediate surface of this gravel and appears to be coeval with local lenticles of light gray clay that indicate a shift from high-energy slope mobilization to low-energy subaqueous sedimentation. A single pollen sample from this clayey layer shows high AP values (76 percent), predominantly *Pinus silvestris. Sphagnum* spores suggest poorly drained or boggy ground near the site, while sedges are absent. The fact that the NAP is essentially all grasses, with a trace of *Artemisia,* indicates open vegetation on drier plateau surfaces nearby.

Size distribution histograms for rocks from all the archeological levels at Torralba reveal an abnormal frequency of large stones in this level, either indicating far more effective frost-shattering than can be found in recent analogues, or that the larger stones were concentrated in the site through human activity.

The presence of stone rings, garlands, and stripes more or less contemporaneous with the earliest archeological level raises the question whether solifluidal transport or sheetwash disturbed the cultural associations of this particular horizon. From the orientation and dispersal of bones, it is obvious that some sliding has taken place, particularly on slopes exceeding 10 degrees. However, the limited rolling or wear of articular bone surfaces, the lack of size sorting of bones or artifacts, and the nearly articulated position of bones of single animals, all argue that, in general, such sliding has not destroyed the validity of cultural associations—independent of orientation.

Other archeological levels lack soil-frost structures and have not been so extensively disturbed. The best occurrences are in semiprimary context.

Most of the Acheulian occupations are concentrated in levels in Member IIc (Lower Gray Colluvium), disconformably deposited atop the "A-Gravel." The A-Gravel is absent in the western sector of the site, where unit IIc rests directly on earlier deposits, the contact distorted by congeliturbation structures. There are several well-stratified subunits and facies in Member IIc that range from gravel layers to unconsolidated, white to light gray or pale brown gritty sands with lenses of fine gravel and sandy silts. Periodic halts in deposition or episodes of erosion interrupt these deposits. This unit is a quasi-horizontal graded valley fill, with abundant fragments of thin-shelled aquatic gastropods (see later discussion) in most finer facies. Current-bedding is visible in some of the fine-sediment subunits. The gravel facies of unit IIc is characterized by angular to subangular shapes, containing some 19 percent pebbles fractured during transport. This suggests very short transport distances but only an intermediate intensity of frost-weathering. There is no evidence of soil-frost structures. Limonitic staining and mottling band the sediments, showing water-table fluctuations. Since these stains do not conform to the lay of the deposits, the water-table changes happened after the site deposits accumulated and even after some micro-faulting took place.

A number of cobbles and boulders, varying in major diameter from 20-55 cm, were probably carried into the site area by Acheulian people, and numerous archeological horizons of variable area are found throughout this unit.

The range of horizontal facies from sandy clays to gritty sands, with some current-bedding and discontinuous rubble bands, combined with the aquatic gastropods, suggests a predominantly fluvial depositional environment. A low- to moderate-energy stream crossed parts of the site, and incorporated some slope rubble during periods of intense overland flow, while ponding was not uncommon farther down valley, at least during the early phases of accumulation. Climate was quite cold, but not as severe as during accumulation of the A-Gravel, and surface denudation was less vigorous.

Pollen spectra attest a cycle of shrinkage and later recovery of a swamp or lake near the site, and continued very cold conditions. Arboreal pollen drops to 36 percent before rising again to its former level. At that point, pine forests must have been reduced to scrubby stands in a largely grassland environment. Other (rare) tree species are those that would fringe water-courses or ponds/lakes nearby. Preservation is unusually good for plant material, and bits of wood as well as other material are preserved. Identifications of macrobotanical remains, by Dr. B. F. Kukachka of the Wood Products Laboratory in Madison, Wisconsin, are mostly of conifers, among which *Pinus silvestris* is predominant—presumably brought to the site by humans—but they additionally include one bit of birch and another of *Salix* or *Populus*, that could have been obtained locally. Chenopod pollen increases with grasses in the middle of the series, when attractive and nutritious pasturage was most abundant. Sedges are represented in the earliest and latest pollen samples in the sequence, while *Artemisia* is always present. Increasing desiccation in the midpart of this Member was likely due to physiological drought during the cold season, not to a total drop in precipitation: The presence of water-lily in one of the grass-rich samples betrays the (perennial) presence of standing water from 1 to 3 m deep.

The final bed of Member II rests on an eroded surface, attaining a thickness of 90 cm in a former topographic hollow in the northeastern part of the site. This "Brown Marl" bed is a compact, light gray to brownish gray marl, intermixed with lime sand or grit. Diffuse limonitic staining as well as reddish-yellow mottling indicate oxidation in a zone of fluctuating watertable. Some cryoturbation festooning is present. A ponded stream channel, spring seep, or the margin of a swampy floodplain is implied. Slope denudation was minimal and the environment was more temperate as well as wetter than during accumulation of unit IIc, but never as benign as it would be during a full interglacial. Archeological materials in the Brown Marl are very localized in their occurrence. Pollen samples show an initial peak of AP (80 percent +), declining thereafter. The last Acheulian occupation at Torralba occurs in this unit.

At Ambrona, however, occupations continue into Units IV (Upper Gray Colluvium and Gray Marls) and V (Rubefied Colluvium) of the Torralba Formation. Unit IV begins with moderate-energy fluvial deposition, becoming increasingly low energy, and attests cold conditions with intensive seasonally concentrated runoff at first. At Ambrona this unit is terminated by gravels indicating a return to higher energy conditions. There follow marly mixed slope and fluvial accumulations, indicating intensive seasonally concentrated runoff under cold conditions (Gritty Gray Marls) and then the Upper Gray Marls, low-energy ponded or lacustrine deposits in more temperate conditions (though temperate, climate was still some 5° C colder than today). Last come the stratified, in part lenticular, deposits of Unit V, resulting from moderate-energy footslope and valley-margin accumulation by surface runoff and frost-assisted gravity transfer, including alluvial fans at Ambrona. Intensive frost-weathering and vigorous denudation took place on higher slopes, with incomplete vegetative mat (very cold). Dr. Thure Cerling noted that small red quartz crystals in the gravels of this unit were so fresh, and their surfaces so free from abrasion, that they could not possibly have travelled far by hydraulic action (personal communication in Toth, *in litt.*). The final Acheulian occupation at Ambrona took place during the first of the moister episodes in this Unit.

There are several distinct Acheulian occupations in the Ambrona deposits just as at Torralba (though they may be fewer in number); since the distributions are still not completely analyzed, they have been grouped into two larger sets in earlier descriptions: those from the Lower Unit and those from the Upper Unit.

Butzer notes that most of the major archeological horizons at both sites are found in seasonally active, valley-margin deposits, in close proximity to permanently wet ground. However, a minority of archeological levels—more at Ambrona than at Torralba—occur within more clayey swamp- or pond-edge sediments themselves, as though shallow water or waterlogged marshy areas were sometimes used for the accumulation of or disposal of archeological residues.

Though Butzer estimates that the accumulation of the Torralba Formation sediments may have taken some 125,000 years, and the Acheulian deposits may date between very roughly 420,000 and 450,000 BP, it must be noted that the deposits and the archeological materials they contain were not accumulated continuously, as Binford (1987) seems to suggest, but rather episodically; long periods of nondeposition and some erosion, and even longer periods when neither artifacts nor animal remains were accumulating in the site deposits, were followed by relatively brief moments of active site use by animals and/or humans, and then by other periods of disuse.

None of the occupations at Torralba is a pristine intact association in true "primary" archeological context, and if earlier papers have not made that sufficiently clear, it has not been our intention to deny it, as some secondary sources seem to suggest (see later discussion).

Size of Samples

If the density of finds at Torralba and Ambrona is not particularly high for well-excavated sites of their age and type, neither is it especially low. The very large size of exposures, coupled with good preservation of organic materials, should suggest that sample sizes of recovered artifacts, bones and other materials of all kinds, are likely to be larger, not smaller, than "average." At Torralba, 2141 bones and 689 stone artifacts were excavated during the 1962-1963 field seasons alone.

I find Villa's (1990:307) observation that this sample size is too small and sparse for reliable statistical analysis puzzling to say the least. It betrays a surprising ignorance of statistics; worse, it is fundamentally illogical, since she finds no such fault with the much smaller samples from the Aridos quarry localities, which together are less than half that size. In fact, from the published evidence, I see no more reason to believe Aridos a convincing intact butchery site than to consider Torralba the same. At Ambrona, Howell's investigations produced vastly larger quantities of varied occupation residues: Over 2085 fragmentary remains of the single taxon *Elephas*, and more than 1400 stone artifacts, have been found in the Lower depositional unit alone to date.

Lithic Artifacts

Stone artifacts are, of course, one principal evidence of human activity at Torralba and Ambrona. Various aspects of the lithic assemblages at these sites are interesting: the raw materials used, the composition of assemblages, the presence of wear traces, spatial associations with other evidence, including conjoinability, and relationships in abundance of specific sets of tools and particular animal species or body parts, are all informative in their respective ways.

Freeman (1991) provides a more detailed discussion of raw material use at Torralba. None of the raw materials used for stone tool manufacture at either site is local. Three basic kinds of stone are represented: cherts/chalcedonous flints, quartzites of variable grain size, and limestones. Although there are outcrops of porous limestone a few hundred meters from either site, they are not really suitable for tool manufacture and were not used. The Triassic clays underlying the site contain no stone raw material. The closest stone sources are suitable limestones a few kilometers from the site; the quartzites used are found no closer than 10 km away, and the flints and cherts would have had to been transported scores of kilometers to the sites. One distinctive and rare kind of flint seems to have been imported from the Jalón drainage, more than 50 km from the site. The Río Ambrona flowing past that site has none of this material in its bed—it could not, for the source is across the divide separating the site from the Ebro drainage, several kilometers downstream on that side. The most probable sources of commoner raw materials are downstream from the Torralba site in the Tajo/Duero drainages. Raw material from any of these sources would have had to be transported upstream to reach the sites, so it must have been imported by humans. At Torralba, aside from the fact that flints are not frequently used to make bifaces, the finer cryptocrystalline materials—the flints and cherts—were not especially chosen for the manufacture of smoothly retouched working edges such as sidescrapers.

From the 1962-1963 excavations at Torralba, there are 689 stone artifacts, of which 63, or about 9 percent of the total, are geologically crushed (rather cryoturbated than rolled) pieces on flakes. Though they are or once were artifacts, their original typology is indeterminate, so they have always been excluded from detailed analysis of the stone artifact collections, leaving 626 identifiable artifacts. The total includes 1 battered polyhedron and 5 hammerstones (1 percent). Thirty cores and discs make up 4.8 percent of the collection. There are 36 or 5.8 percent bifaces, and 212 or 34 percent shaped flake tools. Minimally retouched/utilized flakes, 160 are 25.6 percent, and unretouched so-called "waste" another 159 pieces or 25.4 percent: together they compose 51 percent of the artifacts in the combined collection. When just the shaped tool collection—the 212 flake tools plus 36 bifaces—is considered, bifaces are 14.5 percent of the total for all levels. Scraping tools (60) are 24.2 percent, notches (21) 8.5 percent and denticulates (48) 19.4 percent of the shaped tool series. There are small proportions of burins (5.2 percent) and backed knives (0.8 percent), while perforators and becs are more frequent (10.9 percent). Two points were recovered. About 4 percent of the pieces are raclette-like artifacts with continuous abrupt retouch on much or all of the circumference. Unclassifiable variants (usually multiple-edged, prismatic-sectioned pieces) are quite numerous—10.1 percent of shaped tools.

From my counts, the lithic collection from the 1962-1963 excavations at Ambrona (all units) is more than twice as large: 1520 total pieces. These were apparently not all included in Howell et al.'s earlier (1992) summary. The counts that follow are complete for the years in question: I studied the Ambrona artifacts piece by piece when they were on loan to the University of California in the 1970s.

My records show geological crushing to be much less evident than it was in the Torralba series: most of the 199 pieces with coarse abrupt retouch may well be heavily utilized, rather than cryoturbated. But, since the threshold of differentiation between deliberate, irregular, coarse retouch and geological crushing is hard to draw consistently, they are excluded from the remaining calculations, leaving 1321 undoubted artifacts. The 50 cores make up about 3.8 percent of that total. Minimally utilized flakes are 212 (16.1 percent) and waste flakes another 636 (48.2 percent) of these: Together they constitute just over 64 percent of the collection. The "waste" series included 14 biface trimming flakes and a pseudo-Levallois point. Shaped tools are 391 or 26 percent of the total. The proportion of shaped tools is smaller than at Torralba, and other differences between the two sites also appear. The 47 bifaces (including 3 roughouts) make up 12 percent of the shaped tool collection, scrapers are 36.6 percent (more than at Torralba), notches 13.3 percent and denticulates 14.8 percent. While notches are more numerous and denticulates less so than at Torralba, their summed percentage representation is about the same at the two sites. The proportion of unclassifiable tools is smaller (only 1.5 percent—multiple-edged pieces are rarer), while burins, perforators, and alternate burinating becs (1.2 percent) are about equally well represented in this shaped tool collection.

Despite the opinion of some authors, such figures—particularly the proportion of bifaces and ratios of unretouched or minimally utilized pieces to shaped tools—are not in any way anomalous for well-excavated Acheulian assemblages from stratified contexts. The proportion of bifacial tools is not particularly low, nor is it uniform from occupation to occupation. While in some units at both sites, there are few bifaces or none at all, there are major occupations with more than 15 percent bifaces (Torralba Level 3), and in Torralba Level 2a, the total is nearly twice that (the Level 2a collection is very small). The proportion of waste and minimally retouched pieces would probably be considered low for sites located near contemporary sources of good raw material, but the stone at Torralba (as at Ambrona) was all imported from some distance—some of it from scores of kilometers away, as noted. There is very little evidence for primary flaking or workshop activities at either site, as one might expect from that fact alone. Nor would one expect a great many (but see later) conjoinable pieces at these sites, as compared to the situation at the Aridos or Pinedo quarries, where sources of good stone in reasonably large sizes were readily available locally as cobbles from river terraces—a point that I have tried previously to make, apparently without much effect (Freeman 1991).

While we have called the rather idiosyncratic Torralba artifact assemblage "Late Early Acheulian or Early Middle Acheulian" (*in litt.*), Santonja and Villa consider them typologically later in the Middle Acheulian comparing our better-formed bifaces to the cruder pieces from Pinedo. Pinedo's age is itself in question, though it is respectably old, but even if it were Early Acheulian, the comparison would still not be conclusive. At Pinedo, a quarry-workshop site near Toledo, the biface series consists mostly of abandoned roughouts, not finished pieces, many of them on obviously flawed raw material. Naturally they look crude. An earlier (1987) study by Carbonell et al., also suggests that the Torralba series, though it may overlap in age with Aridos, is later than Pinedo, and possibly than Aridos as well. They provide no new evidence for their assessment.

Wear Traces on Stone

Dr. Nicholas Toth of Indiana University examined the Ambrona artifact collections for traces of wear-polish (*in litt.*). He found that none of the tools from atop and in the "pebble-pavement" in the earlier part of the Lower Unit was suited to study: all had a "frosted" surface lustre that obliterated any use-polish.

Artifacts in clayey and sandy deposits of the Upper Unit (Va and Vb), including the fan sediments, were relatively fresh and 37 pieces were chosen as suitable for analysis. Of the larger flakes and retouched pieces, most had use-wear polishes, and where striations were present they were normally parallel to working edges, suggesting slicing. All wear patterns found are consistent with hide, meat, and (rarely) bone being the material operated on. In only one case was there wear indicative of "heavy" hide working, and no plant polish was observed. Toth concludes that the presence of little "unused" waste suggests minimal on-site flaking, and since microwear patterns are consistent, indicating animal butchery, while other patterns are lacking, the site seems to have been specialized, rather than a base camp or some other general station.

Conjoinable Lithics

The study of conjoinable stone artifacts is an informative addition to the analytical battery of the prehistorian; despite a widespread misapprehension, it was not ignored in our work at Torralba and Ambrona. Dr. Nicholas Toth has had the Ambrona study under way for some time, but my knowledge of his results is too sketchy to include. I do know that there were conjoinable pieces in the 1962-1963 collections from that site: my notes indicate that feature 50D,IV,7a&7b (two fragments of a quartzite chunk) can be rejoined and refit to 50F,IV,13, and that 50F,IV,1, is also attributable to this chunk (but will not join); another pair of refittable pieces is 48E,IV,6 and 48F,IV,6. I presume that Toth may have identified other cases.

For Torralba, my information is relatively complete, since we had the lithics in Chicago for study (and replication) for an extended period. The series includes a relatively small number of conjoinable stone artifacts. Of the 626 classifiable artifacts (excluding congelifracts), 29 are conjoinable fragments. We were quite aware of the potential information to be gained from such pieces, and most of them were detected during the course of excavations. The field identifications were all verified in Chicago. There were only two cases (totalling 6 flakes), where the conjoinability of pieces was first recognized in Chicago. It is possible of course that the collections still contain one or more conjoinable pieces that I missed, but I would not expect their number to be large. Nor would I expect there to be many such pieces in the smaller 1961 collection that I have not examined as closely. The following list does not include the several cases discovered of artifacts that are probably attributable to the same core or chunk of raw material, but could not actually be physically conjoined.

The 29 conjoinable artifacts found in 1962-1963 are from 12 occurrences in 7 levels at Torralba. Their provenience and separation are shown in Table 27-1.

TABLE 27-1 Conjoinable Pieces from Torralba

Occ.	Feature	Material	Description	Separation
I	L9, 37	Chal. flint	Tr. s/scr w/2 flakes	0 (touch)
	L18,1	Chal. flint	2 flakes	6 cm.
Id	D6, 58	Quartzite	1 fl. atop 1 core	0 (touch)
2	I24, Lev	Chal. flint	4 tiny flake frags	< 10 cm
2b	D9, 69	Chal. flint	2 flakes together	2 cm
4	I21,1	Chal. flint	2 util flake frags	0 (touch)
	I18,26	Chal. flint	2 flakes	< 10 cm
	H18,8	Chal. flint	1 burin, 1 ret. fl.	< 10 cm
	I21,1& J24,36	Quartzite	2 complete s/scr	4.2 m
5	F9,39	Quartzite	2 lg. fl atop 2 small	10 cm
7	G15,21	Chal. flint	2 flakes	0 (touch)
	H15,17& G12,44	Chal. flint	2 complete s/s	4.5 m

The data in Table 27-1 are remarkable. Virtually all the conjoinable materials identified in the Torralba collections are pieces that were found with very little lateral separation between them or none at all (the four pieces level-bagged from Occupation 2 were found very close together and placed in a matchbox, but the markers indicating find positions were accidentally disturbed before they could be mapped). The unusually small lateral distance between the pieces would seem to imply that neither during deposition nor afterwards were they affected by any appreciable lateral transport. The separations noted are in fact small in comparison with average distances separating conjoinable finds in other situations where there is no possible question of fluviatile transport, where distributions are universally agreed to be "human-made," and have always been interpreted as such. That would seem to be a datum to bear in mind in evaluating the possibility that long-distance water transport and rolling have altered bone surfaces or materially affected the original distribution of recovered materials at Torralba.

In two cases only at Torralba, fragments of the same original piece were found separated by 4.2-4.5 m. But those cases are unique. Each involves a pair of complete sidescrapers made on two refittable pieces of a single large flake. In both cases, the flake was broken before the final shaping of the individual sidescraper edges took place. With such data, human agency seems the most likely explanation for the separation of the find-spots.

Faunal Samples and MNIs

The Torralba fauna—its makeup, condition, significance, and abundance—has been the subject of some debate, partly because of differences of opinion about identifications and individual estimates provided by the two principal faunal analysts, Emiliano Aguirre and Richard Klein. I believe that a significant part of the disagreement between them can be resolved at this time. A certain amount of disagreement will remain unexplained, particularly where a single feature seems to have been attributed to two different taxa. Even in that case, part of the difference is due to the assignment of a single feature number to two (rarely three or four) pieces found together in a level, in intimate contact.

Sometimes, curation procedures that are beyond the control of Howell or the excavators were the cause of later analytical problems. Materials once excavated were removed (after plaster jacketing, where necessary) for shipment to the Museo Nacional de Ciencias Naturales by workers under Aguirre's direction. Some faunal materials—and this is particularly true for the shafts of ribs—were discarded by that team, as "requiring excessive museum space for their limited scientific interest." All such items were identified and thoroughly examined by Aguirre beforehand. While I have no reason to question his identifications, such pieces will of course not have been available to Klein for his later study. There is some reason to believe that among the bones so treated were some that bore possible marks of human modification.

After arrival at the museum, several of the bigger and more impressive bones—particularly elephant bones—were selected for display. Those pieces were repaired—sometimes separately found fragments of the same bone were rejoined—and their surfaces smoothed where necessary and coated with preservative. Pieces so treated often or usually lost their identifying labels in the process. And, the surface treatment they received obliterated what I had identified as cutmarks in some cases, or made it impossible for Klein to differentiate modern damage from ancient modification. While the number of pieces so affected is not large, most of the information that they might have provided is forever lost. A larger number of plaster-jacketed pieces—some tusks, skulls, mandibles, pelvis and scapula fragments, as well as the bigger and more complete limb bones, were stored in their jackets and remain in them. Consequently, Klein was unable to examine and identify them, and any information they provide about human agency or carnivore action is for the time being inaccessible. A still more important problem has been that most of the pieces have been relocated and relabelled on several occasions during periodic museum reorganization, and an even larger number of (usually smaller) items has become detached from its labels, misplaced, confused with other materials, or outright lost in the process. Last, the 1960s collections are reported to have been partly dispersed due to overlap in function between museums on an intra- (should these remains be regarded as primarily paleontological with tools, or primarily archeological, with bones?) or inter-regional (do they belong in Madrid or in Soria?) scale.

In cases where finds can no longer be identified, we have no recourse except to accept Aguirre's faunal identifications and his, Howell's and my observations recorded in our field and laboratory notes. The discrepancy between Aguirre's counts of taxa and Klein's can be partly explained on this basis. Klein, after all, saw only 1521 (71 percent) of the 2141 bone fragments recovered, and among the bones he could not examine were a substantial part of the largest, most readily identifiable skeletal elements.

That by itself will probably not account for most of the discrepancy. For the 1962-1963 Torralba excavations, Aguirre calculates an estimated minimum of 116 individual animals (112 mammals) for all levels, of which 37 are elephants, 23 equids, 21 red deer, 15 aurochs, 7 *Dama*, 5 rhinoceros, 2 lions, 2 small carnivores, and 4 Aves. (Azzaroli *in litt.* identifies one of the cervid mandibles as *Megaceros* sp.). Klein, in contrast, estimates only 64 individual mammals: 15 horses, 14 elephants, 10 red deer, 10 aurochs, 8 *Dama*, 4 rhinos, 2 lions and 1 lagomorph. Klein then has 48 fewer individuals (excluding the birds) than Aguirre. Another factor helps resolve most of this difference.

Klein's MNI calculations were derived on the basis of counting repetitions of the best represented body part for each taxon in each "level"—surely accepted practice, and the most conservative, justifiable way to proceed. However, when Klein calculated MNIs, he combined remains from the sublevels or spreads discussed earlier with the major horizon with which they were associated: All sublevels of Level 1 were united in his Level 1, and so on. When levels are combined, the MNI count invariably drops, as Klein himself illustrates in his chapter in the forthcoming monograph: uniting all Torralba levels drops his total MNI by almost 50 percent—from 64 to 34! Combining sublevels as he did by itself eliminates from Klein's level–by–level counts 42 animals that would be called different individuals were the subhorizons differentiated, reducing the overall discrepancy between Klein and Aguirre to 7 animals. Since Klein only saw 70 percent of the bones, a difference of this order of magnitude is scarcely cause for alarm. Some unexplainable differences still remain: Klein's list, though shorter, has one more *Dama* than Aguirre's, and a lagomorph (which may be one of the otherwise missing "small carnivores" in Aguirre's list).

Klein originally characterized the mortality profiles for Torralba elephants as catastrophic (*in litt.*) but has later stated that sample size was probably too small for reliable estimation, suggesting that "if the Torralba and Ambrona 'Lower' samples are combined, the case for attritional mortality is especially strong" (Klein 1987: 29). However, if combining remains from different sublevels is likely to be misleading, combining remains from different sites is much more perilous. In fact, when the remains of all bones (not just teeth) from the larger sample of ageable "individuals" obtained from the separated sublevels are examined, the Torralba mortality profiles once more become catastrophic rather than attritional. If that is a correct diagnosis, the observation made by Santonja and Villa (1990:61), that "the mortality profiles . . . cannot be reconciled with Freeman's and Howell's view of the sites" is wrong. (I believe that it is best to reserve judgment about the shape of the age-distribution at Ambrona until the final level distinctions have been established, and occupation contents correlated across the site).

At various times Klein has suggested that even catastrophic profiles might be explained by nonhuman agency, suggesting the drying of water holes or flash-flooding as likely alternatives. However, there is not the least geological or paleoenvironmental evidence for either phenomenon at either site. In the prehistoric environmental settings as they are now understood, truly catastrophic age-profiles would almost certainly imply human agency.

Birds

Bird remains from Torralba and Ambrona have been identified by Antonio Sanchez and E. Aguirre (Sanchez and Aguirre *in litt.*) At Torralba, the four specimens recovered are all water birds: a "wishbone" of *Tadorna ferruginea*, the ruddy shelduck, a scapula of *Mergus serrator*, the red–breasted merganser, a humerus of *Porphyrio porphyrio*, the purple swamphen, and a coracoid from an unidentified anatid. There is no reason to believe that these creatures were captured by humans—such small, light remains may have been dropped nearby by kites or other predators and washed into the site deposits, and none is cut or otherwise altered. At the right season, all could have been found nesting in the reedy edges of lakes or slow–moving streams at Torralba--the merganser would normally be found near more northerly seacoasts, far from Torralba, at other seasons, and the swamphen, a partial migrant, though occasionally reported as far from its southerly range as Norway, would not ordinarily be found in as cold conditions as those at either Torralba or Ambrona during the winter season (Vaurie 1965:138–139, 357–358).

Twelve bird bones were recovered from Ambrona; the provenience label is missing from one of them. In addition to the swamphen and merganser represented at Torralba, the provenienced items are bones of *Anser anser* (greylag), *Anas acuta* (pintail), *Fulica cf. atra* (coot), and *Vanellus vanellus* (lapwing). All but the lapwing are waterfowl, and it too inhabits the banks of ponds and shores as frequently as moist meadows. The coot prefers large, open bodies of water. Like the merganser, the pintail is tolerant of brackish water (Vaurie 1965:116–117, 359–360, 389). Again, there is no evidence that these bones are related to any human activity at the site.

The avifauna tells us something about local environments, but the species list is chronologically uninformative. It is interesting that most bird remains were detected in the course of excavation, even at Ambrona; few specimens were recovered by washing.

Small Fauna

The Torralba deposits did not yield much in the way of small fauna, aside from the often intact remains of tiny fresh-water snails, some specifically pond dwellers, dominated by *Hydrobia* sp., denizens of streams, ponds, marshes, and backwaters. It is notable that this genus is a recent invader of fresh water and is salt tolerant. They and the birds confirm the presence of bodies of water near the site but are otherwise climatically uninformative. My notes also indicate a 1960s identification of a pelobatid (spadefoot) toad from the site, but it is unclear and the material is not mentioned in later references.

At Ambrona, where the 1980s sediments were washed, samples of small animals were recovered in some abundance. They were identified by Drs. C. Sese, B. Sanchiz and I. Doadrio in Madrid. Sese recognized the insectivore *Crocidura* sp., the rodents *Arvicola aff. sapidus*, *Microtus brecciensis* and *Apodemus aff. sylvaticus*, as well as the leporid *Oryctolagus* (Sese *in litt.*). (*Lepus* was said by Aguirre to be represented in the 1960s material.) *Arvicola*, the water vole, is a strong swimmer that prefers to live in cool, humid ground near bodies of water—I would be surprised if *A. sapidus* can be differentiated from the more northern form *A. amphibius* from the material recovered. This surprisingly impoverished fauna suggests a post-Biharian age for the site but is not otherwise very informative.

Sanchiz identified anurids including *Discoglossus pictus*, *Pelobates cultripes*, *Pelodytes punctatus*, *Bufo bufo*, *Bufo calamita*, a *Hyla* (*H. arborea* or *H. meridionalis*), and *Rana perezi*, as well as the water snake cf. *Natrix* (Sanchiz *in litt.*). *Discoglossus* is usually found in bodies of water or their damp grassy banks. *Pelobates*, the spadefoot "toad," lives in dry, sandy ground close to bodies of water (Salvador 1974), excavating galleries in which it can survive long dry or cold periods.

Fish remains were found in considerable numbers, but all may probably represent a single species: *Rutilus arcasii*—its first documented fossil occurrence; less precisely identifiable remains were all attributable to *Rutilus/Chondrostoma* sp. or to indeterminate cyprinids (Doadrio *in litt.*), all of which may very well be from the same species. That in itself is interesting since *R. arcasii* has been found as the exclusive fish colonizing some interior drainage lakes in Spain (Diodario *in litt.*). The species prefers to live in and near the reedy shallows of sluggish or tranquil waters and is absent from turbulent streams or very cold water. The waters of lakes deep enough not to freeze solid may be warm enough for them to survive year-round even in cold climates.

For the number of remains that were recovered by washing, the poverty of small mammal, reptile, and fish taxa is noteworthy. These creatures were all most probably resident at the site during its formation. The species found coincide in showing that the site environment was characterized by lakes, ponds, and marshy ground. As far as refinements of dating are concerned, they are unfortunately banal. There is no indication that any of them were used by people at Ambrona.

Carnivores as Agents of Bone Accumulation

Binford and others have suggested that the accumulations of animal remains at Torralba and Ambrona may be due to natural causes having nothing at all to do with the human presence seemingly attested by the stone tools. The excavators (and later, Shipman) detected traces of animal gnawing on a few bones. Discussions by Klein have reinforced the impression that carnivore remains or coprolites are quite abundant at the sites. Klein characterizes coprolites as "numerous . . . although artifacts are more numerous than coprolites" (1987:18), thus giving the unfortunate impression that there must be many hundreds of large carnivore coprolites at Torralba and Ambrona, when in fact that has never been demonstrated. These observations have suggested to some that animals may be the major agents involved in the bone accumulations.

To the contrary, carnivore remains—bones as well as coprolites—while present are rare at both sites. Even where present, specimens that are apparently coprolites must be further analyzed before their meaning is clear. Most of the fragments considered to be coprolites are not well-formed scats, but fragments of clayey sediment containing small bits of bone. At some mid-Pleistocene sites near Madrid, I have seen small clumps of clay filled with crushed or whole remains of the bones of small mammals and reptiles that are probably fossil pellets of raptorial birds. In the case of true coprolites, only detailed analysis of their contents can determine which carnivore is responsible: Even some amount of decayed "bone-meal" (which may be present in scales of foxes and smaller carnivores) is no guarantee that hyenas are responsible. Furthermore, the feces of several small carnivores contain bone fragments. Klein has certainly identified coprolites at Ambrona—I have seen some of them myself—but I don't think that analysis of the specimens has been thorough enough to show that all the bits of bone-rich clayey sediment from the site were produced by large carnivores, or in particular, hyenas.

Bones of carnivores large enough to have killed the animals represented at either site or to have gotten their jaws around the bones of the larger ungulates to gnaw them are very few indeed, and marks of gnawing at Torralba have been said to be as rare as cutmarks apparently due to human modification. There are just two lion bones at Torralba: one in Occupation 1c and one in Occupation 4 (Klein lists the latter in Level 3). No wolves, no bears, no hyenas—in fact, no other large carnivores at all—are represented at that site. There are, of course, possibly two small (mustelid-sized) carnivores in Aguirre's list, one from Level 4b and one from Level 10, but even if both are carnivores, they are certainly not the bone accumulators at Torralba. Even the Torralba lion, a respectably large cat, could not have dealt with a healthy adult elephant the size of those at Torralba—with shoulder heights verging on 11 to 12 feet—though lions could certainly have killed some of the other animals, and they might very well have—probably did—scavenge from carcasses of animals dead from other causes. How any of the carnivores represented could have managed to remove the appendicular bones of the large elephants, as Klein (1987: 25) suggests to explain their rareness compared to the abundance of innominates, is quite unclear; the imbalance must be due to some other agency, and among the alternative possibilities human activity seems the strongest.

At Ambrona, in the Lower Unit, both hyena and lynx are represented by but a single individual each, while indications of carnivore activity are not abundant at Ambrona, and Klein and Cruz-Uribe identified just three bones as bearing marks of carnivore chewing (Klein 1987). Such figures as these do attest a carnivore presence but are scarcely convincing evidence of a major carnivore role in the accumulation or alteration of the mammal remains from either site.

One might object that marks of carnivore activity could have been obliterated by natural alterations of the bone surfaces during or after their deposition. But if that is the case, as many marks of human alteration could have been obliterated at the same time. Arguments that postulate that a mechanism that is inherently nonselective is responsible for selective destruction of particular kinds of data are inherently fallacious.

Implications of the Surface Condition of Bones

Emiliano Aguirre, in his original study of the faunal remains from Torralba (*in litt.*) said:

> The preservation of the vertebrate remains at Torralba varies from good, even sometimes excellent, to specimens having been altered in various ways, some prior to the process of fossilization and others, clearly subsequent to that process. In respect to the latter situations it is worth noting that there is relatively little breakage attributable to processes—such as mechanical deformation due to tectonic events or other such causes—within the sedimentary body itself. . . . On the other hand, in not a few instances, there are clear evidences of modification to faunal elements as a consequence of post-depositional chemical or biological processes, which hamper the identification of features of interest on a number of pieces.

Superficial decay or degradation of the bone and dendritic patterns produced by invertebrates and roots occur with some frequency, indicating interruptions in the process of sedimentation, deflation, and even periods of atmospheric exposure. He noted that exceptionally, bones were seen to exhibit a uniform polish all over, or all over one flat surface, but observed that "relatively few bones exhibit erosive traces over the entire surface, such as might result from water washing over a fossiliferous horizon, and leading to smoothing of protruding body parts through transport and rolling, or more rarely, aeolian processes" (Aguirre *in litt.*). He goes on to say "the great majority of modifications of bony elements fall into regular patterns" particularly patterns of breakage, incision and percussion, "that can be attributed to cultural activities." Aguirre thus suggested that the bone was in good enough condition, despite surface alteration, so that traces of deliberate cultural modification could still be recognized on some—perhaps many—bones, and in this I concur. From the outset, Aguirre and all other analysts have recognized that surface abrasion exists on a number of specimens from Torralba (and Ambrona). However, Aguirre's assessment of the general state of the bones is much more positive than the later diagnosis by Richard Klein.

Klein (1987:19-21) states that at both Torralba and Ambrona

> intense post-depositional leaching has corroded bone surfaces. . . . The alterations introduced by leaching and corrosion are compounded by the massive fragmentation that occurred during and after burial at both Torralba and Ambrona. . . . It is notable that one-third of the 1779 bones at Torralba and one-sixth of the 4326 bones from Ambrona "Lower" exhibit edge-rounding that Butzer (pers. comm.) suggests occurred during limited fluvial transport on seasonally-activated valley-margins or during net transport of sandy alluvium that partially buried the bones. Many bones that are not conspicuously rounded show a distinctive polish or luster and probably would exhibit abrasion or edge rounding under magnification. . . . Using a hand-held glass on a sample of lustrous Torralba and Ambrona bones Butzer (pers. comm.) found parallel microstriations from abrasion by sand-sized particles on every one (1987:20).

He notes that Shipman and Rose also found "rounding" (under greater magnification) on nearly every specimen they examined from the two sites and goes on to say that "excepting abrasion and corrosion, Cruz-Uribe and I found little other damage on the Torralba and Ambrona bones" (1987:21). The total number of carnivore-chewed pieces they detected was 14 from Torralba and 3 from Ambrona Lower, while the number of possible stone tool cut marks was 22 at Torralba and none from Ambrona Lower. (Klein recognizes, of course, that surface corrosion may have obliterated other traces of both kinds.)

Butzer's observations on this subject are reported in an appendix to his final faunal chapter in the Torralba monograph. It is worth quoting *in extenso*. He reports:

> The conspicuous concentration of archeological materials in such coarser-grained, intermediate energy horizons cautions strongly against diagnosing these as intact, primary associations. Instead, it is probable not only that there has been a measure of pre-depositional dispersal, but that at least some of the archeological micro-horizons are telescoped lag levels. This is strongly supported by my 1981 examination of the Torralba bone in the Madrid museum. Every specimen selected at random under low-power magnification showed systematic, very fine, longitudinal and parallel striations and had a "sandpapered" feel. This systematic striation was noted on all sides of each bone and was strongest on the most-exposed ends. It can only be explained by sand transport below, above and around the bone, resulting from energy conditions adequate to transport sand but mainly insufficient to move large bone; repeated burial and exposure is therefore probable. This is not incompatible with my conclusion that the archeological occurrences may retain their basic associations, i.e. between bone and bone, or bone and artifact, despite some horizontal displacement and changes in orientation. But the problem of telescoping bone and artifacts into "pseudo-floor" lags is more serious than I had anticipated. Trampling and sinking of heavy objects in wet clayey sediments is less problematical than at Ambrona, although it bedevils interpretation of those archeological materials at Torralba that are found in or at the base of clayey deposits. In effect, like all other Paleolithic open-air sites that I have examined since 1961-1963, the best associations at Torralba are, in sediment taphonomic terms, semi-primary (see Butzer 1982:120-122).

Some more or less significant differences in these three observations call for comment. Sometimes they are quite subtle, but the differences have such important consequences for interpretation that it is essential to be quite careful about language. Aguirre's description makes the Torralba fauna sound relatively intact, and relatively informative about cultural behavior. Klein in contrast talks of the "intense pre- and post-depositional destruction that affected the Ambrona bones" (1987:27) (a description that can only be fairly applied to the Ambrona Upper series, where intense leaching has removed most bone).

Butzer's description does not make it clear whether his sample was chosen from all bones or all visibly polished bones, as Klein suggests, but that is of less consequence than the conclusions he derives.

His term "semi-primary" implies limited dispersal of cultural materials prior to burial, after which the buried deposits are subject to some disturbance (Butzer 1982:121). In the depositional unit at Torralba bearing most of the archeological materials, though its sediments deposited under cold conditions in valley-bottom deposits, there is little cryoturbation and transport distances must have been quite short. Surface abrasion of bone could be ascribed to sediments passing around the bones, rather than to lateral movement of the bones in the sediments. The lack of preferential orientations or size-sorting would seem to support this possibility. Archeological associations, as Butzer points out, could survive this degree of disturbance and still be recognizable. Only his conclusion that the depositional environment is one in which different archeological levels might have been telescoped into "pseudo-floor lags" poses any substantial theoretical problem to cultural interpretation.

Butzer's conclusions are borne out by the archeological field observations. The local merging of elsewhere discrete levels shows that even the thinnest, apparently most pristine level might contain materials originally deposited in several separate episodes. But lag deposits have a geo-archeological signature. Ordinarily lag deposits built up over any length of time may be expected to be heterogeneous in content, and different lag deposits should differ in random ways. That is because, as a rule, the depositional conditions were different for each of the discrete "moments" that later telescope to form a single apparent "floor." Ordinarily, the materials in one lag deposit don't differ from those in another in patterned ways, unless the landscape and the conditions of deposition have remained so constant that the local depositional environment has repeatedly caused accumulations of materials of the same size and shape to be dropped in essentially "the same spot." Only then should telescoping of formerly disparate levels produce a horizon (or horizons) whose contents are both internally homogeneous in their characteristics, and different from others in regularly repeated and predictable ways. Such cases are by no means geologically exceptional; nevertheless, careful examination should reveal the essentially "geological" nature of the accumulation (due to similar behavior of items whose sizes or shapes are analogous when waterborne or moved by gravity, etc.). What is more, in the archeological case, the original cultural behavior that produced the residues forming the lag would of course have had to be essentially similar during each episode of accumulation, implying the repeated performance of the same set of activities in the same part of the changing prehistoric landscape (whether this is the actual area excavated or other areas which served as sources to the lag). The evidence of the accumulations called "Occupations" at Torralba and Ambrona runs contrary to such an interpretation.

Another problematic situation he mentions is that of the "sinking of heavy objects in wet, clayey sediments." At Ambrona, there are some situations in clayey sediments in which skeletal remains of several animals were found lying one above another in layer-cake fashion, and in the absence of other evidence, it would be a mistake to interpret these as single cultural accumulations. At Torralba, this is less a potential problem than at Ambrona, since the major accumulation at the base of clays (the clay facies in the north half of Occupation 7) consists principally of the bones from one side of one individual animal, in a somewhat rearranged "near-anatomical" position. Since that individual died but once, the question of whether or not it sank, and at what rate, is immaterial. The large stones in the same horizon that are interpreted as part of this accumulation were pretty evidently positioned in relation to the bones: Again, sinking provides no objection to previous interpretation. I see no reason to believe that all else is a culturally meaningful association, while the stone tools in intimate juxtaposition to the bones are extraneous.

The concentration of accumulations within or at the base of clayey deposits certainly does impose peculiar restraints on interpretation—in some cases it may even rule out explanations in purely geological terms. It is hard to account for differences in the distributions of materials deposited in still-water or marshy sediments, particularly the sorting of large, dense items such as elephant bone, in terms of geological agency. If the accumulations are found at some distance from the edge of a prehistoric lake or bog, and there are no nearby channels in which flow would have sorted them as they were swept along, the discovery of sorting by body part or bone size or shape may well have cultural rather than simply geological significance.

Figure 27-2 shows an example of this sort from Ambrona. In 1980, we found a group of five elephant tusks of different sizes, lying in close proximity in clays (not an isolated example—other tusks were found grouped together not far away in the same deposits). One of the tusks was near vertical in the clayey sediment. There are no faults or other disturbances in the deposits that could account for its attitude. It must have been buried that way, fast enough so that it was not weathered to pieces. Its position may very likely be due to a heavier tusk having sunk more rapidly, trapping the point of the smaller one and pulling it down into the angle it maintained at discovery. While the attitude of this single find may be purely a result of depositional processes, I do not see how any natural agent other than human activity can explain the spatial segregation of the tusks from other bones in these deposits.

The sediments are still-water beds, not stream deposits, and assortment by channeled flow is out of the question. No geological force as far from a contemporary channel would have separated these five tusks from other relatively same-shaped body parts and dumped them all together.

Nonhuman biotic agencies are also improbable agents. The tusks are uninteresting to carnivores, who in any case would scarcely have dragged them all into a separate pile in muck or standing water. As Villa notes, elephants today often pick up and carry about bones of their dead congeners, and anyone who has seen filmed behavior of this sort must admit that it is remarkable. However, they do not sort the bones and dispose of them in piles segregated by body part. Rather, they seem to carry or drag the bones about for a bit, then toss them away apparently at random. Peter Beard (1977) has published scores of photographs of dead elephants, including some astonishing natural accumulations of bone, but in the few cases he shows where bones are segregated by body part (or arranged into tidy localized piles) the hands of humans were responsible.

Butzer's concern about the problematic effects of sinking in clayey deposits is doubtless well placed. On the other hand, such sediments may, in special cases such as the ones just described,

FIGURE 27-2
Group of elephant tusks—one vertical—in Ambrona Lower Unit (1980).

constrain geological interpretation in directions that pave the way toward an understanding of cultural phenomena.

Marks of Human Alteration on Bone

On many of the bones from Torralba and Ambrona, there are marks that I do not believe could have been made by any nonhominid agency. The marks are gross enough in most cases so that surface alteration has not obliterated them or rendered them unrecognizable. Of course, those marks will never pass muster as evidence for hominid alteration if one insists that the bone surface topography must be essentially fresh for the markings to be studied at all. That requirement has been both one of the strengths, and, at the same time, one of the weaknesses, of a recent study of some Torralba and Ambrona bones undertaken by Shipman and Rose (1983).

They subjected replicas of surfaces of some of the smaller Torralba bone fragments to reexamination for microscopic evidence of cutting and gnawing, using the scanning electron microscope. Shipman's criteria for identifying cut-marks are very exacting, and her procedure rigorous. Therefore, there is very little doubt that a bone she identifies as cut actually bears marks that most would find convincing evidence of such alteration. She found such marks on some of the Torralba bone, and I suppose that I should be pleased that there is some support for my belief that hominids altered some of the Torralba bones. While it may seem contrary of me, I have several reservations about the Shipman and Rose study.

Most important, I believe that their description of procedures as published is unreliable, and the estimate of the proportion of sliced bones in the collection they offer is therefore unusable. I don't mean that the marks identified do not exist or were wrongly counted, but that other statements about the study and the size of samples examined incorporate serious errors.

For one thing, Shipman claims to have examined all the Torralba bones. However, there is no way she could have examined the whole collection, since the plaster-jacketed bones Klein was unable to see are still in the same jackets. (Many of the marks I find most convincing are found on those larger bones—major parts of elephant pelvis, whole tusks, mandibles or elephant and bovid crania—under the plaster jackets).

Second, she claims to have found convincing marks of hominid alteration on a total of 12—some 1.2 percent—of the replicas examined from Torralba. The figure cannot possibly be correct.

After Klein's reclassification of the Torralba bone, the museum collections were reorganized, replacing the finds in shelved lots by square, level, and feature number, rather than by species and body part. (The only exception is a lot of 22 bones Klein suspected might be cut and had shelved separately for future study.) Any thorough examination of all bone finds would first require opening every box, locating the label (square and feature number) on each piece, and then identifying it from Klein's inventory of taxa and body parts, a process that by itself would necessitate several days' work. Then the surface of each bone would have had to be completely examined under proper lighting, even on occasion under low magnification. Next, suspect bone surfaces would need replication, in itself a time-consuming process. To examine this bulk of material carefully and replicate the specimens that seemed altered would require a minimum of several weeks' time. This estimate may be approximately doubled because of the shortness of the museum's hours—ordinarily only 4 to 5 hours of access to its warehoused collections are permitted each day.

Shipman spent in all several hours with the collections, not several weeks. In such a limited time it is not possible that she could have had time to examine, let alone replicate, more than a few bones from the Torralba collection. For 12 to be 1.2 percent of the replicas made, Shipman would have had to make a thousand of them. In the short time available, this is an unrealizably high number, even if several replicas were made of any single bone.

It seems to me most probable that, given the time restrictions of her study, Shipman must in fact have spent almost all her time on the two dozen bones Klein had set aside, looking at others only as (or if) time permitted. Perhaps this is actually what Shipman and Rose intended to say. Whatever the explanation, the account they give of sample size and procedures is inconsistent with the nature and size of the collections.

If what Shipman and Rose really examined was just the collection Klein thought might be worked, their sample was doubly constrained by any preconceptions he may have had at the time about the nature of bone working, and by his ability under the less-than-ideal conditions in the museum to distinguish marks on bone. Results of the Shipman and Rose study would then be unintentionally biased, no matter what their remaining procedures.

The Shipman-Rose study provides some information of qualified interest. They did find 12 convincing marks of human alteration on 4 bones of *Paleoloxodon*, 3 of *Equus*, and 1 of *Cervus*, as well as on 1 bone of indeterminate species. In a clear misunderstanding of the evidence, Santonja and Villa state that "the rarity of cutmarks on the bones . . . cannot be reconciled with Freeman's and Howell's view of the sites as places where herds of elephants were killed and butchered" (1990:61). But Shipman and Rose actually said that their study offers "only limited" support, not "no support" for the fact that Torralba (and Ambrona) were butchery sites. There is a real difference. And, there is no reason to suppose that butchering marks need be abundant even in a culturally modified faunal collection. Visible cut-marks may be very rare—even virtually absent—in more recent butchery sites, such as some "buffalo jumps" in the United States, where humans are known to be the principal or the only agent of bone accumulation and/or alteration. As mentioned previously, Shipman and Rose also detected carnivore tooth scratches in comparable frequency. They were characterized as less abundant than might be expected of assemblages where carnivores were the primary agents of bone alteration. Shipman and Rose's conclusions do not correspond to Santonja and Villa's summary.

In sum, despite its problematic nature, the work of Shipman and Rose is nonetheless interesting insofar as it provides some direct evidence of apparent human intervention in the alteration of the Torralba bones. However, theirs is far from the "last word" on the subject. There are other kinds of apparently cultural marks on bones from these two sites that were not considered in their study. By far the most abundant marks that were earlier interpreted as signs of hominid alteration are grosser traces than the fine slicing studied by Shipman. They consist of large-scale scars of gross damage—marks of battering, chopping with a large, sharp, wedge-shaped edge, scraping or abrasion with a smooth, blunter stone edge, and deep slicing, gouging or grooving. Though they occur on bones whose surfaces have also been altered by natural postdepositional phenomena, they have resisted obliteration. Quite comparable coarse marks of hacking are identified as butchering traces at the Casper site (Frison 1965:36–37) and elsewhere and have been interpreted as important evidence about butchering techniques at those sites. Shipman's methods simply ignore all such evidence, which to me seems as obvious and as convincingly indicative of human handiwork as the pristine fine slicemarks she studies. The coarser topography of such marks was the basis for my own field counts of worked bones, in the majority of cases.

In collections from sites where bone surfaces have undergone more than minimal postdepositional alteration, as at Torralba, macroscopic butchery marks may be the only ones that can survive. Most bone under such conditions cannot preserve the diagnostic microtopographic features, the fresh traces of fine slicing, that Shipman's microscopic study relies on. Marks of gross damage certainly merit further investigation, instead of summary dismissal.

I examined the bones from Torralba while they were being excavated, while the surfaces were "fresh," unvarnished, and still unjacketed. It was then still easy to tell fresh excavation damage from ancient alteration. Workers alerted me as they recognized apparent human modification, so I watched many of the surfaces as they were cleaned and excavated not a few myself. In the field, I identified four types of modifications that seemed to be cultural: slicing, hacking with a wedge-shaped edge, scraping or abrasion, and battering or repeated percussion. My notes show 56 sliced surfaces, 6 cases of hackmarks, 1 abraded bone, and 4 battered specimens. There were in addition a number of charred bones, 2 so heavily burnt that I thought it unlikely that grassfire could be responsible. There were also some large bones that had apparently been deliberately flaked while "green" in such ways that carnivore gnawing as responsible agency was out of the question. Those were not counted; we relied on Aguirre to study them (which he did, in a chapter in the forthcoming monograph). Klein saw the collections only after they were jacketed/warehoused/preserved, when it was much harder or impossible to differentiate fresh damage from ancient modification, and so he quite properly excluded several by then "dubious" cases from his counts. Nonetheless, he recorded 22 bones as potentially cut, 4 charred (possibly naturally) and 10 from which flakes had apparently been struck in the "green" state. In fact, the disagreement between Klein's figures and mine is really not serious, considering what had happened to the collections between the excavation and the time he saw them.

In the 1980 excavation in the Ambrona Lower Unit, I found that about 50 percent of the larger bones bore marks suggestive of cultural alteration. A selection of pieces from both sites is illustrated. Figure 27-3 shows an immense elephant left innominate with subparallel grooves attesting extensive scraping. Figure 27-4 shows hacking and slicing on the premaxilla of an elephant skull. In Figure 27-5, an elephant mandible whose ascending ramus was removed, by repeated chopping with a wedge-shaped edge, is shown. Details of the hacking are illustrated in Figure 27-6. The remainder of the ramus was found just behind the mandible (it can be seen in the first photograph) and bore matching scars (Figure 27-7). Despite the evident surface corrosion on these pieces, the marks are still easily identifiable, and in no case do they seem explicable by carnivore activity. None of these pieces would have been replicated by Shipman: their surfaces are too corroded and the marks they bear are not the sort she studies. Three apparently sliced specimens from Torralba are shown in Figures 27-8 to 27-10 and a hacked bone from the same site in Figure 27-11. Only space limits keep me from illustrating a score of other altered bones, including skulls, scapulae, ribs, innominates, and longbones.

Spatial Associations

The discovery of items in close juxtaposition in an archeological level has traditionally been seen as evidence that there is a real relationship between them. While some apparent spatial associations that are detected by eye are misleading, at least in the absence of statistical demonstration that the associations are unlikely to have arisen by chance, other visual associations are quite valid. No one would believe that an association of the bones of the skeleton of a single individual (such as the focal association in the north part of Occupation 7 at Torralba) needs statistical validation. Nor is that an isolated instance.

FIGURE 27-3 Large elephant innominate, showing marks of scraping (Amb 80).

FIGURE 27-4 Elephant skull, hacked on premaxilla (Amb 80).

FIGURE 27-5 Elephant mandible with marks of hacking (to remove ramus?) (Amb 80).

FIGURE 27-6 Detail of hacking, visible despite corrosion, on mandible (Amb 80).

The separate accumulations of elephant tusks in the Ambrona Lower Unit are statistically significant associations. So are the repeated concentrations of bovine horncores and bifaces in squares G15, G12, H12, and I3 in Torralba's Occupation 3 (Figure 27-12).

Still other associations seem so unlikely that even though their probability cannot be directly determined because each is almost unique, their nature and number still persuasively suggest a direct relationship between the animal bones and the implements found at these two sites. In Torralba Occupation 1 we found one particularly striking case: square J12 held an elephant pelvis with a convergent denticulate tucked inside the acetabulum; a limestone battered polyhedron lay just outside the socket (Figure 27-13). In I9 in the same level, a flint utilized flake lay atop an elephant right pyramidal. In level 7, a small flint biface lay next to an elephant radio-ulna in square M12. At Ambrona, in the Lower Unit we repeatedly found bifaces right beside tusks (Square G99: Figure 27-14) or elephant vertebrae (STE 4: 2 vertebrae with two hand-axes; Figure 27-15). Other cases are too numerous to mention. The sheer number of such finds and their coherence with results of the statistical study of frequency relationships (see later discussion) cannot fail to impress a reasonable analyst.

Statistical Analyses

If this were not enough, there is more abundant—and, to my mind, more convincing—evidence that there is a meaningful, culturally mediated relationship between the remains of large animals found at these sites and the artifacts left there by humans. That is the evidence provided by multivariate statistical analyses of relationships between these different kinds of data, analyses that have been carried furthest (and criticized most) at Torralba.

There is only one problem with the results of statistical testing. Most people still really don't understand the tests or their results, and so they will either reject the whole process as less meaningful than the solid, tangible "real" data an excavator digs up, or—even worse—will uncritically accept any and all statistical manipulations as valid, only to reject each in turn in favor of the latest test claimed to have produced contradictory results. It is an unfortunate problem, but one that eventually must vanish

FIGURE 27-7
Ascending ramus
of same mandible
with matching hack-marks.

with education. In fact, used properly and evaluated critically, statistical tests are just so many more among the tools—knives, brushes, and so on—that excavators use to gather data, and their results are just as real and meaningful as any finds made with those tools. Just as one can pick the wrong tool for excavation, using shovels where trowels are called for, so one can use an inappropriate statistical procedure. Not all statistical procedures are equally justifiable. Just as one can excavate badly, producing erroneous information, one can also use statistics inappropriately to produce wrong or misleading information. Not all statistical results are equally reliable.

The use of any statistical test requires that the data to be analyzed be error free, that if the data must be transformed it be done in a justifiable and appropriate way that will neither invalidate the calculations nor hinder interpretation, and that the measures chosen be suited to the kind of data being studied. The tests chosen here produce measures of bivariate relationship—a matrix of correlation coefficients, in this case—and then use those measures as a basis for further computation. Any measure of the strength of a relationship between two variables should remain the same whether or not a third variable is present; that is not the case for some measures, but it is for the coefficients used here. Some variables are unrepresented in some samples. The problem of zeros in the data was handled by treating them as missing values and deleting any pairwise comparison where a zero occurred. Sometimes statistical software packages perform a multivariate test in different ways, producing different solutions from the same data. Obviously, that is undesirable: it ought to be the case that any analyst, using the same data and the same tests, should get the same results. The tests we used are fully replicable.

FIGURE 27-8
Torralba bone showing marks of slicing.

Table 27-2 lists the more abundant artifact types and MNIs for the major species represented in the Torralba occupations. Since use of edge counts in a previous work (Freeman 1978) drew criticism and, more important, caused confusion, the artifact counts used here are of *whole tools* tabulated by level.

Whether edge counts or whole tool counts are used, significant patterned relationships appear in the data. The solutions are not identical. Multiple tools often combine different kinds of edges, but where the particular combination is not abundant enough to be considered a significant "new" type, they are placed into the type of the best-made edge. There are inevitably differences between solutions based on edge counts (which I still consider more meaningful) and those based on whole tool counts, but the differences are less important than the fact that significant patterning is detected no matter which data are used.

The counts were ranked and used to calculate the rank–order correlation coefficient, Spearman's rho. Rho works with ranks of frequencies, rather than the raw frequencies themselves, and ranking is far and away the most mathematically defensible transformation for these data, where it is inappropriate to make assumptions about the underlying shape of the data distributions. In the past, I have transformed frequencies to square roots and used the more common bivariate correlation coefficient, Pearson's r (with larger data sets the transformation had essentially no effect on results); the results were slightly different

FIGURE 27-9 Torralba bone showing apparent slicing and battering.

from those presented here, but both tests coincide in showing relationship between similar sets of variables. There are no significant patterns of replacement or inverse relationship in the data.

The rank–order correlation coefficients in Table 27-3 were used as measures of nearness in a cluster analysis (Figure 27-16). The structure of variability was simple enough so as to be discernable in most of its details on visual inspection of the coefficient matrix, but the dendrogram, based on a single-linkage procedure, shows its characteristics more clearly. Two data categories are really unlike the rest: congelifracts and bovid MNIs. Since the number of cattle is nearly invariant, this is as one would expect. The fact that congelifracts are unlike other data categories is reassuring. The animal species remaining are in fact related to each other, and also to particular stone-tool types. Notches and denticulates stand apart from the rest of the tools, but choppers are related in frequency to equids and cervids, while bifaces, end

FIGURE 27-10 Torralba bone showing apparent slicing.

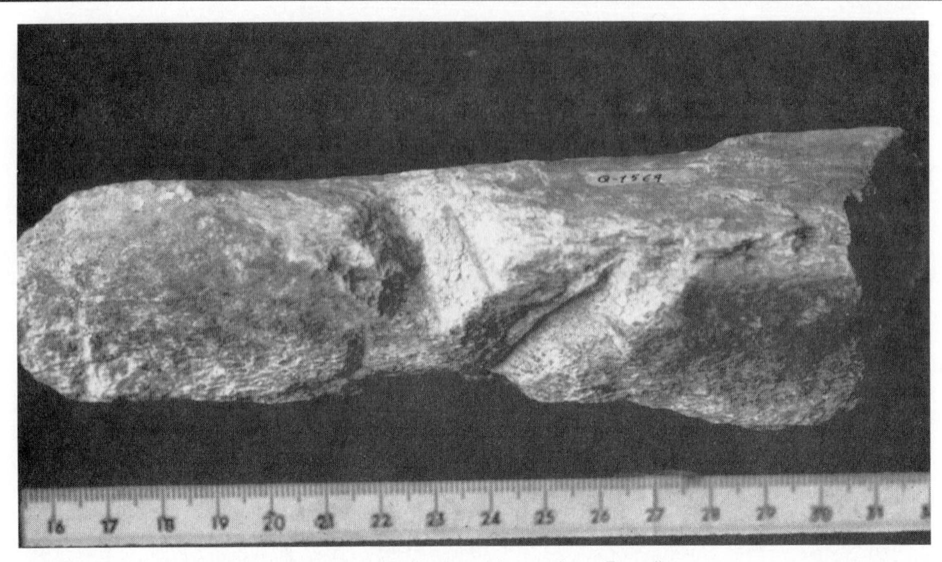

FIGURE 27-11 Fragment of "chopped" elephant bone from Torralba.

and sidescrapers, and cores relate more closely to elephant counts. That is not to say that elephants, cervids, and equids are unrelated to each other—all, and the other tools, form an interrelated group at a more distant level.

This simple test indicates beyond any doubt that there are meaningful relationships between the abundance of particular stone tools and the abundance of particular animal species. That simply would not be the case if it were true as some allege that the human presence at Torralba was essentially unrelated to the presence of animals at the site. But the tests are not in every respect satisfactory explorations of the data. Bovid MNIs, as noted, are small and nearly invariant. There are fewer cervids and equids than one would ideally prefer, and there are a lot of tied ranks. In these respects, the correlation matrix and cluster procedure, though certainly conclusive, leave something to be desired.

FIGURE 27-12 Association of bifaces and horncores, Occ. 3, Torralba.

FIGURE 27-13
Elephant pelvis, with denticulate inside acetabulum; polyhedron nearby. Torralba.

When body part counts are used rather than counts of MNIs, both counts and variability in the faunal categories increase. The picture of relationships is both strengthened and clarified as details are added. At the same time, new dimensions of variability appear that are not adequately depicted in the essentially two-dimensional cluster analysis; fortunately, the related but more elegant principal components analysis can show these relationships quite well.

In the following test I have used the stone artifact frequencies from the previous table, dropped the MNIs, and added the body part counts shown in Table 27-4. Differences in frequency are not so great that raw frequencies and Pearson's r could not have been used, and it might very well have been appropriate to do so, but since nothing is known of nature of the underlying distribution of these data, to be safe the same nonparametric measure of correlation, Spearman's rho, was chosen.

The matrix of bivariate rank-order correlations is given in Table 27-5. Binford (1987) claims his analyses show that the Torralba deposits show a palimpsest of two major patterns: one in which bovid, equid, and cervid remains were deposited with tools while elephant remains were deposited in unrelated fashion; the other in which elephant bones were deposited in association with stone tools, but in which bones were broken into unidentifiable bits by forces other than human agency (1987:66). More detailed examination of frequency relationships including body parts leads him to identify one pattern as potentially due to hominids, only to reject that possibility in the following terms: "No matter how we interpret the patterning, the case for 'activity areas' is very hard to sustain. The elephant carcass material is inversely related to remains of other species, making it difficult to argue that the differences in tools represent tools appropriate to sequential processing steps in the butchering of a single animal" (Binford 1987:90). In his detailed "analysis," in fact, he claims to find in the Torralba data an inverse relationship between frequencies of a kind of pseudo "Mousterian of Acheulian Tradition" tool set including especially bifaces, notches, and denticulates, on the one hand, and those of waste on the other (1987:75); an association of scrapers and choppers—which he also wrongly calls "corescrapers or core axes" (1987:77); elephants varying inversely with other animals (1987:83, 85, 91); and an association between bifaces, sidescrapers, and elephant remains that he explains away as partly related to the paucity of those tool types (sidescrapers are, on the contrary, the next most abundant flake tool category—whether whole tools or edges are counted—in the collection, and only five fewer in number than denticulates), and partly due to the fact that the sidescraper counts are elevated because they are "compound edged tools" (1987:89). While it is true that counts of working edges were used instead of whole tool counts for most flake–tool types in the study (Freeman 1978) that was the source of data reanalyzed by Binford, patterned relationships between

FIGURE 27-14
Associated biface
and elephant tusk, Ambrona.

sidescrapers, other tools, and bones appears just as clearly when whole tool counts are used, as the present study shows. One could go on to contest other "results" of Binford's "analysis" in detail, but it is pointless. No matter what one feels about the logical coherence of his explanations of patterning (there, too, I find much that is questionable), the statistical results on which the arguments are based are worthless, since he used erroneous data, unjustified and unnecessarily convoluted data transformations, and inappropriate analytical procedures—no one whose hand was not guided by Binford could repeat his test and obtain the same results.

In fact, a simple inspection of the correlation matrices in this chapter is enough to show that Binford's claims are wrong. Aside from the association of notches and denticulates, which is only part of a more heterogeneous group he defines, not a single one of his claimed relationships has any validity—a very unfortunate state of affairs, since his results have been uncritically accepted at face value by Villa (1990, 1991—though she interprets them differently) among others.

True, pairs of items that cluster in Binford's solution sometimes also cluster in mine, but not in the ways or for the reasons he specifies, and there is no evidence of any substantial "inverse relationship" between variables. The numerous inverse relationships that so preoccupy Binford are in fact mathematical fictions. They occur neither in a correctly calculated matrix of correlation coefficients—product–moment or rank-order—nor are they at all numerous in an appropriate matrix of component loadings. Any real inverse relationship between variables has to be reflected in one increasing as the other decreases—producing at least a partial inversion of their numbers or rank orders—and that must result in a significant

FIGURE 27-15 Association of vertebrae and hand-axes, Ambrona.

negative correlation. This simply is very unusual in the Torralba data: only one of the small number of negative coefficients (notches vs. elephant feet) reaches significance at the .05 level. It doesn't even happen when the erroneously copied data Binford presents are analyzed correctly.

I can only explain the large number of negative loadings in Binford's tables by assuming that either his "chi-square" transforms were inappropriately calculated from percentages (I suspect this may be the case, since Binford has been so fond of percentages in past), or that he has presented an incomplete solution, which, had he allowed the test to continue to iterate until it reached a unique solution, would have eliminated the negative loadings. (There may be other mathematical explanations for his results, but no one could isolate them from Binford's almost deliberately obtuse procedural description.) Whatever the case, the statistical procedure is—has to be—invalid, as one can determine just by inspection of his data tables.

Our table of rotated factor loadings (Table 27-6) shows that seven factors or components are adequate to account for over 92 percent of the variance in the matrix of correlation coefficients. As is my usual practice, I rotated one more component (as a possible "error component") than the number with eigenvalues of 1.0 or greater. The last component does not principally determine variation in any variable; that is as one would hope. The seventh component only loads highly on geologically crushed pieces. That is also an encouraging sign.

Such tests as these are most justifiable when applied to data about whose structure there are some prior expectations. We had some idea beforehand what the statistical results at Torralba might show. Field observations of spatial associations suggested that cores and scrapers should each be related to elephant tusk, ribs, limbs, vertebrae, scapula, and pelvis (these two were combined in the statistical test), that bifaces and perforators should be related to elephant skull, that denticulates and notches were related (a pattern also incidentally found by Binford), and that both bifaces and denticulates were related to bovid skull (but bovid skull was too infrequent to be used alone in the test). Cervid metapodials and bovid and elephant

TABLE 27-2 Torralba Major Data Categories by Level (Lithics are whole pieces; taxa are Aguirre's MNIs)

	BIFACES	CHOPPERS	CORES	WASTE	SIDESCR
OCC 1	3	5	7	27	13
OCC 1C	1	2	0	2	0
OCC 1D	2	1	2	9	0
OCC 2	2	3	1	28	2
OCC 2A	2	0	0	3	0
OCC 3	10	3	5	27	7
OCC 3A	1	0	1	10	0
OCC 4	0	2	2	25	3
OCC 5	2	0	1	10	1
OCC 7	8	2	4	32	8
OCC 8	3	2	3	93	3
	ENDSCR	**PERF**	**NOTCH**	**DENTIC**	**CONGEL**
OCC 1	4	2	3	6	6
OCC 1C	0	0	0	2	0
OCC 1D	1	2	3	4	1
OCC 2	2	0	4	3	1
OCC 2A	0	0	0	1	2
OCC 3	2	2	4	10	10
OCC 3A	1	0	1	4	3
OCC 4	0	2	0	5	3
OCC 5	0	1	1	2	5
OCC 7	3	7	2	3	7
OCC 8	1	3	1	2	1
	ELEPHAS	***EQUUS***	***BOS***	***CERVUS***	
OCC 1	5	3	1	3	
OCC 1C	1	1	1	1	
OCC 1D	2	1	1	1	
OCC 2	2	1	1	1	
OCC 2A	1	1	1	0	
OCC 3	3	2	2	1	
OCC 3A	1	1	0	0	
OCC 4	3	2	1	1	
OCC 5	1	1	1	1	
OCC 7	6	1	1	2	
OCC 8	2	2	1	2	

TABLE 27-3 Torralba Artifacts and Species Matrix of Spearman Rank Correlation Coefficients (RHO) (Zeros treated as missing data; pairs with any zero member eliminated)

	BIFACES	CHOPPERS	CORES	WASTE	SIDESCR
BIFACES	1.000				
CHOPPERS	0.333	1.000			
CORES	0.843	0.472	1.000		
WASTE	0.687	0.276	0.429	1.000	
SIDESCR	0.677	0.313	0.945	0.309	1.000
ENDSCR	0.543	0.712	0.670	0.255	0.821
PERF	0.532	-0.216	0.437	0.775	0.462
NOTCH	0.363	0.485	0.342	0.025	0.265
DENTIC	0.375	0.452	0.513	0.257	0.633
CONGEL	0.578	0.408	0.561	0.133	0.609
ELEPHAS	0.842	0.356	0.825	0.679	0.917
EQUUS	0.642	0.539	0.723	0.460	0.584
BOS	0.575	0.354	0.417	0.116	0.206
CERVUS	0.489	0.310	0.682	0.615	0.724

	ENDSCR	PERF	NOTCH	DENTIC	CONGEL
ENDSCR	1.000				
PERF	0.000	1.000			
NOTCH	0.423	-0.094	1.000		
DENTIC	0.295	-0.139	0.604	1.000	
CONGEL	0.602	0.030	0.177	0.475	1.000
ELEPHAS	0.854	0.572	0.497	0.637	0.491
EQUUS	0.376	0.021	0.193	0.573	0.292
BOS	0.000	-0.113	0.525	0.530	0.557
CERVUS	0.556	0.560	-0.338	0.051	0.211

	ELEPHAS	*EQUUS*	*BOS*	*CERVUS*
ELEPHAS	1.000			
EQUUS	0.585	1.000		
BOS	0.239	0.332	1.000	
CERVUS	0.582	0.490	-0.245	1.000

foot bones were suspected to be related to waste and minimally utilized pieces, but waste flakes often occurred near skull fragments. (Note that expectations would be different for Ambrona, where other spatial associations were observed with greater frequency.) In my previously published analysis based on Pearson's r, several of these associations were confirmed statistically.

In this test, using a less powerful but more justifiable measure of association, and a slightly different set of data, with fewer collapsed categories, fewer correspondences occur, but the general picture remains the same.

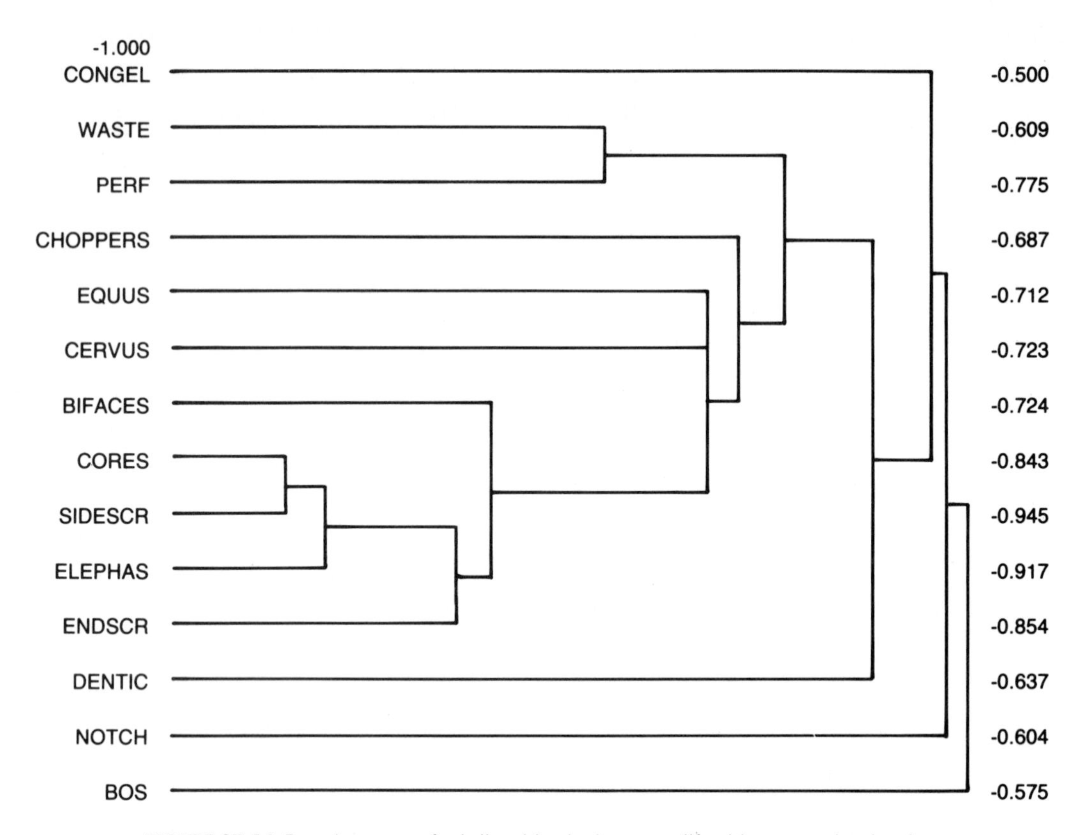

FIGURE 27-16 Dendrogram of relationships between artifact types and animal species, Torralba. Single linage method. Distance measure = Spearman's rho.

The first and largest tendency for variation is associated very strongly with sidescrapers, elephant teeth, tusk, limbs, ribs, feet and vertebrae, scapula/pelvis, and equid teeth and feet, and less strongly with cores, elephant skull, equid skull, and bovid limbs. The second is still less strongly associated with variance in bifaces and endscrapers but strongly determines variation in cervid antler. The third is highly associated with cervid limbs, less strongly with cervid skull, and less still with equid limbs, while equid skull and endscrapers show moderate negative loadings on this factor—that suggests simply that we may be sampling different aspects of the "cultural landscape" in each level, and that the places of discovery of these latter items are different from those of the former (a fact that has no evident geological explanation). Perforators, "waste," and elephant skull fragments are strongly associated with Factor 4. The fifth tendency for variation strongly determines variation in choppers and horse scapula/pelvis, "explaining" to a smaller degree variation in bovid limbs. Notches and denticulates are found alone to be determined by Factor 6.

Using Pearson's r (results not shown) the number of meaningful factors isolated was 7. The associations detected remained essentially similar, but one factor determined most variation in both waste and equid limbs, and elephant foot-bones and cervid limbs were found related to another.

The results of the principal components analysis indicate that there are in fact patterned relationships between stone artifact types and particular animal body parts, for all the major species represented at Torralba. They are nontrivial: a trivial association would be, for example, a single tendency that determined variation in all the variables, which would indicate that sample size was the only operative variable. They do not bear out Villa's (1990:304) claim that at Torralba people butchered "elephant carcass leftovers" rather than whole carcasses—all major elephant body parts are involved in these patterned relationships. The

TABLE 27-4 Torralba Body Parts Count by Level

	ELTTH	ELTSK	ELSKL	ELRIB	ELLMB	ELFET	ELSCPL	ELVRT	EQTTH	EQSKL
OCC 1	16	30	4	29	42	11	17	33	32	5
OCC 1C	0	0	2	0	1	1	0	0	1	0
OCC 1D	0	3	0	0	1	0	0	0	0	0
OCC 2	2	1	0	2	1	1	1	1	5	0
OCC 2A	5	6	2	12	9	4	2	2	9	3
OCC 3	0	0	0	1	0	1	0	0	2	1
OCC 3A	8	14	10	17	24	11	3	11	21	2
OCC 4	4	4	4	5	4	3	1	6	19	3
OCC 5	0	5	1	3	4	0	1	3	2	1
OCC 7	10	18	12	27	23	18	10	22	23	4
OCC 8	5	6	0	4	3	0	2	6	22	2

	EQLMB	EQFET	EQSCPL	BOLMB	CRVNT	CRSKL	CRLMB
OCC 1	14	7	11	9	15	2	2
OCC 1C	0	0	0	0	0	0	0
OCC 1D	1	1	1	1	0	0	2
OCC 2	0	1	0	0	0	0	1
OCC 2A	6	4	8	0	4	1	0
OCC 3	0	0	0	2	0	0	0
OCC 3A	13	6	9	5	2	2	3
OCC 4	4	2	4	0	3	1	1
OCC 5	7	1	1	2	5	1	0
OCC 7	7	6	2	3	5	1	1
OCC 8	18	4	7	6	8	2	4

statistical tests demonstrate that the human presence and the animal remains really cannot be independent of each other. In some cases, at least, they correspond to spatial associations viewed during the course of excavation, and their contents could not have been acted on similarly by natural depositional agencies other than humans, so that no simple explanation of site formation processes that excludes human agency can adequately account for them.

When the results of the level–by–level statistical analysis are evaluated in light of all other evidence from Torralba, the most economical way to account for the presence of the different components, the distinctive clusters of variables associated with each, and the fact that different clusters were frequently found in different areas, is to ascribe them largely to the organization of human activities. That is not to say that all materials from Torralba reflect human behavior, for there are many kinds of data that were not included in the tests, and some that were not adequately explained in terms of the factors isolated. Nor is it to claim that there has been no natural disturbance of the original patterns in the residues. Despite these processes, however, a picture of human activity emerges among the other pictures reflected in the Torralba finds.

TABLE 27-5 Matrix of Spearman Correlation Coefficients - Ranked Data

	BIFACES	CHOPPERS	CORES	WASTE	SIDESCR
BIFACES	1.000				
CHOPPERS	0.333	1.000			
CORES	0.843	0.472	1.000		
WASTE	0.687	0.276	0.429	1.000	
SIDESCR	0.677	0.313	0.945	0.309	1.000
ENDSCR	0.543	0.712	0.670	0.255	0.821
PERF	0.532	-0.216	0.437	0.775	0.462
NOTCH	0.363	0.485	0.342	0.025	0.265
DENTIC	0.375	0.452	0.513	0.257	0.633
CONGEL	0.578	0.408	0.561	0.133	0.609
ELETEETH	0.448	0.224	0.754	0.036	0.975
ELTUSK	0.462	0.334	0.626	0.210	0.783
ELSKULL	0.358	0.000	0.263	0.679	0.632
ELRIBS	-0.074	-0.000	0.317	-0.084	0.595
ELLIMBS	0.322	0.429	0.373	0.194	0.721
ELFEET	0.143	-0.098	0.179	0.253	0.718
ELSCPEL	0.467	0.344	0.731	0.185	0.893
ELVERTS	0.505	0.229	0.767	0.277	0.941
EQUTEETH	0.378	0.179	0.479	0.563	0.709
EQUSKULL	0.056	0.224	0.459	0.178	0.735
EQULIMBS	0.321	0.671	0.339	0.657	0.263
EQUFEET	0.384	0.462	0.646	0.262	0.971
EQUSCPEL	-0.019	0.894	0.385	0.139	0.667
BOSLIMBS	0.064	0.667	0.373	0.606	0.616
CERVANTL	0.702	0.775	0.750	0.609	0.526
CERSKULL	-0.101	0.577	0.198	0.364	0.444
CERLIMBS	-0.258	-0.131	0.048	-0.019	0.287

	ENDSCR	PERF	NOTCH	DENTIC	CONGEL
ENDSCR	1.000				
PERF	0.000	1.000			
NOTCH	0.423	-0.094	1.000		
DENTIC	0.295	-0.139	0.604	1.000	
CONGEL	0.602	0.030	0.177	0.475	1.000
ELTEETH	0.667	0.105	-0.051	0.327	0.761
ELTUSK	0.577	0.395	-0.281	0.140	0.701
ELSKULL	-0.500	1.000	0.211	0.578	0.426
ELRIBS	0.500	0.395	-0.356	0.025	0.168
ELLIMBS	0.462	0.092	-0.330	0.326	0.794
ELFEET	0.379	0.775	-0.811	0.000	0.266
ELSCPEL	0.667	0.500	0.154	0.329	0.528
ELVERTS	0.667	0.500	0.030	0.636	0.715
EQUTEETH	0.500	0.585	-0.104	0.253	0.017
EQUSKULL	0.684	0.462	0.156	0.061	0.012
EQULIMBS	0.112	0.308	-0.344	-0.030	0.206
EQUFEET	0.563	0.585	-0.137	0.290	0.572
EQUSCPEL	0.447	0.339	0.047	0.267	0.103

TABLE 27-5 continued

CERVANTL	0.632	0.105	0.574	0.045	0.236
CERSKULL	–0.275	0.148	–0.000	0.364	–0.218
CERLIMBS	–0.625	-0.059	–0.727	–0.162	–0.314

	ELTEETH	**ELTUSK**	**ELSKULL**	**ELRIBS**	**ELLIMBS**
ELTEETH	1.000				
ELTUSK	1.000	1.000			
ELSKULL	0.410	0.522	1.000		
ELRIBS	0.937	0.910	0.667	1.000	
ELLIMBS	0.901	0.903	0.615	0.934	1.000
ELFEET	0.899	0.899	0.821	0.927	0.881
ELSCPEL	0.991	0.976	0.603	0.908	0.827
ELVERTS	0.918	0.849	0.696	0.850	0.777
EQUTEETH	0.883	0.826	0.782	0.854	0.714
EQUSKULL	0.567	0.606	0.456	0.885	0.596
EQULIMBS	0.493	0.711	0.265	0.162	0.428
EQUFEET	0.991	0.944	0.647	0.946	0.879
EQUSCPEL	0.406	0.687	0.232	0.607	0.717
BOSLIMBS	0.200	0.771	0.200	0.754	0.657
CERVANTL	0.493	0.418	–0.176	0.126	–0.009
CERSKULL	0.396	0.510	0.315	0.289	0.291
CERLIMBS	0.273	0.243	–0.056	0.030	0.160

	ELFEET	**ELSCPEL**	**ELVERTS**	**EQUTEETH**	**EQUSKULL**
ELFEET	1.000				
ELSCPEL	0.882	1.000			
ELVERTS	0.841	0.821	1.000		
EQUTEETH	0.908	0.872	0.886	1.000	
EQUSKULL	0.647	0.676	0.606	0.793	1.000
EQULIMBS	0.667	0.523	0.473	0.595	-0.073
EQUFEET	0.868	0.975	0.860	0.897	0.676
EQUSCPEL	–0.051	0.600	0.378	0.464	0.400
BOSLIMBS	0.316	0.600	0.600	0.824	0.806
CERVANTL	0.410	0.376	0.327	0.523	0.312
CERSKULL	0.296	0.514	0.510	0.577	0.000
CERLIMBS	0.287	0.277	0.154	0.334	–0.632

	EQULIMBS	**EQUFEET**	**EQUSCPEL**	**BOSLIMBS**	**CERVANTL**
EQULIMBS	1.000				
EQUFEET	0.634	1.000			
EQUSCPEL	0.572	0.805	1.000		
BOSLIMBS	0.928	0.794	0.928	1.000	
CERVANTL	0.600	0.266	0.054	0.667	1.000
CERSKULL	0.874	0.588	0.722	0.866	0.291
CERLIMBS	0.647	0.276	0.441	0.410	0.154

	CERSKULL	**CERLIMBS**
CERSKULL	1.000	
CERLIMBS	0.889	1.000

TABLE 27-6 PC Analysis with Rotation Torralba Artifacts and Body Parts

I) LATENT ROOTS (EIGENVALUES)

1	2	3	4	5	
13.576	4.308	3.484	2.718	2.158	
6	**7**	**8**	**9**	**10**	
1.973	1.295	0.893	0.589	0.327	
11	**12**	**13**	**14**	**15**	**16**
0.289	0.209	0.079	0.047	0.010	0.000

II) ROTATED LOADINGS

	1	2	3	4	5	6	7	8
BIFACES	0.337	0.571	-0.133	0.459	-0.102	0.222	0.476	-0.272
CHOPPERS	0.244	0.437	-0.018	-0.093	0.844	0.325	0.195	0.157
CORES	0.607	0.501	-0.025	0.145	0.050	0.301	0.190	-0.489
WASTE	0.185	0.423	0.070	0.864	0.199	0.060	0.072	0.156
SIDESCR	0.909	0.195	0.022	0.081	0.040	0.323	0.049	-0.357
ENDSCR	0.587	0.561	-0.602	-0.284	0.236	0.062	0.146	-0.132
PERF	0.430	-0.063	-0.067	0.920	-0.085	-0.152	-0.148	-0.256
NOTCH	-0.147	0.343	-0.375	0.013	0.130	0.886	-0.083	-0.174
DENTIC	0.308	-0.043	0.049	0.035	0.052	0.913	0.222	0.079
CONGEL	0.519	0.065	-0.221	-0.048	-0.003	0.219	0.860	0.020
ELTEETH	1.005	0.172	0.096	-0.244	-0.181	0.074	0.191	-0.022
ELTUSK	0.960	0.085	0.085	0.029	0.175	-0.181	0.193	0.007
ELSKULL	0.605	-0.420	0.055	0.765	-0.160	0.366	0.018	0.174
ELRIBS	0.965	-0.270	-0.164	-0.083	0.042	-0.198	-0.287	0.076
ELLIMBS	0.902	-0.244	-0.042	-0.050	0.249	-0.066	0.333	0.211
ELFEET	0.922	-0.002	0.102	0.233	-0.355	-0.377	-0.027	0.327
ELSCPEL	0.969	0.095	0.055	0.032	0.057	0.101	-0.006	-0.128
ELVERTS	0.940	0.078	0.048	0.115	-0.068	0.248	0.102	0.027
EQUTEETH	0.890	0.179	0.123	0.287	0.005	0.009	-0.347	0.170
EQUSKULL	0.728	0.027	-0.613	0.098	0.153	-0.014	-0.493	0.030
EQULIMBS	0.445	0.387	0.591	0.286	0.436	-0.242	0.165	0.282
EQUFEET	0.981	-0.040	0.090	0.119	0.251	-0.014	0.052	-0.111
EQUSCPEL	0.537	-0.168	0.181	-0.020	0.875	0.074	-0.074	-0.281
BOSLIMBS	0.612	0.216	0.188	0.220	0.733	-0.221	-0.340	0.093
CERVANTL	0.325	1.006	0.046	0.081	0.138	0.086	-0.082	0.053
CERSKULL	0.403	0.051	0.797	0.093	0.450	0.228	-0.281	0.135
CERLIMBS	0.183	-0.030	1.054	-0.106	0.037	-0.245	-0.089	-0.133

Binford's dubious statistical procedures and errors and his mistranscriptions—perhaps better, "manipulations"—of artifact and bone counts from the site have misled readers about the nature and composition of the Torralba assemblages, and about the relationships between data categories. As Howell noted in a review in the *Journal of Human Evolution* (1989), 14.3 percent (10) of the 70 cells in the matrix Binford supposedly compiled from my earlier published figures are wrong: Even had he used identical tests, he would therefore have obtained different results from mine. The situation is aggravated by questionable transformations of the data, over interpretation of mathematical results that are not statistically significant to begin with, and the use of analytical procedures that I defy anyone (other than Binford) to understand or replicate. There are perfectly appropriate ways of transforming the data for his purposes—simply ranking the raw counts and using a rank-order correlation procedure as has been done here is the simplest and probably the best, while square root or log transformations of all the data and the use of Pearson's r is probably also defensible in this case—and when error-free data, transformed appropriately, are used as input to ordinary principal components analysis and rotation (or related multivariate tests whose results are free of operator bias and equally insensitive to the order of data entry), the results obtained are the ones I have published here and elsewhere, not those Binford presents.

CONCLUSIONS

I hope that I have presented enough information regarding Howell's work at Torralba and Ambrona in the 1960s and later to indicate the significance of those investigations once and for all, and to lay the less well-founded criticisms of our work to rest. I do not imply that our interpretations—specifically, my own—have always been impeccable and infallible. They have certainly not. In my earlier work, I seriously misjudged the extent of cultural elaboration expectable in a Mid-Pleistocene site and underestimated the difficulties in unravelling what cultural information there is from the overlay of other processes—geological, mechanical, chemical, and biological—that may embed and hide it. Nor have I always expressed myself as well as I could have done. Excavations in 1963, conducted under my guidance, while good enough, could nonetheless have been better; I paid too little attention in the 1960s to marks of gnawing or to marks of butchery; it is probably my own fault that no one knows that screens were used at Torralba; I should undoubtedly have indicated more clearly that we knew that carnivores were present at the sites, or that we had taken geological processes such as slopewash and channel flow into consideration; I never stressed enough that our statistical tests only included some of the Torralba data, or that some of the tested variables were not adequately explained. Of course there are animal remains at Torralba and Ambrona that were not manipulated by humans, or even evident to them. That should also have been made clearer.

Questions about the causes of patterning in these residues are not simple black–and–white issues. It is irresponsible and nonscientific to decide that either all the patterning detected must result from human cultural agency or none of it can. There is patterning due to nonhuman agency at the two sites. At the same time, a substantial basis for cultural interpretation can still be recognized at both. The archeological record of hominid activity at Torralba and Ambrona is not pristine and free from pre- and postdepositional distortions. However, even if these obscure the message, they do not obliterate it entirely, and enough remains to tell at least part of a story of hominid-animal interactions at Torralba and Ambrona.

No single kind of evidence tells the whole story. The sediments and fauna pose questions that must be answered with conjoinable stone tools, wear polishes, skeletal dispersal, and patterned regularities discerned statistically. No single line of evidence—sinking in clayey deposits, MNIs, age distributions, tooth-marks, stone tools—tells its own story unambiguously. To decipher what Binford has called the "palimpsest" of Torralba requires assembling and comparing all these multitudinous kinds of information, and trying to reconcile each with the rest. But when that is done, the outlines of a message about human adaptation appears, behind other messages, it is true, but nevertheless still legible.

I hope that these observations will help in some small way to clarify the importance and potential of Torralba and Ambrona and to secure for them the recognition they deserve. It is unfortunately true that we still know all too little about hominid adaptations of the Mid-Pleistocene. That is so in spite of the number of new Mid-Pleistocene sites that have been discovered and carefully excavated since the 1960s. Despite the high quality of excavations at sites like Aridos and Isernia, much more information will be needed before any satisfactory idea of the nature of Mid-Pleistocene adaptations in any region can be derived. Each site we now know is like an irreplaceable piece of a huge and variable picture puzzle most of whose pieces are missing. Each site we know so far has proven to be unique in scale, in scope, and in quality of information; it would be absolutely senseless to discard or ignore any of the pieces we have so laboriously assembled, assuming that it is replicated by any other. The pattern on each and every piece is damaged or obscure—every other European site from this time range presents at least as many problems of interpretation as do Torralba and Ambrona. To progress, we must try to understand every piece in its own terms, and to see how each relates to all the rest.

In this interpretive process, Torralba and Ambrona will continue to play a large part. Few European Mid-Pleistocene sites are nearly as informative as they. In the last analysis, that we owe to the vision, care, and scholarship of F. Clark Howell.

LITERATURE CITED

Beard, P. (1977) *The End of the Game.* Garden City: Doubleday.

Binford, L. R. (1981) *Bones: Ancient Men and Modern Myths.* New York: Academic.

Binford, L. R. (1987) Were there elephant hunters at Torralba? In M. Nitecki and D. Nitecki (eds.): *The Evolution of Human Hunting.* New York: Plenum, pp. 47–105.

Butzer, K. W. (1982) *Archaeology as Human Ecology.* Cambridge: Cambridge University.

Cerralbo, E. Aguilera y Gamboa, Marqués de (1909) *El Alto Jalón . Descubrimientos Arqueológicos.* Madrid: Fortanet.

Cerralbo, E. Aguilera y Gamboa, Marqués de (1913a) Torralba, la plus ancienne station humaine de l'Europe?, *Cong. Int. Anthropol. Archeol. Prehist., Geneve, C. R. XIV Session, pp. 227–290.*

Cerralbo, E. Aguilera y Gamboa, Marqués de (1913b) Torralba, la estación mas antigua de Europa entre las hoy conocidas, *Asoc. Esp. Progr. Cienc.*, Tomo I, Sección 4, pp. 197–210.

Freeman, L. G. (1975) Acheulean sites and stratigraphy in Iberia and the Maghreb. In K. B. Butzer and G. L. Isaac (eds.): *After the Australopithecines.* The Hague: Mouton, pp. 661–743.

Freeman, L. G. (1978)The analysis of some occupation floor distributions from Earlier and Middle Paleolithic sites in Spain. In L. G. Freeman (ed.): *Views of the Past: Essays in Old World Prehistory and Paleoanthropology.* The Hague: Mouton, pp. 57–115.

Freeman, L. G. (1981) The fat of the land: Notes on Paleolithic diet in Iberia. In R. S. Harding and G. Teleki (eds.): *Omnivorous Primates: Gathering and Hunting in Human Evolution.* New York: Columbia University Press, pp. 104–165.

Freeman, L. G. (1991) What mean these stones? Remarks on raw material use in the Spanish Paleolithic. In A. Montet-White and S. Holen (eds.): *Raw Material Economy among Prehistoric Hunter-Gatherers.* Lawrence: University of Kansas Press, pp. 73–125.

Freeman, L. G., and Butzer, K. W. (1966) The Acheulean station of Torralba, Spain: A progress report. *Quaternaria* 8, 9–21.

Frison, G. (ed.) (1974) *The Casper Site.* New York: Academic.

Frison, G. (1991) *Prehistoric Hunters of the High Plains.* New York: Academic.

Howell, F. C. (1989) Review of M. Nitecki and D. Nitecki, The Evolution of Human Hunting. *J. Hum. Evol.* 18:583–594.

Howell, F. C., and Freeman, L. G. (1982) Ambrona, an early stone age site on the Spanish meseta. *L. S. B. Leakey Foundation News* 22:11–13.

Howell, F. C., Butzer, K. W., and Aguirre, E. (1963) Noticia preliminar sobre el emplazamiento Achelense de Torralba, Excav. *Arqueol. Esp.* 10:1-38.

Howell, F. C., Freeman, L. G., Butzer, K. W., and Klein, R. G. (1992) Observations on the Acheulean occupation site of Ambrona (Soria Province, Spain), with particular reference to recent (1980–1983) investigations and the Lower Occupation. In G. Bosinski (ed.): *Die Altest Besiedlung Europas.*

Klein, R. (1987) Reconstructing how early people exploited animals: Problems and prospects. In M. Nitecki and D. Nitecki (eds.): *The Evolution of Human Hunting.* New York: Plenum, pp. 11–43.

Obermaier, H. (1916) *El Hombre Fósil.* Comisión de investigaciones paleontológicas y prehistóricas. Memoria 9.

Obermaier, H. (1925) *El Hombre Fósil.* Comisión de investigaciones paleontológicas y prehistóricas. Memoria 9 (Second Spanish Edition).

Santonja, M., and Villa, P. (1990) The Lower Paleolithic of Spain and Portugal. *J. World Prehist.* 4:45–94.

Shipman, P., and Rose, J. (1983) Evidence of butchery and hominid activities at Torralba and Ambrona: An evaluation using microscopic techniques. *J. Archaeol. Sci.* 10:465–474.

Vaurie, C. (1965) *Birds of the Palearctic Fauna. Non-passeriformes.* London: H. F. & G. Witherby.

Villa, P. (1990) Torralba and Aridos: Elephant exploitation in Middle Pleistocene Spain. *J. Hum. Evol.* 19:299–309.

Villa, P. (1991) Middle Pleistocene prehistory in southwestern Europe. *J. Anthropol. Res.* 47:193–217.

CHAPTER 28

New Archaic Human Fossil Discoveries in China and Their Bearing on Hominid Species Definition During the Middle Pleistocene

Dennis A. Etler
and
Li Tianyuan

INTRODUCTION

The disappearance during World War II of nearly the entire fossil human assemblage recovered at Zhoukoudian during the 1920s and 1930s, while deeply lamented by the international scientific community, was an irreplaceable loss for the Chinese themselves (Jia and Huang 1990). After the war the natural focus of paleoanthropological research in China was therefore to retrieve new fossil human material, be it from Zhoukoudian or other, yet undiscovered sites in China (Jia 1980). Decades would pass with few additional discoveries. Recently, however, more and more material attributable to archaic forms of humanity has been forthcoming from China and today a diverse array of fossil hominid remains has been collected from a number of different sites. This sample rivals in size and surpasses in scope what was lost from Zhoukoudian.

Archaic human material (generally attributed to the taxon *Homo erectus*) discovered in China since the 1950s includes additional cranial, gnathic, and dental remains from Locality 1 at Zhoukoudian; well-known cranial and gnathic specimens from Gongwangling and Chenjiawo, in Lantian county, Shaanxi; dental remains from Yuanmou, Yunnan and various sites in northwestern Hubei and adjacent areas of central China; fragmentary cranial, postcranial, and dental remains from Yiyuan, Shandong; cranial, gnathic, and dental remains from Hexian, Anhui; and two new crushed but relatively intact crania from Yunxian, Hubei. Much of this material is still poorly known in the West. In some ways, the new specimens differ from what was previously discovered and described at Zhoukoudian, expanding to a considerable degree the known range of variation in archaic Chinese hominids.

Of equal significance to the recovery of new archaic human fossils from China has been the remarkable set of discoveries pertaining to later, premodern humans that have come to light over the last two decades. Hominid remains from Dali, Yingkou (Jinniushan), Xujiayao, Chaoxian, and Tongzi have all been discovered within the last 15 years. Coupled with premodern material from Changyang, Dingcun, and Maba, recovered during the 1950s, human paleontologists are now in a better position to assess the status of hominid material from China that is clearly distinct and more progressive than archaic specimens generally attributed to *H. erectus*, but outside the accepted range of variation associated with modern humans. Comparisons with similar material of approximately the same age from Europe (early Neanderthals), India (Narmada), and Africa (Jebel Irhoud, Ngaloba, Florisbad, etc.) will yield insights into the dynamics of human evolution during the late middle and early late Pleistocene.

OVERVIEW OF DATING OF ARCHAIC HUMAN FOSSILS FROM CHINA

The earliest human fossil and artifactual remains recovered in China are with little doubt over 1 million years old (An and Ho 1989; Schick et al. 1991). Fragmentary fossils and scanty archeological remains may push this record back to nearly 2 million years. Evidence for an early Pleistocene human presence in east Asia, while equivocal, is afforded by the magnetostratigraphic dating of the Yuanmou hominid-bearing stratum to the Olduvai polarity subchron by Qian and colleagues (Qian 1985). Additional dentognathic material attributed to *H. erectus* from what are thought to be early Pleistocene cave deposits at Wushan in Sichuan (Huang and Fang 1991) and Liucheng in Guangxi are also of potential significance for documenting early human habitation in China. A better understanding of this material awaits the recovery of more abundant fossil remains and their more secure geochronologic placement.

Regardless of the outcome of research into these finds, the fossil record of archaic humans (i.e., *H. erectus*) in China has been amply supplemented over the last 30 years. If preliminary datings are confirmed by further study, the temporal range of archaic humans in China could be considerable, spanning 700,000 or more years (Figure 28-1).

In this regard a date of approximately 1.15 MYA given to the Lantian Gongwangling cranium (An and Ho 1989) is consistent with its primitive morphology and could set a baseline for the introduction of early humans into east Asia. The Lantian Chenjiawo mandible is reasonably dated to approximately 600 + KYA (Xu 1990), while the bulk of the Zhoukoudian material is probably somewhat younger (~440 KYA) (Zhou and Ho 1990). Abundant archaic dental remains attributed to *H. erectus* from the Nanzhao/Yunxian region seem to date from a time close to the base of the Zhoukoudian sequence (Wu and Dong 1982). The recently described material of *H. erectus* from Qizianshan, Yiyuan county, Shandong is thought to be contemporaneous with the middle layers at Zhoukoudian (Lü et al. 1989) while *H. erectus* material from Hexian in Anhui province is apparently coeval with or younger than the upper layers at Zhoukoudian (see review in Etler 1990). The new hominid crania from Yunxian in Hubei can at present only be broadly dated to the middle Pleistocene (Li and Etler 1992). All of these recent and not so recent discoveries go a long way in helping to elucidate the extent of variation in archaic humans from the middle Pleistocene of east Asia as well as the magnitude of evolutionary transformation that affected them during their long history in this region of the world.

ZHOUKOUDIAN

The famous site at Locality 1 Zhoukoudian (39° 41′ N, 115° 55′ E), 48 km SW of Beijing, which yielded the first remains of "*Sinanthropus pekinensis*" in the 1920s and 1930s, is well documented in English (Chia 1975; Aigner 1981; Binford and Ho 1984; Binford and Stone 1985; and included references) and need not be extensively reviewed here. Nevertheless, three recently published monographs concerning Zhoukoudian make major contributions to our understanding and appreciation of the site both in its historical and more research-oriented contexts. Jia and Huang (1990) present a retrospective on the early, middle, and later years of investigations at Zhoukoudian that offers a rare glimpse of the history of the excavations from the personal perspective of Prof. Jia Lanpo, who has worked the site for over 50 years. The memoir also paints revealing portraits of other famous personalities involved in the work at Zhoukoudian from the 1920s through the 1980s.

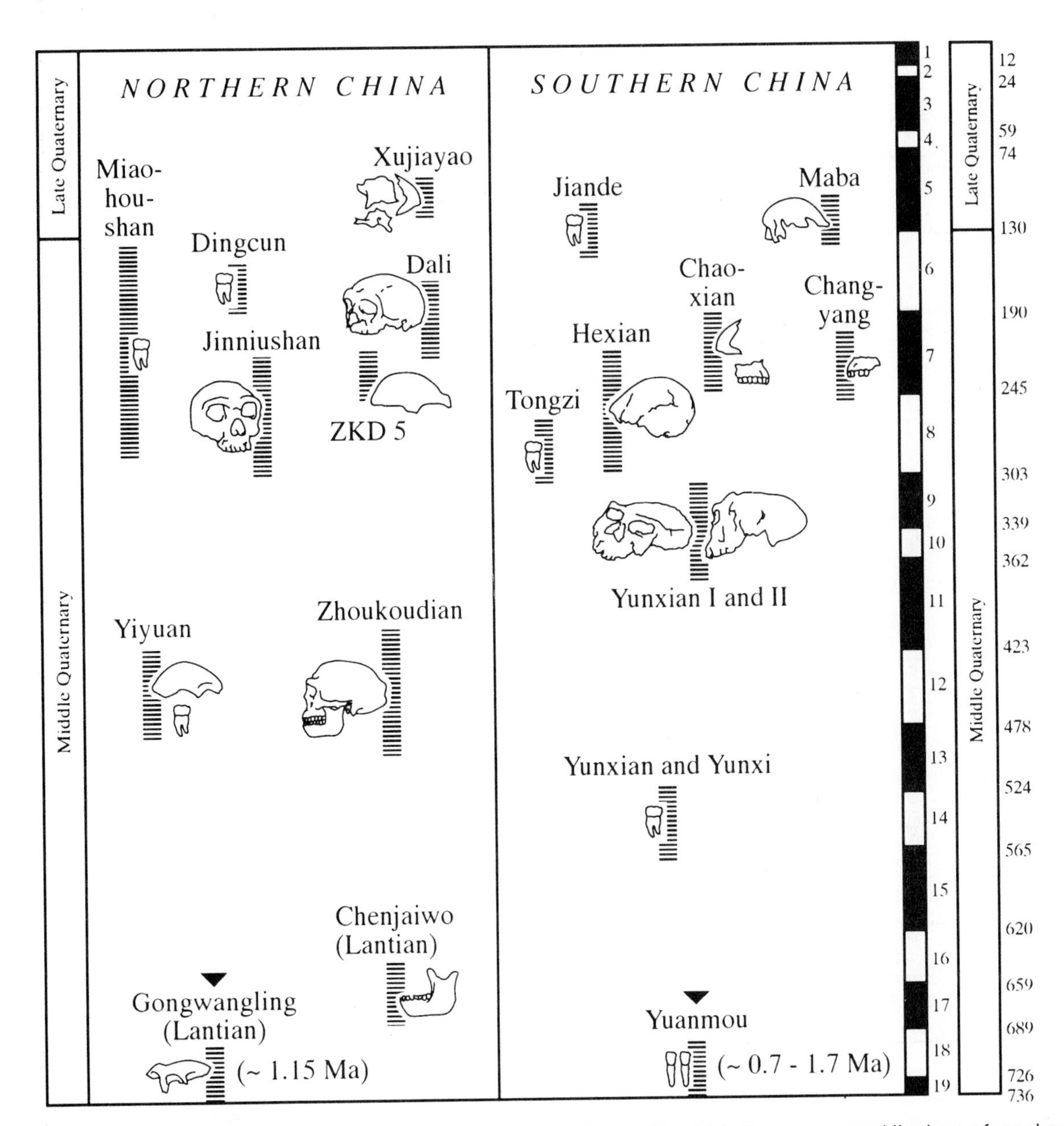

FIGURE 28-1 Temporal Distribution of Middle-Early Late Pleistocene Chinese Hominids. Figures represent the type of specimens recovered from each site. Archaic material from Yuanmou, Gongwangling, Chenjiawo, Yiyuan, Zhoukoudian, Yunxian, and Hexian is generally attributed to *H. erectus.* Material from Changyang, Dali, Jinniushan, Maba and Xujiayao is generally considered pre-modern *H. sapiens.* Chaoxian, Miaohoushan, and Tongzi are thought by some to be *H. erectus* and by others to be pre-modern *H. sapiens.* Dates are tentative and subject to revision. Dating is based on uranium series tests, biostratigraphic correlation, paleomagnetostratigraphy, and correlation of the Chinese loess/palaeosol column with the global oxygen-isotope scale. Late dates assigned to *H. erectus* remains from the upper layers at Zhoukoudian, Locality 1 (ZKD 5) and Hexian by uranium series tests are problematical. There is a possibility they may be older than first thought (see discussion in text).

A posthumous summation of research done by the late Prof. Pei Wenzhong on the lithic artifacts of "*Sinanthropus*" recovered at Locality 1, edited by Zhang Senshui (Pei and Zhang 1985), presents a thorough description of the culture-bearing deposits, a layer by layer inventory of the recovered cultural remains, and a comprehensive analysis of the lithic assemblage.

Finally a monographic treatment of the multidisciplinary researches conducted at Zhoukoudian during the late 1970s and early 1980s (Wu et al. 1985) builds on the foundation laid by Chinese and foreign

TABLE 28-1 Zonation of Depositional Sequence at Zhoukoudian

Age (kya)	Strata	Typical strata inside the cave	Typical strata outside the cave	Climatic characteristics
12	Holocene		River, lake and marsh facies of the Fenzhuang Group	Cool---warm and humid ---dry and cool
(C14) 49 (TL)	Upper Pleistocene series	Upper Cave; The new cave at Locality 4	Diluvial stratum of the Mangniu River containing *Paleoloxodon naumani*; Malan loess; Loess with paleosol of brown earth type	Semi-arid climate of the temperate zone in which the annual average temperature of the cold period may drop down by 5-6 degrees C
230 (U) 256	Middle Pleistocene series; The Zhoukoudian Group	Locality 1 — First section: 1st, 2nd and 3rd strata	The lower gravel stratum — The upper section: the red debris talus stratum; Alluvial stratum of the second terrace (the gravel stratum of the Yangery valley)	Semi-arid climate of the temperate zone
290-310 (TL)		Second section: The 4th stratum		Turning dry
		Third section: The 5th-6th strata	Middle section: yellowish-red loessal soil containing debris	Semi-arid climate of the temperate zone
462 (FT)		Fourth section: The 7th-11th strata		Semi-arid climate of the temperate zone tending towards dry and cold with intermittant warm and humid periods
700		Fifth section: The 12th-13th strata	Lower section: Weathered sand and gravel stratum	Semi-humid climate of the warm temperate zone
(PM)	The Long-gushan Group	The test well at Locality 1 14th-15th strata		Somewhat dry climate of the temperate zone with the upper part turning warm and semi-humid
800 (PM) 1000 (PM)	Lower Pleistocene series	The test well at Locality 1 16th-17th strata; Locality 12		Semi-humid climate of the warm temperate zone
	Pliocene series — Upper part: Red clay layer	Funnel-shaped agglutinated red calcitic clay in the cave	The Donglingzi gravel stratum; Weathering residium of red kaolin type	Semi-humid climate of the north subtropical zone
	Pliocene series — Lower part: Yuling Group	Subterranean deposits at Locality 14 — Upper gravel stratum: Upper section: upper travertine; Middle section: weathered sand and gravel stratum; Lower section: thin silt and fine sand		Humid climate of the subtropical zone

TABLE 28-2 Sedimentary Sequence at Zhoukoudian Locality 1

Stratigraphic layer	Thickness (m)	Lithologic characteristics	Major clay minerals	Vegetation		Mammalian fauna			Hominid fossils
				Pollen	Phase of	Warm climate faunas	Cold climate faunas	Ecological ratio	
1st	1-2	Breccia with yellowish-red clayey silt							
2nd	1.7	Travertine and light-red silt with small quantity of fine breccia	Illite, montmorillonite, kaolinite	AP=12.5% *Betula, Juglans, Alnus,* Chenopodiuceae	Broadleaf forest with grassland	*Rhizomys*		Decrease of forest animals and increase in grassland animals	
3rd	3.6	Breccia with reddish clay and rounded cobbles	Illite, kaolinite, montmorillonite,	*Pinus, Abies, Artemesia, Selaginella*	Coniferous forest		*Megaceros*		Cranium, two mandibular fragments
4th	6.9	Colored ash and light-red clayey silt	Montmorillonite, Illite, kaolinite	*Betula, Salix, Picea,* Cupressaceae	Coniferous and broadleaf mixed forest with grassland	*Elephus* cf. *namadicus*			Cranium, five mandibular fragments, two postcranial fragments
5th	0.15	Travertine				*Ailuropoda, Acinonyx*			One mandibular fragment
6th	7.2	Coarse breccia intercalated with fine breccia	Illite, kaolinite, chlorite	*Quercus, Betula, Corylus, Artemisia*			*Castor, Megaceros*	Forest animals more than grassland animals	
7th	1.5	Grayish-yellow fine silt	Montmorillonite, kaolinite	AP=40% *Pinus, Corylus, Alnus, Betula*	Dediduous broadleaf forest	*Bubulus*			One cranium and one postcranial fragment
8th	2	Travertine and fine breccia with brownish silty clay	Illite, kaolinite, chlorite, montmorillonite,	*Betula, Alnus, Celtis, Symplocas, Corylus*		*la io, Hystrix* cf. *subcristata*			Eight crania, four facial fragments, four mandibular fragments, eight postcrania
9th	5	Breccia with light-red clayey silt	Illite, kaolinite, chlorite	*Pinus, Abies, Betula, Quercus, Salix, Selaginella*	Conifderous broadleaf mixed forest		*Gulo, Trogontherium, Coelodonta*		
10th	0.6	Colored ash and brownish silty clay	Illite, kaolinite, chlorite	AP=33% *Pinus, Betula, Liquidambar, Aralia, Artemisia,* Gramineae, *Typha*	Deciduous broadleaf mixed forest	*Dicerorhinus choukoutienesis*			Two crania and one mandibular fragment
11th	0.8	Breccia and reddish clay	Illite, kaolinite, chlorite	*Pinus, Abies, Artemisia, Selaginella*		*Ursus* cf. *spelaeus*		Grassland animals more than forest animals	One mandibular fragment
12th	1.5	Coarse silt	Illite, kaolinite, chlorite	AP=85% *Pinus, Betula, Fraxinus,* Compositae	Coniferous broadleaf mixed forest	*Dicerorhinus choukoutienesis*			
13th	2.9-4.8	Reddish-brown clay and breccia	Illite, kaolinite, chlorite	Small quantity of *Artemisia, Selaginella, Silensis*			*Megaceros flabellatus*		
14th	7.6	Reddish-brown clay and larger gravels,	Illite, kaolinite, chlorite, montmorillonite,	AP=95.7% *Juglans, Mandshuri*			Mammalian fossils not found at or below this level		
15th	1.7-2.8	Brownish-red sand with gravels, fine sand, silt and clay	Kaolinite, illite, chlorite, montmorillonite	*Alnus, Pinus, Sygodium, Selaginella, Sinensis* (80.2%)	Broadleaf or broadleaf coniferous mixed forest				
16th	1.0-2.5	Gray and grayish-brown clayey silt with laminations	Illite, kaolinite	AP=35% *Quercus, Castanea, Ulmus, Pinus, Artemisia, Selaginella,* Taxodiacea	Coniferous broadleaf mixed forest with grasslands				
17th	0.7	Light-brown, coarse sand with some rounded gravels	Kaolinite, illite	Chenopodieceae, *Typha*					

TABLE 28-3 Archaic human remains from Zhoukoudian (after Weidenreich 1943 and Wu and Dong 1985a). Specimen # refers to the colloquial designation given to the Zhoukoudian remains by the original researchers. Sex and age were attributed by Weidenreich and other primary researchers. Individual refers to the locus from which individual human remains were recovered and the sequential number of the individual with which they are thought to be associated. Year of discovery refers to the year when individual specimens attributed to a single individual were found.

Specimen #	Fossil hominid remains	Sex	Age	Locality			Individual	Year of Discovery
				Locus	Layer	Level		
	Crania and cranial fragments							
I	fragmentary rt. parietal, lf. frontal	M	A	B	4	-	B II	1928
II	calvarium (lacking 2 temporals and occipital)	?	A	D	8-9	-	D I	1929
III	calvarium	M	J	E	11	-	E I	1929
IV	rt. parietal fragment	M	J	G	7	-	G II	1931
V	calvarium	M	A	H	3	-	H III	1934, 1934/1936, 1966
VI	frontal frag., lf. parietal frag., rt. temporal frag.	F	A	I	8-9	22	I I	1936
VII	rt. parietal at mastoid angle	M	AD	I	8-9	22	I II	1936
VII	occipital frag.	F?	I	J	8-9	23	J I	1936
IX	frontal frag., 4 small cranial frags.	M	I	J	8-9	23	J IV?	1936
X	calvarium	M	A	L	8-9	25	L I	1936
XI	calvarium	F	A	L	8-9	25	L II	1936
XII	calvarium	M	A	L	8-9	25	L III	1936
XIII	lf. maxillary frag. (with I2, P3-M3)	M?	A	O	10	29	O I	1937
XIV	lf. maxillary frag. (with P3, M1-M3)	M	A	Uppper Cave	-	-	UC?	1933
	Facial bones and facial bone fragments							
I	frontal process of lf. maxilla	M	A	L	8-9	25	Skull X	1936
II	left malar frag.	M	A	L	8-9	25	Skull X	1936
III	lf. maxillary frag. (with P3-M3)	F	A	L	8-9	25	Skull XI	1936
IV	rt. palate	F	A	L	8-9	25	Skull XI	1936
V	cf. Skull XIII							
VI	cf. Skull XIV							
	Adult mandibles							
I	part of rt. moiety	F	A	A	5	-	A II	1928
II	lf. condyle	M?	A	B	4	-	B II	1928/35
III	3 lf./rt. frags.	M	A	G	7	-	G I	1931
IV	symphysis and rt. corpus	F	A	H	3	-	H I	1934
V	symphyseal frag. and lateral bodies	F?	A	H	3	-	K I	1936
VI	lf. corpus	M	A	K	8-9	24	K I	1936
VII	lf. frag.	M	A	M	8-9	36	M I	1937
VIII	lf. hemi-mandible	F	A	M	8-9	26	M II	1937
IX	lf./rt. frags	F	A	-	10	27	-	1959
	Non-adult mandibles							
I	symphysis and rt. frag.	F	J	B	4	-	B I	1928
II	rt. frag.	M	J	B	4	-	B III	1932/38
III	rt. frag.	F	I	B	4	-	B IV	1932/35
IV	symphysis and rt. frag.	M	J	B	4	-	B V	1928/35
V	rt. ramus frag.	F	J	C	8-9	-	C I	1929
VI	rt. frag.	M	J	F	11	-	F I	1930

TABLE 28-3 cont'd

Specimen #	Fossil hominid remains	Sex	Age	Locality			Individual designation	Year of Discovery
				Locus	Layer	Level		
Femora								
I	proximal half lf. shaft	M	A	C	8-9	-	C III	1929/1936
II	lf. mid-shaft	F	A	J	8-9	23	J II	1936/38
III	rt. prox.imal frag.	M	A	J	8-9	23	J III	1936/38
IV	near complete rt. shaft	M	A	M	8-9	26	M IV	1937/38
V	lf. proximal shaft	M	A	M	8-9	26	M IV	1937/38
VI	lf. mid-shaft (2 pieces)	M	A	M	8-9	26	M I	1937/38
Assorted postcrania								
Tibia I	lf. shaft frag.	-	A	-	-	-	-	1951
Humeri I	lf. distal frag.	M	A	B	4	-	B II	1928/35
II	lf. shaft	M	A	J	8-9	-	J III	1936/38
III	rt. mid-shaft	M	A	-	-	-	-	1951
Clavicle I	lf. shaft	M	A	G	7	-	G II	1931
Lunate I	rt. lunate	M?	A	B	4	-	B II	1928
Teeth								
	2 specimens; lower LP3, Upper RM3							1921/1923
1-147	64 isolated specimens; rest in upper and lower jaws; 52 upper, 82 lower, 13 deciduous							1927-1937
	5 specimens; upper LI1, upper RP3, upper RP4, lower LM1, lower LM2							1949/50, 1951/1953
	lower RP4							1953
	lower LP3 (in jaw)							1959
	lower RP3							1966

workers during the 1920s through the 1960s, yielding a new understanding of the paleoenvironmental, including paleogeographical and paleoclimatological, setting of Zhoukoudian throughout the Quaternary based on the results of modern scientific investigations into the sedimentology, geochronology, palynology, and biostratigraphy of the deposits. Much of this research has been reviewed by the primary investigators in English language publications (Li & Ji 1981; Ren et al. 1981; Ren & Liu 1985; Yang and Mou 1982; Wu and Lin 1983; Liu 1985, 1988; see Aigner 1981 for a review of the literature on Zhoukoudian through the 1960s).

The question of the geologic age of the Zhoukoudian deposition is of obvious importance for understanding the tempo and mode of human evolution in east Asia. The application of modern dating techniques to the various stratigraphic layers at Zhoukoudian Locality 1 (Guo et al. 1980; Huang and Fang 1991; S. Liu et al. 1985; Pei 1980, 1985; Pei and Sun 1981; Qian et al. 1980, 1985; Shen and Jin 1991; Zhao et al. 1980, 1985a, 1985b) and the correlation of the sedimentary sequence at Zhoukoudian Locality 1 with the loess/paleosol column and the oxygen isotope scale as recorded in deep sea cores (Xu and Ouyang 1982; Liu and Ding 1984; Liu 1985, 1988) has resulted in a tentative absolute chronology for the entire deposition (reviewed in Zhou and Ho 1990 and Table 28-1). Studies of the lithologic character of the deposits including analysis of the clay mineral assemblage, the heavy mineral assemblage, the macroscopic structure of fluvially transported sands (Yang et al. 1985; Xie et al. 1985; NWISWC 1985) and the morphogenesis of karstic features within the cave (Wang 1983; Ren et al. 1985; Liu 1988) combined with palynological (Kong et al. 1985) and paleontological (Hu 1985; Hou 1985) studies have led to a much better appreciation of macro- and microclimatic shifts that have occurred during the depositional sequence. This wealth of new data is summarized in Table 28-2.

Aigner (1986, 1987) has been critical of some of these efforts, voicing skepticism as to the efficacy of the employed dating techniques and the utility of the analytic methods underlying interpretations of the zonation of the Zhoukoudian depositional sequence. She believes that evidence for a long depositional record and concomitant climatic oscillations at Zhoukoudian Locality 1 is inconclusive and opts for the hominid-bearing layers being laid down during a relatively short interglacial period (Holsteinian) of the mid-middle Pleistocene. Aigner's long-standing contention that the deposits at Zhoukoudian Locality 1 represent a single interglacial episode (Aigner 1981) rests primarily on an interpretation of the palynological and paleontological record that minimizes the extent of faunal and floral excursions during the depositional sequence. The static nature of the Zhoukoudian Locality 1 fauna and flora is contrasted with the faunal replacement and floral succession seen in Europe during the middle Pleistocene, suggesting to Aigner (1986, 1987) that the Zhoukoudian Locality 1 fauna and flora is time restricted. Recent palynological research at Zhoukoudian (Kong et al. 1985), which augments the initial probes conducted during the 1960s by Hsu (1966) and Sun (1965), however, may document climatic oscillations in keeping with those known elsewhere in the Palearctic during the middle Pleistocene (see Howell 1986: Appendix 2 for a highly informative review of the status of Quaternary research at Zhoukoudian and elsewhere in China as of 1985).

The vast majority of fossil hominids from Locality 1 were recovered from layers 8-9 and 3-4 (Wu and Dong 1985a; see Table 28-3). The remains from the upper layers, in particular Skull V from Locality 1, are thought by some to be more progressive than remains from the lower depositional layers (see below). Zhang (1991), likewise, documents secular changes in dental dimensions between the upper and lower layers at Zhoukoudian as do Pei and Zhang (1985) for various parameters of the lithic assemblage. These results have been used to bolster the idea that a considerable length of time separates the two periods of deposition.

Supporting this interpretation is the dating of the upper layers at Locality 1 to approximately 230 KYA by uranium series tests (Wang 1989) and ESR determinations (Huang and Fang 1991). Shen and Jin (1991), however, have conducted U-series tests on a capping travertine intercalated between layers 1 and 2 from which a pure, densely crystallized calcite was obtained. According to their results, the concordance between the 230Th/234U and 227Th/230U ages shows that the crystals come from a closed system. These findings indicate that the age of the upper layers at Zhoukoudian may be considerably older than indicated by previous dating attempts, on the order of 420+110/-54 KYA. The upper age limit of the hominid bearing strata at Zhoukoudian may, therefore, be either nearly 200,000 years younger than the majority of *H. erectus* remains from Locality 1, which are thought to date from 400-460 KYA (Wang 1989) or, if the dates reported by Shen and Jin are correct, of the same general age as the lower layers. The latter age would agree with Aigner's (1986) interpretation of the depositional sequence at Locality 1.

MORPHOLOGICAL FEATURES OF THE ZHOUKOUDIAN HOMINIDS

Wu and Dong (1985a) review the history of discovery of the hominid remains at Zhoukoudian (see Table 28-3). Between 1921 and 1966 a total of six complete or near complete calvaria, 12 cranial fragments, 15 mandibular pieces, and 157 teeth (including 84 socketed and 73 isolated) were recovered. Postcranial finds include seven femoral pieces, one fragment of a distal tibia, three humeral pieces, one clavicular fragment and one lunate.

The analysis of the morphological characteristics of this assemblage by Weidenreich (1936a, 1936b, 1937, 1939a,b, 1941, 1943) constitutes a remarkable repository of comparative anatomical detail which to date serves as a baseline for interpreting fossil hominids regardless of their spatial or temporal provenance. Given the historical importance of the description of this material for interpreting human fossil remains found subsequently in east Asia, a brief review of the morphological features that distinguish the Zhoukoudian remains from anatomically modern humans will serve as a convenient point of departure for the discussions to follow.

The skull is long and low (basibregmatic height/glabello-occipital length = 0.62) (Stringer 1984). Greatest cranial breadth lies toward the base of the cranium in the temporal region, at the level of the supramastoid crests. The sides of the cranium converge toward the apex. The frontal is relatively flat (frontal angle > 136.5^{0}) (Stringer 1984) and slopes posteriorly. It is characterized by pronounced, projecting brow ridges that are united medially by a distinct glabellar prominence, the whole unitary structure being termed the supraorbital torus. The boundary between the supraorbital torus and the frontal squama is typically marked by a troughlike depression called either the supratoral or posttoral sulcus. There is a significant degree of postorbital constriction that sets the supraorbital structures away from the cranial vault as a supraorbital visor. The frontal is further characterized by a thickening of bone metopically along the median plane generally referred to as a frontal keel. This thickening begins at the level of the frontal eminences and continues posteriorly along the median plane to the obelion region of the parietals where it fades out.

The parietals are characterized by parasagittal depressions on either side of the mid-sagittal keel, anteroposterior abbreviation and a high sagittal arc/chord index indicative of overall longitudinal flatness.

The occipital has a short upper (occipital) and long lower (nuchal) scale. They meet at a sharp angle along which forms a strongly developed transverse occipital torus. The torus spans the entire breadth of the occipital and continues on to the mastoid portion of the temporal. The central point of the torus (incorrectly defined as inion by Weidenreich, cf. Hublin 1986) is coincident with opisthocranion (point of greatest posterior cranial projection). The occipital torus is demarcated by a supratoral sulcus above and a well-defined superior nuchal line below.

The temporal bone has a number of distinctive features relative to the condition generally seen in later premodern and modern *H. sapiens*. The squamosal (superior) border is, as a rule, straight and low versus the highly arched border seen in modern humans. The tympanic plate is very thick, inclined more horizontally than in modern humans and oriented along the coronal plane. At its medial extremity is a large "clublike" eustachian process (processus supratubarius) (Aiello and Dean 1990). The axis of the pyramidal process is oriented parasagittally rather than being perpendicular to the mid-sagittal plane, producing an acute posterior angle with the tympanic. The mandibular fossa is deep and narrow. It is bounded medially by the sphenoid rather than an entoglenoid process and anteriorly by a preglenoid planum. There is no postglenoid process. The external auditory meatus is relatively broad in comparison to modern humans and overhung by a suprameatal tegmen formed by the laterally projecting zygomatic root. The mastoid process is relatively small. The styloid process is not ossified to the cranial base, the position of its vaginal sheath being occupied by a elongate "styloid pit."

The cranial vault bones are thick ($\overline{x}$ = 9.7 mm), nearly twice as thick as the modern human average ($\overline{x}$ = 5.2 mm) (Weidenreich 1943). The vault bones are also characterized by extremely robust ectocranial buttresses including the aforementioned sagittal keel; thickening along the coronal suture which, along with the sagittal keel, forms a distinct cruciate eminence at bregma; a well-developed temporal ridge that forms a prominent angular torus at the parietal mastoid angle; and pronounced occipitomastoid and supramastoid crests that are the terminal extensions of the occipital and angular tori respectively.

The cranium, while robust, has a poorly pneumatized frontal sinus, which tends to be small and limited to the glabellar region. The maxillary sinuses are, however, very large. Average cranial capacity is 1088 cc with a range of 915 cc (juvenile Skull III) to 1225 cc (Skull X). Internally the cranium shows a low division of the middle meningeal artery from its main trunk with the imprint of the posterior branch (ramus temporalis) somewhat larger than that of the anterior branch (ramus frontoparietalis). Sub-branches from both are much less abundant than in modern humans. The cerebral fossae of the occipital are distinctly larger than the cerebellar fossae by a factor of two, opposite the modern condition. Endinion is well separated from inion. There is a prominent Sylvian crest.

What is known of the facial skeleton is also robust. The orbits are deep and broad. There is no lacrimal fossa. The nasal bones are very broad and the bridge of the nose is relatively flat. The nasal aperture is low and broad and lacks an anterior nasal spine. The cheek bones are very deep (i.e., tall) and rotated anteriorly. As in many modern humans, a robust malar tubercle occurs at the inferomedial corner of the maxillary process of the zygomatic along the inferior zygomaxillary border and there is a distinct incurvature, the incisura malaris, medial to the malar tubercle along the inferior margin of the zygomatic process of the maxilla. The modern-looking zygomaticomaxillary features encountered at Zhoukoudian are not seen or are poorly expressed in earlier human specimens from Asia and Africa, and archaic middle Pleistocene hominids from Europe and Africa. The palate is deep, has a rugose surface and a well-expressed palatine torus is often seen along its longitudinal axis. The alveolar process of the maxilla is, however, relatively narrow. There is a marked degree of alveolar prognathism.

The mandibles from Zhoukoudian are considerably more robust than in modern humans. They are distinguished by their large bicondylar breadth, posteriorly inclined symphyseal region, and lack of a chin. The dental arcade is "U"-shaped, long and narrow. The lateral surface of the corpus has strong surface relief with a well-expressed lateral prominence; strong marginal and lateral superior tori and an anterior marginal tuberculum. There are, as a rule, multiple mental foramina. The digastric fossae, situated on the inferior border of the mandibular corpus, are long and narrow. The ascending ramus is very broad with a well-excavated masseteric fossa. The gonial angle is strongly everted. Internally the symphysis has poorly expressed superior and inferior transverse tori and a weak postincisive planum. The inner surface of the corpus has a well-expressed mandibular torus.

The teeth recovered from Locality 1 are larger and more robust than in modern humans. The lower premolar and molar crowns are, however, narrower and lower than in later hominids. Molars follow the serial pattern M2>M1>M3. The occlusal surfaces of the chewing teeth are complexly ridged and crenulated in comparison to modern humans. Upper and lower canines, premolars and molars show remnants of a basal cingulum. Pulp cavities tend to be enlarged (taurodont). The upper central incisors are typically shovel-shaped and have a basal tubercle and elongated longitudinal enamel ridges lingually. The upper canine is large and pyramidal with a well-defined apex, prominent lingual groove, and a mesial lingual ridge. Upper P3 is asymmetrical with a projecting mesiobuccal corner and a large buccal cusp. Upper P4 is smaller than P3 and ovoid in shape with subequal buccal and lingual cusps. The lower canine has a pointed apex but is reduced in size relative to the upper canine and somewhat incisiform. The lower P3 is inflated distolingually yielding an asymmetrical, oblong shape to the crown. The buccal cusp is more salient and larger than the low lingual cusp. Lower P4 has subequal buccal and lingual cusps with the latter shifted mesially. The lower molars follow the dryopithecine pattern and often retain a tuberculum sextum.

The postcrania recovered from Zhoukoudian are basically restricted to diaphysial elements of the long bones. They are characterized by thick cortical walls (hyperostosis), reduced medullary cavities (stenosis), and platymeria. The femora are distinguished by a variably developed linea aspira and low pilaster.

Skull V from bed three at Zhoukoudian has been considered more progressive, in a number of respects, than other cranial specimens recovered at the site. It consists of four major fragments, found separately, that come together to form a relatively complete calvaria. The left temporal and adjoining portions of the parietal and occipital bones were discovered in 1934, a fragment of the right temporal centered around the external auditory meatus was found in 1936 (Weidenreich 1943), while the frontal and much of the remaining portion of the occipital were recovered in 1966 (Qiu et al. 1973).

The reconstructed calvaria shows a suite of consistently progressive features when compared to other fossil human specimens from Locality 1 at Zhoukoudian. These include an arched parietal margin of the temporal, relatively slender supraorbital tori, a reduced occipital torus, a closer approximation between the internal and external occipital protuberances (29.5 mm vs. an average of 35.7 mm in other Zhoukoudian crania) and a thinning of the cranial vault bones (Qiu et al. 1973; Wu and Dong 1985a,b). Many of the

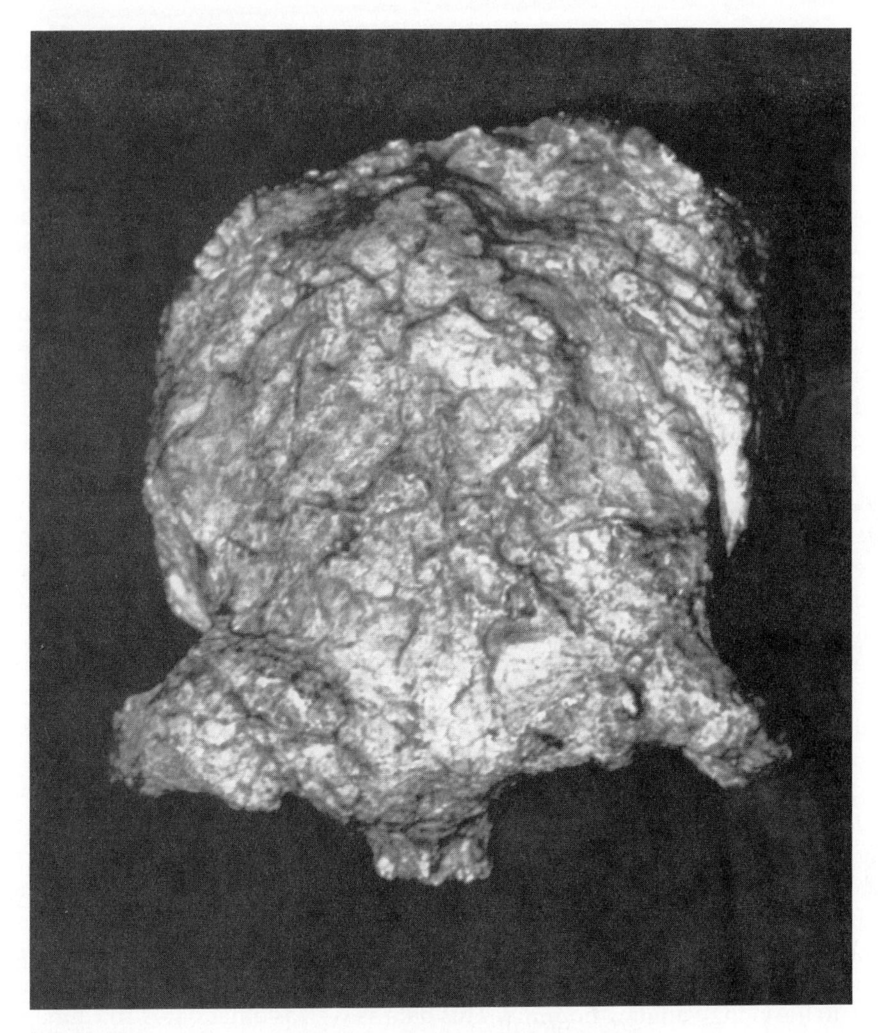

FIGURE 28-2 The Gongwangling, Lantian Calotte, Superior View.

progressive features seen in Skull V have also been noted in the recently discovered Hexian calvaria to be discussed later. The material does not, however, seem to differ fundamentally from other human remains recovered from the middle depositional layers at Zhoukoudian Locality 1.

Although the fossil human assemblage from Locality 1 has come to characterize the taxon "*H. erectus*" in east Asia, the following accounts will show that archaic humans as now known from China encompass considerably more variation than previously realized. Documentation of the extent of polytypic variation in middle Pleistocene east Asian hominids is of great consequence for evaluating the character and affinities of archaic humans known from elsewhere in the Old World.

GONGWANGLING AND CHENJIAWO, LANTIAN COUNTY, SHAANXI

In 1963, a field team of the IVPP conducting a survey of Quaternary deposits in the vicinity of Lantian discovered a fossil locality near Chenjiawo village in Xiehu district, 10 km NW of the countyseat (E 109° 15′, N 34° 14′) (IVPP 1966). Excavations led to the recovery of a fossil hominid mandible. Following this discovery the calvarium, maxillae, and isolated tooth of a single individual were found 16 km east of

Lantian City at Gongwangling (E 109° 29′, N 34° 11′) in 1964. The Chenjiawo and Gongwangling hominids have been reported on extensively in English (Woo 1964a,b, 1965, 1966).

The dating of the Lantian hominid sites has been the subject of some controversy, with calculated ages ranging from 300 KYA for the Chenjiawo deposits (Aigner and Laughlin 1973) to 1.15 MYA for the Gongwangling site (An and Ho 1989). A variety of techniques have been utilized to arrive at these estimates, including faunal correlation (Zhou and Li 1965; Aigner and Laughlin 1973; Zhang et al. 1978), paleomagnetic correlation (Cheng et al. 1978; Ma et al. 1978; An and Ho 1989), loess/deep sea oxygen isotope correlation (Liu and Ding 1984; An and Ho 1989) and amino acid racimization (Li and Lin 1979). Recent advances in dating the Gongwangling remains, however, now make it perhaps the earliest hominid on the Asian mainland (An and Ho 1989; Xu 1990). The Chenjiawo find, although not nearly as ancient, most likely represents an earlier human occurrence in China than do the human remains from Zhoukoudian (An and Ho 1989; Xu 1990).

According to An and Ho (1989), the Chenjiawo column unconformably overlies Eocene sandstones of the Beiluyan Group. The column itself is described as a 50-m thick sequence of middle-late Pleistocene loesses and paleosols that encompass the Brunhes chron and the uppermost portion of the Matuyama chron. Particular importance is given to the presence of a strongly weathered paleosol complex mid-way up the section equated with the S5 paleosol of the type Luochuan loess/paleosol sequence for northern China (Heller and Liu 1982). The S5 paleosol is in turn correlated with stages 13-15 of the O-18 sequence as recorded in deep-sea cores dated radiometrically to between 500,000-620,000 YA. The Chenjiawo mandible was found approximately 5 m below the paleosol complex referred to previously, in another paleosol, correlated with the S6 paleosol at Luochuan, dated to approximately 650,000 YA. Below the fossil-bearing deposits are two more paleosols followed by a well-developed silty loess of negative polarity correlated with the L9 loess of the Luochuan sequence.

The stratigraphic column at Gongwangling records the Brunhes normal polarity chron to a depth of 5 m. Below this level, to the basal gravels, the section is characterized by negative polarity of the Matuyama chron. A strong normal polarity interval is recorded at a depth of 17-20 m, thought to correspond to the Jaramillo subchron. Above and below this interval, at depths of 9-14 m and 24-27 m respectively, are well-developed silty loess layers said to be equivalent to the L9 and L15 marker horizons at Luochuan. The Gongwangling fauna, including the hominid remains, comes from the mid-portion of the proposed L15 equivalent which is dated at Luochuan to between 1.09–1.20 MYA. An and Ho (1989), therefore, assign an age of 1.15 MY to the Gongwangling hominid. This date as well as other dates, in China and elsewhere, that rely on long-range lithologic correlation and uncontrolled paleomagnetic interpolation should, however, be viewed as largely hypothetical and in need of further testing.

THE GONGWANGLING HOMINID

The Gongwangling hominid consists of a partial calvarium including a complete frontal, large portions of the parietals and a right temporal fragment (excluding the mastoid part); most of the left nasal and the root of the right nasal; most of the right maxilla, retaining M2-3, and the corpus and frontal process of the left maxilla. A left upper M2, an antimere to the M2 retained in the right maxilla, was also recovered. The calvarium is heavily fossilized, mottled in color and has a corrugated surface relief possibly due to the acidic nature of the deposits in which it was contained. It has been subject to extensive plastic deformation (Figure 28-2). The relatively small size of the teeth, combined with the degree of tooth wear and sutural closing, led Woo (1966) to conclude that the specimen was a 30+ year-old woman.

The Gongwangling cranium seems to represent an early Asian form of archaic human. It exhibits the major traits of "*Sinanthropus pekinenis*" as defined by Weidenreich in his series of monographs on the Zhoukoudian remains. It deviates from the Zhoukoudian bauplan, however, in a number of ways, perhaps reflective of a more overall primitive construction (the Gongwangling cranium may be nearly 700,000 years older than the majority of archaic human remains from Zhoukoudian).

The frontal appears extremely low and receding although the degree to which this is influenced by damage to the cortical surface of the cranial bones and postdepositional crushing is hard to determine. Concomitant to this is the apparent lack of a well expressed supraorbital sulcus commonly seen in the Zhoukoudian crania. The temporal lines are well expressed, forming a salient ridge. Although the external surface of the frontal has been damaged, a median frontal tuberosity and traces of a frontal keel and a cruciate eminence at bregma are still discernible.

The supraorbital tori are massive and extend laterally to a greater extent than in archaic human specimens from Java or Zhoukoudian. One conventional measure of postorbital constriction, the ratio of minimum frontal breadth to biorbital chord length, is approximately 73.6 in the Gongwangling cranium. This compares to 76.1 in KNM-ER 3733, 72.9 in KNM-ER 3883, and 71.5 in OH 9 (Rightmire 1990). This is a much lower value than recorded at Zhoukoudian or elsewhere in Asia. In this respect Gongwangling is more similar to early forms of *Homo* from Africa rather than later Asian material attributed to *H. erectus*.

Another measure of postorbital constriction, commonly used by Chinese workers, is an index relating total breadth across the supraorbital tori to minimum postorbital breadth measured at the temporal fossae. These dimensions in Lantian are estimated to be 123 mm and 94 mm respectively, yielding an index of 76.4. This index at Zhoukoudian varies between 80.7 and 82.9. It is 91.0 at Hexian and 85.1 at Dali. The Lantian calvarium, therefore, has a degree of postorbital constriction more comparable to that of early specimens of *Homo* from eastern Africa rather than later specimens of archaic and premodern human from Asia.

The robusticity of the supraorbital structures of the Gongwangling frontal surpasses that of most other specimens of archaic Asian hominid. The supraorbital torus, while thick and continuous through glabella, has a depression in its central anterior portion and the brow ridges arch superiorly to a greater extent than typical among Zhoukoudian specimens. They are thickest at mid-point (19 mm) and thin quickly and appreciably laterally (14 mm). The entire torus can hence be divided into three components: two lateral bodies and a glabellar prominence. X-rays show the absence of a frontal sinus (Woo 1966).

The cranial vault is extraordinarily thick (16.0 mm at bregma, 11.5 mm on the temporal squama, 15.0 mm on the frontal squama), surpassing the same measures at Zhoukoudian. Cranial capacity, estimated at 780 cc, is well below the range of variation sampled at Zhoukoudian (Woo 1966).

In contrast to the primitive features of the frontal enumerated above, the pyramidal process of the Gongwangling temporal shows a more modern configuration than seen in the Zhoukoudian sample. The body of the process is slender, approaching the condition seen in modern females; the anterior and posterior surfaces of the pyramid are steep and there is a sharp margo superior (Woo 1966). According to Weidenreich, the condition observed at Zhoukoudian differs considerably. In the Zhoukoudian pyramid the margo superior "is an obtuse or completely rounded corner with both of the adjacent surfaces pressed down toward the floor of the (cranial) fossa" (Weidenreich 1943:67). Weidenreich takes note of the steep-sided petrous pyramid of modern humans, which has a well-developed arcuate eminence along its anterior surface. At Zhoukoudian the anterior surface of the pyramid is flatter and more even, lacking the clear expression of an arcuate eminence. In this character, the Lantian temporal fragment appears to manifest a more modern morphology (Woo 1966). The presence of a more modern-looking temporal morphology in the Gongwangling specimen at first seems incongruous given the overall primitiveness of the remainder of the cranium. Clarke (1977, 1990), however, documents an "advanced" temporal morphology in early members of the genus *Homo* from southern and eastern Africa dating as far back as 1.5 MYA. Among other features, the derived petrous morphology of the African remains (i.e., "the posterior surface of the petrous is vertical and partly undercut by the sigmoid sulcus" [Clarke 1990:729]), relative to the Zhoukoudian and Javan finds (which are said to evince a more archaic anthropoid pyramidal structure) leads Clarke (1990) to question the inclusion of the African specimens in the hypodigm of an Asian delimited *H. erectus*. The morphological description of the Gongwangling petrous

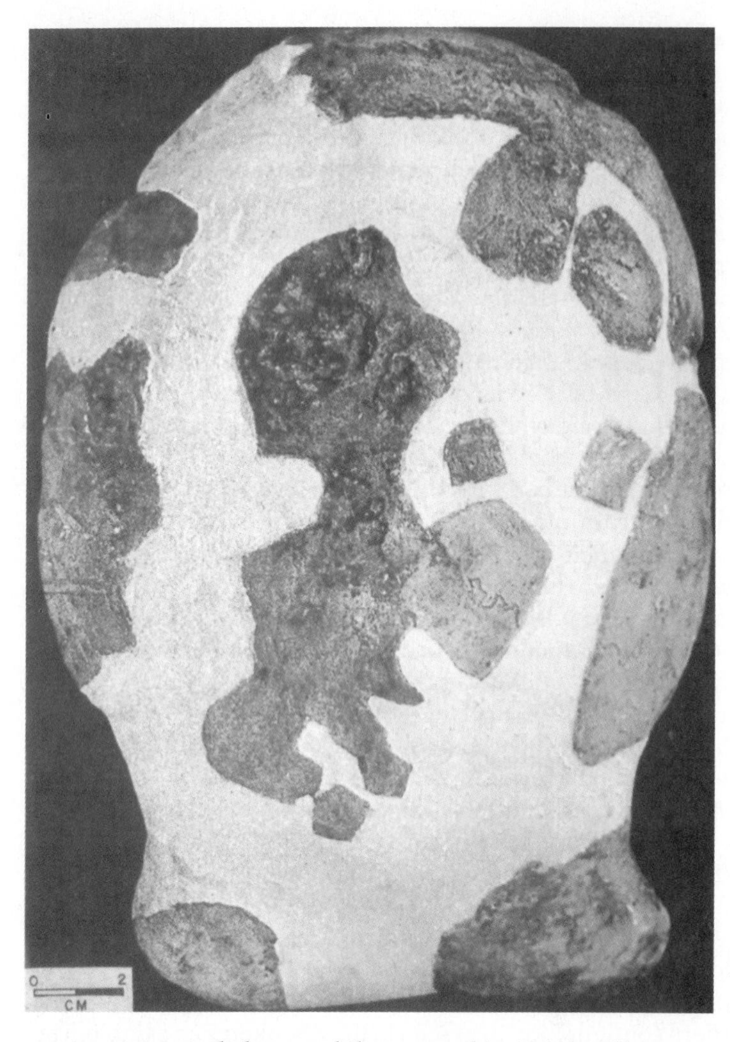

FIGURE 28-3
The Yiyuan Partial Calvaria
(Courtesy Dr. Ho Chuan Kun).

portion presented above, while reported in 1966 by Woo, was only briefly noted in his English-language text of the same year and until now has not drawn any mention in the Western literature. The Gongwangling temporal is not complete enough to ascertain whether it has other modernlike morphologies homologous to those mentioned by Clarke (1990) in early to middle Pleistocene "*erectus*"-like specimens from Africa. It is, however, not inconceivable that early Asian specimens such as Gongwangling may well have retained other modern-looking "derived" traits seen first in early Pleistocene African *Homo* and later lost by "reversal" in subsequent Asian hominids. The presence of such traits in early hominids from Asia may in fact herald the onset of human occupation in this area of the world. Caution must, therefore, be advised when evaluating the presence or absence of such supposedly derived traits in the human fossil record, particularly when potentially polymorphic features are utilized to define distinct hominid taxa.

As regards its facial aspect, the nasal bones of Gongwangling are short and broad. The nasal root is preserved on the right side and the nasal root and a portion of the nasal bridge are preserved on the left side. The frontal processes of both maxillae are also preserved. The entire course of the nasofrontal suture together with the frontomaxillary suture is hence observable. It is horizontally oriented, curving upward slightly toward the midline. The nasal bridge of Gongwangling is slightly higher than in the Zhoukoudian hominid but still quite a bit lower than in modern humans (Woo 1966). Least width of the nasalia is not preserved, but from what remains of the left nasal bone, it can be inferred that least width across the nasal bridge was not appreciably narrower than greatest width across the nasal root (Wu 1966). The nasalia do not narrow along their preserved length. This is similar to the condition at Zhoukoudian and differs from modern humans in which the nasalia tend to narrow at their mid-points (Woo 1966).

The maxillary corpus of the Gongwangling cranium is relatively slender. The maxillary antrum is similar in size and extent to Zhoukoudian specimens and modern humans as discussed by Weidenreich (1943:75, see also, Woo 1966). The anterior face of the maxilla is relatively intact. The clivus is convex as in the Zhoukoudian material, but there is a slightly greater degree of alveolar prognathism (Woo 1966). This differs from the more concave clivus seen in some early Plio-Pleistocene forms of *Homo* from eastern Africa. The junction of the inferior border of the nasal aperture with the alveolar process of the maxilla is marked by a relatively sharp angle, unlike the open gutter seen in the great apes and *Australopithecus*. There is also a clear indication of a small but distinct anterior nasal spine, a feature poorly expressed in archaic human specimens from Zhoukoudian (Woo 1966).

What is known of the Gongwangling mid-face is structurally primitive. The lateral facies of the maxilla has a strongly expressed canine jugum. There is a broad, shallow maxillary sulcus rather than a well-expressed canine fossa inferior to the infraorbital foramen. The right maxillary fragment preserves the root of the zygomatic process, which is positioned low on the maxillary body near the alveolar margin. The lower margin of the zygomaxillary border is oriented obliquely rather than horizontally, differing from the more modern condition seen in later examples of archaic human from China. In all these features the Lantian Gongwangling maxilla is phenetically more similar to early *Homo* from Africa (e.g., KNM-ER 3733 and KNM-WT 15000) rather than later archaic, premodern, and modern human specimens from China.

In summary, the Gongwangling cranium preserves a set of features relating to the construction of the supraorbital torus, the degree of postorbital constriction, the discrete morphology of the temporal, and the structure of the mid-face, highly reminiscent of early east African humans. These features are, however, combined with remarkably thick cranial vault bones and a relatively well-expressed cranial reinforcement system characteristic of later occurring hominids in Asia. Given the new evidence reviewed here supporting a late early Pleistocene age for the Gongwangling remains, a good case can be made that the specimen represents an early incursion of an African-derived hominid into an east Asian milieu.

THE CHENJIAWO HOMINID

The Chenjiawo mandible (PA 102), discovered in July 1963, is thought to belong to an older female (Woo 1964a,b). It is distinguished by the agenesis of M3, a somewhat less vertical symphysis than typical at Zhoukoudian, a greater degree of posterior divergence along the molar row and a disproportion in the height of the corpus, with a shallowing in the area of the mental foramina.

In overall character the Lantian mandible diverges little from the morphological pattern set at Zhoukoudian. Those differences enumerated by Woo (1964a,b) do not have much if any phylogenetic weight. The Lantian mandible in a number of angular and metric variables, however, falls outside the range of variation seen at Zhoukoudian (Woo 1964).

QIZIANSHAN, YIYUAN, SHANDONG

The Yiyuan fossil human site is located in the Tumen karst region of central Shandong 7.5 km NW of Zhifang Village (E 118° 8′, N 36° 14′). To the north, west, and south are metamorphic highlands. The eastern margin of the region is bounded by the upper Wujing Rift System. The areal extent of the Tumen karst is approximately 50 sq km. Within these confines there are over 40 caves of various sizes making it the most concentrated area of karst cave formation in central Shandong. In 1981, a fragmentary human skullcap, several human teeth, a number of stone artifacts, and mammalian fossils were discovered in middle Pleistocene limestone fissure fills and cave deposits of the Xiaya cave group along the foothills of

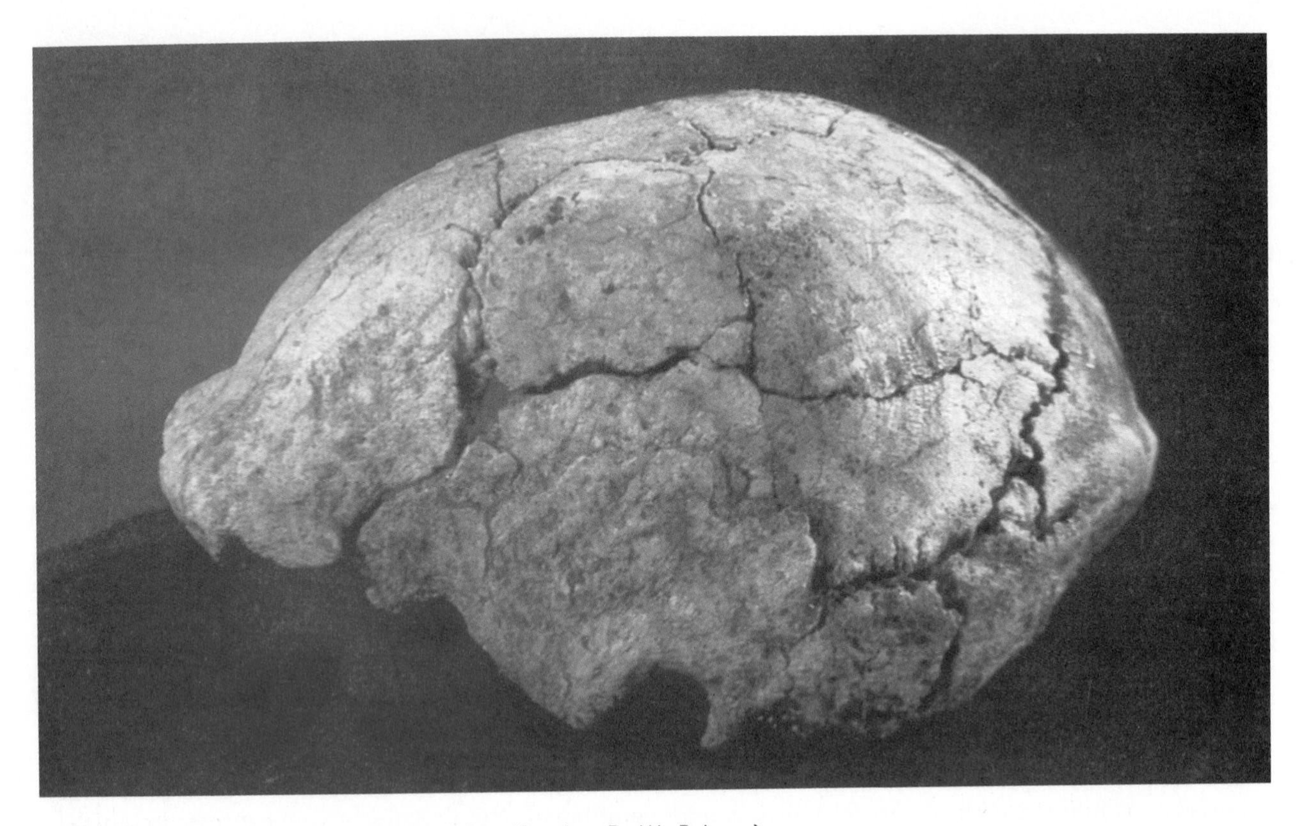

FIGURE 28-4 The Hexian Calvaria, Lateral View (Courtesy Dr. Wu Rukang).

the Qizian Mountains. The hominid material and most of the faunal remains come from a secondary depositional context; no hominid occupation site has yet been found.

The mammalian fauna recovered at Qizianshan comes from the two hominid-bearing localities as well as three other localities at the same stratigraphic level. The fauna include *Megaloceros pachyosteus, Sus lydekkeri, Cervus* sp., Bovinae, *Equus sanmeniensis, Dicerorhinus merki, Hyaena* sp., *Canis variablis, Ursus arctos, U. thibetanus, Panthera tigris, Trogontherium* sp., and *Macaca robustus*, all typical components of the Locality 1 Zhoukoudian fauna. The fossils are relatively fragmentary and show clear signs of fluvial transport. The most abundant material belongs to *U. arctos* (a well-preserved left dentary and 28 isolated teeth), *S. lydekkeri* (two fragmentary mandibles and 30 isolated teeth), *M. pachyosteus* (two antler fragments and 15 upper and lower jaw fragments) and *E. sanmeniensis* (7 cheek teeth, a lower incisor, 1 metacarpal). The material is said to be metrically and morphologically indistinguishable from similar taxa recovered at Zhoukoudian Locality 1 (Lü et al. 1989).

THE YIYUAN HOMINID

The fossil-bearing beds at Locality 1 were discovered in September 1981 when an archeological survey team of the cultural section of the Xiyuan County Library was directed to the site by a local resident. Soon thereafter the left and right parietals of a single individual and several cranial pieces including two supraorbital torus fragments, a portion of a frontal, and a portion of an occipital were recovered. Together with the parietals, the majority of cranial fragments can be reconstructed into a partial calvarium. Differences in the degree of robusticity and coloration of the supraorbital fragments, however, suggest that they represent two individuals. The cranial specimens are crushed and deeply weathered. Subsequent to these initial discoveries, a joint investigation team of the Shandong Provincial Museum, the archeological section of the provincial Cultural Bureau, and the Yiyuan County Cultural Department conducted excavations at two of the fossiliferous localities (Loc. 1 and 3) in November of the same year, recovering five fossilized human teeth and associated vertebrate fossils. In 1982, further work at the site by the provincial museum and the Archeology Department of Peking University led to the discovery of the distal portion of a humerus, a femoral head, one rib, and two fossilized human teeth. The cranial and dental specimens have been recently described in a preliminary fashion by Xu (1986) and more completely by Lü et al. (1989).

The skull fragments (Sh.y. 001) have been reconstructed into a partial calvaria (Figure 28-3). Internally the endocranial surface is smooth and sutures are fully fused. Little of the morphology of the inner surface of the cranium can be discerned. The external surface preserves the coronal suture, which is of simple character and nearly obliterated. A small portion of the sagittal suture is visible at lambda. The lambdoidal suture is more complex. The amount of sutural closing suggests an adult individual. Because of the fragmentary nature of the cranial remains and exfoliation of the external bone shelf, the degree of sagittal keeling is hard to determine. It is clearly discernable, however, anteroposterior to bregma. Temporal lines are indistinct on either parietal. Cranial thickness at bregma is 9 mm, well within the range of variation at Locality 1, Zhoukoudian (7-10) and greater than at Xujiayao (8.5) and Maba (7.0). Thickness in the vicinity of asterion is 13, again close to the Zhoukoudian mean (Lü et al. 1989).

The supraorbital fragments (Sh.y. 002.1-2) consist of portions of the left and right supraorbital tori. Sh.y. 002.1 is a nearly complete left torus that preserves a portion of the temporal fossa at the anterosuperior angle of the greater wing of the left sphenoid. The torus is broken off medially along a line extending from its inner wall toward glabella. Laterally a small portion of a robust zygomatic process is preserved. The torus is both long and thick. It has a medial thickness of 13 mm, a median thickness of 12 mm, and a lateral thickness of 16.5 mm. Compared to other specimens of archaic human from China, the Yiyuan specimen is most similar to tori from Zhoukoudian, which share a similar tendency toward greatest toral thickness laterally. It differs from the Hexian specimens, which show greatest thickness medially (PA 830) or medianly (PA 840). The left toral fragment lacks a supraorbital foramen but preserves a supraorbital notch as at Zhoukoudian and Hexian. What remains of the frontal squama is relatively low and flat. A supratoral sulcus is clearly expressed. Thickness of the squama at the sulcus is 11 mm, close to the condition seen at Zhoukoudian. This measure in Dali is considerably greater (18) due to its lack of development of a well-expressed supratoral sulcus. Although the frontal posterior to the torus is mostly lost, the degree of postorbital constriction can be approximated. It appears to be close to that seen at Zhoukoudian and far more expressed than in Hexian or Dali. The roof of the orbit is only slightly curved (Lü et al. 1989).

The right toral fragment (Sh.y. 002.2) is preserved from its mid-portion to the zygomatic process. It is similar in character to the left fragment. Its median thickness is 12 mm, and its lateral thickness is 14.7 mm. It has a similar coloration to the cranial fragments and may represent the same individual. The left toral fragment is more deeply colored and thicker and is thought to represent another individual. In sum, the preserved features of the cranium show closest affinities to similar remains from Zhoukoudian.

The seven well-fossilized dental specimens include an upper left P3, upper right P3-M1, lower C, and a lower M1 or M2 allowing for a good characterization of the Yiyuan material vis-à-vis other archaic dental remains from China. Based on size, morphology, degree of wear and coloration, they are thought to represent at least two adult individuals. Specimens Sh.y. 003 (RP3), Sh.y. 004 (LP3), and Sh.y. 071 (LP4) clearly represent one individual. In terms of overall morphology, the Yiyuan specimens are closest to those previously recovered from Zhoukoudian, to the extent that detailed comparisons can be made. They differ somewhat more from specimens recovered from Yunxian, Xichuan, and Hexian. Of particular interest is the relatively great breadth of the upper dentition, said to be an unusual feature of the Yiyuan specimens (Lü et al. 1989). In overall character, however, the Yiyuan remains are remarkably similar to those from Zhoukoudian.

LONGTANDONG, HEXIAN COUNTY, ANHUI

The Longtandong site in He county (Hexian), Anhui has produced the most prolific fossil hominid remains unearthed in China since the height of excavations at Zhoukoudian in the 1930s (Huang et al. 1981, 1982; Wu and Dong 1982; see review in Etler 1990). An inventory of recovered specimens includes a nearly complete calvaria (PA 830), the supraorbital portion of a frontal (PA 840), a right parietal (PA 841), a fragment of a left mandibular corpus retaining M2 and M3 (PA 831), a right upper P4 (PA 832), a left upper M2 (PA 833), left lower M1 and M2 of one individual (PA 834[1-2]), a right upper I1 (PA 835), a left upper M1 or M2 (PA 836), a right upper M2 (PA 837), and two left lower M2's (PA 838 and PA 839).

The site was discovered in 1973 when it was blasted open during canal construction. The cave was found to be richly fossiliferous. In 1979 at the request of the Anhui Province Water Conservancy Bureau, a team led by Huang Wanpo was dispatched by the IVPP to investigate the site. Excavations were resumed in January of 1980. In July, the first human fossil, an upper molar, was recovered from previously excavated back dirt within the cave. Later that fall during the third excavation, the calvaria and other human fossils were found.

Longtandong is located on the northern slope of Wangjia Mountain, 30 km northwest of the Yangtze River at an elevation of 23 m asl. (E 118° 20′, N 31° 45′). The deposits within the cave can be divided into 4 layers. The hominid material comes from the second stratigraphic level, which consists of a 0.5–1.0 m tawny clay (Huang et al. 1981, 1982). The Hexian fauna is biogeographically transitional, containing both Palearctic and Subtropical elements. The micro-mammalian component shows considerable faunal mixing with northern, southern, and "alpine" species from the western montane regions of China all sampled (Zheng 1982; Xu and You 1984). Palynological sampling at Longtandong likewise indicates the presence of northern and southern taxa (Huang and Huang 1985).

The Longtandong fauna has been correlated with the upper layers at Zhoukoudian and oxygen isotope Stage 8, which give an age of 240–280 KYA (Xu and You 1984). Recent uranium series dating of bovid teeth associated with the human remains, however, indicates an age of between 150–190 KYA for the hominid-bearing strata, somewhat younger than the upper layers at Zhoukoudian (Chen et al. 1987; Chen and Yuan 1988). As such, the Hexian remains would be the youngest demonstrable evidence of a *H. erectus* in Asia and coeval with remains of more advanced premodern humans now known from China (see Etler 1990). The efficacy of using Uranium-series tests to date enamel recovered from paleontological contexts is, however, questionable as enamel is prone to exchange uranium within its depositional environment (Schwarts 1993). The dating of the upper layers at Zhoukoudian in excess of 400 KYA by Shen and Jin (1991), based on tests conducted on calcite samples, suggests that the true age of the Hexian remains may be considerably older than at first indicated. Given the vagaries of uranium series dating and the contradictory results obtained at Zhoukoudian, it is premature to suggest that the Hexian remains represent a late occurring *H. erectus* contemporaneous with more advanced Chinese hominids such as those recovered at Dali and Yingkou (Jinniushan).

THE HEXIAN HOMINID

The Hexian cranium (Figure 28-4) has been studied in detail by Wu and Dong (1982). When first found, it was severely fragmented. Reconstruction produced a relatively intact specimen that retains all bones of the cranial vault and a substantial part of the cranial base. The vault between the temporal lines anterior to the vertex and posterior to bregma was apparently flattened during fossilization (Wu and Dong 1982). Cranial capacity is estimated to be approximately 1050 cc.

The robust construction of the calvaria, the thick and well-developed supraorbital and occipital tori, the well-expressed temporal and nuchal lines, and the sloping forehead combine to suggest a male specimen. It is judged to be a young adult as both external and internal sutures are open (Wu and Dong 1982).

TABLE 28-4 Comparative cranial metrics: Hexian calvaria and other archaic Asians

Martin No.	Measurement	Hexian	ZKD			Java			Ngandong		
			n	x	SD	n	x	SD	n	x	SD
1(2)	Maximum length: g-op(i)	190	6	197	8.7	4	184	6.3	6	202	9.4
5(1)	Nasion-opisthion line: n-o	131?	3	145	1.5	3	139	5.1	6	154	7.4
8	Maximum breadth: eu-eu	160	4	141	3.2	4	138	7.2	6	146	7.0
8c	Temporo-parietal breadth	145	4	136	2.8	2	129	3.5	4	145	3.6
9	Least frontal breadth: ft-ft	93	6	87	4.1	3	82	3.0	5	103	3.8
11	Biauricular breadth: au-au	144	5	146	4.1	3	126	10.3	5	148	5.0
12	Biasterionic breadth: ast-ast	142	6	114	6.9	4	120	22.2	6	126	1.5
20	Auricular-bregmatic height: po-b	95	4	99	5.3	4	94	5.5	6	109	5.1
23	Max. horizontal circ.: gi(op)	571	4	564	12.3	2	526	0.7	-	-	-
24	Auriculo-bregmatic arc: po-po(b)	291	4	287	15.6	4	266	6.9	6	308	11.5
25	Median sagittal arc II: no	340?	3	330	8.2	2	303	1.4	6	356	13.3
26	Nasion-bregma arc: nb	120?	5	123	5.0	2	104	4.9	6	130	6.1
27	Bregma-lambda arc: bl	110	5	102	9.7	3	93	2.1	6	108	6.0
28	Lambda-opisthion arc: l o	110	3	114	6.9	4	105	5.4	6	118	6.7
29	Nasion-bregma chord: n-b	99?	5	110	5.5	3	93	6.1	6	117	4.0
30	Bregma-lambda chord: b-l	103	5	96	8.6	3	90	2.4	6	102	5.0
31	Lambda-opisthion chord: l-o	83	4	84	3.0	4	79	3.5	6	86	4.8
32a	Frontal profile: m-g-i (angle)	58	5	59	4.3	2	51	5.3	6	62	4.4
32(2)	Inclination of frontal squama: b-g-i (angle)	41	5	43	2.9	2	40	3.2	6	46	1.3
33(4)	Occipital curvature: l-i-o (angle)	101	5	103	3.3	3	109	6.0	6	99	4.3
8/1(2)	Length-breadth index	84	4	73	0.8	4	75	2.7	6	71	2.9
20/1(2)	Length-height (po-b) index	50	4	51	1.8	4	51	3.1	6	54	2.1
20/8	Breadth-height (po-b) index	59	4	71	2.8	4	69	5.1	6	74	3.4
5(1)/25	Sagittal cranial curvature	39?	3	44	0.7	2	46	2.1	6	43	1.8
11/24	Transverse cranial curvature	50	4	51	2.7	3	48	5.0	6	52	3.4
9/8	Transverse fronto-parietal index	58	4	62	2.3	3	61	3.9	-	-	-
12/8	Transverse parieto-occipital index	89	4	81	3.2	4	87	11.9	-	-	-
27/26	Fronto-parietal (arc) index	92?	5	84	7.1	2	90	2.2	6	83	2.7
28/27	Parieto-occipital (arc) index	100	3	120	11.9	3	110	3.1	6	110	8.9
29/26	Frontal curvature	83?	5	90	1.2	2	92	11.2	6	90	1.7
30/27	Parietal curvature	94	5	94	1.1	3	97	0.4	6	95	1.3
31/28	Occipital curvature	76	3	74	1.5	4	76	2.2	6	73	1.8
	Cranial capacity	1025	6	1059	108.0	4	879	86.0	6	1096	86.0

The cranium preserves many characteristics commonly seen in archaic Asian hominids generally attributed to "*H. erectus*." It is low vaulted and elliptical in shape. Opisthocranium falls on the midpoint of the occipital torus. The temporal lines are well expressed and terminate in a distinct torus angularus at the mastoid angle of the parietal. The temporal squama is relatively high and arched superiorly at the parietal margin unlike the condition commonly seen at Zhoukoudian, Locality 1 (with the exception of Skull V) in which the squamosal margin is low and straight. The left squamous portion is 70 mm long and 42 mm high, yielding a length/height index of 60, compared to an average of 49.9 at Zhoukoudian, 64.6 in the Dali cranium, and 65.2 in modern humans. The parietal notch separating the temporal squama from the mastoid region is deeply incised. The Hexian temporal squama, therefore, approximates that of modern humans in shape but preserves the large overall size of archaic hominids. The Hexian cranium preserves other features of the vault commonly seen at Zhoukoudian, including a wide and shallow sulcus processus zygomatici, the configuration of the external auditory meatus, and the overhanging suprameatal tegmen, a posterosuperiorly slanted root of the zygomatic process of the temporal and well-developed supramastoid and mastoid crests. There is also a well-expressed supramastoid sulcus. The digastric groove is broad and shallow, opening up posteriorly, differing from the deeper and more narrow incisura mastoidea seen in modern humans. What is preserved of the mammillary process of the mastoid is relatively small, as in many specimens of archaic Asian hominid, and lies below the level of the supramastoid crest.

The frontal has a low and sloping squamous portion. The frontal angle measures 58°, the glabello-bregmatic angle measures 41°. These values are comparable to those seen at Zhoukoudian but considerably less than at Dali in which the above angles measure 72° and 50° respectively.

The supraorbital torus is wide laterally and thick inferosuperiorly. Although the superior margin of the glabellar region is slightly depressed, the left and right supraorbitals unite into a single body. Greatest thickness is medial and least thickness is lateral. The frontal squama above the supraorbitals has a pronounced "bumplike" protuberance similar to Zhoukoudian (Weidenreich 1943:225), differing in this respect from archaic Javanese specimens, which show a flat frontal squama. There is a distinct supratoral sulcus in the Hexian calotte, but it is much more weakly expressed than in analogous specimens from Zhoukoudian.

The frontal eminences are only slightly expressed, similar in degree to male specimens from Zhoukoudian. A metopic suture is evident from a point level with the frontal eminences to bregma. An unfused metopic suture is also seen in Skull XI from Zhoukoudian.

Least frontal breadth (ft-ft) measures 93 mm, greatest frontal breadth (co-co) 118 mm, yielding a frontal breadth index of 78.8, within the range of other archaic Asians. The degree of postorbital constriction can be gauged by an index that measures maximum postorbital constriction (minimum breadth of the frontal behind the orbits) against the distance between the outer edges of the brow ridges (101 mm and 111 mm respectively in the Hexian specimen)(Wu and Dong 1982). In the Hexian specimen, this index stands at 91 versus a range of 80.7-82.9 in specimens from Zhoukoudian and 85.1 for the Dali cranium. This relatively minor degree of postorbital constriction is a distinguishing feature of the Hexian calvaria.

A sagittal keel is apparent along the median plane of the skull from a position level with the frontal eminences to bregma where it gradually fades out. The sagittal keel at Zhoukoudian is more prominent, extending from a position level with the frontal eminences to the obelion region of the parietals. This robust construction of the vault is further enhanced by the presence of a cruciate eminence at the junction of the coronal and sagittal sutures. The Hexian specimen lacks this latter structure, as well as parasagittal depressions commonly seen in specimens of archaic hominids from Zhoukoudian and Java (Wu and Dong 1982).

Greatest cranial width is low on the skull, in line with the supramastoid crests. The parietal eminences are, nevertheless, well expressed. The torus occipitalis is well developed and there is a clear supratoral sulcus. The occipital angle separating the occipital and nuchal planes is sharp. There is a large intersutural bone on the right side at asterion (Wu and Dong 1982).

As in specimens from Zhoukoudian, the mandibular fossae are deep and narrow and the articular eminences are weakly expressed (Weidenreich 1943:47). The size of the tympanic plate, its position in relation to the mid-sagittal plane of the cranium, and its spatial orientation is likewise similar to Zhoukoudian (Weidenreich 1943:52-57). The angle between the tympanic plate and the petrous process is, however, more acute in the Hexian specimen (30°) than in specimens from Zhoukoudian (50° in Skull III). In Hexian the petrous process is positioned less obliquely in relation to the midsagittal plane than in material from Zhoukoudian (60° vs. 40°), falling within the upper range of variation seen in modern human samples (38°-63°)(Weidenreich 1943). Both a styloid process and vaginal sheath are lacking. The occipitomastoid crest is not as well expressed as at Zhoukoudian (Wu and Dong 1982).

The endocranial surface of the Hexian calotte preserves a low, broad frontal crest. It divides into two branches, separated by a sagittal sulcus along the lower 9 mm of its course. (Wu and Dong 1982). This is similar to the condition in Zhoukoudian Skull III as reported by Weidenreich (1943:32). There is no foramen caecum, confirming the absence of this feature in the Zhoukoudian material as noted by Weidenreich (1943:34).

The ramification of the middle meningeal artery follows the pattern set by *H. erectus pekinensis* in which impressions of the posterior branch (ramus temporalis) are somewhat more robust than the anterior branch (ramus fronto-parietalis) and the subbranches from both are much less abundant than in modern humans.

The internal surface of the occipital shows the typical division of the upper (cerebral) and lower (cerebellar) occipital fossae by a clear cruciate eminence. As in other specimens of archaic Asian humans, the cerebellar fossae are distinctly smaller than the cerebral fossae, the reverse of the condition in modern humans. At Zhoukoudian, the ratio of the area of the cerebral to cerebellar fossae is more or less 2:1, while in Hexian it is closer to 4:3, indicating an expansion of the relative area of the cerebellar fossae in relation to the cerebral fossae and a closer approximation to the modern human condition. There is a 22-mm separation of the internal and external occipital protuberances from each other. This is clearly greater than in Dali (11 mm) but less than in Zhoukoudian (range 27.5–38.0, x = 32.8). In modern humans these structures are coincident (Wu and Dong 1982).

Measurements of the Hexian cranium in comparison to other archaic humans are given in Table 28-4. A number of these measurements are close to those for specimens from Zhoukoudian and Java. These include cranial length (g-op[i]), cranial height (po-b), inclination of the frontal squama (b-g-i angle), horizontal curvature of the cranium (biauricular breadth/auriculobregmatic arc), frontoparietal breadth index (minimum frontal breadth [ft-ft]/maximum cranial breadth [eu-eu]), and parietal and occipital curvature. In a number of features, Hexian differs from material from Java and is more similar to material from Zhoukoudian, that is, measures of biauricular breadth (au-au); cranial circumference, transverse arc of the cranium, sagittal arc of the cranium, inferior inclination of the frontal, occipital angle, and cranial capacity. The only index in which Hexian differs from specimens of *H. erectus* at Zhoukoudian and is more similar to specimens from Java is the parietooccipital breadth index. This index demonstrates the relatively larger breadth of the occiput in the Hexian cranium when compared to crania from Zhoukoudian and the overall spherical shape of the cranium versus the more elliptical shape characteristic of material from Zhoukoudian. Relatively progressive metric features of the Hexian cranium include an increase in minimum frontal breadth, a greater degree of sagittal and frontal curvature, and a relatively low value for the parieto-occipital arc index (indicating that the parietal and occipital arcs are nearly equal to one another versus the greater length of the occipital arc in most archaic human fossils from Asia and the greater length of the parietal arc in modern humans). In addition the Hexian cranium shows a much greater maximum breadth than either the Zhoukoudian or Javanese material and its length/breadth index

FIGURE 28-5 The Yunxian 1 Cranium, Lateral View. (Courtesy Prof. Li Tianyuan)

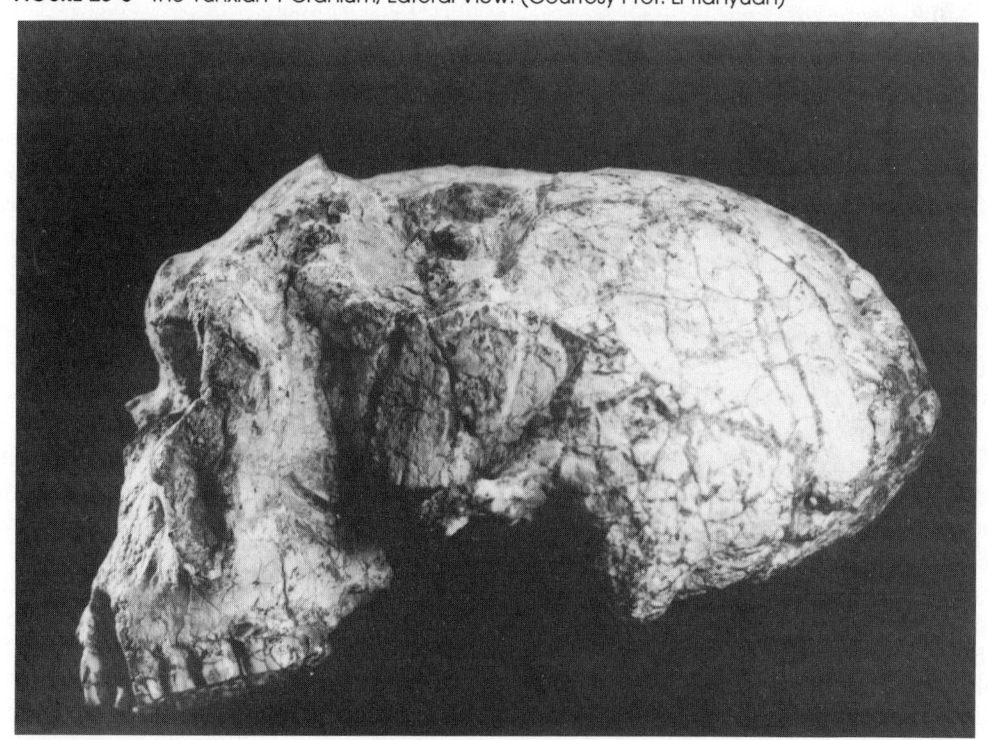

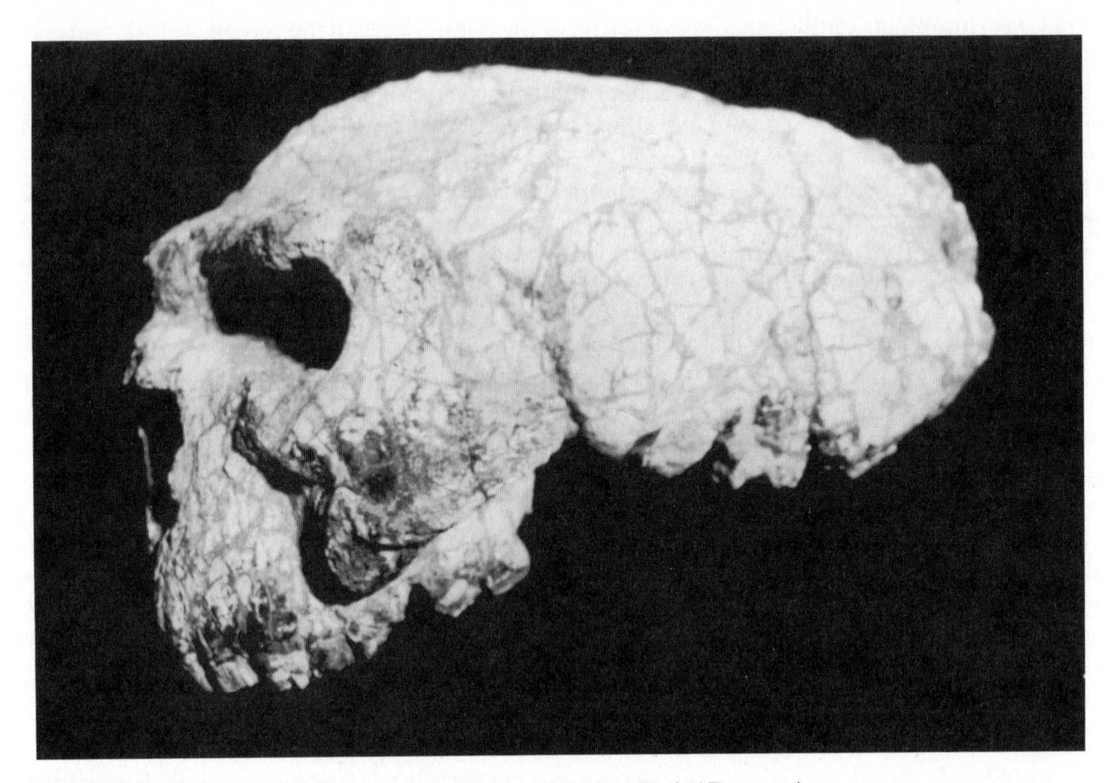

FIGURE 28-6 The Yunxian 2 Cranium, Lateral View. (Courtesy Prof. Li Tianyuan).

falls within the brachycranic range, rather than the dolichocranic range of most of the Zhoukoudian crania or the mesocranic range of the Javanese specimens.

The thickness of the cranial vault bones at various points along the cranium is 7.0 mm at the median point of the frontal squama, 13.5 mm at the left parietal eminence, 18.0 mm at the mastoid angle of left parietal, 18.0 mm at the midpoint of the occipital crest, 6.0 mm at the cerebellar fossa of the occipital, and 10.0 mm at the median point of the temporal squama. These values demonstrate that cranial vault thickness is overall less than in the Zhoukoudian material and much less than in Gongwangling.

Other cranial remains from Hexian include a supraorbital fragment (PA 840) described by Wu (1983). It preserves much of the right torus and a small portion of the frontal squama immediately superior to it. Medially it is broken at glabella. The nasal process of the frontal is preserved as is a small portion of the roof of the ethmoidal sinus. The supraorbital torus is strongly developed, although it is not quite as robust as PA 830. It differs from PA 830 and homologous Zhoukoudian material in that the thickest part of the torus is situated along the supraorbital midline rather than further medially. Although most of the frontal squama is lost, it is clear that postorbital constriction is less than in crania from Zhoukoudian. In like manner, the supratoral sulcus is reduced in both PA 830 and PA 840 differing in this respect from the Zhoukoudian crania.

The right parietal fragment (PA 841) described by Wu (1983) is 40 mm long and 60 mm wide. It preserves the area around the parietal eminence and a small portion posterior to the parieto-temporal suture. The wall of PA 841 is relatively thick, 11 mm in the vicinity of the parietal eminence, well within the Zhoukoudian range (5.0-16 mm).

The left mandibular corpus fragment (PA 831) retains wellworn M2 and M3 in situ. The corpus is extremely robust and particularly thick and is most likely male. There are three mental foramina positioned between P4 and M1. Corporal height between M1 and M2 is 32.0 mm; corporal thickness at the same position is 20.7 mm, yielding a robusticity index of 64.7. This is greater than in all Zhoukoudian and Javanese specimens (both male and female), and the Chenjiawo mandible from Lantian.

The dental remains from Hexian are remarkable for their overall robustness and large size. In nearly all instances (except M3), they approach or exceed the upper limit of the range of metric variation for previously known archaic dental specimens from Asia. Coupled with the large size of dental remains and the extreme thickness of cranial vault bones from the upper Pleistocene site of Xujiayao in Shanxi Province, it is apparent that mere measures of overall size and robusticity do not serve as good phylogenetic indicators among east Asian Pleistocene hominids.

Teeth recovered from Hexian include the in situ left lower M2 and M3 from the PA 831 mandible, a right upper P4 (PA 832), a left upper M2 (PA 833), and left lower M1-M2 (PA 834[1-2]) from the 1980 excavation and a right upper I1 (PA 835), a left upper M1 (PA 836), a right upper M2 (PA 837), and two left lower M2's (PA 838 and PA 839) recovered during the 1981 excavation.

The central incisor is particularly large, exceeding even the Yuanmou incisors in MD length and BL breadth. Its overall size is most comparable to the Krapina Neandertals. The specimen is clearly shovel-shaped with a broad incisive margin and thickened lateral margins which turn inward toward the lingual surface of the tooth. The labial face of the crown has three prominent longitudinal swellings. Lingually there is a pronounced basal tubercle which is confluent with the lateral margins. There is a central lingual fossa, but it is not as well expressed as in the Yuanmou specimens. Four fingerlike projections of various lengths emanate from the basal tubercle and terminate in the central fossa. The root is extremely robust and conical. There are distinct broad and shallow mesial and distal vertical grooves.

The other teeth recovered at Hexian have complexly wrinkled occlusal surfaces but lack developed cingula. The cingula of upper M2 and lower M1 are, however, relatively well expressed. The upper P4 is again very large and robust, well exceeding means for MD length and BL breadth seen for this tooth in samples of *H. erectus* from Asia. The lower M1 and M2 (PA 834 [1-2]) are well worn and probably represent a single middle-aged individual. The upper and lower molars, besides their large size, are morphologically consistent with other archaic dental specimens from China.

In overall character, the Hexian specimens conform to the pattern set at Zhoukoudian, diverging only in certain respects. The cranium is distinguished by its greater breadth, transverse expansion of the frontal squama and reduced postorbital constriction. The supraorbital region also differs from Zhoukoudian, being more similar in many respects to specimens from Sangiran and Ngandong in Java. The morphology of the Hexian temporal is more modern than seen at Zhoukoudian, approximating more or less the pattern seen in later hominid specimens from China. Of particular significance is the large size of the Hexian dentition and the robusticity of the Hexian mandibular remains, which fall well beyond the limits of variation sampled at Zhoukoudian.

OTHER ARCHAIC HUMAN REMAINS FROM CHINA

Archaic dental remains attributed to *H. erectus* are now known from a number of sites in China. These include Yuanmou, Yunnan (Hu 1973); Wushan, Sichuan (Huang and Fang 1991); Nanzhao, Henan; Luonan, Shaanxi; and sites in Yunxian and Yunxi counties, Hubei (Wu and Wu 1982). Similar teeth have also been collected from medicinals warehouses in Xichuan county, Henan. Yunxian and Yunxi are located in northwestern Hubei while Nanzhao and Xichuan are located in southwestern Henan. They are all situated on the western margin of the North China Plain between the Huanghe (Yellow River) and Changjiang (Yangtze River), south of the divide between the northern (Palearctic) and southern (Subtropical) Chinese biogeographic zones.

A total of 25 archaic human teeth have been collected from medicinal warehouses or recovered in situ from the Nanyang region of southwestern Henan and the Yunxian region of northwestern Hubei. Of these 18 have been described in the literature.

Taken as a whole, the teeth present a mix of primitive and progressive features, which is in keeping with the wide variation in dental morphology of specimens recovered from Zhoukoudian Locality 1. In a number of characters the variation encountered at Zhoukoudian is increased when the material documented here is taken into account. The large size of a number of specimens suggests that size alone does not effectively discriminate early hominids from one another. Other isolated teeth from Yuanmou, Yunnan (two upper central incisors); Luonan, Shanxi; and Jianshi, Hubei differ little from the material described here although the Jianshi molars are morphologically aberrant.

QUYUANHEKOU, YUNXIAN, HUBEI

In May and June of 1989, two damaged but relatively complete hominid crania were excavated from middle Pleistocene terrace deposits of the Han River in Yunxian (Yun county), northwestern Hubei province (Li et al. 1991; Li and Etler 1992). The two crania are the most complete specimens of this antiquity ever found on the Asian mainland. Of added significance is their overall morphological character, which differs in significant respects from previously recovered fossil hominids in China. In terms of sheer size the Yunxian crania surpass the range of variation seen at Zhoukoudian. While the overall bauplan of the crania is undoubtedly similar to other archaic east Asians, various aspects of their faces, cranial vaults and basicranial anatomy differ considerably from the pattern established at Zhoukoudian.

The Yunxian crania come from highly calcified upper terrace deposits not far from Qingqu village at the mouth of the Quyuan River. The locality is situated on a east/west trending ridge named Xuetangliangzi (Schoolhouse Ridge) atop a high terrace of the Han River. Its summit is approximately 50 m above river level.

The mammalian fossils associated with the human remains include relatively complete crania, limb bones, and isolated teeth of more than 10 genera and species. Major taxa include *Hystrix* sp., *Ursus* sp., *Hyaena licenti, Tapirus* sp., *Rhinoceros* sp., *Sus scrofa, Cervus* sp., and *Bos* sp.

In overall character, the Yunxian faunal assemblage shows similarities to the middle Pleistocene *Stegodon-Ailuropoda* fauna often seen in southern (and central) China (Han and Xu 1985). All genera and species discovered at the site can be found represented in the latter fauna. It differs primarily in the lack of typical members *Ailuropoda* and *Stegodon* (Li et al. 1991).

The new Yunxian locality is 69 km from the cave site of Longgudong, Meipu, which has yielded archaic human dental remains. The Meipu fauna has a number of archaic elements including the Tertiary relic *Gomphotherium*, the lower Pleistocene *Hyaena licenti* and the lower-middle Pleistocene *Sus xiaozhu* (Xu, C. 1978). At present the Quyuanhekou fauna lacks two of these elements and is unlikely to be older than the Meipu fauna. It is accordingly felt that a middle Pleistocene age for the Quyuanhekou fossil locality is reasonable.

Sixty-eight archaeological specimens have been surface collected or excavated *in situ* at the Quyuanhekou locality. The artifacts consist primarily of amorphous cores and bifacial point tools fashioned on cobbles. The lithic assemblage is similar to those recovered at Lantian and other sites in China that are characterized by large tools made on cobbles and lack small flake tool components. Twenty-four of the artifacts were recovered during survey. According to Li et al. (1991) these include 9 cores, 5 flakes, 1 hammerstone, 5 chopping tools, and 1 small point tool. There were, additionally, 10 blocks that show traces of human alteration. Most of these artifacts appear to have been fabricated on cobbles. Another 44 artifacts were found during inspection and excavation of the site. Of these, 23 were surface collected and 21 were recovered *in situ*. Only this latter material was analyzed by Li et al. (1991). Among the 44 specimens collected at the site were 18 cores, 16 flakes, 4 scrapers, 3 chopping tools, and 3 point tools. Of the cores, 7 (EP 9001-9007) were surface collected, while 11 (EP 9008-9018) were excavated. Ten pieces were made on quartzite. All were single platform cores, fabricated on cobbles. Most were struck on the cortical surface. Three point tools reminiscent of points found at Lantian have been described. All were made on quartz cobbles.

THE YUNXIAN HOMINID

Both crania from Yunxian were skillfully prepared by Prof. Hu Chengzhi of the Geological Museum of China (Figures 28-5 and 28-6). Based on the full eruption of the third molar in EV 9001 and the degree of sutural closure in both EV 9001 and EV 9002, it can be assumed that both crania represent fully adult individuals. Although the crania are largely intact, they have suffered varying degrees of postdepositional fracturing and plastic deformation. EV 9001 is so severely shattered, crushed, and altered that precise craniometric measurements are impossible to obtain. Significant amounts of morphological detail are, however, preserved. EV 9002 is in a much better state of preservation. Its facial and basal aspects have been damaged but are largely preserved.

In most important respects the morphological characteristics of the two crania are similar. They are both large with low cranial vault contours. There is no discernable mid-sagittal keeling on the frontal or along the parietals. There is, likewise, no cruciate eminence at bregma. Other ectocranial features commonly associated with archaic Asian hominids, such as a pronounced angular torus at the mastoid angle of the parietal, and a well-expressed supramastoid crest confluent with the anterolateral portion of the occipital torus are poorly expressed.

While the cranial vaults bear a strong resemblance to other archaic Asians, perhaps their most striking feature is the great cranial breadth and overall large size. Although it is impossible to obtain reliable craniometric measurements of the whole cranium of EV 9001, EV 9002 can be measured with a reasonable degree of accuracy, allowing for estimates of overall cranial length, breadth, and height.

In its current state of preservation, EV 9002 has a length (g-op[i]) of 217 mm, well above the range of variation seen at Zhoukoudian (188-199) and at the upper limit of the range of variation seen at Ngandong (Weidenreich 1943). Even after compensating for some increase in total cranial length due to breakage and subsequent expansion, the cranium is undoubtedly long.

Greatest cranial breadth (eu-eu), estimated to be approximately 170 mm, is positioned low on the vault, just above the level of the auditory meatus in the supramastoid region of the temporal. A cranial breadth of this magnitude is extremely rare. A search of the literature has failed to uncover any equivalently broad fossil hominid cranium. Petralona, with an estimated greatest cranial breadth of 165 mm across the supramastoid crests (Murrill 1981), and Hexian with a breadth of 160 mm, come closest. It seems unlikely that the measurement of greatest cranial breadth in EV 9002 is exaggerated to any large degree by breakage and subsequent displacement. The nuchal plane of the occiput is preserved intact and has not been severely fractured or otherwise altered. There is a network of small matrix filled cracks which covers the lower scale of the occipital, but they are small fractures that could not contribute more than a few millimeters at most to the total measurement.

The whole cranial vault, from a point on the frontal squama well anterior of bregma, has collapsed into the cranial fossa. The curvature of the intact frontal squama is, however, unaltered as is the curvature of the lower portion of the occipital plane, allowing for a rough approximation of the overall cranial vault contour. Another complicating factor is the degree of flattening of the basicranium. It is difficult to gauge whether the small degree of basicranial flexion is due to postdepositional crushing or is reflective of true morphology.

Given the damaged condition of both Yunxian crania, and the fact that the cranial fossae are still filled with a hard calcareous matrix, it is impossible to apply standard techniques to estimate cranial capacity. There can be little doubt, however, that in sheer size the Yunxian crania surpass the condition seen at Zhoukoudian.

The supraorbital tori of EV 9002 are heavily constructed and moderately arched over the orbits. They are joined by a robust glabellar prominence that is slightly indented anteriorly. The tori in EV 9002 do not form a continuous horizontal bony shelf as in specimens from Zhoukoudian but are moderately swept back posterolaterally as in the Broken Hill, Petralona, Sangiran 17, and Dali crania. The tori are thickest medially (18 mm) approaching the thickness seen in Bodo, Petralona, and Dali. The lateral portions of the

TABLE 28-5 Dental dimensions among archaic and pre-modern human fossils

Specimens	Upper central incisor	
	Mesiodistal length	Buccolingual breadth
Zhoukoudian (n=6)	9.8-10.8	7.5-8.1
Other Archaic Chinese (n=7)	8.1-11.7	7.0-9.4
Pre-modern Chinese (n=2)	8.3-10.3	6.4-8.4
Modern Chinese*	7.3-12.8	6.1-8.5

*(n=1401) Wang 1965

Specimens	Upper lateral incisor	
	Mesiodistal length	Buccolingual breadth
Zhoukoudian (n=3)	6.0-8.3	8.0-8.2
Other Archaic Chinese	-	-
Pre-modern Chinese (n=1)	7.0	6.0
Modern Chinese	5.7-8.3	5.2-7.7

Specimens	Upper canine	
	Mesiodistal length	Buccolingual breadth
Zhoukoudian (n=6)	8.5-10.5	9.8-10.6
Other Archaic Chinese (n=2)	9.4	8.7-9.7
Pre-modern Chinese (n=2)	8.2-10.8	8.7-10.4
Modern Chinese	6.9-9.2	6.2-9.9

Specimens	Upper first premolar	
	Mesiodistal length	Buccolingual breadth
Zhoukoudian (n=5)	7.4-9.2	10.5-12.8
Other Archaic Chinese (n=4)	8.5-9.0	11.1-12.8
Pre-modern Chinese (n=2)	7.4-8.5	10.6-11.0
Modern Chinese	5.9-8.9	7.8-11.0

Specimens	Upper second premolar	
	Mesiodistal length	Buccolingual breadth
Zhoukoudian (n=11)	7.2-8.9	10.2-12.5
Other Archaic Chinese (n=6)	7.8-9.0	9.9-13.4
Pre-modern Chinese	-	-
Modern Chinese	5.3-8.0	7.7-10.9

Specimens	Upper first molar	
	Mesiodistal length	Buccolingual breadth
Zhoukoudian (n=6)	10.0-13.1	11.7-13.7
Other Archaic Chinese (n=8)	10.5-12.7	11.1-14.8
Pre-modern Chinese (n=2)	10.8-13.4	12.8-14.0
Modern Chinese	9.1-11.9	10.0-13.0

Specimens	Upper second molar	
	Mesiodistal length	Buccolingual breadth
Zhoukoudian (n=7)	10.2-12.2	12.2-13.4
Other Archaic Chinese (n=4)	10.9-12.5	12.9-15.5
Pre-modern Chinese (n=2)	11.4-12.0	13.7-13.8
Modern Chinese	8.3-11.7	9.3-13.8

Specimens	Upper third molar	
	Mesiodistal length	Buccolingual breadth
Zhoukoudian (n=8)	8.7-10.4	10.4-12.5
Other Archaic Chinese (n=1)	9.5	13.0
Pre-modern Chinese	-	-
Modern Chinese	7.6-11.0	9.4-12.9

Specimens	Lower central incisor	
	Mesiodistal length	Buccolingual breadth
Zhoukoudian (n=7)	6.0-6.8	5.8-6.8
Other Archaic Chinese	-	-
Pre-modern Chinese	-	-
Modern Chinese	4.4-6.3	4.8-6.7

Specimens	Lower lateral incisor	
	Mesiodistal length	Buccolingual breadth
Zhoukoudian (n=10)	6.3-7.2	6.4-7.3
Other Archaic Chinese (n=1)	7.7	8.4
Pre-modern Chinese	-	-
Modern Chinese	5.0-7.0	5.5-7.1

Specimens	Lower canine	
	Mesiodistal length	Buccolingual breadth
Zhoukoudian (n=8)	8.1-9.0	8.2-10.4
Other Archaic Chinese (n=1)	8.3	8.4
Pre-modern Chinese	-	-
Modern Chinese	6.0-8.2	6.9-9.0

Specimens	Lower third molar	
	Mesiodistal length	Buccolingual breadth
Zhoukoudian (n=10)	10.0-13.8	10.0-12.4
Other Archaic Chinese	10.9-11.3	10.7-11.0
Pre-modern Chinese	-	-
Modern Chinese	9.2-13.4	8.9-12.5

Secimens	Lower first premolar	
	Mesiodistal length	Buccolingual breadth
Zhoukoudian (n=15)*	7.9-9.8	8.2-10.8
Other Archaic Chinese (n=4)	7.2-8.1	9.1-9.8
Pre-modern Chinese	-	-
Modern Chinese (range)	5.4-8.1	6.6-9.3

*Including original 1921-1923 specimen recoverd by Zdansky

Specimens	Lower second premolar	
	Mesiodistal length	Buccolingual breadth
Zhoukoudian (n=7)	8.2-9.2	8.0-11.1
Other Archaic Chinese (n=2)	7.2-9.1	9.6-11.2
Pre-modern Chinese	-	-
Modern Chinese	5.5-8.8	6.7-9.8

Specimens	Lower first molar	
	Mesiodistal length	Buccolingual breadth
Zhoukoudian (n=15)	9.9-14.1	10.1-12.8
Other Archaic Chinese (n=6)	10.0-12.8	10.1-13.2
Pre-modern Chinese	11.6	10.8
Modern Chinese	9.9-12.8	9.1-13.0

Specimens	Lower second molar	
	Mesiodistal length	Buccolingual breadth
Zhoukoudian (n=12)	11.3-13.2	11.1-13.0
Other Archaic Chinese (n=4)	11.7-14.3	11.4-13.9
Pre-modern Chinese (n=1)	8.3	10.6
Modern Chinese	8.5-12.9	8.7-12.3

tori are lost, so it is impossible to say whether they thin out to either side. The condition of EV 9001, however, suggests that they do not taper to any appreciable degree laterally.

The frontal is low and flat with a strong posterior slope. The ratio between least and greatest frontal breadth, which expresses the relative degree of transversal expansion of the frontal squama, is 96.5 in EV 9002, virtually identical to Dali (96.3) and considerably greater than at Zhoukoudian. The great breadth across the squama contributes to an overall absolute and relative reduction in postorbital constriction as compared to early *Homo* from Africa. The postorbital constriction index, calculated as the ratio of minimum frontal breadth to biorbital chord length, is 89.1. This compares to 76.1 in KNM-ER 3733, 72.9 in KNM-ER 3883, and 71.5 in OH 9 (Rightmire 1990). The EV 9002 index is, however, very close to Petralona (87.3), Sambungmachan 1 (89.4), and Ngandong 12 (91.1) (Rightmire 1990). It is interesting to note that Rightmire

(1990) records a value of 78.4 for this index in the Broken Hill skull, in keeping with the low values of earlier African crania. Another measure of postorbital constriction, commonly used by Chinese workers, is an index relating total breadth across the supraorbital tori to minimum postorbital breadth measured at the temporal fossae. These dimensions in EV 9002 are 135 mm and 114 mm respectively, yielding an index of 84.4. This index at Zhoukoudian varies between 80.7 and 82.9. It is 91.0 at Hexian and 85.1 at Dali. The Lantian calvarium has a degree of postorbital constriction comparable to that of early specimens of *Homo* from east Africa. These measurements reveal that EV 9002 has a reduced degree of postorbital constriction relative to the condition seen in early Pleistocene hominids and at Zhoukoudian. It is most comparable in this regard to late archaic Asian specimens as known from Hexian and Indonesia as well as archaic hominid specimens from Europe and India (Narmada).

Between the brow ridges and the frontal squama is a gentle bow-shaped ophryonic depression, rather than a prominent post-toral sulcus as seen in specimens from Zhoukoudian (Weidenreich 1943) or the steeply graded continuous squamous-toral junction seen in the Ngandong hominids. The orphryonic region of the Yunxian crania appear most similar to hominid specimens from Sangiran (particularly S-17), Arago, Petralona and Broken Hill.

In its overall character the Yunxian frontal diverges significantly from the pattern established at Zhoukoudian and is closer to the frontal of western archaics (Petralona, Arago, etc.) and premodern humans in China (Dali, Yingkou, Maba, etc.). This is reflected in its supraorbital structure, post-toral morphology, degree of postorbital constriction, and transverse squamal expansion. It retains the ancestral condition in terms of degree of frontal inclination, but in this regard it still falls within the range of variation of other archaic humans known from Asia, Africa, and Europe.

The parietals of EV 9002 have collapsed into the cranial fossa and have been flattened and deformed by postdepositional crushing. It is unclear, therefore, if keeling along the sagittal suture or parasagittal depressions in the obelion region is present or not. There is, however, no indication of a pronounced angular torus at the mastoid angle of the parietal. The parietals are short and flat. In relative proportions the length of the reconstructed parietal sagittal arc of EV 9002 appears to be shorter than both the frontal sagittal arc and occipital sagittal arc, which is consistent with the condition generally seen in archaic Asians. Parietal eminences are indistinct suggesting that there has been little or no overall parietal expansion.

The posterior portion of the skull has a well-expressed occipital torus that forms a smooth rounded transverse projection onto the occipital squama; the occipital and nuchal planes are strongly angled to one another so that midpoint of the occipital torus along the median plane of the cranium and the point of greatest posterior projection of the cranium (opisthocranion) are coincident. The occipital is characterized by a sharply delimited upper (occipital) scale and an expansive lower (nuchal) scale. Both the external occipital protuberance and linear tubercle are indistinct, although they are somewhat more strongly expressed in EV 9001. The great breadth and length of the nuchal attachment area is a distinguishing feature of the Yunxian crania. Fossae for the insertion of the nuchal musculature are, however, shallow and poorly defined, unlike the deeply excavated hollows seen in the Ngandong hominids.

The temporal bone and associated basicranial structures are well preserved in EV 9002. The squamous portion of the temporal is elevated and somewhat arched superiorly, unlike the condition in most archaic Asians in which the parietal margin is low and straight. A similar pattern to that seen in EV 9002 has, however, been noted in crania from the upper layers at Zhoukoudian (Skull V) and Hexian. This condition is best exemplified by the temporal squama length/height index. In EV 9002 maximum length parallel to the Frankfurt plane is 78 mm and vertical height from auriculare to the top of the squama is 45 mm, yielding an index of 57.7. This value is similar to that of crania from the upper layers at Zhoukoudian (Skull V) and Hexian, later premodern humans in China, middle and late Pleistocene hominids from Europe and Africa, and modern humans.

The zygomatic process of the temporal is robust and associated with a broad sulcus processus zygomatici. The auditory meatus is elliptically shaped and recessed below a strong suprameatal tegmen. These latter features are commonly seen traits in archaic Asians. The ventral aspect of the temporal, however, shows a mix of both archaic and progressive characters. The tympanic plate is very strongly developed, extremely thick and elongate (a-p length approximately 43 mm), with a robust "clublike" eustachian process (*processus supratubarius*) at its medial extremity, much as in previously known specimens of archaic Asians attributed to *H. erectus* (Aiello and Dean 1990). On the other hand, the tympanic is vertically emplaced and strongly angled to the mid-sagittal plane (60°). At Zhoukoudian and Ngandong, the tympanic plate is generally more horizontally inclined, tubular, and perpendicular to the mid-sagittal plane. In EV 9002 the petrous portion is oriented sagittally producing a relatively obtuse petrotympanic angle (155°). In other Asian archaics, this angle tends to be close to 90° while in modern humans it is closer to 180°.

Notwithstanding these "progressive" features, the overall robusticity and character of the tympanic region is very archaic. The tympanic is low, with a blunt petrosal spine and elongate, rather than tall, thin, and foreshortened as in later human specimens. EV 9001 and 9002 differ in this respect from Dali, Xujiayao, and Jinniushan, in which the tympanic plates are fundamentally modern in character.

The mandibular fossa is extremely elongate and deeply excavated. It is bounded posteriorly by a distinct postglenoid process set well in front of the tympanic plate and medially by a large entoglenoid process. The sphenoid spine, which in more modern humans contributes to the medial border of the glenoid cavity, is replaced by a deep pit medial to the entoglenoid process. Associated with this morphology is a distinct recessus medialis fossae mandibularis. This feature is defined by Weidenreich (1943:47) as a narrow cleftlike recess between the ventral aspect of the temporal squama and the tympanic plate at the posterior medial part of the mandibular fossa. As at Zhoukoudian "the medial wall of the fossa is not concave conforming with the contour of the whole fossa but rather convex projecting toward the tympanic plate" (Weidenreich 1943:47). The articular eminance, which forms the anterior boundary of the fossa, is a low, mound-shaped surface that is slightly concave mediolaterally. It opens anteriorly onto a broad preglenoid planum much as in other specimens of early and late Asian archaics. In modern *H. sapiens* the tympanic plate, postglenoid process, sphenoid spine, and articular tubercle have coalesced into a well-defined, structurally integrated whole in which the tympanic plate and postglenoid process are fully incorporated into the posterior wall of the mandibular fossa.

A broad incisura mastoidea (digastric fossa in modern humans) sets the massive, inferiorly projecting mastoid process away from a well-developed juxtmastoid eminence that is confluent with a robust occipitomastoid crest much as in other archaic human specimens and Neandertals. Neandertals differ, however, in having relatively reduced mastoid processes. There is no evidence of a fissure separating the tympanic plate from the mastoid process as seen in many archaic Asian specimens. The vaginal process of the styloid is replaced by an elongate pit, indicating that the styloid process was not ossified to the cranial base as is generally the case in more modern humans.

The mastoid process is fully modern in form. It is large and massive with a total projection from porion to the mastoid tip of 30.1 mm. Its longitudinal axis is oriented medially so that the mastoid tips fall beneath the cranial base.

The facial aspect of the two Yunxian crania is remarkably large and massive. In most facial dimensions the two specimens closely approximate the condition seen in large (presumably male) archaic human specimens from Europe and Africa. There are particular similarities between the Yunxian and Petralona, Broken Hill, and Bodo crania in terms of upper facial height (n-pr), interorbital breadth (fmo-fmo), upper facial breadth (ek-ek), zygomaxillary breadth (zm-zm), the height (rhi-ns) and breadth (al-al) of the nasal aperture and subnasal alveolar height (ns-pr).

The mid-face is very broad. Upper facial height is absolutely great and virtually identical to Petralona and Broken Hill. Total facial breadth, however, is estimated to be very great so that the upper facial index would be mesene as in other middle and early late Pleistocene fossil hominids from east Asia, early modern *H. sapiens* specimens from Europe, Africa, and Asia and modern Asian and derivative people.

Structurally the mid-face looks modern. It is characterized by transversely oriented infraorbital plates with flat, vertically emplaced malar facies. A well-expressed canine fossa occurs inferior to the infraorbital foramen giving expression to the fully retracted and orthogonal character of the mid-face. The inferior zygomaxillary border is oriented horizontally and is indented at the zygomaxillary suture. In EV 9001 the zygomatic process of the maxilla is strongly angled at its junction with the maxillary process of the zygomatic. The root of the zygomatic process is anteriorly placed and originates well above the alveolar margin, high on the maxillary body above M1. These features are commonly seen mid-facial characteristics of living Asians and fossil hominids previously known from China. Archaic humans from western Eurasia and Africa (e.g., Arago, Petralona, Broken Hill 1, and Bodo) differ considerably in their facial morphology from the pattern previously described. They presage Neandertals by the development of moderate to extreme mid-facial projection characterized by sagittally oriented infraorbital plates, an obliquely oriented inferior zygomaxillary border and a low horizontal curvature index of the zygomatic. The latter characteristics are also seen to a lesser extent in KNM-ER 3733, the Lantian maxillary fragment, and Sangiran 17. The Yunxian crania, therefore, appear to represent an Asian hominid that retains a dimensionally large face but possesses a mid-facial morphology derived in the direction of modern Asians and divergent from the pattern established in western Eurasia and Africa during the middle Pleistocene.

The teeth of the Yunxian hominids are extremely robust. The labiolingual diameter at the base of the incisor crowns is especially thick. After wear, the incisal edge is worn flat with the labial margin slightly lower than the lingual margin, indicative of heavy mechanical loading of the anterior dentition. The upper central incisors lack features commonly associated with both fossil and living Asians, such as appreciable shoveling, a basal lingual tubercle, or fingerlike projections on the lingual surface of the tooth. The canines lack mesial and distal ridges and do not project beyond the occlusal level of the dental row. Of particular note is the large size of M3 in EV 9001. In EV 9001 the molar size sequence is M3>M2>M1 (in EV 9002 M3 is degenerate and peg-shaped).

In summary, the Yunxian crania retain cranial vault and basicranial character states commonly seen in archaic humans previously attributed to *H. erectus* in association with some features of the tympanic, temporal squama, frontal squama, and supraorbital torus seen in contemporary Eurafrican hominids and later hominids known from China and farther west. They possess large faces but have a mid-facial morphology similar to that of modern Asians and divergent from the pattern established in western Eurasia during the middle Pleistocene and elaborated upon by late Pleistocene Neandertals of Europe and the Near East.

DISCUSSION

New evidence afforded by the recent increase in sampling of east Asian middle Pleistocene hominids dispels the notion that archaic humans in China consist of a fairly homogeneous type, distinct at the species level from other contemporaneous humans. The notion that archaic Asian hominids remained in a condition of stasis throughout their long history (Rightmire 1990) becomes harder to maintain as more fossil material comes to light. A comparison of the Lantian and Hexian remains, which represent the terminal ends of the archaic human lineage in China, amply demonstrates this fact. Although the record is still scanty, there is evidence of change and significant amounts of infraspecific variation in the following features of middle Pleistocene humans in China: (1) Cranial capacity varies from approximately 780 cc to over 1250 cc; (2) dental size, particularly of the molar teeth, varies considerably more than within the relatively homogeneous Zhoukoudian sample (see Table 28-5); (3) supraorbital morphology varies from the more generalized pattern seen at Lantian to the classic pattern seen at Zhoukoudian and elsewhere; (4) variation in the morphology of the temporal and occipital bones, which is, perhaps, concomitant to increases in overall vault size; and (5) the transformation of the primitive hominid face along more modern lines.

The relationship of early African members of the genus *Homo* to archaic Asians has generated considerable comment (Stringer 1984; Andrews 1984; Wood 1984; Clarke 1990). Most of the characters that have previously been used to unite the African and Asian material taxonomically (i.e., projecting supraorbital torus, small cranial capacity, sloping forehead, angled occipital) have been dismissed as primitive for the genus *Homo* and, therefore, not indicative of a close phylogenetic relationship between the early African and later Asian remains. Apparently derived features of Asian *H. erectus* such as a cruciate eminence at bregma, parasagittal depressions, and an angular torus are not seen in the earlier African remains. Clarke, moreover, notes a number of derived features of the petrous portion of the temporal in the early African material not seen in later archaic Asians that tend to unite the former with later members of *H. sapiens*. These considerations have led some to question the attribution of the African material to the same hominid taxon as the Asian material. Evidence detailed in this chapter, however, indicates that both the earlier (Lantian) and later (Hexian) examples of archaic Asian generally attributed to *H. erectus* in China diverge to differing extents from the pattern set at Zhoukoudian. The Lantian cranium, for instance, seems to align well with material attributed to early *H. erectus* in Africa and there is little or no question regarding its attribution to the same taxon as other archaic Asian specimens. It seems quite premature to erect new hominid taxa based on the application of cladistic criteria that are not designed to be applied at what may well be an intraspecific level of classification (Trinkhaus 1990).

While the recovery of new and diverse finds of middle Pleistocene hominid in China has increased our awareness and appreciation of variation within the taxon both in time and space, it cannot be denied that the total morphological pattern of archaic hominids in Asia throughout the Quaternary was remarkably consistent and resistant to fundamental change. This makes the evolution of east Asian populations toward a more modern morphology at approximately 250,000 YA all the more significant.

The importance of the Yunxian crania lies in the fact that prior to their discovery, there was only limited evidence of the facial morphology of archaic people in east Asia. Their discovery allows, for the first time, a thorough evaluation of the distribution of craniofacial features in middle Pleistocene fossil hominids. In this respect the Yunxian crania display a combination of primitive and derived features of the mid-face seen in other later Asian fossil hominids, some early premodern *H. sapiens* material from Africa (e.g., Irhoud 1, Ngaloba, Florisbad, Eliye Springs, and Broken Hill 2) and early modern humans from several regions (e.g., Qafzeh 6, Dar es Soltan 5, Skhul 4, upper Paleolithic Europeans, and north Africans). Archaic humans from Europe and Africa (e.g., Arago, Petralona, Bodo) preserve a different set of primitive versus derived features of the mid-face that trend in the direction of later Neandertals.

While the Yunxian crania display a mid-facial architecture characterized by a "horizontal inferiolateral zygomaticoalveolar margin" and a canine fossa, features which Smith et al. (1989) consider diagnostic of modern humans, they possess the large overall facial dimensions and supraorbital morphology common to archaic human crania from Europe and Africa.

The cranial vault and basicranial structures of the Yunxian crania, on the other hand, preserve many typical *erectus*-like features (Li and Etler 1992). Other features commonly used to define an Asian *H. erectus* as distinct from contemporary forms further to the west in Europe and Africa are not well expressed in the Yunxian crania, suggesting that such characters (i.e., the degree of ectocranial buttressing and the form of the supraorbital tori and ophryonic region of the frontal) are polymorphic rather that species defining autapomorphies.

The combination of features in the Yunxian crania, including a mix of primitive, intermediate, and advanced character states of the cranial vault, cranial base, and face, suggests that they should best be viewed as members of an archaic hominid lineage evolving in a regional context toward an Asian variant of later premodern *H. sapiens*. The Yunxian crania and other remains of archaic east Asians, moreover, possessed a facial structure characteristic of modern humans long before its occurrence in European and African hominids. It has been suggested elsewhere that the cranial vault and basicranial structure characteristic of modern *H. sapiens* makes its debut in Africa during the early middle Pleistocene if not earlier (Rightmire 1990; Clarke 1990). The Yunxian material, therefore, gives strong testimony in support of the idea that human evolution during the middle and upper Pleistocene is characterized by mosaic transitions occurring in regional contexts and that simple models of regional continuity or replacement as explanations for the emergence of modern humans throughout the Old World need to be reevaluated.

CONCLUSIONS

One problem in assessing middle Pleistocene hominid variation has been the tendency to compare specimens that may be from widely differing periods of time. This problem has arisen through a set of historical circumstances associated with the methodological difficulties involved in dating most of the relevant specimens. Many of these specimens were discovered decades ago and can no longer be reasonably dated by modern techniques because the contextual framework within which they were discovered is irretrievably lost. It is probably the case that such important specimens as Kabwe (Broken Hill), Steinheim, and Petralona will never be securely dated. Other equally important crania such as Arago and Ndutu can at present only be broadly assigned to the late middle Pleistocene (Day 1986; Klein 1989). The same problems prevail in China and Java. The situation in China has improved somewhat, as it has in other parts of the world, but problems of reliable dating are still rampant. Many of the employed dating techniques have very high margins of error and the material tested is often times highly subject to contamination from the surrounding environment. It is quite possible that spurious dates, either too old or too young, may enter into the literature, having at times profound influence on the interpretation of the fossil record. This is not to deny the veracity of some key new datings that have recently generated much deserved attention. Many other less secure dates should, however, be viewed with extreme caution.

With these caveats in mind, it can probably be safely assumed that previously known archaic human fossils from China, as exemplified by Zhoukoudian, Yiyuan, Chenjiawo, and Hexian date from a period between 650,000 to 400,000 years ago. This period of human evolution in Europe and Africa is extremely poorly sampled. Early Pleistocene hominid material from Tighenif (Ternifine) in Algeria, which has long been favorably compared to Zhoukoudian (Howell 1960), may predate this period by a considerable margin (Klein 1988). It may well be that no significant hominid material can be assigned to this period of time in Europe or sub-Saharan Africa with the possible exception of the Mauer mandible from Europe (Klein 1988) and Bodo from Ethiopia (White pers. comm. n.d.). Basically it can be argued that there are few diagnostic hominid specimens in either Europe or Africa from the time period in which archaic humans are sampled at Zhoukoudian and elsewhere in Asia. It is, therefore, very difficult to make the necessary comparisons between western and eastern populations during the earlier stretches of the middle Pleistocene.

At present dating of the Yunxian crania remains under study. Recent evidence, however, indicates an earlier age that previously estimated (Yan 1993). The period of archaic human evolution best sampled in Europe and Africa, i.e., between 400,000-250,000 YA thus continues to be poorly sampled in Asia. The young dates assigned to the Upper Layers at Zhoukoudian and Hexian are most likely too recent, as more recent investigations suggest. As regards Europe and Africa there is at present no evidence that would contradict an age of between 400,000 and 250,000 years for the vast majority of specimens. If this is the case, there is no real discrepancy between *H. erectus* as previously known in Asia and "archaic *H. sapiens*" as known in Europe and Africa. In point of fact, these populations most likely do not overlap in time.

Given their completeness and middle Pleistocene or earlier age, the Yunxian crania are the first specimens from China that are directly comparable to archaic hominids from Europe and Africa. Although they are similar in many respects to European and African material of broadly comparable or younger age their principal affinities lie with other archaic Asians with whom they share many features. We feel, therefore, that the term "archaic *H. sapiens*," often applied to Eurafrican specimens to distinguish them from an Asian defined *H. erectus*, is inappropriate. In like manner, the continued use of the term *H. erectus* to refer to some middle Pleistocene hominids but not others is mistaken. All human fossils from the middle Pleistocene should be viewed as members of one species.

There are, however, clear temporal and regional differences between various human groups during the vast stretches of the middle Pleistocene. One taxonomic approach to this problem has been to assign middle Pleistocene hominids from different regions to distinct human species. The taxon *H. heidelbergensis*, for instance, has been applied recently to much of the European and African material previously discussed. The reasoning behind the resurrection of *H. heidelbergensis* was, however, to construct a taxon in opposition to an Asian defined *H. erectus*. It was, therefore, conceived to be cladistically distinct from contemporary humans from further east. If, however, as argued here, there is little or no temporal overlap between classic "*H. erectus*" in Asia and "*H. heidelbergensis*" in Europe and Africa, there is no need to maintain the cladistic distinction. Furthermore, if material such as Yunxian is considered an Asian variant of "*H. heidelbergensis*," there is even less call to accept it as a cladistically defined species. "*H. heidelbergensis*" then becomes merely a gradistic construct interposed between more classic "*H. erectus*" and early premodern *H. sapiens*. In that case "*H. heidelbergensis*" is no different in meaning than advanced or evolved "*H. erectus*." In point of fact, it may be more reflective of our present state of knowledge to simply refer to all middle Pleistocene hominids as archaic humans, with appropriate temporal and spacial divisions as warranted by the material under discussion.

In the previous discussion, we have used the term "premodern *H. sapiens*" to refer to another set of human fossils in China exemplified by Dali, Jinniushan, Maba, and so on that seem to come from a period between 250,000 and 100,000 years ago. We think that this material and similar material from Europe (early Neandertals) and Africa (Jebel Irhoud, Florisbad, Ngaloba, and Omo-Kibish, etc.) should best be considered as representing an evolutionary phase subsequent to late archaic but prior to anatomically modern humans. This is somewhat akin to the so-called Neandertaloid phase of human evolution popularized in the 1960s but with the caveat that the neandertals were the least characteristic, most isolated, and temporally most persistent of this group of fossil hominids, having developed to the extreme a highly derived set of regional characters. Similarly aged material from Africa and Asia, while of the same general premodern character, is seemingly more generalized, although the Asian material displays a set of regional Asian characteristics of its own. African specimens from the late middle and early late Pleistocene are the earliest from this region of the world that begin to display a fully modern facial structure in the same way that contemporary and earlier material in China do (Pope 1991, 1992). This may be evidence that a dispersal out of Asia toward the Levant and north Africa introduced new variation into the human populations of those regions, perhaps serving as a spur toward the development of anatomically modern humans. This possibility is best exemplified by Zuttiyeh (Song and Wolpoff 1990) and Jebel Irhoud 1 (Hublin 1993). Zuttiyeh (the well known Galilee cranial fragments), which consists of much of the upper face and frontal, has features highly reminiscent of material known from Zhoukoudian in China, while the Jebel Irhoud I cranium is extremely similar to Dali in overall facial morphology and dimensionality. It can therefore be argued that during the early late Pleistocene the Levant and North and East Africa were being as influenced by Asia as vice versa.

As the human fossil record becomes more dense and we begin to get a better grasp of both regional and temporal variation, it is very likely that the sort of patterns alluded to here will become clearer. In all probability, ever since the advent of the basal human stock of early Pleistocene East Africa (KNM-ER 3733, etc.) from which all later hominids evolved, there has been no discrete speciation events along our lineage. There are, however, periods of relative stasis and periods of accelerated change. There are aspects of regional continuity as well as the likelihood of important dispersals between contiguous regions, all influencing the course of human evolution during the middle and late Pleistocene. There are also real transitions in the fossil record that have occurred periodically and relatively synchronously throughout the habitable world, the most recent being the transition to anatomically modern humans. Whether these transitions should be recognized taxonomically is a matter of debate. Which transitions warrant taxonomic recognition and which do not is somewhat arbitrary. Given the sparse nature of the middle Pleistocene human fossil record and our lack of understanding of temporal and spacial variation among hominids of this period, it is perhaps premature to even attempt such divisions at a taxonomic level. As a consequence of an overemphasis on such taxonomic quibbling, some very important questions in human evolution have not been adequately addressed. In the furor over contending hypotheses of modern human origins, questions relating to precisely what were the factors mediating the process of human evolution during the middle Pleistocene leading to the eventual rise of modern people have seldom even been asked. The fact that such fundamental questions have not been adequately addressed speaks directly to our ignorance of what was really going on during the middle Pleistocene.

ACKNOWLEDGMENTS

We are grateful to Professor Jia Lanpo, Professor Chen Zhenyu, Professor J. Desmond Clark, Professor F. Clark Howell, and Professor Tim D. White for their invaluable support. We acknowledge the financial support of the Luce Foundation, the L.S.B. Leakey Foundation and the Committee on Scholarly Communication with China that has allowed us to conduct research in each country. Without the warm-hearted friendship of many colleagues in both the United States and China our collaboration would not have been possible.

LITERATURE CITED

Aiello, L., and Dean, C. (1990) *An Introduction to Human Evolutionary Anatomy*. San Diego: Academic Press.

Aigner, J. S. (1981) *Archaeological Remains in Pleistocene China*. Forschungen zur Allgemeinen und Vergleichrenden Archaologie, Band 1. Munich.

Aigner, J. S. (1986) The age of Zhoukoudian Locality 1: The newly proposed O18 correspondences. *Anthropos* (Brno) 23:157-173.

Aigner, J. S. (1987) Corrélations ^{18}O et localité 1 de Chou-kou-tien. *L'Anthropologie* 91:733-748.

Aigner, J. S., and Laughlin, W. S. (1973) The dating of Lantian Man and his significance for analyzing trends in human evolution. *Am. J. Phys. Anthropol.* 39:97-110.

An, Z., and Ho, C. K. (1989) New magnetostratigraphic dates of Lantian *Homo erectus*. *Quat. Res.* 32:213-221.

Andrews, P. (1984) On the characters that define *Homo erectus*. *Courier Forsch. Senck.* 69:167-178.

Binford, L. R., and Ho, C. K. (1985) Taphonomy at a distance: Zhoukoudian, "the cave home of Peking Man"? *Curr. Anthropol.* 26:413-442.

Binford, L. R. and Stone, N. M. (1986) Zhoukoudian: A closer look. *Curr. Anthropol.* 27:453-475.

Chen, T., and Yuan, S. (1988) Uranium-series dating of bones and teeth from Chinese Palaeolithic sites. *Archaeometry* 30:59-76.

Chen, T., Yuan, S., Gao, S., and Hu, Y. (1987) Uranium series dating of fossil bones from Hexian and Chaoxian fossil human sites. *Acta Anthropol. Sin.* 6:249-254 (in Chinese).

Cheng, G., Li, S., and Lin, J. (1978) A research on the ages of the strata of "Lantian Man." In Institute of Vertebrate Palaeontology and Palaeoanthropology, Chinese Academy of Sciences (ed.): *Gurenlei Lunwenji (Collected Papers of Palaeoanthropology)*. Beijing: pp. 151-157 (in Chinese).

Chia, L. (Jia, L.) (1975) *The Cave Home of Peking Man*. Beijing: Foreign Language Press.

Clarke, R. J. (1977) The Cranium of the Swartkrans Hominid, SK 847 and Its Relevance to Human Origins. Ph. D. Dissertation. University of the Witwatersrand, Johannesburg.

Clarke, R. J. (1990) The Ndutu cranium and the origin of *Homo sapiens*. *J. Hum. Evol.* 19:699-736.

Day, M. H. (1986) *Guide to Fossil Man*. Chicago: University of Chicago Press.

Etler, D. A. (1990) A case study of the "erectus"-"sapiens" transition in Asia: Hominid remains from Hexian and Chaoxian counties, Anhui province, China. *Kroeber Anthropol. Papers* 71-72:1-19.

Guo, S., Zhou, S., Meng, W., Zhang, R., Shun, S., Hao, X., Liu, S., Zhang, F., Hu, R., and Liu, J. (1980) The dating of Peking Man by the fission track technique. *Sci. Bull. (Kexue Tongbao)* 25:384 (in Chinese).

Habgood, P. (1989) An investigation into the usefulness of a cladistic approach to the study of the origin of anatomically modern humans. *Hum. Evol.* 4:241-252.

Han, D., and Xu, C. (1985) Pleistocene mammalian faunas of China. In R. Wu and J. W. Olsen (eds.): *Palaeoanthropology and Palaeolithic Archaeology in the People's Republic of China*. Orlando: Academic Press, pp. 267-286.

Heller, F., and Liu, T. (1982) Magnetostratigraphical dating of loess deposits in China. *Nature* 300:431-433.

Hou, L. (1985) Fossil birds from Zhoukoudian Locality 1. In R. Wu et al. (eds.): *Multidisciplinary Study of the Peking Man Site at Zhoukoudian*. Beijing: Science Press, pp. 114-118 (in Chinese).

Howell, F. C. (1960) European and northwest African Middle Pleistocene Pleistocene hominids. *Curr. Anthropol.* 1:195-232.

Howell, F. C. (1986) Variabilité chez *Homo erectus*, et problème de la présence de cette espèce en Europe. *L'Anthropologie* 90:447-481.

Hsu, R. (1966) The climatic condition in north China during the time of *Sinanthropus*. *Scientia Sinica* (English ed.) 25:410-414.

Hu, C. (1985) The history of mammalian fauna of Locality 1 of Zhoukoudian and its recent advance. In R. Wu et al. (eds.): *Multidisciplinary Study of the Peking Man Site at Zhoukoudian*. Beijing: Science Press, pp. 107-113 (in Chinese).

Huang, W., and Fang, Q. (1991) *Wushan Hominid Site*. Beijing: Ocean Press.

Huang, W., Fang, D., and Ye, Y. (1981) Observations on the Homo erectus calvarium discovered at Longtandong, Hexian, Anhui. *Kexue Tongbao* 26:1508-1510 (in Chinese).

Huang, W., Fang, D., and Ye, Y. (1982) Preliminary study on the fossil hominid skull and fauna from Hexian, Anhui. *Vert. PalAs.* 20:248-256 (in Chinese).

Huang, W., and Huang, C. (1985) Mammal fossils and sporopollen compositions at Hexian Man locality and their significance. In *Symposium of National Conference on Quaternary Glacier and Periglacial*. Beijing: Science Press, pp 180-183 (in Chinese).

Hublin, J. J. (1986) Some comments on the diagnostic features of *Homo erectus*. *Anthropos* (Brno) 23:175-185.

Hublin, J. J. (1993) Recent human evolution in north-western Africa. In M. J. Aitken, C. B. Stringer, and P. A. Mellars (eds.): *The Origin of Modern Humans and the Impact of Chronometric Dating*. Princeton, pp.118-131.

IVPP (1966) *The Cenozoic of Lantian*. Beijing: Science Press.

Jia, L. (Chia, L.) (1980) *Early Man in China*. Beijing: Foreign Language Press.

Jia, L. (Chia, L.), and Huang, W. W. (1990) *The Story of Peking Man*. London: Oxford.

Klein, R. G. (1989) *The Human Career: Human Biological and Cultural Origins*. Chicago: Chicago University Press.

Kong, Z., Du, N., Wu, Y., Yu, Q., Yi, M., Ren, Z., Min, Z., Cui, S., Luo, B., Wang, S., and Hu, J. (1985) Vegetational and climatic changes since Paleogene at Zhoukoudian and its adjacent regions. In R. Wu et al. (eds.): *Multidisciplinary Study of the Peking Man Site at Zhoukoudian*. Beijing: Science Press, pp. 119-154 (in Chinese).

Li, R., and Lin, D. (1979) Geochemistry of amino acid of fossil bone from deposits of "Peking Man," "Lantian Man" and "Yuanmou Man" in China. *Sci. Geol. Sin.* 1:56-61 (in Chinese).

Li, T. (1990) *The Study of Early Man*. Wuhan: Wuhan University Press (in Chinese).

Li, T., Wang, Z., Feng, W., Hu, K., and Liu, W. (1991) Investigation and excavation of the fossil place at the outfall of Quyuan River, Yun county, Hubei province. *Jianghan Kaogu* 39:1-14 (in Chinese).

Li, T., and Etler, D. A. (1992) New middle Pleistocene hominid crania from Yunxian in China. *Nature* 357:404-407.

Li, Y., and Ji, H. (1981) Environment during Peking Man's time as viewed from mammalian fossils. *Sci. Bull. (Kexue Tongbao)* (English ed.) 26:170-172.

Liu, T., and Ding, M. (1984) A tentative chronological correlation of early human fossil horizons in China with the loess-deep sea records. *Acta Anthropol. Sin.* 3:93-101 (in Chinese).

Liu, C., Zhu, X., and Ye, S. (1977) A palaeomagnetic study of the cave deposits of Zhoukoudian (Choukoutien), the locality of *Sinanthropus*. *Sci. Geol. Sin.* 1:26-32 (in Chinese).

Liu, S., Zhang, F., Hu, R., Liu, J., Guo, S., Zhou, S., Meng, W., Zhang, P., Sun, S., and Hao, X. (1985) Dating Peking Man site by fission-track method. In R. Wu et al. (eds.): *Multidisciplinary Study of the Peking Man Site at Zhoukoudian*. Beijing: Science Press, pp. 241-245 (in Chinese).

Liu, Z. (1985) Sequence of sediments at Locality 1 in Zhoukoudian and correlation with loess stratigraphy in northern China and with the chronology of deep-sea cores. *Quat. Res.* 23:139-153.

Liu, Z. (1988) Paleoclimatic changes as indicated by the Quaternary karstic cave deposits in China. *Geoarchaeol.* 3:103-115.

Lü, Z., Huang, Y., Li, P., and Meng, Z. (1989) Yiyuan fossil man. *Acta Anthropol. Sin.* 8:301-313 (in Chinese).

Ma, X., Qian, F., Li, P., and Ju, S. (1978) Palaeomagnetic dating of Lantian Man. *Vert. PalAs.* 16:238-243 (in Chinese).

Murrill, R. I. (1981) *Petralona Man*. Springfield: Charles C. Thomas.

NISWC (Northwestern Institute of Soil and Water Conservation, Academia Sinica) (1985) Characteristics of the paleosoil and deposits and their formation at Zhoukoudian area. In R. Wu et al. (eds.): *Multidisciplinary Study of the Peking Man Site at Zhoukoudian*. Beijing: Science Press, pp. 216-238 (in Chinese).

Pei, J. (1980) An application of thermoluminescence dating to the cultural layers of "Peking Man" site. *Quater. Sin.* 5:87-95 (in Chinese).

Pei, J. (1985) Thermoluminescence dating of the Peking Man site and other caves. In R. Wu et al. (eds.): *Multidisciplinary Study of the Peking Man Site at Zhoukoudian*. Beijing: Science Press, pp. 256-260 (in Chinese).

Pei, J., and Sun, J. (1979) Thermoluminescence ages of quartz in ash materials from *Homo erectus pekinensis* site and its geological implication. *Sci. J. (Kexue Tongbao)* 24:849 (in Chinese and English).

Pei, W., and Zhang, S. (1985) A study of the lithic artifacts of *Sinanthropus*. *Palaeontol. Sin.*, New Series D 12:1-277 (in Chinese with English summary).

Pope, G. (1991) Evolution of the zygomaxillary region in the genus *Homo* and its relevance to the origin of modern humans. *J. Hum. Evol.* 21:189-213.

Pope, G. (1992) The craniofacial evidence for the emergence of modern humans in China. *Yearb. Phys. Anthropol.* 35:243-298.

Qian, F. (1985) On the age of "Yuanmou Man"—a discussion with Liu Tungsheng et al. *Acta Anthropol. Sin.* 4(4)324-331.

Qian, F., Zhang, J., and Yin, W. (1980) A magnetostratigraphical study of deposits in "Peking Man Cave." *Sci. Bull. (Kexue Tongbao)* 25:192.

Qian, F., Zhang, J., and Yin, W. (1985) Magnetic stratigraphy from the sediment of west wall and test pit of Locality 1 at Zhoukoudian. In R. Wu et al. (eds.): *Multidisciplinary Study of the Beijing Man Site at Zhoukoudian*. Beijing: Science Press, pp. 251-255 (in Chinese).

Qiu, Z., Gu, Y., Zhang, Y., and Zhang, S. (1973) Newly discovered *Sinanthropus* remains and stone artifacts at Choukoutien. *Vert. PalAs.* 11:109-131 (in Chinese).

Ren, M., and Liu, Z. (1985) Development of the Peking Man's cave in relation to early man at Zhoukoudian, Beijing. In T. Liu (ed.): *Quaternary Geology and Environment of China.* Beijing: China Ocean Press, pp. 186-193.

Ren, M., Liu, Z., Jin, J., Deng, X., Wang, F., Peng, B., Wang, X., and Wang, Z. (1981) Evolution of limestone caves in relation to the life of early man at Zhoukoudian, Beijing. *Scientia Sin.* (English ed.) 24:843-850.

Ren, M., Liu, Z., Jin, J., Deng, X., Wang, F., Peng, B., Wang, X., and Wang, Z. (1985) Development of Zhoukoudian caves and its relation to the life of fossil man. In R. Wu et al. (eds.): *Multidisciplinary Study of the Peking Man Site at Zhoukoudian.* Beijing: Science Press, pp. 155-184 (in Chinese).

Rightmire, G. P. (1990) *The Evolution of* Homo erectus. Cambridge: Cambridge University Press.

Schick, K., Toth, N., Wei, Q., Clark, J. D., and Etler, D. A. (1991) Archeological perspectives in the Nihewan Basin, China. *J. Hum. Evol.* 21:13-26.

Schwartz, H. P. (1993) Uranium-series dating and the origin of modern man. In: M. J. Aitken, C. B. Stringer, and P.A. Mellars (eds.): *The Origin of Modern Humans and the Impact of Chronometric Dating.* Princeton, pp. 12-26.

Shao, X. (1985) *Handbook of Anthropometry.* Shanghai: Cishu Publishing House.

Shen, G., and Jin, L. (1991) Restudy of the upper age limit of Beijing man site. *Acta Anthropol. Sin.* 10:273-277.

Smith, F. H., Falsetti, A. B., and Donnelly, S. M. (1989) Modern human origins. *Yearb. Phys. Anthropol.* 32:35-68.

Song, S., and Wolpoff, M. H. (1990) Zuttiyeh: A new look at an old face. *Acta Anthropol. Sin.* 9:359-370.

Stringer, C. B. (1984) The definition of *Homo erectus* and the existence of the species in Africa and Europe. *Courier Forsch. Senck.* 69:131-144.

Sun, M. (1965) On the pollen/spore assemblage from the *Sinanthropus* beds at Zhoukoudian. *Quaternaria Sinica* 4:84-96 (in Chinese).

Trinkaus, E. (1990) Cladistics and the hominid fossil record. *Am. J. Phys. Anth.* 83:1-11.

Wang, F. (1983) Limestone caves and karst development at Zhoukoudian. *Carsologica Sin.* 1:41-48 (in Chinese).

Wang, H. (1965) *Dental Anatomy and Physiology.* Beijing: People's Press (in Chinese).

Wang, L. (1989) Chronology in Chinese paleoanthropology. In R. Wu, X. Wu, and S. Zhang (eds.): *Early Humankind in China.* Beijing: Science Press, pp. 392-409 (in Chinese).

Weidenreich, F. (1936a) The mandibles of *Sinanthropus pekinensis*: A comparative study. *Palaeontol. Sin.*, Series D, 7:1-162.

Weidenreich, F. (1936b) Observations on the form and proportions of the endocranial casts of *Sinanthropus pekinensis*, other hominids, and the great apes: A comparative study of brain size. *Palaeontol. Sin.*, Series D, 7:1-50.

Weidenreich, F. (1937) The dentition of *Sinanthropus pekinensis*: A comparative odontography of the hominids. *Palaeontol. Sin.*, New Series D, 1:1-180.

Weidenreich, F. (1939a) On the earliest representatives of modern mankind recovered on the soil of East Asia. *Bull. Nat. Hist. Soc. Peking* 13:161-174.

Weidenreich, F. (1939b) Six lectures on *Sinanthropus pekinensis* and related problems. *Bull. Geol. Soc. China* 19:1-110.

Weidenreich, F. (1941) The extremity bones of *Sinanthropus pekinensis. Palaeontol. Sin.*, New Series D 5:1-150.

Weidenreich, F (1943) The skull of *Sinanthropus pekinensis*: A comparative study on a primitive hominid skull. *Palaeontol. Sin.*, New Series D 10:1-484.

Woo, J. (Wu, R.) (1964a) A newly discovered mandible of the *Sinanthropus* type—*Sinanthropus lantianensis. Sci. Sin.* (English edition) 13:801-811.

Woo, J. (Wu, R.) (1964b) Mandible of *Sinanthropus lantianensis. Curr. Anthropol.* 5:98-99.

Woo, J. (Wu, R.) (1965) Preliminary report on a skull of *Sinanthropus lantianensis* of Lantian, Shensi. *Sci. Sin.* (English edition) 14:1032-1035.

Woo, J. (Wu, R.) (1966) The skull of Lantian man. *Curr. Anthropol.* 7:83-86.

Wood, B. A. (1984) The origins of *Homo erectus. Courier Forsch. Senck.* 69:99-112.

Wu, M. (1983) *Homo erectus* from Hexian, Anhui found in 1981. *Acta Anthropol. Sin.* 2:109-115 (in Chinese).

Wu, R. (Woo, J.), and Dong, X. (1980) The teeth from Yunxian, Hubei. *Vert. PalAs.* 18:142-149 (in Chinese).

Wu, R. (Woo, J.) and Dong, X. (1980) The teeth from Yunxian, Hubei. *Verebrata. PalAsiatica.* 18:142-149 (in Chinese).

Wu, R. (Woo, J.), and Dong, X. (1982) Preliminary study of *Homo erectus* remains from Hexian, Anhui. *Acta Anthropol. Sin.* 1:2-13 (in Chinese).

Wu, R. (Woo, J.), and Dong, X. (1985a) Retrospect and prospect of the study of Peking man. In R. Wu et al. (eds.): *Multidisciplinary Study of the Peking Man Site at Zhoukoudian.* Beijing: Science Press, pp. 86-94 (in Chinese).

Wu, R. (Woo, J.), and Dong, X. (1985b) *Homo erectus* in China. In R. Wu and J. W. Olsen (eds.): *Palaeoanthropology and Palaeolithic Archaeology in the People's Republic of China.* Orlando: Academic Press, pp. 79-89.

Wu, R. (Woo, J.) and Lin, S. (1983) Peking Man. *Sci. Am.* 248:78-86.

Wu, R. (Woo, J.), and Lin, S. (1985) Chinese palaeoanthropology: Retrospect and prospect. In R. Wu and J. W. Olsen (eds.): *Palaeoanthropology and Palaeolithic Archaeology in the People's Republic of China.* Orlando: Academic Press, pp. 1-27.

Wu, R. (Woo, J.), and Wu, X. (1982) Human fossil teeth from Xichuan, Henan. *Vertebrata PalAsiatica.* 20:1-9 (in Chinese).

Wu, R., Ren, M., Zhu, X., Yang, Z., Hu, C., Kong, Z., Xie, Y., and Zhao, S. (eds.) (1985) *Multi-disciplinary Study of the Peking Man Site at Zhoukoudian.* Beijing.

Xie, Y., Xing, J., Yu, J., Zhou, B., Huang, Y., Liu, Y., Yang, J., Cui, Z., Li, R., and Sun, X. (1985) The sedimentary environment of the Peking Man period. In R. Wu et al. (eds.):

Multidisciplinary Study of the Peking Man Site at Zhoukoudian. Beijing: Science Press, pp. 185-215 (in Chinese).

Xu, C. (1978) The excavation of the Yunxian *Homo erectus* locality in Hubei. In Institute of Vertebrate Palaeontology and Palaeoanthropology, Chinese Academy of Sciences (ed.): *Gurenli Lunwenji (Collected Papers of Palaeoanthropology)*. Beijing: Science Press, pp. 175-179 (in Chinese).

Xu, S. (1986) Discovery of human fossils at Qizianshan, Yiyuan county, Shandong. *Acta Anthropol. Sin.* 5:398-399 (in Chinese).

Xu, Q. (1990) General trend of the development of the Quaternary stratigraphy using the study of Lantian Quaternary stratigraphy as an example. *J. Stratigraphy* 14:77-80 (in Chinese).

Xu, Q., and Ouyang, L. (1982) Climatic changes during Peking Man's time. *Acta Anthropol. Sin.* 1:89-91 (in Chinese).

Xu, Q., and You, Y. (1984) Hexian fauna: Correlation with deep-sea sediments. *Acta Anthropol. Sin.* 3:62-66 (in Chinese).

Yan, G. (1993) A preliminary study on magnetic stratigraphy of the geological section with the fossil bed of Yunxian *Homo* of Hubei. *Earth Science-Journal of China University of Geosciences* 18:221-226 (in Chinese).

Yang, Z., and Mou, Y. (1982) Latest conceptions of the late Cenozoic strata in Zhoukoudian (Choukoutien). *Sci. Bull. (Kexue Tongbao)* (English ed.) 27:537-543 (in Chinese).

Yang, Z., Mou, Y., Qian, F., Wang, X., Niu, P., Chen, H., Yin, W., and Wei, X. (1985) Study of the late Cenozoic strata at Zhoukoudian. In R. Wu et al. (eds.): *Multidisciplinary Study of the Peking Man Site at Zhoukoudian*. Beijing: Science Press, pp. 1-87 (in Chinese).

Yuan, S., Chen, T., Gao, S., and Hu, Y. (1991) Study on uranium series dating of fossil bones and teeth from Zhoukoudian site. *Acta Anthropol. Sin.* 10:189-193 (in Chinese).

Zhang, Y. (1991) An examination of temporal variation in the hominid dental sample from Zhoukoudian Locality 1. *Acta Anthropol. Sin.* 10:85-95 (in Chinese).

Zhang, Y., Huang, W., Tang, Y., Ji, H., You, Y., Tong, Y., Ding, S., Huang, X., and Zheng, J. (1978) Cenozoic stratigraphy of the Lantian region, Shaanxi. *Mem. Inst. Vert. Paleontol. and Paleoanthropol.*, Series A 14. Beijing: Science Press.

Zhao, S., Xia, M., Zhang, Z., Liu, M., Wang, S., Wu, Q., and Ma, Z. (1980) A study of the age of Peking Man by the uranium series method. *Sci. Bull. (Kexue Tongbao)* 25:182.

Zhao, S., Xia, M., Zhang, Z., Liu, M., Wang, S., Wu, Q., and Ma, Z. (1985a) Uranium series dating of Peking Man site. In R. Wu et al. (eds.): *Multidisciplinary Study of the Peking Man Site at Zhoukoudian*. Beijing: Science Press, pp. 246-250 (in Chinese).

Zhao, S., Pei, J., Gao, S., Liu, S., Qian, F., Qiu, S., and Li, X. (1985b) Study of chronology of Peking Man site. In R. Wu et al. (eds.): *Multidisciplinary Study of the Peking Man Site at Zhoukoudian*. Beijing: Science Press, pp. 239-240 (in Chinese).

Zheng, S. (1982) Character and significance of Hexian man micro-mammalian fauna. *Sci. Bull. (Kexue Tongbao)* 26:683-685 (in Chinese).

Zhou, M., and Ho, C. K. (1990) History of the dating of *Homo erectus* at Zhoukoudian. *Geol. Soc. Am. Special Paper* 242:69-74.

Zhou, M., and Li, C. (1965) Mammalian fossils in association with the mandible of Lantian man at Chen-chia-ou, in Lantian, Shensi. *Vert. PalAs.* 9:377-393 (in Chinese).

CHAPTER 29

Mechanical and Other Perspectives on Neandertal Craniofacial Morphology

Susan C. Antón

INTRODUCTION

Based upon inferences of Neandertal posture and locomotion, Boule and Vallois (1957; Boule 1923) argued that the differences between Neandertals and modern humans were substantial enough to warrant separate specific designations and to remove Neandertals from the *Homo sapiens* lineage entirely. But to what extent and by what method was the initial behavioral inference tested? Anthropologists, as evolutionary biologists, develop functional/adaptational hypotheses concerning the peculiarities of fossil primate morphology. From these inferences of function (activity), we build scenarios of a phylogenetic nature whose robusticity depends largely upon the validity of our initial behavioral inference.

In recent years there has been an increase in the use of biomechanical principles as methods of testing or promoting hypotheses concerning the functional capabilities of fossil hominids (Rak 1986; Demes 1987; Demes and Creel 1988; Antón 1990). This recent move toward quantification of possible activities and their ranges is the logical continuation of the search for functional explanations for fossil morphologies. In many cases these studies have provided some means of bounding the ranges of capability of fossil forms (Demes and Creel 1988; Antón 1990). The ability to bound the results of these studies remains the single most important aspect of improving our ability to distinguish between competing hypotheses of activity in fossil forms.

There is no dearth of literature concerning functional implications of the Neandertal face (e.g., Brace 1963, 1964; Smith 1983; Rak 1986; Demes 1987; Trinkaus 1987). The debate between nonmechanical (sexual selection, cold adaptation, gene drift) and mechanical hypotheses and among the mechanical hypotheses has been a persistent hallmark of this search. The methods of evidence have changed from pure supposition to anecdotal comparative-anatomical evidence, experimental data, stringent biomechanical interpretation, and well-controlled comparative data. This is not to imply that this change of evidence has been unilineal or complete. However, general interest (pure speculation) concerning the functional implications of morphology preceded more refined methods.

This chapter considers some of the history of the Neandertal craniofacial debate and its implications for phylogeny. Particular focus will be placed on the biomechanical perspective so recently popular and new data will be brought to bear on these issues.

NONMECHANICAL HYPOTHESES

In the 1950s cold adaptation was the paradigm behind the evolution of the Neandertal face (Howell 1952, 1957; Coon 1962). Howell made a strong case for such an adaptation by correlating the existence of a classic Neandertal morphology with data on glacial periods. While subsequent evidence has changed the dating and hence the conclusions drawn from some of this material, more involved morphological and physiological evidence has also been used to argue against a cold-adapted Neandertal face (see Mann and Trinkaus 1974; Hylander 1977).

Recent workers have revived the hypothesis in a slightly altered form arguing that the inferred curvature of the external nose was advantageous for moisture retention in a cold and dry climate (Franciscus and Trinkaus 1988; Trinkaus 1989). Coupled with inferences of high activity levels and associated increases in body heat, the nose is suggested to be both a moisture retainer and heat dissipator (Franciscus and Trinkaus 1988; Trinkaus 1989). Laitman and et al. (1992) also suggest that the large Neandertal nose was adapted to the cold/dry environment indicating that Neandertals were nasal rather than oral respirators.

Hublin (1990) considers glacial environments to be important to the production of the derived Neandertal morphology in an indirect way. He suggests that much of Neandertal morphology is attributable to gene drift resulting from glacial isolation of small groups in Europe (*sensu* Howell 1960, 1964; Boaz et al. 1982). He correlates new data on oxygen isotope stages with Neandertal and pre-Neandertal (e.g., Arago) morphologies in a convincing argument. While Hublin (1990) does not rule out direct adaptive roles for some Neandertal morphologies (e.g., limb proportions), he considers their highly derived cranial morphology to be more likely related to gene drift.

Brothwell (1975) hypothesized that differential growth rates resulted in the distinctive Neandertal facial morphology. In particular, he suggested that early Neandertal maturation (epiphyseal closure) followed by a long post-pubertal growth period could explain Neandertal craniofacial and postcranial morphology. More recently Green and Smith (1990; Smith and Green 1991; Smith 1991) have also suggested that developmental differences between Neandertals and modern humans account for morphological differences in their craniofacial skeletons. They specifically identify acceleration of in utero growth rates in basicranial cartilages (midsphenoidal and sphenoccipital synchondroses) as the cause of the distinctive Neandertal facial morphology. They cite other examples of the accelerated growth rate in Neandertals (Dean 1985) as supporting evidence for their suggestion. Heim (1978) suggests that the cause of Neandertal facial morphology is the enlargement of the facial sinuses. The importance of these contributions lie not in their specific arguments but in their explicit recognition of evolutionary changes as developmental phenomena. We rely so often upon the comparison of adult forms that we misplace the notion that evolution proceeds by changing the developmental pattern (Hurov 1991; Thomson 1992). We also incorporate into our hypotheses the notion that the adult form is the goal as opposed to the product of a particular developmental history. And perhaps because functional hypotheses are easier to comprehend and to test, we tend to dismiss developmental/structural hypotheses.

MECHANICAL HYPOTHESES

The majority of mechanical hypotheses have centered around concepts of high masticatory or gravitational forces acting on the prognathic Neandertal face. Little attempt has been made to distinguish functional factors important in the origin of the Neandertal face from the functional effects of Neandertal facial morphology. Most functional interpretations of Neandertal facial morphology are assumed to represent functional causes driving the evolution of the Neandertal craniofacial complex.

Supraorbital torus

The earliest overtly mechanical interpretation of Neandertal cranial form was made by Dubois (1922). He rejected the popular notion that the "simian form" of the Neandertal vault was a primitive retention related to brain organization and sought a mechanical reason for this vault form. Dubois (1922) argued that the long, low vault was in part a consequence of the massive prognathic face. This face was, he inferred, quite heavy and during locomotion required muscular force to act in opposition to its gravitational pull (i.e., large nuchal muscles) and a strong epicranius muscle to keep the thin vault bones from cracking and collapsing (Figure 29-1). The strength of the epicranius muscle encouraged the development of a large Neandertal supraorbital torus. In addition, it exerted force inferiorly onto the vault causing the vault to grow posteriorly rather than superiorly. Dubois (1922) marshalled comparative anatomical data to support his arguments emphasizing comparisons between vault-form in gibbons and larger faced siamangs. However, the most important supposition, the action of the epicranius muscle in supporting the vault bones, remained unsubstantiated.

An alternative reason for Neandertal browridge formation was provided by Tappen (1978). He suggested that the high frequency with which Neandertal brows exhibit vermiculate bone was due to the adaptive advantage of unorganized bone in a high-stress environment. Tappen suggests that vermiculate bone was better able to resist fracture due to its disorganized pattern and therefore mitigated the effects (increased the survivability) of cranial trauma; which he suggests was an important aspect of Neandertal life (*sensu* Boule and Vallois 1957). Oyen et al. (1979) via a histological examination concluded that the appearance of vermiculate bone is a stage in the depositional history of the brow. They tacitly link their argument with the anterior dental loading hypothesis by suggesting that changes in bony deposition in the brow are correlated with changes in masticatory demands; although the nature of these changing demands is not examined further.

Other workers have argued that the presence of a supraorbital torus is a requirement to maintain structural continuity between cranial vault and prognathic face (Moss and Young 1960; Picq and Hylander

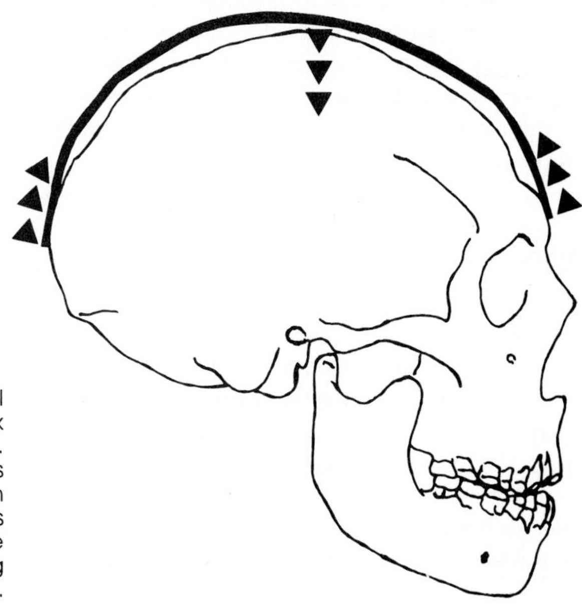

FIGURE 29-1. Lateral view of Neandertal silhouette (after Stewart 1958). The thick black line represents the position of the epicranius muscle. As envisioned by Dubois (1921) the epicranius contracts exerting force anteriorly and posteriorly in the direction of the arrows. This exerts pressure on the supraorbital torus which enlarges. Meanwhile, force is also exerted at the center of the vault inferiorly (see central arrows) restricting SI vault growth. The result is a long, low vault form.

1989; Hylander et al. 1991; Ravosa 1991). Smith and Ranyard (1980; Smith 1991) draw this general conclusion with regard to Neandertals but also conclude that the relative size of the Neandertal torus suggests that a stress-bearing function was also operating. Russell (1985) makes a particularly strong argument in favor of the functional significance of the supraorbital torus. It is important to note here that recent *in vivo* studies found absolutely low strains in the supraorbital region of masticating primates (Hylander et al. 1991). Recorded strains were so far below failure rates that buildup of additional bony mass on the basis of masticatory forces alone seems improbable. Similarly, reanalysis of Endo's (1966) classic work also revealed little support for high strain in this region (Picq and Hylander 1989). The testing of these competing hypotheses by experimental methods clearly indicates that supraorbital tori more likely result from structural continuity requirements than from counteracting masticatory forces.

Dentition

Several authors have examined Neandertal dentitions to unravel the uses to which Neandertals put their teeth. Examination of the La Ferrassie dentition suggested to Boule and Vallois (1957) that Neandertals produced greater anterior movements of the lower jaw than modern humans. They supported this inference by citing the relatively large, shallow Neandertal glenoid fossae. Smith (1976) found that the severity of Neandertal dental attrition was greater than in earlier *Homo* and suggests this was caused by increased functional demands on the dentition and concomitant dental reduction. Such an inference led her to conclude that it was small teeth in Neandertal children that were selected for since small adult teeth were evidently less advantageous.

Alternatively, Puech's (1981) examination of the occlusal wear patterns of La Ferrassie I and II led him to refute a "teeth-as-tools" origin for these patterns. Instead, he suggests that horizontal grinding of meat, plant, and accompanying grit generated the observed wear patterns (*sensu* Boule and Vallois 1957). However, Brace et al. (1981), reporting on comparative evidence of dental wear in gorillas (largely plant and grit diet) and eskimos (high levels of paramasticatory activity), found Neandertal dental wear more closely reflected paramasticatory activities. Indeed Neandertals do show wear patterns that in modern humans are considered indicative of paramasticatory activity (occlusal surface form categories of 5, notched, and 6, rounded, of Molnar 1971).

These authors are thus in agreement that Neandertal dentitions show evidence of some paramasticatory activity (Brace et al. 1981; P. Smith 1976; F. H. Smith 1983; Trinkaus 1987). However, there is little consensus as to whether this activity was more intense than in other modern or archaic hominids. My own experience with California native populations suggests that Neandertals were not unique in their degree of paramasticatory activity and related dental attrition. This is thus not a persuasive argument for craniofacial evolution.

Anterior Dental Loading: Teeth-as-Tools

Overarching examinations have sought to explain the evolution of Neandertal craniofacial morphology via reference to Neandertal masticatory activities—in particular unusual loading of the anterior dentition (Brace 1964; Smith 1983; Rak 1986; Demes 1987; Smith and Paquette 1989). In a detailed discussion of each portion of the Neandertal craniofacial skeleton, Smith (1983) explains what he sees as the "logical" morphological results of Neandertal anterior dental loading. This work (Smith 1983) offers important suppositions to be tested.

Both Rak (1986) and Demes (1987) offer alternative, theoretical, mechanical analyses of forces in the Neandertal craniofacial skeleton. Both assume the presence of either repetitive or absolutely high anterior dental loads. The perspective, as Rak (1986) explains it, is that absolutely high force and torsion were being exerted at the anterior dentition when Neandertals tried to gnaw on objects bigger than their heads. Demes (1987) is careful to point out the hypothetical nature of this type of examination and the need to test the hypotheses it generates.

Most researchers promoting the anterior dental loading hypothesis agree that the geometry of the Neandertal craniofacial skeleton is not mechanically advantageous for producing masticatory force (Smith 1983, 1991; Rak 1986; Trinkaus 1987). Nonetheless, Neandertals are presumed by some to have produced absolutely high anterior forces. Researchers should be careful to differentiate repetitive anterior dental loading from absolutely high bite forces at the anterior dentition. Although Neandertal craniofacial geometry makes the latter improbable (see later discussion), it makes the former only physiologically expensive. Supporting evidence for high anterior force production includes the greater size and degree of wear and extensive enamel chipping of the anterior relative to posterior dentition and degenerative joint disease (DJD) at the temporomandibular joints (TMJ).

PLAUSIBILITY OF THE ANTERIOR DENTAL LOADING HYPOTHESIS

Could Neandertals Produce Absolutely High Anterior Dental Loads?

A simple vector analysis model was used to test whether the assumption of absolutely great anterior dental loads in Neandertals is warranted (Antón 1990). The importance of this study is the attempt to predict the range of muscle force producing capabilities from bony morphology and to use these predictions as input for mechanical models that say something about absolute bite force production. Portions of this analysis are presented here.

Materials

Casts of cranial and mandibular remains of Amud I were used to determine cross-sectional areas, orientations, and moments arms for the Neandertal model (see later discussion). Cranial remains from La Ferrassie, France, were used in determining the area between the zygomatic arch and the vault due to damage in the Amud specimen. These specimens were chosen on the basis of their relatively complete state and their morphological approximation of one another.

Methods

The mandible was modelled as a lever with its fulcrum at the condyle. Because the total muscle force vector (Fm) is positioned posterior to the bite point, in static equilibrium both useful bite force (Fb) and condylar reaction force (Fc) are produced (Hylander 1985:Figure 29-2). To estimate the absolute Fb production in Neandertals it is therefore necessary to understand the relationship between Fb, Fm, and Fc and to be able to estimate Fm.

To estimate muscle force it is necessary to define both the direction and the magnitude of the force for each muscle modelled. The muscles used here (superficial masseter, medial pterygoid, and temporalis) are considered power muscles (Osborn and Baragar 1985). The lateral pterygoid muscle was not used in this model because it has a poor moment arm for producing useful Fb and acts mainly to stabilize the condyle (Hylander 1985).

Experimental studies indicate that the muscles modelled here are functionally heterogenous pinnate muscles (Moller 1966; Herring et al 1979). However, due to the inherent difficulties of determining fiber angle from fossil remains, and most important, the lack of comparative data on fiber angles and their variability in human muscles of mastication, this model assumes simple parallel-fibered muscles. This assumption allows muscular orientations and force vector directions to be inferred from bony morphology (van der Klaauw 1963; Gans and Bock 1965; Gans and Vrees 1987).

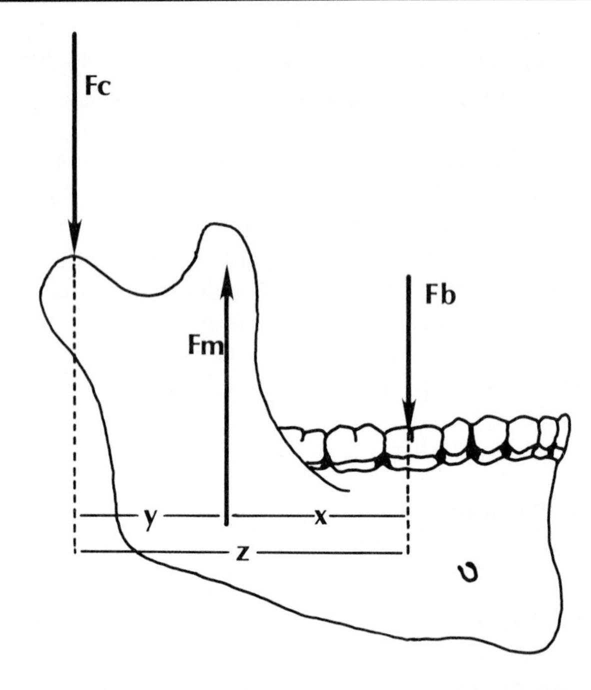

FIGURE 29-2 Lateral projection of external forces acting on the human mandible during biting. Fc, Fm, and Fb are total condylar reaction, muscle and bite forces respectively. In conditions of static equilibrium these forces are related as follows: Fm = (Fb)(z)/y Fb = (Fm)(y)/z Fc = (Fb)(x)/y

Muscle force vector direction was modelled as a single vector positioned as the central fiber (essentially the centroid: Hiiemae 1971) within the body of each muscle. Vector direction was determined by joining the areas of origin and insertion by this central line (Figure 29-3). The greater surface area under the temporal line and the further posterior extension of this line indicates a larger posterior temporalis component in Neandertals than in modern humans. The temporal fossa is similarly elongated posteriorly, providing an increased advantage to some of these posterior fibers over the condition in modern humans.

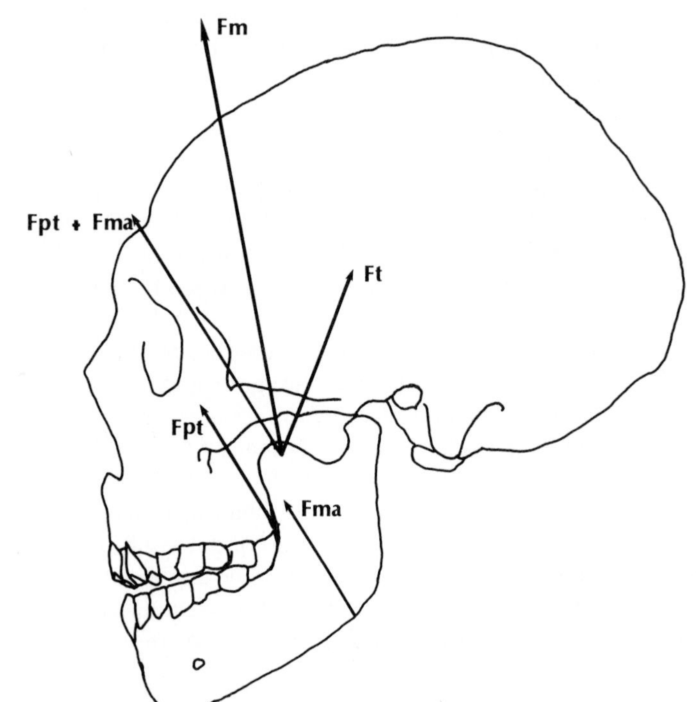

FIGURE 29-3.
Estimated Neandertal muscle force vectors. Vectors are combined right and left force for masseter (Fma), medial pterygoid (Fpt) and temporalis (Ft). Fm = 2010 N. (Silhouette after Trinkaus 1983).

For these reasons the resultant temporalis muscle vector direction has been placed slightly more obliquely than in modern humans (Figure 29-3). This placement is not critical to the determination of maximum occlusal and condylar reaction force magnitudes. The direction would, however, significantly affect the direction of the condylar reaction force vector (Throckmorton 1985). The resultant vectors of the masseter and medial pterygoid muscles are similar to modern humans.

Assuming simple muscle architecture, the maximum force a muscle can exert is equal to the product of its physiological cross-sectional area (i.e., an estimate of the number of muscle fibers firing in unison) and the stress in the muscle (Weijs 1980; Dul et al. 1984). Muscle cross-sectional areas (Table 29-1) were determined as follows: (1) The area enclosed between the zygomatic arch and the cranial vault estimated temporalis; (2) The triangular area formed by the gonial angle and the mylohyoid groove estimated medial pterygoid; (3) The product of the length of the masseteric origin on the zygomatic arch and the distance between the lateral edge of the zygomatic arch and the lateral edge of the mandibular ramus estimated masseter. Cross-sectional areas for these muscles in modern humans are given in Table 29-2.

These measurements were also taken on a modern human skull and compared to physiological cross-sections for these muscles (Table 29-2). Correction factors were determined by assuming a linear relationship between physiological and bony cross-sectional area. Cross-sectional area of the temporalis muscle was overestimated by a value of 6. The medial pterygoid was overestimated by a value of 1.5. The masseter value was not corrected. The Neandertal cross-sectional areas were corrected as previously noted (Table 29-1).

The force and cross-sectional areas reported by Schumacher (1961) for modern humans were used to calculate the stress in each muscle (Table 29-1). Given the close phylogenetic relationship of Neandertals and modern humans, stress was assumed to be the same in both groups.

Individual muscle force magnitude was calculated using the product of the corrected cross-sectional areas for Neandertal muscles and the modern human muscle stresses (Table 29-1). Position, direction, and magnitude of the combined muscle force (Fm) was determined by simple vector analysis after projecting all vectors onto the same plane (Figure 29-3).

External forces at the TMJ were analyzed by determining moments about the mandibular condyle, assuming static equilibrium, in lateral projection with the jaw in closed position and with a fixed center of rotation. Lateral projection analyses derive two of the three variables (Fm, Fb, and Fc) from the third (known) variable. In this model, Neandertal Fb is determined from Fm (see later discussion: Table 29-3). For the system to be in static equilibrium, the summation of the moments around any point is equal to zero (Figure 29-2; Hylander 1975; Smith 1978). Fb is measured perpendicular to the occlusal plane. Forces were calculated for both molar (first molar, M/1) and incisal (lateral incisor, I/2) biting (Table 29-3), and the forces were considered point loads. A frontal projection analysis was performed to consider unilateral biting (Figure 29-4; see Hylander 1985). Bending and twisting moments produced by the positioning of the Fm lateral to the Fb were not considered. As such, Fb is likely to be overestimated as all components of Fm were considered to produce useful Fb.

TABLE 29-1 Estimated Neandertal muscle cross-sectional area and force (for one side)

Muscle	Uncorrected Cross section cm^2	Corrected Cross section cm^2	Human Stress $kgm^{-1}s^{-2}$	Force N
Temporalis	3.2	5.5	$8.4x10^5$	$4.6x10^2$
Masseter	3.7	3.7	$8.4x10^5$	$3.1x10^2$
M. Pterygoid	5.5	3.8	$9.3x10^5$	$3.5x10^2$

TABLE 29-2 Modern Human bony and physiological cross-sectional areas.

Muscle	Bony Cross section cm^2	PhysiologicalCross section cm^2	Correction Factor
Temporalis	2.4×10^1	4.2	6
Masseter	3.4	3.4	-
M. Pterygoid	2.8	1.9	1.5

Results

Corrected muscle cross-sectional areas for Neandertals are slightly larger than those for the modern human muscles (Schumacher 1961:Tables 1,2). Muscle force magnitude estimates are thus slightly larger for Neandertal masseter and temporalis and twice as large for the medial pterygoid estimates than those of Schumacher (1961). As in modern humans, the combined force of the Neandertal masseter and medial pterygoid is slightly greater than that of the temporalis (Carlsöo 1952; Schumacher 1961; Pruim et al. 1980). These estimates are comparable to those of Pruim et al. (1980) and lower than those of Carlsröo (1952). Carlsöo's (1952) larger force determinations are due largely to his use of a higher muscle stress value (1.1 times 10^6 N/m2).

Given the estimates of Fm, bite force production at both molar and incisal bite points is 20-22 percent smaller in Neandertals than in modern humans (Table 29-3). This is despite a more than 15 percent increase in estimated muscle force production capability in Neandertals. However, the relative proportion of Fb at I2 relative to Fb at M1 is nearly identical (approximately 70 percent: Table 29-4).

Neandertal condylar reaction force is substantially greater at both molar and incisal bite points than is bite force. At M1 the Fc is 158 percent of Fb and at I2 Fc is 264 percent of Fb (Table 29-4). In contrast, Fc in modern humans is 74 percent of Fb at M1 and 145 percent at I2. The absolute values of Fc are 30 to 40 percent greater than for modern humans (Table 29-3).

Neandertals could produce less useful Fb per Fm than modern humans at a much greater expense to the TMJ. In Neandertals, bite force is only 39 percent and 28 percent of Fm at M1 and I2, respectively while it is 57 percent and 42 percent of Fm in modern humans (Table 29-4). Likewise, Fc is 61 percent and 72

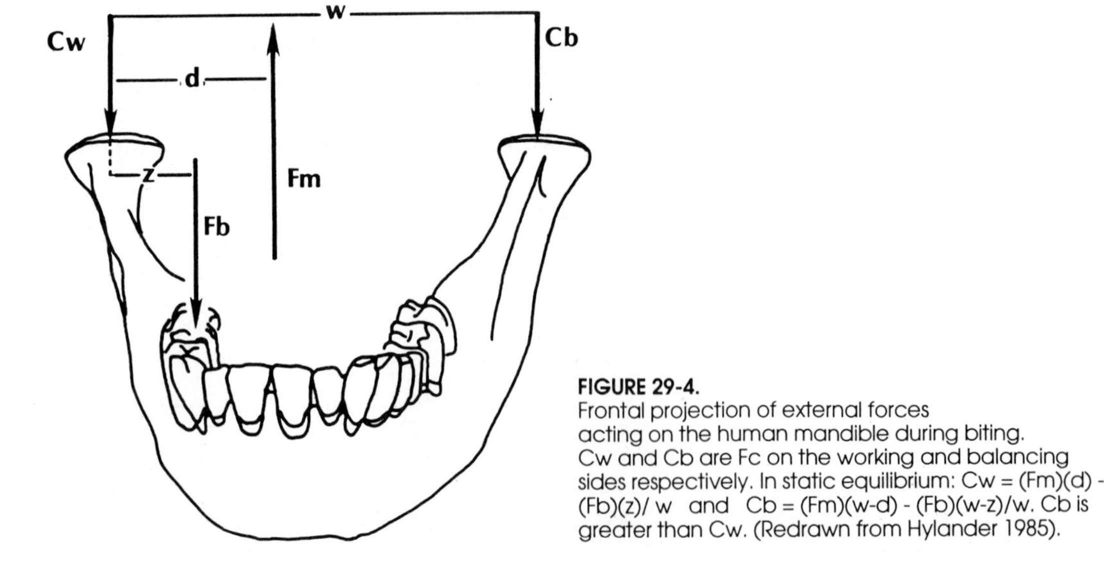

FIGURE 29-4.
Frontal projection of external forces acting on the human mandible during biting. Cw and Cb are Fc on the working and balancing sides respectively. In static equilibrium: Cw = (Fm)(d) - (Fb)(z)/ w and Cb = (Fm)(w-d) - (Fb)(w-z)/w. Cb is greater than Cw. (Redrawn from Hylander 1985).

TABLE 29-3 External forces (in Newtons) at the TMJ in molar and incisal biting determined in lateral projection

	Molar Biting		
	Fb	Fm	Fc
Modern Human Pruim et al. 1980	9.6×10^2	1.6×10^3	7.1×10^2
Neandertal	7.8×10^2	2.0×10^3	1.2×10^3
	Incisal Biting		
	Fb	**Fm**	**Fc**
Modern Human Pruim et al. 1980	7.0×10^2	1.6×10^3	9.9×10^2
Neandertal	5.5×10^2	2.0×10^3	1.5×10^3

percent of Fm at M1 and I2, respectively, in Neandertals. Fc is only 42 percent and 59 percent of Fm in modern humans.

As is true for modern humans, frontal projection analysis indicates that the Neandertal balancing side condyle is more heavily stressed than the working side condyle (Smith 1978: Table 29-5). The greater Fc in Neandertals as opposed to modern humans is reflected in the greater percentage of Fb represented by Fc (Table 29-5). Neandertal Fc is 62 percent of Fb at the working side condyle and 92 percent of Fb at the balancing side during molar biting. During incisal biting Fc is 87 percent of Fb on the working side and 171 percent of Fb on the balancing side. Thus, incisal biting is extremely costly. In modern humans Fc is never more than 71 percent of Fb.

Discussion

Neandertal facial prognathism has been suggested to be the result of the rearrangement of the infraorbital plates and the anterior migration of the tooth row with respect to the mandibular ramus (Rak 1986). The relatively prognathic Neandertal midface results in more anteriorly placed cranial but not mandibular muscular attachments. Thus, masticatory muscular relations with respect to the TMJ are not greatly altered by this facial arrangement. However, the Fb moment arms are elongated resulting in the production of less useful bite force and a very large condylar reaction force. Similar results were found by Grey (1992) using the three-dimensional computer-assisted model of Osborn and Baragar (1985).

Contrary to this, Spencer (1991) found equivalent mechanical advantage ratios (load arm/lever arm ratio) between Neandertals and modern humans. He ascribes this similarity to more anteriorly positioned

TABLE 29-4 Relationships between masticatory forces for modern humans and Neandertals

Fb @ I2	Modern Humans		Neandertals	
Fb @ M1	73%		71%	
	molar	**incisal**	**molar**	**incisal**
Fc/Fb	74%	145%	158%	264%
Fb/Fm	57%	42%	39%	28%
Fc/Fm	42%	59%	61%	72%

TABLE 29-5 Working and balancing side condylar reaction forces as a percentage of useful bite force. (Modern human data from Smith 1978)

	Molar		Incisal	
	Working	Balancing	Working	Balancing
Human	15%	63%	71%	71%
Neandertal	62%	92%	87%	171%

total muscle force resultant relative to the TMJ in Neandertals, due entirely to the more anterior placement of the medial pterygoid origin. The difference between these studies lies largely in their estimation of muscle force vectors. Spencer (1991) and Spencer and Demes (1993) estimate position of each muscle relative to the TMJ on the basis of cranial attachments alone. Grey (1992) and I (1990, this volume) determine vector direction by connecting the mandibular and cranial attachments with a line through their centroids (Hiiemae 1971). This is an important methodological difference when considering Neandertals because the relatively prognathic Neandertal middle face results in more anteriorly placed cranial but not mandibular muscular attachments. Thus the degree of anterior migration of the muscle force direction vectors is exaggerated by using only their cranial positions. As a result, the appearance of equivalent mechanical advantage is found.

The consistent appearance of DJD is most likely due to the proportionally larger Fc produced at the Neandertal TMJ. Even moderate occlusal loading inflicts large reaction forces at the condyles leading, over time, to this degeneration. The physiological restrictions imposed by the production of large Fc make it unlikely that the attrition of the anterior dentition is related to absolutely greater occlusal loading. This absence of large anterior occlusal loads is corroborated by the presences of only minor trauma (enamel microfracture and flaking) in Neandertal anterior teeth (Smith 1983). This conclusion contradicts Smith's (1983) suggestion that the minor trauma was due to absolutely larger teeth that were better able to bear greater loading.

Heavy anterior dental attrition is more likely due to repetitive usage of the anterior dentition in food preparation or other behaviors. Similar wear patterns are observable in prehistoric California groups that exhibit heavy anterior dental wear (personal observation). These patterns exist without the elongated facial geometry typical of the Neandertal face. Such paramasticatory behavior is also suggested by microwear studies of the Neandertal anterior dentition that show labial wear striae indicating the use of the anterior dentition in a viselike grip (Trinkaus 1983).

The assumption that absolutely large anterior occlusal loads contribute to bending moments of the Neandertal face cannot be supported. However, bending moments in the sagittal plane due to increased Fb moment arms are greater than those in modern humans. Additionally, forces generated by continuous use of the anterior dentition over time may be a factor in the evolution of facial morphology (see Hylander 1979).

The unique facial morphology of the Neandertals certainly affected their masticatory force production capabilities. However, the supposition that masticatory forces were the driving forces in the evolution of facial morphology begs the question of the origin of the morphology (see Rak 1986). Force production capabilities of Neandertals are disadvantageous compared to those of less prognathic hominids. Spencer and Demes (1993), who consider Neandertals more efficient than early *Homo*, concur that Neandertals are less efficient than less prognathic hominids (modern humans). They make the important point that Neandertal efficiency is best compared to the efficiency of its immediate forebears; an *already prognathic group*. Facial elongation (with or without the rearrangement of the infraorbital plates typical of the Neandertal face) is not an effective method of counter-balancing high anterior dental loading and sagittal bending moments when compared to forces incurred by less prognathic hominids. Therefore, it seems unlikely that facial prognathism *would have developed* as a direct result of heavy anterior dental loading *unless* the prognathic position of the face was advantageous for some separate reason.

FACIAL FORM AND VAULT FORM

An important avenue in the investigation of the Neandertal face lies in understanding the relationship between facial form and the vault and cranial base. That is, to what degree is Neandertal facial form explainable by and dependent upon the morphology of other cranial regions? This question can be approached by examining the morphological results of "natural experiments" such as intentional artificial vault deformation and congenital vault deformation in modern humans.

Artificial Deformation

To examine the consequences to facial morphology of altering vault morphology, artificially deformed human crania were metrically compared to undeformed crania. The deformed crania included anteroposteriorly (AP; n = 55) and circumferentially (C; n = 30) deformed crania from Peru and circumferentially deformed crania from Vancouver, British Columbia (V; n = 5; Figure 29-5). Anteroposteriorly deformed crania are AP foreshortened with compensatory lateral expansion while the circumferentially deformed crania are AP elongated and mediolaterally narrowed. The Vancouver crania are not as narrow and are superoinferiorly flattened (Figure 29-5). The undeformed sample (N; n = 45) is also from Peru. Older individuals (+40 years) and those with significant antemortem tooth loss were excluded from the samples. All crania are housed at the Phoebe Hearst Museum of Anthropology at UCB (HMA). Standard cranial along with specialized facial measurements were used to evaluate the degree and kind of effect vault deformation has on facial morphology. Specialized measurements were designed to evaluate where in the facial mask tendencies toward broadening or narrowing occur. These included nested sets of smaller, more midline facial breadths and more laterally inclusive units (Figure 29-6). Students' t-tests were used to compare Peruvian samples (Table 29-6).

There is a general tendency toward narrowing in the C and widening in the AP crania (Antón 1989). However, midline facial breadths (nasal, bi-infraorbital, bizygo-orbitale, bimaxillofrontalle, biorbital from the midorbital plane medially, interorbital, anterior ethmoid and palatal) are unaffected by deformation (Table 29-6). More laterally inclusive breadths (orbital, biorbital lateral of the midorbital plane, bisphenoidal wing, bi-superior orbital fissure, bimaxillary, and bizygomatic) account for the changes in facial breadths (Table 29-6).

Clearly, facial dimensions are affected by the presence of a relatively longer or shorter vault with respect to breadth. On average, longer, narrower vaults are associated with relatively narrower faces. This relative narrowness, at least in the artifically deformed cranium, is achieved lateral to the midline of the orbit and in the midface by the reorientation of the parts of the zygomatic bone. A core facial region representing the skeletal unit of a possible functional matrix (*sensu* Moss and Salentijin 1969) thus retains

FIGURE 29-5. Lateral silhouettes of cranial deformation types. A) Anteroposteriorly deformed, B) Undeformed, C) Circumferentially deformed, D) Circumferentially deformed Vancouver crania. Note elongation of cranial vaults in C and D. Note superioinferior shortening in D.

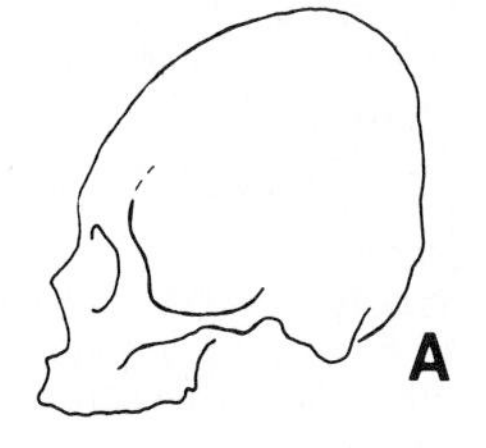

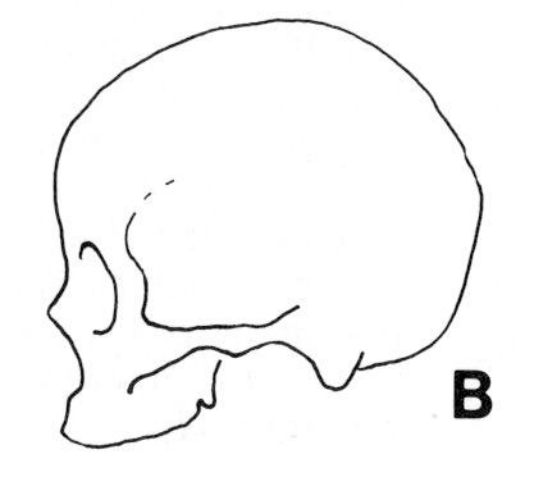

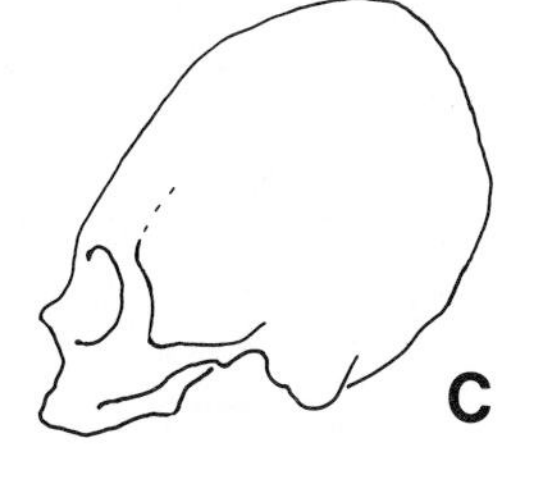

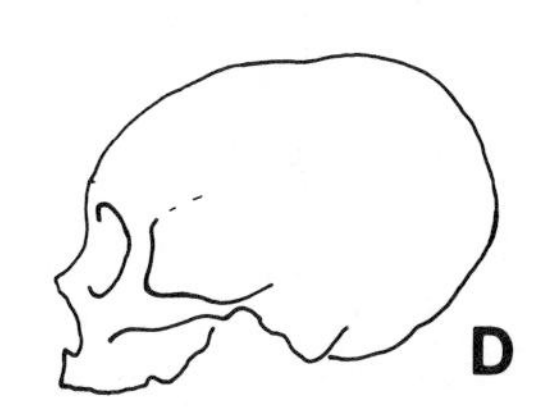

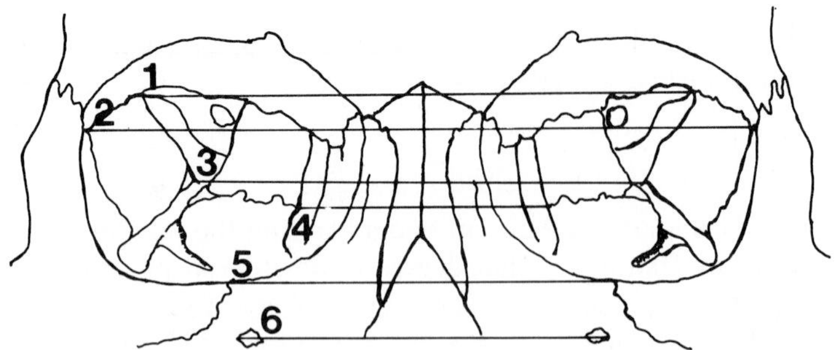

FIGURE 29-6.
Specialized orbital breadth measurements. (1) Bi-superior orbital fissure, (2) Bi-sphenoidal wing, (3) Bi-sphenoid root, (4) Anterior Ethmoid breadth, (5) Bi-zygoorbitale, and (6) Bi-infraorbital breadth.

its integrity. To accomplish this, first, the horizontal portion of the zygomatic is displaced anteriorly relative to the vertical portion. The angle between the two at the inferolateral orbital corner is increased (Figure 29-7). Second, the vertical portion of the zygomatic is reoriented more laterally; in facial view less of its external surface is visible than in lateral view (Figure 29-8). In addition to these reorientations of the zygomatic, the midsagittal plane of the face is positioned more anteriorly than the parasagittal planes. The impression is given that the zygomatics are retreating from the middle face. These conditions are magnified in the superoinferiorly shorter Vancouver circumferential crania. These changes are also affected by the degree of cranial elongation and the degree of intrinsic facial robusticity (i.e., male vs. female). Similar morphologies are seen in non-deformed dolichocephalic individuals (Figure 29-8: see later discussion).

Although these compensations are far less extreme than the morphology of the Neandertal midface, they suggest that some of the attributes of the Neandertal face are related to cranial vault elongation and supero-inferior shortening, specifically in the infraorbital and zygomatic regions.

Modern Human Variability

In Howells's (1973) extensive metric analysis of modern human cranial variability, he considered "facial forwardness"—the relationship between medial and lateral facial (particularly orbital) components. Neandertals score very high in this factor (Howells 1974) as recognized morphologically by most investigators (e.g., Smith 1991; Trinkaus 1987). The relationship between the projecting medial face and retreating lateral face is again tacitly linked to vault form. In this case, the highest scoring modern human groups are the Inugsuk Eskimos from Greenland, a strongly dolichocephalic group (Jorgensen 1953). Again these factors are far less developed than in Neandertals—some one to two standard deviations

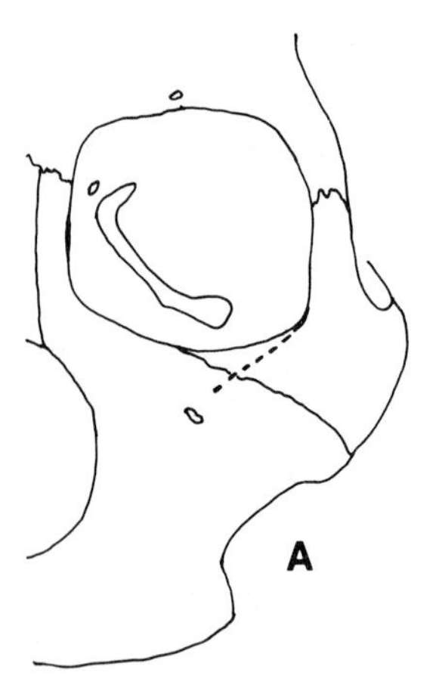

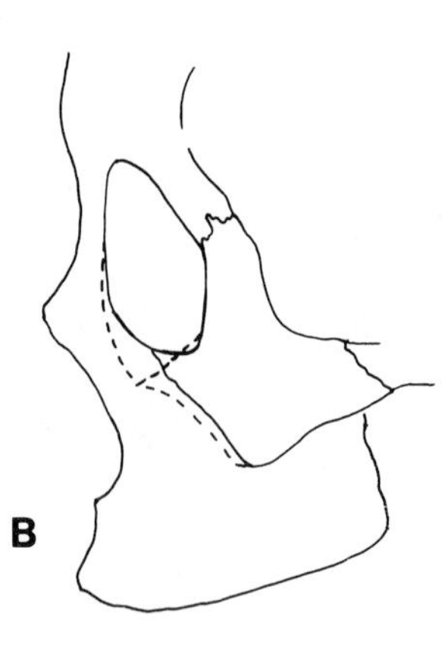

FIGURE 29-7.
Schematic representationof the relationship beween vertical and horizontal zygomatic compoents. Solid lines indicate position in undeformed crania, dashed lines represent circumferentially deformed crania.
(A) Anterior view: Note that the angle formed by the zygomatic at the inferolateral orbital corner is nearly 90 degrees in undeformed crania but significantly more obtuse in circumferentially deformed crania. This is due to inferior migration of the horizontal portion of the zygomatic.
(B) Lateral view: The reorientation of the horizontal zygomatic includes both inferior and anterior migration in the circumferentially deformed crania.

TABLE 29-6 Facial measurements of artificially deformed Peruvian samples.

	N			C			AP		
	n	**mean**	**sd**	**n**	**mean**	**sd**	**n**	**mean**	**sd**
Measurements[1]									
Orbital breadth[2]	45	37.7	1.5	30	36.4	2.0	55	38.5	1.5
Biorbital breadth[2]	45	95.1	3.6	30	92.7	4.3	55	97.6	3.4
Biorbital 1/4	35	40.2	2.2	22	40.5	2.9	38	40.9	2.3
Biorbital 1/2[3]	35	58.9	2.6	22	58.5	3,4	38	60.2	2.7
Biorbital 3/4[3]	35	77.6	3.2	22	76.5	3.9	38	79.5	3.2
Bimaxillary br[2]	43	90.1	4.0	26	87.8	5.4	55	91.7	1.5
Bizygomatic br[2]	39	130.0	5.7	24	128	7.1	48	134.2	6.9
Zygomatic angle[2]	35	96.1	5.0	22	100.6	1.5	38	93.0	5.4
Bi-orbital fissure[3]	35	47.9	3.1	21	46.4	3.3	37	49.9	2.5
Bi-sphenoidwing[3]	35	78.8	3.7	21	78.6	4.2	38	82.0	3.5
Bi-sphenoidroot	35	36.8	2.8	22	36.7	2.7	37	38.2	3.2
Interorbital br	35	21.7	1.7	22	22.1	2.8	38	22.2	2.0
Anterior ethmoid	32	28.6	3.0	19	28.9	3.6	38	28.8	2.8
Bimaxillofrontale	35	16.8	1.8	22	17.8	2.5	38	16.9	1.9
Bi-infraorbital br	35	55.2	3.2	22	55.7	4.3	38	55.2	3.4
Bi-zygoorbitale	35	56.4	4.7	22	56.5	4.8	38	54.9	5.3
Nasal breadth	45	23.3	1.8	26	23.8	1.4	55	22.7	2.7
Palatal breadth	29	37.2	2.7	16	35.6	3.1	35	37.1	2.2
Palatal length[4]	30	46.5	3.4	15	42.8	2.4	35	46.8	3.2
Orbital height[4]	45	34.8	1.9	30	36.4	1.8	55	35.4	1.9
N-Pr height[3]	45	63.7	4.2	28	61.6	5.5	55	66.4	3.9
Nasal height	45	48.4	2.7	30	48.1	3.5	55	49.4	2.7

[1]Standardized measurements follow Antón 1989. Customized measurements are defined in Figure six. Biorbital 1/4,1/2, and 3/4 are breadths taken from 1/4, 1/2, and 3/4 the orbital breadth distance from the medial orbital rim. Bimaxillofrontale is the distance between the left and right maxillofrontale.

[2]All pairwise comparisons significant at ≤ 0.05 level.

[3]AP group significantly different from all others.

[4]C group significantly different from all others.

below the Neandertal factor scores. However, they suggest a link between these key aspects of facial and vault form in hominids.

Visual examination of dolichocephalic and brachycephalic crania from a variety of geographic regions housed in the Atkinson collection, University of the Pacific (SRA), confirms the relationship between lateral facial recession and cranial elongation. The relationship is particularly strong in SI shortened individuals with strongly AP oriented cranial base axes (Figure 29-8).

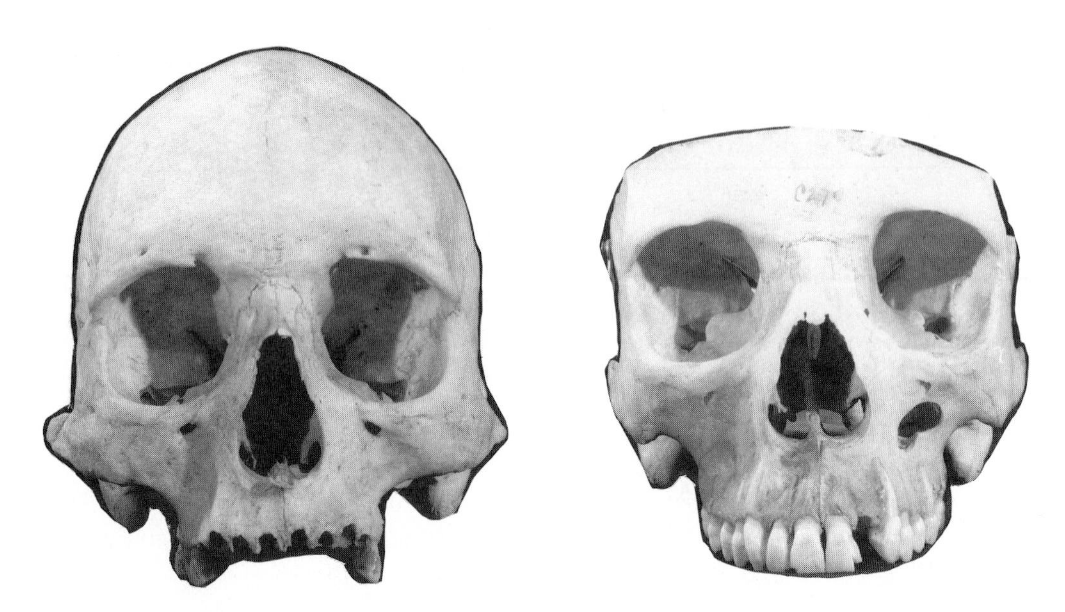

FIGURE 29-8. Facial view of dolichocephalic individual (SRA C-210) demonstrating lateral orientation of the zygomatic. Note how the portion of the zygomatic lateral of the zygomaticofrontal-zygomaticomaxillary line is oriented strongly laterally and is thus barely visible in this view.

FIGURE 29-9. Endocranial view of a dolichocephalic (left; SRA C-210) and scaphocephalic (right; HMA) individual. Note the strong AP orientation of the cranial base axis in the dolichocephalic individual versus the SI heightening in the scaphocephalic. The increased SI height of the scaphocephalic is markedly demonstrated in the steep relationship between the orbital plates of the frontal and the cribriform plate of the ethmoid.

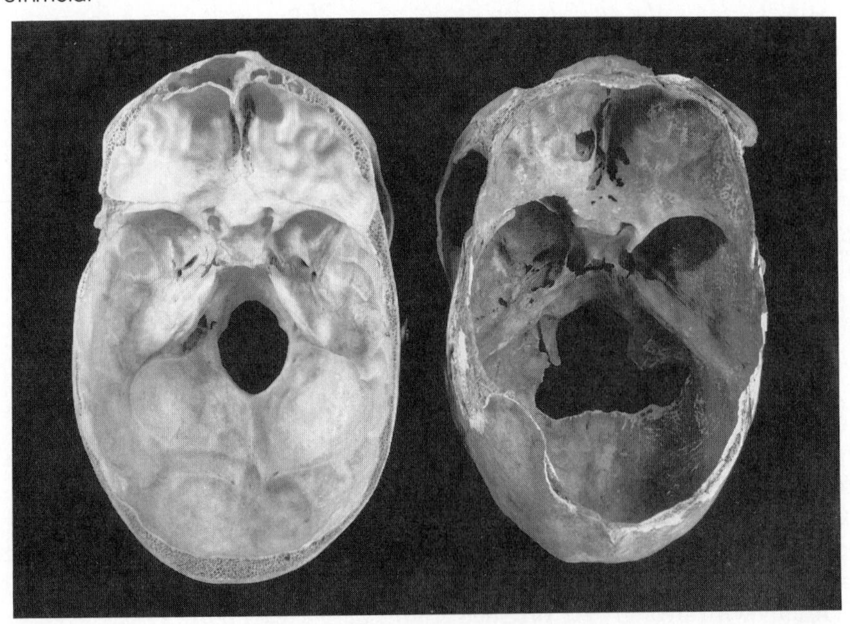

Congenital Vault Deformation

Premature sutural synostosis provides an important perspective on the relationship between vault form and facial morphology because it allows us to consider the impact of AP elongation and SI height increases. Sagittal synostosis is the most frequently seen simple synostosis (1:2000; Cohen 1986) and increases both AP vault length and SI height to some degree (Cohen 1986; Kohn et al. 1992). Given statements in the literature concerning the AP elongation resulting from sagittal fusion, zygomatic compensation similar to that seen in dolichocephalic nondeformed and circumferentially deformed individuals was anticipated.

Visual examination of 7 sagittally synostosed crania housed in the SRA and the HMA yielded some surprising results (Table 29-7). Two of the seven showed no AP elongation and no zygomatic compensation. Three more showed dramatic SI height increases especially in the cranial base (Figure 29-9) and mild AP elongation of the vault but not the cranial base. No zygomatic compensation was evident. Two individuals showed the expected zygomatic reorientation. These individuals exhibit AP elongation of both vault and base and are SI shortened particularly around bregma.

The skeletal evidence does not confirm the conventional clinical wisdom that sagittal synostosis predominantly causes AP elongation. SI heightening is most dramatic and most frequently seen in this sample. This increased height has significant impact on the inclination of the cranial base axis (Figure 29-9).

A long, low vault form is again implicated in the form of the zygomatic/lateral facial region. Sagittal synostosis examples suggest that the requirement of a low vault is related to the axis of the cranial base. Indeed, Howell (1951) noted this connection between vault and base form in Neandertals. AP elongation of both the vault and base is critical in producing the form of the zygomatic region. This is corroborated by artificial deformation studies that have shown that the C group has more obtuse anterior cranial base angles (N-S-Ba, Planum, and Orbital) than the undeformed base (Table 29-8; Antón 1989). Further, the Vancouver crania, which are the most strongly SI restricted, show the most obtuse cranial base angles (Table 29-8). Howell's (1951) data indicate Neandertal cranial base (planum) angles to be similarly flattened with values between 123° and 135°.

Discussion

Modern human data strongly suggest that the form and position of the lateral orbital/zygomatic components of the face are related to AP elongation and SI shortening of the cranial base and vault. By pointing out this structural relationship I am not arguing for Neandertals being phylogenetically more or less like modern humans. Nor am I denying that the Neandertals possess a remarkably derived cranial morphology. Rather, I believe it is important to recognize and understand the structural interdependencies of the cranial system. Evolution of any system entails both changes in morphology that are selected for and others that are only secondary occurrences. Understanding these morphological interrelations allows us to

TABLE 29-7 Sagittally synostosed individuals

	Vault Elongation?	Cranial Base Elongation?	Zygomatics Reoriented?
SRA			
D-165	Slight	Anteroposterior	Yes
D-293	Slight	Anteroposterior	Yes
D-226	Slight	Superoinferior	No
HMA			
12-3933	None	None	No
12-5524	None	None	No
12-5516	Slight	Superoinferior	No
12-6246	Slight	Superoinferior	No

TABLE 29-8 Cranial base angles for artificially deformed crania

	N n = 45		C n = 30		V n = 3	
Measurements[1]	mean	sd	mean	sd	mean	sd
Orbital angle	147.9	6.3	153.6	6.7	160.7	5.3
Planum angle	118.8	10.5	127.2	10.1	133.0	7.1
N-S-Ba angle	132.9	4.9	135.8	5.8	146.0	5.0

[1]Measurement definitions follow Antón (1989).

avoid developing complicated hypotheses concerning the selective advantages of structures that do not need to be explained and did not function in this manner.

This structural relationship coupled with mechanical analysis demonstrating the relative inefficiency of the Neandertal bite force production system most positively does not support the anterior dental loading hypothesis as a viable origin of the Neandertal face. I would agree with Trinkaus (1987) that Neandertal facial morphology certainly had implications for the biomechanics of their masticatory system. However, I believe it is much more likely that the origin of the derived Neandertal morphology is, as Hublin (1990) suggests, the result of a genetic drift/founder effect phenomenon. That aspects of the cranial and postcranial skeleton exhibit morphologies with certain selective advantages is no doubt the case. However, it is our responsibility to sort out those features that are the result of populational/structural causes and those features that have been modified by selectional forces. In short, the Neandertal picture is vastly more complicated than can be accounted for by a single functional hypothesis. The morphologies of these individuals must be evaluated in a temporal and climatic context as well as with reference to what we know about human craniofacial growth and development.

SUMMARY

Preliminary mechanical analysis of bite force producing capabilities of Neandertals indicates that Neandertals produced absolutely lower bite force values at the anterior dentition and far greater percentages of condylar reaction forces compared to modern humans. Degenerative joint disease should be expected in the Neandertal TMJ, even in individuals seeing loads only as great as or less than average modern human bite forces. It is likely that the level of condylar reaction forces provided an upper limit for bite force production.

The attrition of the Neandertal anterior dentition is not likely to be related to absolutely greater occlusal loads but to consistent usage of the anterior dentition through time. This use might have entailed paramasticatory behaviors such as food, tool, or leather preparation. In sum, the central role that the anterior dental loading hypothesis has been given in the origin and evolution of the Neandertal face appears unwarranted.

At least some of the relationships within the Neandertal face appear to be influenced by the long, low Neandertal vault/base form. These include, in particular, the relatively posterior positioning of the lateral relative to medial facial (particularly orbital) components, the strongly anteroposteriorly oriented zygomatic processes and the more sagittally oriented infraorbital plates. This is not to suggest necessarily that cranial elongation was the driving force in Neandertal facial evolution; only that there is interdependence in the two forms.

It is considered more plausible that the highly derived Neandertal cranial morphology is attributable to a populational variant experiencing gene drift (e.g., Hublin 1990).

ACKNOWLEDGMENTS

I am grateful to the editors for having encouraged my participation in this Festschrift for FCH. The staff of the SRA, HMA, and LHES provided access to specimens considered in this study. Curation of the HMA collections under an NSF grant greatly facilitated the research accessibility of these collections. Drs. M. Koehl, S. M. Swartz and W. L. Hylander, D. Pentcheff, and S. Worchester provided valuable comments on the mechanical analysis. Discussions with G. D. Richards and J.-J. Hublin proved invaluable as did Richard's comments on the manuscript. Finally, thanks 1 x 10^6 to Clark for having inspired us all by his own example to think broadly and pursue many avenues of interest. These disparate areas have made me constantly aware of wonder and more often than not have dovetailed, sometimes years later, to provide me with new perspectives.

LITERATURE CITED

Antón, S. C. (1989) Intentional cranial vault deformation and induced changes of the cranial base and face. *Am. J. Phys. Anthropol.* 79:253-267.

Antón, S. C. (1990) Neandertals and the anterior dental loading hypothesis: A biomechanical evaluation of bite force production. *Kroeber Anthropol. Soc. Pap.* 71-72:67-76.

Boaz, N. T., Ninkovich, D., and Rossignol-Strick, M. (1982) Paleoclimatic setting for *Homo sapiens neanderthalensis. Naturwissenschaften* 69:29-33.

Boule, M. (1923) *Les Hommes Fossiles: Elements de Paleontologie Humaine,* 2nd ed. Paris: Masson et Cie.

Boule, M., and Vallois, H. V. (1957) *Fossil Men.* New York: Dryden Press.

Brace, C. L. (1963) Structural reduction in evolution. *Am. Nat.* 93:39-49.

Brace, C. L. (1964) The fate of the "Classic" Neandertals: A consideration of hominid catastrophism. *Curr. Anthropol.* 5:3-9.

Brace, C. L., Ryan, A. S., and Smith, B. H. (1981) Comment to Puech 1981. *Curr. Anthropol.* 22:426-430.

Brothwell, D. R. (1975) Adaptive growth rate changes as a possible explanation for the distinctiveness of the Neanderthalers. *J. Archaeol. Sci.* 2:161-163.

Carlsöo, S. (1952) Nervous coordination and mechanical function of the mandibular elevators. *Acta Odontol. Scand. Suppl.* 11:1-129.

Cohen, M. M., Jr. (ed.) (1986) *Craniosynostosis: Diagnosis Evaluation and Management.* New York: Raven Press.

Coon, C. S. (1962) *The Origin of Races.* New York: Knopf.

Dean, M. C. (1985) Root cone angle of permanent mandibular teeth of modern man and certain fossil hominids. *Am. J. Phys. Anthropol.* 68:233-238.

Demes, B. (1987) Another look at an old face: biomechanics of the Neandertal facial skeleton reconsidered. *J. Hum. Evol.* 16:297-304.

Demes, B., and Creel, N. (1988) Bite force, diet and cranial morphology of fossil hominids. *J. Hum. Evol.* 17:657-670.

Dubois, E. (1922) On the cranial form of *Homo neanderthalensis* and *Pithecanthropus erectus* determined by mechanical factors. *Koninkl. Akad. Wetensch. Amsterdam* 24:313-332.

Dul, J. G., Johnson, E., Shiavi, R., and Townsend, M. A. (1984) Muscular synergism—II. A minimum-fatigue criterion for load sharing between synergistic muscles. *J. Biomech.* 9:675-684.

Endo, B. (1966) Experimental studies on the mechanical significance of the human facial skeleton. *J. Fac. Sci. Univ. Tokyo* Sec. V, 3:1-106.

Franciscus, R. G., and Trinkaus, E. (1988) The Neandertal nose (abstract). *Am. J. Phys. Anthropol.* 75:209-210.

Gans, C., and Bock, W. J. (1965) IV. The functional significance of muscle architecture: A theoretical analysis. *Ergeb. Anat. Entwicklungsgesch.* 38:115-142.

Gans, C., and de Vrees, F. (1987) Functional bases of fiber length and angulation in muscle. *J. Morphol.* 192:63-85.

Green, M., and Smith, F. H. (1990) Neandertal craniofacial growth (abstract). *Am. J. Phys. Anthropol.* 81:232.

Grey, P. E. (1992) Preliminary application of a mathematical model to hominid mastication (abstract). *Am. J. Phys. Anthropol.* Suppl. 14:84-85.

Heim, J-L.(1978) Contribution du massif facial a la morphogenèse du crâne Néanderthalien. In J. Piveteau (ed.): *Les Origines Humaines et les Epoques de l'Intelligence.* pp. 183-215.

Herring, S., Grimm, A. F., and Grimm, B. R. (1979) Functional heterogeneity in a multipinnate muscle. *Am. J. Anat.* 154:563-576.

Hiiemae, K. M. (1971) The structure and function of the jaw muscles in the Rat (*Rattus norvegicus* L.) III. The mechanics of the muscles. *Zool. J. Linn. Soc.* 50:111-132.

Howell, F. C. (1951) The place of Neandertal man in human evolution. *Am. J. Phys. Anthropol.* 9:379-416.

Howell, F. C. (1952) Pleistocene glacial ecology and the evolution of "Classic Neandertal" man. *Southw. J. Anthropol.* 8:377-410.

Howell, F. C. (1957) II. The evolutionary significance of variation and varieties of "Neandertal" man. *Quart. Rev. Biol.* 32:330-347.

Howell, F. C. (1960) European and North West African Middle Pleistocene hominids. *Curr. Anthropol.* 1:195-232.

Howell, F. C. (1964) Reply to Brace. *Curr. Anthropol.* 5:25-26.

Howells, W. W. (1973) Cranial variation in man: A study by multivariate analysis of patterns of difference among recent populations. *Peabody Mus. Pap.* 67.

Howells, W. W. (1974) Neandertal man: Facts and figures. *Yearb. Phys. Anthropol.* 18:7-18.

Hublin, J. J. (1990) Le peuplements paléolithiques de L'Europe: Un point de vue paléobiogéographique. *Mem. Mus. Prehist. France* 3:29-37.

Hurov, J. R. (1991) Rethinking primate locomotion: What can we learn from development? *J. Motor Behav.* 23:211-218.

Hylander, W. L. (1975) The human mandible: Lever or link? *Am. J. Phys. Anthropol.* 43:227-242.

Hylander, W. L. (1977) The adaptive significance of Eskimo craniofacial morphology. In A. A. Dahlberg and T. M. Graber (eds.): *Orofacial Growth and Development.* Paris: Mouton. pp. 129-170.

Hylander, W. L. (1979) Functional anatomy. In B. G. Sarnat and D. M. Laskin (eds.): *The Temporomandibular Joint: A Biological Basis for Clinical Practice.* Springfield: Charles C. Thomas. pp. 85-113.

Hylander, W. L. (1985) Mandibular function and temporomandibular joint loading. In D. S. Carlson, J. A. McNamara, and K. A. Ribbens (eds.): *Developmental Aspects of Temporomandibular Joint Disorders.* Ann Arbor: University of Michigan Press. pp. 19-35.

Hylander, W. L., Picq, P. G., and Johnson, K. R. (1991) Function of the supraorbital region of primates. *Arch. Oral Biol.* 36:273-281.

Jorgensen, J. B. (1953) The Eskimo skeleton: Contributions to the physical anthropology of the aboriginal *Greenlanders. Meddel. Grønland* 146:1-158.

Kohn, L. A. P., Vannier, M. W., Marsh, J. L., and Cheverud, J. M. (1992) The effect of premature suture closure on craniofacial morphology (abstract). *Am. J. Phys. Anthropol.* Suppl. 14:101-102.

Laitman, J. T., Reidenberg, J. S., Friedland, D. R., and Gannon, P. J. (1992) The demise of the Neandertals: The respiratory specialization hypothesis (abstract). *Am. J. Phys. Anthropol.* Suppl. 14:104.

Mann, A., and Trinkaus, E (1974) Neandertal and Neandertal-like fossils from the Upper Pleistocene. *Yearb. Phys. Anthropol.* 17:169-193.

Moller, E. (1966) The chewing apparatus: An electromyographic study of the action of the muscles of mastication and its correlation to facial morphology. *Acta Physiol. Scand.* 69, *Suppl.* 280:1-229.

Molnar, S. J. (1971) Human tooth wear, tooth function and cultural variability. *Am. J. Phys. Anthropol.* 34:175-189.

hMoss, M. L., and Salentijin, L. (1969) The capsular matrix. *Am. J. Orthodont.* 56:474-490.

Moss, M. L., and Young, R. (1960) A functional approach to craniology. *Am. J. Phys. Anthropol.* 18:281-292.

Osborn, J. W., and Baragar, F. A. (1985) Predicted pattern of human muscle activity during clenching derived from a computer assisted model: Symmetric vertical bite forces. *J. Biomech.* 18:599-612.

Oyen, O. J., Rice, R. W., and Cannon, S. (1979) Browridge structure and function in extant primates and Neandertals. *Am. J. Phys. Anthropol.* 51:83-96.

Picq, P. G., and Hylander, W. L. (1989) Endo's stress analysis of the primate skull and the functional significance of the supraorbital region. *Am. J. Phys. Anthropol.* 79:393-398.

Puech, P.-F. (1981) Toothwear in La Ferassie man. *Curr. Anthropol.* 22:424-425.

Pruim, G. J., de Jongh, H. J., and Ten Bosch, J. J. (1980) Forces acting on the mandible during bilateral static bite at different bite force levels. *J. Biomech.* 13:755-763.

Rak, Y.(1986) The Neandertal face: A new look at an old face. *J. Hum. Evol.* 15:151-164.

Ravosa, M. J. (1991) Interspecific perspective on mechanical and nonmechanical models of primate circumorbital morphology. *Am. J. Phys. Anthropol.* 86:369-396.

Russell, M. D. (1985) The supraorbital torus: "A most remarkable peculiarity." *Curr. Anthropol.* 26:337-360.

Schumacher, G. H. (1961) *Funktionelle Morphologie der Kaumuskulatur.* Jena: Gustav Fischer.

Smith, F. H. (1983) Behavioral interpretation of changes in craniofacial morphology across the archaic/modern *Homo sapiens* transition. In E. Trinkaus (ed.): *The Mousterian Legacy: Human Biocultural Change in the Upper Pleistocene.* British Archaeological Reports, International Series 164:141-163.

Smith, F. H. (1991) The Neandertals: Evolutionary dead ends or ancestors of modern people. *J. Anthropol. Res.* 47:219-238.

Smith, F. H., and Green, M. (1991) Heterchrony, life history and Neandertal morphology (abstract). *Am. J. Phys. Anthropol.* Suppl. 12:164.

Smith, F. H., and Paquette, S. P. (1989) The adaptive basis of Neandertal facial form, with some thoughts on the nature of modern human origins. In E. Trinkaus (ed.): *The Emergence of Modern Humans: Biocultural Adaptations in the Later Pleistocene.* New York: Cambridge University Press. pp. 181-276.

Smith, F. H. and Ranyard, G. C. (1980) Evolution of the supraorbital region in upper Pleistocene fossil hominids from south Central Europe. *Am. J. Phys. Anthropol.* 53:589-610.

Smith, P. (1976) Dental pathology in fossil hominids: What did Neandertals do with their teeth? *Curr. Anthropol.* 17:149-151.

Smith, R. J. (1978) Mandibular biomechanics and temporomandibular joint function in Primates. *Am. J. Phys. Anthropol.* 49:341-350.

Spencer, M. A. (1991) Neandertal facial mechanics: A new perspective (abstract). *Am. J. Phys. Ant ropol.* Suppl. 12:166.

Spencer, M. A., and Demes, B. (1993) Biomechanical analysis of masticatory configuration in Neandertals and Inuits. *Am. J. Phys. Anthropol.* 91:1-20.

Tappen, N. C. (1978) The vermiculate surface pattern of brow ridges in Neandertal and modern human crania. *Am. J. Phys. Anthropol.* 49:1-10.

Thomson, K. S. (1992) Macroevolution: The morphological problem. *Am. Zool.* 32:106-112.

Throckmorton, G. S. (1985) Quantitative calculations of temporomandibular joint reaction forces—II. The importance of the direction of the jaw muscle forces. *J. Biomech.* 18:451-461.

Trinkaus, E. (1983) *The Shanidar Neandertals.* New York: Academic Press.

Trinkaus, E. (1987) The Neandertal face: Evolutionary and functional perspectives on a recent hominid face. *J. Hum. Evol.* 16:429-443.

Trinkaus, E. (1989) The upper Pleistocene transition. In E. Trinkaus (ed.): *The Emergence of Modern Humans: Biocultural Adaptations in the Later Pleistocene.* New York: Cambridge University Press, pp. 42-66.

van der Klaauw, C. J. (1963) Projections, deepenings and undulations of the surface of the skull in relation to the attachment of muscles. *Verhand. Koninkl. Nederl. Akad. Wetensch.* Natuur. 55:1-270.

Weijs, W. (1980) Biomechanical models and the analysis of form: A study of the mammalian masticatory apparatus. *Am. Zool.* 20:707-719.

CHAPTER 30

Reaganomics and the Fate of the Progressive Neandertals

Robert S. Corruccini

HISTORICAL CYCLES

Many thoughts, themes, and ideologies have revolved, waxed, and waned in paleoanthropology. Since the early 1980s the idea has energetically reemerged that modern *Homo sapiens* had an early origination, coexisting for a long time as a distinct species alongside Neandertals. This idea, traditionally known as the "presapiens" theory of modern human origins (Howell 1957; Vallois 1958), has come and gone in anthropological fashion periodically through the last 110 years. The current surge of interest in the idea (e.g., Smith and Spencer 1984; Stringer and Andrews 1988; Trinkaus 1989) has been referred to as the "recent African origin" and "mitochondrial Eve" model. The idea's new proponents have not always been responsible to the analogy with the earlier historical schools of thought, and I believe Spencer's (1984) analogy of "new wine in an old bottle" applies well to the relation between the current "out-of-Africa" and the timeworn presapiens theories (Corruccini 1992). Both theories posit a long CroMagnon-like lineage of perhaps even greater than 200,000 YR antiquity (Brauer 1984a) relegating much of pre-Wurm, non-Western European hominids of the later Pleistocene to the ancestry only of Neandertals rather than of modern humans or of both. Brauer's (1984b) title was honest in directly linking the two theories' names.

Support for the "out-of-Africa" model is derived from the supposed mitochondrial DNA evidence pointing to a single African common ancestry of living humans, plus fossils of very early anatomically modern *Homo sapiens* (hereafter "amhs") in South Africa (see reviews in Smith et al. 1989; Trinkaus 1986, 1989; Smith 1984; Stringer 1989; Stringer and Andrews 1988; Spencer 1984; Wolpoff et al. 1984; Mellars 1988; Clark and Lindley 1989; Gould 1988a,b). To some those fossil amhs strengthen the interpretation that intermediary forms between Neandertals and amhs are lacking, whereas a sudden peripheral appearance of amhs is consistent with current thinking about punctuated equilibria. According to that thinking the earliest amhs of 200,000 YR in age should be "essentially us" (Gould 1988b) when they appear, rather than gradually approaching the human tendency. Thus seemingly mosaic specimens are explained as either robust moderns or gracile Neandertals (Stringer 1989). This out-of-Africa model postulates archaic or Neandertal humans existing elsewhere while early evolution occurred of anatomical modernity in south or central Africa. While one version of the model admits to possible gradual change, intermediate forms, and interbreeding with Neandertals upon contact, stricter views deny gradualism or interbreeding in keeping with the presapiens legacy.

Some logical flaws in this reasoning are well known. First, complete (hence diagnostic) and context-rich later Pleistocene remains (up to c. 40,000 BP) are seemingly always Neandertal-like and archaic to an appreciable degree (including Qafzeh and Skhul: Corruccini 1992). The earliest putative amhs are always the most fragmentary (Klasies, Border Cave, Mumba, and even the resurrection of Kanjera: Brauer 1984a). The pattern is statistically strong enough to require an explanation that those presapiens populations were of lower density and/or not practicing interment (whereas the Neandertals probably were), far from consistent with any technological superiority or behavioral/psychological equivalence to modern humans of the early amhs.

According to Brauer (1984a) a hominid population typified by the Kabwe cranium could have served as the ancestor of true *Homo sapiens*. While this strain of hominid is indeed geographically and temporally available for such ancestry, to suddenly evolve amhs from the massive Kabwe face is a more "catastrophic" (Brace 1964) remodeling of anatomy than to transform late Classic Neandertals into CroMagnons (this is craniometrically demonstrable: Corruccini 1974) even though the latter change (justifiably, in my opinion) cannot be accepted by the out-of-Africa proponents. It would be much easier to transform an amhs out of a similarly early Zuttiyeh-like frontal bone.

The later Levantine presapiens (Qafzeh and Skhul) carry quite typical Mousterian (i.e., Neandertal) industries, a longstanding paradox (Klein 1992). Clark and Lindley (1989), Klein (1992), and Marshack (1989) review the unchanging evidence that Levantine stone tools and indications of symbolic behavior fit the model of fluid transitional variability to Upper Paleolithic technology, signalling a very late attainment of truly modern human lifeways, such that replacement of Western Asian Neandertals by African presapiens did not incur technological replacement.

Rising skepticism afflicts the mtDNA clock. Ward (1991) and Valencia and Ward (1991) find mtDNA too diverse within populations to provide credible dates between populations. They find mate separations of 50,000-75,000 YR and Asian subpopulations separated by 110,000-140,000 YR, weakening credibility of the calibration of lineage antiquity. Other critical shortcomings of molecular and population genetic assumptions and of analysis, according to Pickford's (1991) review, render any mtDNA conclusions virtually useless to ruminations about human evolution. The mtDNA labs need to go back to the drawing board and make up the baseline work that was missing at the premature announcement of the Eve hypothesis. Hoelzer et al. (1992) and Melnick and Hoelzer (1992) raise other objections concerning lineage fidelity and evolutionary rate of mtDNA.

Why has so much new and seriously contemplated thinking arisen out of so little new fossil material? (All workers agree that sparse new fossil data have been recovered in the last 20 years.) By now we nearly all accept that scientific thinking and the preeminence of certain paradigms are embedded in and influenced by the society (and its norms) within which the scientific views are taking place (e.g., Gould 1981). It would perhaps be not too radical to assert that the cyclical nature of Neandertal evolutionary views could be thus affected. One among many possible avenues to investigate this suspicion is to consider scientific views against the background of twentieth-century Western (primarily Anglo-American) nationalism and chauvinism related to prevailing political conservatism.

I have very arbitrarily and subjectively charted such fluctuations over time in Figure 30-1. Generally moderate turn-of-the-century tendencies led toward ever heightened imperial jealousies immediately prior to World War I, while attitudes hardened among the victors for a while in the aftermath of a bitter victory. The 1929 depression led to rapid gains on the part of leftist constituencies. A new rise in conservative political fortunes following World War II begins around 1950. In the 1960s, however, powerful liberal ideologies briefly hold upper sway as the civil rights movement and economic growth make a great new society seem just around the corner. Social conservatism slowly climbs back in the 1970s, but who could have foretold its accelerated rise in the 1980s as Reaganomics and Thatcherism opened the gates to a new era of reckless capitalism, the dismantling of a noble English university system, strong (neo Cyril Burt) revival of respectability of biogenetic determinism in psychology and sociology (Degler 1991), and, among many signs of the times, the most poignant for me, the American Presidential Medal of Freedom being awarded to the judge who executed the Rosenbergs, all things which would have been unthinkable in the 1960s. There is currently the hint of some reversal in conservative political fortunes.

What is the relationship of all this to Neandertal evolutionary views? Borrowing heavily from Spencer's (1984) treatment, we see that moderate academic receptiveness to some sort of Neandertal ancestry takes a nosedive in the 1914-1924 period. It is no coincidence that the pro-Neandertal thinking was more prevalent in Central to Eastern European scholarly circles than those of France and England from 1890-1910, but the nationalities involved were part of the losing Central Powers of World War I and "history is written by the winners." After the situation stabilizes, the resistance to Neandertal theories diminishes somewhat, punctuated by the circum-1950 downfall of Piltdown. The 1950s see a revival of a retooled presapiens theory (Vallois 1949), but this is moderated by the conciliation provided by the Progressive Neandertal idea (Howell 1951). Surely Neandertals were seen in their best light during the 1960s when most textbooks began deriding their clumsy stereotype and asserted their behavioral similarity to modern people. The tendency to emphasize a fashionable bond of all humanity underlay the rehabilitation of Neandertals and this was by no means subconscious; Howells (1974) confirms that it seemed akin to racism to acknowledge fundamental differentness of Neandertals in the 1960s, when description of the profoundness of racial diversity throughout the bioanthropological profession was being retracted.

The abrupt downfall of Neandertal relatedness to modern humanity characterizing the 1980s has been conspicuously accompanied by the return of views that were derided in the 1960s as antiquated holdovers of the biases of an earlier century. Now there are serious discussions (at least in the popular science literature) of social and biological incompatibility of speechless and uninnovative Neandertal

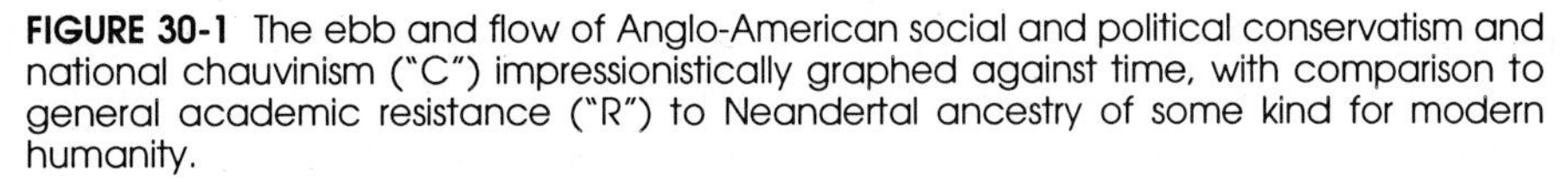

FIGURE 30-1 The ebb and flow of Anglo-American social and political conservatism and national chauvinism ("C") impressionistically graphed against time, with comparison to general academic resistance ("R") to Neandertal ancestry of some kind for modern humanity.

populations upon their contact with relentlessly marching moderns, and the inevitability of genocide as a natural human consequence of contact. The parallel between current Western eradication of "inefficient" social programs since they hinder economic growth, and the heartless elimination of benign yet bumbling Neandertals since they would have hindered the westward march of competitive CroMagnons, cannot be missed.

Thus I postulate a significantly interrelated time series between the fictive coefficients "C" and "R" in Figure 30-1. This is not to say that society is a conscious determinant of views (but rather a subconscious influence), and I also do not wish to imply that the academicians on both sides of the issue are not predominantly "knee-jerk" political liberals like myself and most others in academe, nor that there has not always been a diversity of views over time.

A RELATION THAT INSPIRES SUSPICION

As with earlier periodic popularizations of the idea, the new vigor of renovated presapiens thinking relies too heavily on fragmentary (hence undiagnostic) remains of relatively uncertain provenience or dating. These are now Klasies River Mouth and Border Cave fragments as well as such things as isolated teeth from Mumba; previously they have been Fontechevade, Swanscombe, Galley Hill, and even Piltdown in the earlier runs of the presapiens legend. The emphasis on unusually poorly preserved or dubiously contextualized and recovered remains thus is a century-long tendency in theories of this nature. Ironically White (1987) raises the possibility of the Klasies remains being results of cannibalism; one wonders if it was the inferior archaics who were eating them.

Perhaps this presapiens tendency to lean toward seeing modern characteristics in very fragmentary remains represents a natural human proclivity to deny ancestral status to forms unlike ourselves. Humans have always been uneasy with the Darwinian concept when applied to themselves. Whereas Brace (1964) applied the concept of catastrophism to Neandertal evolutionary discontinuities with amhs, I believe "homunculism" might be the more appropriate analogy. By this analogy people have strived continuously for decades to find ever earlier traces of *Homo* that preserve essentially human characteristics, but at a smaller size, at the same time that the fossil record is dominated by more primitive or apelike hominids. The latter with their unseemly anthropoid or Neandertal features need not now be admitted into our ancestry because of the contemporaneity of the homunculus.

Figure 30-2 illustrates the foundations of my skepticism concerning the randomness of fossil preservation of presapiens as opposed to less humanlike hominids. The relationship between, on the one hand, (X axis) completeness of cranial fossils plus quality of data concerning their geologic provenience, and on the other hand (Y axis) the Neandertal apomorphic or archaic plesiomorphous diagnosis of morphology as opposed to amhs apomorphy, really cannot be missed. Statistically speaking, there is miniscule probability that these qualities have been sampled from a parent population where they are randomly related. The lower bound of the 95 percent confidence limits of the correlation, even when allowing for the nonparametric nature of the data, is well above 50 percent shared standard deviation between the two quantities. The more complete, visible, and diagnostic a cranium is, the more likely it is to be considered very archaic or specifically (Classic) Neandertal in morphology. Just why are many key specimens underpinning the recent African evolution of amhs (primarily Group A in Figure 30-2) so incomplete?

If we seek naturalistic explanations, as opposed to human psychological ones, it may be that the early amhs paradoxically did not practice inhumation whereas populations representing certain archaic strains, possibly the rhodesioid and particularly European Classic Neandertals, did engage in this emphatically human behavior. Aurignacian remains similarly seem rarely to have been carefully interred (see Smith 1984:142).

It could be that early South African amhs were a rare and peripheral population for a long time and therefore simply not as numerous as the more northerly archaics and Neandertals. Assuming (perhaps hazardously) that fossilization rate bears some vague relation to population density per unit of time, the

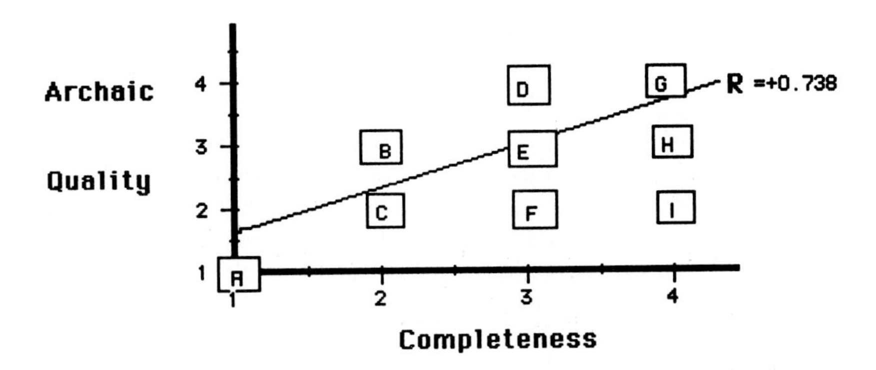

FIGURE 30-2 Quality of fossil information (horizontal axis) plotted against the degree of confident diagnosis of nonmodern morphology (vertical axis) of crania from the early part of later Pleistocene time up until the advent c. 40,000 BP of Upper Paleolithic technology. Both axes have been quantified as four-stage scales. For the horizontal axis completeness of cranial preservation (mandible not included) and sufficiency of data concerning the geologic provenience of hominid fossils are as follows: (1) extremely fragmentary pieces and/or stratigraphic provenience practically unknown; (2) part of calotte only or face only or else substantial portion of cranium but necessitating extensive reconstruction; (3) substantial parts of face and calotte exist, or ample site context data; (4) nearly complete crania. For the vertical axis the degree of archaic/Neandertal as opposed to modern morphology is: (1) amhs diagnosis confidently upheld by at least some; (2) indications at least of some amhs apomorphies in association with other more archaic traits; (3) generalized archaic or "Progressive" Neandertal with possible hint of advanced features; (4) extreme archaic or Classic Neandertal. The squares encompass these clouds of points: (A) Border Cave, Fontechevade, Klasies, Kanjera, Piltdown. (B) Biache, Engis, Eyasi, Hanofersand, Sala, Salzgitter-Lebenstedt, Zuttiyeh. (C) Omo 1. (D) Hopefield, Feldhofer Cave. (E) Bodo, La Chaise, Ehringsdorf, Florisbad, Krapina, Mapa, Ndutu, Omo 2, Saldanha, Steinheim, Vindija. (F) Swanscombe. (G) Amud, La Chapelle, Circeo, Dali, La Ferrassie, Kabwe, Le Moustier, La Quina, Shanidar, Spy, St. Cesaire, Tabun, Teshik-Tash. (H) Gibraltar, Ighoud, Ngaloba, Saccopastore, Skhul. (I) Qafzeh. Sites with more than one hominid cranium are considered as a whole.

presapiens may have simply "kept their heads down" demographically until embarking on a precipitous population growth in conjunction with northward migration. Why this emergence, why it is not signalled by any sort of technological breakthrough, and why the amhs displaced endemically climatically adapted populations so rapidly once they began their odyssey, are not explained.

Least complimentary to us as scientists, the last explanation is that the relationship indicated in Figure 30-2 takes place solely in our subjective minds. In other words, due to homunculism or to the drive for fame emanating from discovering earlier human emergence, one subconsciously leans toward modern diagnosis of remains sufficiently fragmentary to permit such a possibility. I submit that this is precisely why we need metrical analyses of such fragments, to control the subjectivity of the mind and eye. Nearly every single adequately complete hominid fossil up until 40,000 BP retains at least some clearly archaic morphology; it is only the scraps that underlie the vision of very early anatomical modernity. Could the latter entail wishful thinking based upon materials too undiagnostic as to discourage the wishful thoughts? It is also worthwhile to point out that in its periodic rise and fall the presapiens theory has relied upon many fossils that subsequently turned out to be less amhs in morphology, later in time, or both. A purely historical prediction would be for the same fate to eventually befall the collection of fossils currently underpinning the recent African origins thinking, when they have been more intensively scrutinized. Klein's (this volume) comments on Border Cave 1's provenience may be consistent with this.

One is tempted to remind the reader that the only thing we learn from history is that people don't learn from history.

THE FORGOTTEN SKHUL SPECIMENS

A continually curious tendency in the Neandertal-amhs conundrum is the use of Skhul V to characterize early amhs in the Middle East while paying scant comparative attention to the other remains from Skhul. This point was raised by Corruccini (1974:100) in the most comprehensive calvariometric study of fossil hominids. Crania IV and IX from Skhul are not vastly less complete (although more crushed), and V for that matter required substantial midface reconstruction.

Has the disproportionate attention to V somehow grown from its more modern appearance among an actually quite variable population? Both IV and IX, and other Skhul fragments, seem to have a relatively low position of maximum breadth on the parietals. In fact, Howell (1957) drew attention to Skhul VI and VII as the most Neandertal-like of that collection, but these are too crushed to make any craniometrical comparisons (C. Stringer, pers. comm., says the same of Skhul IV). McCown and Keith (1939) chose V as a sort of type specimen because it was the tallest individual and (only in the sense of the entire skeleton) most complete, and also because it was very distinct from Neandertals thus "foreshadowing" the CroMagnons. However, IV was said to be the "most representative of the Skhul type of ancient Palestinian." They documented the sharp occipital angulation and the long flat face of IV in contrast to V. In Skhul IX, Neandertal features predominate over "Neanthropic" ones to the contrary of IV according to McCown and Keith, and IX is the most massive prompting some positive comparison with La Chapelle. Otherwise McCown and Keith (1939) and Howell (in 1951 and 1957) emphasized the "unexpectedly great" variation within Skhul individuals, "overlapping the Neandertal range" when viewed as a whole. It is cautioned that some of the variation could be due to deformation and reconstruction.

At that time there seemed to be no evidence for other than a "homogeneous deposit" signifying a single population in the Skhul burial level; there seems to be scant revision as yet of that impression. A few workers have subsequently been alive to the implications of Skhul crania other than V in formulations about modern human origins (Coon 1962; Brose and Wolpoff 1971; Wolpoff and Caspari 1990).

I undertook a new analysis of size/shape patterning in later Pleistocene hominid crania (Corruccini 1992), with focus on how Skhul IV and IX reflect on the affinities assumed for Skhul V. The 27 craniometric variables used were just those measurements that could be estimated or have been published on all three of these Skhul crania. Methods included simple distances based on raw dimensions ("size"), distances following Mosimann-style shape vector alteration of the raw dimensions, and Generalized distances.

The comparative samples were (1) a broadly defined Neandertal grouping (Western European Classic, Middle Eastern, and earlier Pre- or Progressive Neandertals) consisting of La Chapelle, La Ferrassie, Le Moustier, Spy I and II, La Quina, Neandertal, Circeo, Gibraltar, Teshik-Tash, Steinheim, Ehringsdorf, Saccopastore 1 and 2, Krapina C, Shanidar, Tabun I, Amud, and Zuttiyeh; (2) putative early African and Levant precocious anatomical moderns (so defined by at least one proponent [Smith et al. 1989] or indicated as "transitional" from archaic morphology, and for which could be found adequate data) including the Skhul material, Florisbad, Djebel Ighoud 1 and 2, Qafzeh 6, and Omo 2 (this will be referred to as "FIQO"); and (3) True Upper Paleolithic Eastern European crania (6 Predmost, 2 Oberkassel, Mladec (Lautsch), Brno, the admittedly dubiously included Brux, and Dolni Vestonice).

Internally consistent results produced clusters quite recognizable within existing frameworks of relationships and similarity. Classic (Western European Wurmian) Neandertals plus Shanidar 1 clustered, then joined a group of 5 Middle Eastern and other less classic (perhaps "progressive"?) Neandertals: Tabun 1, Amud, Zuttiyeh, Gibraltar, and the admittedly juvenile Teshik-Tash. Skhul V joined earlier and more generalized European specimens (Steinheim, Ehringsdorf) and then the broadly constituted Neandertal cluster at lower levels of distance.

Skhul IV and IX, meanwhile, although sharing a relatively small distance from V, joined the FIQO cluster of supposedly early African/Levantine amhs, and then this increasingly heterogeneous Neandertal grouping. The Upper Paleolithic true amhs accurately clustered together without exception, relatively far removed from these other groupings. Thus, the picture is one of overriding affinity among all the crania earlier than the European Upper Paleolithic, whether they be considered Classic Neandertal, Progressive Neandertal, amhs, presapiens, or whatever. Strongly implied is that, despite much recent discounting, the old orthodoxy remains accurate that true (craniofacial) anatomical modernity still is uniquely associated with the appearance of the Upper Paleolithic.

These measurements sufficed to re-emphasize that the Classic Neandertals are distinct from the variety of other hominids that have acquired the Neandertal tag, such as pre-Wurm or non-Western European specimens (Howell 1951; Trinkaus 1991). The FIQO sample construed to represent early non-European amhs is nearer to Neandertals than to the true amhs of the Upper Paleolithic. Skhul crania IV and IX, in contrast to the celebrated V, are somewhat more within the earlier neandertaloid than the later, more modern range.

Other similarities conform to the older idea that all pre-Wurm forms could be intermediate and ancestral to both the divergent lineages of Neandertals and amhs (this is the Preneandertal theory); the FIQO "presapiens" assemblage of crania, plus Skhul, could better be called preneandertal since they are metrically closer to Neandertals than to true (Upper Paleolithic) amhs.

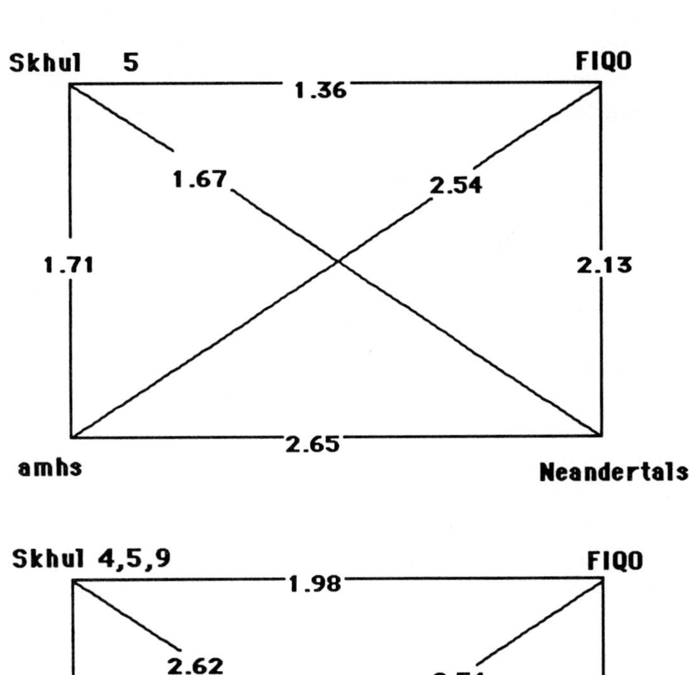

FIGURE 30-3
The Generalized Distance (D) based on 27 craniometric variables represented as a rectangle when only Skhul V is considered (top) versus the distances when Skhul IV, V, and IX are combined as a sample (bottom). To give an accurate picture of actual distance relationships, the corner for Skhul V alone in the top part would have to be folded substantially through the vertical axis toward the opposite corner. FIQO refers to the Florisbad, Ighoud 1 and 2, Qafzeh 6, and Omo II crania.

The FIQO, Neandertal and amhs samples form a roughly triangular arrangement of averaged individual Generalized distances. Skhul V is very nearly in the center of this triangle. Skhul IV plus IX on average are further removed from both Neandertals and amhs than is the very neutral Skhul V, yet they are much more variable in these distances. Figure 30-3 depicts these patterns.

Similar allusions to the variability within Middle Eastern sites are now increasing (these are the source of the blip in "R" at the extreme right of Figure 30-1). Arensburg (1991) denies that any of the Levantine specimens are true Neandertals, and Trinkaus (1991) supports the reasoning that only very late, isolated Western European specimens fit the Neandertal stereotype; both authors refer to clinal distributions of characters (as did Corruccini 1974). Arensburg (1991) reports "the morphological variability within sites (e.g., Qafzeh VI vs. Qafzeh IX, Tabun C vs. Tabun B, Skhul V vs. Skhul IX) makes a separation between Neandertals and *Homo sapiens* in these localities impossible and demonstrates a great range of variation within a unique population." Vandermeersch (1991) seemingly agrees regarding characteristics of Tabun 2 in contrast to Tabun 1. Rak (1991) refers to Shanidar II-IV as "less differentiated" Neandertals coexisting with the "classic" Shanidar I and V. Wolpoff and Caspari (1990), in addition to confirmatory reference to within-Skhul variations, find Qafzeh 3 more archaic than 6 and 9. Simmons (1991) finds major aspects of similarity between the Ighoud and Qafzeh frontal bones while Simmons et al. (1991) allude to substantial within-site and between-site differences at both Skhul and Qafzeh.

According to the most strictly interpreted sudden-replacement model of amhs origins (Stringer and Andrews 1988), the Skhul population should represent a distinct species from Neandertals (whether the latter are broadly or narrowly defined) that is indistinguishable from modern humans (unless Neandertal gene flow is incorporated). Taken all together, however, the previous results raise perplexing questions about the interpretation of two supposedly long-separated species. The variation within Skhul alone spans much of the phenetic spectrum between archaic (or "neandertaloid") and modern cranial form, while the remainder of specimens considered here fall not into two well-defined clusters but three fuzzy ones. Such results are hardly unprecedented. Howells (1989) and Corruccini (1974) find Skhul V no better than intermediate between moderns and Neandertals or slightly nearer the latter. Indeed, ironically Stringer (1974) and Andrews and Williams (1973), major proponents of early African amhs origins, demonstrate a clearly intermediate affinity for Skhul V between Neandertals and later amhs in their unrelated earlier craniometric analyses. Stringer's (1991 pers. comm.) recent interpretation of long-term craniometric study again indicates intermediacy of Skhul and Qafzeh crania between Upper Paleolithic and Neandertal samples. Upon craniometric reexamination (Corruccini 1992 and references therein), the Border Cave cranium, so central to the course of out-of-Africa thinking, can support no special relationship to living African *Homo sapiens*. In fact, it may be as similar to archaic and earlier Late Pleistocene hominids as it is to modern people (Van Vark et al. 1989). Klein (this volume) alludes to some additional evidence that the cranium also may not derive from the older levels of the cave; thus it has become both later and less morphologically advanced than once promoted. This does contradict much of the specific language encountered in support of the punctuated and African nature of the emergence of the earliest anatomical moderns.

WHAT EVER HAPPENED TO THE PROGRESSIVE NEANDERTALS?

In many ways the newly emergent non-Eurocentric, competing out-of-Africa versus "multiregional continuity" hypotheses are geographically generalized recirculations of the old presapiens versus Neandertal theories of human origins, respectively (Spencer 1984; Smith et al. 1989). What are the prospects of likewise retooling and resurrecting the old middle ground of the "Preneandertal" or "Progressive Neandertal" theory (Howell 1951, 1957) as a phylogenetically more complex (cf. Trinkaus 1986; Foley and Lahr 1992) alternative to the two newer models that are increasingly being mentioned as the two sole alternatives? The key feature of the Progressive Neandertal model is that it allows for *some* Neandertal ancestry of amhs but rules out Classic Neandertals as ancestors. The "Progressive" (let us call them pre-Wurm and/or non-Western European) Neandertal-like populations are the only confidently preserved predecessors both of amhs and of later glacially isolated extreme Western European Neandertals; anatomical modernity according to this can be first traced to hominids of the general middle ground of North Africa, Western Asia, and Southeastern Europe c. 40,000+ YA—again an old idea (Birdsell 1963) but one I believe deserves much more consideration among the other revivals currently provoking debate (see Figure 30-4). A revived Preneandertal formulation would incorporate a gradualistic version of the single origin model but deny its earliness and exclusively South African Eden. Brauer's (1984a,b) version of African origins and some others bear close resemblance to the Progressive Neandertal model, but often without invoking its name.

In one of his few recent references to problems of later Pleistocene hominid evolution Howell (1984) distinguishes the plesiomorphous features of earlier and non-Western European Neandertal-like fossils from the apomorphic (i.e., Classic, Wurmian, Western European) Neandertal features. Clearly some terminological refinement is needed to deal with the situation, and the "Preneandertals" would perhaps better be called Progressive Archaics rather than any sort of Neandertal (despite "incipient" Neandertalization of certain features such as the nasal aperture: Howell 1957) to reserve the Neandertal label only for

FIGURE 30-4 A subjective view crosscutting location, time, and morphology in Upper Pleistocene crania illustrating interaction among evident geo-, chrono- and morpho-clines. There is a cline of increasing Classic Neandertal specialization in the western "Wurmian" zone, a reverse cline in primitive holdover traits among earliest amhs, and a particularly marked resultant cline in the morphological gap and suddenness of change between latest Neandertal (broad sense) and earliest appearing Upper Paleolithic crania.

```
DECREASING
LONGITUDE
<------------------------------------------------------------------
WESTWARD

                (WESTERN)     (CENTRAL)     (EASTERN       (SOUTH & EASTERN
                 EUROPE)       EUROPE)       EUROPE)        MEDITERRANEAN)

INCREASING
MODERNITY
   /|\          SOLUTRE
    |           CRO-MAGNON
    |                        PREDMOST>                     KIIK-KOBA
    |                                     <MLADEC                         EARLY
                                                           QAFZEH         AMHS-LIKE
-------------------------------------------------------::::::::::------------------
                                                           SKHUL          LATE
    |                                     KRAPINA          TABUN          NEANDERTAL
    |                        HANOFER-
    |           LA QUINA      SAND                         AMUD
   \|/          ST CESAIRE
INCREASING
NEANDERTAL APOMORPHY
                                                                          INCREASING
------------------------------------------------------------->
WITHIN-POPULATION
                                                                          VARIABILITY
```

the true Classics of later Western Europe. Specimens such as Florisbad, Ngaloba, Omo Kibish, Ighoud, Ehringsdorf, and Saccopastore are these late progressive archaic/early incipient Neandertals. Indeed, the lateral profile of Omo II or of Ngaloba compared with Ehringsdorf suggests a similarity of grade that transcends geographic limitations, much like the earlier Kabwe-Petralona continuity (and this is reflected in relatively high calvarial shape similarity: Corruccini 1974, 1992).

Krapina and most of the Levant assemblage, most notably Skhul, Tabun, and Qafzeh, are of a different nature with more of a volatile mixture of Classic Neandertal plus amhs apomorphies rather than a more uniformly plesiomorphous craniofacial morphology, and perhaps the Progressive Neandertal label is more appropriate for these although the suggestion of mixed apomorphies is more conformable to the traditional late (c. 50,000 YR) date for these sites than to the recent claims for c. 100,000 YR (e.g., Schwarcz et al. 1988; Klein 1992). Stringer (1989) maintains that if Klasies and Border Cave are considered too uncertain for basing expansive phylogenies then the (much more decisively transitional-mosaic) Ighoud crania serve as the best evidence for African origin of amhs! This rather directly renders a Progressive Neandertal interpretation.

Howells' (1989) expansive multivariate cranial study of modern humanity shows that intraregional similarities between fossil and modern hominids are not evident and that all modern populations are fairly cohesive relative to Pleistocene forerunners, suggesting recency of "racial" formation. Such results defy the multiregional continuity idea resurrected and adapted from Coon (1962) and the 200+ KYA origination of current human varieties suggested by mtDNA. It is unfortunate that the multiregional continuity idea unavoidably implies very ancient, deeply rooted, and therefore potentially quite biologically divergent modern races. Kamminga and Wright's (1988) and Van Vark and Dijkema's (1989) recent craniometric reconsiderations of Zhoukoudian Upper Cave hominids, furthermore, show that there is no "mongoloid" quality whatsoever to those hominids (but some vague similarity to Ainu and Australian aboriginals), denying the great antiquity of Asian races. It is interesting here to recall also the variable nature of historical attributions of early Upper Paleolithic humans within geographic areas such as Chancelade with Eskimos versus Grimaldi with Africans, and so on, and Weidenreich's analogy respectively among the three Upper Cave Zhoukoudian crania with Chinese, Melanesians, and Eskimos.

SUMMARY

The out-of-Africa or "Noah's Ark" model for early origins of modern *Homo sapiens* is a revival of the two-lineage model for explaining Neandertal and modern and for disallowing mixed craniofacial morphologies. It continues strong historical precedents associated with the presapiens models of recent human evolution: revival in times of notable Anglo-American politically conservative upswings, and heavy reliance on the most incomplete of available specimens of the late Middle and early Late Pleistocene.

The Recent African Evolution school of thought typically presents Skhul as one site documenting early origination and conquering radiation of amhs. However, only Skhul V is usually considered in the comparative craniology of the question. Omission of other substantial Skhul crania leads to failure to perceive continuous gradations of Neandertal-to-modern variations throughout the Near East, where much of the variation is thus represented within this one site. Substantial craniometric diversity within the Skhul remains parallels the problem of gradational variation among other Middle Eastern (especially Tabun, Qafzeh) and other Upper Pleistocene sites, raising anew the age-old difficulties with the presapiens interpretation of Neandertals (broad sense) and amhs as two long-separate species. Skhul, Qafzeh, Ighoud, and other crania as a group are quite phenetically intermediate between Upper Paleolithic samples and Neandertals—much more consistent with the old "Progressive" Neandertal thinking than with the presapiens and multiregional continuity revivals currently generating copious discussion.

LITERATURE CITED

Andrews, P., and Williams, D. B. (1973) The use of principal components analysis in physical anthropology. *Am. J. Phys. Anthropol.* 39:291-303.

Arensburg, B. (1991) From sapiens to neandertals: Rethinking the Middle East (abstract). *Am. J. Phys. Anthropol. Suppl.* 12:44.

Birdsell, J. B. (1963) The origin of human races. *Quart. Rev. Biol.* 38:178-185.

Brace, C. L. (1964) The fate of the "Classic" Neanderthals: A consideration of hominid catastrophism. *Curr. Anthropol.* 5:3-43.

Brauer, G. (1984a) A craniological approach to the origin of anatomically modern *Homo sapiens* in Africa and implications for the appearance of modern Europeans. In F. H. Smith and F. Spencer (eds.): *The Origins of Modern Humans.* New York: Alan R. Liss, pp. 327-410.

Brauer, G. (1984b) Presapiens-hypothese oder Afro-europaische sapiens-hypothese? *Zeit. Morph. Anthropol.* 75:1-25.

Brose, D. S., and Wolpoff, M. H. (1971) Early Upper Paleolithic man and late Middle Paleolithic tools. *Am. Anthropol.* 73:1156-1194.

Clark, G. A., and Lindley, J. M. (1989) Modern human origins in the Levant and Western Asia: The fossil and archeological evidence. *Am. Anthropol.* 91:962-985.

Coon, C. S. (1962) *The Origin of Races.* New York: Knopf.

Corruccini, R. S. (1974) Calvarial shape relationships between fossil hominids. *Yearb. Phys. Anthropol. 18 :89-109.*

Corruccini, R. S. (1992) Metrical reconsideration of the Skhul IV and IX and Border Cave 1 crania in the context of modern human origins. *Am. J. Phys. Anthropol.* 87:433-445.

Degler, C. N. (1991) *In Search of Human Nature. The Decline and Revival of Darwinism in American Social Thought.* New York: Oxford University Press.

Foley, R. A., and Lahr, M. M. (1992) Beyond "out of Africa": Reassessing the origins of *Homo sapiens. J. Hum. Evol.* 22:523-529.

Gould, S. J. (1981) *The Mismeasure of Man.* Norton: New York.

Gould, S. J. (1988a) Honorable men and women. *Nat. Hist.* 97(3):16-20.

Gould, S. J. (1988b) A novel notion of neanderthal. *Nat. Hist.* 97(6):16-21.

Hoelzer, G., Hoelzer, M., Rosenblum, L., and Melnick, D. (1992) A molecular phylogeny of Asian macaques using mitochondrial DNA (abstract). *Am. J. Phys. Anthropol. Suppl.* 14:91.

Howell, F. C. (1951) The place of Neanderthal man in human evolution. *Am. J. Phys. Anthropol.* 9:379-416.

Howell, F. C. (1957) The evolutionary significance of variation and varieties of "Neanderthal" man. *Quart. Rev. Biol.* 32:330-347.

Howell, F. C. (1984) Introduction. In F. H. Smith and F. Spencer (eds.): *The Origins of Modern Humans.* New York: Alan R. Liss, pp. viii-xxii.

Howells, W. W. (1974) Neanderthal man: Facts and figures. *Yearb. Phys. Anthropol.* 18:7-18.

Howells, W. W. (1989) Skull shapes and the map: Craniometric analyses in the dispersion of modern *Homo. Pap. Peabody Mus.* Archaeol. Ethnol. 79.

Kamminga, J., and Wright, R. V. S. (1988) The Upper Cave at Zhoukoudian and the origins of the Mongoloids. *J. Hum. Evol.* 17:739-767.

Klein, R. G. (1992) The archeology of modern human origins. *Evol. Anthropol.* 1:5-14.

McCown, T. D., and Keith, A. (1939) *The Stone Age of Mount Carmel. II. The Fossil Human Remains from the Levalloiso-Mousterian.* Oxford: Clarendon.

Marshack, A. (1989) Evolution of the human capacity: The symbolic evidence. *Yearb. Phys. Anthropol.* 32:1-34.

Mellars, P. (1988) The origins and dispersal of modern humans. *Curr. Anthropol.* 29:186-188.

Melnick, D., and Hoelzer, G. (1992) What in the study of primate evolution is mtDNA good for? (abstract). *Am. J. Phys. Anthropol. Suppl.* 14:122.

Pickford, M. (1991) Paradise lost: Mitochondrial Eve refuted. *Hum. Evol.* 6:263-268.

Rak, Y. (1991) A model for morphologic and taxonomic variation in Neandertals and early *Homo Sapiens* (abstract). *Am. J. Phys. Anthropol. Suppl.* 12:147-148.

Schwarcz, H. P., Grun, R., Vandermeersch, B., Bar-Yosef, O., Valladas, H., and Tchernov, E. (1988) ESR dates for the hominid burial site of Qafzeh in Israel. *J. Hum. Evol.* 17:733-737.

Simmons, T. (1991) North African and Levantine hominid affinities: Frontal squama and browridge morphometry (abstract). *Am. J. Phys. Anthropol. Suppl.* 12:162.

Simmons, T., Falsetti, A. B., and Smith, F. H. (1991) Frontal bone morphometrics of southwest Asian Pleistocene hominids. *J. Hum. Evol.* 20:249-269.

Smith, F. H. (1984) Fossil hominids from the upper Pleistocene of Central Europe and the origin of modern Europeans. In F. H. Smith and F. Spencer (eds.): *The Origin of Modern Humans.* New York: Alan R. Liss, pp. 137-210.

Smith, F. H., Falsetti, A. B., and Donnelly, S. M. (1989) Modern human origins. *Yearb. Phys. Anthropol. 32:35-68.*

Smith, F. H., and Spencer, F. (eds.) (1984) *The Origins of Modern Humans.* New York: Alan R. Liss.

Spencer, F. (1984) The Neanderthals and their evolutionary sequence: A brief historical survey. In F. H. Smith and F. Spencer (eds.): *The Origins of Modern Humans.* New York: Alan R. Liss, pp. 1-50.

Stringer, C. B. (1974) Population relationships of later Pleistocene hominids: A multivariate study of available crania. *J. Archaeol. Sci.* 1:317-342.

Stringer, C. B. (1989) Documenting the origin of modern humans. In E. Trinkaus (ed.): *The Emergence of Modern Humans. Biocultural Adaptations in the Later Pleistocene.* Cambridge: Cambridge University Press, pp. 67-96.

Stringer, C. B. (1991) Personal communication of talk, 1991 Chicago Systematics meeting.

Stringer, C. B., and Andrews, P. (1988) Genetic and fossil evidence for the origin of modern humans. *Science* 239:1263-1268.

Trinkaus, E. (1986) The Neandertals and modern human origins. *Ann. Rev. Anthropol.* 15:193-218.
Trinkaus, E. (ed.) (1989) *The Emergence of Modern Humans. Biocultural Adaptations in the Later Pleistocene.* Cambridge: Cambridge University Press.
Trinkaus, E. (1991) Would the real Neandertal please stand up—the search for Neandertal apomorphies (abstract). *Am. J. Phys. Anthropol. Suppl.* 12:174-175.
Valencia, D., and Ward, R. H. (1991) Using mitochrondrial sequence data to estimate evolutionary distance between mates (abstract). *Am. J. Phys. Anthropol. Suppl.* 12:176-177.
Vallois, H. V. (1949) The Fontechevade fossil man. *Am. J. Phys. Anthropol.* 7:339-362.
Vallois, H. V. (1958) La Grotte de Fontechevade. *Arch. Inst. Paleont. Hum.* 29(2).
Vandermeersch, B. (1991) Contemporaneity of *Homo sapiens* and Neandertals in the Near East? (abstract). *Am. J. Phys. Anthropol. Suppl.* 12:177.
Van Vark, G. N., Bilsborough, A., and Dijkema, J. (1989) A further study of the morphological affinities of the Border Cave 1 cranium, with special reference to the origin of modern man. *Anthropol. Prehist.* 100:43-56.
Van Vark, G. N., and Dijkema, J. (1989) Some notes on the origin of the Chinese people. *Homo* 39:143-148.
Ward, R. H. (1991) On the age of our mitochondrial ancestors: evidence for deep lineages in "mongoloid populations" (abstract). *Am. J. Phys. Anthropol. Suppl.* 12:180-181.
White, T. D. (1987) Cannibals at Klasies? *Sagittarius* 2:6-9.
Wolpoff, M. H., and Caspari, R. (1990) On middle paleolithic/middle stone age hominid taxonomy. *Curr. Anthropol.* 31:394-395.
Wolpoff, M. H., Wu, X., and Thorne, A. G. (1984) Modern *Homo sapiens* origins: A general theory of hominid evolution involving the fossil evidence from east Asia. In F. H. Smith and F. Spencer (eds.): *The Origins of Modern Humans.* New York: Alan R. Liss, pp. 411-483.

INDEX

E

F

G

H

I

J

K